W9-CSI-811

Engineering Properties of Steel

Philip D. Harvey
Editor

AMERICAN SOCIETY FOR METALS
METALS PARK, OHIO 44073

First printing, December 1982
Second printing, July 1985
Third printing, August 1992
Fourth printing, May 1994
Fifth printing, April 1995
Sixth printing, May 1999

Library of Congress Cataloging in Publication Data

Engineering properties of steel.

Includes bibliographical references and index.
1. Steel. I. Harvey, Philip D. II. American Society for Metals.
TA472.E63 1982 620.1′7 82-8829
ISBN 0-87170-144-8 AACR2

SAN 204-7586

PRINTED IN THE UNITED STATES OF AMERICA

Preface

When a manufacturing company considers making a new product or improving an old one, the materials engineer or the design engineer must examine both the technical and economic feasibility of available materials. *Only* if the materials are properly selected to perform satisfactorily in service can parts and components meet major technical requirements.

Economics of production are determined, in large part, by the cost of the materials, plus the cost of converting them into the finished product. Both costs are governed by the materials chosen. Thus, proper knowledge of the mechanical and physical properties, fabrication characteristics, and applications of materials is essential for the economic and technical success of the product.

Engineering Properties of Steel presents the technical data needed for materials selection. The extensive properties data are presented in a ready-reference format which makes the information easy to find.

Because the materials selected are usually a compromise of properties, costs, availability, and suitability, engineers must frequently survey a large number of steels. *Engineering Properties of Steel* provides the information to make such surveys quickly and efficiently.

Mechanical properties refer to the response of the material to applied forces or loads, and indicate the strength and shape integrity of the material. Physical properties or characteristics indicate how a material is affected by non-mechanical variables such as thermal energy, electrical energy, magnetic fields, chemicals, and general or specific environments.

Engineering Properties of Steel presents the chemical composition, mechanical properties, physical properties, general characteristics and uses, and machining data for most steels used in industrial applications. Data are given for carbon, alloy, stainless and heat resisting, tool, ASTM structural, and maraging steels. For properties of nonferrous metals and alloys, the reader can consult the *Metals Handbook,* 9th ed., Vol 2 and 3, published by the American Society for Metals. The data listed for each steel should be regarded as average or nominal and adjustments made to suit the prevailing conditions.

Where applicable, data for the steels are arranged in numerical order according to the AISI/SAE identification number. The ASTM structural steels are given in numerical order using the ASTM standard specification number. Maraging steels are given in the order of nominal yield strength. For the AISI/SAE steels, the designations of similar steels as listed by United States and foreign standards specification organizations are given.

Our wholehearted thanks are extended to the many steel companies in the United States and Canada who provided us with their technical bulletins from which much of the data was adapted. We also gratefully acknowledge the following as sources of reference and data: Society of Automotive Engineers, American Iron and Steel Institute, American Society for Testing and Materials, and the Machinability Data Center of Metcut Research Associates, Inc.

Philip D. Harvey
Editor

Contents

Carbon Steels

Carbon steels are designated by an identical AISI or SAE four-digit number, in which the last two numbers indicate the approximate middle of the carbon range. AISI 1010, for example, has a carbon content of 0.08 to 0.13%. The initial two digits of the AISI/SAE numbers are also significant. The number 10 indicates nonresulfurized grades. Similarly, grade numbers beginning with 11 denote resulfurized grades. Resulfurized and rephosphorized grades are indicated by the number 12, and nonresulfurized grades with a maximum manganese content of more than 1.00%, by the number 15.

An L between the second and third digits denotes a leaded steel, while B denotes a boron steel. The suffix H after the four digits indicates a steel produced to prescribed hardenability limits.

Types

During the manufacture of steel, oxygen is used to remove excess carbon from the molten metal and to attain the carbon content desired for the finished steel. Carbon and oxygen combine to form carbon monoxide, which bubbles to the surface of the melt. If the excess oxygen is not removed, the gaseous product continues to evolve as the steel solidifies.

If little or no gas evolves during solidification, the steel is deoxidized. This deoxidized steel is termed killed, because it lies quietly in the ingot mold. Increasing degrees of gas evolution result in semikilled, rimmed, and capped steels.

Killed steels are characterized by good uniformity of chemical composition and mechanical properties. They are generally specified when a homogeneous structure is desired. Any steel chosen for rods with more than 0.23% carbon is usually killed steel.

Semikilled steels have a structure which is intermediate in homogeneity between killed and rimmed steels. Carbon contents of these steels are usually up to 0.24%.

Rimmed steels are produced by carefully controlling the addition of small quantities of deoxidizers to the ladle. Rimming is caused by gases continuing to evolve after the ingot is teemed and as the steel solidifies. This action results in a rim or case of essentially carbon-free ferrite on the ingot with a minimum of blowholes and oxide inclusions. Rimmed steel products, particularly cold rolled sheet, can be used to advantage whenever extensive cold forming or superior surface appearance is required. Because carbon contents in excess of 0.25% or manganese contents greater than 0.60% prevent proper rimming action, only low-carbon steels are made as rimmed steels.

Capped steels, although similar to rimmed steels, differ in that the rimming action is stopped at a specified point during the solidification process. A capped steel has a low-carbon rim typical of a rimmed steel, but the uniformity of composition and mechanical properties in the center that might be expected from a killed steel ingot. This combination of properties makes capped steels particularly well suited for applications involving cold forming or cold heading.

Selection

The feasibility of using a carbon steel is determined by assessing whether or not it is metallurgically suitable for the application. Evaluations may include tensile and fatigue strengths, impact resistance, size of the part compared with the need for through hardenability, fabricability, ductility and/or machinability, potential of heat treating, service temperature of the part, and corrosion resistance.

When the desired characteristics can be obtained with a plain carbon grade, most users select this less costly steel. If critical strength requirements or other specified needs are beyond the inherent capabilities of carbon steel, then alloy steel is the obvious choice. One consideration worth noting is that most carbon steels are not through hardening on heat treatment, except in relatively thin sections.

The selection of a specific grade of steel is a more complex decision. Grade selection should be undertaken in collaboration with the supplier, who can call upon the expertise of its metallurgists and broad practical experience, as well as the experience of other customers who have put carbon or alloy steel sheet and bar to the same or similar use.

Classification

The numerical classifications of low-, medium-, and high-carbon steels may vary with individual mills, but they generally fall into the following approximate ranges.

Low-carbon steels (AISI 1005 to 1026) have lower carbon content by definition. Because ease of formability is directly related to hardness (carbon content) of the sheet or bar, these softer steels are more ductile—better for operations involving some degree of cold forming, drawing, bending, punching, or swaging.

Low-carbon steels are used for cold heading, a major process for fastener production, and deep drawing. They are also commonly found in machined and welded components. Because they are relatively inexpensive, the low-carbon steels are a popular choice when great strength is not needed. Tensile strengths range from 300 to 440 MPa (43 to 64 ksi).

Although ductile, low-carbon steels are not very resistant to wear. However, wear resistance can be obtained in low-carbon grades by increasing the carbon content of the surface (carburizing). This is done by heating at elevated temperature in a controlled, carbon-rich atmosphere. Low-carbon steels that are commonly carburized include AISI 1015, 1018, 1020, and 1117.

Medium-carbon steels (AISI 1029 to 1053) are often selected where higher strength is required. These can be thermally treated for even greater strength. Tensile strengths range from 470 to 620 MPa (68 to 90 ksi). Used in larger parts and forgings, these bar grades are also among the most widely used steels for machined components. AISI 1040 and its modifications are specified extensively in the automotive industry for bolts, connecting rods, crankshafts, and tubing. AISI 1050 is frequently used for axles, gears, and heavier forgings, where slightly higher hardness and wear resistance are required.

High-carbon steels (AISI 1055 to 1095) are specified when the highest available strength is needed in the carbon range. Because of superior surface hardness, they provide better wear resistance than the plain carbon steels. When properly heat treated these grades are highly wear resistant. In addition, high-carbon steels offer longer service life and maximum response to heat treatment. Tensile strengths range from 650 to 830 MPa (94 to 120 ksi).

Ductility of the high-carbon steels is lower than that of the low- and medium-carbon steels, but cold formability of high-carbon steels can be improved through thermal treatment. However, this treatment reduces strength and hardness.

None of the 10XX series grades contain more than 1% manganese. For greater hardenability, the 15XX series grades (AISI 1513 to 1572) are available with maximum manganese contents of 1.05 to 1.65%.

Free-machining grades are intended for use where improved machinability is desired, compared to that for carbon steels of similar carbon and manganese contents. Machinability refers to the effects of hardness, strength, ductility, grain size, microstructure, and chemical composition on cutting tool wear, chip formation, ease of metal removal, and surface finish quality of the steel being cut.

Free-machining steels contain one or more additives which enhance machining characteristics and lower machining costs. The lower costs result from either the increased production through greater machining speeds and improved tool life, or the elimination of secondary operations through an improvement in surface finish. The addition of bismuth, selenium, or tellurium enhances the machinability of free-machining steels. Sulfur and phosphorus additions cause a reduction in cold forming properties, weldability, and forging characteristics. Lead additions, however, have very little effect on those characteristics.

The resulfurized 11XX series grades (AISI 1108 to 1151) provide improved machinability with increased sulfur content in amounts up to 0.33% in some grades. The 12XX series grades (AISI 1211 to 1215) are both resulfurized and rephosphorized to increase chip control.

Leaded carbon steels in the free-machining grades contain from 0.15 to 0.35% lead. Serving as an internal lubricant, lead reduces friction and the buildup of heat between the cutting edge of the tool and the work. Leaded steels are most commonly used when a large amount of machining is necessary to produce a finished part. An increase of 25% or more in productivity may result from the use of leaded steels.

Hardenability

Hardenability in plain carbon steels can be improved through the use of additives or by controlling the chemical composition and steelmaking practice (H-steels). Boron is added, usually to medium-carbon grades, to increase the depth of hardening when quenched. H-steels are steels that can be ordered to specific and finite hardenability ranges. These range limits, upper and lower limits from the center or core to the surface, are called H-bands. The H-steel classification applies mainly to alloy steels, but also includes a number of carbon grades. For more information refer to the alloy steel introduction which includes an explanation on the use of hardenability bands.

Quality Designations

There are four quality designations a customer may specify for hot and cold rolled sheet: commercial quality (CQ), drawing quality (DQ), drawing quality special killed (DQSK), and structural quality (SQ). Hot rolled band is available in CQ, DQ, and DQSK. ASTM A568 gives the general requirements for hot rolled sheet and strip and cold rolled sheet carbon steel, and ASTM A29 provides the general requirements for hot rolled and cold finished carbon steel bars.

Commercial quality sheet and bands are usually produced from rimmed, capped, or semikilled steel. Because of the segregation that occurs in the solidification of these types of steels, CQ materials are less ductile than either DQ or DQSK materials and can be expected to show wider variations in mechanical properties and chemical composition. In addition, some CQ sheet may be subject to loss of ductility with time; bands, however, are not subject to strain aging because they are not processed. If sheet steel must be essentially free from stretcher strains during fabrication, it should be roller leveled just before the forming operation.

Drawing quality material is produced from specially selected steel, then carefully processed to result in more uniform drawing properties than commercial quality material. Drawing quality sheet and bands are more ductile and more uniform in chemical composition than CQ material. Parts that are too difficult for the forming properties of CQ sheet are made from DQ sheet.

Because of aging, time delays between drawing stages and heat treatment and other processing operations may adversely affect the properties of DQ steel. Therefore, these operations must be carefully controlled either before or during fabrication to provide the required performance.

Drawing quality sheet is subject to coil breaks, stretcher strains, and fluting in the as-rolled condition. If DQ sheet is to remain essentially free of stretcher strains through fabrication, it should be roller leveled by the customer just before forming.

Drawing quality special-killed materials are rolled from aluminum-killed steel which is produced by special steelmaking and processing practices. Because these steels offer forming characteristics superior to commercial and drawing quality materials, they are excellent for use in severe drawing operations. DQSK steels are recommended for processes in which delays between draws would detrimentally affect the drawing performance of CQ and DQ materials, or where roller leveling equipment is not available.

Structural quality sheet is ordered when specific mechanical or structural properties are required. Orders usually specify hardness or tensile properties. When indicating mechanical properties, the customer should carefully consider compatibility of the properties with forming requirements. ASTM A570 is the standard specification for structural quality hot rolled sheet, and ASTM A611 is the standard specification for structural quality cold rolled sheet.

REFERENCES

1. J & L Cold Finished Bars, Jones & Laughlin Steel Corporation, Pittsburgh, PA
2. *Modern Steels and Their Properties*, Handbook 3310, Bethlehem Steel Corporation, Bethlehem, PA, March 1980
3. Ryerson Stocks and Services, Joseph T. Ryerson & Sons, 1980
4. *Metals Handbook*, 9th ed., Vol 1, American Society for Metals, 1978
5. Mechanical Properties of Alloy Steel, Adv 1099, Republic Steel Corporation, Cleveland, OH, 1979
6. *Metals Handbook*, 8th ed., Vol 1, American Society for Metals, 1961
7. *Handbook of Spring Design*, Spring Manufacturers Institute, Oak Brook, IL, 1981
8. *Machining Data Handbook*, 3rd ed., Metcut Research Associates, Cincinnati, OH, 1980
9. Cold Finished Steel Bars: Selection and Uses, Stelco, Inc., Toronto, Ontario, Canada
10. *Steel Products Manual: Wire and Rods, Carbon Steel*, American Iron and Steel Institute, Washington, D.C., Sept 1981
11. *Metals Handbook*, 9th ed., Vol 4, American Society for Metals, 1981
12. *Steel Products Manual*: *Alloy Carbon and High Strength Low Alloy Steels,* American Iron and Steel Institute, Washington, D.C., Aug 1977 (Revised April 1981)

Composition of Carbon Steels (Ref 12)

AISI No.	UNS No.	C	Composition(a), % Mn	P	S
Nonresulfurized grades, manganese 1.00% max					
1005 (b)	G10050	0.06 max	0.35 max	0.040 max	0.050 max
1006 (b)	G10060	0.08 max	0.25-0.40	0.040 max	0.050 max
1008	G10080	0.10 max	0.30-0.50	0.040 max	0.050 max
1010	G10100	0.08-0.13	0.30-0.60	0.040 max	0.050 max
1012	G10120	0.10-0.15	0.30-0.60	0.040 max	0.050 max
1015	G10150	0.13-0.18	0.30-0.60	0.040 max	0.050 max
1016	G10160	0.13-0.18	0.60-0.90	0.040 max	0.050 max
1017	G10170	0.15-0.20	0.30-0.60	0.040 max	0.050 max
1018	G10180	0.15-0.20	0.60-0.90	0.040 max	0.050 max
1019	G10190	0.15-0.20	0.70-1.00	0.040 max	0.050 max
1020	G10200	0.18-0.23	0.30-0.60	0.040 max	0.050 max
1021	G10210	0.18-0.23	0.60-0.90	0.040 max	0.050 max
1022	G10220	0.18-0.23	0.70-1.00	0.040 max	0.050 max
1023	G10230	0.20-0.25	0.30-0.60	0.040 max	0.050 max
1025	G10250	0.22-0.28	0.30-0.60	0.040 max	0.050 max
1026	G10260	0.22-0.28	0.60-0.90	0.040 max	0.050 max
1029	G10290	0.25-0.31	0.60-0.90	0.040 max	0.050 max
1030	G10300	0.28-0.34	0.60-0.90	0.040 max	0.050 max
1034	G10340	0.32-0.38	0.50-0.80	0.040 max	0.050 max
1035	G10350	0.32-0.38	0.60-0.90	0.040 max	0.050 max
1037	G10370	0.32-0.38	0.70-1.00	0.040 max	0.050 max
1038	G10380	0.35-0.42	0.60-0.90	0.040 max	0.050 max
1039	G10390	0.37-0.44	0.70-1.00	0.040 max	0.050 max
1040	G10400	0.37-0.44	0.60-0.90	0.040 max	0.050 max
1042	G10420	0.40-0.47	0.60-0.90	0.040 max	0.050 max
1043	G10430	0.40-0.47	0.70-1.00	0.040 max	0.050 max
1044	G10440	0.43-0.50	0.30-0.60	0.040 max	0.050 max
1045	G10450	0.43-0.50	0.60-0.90	0.040 max	0.050 max
1046	G10460	0.43-0.50	0.70-1.00	0.040 max	0.050 max
1049	G10490	0.46-0.53	0.60-0.90	0.040 max	0.050 max
1050	G10500	0.48-0.55	0.60-0.90	0.040 max	0.050 max
1053	G10530	0.48-0.55	0.70-1.00	0.040 max	0.050 max
1055	G10550	0.50-0.60	0.60-0.90	0.040 max	0.050 max
1059 (b)	G10590	0.55-0.65	0.50-0.80	0.040 max	0.050 max
1060	G10600	0.55-0.65	0.60-0.90	0.040 max	0.050 max
1064 (b)	G10640	0.60-0.70	0.50-0.80	0.040 max	0.050 max
1065 (b)	G10650	0.60-0.70	0.60-0.90	0.040 max	0.050 max
1069 (b)	G10690	0.65-0.75	0.40-0.70	0.040 max	0.050 max
1070	G10700	0.65-0.75	0.60-0.90	0.040 max	0.050 max
1074	G10740	0.70-0.80	0.50-0.80	0.040 max	0.050 max
1078	G10780	0.72-0.85	0.30-0.60	0.040 max	0.050 max
1080	G10800	0.75-0.88	0.60-0.90	0.040 max	0.050 max
1084	G10840	0.80-0.93	0.60-0.90	0.040 max	0.050 max
1086 (b)	G10860	0.80-0.93	0.30-0.50	0.040 max	0.050 max
1090	G10900	0.85-0.98	0.60-0.90	0.040 max	0.050 max
1095	G10950	0.90-1.03	0.30-0.50	0.040 max	0.050 max
Nonresulfurized grades, manganese greater than 1.00% max					
1513	G15130	0.10-0.16	1.10-1.40	0.040 max	0.050 max
1522	G15220	0.18-0.24	1.10-1.40	0.040 max	0.050 max
1524	G15240	0.19-0.25	1.35-1.65	0.040 max	0.050 max
1526	G15260	0.22-0.29	1.10-1.40	0.040 max	0.050 max
1527	G15270	0.22-0.29	1.20-1.50	0.040 max	0.050 max
1541	G15410	0.36-0.44	1.35-1.65	0.040 max	0.050 max
1547	G15470	0.43-0.51	1.35-1.65	0.040 max	0.050 max
1548	G15480	0.44-0.52	1.10-1.40	0.040 max	0.050 max
1551	G15510	0.45-0.56	0.85-1.15	0.040 max	0.050 max
1552	G15520	0.47-0.55	1.20-1.50	0.040 max	0.050 max
1561	G15610	0.55-0.65	0.75-1.05	0.040 max	0.050 max
1566	G15660	0.60-0.71	0.85-1.15	0.040 max	0.050 max
Free-machining grades, resulfurized					
1108	G11080	0.08-0.13	0.50-0.80	0.040 max	0.08-0.13
1109	G11090	0.08-0.13	0.60-0.90	0.040 max	0.08-0.13
1110	G11100	0.08-0.13	0.30-0.60	0.040 max	0.08-0.13
1116	G11160	0.14-0.20	1.10-1.40	0.040 max	0.16-0.23
1117	G11170	0.14-0.20	1.00-1.30	0.040 max	0.08-0.13
1118	G11180	0.14-0.20	1.30-1.60	0.040 max	0.08-0.13
1119	G11190	0.14-0.20	1.00-1.30	0.040 max	0.24-0.33
1137	G11370	0.32-0.39	1.35-1.65	0.040 max	0.08-0.13
1139	G11390	0.35-0.43	1.35-1.65	0.040 max	0.13-0.20
1140	G11400	0.37-0.44	0.70-1.00	0.040 max	0.08-0.13
1141	G11410	0.37-0.45	1.35-1.65	0.040 max	0.08-0.13
1144	G11440	0.40-0.48	1.35-1.65	0.040 max	0.24-0.33
1146	G11460	0.42-0.49	0.70-1.00	0.040 max	0.08-0.13
1151	G11510	0.48-0.55	0.70-1.00	0.040 max	0.08-0.13
Free-machining grades, resulfurized and rephosphorized					
1211	G12110	0.13 max	0.60-0.90	0.07-0.12	0.10-0.15
1212	G12120	0.13 max	0.70-1.00	0.07-0.12	0.16-0.23
1213	G12130	0.13 max	0.70-1.00	0.07-0.12	0.24-0.33
1215	G12150	0.09 max	0.75-1.05	0.04-0.09	0.26-0.35
12L14 (c)	G12144	0.15 max	0.85-1.15	0.04-0.09	0.26-0.35

(a) The following notes refer to boron, copper, lead and silicon additions: Boron: standard killed carbon steels, which are generally fine grain, may be produced with a boron treatment addition to improve hardenability. Such steels are produced to a range of 0.0005 to 0.003% B. These steels are identified by inserting the letter B between the second and third numerals of the AISI or SAE number, such as 10B46. Copper: when copper is required, 0.20% min is generally specified. Lead: standard carbon steels can be produced with a lead range of 0.15 to 0.35% to improve machinability. Such steels are identified by inserting the letter L between the second and third numerals of the AISI or SAE number, such as 12L15 and 10L45. Silicon: it is not common practice to produce the 12XX series of resulfurized and rephosphorized steels to specified limits for silicon because of its adverse effect on machinability. When silicon ranges or limits are required for resulfurized or nonresulfurized steels, however, these values apply: a range of 0.08% for maximum silicon contents up to 0.15% inclusive, a range of 0.10% for maximum silicon contents over 0.15 to 0.20% inclusive, a range of 0.15% for maximum silicon contents over 0.20 to 0.30% inclusive, and a range of 0.20% for maximum silicon contents over 0.30 to 0.60% inclusive. Example: maximum silicon content is 0.25%, range is 0.10 to 0.25%. (b) Standard grades for wire rod and wire only. (c) 0.15 to 0.35% lead

Estimated Mechanical Properties and Machinability of Carbon Steel Bar (Ref 1)

All values are estimated minimum values; all SAE 1100 series steels are rated on the basis of 0.10% max silicon or coarse-grain melting practice; the mechanical properties shown are expected minimums for the sizes ranging from 19 to 31.8 mm (0.75 to 1.25 in.)

AISI grade	Type of processing	Tensile strength MPa	Tensile strength ksi	Yield strength MPa	Yield strength ksi	Elongation(a), %	Reduction in area, %	Hardness, HB	Machinability rating(b)
1006	Hot rolled	295	43	165	24	30	55	86	...
	Cold drawn	330	48	285	41	20	45	95	50
1008	Hot rolled	305	44	170	24.5	30	55	86	...
	Cold drawn	340	49	285	41.5	20	45	95	55
1010	Hot rolled	325	47	180	26	28	50	95	...
	Cold drawn	365	53	305	44	20	40	105	55
1012	Hot rolled	330	48	185	26.5	28	50	95	...
	Cold drawn	370	54	310	45	19	40	105	55
1015	Hot rolled	345	50	190	27.5	28	50	101	...
	Cold drawn	385	56	325	47	18	40	111	60
1016	Hot rolled	380	55	205	30	25	50	111	...
	Cold drawn	420	61	350	51	18	40	121	70
1017	Hot rolled	365	53	200	29	26	50	105	...
	Cold drawn	405	59	340	49	18	40	116	65
1018	Hot rolled	400	58	220	32	25	50	116	...
	Cold drawn	440	64	370	54	15	40	126	70
1019	Hot rolled	405	59	225	32.5	25	50	116	...
	Cold drawn	455	66	380	55	15	40	131	70
1020	Hot rolled	380	55	205	30	25	50	111	...
	Cold drawn	420	61	350	51	15	40	121	65
1021	Hot rolled	420	61	230	33	24	48	116	...
	Cold drawn	470	68	395	57	15	40	131	70
1022	Hot rolled	425	62	235	34	23	47	121	...
	Cold drawn	475	69	400	58	15	40	137	70
1023	Hot rolled	385	56	215	31	25	50	111	...
	Cold drawn	425	62	360	52.5	15	40	121	65
1524	Hot rolled	510	74	285	41	20	42	149	...
	Cold drawn	565	82	475	69	12	35	163	60
1025	Hot rolled	400	58	220	32	25	50	116	...
	Cold drawn	440	64	370	54	15	40	126	65
1026	Hot rolled	440	64	240	35	24	49	126	...
	Cold drawn	490	71	415	60	15	40	143	75
1527	Hot rolled	515	75	285	41	18	40	149	...
	Cold drawn	570	83	485	70	12	35	163	65
1030	Hot rolled	470	68	260	37.5	20	42	137	...
	Cold drawn	525	76	440	64	12	35	149	70
1035	Hot rolled	495	72	270	39.5	18	40	143	...
	Cold drawn	550	80	460	67	12	35	163	65
1536	Hot rolled	570	83	315	45.5	16	40	163	...
	Cold drawn	635	92	535	77.5	12	35	187	55
1037	Hot rolled	510	74	280	40.5	18	40	143	...
	Cold drawn	565	82	475	69	12	35	167	65
1038	Hot rolled	515	75	285	41	18	40	149	...
	Cold drawn	570	83	485	70	12	35	163	65
1039	Hot rolled	545	79	300	43.5	16	40	156	...
	Cold drawn	605	88	510	74	12	35	179	60
1040	Hot rolled	525	76	290	42	18	40	149	...
	Cold drawn	585	85	490	71	12	35	170	60
1541	Hot rolled	635	92	350	51	15	40	187	...
	Cold drawn	705	102.5	600	87	10	30	207	45
	Annealed, cold drawn	650	94	550	80	10	45	184	60
1042	Hot rolled	550	80	305	44	16	40	163	...
	Cold drawn	615	89	515	75	12	35	179	60
	Normalized, cold drawn	585	85	505	73	12	45	179	70
1043	Hot rolled	565	82	310	45	16	40	163	...
	Cold drawn	625	91	530	77	12	35	179	60
	Normalized, cold drawn	600	87	515	75	12	45	179	70
1044	Hot rolled	550	80	305	44	16	40	163	...
1045	Hot rolled	565	82	310	45	16	40	163	...
	Cold drawn	625	91	530	77	12	35	179	55
	Annealed, cold drawn	585	85	505	73	12	45	170	65
1046	Hot rolled	585	85	325	47	15	40	170	...
	Cold drawn	650	94	545	79	12	35	187	55
	Annealed, cold drawn	620	90	515	75	12	45	179	65
1547	Hot rolled	650	94	360	52	15	30	192	...
	Cold drawn	710	103	605	88	10	28	207	40
	Annealed, cold drawn	655	95	585	85	10	35	187	45
1548	Hot rolled	660	96	365	53	14	33	197	...
	Cold drawn	735	106.5	615	89.5	10	28	217	45
	Annealed, cold drawn	645	93.5	540	78.5	10	35	192	50
1049	Hot rolled	600	87	330	48	15	35	179	...
	Cold drawn	670	97	560	81.5	10	30	197	45
	Annealed, cold drawn	635	92	530	77	10	40	187	55
1050	Hot rolled	620	90	340	49.5	15	35	179	...
	Cold drawn	690	100	580	84	10	30	197	45
	Annealed, cold drawn	655	95	550	80	10	40	189	55

(continued)

(a) In 50 mm (2 in.). (b) Based on cold drawn AISI 1212 steel as 100% average machinability

Estimated Mechanical Properties and Machinability of Carbon Steel Bar (Ref 1) (continued)

All values are estimated minimum values; all SAE 1100 series steels are rated on the basis of 0.10% max silicon or coarse-grain melting practice; the mechanical properties shown are expected minimums for the sizes ranging from 19 to 31.8 mm (0.75 to 1.25 in.)

AISI grade	Type of processing	Tensile strength MPa	Tensile strength ksi	Yield strength MPa	Yield strength ksi	Elongation(a), %	Reduction in area, %	Hardness, HB	Machinability rating(b)
1552	Hot rolled	745	108	410	59.5	12	30	217	...
	Annealed, cold drawn	675	98	570	83	10	40	193	50
1055	Hot rolled	650	94	355	51.5	12	30	192	...
	Annealed, cold drawn	660	96	560	81	10	40	197	55
1060	Hot rolled	675	98	370	54	12	30	201	...
	Spheroidized annealed, cold drawn	620	90	485	70	10	45	183	60
1064	Hot rolled	670	97	370	53.5	12	30	201	...
	Spheroidized annealed, cold drawn	615	89	475	69	10	45	183	60
1065	Hot rolled	690	100	380	55	12	30	207	...
	Spheroidized annealed, cold drawn	635	92	490	71	10	45	187	60
1070	Hot rolled	705	102	385	56	12	30	212	...
	Spheroidized annealed, cold drawn	640	93	495	72	10	45	192	55
1074	Hot rolled	725	105	400	58	12	30	217	...
	Spheroidized annealed, cold drawn	650	94	505	73	10	40	192	55
1078	Hot rolled	690	100	380	55	12	30	207	...
	Spheroidized annealed, cold drawn	650	94	500	72.5	10	40	192	55
1080	Hot rolled	770	112	425	61.5	10	25	229	...
	Spheroidized annealed, cold drawn	675	98	515	75	10	40	192	45
1084	Hot rolled	820	119	450	65.5	10	25	241	...
	Spheroidized annealed, cold drawn	690	100	530	77	10	40	192	45
1085	Hot rolled	835	121	460	66.5	10	25	248	...
	Spheroidized annealed, cold drawn	695	100.5	540	78	10	40	192	45
1086	Hot rolled	770	112	425	61.5	10	25	229	...
	Spheroidized annealed, cold drawn	670	97	510	74	10	40	192	45
1090	Hot rolled	840	122	460	67	10	25	248	...
	Spheroidized annealed, cold drawn	695	101	540	78	10	40	197	45
1095	Hot rolled	825	120	455	66	10	25	248	...
	Spheroidized annealed, cold drawn	680	99	525	76	10	40	197	45
1211	Hot rolled	380	55	230	33	25	45	121	...
	Cold drawn	515	75	400	58	10	35	163	95
1212	Hot rolled	385	56	230	33.5	25	45	121	...
	Cold drawn	540	78	415	60	10	35	167	100
1213	Hot rolled	385	56	230	33.5	25	45	121	...
	Cold drawn	540	78	415	60	10	35	167	135
12L14	Hot rolled	395	57	235	34	22	45	121	...
	Cold drawn	540	78	415	60	10	35	163	160
1108	Hot rolled	345	50	190	27.5	30	50	101	...
	Cold drawn	385	56	325	47	20	40	121	80
1109	Hot rolled	345	50	190	27.5	30	50	101	...
	Cold drawn	385	56	325	47	20	40	121	80
1117	Hot rolled	425	62	235	34	23	47	121	...
	Cold drawn	475	69	400	58	15	40	137	90
1118	Hot rolled	450	65	250	36	23	47	131	...
	Cold drawn	495	72	420	61	15	40	143	85
1119	Hot rolled	425	62	235	34	23	47	121	...
	Cold drawn	475	69	400	58	15	40	137	100
1132	Hot rolled	570	83	315	45.5	16	40	167	...
	Cold drawn	635	92	530	77	12	35	183	75
1137	Hot rolled	605	88	330	48	15	35	179	...
	Cold drawn	675	98	565	82	10	30	197	70
1140	Hot rolled	545	79	300	43.5	16	40	156	...
	Cold drawn	605	88	510	74	12	35	170	70
1141	Hot rolled	650	94	355	51.5	15	35	187	...
	Cold drawn	725	105.1	605	88	10	30	212	70
1144	Hot rolled	670	97	365	53	15	35	197	...
	Cold drawn	745	108	620	90	10	30	217	80
1145	Hot rolled	585	85	325	47	15	40	170	...
	Cold drawn	650	94	550	80	12	35	187	65
1146	Hot rolled	585	85	325	47	15	40	170	...
	Cold drawn	650	94	550	80	12	35	187	70
1151	Hot rolled	635	92	350	50.5	15	35	187	...
	Cold drawn	705	102	595	86	10	30	207	65

(a) In 50 mm (2 in.). (b) Based on cold drawn AISI 1212 steel as 100% average machinability

AISI 1005, 1006, 1008

AISI 1005, 1006, 1008: Chemical Composition

AISI grade	C max	Mn	P max	S max
	Chemical composition, %			
1005 (a)	0.06	0.35 max	0.040	0.050
1006 (a)	0.08	0.25-0.40	0.040	0.050
1006	0.08	0.25-0.45	0.040	0.050
1008	0.10	0.30-0.50	0.040	0.050

(a) Standard grades for wire rod and wire only

Characteristics. AISI grades 1005, 1006, and 1008 are usually produced as rimmed, capped, semikilled, and fully killed steels. Rimmed steels of these grades have exceptionally good cold formability. The soft rim enhances drawability and surface finish. Because they have relatively low tensile strengths, these grades are used where strength is of minor importance. Aluminum killed steels are very soft, mild steels of low strength and high ductility, which are easy to forge or to form and draw cold. Grades 1005, 1006, and 1008 are not heat treatable in the usual manner (by quenching and tempering), but can be hardened or strengthened by cold working. These steels can be restored to their initial soft condition by annealing. Weldability (spot, projection, butt, and fusion) and brazeability are excellent.

Typical Uses. Parts made from grades 1005, 1006, and 1008 include extruded, cold headed, cold upset, and cold pressed parts and forms, and a large variety of parts requiring severe bending and welding. For more severe drawing operations, fully killed steel should be used to eliminate or minimize the strain-aging behavior of rimmed steels. AISI grade 1006 steel can be used in magnet core applications.

AISI 1005: Similar Steels (U.S. and/or Foreign). UNS G10050; ASTM A29, A510; MIL SPEC MIL-S-11310 (CS1005); SAE J403, J412

AISI 1006: Similar Steels (U.S. and/or Foreign). UNS G10060; ASME 5041; ASTM A29, A510, A545; FED QQ-W-461; MIL SPEC MIL-S-11310 (CS1006); SAE J403, J412, J414

AISI 1008: Similar Steels (U.S. and/or Foreign). UNS G10080; ASTM A29, A108, A510, A519, A545, A549, A575, A576; FED QQ-S-637 (C1008), QQ-S-698 (C1008); MIL SPEC MIL-S-11310 (CS1008); SAE J403, J412, J414; (W. Ger.) DIN 1.0204; (Ital.) UNI CB 10 FU

AISI 1008: Approximate Critical Points

Transformation point	Temperature(a) °C	°F
Ac_1	730	1350
Ac_3	875	1605
Ar_3	855	1570
Ar_1	680	1255

(a) On heating or cooling at 28 °C (50 °F) per hour

Physical Properties

AISI 1005, 1006, 1008: Average Coefficients of Linear Thermal Expansion (Ref 4)

Temperature range °C	°F	Coefficient μm/m·K	μin./in.·°F
0-100	32-212	12.6	7.0
0-200	32-390	13.1	7.3
0-300	32-570	13.5	7.5
0-400	32-750	13.7	7.6
0-500	32-930	14.2	7.9
0-600	32-1110	14.6	8.1
0-700	32-1290	14.9	8.3
0-800	32-1470	16.6	9.2
0-1000	32-1830	13.7	7.6

Material composition: 0.06% carbon, 0.38% manganese, 0.55% nickel, 0.02% chromium, 0.03% molybdenum

AISI 1005, 1006, 1008: Mean Apparent Specific Heat (Ref 4)

Temperature range °C	°F	Material A(a) J/kg·K	Material A(a) Btu/lb·°F	Material B(b) J/kg·K	Material B(b) Btu/lb·°F
50-100	120-212	481	0.115	481	0.115
150-200	300-390	519	0.124	523	0.125
200-250	390-480	536	0.128	544	0.130
250-300	480-570	553	0.132	557	0.133
300-350	570-660	574	0.137	569	0.136
350-400	660-750	595	0.142	595	0.142
450-500	840-930	662	0.158	662	0.158
550-600	1020-1110	754	0.180	858	0.205
650-700	1200-1290	867	0.207	1139	0.272
700-750	1290-1380	1105	0.264	959	0.229
750-800	1380-1470	875	0.209	816	0.195
850-900	1560-1650	846	0.202	...	...

(a) Material composition: 0.06% carbon, 0.38% manganese. (b) Material composition: 0.08% carbon, 0.31% manganese

AISI 1008: Electrical Resistivity and Thermal Conductivity

Temperature °C	°F	Electrical resistivity, μΩ·m (a)	(b)	Thermal conductivity W/m·K (b)	W/m·K (c)	Btu/ft·h·°F (b)	Btu/ft·h·°F (c)
0	32	...	...	62.8	65.2	36.3	37.7
20	70	0.130	0.142	...	...	...	...
100	212	0.178	0.190	57.6	60.2	33.3	34.8
200	390	0.252	0.263	53.1	54.7	30.7	31.6
300	570	...	...	49.1	...	28.4	...
400	750	0.448	0.458	45.7	45.2	26.4	26.1
500	930	...	...	40.8	...	23.6	...
600	1110	0.725	0.734	36.7	36.3	21.2	21.0
700	1290	0.898	0.905	33.2	...	19.2	...
800	1470	1.073	1.081	28.4	28.4	16.4	16.4
900	1650	1.124	1.130	...	...	...	...
1000	1830	1.160	1.165	27.3	27.3	15.8	15.8
1100	2010	1.189	1.193	...	...	...	...
1200	2190	1.126	1.220	29.9	...	17.3	...
1300	2370	1.241	1.244	...	...	...	...

(a) Material composition: 0.06% carbon, 0.38% manganese. (b) Material composition: 0.08% carbon, 0.31% manganese. (c) Material composition: 0.06% carbon, 0.4% manganese

AISI 1005, 1006, 1008: Density (Ref 4)

7.872 g/cm³ .... 0.2844 lb/in.³

Material composition: 0.06% carbon, 0.01% silicon, 0.38% manganese; annealed at 925 °C (1700 °F)

Mechanical Properties

AISI 1006, 1008: Tensile Properties and Machinability

Product form	Size round or thickness, mm	Size round or thickness, in.	Tensile strength, MPa	Tensile strength, ksi	Yield strength, MPa	Yield strength, ksi	Elongation(a), %	Reduction in area, %	Hardness, HB	Average machinability rating(b)
AISI 1006										
Hot rolled bar (Ref 1)	19-32	0.75-1.25	295	43	165	24	30	55	86	...
Cold drawn bar (Ref 1)	19-32	0.75-1.25	330	48	285	41	20	45	95	50
CQ, DQ, and DQSK sheet (Ref 11)	1.6-5.8	0.064-0.229	305-360	44-52	180-240	26-35	32-48	...	...	...
AISI 1008										
Hot rolled bar (Ref 1)	19-32	0.75-1.25	305	44	170	24.5	30	55	86	...
Cold drawn bar (Ref 1)	19-32	0.75-1.25	340	49	285	41.5	20	45	95	55
CQ, DQ, and DQSK sheet (Ref 11)	1.6-5.8	0.064-0.229	303-358	44-52	180-240	26-35	32-48	...	...	...

(a) In 50 mm (2 in.). (b) Based on AISI 1212 steel as 100% average machinability

Machining Data (Ref 8)

AISI 1005, 1006, 1008: Turning (Single Point and Box Tools)

Depth of cut		M2 and M3 high speed steel				Uncoated carbide						Coated carbide			
		Speed		Feed		Speed, brazed		Speed, inserted		Feed		Speed		Feed	
mm	in.	m/min	ft/min	mm/rev	in./rev	m/min	ft/min	m/min	ft/min	mm/rev	in./rev	m/min	ft/min	mm/rev	in./rev
Hardness, 85 to 125 HB															
1	0.040	56	185	0.18	0.007	165(a)	535(a)	215(a)	700(a)	0.18	0.007	320(b)	1050(b)	0.18	0.007
4	0.150	44	145	0.40	0.015	135(c)	435(c)	165(c)	540(c)	0.50	0.020	215(d)	700(d)	0.40	0.015
8	0.300	35	115	0.50	0.020	105(c)	340(c)	130(c)	420(c)	0.75	0.030	170(d)	550(d)	0.50	0.020
16	0.625	27	90	0.75	0.030	81(c)	265(c)	100(c)	330(c)	1.00	0.040	...	...	...	...
Hardness, 125 to 175 HB															
1	0.040	46	150	0.18	0.007	150(a)	485(a)	195(a)	640(a)	0.18	0.007	290(b)	950(b)	0.18	0.007
4	0.150	38	125	0.40	0.015	125(c)	410(c)	150(c)	500(c)	0.50	0.020	190(d)	625(d)	0.40	0.015
8	0.300	30	100	0.50	0.020	100(c)	320(c)	120(c)	390(c)	0.75	0.030	150(d)	500(d)	0.50	0.020
16	0.625	24	80	0.75	0.030	75(c)	245(c)	95(c)	305(c)	1.00	0.040	...	...	...	...
Hardness, 175 to 225 HB															
1	0.040	44	145	0.18	0.007	140(a)	460(a)	175(a)	570(a)	0.18	0.007	260(b)	850(b)	0.18	0.007
4	0.150	35	115	0.40	0.015	115(c)	385(c)	135(c)	450(c)	0.50	0.020	170(d)	550(d)	0.40	0.015
8	0.300	29	95	0.50	0.020	90(c)	300(c)	105(c)	350(c)	0.75	0.030	135(d)	450(d)	0.50	0.020
16	0.625	23	75	0.75	0.030	72(c)	235(c)	81(c)	265(c)	1.00	0.040	...	...	...	...

(a) Carbide tool material: C-7. (b) Carbide tool material: CC-7. (c) Carbide tool material: C-6. (d) Carbide tool material: CC-6

AISI 1005, 1006, 1008: Turning (Cutoff and Form Tools)

Tool material	Speed, m/min (ft/min)	Feed per revolution for cutoff tool width of:						Feed per revolution for form tool wodth of:							
		1.5 mm (0.062 in.)		3 mm (0.125 in.)		6 mm (0.25 in.)		12 mm (0.5 in.)		18 mm (0.75 in.)		25 mm (1 in.)		50 mm (2 in.)	
		mm	in.	mm	in.	mm	in.	mm	in.	mm	in.	mm	in.	mm	in.
Hardness, 85 to 125 HB															
M2 and M3 high speed steel	43 (140)	0.038	0.0015	0.050	0.002	0.061	0.0024	0.046	0.0018	0.041	0.0016	0.036	0.0014	0.028	0.0011
C-6 carbide	135 (450)	0.038	0.0015	0.050	0.002	0.061	0.0024	0.046	0.0018	0.041	0.0016	0.036	0.0014	0.028	0.0011
Hardness, 125 to 175 HB															
M2 and M3 high speed steel	37 (120)	0.038	0.0015	0.050	0.002	0.061	0.0024	0.046	0.0018	0.041	0.0016	0.036	0.0014	0.028	0.0011
C-6 carbide	120 (390)	0.038	0.0015	0.050	0.002	0.061	0.0024	0.046	0.0018	0.041	0.0016	0.036	0.0014	0.028	0.0011
Hardness, 175 to 225 HB															
M2 and M3 high speed steel	30 (100)	0.033	0.0013	0.046	0.0018	0.056	0.0022	0.041	0.0016	0.036	0.0014	0.030	0.0012	0.023	0.0009
C-6 carbide	100 (325)	0.033	0.0013	0.046	0.0018	0.056	0.0022	0.041	0.0016	0.036	0.0014	0.030	0.0012	0.023	0.0009

Machining Data (Ref 8) (continued)

AISI 1005, 1006, 1008: Face Milling

Depth of cut		M2 and M7 high speed steel				Uncoated carbide						Coated carbide			
		Speed		Feed/tooth		Speed, brazed		Speed, inserted		Feed/tooth		Speed		Feed/tooth	
mm	in.	m/min	ft/min	mm	in.	m/min	ft/min	m/min	ft/min	mm	in.	m/min	ft/min	mm	in.
Hardness, 85 to 125 HB															
1	0.040	70	230	0.20	0.008	220(a)	725(a)	245(a)	800(a)	0.20	0.008	365(b)	1200(b)	0.20	0.008
4	0.150	53	175	0.30	0.012	160(a)	525(a)	190(a)	625(a)	0.30	0.012	245(b)	810(b)	0.30	0.012
8	0.300	41	135	0.40	0.016	120(c)	400(c)	150(c)	490(c)	0.40	0.016	195(d)	635(d)	0.40	0.016
Hardness, 125 to 175 HB															
1	0.040	64	210	0.20	0.008	205(a)	665(a)	220(a)	725(a)	0.20	0.008	330(b)	1075(b)	0.20	0.008
4	0.150	49	160	0.30	0.012	150(a)	500(a)	170(a)	560(a)	0.30	0.012	225(b)	730(b)	0.30	0.012
8	0.300	38	125	0.40	0.016	110(c)	355(c)	135(c)	435(c)	0.40	0.016	170(d)	565(d)	0.40	0.016
Hardness, 175 to 225 HB															
1	0.040	58	190	0.20	0.008	170(a)	550(a)	180(a)	590(a)	0.20	0.008	270(b)	885(b)	0.20	0.008
4	0.150	43	14	0.30	0.012	135(a)	450(a)	150(a)	490(a)	0.30	0.012	195(b)	635(b)	0.30	0.012
8	0.300	34	110	0.40	0.016	95(c)	310(c)	115(c)	380(c)	0.40	0.016	150(d)	495(d)	0.40	0.016

(a) Carbide tool material: C-6. (b) Carbide tool material: CC-6. (c) Carbide tool material: C-5. (d) Carbide tool material: CC-5

AISI 1005, 1006, 1008: Drilling

Tool material	Speed		Feed per revolution for nominal hole diameter of:											
			1.5 mm (0.062 in.)		3 mm (0.125 in.)		6 mm (0.25 in.)		12 mm (0.5 in.)		18 mm (0.75 in.)		25 mm (1 in.)	
	m/min	ft/min	mm	in.	mm	in.	mm	in.	mm	in.	mm	in.	mm	in.
Hardneess, 85 to 125 HB														
M10, M7, and M1 high speed steel	24	80	0.025	0.001	...	...	...	...	...	...	...	...	...	...
	29	95	...	...	0.075	0.003	0.15	0.006	0.25	0.010	0.40	0.015	0.50	0.020
Hardness, 125 to 175 HB														
M10, M7, and M1 high speed steel	23	75	0.025	0.001	...	...	...	...	...	...	...	...	...	...
	26	85	...	...	0.075	0.003	0.13	0.005	0.23	0.009	0.30	0.012	0.40	0.015
Hardness, 175 to 225 HB														
M10, M7, and M1 high speed steel	21	70	0.025	0.001	...	...	...	...	...	...	...	...	...	...
	24	80	...	...	0.075	0.003	0.13	0.005	0.23	0.009	0.30	0.012	0.36	0.014

AISI 1005, 1006, 1008: Planing

Tool material	Depth of cut		Speed		Feed/stroke	
	mm	in.	m/min	ft/min	mm	in.
Hardness, 85 to 125 HB						
M2 and M3 high speed steel	0.1	0.005	18	60	(a)	(a)
	2.5	0.100	24	80	1.25	0.050
	12	0.500	15	50	1.50	0.060
C-6 carbide	0.1	0.005	90	300	(a)	(a)
	2.5	0.100	90	300	2.05	0.080
	12	0.500	90	300	1.50	0.060
Hardness, 125 to 175 HB						
M2 and M3 high speed steel	0.1	0.005	15	50	(a)	(a)
	2.5	0.100	21	70	1.25	0.050
	12	0.500	14	45	1.50	0.060
C-6 carbide	0.1	0.005	90	300	(a)	(a)
	2.5	0.100	90	300	2.05	0.080
	12	0.500	90	300	1.50	0.060
Hardness, 175 to 225 HB						
M2 and M3 high speed steel	0.1	0.005	14	45	(a)	(a)
	2.5	0.100	20	65	1.25	0.050
	12	0.500	12	40	1.50	0.060
C-6 carbide	0.1	0.005	84	275	(a)	(a)
	2.5	0.100	90	300	2.05	0.080
	12	0.500	84	275	1.50	0.060

(a) Feed is 75% the width of the square nose finishing tool

AISI 1005, 1006, 1008: End Milling (Profiling)

Tool material	Depth of cut mm	Depth of cut in.	Speed m/min	Speed ft/min	Feed per tooth for cutter diameter of: 10 mm (0.375 in.) mm	in.	12 mm (0.5 in.) mm	in.	18 mm (0.75 in.) mm	in.	25-50 mm (1-2 in.) mm	in.
Hardness, 85 to 125 HB												
M2, M3, and M7 high speed steel	0.5	0.020	67	220	0.025	0.001	0.050	0.002	0.102	0.004	0.13	0.005
	1.5	0.060	52	170	0.050	0.002	0.075	0.003	0.13	0.005	0.15	0.006
	diam/4	diam/4	46	150	0.025	0.001	0.050	0.002	0.102	0.004	0.13	0.005
	diam/2	diam/2	40	130	0.018	0.0007	0.025	0.001	0.075	0.003	0.102	0.004
C-5 carbide	0.5	0.020	170	550	0.038	0.0015	0.089	0.0035	0.15	0.006	0.18	0.007
	1.5	0.060	130	420	0.063	0.0025	0.102	0.004	0.15	0.006	0.20	0.008
	diam/4	diam/4	110	360	0.050	0.002	0.075	0.003	0.13	0.005	0.15	0.006
	diam/2	diam/2	100	335	0.038	0.0015	0.050	0.002	0.102	0.004	0.13	0.005
Hardness, 125 to 175 HB												
M2, M3, and M7 high speed steel	0.5	0.020	64	210	0.025	0.001	0.050	0.002	0.102	0.004	0.13	0.005
	1.5	0.060	49	160	0.050	0.002	0.075	0.003	0.13	0.005	0.15	0.006
	diam/4	diam/4	43	140	0.025	0.001	0.050	0.002	0.102	0.004	0.13	0.005
	diam/2	diam/2	37	120	0.018	0.0007	0.025	0.001	0.075	0.003	0.102	0.004
C-5 carbide	0.5	0.020	160	520	0.038	0.0015	0.089	0.0035	0.15	0.006	0.18	0.007
	1.5	0.060	120	400	0.063	0.0025	0.102	0.004	0.15	0.006	0.20	0.008
	diam/4	diam/4	105	350	0.050	0.002	0.075	0.003	0.13	0.005	0.15	0.006
	diam/2	diam/2	100	320	0.038	0.0015	0.050	0.002	0.102	0.004	0.13	0.005
Hardness, 175 to 225 HB												
M2, M3, and M7 high speed steel	0.5	0.020	52	170	0.025	0.001	0.050	0.002	0.075	0.003	0.102	0.004
	1.5	0.060	38	125	0.050	0.002	0.075	0.003	0.102	0.004	0.13	0.005
	diam/4	diam/4	34	110	0.025	0.001	0.050	0.002	0.075	0.003	0.102	0.004
	diam/2	diam/2	29	95	0.018	0.0007	0.025	0.001	0.050	0.002	0.075	0.003
C-5 carbide	0.5	0.020	150	485	0.038	0.0015	0.075	0.003	0.13	0.005	0.15	0.006
	1.5	0.060	115	375	0.063	0.0025	0.102	0.004	0.15	0.006	0.18	0.007
	diam/4	diam/4	100	330	0.050	0.002	0.075	0.003	0.13	0.005	0.15	0.006
	diam/2	diam/2	95	310	0.038	0.0015	0.050	0.002	0.102	0.004	0.13	0.005

AISI 1005, 1006, 1008: Reaming

Based on 4 flutes for 3- and 6-mm (0.125 and 0.25 in.) reamers, 6 flutes for 12-mm (0.5-in.) reamers, and 8 flutes for 25-mm (1-in.) and larger reamers

Tool material	Speed m/min	Speed ft/min	Feed per revolution for reamer diameter of: 3 mm (0.125 in.) mm	in.	6 mm (0.25 in.) mm	in.	12 mm (0.5 in.) mm	in.	25 mm (1 in.) mm	in.	35 mm (1.5 in.) mm	in.	50 mm (2 in.) mm	in.
Hardness, 85 to 125 HB														
Roughing														
M1, M2, and M7 high speed steel	35	115	0.102	0.004	0.18	0.007	0.30	0.012	0.50	0.020	0.65	0.025	0.75	0.030
C-2 carbide	37	120	0.102	0.004	0.18	0.007	0.30	0.012	0.50	0.020	0.65	0.025	0.75	0.030
Finishing														
M1, M2, and M7 high speed steel	14	45	0.15	0.006	0.25	0.010	0.40	0.015	0.65	0.025	0.75	0.030	0.90	0.035
C-2 carbide	17	55	0.15	0.006	0.25	0.010	0.40	0.015	0.65	0.025	0.75	0.030	0.90	0.035
Hardness, 125 to 175 HB														
Roughing														
M1, M2, and M7 high speed steel	32	105	0.102	0.004	0.18	0.007	0.30	0.012	0.50	0.020	0.65	0.025	0.75	0.030
C-2 carbide	35	115	0.102	0.004	0.18	0.007	0.30	0.012	0.50	0.020	0.65	0.025	0.75	0.030
Finishing														
M1, M2, and M7 high speed steel	12	40	0.15	0.006	0.25	0.010	0.40	0.015	0.65	0.025	0.75	0.030	0.90	0.035
C-2 carbide	15	50	0.15	0.006	0.25	0.010	0.40	0.015	0.65	0.025	0.75	0.030	0.90	0.035
Hardness, 175 to 225 HB														
Roughing														
M1, M2, and M7 high speed steel	29	95	0.102	0.004	0.18	0.007	0.30	0.012	0.50	0.020	0.65	0.025	0.75	0.030
C-2 carbide	34	110	0.102	0.004	0.18	0.007	0.30	0.012	0.50	0.020	0.65	0.025	0.75	0.030
Finishing														
M1, M2, and M7 high speed steel	11	35	0.13	0.005	0.20	0.008	0.30	0.012	0.50	0.020	0.65	0.025	0.75	0.030
C-2 carbide	14	45	0.13	0.005	0.20	0.008	0.30	0.012	0.50	0.020	0.65	0.025	0.75	0.030

Machining Data (Ref 8) (continued)

AISI 1005, 1006, 1008: Boring

Depth of cut mm	Depth of cut in.	M2 and M3 high speed steel Speed m/min	Speed ft/min	Feed mm/rev	Feed in./rev	Uncoated carbide Speed, brazed m/min	Speed, brazed ft/min	Speed, inserted m/min	Speed, inserted ft/min	Feed mm/rev	Feed in./rev	Coated carbide Speed m/min	Speed ft/min	Feed mm/rev	Feed in./rev
Hardness, 85 to 125 HB															
0.25	0.010	56	185	0.075	0.003	160(a)	525(a)	185(a)	615(a)	0.075	0.003	280(b)	920(b)	0.075	0.003
1.25	0.050	46	150	0.13	0.005	125(c)	415(c)	150(c)	490(c)	0.13	0.005	225(d)	735(d)	0.13	0.005
2.5	0.100	35	115	0.30	0.012	100(e)	320(e)	115(e)	375(e)	0.40	0.015	150(f)	490(f)	0.30	0.012
Hardness, 125 to 175 HB															
0.25	0.010	46	150	0.075	0.003	145(a)	475(a)	170(a)	560(a)	0.075	0.003	255(b)	830(b)	0.075	0.003
1.25	0.050	37	120	0.13	0.005	115(c)	380(c)	135(c)	450(c)	0.13	0.005	205(d)	665(d)	0.13	0.005
2.5	0.100	30	100	0.30	0.012	90(e)	300(e)	105(e)	350(e)	0.40	0.015	135(f)	440(f)	0.30	0.012
Hardness, 175 to 225 HB															
0.25	0.010	44	145	0.075	0.003	130(c)	425(c)	150(c)	500(c)	0.075	0.003	230(d)	750(d)	0.075	0.003
1.25	0.050	35	115	0.13	0.005	105(e)	340(e)	120(e)	400(e)	0.13	0.005	185(f)	600(f)	0.13	0.005
2.5	0.100	27	90	0.30	0.012	82(e)	270(e)	95(e)	315(e)	0.40	0.015	115(f)	385(f)	0.30	0.012

(a) Carbide tool material: C-8. (b) Carbide tool material: CC-8. (c) Carbide tool material: C-7. (d) Carbide tool material: CC-7. (e) Carbide tool material: C-6. (f) Carbide tool material: CC-6

AISI 1010, 1012

AISI 1010, 1012: Chemical Composition

AISI grade	C	Mn	P max	S max
1010	0.08-0.13	0.30-0.60	0.040	0.050
1012	0.10-0.15	0.30-0.60	0.040	0.050

(Chemical composition, %)

Characteristics. AISI grades 1010 and 1012 are widely used for low-strength applications because of good formability and fair response to moderate machining, particularly when following cold drawing. These steels are less expensive than grade 1008 and can be used when the requirements for drawing and forming are less exacting. Grades 1010 and 1012 can be hardened by cyaniding.

Typical Uses. AISI 1010 and 1012 steels are easily cold formed by heading, extruding, upsetting, bending, and other deforming processes. End uses for wires of these grades include electroplated products, such as racks, storage bins, shopping carts, fan guards, and jewelry, and unplated products, such as wires, staples, hardware items, welded wire fabric, and barbed wire. Wires of this carbon range are also used for shielded-metal-arc welding rods. AISI 1010 wire is suitable for magnet core applications.

AISI 1010: Similar Steels (U.S. and/or Foreign). UNS G10100; AMS 5040, 5042, 5044, 5047, 5053; ASTM A29, A108, A510, A519, A545, A549, A575, A576; MIL SPEC MIL-S-11310 (CS1010); SAE J403, J412, J414; (W. Ger.) DIN 1.1121; (Fr.) AFNOR XC 10; (Jap.) JIS S 10 C, S 12 C, S 9 CK

AISI 1012: Similar Steels (U.S. and/or Foreign). UNS G10120; ASTM A29, A510, A519, A545, A549, A575, A576; MIL SPEC MIL-S-11310 (CS1012); SAE J403, J412, J414

AISI 1010, 1012: Approximate Critical Points

Transformation point	Temperature(a) °C	°F
Ac_1	725	1335
Ac_3	875	1610
Ar_3	850	1560
Ar_1	680	1260

(a) On heating or cooling at 28 °C (50 °F) per hour

Physical Properties

AISI 1010: Average Coefficients of Linear Thermal Expansion (Ref 4)

Temperature range °C	°F	Coefficient μm/m·K	μin./in.·°F
0-100	32-212	12.2	6.8
0-200	32-390	13.0	7.2
0-300	32-570	13.5	7.5
0-400	32-750	13.8	7.7
0-500	32-930	14.2	7.9
0-600	32-1110	14.6	8.1
0-700	32-1290	14.9	8.3

AISI 1010, 1012: Thermal Treatment Temperatures

Treatment	Temperature range °C	°F
Annealing	540-730	1000-1350
Normalizing	900-955	1650-1750
Carburizing	900-925	1650-1700
Quenching	900-925	1650-1700

Physical Properties (continued)

AISI 1010: Mean Apparent Specific Heat (Ref 4)

Temperature range °C	°F	Specific heat J/kg·K	Btu/lb·°F
50-100	120-212	448	0.107
150-200	300-390	498	0.119
200-250	390-480	519	0.124
250-300	480-570	536	0.128
300-350	570-660	565	0.135
350-400	660-750	590	0.141
450-500	840-930	649	0.155
550-600	1020-1110	729	0.174
650-700	1200-1290	825	0.197

Machining Data

For machining data on AISI grades 1010 and 1012, refer to the preceding machining tables for AISI grades 1005, 1006, and 1008.

Mechanical Properties

AISI 1010, 1012: Tensile Properties and Machinability (Ref 1)

AISI grade	Condition	Size round or thickness mm	in.	Tensile strength MPa	ksi	Yield strength MPa	ksi	Elongation(a), %	Reduction in area, %	Hardness, HB	Average machinability rating(b)
1010	Hot rolled bar	19-32	0.75-1.25	325	47	180	26	28	50	95	...
	Cold drawn bar	19-32	0.75-1.25	365	53	305	44	20	40	105	55
	CQ sheet	1.6-5.8	0.064-0.229	310-360	45-52	180-240	26-35	32-48	...	...	...
1012	Hot rolled bar	19-32	0.750-1.250	330	48	185	26.5	28	50	95	...
	Cold drawn bar	19-32	0.750-1.250	370	54	310	45	19	20	105	55
	CQ sheet	1.6-5.8	0.064-0.229	310-360	45-52	180-240	26-35	32-48	...	...	...

(a) In 50 mm (2 in.). (b) Based on AISI 1212 steel as 100% average machinability

AISI 1015

AISI 1015: Chemical Composition

AISI grade	C	Mn	P max	S max
1015	0.13-0.18	0.30-0.60	0.040	0.050

Characteristics. AISI 1015 steel has better machinability and carburizing characteristics than the more formable extra-low carbon grades. It is a carburizing grade, and can be strengthened by cold working or surface hardened by carburizing or cyaniding. Hot rolled sheet has excellent weldability with either the arc or resistance method. The killed steel is best suited for arc welding. Rimmed and capped steel can be welded with proper welding techniques.

Typical Uses. Grade 1015 steel is used in the as-rolled condition where ease of forming and joining is important. In general, applications for grade 1015 are similar to those for grades 1010 and 1012.

AISI 1015: Similar Steels (U.S. and/or Foreign). UNS G10150; AMS 5060; ASTM A29, A108, A510, A519, A545, A549, A575, A576, A659; FED QQ-S-698 (C1015); MIL SPEC MIL-S-16974; SAE J403, J412, J414; (W. Ger.) DIN 1.1141; (Fr.) AFNOR XC 15, XC 18; (Jap.) JIS S 15 C, S 17 C, S 15 CK; (Swed.) SS_{14} 1370

AISI 1015: Approximate Critical Points (Ref 2)

Transformation point	Temperature(a) °C	°F
Ac_1	755	1390
Ac_3	850	1560
Ar_3	820	1510
Ar_1	755	1390

(a) On heating or cooling at 28 °C (50 °F) per hour

Machining Data

For machining data on AISI grade 1015, refer to the preceding machining tables for AISI grades 1005, 1006, and 1008.

Physical Properties

AISI 1015: Average Coefficients of Linear Thermal Expansion (Ref 4)

Temperature range °C	°F	Coefficient µm/m·K	µin./in.·°F
0-100	32-212	11.9	6.6
0-200	32-390	12.4	6.9
0-300	32-570	13.0	7.2
0-400	32-750	13.5	7.5
0-500	32-930	14.2	7.9

AISI 1015: Thermal Treatment Temperatures

Treatment	Temperat °C
Annealing	870-900
Process annealing	540-730
Normalizing	900-925
Carburizing	900-925
Carburizing, low quench	760-790
Quenching	900-925

Mechanical Properties

AISI 1015: Tensile Properties

Condition or treatment	Size round mm	in.	Tensile strength MPa	ksi	Yield strength MPa	ksi	Elongation(a), %	Reduction in area, %	Hardness, HB	Izod impact energy J	ft·lb
Hot rolled (Ref 1)	19-32	0.75-1.25	345	50	190	27.5	28	50	101	...	...
Cold drawn (Ref 1)	19-32	0.75-1.25	385	56	325	47	18	40	111	...	...
As rolled (Ref 4)	...	...	420	61	315	46	39	61	126	110	82
Normalized at 925 °C (1700 °F) (Ref 4)	...	...	425	62	325	47	37	70	121	115	85
Annealed at 870 °C (1600 °F) (Ref 4)	...	...	385	56	285	41	37	70	111	115	85

(a) In 50 mm (2 in.). (b) Average machinability rating of 60% based on AISI 1212 steel as 100% average machinability

AISI 1015: Mass Effect on Mechanical Properties (Ref 2)

Condition or treatment	Size round mm	in.	Tensile strength MPa	ksi	Yield strength MPa	ksi	Elongation(a), %	Reduction in area, %	Hardness, HB
Annealed (heated to 870 °C or 1600 °F, furnace cooled 17 °C or 30 °F per hour to 725 °C or 1340 °F, cooled in air)	25	1	385	56	284	41.25	37.0	69.7	111
Normalized (heated to 925 °C or 1700 °F, cooled in air)	13	0.5	436	63.25	330	48	38.6	71	126
	25	1	424	61.5	325	47	37	69.6	121
	50	2	415	60	307	44.5	37.5	69.2	116
	100	4	409	59.25	288	41.8	36.5	67.8	116
Mock carburized at 915 °C (1675 °F) for 8 h; reheated to 775 °C (1425 °F); quenched in water; tempered at 175 °C (350 °F)	13	0.5	733	106.25	415	60	15.0	32.9	217
	25	1	521	75.5	305	44	30	69	156
	50	2	488	70.75	285	41.38	32	70.4	131
	100	4	464	67.25	270	39	30.5	69.5	121

(a) In 50 mm (2 in.)

AISI 1015: Effect of the Mass on Hardness at Selected Points (Ref 2)

Size round mm	in.	As-quenched hardness after quenching in water at: Surface	½ radius	Center
13	0.5	36.5 HRC	23 HRC	22 HRC
25	1	99 HRB	91 HRB	90 HRB
50	2	98 HRB	84 HRB	82 HRB
100	4	97 HRB	80 HRB	78 HRB

AISI 1016, 1017, 1018, 1019

AISI 1016, 1017, 1018, 1019: Chemical Composition

AISI grade	Chemical composition, % C	Mn	P max	S max
1016	0.13-0.18	0.60-0.90	0.040	0.050
1017	0.15-0.20	0.30-0.60	0.040	0.050
1018	0.15-0.20	0.60-0.90	0.040	0.050
1019	0.15-0.20	0.70-1.00	0.040	0.050

Characteristics. AISI grades 1016, 1017, 1018, and 1019 are supplied as killed, semikilled, rimmed, and capped steels. These medium-low carbon steels have good weldability and slightly better machinability than the lower carbon grades. They are relatively soft and strengthen with cold work during forming or drawing. They can also be case hardened. The higher carbon and manganese contents result in slightly increased strength.

Typical Uses. Grade 1016, 1017, 1018, and 1019 steels are most widely used in cold forming operations such as heading, upsetting, and extrusion. They are used in the as-rolled condition as bar, shapes, sheet, and strip where ease of forming and joining, as well as stiffness, are important.

AISI 1016: Similar Steels (U.S. and/or Foreign). UNS G10160; ASTM A29, A108, A510, A513, A545, A548, A549, A576, A659; MIL SPEC MIL-S-866; SAE J403, J412, J414; (W. Ger.) DIN 1.0419

AISI 1017: Similar Steels (U.S. and/or Foreign). UNS G10170; ASTM A29, A108, A510, A513 , A519 , A544, A549 , A575, A576 , A659; MIL SPEC MIL-S-11310 (CS1017); SAE J403, J412 , J414 ; (W. Ger.) DIN 1.1141; (Fr.) AFNOR XC 15, XC 18; (Jap.) JIS S 15 C, S 17 C, S 15 CK; (Swed.) SS_{14} 1370

AISI 1018: Similar Steels (U.S. and/or Foreign). UNS G10180; AMS 5069; ASTM A29, A108, A510, A513, A519, A544, A545, A548, A549, A576, A659; MIL SPEC MIL-S-11310 (CS1018); SAE J403, J412, J414

AISI 1019: Similar Steels (U.S. and/or Foreign). UNS G10190; ASTM A29, A510, A513, A519, A545, A548, A576; SAE J403, J412, J414

AISI 1016, 1018: Approximate Critical Points

AISI grade	Transformation point	Temperature(a) °C	°F
1016	Ac_1	730	1350
	Ac_3	850	1560
	Ar_3	825	1520
	Ar_1	680	1260
1018	Ac_1	730	1350
	Ac_3	840	1545
	Ar_3	820	1505
	Ar_1	680	1255

(a) On heating or cooling at 28 °C (50 °F) per hour

Mechanical Properties

AISI 1016, 1017, 1018, 1019: Tensile Properties and Machinability

Condition or treatment	Size round mm	in.	Tensile strength MPa	ksi	Yield strength MPa	ksi	Elongation(a), %	Reduction in area, %	Hardness, HB	Average machinability rating(b)
AISI 1016										
Hot rolled (Ref 1)	19-32	0.75-1.25	380	55	205	30	25	50	111	...
Cold drawn (Ref 1)	19-32	0.75-1.25	420	61	350	51	18	40	121	70
AISI 1017										
Hot rolled (Ref 1)	19-32	0.75-1.25	365	53	200	29	26	50	105	...
Cold drawn (Ref 1)	19-32	0.75-1.25	405	59	340	49	18	40	116	65
AISI 1018										
Hot rolled (Ref 1)	19-32	0.75-1.25	400	58	220	32	25	50	116	...
Hot rolled, quenched, and tempered (Ref 3)	19-32	0.75-1.25	475	69	275	40	38	62	143	52
Cold drawn (Ref 1)	19-32	0.75-1.25	440	64	370	54	15	40	126	70
Cold drawn, quenched, and tempered (Ref 3)	19-32	0.75-1.25	565	82	485	70	20	57	163	65
As cold drawn (Ref 1)	16-22	0.63-0.88	485	70	415	60	18	40	143	...
	22-32	0.88-1.25	450	65	380	55	16	40	131	...
	32-50	1.25-2	415	60	345	50	15	35	121	...
	50-76	2-3	380	55	310	45	15	35	111	...
Cold drawn, high temperature, stress relieved (Ref 1)	16-22	0.63-0.88	450	65	310	45	20	45	131	...
	22-32	0.88-1.25	415	60	310	45	20	45	121	...
	32-50	1.25-2	380	55	310	45	16	40	111	...
	50-76	2-3	345	50	275	40	15	40	101	...
Carburized at 925 °C (1700 °F); cooled in box, reheated to 775 °C (1425 °F), water quenched, tempered at 175 °C (350 °F), core properties (Ref 1)	19-32	0.75-1.25	634	92	386	56	27	48	197	...
AISI 1019										
Hot rolled (Ref 1)	19-32	0.75-1.25	407	59	224	32	25	50	116	...
Cold drawn (Ref 1)	19-32	0.75-1.25	455	66	379	55	15	40	131	70

(a) In 50 mm (2 in.). (b) Based on AISI 1212 steel as 100% average machinability

Physical Properties

AISI 1016, 1017, 1018, 1019: Thermal Treatment Temperatures

Treatment	Temperature range °C	°F
Forging	1290-1120	2350-2050
Annealing	850-900	1575-1650
Normalizing	900-955	1650-1750
Quenching	870-900	1600-1650

Machining Data (Ref 8)

AISI 1016, 1017, 1018, 1019: Turning (Single Point and Box Tools)

Depth of cut mm	in.	M2 and M3 high speed steel Speed m/min	ft/min	Feed mm/rev	in./rev	Uncoated carbide Speed, brazed m/min	ft/min	Speed, inserted m/min	ft/min	Feed mm/rev	in./rev	Coated carbide Speed m/min	ft/min	Feed mm/rev	in./rev
Hardness, 85 to 125 HB															
1	0.040	55	180	0.18	0.007	165(a)	540(a)	210(a)	690(a)	0.18	0.007	310(b)	1025(b)	0.18	0.007
4	0.150	43	140	0.40	0.015	130(c)	425(c)	160(c)	530(c)	0.50	0.020	205(d)	675(d)	0.40	0.015
8	0.300	34	110	0.50	0.020	100(c)	330(c)	125(c)	410(c)	0.75	0.030	160(d)	525(d)	0.50	0.020
16	0.625	26	85	0.75	0.030	78(c)	255(c)	100(c)	320(c)	1.00	0.040	...	...	...	...
Hardness, 125 to 175 HB															
1	0.040	44	145	0.18	0.007	145(a)	475(a)	190(a)	630(a)	0.18	0.007	290(b)	950(b)	0.18	0.007
4	0.150	37	120	0.40	0.015	120(c)	400(c)	150(c)	490(c)	0.50	0.020	190(d)	625(d)	0.40	0.015
8	0.300	29	95	0.50	0.020	95(c)	310(c)	115(c)	380(c)	0.75	0.030	150(d)	500(d)	0.50	0.020
16	0.625	23	75	0.75	0.030	72(c)	235(c)	90(c)	300(c)	1.00	0.040	...	...	...	...

(a) Carbide tool material: C-7. (b) Carbide tool material: CC-7. (c) Carbide tool material: C-6. (d) Carbide tool material: CC-6

AISI 1016, 1017, 1018, 1019: Turning (Cutoff and Farm Tools)

Tool material	Speed, m/min (ft/min)	Feed per revolution for cutoff tool width of: 1.5 mm (0.026 in.) mm	in.	3 mm (0.125 in.) mm	in.	6 mm (0.25 in.) mm	in.	Feed per revolution for form tool width of: 12 mm (0.5 in.) mm	in.	18 mm (0.75 in.) mm	in.	25 mm (1 in.) mm	in.	50 mm (2 in.) mm	in.
Hardness, 85 to 125 HB															
M2 and M3 high speed steel	46 (150)	0.041	0.0016	0.048	0.0019	0.058	0.0023	0.048	0.0019	0.043	0.0017	0.041	0.0016	0.030	0.0012
C-6 carbide	150 (500)	0.041	0.0016	0.048	0.0019	0.058	0.0023	0.048	0.0019	0.043	0.0017	0.041	0.0016	0.030	0.0012
Hardness, 125 to 175 HB															
M2 and M3 high speed steel	40 (130)	0.041	0.0016	0.043	0.0017	0.053	0.0021	0.043	0.0017	0.038	0.0015	0.036	0.0014	0.025	0.0010
C-6 carbide	120 (400)	0.041	0.0016	0.043	0.0017	0.053	0.0021	0.043	0.0017	0.038	0.0015	0.036	0.0014	0.025	0.0010

AISI 1016, 1017, 1018, 1019: Face Milling

Depth of cut mm	in.	M2 and M7 high speed steel Speed m/min	ft/min	Feed/tooth mm	in.	Uncoated carbide Speed, brazed m/min	ft/min	Speed, inserted m/min	ft/min	Feed/tooth mm	in.	Coated carbide Speed m/min	ft/min	Feed/tooth mm	in.
Hardness, 85 to 125 HB															
1	0.040	72	235	0.20	0.008	225(a)	735(a)	245(a)	810(a)	0.20	0.008	365(b)	1200(b)	0.20	0.008
4	0.150	55	180	0.30	0.012	165(a)	535(a)	195(a)	635(a)	0.30	0.012	250(b)	825(b)	0.30	0.012
8	0.300	43	140	0.40	0.016	125(c)	405(c)	150(c)	495(c)	0.40	0.016	195(d)	645(d)	0.40	0.016
Hardness, 125 to 175 HB															
1	0.040	66	215	0.20	0.008	205(a)	675(a)	225(a)	735(a)	0.20	0.008	335(b)	1100(b)	0.20	0.008
4	0.150	50	165	0.30	0.012	155(a)	510(a)	175(a)	570(a)	0.30	0.012	225(b)	740(b)	0.30	0.012
8	0.300	40	130	0.40	0.016	110(c)	365(c)	135(c)	445(c)	0.40	0.016	175(d)	575(d)	0.40	0.016

(a) Carbide tool material: C-6. (b) Carbide tool material: CC-6. (c) Carbide tool material: C-5. (d) Carbide tool material: CC-5

AISI 1016, 1017, 1018, 1019: End Milling (Profiling)

Tool material	Depth of cut mm	Depth of cut in.	Speed m/min	Speed ft/min	Feed per tooth for cutter diameter of: 10 mm (0.375 in.) mm	10 mm (0.375 in.) in.	12 mm(0.5 in.) mm	12 mm(0.5 in.) in.	18 mm (0.75 in.) mm	18 mm (0.75 in.) in.	25-50 mm (1-2 in.) mm	25-50 mm (1-2 in.) in.
Hardness, 85 to 125 HB												
M2, M3, and M7 high speed steel	0.5	0.020	58	190	0.025	0.001	0.050	0.002	0.102	0.004	0.13	0.005
	1.5	0.060	44	145	0.050	0.002	0.075	0.003	0.13	0.005	0.15	0.006
	diam/4	diam/4	40	130	0.025	0.001	0.050	0.002	0.102	0.004	0.13	0.005
	diam/2	diam/2	35	115	0.018	0.0007	0.025	0.001	0.075	0.003	0.102	0.004
C-5 carbide	0.5	0.020	170	550	0.038	0.0015	0.089	0.0035	0.15	0.006	0.18	0.007
	1.5	0.060	130	425	0.063	0.0025	0.102	0.004	0.15	0.006	0.20	0.008
	diam/4	diam/4	110	365	0.050	0.002	0.075	0.0030	0.13	0.005	0.15	0.006
	diam/2	diam/2	105	340	0.038	0.0015	0.050	0.002	0.102	0.004	0.13	0.005
Hardness, 125 to 175 HB												
M2, M3, and M7 high speed steel	0.5	0.020	64	210	0.025	0.001	0.050	0.002	0.102	0.004	0.13	0.005
	1.5	0.060	49	160	0.050	0.002	0.075	0.003	0.13	0.005	0.15	0.006
	diam/4	diam/4	43	140	0.025	0.001	0.050	0.002	0.102	0.004	0.13	0.005
	diam/2	diam/2	37	120	0.018	0.0007	0.025	0.001	0.075	0.003	0.102	0.004
C-5 carbide	0.5	0.020	160	520	0.038	0.0015	0.089	0.0035	0.15	0.006	0.18	0.007
	1.5	0.060	120	400	0.063	0.0025	0.102	0.004	0.15	0.006	0.20	0.008
	diam/4	diam/4	105	350	0.050	0.002	0.075	0.003	0.13	0.005	0.15	0.006
	diam/2	diam/2	100	320	0.038	0.0015	0.050	0.002	0.102	0.004	0.13	0.005

AISI 1016, 1017, 1018, 1019: Boring

Depth of cut mm	Depth of cut in.	M2 and M3 high speed steel: Speed m/min	Speed ft/min	Feed mm/rev	Feed in./rev	Uncoated carbide: Speed, brazed m/min	Speed, brazed ft/min	Speed, inserted m/min	Speed, inserted ft/min	Feed mm/rev	Feed in./rev	Coated carbide: Speed m/min	Speed ft/min	Feed mm/rev	Feed in./rev
Hardness, 85 to 125 HB															
0.25	0.010	55	180	0.075	0.003	160(a)	520(a)	185(a)	610(a)	0.075	0.003	275(b)	900(b)	0.075	0.003
1.25	0.050	44	145	0.130	0.005	125(c)	410(c)	150(c)	485(c)	0.130	0.005	220(d)	715(d)	0.130	0.005
2.5	0.100	34	110	0.300	0.012	95(e)	315(e)	115(e)	315(e)	0.400	0.015	145(f)	475(f)	0.300	0.012
Hardness, 125 to 175 HB															
0.25	0.010	44	145	0.075	0.003	140(a)	465(a)	170(a)	550(a)	0.075	0.003	255(b)	830(b)	0.075	0.003
1.25	0.050	35	115	0.130	0.005	115(c)	375(c)	135(c)	440(c)	0.130	0.005	205(d)	665(d)	0.130	0.005
2.5	0.100	29	95	0.300	0.012	90(e)	295(e)	105(e)	345(e)	0.400	0.015	135(f)	440(f)	0.300	0.012

(a) Carbide tool material: C-8. (b) Carbide tool material: CC-8. (c) Carbide tool material: C-7. (d) Carbide tool material: CC-7. (e) Carbide tool material: C-6. (f) Carbide tool material: CC-6

AISI 1016, 1017, 1018, 1019: Reaming

Based on 4 flutes for 3- and 6-mm (0.125- and 0.25-in.) reamers, 6 flutes for 12-mm (0.5-in.) reamers, and 8 flutes for 25-mm (1-in.) and larger reamers

Tool material	Speed m/min	Speed ft/min	Feed per revolution for reamer diameter of: 3 mm (0.125 in.) mm	3 mm (0.125 in.) in.	6 mm (0.25 in.) mm	6 mm (0.25 in.) in.	12 mm (0.5 in.) mm	12 mm (0.5 in.) in.	25 mm (1 in.) mm	25 mm (1 in.) in.	35 mm (1.5 in.) mm	35 mm (1.5 in.) in.	50 mm (2 in.) mm	50 mm (2 in.) in.
Hardness, 85 to 125 HB														
Roughing														
M1, M2, and M7 high speed steel	35	115	0.102	0.004	0.18	0.007	0.30	0.012	0.50	0.020	0.65	0.025	0.75	0.030
C-2 carbide	35	120	0.102	0.004	0.18	0.007	0.30	0.012	0.50	0.020	0.65	0.025	0.75	0.030
Finishing														
M1, M2, and M7 high speed steel	14	45	0.15	0.006	0.25	0.010	0.40	0.015	0.65	0.025	0.75	0.030	0.90	0.035
C-2 carbide	17	55	0.15	0.006	0.25	0.010	0.40	0.015	0.65	0.025	0.75	0.030	0.90	0.035
Hardness, 125 to 175 HB														
Roughing														
M1, M2, and M7 high speed steel	32	105	0.102	0.004	0.18	0.007	0.30	0.012	0.50	0.020	0.65	0.025	0.75	0.030
C-2 carbide	35	115	0.102	0.004	0.18	0.007	0.30	0.012	0.50	0.020	0.65	0.025	0.75	0.030
Finishing														
M1, M2, and M7 high speed steel	12	40	0.15	0.006	0.25	0.010	0.40	0.015	0.65	0.025	0.75	0.030	0.90	0.035
C-2 carbide	15	50	0.15	0.006	0.25	0.010	0.40	0.015	0.65	0.025	0.75	0.030	0.90	0.035

Machining Data (Ref 8) (continued)

AISI 1016, 1017, 1018, 1019: Planing

Tool material	Depth of cut mm	Depth of cut in.	Speed m/min	Speed ft/min	Feed/stroke mm	Feed/stroke in.
Hardness, 85 to 125 HB						
M2 and M3 high speed steel	0.1	0.005	17	55	(a)	(a)
	2.5	0.100	23	75	1.25	0.050
	12	0.500	14	45	1.50	0.060
C-6 carbide	0.1	0.005	90	300	(a)	(a)
	2.5	0.100	90	300	2.05	0.080
	12	0.500	90	300	1.50	0.060
Hardness, 125 to 175 HB						
M2 and M3 high speed steel	0.1	0.005	14	45	(a)	(a)
	2.5	0.100	20	65	1.25	0.050
	12	0.500	12	40	1.50	0.060
C-6 carbide	0.1	0.005	84	275	(a)	(a)
	2.5	0.100	90	300	2.05	0.080
	12	0.500	84	275	1.50	0.060

(a) Feed is 75% the width of the square nose finishing tool

AISI 1016, 1017, 1018, 1019: Drilling

Tool material	Speed m/min	Speed ft/min	Feed per revolution for nominal hole diameter of: 1.5 mm (0.062 in.) mm	1.5 mm in.	3 mm (0.125 in.) mm	3 mm in.	6 mm (0.25 in.) mm	6 mm in.	12 mm (0.5 in.) mm	12 mm in.	18 mm (0.75 in.) mm	18 mm in.	25 mm (1 in.) mm	25 mm in.
Hardness, 85 to 125 HB														
M10, M7, and M1 high speed steel	24	80	0.025	0.001	...	...	...	...	...	...	...	...	...	...
	29	95	...	...	0.075	0.003	0.15	0.006	0.25	0.010	0.40	0.015	0.50	0.020
Hardness, 125 to 175 HB														
M10, M7, and M1 high speed steel	23	75	0.025	0.001	...	...	...	...	...	...	...	...	...	...
	26	85	...	...	0.075	0.003	0.13	0.005	0.23	0.009	0.30	0.012	0.40	0.015

AISI 1020

AISI 1020: Chemical Composition

AISI grade	C	Mn	P max	S max
1020	0.18-0.23	0.30-0.60	0.040	0.050

Characteristics. AISI grade 1020 is stronger and less easily formed than grade 1018. This steel responds well to cold work and heat treating. To minimize soft spots in the case, when case hardened, orders should specify that the steel be "normal" in the McQuaid-Ehn test. Weldability of grade 1020 is fair.

Typical Uses. AISI grade 1020 is suitable for parts in the case hardened condition where core strength is not critical, and for shafts of larger cross section that are not highly stressed. Other uses include lightly stressed gears with hard wearing surfaces and case hardened pins and chains.

AISI 1020: Similar Steels (U.S. and/or Foreign). UNS G10200; AMS 5032, 5045; ASTM A29, A108, A510, A519, A544, A575, A576, A659; MIL SPEC MIL-S-11310 (CS1020); SAE J403, J412, J414; (W. Ger.) DIN 1.0402; (Fr.) AFNOR CC 20; (Ital.) UNI C 20; (Swed.) SS_{14} 1450; (U.K.) B.S. 040 A 20, 070 M 20

AISI 1020: Approximate Critical Points

Transformation point	Temperature(a) °C	°F
Ac_1	725	1335
Ac_3	845	1555
Ar_3	815	1500
Ar_1	680	1260

(a) On heating or cooling at 28 °C (50 °F) per hour

Physical Properties

AISI 1020: Average Coefficients of Linear Thermal Expansion (Ref 4)

Temperature range °C	°F	Coefficient µm/m · K	µin./in. · °F
20-100	68-212	11.7	6.5
20-200	68-390	12.1	6.7
20-300	68-570	12.8	7.1
20-400	68-750	13.3	7.4
20-500	68-930	13.9	7.7
20-600	68-1110	14.4	8.0
20-700	68-1290	14.8	8.2

AISI 1020: Thermal Treatment Temperatures

Treatment	Temperature range °C	°F
Forging	1260-760	2300-1400
Annealing	870-900	1600-1650
Process annealing	540-730	1000-1350
Normalizing	900-955	1650-1750
Carburizing	900-925	1650-1700
Quenching	870-915	1600-1675

Mechanical Properties

AISI 1020: Tensile Properties and Machinability

Condition or treatment	Size round mm	Size round in.	Tensile strength MPa	Tensile strength ksi	Yield strength MPa	Yield strength ksi	Elongation(a), %	Reduction in area, %	Hardness, HB	Average machinability rating(b)
Hot rolled (Ref 1)	19-32	0.75-1.25	380	55	205	30	25	50	111	...
Cold rolled (Ref 1)	19-32	0.75-1.25	420	61	350	51	15	40	121	65
Hot rolled, quenched and tempered, 0.2% offset (Ref 3)	19-32	0.75-1.25	475	69	275	40	38	62	163	52
As rolled (Ref 12)	...	...	450	65	330	48	36	59	143	...
Normalized at 870 °C (1600 °F) (Ref 12)	...	...	440	64	345	50	36	68	131	...
Annealed at 870 °C (1600 °F) (Ref 12)	...	...	505	73	360	52	36	66	111	...

(a) In 50 mm (2 in.). (b) Based on AISI 1212 steel as 100% average machinability

AISI 1020: Mass Effect on Mechanical Properties (Ref 2)

Condition or treatment	Size round mm	Size round in.	Tensile strength MPa	Tensile strength ksi	Yield strength MPa	Yield strength ksi	Elongation(a), %	Reduction in area, %	Hardness, HB	Izod impact energy J	Izod impact energy ft·lb
As rolled	25	1	472	68.5	384	55.75	32.0	66.5	137	...	...
Annealed (heated to 870 °C or 1600 °F, furnace cooled, 17 °C or 30 °F per hour to 700 °C or 1290 °F, cooled in air)	25	1	395	57.25	295	42.75	36.5	66	111	...	...
Normalized (heated to 925 °C or 1700 °F, cooled in air)	13	0.5	445	64.5	346	50.25	39.3	69.1	131	...	...
	25	1	440	64	346	50.25	35.8	67.9	131	...	...
	50	2	438	63.5	319	46.25	35.5	65.5	126	...	...
	100	4	415	60	281	40.75	36.0	66.6	121	...	...
Mock carburized at 915 °C (1675 °F) for 8 h; reheated to 775 °C (1425 °F); quenched in water; tempered at 175 °C (350 °F)	13	0.5	889	129	495	72	11.4	29.4	255	55	40
	25	1	600	87	370	54	23	64.2	179	110	80
	50	2	521	75.5	302	43.75	31.3	67.9	156	135	98
	100	4	491	71.25	290	42	33	67.6	143	130	97

(a) In 50 mm (2 in.)

AISI 1020: Effect of the Mass on Hardness at Selected Points (Ref 2)

Size round mm	Size round in.	As-quenched hardness after quenching in water at: Surface	½ radius	Center
13	0.5	40.5 HRC	30 HRC	28 HRC
25	1	29.5 HRC	96 HRB	93 HRB
50	2	95 HRB	85 HRB	83 HRB
100	4	94 HRB	78 HRB	77 HRB

AISI 1020: Impact Properties (Ref 6)

Temperature °C	Temperature °F	Energy J	Energy ft·lb
−30	−25	16.9	12.5
−18	0	18	13
−3	25	20	15
10	50	24	18
38	100	41	30
65	150	54	40
95	200	61	45
150	300	68	50

AISI 1020: Izod Impact Properties (Ref 12)

Condition or treatment	Energy J	Energy ft·lb
As rolled	85	64
Normalized at 870 °C (1600 °F)	120	87
Annealed at 870 °C (1600 °F)	125	91

AISI 1020: Recommended Condition to Provide Maximum Drawability (Ref 6)

Punch nose radius, r(a)	Condition, HRB, for cup reduction of: 20%	30%	40%	50%
2t	60	65(b)(c)	65(b)(c)(d)	...
4t	65	60	65(b)(c)	...
8t	70	70	60	...
16t	70	65	60	65(b)(c)(d)
32t	70	70	65	65(b)(c)

Drawing limitations refer to cold rolled and mill annealed sheet steel; simple round cups, no flanges, not ironed, 20-gage (0.91-mm or 0.036-in.) sheet steel. (a) In relation to stock thickness, t. (b) Aluminum killed. (c) Fine grained. (d) Spheroidized

Machining Data

For machining data on AISI grade 1020, refer to the preceding machining tables for AISI grades 1005, 1006, and 1008.

AISI 1021, 1022

AISI 1021, 1022: Chemical Composition

AISI grade	C	Mn	P max	S max
	Chemical composition, %			
1021	0.18-0.23	0.60-0.90	0.040	0.050
1022	0.18-0.23	0.70-1.00	0.040	0.050

Characteristics. AISI grades 1021 and 1022 exhibit fair machinability, fair to good weldability, and excellent cold formability. Cold drawing markedly increases the tensile strength of grade 1022. The higher manganese content of 1022 steel makes this steel more likely to meet required specifications for heat treatment and better adapted for use in larger sections than grade 1020 or 1021 steels.

Typical Uses. AISI grades 1021 and 1022 are used in cold heading and extruding. When case hardened, 1022 steel has high core strength. This grade is used in producing low-strength fasteners, die pins, motor spindles, and recessed-head screws.

AISI 1021: Similar Steels (U.S. and/or Foreign). UNS G10210; ASTM A29, A510, A519, A545, A548, A576, A659; SAE J403, J412, J414

AISI 1022: Similar Steels (U.S. and/or Foreign). UNS G10220; AMS 5070; ASTM A29, A510, A519, A544, A545, A548, A576; MIL SPEC MIL-S-11310 (CS1022); SAE J403, J412, J414; (W. Ger.) DIN 1.1133; (Ital.) UNI G 22 Mn 3; (Jap.) JIS SMnC 21

AISI 1022: Approximate Critical Points (Ref 2)

Transformation point	Temperature(a) °C	°F
Ac_1	740	1360
Ac_3	830	1530
Ar_3	780	1440
Ar_1	705	1300

(a) On heating or cooling at 28 °C (50 °F) per hour

Mechanical Properties

AISI 1021, 1022: Tensile Properties and Machinability

Condition or treatment	Size round mm	in.	Tensile strength MPa	ksi	Yield strength MPa	ksi	Elongation(a), %	Reduction in area, %	Hardness, HB	Average machinability rating(b)
AISI 1021										
Hot rolled (Ref 1)	19-32	0.75-1.25	420	61	230	33	24	48	116	...
Cold drawn (Ref 1)	19-32	0.75-1.25	470	68	395	57	15	40	121	70
AISI 1022										
Hot rolled (Ref 1)	19-32	0.75-1.25	425	62	235	34	23	47	121	...
Cold drawn (Ref 1)	19-32	0.75-1.25	475	69	400	58	15	40	137	70
As rolled (Ref 12)	...	...	505	73	360	52	35	67	149	...
Normalized at 925 °C (1700 °F) (Ref 12)	...	...	485	70	360	52	34	68	143	...
Annealed at 870 °C (1600 °F) (Ref 12)	...	...	450	65	315	46	35	64	137	...

(a) In 50 mm (2 in.). (b) Based on AISI 1212 steel as 100% average machinability

AISI 1022: Approximate Core Properties (Ref 5)

Heat treatment of test specimens: 1 normalized at 925 °C (1700 °F) in 32 mm (1.25 in.) rounds; 2 machined to 25 or 13 mm (1 or 0.50 in.); 3 pseudocarburized at 925 °C (1700 °F) for 8 h; 4 box cooled to room temperature; 5 reheated to temperatures shown below and oil quenched; 6 tempered at 150 °C (300 °F); 7 tested in 12.8 mm (0.505 in.) rounds; tests were conducted using test specimens machined to English units

Reheat temperature °C	°F	Yield strength(a) MPa	ksi	Tensile strength MPa	ksi	Elongation(b), %	Reduction in area, %	Hardness, HB
Heat treated in 1 in. (25 mm) rounds								
775	1425	355	51.5	560	81.5	31	71	163
800	1475	350	51	555	80.5	31.5	71	163
830	1525	360	52	570	83	31	70.5	174
(c)	(c)	420	61.0	585	85	29.5	70.5	179
Heat treated in 0.540 in. (13.7 mm) rounds								
775	1425	380	55	550	80	30	68.5	170
800	1475	355	51.5	560	81	30	70.5	170
830	1525	400	58	565	82	29.5	72.5	179
(c)	(c)	415	60	570	83	30	71	179

(a) 0.2% offset. (b) In 50 mm (2 in.). (c) Quenched from step 3 of heat treatment

Mechanical Properties (continued)

AISI 1021, 1022: Mass Effect on Mechanical Properties (Ref 2)

Condition or treatment	Size round mm	Size round in.	Tensile strength MPa	Tensile strength ksi	Yield strength MPa	Yield strength ksi	Elongation(a), %	Reduction in area, %	Hardness, HB
Cold rolled	25	1	484	70.25	360	52.25	33	65.2	137
Annealed (heated to 870 °C or 1600 °F, furnace cooled 17 °C or 30 °F per hour to 675 °C or 1250 °F, cooled in air)	25	1	450	65.25	315	46	35	63.5	137
Normalized (heated to 925 °C or 1700 °F, cooled in air)	13	0.5	486	70.5	365	53	35.7	68.3	143
	25	1	483	70	360	52	34	67.5	143
	50	2	474	68.75	330	48	34	66.6	137
	100	4	464	67.25	310	45	33.8	63.9	131
Mock carburized at 910 °C (1675 °F) for 8 h; reheated to 775 °C (1425 °F); quenched in water; tempered at 175 °C (350 °F)	13	0.5	931	135	515	75	13.6	24.3	262
	25	1	615	89	380	55	25.5	57.3	179
	50	2	565	82	346	50.25	30	69.6	163
	100	4	510	74	293	42.5	32.5	71.1	149

(a) In 50 mm (2 in.)

AISI 1021, 1022: Effect of the Mass on Hardness at Selected Points (Ref 2)

Size round mm	Size round in.	As-quenched hardness after quenching in water at: Surface	½ radius	Center
13	0.5	45 HRC	29 HRC	27 HRC
25	1	41 HRC	95 HRB	92 HRB
50	2	38 HRC	88 HRB	84 HRB
100	4	34 HRC	84 HRB	81 HRB

AISI 1021, 1022: Izod Impact Properties (Ref 12)

Condition or treatment	Energy J	Energy ft·lb
As rolled	81	60
Normalized at 925 °C (1700 °F)	118	87
Annealed at 870 °C (1600 °F)	121	89

Physical Properties

AISI 1021, 1022: Thermal Treatment Temperatures (Ref 4)

Treatment	Temperature range °C	°F
Forging	740 max	1360 max
Annealing	870-900	1600-1650
Normalizing	900-955	1650-1750
Carburizing	900-925	1650-1700

AISI 1021, 1022: Density (Ref 4)

7.858 g/cm^3 .... 0.2839 lb/in.^{3}

Machining Data

For machining data on AISI grades 1021 and 1022, refer to the preceding machining tables for AISI grades 1016, 1017, 1018, and 1019.

AISI 1023, 1025, 1026

AISI 1023, 1025, 1026: Chemical Composition

AISI grade	C	Mn	P max	S max
	Chemical composition, %			
1023	0.20-0.25	0.30-0.60	0.040	0.050
1025	0.22-0.28	0.30-0.60	0.040	0.050
1026	0.22-0.28	0.60-0.90	0.040	0.050

Characteristics. Similarly to grade 1020, AISI grades 1023, 1025, and 1026 respond favorably to heat treating and cold work. Machinability and weldability are fair. Cold formability of these steels is good to excellent.

Typical Uses. AISI grades 1023, 1025, and 1026 have uses similar to those described for grade 1020 steel.

AISI 1023: Similar Steels (U.S. and/or Foreign). UNS G10230; ASTM A29, A510, A575, A576, A659; SAE J403, J412, J414; (W. Ger.) DIN 1.1151; (Fr.) AFNOR XC 18 S, XC 25; (Jap.) JIS S 20 C, S 22 C, S 20 CK

AISI 1025: Similar Steels (U.S. and/or Foreign). UNS G10250; AMS 5075, 5077; ASTM A29, A510, A512, A519, A575, A576; FED QQ-S-700 (C1025); MIL SPEC MIL-S-11310 (CS1025); SAE J403, J412, J414; (W. Ger.) DIN 1.1158; (Jap.) JIS S 25 C, S 28 C

AISI 1026: Similar Steels (U.S. and/or Foreign). UNS G10260; ASTM A29, A273, A510, A519, A545, A576; SAE J403, J412, J414

AISI 1025: Approximate Critical Points (Ref 2)

Transformation point	Temperature(a) °C	°F
Ac_1	735	1355
Ac_3	830	1530
Ar_3	780	1440
Ar_1	695	1285

(a) On heating or cooling at 28 °C (50 °F) per hour

Physical Properties

AISI 1023, 1025, 1026: Mean Apparent Specific Heat (Ref 4)

Temperature range °C	°F	Specific heat J/kg·K	Btu/lb·°F
50-100	120-212	486	0.116
150-200	300-390	519	0.124
200-250	390-480	532	0.127
250-300	480-570	557	0.133
300-350	570-660	574	0.137
350-400	660-750	599	0.143
450-500	840-930	662	0.158
550-600	1020-1110	749	0.179
650-700	1200-1290	846	0.202
700-750	1290-1380	1432	0.342
750-800	1380-1470	950	0.227

Material composition: 0.23% carbon, 0.635% manganese

AISI 1023, 1025, 1026: Average Coefficients of Linear Thermal Expansion (Ref 4)

Temperature range °C	°F	Coefficient μm/m·K	μin./in.·°F
20-100	68-212	12.1	6.7
20-200	68-390	12.8	7.1
20-300	68-570	13.3	7.4
20-400	68-750	13.9	7.7
20-500	68-930	14.4	8.0
20-600	68-1110	14.8	8.2

AISI 1023, 1025, 1026: Electrical Resistivity and Thermal Conductivity (Ref 4)

Temperature °C	°F	Electrical resistivity, μΩ·m	Thermal conductivity W/m·K	Btu/ft·h·°F
0	32	...	51.9	30.0
20	68	0.169	...	...
100	212	0.219	51.2	29.6
200	390	0.292	49.0	28.3
300	570	...	46.0	26.7
400	750	0.487	42.7	24.7
500	930	...	39.4	22.8
600	1110	0.758	35.6	20.6
700	1290	0.925	31.8	18.4
800	1470	1.094	26.1	15.1
900	1650	1.136	...	...
1000	1830	1.167	27.2	15.7
1100	2010	1.207	...	...
1200	2190	1.219	29.8	17.2
1300	2370	1.239	...	...

Material composition: 0.23% carbon, 0.635% manganese

AISI 1023, 1025, 1026: Density (Ref 4)

7.858 g/cm³ .... 0.2839 lb/in.³

Material composition: 0.23% carbon, 0.635% manganese, 0.11% silicon; annealed at 925 °C (1700 °F)

Machining Data

For machining data on AISI grades 1023, 1025, and 1026, refer to the preceding machining tables for AISI 1005, 1006, and 1008.

Mechanical Properties

AISI 1023, 1025, 1026: Tensile Properties and Machinability

Condition or treatment	Size round mm	Size round in.	Tensile strength MPa	Tensile strength ksi	Yield strength MPa	Yield strength ksi	Elongation(a), %	Reduction in area, %	Hardness, HB	Average machinability rating(b)
AISI 1023										
Hot rolled (Ref 1)	19-32	0.75-1.25	385	56	215	31	25	50	111	...
Cold drawn (Ref 1)	19-32	0.75-1.25	425	62	360	52.5	15	40	121	65
AISI 1025										
Hot rolled (Ref 1)	19-32	0.75-1.25	400	58	220	32	25	50	116	...
Cold drawn (Ref 1)	19-32	0.75-1.25	440	64	370	54	15	40	126	65
Turned, ground, and polished (Ref 3)	19-32	0.75-1.25	460	67	310	45	36	56	143	50
As drawn (Ref 1)	16-22	0.63-0.88	485	70	415	60	18	40	143	...
	22-32	0.88-1.25	450	65	380	55	16	40	131	...
	32-50	1.25-2	415	60	345	50	15	35	121	...
	50-75	2-3	380	55	310	45	15	35	111	...
Cold drawn, high temperature, stress relieved (Ref 1)	16-22	0.63-0.88	450	65	310	45	20	45	131	...
	22-32	0.88-1.25	415	60	310	45	20	45	121	...
	32-50	1.25-2	380	55	310	45	16	40	111	...
	50-75	2-3	345	50	275	40	15	40	101	...
AISI 1026										
Hot rolled (Ref 1)	19-32	0.75-1.25	440	64	240	35	24	49	126	...
Cold drawn (Ref 1)	19-32	0.75-1.25	490	71	415	60	15	40	143	75

(a) In 50 mm (2 in.). (b) Based on AISI 1212 steel as 100% average machinability

AISI 1030, 1034, 1035

AISI 1030, 1034, 1035: Chemical Composition

AISI grade	C	Mn	P max	S max
1030	0.28-0.34	0.60-0.90	0.040	0.050
1034	0.32-0.38	0.50-0.80	0.040	0.050
1035	0.32-0.38	0.60-0.90	0.040	0.050

Characteristics. AISI grades 1030, 1034, and 1035 are water-hardening steels suitable for small parts of moderate strength. Grades 1034 and 1035 have a slightly greater hardenability and strength. All have moderate strength and hardness in the as-rolled condition, and can be strengthened and hardened by cold work. Machinability is only fair, as is weldability. Precautions must be taken when welding to avoid cracking from cooling too rapidly.

Typical Uses. Grades 1030, 1034, and 1035 are used in manufacturing levers, bolts, studs, nuts, and similar parts which are headed, upset, or extruded. Applications for wires of these grades include nails requiring high shank rigidity or strength, and quench-hardened nails.

AISI 1030: Similar Steels (U.S. and/or Foreign). UNS G10300; ASTM A29, A108, A510, A512, A519, A544, A545, A546, A576, A682; FED QQ-S-635 (C1030), QQ-S-700 (C1030); MIL SPEC MIL-S-11310 (CS1030); SAE J403, J412, J414; (W. Ger.) DIN 1.1172; (Ital.) UNI CB 35

AISI 1034: Similar Steels (U.S. and/or Foreign). UNS G10340; ASTM A29, A181; SAE J118, J412; (W. Ger.) DIN 1.1181; (Fr.) AFNOR XC 32, XC 35, XC 38; (Jap.) JIS S 38 C

AISI 1035: Similar Steels (U.S. and/or Foreign). UNS G10350; AMS 5080, 5082; ASTM A29, A108, A510, A519, A544, A545, A546, A576, A682; FED QQ-S-635 (C1035), QQ-S-700 (C1035); SAE J403, J412, J414; (W. Ger.) DIN 1.0501; (Fr.) AFNOR CC 35; (Ital.) UNI C 35; (Swed.) SS_{14} 1550; (U.K.) B.S. 060 A 35, 080 A 32, 080 A 35, 080 A 37, 080 M 36

AISI 1030, 1035: Approximate Critical Points

AISI grade	Transformation point	Temperature(a) °C	Temperature(a) °F
1030	Ac_1	725	1340
	Ac_3	815	1495
	Ar_3	790	1450
	Ar_1	675	1250
1035	Ac_1	730	1345
	Ac_3	800	1475
	Ar_3	790	1455
	Ar_1	690	1275

(a) On heating or cooling at 28 °C (50 °F) per hour

Physical Properties

AISI 1030, 1035: Average Coefficients of Linear Thermal Expansion (Ref 4)

AISI grade	Temperature range °C	°F	Coefficient µm/m·K	µin./in.·°F
1030	15-75	59-165	11.7	6.5
1035	20-100	68-212	11.0	6.1
	20-200	68-390	11.9	6.6
	20-300	68-570	12.6	7.0
	20-400	68-750	13.3	7.4
	20-500	68-930	13.9	7.7
	20-600	68-1110	14.4	8.0
	20-700	68-1290	13.9	8.2

Mechanical Properties

AISI 1030: Mass Effect on Mechanical Properties (Ref 2)

Condition or treatment	Size round mm	in.	Tensile strength MPa	ksi	Yield strength MPa	ksi	Elongation(a), %	Reduction in area, %	Hardness, HB
Annealed (heated to 845 °C or 1550 °F, furnace cooled 11 °C or 20 °F per hour to 650 °C or 1200 °F, cooled in air)	25	1	460	67	345	50	31.2	57.9	126
Normalized (heated to 925 °C or 1700 °F, cooled in air)	13	0.5	540	78	345	50	32.1	61.1	156
	25	1	525	76	345	50	32.0	60.8	149
	50	2	510	74	345	50	29.5	58.9	137
	100	4	495	72	325	47	29.7	56.2	137
Water quenched from 870 °C (1600 °F), tempered at 540 °C (1000 °F)	13	0.5	635	92	515	75	28.2	58.0	187
	25	1	605	88	470	68	28.0	68.6	179
	50	2	595	86	440	64	28.2	65.8	170
	100	4	560	81	380	55	32.0	68.2	163
Water quenched from 870 °C (1600 °F), tempered at 595 °C (1100 °F)	13	0.5	605	88	440	64	28.9	69.7	179
	25	1	585	85	435	63	29.0	70.8	170
	50	2	580	84	395	57	29.0	69.1	167
	100	4	550	80	370	54	32.0	68.5	163
Water quenched from 870 °C (1600 °F), tempered at 650 °C (1200 °F)	13	0.5	595	86	425	62	29.9	70.5	174
	25	1	580	84	425	62	28.5	71.4	170
	50	2	550	80	395	57	30.2	70.9	156
	100	4	510	74	345	50	34.2	71.0	149

(a) In 50 mm (2 in.). (b) Based on AISI 1212 steel as 100% average machinability

AISI 1030: Effect of the Mass on Hardness at Selected Points (Ref 2)

Size round mm	in.	As-quenched hardness after quenching in water at: Surface	½ radius	Center
13	0.5	50 HRC	50 HRC	23 HRC
25	1	46 HRC	23 HRC	21 HRC
50	2	30 HRC	93 HRB	90 HRB
100	4	97 HRB	88 HRB	85 HRB

AISI 1030, 1034, 1035: Izod Impact Properties (Ref 12)

Condition or treatment	Energy J	ft·lb
As rolled	75	55
Normalized at 925 °C (1700 °F)	94	69
Annealed at 845 °C (1550 °F)	69	51

AISI 1030: Recommended Condition to Provide Maximum Drawability (Ref 6)

Punch nose radius, r(a)	Condition, HRB, for cup reduction of: 20%	30%	40%	50%
2t	65(b)(c)	65(b)(c)	65(b)(c)(d)	...
4t	60	60	65(b)(c)(d)	...
8t	65	65	65(b)(c)(d)	...
16t	60	60	65(b)(c)	...
32t	70	65	65(b)(c)	65 (b)(c)(d)

Drawing limitations refer to cold rolled and mill annealed sheet steel; simple round cups, no flanges, not ironed, 20-gage (0.91-mm or 0.36-in.) sheet steel. (a) In relation to stock thickness, *t*. (b) Aluminum killed. (c) Fine grained. (d) Spheroidized

Mechanical Properties (continued)

AISI 1030, 1035: Tensile Properties and Machinability

Condition or treatment	Size round mm	Size round in.	Tensile strength MPa	Tensile strength ksi	Yield strength MPa	Yield strength ksi	Elongation(a), %	Reduction in area, %	Hardness, HB	Average machinability rating(b)
AISI 1030										
Hot rolled (Ref 1)	19-32	0.75-1.25	470	68	260	38	20	42	137	...
Cold drawn (Ref 1)	19-32	0.75-1.25	525	76	440	64	12	35	149	70
As rolled (Ref 12)	...	...	550	80	345	50	32	57	179	...
Normalized at 925 °C (1700 °F) (Ref 12)	...	...	525	76	345	50	32	61	149	...
Annealed at 845 °C (1550 °F) (Ref 12)	...	...	460	67	345	50	31	58	126	...
AISI 1035										
Hot rolled (Ref 1)	19-32	0.75-1.25	495	72	275	40	18	40	143	...
As rolled (Ref 1)	19-32	0.75-1.25	585	85	370	54	30	53	183	65
Water quenched from 845 °C (1550 °F), tempered at 540 °C (1000 °F) (Ref 3)	19-32	0.75-1.25	710	103	615	89	16	40	207	...
Cold drawn (Ref 1)	19-32	0.75-1.25	550	80	460	67	12	35	163	65
As cold drawn (Ref 1)	16-22	0.63-0.88	585	85	515	75	13	35	170	...
	22-32	0.88-1.25	550	80	485	70	12	35	163	...
	32-50	1.25-2	515	75	450	65	12	35	149	...
	50-75	2-3	485	70	415	60	10	30	143	...
Cold drawn, low temperature, stress relieved (Ref 1)	16-22	0.63-0.88	620	90	550	80	13	35	179	...
	22-32	0.88-1.25	585	85	515	75	12	35	170	...
	32-50	1.25-2	550	80	485	70	12	35	163	...
	50-75	2-3	515	75	450	65	10	30	149	...
Cold drawn, high temperature, stress relieved (Ref 1)	16-22	0.63-0.88	550	80	415	60	16	45	163	...
	22-32	0.88-1.25	515	75	415	60	15	45	149	...
	32-50	1.25-2	485	70	415	60	15	40	143	...
	50-75	2-3	450	65	380	55	12	35	131	...

(a) In 50 mm (2 in.). (b) Based on AISI 1212 steel as 100% average machinability

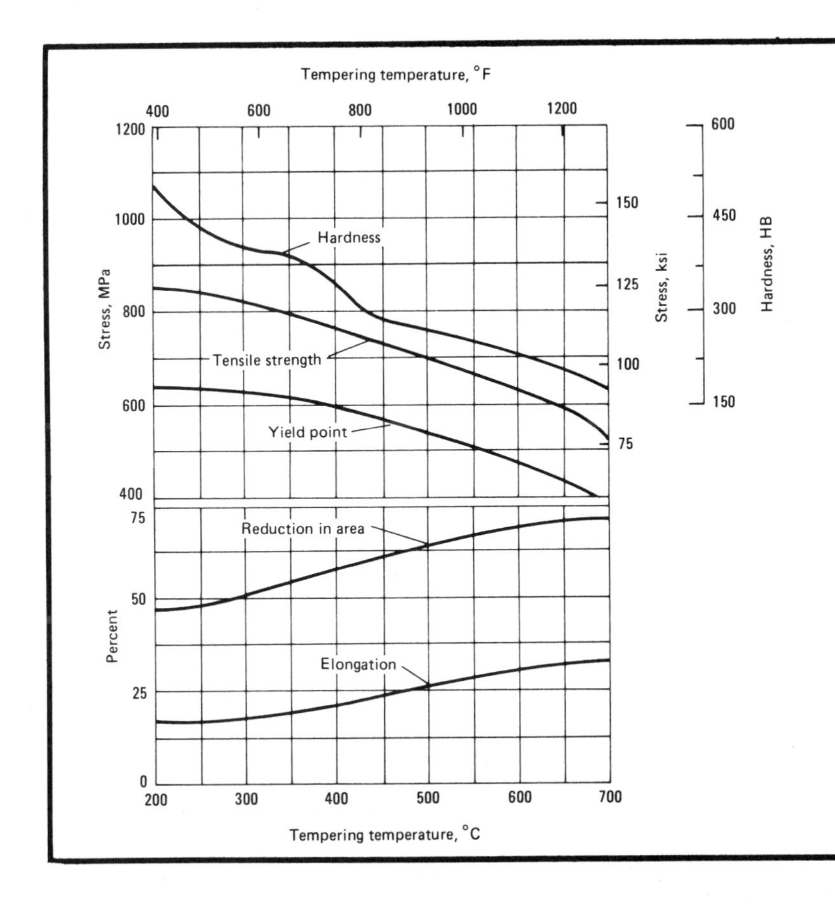

AISI 1030: Effect of Tempering Temperature on Tensile Properties. Normalized at 925 °C (1700 °F); reheated to 870 °C (1600 °F); quenched in water. Specimens were treated in 25-mm (1.0-in.) diam and machined to 12.8-mm (0.505-in.) diam for testing. Tests were conducted using test specimens machined to English units. As-quenched hardness was 514 HB. Elongation was measured in 50 mm (2 in.). (Ref 2)

Machining Data (Ref 8)

AISI 1030, 1034, 1035: Turning (Single Point and Box Tools)

Depth of cut mm	Depth of cut in.	M2 and M3 high speed steel Speed m/min	Speed ft/min	Feed mm/rev	Feed in./rev	Uncoated carbide Speed, brazed m/min	Speed, brazed ft/min	Speed, inserted m/min	Speed, inserted ft/min	Feed mm/rev	Feed in./rev	Coated carbide Speed m/min	Speed ft/min	Feed mm/rev	Feed in./rev
Hardness, 125 to 175 HB															
1	0.040	43	140	0.18	0.007	140(a)	465(a)	180(a)	590(a)	0.18	0.007	280(b)	925(b)	0.18	0.007
4	0.150	35	115	0.40	0.015	110(c)	360(c)	140(c)	460(c)	0.50	0.020	185(d)	600(d)	0.40	0.015
8	0.300	27	90	0.50	0.020	85(c)	280(c)	110(c)	360(c)	0.75	0.030	145(d)	475(d)	0.50	0.020
16	0.625	11	70	0.75	0.030	67(c)	220(c)	85(c)	280(c)	1.00	0.040	...	...	...	...
Hardness, 175 to 225 HB															
1	0.040	40	130	0.18	0.007	130(a)	430(a)	160(a)	525(a)	0.18	0.007	240(b)	785(b)	0.18	0.007
4	0.150	30	100	0.40	0.015	100(c)	325(c)	125(c)	410(c)	0.50	0.020	160(d)	525(d)	0.40	0.015
8	0.300	26	85	0.50	0.020	78(c)	255(c)	100(c)	320(c)	0.75	0.030	125(d)	415(d)	0.50	0.020
16	0.625	20	65	0.75	0.030	60(c)	200(c)	78(c)	255(c)	1.00	0.040	...	...	...	...

(a) Carbide tool material: C-7. (b) Carbide tool material: CC-7. (c) Carbide tool material: C-6. (d) Carbide tool material: CC-6

AISI 1030, 1034, 1035: Turning (Cutoff and Form Tools)

Tool material	Speed, m/min (ft/min)	Cutoff 1.5 mm (0.062 in.) mm	in.	Cutoff 3 mm (0.125 in.) mm	in.	Cutoff 6 mm (0.25 in.) mm	in.	Form 12 mm (0.5 in.) mm	in.	Form 18 mm (0.75 in.) mm	in.	Form 25 mm (1 in.) mm	in.	Form 50 mm (2 in.) mm	in.
Hardness, 125 to 175 HB															
M2 and M3 high speed steel	35 (115)	0.036	0.0014	0.043	0.0017	0.056	0.0022	0.043	0.0017	0.038	0.0015	0.036	0.0014	0.028	0.0011
C-6 carbide	115 (370)	0.036	0.0014	0.043	0.0017	0.056	0.0022	0.043	0.0017	0.038	0.0015	0.036	0.0014	0.028	0.0011
Hardness, 175 to 225 HB															
M2 and M3 high speed steel	29 (95)	0.036	0.0014	0.043	0.0017	0.056	0.0022	0.043	0.0017	0.038	0.0015	0.036	0.0014	0.028	0.0011
C-6 carbide	95 (305)	0.036	0.0014	0.043	0.0017	0.056	0.0022	0.043	0.0017	0.038	0.0015	0.036	0.0014	0.028	0.0011

Feed per revolution for cutoff tool width of: 1.5, 3, 6 mm; Feed per revolution for form tool width of: 12, 18, 25, 50 mm.

AISI 1030, 1034, 1035: Face Milling

Depth of cut mm	Depth of cut in.	M2 and M7 high speed steel Speed m/min	Speed ft/min	Feed/tooth mm	Feed/tooth in.	Uncoated carbide Speed, brazed m/min	Speed, brazed ft/min	Speed, inserted m/min	Speed, inserted ft/min	Feed/tooth mm	Feed/tooth in.	Coated carbide Speed m/min	Speed ft/min	Feed/tooth mm	Feed/tooth in.
Hardness, 125 to 175 HB															
1	0.040	55	180	0.20	0.008	190(a)	625(a)	215(a)	700(a)	0.20	0.008	320(b)	1050(b)	0.20	0.008
4	0.150	46	150	0.30	0.012	145(a)	475(a)	165(a)	540(a)	0.30	0.012	215(b)	700(b)	0.30	0.012
8	0.300	35	115	0.40	0.016	105(c)	345(c)	130(c)	420(c)	0.40	0.016	165(d)	545(d)	0.40	0.016
Hardness, 175 to 225 HB															
1	0.040	49	160	0.20	0.008	160(a)	525(a)	170(a)	560(a)	0.20	0.008	255(b)	840(b)	0.20	0.008
4	0.150	38	125	0.30	0.012	125(a)	410(a)	145(a)	480(a)	0.30	0.012	190(b)	625(b)	0.30	0.012
8	0.300	30	100	0.40	0.016	95(c)	310(c)	115(c)	375(c)	0.40	0.016	150(d)	485(d)	0.40	0.016

(a) Carbide tool material: C-6. (b) Carbide tool material: CC-6. (c) Carbide tool material: C-5. (d) Carbide tool material: CC-5

AISI 1030, 1034, 1035: Drilling

Tool material	Speed m/min	Speed ft/min	1.5 mm (0.062 in.) mm	in.	3 mm (0.125 in.) mm	in.	6 mm (0.25 in.) mm	in.	12 mm (0.5 in.) mm	in.	18 mm (0.75 in.) mm	in.	25 mm (1 in.) mm	in.
Hardness, 125 to 175 HB														
M10, M7, and M1 high speed steel	21	70	0.025	0.001	...	...	...	...	...	...	...	...	...	...
	26	85	...	...	0.075	0.003	0.13	0.005	0.23	0.009	0.30	0.012	0.45	0.018
Hardness, 175 to 225 HB														
M10, M7, and M1 high speed steel	20	65	0.025	0.001	...	...	...	...	...	...	...	...	...	...
	23	75	...	...	0.075	0.003	0.13	0.005	0.23	0.009	0.30	0.012	0.45	0.018

Feed per revolution for nominal hole diameter of: 1.5 mm to 25 mm.

Machining Data (Ref 8) (continued)

AISI 1030, 1034, 1035: End Milling (Profiling)

Tool material	Depth of cut mm	Depth of cut in.	Speed m/min	Speed ft/min	Feed per tooth for cutter diameter fo: 10 mm (0.375 in.) mm	10 mm (0.375 in.) in.	12 mm (0.5 in.) mm	12 mm (0.5 in.) in.	18 mm (0.75 in.) mm	18 mm (0.75 in.) in.	25-50 mm (1-2 in.) mm	25-50 mm (1-2 in.) in.
Hardness, 125 to 175 HB												
M2, M3, and M7 high speed steel	0.5	0.020	60	200	0.025	0.001	0.050	0.002	0.102	0.004	0.13	0.005
	1.5	0.060	46	150	0.050	0.002	0.075	0.003	0.13	0.005	0.15	0.006
	diam/4	diam/4	43	140	0.025	0.001	0.050	0.002	0.102	0.004	0.13	0.005
	diam/2	diam/2	37	120	0.018	0.0007	0.025	0.001	0.075	0.003	0.102	0.004
C-5 carbide	0.5	0.020	160	520	0.038	0.0015	0.089	0.0035	0.13	0.005	0.18	0.007
	1.5	0.060	120	400	0.063	0.0025	0.102	0.004	0.15	0.006	0.20	0.008
	diam/4	diam/4	105	350	0.050	0.002	0.075	0.003	0.13	0.005	0.15	0.006
	diam/2	diam/2	100	320	0.038	0.0015	0.050	0.002	0.102	0.004	0.13	0.005
Hardness, 175 to 225 HB												
M2, M3, and M7 high speed steel	0.5	0.020	49	160	0.025	0.001	0.050	0.002	0.075	0.003	0.102	0.004
	1.5	0.060	37	120	0.050	0.002	0.075	0.003	0.102	0.004	0.13	0.005
	diam/4	diam/4	32	105	0.025	0.001	0.050	0.002	0.075	0.003	0.102	0.004
	diam/2	diam/2	27	90	0.018	0.0007	0.025	0.001	0.050	0.002	0.075	0.003
C-5 carbide	0.5	0.020	150	485	0.038	0.0015	0.075	0.003	0.13	0.005	0.15	0.006
	1.5	0.060	115	370	0.063	0.0025	0.102	0.004	0.15	0.006	0.18	0.007
	diam/4	diam/4	100	325	0.050	0.002	0.075	0.003	0.13	0.005	0.15	0.006
	diam/2	diam/2	90	300	0.038	0.0015	0.050	0.002	0.102	0.004	0.13	0.005

AISI 1030, 1034, 1035: Boring

Depth of cut mm	Depth of cut in.	M2 and M3 high speed steel Speed m/min	Speed ft/min	Feed mm/rev	Feed in./rev	Uncoated carbide Speed, brazed m/min	Speed, brazed ft/min	Speed, inserted m/min	Speed, inserted ft/min	Feed mm/rev	Feed in./rev	Coated carbide Speed m/min	Speed ft/min	Feed mm/rev	Feed in./rev
Hardness, 125 to 175 HB															
0.25	0.010	43	140	0.075	0.003	135(a)	445(a)	160(a)	525(a)	0.075	0.003	210(b)	690(b)	0.075	0.003
1.25	0.050	34	110	0.13	0.005	105(c)	350(c)	125(c)	415(c)	0.13	0.005	170(d)	550(d)	0.13	0.005
2.5	0.100	27	90	0.30	0.012	84(e)	275(e)	100(e)	325(e)	0.40	0.015	115(f)	370(f)	0.30	0.012
Hardness, 175 to 225 HB															
0.25	0.010	37	120	0.075	0.003	120(c)	390(c)	140(c)	460(c)	0.075	0.003	245(d)	810(d)	0.075	0.003
1.25	0.050	29	95	0.13	0.005	95(e)	315(e)	115(e)	370(e)	0.13	0.005	200(f)	650(f)	0.13	0.005
2.5	0.100	24	80	0.30	0.012	84(e)	275(e)	100(e)	325(e)	0.40	0.015	115(f)	370(f)	0.30	0.012

(a) Carbide tool material: C-8. (b) Carbide tool material: CC-8. (c) Carbide tool material: C-7. (d) Carbide tool material: CC-7. (e) Carbide tool material: C-6. (f) Carbide tool material: CC-6

AISI 1030, 1034, 1035: Planing

Tool material	Depth of cut mm	Depth of cut in.	Speed m/min	Speed ft/min	Feed/stroke mm	Feed/stroke in.
Hardness, 125 to 175 HB						
M2 and M3 high speed steel	0.1	0.005	14	45	(a)	(a)
	2.5	0.100	20	65	1.25	0.050
	12	0.500	12	40	1.50	0.060
C-6 carbide	0.1	0.005	76	250	(a)	(a)
	2.5	0.100	84	275	2.05	0.080
	12	0.500	69	225	1.50	0.060
Hardness, 175 to 225 HB						
M2 and M3 high speed steel	0.1	0.005	12	40	(a)	(a)
	2.5	0.100	18	60	1.25	0.050
	12	0.500	11	35	1.50	0.060
C-6 carbide	0.1	0.005	70	230	(a)	(a)
	2.5	0.100	76	250	2.05	0.080
	12	0.500	60	200	1.50	0.060

(a) Feed is 75% the width of the square nose finishing tool

Machining Data (Ref 8) (continued)

AISI 1030, 1034, 1035: Reaming

Based on 4 flutes for 3- and 6-mm (0.125- and 0.25-in.) reamers, 6 flutes for 12-mm (0.5-in.) reamers, and 8 flutes for 25-mm (1-in.) and larger reamers

Tool material	Speed		Feed per revolution for reamer diameter of: 3 mm (0.125 in.)		6 mm (0.25 in.)		12 mm (0.5 in.)		25 mm (1 in.)		35 mm (1.5 in.)		50 mm (2 in.)	
	m/min	ft/min	mm	in.	mm	in.	mm	in.	mm	in.	mm	in.	mm	in.
Hardness, 125 to 175 HB														
Roughing														
M1, M2, and M7 high speed steel	26	85	0.102	0.004	0.18	0.007	0.30	0.012	0.50	0.020	0.65	0.025	0.75	0.030
C-2 carbide	30	100	0.102	0.004	0.18	0.007	0.30	0.012	0.50	0.020	0.65	0.025	0.75	0.030
Finishing														
M1, M2, and M7 high speed steel	12	40	0.15	0.006	0.25	0.010	0.40	0.015	0.65	0.025	0.75	0.030	0.90	0.035
C-2 carbide	15	50	0.15	0.006	0.25	0.010	0.40	0.015	0.65	0.025	0.75	0.030	0.90	0.035
Hardness, 175 to 225 HB														
Roughing														
M1, M2, and M7 high speed steel	23	75	0.102	0.004	0.18	0.007	0.30	0.012	0.50	0.020	0.65	0.025	0.75	0.030
C-2 carbide	27	90	0.102	0.004	0.18	0.007	0.30	0.012	0.50	0.020	0.65	0.025	0.75	0.030
Finishing														
M10, M2, and M7 high speed steel	11	35	0.13	0.005	0.20	0.008	0.30	0.012	0.50	0.020	0.65	0.025	0.75	0.030
C-2 carbide	14	45	0.13	0.005	0.20	0.008	0.30	0.012	0.50	0.020	0.65	0.025	0.75	0.030

AISI 1038, 1038H, 1039, 1040

AISI 1038, 1038H, 1039, 1040: Chemical Composition

AISI grade	Chemical composition, % C	Mn	P max	S max
1038	0.35-0.42	0.60-0.90	0.040	0.050
1038H	0.34-0.43	0.50-1.00	0.040	0.050
1039	0.37-0.44	0.70-1.00	0.040	0.050
1040	0.37-0.44	0.60-0.90	0.040	0.050

Characteristics. AISI grades 1038, 1038H, 1039, and 1040 are medium-high carbon steels that can be strengthened by heat treating after forming. Machinability and weldability are fair.

Typical Uses. Machine, plow, and carriage bolts, tie wire, cylinder head studs, and machined parts are made from grade 1038, 1038H, 1039, and 1040 steels. These grades are also used for U-bolts, concrete reinforcing rods, forgings, and noncritical springs. Grade 5, SAE J429 fasteners are made from grade 1038 steel.

AISI 1038: Similar Steels (U.S. and/or Foreign). UNS G10380; ASTM A29, A510, A544, A545, A546, A576; SAE J403, J412, J414; (W. Ger.) DIN 1.1176; (Fr.) AFNOR XC 38 TS

AISI 1038H: Similar Steels (U.S. and/or Foreign). UNS H10380; ASTM A29; SAE J776; (W. Ger.) DIN 1.1176; (Fr.) AFNOR XC 38 TS

AISI 1039: Similar Steels (U.S. and/or Foreign). UNS G10390; ASTM A29, A510, A546, A576; SAE J403, J412, J414; (W. Ger.) DIN 1.1157; (Fr.) AFNOR 35 M 5; (U.K.) B.S. 120 M 36, 150 M 36, CDS 105/106

AISI 1040: Similar Steels (U.S. and/or Foreign). UNS G10400; ASTM A29, A108, A510, A519, A546, A576, A682; MIL SPEC MIL-S-11310 (CS1040); SAE J403, J412, J414; (W. Ger.) DIN 1.1186; (Jap.) JIS S 40 C; (U.K.) B.S. 080 A 40, 2 S 93

AISI 1040: Approximate Critical Points

Transformation point	Temperature(a) °C	°F
Ac_1	725	1340
Ac_3	795	1460
Ar_3	755	1395
Ar_1	670	1240

(a) On heating or cooling at 28 °C (50 °F) per hour

Machining Data

For machining data on AISI grades 1038, 1038H, 1039, and 1040, refer to the preceding machining tables for AISI grades 1030, 1034, and 1035.

Physical Properties

AISI 1038, 1038H, 1039, 1040: Electrical Resistivity and Thermal Conductivity (Ref 4)

Temperature °C	Temperature °F	Electrical resistivity, μΩ·m	Thermal conductivity W/m·K	Thermal conductivity Btu/ft·h·°F
0	32	...	51.9	30.0
20	68	0.171	...	...
100	212	0.221	50.7	29.3
200	390	0.296	48.1	27.8
300	570	...	45.7	26.4
400	750	0.493	41.7	24.1
500	930	...	38.2	22.1
600	1110	0.763	33.9	19.6
700	1290	0.932	30.1	17.4
800	1470	1.111	24.7	14.3
900	1650	1.149	...	...
1000	1830	1.179	32.9	19.0
1100	2010	1.207	...	...
1200	2190	1.230	29.8	17.2

Material composition: 0.415% carbon, 0.643% manganese

AISI 1038, 1038H, 1039, 1040: Mean Apparent Specific Heat (Ref 4)

Temperature range °C	Temperature range °F	Specific heat J/kg·K	Specific heat Btu/lb·°F
50-100	120-212	486	0.116
150-200	300-390	515	0.123
200-250	390-480	528	0.126
250-300	480-570	548	0.131
300-350	570-660	569	0.136
350-400	660-750	586	0.140
450-500	840-930	649	0.155
550-600	1020-1110	708	0.169
650-700	1200-1290	770	0.184
700-750	1290-1380	1583	0.378
750-800	1380-1470	624	0.149
850-900	1560-1650	548	0.131

Material composition: 0.415% carbon, 0.643% manganese

AISI 1040: Average Coefficients of Linear Thermal Expansion (Ref 4)

Temperature range °C	Temperature range °F	Coefficient μm/m·K	Coefficient μin./in.·°F
Material A(a)			
20-100	68-212	11.3	6.3
20-200	68-390	12.1	6.7
20-300	68-570	12.2	6.8
20-400	68-750	13.3	7.4
20-500	68-930	13.9	7.7
20-600	68-1110	14.2	7.9
20-700	68-1290	14.8	8.2
Material B(b)			
0-100	32-212	11.2	6.2
0-200	32-390	12.1	6.7
0-300	32-570	13.0	7.2
0-400	32-750	13.5	7.5
0-500	32-930	14.0	7.8
0-600	32-1110	14.6	8.1
0-700	32-1290	14.8	8.2

(a) Material composition: 0.40% carbon, 0.11% manganese, 0.01% phosphorus, 0.03% sulfur, 0.03% silicon, 0.03% copper. (b) Material composition: 0.42% carbon, 0.64% manganese, 0.031% phosphorus, 0.029% sulfur, 0.11% silicon, 0.06% nickel, 0.12% copper, 0.006% aluminum, 0.033% arsenic

AISI 1040: Thermal Treatment Temperatures

Treatment	Temperature range °C	Temperature range °F
Forging, start	1290-1150	2350-2100
Forging, finish	1010-870	1850-1600
Annealing	790-870	1450-1600
Normalizing	885-915	1625-1675
Quenching	830-855	1525-1575

AISI 1038, 1038H, 1039, 1040: Density (Ref 4)

7.845 g/cm³ .... 0.2834 lb/in.³

Material composition: 0.435% carbon, 0.69% manganese, 0.20% silicon; annealed at 860 °C (1580 °F)

Mechanical Properties

AISI 1038, 1039, 1040: Tensile Properties

Condition or treatment	Size round mm	Size round in.	Tensile strength MPa	Tensile strength ksi	Yield strength MPa	Yield strength ksi	Elongation(a), %	Reduction in area, %	Hardness, HB
AISI 1038									
Hot rolled (Ref 1)	19-32	0.75-1.25	515	75	285	41	18	40	149
Cold drawn(b) (Ref 1)	19-32	0.75-1.25	570	83	485	70	12	35	163
AISI 1039									
Hot rolled (Ref 1)	19-32	0.75-1.25	545	79	300	43.5	16	40	156
Cold drawn(c) (Ref 1)	19-32	0.75-1.25	605	88	510	74	12	35	179
AISI 1040									
Hot rolled (Ref 1)	19-32	0.75-1.25	525	76	290	42	18	40	149
Cold drawn(c) (Ref 1)	19-32	0.75-1.25	585	85	490	71	12	35	170
As cold drawn (Ref 1)	16-22	0.63-0.88	620	90	550	80	12	35	170
	22-32	0.88-1.25	585	85	515	75	12	35	170
	32-50	1.25-2	550	80	485	70	10	30	163
	50-75	2-3	515	75	450	65	10	30	149
Cold drawn, low temperature, stress relieved (Ref 1)	16-22	0.63-0.88	655	95	585	85	12	35	187
	22-32	0.88-1.25	620	90	550	80	12	35	178
	32-50	1.25-2	585	85	515	75	10	30	170
	50-75	2-3	550	80	485	70	10	30	163
Cold drawn, high temperature, stress relieved (Ref 1)	16-22	0.63-0.88	585	85	450	65	15	45	170
	22-32	0.88-1.25	550	80	450	65	15	45	163
	32-50	1.25-2	515	75	415	60	15	40	149
	50-75	2-3	485	70	450	65	12	35	143
As rolled (Ref 12)	...	...	620	90	415	60	25	50	201
Normalized at 900 °C (1650 °F) (Ref 12)	...	...	595	86	370	54	28	55	170
Annealed at 790 °C (1450 °F) (Ref 12)	...	...	515	75	350	51	30	57	149

(a) In 50 mm (2 in.). (b) Average machinability rating of 65% based on AISI 1212 steel as 100% average machinability. (c) Average machinability rating of 60% based on AISI 1212 steel as 100% average machinability

Mechanical Properties (continued)

AISI 1040: Mass Effect on Mechanical Properties (Ref 2)

Condition or treatment	Size round mm	Size round in.	Tensile strength MPa	Tensile strength ksi	Yield strength MPa	Yield strength ksi	Elongation(a), %	Reduction in area, %	Hardness, HB
Annealed (heated to 790 °C or 1450 °F, furnace cooled 11 °C or 20 °F per hour to 650 °C or 1200 °F, cooled in air)	25	1	519	75.25	353	51.25	30.2	57.2	149
Normalized (heated to 900 °C or 1650 °F, cooled in air)	13	0.5	608	88.25	403	58.50	30.0	56.5	183
	25	1	590	85.50	374	54.25	28.0	54.9	170
	50	2	581	84.25	365	53.00	28.0	53.3	167
	100	4	583	84.50	340	49.25	27.0	51.8	167
Oil quenched from 855 °C (1575 °F), tempered at 540 °C (1000 °F)	13	0.5	722	104.75	500	72.50	27.0	62.0	217
	25	1	664	96.25	469	68.00	26.5	61.1	197
	50	2	635	92.25	412	59.75	27.0	59.7	187
	100	4	621	90.00	396	57.50	27.0	60.3	179
Oil quenched from 855 °C (1575 °F), tempered at 595 °C (1100 °F)	13	0.5	693	100.50	479	69.50	27.0	65.2	207
	25	1	631	91.50	443	64.25	28.2	63.5	187
	50	2	598	86.75	392	56.87	28.0	62.5	174
	100	4	571	82.75	360	52.25	30.0	61.6	170
Oil quenched from 855 °C (1575 °F), tempered at 650 °C (1200 °F)	13	0.5	655	95.00	459	66.62	28.9	65.4	197
	25	1	588	85.25	415	60.25	30.0	67.4	170
	50	2	569	82.50	376	54.50	31.0	66.4	167
	100	4	543	78.75	345	50.00	31.2	64.5	156
Water quenched from 845 °C (1550 °F), tempered at 540 °C (1000 °F)	13	0.5	752	109.00	562	81.5	23.8	61.5	223
	25	1	743	107.75	541	78.5	23.2	62.6	217
	50	2	702	101.75	479	69.5	24.7	63.6	207
	100	4	683	99.00	440	63.8	24.7	60.2	201
Water quenched from 845 °C (1550 °F), tempered at 595 °C (1100 °F)	13	0.5	698	101.25	490	71.0	26.4	65.2	212
	25	1	689	100.00	479	69.5	26.0	65.0	207
	50	2	655	95.00	469	68.0	29.0	69.2	197
	100	4	650	94.25	407	59.1	27.0	63.4	192
Water quenched from 845 °C (1550 °F), tempered at 650 °C (1200 °F)	13	0.5	662	96.00	476	69	27.7	66.6	201
	25	1	645	93.50	469	68	27.0	67.9	197
	50	2	614	89.00	413	59.9	28.7	69.0	183
	100	4	586	85.00	378	54.8	30.2	67.2	170

(a) In 50 mm (2 in.)

AISI 1040: Effect of the Mass on Hardness at Selected Points (Ref 2)

Size round mm	Size round in.	As-quenched hardness at: Surface	½ radius	Center
After quenching in oil				
13	0.5	28 HRC	22 HRC	21 HRC
25	1	23 HRC	21 HRC	18 HRC
50	2	93 HRB	92 HRB	91 HRB
100	4	91 HRB	91 HRB	89 HRB
After quenching in water				
13	0.5	54 HRC	53 HRC	53 HRC
25	1	50 HRC	22 HRC	18 HRC
50	2	50 HRC	97 HRB	95 HRB
100	4	98 HRB	96 HRB	95 HRB

AISI 1038H: End-Quench Hardenability Limits (Ref 2)

Distance from quenched end, 1/16 in.	Hardness, HRC max	Hardness, HRC min
1	58	51
2	55	34
3	49	26
4	37	23
5	30	22
6	28	21
7	27	...
8	26	...
10	25	...
12	24	...
14	23	...
16	21	...

Mechanical Properties (continued)

AISI 1040: Recommended Condition to Provide Maximum Drawability (Ref 6)

Punch nose radius, r(a)	Condition, HRB, for cup reduction of: 20%	30%	40%
2t	65(b)(c)(d)	...	...
4t	60	65(b)(c)	...
8t	60	60	...
16t	65(b)(c)	65(b)(c)	...
32t	60	65(b)(c)	65(b)(c)(d)

Drawing limitations refer to cold rolled and mill annealed sheet steel; simple round cups, no flanges, not ironed, 20-gage (0.91-mm or 0.036-in.) sheet steel. (a) In relation to stock thickness, *t*. (b) Aluminum killed. (c) Fine grained. (d) Spheroidized

AISI 1038, 1038H, 1039, 1040: Izod Impact Properties (Ref 12)

Condition or treatment	Energy J	ft·lb
As rolled	49	36
Normalized at 900 °C (1650 °F)	65	48
Annealed at 790 °C (1450 °F)	45	33

AISI 1040: Effect of Tempering Temperature on Tensile Properties After Quenching in Water.

Normalized at 900 °C (1650 °F); reheated to 845 °C (1550 °F); quenched in water. Specimens were treated in 25-mm (1.0-in.) diam and machined to 12.8-mm (0.505-in.) diam for testing. Tests were conducted using test specimens machined to English units. As-quenched hardness was 534 HB. Elongation was measured in 50 mm (2 in.). (Ref 2)

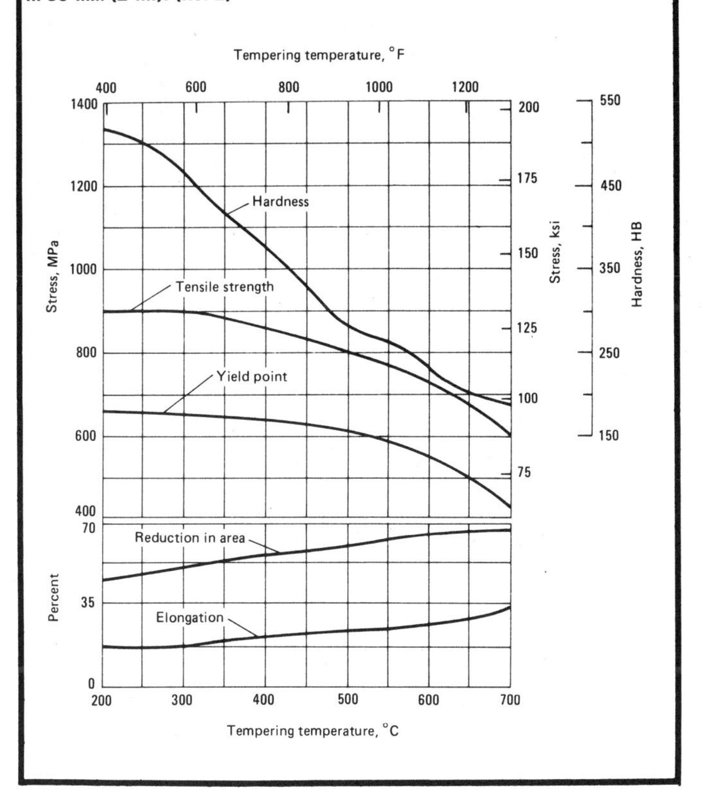

AISI 1040: Effect of Tempering Temperature on Tensile Properties After Quenching in Oil.

Normalized at 900 °C (1650 °F); reheated to 855 °C (1575 °F); quenched in oil. Specimens were treated in 25-mm (1.0-in.) diam and machined to 12.8-mm (0.505-in.) diam for testing. Tests were conducted using test specimens machined to English units. As-quenched hardness was 269 HB. Elongation was measured in 50 mm (2 in.). (Ref 2)

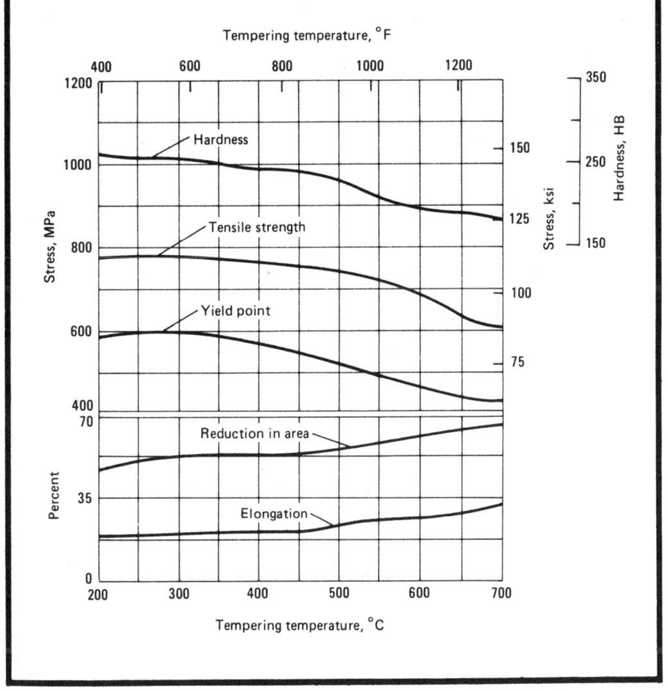

AISI 1044, 1045, 1045H, 1046

AISI 1044, 1045, 1045H, 1046: Chemical Composition

AISI grade	Chemical composition, % C	Mn	P max	S max
1044	0.43-0.50	0.30-0.60	0.040	0.050
1045	0.43-0.50	0.60-0.90	0.040	0.050
1045H	0.42-0.51	0.50-1.00	0.040	0.050
1046	0.43-0.50	0.70-1.00	0.040	0.050

Characteristics. AISI grades 1044, 1045, 1045H, and 1046 are used when greater strength and hardness are desired in the as-rolled condition. These medium-carbon steels can be hammer forged. They respond to heat treatment, and flame and induction hardening, but are not recommended for carburizing or cyaniding. When special practices are employed, weldability is fair. Die forging and hot upsetting of these steels are good to excellent.

Typical Uses. AISI grades 1044, 1045, 1045H, and 1046 are used for gears, shafts, axles, bolts, studs, and machine parts.

AISI 1044: Similar Steels (U.S. and/or Foreign). UNS G10440; ASTM A29, A510, A575, A576; SAE J403, J412, J414

AISI 1045: Similar Steels (U.S. and/or Foreign). UNS G10450; ASTM A29, A510, A519, A576, A682; FED QQ-S-635 (C1045), QQ-S-700 (C1045); SAE J403, J412, J414; (W. Ger.) DIN 1.1191; (Fr.) AFNOR XC 42, XC 42 TS, XC 45, XC 48; (Jap.) JIS S 45 C, S 48 C; (Swed.) SS_{14} 1672

AISI 1045H: Similar Steels (U.S. and/or Foreign). UNS H10450; ASTM A29; SAE J776; (W. Ger.) DIN 1.1191; (Fr.) AFNOR XC 42, XC 42 TS, XC 45, XC 48; (Jap.) JIS S 45 C, S 48 C; (Swed.) SS_{14} 1672

AISI 1046: Similar Steels (U.S. and/or Foreign). UNS G10460; ASTM A29, A510, A576; SAE J403, J412, J414

AISI 1044, 1045, 1045H, 1046: Approximate Critical Points

Transformation point	Temperature(a) °C	°F
Ac_1	725	1340
Ac_3	780	1435
Ar_3	750	1385
Ar_1	680	1260

(a) On heating or cooling at 28 °C (50 °F) per hour

Physical Properties

AISI 1045: Average Coefficients of Linear Thermal Expansion (Ref 4)

Temperature range °C	°F	Coefficient μm/m·K	μin./in.·°F
0-100	32-212	11.5	6.4
0-200	32-390	12.2	6.8
0-300	32-570	13.0	7.2
0-400	32-750	13.7	7.6
0-500	32-930	14.0	7.8
0-600	32-1110	14.6	8.1
0-700	32-1290	15.1	8.4
25-100	75-212	11.2	6.2
25-200	75-390	11.9	6.6
25-300	75-570	12.6	7.0
25-400	75-750	13.5	7.5
25-500	75-930	14.0	7.8
25-600	75-1110	14.4	8.0
25-700	75-1290	14.8	8.2

AISI 1045: Thermal Treatment Temperatures

Treatment	Temperature range °C	°F
Forging	1245 max	2275 max
Annealing	790-870	1450-1600
Normalizing	830-915	1525-1675
Austenitizing	800-845	1475-1550

Machining Data

For machining data on AISI grades 1044, 1045, 1045H, and 1046, refer to the preceding machining tables for AISI grades 1030, 1034, and 1035.

Mechanical Properties

AISI 1045H: End-Quench Hardenability Limits (Ref 2)

Distance from quenched end, 1/16 in.	Hardness, HRC max	min
1	62	55
2	59	42
3	52	31
4	38	28
5	33	26
6	32	25
7	31	25
8	30	24
10	29	22
12	28	21
14	27	20
16	26	...

5, 1046: Mechanical Properties

atment	Size round mm	in.	Tensile strength MPa	ksi	Yield strength MPa	ksi	Elongation(a), %	Reduction in area, %	Hardness, HB	Average machinability rating(b)
hot										
....	19-32	0.75-1.25	620	90	415	60	26	50	...	64
....	19-32	0.75-1.25	550	80	310	45	16	40	163	...
....	19-32	0.75-1.25	565	82	310	45	16	40	163	...
)...............	19-32	0.75-1.25	625	91	530	77	12	35	179	55
aled										
....	19-32	0.75-1.25	585	85	505	73	12	45	170	65
nd										
....	19-32	0.75-1.25	675	98	405	59	24	45	212	56
f 1)	16-22	0.63-0.88	655	95	585	85	12	35	187	...
	22-32	0.88-1.25	620	90	550	80	11	30	179	...
	32-50	1.25-2	585	85	515	75	10	30	170	...
	50-75	2-3	515	75	485	70	10	30	163	...
Cold drawn, low temperature, stress relieved (Ref 1)	16-22	0.63-0.88	690	100	620	90	12	35	197	...
	22-32	0.88-1.25	655	95	585	85	11	30	187	...
	32-50	1.25-2	620	90	550	80	10	30	179	...
	50-75	2-3	585	85	515	75	10	25	170	...
Cold drawn, high temperature, stress relieved (Ref 1)	16-22	0.63-0.88	655	95	515	75	15	45	187	...
	22-32	0.88-1.25	620	90	515	75	15	40	179	...
	32-50	1.25-2	585	85	485	70	15	40	170	...
	50-75	2-3	550	80	450	65	12	35	163	...
AISI 1046										
Hot rolled (Ref 1)	19-32	0.75-1.25	585	85	325	47	15	40	170	...
Cold drawn (Ref 1)	19-32	0.75-1.25	650	94	545	79	12	35	187	55
Cold drawn, annealed (Ref 1)	19-32	0.75-1.25	620	90	515	75	12	45	179	65

(a) In 50 mm (2 in.). (b) Based on AISI 1212 steel as 100% average machinability

AISI 1050

AISI 1050: Chemical Composition

AISI grade	C	Mn	P max	S max
1050	0.48-0.55	0.060-0.090	0.040	0.050

Characteristics and Typical Uses. Strain-tempered AISI grade 1050 steel bar offers a minimum yield strength of 690 MPa (100 ksi). For applications such as piston rods, grade 1050 can be flame hardened to a hardness of 56 to 60 HRC. This medium-carbon steel is also used as a spring material from cold rolled carbon strip in the annealed or tempered condition.

AISI 1050: Similar Steels (U.S. and/or Foreign). UNS G10500; AMS 5085; ASTM A29, A510, A519, A576, A682; FED QQ-S-635 (C1050), QQ-S-700 (C1050); MIL SPEC MIL-S-16974; SAE J403, J412, J414; (W. Ger.) DIN 1.1210; (Jap.) JIS S 53 C, S 55 C

AISI 1050: Approximate Critical Points

Transformation point	Temperature(a) °C	°F
Ac_1	730	1340
Ac_3	770	1415
Ar_3	740	1365
Ar_1	680	1260

(a) On heating or cooling at 28 °C (50 °F) per hour

Machining Data

For machining data on AISI grade 1050, refer to the preceding machining tables for AISI grades 1030, 1034, and 1035.

Mechanical Properties

AISI 1050: Tensile Properties

Condition or treatment	Size round mm	Size round in.	Tensile strength MPa	Tensile strength ksi	Yield strength MPa	Yield strength ksi	Elongation(a), %	Reduction in area, %	Hardness, HB
Hot rolled (Ref 1)	19-32	0.75-1.25	620	90	345	50	15	35	179
Cold drawn(b) (Ref 1)	19-32	0.75-1.25	690	100	580	84	10	30	197
Cold drawn, annealed(c) (Ref 1)	19-32	0.75-1.25	655	95	550	80	10	40	189
As cold drawn bar (Ref 1)	16-22	0.63-0.88	690	100	620	90	11	35	197
	22-32	0.88-1.25	655	95	585	85	11	30	187
	32-50	1.25-2	620	90	550	80	10	30	179
	50-75	2-3	585	85	515	75	10	30	170
Cold drawn, low temperature, stress relieved (Ref 1)	16-22	0.63-0.88	725	105	655	95	11	35	212
	22-32	0.88-1.25	690	100	620	90	11	30	197
	32-50	1.25-2	655	95	585	85	10	30	187
	50-75	2-3	620	90	550	80	10	25	179
Cold drawn, high temperature, stress relieved (Ref 1)	16-22	0.63-0.88	655	95	515	75	15	45	187
	22-32	0.88-1.25	620	90	515	75	15	40	179
	32-50	1.25-2	585	85	485	70	15	40	170
	50-75	2-3	550	80	450	65	12	35	163
As rolled (Ref 12)	...	...	725	105	415	60	20	40	229
Normalized at 900 °C (1650 °F) (Ref 12)	...	...	752	109	425	62	20	39	217
Annealed at 790 °C (1450 °F) (Ref 12)	...	...	635	92	365	53 24		40	187

(a) In 50 mm (2 in.). (b) Average machinability rating of 45% based on AISI 1212 steel as 100% average machinability. (c) Average machinability rating of 55% based on AISI 1212 steel as 100% average machinability

AISI 1050: Effect of Tempering Temperature on Tensile Properties After Quenching in Water. Normalized at 900 °C (1650 °F); reheated to 830 °C (1525 °F); quenched in water. Specimens were treated in 25-mm (1.0-in.) diam and machined to 12.8-mm (0.505-in.) diam for testing. Tests were conducted using test specimens machined to English units. As-quenched hardness was 601 HB. Elongation was measured in 50 mm (2 in.). (Ref 2)

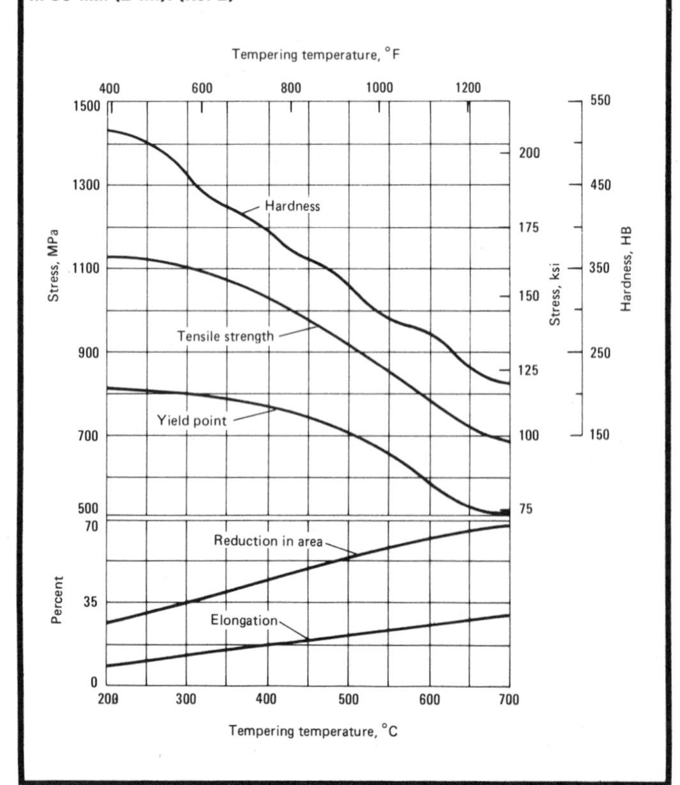

AISI 1050: Effect of Tempering Temperature on Tensile Properties After Quenching in Oil. Normalized at 900 °C (1650 °F); reheated to 845 °C (1550 °F); quenched in oil. Specimens were treated in 25-mm (1.0-in.) diam and machined to 12.8-mm (0.505-in.) diam for testing. Tests were conducted using test specimens machined to English units. As-quenched hardness was 321 HB. Elongation was measured in 50 mm (2 in.). (Ref 2)

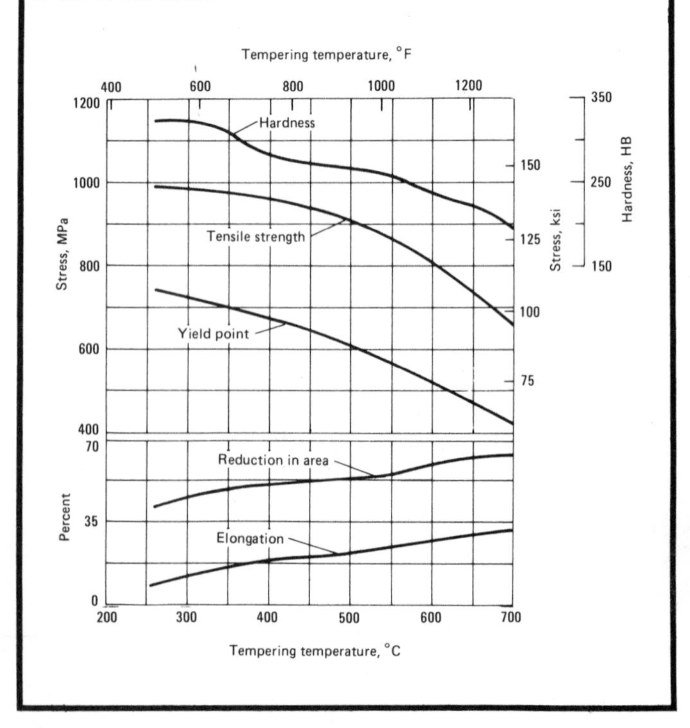

Mechanical Properties (continued)

AISI 1050: Mass Effect on Mechanical Properties (Ref 2)

Condition or treatment	Size round mm	Size round in.	Tensile strength MPa	Tensile strength ksi	Yield strength MPa	Yield strength ksi	Elongation(a), %	Reduction in area, %	Hardness, HB
Annealed (heated to 790 °C or 1450 °F, furnace cooled 10 °C or 20 °F per hour to 650 °C or 1200 °F, cooled in air)	25	1	636	92.2	365	53.0	24.0	40.0	187
Normalized (heated to 900 °C or 1650 °F, cooled in air)	13	0.5	769	111.5	431	62.5	21.5	45.1	223
	25	1	748	108.5	427	62.0	20.0	39.4	217
	50	2	732	106.2	402	58.3	20.0	38.8	212
	100	4	689	100.0	386	56.0	21.7	41.6	201
Oil quenched from 845 °C (1550 °F), tempered at 540 °C (1000 °F)	13	0.5	914	132.5	603	87.5	20.7	52.9	262
	25	1	852	123.5	524	76.0	20.2	53.3	248
	50	2	845	122.5	516	74.8	19.7	51.4	248
	100	4	834	121.0	476	69.0	19.7	48.0	241
Oil quenched from 845 °C (1550 °F), tempered at 595 °C (1100 °F)	13	0.5	841	122.0	558	81.0	22.8	58.1	248
	25	1	786	114.0	486	70.5	23.5	57.6	223
	50	2	772	112.0	469	68.0	23.0	55.6	223
	100	4	696	101.0	405	58.7	25.2	54.5	207
Oil quenched from 845 °C (1550 °F), tempered at 650 °C (1200 °F)	13	0.5	776	112.5	510	74.0	24.6	61.8	229
	25	1	731	106.0	443	64.2	24.7	60.5	217
	50	2	724	105.0	441	64.0	25.0	59.1	217
	100	4	667	96.7	384	55.7	25.5	59.6	197
Water quenched from 830 °C (1525 °F), tempered at 540 °C (1000 °F)	13	0.5	924	134.0	683	99.0	20.0	54.4	269
	25	1	905	131.2	636	92.2	20.0	55.2	262
	50	2	893	129.5	580	84.1	20.7	56.6	255
	100	4	846	122.7	539	78.2	21.5	55.3	248
Water quenched from 830 °C (1525 °F), tempered at 595 °C (1100 °F)	13	0.5	820	119.0	607	88.0	21.7	59.9	241
	25	1	814	118.0	552	80.0	22.5	59.9	241
	50	2	808	117.2	543	78.7	23.0	61.0	235
	100	4	774	112.2	470	68.2	23.7	55.5	229
Water quenched from 830 °C (1525 °F), tempered at 650 °C (1200 °F)	13	0.5	758	110.0	593	86.0	24.8	60.6	229
	25	1	752	109.0	527	76.5	23.7	61.2	229
	50	2	743	107.7	472	68.5	24.7	61.0	223
	100	4	721	104.5	450	65.2	25.2	60.8	217

(a) In 50 mm (2 in.)

AISI 1050: Effect of the Mass on Hardness at Selected Points (Ref 2)

Size round mm	Size round in.	As-quenched hardness at: Surface	½ radius	Center
After quenching in oil				
13	0.5	57 HRC	37 HRC	34 HRC
25	1	33 HRC	30 HRC	26 HRC
50	2	27 HRC	25 HRC	21 HRC
100	4	98 HRB	95 HRB	91 HRB
After quenching in water				
13	0.5	64 HRC	59 HRC	57 HRC
25	1	60 HRC	35 HRC	33 HRC
50	2	50 HRC	32 HRC	26 HRC
100	4	33 HRC	27 HRC	20 HRC

AISI 1050: Izod Impact Properties (Ref 12)

Condition or treatment	Energy J	Energy ft·lb
As rolled	31	23
Normalized at 900 °C (1650 °F)	27	20
Annealed at 790 °C (1450 °F)	18	13

AISI 1055

AISI 1055: Chemical Composition

AISI grade	Chemical composition, % C	Mn	P max	S max
1055	0.50-0.60	0.60-0.90	0.040	0.050

Characteristics. When heat treated, AISI 1055 steel yields a high surface hardness, combined with relatively good toughness. This grade also has good forging characteristics. It is shallow hardening, however, and useful section size is limited. Parts made from grade 1055 steel requiring strength are oil quenched; parts requiring high hardness are water quenched.

Typical Uses. Applications for AISI grade 1055 steel include battering tools, hot upset forging dies, ring-rolling tools, wear-resistant parts, hand tools, and parts for agricultural implements which require high strength at low cost.

AISI 1055: Similar Steels (U.S. and/or Foreign). UNS G10550; ASTM A29, A510, A576, A682; FED QQ-S-700 (C1055); SAE J403, J412, J414; (W. Ger.) DIN 1.1209

AISI 1055: Approximate Critical Points

Transformation point	Temperature(a) °C	°F
Ac_1	725	1340
Ac_3	755	1390
Ar_3	730	1350
Ar_1	680	1260

(a) On heating or cooling at 28 °C (50 °F) per hour

Mechanical Properties

AISI 1055: Tensile Properties and Machinability (Ref 1)

Condition or treatment	Size round mm	in.	Tensile strength MPa	ksi	Yield strength MPa	ksi	Elongation(a), %	Reduction in area, %	Hardness, HB	Average machinability rating(b)
Hot rolled	19-32	0.75-1.25	650	94	355	51.5	12	30	192	...
Cold drawn, annealed	19-32	0.75-1.25	660	96	560	81.0	10	40	197	55

(a) In 50 mm (2 in.). (b) Based on AISI 1212 steel as 100% average machinability

Physical Properties

AISI 1055: Average Coefficients of Linear Thermal Expansion (Ref 4)

Temperature range °C	°F	Coefficient µm/m·K	µin./in.·°F
20-100	68-212	11.0	6.1
20-200	68-390	12.1	6.7
20-300	68-570	12.4	6.9
20-400	68-750	13.3	7.4
20-500	68-930	13.9	7.7
20-600	68-1110	14.4	8.0
20-700	68-1290	14.8	8.2

AISI 1055: Thermal Treatment Temperatures

Treatment	Temperature range °C	°F
Annealing	815-855	1500-1575
Normalizing	845-900	1550-1650
Austenitizing	790-845	1450-1550

Machining Data

For machining data on AISI grade 1055, refer to the preceding machining tables for AISI grade 1030.

AISI 1059, 1060

AISI 1059, 1060: Chemical Composition

AISI grade	Chemical composition, % C	Mn	P max	S max
1059	0.55-0.65	0.50-0.80	0.040	0.050
1060	0.55-0.65	0.60-0.90	0.040	0.050

Characteristics and Typical Uses. AISI grade 1059 and 1060 steels have characteristics and uses similar to those described for AISI grade 1055.

AISI 1059: Similar Steels (U.S. and/or Foreign). UNS G10590; ASTM A29; SAE J118, J412

AISI 1060: Similar Steels (U.S. and/or Foreign). UNS G10600; AMS 7240; ASTM A29, A510, A576, A682; MIL SPEC MIL-S-16974; SAE J403, J412, J414; (W. Ger.) DIN 1.0601; (Fr.) AFNOR CC 55; (Ital.) UNI C 60; (U.K.) B.S. 060 A 62

AISI 1060: Approximate Critical Points

Transformation point	Temperature(a) °C	°F
Ac_1	725	1340
Ac_3	745	1375
Ar_3	725	1340
Ar_1	685	1265

(a) On heating or cooling at 28 °C (50 °F) per hour

Physical Properties

AISI 1060: Average Coefficients of Linear Thermal Expansion (Ref 4)

Temperature range °C	°F	Coefficient µm/m · K	µin./in. · °F
20-100	68-212	11.0	6.1
20-200	68-390	11.5	6.4
20-300	68-570	12.2	6.8
20-400	68-750	13.1	7.3
20-500	68-930	13.7	7.6
20-600	68-1110	14.0	7.8
20-700	68-1290	14.4	8.0

AISI 1059, 1060: Thermal Treatment Temperatures

Treatment	Temperature range °C	°F
Annealing	790-845	1450-1550
Normalizing	845-900	1550-1650
Quenching	790-845	1450-1550

Mechanical Properties

AISI 1060: Tensile Properties

Treatment or condition	Tensile strength MPa	ksi	Yield strength MPa	ksi	Elongation(a), %	Reduction in area, %	Hardness, HB
Hot rolled (Ref 1)(b)	660	96	370	54	12	30	201
Cold drawn, spheroidized annealed (Ref 1)(b)	620	90	485	70	10	45	183
As rolled (Ref 12)	814	118	485	70	17	34	241
Normalized at 900 °C (1650 °F) (Ref 12)	779	113	420	61	18	32	229
Annealed at 790 °C (1450 °F) (Ref 12)	625	91	370	54	22	38	179

(a) In 50 mm (2 in.). (b) Test specimens were 19- to 32-mm (0.75 to 1.25-in.) rounds

AISI 1060: Effect of Tempering Temperature on Tensile Properties. Normalized at 900 °C (1650 °F); reheated to 845 °C (1550 °F); quenched in oil. Specimens were treated in 25-mm (1.0-in.) diam and machined to 12.8-mm (0.505-in.) diam for testing. Tests were conducted using test specimens machined to English units. As-quenched hardness was 321 HB. Elongation was measured in 50 mm (2 in.). (Ref 2)

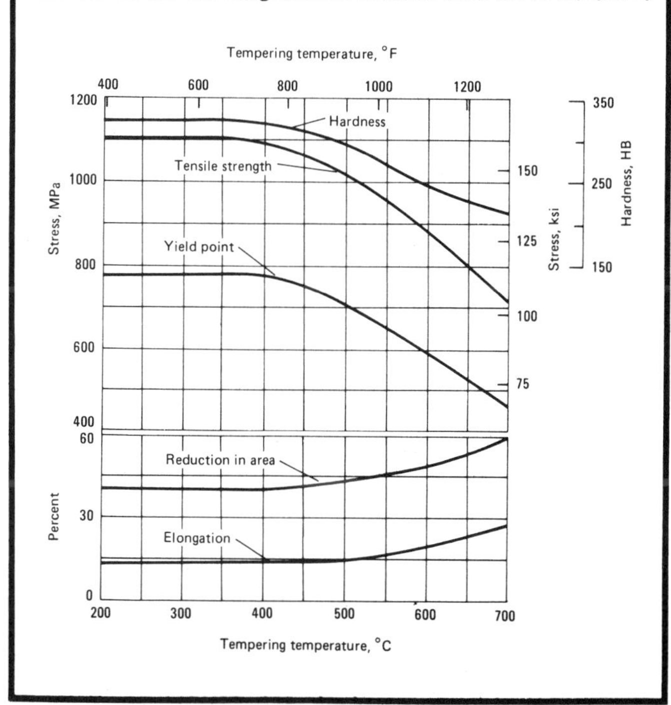

AISI 1059, 1060: Recommended Condition to Provide Maximum Drawability (Ref 6)

Punch nose radius, r(a)	Condition, HRB, for cup reduction of: 20%	30%
2t	...	...
4t	65(b)(c)(d)	...
8t	65(b)(c)(d)	...
16t	65(b)(c)(d)	...
32t	65(b)(c)	65(b)(c)(d)

Drawing limitations refer to cold rolled and mill annealed sheet steel; simple round cups, no flanges, not ironed 20-gage (0.91-mm or 0.036-in.) sheet steel; omission of entry indicates that the draw is usually too severe to be completed successfully. (a) In relation to stock thickness, *t*. (b) Aluminum killed. (c) Fine grained. (d) Spheroidized

AISI 1059, 1060: Izod Impact Properties (Ref 12)

Condition or treatment	Energy J	ft · lb
As rolled	18	13
Normalized at 900 °C (1650 °F)	14	10
Annealed at 790 °C (1450 °F)	11	8

Mechanical Properties (continued)

AISI 1060: Mass Effect on Mechanical Properties (Ref 2)

Condition or treatment	Size round mm	Size round in.	Tensile strength MPa	Tensile strength ksi	Yield strength MPa	Yield strength ksi	Elongation(a), %	Reduction in area, %	Hardness, HB
Annealed (heated to 790 °C or 1450 °F, furnace cooled 11 °C or 20 °F per hour to 650 °C or 1200 °F, cooled in air)	25	1	625	91	372	54.0	22.5	38.2	179
Normalized (heated to 900 °C or 1650 °F, cooled in air)	13	0.5	779	113	427	62.0	20.4	40.6	229
	25	1	772	112	421	61.0	18.0	37.2	229
	50	2	758	110	396	57.5	17.7	34.0	223
	100	4	745	108	353	51.2	18.0	31.3	223
Oil quenched from 845 °C (1550 °F), tempered at 480 °C (900 °F)	13	0.5	1025	149	677	98.2	15.1	46.0	302
	25	1	1005	146	641	93.0	16.2	44.0	293
	50	2	986	143	617	89.5	16.5	46.2	285
	100	4	931	135	518	75.2	18.2	44.8	269
Oil quenched from 845 °C (1550 °F), tempered at 540 °C (1000 °F)	13	0.5	965	140	634	92.0	19.6	52.1	277
	25	1	938	136	591	85.7	17.7	48.0	269
	50	2	917	133	546	79.2	18.5	50.3	262
	100	4	855	124	456	66.2	20.0	48.0	248
Oil quenched from 845 °C (1550 °F), tempered at 595 °C (1100 °F)	13	0.5	910	132	569	82.5	20.7	53.5	262
	25	1	883	128	545	79.0	20.0	51.7	255
	50	2	862	125	527	76.5	20.2	53.3	248
	100	4	820	119	427	62.0	21.5	49.4	241

(a) In 50 mm (2 in.)

AISI 1060: Effect of the Mass on Hardness at Selected Points

Size round mm	Size round in.	As-quenched hardness after quenching in oil at: Surface	½ radius	Center
13	0.5	59 HRC	37 HRC	35 HRC
25	1	34 HRC	32 HRC	30 HRC
50	2	30.5 HRC	27.5 HRC	25 HRC
100	4	29 HRC	26 HRC	24 HRC

Machining Data (Ref 8)

AISI 1059, 1060: Turning (Single Point and Box Tools)

Depth of cut mm	Depth of cut in.	High speed steel Speed m/min	Speed ft/min	Feed mm/rev	Feed in./rev	Uncoated carbide Speed, brazed m/min	Speed, brazed ft/min	Speed, inserted m/min	Speed, inserted ft/min	Feed mm/rev	Feed in./rev	Coated carbide Speed m/min	Speed ft/min	Feed mm/rev	Feed in./rev
Hardness, 175 to 225 HB															
1	0.040	37(a)	120(a)	0.18	0.007	130(b)	420(b)	155(b)	510(b)	0.18	0.007	230(c)	750(c)	0.18	0.007
4	0.150	27(a)	90(a)	0.40	0.015	95(d)	315(d)	120(d)	390(d)	0.50	0.020	150(e)	500(e)	0.40	0.015
8	0.300	21(a)	70(a)	0.50	0.020	76(d)	250(d)	95(d)	310(d)	0.75	0.030	120(e)	400(e)	0.50	0.020
16	0.625	15(a)	50(a)	0.75	0.030	58(d)	190(d)	73(d)	240(d)	1.00	0.040	...	...	...	...
Hardness, 225 to 275 HB															
1	0.040	30(a)	100(a)	0.18	0.007	115(b)	375(b)	140(b)	460(b)	0.18	0.007	215(c)	700(c)	0.18	0.007
4	0.150	24(a)	80(a)	0.40	0.015	88(d)	290(d)	105(d)	350(d)	0.50	0.020	145(e)	475(e)	0.40	0.015
8	0.300	18(a)	60(a)	0.50	0.020	69(d)	225(d)	84(d)	275(d)	0.75	0.030	115(e)	375(e)	0.50	0.020
16	0.625	14(a)	45(a)	0.75	0.030	49(d)	160(d)	64(d)	210(d)	1.00	0.040	...	...	...	...
Hardness, 275 to 325 HB															
1	0.040	24(f)	80(f)	0.18	0.007	110(b)	360(b)	130(b)	420(b)	0.18	0.007	205(c)	675(c)	0.18	0.007
4	0.150	18(f)	60(f)	0.40	0.015	84(d)	275(d)	100(d)	325(d)	0.40	0.015	135(e)	450(e)	0.40	0.015
8	0.300	15(f)	50(f)	0.50	0.020	60(d)	200(d)	76(d)	250(d)	0.65	0.025	105(e)	350(e)	0.50	0.020
Hardness 325 to 375 HB															
1	0.040	18(f)	60(f)	0.18	0.007	95(b)	305(b)	110(b)	360(b)	0.18	0.007	170(c)	550(c)	0.18	0.007
4	0.150	14(f)	45(f)	0.40	0.015	72(d)	235(d)	85(d)	280(d)	0.40	0.015	115(e)	375(e)	0.40	0.015
8	0.300	11(f)	35(f)	0.50	0.020	53(d)	175(d)	67(d)	220(d)	0.65	0.025	90(e)	300(e)	0.50	0.020

(a) High speed steel tool material: M2 or M3. (b) Carbide tool material: C-7. (c) Carbide tool material: CC-7. (d) Carbide tool material: C-6. (e) Carbide tool material: CC-6. (f) Any premium high speed steel tool material (T15, M33, M41 to M47)

AISI 1059, 1060: Turning (Cutoff and Form Tools)

Tool material	Speed, m/min (ft/min)	Feed per revolution for cutoff tool width of: 1.5 mm (0.062 in.)		3 mm (0.125 in.)		6 mm (0.25 in.)		Feed per revolution for form tool width of: 12 mm (0.5 in.)		18 mm (0.75 in.)		25 mm (1 in.)		50 mm (2 in.)	
		mm	in.	mm	in.	mm	in.	mm	in.	mm	in.	mm	in.	mm	in.
Hardness, 175 to 225 HB															
M2 and M3 high speed steel	24 (80)	0.025	0.001	0.030	0.0012	0.038	0.0015	0.033	0.0013	0.028	0.0011	0.025	0.001	0.020	0.0008
C-6 carbide	78 (255)	0.025	0.001	0.030	0.0012	0.038	0.0015	0.033	0.0013	0.028	0.0011	0.025	0.001	0.020	0.0008
Hardness, 225 to 275 HB															
M2 and M3 high speed steel	18 (60)	0.025	0.001	0.030	0.0012	0.038	0.0015	0.033	0.0013	0.028	0.0011	0.025	0.001	0.020	0.0008
C-6 carbide	58 (190)	0.025	0.001	0.030	0.0012	0.038	0.0015	0.033	0.0013	0.028	0.0011	0.025	0.001	0.020	0.0008
Hardness, 275 to 325 HB															
High speed steel(a)	14 (45)	0.023	0.0009	0.028	0.0011	0.036	0.0014	0.030	0.0012	0.025	0.001	0.023	0.0009	0.018	0.0007
C-6 carbide	44 (145)	0.023	0.0009	0.028	0.0011	0.036	0.0014	0.030	0.0012	0.025	0.001	0.023	0.0009	0.018	0.0007
Hardness, 325 to 375 HB															
High speed steel(a)	11 (35)	0.020	0.0008	0.025	0.001	0.033	0.0013	0.028	0.0011	0.023	0.0009	0.020	0.0008	0.015	0.0006
C-6 carbide	35 (115)	0.020	0.0008	0.025	0.001	0.033	0.0013	0.028	0.0011	0.023	0.0009	0.020	0.0008	0.015	0.0006

(a) Any premium high speed steel tool material (T5, M33, M41 to M47)

AISI 1059, 1060: Drilling

Tool material	Speed m/min	Speed ft/min	Feed per revolution for nominal hole diameter of: 1.5 mm (0.062 in.)		3 mm (0.125 in.)		6 mm (0.25 in.)		12 mm (0.5 in.)		18 mm (0.75 in.)		25 mm (1 in.)	
			mm	in.	mm	in.	mm	in.	mm	in.	mm	in.	mm	in.
Hardness, 175 to 225 HB														
M10, M7, and M1 high speed steel	14	45	0.025	0.001	...	...	...	...	...	...	...	...	...	...
	20	65	...	...	0.075	0.003	0.13	0.005	0.23	0.009	0.30	0.012	0.45	0.018
Hardness, 225 to 275 HB														
M10, M7, and M1 high speed steel	17	55	0.025	0.001	0.050	0.002	0.102	0.004	0.18	0.007	0.25	0.010	0.40	0.015
Hardness, 275 to 325 HB														
M10, M7, and M1 high speed steel	14	45	...	...	0.050	0.002	0.102	0.004	0.18	0.007	0.25	0.010	0.30	0.012
Hardness, 325 to 375 HB														
M10, M7, and M1 high speed steel	12	40	...	...	0.050	0.002	0.075	0.003	0.15	0.006	0.23	0.009	0.28	0.011

AISI 1059, 1060: Planing

Tool material	Depth of cut mm	Depth of cut in.	Speed m/min	Speed ft/min	Feed/stroke mm	Feed/stroke in.
Hardness, 175 to 225 HB						
M2 and M3 high speed steel	0.1	0.005	9	30	(a)	(a)
	2.5	0.100	14	45	0.75	0.030
	12	0.500	8	25	1.15	0.045
C-6 carbide	0.1	0.005	58	190	(a)	(a)
	2.5	0.100	64	210	1.50	0.050
	12	0.500	50	165	1.25	0.050
Hardness, 225 to 275 HB						
M2 and M3 high speed steel	0.1	0.005	8	25	(a)	(a)
	2.5	0.100	12	40	0.75	0.030
	12	0.500	6	20	1.15	0.045
C-6 carbide	0.1	0.005	50	165	(a)	(a)
	2.5	0.100	58	190	1.50	0.060
	12	0.500	40	130	1.25	0.050
Hardness, 275 to 325 HB						
M2 and M3 high speed steel	0.1	0.005	8	25	(a)	(a)
	2.5	0.100	11	35	0.50	0.020
	12	0.500	6	20	0.75	0.030
C-6 carbide	0.1	0.005	43	140	(a)	(a)
	2.5	0.100	50	165	1.50	0.060
	12	0.500	37	120	1.25	0.050
Hardness, 225 to 375 HB						
M2 and M3 high speed steel	0.1	0.005	6	20	(a)	(a)
	2.5	0.100	9	30	0.50	0.020
	12	0.500	6	20	0.75	0.030
C-6 carbide	0.1	0.005	40	130	(a)	(a)
	2.5	0.100	46	150	1.50	0.060
	12	0.500	34	110	1.25	0.050

(a) Feed is 75% the width of the square nose finishing tool

Machining Data (Ref 8) (continued)

AISI 1059, 1060: End Milling (Profiling)

Tool material	Depth of cut mm	Depth of cut in.	Speed m/min	Speed ft/min	Feed per tooth for cutter diameter of: 10 mm (0.375 in.) mm	10 mm (0.375 in.) in.	12 mm (0.5 in.) mm	12 mm (0.5 in.) in.	18 mm (0.75 in.) mm	18 mm (0.75 in.) in.	25-50 mm (1-2 in.) mm	25-50 mm (1-2 in.) in.
Hardness, 175 to 225 HB												
M2, M3, and M7 high speed steel	0.5	0.020	46	150	0.025	0.001	0.050	0.002	0.075	0.003	0.102	0.004
	1.5	0.060	34	110	0.050	0.002	0.075	0.003	0.102	0.004	0.13	0.005
	diam/4	diam/4	29	95	0.025	0.001	0.050	0.002	0.075	0.003	0.102	0.004
	diam/2	diam/2	24	80	0.018	0.0007	0.025	0.001	0.050	0.002	0.075	0.003
C-5 carbide	0.5	0.020	145	475	0.038	0.0015	0.075	0.003	0.13	0.005	0.15	0.006
	1.5	0.060	115	370	0.063	0.0025	0.102	0.004	0.15	0.006	0.18	0.007
	diam/4	diam/4	100	320	0.050	0.002	0.075	0.003	0.18	0.005	0.15	0.006
	diam/2	diam/2	90	300	0.038	0.0015	0.050	0.002	0.102	0.004	0.13	0.005
Hardness, 225 to 275 HB												
M2, M3, and M7 high speed steel	0.5	0.020	32	105	0.025	0.001	0.050	0.002	0.075	0.003	0.102	0.004
	1.5	0.060	24	80	0.050	0.002	0.075	0.003	0.102	0.004	0.13	0.005
	diam/4	diam/4	21	70	0.025	0.001	0.050	0.002	0.075	0.003	0.102	0.004
	diam/2	diam/2	18	60	0.018	0.0007	0.025	0.001	0.050	0.002	0.075	0.003
C-5 carbide	0.5	0.020	130	425	0.025	0.001	0.050	0.002	0.102	0.004	0.13	0.005
	1.5	0.060	100	325	0.050	0.002	0.075	0.003	0.13	0.005	0.18	0.007
	diam/4	diam/4	85	280	0.038	0.0015	0.063	0.0025	0.102	0.004	0.13	0.005
	diam/2	diam/2	79	260	0.025	0.001	0.050	0.002	0.075	0.003	0.102	0.004
Hardness, 275 to 325 HB												
M2, M3, and M7 high speed steel	0.5	0.020	27	90	0.018	0.0007	0.038	0.0015	0.075	0.003	0.102	0.004
	1.5	0.060	20	65	0.025	0.001	0.050	0.002	0.102	0.004	0.13	0.005
	diam/4	diam/4	18	60	0.018	0.0007	0.038	0.0015	0.075	0.003	0.102	0.004
	diam/2	diam/2	15	50	0.013	0.0005	0.025	0.001	0.050	0.002	0.075	0.003
C-5 carbide	0.5	0.020	110	360	0.025	0.001	0.050	0.002	0.102	0.004	0.13	0.005
	1.5	0.060	84	275	0.050	0.002	0.075	0.003	0.13	0.005	0.15	0.006
	diam/4	diam/4	72	235	0.038	0.0015	0.063	0.0025	0.102	0.004	0.13	0.005
	diam/2	diam/2	67	220	0.025	0.001	0.050	0.002	0.075	0.003	0.102	0.004
Hardness, 325 to 375 HB												
M2, M3, and M7 high speed steel	0.5	0.020	20	65	0.013	0.0005	0.038	0.0015	0.075	0.003	0.102	0.004
	1.5	0.060	17	55	0.013	0.0005	0.038	0.0015	0.102	0.004	0.13	0.005
	diam/4	diam/4	15	50	0.013	0.0005	0.038	0.0015	0.075	0.003	0.102	0.004
	diam/2	diam/2	12	40	0.013	0.0005	0.025	0.001	0.050	0.002	0.075	0.003
C-5 carbide	0.5	0.020	76	250	0.025	0.001	0.038	0.0015	0.102	0.003	0.13	0.005
	1.5	0.060	58	190	0.038	0.0015	0.075	0.003	0.13	0.005	0.15	0.006
	diam/4	diam/4	50	165	0.038	0.0015	0.050	0.002	0.075	0.003	0.102	0.004
	diam/2	diam/2	46	150	0.025	0.001	0.038	0.0015	0.050	0.002	0.075	0.003

AISI 1059, 1060: Face Milling

Depth of cut mm	Depth of cut in.	High speed steel: Speed m/min	Speed ft/min	Feed/tooth mm	Feed/tooth in.	Uncoated carbide: Speed, brazed m/min	Speed, brazed ft/min	Speed, inserted m/min	Speed, inserted ft/min	Feed/tooth mm	Feed/tooth in.	Coated carbide: Speed m/min	Speed ft/min	Feed/tooth mm	Feed/tooth in.
Hardness, 175 to 225 HB															
1	0.040	44(a)	145(a)	0.20	0.008	150(b)	500(b)	160(b)	530(b)	0.20	0.008	245(c)	800(c)	0.20	0.008
4	0.150	34(a)	110(a)	0.30	0.012	120(b)	390(b)	135(b)	450(b)	0.30	0.012	180(c)	585(c)	0.30	0.012
8	0.300	26(a)	85(a)	0.40	0.016	87(d)	285(d)	105(d)	350(d)	0.40	0.016	140(e)	455(e)	0.40	0.016
Hardness, 225 to 275 HB															
1	0.040	35(a)	115(a)	0.15	0.006	130(b)	425(b)	130(b)	480(b)	0.18	0.007	220(c)	725(c)	0.18	0.007
4	0.150	27(a)	90(a)	0.25	0.010	110(b)	360(b)	110(b)	400(b)	0.25	0.010	160(c)	520(c)	0.25	0.010
8	0.300	21(a)	70(a)	0.36	0.014	78(d)	255(d)	78(d)	310(d)	0.36	0.014	125(e)	405(e)	0.36	0.014
Hardness, 275 to 325 HB															
1	0.040	30(f)	100(f)	0.15	0.006	120(b)	390(b)	130(b)	420(b)	0.15	0.006	190(c)	625(c)	0.13	0.005
4	0.150	24(f)	80(f)	0.23	0.009	90(b)	300(b)	100(b)	330(b)	0.20	0.008	130(c)	430(c)	0.18	0.007
8	0.300	18(f)	60(f)	0.30	0.012	64(d)	210(d)	78(d)	255(d)	0.25	0.010	100(e)	330(e)	0.23	0.009
Hardness, 325 to 375 HB															
1	0.040	21(f)	70(f)	0.13	0.005	95(b)	310(b)	100(b)	330(b)	0.13	0.005	150(c)	500(c)	0.102	0.004
4	0.150	15(f)	50(f)	0.20	0.008	76(b)	250(b)	88(b)	290(b)	0.18	0.007	115(c)	375(c)	0.15	0.006
8	0.300	2(f)	40(f)	0.25	0.010	56(d)	185(d)	69(d)	225(d)	0.23	0.009	88(e)	290(e)	0.20	0.008

(a) High speed steel tool material: M2 or M3. (b) Carbide tool material: C-6. (c) Carbide tool material: CC-6. (d) Carbide tool material: C-5. (e) Carbide tool material: CC-5. (f) Any premium high speed steel tool material (T15, M33, M41 to M47)

Machining Data (Ref 8) (continued)

AISI 1059, 1060: Boring

Depth of cut mm	Depth of cut in.	High speed steel Speed m/min	High speed steel Speed ft/min	High speed steel Feed mm/rev	High speed steel Feed in./rev	Uncoated carbide Speed, brazed m/min	Uncoated carbide Speed, brazed ft/min	Uncoated carbide Speed, inserted m/min	Uncoated carbide Speed, inserted ft/min	Uncoated carbide Feed mm/rev	Uncoated carbide Feed in./rev	Coated carbide Speed m/min	Coated carbide Speed ft/min	Coated carbide Feed mm/rev	Coated carbide Feed in./rev
Hardness, 175 to 225 HB															
0.25	0.010	37(a)	120(a)	0.075	0.003	115(b)	380(b)	135(b)	450(b)	0.075	0.003	200(c)	655(c)	0.075	0.003
1.25	0.050	29(a)	95(a)	0.130	0.005	95(d)	305(d)	110(d)	360(d)	0.130	0.005	160(e)	525(e)	0.130	0.005
2.5	0.100	21(a)	70(a)	0.300	0.012	72(d)	235(d)	84(d)	275(d)	0.400	0.015	105(e)	350(e)	0.300	0.012
Hardness, 225 to 275 HB															
0.25	0.010	30(a)	100(a)	0.075	0.003	105(b)	340(b)	120(b)	400(b)	0.075	0.003	185(c)	610(c)	0.075	0.003
1.25	0.050	24(a)	80(a)	0.130	0.005	84(d)	275(d)	100(d)	320(d)	0.130	0.005	150(e)	490(e)	0.130	0.005
2.5	0.100	20(a)	65(a)	0.300	0.012	66(d)	215(d)	76(d)	250(d)	0.400	0.015	100(e)	330(e)	0.300	0.012
Hardness, 275 to 325 HB															
0.25	0.010	24(f)	80(f)	0.075	0.003	95(b)	315(b)	115(b)	370(b)	0.075	0.003	180(c)	595(c)	0.075	0.003
1.25	0.050	20(f)	65(f)	0.130	0.005	76(d)	250(d)	90(d)	295(d)	0.130	0.005	145(e)	475(e)	0.130	0.005
2.5	0.100	15(f)	50(f)	0.300	0.012	58(d)	190(d)	69(d)	225(d)	0.300	0.012	95(e)	315(e)	0.300	0.012
Hardness, 325 to 375 HB															
0.25	0.010	20(f)	65(f)	0.075	0.003	66(b)	215(b)	78(b)	255(b)	0.075	0.003	145(c)	480(c)	0.075	0.003
1.25	0.050	15(f)	50(f)	0.130	0.005	53(d)	175(d)	62(d)	205(d)	0.130	0.005	115(e)	385(e)	0.130	0.005
2.5	0.100	12(f)	40(f)	0.300	0.012	50(d)	165(d)	59(d)	195(d)	0.300	0.012	79(e)	260(e)	0.300	0.012

(a) High speed steel tool material: M2 or M3. (b) Carbide tool material: C-7. (c) Carbide tool material: CC-7. (d) Carbide tool material: C-6. (e) Carbide tool material: CC-6. (f) Any premium high speed steel tool material (T15, M33, M41 to M47)

AISI 1059, 1060: Reaming

Based on 4 flutes for 3- and 6-mm (0.125- and 0.25-in.) reamers, 6 flutes for 12-mm (0.5-in.) reamers, and 8 flutes for 25-mm (1-in.) and larger reamers

Tool material	Speed m/min	Speed ft/min	3 mm (0.125 in.) mm	3 mm (0.125 in.) in.	6 mm (0.25 in.) mm	6 mm (0.25 in.) in.	12 mm (0.5 in.) mm	12 mm (0.5 in.) in.	25 mm (1 in.) mm	25 mm (1 in.) in.	35 mm (1.5 in.) mm	35 mm (1.5 in.) in.	50 mm (2 in.) mm	50 mm (2 in.) in.
			Feed per revolution for reamer diameter of:											
Hardness, 175 to 225 HB														
Roughing														
M1, M2, and M7 high speed steel	20	65	0.075	0.003	0.15	0.006	0.20	0.008	0.30	0.012	0.40	0.015	0.50	0.020
C-2 carbide	24	80	0.102	0.004	0.15	0.006	0.20	0.008	0.30	0.012	0.40	0.015	0.50	0.020
Finishing														
M1, M2, and M7 high speed steel	11	35	0.13	0.005	0.20	0.008	0.30	0.012	0.50	0.020	0.65	0.025	0.75	0.030
C-2 carbide	14	45	0.13	0.005	0.20	0.008	0.30	0.012	0.50	0.020	0.65	0.025	0.75	0.030
Hardness, 225 to 275 HB														
Roughing														
M1, M2, and M7 high speed steel	17	55	0.075	0.003	0.15	0.006	0.20	0.008	0.30	0.012	0.40	0.015	0.50	0.020
C-2 carbide	21	70	0.102	0.004	0.15	0.006	0.20	0.008	0.30	0.012	0.40	0.015	0.50	0.020
Finishing														
M1, M2, and M7 high speed steel	9	30	0.102	0.004	0.18	0.007	0.25	0.010	0.40	0.015	0.50	0.020	0.65	0.025
C-2 carbide	12	40	0.102	0.004	0.18	0.007	0.25	0.010	0.40	0.015	0.50	0.020	0.65	0.025
Hardness, 275 to 325 HB														
Roughing														
M1, M2, and M7 high speed steel	14	45	0.075	0.003	0.102	0.004	0.15	0.006	0.25	0.010	0.30	0.012	0.40	0.015
C-2 carbide	18	60	0.102	0.004	0.13	0.005	0.15	0.006	0.25	0.010	0.30	0.012	0.40	0.015
Finishing														
M1, M2, and M7 high speed steel	8	25	0.075	0.003	0.15	0.006	0.20	0.008	0.30	0.012	0.40	0.015	0.50	0.020
C-2 carbide	11	35	0.102	0.004	0.15	0.006	0.20	0.008	0.30	0.012	0.40	0.015	0.50	0.020
Hardness, 325 to 375 HB														
Roughing														
M1, M2, and M7 high speed steel	11	35	0.05	0.002	0.102	0.004	0.13	0.005	0.20	0.008	0.25	0.010	0.30	0.012
C-2 carbide	15	50	0.102	0.004	0.15	0.006	0.20	0.008	0.25	0.010	0.30	0.012	0.40	0.015
Finishing														
M1, M2, and M7 high speed steel	8	25	0.075	0.003	0.13	0.005	0.15	0.006	0.25	0.010	0.30	0.012	0.40	0.015
C-2 carbide	11	35	0.102	0.004	0.15	0.006	0.20	0.008	0.25	0.010	0.30	0.012	0.40	0.015

AISI 1064, 1065, 1070, 1074

AISI 1064, 1065, 1070, 1074: Chemical Composition

AISI grade	Chemical composition, % C	Mn	P max	S max
1064	0.60-0.70	0.50-0.80	0.040	0.050
1065	0.60-0.70	0.60-0.80	0.040	0.050
1070	0.65-0.75	0.60-0.90	0.040	0.050
1074	0.70-0.80	0.50-0.80	0.040	0.050

Characteristics and Typical Uses. AISI grade 1064, 1065, 1070, and 1074 are made into wire which is high in tensile strength and heat treatable. Cold forming of these grades is generally limited to coiling in smaller diameters for coil-type springs, or sinusoidal forming for springs used in upholstered furniture and automotive seats and backs. Material for mechanical and upholstery springs is ordered by tensile strength and mechanical properties, rather than by grade. Thus, the producer provides a steel that will meet the requirements.

These high-carbon steels are produced as cold rolled strip for flat springs in the annealed or tempered condition. Oil tempered wire and strip are suitable for springs subject to high and frequent stress. For springs subject to static loads or relatively infrequent stress cycles, however, hard drawn or cold rolled steels are generally used.

High-carbon spring wire, which is heat treatable after forming, is used when the end product is quenched and tempered after forming. Other applications include wire that may be subject to severe forming or for which high hardness is required. This spring wire is available as untempered, spheroidized annealed in process, and as spheroidized annealed at finished size. Composition of the wire is critically important to assure uniform response to subsequent heat treatment.

AISI 1064: Similar Steels (U.S. and/or Foreign). UNS G10640; ASTM A26, A29, A57, A230, A682; SAE J403, J412, J414; (W. Ger.) DIN 1.1221; (Fr.) AFNOR XC 60, XC 65; (Swed.) SS_{14} 1665, 1678

AISI 1065: Similar Steels (U.S. and/or Foreign). UNS G10650; ASTM A29, A229, A682; FED QQ-S-700 (C1065); MIL SPEC MIL-S-46049, MIL-S-46409; SAE J403, J412, J414; (W. Ger.) DIN 1.1230

AISI 1070: Similar Steels (U.S. and/or Foreign). UNS G10700; AMS 5115; ASTM A29, A510, A576, A682; MIL SPEC MIL-S-11713 (2); SAE J403, J412, J414; (W. Ger.) DIN 1.1231; (Fr.) AFNOR XC 68; (Swed.) SS_{14} 1770, 1778

AISI 1074: Similar Steels (U.S. and/or Foreign). UNS G10740; AMS 5120 D; ASTM A29, A682; FED QQ-S-700 (C1074); MIL SPEC MIL-S-46049; SAE J403, J412, J414

AISI 1070: Approximate Critical Points

Transformation point	Temperature(a) °C	°F
Ac_1	725	1340
Ac_3	730	1350
Ar_3	710	1310
Ar_1	690	1275

(a) On heating or cooling at 28 °C (50 °F) per hour

Mechanical Properties

AISI 1064, 1065, 1070, 1074: Tensile Properties and Machinability

Condition or treatment	Size round mm	in.	Tensile strength MPa	ksi	Yield strength MPa	ksi	Elongation(a), %	Reduction in area, %	Hardness, HB	Average machinability rating(b)
AISI 1064										
Hot rolled (Ref 1)	19-32	0.75-1.25	670	97	370	54	12	30	201	...
Cold drawn, spheroidized annealed (Ref 1)	19-32	0.75-1.25	615	89	475	69	10	45	183	60
AISI 1065										
Hot rolled (Ref 1)	19-32	0.75-1.25	690	100	380	55	12	30	207	...
Cold drawn, spheroidized annealed (Ref 1)	19-32	0.75-1.25	635	92	490	71	10	45	187	60
AISI 1070										
Hot rolled (Ref 1)	19-32	0.75-1.25	703	102	385	56	12	30	212	...
Cold drawn, spheroidized annealed (Ref 1)	19-32	0.75-1.25	640	93	495	72	10	45	192	55
AISI 1074										
Hot rolled (Ref 1)	19-32	0.75-1.25	724	105	400	58	12	30	217	...
Cold drawn, spheroidized annealed (Ref 1)	19-32	0.75-1.25	650	94	505	73	10	40	192	55
Cold rolled strip (Ref 7)	...	...	1105-2205	160-320	...	...	...	...	...	(c)

(a) In 50 mm (2 in.). (b) Based on AISI 1212 steel as 100% average machinability. (c) Hardness of 85 HRB in the annealed condition, 38 to 50 HRC in the tempered condition

Machining Data

For machining data on AISI grade 1064, 1065, 1070, and 1074 steels, refer to the preceding machining tables for AISI grades 1059 and 1060.

Physical Properties

AISI 1070: Average Coefficients of Linear Thermal Expansion (Ref 4)

Temperature range		Coefficient	
°C	°F	μm/m · K	μin./in. · °F
0-100	32-212	11.6	6.4
0-200	32-390	12.6	7.0
0-300	32-570	13.3	7.4
0-400	32-750	14.0	7.8

AISI 1074: Thermal Treatment Temperatures

Treatment	Temperature range	
	°C	°F
Annealing	790-845	1450-1550
Normalizing	845-900	1550-1650
Hardening	855-885	1575-1625

AISI 1080

AISI 1080: Chemical Composition

AISI grade	Chemical composition, %			
	C	Mn	P max	S max
1080	0.75-0.88	0.60-0.90	0.040	0.050

Characteristics and Typical Uses. AISI grade 1080 steel has characteristics and uses similar to those described for AISI grades 1055 and 1074.

AISI 1080: Similar Steels (U.S. and/or Foreign). UNS G10800; AMS 5110; ASTM A29, A510, A576, A682; FED QQ-S-700 (C1080); MIL SPEC MIL-S-16974; SAE J403, J412, J414

AISI 1080: Approximate Critical Points

Transformation point	Temperature(a)	
	°C	°F
Ac_1	730	1345
Ac_3	735	1355
Ar_3	700	1290
Ar_1	695	1280
M_s	215	415

(a) On heating or cooling at 28 °C (50 °F) per hour

Physical Properties

AISI 1080: Electrical Resistivity and Thermal Conductivity (Ref 4)

Temperature		Electrical resistivity, μΩ·m	Thermal conductivity	
°C	°F		W/m·K	Btu/ft·h·°F
0	32	···	47.7	27.6
20	68	0.180	···	···
100	212	0.232	48.1	27.8
200	390	0.308	45.2	26.1
300	570	···	41.3	23.9
400	750	0.505	38.1	22.0
500	930	···	35.1	20.3
600	1110	0.772	32.7	18.9
700	1290	0.935	30.1	17.4
800	1470	1.129	24.4	14.1
900	1650	1.164	···	···
1000	1830	1.191	26.8	15.5
1100	2010	1.214	···	···
1200	2190	1.234	30.1	17.4
1250	2270	1.246	···	···

Material composition: 0.08% carbon, 0.32% manganese

AISI 1080: Thermal Treatment Temperatures

Quenching medium: oil

Treatment	Temperature range	
	°C	°F
Annealing	760-815	1400-1500
Normalizing	845-900	1550-1650
Hardening	855-885	1575-1625
Tempering	(a)	(a)

(a) Temper to desired hardness

AISI 1080: Mean Apparent Specific Heat (Ref 4)

Temperature range		Specific heat	
°C	°F	J/kg·K	Btu/lb·°F
50-100	120-212	490	0.117
150-200	300-390	531	0.127
200-250	390-480	548	0.131
250-300	480-570	565	0.135
300-350	570-660	586	0.140
350-400	660-750	607	0.145
450-500	840-930	670	0.160
550-600	1020-1110	712	0.170
650-700	1200-1290	770	0.184
700-750	1290-1380	2080	0.497
750-800	1380-1470	615	0.147

Material composition: 0.80% carbon, 0.32% manganese

AISI 1080: Average Coefficients of Linear Thermal Expansion (Ref 4)

Temperature range		Coefficient	
°C	°F	μm/m·K	μin./in.·°F
0-100	32-212	11.0	6.1
0-200	32-390	11.5	6.4
0-300	32-570	12.2	6.8
0-400	32-750	13.1	7.3
0-500	32-930	13.7	7.6
0-600	32-1110	14.0	7.8
0-700	32-1290	14.6	8.1

Mechanical Properties

AISI 1080: Mass Effect on Mechanical Properties (Ref 2)

Condition or treatment	Size round mm	Size round in.	Tensile strength MPa	Tensile strength ksi	Yield strength MPa	Yield strength ksi	Elongation(a), %	Reduction in area, %	Hardness, HB
Annealed (heated to 790 °C or 1450 °F, furnace cooled 11 °C or 20 °F per hour to 650 °C or 1200 °F, cooled in air)	25	1	615	89	370	54	24.7	45.0	174
Normalized (heated to 900 °C or 1650 °F, cooled in air)	13	0.5	1035	150	550	80	12.4	27.7	293
	25	1	1005	146	525	76	11.0	20.6	293
	50	2	972	141	485	70	10.7	17.0	285
	100	4	931	135	440	64	10.7	15.5	269
Oil quenched from 815 °C (1500 °F), tempered at 480 °C (900 °F)	13	0.5	1270	184	869	126	12.1	34.4	363
	25	1	1255	182	772	112	13.0	35.8	352
	50	2	1240	180	758	110	12.7	37.3	352
	100	4	1180	171	717	104	11.7	28.6	341
Oil quenched from 815 °C (1500 °F), tempered at 540 °C (1000 °F)	13	0.5	1165	169	841	122	15.0	38.6	341
	25	1	1145	166	717	104	15.0	37.6	331
	50	2	1130	164	710	103	15.2	38.0	321
	100	4	1080	157	620	90	11.5	24.4	311
Oil quenched from 815 °C (1500 °F), tempered at 595 °C (1100 °F)	13	0.5	1050	152	738	107	17.0	43.6	302
	25	1	1035	150	670	97	16.5	40.3	302
	50	2	965	140	605	88	17.7	42.2	277
	100	4	924	134	515	75	15.7	33.1	269

(a) In 50 mm (2 in.)

AISI 1080: Effect of the Mass on Hardness at Selected Points

Size round mm	Size round in.	As-quenched hardness after quenching in oil at: Surface	½ radius	Center
13	0.5	60 HRC	43 HRC	40 HRC
25	1	45 HRC	42 HRC	39 HRC
50	2	43 HRC	40 HRC	37 HRC
100	4	39 HRC	37 HRC	32 HRC

AISI 1080: Izod Impact Properties (Ref 12)

Condition or treatment	Energy J	Energy ft·lb
As rolled	7	5
Normalized at 900 °C (1650 °F)	7	5
Annealed at 790 °C (1450 °F)	7	5

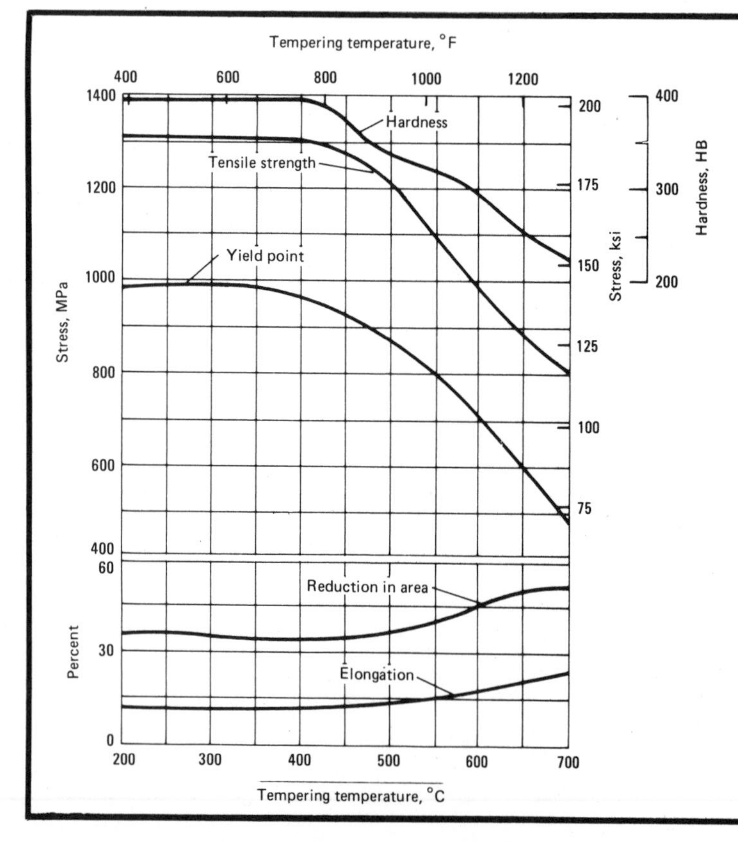

AISI 1080: Effect of Tempering Temperature on Tensile Properties. Normalized at 900 °C (1650 °F); reheated to 815 °C (1500 °F); quenched in oil. Specimens were treated in 25-mm (1.0-in.) diam and machined to 12.8-mm (0.505-in.) diam for testing. Tests were conducted using test specimens machined to English units. As-quenched hardness was 388 HB. Elongation was measured in 50 mm (2 in.). (Ref 2)

Mechanical Properties (continued)

AISI 1080: Tensile Properties

Condition or treatment	Size round mm	Size round in.	Tensile strength MPa	Tensile strength ksi	Yield strength MPa	Yield strength ksi	Elongation(a), %	Reduction in area, %	Hardness, HB
Hot rolled (Ref 1)	19-32	0.75-1.25	772	112	425	62	10	25	229
Cold drawn, spheroidized annealed(b) (Ref 1)	19-32	0.75-1.25	675	98	515	75	10	40	192
As rolled (Ref 12)	...	...	965	140	585	85	12	17	293
Normalized at 900 °C (1650 °F) (Ref 12)	...	...	1200	174	525	76	11	21	293
Annealed at 790 °C (1450 °F) (Ref 12)	...	...	615	89	380	55	25	45	174

(a) In 50 mm (2 in.). (b) Average machinability rating of 45% based on AISI 1212 steel as 100% average machinability

Machining Data

For machining data on AISI grade 1080, refer to the preceding machining tables for AISI grades 1059 and 1060.

AISI 1090, 1095

AISI 1090, 1095: Chemical Composition

AISI grade	C	Mn	P max	S max
1090	0.85-0.98	0.60-0.90	0.040	0.050
1095	0.90-1.03	0.30-0.50	0.040	0.050

(Chemical composition, %)

Characteristics. The high-carbon grade 1090 and 1095 steels provide maximum surface hardness with improved wear resistance and high strength. There is, however, a loss of toughness. Because cold forming methods are generally not practical with these steels, uses are limited to flat stampings and springs coiled from small-diameter wire.

Typical Uses. Applications for grade 1090 and 1095 steels include edge tools, wear-resistant parts, high-stress flat springs, hot coiled springs, plow beams, plow shares, scraper blades, discs, mower knives, and harrow teeth.

AISI 1090: Similar Steels (U.S. and/or Foreign). UNS G10900; AMS 5112; ASTM A29, A510, A576; SAE J403, J412, J414; (W. Ger.) DIN 1.1273

AISI 1095: Similar Steels (U.S. and/or Foreign). UNS G10950; AMS 5121, 5122, 5132, 7304; ASTM A29, A510, A576, A682; FED QQ-S-700 (C1095); MIL SPEC MIL-S-16788 (C10); SAE J403, J412, J414; (W. Ger.) DIN 1.1274; (Jap.) JIS SUP 4; (Swed.) SS_{14} 1870; (U.K.) B.S. 060 A 96, EN 44 B

AISI 1090, 1095: Approximate Critical Points

AISI grade	Transformation point	Temperature(a) °C	Temperature(a) °F
1090	Ac_1	730	1345
	Ac_3	745	1370
	Ar_3	700	1290
	Ar_1	690	1270
1095	Ac_1	730	1350
	Ac_3	770	1415
	Ar_3	725	1340
	Ar_1	700	1290

(a) On heating or cooling at 28 °C (50 °F) per hour

Physical Properties

AISI 1090, 1095: Thermal Treatment Temperatures

Treatment	Temperature range °C	Temperature range °F
Annealing(a)	760-790	1400-1450
Normalizing	845-900	1550-1650
Hardening(b)	855-885	1575-1625
Induction hardening(c)	790-815	1450-1500

(a) Spheroidal structures, which are often required for machining purposes, should be cooled very slowly, or be isothermally transformed, to produce the desired structure. (b) For AISI 1090 only, the quenching medium is oil. (c) In addition to the quenching medium of oil, suitable for AISI 1090, this temperature range may be employed for water or brine when quenching 1095 steel

AISI 1095: Average Coefficients of Linear Thermal Expansion (Ref 4)

Temperature range °C	Temperature range °F	Coefficient μm/m·K	Coefficient μin./in.·°F
0-100	32-212	11.0	6.1
0-200	32-390	11.7	6.5
0-300	32-570	12.4	6.9
0-400	32-750	13.1	7.3
0-500	32-930	13.5	7.5
0-600	32-1110	14.2	7.9
0-700	32-1290	14.6	8.1

Mechanical Properties

AISI 1095: Mass Effect on Mechanical Properties (Ref 2)

Condition or treatment	Size round mm	Size round in.	Tensile strength MPa	Tensile strength ksi	Yield strength MPa	Yield strength ksi	Elongation(a), %	Reduction in area, %	Hardness, HB
Annealed (heated to 790 °C or 1450 °F, furnace cooled 11 °C or 20 °F per hour to 665 °C or 1215 °F, cooled in air)	25	1	655	95	380	55	13.0	21.0	192
Normalized (heated to 900 °C or 1650 °F, cooled in air)	13	0.5	1040	151	550	80	12.3	27.7	302
	25	1	1015	147	495	72	9.5	13.5	293
	50	2	910	132	400	58	9.2	13.4	269
	100	4	882	128	395	57	10.0	13.9	255
Oil quenched from 800 °C (1475 °F), tempered at 480 °C (900 °F)	13	0.5	1270	184	800	116	12.8	35.5	363
	25	1	1215	176	703	102	10.0	23.4	352
	50	2	1160	168	675	98	12.0	29.8	331
	100	4	1140	165	640	93	12.2	17.3	331
Oil quenched from 800 °C (1475 °F), tempered at 540 °C (1000 °F)	13	0.5	1145	166	703	102	15.7	40.0	331
	25	1	1105	160	655	95	13.2	32.4	321
	50	2	1040	151	635	92	13.7	31.4	311
	100	4	1020	148	550	80	11.7	22.1	302
Oil quenched from 800 °C (1475 °F), tempered at 595 °C (1100 °F)	13	0.5	979	142	600	87	17.4	42.8	293
	25	1	965	140	545	79	17.2	38.8	277
	50	2	924	134	530	77	18.7	43.4	269
	100	4	896	130	455	66	17.2	34.4	262

(a) In 50 mm (2 in.)

AISI 1095: Effect of Tempering Temperature on Tensile Properties.

Normalized at 900 °C (1650 °F); reheated to 800 °C (1475 °F); quenched in oil. Specimens were treated in 25-mm (1.0-in.) diam and machined to 12.8-mm (0.505-in.) diam for testing. Tests were conducted using test specimens machined to English units. As-quenched hardness was 401 HB. Elongation was measured in 50 mm (2 in.). (Ref 2)

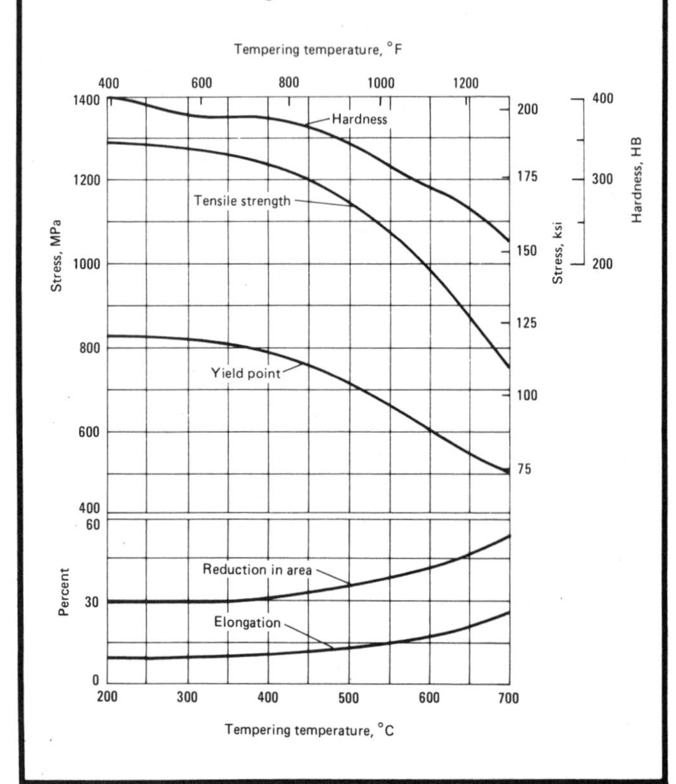

AISI 1095: Effect of the Mass on Hardness at Selected Points (Ref 2)

Size round mm	Size round in.	As-quenched hardness after quenching in oil at: Surface	½ radius	Center
13	0.5	60 HRC	44 HRC	41 HRC
25	1	46 HRC	42 HRC	40 HRC
50	2	43 HRC	40 HRC	37 HRC
100	4	40 HRC	37 HRC	30 HRC

AISI 1095: Izod Impact Properties (Ref 12)

Condition or treatment	Energy J	Energy ft·lb
As rolled	4	3
Normalized at 900 °C (1650 °F)	5	4
Annealed at 790 °C (1450 °F)	3	2

Mechanical Properties (continued)

AISI 1090, 1095: Tensile Properties

Condition or treatment	Size round mm	Size round in.	Tensile strength MPa	Tensile strength ksi	Yield strength MPa	Yield strength ksi	Elongation(a), %	Reduction in area, %	Hardness, HB
AISI 1090									
Hot rolled (Ref 1)	19-32	0.75-1.25	841	122	460	67	10	25	248
Cold drawn, spheroidized annealed(b) (Ref 1)	19-32	0.75-1.25	696	101	540	78	10	40	197
AISI 1095									
Hot rolled (Ref 1)	19-32	0.75-1.25	827	120	455	66	10	25	248
Cold drawn, spheroidized annealed(b) (Ref 1)	19-32	0.75-1.25	685	99	525	76	10	40	197
As rolled (Ref 12)	...	...	965	140	570	83	9	18	293
Normalized at 900 °C (1650 °F) (Ref 12)	...	...	1015	147	505	73	10	14	293
Annealed at 790 °C (1450 °F) (Ref 12)	...	...	665	95	380	55	13	21	192

(a) In 50 mm (2 in.). (b) Average machinability rating of 45% based on AISI 1212 steel as 100% average machinability

Machining Data

For machining data on AISI grades 1090 and 1095, refer to the preceding machining tables for AISI grades 1059 and 1060.

AISI 1108, 1109

AISI 1108, 1109: Chemical Composition

AISI grade	C	Mn	P max	S max
1108	0.08-0.13	0.50-0.80	0.040	0.08-0.13
1109	0.08-0.13	0.60-0.90	0.040	0.08-0.13

Characteristics. Machinability of AISI grades 1108 and 1109 is better than that of carbon steels with similar carbon and manganese content. Grade 1109 can be carburized to increase surface hardness. Cold forming properties, weldability, and forging characteristics of 1108 and 1109 steels are inferior to similar carbon steels.

Typical Uses. AISI 1108 and 1109 steels are used for screw machine parts requiring considerable machining.

AISI 1108: Similar Steels (U.S. and/or Foreign). UNS G11080; ASTM A29; SAE J403, J412; (W. Ger.) DIN 1.0700

AISI 1109: Similar Steels (U.S. and/or Foreign). UNS G11090; ASTM A29; FED QQ-S-637 (C1109); SAE J403, J412, J414; (W. Ger.) DIN 1.0702; (Jap.) JIS SUM 11, SUM 12

Mechanical Properties

AISI 1108: Tensile Properties and Machinability (Ref 1)

Condition	Size round mm	Size round in.	Tensile strength MPa	Tensile strength ksi	Yield strength MPa	Yield strength ksi	Elongation(a), %	Reduction in area, %	Hardness, HB	Average machinability rating(b)
Hot rolled	19-32	0.75-1.25	345	50	195	28	30	50	101	...
Cold drawn	19-32	0.75-1.25	385	56	325	47	20	40	121	80

(a) In 50 mm (2 in.). (b) Based on AISI 1212 steel as 100% average machinability

Physical Properties

AISI 1109: Thermal Treatment for Case Hardening. Carburize at 900 to 925 °C (1650 to 1700 °F), cool in water or oil; reheat at 760 to 790 °C (1400 to 1450 °F), cool in water or caustic (3% sodium hydroxide, NaOH); temper at 120 to 205 °C (250 to 400 °F) to desired hardness

Machining Data (Ref 8)

AISI 1108, 1109: Turning (Single Point and Box Tools)

Depth of cut		M2 and M3 high speed steel				Uncoated carbide						Coated carbide			
		Speed		Feed		Speed, brazed		Speed, inserted		Feed		Speed		Feed	
mm	in.	m/min	ft/min	mm/rev	in./rev	m/min	ft/min	m/min	ft/min	mm/rev	in./rev	m/min	ft/min	mm/rev	in./rev
Hot rolled or annealed, with hardness of 100 to 150 HB															
1	0.040	55	180	0.20	0.008	185(a)	600(a)	215(a)	700(a)	0.18	0.007	320(b)	1050(b)	0.18	0.007
4	0.150	41	135	0.40	0.015	135(c)	450(c)	160(c)	525(c)	0.50	0.020	215(d)	700(d)	0.40	0.015
8	0.300	34	110	0.50	0.020	110(c)	360(c)	130(c)	420(c)	0.75	0.030	170(d)	550(d)	0.50	0.020
16	0.625	26	85	0.75	0.030	85(c)	280(c)	100(c)	330(c)	1.00	0.040	...	...	...	...
Cold drawn, with hardness of 150 to 200 HB															
1	0.040	58	190	0.20	0.008	185(a)	615(a)	220(a)	725(a)	0.18	0.007	335(b)	1100(b)	0.18	0.007
4	0.150	44	145	0.40	0.015	145(c)	475(c)	170(c)	550(c)	0.50	0.020	215(d)	700(d)	0.40	0.015
8	0.300	34	110	0.50	0.020	115(c)	375(c)	130(c)	430(c)	0.75	0.030	175(d)	575(d)	0.50	0.020
16	0.625	27	90	0.75	0.030	88(c)	290(c)	105(c)	340(c)	1.00	0.040	...	...	...	...

(a) Carbide tool material: C-7. (b) Carbide tool material: CC-7. (c) Carbide tool material: C-6. (d) Carbide tool material: CC-6

AISI 1108, 1109: Turning (Cutoff and Form Tools)

Tool material	Speed, m/min (ft/min)	Feed per revolution for cutoff tool width of:						Feed per revolution for form tool width of:							
		1.5 mm (0.062 in.)		3 mm (0.125 in.)		6 mm (0.25 in.)		12 mm (0.5 in.)		18 mm (0.75 in.)		25 mm (1 in.)		50 mm (2 in.)	
		mm	in.	mm	in.	mm	in.	mm	in.	mm	in.	mm	in.	mm	in.
Hot rolled or annealed, with hardness of 100 to 150 HB															
M2 and M3 high speed steel	40 (130)	0.041	0.0016	0.050	0.0020	0.061	0.0024	0.050	0.0020	0.046	0.0018	0.043	0.0017	0.030	0.0012
C-6 carbide	120 (400)	0.041	0.0016	0.050	0.0020	0.061	0.0024	0.050	0.0020	0.046	0.0018	0.043	0.0017	0.030	0.0012
Cold drawn, with hardness of 150 to 200 HB															
M2 and M3 high speed steel	44 (145)	0.041	0.0016	0.050	0.0020	0.061	0.0024	0.050	0.0020	0.046	0.0018	0.043	0.0017	0.030	0.0012
C-6 carbide	135 (450)	0.041	0.0016	0.050	0.0020	0.061	0.0024	0.050	0.0020	0.046	0.0018	0.043	0.0017	0.030	0.0012

AISI 1108, 1109: Drilling

Tool material	Speed		Feed per revolution for nominal hole diameter of:											
			1.5 mm (0.062 in.)		3 mm (0.125 in.)		6 mm (0.25 in.)		12 mm (0.5 in.)		18 mm (0.75 in.)		25 mm (1 in.)	
	m/min	ft/min	mm	in.	mm	in.	mm	in.	mm	in.	mm	in.	mm	in.
Hot rolled or annealed, with hardness of 100 to 150 HB														
M10, M7, and M1 high speed steel	21	70	0.025	0.001	...	...	...	...	...	...	...	...	...	...
	29	95	...	...	0.075	0.003	0.13	0.005	0.30	0.012	0.45	0.018	0.55	0.022
Cold drawn, with hardness of 150 to 200 HB														
M10, M7, and M1 high speed steel	21	70	0.025	0.001	...	...	...	...	...	...	...	...	...	...
	30	100	...	...	0.075	0.003	0.13	0.005	0.30	0.012	0.45	0.018	0.55	0.022

AISI 1059, 1060: Planing

Tool material	Depth of cut		Speed		Feed/stroke	
	mm	in.	m/min	ft/min	mm	in.
Hot rooled or annealed, with hardness of 100 to 150 HB						
M2 and M3 high speed steel	0.1	0.005	17	55	(a)	(a)
	2.5	0.100	27	90	1.25	0.050
	12	0.500	14	45	1.50	0.060
C-6 carbide	0.1	0.005	90	300	(a)	(a)
	2.5	0.100	90	300	2.05	0.080
	12	0.500	90	300	1.50	0.060

Tool material	Depth of cut		Speed		Feed/stroke	
	mm	in.	m/min	ft/min	mm	in.
Cold drawn, with hardness of 150 to 200 HB						
M2 and M3 high speed steel	0.1	0.005	20	65	(a)	(a)
	2.5	0.100	30	100	1.25	0.050
	12	0.500	15	50	1.50	0.060
C-6 carbide	0.1	0.005	90	300	(a)	(a)
	2.5	0.100	90	300	2.05	0.080
	12	0.500	90	300	1.50	0.060

(a) Feed is 75% the width of the square nose finishing tool

Machining Data (Ref 8) (continued)

AISI 1108, 1109: Face Milling

Depth of cut		M2 and M7 high speed steel				Uncoated carbide						Coated carbide			
		Speed		Feed/tooth		Speed, brazed		Speed, inserted		Feed/tooth		Speed		Feed/tooth	
mm	in.	m/min	ft/min	mm	in.	m/min	ft/min	m/min	ft/min	mm	in.	m/min	ft/min	mm	in.
Hot rolled or annealed, with hardness of 100 to 150 HB															
1	0.040	72	235	0.20	0.008	200(a)	655(a)	215(a)	700(a)	0.20	0.008	320(b)	1050(b)	0.20	0.008
4	0.150	55	180	0.30	0.012	150(a)	495(a)	165(a)	545(a)	0.30	0.012	215(b)	710(b)	0.30	0.012
8	0.300	43	140	0.40	0.016	105(c)	345(c)	130(c)	425(c)	0.40	0.016	170(d)	550(d)	0.40	0.016
Cold drawn, with hardness of 150 to 200 HB															
1	0.040	69	225	0.20	0.008	185(a)	600(a)	195(a)	640(a)	0.20	0.008	290(b)	950(b)	0.20	0.008
4	0.150	52	170	0.30	0.012	135(a)	450(a)	150(a)	495(a)	0.30	0.012	195(b)	645(b)	0.30	0.012
8	0.300	40	130	0.40	0.016	95(c)	315(c)	115(c)	385(c)	0.40	0.016	150(d)	500(d)	0.40	0.016

(a) Carbide tool material: C-6. (b) Carbide tool material: CC-6. (c) Carbide tool material: C-5. (d) Carbide tool material: CC-5

AISI 1108, 1109: End Milling (Profiling)

Tool material	Depth of cut		Speed		Feed per tooth for cutter diameter of:							
					10 mm (0.375 in.)		12 mm(0.5 in.)		18 mm (0.75 in.)		25-50 mm (1-2 in.)	
	mm	in.	m/min	ft/min	mm	in.	mm	in.	mm	in.	mm	in.
Hot rolled or annealed, with hardness of 100 to 150 HB												
M2, M3, and M7 high speed steel	0.5	0.020	69	275	0.025	0.001	0.050	0.002	0.102	0.004	0.13	0.005
	1.5	0.060	53	175	0.050	0.002	0.075	0.003	0.13	0.005	0.15	0.006
	diam/4	diam/4	47	155	0.025	0.001	0.050	0.002	0.102	0.004	0.13	0.005
	diam/2	diam/2	41	135	0.018	0.0007	0.025	0.001	0.075	0.003	0.102	0.004
C-5 carbide	0.5	0.020	185	600	0.050	0.002	0.075	0.003	0.13	0.005	0.18	0.007
	1.5	0.060	135	450	0.075	0.003	0.102	0.004	0.15	0.006	0.20	0.008
	diam/4	diam/4	120	390	0.050	0.002	0.075	0.003	0.13	0.005	0.18	0.007
	diam/2	diam/2	105	340	0.038	0.0015	0.050	0.002	0.102	0.004	0.15	0.006
Cold drawn, with hardness of 150 to 200 HB												
M2, M3, and M7 high speed steel	0.5	0.020	58	190	0.025	0.001	0.050	0.002	0.102	0.004	0.13	0.005
	1.5	0.060	44	145	0.050	0.002	0.075	0.003	0.13	0.005	0.15	0.006
	diam/4	diam/4	40	130	0.025	0.001	0.050	0.002	0.102	0.004	0.13	0.005
	diam/2	diam/2	35	115	0.018	0.0007	0.025	0.001	0.075	0.003	0.102	0.004
C-5 carbide	0.5	0.020	175	570	0.050	0.002	0.075	0.003	0.13	0.005	0.18	0.007
	1.5	0.060	135	440	0.075	0.003	0.102	0.004	0.15	0.006	0.20	0.008
	diam/4	diam/4	115	385	0.050	0.002	0.075	0.003	0.13	0.005	0.18	0.007
	diam/2	diam/2	105	340	0.038	0.0015	0.050	0.002	0.102	0.004	0.15	0.006

AISI 1108, 1109: Reaming

Based on 4 flutes for 3- and 6-mm (0.125- and 0.25-in.) reamers, 6 flutes for 12-mm (0.5-in.) reamers, and 8 flutes for 25-mm (1-in.) and larger reamers

Tool material	Speed		Feed per revolution for reamer diameter of:											
			3 mm (0.125 in.)		6 mm (0.25 in.)		12 mm (0.5 in.)		25 mm (1 in.)		35 mm (1.5 in.)		50 mm (2 in.)	
	m/min	ft/min	mm	in.	mm	in.	mm	in.	mm	in.	mm	in.	mm	in.
Hot rolled or annealed, with hardness of 100 to 150 HB														
Roughing														
M1, M2, and M7 high speed steel	35	115	0.13	0.005	0.20	0.008	0.30	0.012	0.50	0.020	0.65	0.025	0.75	0.030
C-2 carbide	40	130	0.15	0.006	0.25	0.010	0.40	0.015	0.65	0.025	0.75	0.030	0.90	0.035
Finishing														
M1, M2, and M7 high speed steel	18	60	0.15	0.006	0.25	0.010	0.40	0.015	0.65	0.025	0.75	0.030	0.90	0.035
C-2 carbide	20	65	0.15	0.006	0.25	0.010	0.40	0.015	0.65	0.025	0.75	0.030	0.90	0.035
Cold drawn, with hardness of 150 to 200 HB														
Roughing														
M1, M2, and M7 high speed steel	37	120	0.13	0.005	0.20	0.008	0.30	0.012	0.50	0.020	0.65	0.025	0.75	0.030
C-2 carbide	43	140	0.15	0.006	0.25	0.010	0.40	0.015	0.65	0.025	0.75	0.030	0.90	0.035
Finishing														
M1, M2, and M7 high speed steel	20	65	0.15	0.006	0.25	0.010	0.40	0.015	0.65	0.025	0.75	0.030	0.90	0.035
C-2 carbide	21	70	0.15	0.006	0.25	0.010	0.40	0.015	0.65	0.025	0.75	0.030	0.90	0.035

Machining Data (Ref 8) (continued)

AISI 1108, 1109: Boring

Depth of cut		M2 and M3 high speed steel				Uncoated carbide						Coated carbide			
		Speed		Feed		Speed, brazed		Speed, inserted		Feed		Speed		Feed	
mm	in.	m/min	ft/min	mm/rev	in./rev	m/min	ft/min	m/min	ft/min	mm/rev	in./rev	m/min	ft/min	mm/rev	in./rev
Hot rolled or annealed, with hardness of 100 to 150 HB															
0.25	0.010	67	220	0.102	0.004	160(a)	520(a)	185(a)	615(a)	0.075	0.003	280(b)	920(b)	0.075	0.003
1.25	0.050	44	145	0.15	0.006	130(c)	425(c)	150(c)	500(c)	0.13	0.005	225(d)	735(d)	0.13	0.005
2.50	0.100	34	110	0.30	0.012	95(e)	315(e)	115(e)	370(e)	0.40	0.015	150(f)	500(f)	0.30	0.012
Cold drawn, with hardness of 150 to 200 HB															
0.25	0.010	70	230	0.102	0.004	165(a)	545(a)	195(a)	640(a)	0.075	0.003	295(b)	960(b)	0.075	0.003
1.25	0.050	46	150	0.15	0.006	135(c)	435(c)	155(c)	510(c)	0.13	0.005	235(d)	770(d)	0.13	0.005
2.50	0.100	35	115	0.30	0.012	100(e)	330(e)	115(e)	385(e)	0.40	0.015	150(f)	490(f)	0.30	0.012

(a) Carbide tool material: C-8. (b) Carbide tool material: CC-8. (c) Carbide tool material: C-7. (d) Carbide tool material: CC-7. (e) Carbide tool material: C-6. (f) Carbide tool material: CC-6

AISI 1116, 1117, 1118, 1119

AISI 1116, 1117, 1118, 1119: Chemical Composition

AISI grade	Chemical composition, % C	Mn	P max	S
1116	0.14-0.20	1.10-1.40	0.040	0.16-0.23
1117	0.14-0.20	1.00-1.30	0.040	0.08-0.13
71118	0.14-0.20	1.30-1.60	0.040	0.08-0.13
1119	0.14-0.20	1.00-1.30	0.040	0.24-0.33

Characteristics. AISI grades 1116, 1117, 1118, and 1119 carry more manganese than the lower-carbon grades for better hardenability. In most instances this composition permits oil quenching after case hardening heat treatments. These steels are available with added lead to enhance machinability. They are more ductile than the lower manganese grades and will withstand moderately severe cold deformation.

Typical Uses. AISI grades 1116, 1117, 1118, and 1119 are used where good machinability and case hardening are required. Applications include medium duty shafts, studs, pins, distributor cams, cam shafts, and universal joints.

Grades 1116, 1117, and 1119 can be satisfactorily hardened superficially in activated baths, and subsequently heat treated. Because of greater hardenability, grade 1118 is better for solid parts or heavier walls. For more drastic quenching or lighter sections, grade 1117 is preferred.

AISI 1116: Similar Steels (U.S. and/or Foreign). UNS G11160; ASTM A29; FED QQ-S-637 (C1116); MIL SPEC MIL-S-20166 (CS1116); SAE J412

AISI 1117: Similar Steels (U.S. and/or Foreign). UNS G11170; ASTM A29, A107, A108; FED QQ-S-637 (C1117); MIL SPEC MIL-S-18411; SAE J403, J412, J414

AISI 1118: Similar Steels (U.S. and/or Foreign). UNS G11180; ASTM A29, A107, A108; FED QQ-S-637 (C1118); SAE J403, J412, J414

AISI 1119: Similar Steels (U.S. and/or Foreign). UNS G11190; ASTM 29; FED QQ-S-637 (C1119); SAE J403, J412, J414

AISI 1117, 1118: Approximate Critical Points

AISI grade	Transformation point	Temperature(a) °C	°F
1117	Ac_1	730	1350
	Ac_3	845	1550
	Ar_3	790	1450
	Ar_1	675	1245
1118	Ac_1	730	1345
	Ac_3	825	1520
	Ar_3	815	1495
	Ar_1	675	1245

(a) On heating or cooling at 28 °C (50 °F) per hour

Physical Properties

AISI 1117: Thermal Treatment Temperatures

Treatment	Temperature range °C	°F
Annealing	855-885	1575-1625
Normalizing	875-955	1605-1750
Carburizing(a)	900-955	1650-1750
Reheating(a)	790-870	1450-1600
Carbonitriding(b)	790-900	1450-1650
Tempering(c)	120-205	250-400

(a) Cool in water or oil. For grade 1118, use the same temperatures, but cool in oil. (b) Cool in oil. (c) Grades 1117 and 1118 are tempered to desired hardness

Mechanical Properties

AISI 1116, 1117, 11L17, 1118, 1119: Tensile Properties

Condition or treatment	Size round mm	Size round in.	Tensile strength MPa	Tensile strength ksi	Yield strength MPa	Yield strength ksi	Elongation(a), %	Reduction in area, %	Hardness, HB
AISI 1116									
Hot rolled (Ref 1)	...	...	440	64	...	...	33	41	...
Cold drawn (Ref 1)	...	...	530	77	460	67	17	...	...
AISI 1117									
Hot rolled (Ref 1)	19-32	0.75-1.25	425	62	235	34	23	47	121
Cold drawn(b) (Ref 1)	19-32	0.75-1.25	475	69	400	58	15	40	137
As rolled (Ref 12)	...	...	490	71	305	44	33	63	143
Normalized at 900 °C (1650 °F) (Ref 12)	...	...	465	68	305	44	34	54	137
Annealed at 855 °C (1575 °F) (Ref 12)	...	...	425	62	285	41	33	58	121
Cold finished (Ref 3)	19-32	0.75-1.25	595	86	515	75	22	52	170
AISI 11L17									
Cold finished (Ref 3)	19-32	0.75-1.25	595	86	515	75	22	53	179
AISI 1117, 1118									
As cold drawn (Ref 1)	16-22	0.63-0.88	515	75	450	65	15	40	149
	22-32	0.88-1.25	485	70	415	60	15	40	143
	32-50	1.25-2	450	65	380	55	13	35	131
	50-75	2-3	415	60	345	50	12	30	121
Cold drawn, low temperature, stress relieved (Ref 1)	16-22	0.63-0.88	550	80	485	70	15	40	163
	22-32	0.88-1.25	515	75	450	65	15	40	149
	32-50	1.25-2	485	70	415	60	13	35	143
	50-75	2-3	450	65	380	55	12	35	131
Cold drawn, high temperature, stress relieved (Ref 1)	16-22	0.63-0.88	485	70	345	50	18	45	143
	22-32	0.88-1.25	450	65	345	50	16	45	131
	32-50	1.25-2	415	60	345	50	15	40	121
	50-75	2-3	380	55	310	45	15	40	111
Hot rolled (Ref 1)	19-32	0.75-1.25	400	58	220	32	25	50	116
AISI 1118									
Cold drawn(c) (Ref 1)	19-32	0.75-1.25	440	64	370	54	15	40	126
As rolled (Ref 12)	...	...	525	76	315	46	32	70	149
Normalized at 925 °C (1700 °F) (Ref 12)	...	...	475	69	315	46	34	66	143
Annealed at 790 °C (1450 °F) (Ref 12)	...	...	450	65	285	41	34	67	131

(a) In 50 mm (2 in.). (b) Average machinability rating of 90% based on AISI 1212 steel as 100% average machinability. (c) Average machinability rating of 70% based on AISI 1212 steel as 100% average machinability

AISI 1117: Mass Effect on Mechanical Properties (Ref 2)

Condition or treatment	Size round mm	Size round in.	Tensile strength MPa	Tensile strength ksi	Yield strength MPa	Yield strength ksi	Elongation(a), %	Reduction in area, %	Hardness, HB
As rolled	25	1	483	70.00	345	50.25	33.5	61.0	149
Annealed (heated to 855 °C or 1575 °F, furnace cooled 17 °C or 30 °F per hour to 700 °C or 1290 °F, cooled in air)	25	1	430	62.30	279	40.50	33.0	58.0	121
Normalized (heated to 900 °C or 1650 °F, cooled in air)	13	0.5	481	69.75	310	45.00	34.3	61.0	143
	25	1	467	67.75	303	44.00	33.5	63.8	137
	50	2	462	67.00	286	41.50	33.5	64.7	137
	100	4	440	63.75	241	35.00	34.3	64.7	126
Mock carburized at 925 °C (1700 °F) for 8 h; reheated to 790 °C (1450 °F); quenched in water; tempered at 175 °C (350 °F)	13	0.5	860	124.75	459	66.50	9.7	18.4	235
	25	1	617	89.50	348	50.50	22.3	48.8	183
	50	2	538	78.00	329	47.75	26.3	65.7	156
	100	4	515	74.75	295	42.75	27.3	62.6	149

(a) In 50 mm (2 in.)

AISI 1118: Effect of the Mass on Hardness at Selected Points (Ref 2)

Size round mm	Size round in.	As-quenched hardness after quenching in water at: Surface	½ radius	Center
13	0.5	43 HRC	36 HRC	33 HRC
25	1	36 HRC	99 HRB	96.5 HRB
50	2	34 HRC	91 HRB	87 HRB
100	4	32 HRC	84 HRB	82 HRB

AISI 1117, 1118: Izod Impact Properties (Ref 12)

AISI grade	Condition or treatment	Energy J	Energy ft·lb
1117	As rolled	81	60
	Normalized at 900 °C (1650 °F)	85	63
	Annealed at 855 °C (1575 °F)	94	69
1118	As rolled	108	80
	Normalized at 925 °C (1700 °F)	103	76
	Annealed at 790 °C (1450 °F)	107	79

Mechanical Properties (continued)

AISI 1118: Mass Effect on Mechanical Properties (Ref 2)

Condition or treatment	Size round mm	Size round in.	Tensile strength MPa	Tensile strength ksi	Yield strength MPa	Yield strength ksi	Elongation(a), %	Reduction in area, %	Hardness, HB
As rolled	25	1	486	70.50	355	51.50	32.3	63.0	143
Annealed (heated to 790 °C or 1450 °F, furnace cooled 17 °C or 30 °F per hour to 605 °C or 1125 °F, cooled in air)	25	1	450	65.30	285	41.30	34.5	67.0	131
Normalized (heated to 925 °C or 1700 °F, cooled in air)	13	0.5	502	72.75	330	47.80	33.3	62.8	156
	25	1	477	69.25	319	46.25	33.5	65.9	143
	50	2	472	68.50	298	43.25	33.0	67.7	137
	100	4	457	66.25	260	37.75	34.0	67.4	131
Mock carburized at 925 °C (1700 °F) for 8 h; reheated to 790 °C (1450 °F); quenched in water; tempered at 175 °C (350 °F)	13	0.5	996	144.50	621	90.00	13.2	30.8	285
	25	1	707	102.50	409	59.25	19.0	48.9	207
	50	2	567	82.25	330	47.87	27.3	65.5	167
	100	4	531	77.00	310	45.00	31.0	67.4	156

(a) In 50 mm (2 in.)

AISI 1117: Effect of the Mass on Hardness at Selected Points (Ref 2)

Size round mm	Size round in.	As-quenched hardness after quenching in water at: Surface	½ radius	Center
13	0.5	42 HRC	34.5 HRC	29.5 HRC
25	1	37 HRC	96 HRB	93 HRB
50	2	33 HRC	90 HRB	86 HRB
100	4	32 HRC	83 HRB	81 HRB

Machining Data (Ref 8)

AISI 1116, 1117, 1118, 1119: Face Milling

Depth of cut mm	Depth of cut in.	M2 and M7 high speed steel Speed m/min	Speed ft/min	Feed/tooth mm	Feed/tooth in.	Uncoated carbide Speed, brazed m/min	Speed, brazed ft/min	Speed, inserted m/min	Speed, inserted ft/min	Feed/tooth mm	Feed/tooth in.	Coated carbide Speed m/min	Speed ft/min	Feed/tooth mm	Feed/tooth in.
Hot rolled or annealed, with hardness of 100 to 150 HB															
1	0.040	79	260	0.20	0.008	225(a)	730(a)	245(a)	800(a)	0.20	0.008	365(b)	1200(b)	0.20	0.008
4	0.150	60	200	0.30	0.012	170(a)	550(a)	185(a)	605(a)	0.30	0.012	240(b)	785(b)	0.30	0.012
8	0.300	47	155	0.40	0.016	115(c)	385(c)	145(c)	470(c)	0.40	0.016	185(d)	610(d)	0.40	0.016
Cold drawn, with hardness of 150 to 200 HB															
1	0.040	76	250	0.20	0.008	205(a)	665(a)	225(a)	730(a)	0.20	0.008	335(b)	1100(b)	0.20	0.008
4	0.150	58	190	0.30	0.012	150(a)	500(a)	170(a)	550(a)	0.30	0.012	220(b)	715(b)	0.30	0.012
8	0.300	46	150	0.40	0.016	105(c)	350(c)	130(c)	430(c)	0.40	0.016	170(d)	560(d)	0.40	0.016

(a) Carbide tool material: C-6. (b) Carbide tool material: CC-6. (c) Carbide tool material: C-5. (d) Carbide tool material: CC-5

AISI 1116, 1117, 1118, 1119: Drilling

Tool material	Speed m/min	Speed ft/min	Feed per revolution for nominal hole diameter of: 1.5 mm (0.062 in.) mm	1.5 mm in.	3 mm (0.125 in.) mm	3 mm in.	6 mm (0.25 in.) mm	6 mm in.	12 mm (0.5 in.) mm	12 mm in.	18 mm (0.75 in.) mm	18 mm in.	25 mm (1 in.) mm	25 mm in.
Hot rolled or annealed, with hardness of 100 to 150 HB														
M10, M7, and M1 high speed steel	21	70	0.025	0.001	...	...	...	...	...	...	...	...	...	...
	32	105	...	...	0.075	0.003	0.13	0.005	0.30	0.012	0.45	0.018	0.55	0.022
Cold drawn, with hardness of 150 to 200 HB														
M10, M7, and M1 high speed steel	21	70	0.025	0.001	...	...	...	...	...	...	...	...	...	...
	34	110	...	...	0.075	0.003	0.13	0.005	0.30	0.012	0.45	0.018	0.55	0.022

Machining Data (Ref 8) (continued)

AISI 1116, 1117, 1118, 1119: Turning (Single Point and Box Tools)

Depth of cut		M2 and M3 high speed steel				Uncoated carbide						Coated carbide			
		Speed		Feed		Speed, brazed		Speed, inserted		Feed		Speed		Feed	
mm	in.	m/min	ft/min	mm/rev	in./rev	m/min	ft/min	m/min	ft/min	mm/rev	in./rev	m/min	ft/min	mm/rev	in./rev
Hot rolled or annealed, with hardness of 100 to 150 HB															
1	0.040	60	200	0.18	0.007	205(a)	670(a)	240(a)	790(a)	0.18	0.007	365(b)	1200(b)	0.18	0.007
4	0.150	45	150	0.40	0.015	155(c)	510(c)	185(c)	600(c)	0.50	0.020	235(d)	775(d)	0.40	0.015
8	0.300	37	120	0.50	0.020	120(c)	400(c)	145(c)	475(c)	0.75	0.030	190(d)	625(d)	0.50	0.020
16	0.625	27	90	0.75	0.030	100(c)	320(c)	115(c)	370(c)	1.00	0.040	...	...	...	...
Cold drawn, with hardness of 150 to 200 HB															
1	0.040	64	210	0.18	0.007	205(a)	680(a)	250(a)	820(a)	0.18	0.007	375(b)	1225(b)	0.18	0.007
4	0.150	49	160	0.40	0.015	160(c)	520(c)	190(c)	625(c)	0.50	0.020	245(d)	800(d)	0.40	0.015
8	0.300	38	125	0.50	0.020	125(c)	410(c)	150(c)	495(c)	0.75	0.030	200(d)	650(d)	0.50	0.020
16	0.625	30	100	0.75	0.030	100(c)	330(c)	115(c)	385(c)	1.00	0.040	...	...	...	...

(a) Carbide tool material: C-7. (b) Carbide tool material: CC-7. (c) Carbide tool material: C-6. (d) Carbide tool material: CC-6

AISI 1116, 1117, 1118, 1119: Turning (Cutoff and Form Tools)

Tool material	Speed, m/min (ft/min)	Feed per revolution for cutoff tool width of:						Feed per revolution for form tool width of:							
		1.5 mm (0.062 in.)		3 mm (0.125 in.)		6 mm (0.25 in.)		12 mm (0.5 in.)		18 mm (0.75 in.)		25 mm (1 in.)		50 mm (2 in.)	
		mm	in.	mm	in.	mm	in.	mm	in.	mm	in.	mm	in.	mm	in.
Hot rolled or annealed, with hardness of 100 to 150 HB															
M2 and M3 high speed steel	43 (140)	0.048	0.0019	0.058	0.0023	0.071	0.0028	0.058	0.0023	0.050	0.0020	0.048	0.0019	0.036	0.0014
C-6 carbide	135 (450)	0.048	0.0019	0.058	0.0023	0.071	0.0028	0.058	0.0023	0.050	0.0020	0.048	0.0019	0.036	0.0014
Cold drawn, with hardness of 150 to 200 HB															
M2 and M3 high speed steel	47 (155)	0.048	0.0019	0.058	0.0023	0.071	0.0028	0.058	0.0023	0.050	0.0020	0.048	0.0019	0.036	0.0014
C-6 carbide	150 (500)	0.048	0.0019	0.058	0.0023	0.071	0.0028	0.058	0.0023	0.050	0.0020	0.048	0.0019	0.036	0.0014

AISI 1116, 1117, 1118, 1119: Reaming

Based on 4 flutes for 3- and 6-mm (0.125- and 0.25-in.) reamers, 6 flutes for 12-mm (0.5-in.) reamers, and 8 flutes for 25-mm (1-in.) and larger reamers

Tool material	Speed		Feed per revolution for reamer diameter of:											
			3 mm (0.125 in.)		6 mm (0.25 in.)		12 mm (0.5 in.)		25 mm (1 in.)		35 mm (1.5 in.)		50 mm (2 in.)	
	m/min	ft/min	mm	in.	mm	in.	mm	in.	mm	in.	mm	in.	mm	in.
Hot rolled or annealed, with hardness of 100 to 150 HB														
Roughing														
M1, M2, and M7 high speed steel	37	120	0.13	0.005	0.20	0.008	0.30	0.012	0.50	0.020	0.65	0.025	0.75	0.030
C-2 carbide	43	140	0.15	0.006	0.25	0.010	0.40	0.015	0.65	0.025	0.75	0.030	0.90	0.035
Finishing														
M1, M2, and M7 high speed steel	18	60	0.15	0.006	0.25	0.010	0.40	0.015	0.65	0.025	0.75	0.030	0.90	0.035
C-2 carbide	21	70	0.15	0.006	0.25	0.010	0.40	0.015	0.65	0.025	0.75	0.030	0.90	0.035
Cold drawn, with hardness of 150 to 200 HB														
Roughing														
M1, M2, and M7 high speed steel	40	130	0.13	0.005	0.20	0.008	0.30	0.012	0.50	0.020	0.65	0.025	0.75	0.030
C-2 carbide	46	150	0.15	0.006	0.25	0.010	0.40	0.015	0.65	0.025	0.75	0.030	0.90	0.035
Finishing														
M1, M2, and M7 high speed steel	20	65	0.15	0.006	0.25	0.010	0.40	0.015	0.65	0.025	0.75	0.030	0.90	0.035
C-2 carbide	21	70	0.15	0.006	0.25	0.010	0.40	0.015	0.65	0.025	0.75	0.030	0.90	0.035

Machining Data (Ref 8) (continued)

AISI 1116, 1117, 1118, 1119: End Milling (Profiling)

Tool material	Depth of cut mm	Depth of cut in.	Speed m/min	Speed ft/min	Feed per tooth for cutter diameter of: 10 mm (0.375 in.) mm	in.	12 mm (0.5 in.) mm	in.	18 mm (0.75 in.) mm	in.	25-50 mm (1-2 in.) mm	in.
Hot rolled or annealed, with hardness of 100 to 150 HB												
M2, M3, and M7 high speed steel	0.5	.020	72	235	0.025	0.001	0.050	0.002	0.102	0.004	0.13	0.005
	1.5	0.060	55	180	0.050	0.002	0.075	0.003	0.13	0.005	0.15	0.006
	diam/4	diam/4	49	160	0.025	0.001	0.050	0.002	0.102	0.004	0.13	0.005
	diam/2	diam/2	43	140	0.018	0.0007	0.025	0.001	0.075	0.003	0.102	0.004
C-5 carbide	0.5	0.020	190	620	0.050	0.002	0.075	0.003	0.13	0.005	0.18	0.007
	1.5	0.060	145	475	0.075	0.003	0.102	0.004	0.15	0.006	0.20	0.008
	diam/4	diam/4	130	420	0.050	0.002	0.075	0.003	0.13	0.005	0.15	0.006
	diam/2	diam/2	110	360	0.038	0.0015	0.015	0.002	0.102	0.004	0.13	0.005
Cold drawn, with hardness of 150 to 200 HB												
M2, M3, and M7 high speed steel	0.5	0.020	60	200	0.025	0.001	0.050	0.002	0.102	0.004	0.13	0.005
	1.5	0.060	46	150	0.050	0.002	0.075	0.003	0.13	0.005	0.15	0.006
	diam/4	diam/4	40	130	0.025	0.001	0.050	0.002	0.102	0.005	0.13	0.005
	diam/2	diam/2	35	115	0.018	0.0007	0.025	0.001	0.075	0.003	0.102	0.004
C-5 carbide	0.5	0.020	180	585	0.050	0.002	0.075	0.003	0.13	0.005	0.18	0.007
	1.5	0.060	135	450	0.075	0.003	0.102	0.004	0.15	0.006	0.20	0.008
	diam/4	diam/4	120	390	0.050	0.002	0.075	0.003	0.13	0.005	0.15	0.006
	diam/2	diam/2	110	360	0.038	0.0015	0.050	0.002	0.102	0.004	0.13	0.005

AISI 1116, 1117, 1118, 1119: Boring

Depth of cut mm	in.	M2 and M3 high speed steel Speed m/min	ft/min	Feed mm/rev	in./rev	Uncoated carbide Speed, brazed m/min	ft/min	Speed, inserted m/min	ft/min	Feed mm/rev	in./rev	Coated Carbide Speed m/min	ft/min	Feed mm/rev	in./rev
Hot rolled or annealed, with hardness of 100 to 150 HB															
0.25	0.010	73	240	0.075	0.003	245(a)	810(a)	290(a)	950(a)	0.075	0.003	365(b)	1200(b)	0.075	0.003
1.25	0.050	49	160	0.13	0.005	165(c)	540(c)	190(c)	630(c)	0.13	0.005	295(d)	960(d)	0.13	0.005
2.5	0.100	38	125	0.25	0.010	125(e)	410(e)	145(e)	480(e)	0.40	0.015	190(f)	620(f)	0.25	0.010
Cold drawn, with hardness of 150 to 200 HB															
0.25	0.010	76	250	0.075	0.003	225(a)	730(a)	260(a)	860(a)	0.075	0.003	330(b)	1075(b)	0.075	0.003
1.25	0.050	52	170	0.13	0.005	150(c)	490(c)	175(c)	575(c)	0.13	0.005	260(d)	860(d)	0.13	0.005
2.5	0.100	40	130	0.30	0.012	115(e)	375(e)	135(e)	440(e)	0.40	0.015	170(f)	560(f)	0.30	0.012

(a) Carbide tool material: C-8. (b) Carbide tool material: CC-8. (c) Carbide tool material: C-7. (d) Carbide tool material: CC-7. (e) Carbide tool material: C-6. (f) Carbide tool material: CC-6

AISI 1116, 1117, 1118, 1119: Planing

Tool material	Depth of cut mm	in.	Speed m/min	ft/min	Feed/stroke mm	in.
Hot rolled or annealed, with hardness of 100 to 150 HB						
M2 and M3 high speed steel	0.1	0.005	18	60	(a)	(a)
	2.5	0.100	29	95	1.25	0.050
	12	0.500	15	50	1.50	0.060
C-6 carbide	0.1	0.005	90	300	(a)	(a)
	2.5	0.100	90	300	2.05	0.080
	12	0.500	90	300	1.50	0.060
Cold drawn, with hardness of 150 to 200 HB						
M2 and M3 high speed steel	0.1	0.005	20	65	(a)	(a)
	2.5	0.100	32	105	1.25	0.050
	12	0.500	17	55	1.50	0.060
C-6 carbide	0.1	0.005	90	300	(a)	(a)
	2.5	0.100	90	300	2.05	0.080
	12	0.500	90	300	1.50	0.060

(a) Feed is 75% the width of the square nose finishing tool

AISI 1132

AISI 1132: Chemical Composition

AISI grade	C	Chemical composition, % Mn	P max	S
1132	0.27-0.34	1.35-1.65	0.04	0.08-0.13

Characteristics. AISI grade 1132 is a high manganese, medium-carbon resulphurized steel with medium strength and good machinability. It is used where a large amount of machining is necessary, or where threads, splines, and other operations with special tools are required. This steel can be selectively hardened by induction or flame heating, but quench cracking is possible.

Typical Uses. Applications of AISI 1132 steel include nuts, bolts, and studs with machined threads.

AISI 1132: Similar Steels (U.S. and/or Foreign). UNS G11320; ASTM A29; FED QQ-S-637 (C1132); SAE J403, J412, J414

Mechanical Properties

AISI 1132: Tensile Properties

Condition or treatment	Size round mm	Size round in.	Tensile strength MPa	Tensile strength ksi	Yield strength MPa	Yield strength ksi	Elongation(a), %	Reduction in area, %	Hardness, HB
Hot rolled	19-32	0.75-1.25	570	83	315	46	16	40	167
Cold drawn(b)	19-32	0.75-1.25	635	92	530	77	12	35	183
Hot rolled and cold drawn	<16	<0.63	640-779	93-113	515-655	75-95	14-24	38-58	187-229
	16-38	0.63-1.5	605-760	88-110	485-620	70-90	13-23	35-60	179-217
	38-89	1.5-3.5	550-690	80-100	450-585	65-85	13-23	30-55	164-202

(a) In 50 mm (2 in.). (b) Average machinability rating of 75% based on AISI 1212 steel as 100% average machinability

Machining Data (Ref 8)

AISI 1132: Turning (Single Point and Box Tools)

Depth of cut mm	Depth of cut in.	M2 and M3 high speed steel Speed m/min	Speed ft/min	Feed mm/rev	Feed in./rev	Uncoated carbide Speed, brazed m/min	Speed, brazed ft/min	Speed, inserted m/min	Speed, inserted ft/min	Feed mm/rev	Feed in./rev	Coated carbide Speed m/min	Speed ft/min	Feed mm/rev	Feed in./rev
Hardness, 175 to 225 HB															
1	0.040	52	170	0.20	0.008	165(a)	540(a)	200(a)	660(a)	0.18	0.007	305(b)	1000(b)	0.18	0.007
4	0.150	40	130	0.40	0.015	125(c)	410(c)	150(c)	500(c)	0.50	0.020	200(d)	650(d)	0.40	0.015
8	0.300	30	100	0.50	0.020	100(c)	325(c)	120(c)	400(c)	0.75	0.030	160(d)	525(d)	0.50	0.020
16	0.625	24	80	0.75	0.030	78(c)	255(c)	95(c)	310(c)	1.00	0.040	...	...	...	...
Hardness, 225 to 275 HB															
1	0.040	35	115	0.18	0.007	120(a)	390(a)	150(a)	490(a)	0.18	0.007	230(b)	750(b)	0.18	0.007
4	0.150	27	90	0.40	0.015	90(c)	300(c)	115(c)	375(c)	0.40	0.015	150(d)	500(d)	0.40	0.015
8	0.300	21	70	0.50	0.020	72(c)	235(c)	90(c)	295(c)	0.50	0.020	120(d)	400(d)	0.50	0.020

(a) Carbide tool material: C-7. (b) Carbide tool material: CC-7. (c) Carbide tool material: C-6. (d) Carbide tool material: CC-6

AISI 1132: Drilling

Tool material	Speed m/min	Speed ft/min	Feed per revolution for nominal hole diameter of: 1.5 mm (0.062 in.) mm	1.5 mm in.	3 mm (0.125 in.) mm	3 mm in.	6 mm (0.25 in.) mm	6 mm in.	12 mm (0.5 in.) mm	12 mm in.	18 mm (0.75 in.) mm	18 mm in.	25 mm (1 in.) mm	25 mm in.
Hardness, 175 to 225 HB														
M10, M7, and M1 high speed steel	18	60	0.025	0.001	...	...	...	...	...	...	...	...	...	...
	24	80	...	...	0.075	0.003	0.13	0.005	0.30	0.012	0.45	0.018	0.55	0.022
Hardness, 275 to 325 HB														
M10, M7, and M1 high speed steel	21	70	...	...	0.075	0.003	0.13	0.005	0.25	0.010	0.40	0.015	0.45	0.018

Machining Data (Ref 8) (continued)

AISI 1132: Boring

Depth of cut mm	Depth of cut in.	High speed steel Speed m/min	ft/min	Feed mm/rev	in./rev	Uncoated carbide Speed, brazed m/min	ft/min	Speed, inserted m/min	ft/min	Feed mm/rev	in./rev	Coated carbide Speed m/min	ft/min	Feed mm/rev	in./rev
Hardness, 175 to 225 HB															
0.25	0.010	62(a)	205(a)	0.102	0.004	150(b)	490(b)	175(b)	575(b)	0.075	0.003	265(c)	875(c)	0.075	0.003
1.25	0.050	41(a)	135(a)	0.15	0.006	120(d)	390(d)	140(d)	460(d)	0.13	0.005	215(e)	700(e)	0.13	0.005
2.5	0.100	32(a)	105(a)	0.30	0.012	90(f)	300(f)	105(f)	350(f)	0.40	0.015	140(g)	455(g)	0.30	0.012
Hardness, 275 to 325 HB															
0.25	0.010	43(h)	140(h)	0.075	0.003	110(d)	365(d)	130(d)	430(d)	0.075	0.003	200(e)	655(e)	0.075	0.003
1.25	0.050	27(h)	90(h)	0.13	0.005	90(f)	295(f)	105(f)	345(f)	0.13	0.005	160(g)	525(g)	0.13	0.005
2.5	0.100	21(h)	70(h)	0.30	0.012	67(f)	220(f)	79(d)	260(d)	0.30	0.012	105(g)	350(g)	0.30	0.012

(a) High speed steel tool material: M2 or M3. (b) Carbide tool material: C-8. (c) Carbide tool material: CC-8. (d) Carbide tool material: C-7. (e) Carbide tool material: CC-7. (f) Carbide tool material: C-6. (g) Carbide tool material: CC-6. (h) Any premium high speed steel tool material (T15, M33, M41 to M47)

AISI 1132: End Milling (Profiling)

Tool material	Depth of cut mm	in.	Speed m/min	ft/min	Feed per tooth for cutter diameter of: 10 mm (0.375 in.) mm	in.	12 mm (0.5 in.) mm	in.	18 mm (0.75 in.) mm	in.	25-50 mm (1-2 in.) mm	in.
Hardness, 175 to 225 HB												
M2, M3, and M7 high speed steel	0.5	0.020	67	220	0.025	0.001	0.050	0.002	0.102	0.004	0.13	0.005
	1.5	0.060	52	170	0.050	0.002	0.075	0.003	0.13	0.005	0.15	0.006
	diam/4	diam/4	46	150	0.025	0.001	0.050	0.002	0.102	0.004	0.13	0.005
	diam/2	diam/2	40	130	0.018	0.0007	0.025	0.001	0.075	0.003	0.102	0.004
C-5 carbide	0.5	0.020	160	525	0.050	0.002	0.075	0.003	0.13	0.005	0.18	0.007
	1.5	0.060	120	400	0.075	0.003	0.102	0.004	0.15	0.006	0.20	0.008
	diam/4	diam/4	105	350	0.050	0.002	0.075	0.003	0.13	0.005	0.18	0.007
	diam/2	diam/2	100	325	0.038	0.0015	0.050	0.002	0.102	0.004	0.15	0.006
Hardness, 275 to 325 HB												
M2, M3, and M7 high speed steel	0.5	0.020	50	165	0.018	0.0007	0.050	0.002	0.102	0.004	0.13	0.005
	1.5	0.060	38	125	0.025	0.001	0.075	0.003	0.13	0.005	0.15	0.006
	diam/4	diam/4	34	110	0.018	0.0007	0.050	0.002	0.102	0.004	0.13	0.005
	diam/2	diam/2	30	110	0.013	0.0005	0.025	0.001	0.075	0.003	0.102	0.004
C-5 carbide	0.5	0.020	115	380	0.050	0.002	0.075	0.003	0.102	0.004	0.15	0.006
	1.5	0.060	88	290	0.075	0.003	0.102	0.004	0.13	0.005	0.20	0.008
	diam/4	diam/4	76	250	0.050	0.002	0.075	0.003	0.102	0.004	0.13	0.005
	diam/2	diam/2	70	230	0.038	0.0015	0.050	0.002	0.075	0.003	0.102	0.004

AISI 1132: Reaming

Based on 4 flutes for 3- and 6-mm (0.125- and 0.25-in.) reamers, 6 flutes for 12-mm (0.5-in.) reamers, and 8 flutes for 25-mm (1-in.) and larger reamers

Tool material	Speed m/min	ft/min	Feed per revolution for reamer diameter of: 3 mm (0.125 in.) mm	in.	6 mm (0.25 in.) mm	in.	12 mm (0.5 in.) mm	in.	25 mm (1 in.) mm	in.	35 mm (1.5 in.) mm	in.	50 mm (2 in.) mm	in.
Hardness, 175 to 225 HB														
Roughing														
M1, M2, and M7 high speed steel	34	110	0.13	0.005	0.20	0.008	0.30	0.012	0.50	0.020	0.65	0.025	0.75	0.030
C-2 carbide	38	125	0.15	0.006	0.25	0.010	0.40	0.015	0.65	0.025	0.75	0.030	0.90	0.035
Finishing														
M1, M2, and M7 high speed steel	17	55	0.15	0.006	0.25	0.010	0.40	0.015	0.65	0.025	0.75	0.030	0.90	0.035
C-2 carbide	20	65	0.15	0.006	0.25	0.010	0.40	0.015	0.65	0.025	0.75	0.030	0.90	0.035
Hardness, 275 to 325 HB														
Roughing														
M1, M2, and M7 high speed steel	23	75	0.102	0.004	0.15	0.006	0.25	0.010	0.40	0.015	0.50	0.020	0.65	0.025
C-2 carbide	27	90	0.13	0.005	0.18	0.007	0.30	0.012	0.50	0.020	0.65	0.025	0.75	0.030
Finishing														
M1, M2, and M7 high speed steel	11	35	0.102	0.004	0.20	0.008	0.30	0.012	0.50	0.020	0.65	0.025	0.75	0.030
C-2 carbide	14	45	0.13	0.005	0.20	0.008	0.30	0.012	0.50	0.020	0.65	0.025	0.75	0.030

Machining Data (Ref 8) (continued)

AISI 1132: Turning (Cutoff and Form Tools)

Tool material	Speed, m/min (ft/min)	Feed per revolution for cutoff tool width of: 1.5 mm (0.062 in.) mm	in.	3 mm (0.125 in.) mm	in.	6 mm (0.25 in.) mm	in.	Feed per revolution for form tool width of: 12 mm (0.5 in.) mm	in.	18 mm (0.75 in.) mm	in.	25 mm (1 in.) mm	in.	50 mm (2 in.) mm	in.
Hardness, 175 to 225 HB															
M2 and M3 high speed steel	37 (120)	0.038	0.0015	0.048	0.0019	0.056	0.0022	0.048	0.0019	0.043	0.0017	0.038	0.0015	0.030	0.0012
C-6 carbide	120 (390)	0.038	0.0015	0.048	0.0019	0.056	0.0022	0.048	0.0019	0.043	0.0017	0.038	0.0015	0.030	0.0012
Hardness, 275 to 325 HB															
High speed steel(a)	24 (80)	0.030	0.0012	0.041	0.0016	0.048	0.0019	0.041	0.0016	0.036	0.0014	0.030	0.0012	0.025	0.0010
C-6 carbide	78 (225)	0.030	0.0012	0.041	0.0016	0.048	0.0019	0.041	0.0016	0.036	0.0014	0.030	0.0012	0.025	0.0010

(a) Any premium high speed steel tool material (T15, M33, M41 to M47)

AISI 1132: Face Milling

Depth of cut mm	in.	High speed steel Speed m/min	ft/min	Feed/tooth mm	in.	Uncoated carbide Speed, brazed m/min	ft/min	Speed, inserted m/min	ft/min	Feed/tooth mm	in.	Coated carbide Speed m/min	ft/min	Feed/tooth mm	in.
Hardness, 175 to 225 HB															
1	0.040	69(a)	225(a)	0.20	0.008	180(b)	590(b)	190(b)	620(b)	0.20	0.008	280(c)	935(c)	0.20	0.008
4	0.150	50(a)	165(a)	0.30	0.012	170(b)	550(b)	170(b)	560(b)	0.30	0.012	225(c)	730(c)	0.30	0.012
8	0.300	38(a)	125(a)	0.40	0.016	110(d)	355(d)	135(d)	435(d)	0.40	0.016	170(e)	565(e)	0.40	0.016
Hardness, 275 to 325 HB															
1	0.040	47(f)	155(f)	0.15	0.006	135(b)	450(b)	145(b)	475(b)	0.15	0.006	220(c)	715(c)	0.13	0.005
4	0.150	35(f)	115(f)	0.23	0.009	120(b)	400(b)	130(b)	430(b)	0.20	0.008	170(c)	560(c)	0.18	0.007
8	0.300	27(f)	90(f)	0.30	0.012	84(d)	275(d)	100(d)	335(d)	0.25	0.010	135(e)	435(e)	0.23	0.009

(a) High speed steel tool material: M2 or M7. (b) Carbide tool material: C-6. (c) Carbide tool material: CC-6. (d) Carbide tool material: C-5. (e) Carbide tool material: CC-5. (f) Any premium high speed steel tool material (T15, M33, M41 to M47)

AISI 1132: Planing

Tool material	Depth of cut mm	in.	Speed m/min	ft/min	Feed/stroke mm	in.
Hardness, 175 to 225 HB						
M2 and M3 high speed steel	0.1	0.005	12	40	(a)	(a)
	2.5	0.100	21	70	1.25	0.050
	12	0.500	11	35	1.50	0.060
C-6 carbide	0.1	0.005	90	300	(a)	(a)
	2.5	0.100	90	300	2.05	0.080
	12	0.500	76	250	1.50	0.060
Hardness, 275 to 325 HB						
M2 and M3 high speed steel	0.1	0.005	11	35	(a)	(a)
	2.5	0.100	14	45	0.75	0.030
	12	0.500	9	30	1.15	0.045
C-6 carbide	0.1	0.005	76	250	(a)	(a)
	2.5	0.100	76	250	1.50	0.060
	12	0.500	60	200	1.25	0.050

(a) Feed is 75% the width of the square nose finishing tool

AISI 1137

AISI 1137: Chemical Composition

AISI grade	C	Mn	P max	S
	Chemical composition, %			
1137	0.32-0.39	1.35-1.65	0.040	0.08-0.13

Characteristics. AISI 1137 is a high-manganese, medium-carbon, resulfurized steel with medium strength and good machinability. It can be flame or induction hardened, but quench cracking is possible. This steel is used where threads, splines, and other operations with special tools are required.

Typical Uses. Applications of AISI 1137 steel include spline shafts, and studs, bolts, and nuts with machined threads.

AISI 1137: Similar Steels (U.S. and/or Foreign). UNS G11370; AMS 5024 C; ASTM A29, A108, A311; FED QQ-S-637 (C1137); SAE J403, J412, J414

AISI 1137: Approximate Critical Points

Transformation point	Temperature(a) °C	°F
Ac_1	715	1315
Ac_3	770	1420
Ar_3	740	1360
Ar_1	660	1220

(a) On heating or cooling at 28 °C (50 °F) per hour

Machining Data

For machining data on AISI grade 1137, refer to the preceding machining tables for AISI grade 1132.

Physical Properties

AISI 1137: Thermal Treatment Temperatures

Treatment	Temperature range °C	°F
Normalizing	860-940	1580-1725
Austenitizing	830-855	1525-1570

Mechanical Properties

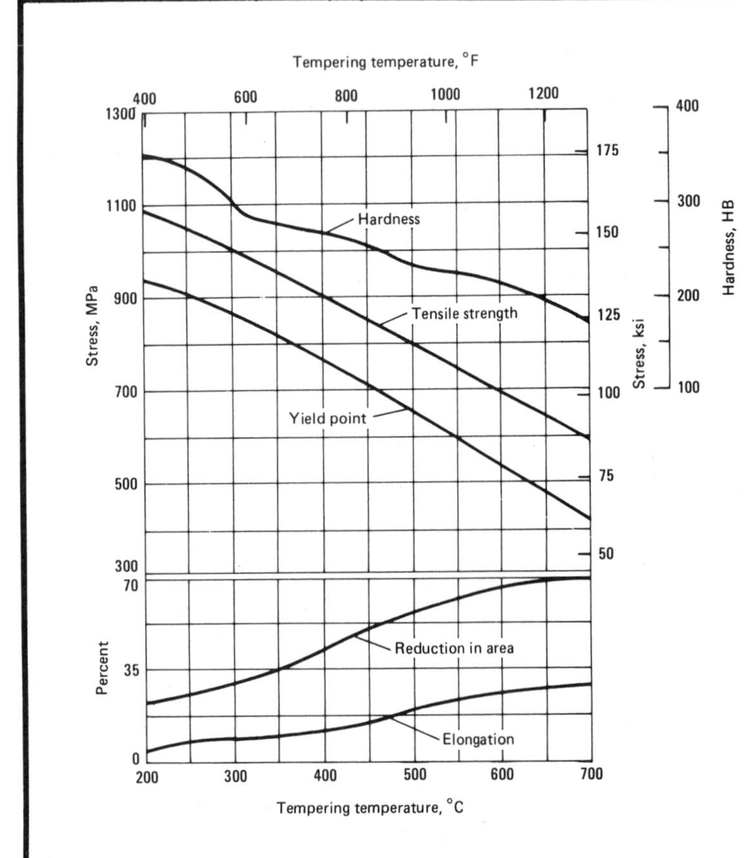

AISI 1137: Effect of Tempering Temperature on Tensile Properties After Quenching in Oil. Normalized at 900 °C (1650 °F); reheated to 855 °C (1575 °F); quenched in oil. Specimens were treated in 25-mm (1.0-in.) diam and machined to 12.8-mm (0.505-in.) diam for testing. Tests were conducted using test specimens machined to English units. As-quenched hardness was 363 HB. Elongation was measured in 50 mm (2 in.). (Ref 2)

Mechanical Properties (continued)

AISI 1137: Mass Effect on Mechanical Properties (Ref 2)

Condition or treatment	Size round mm	Size round in.	Tensile strength MPa	Tensile strength ksi	Yield strength MPa	Yield strength ksi	Elongation(a), %	Reduction in area, %	Hardness, HB
Water quenched from 845 °C (1550 °F), tempered at 540 °C (1000 °F)	13	0.5	896	130	772	112	17.1	51.3	262
	25	1	841	122	675	98	16.9	51.2	248
	50	2	758	110	490	71	20.8	56.1	229
	100	4	745	108	475	69	20.3	52.1	223
Water quenched from 845 °C (1550 °F), tempered at 595 °C (1100 °F)	13	0.5	772	112	655	95	21.4	57.6	229
	25	1	745	108	605	88	21.0	59.2	223
	50	2	724	105	525	76	22.0	61.7	217
	100	4	675	98	420	61	23.5	60.9	201
Water quenched from 845 °C (1550 °F), tempered at 650 °C (1200 °F)	13	0.5	724	105	615	89	23.9	61.2	223
	25	1	703	102	565	82	22.3	58.8	217
	50	2	675	98	460	67	24.0	64.1	201
	100	4	660	96	415	60	24.0	63.5	197
Annealed (heated to 790 °C or 1450 °F, furnace cooled 11 °C or 20 °F per hour to 610 °C or 1130 °F, cooled in air)	25	1	585	85	345	50	27.0	54.0	174
Nmalized (heated to 900 °C or 1650 °F, cooled in air)	13	0.5	675	98	400	58	25.0	58.5	201
	25	1	670	97	400	58	22.5	48.5	197
	50	2	660	96	340	49	21.8	51.6	197
	100	4	650	94	330	48	23.3	51.0	192
Oil quenched from 855 °C (1575 °F), tempered at 540 °C (1000 °F)	13	0.5	883	128	689	100	18.2	55.8	255
	25	1	745	108	525	76	21.3	56.0	223
	50	2	724	105	435	63	23.0	56.2	217
	100	4	689	100	405	59	22.3	55.5	201
Oil quenched from 855 °C (1575 °F), tempered at 595 °C (1100 °F)	13	0.5	772	112	620	90	21.8	61.0	229
	25	1	696	101	475	69	23.5	60.1	207
	50	2	675	98	425	62	23.0	57.8	207
	100	4	655	95	395	57	24.5	59.5	192
Oil quenched from 855 °C (1575 °F), tempered at 650 °C (1200 °F)	13	0.5	717	104	550	80	24.6	63.6	217
	25	1	675	98	475	69	23.5	60.8	201
	50	2	670	97	395	57	25.0	64.1	197
	100	4	650	94	385	56	24.0	61.1	192

(a) In 50 mm (2 in.)

AISI 1137: Effect of the Mass on Hardness at Selected Points (Ref 2)

Size round mm	Size round in.	As-quenched hardness at: Surface	½ radius	Center
After quenching in water				
13	0.5	57 HRC	53 HRC	50 HRC
25	1	56 HRC	50 HRC	45 HRC
50	2	52 HRC	35 HRC	24 HRC
100	4	48 HRC	23 HRC	20 HRC
After quenching in oil				
13	0.5	48 HRC	43 HRC	42 HRC
25	1	34 HRC	28 HRC	23 HRC
50	2	28 HRC	22 HRC	18 HRC
100	4	21 HRC	18 HRC	16 HRC

AISI 1137: Izod Impact Properties (Ref 12)

Condition or treatment	Energy J	Energy ft·lb
As rolled	83	61
Normalized at 900 °C (1650 °F)	64	47
Annealed at 790 °C (1450 °F)	50	37

Mechanical Properties (continued)

AISI 1137: Tensile Properties

Treatment or condition	Size round mm	Size round in.	Tensile strength MPa	Tensile strength ksi	Yield strength MPa	Yield strength ksi	Elongation(a), %	Reduction in area, %	Hardness, HB
Hot rolled (Ref 1)	19-32	0.75-1.25	605	88	330	48	15	35	179
Cold drawn(b) (Ref 1)	19-32	0.75-1.25	675	98	565	82	10	30	197
As cold drawn (Ref 1)	16-22	0.63-0.88	690	100	620	90	11	35	197
	22-32	0.88-1.25	655	95	585	85	11	30	187
	32-50	1.25-2	620	90	550	80	10	30	179
	50-75	2-3	585	85	515	75	10	30	170
Cold drawn, low temperature, stress relieved (Ref 1)	16-22	0.63-0.88	725	105	655	95	11	35	212
	22-32	0.88-1.25	690	100	620	90	11	30	197
	32-50	1.25-2	655	95	585	85	10	30	187
	50-75	2-3	620	90	550	80	10	25	179
Cold drawn, high temperature, stress relieved (Ref 1)	16-22	0.63-0.88	655	95	515	75	15	45	187
	22-32	0.88-1.25	620	90	515	75	15	40	179
	32-50	1.25-2	585	85	485	70	15	40	170
	50-75	2-3	550	80	450	65	12	35	163
As rolled (Ref 12)	...	...	625	91	380	55	28	61	192
Normalized at 900 °C (1650 °F) (Ref 12)	...	...	670	97	400	58	22	49	197
Annealed at 790 °C (1450 °F) (Ref 12)	...	...	585	85	345	50	27	54	174

(a) In 50 mm (2 in.). (b) Average machinability rating of 70% based on AISI 1212 steel as 100% average machinability

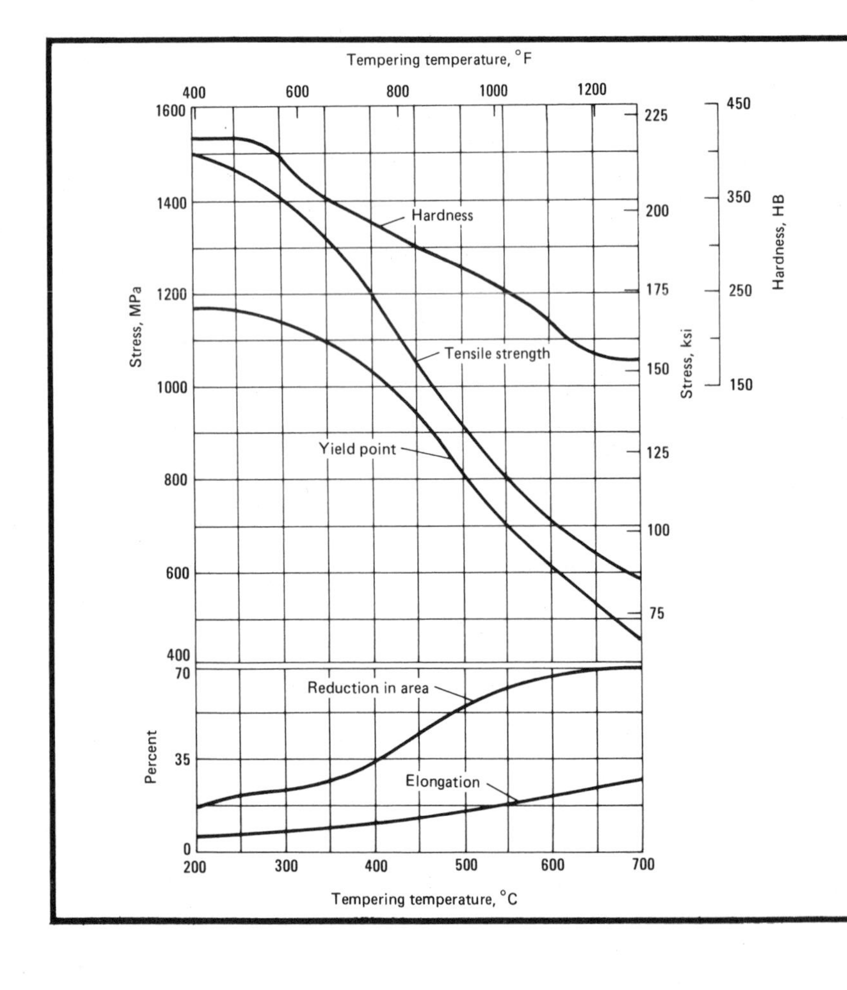

AISI 1137: Effect of Tempering Temperature on Tensile Properties After Quenching in Water. Normalized at 900 °C (1650 °F); reheated to 845 °C (1550 °F); quenched in water. Specimens were treated in 25-mm (1.0-in.) diam and machined to 12.8-mm (0.505-in.) diam for testing. Tests were conducted using test specimens machined to English units. As-quenched hardness was 415 HB. Elongation was measured in 50 mm (2 in.). (Ref 2)

AISI 1141, 1144

AISI 1141, 1144: Chemical Composition

AISI grade	C	Chemical composition, % Mn	P max	S
1141	0.37-0.45	1.35-1.65	0.040	0.08-0.13
1144	0.40-0.48	1.35-1.65	0.040	0.24-0.33

Characteristics. AISI grades 1141 and 1144 are oil-hardening steels with good heat treating response. Many end products made from these steels are cyanided or carbonitrided. Grades 1141 and 1144 can also be flame and induction hardened, and oil quenched after case hardening. The good machinability of these steels makes them suitable for use in the hot rolled, cold drawn, or heat treated condition. The addition of 0.15 to 0.35% lead further increases machinability.

Grade 1144 steel is available in a severely cold worked condition with increased in-the-bar strength. Mechanical properties in this condition are: 860-MPa (125-ksi) tensile strength; 690-MPa (100-ksi) yield strength; hardness of 269 HB; machinability rating of 83% (Ref 9). No heat treatment is necessary. Furthermore, being already stress relieved, grade 1144 bar machines readily with minimum distortion, which reduces or eliminates straightening operations.

Most AISI 1141 and 1144 bars are cold finished. They may be further turned and ground, or ground and polished, for use in screw machine operations.

Typical Uses. Typical end products made from grade 1141 and 1144 steels include cold drawn or finished bar, cold punched nuts, split rivets, machine screws, and wood screws.

AISI 1141: Similar Steels (U.S. and/or Foreign). UNS G11410; ASTM A29, A107, A108, A311; FED QQ-S-637; SAE J403, J412, J414

AISI 1144: Similar Steels (U.S. and/or Foreign). UNS G11440; ASTM A29, A108, A311; FED QQ-S-637 (C1144); SAE J403, J412, J414

AISI 1141, 1144: Approximate Critical Points

AISI grade	Transformation point	Temperature(a) °C	°F
1141	Ac_1	720	1330
	Ac_3	780	1435
	Ar_3	665	1230
	Ar_1	645	1190
1144	Ac_1	725	1335
	Ac_3	760	1400
	Ar_3	695	1285
	Ar_1	650	1200

(a) On heating or cooling at 28 °C (50 °F) per hour

Machining Data

For machining data on AISI grades 1141 and 1144, refer to the preceding machining tables for AISI grade 1132.

Physical Properties

AISI 1141, 1144: Thermal Treatment Temperatures (Ref 12)

Treatment	Temperature range °C	°F
Normalizing	855-910	1570-1670
Austenitizing	800-845	1475-1550

Mechanical Properties

AISI 1141: Tensile Properties

Condition or treatment	Size round mm	in.	Tensile strength MPa	ksi	Yield strength MPa	ksi	Elongation(a), %	Reduction in area, %	Hardness, HB
Hot rolled (Ref 1)	19-32	0.75-1.25	650	94	360	52	15	35	187
Cold drawn(b) (Ref 1)	19-32	0.75-1.25	725	105	605	88	10	30	212
As cold drawn (Ref 1)	16-22	0.63-0.88	725	105	655	95	11	30	212
	22-32	0.88-1.25	690	100	620	90	10	30	197
	32-50	1.25-2	655	95	585	85	10	30	187
	50-75	2-3	620	90	550	80	10	20	179
Cold drawn, low temperature, stress relieved (Ref 1)	16-22	0.63-0.88	760	110	690	100	11	30	223
	22-32	0.88-1.25	725	105	655	95	10	30	212
	32-50	1.25-2	690	100	620	90	10	25	197
	50-75	2-3	655	95	585	85	10	20	187
Cold drawn, high temperature, stress relieved (Ref 1)	16-22	0.63-0.88	690	100	550	80	15	40	197
	22-32	0.88-1.25	655	95	550	80	15	40	187
	32-50	1.25-2	620	90	515	75	15	40	179
	50-75	2-3	585	85	485	70	12	30	170
Cold finished(c) (Ref 3)	19-38	0.75-1.50	772	112	655	95	16	40	223
As rolled (Ref 12)	...	...	675	98	360	52	22	38	192
Normalized at 900 °C (1650 °F)	...	...	710	103	405	59	23	56	201
Annealed at 815 °C (1500 °F)	...	...	600	87	350	51	26	49	163

(a) In 50 mm (2 in.). (b) Average machinability rating of 70% based on AISI 1212 steel as 100% average machinability. (c) Average machinability rating of 78% based on AISI steel as 100% average machinability

Mechanical Properties (continued)

AISI 1144: Tensile Properties

Condition or treatment	Size round mm	Size round in.	Tensile strength MPa	Tensile strength ksi	Yield strength MPa	Yield strength ksi	Elongation(a), %	Reduction in area, %	Hardness, HB
Hot rolled (Ref 1)	19-32	0.75-1.25	670	97	365	53	15	35	197
Cold drawn(b) (Ref 1)	19-32	0.75-1.25	745	108	620	90	10	30	217
As cold drawn (Ref 1)	16-22	0.63-0.88	760	110	690	100	10	30	223
	22-32	0.88-1.25	725	105	655	95	10	30	212
	32-50	1.25-2	690	100	620	90	10	25	197
	50-75	2-3	655	95	585	85	10	20	187
Cold drawn, low temperature, stress relieved (Ref 1)	16-22	0.63-0.88	795	115	725	105	10	30	229
	22-32	0.88-1.25	760	110	690	100	10	30	223
	32-50	1.25-2	725	105	655	95	10	25	212
	50-75	2-3	690	100	620	90	10	20	197
Cold drawn, high temperature, stress relieved (Ref 1)	16-22	0.63-0.88	725	105	585	85	15	40	212
	22-32	0.88-1.25	690	100	585	85	15	40	197
	32-50	1.25-2	655	95	550	80	15	35	187
	50-75	2-3	620	90	515	75	12	30	179
Cold finished(b) (Ref 9)	19-38	0.75-1.50	786	114	670	97	14	36	235
As rolled (Ref 12)	...	...	705	102	420	61	21	41	212
Normalized at 900 °C (1650 °F)	...	...	670	97	400	58	21	40	197
Annealed at 790 °C (1450 °F)	...	...	585	85	345	50	25	41	167

(a) In 50 mm (2 in.). (b) Average machinability rating of 80% based on AISI 1212 steel as 100% average machinability

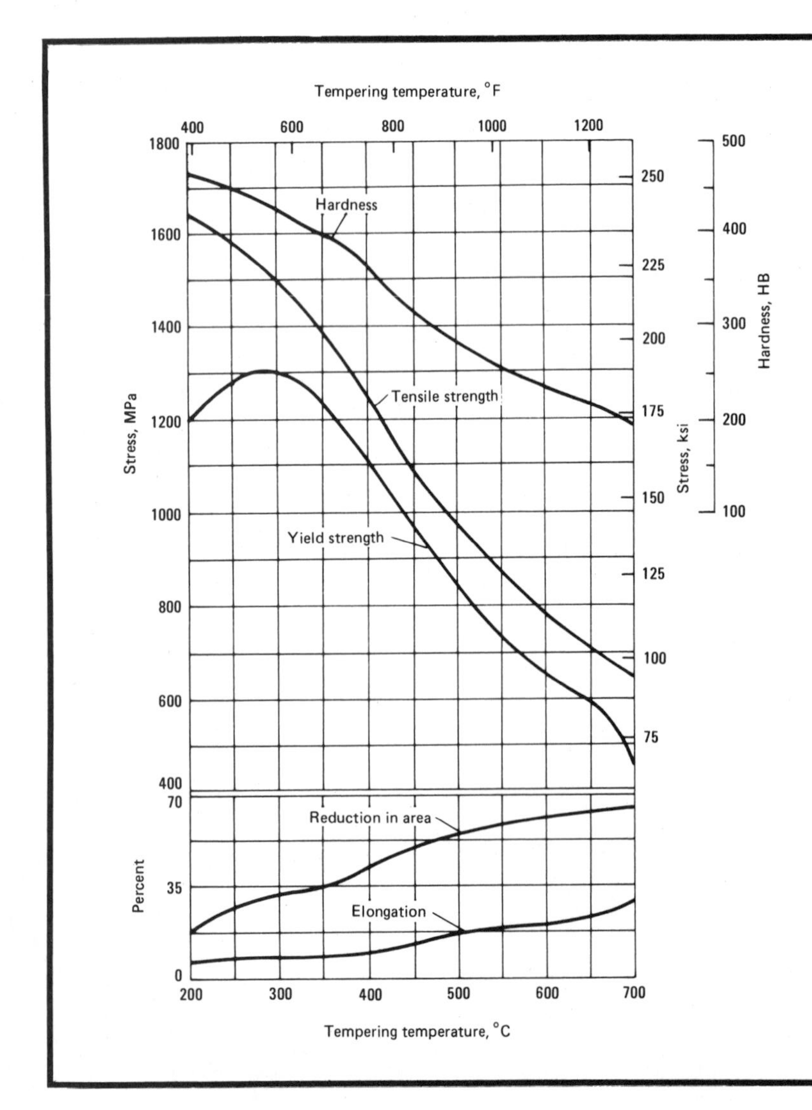

AISI 1141: Effect of Tempering Temperature on Tensile Properties. Normalized at 855 °C (1575 °F); reheated to 815 °C (1500 °F); quenched in oil. Specimens were treated in 13.5-mm (0.530-in.) diam and machined to 12.8-mm (0.505-in.) diam for testing. Tests were conducted using test specimens machined to English units. As-quenched hardness was 495 HB. Elongation was measured in 50 mm (2 in.). (Ref 2)

Mechanical Properties (continued)

AISI 1141, 1144: Mass Effect on Mechanical Properties (Ref 2)

Condition or treatment	Size round mm	Size round in.	Tensile strength MPa	Tensile strength ksi	Yield strength MPa	Yield strength ksi	Elongation(a), %	Reduction in area, %	Hardness, HB
AISI 1141									
Annealed (heated to 815 °C or 1500 °F, furnace cooled 11 °C or 20 °F per hour to 480 °C or 900 °F, cooled in air)	25	1	600	87	350	51	25.5	49.3	163
Normalized (heated to 900 °C or 1650 °F, cooled in air)	13	0.5	731	106	425	62	22.7	57.8	207
	25	1	703	102	405	59	22.7	55.5	201
	50	2	696	101	395	57	22.5	55.8	201
	100	4	689	100	380	55	21.7	49.3	201
Oil quenched from 815 °C (1500 °F), tempered at 540 °C (1000 °F)	13	0.5	896	130	758	110	18.7	57.1	262
	25	1	758	110	515	75	23.5	58.7	229
	50	2	745	108	515	75	21.8	57.2	217
	100	4	738	107	460	67	20.8	54.3	212
Oil quenched from 815 °C (1500 °F), tempered at 595 °C (1100 °F)	13	0.5	800	116	660	96	20.7	60.6	235
	25	1	710	103	485	70	23.8	62.2	207
	50	2	696	101	475	69	24	62.5	201
	100	4	689	100	420	61	23.5	59.1	197
Oil quenched from 815 °C (1500 °F), tempered at 650 °C (1200 °F)	13	0.5	724	105	600	87	23.5	63.8	217
	25	1	660	96	485	70	24.8	64.1	197
	50	2	660	96	450	65	25.2	65.1	192
	100	4	655	95	415	60	25.2	63.0	183
AISI 1144									
Annealed (heated to 790 °C or 1450 °F, furnace cooled 11 °C or 20 °F per hour to 620 °C or 1150 °F, cooled in air)	25	1	585	85	345	50	25.0	41.3	167
Normalized (heated to 900 °C or 1650 °F, cooled in air)	13	0.5	675	98	415	60	24.6	51.0	201
	25	1	670	97	400	58	21.0	40.4	197
	50	2	660	96	370	54	21.5	45.0	192
	100	4	650	94	360	52	21.5	42.7	192
Oil quenched from 845 °C (1550 °F), tempered at 540 °C (1000 °F)	13	0.5	786	114	545	79	20.4	52.1	235
	25	1	745	108	505	73	19.3	46.0	223
	50	2	724	105	470	68	20.5	49.6	212
	100	4	703	102	435	63	21.5	50.0	207
Oil quenched from 845 °C (1550 °F), tempered at 595 °C (1100 °F)	13	0.5	717	104	490	71	20.7	51.2	217
	25	1	710	103	470	68	21.5	51.4	212
	50	2	696	101	450	65	23.3	56.5	207
	100	4	650	94	400	58	23.8	54.4	192
Oil quenched from 845 °C (1550 °F), tempered at 650 °C (1200 °F)	13	0.5	675	98	475	69	23.2	55.2	201
	25	1	670	97	470	68	23.0	52.4	201
	50	2	650	94	425	62	24	57.7	192
	100	4	615	89	370	54	25.8	57.7	183

(a) In 50 mm (2 in.)

AISI 1141, 1144: Effect of the Mass on Hardness at Selected Points (Ref 2)

AISI grade	Size round mm	Size round in.	As-quenched hardness after quenching in oil at: Surface	½ radius	Center
1141	13	0.5	52 HRC	49 HRC	46 HRC
	25	1	48 HRC	43 HRC	38 HRC
	50	2	36 HRC	28 HRC	22 HRC
	100	4	27 HRC	22 HRC	18 HRC
1144	13	0.5	39 HRC	32 HRC	28 HRC
	25	1	36 HRC	29 HRC	24 HRC
	50	2	30 HRC	27 HRC	22 HRC
	100	4	27 HRC	98 HRB	97 HRB

AISI 1141, 1144: Izod Impact Properties (Ref 12)

AISI grade	Condition or treatment	Energy J	Energy ft·lb
1141	As rolled	11	8
	Normalized at 900 °C (1650 °F)	53	39
	Annealed at 815 °C (1500 °F)	34	25
1144	As rolled	53	39
	Normalized at 900 °C (1650 °F)	43	32
	Annealed at 790 °C (1450 °F)	65	48

Mechanical Properties (continued)

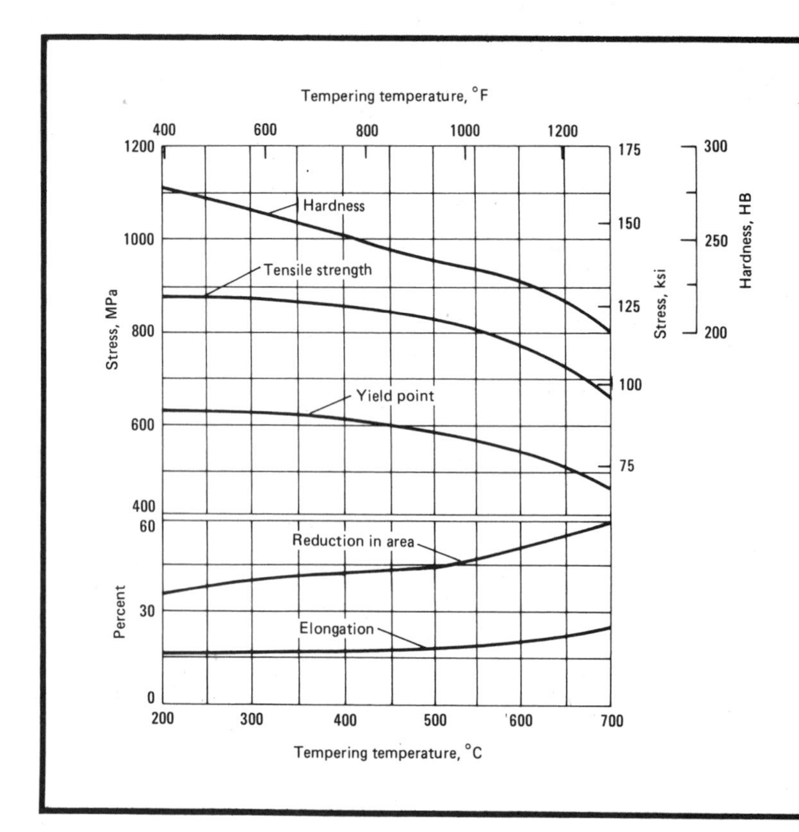

AISI 1144: Effect of Tempering Temperature on Tensile Properties. Normalized at 900 °C (1650 °F); reheated to 845 °C (1550 °F); quenched in oil. Specimens were treated in 25-mm (1.0-in.) diam and machined to 12.8-mm (0.505-in.) diam for testing. Tests were conducted using test specimens machined to English units. As-quenched hardness was 285 HB. Elongation was measured in 50 mm (2 in.). (Ref 2)

AISI 1211, 1212, 1213, 1215

AISI 1211, 1212, 1213, 1215: Chemical Composition

AISI grade	C max	Chemical composition, % Mn	P	S
1211	0.13	0.60-0.90	0.07-0.12	0.10-0.15
1212	0.13	0.70-1.00	0.07-0.12	0.16-0.23
1213	0.13	0.70-1.00	0.07-0.12	0.24-0.33
1215	0.09	0.75-1.05	0.04-0.09	0.26-0.35

Characterisitics. The 12XX series steels are both rephosphorized and resulfurized. Phosphorus, which is soluble in iron, promotes chip breakage in machining operations because of increased hardness and brittleness. Similar to carbon, excessive phosphorus can raise strength and hardness levels adversely high, thereby impairing machinability. Thus the chemical composition of 12XX steels is limited to phosphorus ranges of either 0.04 to 0.09% or 0.07 to 0.12%, and a maximum carbon content of 0.13%.

These free-machining grades have machinability superior to that of grades with similar carbon ranges which have not been resulfurized and rephosphorized. AISI grade 1212 is now employed as the standard for rating the machinability of other steels. Availability of grade 1112, formerly used for rating, is limited.

Typical Uses. AISI grade 1211, 1212, 1213, and 1215 steels are frequently rolled into rods, then drawn into wire. This wire is straightened, cut to length, and used in screw machine operations.

AISI 1211: Similar Steels (U.S. and/or Foreign). UNS G12110; ASTM A29, A108; FED QQ-S-637 (C1211); SAE J403

AISI 1212: Similar Steels (U.S. and/or Foreign). UNS G12120; AMS 5010 D; ASTM A29, A108; FED QQ-S-637 (C1212); SAE J403; (W. Ger.) DIN 1.0711; (Ital.) UNI 10 S 20; (Jap.) JIS SUM 21

AISI 1213: Similar Steels (U.S. and/or Foreign). UNS G12130; ASTM A29, A108; FED QQ-S-637 (C1913); SAE J403; (W. Ger.) DIN 1.0715; (Ital.) UNI 9 SMn 23; (Jap.) JIS SUM 22; (U.K.) B.S. 220 M07

AISI 1215: Similar Steels (U.S. and/or Foreign). UNS G12150; ASTM A29, A108; FED QQ-S-637; SAE J412

Mechanical Properties

AISI 1211, 1212, 1213, 1215: Tensile Properties and Machinability

Condition or treatment	Size round mm	Size round in.	Tensile strength MPa	Tensile strength ksi	Yield strength MPa	Yield strength ksi	Elongation(a), %	Reduction in area, %	Hardness, HB	Average machinability rating(b)
AISI 1211										
Hot rolled (Ref 1)	19-38	0.75-1.5	380	55	230	33	25	45	121	...
Cold drawn (Ref 1)	19-38	0.75-1.5	515	75	400	58	10	35	163	95
AISI 1212										
Hot rolled (Ref 1)	19-38	0.75-1.5	385	56	240	35	25	45	121	...
Cold drawn (Ref 1)	19-38	0.75-1.5	540	78	415	60	10	35	167	100
AISI 1213										
Hot rolled (Ref 1)	19-38	0.75-1.5	385	56	240	35	25	45	121	...
Cold drawn (Ref 1)	19-38	0.75-1.5	540	78	415	60	10	35	167	135
AISI 1215										
Drawn, ground, and polished (Ref 3)	19-38	0.75-1.5	525	76	470	68	...	...	...	...
Cold drawn (Ref 9)	19-38	0.75-1.5	540	78	415	60	10	35	167	135
Turned and polished (Ref 9)	19-38	0.75-1.5	385	56	230	33	25	45	121	...

(a) In 50 mm (2 in.). (b) Based on AISI 1212 steel as 100% average machinability

Machining Data (Ref 8)

For machining data on AISI grades 1211 and 1212, refer to the preceding machining tables for AISI grades 1116, 1117, 1118, and 1119

AISI 1213, 1215: Turning (Single Point and Box Tools)

Depth of cut mm	Depth of cut in.	M2 and M3 high speed steel Speed m/min	Speed ft/min	Feed mm/rev	Feed in./rev	Uncoated carbide Speed, brazed m/min	Speed, brazed ft/min	Speed, inserted m/min	Speed, inserted ft/min	Feed mm/rev	Feed in./rev	Coated carbide Speed m/min	Speed ft/min	Feed mm/rev	Feed in./rev
Hot rolled or annealed, with hardness of 100 to 150 HB															
1	0.040	90	295	0.20	0.008	220(a)	725(a)	260(a)	860(a)	0.18	0.007	395(b)	1300(b)	0.18	0.007
4	0.150	69	225	0.40	0.015	170(c)	550(c)	200(c)	650(c)	0.50	0.020	260(d)	850(d)	0.40	0.015
8	0.300	53	175	0.50	0.020	125(c)	415(c)	155(c)	510(c)	0.75	0.030	205(d)	675(d)	0.50	0.020
16	0.625	43	140	0.75	0.030	105(c)	340(c)	120(c)	400(c)	1.00	0.040	...	...	...	...
Cold drawn, with hardness of 150 to 200 HB															
1	0.040	90	300	0.20	0.008	240(a)	790(a)	275(a)	900(a)	0.18	0.007	410(b)	1350(b)	0.18	0.007
4	0.150	70	230	0.40	0.015	185(c)	600(c)	215(c)	700(c)	0.50	0.020	275(d)	900(d)	0.40	0.015
8	0.300	55	180	0.50	0.020	145(c)	475(c)	170(c)	550(c)	0.75	0.030	220(d)	725(d)	0.50	0.020
16	0.625	43	140	0.75	0.030	115(c)	370(c)	125(c)	415(c)	1.00	0.040	...	...	...	...

(a) Carbide tool material: C-7. (b) Carbide tool material: CC-7. (c) Carbide tool material: C-6. (d) Carbide tool material: CC-6

AISI 1213, 1215: End Milling (Profiling)

Tool material	Depth of cut mm	Depth of cut in.	Speed m/min	Speed ft/min	Feed per tooth for cutter diameter of: 10 mm (0.375 in.) mm	10 mm in.	12 mm (0.5 in.) mm	12 mm in.	18 mm (0.75 in.) mm	18 mm in.	25-50 mm (1-2 in.) mm	25-50 mm in.
Hot rolled or annealed, with hardness of 100 to 150 HB												
M2, M3, and M7 high speed steel	0.5	0.020	79	260	0.025	0.001	0.050	0.002	0.102	0.004	0.15	0.006
	1.5	0.060	60	200	0.050	0.002	0.075	0.003	0.13	0.005	0.18	0.007
	diam/4	diam/4	53	175	0.025	0.001	0.050	0.002	0.102	0.004	0.13	0.005
	diam/2	diam/2	46	150	0.018	0.0007	0.025	0.001	0.075	0.003	0.102	0.004
C-6 carbide	0.5	0.020	205	680	0.050	0.002	0.102	0.004	0.15	0.006	0.20	0.008
	1.5	0.060	160	520	0.075	0.003	0.13	0.005	0.18	0.007	0.23	0.009
	diam/4	diam/4	135	440	0.050	0.002	0.075	0.003	0.15	0.006	0.18	0.007
	diam/2	diam/2	125	410	0.038	0.0015	0.050	0.002	0.13	0.005	0.15	0.006
Cold drawn, with hardness of 150 to 200 HB												
M2, M3, and M7 high speed steel	0.5	0.020	70	230	0.025	0.001	0.050	0.002	0.102	0.004	0.15	0.006
	1.5	0.060	55	180	0.050	0.002	0.075	0.003	0.13	0.005	0.18	0.007
	diam/4	diam/4	49	160	0.025	0.001	0.050	0.002	0.102	0.004	0.13	0.005
	diam/2	diam/2	43	140	0.018	0.0007	0.025	0.001	0.075	0.003	0.102	0.004
C-6 carbide	0.5	0.020	190	620	0.050	0.002	0.102	0.004	0.15	0.006	0.18	0.007
	1.5	0.060	145	475	0.075	0.003	0.13	0.005	0.18	0.007	0.20	0.008
	diam/4	diam/4	120	400	0.050	0.002	0.075	0.003	0.15	0.006	0.18	0.007
	diam/2	diam/2	115	375	0.038	0.0015	0.050	0.002	0.13	0.005	0.15	0.006

Machining Data (Ref 8) (continued)

AISI 1213, 1215: Turning (Cutoff and Form Tools)

Tool material	Speed, m/min (ft/min)	Cutoff 1.5 mm (0.062 in.) mm	in.	Cutoff 3 mm (0.125 in.) mm	in.	Cutoff 6 mm (0.25 in.) mm	in.	Form 12 mm (0.5 in.) mm	in.	Form 18 mm (0.75 in.) mm	in.	Form 25 mm (1 in.) mm	in.	Form 50 mm (2 in.) mm	in.
Hot rolled or annealed, with hardness of 100 to 150 HB															
M2 and M3 high speed steel	59 (195)	0.075	0.0030	0.089	0.0035	0.102	0.0040	0.075	0.0030	0.067	0.0027	0.063	0.0025	0.046	0.0018
C-6 carbide	170 (550)	0.075	0.0030	0.089	0.0035	0.102	0.0040	0.075	0.0030	0.067	0.0027	0.063	0.0025	0.046	0.0018
Cold drawn, with hardness of 150 to 200 HB															
M2 and M3 high speed steel	64 (210)	0.075	0.0030	0.089	0.0035	0.102	0.0040	0.075	0.0030	0.067	0.0027	0.063	0.0025	0.046	0.0018
C-6 carbide	185 (600)	0.075	0.0030	0.089	0.0035	0.102	0.0040	0.075	0.0030	0.067	0.0027	0.063	0.0025	0.046	0.0018

Feed per revolution for cutoff tool width of: 1.5, 3, 6 mm; Feed per revolution for form tool width of: 12, 18, 25, 50 mm.

AISI 1213, 1215: Face Milling

Depth of cut mm	in.	M2 and M7 high speed steel Speed m/min	ft/min	Feed/tooth mm	in.	Uncoated carbide Speed, brazed m/min	ft/min	Speed, inserted m/min	ft/min	Feed/tooth mm	in.	Coated carbide Speed m/min	ft/min	Feed/tooth mm	in.
Hot rolled or annealed, with hardness of 100 to 150 HB															
1	0.040	115	375	0.20	0.008	245(a)	800(a)	270(a)	880(a)	0.20	0.008	405(b)	1325(b)	0.20	0.008
4	0.150	88	290	0.30	0.012	185(a)	605(a)	205(a)	665(a)	0.30	0.012	265(b)	865(b)	0.30	0.012
8	0.300	69	225	0.40	0.016	130(c)	420(c)	155(c)	515(c)	0.40	0.016	205(d)	670(d)	0.40	0.016
Cold drawn, with hardness of 150 to 200 HB															
1	0.040	110	360	0.20	0.008	225(a)	730(a)	245(a)	800(a)	0.20	0.008	365(b)	1200(b)	0.20	0.008
4	0.150	84	275	0.30	0.012	175(a)	575(a)	185(a)	615(a)	0.30	0.012	245(b)	800(b)	0.30	0.012
8	0.300	66	215	0.40	0.016	120(c)	390(c)	145(c)	480(c)	0.40	0.016	190(d)	625(d)	0.40	0.016

(a) Carbide tool material: C-6. (b) Carbide tool material: CC-6. (c) Carbide tool material: C-5. (d) Carbide tool material: CC-5

AISI 1213, 1215: Boring

Depth of cut mm	in.	M2 and M3 high speed steel Speed m/min	ft/min	Feed mm/rev	in./rev	Uncoated carbide Speed, brazed m/min	ft/min	Speed, inserted m/min	ft/min	Feed mm/rev	in./rev	Coated carbide Speed m/min	ft/min	Feed mm/rev	in./rev
Hot rolled or annealed, with hardness of 100 to 150 HB															
0.25	0.010	105	350	0.102	0.004	195(a)	640(a)	230(a)	750(a)	0.075	0.003	350(b)	1150(b)	0.075	0.003
1.25	0.050	72	235	0.15	0.006	155(c)	510(c)	185(c)	600(c)	0.13	0.005	275(d)	910(d)	0.13	0.005
2.5	0.100	55	180	0.30	0.012	120(e)	390(e)	140(e)	460(e)	0.40	0.015	220(f)	720(f)	0.40	0.015
Cold drawn, with hardness of 150 to 200 HB															
0.25	0.010	110	360	0.102	0.004	205(a)	670(a)	240(a)	790(a)	0.075	0.003	360(b)	1180(b)	0.075	0.003
1.25	0.050	73	240	0.15	0.006	165(c)	535(c)	190(c)	630(c)	0.13	0.005	290(d)	945(d)	0.13	0.005
2.5	0.100	56	185	0.30	0.012	130(e)	425(e)	150(e)	500(e)	0.40	0.015	190(f)	630(f)	0.30	0.012

(a) Carbide tool material: C-8. (b) Carbide tool material: CC-8. (c) Carbide tool material: C-7. (d) Carbide tool material: CC-7. (e) Carbide tool material: C-6. (f) Carbide tool material: CC-6

AISI 1213, 1215: Planing

Tool material	Depth of cut mm	in.	Speed m/min	ft/min	Feed/stroke mm	in.
Hot rolled or annealed, with hardness of 100 to 150 HB						
M2 and M3 high speed steel	0.1	0.005	20	65	(a)	(a)
	2.5	0.100	32	105	1.25	0.050
	12	0.500	17	55	1.50	0.060
C-6 carbide	0.1	0.005	90	300	(a)	(a)
	2.5	0.100	90	300	2.05	0.080
	12	0.500	90	300	1.50	0.060
Cold drawn, with hardness of 150 to 200 HB						
M2 and M3 high speed steel	0.1	0.005	21	70	(a)	(a)
	2.5	0.100	37	120	1.25	0.050
	12	0.500	18	60	1.50	0.060
C-6 carbide	0.1	0.005	90	300	(a)	(a)
	2.5	0.100	90	300	2.05	0.080
	12	0.500	60	200	1.50	0.060

(a) Feed is 75% the width of the square nose finishing tool

Machining Data (Ref 8) (continued)

AISI 1213, 1215: Drilling

Tool material	Speed m/min	Speed ft/min	Feed per revolution for nominal hole diameter of: 1.5 mm (0.062 in.) mm	1.5 mm (0.062 in.) in.	3 mm (0.125 in.) mm	3 mm (0.125 in.) in.	6 mm (0.25 in.) mm	6 mm (0.25 in.) in.	12 mm (0.5 in.) mm	12 mm (0.5 in.) in.	18 mm (0.75 in.) mm	18 mm (0.75 in.) in.	25 mm (1 in.) mm	25 mm (1 in.) in.
Hot rolled or annealed, with hardness of 100 to 150 HB														
M10, M7, and M1 high speed steel	27	90	0.025	0.001	...	...	...	...	...	...	...	...	...	...
	38	125	...	...	0.075	0.003	0.13	0.005	0.30	0.012	0.45	0.018	0.55	0.022
Cold drawn, with hardness of 150 to 200 HB														
M10, M7, and M1 high speed steel	30	100	0.025	0.001	...	...	...	...	...	...	...	...	...	...
	40	130	...	...	0.075	0.003	0.13	0.005	0.30	0.012	0.45	0.018	0.55	0.022

AISI 1213, 1215: Reaming

Based on 4 flutes for 3- and 6-mm (0.125- and 0.25-in.) reamers, 6 flutes for 12-mm (0.5-in.) reamers, and 8 flutes for 25-mm (1-in.) and larger reamers

Tool material	Speed m/min	Speed ft/min	Feed per revolution for reamer diameter of: 3 mm (0.125 in.) mm	3 mm (0.125 in.) in.	6 mm (0.25 in.) mm	6 mm (0.25 in.) in.	12 mm (0.5 in.) mm	12 mm (0.5 in.) in.	25 mm (1 in.) mm	25 mm (1 in.) in.	35 mm (1.5 in.) mm	35 mm (1.5 in.) in.	50 mm (2 in.) mm	50 mm (2 in.) in.
Hot rolled or annealed, with hardness of 100 to 150 HB														
Roughing														
M1, M2, and M7 high speed steel	52	170	0.15	0.006	0.25	0.010	0.40	0.015	0.65	0.025	0.75	0.030	0.90	0.035
C-2 carbide	60	200	0.15	0.006	0.25	0.010	0.40	0.015	0.65	0.025	0.75	0.030	0.90	0.035
Finishing														
M1, M2, and M7 high speed steel	20	55	0.20	0.008	0.30	0.012	0.45	0.018	0.75	0.030	0.90	0.035	1.00	0.040
C-2 carbide	21	70	0.20	0.008	0.30	0.012	0.45	0.018	0.75	0.030	0.90	0.035	1.00	0.040
Cold drawn, with hardness of 150 to 200 HB														
Roughing														
M1, M2, and M7 high speed steel	55	180	0.15	0.006	0.25	0.010	0.40	0.015	0.65	0.025	0.75	0.030	0.90	0.035
C-2 carbide	67	220	0.15	0.006	0.25	0.010	0.40	0.015	0.65	0.025	0.75	0.030	0.90	0.035
Finishing														
M1, M2, and M7 high speed steel	18	60	0.20	0.008	0.30	0.012	0.45	0.018	0.75	0.030	0.90	0.035	1.00	0.040
C-2 carbide	21	70	0.20	0.008	0.30	0.012	0.45	0.018	0.75	0.030	0.90	0.035	1.00	0.040

AISI 12L13, 12L14, 12L15

AISI 12L13, 12L14, 12L15: Chemical Composition

AISI grade	C max	Chemical composition, % Mn	P	S	Pb
12L13	0.13	0.70-1.00	0.07-0.12	0.24-0.33	0.15-0.35
12L14	0.15	0.85-1.15	0.04-0.09	0.26-0.35	0.15-0.35
12L15	0.09	0.75-1.05	0.04-0.09	0.26-0.35	0.15-0.35

Characteristics. The lead additions in AISI 12L13, 12L14, and 12L15 steels augment the effect of sulfur, permitting increased machining speeds and better finishes. Lead, which is soluble in steel, disperses microscopically in the rolled product. These lead particles act as a lubricant, helping to prevent tool buildup during machining and serving as chip breakers in a manner similar to that of sulfide inclusions.

Typical Uses. Economic reasons usually limit leaded resulfurized steels to use for high-speed screw machine products where the superior machining characteristics of the steel can be fully utilized.

AISI 12L13: Similar Steels (U.S. and/or Foreign). UNS G12134; ASTM A29; FED QQ-S-633 (C12L13); MIL SPEC MIL-S-18411; SAE J403, J412; (W. Ger.) DIN 1.0718; (Ital.) UNI 9 SMnPb 23; (Jap.) JIS SUM 22 L, SUM 23 L, SUM 24 L; (Swed.) SS_{14} 1914

AISI 12L14: Similar Steels (U.S. and/or Foreign). UNS G12144; ASTM A29, A108; SAE J403, J412, J414; (W. Ger.) DIN 1.0718; (Ital.) UNI 9 SMnPb 23; (Jap.) JIS SUM 22 L, SUM 24 L; (Swed.) SS_{14} 1914

Mechanical Properties (Ref 8)

AISI 12L14: Tensile Properties and Machinability

Condition or treatment	Size round mm	Size round in.	Tensile strength MPa	Tensile strength ksi	Yield strength MPa	Yield strength ksi	Elongation(a), %	Reduction in area, %	Hardness, HB	Average machinability rating(b)
Hot rolled (Ref 1)	19-38	0.75-1.5	395	57	235	34	22	45	121	...
Cold drawn (Ref 1)	19-38	0.75-1.5	540	78	415	60	10	35	163	160

(a) In 50 mm (2 in.) (b) Based on AISI 1212 steel as 100% average machinability

Machining Data (Ref 8)

AISI 12L13, 12L14, 12L15: Turning (Single Point and Box Tools)

Depth of cut mm	Depth of cut in.	M2 and M3 high speed steel Speed m/min	Speed ft/min	Feed mm/rev	Feed in./rev	Uncoated carbide Speed, brazed m/min	Speed, brazed ft/min	Speed, inserted m/min	Speed, inserted ft/min	Feed mm/rev	Feed in./rev	Coated carbide Speed m/min	Speed ft/min	Feed mm/rev	Feed in./rev
Hardness, 100 to 150 HB															
1	0.040	105	340	0.23	0.009	290(a)	950(a)	305(a)	1000(a)	0.18	0.007	470(b)	1550(b)	0.18	0.007
4	0.150	79	260	0.40	0.015	220(c)	725(c)	245(c)	800(c)	0.50	0.020	310(d)	1025(d)	0.40	0.015
8	0.300	64	210	0.50	0.020	175(c)	575(c)	190(c)	620(c)	0.75	0.030	250(d)	825(d)	0.50	0.020
16	0.625	49	160	0.75	0.030	135(c)	450(c)	150(c)	490(c)	1.00	0.040	...	...	...	...
Hardness, 150 to 200 HB															
1	0.040	105	350	0.23	0.009	295(a)	975(a)	320(a)	1050(a)	0.18	0.007	460(b)	1500(b)	0.18	0.007
4	0.150	82	270	0.40	0.015	230(c)	750(c)	255(c)	840(c)	0.50	0.020	305(d)	1000(d)	0.40	0.015
8	0.300	64	210	0.50	0.020	180(c)	590(c)	195(c)	640(c)	0.75	0.030	245(d)	800(d)	0.50	0.020
16	0.625	52	170	0.75	0.030	140(c)	460(c)	145(c)	475(c)	1.00	0.040	...	...	...	...
Hardness, 200 to 250 HB															
1	0.040	79	260	0.23	0.009	260(a)	860(a)	290(a)	950(a)	0.18	0.007	425(b)	1400(b)	0.18	0.007
4	0.150	59	195	0.40	0.015	200(c)	650(c)	220(c)	725(c)	0.50	0.020	280(d)	925(d)	0.40	0.015
8	0.300	47	155	0.50	0.020	160(c)	520(c)	175(c)	575(c)	0.75	0.030	230(d)	750(d)	0.50	0.020
16	0.625	37	120	0.75	0.030	120(c)	400(c)	135(c)	450(c)	1.00	0.040	...	...	...	...

(a) Carbide tool material: C-7. (b) Carbide tool material: CC-7. (c) Carbide tool material: C-6. (d) Carbide tool material: CC-6

AISI 12L13, 12L14, 12L15: Planing

Tool material	Depth of cut mm	Depth of cut in.	Speed m/min	Speed ft/min	Feed/stroke mm	Feed/stroke in.
Hardness, 100 to 150 HB						
M2 and M3 high speed steel	0.1	0.005	21	70	(a)	(a)
	2.5	0.100	37	120	1.25	0.050
	12	0.500	18	60	1.50	0.060
C-6 carbide	0.1	0.005	90	300	(a)	(a)
	2.5	0.100	90	300	2.05	0.080
	12	0.500	90	300	1.50	0.060
Hardness, 150 to 200 HB						
M2 and M3 high speed steel	0.1	0.005	23	75	(a)	(a)
	2.5	0.100	28	125	1.25	0.050
	12	0.500	20	65	1.50	0.060
C-6 carbide	0.1	0.005	90	300	(a)	(a)
	2.5	0.100	90	300	2.05	0.080
	12	0.500	90	300	1.50	0.060
Hardness, 200 to 250 HB						
M2 and M3 high speed steel	0.1	0.005	15	50	(a)	(a)
	2.5	0.100	24	80	1.00	0.040
	12	0.500	17	55	1.25	0.050
C-6 carbide	0.1	0.005	90	300	(a)	(a)
	2.5	0.100	90	300	1.50	0.060
	12	0.500	76	250	1.25	0.050

(a) Feed is 75% the width of the square nose finishing tool

Machining Data (Ref 8) (continued)

AISI 12L13, 12L14, 12L15: Reaming

Based on 4 flutes for 3- and 6-mm (0.125- and 0.25-in.) reamers, 6 flutes for 12-mm (0.5-in.) reamers, and 8 flutes for 25-mm (1-in.) and larger reamers

Tool material	Speed		Feed per revolution for reamer diameter of:											
			3 mm (0.125 in.)		6 mm (0.25 in.)		12 mm (0.5 in.)		25 mm (1 in.)		35 mm (1.5 in.)		50 mm (2 in.)	
	m/min	ft/min	mm	in.	mm	in.	mm	in.	mm	in.	mm	in.	mm	in.
Hardness, 100 to 150 HB														
Roughing														
M1, M2, and M7 high speed steel	69	225	0.15	0.006	0.25	0.010	0.40	0.015	0.65	0.025	0.75	0.030	0.90	0.035
C-2 carbide	79	260	0.15	0.006	0.25	0.010	0.40	0.015	0.65	0.025	0.75	0.030	0.90	0.035
Finishing														
M1, M2, and M7 high speed steel	21	70	0.20	0.008	0.30	0.012	0.45	0.018	0.75	0.030	0.90	0.035	1.00	0.040
C-2 carbide	23	75	0.20	0.008	0.30	0.012	0.45	0.018	0.75	0.030	0.90	0.035	1.00	0.040
Hardness, 150 to 200 HB														
Roughing														
M1, M2, and M7 high speed steel	66	215	0.15	0.006	0.25	0.010	0.40	0.015	0.65	0.025	0.75	0.030	0.90	0.035
C-2 carbide	76	250	0.15	0.006	0.25	0.010	0.40	0.015	0.65	0.025	0.75	0.030	0.90	0.035
Finishing														
M1, M2, and M7 high speed steel	20	65	0.20	0.008	0.30	0.012	0.45	0.018	0.75	0.030	0.90	0.035	1.00	0.040
C-2 carbide	21	70	0.20	0.008	0.30	0.012	0.45	0.018	0.75	0.030	0.90	0.035	1.00	0.040
Hardness, 200 to 250 HB														
Roughing														
M1, M2, and M7 high speed steel	60	200	0.13	0.005	0.20	0.008	0.30	0.012	0.50	0.020	0.65	0.025	0.75	0.030
C-2 carbide	69	225	0.15	0.006	0.25	0.010	0.40	0.015	0.65	0.025	0.75	0.030	0.90	0.035
Finishing														
M1, M2, and M7 high speed steel	18	60	0.15	0.006	0.25	0.010	0.40	0.015	0.65	0.025	0.75	0.030	0.90	0.035
C-2 carbide	20	65	0.15	0.006	0.25	0.010	0.40	0.015	0.65	0.025	0.75	0.030	0.90	0.035

AISI 12L13, 12L14, 12L15: End Milling (Profiling)

Tool material	Depth of cut		Speed		Feed per tooth for cutter diameter of:							
					10 mm (0.375 in.)		12 mm(0.5 in.)		18 mm (0.75 in.)		25-50 mm (1-2 in.)	
	mm	in.	m/min	ft/min	mm	in.	mm	in.	mm	in.	mm	in.
Hardness, 100 to 150 HB												
M2, M3, and M7 high speed steel	0.5	0.020	82	270	0.025	0.001	0.050	0.002	0.102	0.004	0.15	0.006
	1.5	0.060	60	200	0.050	0.002	0.075	0.003	0.13	0.005	0.18	0.007
	diam/4	diam/4	55	180	0.025	0.001	0.050	0.002	0.102	0.004	0.13	0.005
	diam/2	diam/2	49	160	0.018	0.0007	0.025	0.001	0.075	0.003	0.102	0.004
C-5 carbide	0.5	0.020	215	710	0.050	0.002	0.102	0.004	0.15	0.006	0.20	0.008
	1.5	0.060	165	540	0.075	0.003	0.13	0.005	0.18	0.007	0.23	0.009
	diam/4	diam/4	140	460	0.050	0.002	0.102	0.004	0.15	0.006	0.20	0.008
	diam/2	diam/2	130	430	0.038	0.0015	0.075	0.003	0.13	0.005	0.18	0.007
Hardness, 150 to 200 HB												
M2, M3, and M7 high speed steel	0.5	0.020	73	240	0.025	0.001	0.050	0.002	0.102	0.004	0.15	0.006
	1.5	0.060	56	185	0.050	0.002	0.075	0.003	0.13	0.005	0.18	0.007
	diam/4	diam/4	50	165	0.025	0.001	0.050	0.002	0.102	0.004	0.13	0.005
	diam/2	diam/2	44	145	0.018	0.0007	0.025	0.001	0.075	0.003	0.102	0.004
C-5 carbide	0.5	0.020	190	625	0.050	0.002	0.102	0.004	0.15	0.006	0.20	0.008
	1.5	0.060	145	480	0.075	0.003	0.13	0.005	0.18	0.007	0.23	0.009
	diam/4	diam/4	120	400	0.050	0.002	0.102	0.004	0.15	0.006	0.20	0.008
	diam/2	diam/2	115	375	0.038	0.0015	0.075	0.003	0.13	0.005	0.18	0.007
Hardness, 200 to 250 HB												
M2, M3, and M7 high speed steel	0.5	0.020	69	225	0.025	0.001	0.050	0.002	0.102	0.004	0.13	0.005
	1.5	0.060	52	170	0.050	0.002	0.075	0.003	0.13	0.005	0.15	0.006
	diam/4	diam/4	47	155	0.025	0.001	0.050	0.002	0.102	0.004	0.13	0.005
	diam/2	diam/2	41	135	0.018	0.0007	0.025	0.001	0.075	0.003	0.102	0.004
C-5 carbide	0.5	0.020	170	550	0.050	0.002	0.075	0.003	0.13	0.005	0.18	0.007
	1.5	0.060	130	420	0.075	0.003	0.102	0.004	0.15	0.006	0.20	0.008
	diam/4	diam/4	110	360	0.050	0.002	0.075	0.003	0.13	0.005	0.18	0.007
	diam/2	diam/2	100	335	0.038	0.0015	0.050	0.002	0.102	0.004	0.15	0.006

Machining Data (Ref 8) (continued)

AISI 12L13, 12L14, 12L15: Turning (Cutoff and Form Tools)

Tool material	Speed, m/min (ft/min)	Feed per revolution for cutoff tool width of: 1.5 mm (0.062 in.)		3 mm (0.125 in.)		6 mm (0.25 in.)		Feed per revolution for form tool width of: 12 mm (0.5 in.)		18 mm (0.75 in.)		25 mm (1 in.)		50 mm (2 in.)	
		mm	in.	mm	in.	mm	in.	mm	in.	mm	in.	mm	in.	mm	in.
Hardness, 100 to 150 HB															
M2 and M3 high speed steel	85 (280)	0.084	0.0033	0.097	0.0038	0.112	0.0044	0.089	0.0035	0.084	0.0033	0.075	0.0030	0.053	0.0021
C-6 carbide	185 (600)	0.084	0.0033	0.097	0.0038	0.112	0.0044	0.089	0.0035	0.084	0.0033	0.075	0.0030	0.053	0.0021
Hardness, 150 to 200 HB															
M2 and M3 high speed steel	76 (250)	0.084	0.0033	0.097	0.0038	0.112	0.0044	0.089	0.0035	0.084	0.0033	0.075	0.0030	0.053	0.0021
C-6 carbide	175 (575)	0.084	0.0033	0.097	0.0038	0.112	0.0044	0.089	0.0035	0.084	0.0033	0.075	0.0030	0.053	0.0021
Hardness, 200 to 250 HB															
M2 and M3 high speed steel	69 (225)	0.075	0.0030	0.089	0.0035	0.104	0.0041	0.081	0.0032	0.075	0.0030	0.067	0.0027	0.046	0.0018
C-6 carbide	170 (550)	0.075	0.0030	0.089	0.0035	0.104	0.0041	0.081	0.0032	0.075	0.0030	0.067	0.0027	0.046	0.0018

AISI 12L13, 12L14, 12L15: Boring

Depth of cut		M2 and M3 high speed steel: Speed		Feed		Uncoated carbide: Speed, brazed		Speed, inserted		Feed		Coated carbide: Speed		Feed	
mm	in.	m/min	ft/min	mm/rev	in./rev	m/min	ft/min	m/min	ft/min	mm/rev	in./rev	m/min	ft/min	mm/rev	in./rev
Hardness, 100 to 150 HB															
0.25	0.010	105	350	0.13	0.005	225(a)	740(a)	265(a)	875(a)	0.075	0.003	420(b)	1375(b)	0.075	0.003
1.25	0.050	82	270	0.18	0.007	150(c)	490(c)	215(c)	700(c)	0.13	0.005	335(d)	1100(d)	0.13	0.005
2.5	0.100	64	210	0.30	0.012	115(c)	385(c)	170(c)	550(c)	0.40	0.015	215(d)	700(d)	0.30	0.012
Hardness, 150 to 200 HB															
0.25	0.010	105	350	0.13	0.005	240(a)	780(a)	280(a)	920(a)	0.075	0.003	395(b)	1300(b)	0.075	0.003
1.25	0.050	85	280	0.18	0.007	190(c)	625(c)	225(c)	735(c)	0.13	0.005	320(d)	1050(d)	0.13	0.005
2.5	0.100	67	220	0.30	0.012	150(c)	500(c)	180(c)	590(c)	0.40	0.015	215(d)	700(d)	0.30	0.012
Hardness, 200 to 250 HB															
0.25	0.010	79	260	0.13	0.005	215(a)	700(a)	255(a)	830(a)	0.075	0.003	375(b)	1225(b)	0.075	0.003
1.25	0.050	64	210	0.18	0.007	170(c)	565(c)	205(c)	665(c)	0.13	0.005	300(d)	980(d)	0.13	0.005
2.5	0.100	47	155	0.30	0.012	135(c)	435(c)	155(c)	510(c)	0.40	0.015	200(d)	650(d)	0.30	0.012

(a) Carbide tool material: C-7. (b) Carbide tool material: CC-7. (c) Carbide tool material: C-6. (d) Carbide tool material: CC-6

AISI 12L13, 12L14, 12L15: Face Milling

Depth of cut		M2 and M7 high speed steel: Speed		Feed/tooth		Uncoated carbide: Speed, brazed		Speed, inserted		Feed/tooth		Coated carbide: Speed		Feed/tooth	
mm	in.	m/min	ft/min	mm	in.	m/min	ft/min	m/min	ft/min	mm	in.	m/min	ft/min	mm	in.
Hardness, 100 to 150 HB															
1	0.040	100	325	0.20	0.008	255(a)	840(a)	275(a)	900(a)	0.20	0.008	410(b)	1350(b)	0.20	0.008
4	0.150	79	260	0.30	0.012	215(a)	700(a)	220(a)	725(a)	0.30	0.012	285(b)	940(b)	0.30	0.012
8	0.300	60	200	0.40	0.016	140(c)	460(c)	170(c)	565(c)	0.40	0.016	225(d)	735(d)	0.40	0.016
Hardness, 150 to 200 HB															
1	0.040	88	290	0.20	0.008	230(a)	750(a)	245(a)	800(a)	0.20	0.008	365(b)	1200(b)	0.20	0.008
4	0.150	72	235	0.30	0.012	185(a)	600(a)	200(a)	650(a)	0.30	0.012	260(b)	845(b)	0.30	0.012
8	0.300	55	180	0.40	0.016	120(c)	400(c)	150(c)	500(c)	0.40	0.016	200(d)	650(d)	0.40	0.016
Hardness, 200 to 250 HB															
1	0.040	58	190	0.20	0.008	200(a)	650(a)	215(a)	700(a)	0.20	0.008	320(b)	1050(b)	0.18	0.007
4	0.150	46	150	0.30	0.012	150(a)	500(a)	170(a)	550(a)	0.30	0.012	220(b)	715(b)	0.25	0.010
8	0.300	35	115	0.40	0.016	105(c)	350(c)	130(c)	430(c)	0.40	0.016	170(d)	560(d)	0.36	0.014

(a) Carbide tool material: C-6. (b) Carbide tool material: CC-6. (c) Carbide tool material: C-5. (d) Carbide tool material: CC-5

Machining Data (Ref 8) (continued)

AISI 12L13, 12L14, 12L15: Drilling

Tool material	Speed m/min	Speed ft/min	Feed per revolution for nominal hole diameter of: 1.5 mm (0.062 in.) mm	1.5 mm in.	3 mm (0.125 in.) mm	3 mm in.	6 mm (0.25 in.) mm	6 mm in.	12 mm (0.5 in.) mm	12 mm in.	18 mm (0.75 in.) mm	18 mm in.	25 mm (1 in.) mm	25 mm in.
Hardness, 100 to 150 HB														
M10, M7, and M1 high speed steel	24	80	0.025	0.001	...	...	...	...	...	...	...	...	...	...
	46	150	...	...	0.075	0.003	0.13	0.005	0.30	0.012	0.45	0.018	0.55	0.022
Hardness, 150 to 200 HB														
M10, M7, and M1 high speed steel	24	80	0.025	0.001	...	...	...	...	...	...	...	...	...	...
	43	140	...	...	0.075	0.003	0.13	0.005	0.30	0.012	0.45	0.018	0.55	0.022
Hardness, 200 to 250 HB														
M10, M7, and M1 high speed steel	18	60	0.025	0.001	...	...	...	...	...	...	...	...	...	...
	37	120	...	...	0.075	0.003	0.13	0.005	0.30	0.012	0.45	0.018	0.55	0.022

AISI 1547, 1548, 15B48H

AISI 1547, 1548, 15B48: Chemical Composition

AISI grade	Chemical composition, % C	Mn	P max	S max
1547	0.43-0.51	1.35-1.65	0.040	0.050
1548	0.44-0.52	1.10-1.40	0.040	0.050
15B48H(a)	0.43-0.55	1.10-1.15	0.040	0.050

(a) This grade can be expected to contain 0.0005 to 0.0030% B

Characteristics and Typical Uses. The 15XX steels contain a minimum of 1.00% manganese. In addition, most grades are available with an expected boron addition of 0.0005 to 0.0030% and, as H-steels when consistent and exacting heat treating procedures are involved. Steels in the 15XX series can meet SAE J429j, grade 8.1 specifications.

The lower carbon steels in the 15XX series, those with 0.010 to 0.29% carbon, are used for carburizing applications which require greater case hardenability than regular carbon steels provide. These lower carbon grades are also suitable for heavier sections or for thin sections where oil quenching is required.

The medium-carbon steels in this series, those with a carbon content of 0.30% to 0.35%, may be flame or induction hardened. They are suitable for forgings. Selection of steels for forgings is governed by section size and mechanical properties desired after heat treatment. The medium-carbon 15XX steels are widely used for parts machined from bar stock. Depending on the application and the level of properties needed, they are suitable with or without heat treatment.

The higher carbon steels of the 15XX series are used where the increased carbon is needed to improve wear characteristics, and where greater strength levels are required than those attainable with the lower or regular carbon grades.

The H-steels and boron-treated steels are used in the production of rods and wires, which serve as substitutes for alloy and higher carbon steels in the manufacture of heat treated bolts and other cold formed parts. These parts are quenched and tempered after forming to gain the strength and hardness characteristics inherent in boron-treated grades.

AISI 1547: Similar Steels (U.S. and/or Foreign). UNS G15470; ASTM A29, A510, A576; SAE J403, J412, J414

AISI 1548: Similar Steels (U.S. and/or Foreign). UNS G15480; ASTM A29, A510, A576; SAE J403, J412, J414; (W. Ger.) DIN 1.1226

AISI 15B48H: Similar Steels (U.S. and/or Foreign). UNS H15481; ASTM A29, A304

Physical Properties

AISI 1547, 1548, 15B48H: Thermal Treatment Temperatures

Treatment	Temperature °C	°F
Normalizing	870	1600
Austenitizing	845	1550

Mechanical Properties

AISI 1547, 1548: Tensile Properties and Machinability (Ref 1)

Condition or treatment	Size round mm	Size round in.	Tensile trength MPa	Tensile trength ksi	Yield strength MPa	Yield strength ksi	Elongation(a), %	Reduction in area, %	Hardness, HB	Average machinability rating(b)
AISI 1547										
Hot rolled	19-32	0.75-1.25	650	94	360	52	15	30	192	...
Cold drawn	19-32	0.75-1.25	710	103	605	88	10	28	207	40
Annealed, cold drawn	19-32	0.75-1.25	655	95	585	85	10	35	187	45
AISI 1548										
Hot rolled	19-32	0.75-1.25	660	96	365	53	14	33	197	...
Cold drawn	19-32	0.75-1.25	731	106	620	90	10	28	217	45
Annealed, cold drawn	19-32	0.75-1.25	650	94	540	78	10	35	192	50

(a) In 50 mm (2 in.). (b) Based on AISI 1212 steel as 100% average machinability

AISI 15B48H: End-Quench Hardenability Limits and Hardenability Band. Normalize at 870 °C (1600 °F) (for forged or rolled specimens only); austenitize at 845 °C (1550 °F); heat treating temperatures recommended by SAE. (Ref 10)

Distance from quenched end, 1/16 in.	Hardness, HRC max	Hardness, HRC min
1	63	56
2	62	56
3	62	55
4	61	54
5	60	53
6	59	52
7	58	42
8	57	34
9	56	31
10	55	30
11	53	29
12	51	28
13	48	27
14	45	27
15	41	26
16	38	26
18	34	25
20	32	24
22	31	23
24	30	22
26	29	21
28	29	20
30	28	...
32	28	...

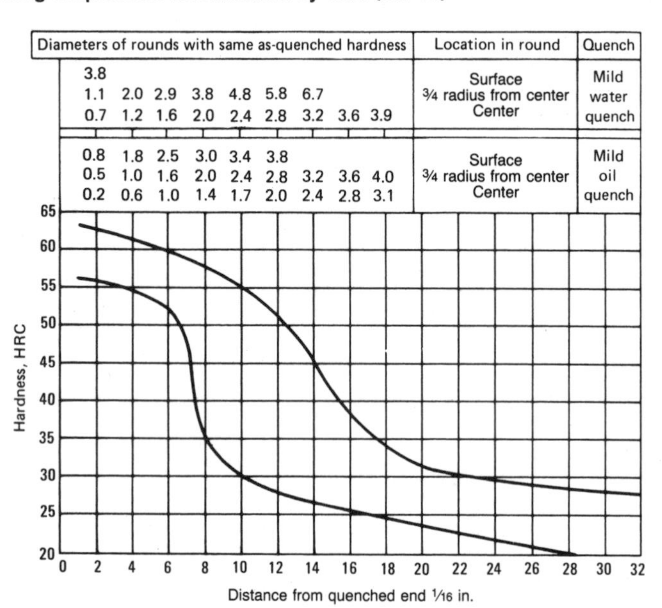

Machining Data (Ref 8)

AISI 1547, 1548, 15B48H: Turning (Single Point and Box Tools)

Depth of cut mm	Depth of cut in.	M2 and M3 high speed steel Speed m/min	Speed ft/min	Feed mm/rev	Feed in./rev	Uncoated carbide Speed, brazed m/min	Speed, brazed ft/min	Speed, inserted m/min	Speed, inserted ft/min	Feed mm/rev	Feed in./rev	Coated carbide Speed m/min	Speed ft/min	Feed mm/rev	Feed in./rev
Hardness, 175 to 225 HB															
1	0.040	38	125	0.18	0.007	130(a)	425(a)	160(a)	520(a)	0.18	0.007	230(b)	750(b)	0.18	0.007
4	0.150	29	95	0.40	0.015	100(c)	320(c)	125(c)	405(c)	0.50	0.020	150(d)	500(d)	0.40	0.015
8	0.300	23	75	0.50	0.020	76(c)	250(c)	95(c)	315(c)	0.75	0.030	120(d)	400(d)	0.50	0.020
16	0.625	17	55	0.75	0.030	59(c)	195(c)	76(c)	250(c)	1.00	0.040	...	...	...	...
Hardness, 225 to 275 HB															
1	0.040	34	110	0.18	0.007	115(a)	375(a)	150(a)	485(a)	0.18	0.007	220(b)	725(b)	0.18	0.007
4	0.150	23	75	0.40	0.015	90(c)	300(c)	115(c)	375(c)	0.50	0.020	145(d)	475(d)	0.40	0.015
8	0.300	20	65	0.50	0.020	70(c)	230(c)	90(c)	295(c)	0.75	0.030	115(d)	375(d)	0.50	0.020
16	0.625	15	50	0.75	0.030	53(c)	175(c)	69(c)	225(c)	1.00	0.040	...	...	...	...

(a) Carbide tool material: C-7. (b) Carbide tool material: CC-7. (c) Carbide tool material: C-6. (d) Carbide tool material: CC-6

Machining Data (Ref 8) (continued)

AISI 1547, 1548, 15B48H: Turning (Cutoff and Form Tools)

Tool material	Speed, m/min (ft/min)	Feed per revolution for cutoff tool width of: 1.5 mm (0.062 in.) mm	in.	3 mm (0.125 in.) mm	in.	6 mm (0.25 in.) mm	in.	Feed per revolution for form tool width of: 12 mm (0.5 in.) mm	in.	18 mm (0.75 in.) mm	in.	25 mm (1 in.) mm	in.	50 mm (2 in.) mm	in.
Hardness, 175 to 225 HB															
M2 and M3 high speed steel	27 (90)	0.028	0.0011	0.033	0.0013	0.041	0.0016	0.036	0.0014	0.030	0.0012	0.028	0.0011	0.020	0.0008
C-6 carbide	88 (290)	0.028	0.0011	0.033	0.0013	0.041	0.0016	0.036	0.0014	0.030	0.0012	0.028	0.0011	0.020	0.0008
Hardness, 225 to 275 HB															
M2 and M3 high speed steel	21 (70)	0.025	0.0010	0.030	0.0012	0.038	0.0015	0.033	0.0013	0.028	0.0011	0.025	0.0010	0.018	0.0007
C-6 carbide	69 (225)	0.025	0.0010	0.030	0.0012	0.038	0.0015	0.033	0.0013	0.028	0.0011	0.025	0.0010	0.018	0.0007

AISI 1547, 1548, 15B48H: Face Milling

Depth of cut mm	in.	M2 and M7 High speed steel Speed m/min	ft/min	Feed/tooth mm	in.	Uncoated carbide Speed, brazed m/min	ft/min	Speed, inserted m/min	ft/min	Feed/tooth mm	in.	Coated carbide Speed m/min	ft/min	Feed/tooth mm	in.
Hardness, 175 to 225 HB															
1	0.040	46	150	0.20	0.008	155(a)	515(a)	170(a)	550(a)	0.20	0.008	250(b)	825(b)	0.20	0.008
4	0.150	35	115	0.30	0.012	120(a)	400(a)	145(a)	470(a)	0.30	0.012	185(b)	610(b)	0.30	0.012
8	0.300	27	90	0.40	0.016	90(c)	300(c)	110(c)	365(c)	0.40	0.016	145(d)	475(d)	0.40	0.016
Hardness, 225 to 275 HB															
1	0.040	37	120	0.15	0.006	135(a)	440(a)	150(a)	490(a)	0.18	0.007	225(b)	735(b)	0.18	0.007
4	0.150	29	95	0.25	0.010	110(a)	365(a)	125(a)	405(a)	0.25	0.010	160(b)	525(b)	0.25	0.010
8	0.300	23	75	0.36	0.014	79(c)	260(c)	95(c)	315(c)	0.36	0.014	125(d)	410(d)	0.36	0.014

(a) Carbide tool material: C-6. (b) Carbide tool material: CC-6. (c) Carbide tool material: C-5. (d) Carbide tool material: CC-5

AISI 1547, 1548, 15B48H: Drilling

Tool material	Speed m/min	ft/min	Feed per revolution for nominal hole diameter of: 1.5 mm (0.062 in.) mm	in.	3 mm (0.125 in.) mm	in.	6 mm (0.25 in.) mm	in.	12 mm (0.5 in.) mm	in.	18 mm (0.75 in.) mm	in.	25 mm (1 in.) mm	in.
Hardness, 175 to 225 HB														
M10, M7, and M1 high speed steel	17	55	0.025	0.001	...	...	...	...	...	...	...	...	...	...
	21	70	...	...	0.075	0.003	0.130	0.005	0.23	0.009	0.30	0.012	0.45	0.018
Hardness, 225 to 275 HB														
M10, M7, and M1 high speed steel	18	60	0.025	0.001	0.050	0.002	0.102	0.004	0.18	0.007	0.25	0.010	0.40	0.015

AISI 1547, 1548, 15B48H: Planing

Tool material	Depth of cut mm	in.	Speed m/min	ft/min	Feed/stroke mm	in.
Hardness, 175 to 225 HB						
M2 and M3 high speed steel	0.1	0.005	11	35	(a)	(a)
	2.5	0.100	17	55	1.25	0.050
	12	0.500	9	30	1.50	0.060
C-6 carbide	0.1	0.005	60	200	(a)	(a)
	2.5	0.100	69	225	2.05	0.080
	12	0.500	60	200	1.50	0.060
Hardness, 225 to 275 HB						
M2 and M3 high speed steel	0.1	0.005	9	30	(a)	(a)
	2.5	0.100	14	45	0.75	0.030
	12	0.500	8	25	1.15	0.045
C-6 carbide	0.1	0.005	58	190	(a)	(a)
	2.5	0.100	64	210	1.50	0.060
	12	0.500	50	165	1.25	0.050

(a) Feed is 75% the width of the square nose finishing tool

Machining Data (Ref 8) (continued)

AISI 1547, 1548, 15B48H: End Milling (Profiling)

Tool material	Depth of cut mm	Depth of cut in.	Speed m/min	Speed ft/min	Feed per tooth for cutter diameter of: 10 mm (0.375 in.) mm	in.	12 mm (0.5 in.) mm	in.	18 mm (0.75 in.) mm	in.	25-50 mm (1-2 in.) mm	in.
Hardness, 175 to 225 HB												
M2, M3, and M7 high speed steel	0.5	0.020	49	160	0.025	0.001	0.050	0.002	0.075	0.003	0.102	0.004
	1.5	0.060	37	120	0.050	0.002	0.075	0.003	0.102	0.004	0.13	0.005
	diam/4	diam/4	32	105	0.025	0.001	0.050	0.002	0.075	0.003	0.102	0.004
	diam/2	diam/2	27	90	0.018	0.0007	0.025	0.001	0.050	0.002	0.075	0.003
C-5 carbide	0.5	0.020	150	485	0.038	0.0015	0.075	0.003	0.13	0.005	0.15	0.006
	1.5	0.060	115	370	0.063	0.0025	0.102	0.004	0.15	0.006	0.18	0.007
	diam/4	diam/4	100	325	0.050	0.002	0.075	0.003	0.13	0.005	0.15	0.006
	diam/2	diam/2	90	300	0.038	0.0015	0.050	0.002	0.102	0.004	0.13	0.005
Hardness, 225 to 275 HB												
M2, M3, and M7 high speed steel	0.5	0.020	34	110	0.025	0.001	0.050	0.002	0.075	0.003	0.102	0.004
	1.5	0.060	26	85	0.050	0.002	0.075	0.003	0.102	0.004	0.13	0.005
	diam/4	diam/4	23	75	0.025	0.001	0.050	0.002	0.075	0.003	0.102	0.004
	diam/2	diam/2	20	65	0.018	0.0007	0.025	0.001	0.050	0.002	0.075	0.003
C-5 carbide	0.5	0.020	130	430	0.025	0.001	0.050	0.002	0.102	0.004	0.13	0.005
	1.5	0.060	100	330	0.050	0.002	0.075	0.003	0.13	0.005	0.18	0.007
	diam/4	diam/4	85	280	0.038	0.0015	0.063	0.0025	0.102	0.004	0.13	0.005
	diam/2	diam/2	81	265	0.025	0.001	0.050	0.002	0.075	0.003	0.102	0.004

AISI 1547, 1548, 15B48H: Boring

Depth of cut mm	in.	M2 and M3 high speed steel: Speed m/min	ft/min	Feed mm/rev	in./rev	Uncoated carbide: Speed, brazed m/min	ft/min	Speed, inserted m/min	ft/min	Feed mm/rev	in./rev	Coated carbide: Speed m/min	ft/min	Feed mm/rev	in./rev
Hardness, 175 to 225 HB															
0.25	0.010	38	125	0.075	0.003	115(a)	385(a)	140(a)	455(a)	0.075	0.003	200(b)	655(b)	0.075	0.003
1.25	0.050	30	100	0.13	0.005	95(c)	310(c)	110(c)	365(c)	0.13	0.005	160(d)	510(d)	0.13	0.005
2.5	0.100	23	75	0.30	0.012	73(c)	240(c)	87(c)	285(c)	0.40	0.015	105(d)	350(d)	0.30	0.012
Hardness, 225 to 275 HB															
0.25	0.010	35	115	0.075	0.003	110(a)	360(a)	130(a)	425(a)	0.075	0.003	195(b)	635(b)	0.075	0.003
1.25	0.050	27	90	0.13	0.005	88(c)	290(c)	105(c)	340(c)	0.13	0.005	155(d)	510(d)	0.13	0.005
2.5	0.100	18	60	0.30	0.012	67(c)	220(c)	79(c)	260(c)	0.40	0.015	100(d)	330(d)	0.30	0.012

(a) Carbide tool material: C-7. (b) Carbide tool material: CC-7. (c) Carbide tool material: C-6. (d) Carbide tool material: CC-6

AISI 1547, 1548, 15B48H: Reaming

Based on 4 flutes for 3- and 6-mm (0.125- and 2.50-in.) reamers, 6 flutes for 12-mm (0.5-in.) reamers, and 8 flutes for 25-mm (1-in.) and larger reamers

Tool material	Speed m/min	Speed ft/min	Feed per revolution for reamer diameter of: 3 mm (0.125 in.) mm	in.	6 mm (0.25 in.) mm	in.	12 mm (0.5 in.) mm	in.	25 mm (1 in.) mm	in.	35 mm (1.5 in.) mm	in.	50 mm (2 in.) mm	in.
Hardness, 175 to 225 HB														
Roughing														
M1, M2, and M7 high speed steel	23	75	0.102	0.004	0.18	0.007	0.30	0.012	0.50	0.020	0.65	0.025	0.75	0.030
C-2 carbide	27	90	0.102	0.004	0.18	0.007	0.30	0.012	0.50	0.020	0.65	0.025	0.75	0.030
Finishing														
M1, M2, and M7 high speed steel	11	35	0.13	0.005	0.20	0.008	0.30	0.012	0.50	0.020	0.65	0.025	0.75	0.030
C-2 carbide	14	45	0.13	0.005	0.20	0.008	0.30	0.012	0.50	0.020	0.65	0.025	0.75	0.030
Hardness, 225 to 275 HB														
Roughing														
M1, M2, and M7 high speed steel	20	65	0.075	0.003	0.15	0.006	0.25	0.010	0.40	0.015	0.50	0.020	0.65	0.025
C-2 carbide	24	80	0.102	0.004	0.15	0.006	0.25	0.010	0.40	0.015	0.50	0.020	0.65	0.025
Finishing														
M1, M2, and M7 high speed steel	9	30	0.102	0.004	0.18	0.007	0.25	0.010	0.40	0.015	0.50	0.020	0.65	0.025
C-2 carbide	12	40	0.102	0.004	0.18	0.007	0.25	0.010	0.40	0.015	0.50	0.020	0.65	0.025

Alloy Steels

Steel is considered an alloy steel when the maximum content range of alloying elements exceeds one or more of the following limits: 1.65% manganese, 0.60% silicon, or 0.60% copper. Also included in the recognized field of alloy steels are steels with a specified or required range or minimum quantity of the following elements: aluminum, boron, chromium (up to 3.99%), cobalt, niobium, molybdenum, nickel, titanium, tungsten, vanadium, zirconium, or any other alloying element added to obtain a desired alloying effect (Ref 1).

Alloy sheet and strip are available as hot rolled and cold rolled steel in coils and cut lengths. Other product forms of alloy steel include hot rolled plate, and hot rolled, cold rolled, and cold drawn bar, rod, and wire.

Basic open-hearth, basic oxygen, or electric furnace processes are used to manufacture alloy steel. To provide optimum degasification, cleanliness, chemical composition control, soundness, and mechanical properties, vacuum melting may be employed. Vacuum-melted steels include those produced by vacuum induction melting (VIM), vacuum arc remelting (VAR), electroslag remelting (ESR), and vacuum carbon deoxidation (VCD). Alloy steel can also be produced by the argon-oxygen decarburization (AOD) process.

Steel melting practices used in the production of alloy steel employ certain deoxidizing elements which act with varying intensities. Silicon and aluminum are the most common, but vanadium, titanium, and zirconium are also used as deoxidizing agents. The deoxidation practice in the manufacture of killed steels, including degassing and the choice and amounts of specific deoxidizers, varies among producers.

Alloy steels are generally more sensitive to thermal and mechanical operations than carbon or high-strength, low-alloy steels. To ensure the most satisfactory results, consumers normally consult steel producers regarding the working, machining, heat-treating, or other operations to be employed in fabricating the steel, the mechanical properties to be obtained, and the conditions of service intended for the finished articles. The producers should also be informed of the operations that the steel will be subject to during fabrication.

AISI-SAE Designations

The most widely used system for designating alloy steels combines the designations of the American Iron and Steel Institute (AISI) and the Society of Automotive Engineers (SAE). Although two separate systems, they are nearly identical and are carefully coordinated by the two groups. The numerical designations are summarized in the table listing the AISI-SAE system of designations.

The first two digits in the designation number indicate the major alloying element or elements; the last two digits indicate the carbon content in hundredths of a percent. The letter B between the second and third digits denotes boron steel to which a minimum of 0.0005% boron has been added.

Hardening Characteristics

Alloy steels, most frequently plate and bar, are either surface hardenable or through hardenable.

Surface-hardenable alloys, usually with a nominal carbon content of up to 0.30%, are specified when a tough core, wear resistance, and case hardness to a controlled depth are necessary. Frequently these grades are selected with later surface treatment, such as nitriding or carburizing, in mind.

Gears need high wear resistance at the surface and hardness in sufficient depth to prevent crushing of the teeth, yet must be resilient deep inside to resist tooth breakage. Surface hardening achieves these characteristics. Shallow hardening is also necessary in many shock applications where a softer core is essential to avoid fracture.

Through-hardening grades, usually quenched and tempered steels with a carbon content of 0.35% and above, are used when maximum hardness and strength must extend deep within the part, even from surface to center, to provide resistance to critical stresses.

As with some carbon grades, there are also alloy steels specified to hardenability band limits. These limits define a predictable range in which specific Rockwell C (HRC) hardness values can be obtained after quenching. These steels, identified by the letter suffix H, are available in many alloy grades. H-steels require a slight modification in the chemical composition of the conventional grade bearing the same AISI numerical designation.

Use of Hardenability Bands. Hardenability band graphs for H-steels are given in the following text for convenience in estimating the hardness values obtainable at various locations on the end-quench test bar and in comparing various H-steels. The graphs also permit specification of maximum and minimum hardenability limits at distances in increments of 1/16 in. from the quenched end.

For specification purposes, tables on end-quench hardenability limits present maximum and minimum values of Rockwell C hardness (HRC). Because they are invalid, values below 20 HRC are not included.

The hardenability band graphs accompanying these tables provide five possible interpretations of the tabulated data points:

- The minimum and maximum hardness values at any desired distance
- The minimum and maximum distance at which any desired hardness value occurs
- Two maximum hardness values at two desired distances
- Two minimum hardness values at two desired distances
- Any minimum hardness plus any maximum hardness

Because the section size required is not always the same as that of the standard test specimen, bar diameters have been plotted for three positions on the end-quench hardenability specimen: surface, ¾ radius from center, and center. The diameters of rounds with the same as-quenched hardness are given for two modes of quenching above each H-band.

The diameter of the standard hardenability test specimen was 1 in. Tests were conducted using test specimens machined to English units.

Referring to the hardenability band for AISI 1330H, the data above the H-band provide the approximate diameter of a bar with the same as-quenched hardness as the test specimen, at a specified distance from the quenched end. For example, at 2/16 in. the hardness of the test sample averaged 47 to 56 HRC. For a 1330H bar with a mild oil quench, this hardness range will occur at the center of a 0.2-in. diam bar, at ¾ radius from center in a 0.5-in. diam bar, and at the surface in a 0.8-in. diam bar.

Quality Descriptors

Steel quality descriptors indicate conditions such as the degree of internal soundness and relative freedom from detrimental surface imperfections. In alloy steel, quality also relates to the general suitability of the steel involved to make the desired part.

Regular quality alloy sheet and strip steels are intended primarily for general or miscellaneous applications where moderate forming and/or bending is required. Regular quality bar is killed steel, usually produced as fine-grain material.

In regular quality steels normal surface imperfections are not objectionable and good finish is not a prime requirement. This quality is not produced to internal cleanliness limits, but can be produced to standard or modified chemical ranges, grain sizes, and/or mechanical property limits.

Drawing quality alloy sheet and strip steels are intended primarily for applications involving moderate to severe cold plastic deformation, such as deep drawn or severely formed parts. Alloy steel plate of drawing quality is specially produced for the fabrication of parts where the consumer's operations involve cold deep drawing and both hot and cold drawing.

Drawing quality sheet, strip, and plate are supplied from steel produced by closely controlled steelmaking practices. They are subject to mill testing and inspection designed to ensure internal soundness and freedom from injurious surface imperfections. Special annealing practices are generally required to obtain optimum forming and drawing characteristics of these steels.

Structural quality alloy sheet and strip steels meet the requirements of regular quality steels and are also produced to specific mechanical property requirements. These requirements include tension, bend, hardness, or other commonly accepted mechanical tests, which may be specified or required.

Structural quality alloy steel plate is intended for general structural applications, such as buildings, bridges, and transportation equipment; machined parts; and miscellaneous end uses in the hot rolled or thermally treated conditions.

Aircraft quality alloy steel is intended for highly stressed parts of aircraft, missiles, and similar applications involving stringent requirements.

Aircraft quality alloy steel requires exacting steelmaking, conditioning, and processing controls. Electric furnace melting and vacuum degassing are sometimes required. Primary requirements for this quality are internal soundness, uniformity of chemical composition, cleanliness, a fine, austenitic grain size, and good surface. Magnetic particle testing requirements of AMS 2301 (published by SAE) are sometimes specified for parts made of aircraft quality alloy steel.

Aircraft structural quality alloy sheet and strip steels meet the stringent requirements of aircraft quality, plus specific mechanical property limits or ranges.

Bearing quality alloy steel sheet, strip, and bar are intended for rolling-element bearing components. This quality is usually specified in the AISI-SAE standard alloy carburizing grades and the AISI-SAE high-alloy chromium grades. Alloy steel bar can be produced in accordance with ASTM A295, A485, or A534.

Production of bearing quality alloy steels entails highly specialized steelmaking and processing practices. The steels are often subjected to one or more tests or acceptance limits in the intermediate or final stages of processing.

Saw quality alloy steel sheet and strip are intended for applications involving the fabrication of metal-cutting saws. Internal cleanliness, soundness, good surface, and weldability are primary requirements. Another frequent requirement is dimensional control.

Pressure vessel quality alloy steel plate is intended for application in pressure vessels. This plate may be supplied to meet specifications for ultra-sonic testing requirements.

Axle shaft quality describes hot rolled steel bar intended for the manufacture of power-driven axle shafts for automotive or truck applications. This bar is not machined all over, or has less than recommended stock removal allowance for the proper cleanup of normal surface imperfections.

Cold heading quality describes alloy steel bar required for special applications which involve cold plastic deformation by operations such as upsetting, heading or forging. Most cold heading alloy steels are low- and medium-carbon grades.

Special cold heading quality applies to alloy steel bar required for applications which involve severe plastic deformation, where slight surface imperfections may cause the splitting of a part. Closely controlled steelmaking practices provide the uniform chemical composition and internal soundness that is essential for bar of this quality. Special processing, which may include grinding or another equivalent surface preparation, is applied in intermediate stages to remove detrimental surface imperfections.

AISI-SAE Standard Alloy Steels: Chemical Composition (Ref 1)

Small quantities of certain elements not specified or required are present; maximum contents of incidental elements are: 0.35% copper, 0.25% nickel, 0.20% chromium, and 0.06% molybdenum; standard alloy steels may also be produced with 0.15 to 0.35% lead and are identified with an L inserted between the second and third numerals of the AISI-SAE number (for example, 41L40)

AISI-SAE No.	UNS No.	Chemical composition, % C	Mn	P max	S max	Si	Ni	Cr	Mo
1330	G13300	0.28-0.33	1.60-1.90	0.035	0.040	0.15-0.35	...	...	...
1335	G13350	0.33-0.38	1.60-1.90	0.035	0.040	0.15-0.35	...	...	...
1340	G13400	0.38-0.43	1.60-1.90	0.035	0.040	0.15-0.35	...	...	...
1345	G13450	0.43-0.48	1.60-1.90	0.035	0.040	0.15-0.35	...	...	...
4023	G40230	0.20-0.25	0.70-0.90	0.035	0.040	0.15-0.35	...	...	0.20-0.30
4024	G40240	0.20-0.25	0.70-0.90	0.035	0.035-0.050	0.15-0.35	...	...	0.20-0.30
4027	G40270	0.25-0.30	0.70-0.90	0.035	0.040	0.15-0.35	...	...	0.20-0.30
4028	G40280	0.25-0.30	0.70-0.90	0.035	0.035-0.050	0.15-0.35	...	...	0.20-0.30
4037	G40370	0.35-0.40	0.70-0.90	0.035	0.040	0.15-0.35	...	...	0.20-0.30
4047	G40470	0.45-0.50	0.70-0.90	0.035	0.040	0.15-0.35	...	...	0.20-0.30
4118	G41180	0.18-0.23	0.70-0.90	0.035	0.040	0.15-0.35	...	0.40-0.60	0.08-0.15
4130	G41300	0.28-0.33	0.40-0.60	0.035	0.040	0.15-0.35	...	0.80-1.10	0.15-0.25
4137	G41370	0.35-0.40	0.70-0.90	0.035	0.040	0.15-0.35	...	0.80-1.10	0.15-0.25
4140	G41400	0.38-0.43	0.75-1.00	0.035	0.040	0.15-0.35	...	0.80-1.10	0.15-0.25
4142	G41420	0.40-0.45	0.75-1.00	0.035	0.040	0.15-0.35	...	0.80-1.10	0.15-0.25
4145	G41450	0.43-0.48	0.75-1.00	0.035	0.040	0.15-0.35	...	0.80-1.10	0.15-0.25
4147	G41470	0.45-0.50	0.75-1.00	0.035	0.040	0.15-0.35	...	0.80-1.10	0.15-0.25
4150	G41500	0.48-0.53	0.75-1.00	0.035	0.040	0.15-0.35	...	0.80-1.10	0.15-0.25
4161	G41610	0.56-0.64	0.75-1.00	0.035	0.040	0.15-0.35	...	0.70-0.90	0.25-0.35
4320	G43200	0.17-0.22	0.45-0.65	0.035	0.040	0.15-0.35	1.65-2.00	0.40-0.60	0.20-0.30
4340	G43400	0.38-0.43	0.60-0.80	0.035	0.040	0.15-0.35	1.65-2.00	0.70-0.90	0.20-0.30
E4340 (a)	G43406	0.38-0.43	0.65-0.85	0.025	0.025	0.15-0.35	1.65-2.00	0.70-0.90	0.20-0.30
4615	G46150	0.13-0.18	0.45-0.65	0.035	0.040	0.15-0.35	1.65-2.00	...	0.20-0.30
4620	G46200	0.17-0.22	0.45-0.65	0.035	0.040	0.15-0.35	1.65-2.00	...	0.20-0.30
4626	G46260	0.24-0.29	0.45-0.65	0.035	0.040	0.15-0.35	0.70-1.00	...	0.15-0.25
4720	G47200	0.17-0.22	0.50-0.70	0.035	0.040	0.15-0.35	0.90-1.20	0.35-0.55	0.15-0.25
4815	G48150	0.13-0.18	0.40-0.60	0.035	0.040	0.15-0.35	3.25-3.75	...	0.20-0.30
4817	G48170	0.15-0.20	0.40-0.60	0.035	0.040	0.15-0.35	3.25-3.75	...	0.20-0.30
4820	G48200	0.18-0.23	0.50-0.70	0.035	0.040	0.15-0.35	3.25-3.75	...	0.20-0.30
5117	G51170	0.15-0.20	0.70-0.90	0.035	0.040	0.15-0.35	...	0.70-0.90	...
5120	G51200	0.17-0.22	0.70-0.90	0.035	0.040	0.15-0.35	...	0.70-0.90	...
5130	G51300	0.28-0.33	0.70-0.90	0.035	0.040	0.15-0.35	...	0.80-1.10	...
5132	G51320	0.30-0.35	0.60-0.80	0.035	0.040	0.15-0.35	...	0.75-1.00	...
5135	G51350	0.33-0.38	0.60-0.80	0.035	0.040	0.15-0.35	...	0.80-1.05	...
5140	G51400	0.38-0.43	0.70-0.90	0.035	0.040	0.15-0.35	...	0.70-0.90	...
5150	G51500	0.48-0.53	0.70-0.90	0.035	0.040	0.15-0.35	...	0.70-0.90	...
5155	G51550	0.51-0.59	0.70-0.90	0.035	0.040	0.15-0.35	...	0.70-0.90	...
5160	G51600	0.56-0.64	0.75-1.00	0.035	0.040	0.15-0.35	...	0.70-0.90	...
E51100 (a)	G51986	0.98-1.10	0.25-0.45	0.025	0.025	0.15-0.35	...	0.90-1.15	...
E52100 (a)	G52986	0.98-1.10	0.25-0.45	0.025	0.025	0.15-0.35	...	1.30-1.60	...
6118 (b)	G61180	0.16-0.21	0.50-0.70	0.035	0.040	0.15-0.35	...	0.50-0.70	...
6150 (c)	G61500	0.48-0.53	0.70-0.90	0.035	0.040	0.15-0.35	...	0.80-1.10	...
8615	G86150	0.13-0.18	0.70-0.90	0.035	0.040	0.15-0.35	0.40-0.70	0.40-0.60	0.15-0.25
8617	G86170	0.15-0.20	0.70-0.90	0.035	0.040	0.15-0.35	0.40-0.70	0.40-0.60	0.15-0.25
8620	G86200	0.18-0.23	0.70-0.90	0.035	0.040	0.15-0.35	0.40-0.70	0.40-0.60	0.15-0.25
8622	G86220	0.20-0.25	0.70-0.90	0.035	0.040	0.15-0.35	0.40-0.70	0.40-0.60	0.15-0.25
8625	G86250	0.23-0.28	0.70-0.90	0.035	0.040	0.15-0.35	0.40-0.70	0.40-0.60	0.15-0.25
8627	G86270	0.25-0.30	0.70-0.90	0.035	0.040	0.15-0.35	0.40-0.70	0.40-0.60	0.15-0.25
8630	G86300	0.28-0.33	0.70-0.90	0.035	0.040	0.15-0.35	0.40-0.70	0.40-0.60	0.15-0.25
8637	G86370	0.35-0.40	0.75-1.00	0.035	0.040	0.15-0.35	0.40-0.70	0.40-0.60	0.15-0.25
8640	G86400	0.38-0.43	0.75-1.00	0.035	0.040	0.15-0.35	0.40-0.70	0.40-0.60	0.15-0.25
8642	G86420	0.40-0.45	0.75-1.00	0.035	0.040	0.15-0.35	0.40-0.70	0.40-0.60	0.15-0.25
8645	G86450	0.43-0.48	0.75-1.00	0.035	0.040	0.15-0.35	0.40-0.70	0.40-0.60	0.15-0.25
8655	G86550	0.51-0.59	0.75-1.00	0.035	0.040	0.15-0.35	0.40-0.70	0.40-0.60	0.15-0.25
8720	G87200	0.18-0.23	0.70-0.90	0.035	0.040	0.15-0.35	0.40-0.70	0.40-0.60	0.20-0.30
8740	G87400	0.38-0.43	0.75-1.00	0.035	0.040	0.15-0.35	0.40-0.70	0.40-0.60	0.20-0.30
8822	G88220	0.20-0.25	0.75-1.00	0.035	0.040	0.15-0.35	0.40-0.70	0.40-0.60	0.30-0.40
9260	G92600	0.56-0.64	0.75-1.00	0.035	0.040	1.80-2.20	...	...	...
Standard boron grades(d)									
50B44	G50441	0.43-0.48	0.75-1.00	0.035	0.040	0.15-0.35	...	0.40-0.60	...
50B46	G50461	0.44-0.49	0.75-1.00	0.035	0.040	0.15-0.35	...	0.20-0.35	...
50B50	G50501	0.48-0.53	0.75-1.00	0.035	0.040	0.15-0.35	...	0.40-0.60	...
50B60	G50601	0.56-0.64	0.75-1.00	0.035	0.040	0.15-0.35	...	0.40-0.60	...
51B60	G51601	0.56-0.64	0.75-1.00	0.035	0.040	0.15-0.35	...	0.70-0.90	...
81B45	G81451	0.43-0.48	0.75-1.00	0.035	0.040	0.15-0.35	0.20-0.40	0.35-0.55	0.08-0.15
94B17	G94171	0.15-0.20	0.75-1.00	0.035	0.040	0.15-0.35	0.30-0.60	0.30-0.50	0.08-0.15
94B30	G94301	0.28-0.33	0.75-1.00	0.035	0.040	0.15-0.35	0.30-0.60	0.30-0.50	0.08-0.15

(a) Electric furnace steel. (b) Includes 0.10 to 0.15% vanadium. (c) Includes 0.15% minimum vanadium. (d) 0.0005 to 0.003% boron

Alloy Steels: AISI-SAE System of Designations (Ref 6)

XX in the last two digits of the designation indicates that the carbon content (in hundredths of a percent) is to be inserted; in the chromium steels with five digits in the designation, the last three digits indicate carbon content

Numerals and digits	Type of steel and nominal alloy content
Manganese steels	
13XX	Mn 1.75
Nickel steels	
23XX	Ni 3.50
25XX	Ni 5.00
Nickel-chromium steels	
31XX	Ni 1.25; Cr 0.65 and 0.80
32XX	Ni 1.75; Cr 1.07
33XX	Ni 3.50; Cr 1.50 and 1.57
34XX	Ni 3.00; Cr 0.77
Nickel-molybdenum steels	
46XX	Ni 0.85 and 1.82; Mo 0.20 and 0.25
48XX	Ni 3.50; Mo 0.25
Molybdenum steels	
40XX	Mo 0.20 and 0.25
44XX	Mo 0.40 and 0.52
Chromium steels	
50XX	Cr 0.27, 0.40, 0.50 and 0.65
51XX	Cr 0.80, 0.87, 0.92, 0.95, 1.00 and 1.05
50XXX	Cr 0.50 } C 1.00 min
51XXX	Cr 1.02 } C 1.00 min
52XXX	Cr 1.45 } C 1.00 min
Chromium-molybdenum steels	
41XX	Cr 0.50, 0.80 and 0.95; Mo 0.12, 0.20, 0.25 and 0.30
Nickel-chromium-molybdenum steels	
43XX	Ni 1.82; Cr 0.50 and 0.80; Mo 0.25
43BVXX	Ni 1.82; Cr 0.50; Mo 0.12 and 0.25; V 0.03 min
47XX	Ni 1.05; Cr 0.45; Mo 0.20 and 0.35
81XX	Ni 0.30; Cr 0.40; Mo 0.12
86XX	Ni 0.55; Cr 0.50; Mo 0.20
87XX	Ni 0.55; Cr 0.50; Mo 0.25
88XX	Ni 0.55; Cr 0.50; Mo 0.35
93XX	Ni 3.25; Cr 1.20; Mo 0.12
94XX	Ni 0.45; Cr 0.40; Mo 0.12
97XX	Ni 0.55; Cr 0.20; Mo 0.20
98XX	Ni 1.00; Cr 0.80; Mo 0.25
Chromium-vanadium steels	
61XX	Cr 0.60, 0.80 and 0.95; V 0.10 and 0.15 min
Chromium-molybdenum-aluminum steels	
71XX	Cr 1.60; Mo 0.35; Al 1.13
Tungsten-chromium steels	
72XX	W 1.75; Cr 0.75
Silicon-manganese steels	
92XX	Si 1.40 and 2.00; Mn 0.65, 0.82 and 0.85; Cr 0.00 and 0.65

REFERENCES

1. Alloy, Carbon, and High Strength Low Alloy Steels, Steel Products Manual, American Iron and Steel Institute, Washington, D.C., Aug 1977 (Revised April 1981)
2. Mechanical Properties of Alloy Steel, Adv. 1303R2, Republic Steel Corporation
3. *Machining Data Handbook,* 3rd ed., Vol 1, Metcut Research Associates, Inc., Cincinnati, OH, 1980
4. Modern Steels and Their Properties, Handbook 3310, Bethlehem Steel Corporation, Bethlehem, PA, March 1980
5. *Metals Handbook,* 9th ed., Vol 4, American Society for Metals, 1981
6. *Metals Handbook,* 9th ed., Vol 1, American Society for Metals, 1978
7. J & L Cold Finished Bars, Jones & Laughlin Steel Company
8. Crucible Data Sheet, AISI 4340, Crucible Specialty Metals Division, Colt Industries, Syracuse, NY
9. Physical Properties of Nickel Alloy Steels, Nickel Alloy Steels Data Book, Section 7, Bulletin A, The International Nickel Company, Inc., New York, NY, 1968
10. Carburized Nickel Alloy Steels, 2-B, The International Nickel Company, Inc., New York, NY, 1975
11. Technical Data, Tool Steels, Carpenter Presto Steel, Carpenter Technology Corporation, 1981
12. C.V. Darragh, J.J. Donze and C.P. Weigel, New Bearing Steel Combines High Hardenability, Low Cost, *Metal Progress,* Vol 100 (No. 6), Dec 1971, p 74-75
13. Crucible Data Sheet, AISI 8620, Crucible Specialty Metals Division, Colt Industries, Syracuse, NY
14. Quenched and Tempered Nickel Alloy Steels, Nickel Alloy Steels Data Book, Section 2, Bulletin A, The International Nickel Company, Inc., New York, NY, 1965
15. Cold Finished Steel Bars: Selection and Uses, Stelco Inc., Toronto, Ontario, Canada, 1981
16. Nickel Alloy Nitriding Steels, Nickel Alloy Steels Data Book, Section 4, Bulletin A, The International Nickel Company, Inc., New York, NY, 1968
17. Low Temperature Properties of Nickel Alloy Steels, 4-C, The International Nickel Company, Inc., New York, NY, 1975

AISI 1330, 1330H

AISI 1330, 1330H: Chemical Composition

AISI grade	C	Mn	P max	S max	Si
1330	0.28-0.33	1.60-1.90	0.035	0.040	0.15-0.30
1330H	0.27-0.33	1.45-2.05	0.035	0.040	0.15-0.30

Chemical composition, %

Characteristics and Typical Uses. AISI 1330 and 1330H grade steels are available as hot rolled and cold finished bar. Typical applications for hot rolled bar include construction and machining into components. Cold finished bar is generally suitable for heat treatment, machining into components, or, in the as-finished condition, constructional applications or similar purposes.

AISI 1330: Similar Steels (U.S. and/or Foreign). UNS G13300; ASTM A322, A331, A519; MIL SPEC MIL-S-16974; SAE J404, J412, J770; (W. Ger.) DIN 1.1165; (Jap.) JIS SMn 1 H, SCMn 2

AISI 1330H: Similar Steels (U.S. and/or Foreign). UNS H13300; ASTM A304; SAE J1268; (W. Ger.) DIN 1.1165; (Jap.) JIS SMn 1 H, SCMn 2

AISI 1330, 1330H: Approximate Critical Points

Transformation point	Temperature(a) °C	°F
Ac_1	720	1325
Ac_3	800	1470
Ar_3	730	1350
Ar_1	630	1170
M_s	355	675

(a) On heating or cooling at 28 °C (50 °F) per hour

Physical Properties

AISI 1330, 1330H: Thermal Treatment Temperatures

Quenching medium: water or oil

Treatment	Temperature range °C	°F
Forging	1230 max	2250 max
Annealing	845-900	1550-1650
Normalizing	870-925	1600-1700
Hardening	830-855	1525-1575
Tempering	(a)	(a)

(a) To desired hardness

Mechanical Properties

AISI 1330H: End-Quench Hardenability Limits and Hardenability Band. Heat treating temperatures recommended by SAE: normalize at 900 °C (1650 °F), for forged or rolled specimens only; austenitize at 870 °C (1600 °F). (Ref 1)

Distance from quenched end, 1/16 in.	Hardness, HRC max	Hardness, HRC min
1	56	49
2	56	47
3	55	44
4	53	40
5	52	35
6	50	31
7	48	28
8	45	26
9	43	25
10	42	23
11	40	22
12	39	21
13	38	20
14	37	...
15	36	...
16	35	...
18	34	...
20	33	...
22	32	...
24	31	...
26	31	...
28	31	...
30	30	...
32	30	...

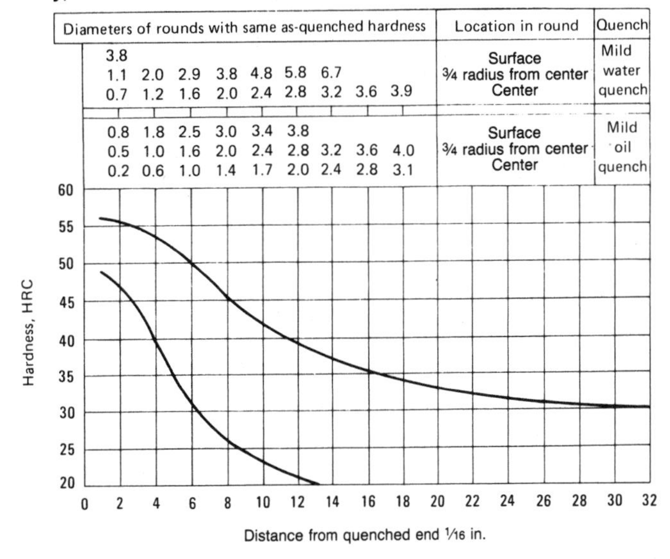

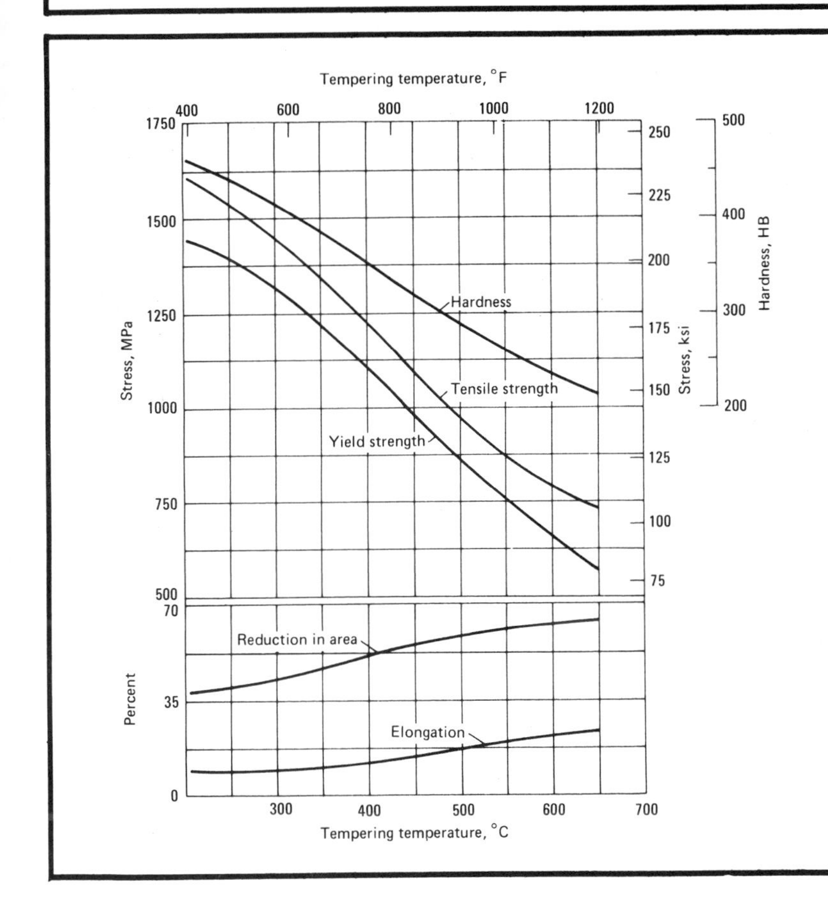

AISI 1330: Effect of Tempering Temperature on Tensile Properties. Normalized at 900 °C (1650 °F); water quenched from 870 °C (1600 °F); tempered at 56 °C (100 °F) intervals in 13.7-mm (0.540-in.) rounds. Tested in 12.8-mm (0.505-in.) rounds. Tests were conducted using specimens machined to English units. Elongation was measured in 50 mm (2 in.). (Ref 2)

AISI 1330, 1330H: Hardness and Machinability (Ref 7)

Condition	Hardness range, HB	Average machinability rating(a)
Annealed and cold drawn	179-335	55

(a) Based on AISI 1212 steel as 100% average machinability

Machining Data (Ref 3)

AISI 1330, 1330H: Turning (Cutoff and Form Tools)

Tool material	Speed, m/min (ft/min)	Feed per revolution for cutoff tool width of: 1.5 mm (0.062 in.)		3 mm (0.125 in.)		6 mm (0.25 in.)		Feed per revolution for form tool width of: 12 mm (0.5 in.)		18 mm (0.75 in.)		25 mm (1 in.)		50 mm (2 in.)	
		mm	in.	mm	in.	mm	in.	mm	in.	mm	in.	mm	in.	mm	in.
Hardness, 175 to 225 HB															
M2 and M3 high speed steel	29 (95)	0.038	0.0015	0.046	0.0018	0.056	0.0022	0.046	0.0018	0.041	0.0016	0.038	0.0015	0.028	0.0011
C-6 carbide	95 (305)	0.038	0.0015	0.046	0.0018	0.056	0.0022	0.046	0.0018	0.041	0.0016	0.038	0.0015	0.028	0.0011
Hardness, 225 to 275 HB															
M2 and M3 high speed steel	23 (75)	0.036	0.0014	0.043	0.0017	0.053	0.0021	0.043	0.0017	0.038	0.0015	0.036	0.0014	0.025	0.0010
C-6 carbide	76 (250)	0.036	0.0014	0.043	0.0017	0.053	0.0021	0.043	0.0017	0.038	0.0015	0.036	0.0014	0.025	0.0010
Hardness, 275 to 325 HB															
M2 and M3 high speed steel	17 (55)	0.033	0.0013	0.041	0.0016	0.050	0.0020	0.041	0.0016	0.036	0.0014	0.033	0.0013	0.023	0.0009
C-6 carbide	53 (175)	0.033	0.0013	0.041	0.0016	0.050	0.0020	0.041	0.0016	0.036	0.0014	0.033	0.0013	0.023	0.0009
Hardness, 325 to 375 HB															
Any premium high speed steel (T15, M33, M41-M47)	14 (45)	0.030	0.0012	0.038	0.0015	0.048	0.0019	0.038	0.0015	0.033	0.0013	0.030	0.0012	0.020	0.0008
C-6 carbide	44 (145)	0.030	0.0012	0.038	0.0015	0.048	0.0019	0.038	0.0015	0.033	0.0013	0.030	0.0012	0.020	0.0008
Hardness, 375 to 425 HB															
Any premium high speed steel (T15, M33, M41-M47)	11 (35)	0.028	0.0011	0.036	0.0014	0.046	0.0018	0.036	0.0014	0.030	0.0012	0.028	0.0011	0.018	0.0007
C-6 carbide	35 (115)	0.028	0.0011	0.036	0.0014	0.046	0.0018	0.036	0.0014	0.030	0.0012	0.028	0.0011	0.018	0.0007

AISI 1330, 1330H: Planing

Tool material	Depth of cut mm	in.	Speed m/min	ft/min	Feed/stroke mm	in.
Hardness, 175 to 225 HB						
M2 and M3 high speed steel	0.1	0.005	11	35	(a)	(a)
	2.5	0.100	15	50	1.25	0.050
	12	0.500	9	30	1.50	0.060
C-6 carbide	0.1	0.005	67	220	(a)	(a)
	2.5	0.100	73	240	2.05	0.080
	12	0.500	60	200	1.50	0.060
Hardness, 225 to 275 HB						
M2 and M3 high speed steel	0.1	0.005	9	30	(a)	(a)
	2.5	0.100	14	45	0.75	0.030
	12	0.500	8	25	1.15	0.045
C-6 carbide	0.1	0.005	52	170	(a)	(a)
	2.5	0.100	60	200	1.50	0.060
	12	0.500	49	160	1.25	0.050
Hardness, 275 to 325 HB						
M2 and M3 high speed steel	0.1	0.005	8	25	(a)	(a)
	2.5	0.100	12	40	0.75	0.030
	12	0.500	6	20	1.15	0.045
C-6 carbide	0.1	0.005	46	150	(a)	(a)
	2.5	0.100	55	180	1.50	0.060
	12	0.500	43	140	1.25	0.050

(a) Feed is 75% the width of the square nose finishing tool

AISI 1330, 1330H: End Milling (Profiling)

Tool material	Depth of cut mm	Depth of cut in.	Speed m/min	Speed ft/min	Feed per tooth for cutter diameter of: 10 mm (0.375 in.) mm	10 mm (0.375 in.) in.	12 mm (0.5 in.) mm	12 mm (0.5 in.) in.	18 mm (0.75 in.) mm	18 mm (0.75 in.) in.	25-50 mm (1-2 in.) mm	25-50 mm (1-2 in.) in.
Hardness, 175 to 225 HB												
M2, M3, and M7 high speed steel	0.5	0.020	37	120	0.025	0.001	0.050	0.002	0.075	0.003	0.102	0.004
	1.5	0.060	27	90	0.050	0.002	0.075	0.003	0.102	0.004	0.13	0.005
	diam/4	diam/4	24	80	0.038	0.0015	0.050	0.002	0.075	0.003	0.102	0.004
	diam/2	diam/2	21	70	0.025	0.001	0.038	0.0015	0.050	0.002	0.075	0.003
C-5 carbide	0.5	0.020	140	455	0.038	0.0015	0.075	0.003	0.13	0.005	0.15	0.006
	1.5	0.060	105	350	0.063	0.0025	0.102	0.004	0.15	0.006	0.18	0.007
	diam/4	diam/4	90	295	0.050	0.002	0.075	0.003	0.13	0.005	0.15	0.006
	diam/2	diam/2	82	275	0.038	0.0015	0.050	0.002	0.102	0.004	0.13	0.005
Hardness, 225 to 275 HB												
M2, M3, and M7 high speed steel	0.5	0.020	32	105	0.025	0.001	0.050	0.002	0.075	0.003	0.102	0.004
	1.5	0.060	24	80	0.050	0.002	0.075	0.003	0.102	0.004	0.13	0.005
	diam/4	diam/4	21	70	0.038	0.0015	0.050	0.002	0.075	0.003	0.102	0.004
	diam/2	diam/2	18	60	0.025	0.001	0.038	0.0015	0.050	0.002	0.075	0.003
C-5 carbide	0.5	0.020	120	390	0.025	0.001	0.050	0.002	0.102	0.004	0.13	0.005
	1.5	0.060	90	300	0.050	0.002	0.075	0.003	0.13	0.005	0.18	0.007
	diam/4	diam/4	78	255	0.038	0.0015	0.075	0.003	0.102	0.004	0.13	0.005
	diam/2	diam/2	72	235	0.025	0.001	0.050	0.002	0.075	0.003	0.102	0.004
Hardness, 275 to 325 HB												
M2, M3, and M7 high speed steel	0.5	0.020	26	85	0.018	0.0007	0.038	0.0015	0.075	0.003	0.102	0.004
	1.5	0.060	20	65	0.025	0.001	0.050	0.002	0.102	0.004	0.13	0.005
	diam/4	diam/4	17	55	0.018	0.0007	0.038	0.0015	0.075	0.003	0.102	0.004
	diam/2	diam/2	15	50	0.013	0.0005	0.025	0.001	0.050	0.002	0.075	0.003
C-5 carbide	0.5	0.020	95	310	0.025	0.001	0.050	0.002	0.102	0.004	0.13	0.005
	1.5	0.060	72	235	0.050	0.002	0.075	0.003	0.13	0.005	0.15	0.006
	diam/4	diam/4	62	205	0.038	0.0015	0.050	0.002	0.102	0.004	0.13	0.005
	diam/2	diam/2	58	190	0.025	0.001	0.038	0.0015	0.075	0.003	0.102	0.004
Hardness, 325 to 375 HB												
M2, M3, and M7 high speed steel	0.5	0.020	20	65	0.013	0.0005	0.038	0.0015	0.075	0.003	0.102	0.004
	1.5	0.060	17	55	0.013	0.0005	0.038	0.0015	0.102	0.004	0.13	0.005
	diam/4	diam/4	15	50	0.013	0.0005	0.038	0.0015	0.075	0.003	0.102	0.004
	diam/2	diam/2	12	40	0.013	0.0005	0.025	0.001	0.050	0.002	0.075	0.003
C-5 carbide	0.5	0.020	79	260	0.025	0.001	0.038	0.0015	0.102	0.003	0.13	0.005
	1.5	0.060	60	200	0.038	0.0015	0.075	0.003	0.13	0.005	0.15	0.006
	diam/4	diam/4	52	170	0.038	0.0015	0.050	0.002	0.102	0.004	0.13	0.005
	diam/2	diam/2	49	160	0.025	0.001	0.050	0.002	0.075	0.003	0.102	0.004
Hardness, 375 to 425 HB												
Any premium high speed steel (T15, M33, M41-M47)	0.5	0.020	17	55	0.013	0.0005	0.018	0.0007	0.025	0.001	0.050	0.002
	1.5	0.060	14	45	0.013	0.0005	0.025	0.001	0.050	0.002	0.075	0.003
	diam/4	diam/4	12	40	0.013	0.0005	0.013	0.0005	0.050	0.002	0.063	0.0025
	diam/2	diam/2	11	35	...	...	...	...	0.038	0.0015	0.050	0.002
C-5 carbide	0.5	0.020	64	210	0.025	0.001	0.025	0.001	0.050	0.002	0.075	0.003
	1.5	0.060	49	160	0.038	0.0015	0.050	0.002	0.075	0.003	0.102	0.004
	diam/4	diam/4	41	135	0.038	0.0015	0.038	0.0015	0.063	0.0025	0.089	0.0035
	diam/2	diam/2	38	125	0.001	0.001	0.025	0.001	0.050	0.002	0.075	0.003

AISI 1330, 1330H: Face Milling

Depth of cut mm	Depth of cut in.	High speed steel: Speed m/min	Speed ft/min	Feed/tooth mm	Feed/tooth in.	Uncoated carbide: Speed, brazed m/min	Speed, brazed ft/min	Speed, inserted m/min	Speed, inserted ft/min	Feed/tooth mm	Feed/tooth in.	Coated carbide: Speed m/min	Speed ft/min	Feed/tooth mm	Feed/tooth in.
Hardness, 175 to 225 HB															
1	0.040	52(a)	170(a)	0.20	0.008	135(b)	450(b)	170(b)	550(b)	0.20	0.008	250(c)	825(c)	0.20	0.008
4	0.150	40(a)	130(a)	0.30	0.012	115(b)	370(b)	130(b)	425(b)	0.30	0.012	170(c)	550(c)	0.30	0.012
8	0.300	34(a)	110(a)	0.40	0.016	79(d)	260(d)	100(d)	320(d)	0.40	0.016	120(e)	400(e)	0.40	0.016
Hardness, 225 to 275 HB															
1	0.040	44(a)	145(a)	0.15	0.006	125(b)	405(b)	150(b)	495(b)	0.18	0.007	230(c)	750(c)	0.18	0.007
4	0.150	32(a)	105(a)	0.25	0.010	95(b)	315(b)	115(b)	385(b)	0.25	0.010	150(c)	500(c)	0.25	0.010
8	0.300	24(a)	80(a)	0.36	0.014	75(d)	245(d)	90(d)	300(d)	0.36	0.014	120(e)	400(e)	0.36	0.014
Hardness, 275 to 325 HB															
1	0.040	30(f)	100(f)	0.15	0.006	115(b)	375(b)	140(b)	460(b)	0.15	0.006	215(c)	700(c)	0.13	0.005
4	0.150	24(f)	80(f)	0.23	0.009	90(b)	300(b)	110(b)	365(b)	0.20	0.008	145(c)	475(c)	0.18	0.007
8	0.300	18(f)	60(f)	0.30	0.012	72(d)	235(d)	87(d)	285(d)	0.25	0.010	115(e)	375(e)	0.23	0.009

(a) High speed steel tool material: M2 or M7. (b) Carbide tool material: C-6. (c) Carbide tool material: CC-6. (d) Carbide tool material: C-5. (e) Carbide tool material: CC-5. (f) Any premium high speed steel tool material (T15, M33, M41 to M47)

Machining Data (Ref 3) (continued)

AISI 1330, 1330H: Reaming

Based on 4 flutes for 3- and 6-mm (0.125- and 0.25-in.) reamers, 6 flutes for 12-mm (0.5-in.) reamers, and 8 flutes for 25-mm (1-in.) and larger reamers

Tool material	Speed		Feed per revolution for reamer diameter of: 3 mm (0.125 in.)		6 mm (0.25 in.)		12 mm (0.5 in.)		25 mm (1 in.)		35 mm (1.5 in.)		50 mm (2 in.)	
	m/min	ft/min	mm	in.	mm	in.	mm	in.	mm	in.	mm	in.	mm	in.
Hardness, 175 to 225 HB														
Roughing														
M1, M2, and M7 high speed steel	21	70	0.102	0.004	0.18	0.007	0.30	0.012	0.50	0.020	0.65	0.025	0.75	0.030
C-2 carbide	26	85	0.102	0.004	0.18	0.007	0.30	0.012	0.50	0.020	0.65	0.025	0.75	0.030
Finishing														
M1, M2, and M7 high speed steel	11	35	0.13	0.005	0.20	0.008	0.30	0.012	0.50	0.020	0.65	0.025	0.75	0.030
C-2 carbide	14	45	0.13	0.005	0.20	0.008	0.30	0.012	0.50	0.020	0.65	0.025	0.75	0.030
Hardness, 225 to 275 HB														
Roughing														
M1, M2, and M7 high speed steel	18	60	0.075	0.003	0.15	0.006	0.25	0.010	0.40	0.015	0.50	0.020	0.65	0.025
C-2 carbide	23	75	0.102	0.004	0.15	0.006	0.25	0.010	0.40	0.015	0.50	0.020	0.65	0.025
Finishing														
M1, M2, and M7 high speed steel	9	30	0.102	0.004	0.18	0.007	0.25	0.010	0.40	0.015	0.50	0.020	0.65	0.025
C-2 carbide	12	40	0.102	0.004	0.18	0.007	0.25	0.010	0.40	0.015	0.50	0.020	0.65	0.025
Hardness, 275 to 325 HB														
Roughing														
M1, M2, and M7 high speed steel	15	50	0.075	0.003	0.13	0.005	0.20	0.008	0.30	0.012	0.40	0.015	0.50	0.020
C-2 carbide	20	65	0.102	0.004	0.15	0.006	0.20	0.008	0.30	0.012	0.40	0.015	0.50	0.020
Finishing														
M1, M2, and M7 high speed steel	8	25	0.075	0.003	0.15	0.006	0.20	0.008	0.30	0.012	0.40	0.015	0.50	0.020
C-2 carbide	11	35	0.102	0.004	0.15	0.006	0.20	0.008	0.30	0.012	0.40	0.015	0.50	0.020
Hardness, 325 to 375 HB														
Roughing														
M1, M2, and M7 high speed steel	11	35	0.05	0.002	0.102	0.004	0.13	0.005	0.20	0.008	0.25	0.010	0.30	0.012
C-2 carbide	15	50	0.102	0.004	0.15	0.006	0.20	0.008	0.25	0.010	0.30	0.012	0.40	0.015
Finishing														
M1, M2, and M7 high speed steel	8	25	0.050	0.002	0.13	0.005	0.15	0.006	0.25	0.010	0.30	0.012	0.40	0.015
C-2 carbide	11	35	0.102	0.004	0.15	0.006	0.20	0.008	0.25	0.010	0.30	0.012	0.40	0.015
Hardness, 375 to 425 HB														
Roughing														
Any premium high speed steel (T15, M33, M41-M47)	8	25	0.050	0.002	0.102	0.004	0.13	0.005	0.15	0.006	0.18	0.007	0.20	0.008
C-2 carbide	12	40	0.102	0.004	0.15	0.006	0.20	0.008	0.25	0.010	0.28	0.011	0.30	0.012
Finishing														
Any premium high speed steel (T15, M33, M41-M47)	6	20	0.050	0.002	0.102	0.004	0.13	0.005	0.15	0.006	0.18	0.007	0.20	0.008
C-2 carbide	9	30	0.102	0.004	0.15	0.006	0.20	0.008	0.25	0.010	0.28	0.011	0.30	0.012

AISI 1330, 1330H: Drilling

High speed steel tool material	Speed		Feed per revolution for nominal hole diameter of: 1.5 mm (0.062 in.)		3 mm (0.125 in.)		6 mm (0.25 in.)		12 mm (0.5 in.)		18 mm (0.75 in.)		25 mm (1 in.)	
	m/min	ft/min	mm	in.	mm	in.	mm	in.	mm	in.	mm	in.	mm	in.
Hardness, 175 to 225 HB														
M10, M7, and M1	21	70	0.025	0.001	0.075	0.003	0.15	0.006	0.25	0.010	0.36	0.014	0.40	0.016
Hardness, 225 to 275 HB														
M10, M7, and M1	18	60	0.025	0.001	0.075	0.003	0.102	0.004	0.18	0.007	0.25	0.010	0.30	0.012
Hardness, 275 to 325 HB														
M10, M7, and M1	15	50	...	...	0.050	0.002	0.102	0.004	0.15	0.006	0.20	0.008	0.23	0.009

Machining Data (Ref 3) (continued)

AISI 1330, 1330H: Turning (Single Point and Box Tools)

Depth of cut		High speed steel				Uncoated carbide						Coated carbide			
		Speed		Feed		Speed, brazed		Speed, inserted		Feed		Speed		Feed	
mm	in.	m/min	ft/min	mm/rev	in./rev	m/min	ft/min	m/min	ft/min	mm/rev	in./rev	m/min	ft/min	mm/rev	in./rev
Hardness, 175 to 225 HB															
1	0.040	41(a)	135(a)	0.18	0.007	115(b)	375(b)	150(b)	500(b)	0.18	0.007	200(c)	650(c)	0.18	0.007
4	0.150	32(a)	105(a)	0.40	0.015	90(d)	300(d)	120(d)	400(d)	0.40	0.015	160(e)	525(e)	0.40	0.015
8	0.300	24(a)	80(a)	0.50	0.020	72(d)	235(d)	95(d)	315(d)	0.50	0.020	120(e)	400(e)	0.50	0.020
16	0.625	20(a)	65(a)	0.75	0.030	56(d)	185(d)	76(d)	250(d)	1.00	0.040	...	...	...	...
Hardness, 225 to 275 HB															
1	0.040	35(a)	115(a)	0.18	0.007	115(b)	375(b)	135(b)	450(b)	0.18	0.007	185(c)	600(c)	0.18	0.007
4	0.150	27(a)	90(a)	0.40	0.015	85(d)	280(d)	105(d)	350(d)	0.40	0.015	135(e)	450(e)	0.40	0.015
8	0.300	21(a)	70(a)	0.50	0.020	69(d)	225(d)	84(d)	275(d)	0.50	0.020	84(e)	275(e)	0.50	0.020
16	0.625	17(a)	55(a)	0.75	0.030	52(d)	170(d)	66(d)	215(d)	0.75	0.030	...	...	...	...
Hardness, 275 to 325 HB															
1	0.040	29(f)	95(f)	0.18	0.007	105(b)	340(b)	130(b)	420(b)	0.18	0.007	170(c)	550(c)	0.18	0.007
4	0.150	23(f)	75(f)	0.40	0.015	79(d)	260(d)	100(d)	330(d)	0.40	0.015	130(e)	425(e)	0.40	0.015
8	0.300	18(f)	60(f)	0.50	0.020	60(d)	200(d)	79(d)	260(d)	0.50	0.020	105(e)	350(e)	0.50	0.020

(a) High speed steel tool material: M2 or M3. (b) Carbide tool material: C-7. (c) Carbide tool material: CC-7. (d) Carbide tool material: C-6. (e) Carbide tool material: CC-6. (f) Any premium high speed steel tool material (T15, M33, M41 to M47)

AISI 1330, 1330H: Boring

Depth of cut		High speed steel				Uncoated carbide						Coated carbide			
		Speed		Feed		Speed, brazed		Speed, inserted		Feed		Speed		Feed	
mm	in.	m/min	ft/min	mm/rev	in./rev	m/min	ft/min	m/min	ft/min	mm/rev	in./rev	m/min	ft/min	mm/rev	in./rev
Hardness, 175 to 225 HB															
0.25	0.010	43(a)	140(a)	0.075	0.003	115(b)	375(b)	135(b)	440(b)	0.075	0.003	175(c)	570(c)	0.075	0.003
1.25	0.050	34(a)	110(a)	0.13	0.005	90(d)	300(d)	105(d)	350(d)	0.13	0.005	140(e)	455(e)	0.13	0.005
2.5	0.100	26(a)	85(a)	0.30	0.012	73(d)	240(d)	85(d)	280(d)	0.30	0.012	115(e)	370(e)	0.30	0.012
Hardness, 225 to 275 HB															
0.25	0.010	34(a)	110(a)	0.075	0.003	100(b)	335(b)	120(b)	395(b)	0.075	0.003	160(c)	525(c)	0.075	0.003
1.25	0.050	27(a)	90(a)	0.13	0.005	81(d)	265(d)	95(d)	315(d)	0.13	0.005	130(e)	420(e)	0.13	0.005
2.5	0.100	21(a)	70(a)	0.30	0.012	64(d)	210(d)	75(d)	245(d)	0.30	0.012	95(e)	315(e)	0.30	0.012
Hardness, 275 to 325 HB															
0.25	0.010	29(f)	95(f)	0.075	0.003	95(b)	315(b)	70(b)	370(b)	0.075	0.003	145(c)	480(c)	0.075	0.003
1.25	0.050	23(f)	75(f)	0.13	0.005	76(d)	250(d)	90(d)	295(d)	0.13	0.005	115(e)	385(e)	0.13	0.005
2.5	0.100	18(f)	60(f)	0.30	0.012	59(d)	195(d)	70(d)	230(d)	0.30	0.012	90(e)	300(e)	0.30	0.012

(a) High speed steel tool material: M2 or M3. (b) Carbide tool material: C-7. (c) Carbide tool material: CC-7. (d) Carbide tool material: C-6. (e) Carbide tool material: CC-6. (f) Any premium high speed steel tool material (T15, M33, M41 to M47)

AISI 1335, 1335H

AISI 1335, 1335H: Chemical Composition

AISI grade	C	Mn	P max	S max	Si
	Chemical composition, %				
1335	0.33-0.38	1.60-1.90	0.035	0.040	0.15-0.30
1335H	0.32-0.38	1.45-2.05	0.035	0.040	0.15-0.30

Characteristics and Typical Uses. AISI grades 1335 and 1335H are available as hot rolled and cold finished bar for constructional applications or machining into components. Also produced as cold heading quality wire, these grades are used for high-strength bolts, screws, cap screws, and socket- and recessed-head screws that can be heat treated to required mechanical properties.

AISI 1335: Similar Steels (U.S. and/or Foreign). UNS G13350; ASTM A322, A331, A519, A547; MIL SPEC MIL-S-16974; SAE J404, J412, J770; (W. Ger.) DIN 1.1167; (Fr.) AFNOR 40 M 5; (Jap.) JIS SMn 2 H, SMn 2, SCMn 3; (Swed.) SS_{14} 2120

AISI 1335H: Similar Steels (U.S. and/or Foreign). UNS H13350; ASTM A304; SAE J1268; (W. Ger.) DIN 1.1167; (Fr.) AFNOR 40 M 5; (Jap.) JIS SMn 2 H, SMn 2, SCMn 3; (Swed.) SS_{14} 2120

AISI 1335, 1335H: Approximate Critical Points

Transformation point	Temperature(a) °C	Temperature(a) °F
Ac_1	720	1330
Ac_3	780	1440
Ar_3	725	1340
Ar_1	625	1160
M_s	340	640

(a) On heating or cooling at 28 °C (50 °F) per hour

Physical Properties

AISI 1335, 1335H: Thermal Treatment Temperatures

Quenching medium: oil or water

Treatment	Temperature range °C	°F
Annealing	845-900	1550-1650
Normalizing	870-925	1600-1700
Hardening	665-690	1225-1275
Tempering	(a)	(a)

(a) To desired hardness

Machining Data

For machining data on AISI grade 1335 and 1335H, refer to the preceding machining tables for AISI grades 1330 and 1330H.

Mechanical Properties

AISI 1335H: End-Quench Hardenability Limits and Hardenability Band.

Heat treating temperatures recommended by SAE: normalize at 870 °C (1600 °F), for forged or rolled specimens only; austenitize at 845 °C (1550 °F). (Ref 1)

Distance from quenched end, 1/16 in.	Hardness, HRC max	Hardness, HRC min
1	58	51
2	57	49
3	56	47
4	55	44
5	54	38
6	52	34
7	50	31
8	48	29
9	46	27
10	44	26
11	42	25
12	41	24
13	40	23
14	39	22
15	38	22
16	37	21
18	35	20
20	34	...
22	33	...
24	32	...
26	31	...
28	31	...
30	30	...
32	30	...

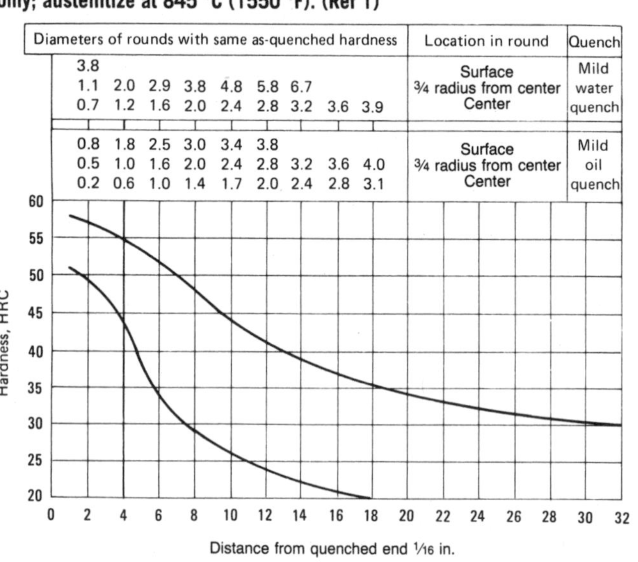

AISI 1335, 1335H: Hardness and Machinability (Ref 7)

Condition	Hardness range, HB	Average machinability rating(a)
Annealed and cold drawn	179-235	55

(a) Based on AISI 1212 steel as 100% average machinability

AISI 1340, 1340H

AISI 1340, 1340H: Chemical Composition

AISI grade	C	Chemical composition, % Mn	P max	S max	Si
1340	0.38-0.43	1.60-1.90	0.035	0.040	0.15-0.30
1340H	0.37-0.44	1.45-2.05	0.035	0.040	0.15-0.30

Characteristics and Typical Uses. AISI grades 1340 and 1340H are generally available as hot rolled and cold finished bar for constructional applications and machining into components. Also available as cold heading quality wire, these steels are used for high-strength bolts, screws, cap screws, and socket- and recessed-head screws that can be heat treated to meet mechanical property requirements.

AISI 1340, 1340H: Approximate Critical Points

Transformation point	Temperature(a) °C	°F	Transformation point	Temperature(a) °C	°F
Ac_1	715	1320	Ar_1	620	1150
Ac_3	775	1430	M_s	320	610
Ar_3	720	1330			

(a) On heating or cooling at 28 °C (50 °F) per hour

AISI 1340: Similar Steels (U.S. and/or Foreign). UNS G13400; ASTM A322, A331, A519, A547; MIL SPEC MIL-S-16974; SAE J404, J412, J770

AISI 1340H: Similar Steels (U.S. and/or Foreign). UNS H13400; ASTM A304; SAE J1268; (W. Ger.) DIN 1.5069

Physical Properties

AISI 1340, 1340H: Thermal Treatment Temperatures

Quenching medium: oil

Treatment	Temperature range °C	°F
Forging	1230 max	2250 max
Annealing	845-900	1550-1650
Normalizing	870-925	1600-1700
Hardening	815-845	1500-1550
Tempering	(a)	(a)

(a) To desired hardness

Mechanical Properties

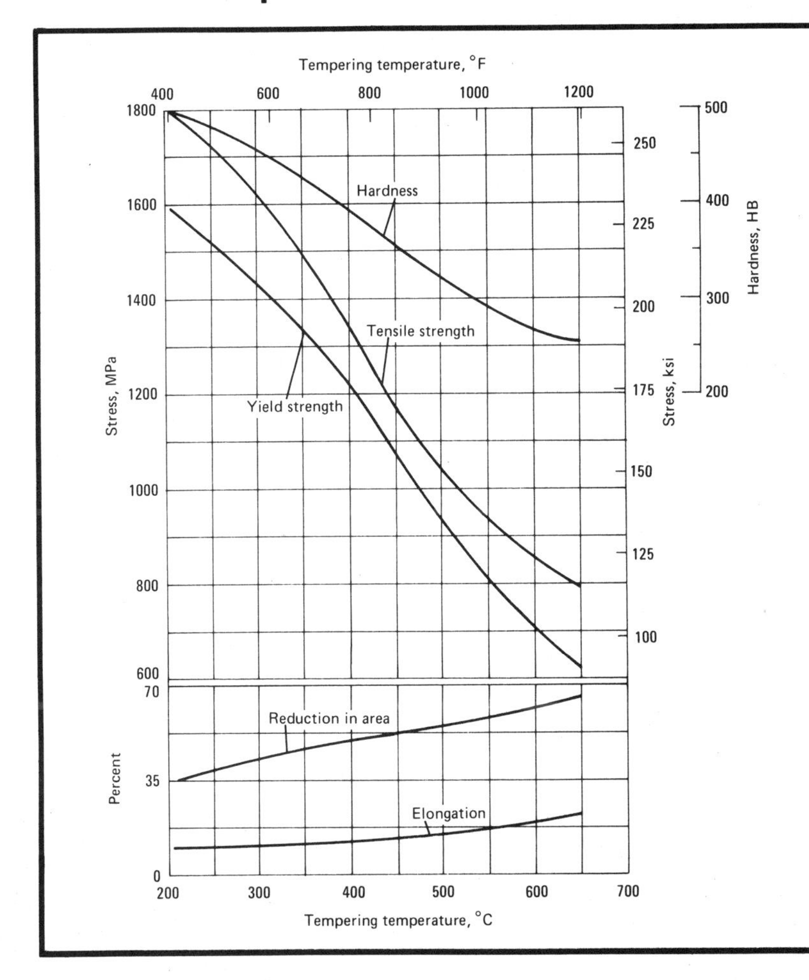

AISI 1340: Effect of Tempering Temperature on Tensile Properties. Normalized at 870 °C (1600 °F); oil quenched from 845 °C (1550 °F); tempered at 56 °C (100 °F) intervals in 13.7-mm (0.540-in.) rounds; tested in 12.8-mm (0.505-in.) rounds. Tests were conducted using specimens machined to English units. Elongation was measured in 50 mm (2 in.). (Ref 2)

Mechanical Properties (continued)

AISI 1340H: End-Quench Hardenability Limits and Hardenability Band. Heat treating temperatures recommended by SAE: normalize at 870 °C (1600 °F), for forged or rolled specimens only; austenitize at 845 °C (1550 °F). (Ref 1)

Distance from quenched end, 1/16 in.	Hardness, HRC max	Hardness, HRC min
1	60	53
2	60	52
3	59	51
4	58	49
5	57	46
6	56	40
7	55	35
8	54	33
9	52	31
10	51	29
11	50	28
12	48	27
13	46	26
14	44	25
15	42	25
16	41	24
18	39	23
20	38	23
22	37	22
24	36	22
26	35	21
28	35	21
30	34	20
32	34	20

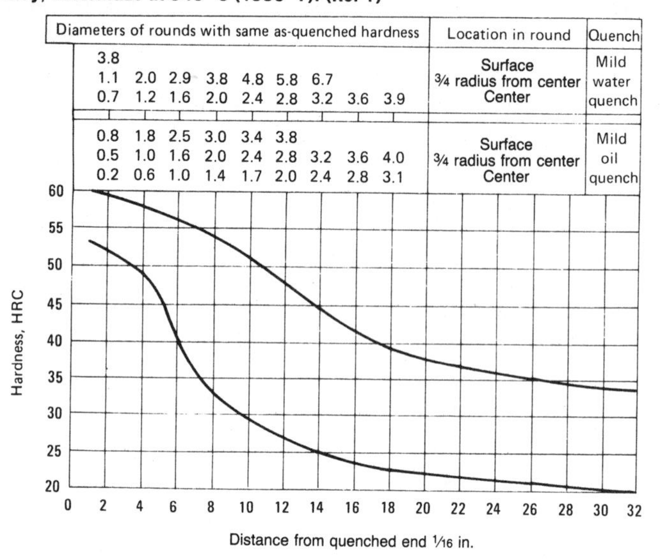

AISI 1340, 1340H: Mass Effect on Mechanical Properties (Ref 4)

Condition or treatment	Size round mm	Size round in.	Tensile strength MPa	Tensile strength ksi	Yield strength MPa	Yield strength ksi	Elongation(a), %	Reduction in area, %	Hardness, HB
Annealed (heated to 800 °C or 1475 °F; furnace cooled 11 °C or 20 °F per hour to 600 °C or 1110 °F; cooled in air)	25	1	703	102	434	63	25.5	57.3	207
Normalized (heated to 870 °C or 1600 °F; cooled in air)	13	0.5	910	132	565	82	20.0	51.0	269
	25	1	834	121	558	81	22.0	62.9	248
	50	2	827	120	524	76	23.5	61.0	235
	100	4	827	120	496	72	21.7	59.2	235
Oil quenched from 830 °C (1525 °F); tempered at 540 °C (1000 °F)	13	0.5	979	142	910	132	18.8	55.2	285
	25	1	951	138	834	121	19.2	57.4	285
	50	2	827	120	579	84	21.2	60.7	248
	100	4	800	116	572	83	21.7	57.9	241
Oil quenched from 830 °C (1525 °F); tempered at 595 °C (1100 °F)	13	0.5	876	127	814	118	21.0	57.9	255
	25	1	814	118	676	98	21.7	60.1	241
	50	2	752	109	565	82	24.7	64.3	217
	100	4	710	103	490	71	25.5	64.5	217
Oil quenched from 830 °C (1525 °F); tempered at 650 °C (1200 °F)	13	0.5	814	118	745	108	22.1	59.5	241
	25	1	772	112	662	96	23.2	62.4	229
	50	2	731	106	552	80	25.5	66.2	217
	100	4	703	102	496	72	26.0	64.8	212

(a) In 50 mm (2 in.)

AISI 1340, 1340H: Effect of the Mass on Hardness at Selected Points (Ref 4)

Size round mm	Size round in.	As-quenched hardness after quenching in oil at: Surface	½ radius	Center
13	0.5	58 HRC	57 HRC	57 HRC
25	1	57 HRC	56 HRC	50 HRC
50	2	39 HRC	34 HRC	32 HRC
100	4	32 HRC	30 HRC	26 HRC

AISI 1340, 1340H: Hardness and Machinability (Ref 7)

Condition	Hardness range, HB	Average machinability rating(a)
Annealed and cold drawn	183-241	50

(a) Based on AISI 1212 steel as 100% average machinability

Mechanical Properties (continued)

AISI 1340, 1340H: Tensile Properties (Ref 5)

Condition or treatment	Tensile strength MPa	Tensile strength ksi	Yield strength MPa	Yield strength ksi	Elongation(a) %	Reduction in area, %	Hardness, HB	Izod impact energy J	Izod impact energy ft·lb
Normalized at 870 °C (1600 °F)	834	121	558	81	22.0	63	248	54	40
Annealed at 800 °C (1475 °F)	703	102	434	63	25.5	57	207	46	34

(a) In 50 mm (2 in.)

Machining Data (Ref 3)

For the following machining data on turning (cutoff and form tools), end milling (profiling), and reaming of AISI grades 1340 and 1340H, refer to the preceding machining tables for AISI grades 1330 and 1330H.

AISI 1340, 1340H: Face Milling

Depth of cut		High speed steel				Uncoated carbide						Coated carbide			
		Speed		Feed/tooth		Speed, brazed		Speed, inserted		Feed/tooth		Speed		Feed/tooth	
mm	in.	m/min	ft/min	mm	in.	m/min	ft/min	m/min	ft/min	mm	in.	m/min	ft/min	mm	in.
Hardness, 175 to 225 HB															
1	0.040	53(a)	175(a)	0.20	0.008	140(b)	460(b)	170(b)	560(b)	0.20	0.008	250(c)	825(c)	0.20	0.008
4	0.150	41(a)	135(a)	0.30	0.012	115(b)	380(b)	135(b)	440(b)	0.30	0.012	175(c)	570(c)	0.30	0.012
8	0.300	35(a)	115(a)	0.40	0.016	85(d)	280(d)	105(d)	345(d)	0.40	0.016	135(e)	450(e)	0.40	0.016
Hardness, 225 to 275 HB															
1	0.040	46(a)	150(a)	0.15	0.006	125(b)	415(b)	155(b)	510(b)	0.18	0.007	235(c)	765(c)	0.18	0.007
4	0.150	34(a)	110(a)	0.25	0.010	100(b)	330(b)	120(b)	400(b)	0.25	0.010	160(c)	520(c)	0.25	0.010
8	0.300	26(a)	85(a)	0.36	0.014	79(d)	260(d)	95(d)	315(d)	0.36	0.014	125(e)	410(e)	0.36	0.014
Hardness, 275 to 325 HB															
1	0.040	32(f)	105(f)	0.15	0.006	120(b)	400(b)	145(b)	485(b)	0.15	0.006	220(c)	725(c)	0.13	0.005
4	0.150	26(f)	85(f)	0.23	0.009	95(b)	305(b)	115(b)	375(b)	0.20	0.008	150(c)	485(c)	0.18	0.007
8	0.300	20(f)	65(f)	0.30	0.012	72(d)	235(d)	88(d)	290(d)	0.25	0.010	115(e)	375(e)	0.23	0.009
Hardness, 325 to 375 HB															
1	0.040	24(f)	80(f)	0.13	0.005	105(b)	345(b)	130(b)	420(b)	0.13	0.005	190(c)	630(c)	0.102	0.004
4	0.150	18(f)	60(f)	0.20	0.008	81(b)	265(b)	100(b)	325(b)	0.18	0.007	130(c)	420(c)	0.15	0.006
8	0.300	15(f)	50(f)	0.25	0.010	62(d)	205(d)	76(d)	250(d)	0.23	0.009	100(e)	325(e)	0.20	0.008
Hardness, 375 to 425 HB															
1	0.040	18(f)	60(f)	0.102	0.004	81(b)	265(b)	100(b)	325(b)	0.102	0.004	150(c)	485(c)	0.075	0.003
4	0.150	15(f)	50(f)	0.15	0.006	66(b)	215(b)	81(b)	265(b)	0.15	0.006	105(c)	345(c)	0.13	0.005
8	0.300	12(f)	40(f)	0.20	0.008	52(b)	170(b)	62(b)	205(b)	0.20	0.008	81(c)	265(c)	0.18	0.007

(a) High speed steel tool material: M2 or M3. (b) Carbide tool material: C-6. (c) Carbide tool material: CC-6. (d) Carbide tool material: C-5. (e) Carbide tool material: CC-5. (f) Any premium high speed steel tool material (T15, M33, M41 to M47)

AISI 1340, 1340H: Drilling

Tool material	Speed		Feed per revolution for nominal hole diameter of:											
			1.5 mm (0.062 in.)		3 mm (0.125 in.)		6 mm (0.25 in.)		12 mm (0.5 in.)		18 mm (0.75 in.)		25 mm (1 in.)	
	m/min	ft/min	mm	in.	mm	in.	mm	in.	mm	in.	mm	in.	mm	in.
Hardness, 175 to 225 HB														
M10, M7, and M1 high speed steel	20	65	0.025	0.001	0.075	0.003	0.15	0.006	0.25	0.010	0.36	0.014	0.40	0.016
Hardness, 225 to 275 HB														
M10, M7, and M1 high speed steel	17	55	0.025	0.001	0.075	0.003	0.102	0.004	0.18	0.007	0.25	0.010	0.30	0.012
Hardness, 275 to 325 HB														
M10, M7, and M1 high speed steel	14	45	...	...	0.050	0.002	0.102	0.004	0.15	0.006	0.20	0.008	0.23	0.009
Hardness, 325 to 375 HB														
M10, M7, and M1 high speed steel	11	35	...	...	0.050	0.002	0.075	0.003	0.13	0.005	0.20	0.008	0.23	0.009
Hardness, 375 to 425 HB														
Any premium high speed steel (T15, M33, M41-M47)	8	25	...	...	0.050	0.002	0.075	0.003	0.102	0.004	0.15	0.006	0.20	0.008

Machining Data (Ref 3) (continued)

AISI 1340, 1340H: Turning (Single Point and Box Tools)

Depth of cut		High speed steel				Uncoated carbide						Coated carbide			
		Speed		Feed		Speed, brazed		Speed, inserted		Feed		Speed		Feed	
mm	in.	m/min	ft/min	mm/rev	in./rev	m/min	ft/min	m/min	ft/min	mm/rev	in./rev	m/min	ft/min	mm/rev	in./rev
Hardness, 175 to 225 HB															
1	0.040	41(a)	135(a)	0.18	0.007	115(b)	375(b)	150(b)	500(b)	0.18	0.007	200(c)	650(c)	0.18	0.007
4	0.150	32(a)	105(a)	0.40	0.015	90(d)	300(d)	120(d)	400(d)	0.50	0.020	160(e)	525(e)	0.40	0.015
8	0.300	24(a)	80(a)	0.50	0.020	73(d)	240(d)	95(d)	315(d)	0.75	0.030	120(e)	400(e)	0.50	0.020
16	0.625	20(a)	65(a)	0.75	0.030	58(d)	190(d)	76(d)	250(d)	1.00	0.040	...	...	...	...
Hardness, 225 to 275 HB															
1	0.040	35(a)	115(a)	0.18	0.007	105(b)	350(b)	140(b)	465(b)	0.18	0.007	185(c)	600(c)	0.18	0.007
4	0.150	27(a)	90(a)	0.40	0.015	85(d)	280(d)	110(d)	365(d)	0.50	0.020	145(e)	475(e)	0.40	0.015
8	0.300	21(a)	70(a)	0.50	0.020	67(d)	220(d)	87(d)	285(d)	0.75	0.030	115(e)	375(e)	0.50	0.020
16	0.625	17(a)	55(a)	0.75	0.030	52(d)	170(d)	69(d)	225(d)	1.00	0.040	...	...	...	...
Hardness, 275 to 325 HB															
1	0.040	27(f)	90(f)	0.18	0.007	100(b)	330(b)	135(b)	440(b)	0.18	0.007	175(c)	575(c)	0.18	0.007
4	0.150	21(f)	70(f)	0.40	0.015	79(d)	260(d)	105(d)	340(d)	0.40	0.015	135(e)	450(e)	0.40	0.015
8	0.300	17(f)	55(f)	0.50	0.020	60(d)	200(d)	82(d)	270(d)	0.50	0.020	105(e)	350(e)	0.50	0.020
Hardness, 325 to 375 HB															
1	0.040	21(f)	70(f)	0.13	0.005	84(b)	275(b)	115(b)	380(b)	0.18	0.007	150(c)	500(c)	0.18	0.007
4	0.150	17(f)	55(f)	0.25	0.010	66(d)	215(d)	90(d)	300(d)	0.40	0.015	120(e)	400(e)	0.40	0.015
8	0.300	12(f)	40(f)	0.40	0.015	52(d)	170(d)	72(d)	235(d)	0.50	0.020	90(e)	300(e)	0.50	0.020
Hardness, 375 to 425 HB															
1	0.040	18(f)	60(f)	0.13	0.005	69(b)	225(b)	90(b)	300(b)	0.18	0.007	120(c)	400(c)	0.18	0.007
4	0.150	14(f)	45(f)	0.25	0.010	53(d)	175(d)	73(d)	240(d)	0.40	0.015	90(e)	300(e)	0.40	0.015
8	0.300	11(f)	35(f)	0.40	0.015	43(d)	140(d)	58(d)	190(d)	0.50	0.020	76(e)	250(e)	0.50	0.020

(a) High speed steel tool material: M2 or M3. (b) Carbide tool material: C-7. (c) Carbide tool material: CC-7. (d) Carbide tool material: C-6. (e) Carbide tool material: CC-6. (f) Any premium high speed steel tool material (T15, M33, M41 to M47)

AISI 1340, 1340H: Boring

Depth of cut		High speed steel				Uncoated carbide						Coated carbide			
		Speed		Feed		Speed, brazed		Speed, inserted		Feed		Speed		Feed	
mm	in.	m/min	ft/min	mm/rev	in./rev	m/min	ft/min	m/min	ft/min	mm/rev	in./rev	m/min	ft/min	mm/rev	in./rev
Hardness, 175 to 225 HB															
0.25	0.010	43(a)	140(a)	0.075	0.003	115(b)	375(b)	135(b)	440(b)	0.075	0.003	175(c)	570(c)	0.075	0.003
1.25	0.050	34(a)	110(a)	0.13	0.005	90(d)	300(d)	105(d)	350(d)	0.13	0.005	140(e)	455(e)	0.13	0.005
2.5	0.100	26(a)	85(a)	0.30	0.012	73(d)	240(d)	85(d)	280(d)	0.40	0.015	110(e)	365(e)	0.30	0.012
Hardness, 225 to 275 HB															
0.25	0.010	34(a)	110(a)	0.075	0.003	105(b)	345(b)	125(b)	405(b)	0.075	0.003	160(c)	525(c)	0.075	0.003
1.25	0.050	27(a)	90(a)	0.13	0.005	84(d)	275(d)	100(d)	325(d)	0.13	0.005	130(e)	420(e)	0.13	0.005
2.5	0.100	21(a)	70(a)	0.30	0.012	66(d)	215(d)	78(d)	255(d)	0.40	0.015	100(e)	330(e)	0.30	0.012
Hardness, 275 to 325 HB															
0.25	0.010	27(f)	90(f)	0.075	0.003	100(b)	330(b)	120(b)	390(b)	0.075	0.003	150(c)	500(c)	0.075	0.003
1.25	0.050	21(f)	70(f)	0.13	0.005	81(d)	265(d)	95(d)	310(d)	0.13	0.005	120(e)	400(e)	0.13	0.005
2.5	0.100	17(f)	55(f)	0.30	0.012	62(d)	205(d)	73(d)	240(d)	0.30	0.012	95(e)	315(e)	0.30	0.012
Hardness, 325 to 375 HB															
0.25	0.010	21(f)	70(f)	0.050	0.002	85(b)	280(b)	100(b)	330(b)	0.075	0.003	135(c)	435(c)	0.075	0.003
1.25	0.050	17(f)	55(f)	0.102	0.004	69(d)	225(d)	81(d)	265(d)	0.13	0.005	105(e)	350(e)	0.13	0.005
2.5	0.100	14(f)	45(f)	0.20	0.008	55(d)	180(d)	64(d)	210(d)	0.30	0.012	85(e)	280(e)	0.30	0.012
Hardness, 375 to 425 HB															
0.25	0.010	18(f)	60(f)	0.050	0.002	67(b)	220(b)	79(b)	260(b)	0.075	0.003	105(c)	350(c)	0.075	0.003
1.25	0.050	15(f)	50(f)	0.102	0.004	55(d)	180(d)	64(d)	210(d)	0.13	0.005	85(e)	280(e)	0.13	0.005
2.5	0.100	11(f)	35(f)	0.20	0.008	44(d)	145(d)	52(d)	170(d)	0.30	0.012	64(e)	210(e)	0.30	0.012

(a) High speed steel tool material: M2 or M3. (b) Carbide tool material: C-7. (c) Carbide tool material: CC-7. (d) Carbide tool material: C-6. (e) Carbide tool material: CC-6. (f) Any premium high speed steel tool material (T15, M33, M41 to M47)

AISI 1340, 1340H: Planing

Tool material	Depth of cut mm	in.	Speed m/min	ft/min	Feed/stroke mm	in.
Hardness, 175 to 225 HB						
M2 and M3 high speed steel	0.1	0.005	11	35	(a)	(a)
	2.5	0.100	15	50	1.25	0.050
	12	0.500	9	30	1.50	0.060
C-6 carbide	0.1	0.005	67	220	(a)	(a)
	2.5	0.100	73	240	2.05	0.080
	12	0.500	60	200	1.50	0.060
Hardness, 225 to 275 HB						
M2 and M3 high speed steel	0.1	0.005	9	30	(a)	(a)
	2.5	0.100	14	45	0.75	0.030
	12	0.500	8	25	1.15	0.045
C-6 carbide	0.1	0.005	52	170	(a)	(a)
	2.5	0.100	60	200	1.50	0.060
	12	0.500	49	160	1.25	0.050
Hardness, 275 to 325 HB						
M2 and M3 high speed steel	0.1	0.005	8	25	(a)	(a)
	2.5	0.100	12	40	0.75	0.030
	12	0.500	6	20	1.15	0.045
C-6 carbide	0.1	0.005	46	150	(a)	(a)
	2.5	0.100	55	180	1.50	0.060
	12	0.500	43	140	1.25	0.050
Hardness, 325 to 375 HB						
M2 and M3 high speed steel	0.1	0.005	6	20	(a)	(a)
	2.5	0.100	11	35	0.75	0.030
	12	0.500	5	15	1.15	0.045
C-6 carbide	0.1	0.005	38	125	(a)	(a)
	2.5	0.100	52	170	1.50	0.060
	12	0.500	38	125	1.25	0.050

(a) Feed is 75% the width of the square nose finishing tool

AISI 1345, 1345H

AISI 1345, 1345H: Chemical Composition

AISI grade	C	Mn	P max	S max	Si
1345	0.43-0.48	1.60-1.90	0.035	0.040	0.15-0.30
1345H	0.42-0.49	1.45-2.05	0.035	0.040	0.15-0.30

Characteristics and Typical Uses. AISI grades 1345 and 1345H have characteristics and typical uses similar to AISI grades 1340 and 1340H. However, the increased carbon content does provide slightly higher maximum hardenability.

AISI 1345: Similar Steels (U.S. and/or Foreign). UNS G13450; ASTM A322; A331, A519; SAE J404, J412, J770; (W. Ger.) DIN 1.0912; (U.K.) B.S. 2 S 516, 2 S 517

AISI 1345H: Similar Steels (U.S. and/or Foreign). UNS H13450; ASTM A304; SAE J1268; (W. Ger.) DIN 1.0912; (U.K.) B.S. 2 S 516, 2 S 517

Mechanical Properties

AISI 1345, 1345H: Hardness and Machinability (Ref 7)

Condition	Hardness range, HB	Average machinability rating(a)
Annealed and cold drawn	183-241	45

(a) Based on AISI 1212 steel as 100% average machinability

AISI 1345, 1345H: Approximate Critical Points

Transformation point	Temperature(a) °C	°F	Transformation point	Temperature(a) °C	°F
Ac_1	715	1315	Ar_3	705	1300
Ac_3	765	1410	Ar_1	625	1160

(a) On heating or cooling at 28 °C (50 °F) per hour

Physical Properties

AISI 1345, 1345H: Thermal Treatment Temperatures

Quenching medium: oil

Treatment	Temperature range °C	°F
Normalizing	845-900	1550-1650
Annealing	870-925	1600-1700
Hardening	815-845	1500-1550

Machining Data

For machining data on AISI grades 1345 and 1345H, refer to the preceding machining tables for AISI grades 1340 and 1340H.

Mechanical Properties (continued)

AISI 1345H: End-Quench Hardenability Limits and Hardenability Band. Heat treating temperatures recommended by SAE: normalize at 870 °C (1600 °F), for forged or rolled specimens only; austenitize at 845 °C (1550 °F). (Ref 1)

Distance from quenched end, 1/16 in.	Hardness, HRC max	Hardness, HRC min
1	63	56
2	63	56
3	62	55
4	61	54
5	61	51
6	60	44
7	60	38
8	59	35
9	58	33
10	57	32
11	56	31
12	55	30
13	54	29
14	53	29
15	52	28
16	51	28
18	49	27
20	47	27
22	45	26
24	44	26
26	47	25
28	46	25
30	45	24
32	45	24

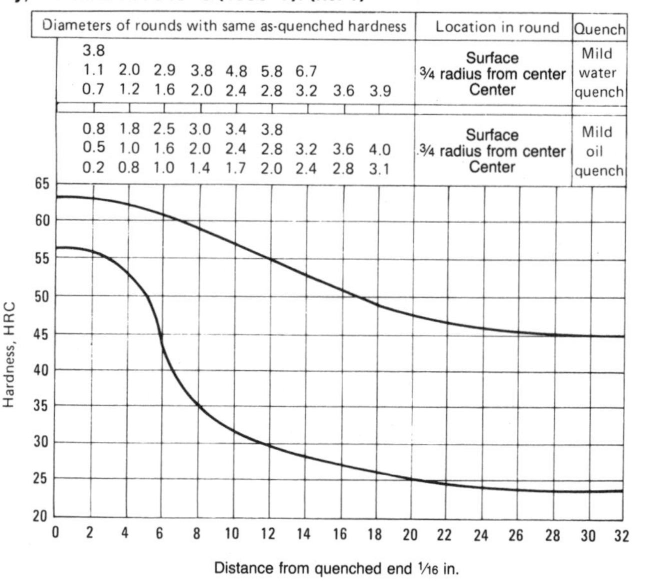

AISI A2317

AISI A2317: Chemical Composition

AISI grade	C	Mn	P max	S max	Si	Ni
A2317	0.15-0.20	0.40-0.60	0.040	0.040	0.20-0.35	3.25-3.75

(Chemical composition, %)

Characteristics and Typical Uses. AISI grade A2317 is a high-nickel, carburizing steel used for relatively small diameter (130- to 200-mm or 5- to 8-in.) machinery gears. Because of the nickel content, applications of AISI A2317 are limited.

AISI A2317: Similar Steels (U.S. and/or Foreign). SAE 2317

AISI A2317: Approximate Critical Points

Transformation point	Temperature(a) °C	Temperature(a) °F	Transformation point	Temperature(a) °C	Temperature(a) °F
Ac_1	700	1295	Ar_1	600	1110
Ac_3	775	1425	M_s	385	725
Ar_3	725	1340			

(a) On heating or cooling at 28 °C (50 °F) per hour

Physical Properties

AISI A2317: Mean Apparent Specific Heat (Ref 9)

Normalized at 925 °C (1700 °F), tempered at 595 °C (1100 °F)

Temperature range °C	Temperature range °F	Specific heat J/kg·K	Specific heat Btu/lb·°F
−100-27	−150-80	364	0.087
27-540	80-1000	595	0.142

AISI A2317: Thermal Treatment Temperatures

Treatment	Temperature range °C	Temperature range °F
Forging	1260 max	2300 max
Normalizing	870-925	1600-1700
Carburizing	925	1700

AISI A2317: Average Coefficients of Linear Thermal Expansion (Ref 9)

Austenitized at 870 °C (1600 °F), tempered at 595 °C (1100 °F) for 2 h

Temperature range °C	°F	Coefficient µm/m·K	µin./in.·°F
−195-21	−320-70	8.6	4.8
−130-21	−200-70	10.1	5.6
−73-21	−100-70	10.4	5.8
−73-93	−100-200	11.0	6.1
21-93	70-200	11.9	6.6
21-150	70-300	12.1	6.7
21-205	70-400	12.2	6.8
21-260	70-500	12.6	7.0
21-315	70-600	12.8	7.1
21-370	70-700	13.0	7.2
21-425	70-800	13.3	7.4
21-480	70-900	13.3	7.4
21-540	70-1000	13.5	7.5

AISI A2317: Thermal Conductivity (Ref 9)

Austenitized at 900 °C (1650 °F), tempered at 595 °C (1100 °F) for 1 h

Temperature °C	°F	Thermal conductivity W/m·K	Btu/ft·h·°F
−130	−200	28.9	16.7
−73	−100	33.7	19.5
−18	0	35.3	20.4
38	100	37.4	21.6
93	200	38.9	22.5
150	300	40.3	23.3
205	400	41.2	23.8
260	500	41.2	23.8
315	600	40.3	23.3
370	700	39.6	22.9
425	800	38.2	22.1
480	900	36.0	20.8
540	1000	34.6	20.0

Mechanical Properties

AISI A2317: Approximate Core Mechanical Properties (Ref 2)

Heat treatment of test specimens: (1) normalized at 925 °C (1700 °F) in 32-mm (1.25-in.) rounds; (2) machined in 25- or 13.7-mm (1- or 0.540-in.) rounds; (3) pseudocarburized at 925 °C (1700 °F) for 8 h; (4) box cooled to room temperature; (5) reheated to temperatures given in table and oil quenched; (6) tempered at 150 °C (300 °F); (7) tested in 12.8-mm (0.505-in.) rounds; tests were conducted using test specimens machined to English units

Reheat temperature °C	°F	Tensile strength MPa	ksi	Yield strength(a) MPa	ksi	Elongation(b), %	Reduction in area, %	Hardness, HB
Heat treated in 25-mm (1-in.) rounds								
775	1425	896	130	689	100	18	55	277
800	1475	951	138	752	109	18	52	285
830	1525	965	140	772	112	20	54	293
(c)	(c)	1000	145	793	115	15	50	302
Heat treated in 13.7-mm (0.540-in.) rounds								
775	1425	910	132	710	103	15	50	277
800	1475	958	139	758	110	15	50	293
830	1525	979	142	793	115	16	55	302
(c)	(c)	1041	151	862	125	14	44	321

(a) 0.2% offset. (b) In 50 mm (2 in.). (c) Quenched from step 3

Machining Data (Ref 3)

AISI A2317: Planing

Tool material	Depth of cut mm	in.	Speed m/min	ft/min	Feed/stroke mm	in.
Hardness, 175 to 225 HB						
M2 and M3 high speed steel	0.1	0.005	11	35	(a)	(a)
	2.5	0.100	17	55	1.25	0.050
	12	0.500	9	30	1.50	0.060
C-6 carbide	0.1	0.005	70	230	(a)	(a)
	2.5	0.100	76	250	2.05	0.080
	12	0.500	69	225	1.50	0.060
Hardness, 225 to 275 HB						
M2 and M3 high speed steel	0.1	0.005	9	30	(a)	(a)
	2.5	0.100	15	50	0.75	0.030
	12	0.500	8	25	1.15	0.045
C-6 carbide	0.1	0.005	53	175	(a)	(a)
	2.5	0.100	64	210	1.50	0.060
	12	0.500	53	175	1.25	0.050
Hardness, 275 to 325 HB						
M2 and M3 high speed steel	0.1	0.005	8	25	(a)	(a)
	2.5	0.100	14	45	0.75	0.030
	12	0.500	8	25	1.15	0.045
C-6 carbide	0.1	0.005	49	160	(a)	(a)
	2.5	0.100	60	200	1.50	0.060
	12	0.500	46	150	1.25	0.050
Hardness, 325 to 375 HB						
M2 and M3 high speed steel	0.1	0.005	6	20	(a)	(a)
	2.5	0.100	11	35	0.75	0.030
	12	0.500	6	20	1.15	0.045
C-6 carbide	0.1	0.005	38	125	(a)	(a)
	2.5	0.100	53	175	1.50	0.060
	12	0.500	38	125	1.25	0.050

(a) Feed is 75% the width of the square nose finishing tool

Machining Data (Ref 3) (continued)

AISI A2317: Turning (Single Point and Box Tools)

Depth of cut		High speed steel				Uncoated carbide						Coated carbide			
		Speed		Feed		Speed, brazed		Speed, inserted		Feed		Speed		Feed	
mm	in.	m/min	ft/min	mm/rev	in./rev	m/min	ft/min	m/min	ft/min	mm/rev	in./rev	m/min	ft/min	mm/rev	in./rev
Hardness, 175 to 225 HB															
1	0.040	41(a)	135(a)	0.18	0.007	135(b)	450(b)	170(b)	560(b)	0.18	0.007	220(c)	725(c)	0.18	0.007
4	0.150	32(a)	105(a)	0.40	0.015	105(d)	350(d)	135(d)	440(d)	0.50	0.020	175(e)	575(e)	0.40	0.015
8	0.300	24(a)	80(a)	0.50	0.020	84(d)	275(d)	105(d)	350(d)	0.75	0.030	135(e)	450(e)	0.50	0.020
16	0.625	20(a)	65(a)	0.75	0.030	66(d)	215(d)	84(d)	275(d)	1.00	0.040	...	...	...	...
Hardness, 225 to 275 HB															
1	0.040	34(a)	110(a)	0.18	0.007	120(b)	400(b)	150(b)	500(b)	0.18	0.007	200(c)	650(c)	0.18	0.007
4	0.150	26(a)	85(a)	0.40	0.015	95(d)	310(d)	120(d)	390(d)	0.50	0.020	150(e)	500(e)	0.40	0.015
8	0.300	20(a)	65(a)	0.50	0.020	73(d)	240(d)	95(d)	310(d)	0.75	0.030	120(e)	400(e)	0.50	0.020
16	0.625	15(a)	50(a)	0.75	0.030	58(d)	190(d)	73(d)	240(d)	1.00	0.040	...	...	...	...
Hardness, 275 to 325 HB															
1	0.040	29(f)	95(f)	0.18	0.007	150(b)	350(b)	135(b)	450(b)	0.18	0.007	175(c)	575(c)	0.18	0.007
4	0.150	23(f)	75(f)	0.40	0.015	84(d)	275(d)	105(d)	350(d)	0.40	0.015	135(e)	450(e)	0.40	0.015
8	0.300	18(f)	60(f)	0.50	0.020	67(d)	220(d)	84(d)	275(d)	0.50	0.020	105(e)	350(e)	0.50	0.020
Hardness, 325 to 375 HB															
1	0.040	24(f)	80(f)	0.13	0.005	88(b)	290(b)	115(b)	375(b)	0.18	0.007	145(c)	475(c)	0.18	0.007
4	0.150	18(f)	60(f)	0.25	0.010	69(d)	225(d)	88(d)	290(d)	0.40	0.015	115(e)	375(e)	0.40	0.015
8	0.300	14(f)	45(f)	0.40	0.015	55(d)	180(d)	70(d)	230(d)	0.50	0.020	90(e)	300(e)	0.50	0.020
Hardness, 375 to 425 HB															
1	0.040	20(f)	65(f)	0.13	0.005	73(b)	240(b)	88(b)	290(b)	0.18	0.007	115(c)	375(c)	0.18	0.007
4	0.150	15(f)	50(f)	0.25	0.010	58(d)	190(d)	70(d)	230(d)	0.40	0.015	90(e)	300(e)	0.40	0.015
8	0.300	12(f)	40(f)	0.40	0.015	46(d)	150(d)	55(d)	180(d)	0.50	0.020	69(e)	225(e)	0.50	0.020

(a) High speed steel tool material: M2 or M3. (b) Carbide tool material: C-7. (c) Carbide tool material: CC-7. (d) Carbide tool material: C-6. (e) Carbide tool material: CC-6. (f) Any premium high speed steel tool material (T15, M33, M41 to M47)

AISI A2317: Turning (Cutoff and Form Tools)

Tool material	Speed, m/min (ft/min)	Feed per revolution for cutoff tool width of: 1.5 mm (0.062 in.)		3 mm (0.125 in.)		6 mm (0.25 in.)		Feed per revolution for form tool width of: 12 mm (0.5 in.)		18 mm (0.75 in.)		25 mm (1 in.)		50 mm (2 in.)	
		mm	in.	mm	in.	mm	in.	mm	in.	mm	in.	mm	in.	mm	in.
Hardness, 175 to 225 HB															
M2 and M3 high speed steel	30 (100)	0.038	0.0015	0.046	0.0018	0.053	0.0021	0.046	0.0018	0.041	0.0016	0.038	0.0015	0.028	0.0011
C-6 carbide	100 (325)	0.038	0.0015	0.046	0.0018	0.053	0.0021	0.046	0.0018	0.041	0.0016	0.038	0.0015	0.028	0.0011
Hardness, 225 to 275 HB															
M2 and M3 high speed steel	24 (80)	0.036	0.0014	0.043	0.0017	0.050	0.0020	0.043	0.0017	0.038	0.0015	0.036	0.0014	0.025	0.0010
C-6 carbide	78 (255)	0.036	0.0014	0.043	0.0017	0.050	0.0020	0.043	0.0017	0.038	0.0015	0.036	0.0014	0.025	0.0010
Hardness, 275 to 325 HB															
M2 and M3 high speed steel	18 (60)	0.033	0.0013	0.041	0.0016	0.048	0.0019	0.041	0.0016	0.036	0.0014	0.033	0.0013	0.023	0.0009
C-6 carbide	58 (190)	0.033	0.0013	0.041	0.0016	0.048	0.0019	0.041	0.0016	0.036	0.0014	0.033	0.0013	0.023	0.0009
Hardness, 325 to 375 HB															
Any premium high speed steel (T15, M33, M41-M47)	15 (50)	0.030	0.0012	0.038	0.0015	0.046	0.0018	0.038	0.0015	0.033	0.0013	0.030	0.0012	0.020	0.0008
C-6 carbide	46 (150)	0.030	0.0012	0.038	0.0015	0.046	0.0018	0.038	0.0015	0.033	0.0013	0.030	0.0012	0.020	0.0008
Hardness, 375 to 425 HB															
Any premium high speed steel (T15, M33, M41-M47)	12 (40)	0.028	0.0011	0.036	0.0014	0.043	0.0017	0.036	0.0014	0.030	0.0012	0.028	0.0011	0.018	0.0007
C-6 carbide	38 (125)	0.028	0.0011	0.036	0.0014	0.043	0.0017	0.036	0.0014	0.030	0.0012	0.028	0.0011	0.018	0.0007

Machining Data (Ref 3) (continued)

AISI A2317: Drilling

Tool material	Speed m/min	Speed ft/min	Feed per revolution for nominal hole diameter of: 1.5 mm (0.062 in.) mm	1.5 mm (0.062 in.) in.	3 mm (0.125 in.) mm	3 mm (0.125 in.) in.	6 mm (0.25 in.) mm	6 mm (0.25 in.) in.	12 mm (0.5 in.) mm	12 mm (0.5 in.) in.	18 mm (0.75 in.) mm	18 mm (0.75 in.) in.	25 mm (1 in.) mm	25 mm (1 in.) in.
Hardness, 175 to 225 HB														
M10, M7, and M1 high speed steel	21	70	0.025	0.001	0.075	0.003	0.15	0.006	0.25	0.010	0.33	0.013	0.40	0.016
Hardness, 225 to 275 HB														
M10, M7, and M1 high speed steel	18	60	0.025	0.001	0.075	0.003	0.102	0.004	0.18	0.007	0.25	0.010	0.30	0.012
Hardness, 275 to 325 HB														
M10, M7, and M1 high speed steel	15	50	...	...	0.050	0.002	0.102	0.004	0.18	0.007	0.20	0.008	0.25	0.010
Hardness, 325 to 375 HB														
M10, M7, and M1 high speed steel	12	40	...	...	0.050	0.002	0.075	0.003	0.13	0.005	0.20	0.008	0.23	0.009
Hardness, 375 to 425 HB														
Any premium high speed steel (T15, M33, M41-M47)	9	30	...	...	0.050	0.002	0.075	0.003	0.13	0.005	0.15	0.006	0.20	0.008

AISI A2317: End Milling (Profiling)

Tool material	Depth of cut mm	Depth of cut in.	Speed m/min	Speed ft/min	Feed per tooth for cutter diameter of: 10 mm (0.375 in.) mm	10 mm (0.375 in.) in.	12 mm (0.5 in.) mm	12 mm (0.5 in.) in.	18 mm (0.75 in.) mm	18 mm (0.75 in.) in.	25-50 mm (1-2 in.) mm	25-50 mm (1-2 in.) in.
Hardness, 175 to 225 HB												
M2, M3, and M7 high speed steel	0.5	0.020	37	120	0.025	0.001	0.050	0.002	0.075	0.003	0.102	0.004
	1.5	0.060	27	90	0.050	0.002	0.075	0.003	0.102	0.004	0.13	0.005
	diam/4	diam/4	24	80	0.038	0.0015	0.050	0.002	0.075	0.003	0.102	0.004
	diam/2	diam/2	21	70	0.025	0.001	0.038	0.0015	0.050	0.002	0.075	0.003
C-5 carbide	0.5	0.020	145	475	0.038	0.0015	0.075	0.003	0.13	0.005	0.15	0.006
	1.5	0.060	115	370	0.063	0.0025	0.102	0.004	0.15	0.006	0.18	0.007
	diam/4	diam/4	100	320	0.050	0.002	0.075	0.003	0.13	0.005	0.15	0.006
	diam/2	diam/2	90	300	0.038	0.0015	0.050	0.002	0.102	0.004	0.13	0.005
Hardness, 225 to 275 HB												
M2, M3, and M7 high speed steel	0.5	0.020	34	110	0.025	0.001	0.050	0.002	0.075	0.003	0.102	0.004
	1.5	0.060	26	85	0.050	0.002	0.075	0.003	0.102	0.004	0.13	0.005
	diam/4	diam/4	23	75	0.038	0.0015	0.050	0.002	0.075	0.003	0.102	0.004
	diam/2	diam/2	20	65	0.025	0.001	0.038	0.0015	0.050	0.002	0.075	0.003
C-5 carbide	0.5	0.020	130	420	0.025	0.001	0.050	0.002	0.102	0.004	0.13	0.005
	1.5	0.060	100	325	0.050	0.002	0.075	0.003	0.13	0.005	0.18	0.007
	diam/4	diam/4	85	280	0.038	0.0015	0.050	0.002	0.102	0.004	0.13	0.005
	diam/2	diam/2	79	260	0.025	0.001	0.038	0.0015	0.075	0.003	0.102	0.004
Hardness, 275 to 325 HB												
M2, M3, and M7 high speed steel	0.5	0.020	26	85	0.018	0.0007	0.038	0.0015	0.075	0.003	0.102	0.004
	1.5	0.060	20	65	0.025	0.001	0.050	0.002	0.102	0.004	0.13	0.005
	diam/4	diam/4	17	55	0.018	0.0007	0.038	0.0015	0.075	0.003	0.102	0.004
	diam/2	diam/2	15	50	0.013	0.0005	0.025	0.001	0.050	0.002	0.075	0.003
C-5 carbide	0.5	0.020	100	325	0.025	0.001	0.050	0.002	0.102	0.004	0.13	0.005
	1.5	0.060	75	245	0.050	0.002	0.075	0.003	0.13	0.005	0.15	0.006
	diam/4	diam/4	64	210	0.038	0.0015	0.050	0.002	0.102	0.004	0.13	0.005
	diam/2	diam/2	60	200	0.025	0.001	0.038	0.0015	0.075	0.003	0.102	0.004
Hardness, 325 to 375 HB												
M2, M3, and M7 high speed steel	0.5	0.020	20	65	0.013	0.0005	0.038	0.0015	0.075	0.003	0.102	0.004
	1.5	0.060	17	55	0.013	0.0005	0.038	0.0015	0.102	0.004	0.13	0.005
	diam/4	diam/4	15	50	0.013	0.0005	0.038	0.0015	0.075	0.003	0.102	0.004
	diam/2	diam/2	12	40	0.013	0.0005	0.025	0.001	0.050	0.002	0.075	0.003
C-5 carbide	0.5	0.020	79	260	0.025	0.001	0.038	0.0015	0.075	0.003	0.13	0.005
	1.5	0.060	60	200	0.038	0.0015	0.075	0.003	0.13	0.005	0.15	0.006
	diam/4	diam/4	52	170	0.038	0.0015	0.050	0.002	0.102	0.004	0.13	0.005
	diam/2	diam/2	49	160	0.025	0.001	0.025	0.001	0.075	0.003	0.102	0.004

(continued)

Machining Data (Ref 3) (continued)

AISI A2317: End Milling (Profiling) (continued)

Tool material	Depth of cut mm	Depth of cut in.	Speed m/min	Speed ft/min	Feed per tooth for cutter diameter of: 10 mm (0.375 in.) mm	10 mm (0.375 in.) in.	12 mm (0.5 in.) mm	12 mm (0.5 in.) in.	18 mm (0.75 in.) mm	18 mm (0.75 in.) in.	25-50 mm (1-2 in.) mm	25-50 mm (1-2 in.) in.
Hardness, 375 to 425 HB												
Any premium high speed steel (T15, M33, M41-M47)	0.5	0.020	17	55	0.013	0.0005	0.018	0.0007	0.025	0.001	0.050	0.002
	1.5	0.060	14	45	0.013	0.0005	0.025	0.001	0.050	0.002	0.075	0.003
	diam/4	diam/4	12	40	0.013	0.0005	0.013	0.0005	0.050	0.002	0.063	0.0025
	diam/2	diam/2	11	35	...	...	...	...	0.038	0.0015	0.050	0.002
C-5 carbide	0.5	0.020	64	210	0.025	0.001	0.025	0.001	0.050	0.002	0.075	0.003
	1.5	0.060	49	160	0.038	0.0015	0.050	0.002	0.075	0.003	0.102	0.004
	diam/4	diam/4	41	135	0.038	0.0015	0.038	0.0015	0.063	0.0025	0.089	0.0035
	diam/2	diam/2	38	125	0.025	0.001	0.025	0.001	0.050	0.002	0.075	0.003

AISI A2317: Reaming

Based on 4 flutes for 3- and 6-mm (0.125- and 0.25-in.) reamers, 6 flutes for 12-mm (0.5-in.) reamers, and 8 flutes for 25-mm (1-in.) and larger reamers

Tool material	Speed m/min	Speed ft/min	Feed per revolution for reamer diameter of: 3 mm (0.125 in.) mm	3 mm (0.125 in.) in.	6 mm (0.25 in.) mm	6 mm (0.25 in.) in.	12 mm (0.5 in.) mm	12 mm (0.5 in.) in.	25 mm (1 in.) mm	25 mm (1 in.) in.	35 mm (1.5 in.) mm	35 mm (1.5 in.) in.	50 mm (2 in.) mm	50 mm (2 in.) in.
Hardness, 175 to 225 HB														
Roughing														
M1, M2, and M7 high speed steel	26	85	0.102	0.004	0.18	0.007	0.30	0.012	0.50	0.020	0.65	0.025	0.75	0.030
C-2 carbide	37	120	0.102	0.004	0.18	0.007	0.30	0.012	0.50	0.020	0.65	0.025	0.75	0.030
Finishing														
M1, M2, and M7 high speed steel	11	35	0.13	0.005	0.20	0.008	0.30	0.012	0.50	0.020	0.65	0.025	0.75	0.030
C-2 carbide	14	45	0.13	0.005	0.20	0.008	0.30	0.012	0.50	0.020	0.65	0.025	0.75	0.030
Hardness, 225 to 275 HB														
Roughing														
M1, M2, and M7 high speed steel	23	75	0.075	0.003	0.15	0.006	0.25	0.010	0.40	0.015	0.50	0.020	0.65	0.025
C-2 carbide	27	90	0.102	0.004	0.15	0.006	0.25	0.010	0.40	0.015	0.50	0.020	0.65	0.025
Finishing														
M1, M2, and M7 high speed steel	9	30	0.102	0.004	0.20	0.008	0.30	0.012	0.50	0.020	0.65	0.025	0.75	0.030
C-2 carbide	12	40	0.102	0.004	0.20	0.008	0.30	0.012	0.50	0.020	0.65	0.025	0.75	0.030
Hardness, 275 to 325 HB														
Roughing														
M1, M2, and M7 high speed steel	20	65	0.075	0.003	0.13	0.005	0.20	0.008	0.30	0.012	0.40	0.015	0.50	0.020
C-2 carbide	24	80	0.102	0.004	0.15	0.006	0.20	0.008	0.30	0.012	0.40	0.015	0.50	0.020
Finishing														
M1, M2, and M7 high speed steel	8	25	0.102	0.004	0.15	0.006	0.25	0.010	0.40	0.015	0.50	0.020	0.65	0.025
C-2 carbide	11	35	0.102	0.004	0.15	0.006	0.25	0.010	0.40	0.015	0.50	0.020	0.65	0.025
Hardness, 325 to 375 HB														
Roughing														
M1, M2, and M7 high speed steel	15	50	0.050	0.002	0.102	0.004	0.15	0.006	0.20	0.008	0.25	0.010	0.30	0.012
C-2 carbide	20	65	0.102	0.004	0.15	0.006	0.20	0.008	0.25	0.010	0.30	0.012	0.40	0.015
Finishing														
M1, M2, and M7 high speed steel	8	25	0.050	0.002	0.13	0.005	0.15	0.006	0.25	0.010	0.30	0.012	0.40	0.015
C-2 carbide	11	35	0.102	0.004	0.15	0.006	0.20	0.008	0.25	0.010	0.30	0.012	0.40	0.015
Hardness, 375 to 425 HB														
Roughing														
Any premium high speed steel (T15, M33, M41-M47)	9	30	0.050	0.002	0.102	0.004	0.13	0.005	0.15	0.006	0.18	0.007	0.20	0.008
C-2 carbide	14	45	0.102	0.004	0.15	0.006	0.20	0.008	0.25	0.010	0.28	0.011	0.30	0.012
Finishing														
Any premium high speed steel (T15, M33, M41-M47)	6	20	0.050	0.002	0.102	0.004	0.13	0.005	0.15	0.006	0.18	0.007	0.20	0.008
C-2 carbide	9	30	0.102	0.004	0.15	0.006	0.20	0.008	0.25	0.010	0.28	0.011	0.30	0.012

Machining Data (Ref 3) (continued)

AISI A2317: Boring

Depth of cut		High speed steel				Uncoated carbide						Coated carbide			
		Speed		Feed		Speed, brazed		Speed, inserted		Feed		Speed		Feed	
mm	in.	m/min	ft/min	mm/rev	in./rev	m/min	ft/min	m/min	ft/min	mm/rev	in./rev	m/min	ft/min	mm/rev	in./rev
Hardness, 175 to 225 HB															
0.25	0.010	43(a)	140(a)	0.075	0.003	125(b)	415(b)	150(b)	490(b)	0.075	0.003	195(c)	640(c)	0.075	0.003
1.25	0.050	34(a)	110(a)	0.13	0.005	100(d)	330(d)	120(d)	390(d)	0.13	0.005	155(e)	510(e)	0.13	0.005
2.5	0.100	26(a)	85(a)	0.30	0.012	81(d)	265(d)	95(d)	310(d)	0.40	0.015	120(e)	400(e)	0.30	0.012
Hardness, 225 to 275 HB															
0.25	0.010	34(a)	110(a)	0.075	0.003	115(b)	375(b)	135(b)	440(b)	0.075	0.003	175(c)	570(c)	0.075	0.003
1.25	0.050	27(a)	90(a)	0.13	0.005	90(d)	300(d)	105(d)	350(d)	0.13	0.005	140(e)	455(e)	0.13	0.005
2.5	0.100	21(a)	70(a)	0.30	0.012	72(d)	235(d)	84(d)	275(d)	0.40	0.015	105(e)	350(e)	0.30	0.012
Hardness, 275 to 325 HB															
0.25	0.010	29(f)	95(f)	0.075	0.003	100(b)	335(b)	120(b)	395(b)	0.075	0.003	150(c)	500(c)	0.075	0.003
1.25	0.050	23(f)	75(f)	0.13	0.005	81(d)	265(d)	95(d)	315(d)	0.13	0.005	120(e)	400(e)	0.13	0.005
2.5	0.100	18(f)	60(f)	0.30	0.012	64(d)	210(d)	75(d)	245(d)	0.30	0.012	95(e)	315(e)	0.30	0.012
Hardness, 325 to 375 HB															
0.25	0.010	24(f)	80(f)	0.050	0.002	84(b)	275(b)	100(b)	325(b)	0.075	0.003	125(c)	410(c)	0.075	0.003
1.25	0.050	20(f)	65(f)	0.102	0.004	67(d)	220(d)	79(d)	260(d)	0.13	0.005	100(e)	330(e)	0.13	0.005
2.5	0.100	15(f)	50(f)	0.20	0.008	53(d)	175(d)	62(d)	205(d)	0.30	0.012	79(e)	260(e)	0.30	0.012
Hardness, 375 to 425 HB															
0.25	0.010	18(f)	60(f)	0.050	0.002	64(b)	210(b)	76(b)	250(b)	0.075	0.003	100(c)	325(c)	0.075	0.003
1.25	0.050	15(f)	50(f)	0.102	0.004	52(d)	170(d)	60(d)	200(d)	0.13	0.005	79(e)	260(e)	0.13	0.005
2.5	0.100	12(f)	40(f)	0.20	0.008	41(d)	135(d)	49(d)	160(d)	0.30	0.012	64(e)	210(e)	0.30	0.012

(a) High speed steel tool material: M2 or M3. (b) Carbide tool material: C-7. (c) Carbide tool material: CC-7. (d) Carbide tool material: C-6. (e) Carbide tool material: CC-6. (f) Any premium high speed steel tool material (T15, M33, M41 to M47)

AISI A2317: Face Milling

Depth of cut		High speed steel				Uncoated carbide						Coated carbide			
		Speed		Feed/tooth		Speed, brazed		Speed, inserted		Feed/tooth		Speed		Feed/tooth	
mm	in.	m/min	ft/min	mm	in.	m/min	ft/min	m/min	ft/min	mm	in.	m/min	ft/min	mm	in.
Hardness, 175 to 225 HB															
1	0.040	55(a)	180(a)	0.20	0.008	160(b)	530(b)	175(b)	580(b)	0.20	0.008	265(c)	870(c)	0.20	0.008
4	0.150	43(a)	140(a)	0.30	0.012	125(b)	405(b)	135(b)	445(b)	0.30	0.012	205(c)	665(c)	0.30	0.012
8	0.300	34(a)	110(a)	0.40	0.016	85(d)	280(d)	105(d)	345(d)	0.40	0.016	155(e)	515(e)	0.40	0.016
Hardness, 225 to 275 HB															
1	0.040	46(a)	150(a)	0.15	0.006	140(b)	460(b)	155(b)	510(b)	0.18	0.007	240(c)	765(c)	0.18	0.007
4	0.150	34(a)	110(a)	0.25	0.010	105(b)	350(b)	120(b)	390(b)	0.25	0.010	180(c)	485(c)	0.25	0.010
8	0.300	26(a)	85(a)	0.36	0.014	76(d)	250(d)	95(d)	305(d)	0.36	0.014	140(e)	455(e)	0.36	0.014
Hardness, 275 to 325 HB															
1	0.040	32(f)	105(f)	0.15	0.006	120(b)	400(b)	150(b)	485(b)	0.15	0.006	190(c)	630(c)	0.13	0.005
4	0.150	26(f)	85(f)	0.23	0.009	100(b)	320(b)	115(b)	375(b)	0.20	0.008	150(c)	485(c)	0.18	0.007
8	0.300	20(f)	65(f)	0.30	0.012	73(d)	240(d)	90(d)	295(d)	0.25	0.010	115(e)	375(e)	0.23	0.009
Hardness, 325 to 375 HB															
1	0.040	26(f)	85(f)	0.13	0.005	95(b)	310(b)	100(b)	330(b)	0.13	0.005	150(c)	500(c)	0.102	0.004
4	0.150	20(f)	65(f)	0.20	0.008	75(b)	245(b)	82(b)	270(b)	0.18	0.007	105(c)	350(c)	0.15	0.006
8	0.300	15(f)	50(f)	0.25	0.010	52(d)	170(d)	64(d)	210(d)	0.23	0.009	84(e)	275(e)	0.20	0.008
Hardness, 375 to 425 HB															
1	0.040	20(f)	65(f)	0.102	0.004	79(b)	260(b)	87(b)	285(b)	0.102	0.004	130(c)	425(c)	0.075	0.003
4	0.150	15(f)	50(f)	0.15	0.006	59(b)	195(b)	66(b)	215(b)	0.15	0.006	84(c)	275(c)	0.13	0.005
8	0.300	12(f)	40(f)	0.20	0.008	41(d)	135(d)	50(d)	165(d)	0.20	0.008	60(e)	200(e)	0.18	0.007

(a) High speed steel tool material: M2 or M7. (b) Carbide tool material: C-6. (c) Carbide tool material: CC-6. (d) Carbide tool material: C-5. (e) Carbide tool material: CC-5. (f) Any premium high speed steel tool material (T15, M33, M41 to M47)

AISI A2515

AISI A2515: Chemical Composition

AISI grade	C	Mn	P max	S max	Si	Ni
	Chemical composition, %					
A2515	0.12-0.17	0.40-0.60	0.040	0.040	0.20-0.35	4.75-5.25

Characteristics and Typical Uses. AISI grade A2515 is a relatively high-nickel, carburizing steel used for machine parts which require a case of medium thickness. Because of the nickel content, applications are limited.

AISI A2515: Similar Steels (U.S. and/or Foreign). SAE 2515; (W. Ger.) DIN 1.5680; (Fr.) AFNOR Z 18 N 5

AISI A2515: Approximate Critical Points

Transformation point	Temperature(a) °C	Temperature(a) °F	Transformation point	Temperature(a) °C	Temperature(a) °F
Ac_1	675	1250	Ar_1	550	1020
Ac_3	760	1400	M_s	365	690
Ar_3	640	1180			

(a) On heating or cooling at 28 °C (50 °F) per hour

Machining Data

For machining data on AISI grade A2515, refer to the preceding machining tables for AISI grade A2317.

Physical Properties

AISI A2515: Average Coefficients of Linear Thermal Expansion (Ref 9)

Austenitized at 870 °C (1600 °F), tempered at 595 °C (1100 °F) for 2 h

Temperature range °C	Temperature range °F	Coefficient µm/m·K	Coefficient µin./in.·°F
−195-21	−320-70	8.8	4.9
−130-21	−200-70	10.4	5.8
−75-21	−100-70	10.4	5.8
−75-93	−100-200	11.2	6.2
21-93	70-200	11.9	6.6
21-150	70-300	12.1	6.7
21-205	70-400	12.2	6.8
21-260	70-500	12.4	6.9
21-315	70-600	12.8	7.1
21-370	70-700	13.0	7.2
21-425	70-800	13.1	7.3
21-480	70-900	13.1	7.3
21-540	70-1000	13.3	7.4

AISI A2515: Thermal Conductivity (Ref 9)

Austenitized at 870 °C (1600 °F), tempered at 595 °C (1100 °F) for 1 h

Temperature °C	Temperature °F	Thermal conductivity W/m·K	Thermal conductivity Btu/ft·h·°F
−130	−200	26.0	15.0
−73	−100	29.6	17.1
−18	0	33.9	19.6
38	100	34.6	20.0
93	200	36.0	20.8
150	300	37.5	21.7
205	400	37.7	21.8
260	500	38.2	22.1
315	600	37.9	21.9
370	700	37.5	21.7
425	800	36.8	21.3
480	900	36.0	20.8
540	1000	34.6	20.0

AISI A2515: Mean Apparent Specific Heat (Ref 9)

Normalized at 925 °C (1700 °F), tempered at 595 °C (1100 °F)

Temperature range °C	Temperature range °F	Specific heat J/kg·K	Specific heat Btu/lb·°F
−100-27	−150-80	335	0.080
27-540	80-1000	615	0.147

AISI A2515: Thermal Treatment Temperatures

Treatment	Temperature range °C	Temperature range °F
Forging	1205 max	2200 max
Normalizing	870-925	1600-1700
Carburizing	925	1700

Mechanical Properties

AISI A2515: Effect of Case Depth on Maximum Specific Compressive Stress Based on Static Loading (Ref 10)

Heat treatment: carburized at 900 °C (1650 °F); oil quenched at 775 °C (1425 °F); tempered 145 °C (290 °F) for 1 h

Test conditions: 50-mm (2-in.) test rolls with a 25-mm (1-in.) face, surfaces were ground and lapped

Failure patterns: 7.6-mm (0.30-in.) case failed at core/case interface where shear stresses extended below the case; 1-mm (0.040-in.) case gave best results; 1.4-mm (0.055-in.) case, possibly embrittled by the conditions of heat treating, shattered around the fatigue pits; tests were conducted using test specimens machined to English units

Case depth mm	Case depth in.	Hardness, HRC Core	Hardness, HRC Case	Maximum specific compressive stress: 10^6 cycles MPa	10^6 cycles ksi	10^7 cycles MPa	10^7 cycles ksi	10^8 cycles MPa	10^8 cycles ksi
0.76	0.030	38	60	2041	296	1862	270	1703	247
1.0	0.040	37	56	2496	362	2213	321	2034	295
1.4	0.055	37	56	1717	249	1558	226	1427	207

Mechanical Properties (continued)

AISI A2515: Approximate Core Mechanical Properties (Ref 2)

Heat treatment of test specimens: (1) normalized at 925 °C (1700 °F) in 32-mm (1.25-in.) rounds; (2) machined in 25- or 13.7-mm (1- or 0.540-in.) rounds; (3) pseudocarburized at 925 °C (1700 °F) for 8 h; (4) box cooled to room temperature; (5) reheated to temperatures given in table and oil-quenched; (6) tempered at 150 °C (300 °F); (7) tested in 12.8-mm (0.505-in.) rounds; tests were conducted using test specimens machined to English units

Reheat temperature		Tensile strength		Yield strength(a)		Elongation(b),	Reduction	Hardness,
°C	°F	MPa	ksi	MPa	ksi	%	in area, %	HB
Heat treated in 25-mm (1-in.) rounds								
775	1425	1089	158	924	134	15.5	50	321
800	1475	1096	159	924	134	16	52	331
830	1525	1096	159	931	135	16	52	331
(c)	(c)	1124	163	951	138	16	51	341
Heat treated in 13.7-mm (0.540-in.) rounds								
775	1425	1145	166	979	142	15.5	50	341
800	1475	1179	171	1007	146	16	52	352
830	1525	1186	172	1020	148	16	52	352
(c)	(c)	1200	174	1034	150	15.5	50	365

(a) 0.2% offset. (b) In 50 mm (2 in.). (c) Quenched from step 3

AISI A2515: Permissible Compressive Stress

Case depth of 1.1 to 1.5 mm (0.045 to 0.060 in.) and a minimum case hardness of 60 HRC

Data point	Compressive stress MPa	ksi
Mean	1795	260
Maximum	2135	310

AISI 3140

AISI 3140: Chemical Composition

AISI grade	C	Mn	P max	S max	Si	Ni	Cr
	Chemical composition, %						
3140	0.38-0.43	0.70-0.90	0.040	0.040	0.20-0.35	1.10-1.40	0.55-0.75

Characteristics and Typical Uses. AISI grade 3140 steel is available in hot rolled and cold finished bar for machining into components or applications in the as-finished condition. This grade is also used for forgings.

AISI 3140: Similar Steels (U.S. and/or Foreign). UNS G31400; ASTM A519, A711, A322, A331; MIL SPEC MIL-S-16974; SAE J778; (W. Ger.) DIN 1.5711; (U.K.) B.S. 640 M 40

AISI 3140: Approximate Critical Points

Transformation point	Temperature(a) °C	°F	Transformation point	Temperature(a) °C	°F
Ac_1	735	1355	Ar_1	660	1220
Ac_3	765	1410	M_s	310	590
Ar_3	720	1330			

(a) On heating or cooling at 28 °C (50 °F) per hour

Physical Properties

AISI 3140: Average Coefficients of Linear Thermal Expansion (Ref 6)

Temperature range °C	°F	Coefficient μm/m·K	μin./in.·°F
0-100	32-212	11.8	6.55
0-200	32-390	12.2	6.80
0-300	32-570	12.9	7.15
0-400	32-750	13.2	7.35
0-500	32-930	14.0	7.75

AISI 3140: Thermal Treatment Temperatures

Quenching medium: oil

Treatment	Temperature range °C	°F
Forging	1230 max	2250 max
Annealing	790-845	1450-1550
Normalizing	845-900	1550-1650
Hardening	830-855	1525-1575

Machining Data

For machining data on AISI grade 3140, refer to the preceding machining tables for AISI grades 1340 and 1340H.

Mechanical Properties

AISI 3140: Low-Temperature Impact Properties (Ref 17)

Treatment	Hardness, HRC	Charpy impact value at temperature of: −185 °C (−300 °F) MPa	ksi	−130 °C (−200 °F) MPa	ksi	−73 °C (−100 °C) MPa	ksi	−18 °C (0 °F) MPa	ksi	38 °C (100 °F) MPa	ksi	Transition temperature(a) °C	°F
Quenched from 845 °C (1550 °F); tempered at 425 °C (800 °F)	43	69	10	103	15	145	21	179	26	207	30	−120	−185
Quenched from 845 °C (1550 °F); tempered at 540 °C (1000 °F)	30	90	13	110	16	241	35	310	45	331	48	−130	−200
Quenched from 845 °C (1550 °F); tempered at 650 °C (1200 °F)	25	97	14	172	25	379	55	434	63	455	66	−130	−200

(a) 50% brittle

AISI 3140: Effect of the Mass After Tempering at 540 °C (1000 °F).

Normalized at 870 °C (1600 °F) in oversize rounds; oil quenched from 845 °C (1550 °F) in sizes shown; tempered at 540 °C (1000 °F); tested in 12.8-mm (0.505-in.) rounds. Tests from 38.1-mm (1.50-in.) diam bar and over were taken at half-radius position. Tests were conducted using specimens machined to English units. Elongation was measured in 50 mm (2 in.). (Ref 2)

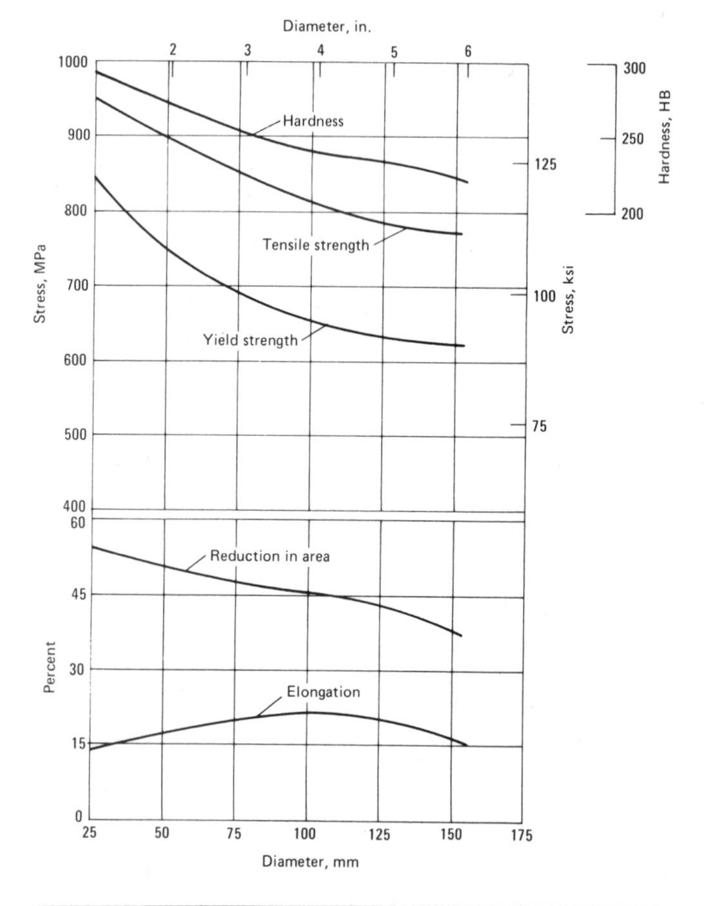

AISI 3140: Effect of the Mass After Tempering at 650 °C (1200 °F).

Normalized at 870 °C (1600 °F) in oversize rounds; oil quenched from 845 °C (1550 °F) in sizes shown; tempered at 650 °C (1200 °F); tested in 12.8-mm (0.505-in.) rounds. Tests from 38.1-mm (1.50-in.) diam bar and over were taken at half-radius position. Tests were conducted using specimens machined to English units. Elongation was measured in 50 mm (2 in.). (Ref 2)

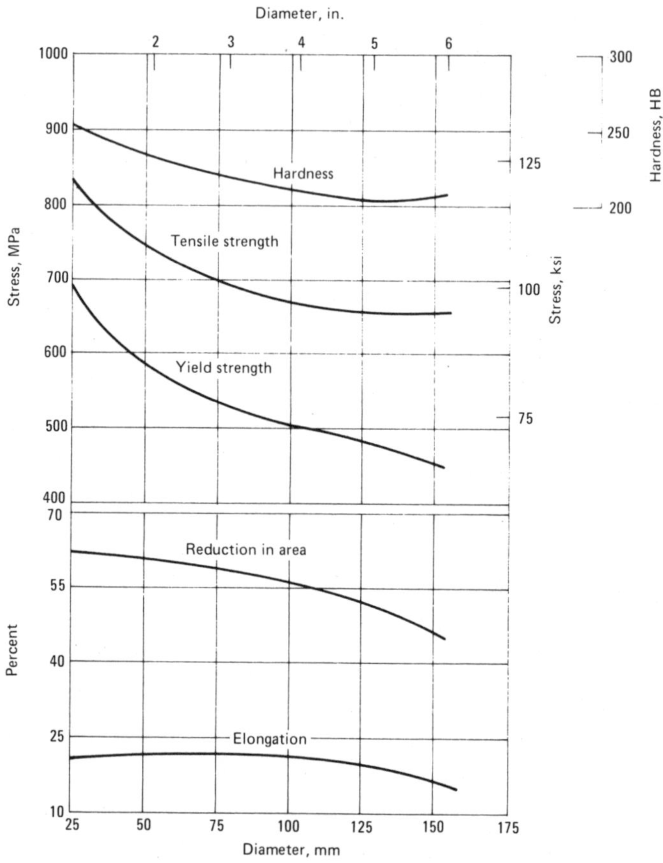

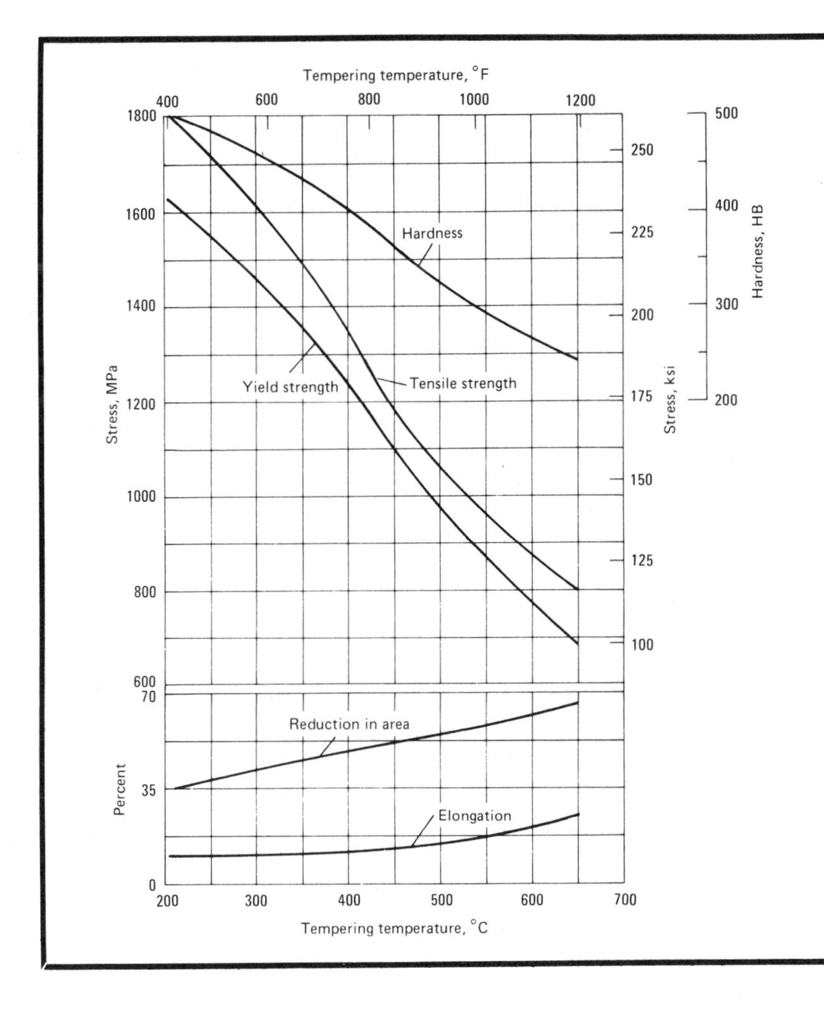

AISI 3140: Effect of Tempering Temperature on Tensile Properties. Normalized at 870 °C (1600 °F); oil quenched from 845 °C (1550 °F); tempered at 56 °C (100 °F) intervals in 13.7-mm (0.540-in.) rounds; tested in 12.8-mm (0.505-in.) rounds. Tests were conducted using test specimens machined to English units. Elongation was measured in 50 mm (2 in.). (Ref 2)

AISI 3140: Tensile Properties (Ref 5)

Condition or treatment	Tensile strength MPa	Tensile strength ksi	Yield strength MPa	Yield strength ksi	Elongation(a), %	Reduction in area, %	Hardness, HB	Izod impact energy J	Izod impact energy ft·lb
Normalized at 870 °C (1600 °F)	889	129	600	87	19.7	57	262	54	40
Annealed at 815 °C (1500 °F)	690	100	420	61	24.5	51	197	46	34

(a) In 50 mm (2 in.)

AISI 4023

AISI 4023: Chemical Composition

AISI grade	C	Mn	P max	S max	Si	Mo
	Chemical composition, %					
4023	0.20-0.25	0.70-0.90	0.035	0.040	0.15-0.30	0.20-0.30

Characteristics and Typical Uses. AISI 4023 is a carburizing grade, alloy steel with low case and core hardenability. This grade is available in bar and rod as hot rolled and cold finished steels. Bearing quality steel billets, another application of grade 4023, are used for rolling or forging, and for tubes, bar, rod, and wire which are employed in manufacturing rolling-element, anti-friction bearings.

AISI 4023: Similar Steels (U.S. and/or Foreign). UNS G40230; ASTM A322, A331, A519, A534; SAE J404, J412, J770

AISI 4023: Approximate Critical Points

Transformation point	Temperature(a) °C	Temperature(a) °F
Ac_1	730	1350
Ac_3	840	1540
Ar_3	780	1440
Ar_1	670	1240
M_s	405	775

(a) On heating or cooling at 28 °C (50 °F) per hour

Physical Properties

AISI 4023: Thermal Treatment Temperatures

Treatment	Temperature range °C	°F
Forging	1230 max	2250 max
Normalizing	900-955	1650-1750
Carburizing	900-925	1650-1700
Tempering	120-175	250-350

Machining Data

For machining data on AISI grade 4023, refer to the preceding machining tables for AISI grade A2317.

Mechanical Properties

AISI 4023: Approximate Core Mechanical Properties (Ref 2)

Heat treatment of test specimens: (1) normalized at 925 °C (1700 °F) in 32-mm (1.25-in.) rounds; (2) machined to 25-mm (1-in.) or 13.7-mm (0.540-in.) rounds; (3) pseudocarburized at 925 °C (1700 °F) for 8 h; (4) box cooled to room temperature; (5) reheated to temperatures given in the table and oil quenched; (6) tempered at 150 °C (300 °F); (7) tested in 12.8-mm (0.505-in.) rounds; tests were conducted using test specimens machined to English units

Reheat temperature °C	°F	Tensile strength MPa	ksi	Yield strength(a) MPa	ksi	Elongation(b), %	Reduction in area, %	Hardness, HB
Heat treated in 25-mm (1-in.) rounds								
775	1425	724	105	415	60	19	48	223
800	1475	745	108	440	64	21	54	229
855	1575	786	114	495	72	22	55	248
(c)	(c)	827	120	585	85	20	53	255
Heat treated in 13.7-mm (0.540-in.) rounds								
775	1425	931	135	620	90	14	40	285
800	1475	965	140	655	95	15	47	293
855	1575	986	143	724	105	16	49	321
(c)	(c)	1050	152	786	114	15	43	331

(a) 0.2% offset. (b) In 50 mm (2 in.). (c) Quenched from step 3

AISI 4023: Hardness and Machinability (Ref 7)

Condition	Hardness range, HB	Average machinability rating(a)
Hot rolled and cold drawn	156-207	70

(a) Based on AISI 1212 steel as 100% average machinability

AISI 4024

AISI 4024: Chemical Composition

AISI grade	C	Mn	P max	S	Si	Mo
	Chemical composition, %					
4024	0.20-0.25	0.70-0.90	0.035	0.035-0.050	0.15-0.30	0.20-0.30

AISI 4024: Similar Steels (U.S. and/or Foreign). UNS G40240; ASTM A322, A331, A519; SAE J404, J412, J770

Characteristics and Typical Uses. AISI grade 4024 is a low-hardenability, carburizing steel which is used for machined parts. It is available in hot rolled and cold finished bar, and seamless tubing. This steel is resulfurized to give better machinability at relatively high hardness.

Mechanical Properties

AISI 4024: Hardness and Machinability (Ref 7)

Condition	Hardness range, HB	Average machinability rating(a)
Hot rolled and cold drawn	156-207	75

(a) Based on AISI 1212 steel as 100% average machinability

Physical Properties

AISI 4024: Thermal Treatment Temperatures

Quenching medium: oil

Treatment	Temperature range °C	°F
Normalizing	(a)	(a)
Carburizing	900-925	1650-1700

(a) At least as high as the carburizing temperature, followed by air cooling

Machinability Data

For machining data on AISI grade 4024, refer to the preceding machining tables for AISI grade A2317.

AISI 4027, 4027H, 4028, 4028H

AISI 4027, 4027H, 4028, 4028H: Chemical Composition

AISI grade	C	Mn	P max	S	Si	Mo
			Chemical composition, %			
4027	0.25-0.30	0.70-0.90	0.035	0.040 max	0.15-0.30	0.20-0.30
4027H	0.24-0.30	0.60-1.00	0.035	0.040 max	0.15-0.30	0.20-0.30
4028	0.25-0.30	0.70-0.90	0.035	0.035-0.050	0.15-0.30	0.20-0.30
4028H	0.24-0.30	0.60-1.00	0.035	0.035-0.050	0.15-0.30	0.20-0.30

Characteristics and Typical Uses. AISI grades 4027, 4027H, 4028, and 4028H are carburizing steels with low to medium hardenability in the core. These steels are available as hot rolled and cold finished bar for machining parts or use in the as-finished condition. Grade 4028 is resulfurized to give improved machinability at relatively high hardness.

AISI 4027: Similar Steels (U.S. and/or Foreign). UNS G40270; ASTM A322, A331, A519; SAE J404, J412, J770

AISI 4027H: Similar Steels (U.S. and/or Foreign). UNS H40270; ASTM A304; SAE J1268

AISI 4028: Similar Steels (U.S. and/or Foreign). UNS G40280; ASTM A322, A331, A519; SAE J404, J412, J770

AISI 4028H: Similar Steels (U.S. and/or Foreign). UNS H40280; ASTM A304; SAE J1268

AISI 4027, 4027H, 4028, 4028H: Approximate Critical Points

Transformation point	Temperature(a) °C	°F	Transformation point	Temperature(a) °C	°F
Ac_1	725	1340	Ar_1	670	1240
Ac_3	805	1485	M_s	400	755
Ar_3	760	1400			

(a) On heating or cooling at 28 °C (50 °F) per hour

Physical Properties

AISI 4027, 4027H, 4028, 4028H: Thermal Treatment Temperatures

Quenching medium: oil

Treatment	Temperature range °C	°F
Normalizing	905	1660
Carburizing	900-925	1650-1700

Machining Data

For machining data on AISI grades 4027, 4027H, 4028, and 4028H, refer to the preceding machining tables for AISI grades 1330 and 1330H.

Mechanical Properties

AISI 4027H, 4028H: End-Quench Hardenability Limits and Hardenability Band. Heat treating temperatures recommended by SAE: normalize at 900 °C (1650 °F), for forged or rolled specimens only; austenitize at 870 °C (1600 °F). (Ref 1)

Distance from quenched end, 1/16 in.	Hardness, HRC max	min
1	52	45
2	50	40
3	46	31
4	40	25
5	34	22
6	30	20
7	28	...
8	26	...
9	25	...
10	25	...
11	24	...
12	23	...
13	23	...
14	22	...
15	22	...
16	21	...
18	21	...
20	20	...

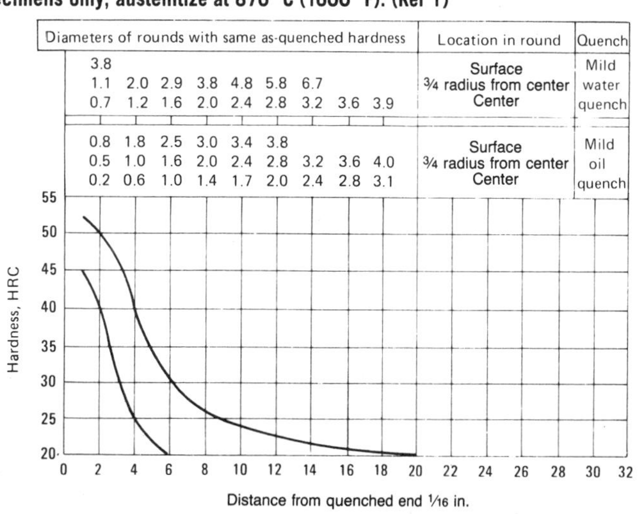

Mechanical Properties (continued)

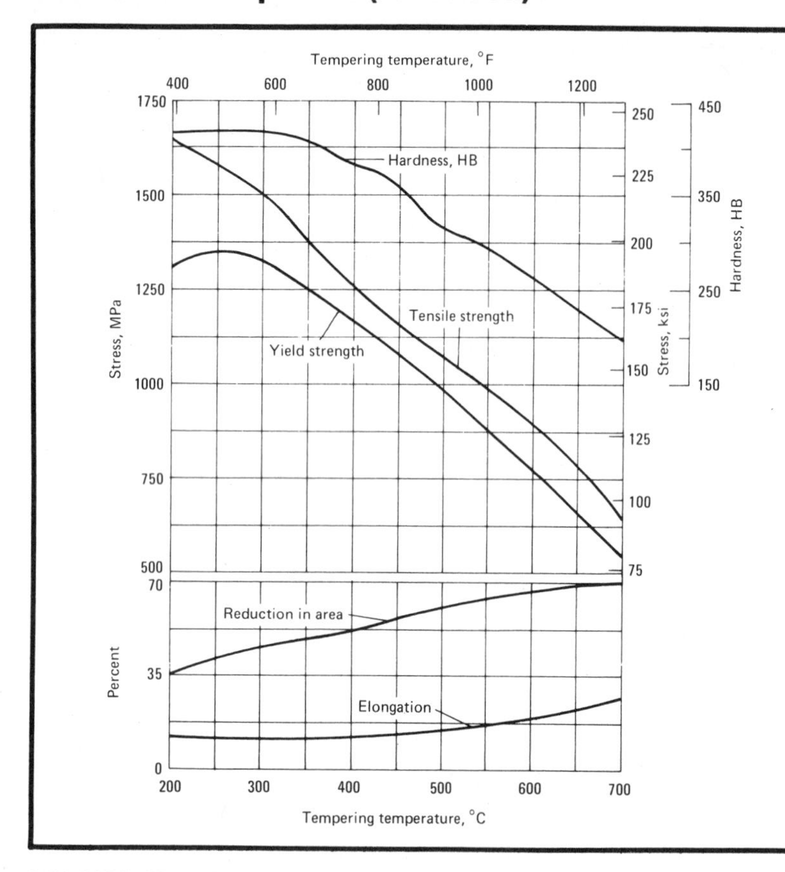

AISI 4027: Effect of Tempering Temperature on Tensile Properties. Normalized at 905 °C (1660 °F); reheated to 865 °C (1585 °F); quenched in water. Specimens were treated in 14.4-mm (0.565-in.) diam and machined to 12.8-mm (0.505-in.) diam for testing. Tests were conducted using specimens machined to English units. As-quenched hardness of 477 HB; elongation measured in 50 mm (2 in.); yield strength at 0.2% offset. (Ref 4)

AISI 4027: Mass Effect on Mechanical Properties (Ref 4)

Condition or treatment	Size round mm	Size round in.	Tensile strength MPa	Tensile strength ksi	Yield strength MPa	Yield strength ksi	Elongation(a), %	Reduction in area, %	Hardness, HB
Annealed (heated to 865 °C or 1585 °F; furnace cooled 11 °C or 20 °F per hour to 425 °C or 800 °F; cooled in air)	25	1	515	75	325	47	30.0	52.9	143
Normalized (heated to 905 °C or 1660 °F; cooled in air)	14	0.57	650	94	425	62	25.5	60.2	179
	25	1	640	93	420	61	25.8	60.2	179
	50	2	595	86	385	56	27.7	57.1	163
	100	4	565	82	350	51	28.3	55.9	156
Water quenched from 865 °C (1585 °F); tempered at 480 °C (900 °F)	14	0.57	1075	156	986	143	15.8	58.4	321
	25	1	1035	150	917	133	16.0	57.8	311
	50	2	786	114	615	89	22.0	66.6	229
	100	4	696	101	540	78	25.0	68.3	201
Water quenched from 865 °C (1585 °F); tempered at 540 °C (1000 °F)	14	0.57	993	144	896	130	17.7	61.3	302
	25	1	958	139	841	122	18.8	60.1	285
	50	2	765	111	585	85	23.7	67.2	223
	100	4	689	100	510	74	25.2	67.4	201
Water quenched from 865 °C (1585 °F); tempered at 595 °C (1100 °F)	14	0.57	896	130	800	116	20.0	64.5	262
	25	1	786	114	640	93	23.0	67.6	229
	50	2	717	104	550	80	24.8	68.3	212
	100	4	655	95	490	71	26.6	68.0	192

(a) In 50 mm (2 in.)

AISI 4027: Effect of the Mass on Hardness at Selected Points (Ref 4)

Size round mm	Size round in.	As-quenched hardness after quenching in water at: Surface	½ radius	Center
14	0.57	50 HRC	50 HRC	50 HRC
25	1	50 HRC	47 HRC	44 HRC
50	2	47 HRC	27 HRC	27 HRC
100	4	83 HRB	77 HRB	75 HRB

AISI 4027, 4028: Hardness and Machinability (Ref 7)

AISI grade	Condition	Hardness range, HB	Average machinability rating(a)
4027	Annealed and cold drawn	167-212	70
4028	Annealed and cold drawn	167-212	75

AISI 4037, 4037H

AISI 4037, 4037H: Chemical Composition

AISI grade	C	Mn	P max	S max	Si	Mo
	Chemical composition, %					
4037	0.35-0.40	0.70-0.90	0.035	0.040	0.15-0.30	0.20-0.30
4037H	0.34-0.41	0.60-1.00	0.035	0.040	0.15-0.30	0.20-0.30

Characteristics. AISI grade 4037 and 4037H steels have 0.20 to 0.30% molybdenum. They are available as hot rolled and cold finished bar and cold heading quality wire.

Typical Uses. Wire is drawn from rods rolled of AISI 4037 and 4037H steels for use in manufacturing high-strength screws, bolts, cap screws, socket-head screws, and self-tapping screws. These parts can be heat treated to meet the mechanical property requirements of SAE grade 8 alloy steel fasteners. Grades 4037 and 4037H are also used for machined parts.

AISI 4037: Similar Steels (U.S. and/or Foreign). UNS G40370; ASTM A322, A331, A519, A547; SAE J404, J412, J770

AISI 4037H: Similar Steels (U.S. and/or Foreign). UNS H40370; ASTM A304; SAE J1268

AISI 4037, 4037H: Approximate Critical Points

Transformation point	Temperature(a) °C	°F
Ac_1	725	1340
Ac_3	815	1495
Ar_3	755	1390
Ar_1	655	1210
M_s	365	690

(a) On heating or cooling at 28 °C (50 °F) per hour

Physical Properties

AISI 4037, 4037H: Thermal Treatment Temperatures (Ref 2)

Quenching medium: oil

Treatment	Temperature range °C	°F
Forging	1230 max	2250 max
Annealing(a)	830-925	1525-1700
Normalizing(b)	855-955	1575-1750
Hardening	830-855	1525-1575

(a) Maximum hardness of 183 HB. (b) Hardness of approximately 235 HB

AISI 4037, 4037H: Thermal Conductivity (Ref 6)

Temperature °C	°F	Thermal conductivity W/m·K	Btu/ft·h·°F
100	212	48.1	27.8
200	390	45.3	26.2
400	750	39.4	22.8
600	1110	33.9	19.6

Mechanical Properties

AISI 4037H: End-Quench Hardenability Limits and Hardenability Band.

Heat treating temperatures recommended by SAE: normalize at 870 °C (1600 °F), for forged or rolled specimens only; austenitize at 845 °C (1550 °F). (Ref 1)

Distance from quenched end, 1/16 in.	Hardness, HRC max	min
1	59	52
2	57	49
3	54	42
4	51	35
5	45	30
6	38	26
7	34	23
8	32	22
9	30	21
10	29	20
11	28	...
12	27	...
13	26	...
14	26	...
15	26	...
16	25	...
18	25	...
20	25	...
22	25	...
24	24	...
26	24	...
28	24	...
30	23	...
32	23	...

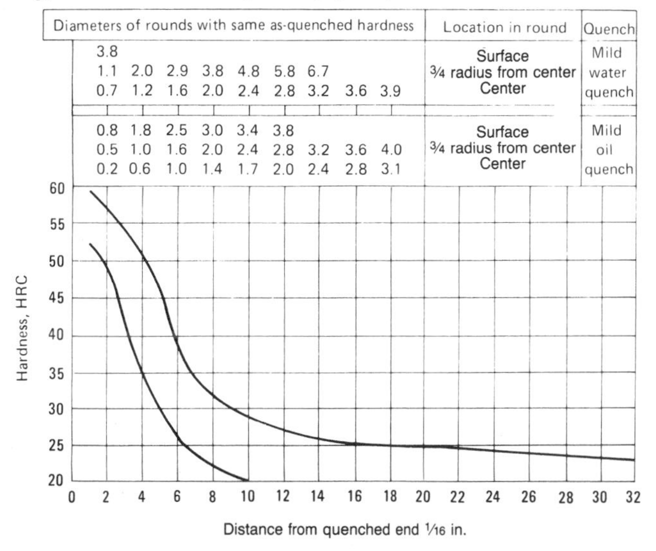

Mechanical Properties (continued)

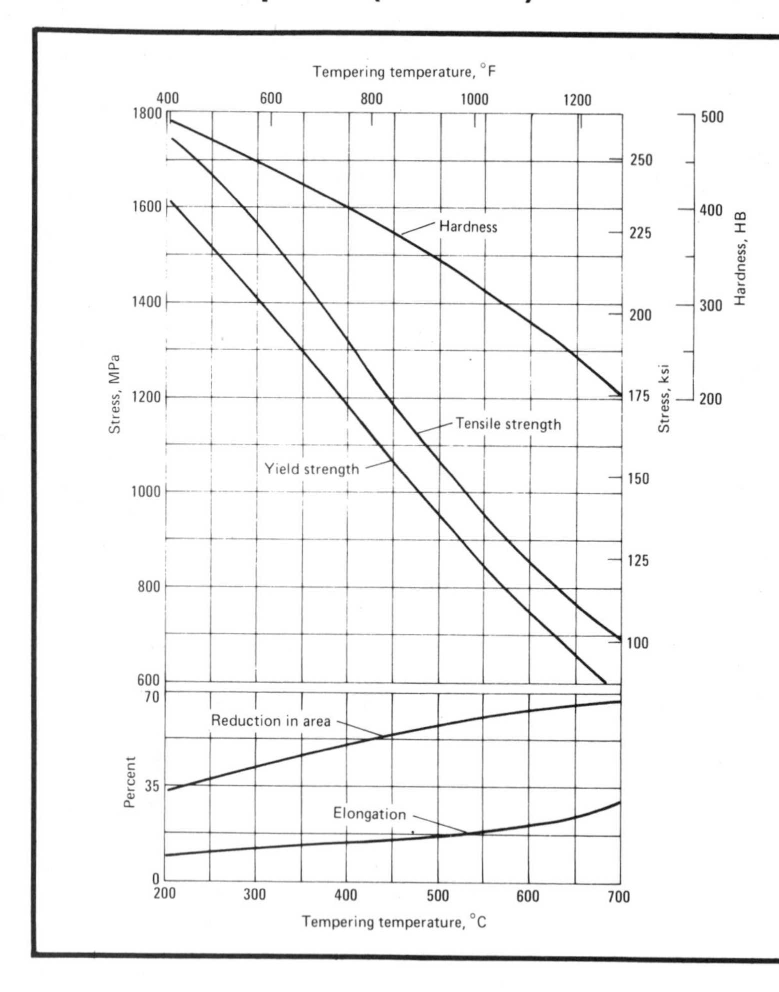

AISI 4037: Effect of Tempering Temperature on Tensile Properties. Normalized at 925 °C (1700 °F); oil quenched from 845 °C (1550 °F); tempered at 56 °C (100 °F) intervals in 13.5-mm (0.530-in.) rounds; tested in 12.8-mm (0.505-in.) rounds. Tests were conducted using specimens machined to English units. Elongation was measured in 50 mm (2 in.). (Ref 2)

AISI 4037, 4037H: Hardness and Machinability (Ref 7)

Condition	Hardness range, HB	Average machinability rating(a)
Annealed and cold drawn	174-217	70

(a) Based on AISI 1212 steel as 100% average machinability

Machining Data

For machining data on AISI grades 4037 and 4037H, refer to the preceding machining tables for AISI grades 1330 and 1330H.

AISI 4042, 4042H

AISI 4042, 4042H: Chemical Composition

AISI grade	C	Mn	Chemical composition, % P max	S max	Si	Mo
4042	0.40-0.45	0.70-0.90	0.035	0.040	0.15-0.30	0.20-0.30
4042H	0.39-0.46	0.60-1.00	0.035	0.040	0.15-0.30	0.20-0.30

Characteristics and Typical Uses. AISI 4042 and 4042H steels are available as hot rolled and cold finished bar which are generally suitable for heat treatment, machining into components, or, in the as-finished condition, constructional purposes. Grades 4042 and 4042H are also used for medium to large parts requiring a high degree of strength and toughness. These steels have medium core hardenability.

AISI 4042: Similar Steels (U.S. and/or Foreign). UNS G40420; ASTM A322, A331, A519; SAE J404, J412, J770

AISI 4042H: Similar Steels (U.S. and/or Foreign). UNS H40420; ASTM A304; SAE J1268

AISI 4042, 4042H: Approximate Critical Points

Transformation point	Temperature(a) °C	°F
Ac_1	725	1340
Ac_3	795	1460
Ar_3	730	1350
Ar_1	655	1210
M_s	345	650

(a) On heating or cooling at 28 °C (50 °F) per hour

Physical Properties

AISI 4042, 4042H: Thermal Treatment Temperatures

Quenching medium: oil

Treatment	Temperature range °C	°F
Forging	1205 max	2200 max
Annealing(a)	815-870	1500-1600
Normalizing(b)	845-900	1550-1650
Hardening	830-855	1525-1575

(a) Maximum hardness of 192 HB. (b) Hardness of approximately 235 HB

Machining Data

For machining data on AISI grades 4042 and 4042H, refer to the preceding machining tables for AISI grades 1340 and 1340H.

Mechanical Properties

AISI 4042, 4042H: Hardness and Machinability (Ref 7)

Condition	Hardness range, HB	Average machinability rating(a)
Annealed and cold drawn	179-229	65

(a) Based on AISI 1212 as 100% average machinability

AISI 4042: Effect of Tempering Temperature on Tensile Properties. Normalized at 870 °C (1600 °F); oil quenched from 845 °C (1550 °F); tempered at 38 °C (100 °F) in 13.7-mm (0.540-in.) rounds; tested in 12.8-mm (0.505-in.) rounds. Tests were conducted using specimens machined to English units. Elongation was measured in 50 mm (2 in.). (Ref 2)

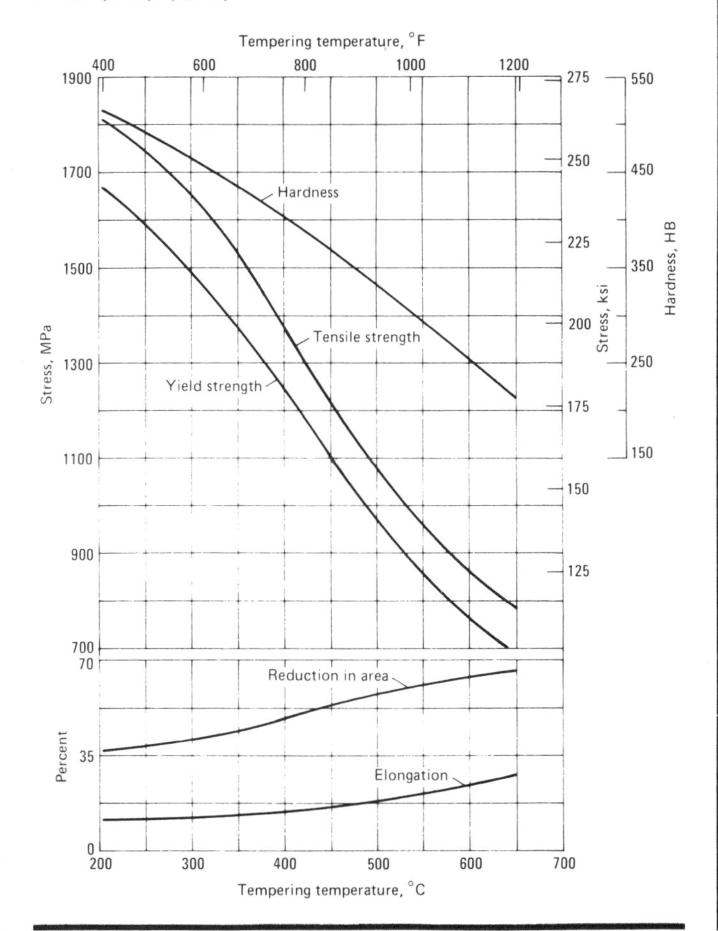

AISI 4042H: End-Quench Hardenability Limits and Hardenability Band. Heat treating temperatures recommended by SAE: normalize at 870 °C (1600 °F), for forged or rolled specimens only; austenitize at 845 °C (1550 °F). (Ref 1)

Distance from quenched end, 1/16 in.	Hardness, HRC max	min
1	62	55
2	60	52
3	58	48
4	55	40
5	50	33
6	45	29
7	39	27
8	36	26
9	34	25
10	33	24
11	32	24
12	31	23
13	30	23
14	30	23
15	29	22
16	29	22
18	28	22
20	28	21
22	28	20
24	27	20
26	27	...
28	27	...
30	26	...
32	26	...

Diameters of rounds with same as-quenched hardness	Location in round	Quench
3.8	Surface	Mild water quench
1.1 2.0 2.9 3.8 4.8 5.8 6.7	3/4 radius from center	
0.7 1.2 1.6 2.0 2.4 2.8 3.2 3.6 3.9	Center	
0.8 1.8 2.5 3.0 3.4 3.8	Surface	Mild oil quench
0.5 1.0 1.6 2.0 2.4 2.8 3.2 3.6 4.0	3/4 radius from center	
0.2 0.6 1.0 1.4 1.7 2.0 2.4 2.8 3.1	Center	

Hardness, HRC: 65 60 55 50 45 40 35 30 25 20

Distance from quenched end: 0 2 4 6 8 10 12 14 16 18 20 22 24 26 28 30 32

Distance from quenched end 1/16 in.

AISI 4047, 4047H

AISI 4047, 4047H: Chemical Composition

AISI grade	Chemical composition, % C	Mn	P max	S max	Si	Mo
4047	0.45-0.50	0.70-0.90	0.035	0.040	0.15-0.30	0.20-0.30
4047H	0.44-0.51	0.60-1.00	0.035	0.040	0.15-0.30	0.20-0.30

Characteristics and Typical Uses. AISI grades 4047 and 4047H are directly hardenable alloy steels with low core hardenability. They are available as hot rolled and cold finished bar suitable for heat treatment, machining into parts, or, in the as-finished condition, constructional purposes. Other uses include medium to large parts requiring a high degree of strength and toughness, and average-size automotive parts.

AISI 4047: Similar Steels (U.S. and/or Foreign). UNS G40470; ASTM A322, A331, A519; SAE J404, J412, J770

AISI 4047H: Similar Steels (U.S. and/or Foreign). UNS H40470; ASTM A304; SAE J1268

AISI 4047, 4047H: Approximate Critical Points

Transformation point	Temperature(a) °C	°F
Ac_1	725	1340
Ac_3	780	1440
Ar_3	720	1330
Ar_1	650	1200
M_s	325	615

(a) On heating or cooling at 28 °C (50 °F) per hour

Mechanical Properties

AISI 4047H: End-Quench Hardenability Limits and Hardenability Band.
Heat treating temperatures recommended by SAE: normalize at 870 °C (1600 °F), for forged or rolled specimens only; austenitize at 845 °C (1550 °F). (Ref 1)

Distance from quenched end, 1/16 in.	Hardness, HRC max	min
1	64	57
2	62	55
3	60	50
4	58	42
5	55	35
6	52	32
7	47	30
8	43	28
9	40	28
10	38	27
11	37	26
12	35	26
13	34	25
14	33	25
15	33	25
16	32	25
18	31	24
20	30	24
22	30	23
24	30	23
26	30	22
28	29	22
30	29	21
32	29	21

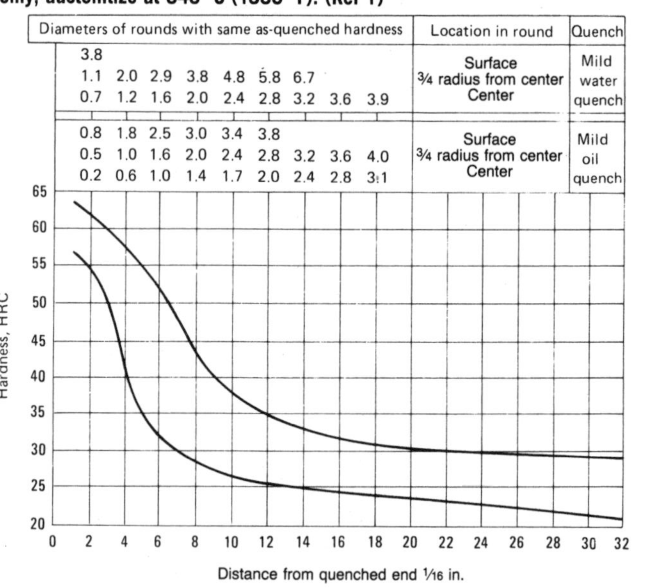

AISI 4047, 4047H: Hardness and Machinability (Ref 7)

Condition	Hardness range, HB	Average machinability rating(a)
Annealed and cold drawn	179-229	65

(a) Based on AISI 1212 steel as 100% average machinability

Machining Data

For machining data on AISI grades 4047 and 4047H, refer to the preceding machining tables for AISI grades 1340 and 1340H.

Physical Properties

AISI 4047, 4047H: Thermal Treatment Temperatures

Quenching medium: oil

Treatment	Temperature range °C	°F
Forging	1230 max	2250 max
Annealing	790-845	1450-1550
Hardening	815-855	1500-1575
Tempering	(a)	(a)

(a) To desired hardness

AISI 4118, 4118H

AISI 4118, 4118H: Chemical Composition

AISI grade	C	Mn	P max	S max	Si	Cr	Mo
	Chemical composition, %						
4118	0.18-0.23	0.70-0.90	0.035	0.040	0.15-0.30	0.40-0.60	0.08-0.15
4118H	0.17-0.23	0.60-1.00	0.035	0.040	0.15-0.30	0.30-0.70	0.08-0.15

Characteristics and Typical Uses. AISI grades 4118 and 4118H are carburizing steels with low to medium core hardenability and intermediate case hardenability. These steels are used for automotive gears, piston pins, ball studs, and roller bearings. Oil quenching will usually produce satisfactory case hardness.

AISI 4118: Similar Steels (U.S. and/or Foreign). UNS G41180; ASTM A322, A331, A505, A519; SAE J404, J412, J770

AISI 4118H: Similar Steels (U.S. and/or Foreign). UNS H41180; ASTM A304; SAE J1268

AISI 4118, 4118H: Approximate Critical Points

Transformation point	Temperature(a) °C	Temperature(a) °F	Transformation point	Temperature(a) °C	Temperature(a) °F
Ac_1	750	1380	Ar_3	775	1430
Ac_3	825	1520	Ar_1	680	1260

(a) On heating or cooling at 28 °C (50 °F) per hour

Physical Properties

AISI 4118, 4118H: Thermal Treatment Temperatures

Quenching medium: oil

Treatment	Temperature range °C	Temperature range °F
Forging	1230 max	2250 max
Carburizing	885-925	1625-1700
Hardening	830	1525
Tempering	120-185	250-365

Machining Data

For machining data on AISI grades 4118 and 4118H, refer to the preceding machining tables for AISI grade A2317.

Mechanical Properties

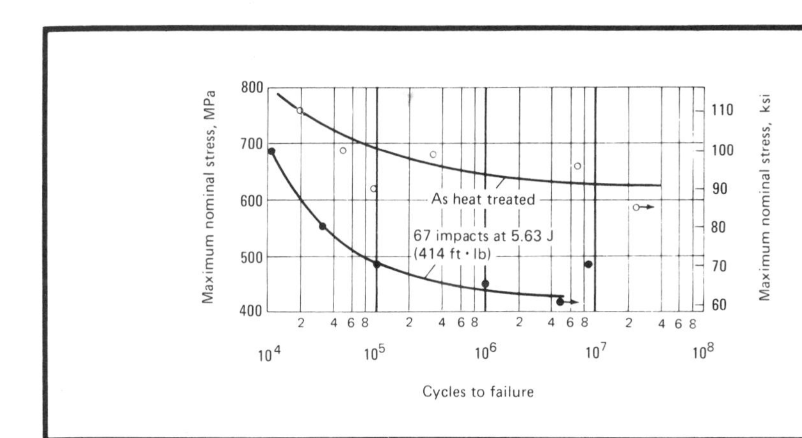

AISI 4118, 4118H: Bending Fatigue Properties (Ref 10)

AISI 4118, 4118H: Hardness and Machinability (Ref 7)

Condition	Hardness range, HB	Average machinability rating(a)
Hot rolled and cold drawn	170-207	60

(a) Based on AISI 1212 steel as 100% average machinability

AISI 4118, 4118H: Fatigue Limit Data (Ref 10)

Effective case depth mm	Effective case depth in.	Peak case hardness, HRC	Nominal fatigue limit MPa	Nominal fatigue limit ksi
1.02	0.040	64	758	110
0.33	0.013	63	772	112

Mechanical Properties (continued)

AISI 4118, 4118H: Bending Fatigue Behavior

Carburized to effective case depth of 0.33 mm or 0.013 in. (○) and 1.0 mm or 0.040 in. (●). (Ref 10)

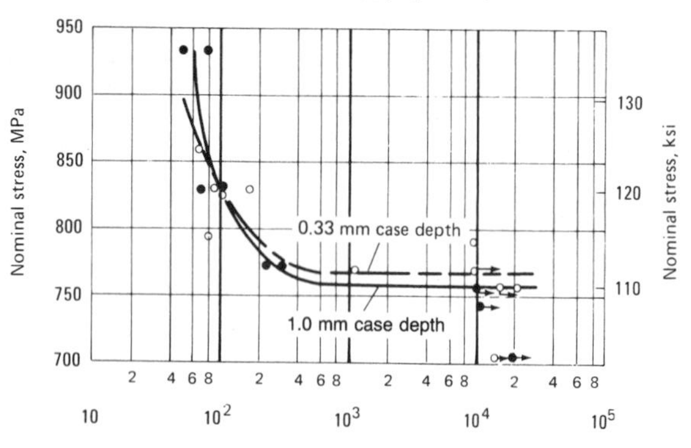

Carburized with effective case depth of 0.31 to 0.46 mm (0.012 to 0.018 in.). (Ref 10)

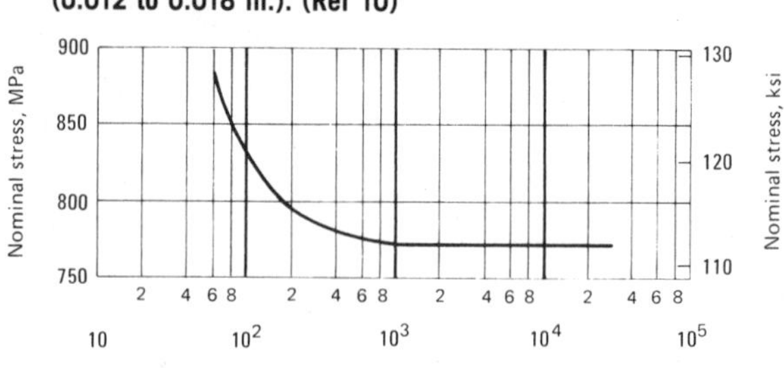

Carburized with effective case depth of 0.94 to 1.2 mm (0.037 to 0.046 in.). (Ref 10)

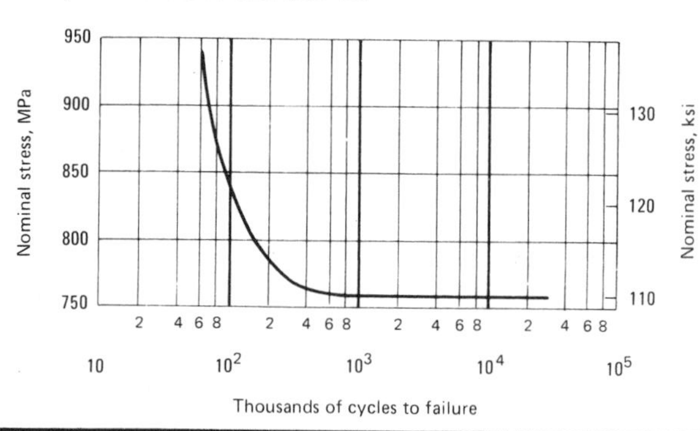

AISI 4118, 4118H: Mass Effect on Mechanical Properties (Ref 4)

Condition or treatment	Size round mm	Size round in.	Tensile strength MPa	Tensile strength ksi	Yield strength MPa	Yield strength ksi	Elongation(a), %	Reduction in area, %	Hardness, HB
Annealed (heated to 870 °C or 1600 °F; furnace cooled 11 °C or 20 °F per hour to 620 °C or 1150 °F; cooled in air)	25	1	517	75.0	365	53.0	33.0	63.7	137
Normalized (heated to 910 °C or 1670 °F; cooled in air)	14	0.57	586	85.0	393	57.0	31.5	70.1	170
	25	1	583	84.5	386	56.0	32.0	71.0	156
	50	2	534	77.5	376	54.5	34.0	74.4	143
	100	4	521	75.5	341	49.5	34.0	71.2	137
Mock carburized at 925 °C (1700 °F) for 8 h; reheated to 830 °C (1525 °F); quenched in oil; tempered at 150 °C (300 °F)	14	0.57	986	143.0	645	93.5	17.5	41.3	293
	25	1	820	119.0	445	64.5	21.0	37.5	241
	50	2	669	97.0	317	46.0	26.5	56.3	201
	100	4	641	93.0	300	43.5	28.0	61.3	192
Mock carburized at 925 °C (1700 °F) for 8 h; reheated to 830 °C (1525 °F); quenched in oil; tempered at 230 °C (450 °F)	14	0.57	951	138.0	617	89.5	17.5	41.9	277
	25	1	793	115.0	441	64.0	22.0	49.0	235
	50	2	645	93.5	314	45.5	28.0	62.0	192
	100	4	617	89.5	296	43.0	28.5	63.5	187

(a) In 50 mm (2 in.)

AISI 4118, 4118H: Effect of the Mass on Hardness at Selected Points (Ref 4)

Size round mm	Size round in.	As-quenched hardness after quenching in oil at: Surface	½ radius	Center
14	0.57	33 HRC	33 HRC	33 HRC
25	1	22 HRC	20 HRC	20 HRC
50	2	88 HRB	88 HRB	87 HRB
100	4	87 HRB	87 HRB	85 HRB

Mechanical Properties (continued)

AISI 4118, 4118H: Core Properties of Specimens Treated for Maximum Case Hardness and Core Toughness

Treatment	Case depth, mm (in.)	Case hardness, HRC	Core properties: Tensile strength, MPa (ksi)	Yield strength(a), MPa (ksi)	Elongation(b), %	Reduction in area, %	Hardness, HB
Recommended practice for maximum case hardness							
Direct quenched from pot: carburized at 925 °C (1700 °F) for 8 h; quenched in agitated oil; tempered at 150 °C (300 °F)	1.6 (0.063)	61	1225 (178)	903 (131)	9	42.3	352
Single quenched and tempered (for good case and core properties): carburized at 925 °C (1700 °F) for 8 h; pot cooled; reheated to 830 °C (1525 °F); quenched in agitated oil; tempered at 150 °C (300 °F)	1.2 (0.047)	62	986 (143)	650 (94)	18	41.3	293
Double quenched and tempered (for maximum refinement of case and core): carburized at 925 °C (1700 °F) for 8 h; pot cooled; reheated to 830 °C (1525 °F); quenched in agitated oil; reheated to 800 °C (1475 °F); quenched in agitated oil; tempered at 150 °C (300 °F)	1.2 (0.047)	62	869 (126)	440 (64)	21	42.4	241
Recommended practice for maximum core toughness							
Direct quenched from pot: carburized at 925 °C (1700 °F) for 8 h; quenched in agitated oil; tempered at 230 °C (450 °F)	1.6 (0.063)	57	1220 (177)	896 (130)	13	48	341
Single quenched and tempered (for good case and core properties): carburized at 925 °C (1700 °F) for 8 h; pot cooled; reheated to 830 °C (1525 °F); quenched in agitated oil; tempered at 230 °C (450 °F)	1.2 (0.047)	56	951 (138)	620 (90)	18	41.9	277
Double quenched and tempered (for maximum refinement of case and core): carburized at 925 °C (1700 °F) for 8 h; pot cooled; reheated to 830 °C (1525 °F); quenched in agitated oil; reheated to 800 °C (1475 °F); quenched in agitated oil; tempered at 230 °C (450 °F)	1.2 (0.047)	56	827 (120)	435 (63)	22	48.9	229

(a) 0.2% offset. (b) In 50 mm (2 in.)

AISI 4118H: End-Quench Hardenability Limits and Hardenability Band. Heat treating temperatures recommended by SAE: normalize at 925 °C (1700 °F), for forged or rolled specimens only; austenitize at 925 °C (1700 °F). (Ref 1)

Distance from quenched end, 1/16 in.	Hardness, HRC max	Hardness, HRC min
1	48	41
2	48	36
3	41	27
4	35	23
5	31	20
6	28	...
7	27	...
8	25	...
9	24	...
10	23	...
11	22	...
12	21	...
13	21	...
14	20	...

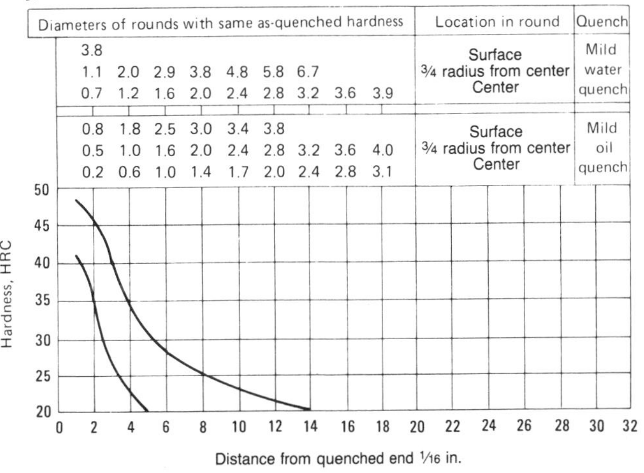

AISI 4130, 4130H

AISI 4130, 4130H: Chemical Composition

AISI grade	Chemical composition, % C	Mn	P max	S max	Si	Cr	Mo
4130	0.28-0.33	0.40-0.60	0.035	0.040	0.15-0.30	0.80-1.10	0.15-0.25
4130H	0.27-0.33	0.30-0.70	0.035	0.040	0.15-0.30	0.75-1.20	0.15-0.25

Characteristics and Typical Uses. AISI grades 4130 and 4130H are medium-carbon, chromium-molybdenum steels. Available as hot rolled and cold finished bar and seamless tube, these steels are intended for general purpose applications. Variations in heat treatment can obtain a broad range of strength and toughness. These steels have good hardenability, strength, wear resistance, toughness, and ductility. In the heat treated condition, they offer good strength and toughness for moderately stressed parts. Grades 4130 and 4130H are available in forging quality, aircraft quality, and premium aircraft quality steels.

AISI 4130: Similar Steels (U.S. and/or Foreign). UNS G41300; AMS 6350, 6356, 6360, 6361, 6362, 6370, 6371, 6373; ASTM A322, A331, A505, A513, A519, A646; MIL SPEC MIL-S-16974; SAE J404, J412, J770; (W. Ger.) DIN 1.7218; (Fr.) AFNOR 25 CD 4 (S); (Ital.) UNI 25 CrMo 4, 25 CrMo 4 KB; (Jap.) JIS SCM 2, SCCrM 1; (Swed.) SS_{14} 2225; (U.K.) B.S. CDS 110

AISI 4130H: Similar Steels (U.S. and/or Foreign). UNS H41300; ASTM A304; SAE J1268; (W. Ger.) DIN 1.7218; (Fr.) AFNOR 25 CD 4 (S); (Ital.) UNI 25 CrMo 4, 25 CrMo 4 KB; (Jap.) JIS SCM 2, SCCrM 1; (Swed.) SS_{14} 2225; (U.K.) B.S. CDS 110

AISI 4130, 4130H: Approximate Critical Points

Transformation point	Temperature(a) °C	°F	Transformation point	Temperature(a) °C	°F
Ac_1	755	1395	Ar_1	695	1280
Ac_3	810	1490	M_s	365	685
Ar_3	755	1390			

(a) On heating or cooling at 28 °C (50 °F) per hour

Physical Properties

AISI 4130, 4130H: Mean Apparent Specific Heat (Ref 6)

Temperature range °C	°F	Specific heat J/kg·K	Btu/lb·°F
50-100	120-212	477	0.114
150-200	300-390	523	0.125
250-300	480-570	544	0.130
350-400	660-750	607	0.145
450-500	840-930	657	0.157
550-600	1020-1110	741	0.177
650-700	1200-1290	829	0.198
750-800	1380-1470	837	0.200

AISI 4130, 4130H: Electrical Resistivity (Ref 6)

Temperature °C	°F	Electrical resistivity, μΩ·m
20	68	0.223
100	212	0.271
200	390	0.342
400	750	0.529
600	1110	0.786
800	1470	1.103
1000	1830	1.171
1200	2190	1.222

AISI 4130, 4130H: Thermal Conductivity (Ref 6)

Temperature °C	°F	Thermal conductivity W/m·K	Btu/ft·h·°F
100	212	42.7	24.7
300	570	40.7	23.5
500	930	37.2	21.5
700	1290	31.0	17.9
1000	1830	28.0	16.2
1200	2190	30.1	17.4

AISI 4130, 4130H: Thermal Treatment Temperatures

Quenching medium: oil or water

Treatment	Temperature range °C	°F
Forging	1230 max	2250 max
Annealing(a)	790-845	1450-1550
Normalizing(b)	870-925	1600-1700
Hardening	815-870	1500-1600
Tempering	(c)	(c)

(a) Maximum hardness of 174 HB. (b) Hardness of approximately 255 HB. (c) To desired hardness

Machining Data

For machining data on AISI grades 4130 and 4130H, refer to the preceding machining tables for AISI grades 1330 and 1330H.

Mechanical Properties

AISI 4130H: End-Quench Hardenability Limits and Hardenability Band.
Heat treating temperatures recommended by SAE: normalize at 845 °C (1550 °F), for forged or rolled specimens only; austenitize at 815 °C (1500 °F). (Ref 1)

Distance from quenched end, 1/16 in.	Hardness, HRC max	Hardness, HRC min
1	56	49
2	55	46
3	53	42
4	51	38
5	49	34
6	47	31
7	44	29
8	42	27
9	40	26
10	38	26
11	36	25
12	35	25
13	34	24
14	34	24
15	33	23
16	33	23
18	32	22
20	32	21
22	32	20
24	31	...
26	31	...
28	30	...
30	30	...
32	29	...

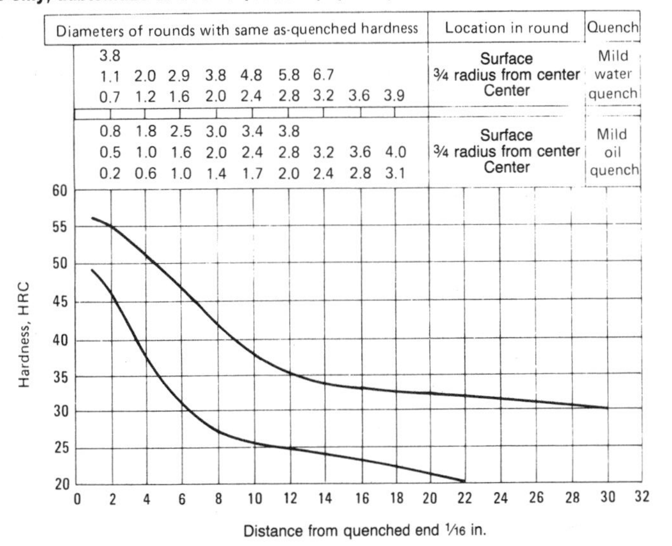

AISI 4130, 4130H: Mass Effect on Mechanical Properties (Ref 4)

Condition or treatment	Size round mm	Size round in.	Tensile strength MPa	Tensile strength ksi	Yield strength MPa	Yield strength ksi	Elongation(a), %	Reduction in area, %	Hardness, HB
Annealed (heated to 865 °C or 1585 °F; furnace cooled 11 °C or 20 °F per hour to 680 °C or 1255 °F; cooled in air)	25	0.5	560	81	460	67	21.5	59.6	217
Normalized (heated to 870 °C or 1600 °F; cooled in air)	13	0.5	731	106	460	67	25.1	59.6	217
	25	1	670	97	435	63	25.5	59.5	197
	50	2	615	89	425	62	28.2	65.4	167
	100	4	615	89	400	58	27.0	61.2	163
Water quenched from 855 °C (1575 °F); tempered at 480 °C (900 °F)	13	0.5	1145	166	1110	161	16.4	61.0	331
	25	1	1110	161	951	138	14.7	54.4	321
	50	2	917	133	758	110	19.0	63.0	269
	100	4	841	122	655	95	20.5	63.6	241
Water quenched from 855 °C (1575 °F); tempered at 540 °C (1000 °F)	13	0.5	1040	151	979	142	18.1	63.9	302
	25	1	993	144	896	130	18.5	61.8	293
	50	2	841	122	685	99	21.2	66.3	241
	100	4	800	116	635	92	21.5	63.5	235
Water quenched from 855 °C (1575 °F); tempered at 595 °C (1100 °F)	13	0.5	917	133	841	122	20.7	69.0	269
	25	1	883	128	779	113	21.2	67.5	262
	50	2	786	114	635	92	21.7	67.7	229
	100	4	703	102	540	78	24.5	69.2	197

(a) In 50 mm (2 in.)

AISI 4130, 4130H: Effect of the Mass on Hardness at Selected Points (Ref 4)

Size round mm	Size round in.	As-quenched hardness after quenching in water at: Surface	½ radius	Center
13	0.5	51 HRC	50 HRC	50 HRC
25	1	51 HRC	50 HRC	44 HRC
50	2	47 HRC	32 HRC	31 HRC
100	4	45.5 HRC	25 HRC	24.5 HRC

AISI 4130, 4130H: Hardness and Machinability (Ref 7)

Condition	Hardness range, HB	Average machinability rating(a)
Annealed and cold drawn	187-229	70

(a) Based on AISI 1212 steel as 100% average machinability

Mechanical Properties (continued)

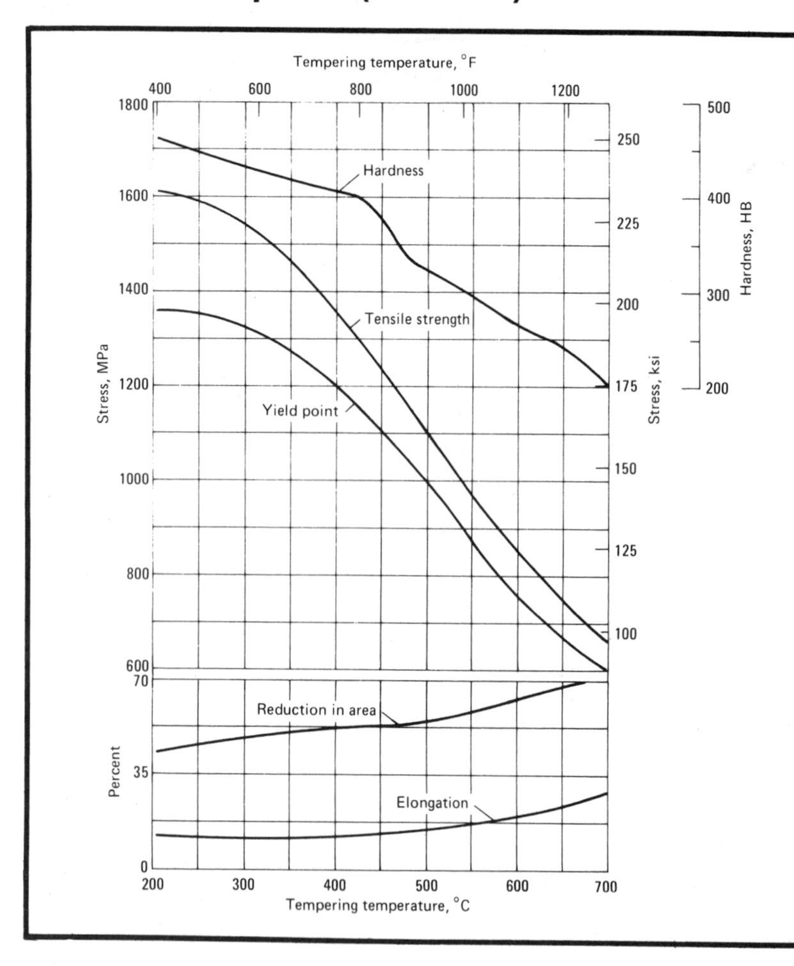

AISI 4130: Effect of Tempering Temperature on Tensile Properties. Normalized at 870 °C (1600 °F); reheated to 855 °C (1575 °F); quenched in water. Specimens were treated in 13.5-mm (0.530-in.) diam and machined to 12.8-mm (0.505-in.) diam for testing. Tests were conducted using specimens machined to English units. As-quenched hardness of 495 HB; elongation measured in 50 mm (2 in.). (Ref 4)

AISI 4130, 4130H: Tensile Properties (Ref 5)

Condition or treatment	Tensile strength MPa	Tensile strength ksi	Yield strength MPa	Yield strength ksi	Elongation(a), %	Reduction in area, %	Hardness, HB	Izod impact energy J	Izod impact energy ft·lb
Normalized at 870 °C (1600 °F)	670	97	435	63	25.5	60	197	87	64
Annealed at 865 °C (1585 °F)	560	81	360	52	28.2	56	156	62	46

(a) In 50 mm (2 in.)

AISI 4135, 4135H

AISI 4135, 4135H: Chemical Composition

AISI grade	C	Mn	P max	S max	Si	Cr	Mo
	Chemical composition, %						
4135	0.33-0.38	0.70-0.90	0.035	0.040	0.15-0.30	0.80-1.10	0.15-0.25
4135H	0.32-0.38	0.60-1.00	0.035	0.040	0.15-0.30	0.75-1.20	0.15-0.25

Characteristics and Typical Uses. AISI grade 4135 and 4135H steels are used in the following applications: forgings for thin-walled pressure vessels; wire for production of high-strength bolts, screws, and recessed- and socket-head screws; and nitrided parts.

AISI 4135, 4135H: Approximate Critical Points

Transformation point	Temperature(a) °C	Temperature(a) °F	Transformation point	Temperature(a) °C	Temperature(a) °F
Ac_1	755	1390	Ar_1	695	1280
Ac_3	805	1485	M_s	340	640
Ar_3	750	1380			

(a) On heating or cooling at 28 °C (50 °F) per hour

AISI 4135: Similar Steels (U.S. and/or Foreign). UNS G41350; AMS 6365 C, 6372 C; ASTM A711, A355, A519; MIL SPEC MIL-S-16974, MIL-S-18733; SAE J404, J412, J770; (W. Ger.) DIN 1.7220; (Fr.) AFNOR 35 CD 4, 35 CD 4 TS; (Ital.) UNI 35 CrMo 4, 35 CrMo 4 F, 34 CrMo 4 KB; (Jap.) JIS SCM 1, SCCrM 3; (Swed.) SS_{14} 2234; (U.K.) B.S. 708 A 37

AISI 4135H: Similar Steels (U.S. and/or Foreign). UNS H41350; ASTM A304; SAE J1268; (W. Ger.) DIN 1.7220; (Fr.) AFNOR 35 CD 4, 35 CD 4 TS; (Ital.) UNI 35 CrMo 4, 35 CrMo 4 F, 34 CrMo 4 KB; (Jap.) JIS SCM 1, SCCrM 3; (Swed.) SS_{14} 2234; (U.K.) B.S. 708 A 37

Physical Properties

AISI 4135, 4135H: Thermal Treatment Temperatures

Quenching medium: oil

Treatment	Temperature range °C	Temperature range °F
Forging	1205 max	2200 max
Annealing	790-845	1450-1550
Hardening	845-870	1550-1600
Tempering	(a)	(a)

(a) To desired hardness

Mechanical Properties

AISI 4135H: End-Quench Hardenability Limits and Hardenability Band.

Heat treating temperatures recommended by SAE: normalize at 870 °C (1600 °F), for forged or rolled specimens only; austenitize at 845 °C (1550 °F). (Ref 1)

Distance from quenched end, 1/16 in.	Hardness, HRC max	Hardness, HRC min
1	58	51
2	58	50
3	57	49
4	56	48
5	56	47
6	55	45
7	54	42
8	53	40
9	52	38
10	51	36
11	50	34
12	49	33
13	48	32
14	47	31
15	46	30
16	45	30
18	44	29
20	42	28
22	41	27
24	40	27
26	39	27
28	38	26
30	38	26
32	37	26

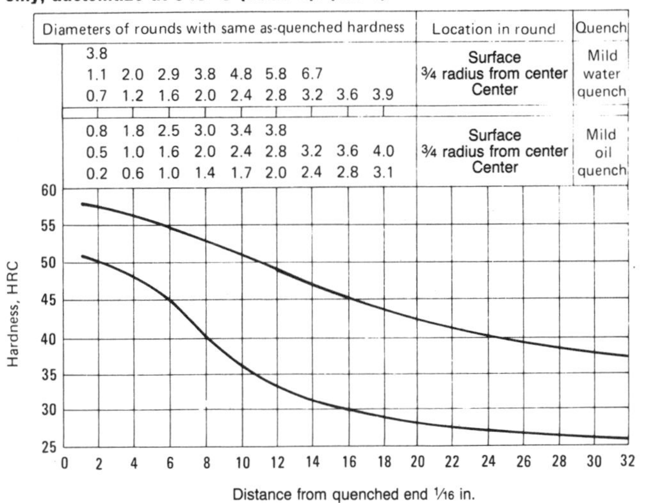

AISI 4135, 4135H: Hardness and Machinability (Ref 7)

Condition	Hardness range, HB	Average machinability rating(a)
Annealed and cold drawn	187-229	70

(a) Based on AISI 1212 steel as 100% average machinability

Machining Data

For machining data on AISI grades 4135 and 4135H, refer to the preceding machining tables for AISI grades 1330 and 1330H.

AISI 4137, 4137H

AISI 4137, 4137H: Chemical Composition

AISI grade	C	Mn	P max	S max	Si	Cr	Mo
	Chemical composition, %						
4137	0.35-0.40	0.70-0.90	0.035	0.040	0.15-0.30	0.80-1.10	0.15-0.25
4137H	0.34-0.41	0.60-1.00	0.035	0.040	0.15-0.30	0.75-1.21	0.15-0.25

Characteristics and Typical Uses. AISI grade 4137 and 4137H steels are available as hot rolled and cold rolled sheet and strip, hot rolled and cold finished bar, and cold heading quality wire. These grades are generally suitable for heat treatment, machining into components, or, in the as-finished condition, constructional applications and similar uses.

AISI 4137: Similar Steels (U.S. and/or Foreign). UNS G41370; ASTM A322, A331, A505, A519, A547; SAE J404, J412, J770; (W. Ger.) DIN 1.7225; (Fr.) AFNOR 40 CD 4, 42 CD 4; (Ital.) UNI G 40 CrMo 4, 38 CrMo 4 KB, 40 CrMo 4; (Jap.) JIS SCM 4 H, SCM 4; (Swed.) SS_{14} 2244; (U.K.) B.S. 708 A 42, 708 M 40, 709 M 40

AISI 4137H: Similar Steels (U.S. and/or Foreign). UNS H41370; ASTM A304; SAE J1268; (W. Ger.) DIN 1.7225; (Fr.) AFNOR 40 CD 4, 42 CD 4; (Ital.) UNI G 40 CrMo 4, 40 CrMo 4, 38 CrMo 4 KB; (Jap.) JIS SCM 4 H, SCM 4; (Swed.) SS_{14} 2244; (U.K.) B.S. 708 A 42, 708 A 40, 709 A 40

Physical Properties

AISI 4137, 4137H: Average Coefficients of Linear Thermal Expansion (Ref 6)

Temperature range °C	°F	Coefficient µm/m·K	µin./in.·°F
0-100	32-212	11.2	6.2
0-200	32-390	11.7	6.5
0-300	32-570	12.4	6.9
0-400	32-750	13.0	7.2
0-500	32-930	13.5	7.5

AISI 4137, 4137H: Thermal Treatment Temperatures

Quenching medium: oil

Treatment	Temperature range °C	°F
Annealing	790-845	1450-1550
Hardening	845-900	1550-1650
Tempering	(a)	(a)

(a) To desired hardness

Mechanical Properties

AISI 4137H: End-Quench Hardenability Limits and Hardenability Band.
Heat treating temperatures recommended by SAE: normalize at 870 °C (1600 °F), for forged or rolled specimens only; austenitize at 845 °C (1550 °F). (Ref 1)

Distance from quenched end, 1/16 in.	Hardness, HRC max	min
1	59	52
2	59	51
3	58	50
4	58	49
5	57	49
6	57	48
7	56	45
8	55	43
9	55	40
10	54	39
11	53	37
12	52	36
13	51	35
14	50	34
15	49	33
16	48	33
18	46	32
20	45	31
22	44	30
24	43	30
26	42	30
28	42	29
30	41	29
32	41	29

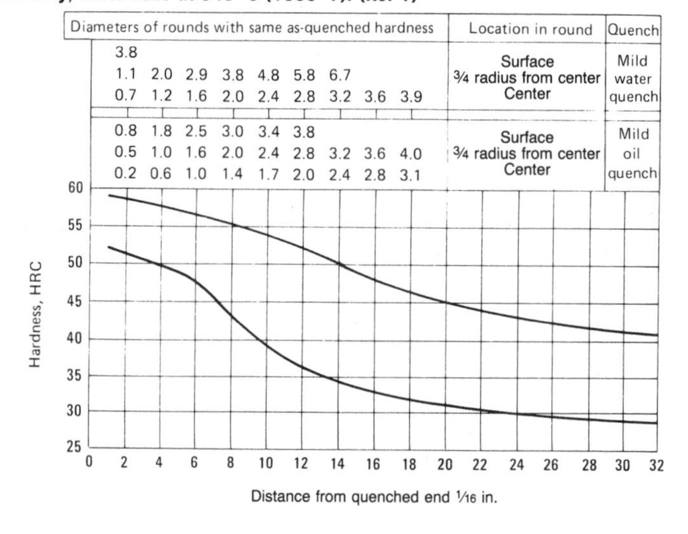

Mechanical Properties (continued)

AISI 4137, 4137H: Hardness and Machinability (Ref 7)

Condition	Hardness range, HB	Average machinability rating(a)
Annealed and cold drawn	187-229	70

(a) Based on AISI 1212 steel as 100% average machinability

Machining Data

For machining data on AISI grades 4137 and 4137H, refer to the preceding machining tables for AISI grades 1330 and 1330H.

AISI 4140, 4140H

AISI 4140, 4140H: Chemical Composition

AISI grade	Chemical composition, %						
	C	Mn	P max	S max	Si	Cr	Mo
4140	0.38-0.43	0.75-1.10	0.035	0.040	0.15-0.30	0.80-1.10	0.15-0.25
4140H	0.37-0.44	0.65-1.10	0.035	0.040	0.15-0.30	0.75-1.20	0.15-0.25

Characteristics. AISI grades 4140 and 4140H are chromium-molybdenum, medium-carbon steels with high hardenability and good fatigue, abrasion, and impact resistance. These steels can be successfully nitrided for maximum wear and abrasion resistance. They are deep-hardening alloys suitable for severe service characterized by fatigue, abrasion, impact, high-temperature stresses, or combinations of such stresses in small and large sections. When fully hardened, grades 4140 and 4140H demonstrate the outstanding property of relatively high impact strength at high-hardness, tensile strength.

AISI 4140 and 4140H are available as hot and cold rolled sheet and strip; hot rolled and cold finished bar; cold heading quality, aircraft quality, and premium aircraft quality bar; and seamless tube.

Typical Uses. AISI grades 4140 and 4140H are used for small gears, pinions, and ball studs, and for high-strength bolts, cap screws, and socket- and recessed-head screws. These products can be heat treated after machining or forming.

AISI 4140: Similar Steels (U.S. and/or Foreign). UNS G41400; AMS 6381, 6382, 6390, 6395; ASTM A322, A331, A505, A519, A547, A646; MIL SPEC MIL-S-16974; SAE J404, J412, J770; (W. Ger.) DIN 1.7225; (Fr.) AFNOR 40 CD 4, 42 CD 4; (Ital.) UNI 40 CrMo 4, G 40 CrMo 4, 38 CrMo 4 KB; (Jap.) JIS SCM 4 H, SCM 4; (Swed.) SS_{14} 2244; (U.K.) B.S. 708 A 42, 708 M 40, 709 M 40

AISI 4140H: Similar Steels (U.S. and/or Foreign). UNS H41400; ASTM A304; SAE J1268; (W. Ger.) DIN 1.7225; (Fr.) AFNOR 40 CD 4, 42 CD 4; (Ital.) UNI G 40 CrMo 4, 40 CrMo 4, 38 CrMo 4 KB; (Jap.) JIS SCM 4 H, SCM 4; (Swed.) SS_{14} 2244; (U.K.) B.S. 708 A 42, 708 M 40, 709 M 40

AISI 4140, 4140H: Approximate Critical Points

Transformation point	Temperature(a) °C	Temperature(a) °F	Transformation point	Temperature(a) °C	Temperature(a) °F
Ac_1	730	1350	Ar_1	680	1255
Ac_3	805	1480	M_s	315	595
Ar_3	745	1370			

(a) On heating or cooling at 28 °C (50 °F) per hour

Physical Properties

AISI 4140, 4140H: Average Coefficients of Linear Thermal Expansion (Ref 6)

Temperature range °C	Temperature range °F	Coefficient μm/m·K	Coefficient μin./in.·°F
20-100	68-212	12.2	6.8
20-200	68-390	12.6	7.0
20-400	68-750	13.7	7.6
20-600	68-1110	14.6	8.1

AISI 4140, 4140H: Electrical Resistivity and Thermal Conductivity (Ref 6)

Temperature range °C	Temperature range °F	Electrical resistivity, μΩ·m	Thermal conductivity W/m·K	Thermal conductivity Btu/ft·h·°F
100	212	0.263	42.6	24.6
200	390	0.326	42.2	24.4
400	750	0.475	37.7	21.8
600	1110	0.646	33.0	19.1

AISI 4140, 4140H: Mean Apparent Specific Heat (Ref 6)

Temperature range °C	Temperature range °F	Specific heat J/kg·K	Specific heat Btu/lb·°F
150-200	300-390	473	0.113
350-400	660-750	519	0.124
550-600	1020-1110	561	0.134

AISI 4140, 4140H: Thermal Treatment Temperatures

Quenching medium: oil

Treatment	Temperature range °C	Temperature range °F
Forging	1205-980	2200-1800
Annealing(a)	815-870	1500-1600
Normalizing(b)	845-900	1550-1650
Hardening	830-855	1525-1575
Tempering	(c)	(c)

(a) Maximum hardness of 197 HB. (b) Hardness of approximately 311 HB. (c) To desired hardness

Mechanical Properties

AISI 4140H: End-Quench Hardenability Limits and Hardenability Band.
Heat treating temperatures recommended by SAE: normalize at 870 °C (1600 °F), for forged or rolled specimen only; austenitize at 845 °C (1550 °F). (Ref 1)

Distance from quenched end, 1/16 in.	Hardness, HRC max	Hardness, HRC min
1	60	53
2	60	53
3	60	52
4	59	51
5	59	51
6	58	50
7	58	48
8	57	47
9	57	44
10	56	42
11	56	40
12	55	39
13	55	38
14	54	37
15	54	36
16	53	35
18	52	34
20	51	33
22	49	33
24	48	32
26	47	32
28	46	31
30	45	31
32	44	30

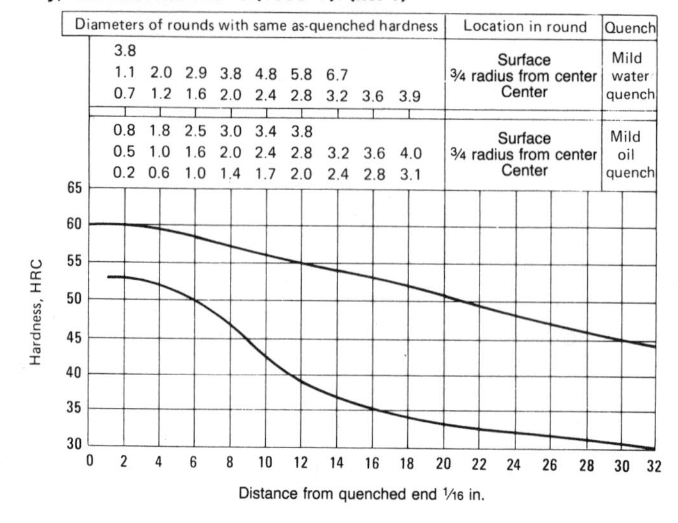

AISI 4140, 4140H: Mass Effect on Mechanical Properties (Ref 4)

Condition or treatment	Size round mm	Size round in.	Tensile strength MPa	Tensile strength ksi	Yield strength MPa	Yield strength ksi	Elongation(a), %	Reduction in area, %	Hardness, HB
Annealed (heated to 815 °C or 1500 °F; furnace cooled 11 °C or 20 °F per hour to 665 °C or 1230 °F; cooled in air)	25	1	655	95	415	60	25.7	56.9	197
Normalized (heated to 870 °C or 1600 °F; cooled in air)	13	0.5	1020	148	675	98	17.8	48.2	302
	25	1	1020	148	655	95	17.7	46.8	302
	50	2	972	141	635	92	16.5	48.1	285
	100	4	814	118	485	70	22.2	57.4	241
Oil quenched from 845 °C (1550 °F); tempered at 540 °C (1000 °F)	13	0.5	1185	172	1110	161	15.4	55.7	341
	25	1	1075	156	986	143	15.5	56.9	311
	50	2	965	140	800	116	17.5	59.8	285
	100	4	883	128	685	99	19.2	60.4	277
Oil quenched from 845 °C (1550 °F); tempered at 595 °C (1100 °F)	13	0.5	1090	158	1025	149	18.1	59.4	321
	25	1	965	140	931	135	19.5	62.3	285
	50	2	883	128	710	103	21.7	65.0	262
	100	4	807	117	600	87	21.5	62.1	235
Oil quenched from 845 °C (1550 °F); tempered at 650 °C (1200 °F)	13	0.5	938	136	889	129	19.9	62.3	277
	25	1	917	133	841	122	21.0	65.0	269
	50	2	841	122	675	98	23.2	65.8	241
	100	4	772	112	580	84	23.2	64.9	229

(a) In 50 mm (2 in.)

AISI 4140, 4140H: Effect of the Mass on Hardness at Selected Points (Ref 4)

Size round mm	Size round in.	As-quenched hardness after quenching in oil at: Surface	½ radius	Center
13	0.5	57 HRC	56 HRC	55 HRC
25	1	55 HRC	55 HRC	50 HRC
50	2	49 HRC	43 HRC	38 HRC
100	4	36 HRC	34.5 HRC	34 HRC

AISI 4140, 4140H: Hardness and Machinability (Ref 7)

Condition	Hardness range, HB	Average machinability rating(a)
Annealed and cold drawn	187-229	65

(a) Based on AISI 1212 steel as 100% average machinability

Mechanical Properties (continued)

AISI 4140, 4140H: Typical Room-Temperature Tensile Properties of Annealed, Normalized, and Hardened Bar (Ref 8)

Test specimens were oil-hardened and tempered, 25-mm (1-in.) rounds

Condition or treatment	Tempering temperature °C	°F	Tensile strength MPa	ksi	Yield strength MPa	ksi	Elongation(a), %	Reduction in area, %	Hardness, HB	Izod impact energy J	ft·lb
Annealed	...	...	655	95	485	70	26.0	52	185-200	...	...
Heat treated, oil-quenched, and tempered	205	400	1795	260	1515	220	8.0	32	520	14	10
	315	600	1585	230	1345	195	11.0	40	455	9	7
	425	800	1380	200	1170	170	15.0	48	385	22	16
	540	1000	1140	165	965	140	18.0	56	320	61	45
	650	1200	931	135	793	115	22.0	63	255	110	80
Normalized at 870 °C (1600 °F); reheated to 845 °C (1550 °F); quenched in oil; tempered	205	400	1965	285	1735	252	11.0	42	578	15	11
	260	500	1860	270	1655	240	11.0	44	534	11	8
	315	600	1725	250	1572	228	11.5	46	495	9	7
	370	700	1595	231	1460	212	12.5	48	461	15	11
	425	800	1450	210	1345	195	15.0	50	429	28	21
	480	900	1295	188	1205	175	16.0	52	388	46	34
	540	1000	1150	167	1050	152	17.5	55	341	65	48
	595	1100	1020	148	910	132	19.0	58	311	94	69
	650	1200	896	130	786	114	21.0	61	277	115	83
	705	1300	807	117	689	100	23.0	65	235	135	100

(a) In 50 mm (2 in.)

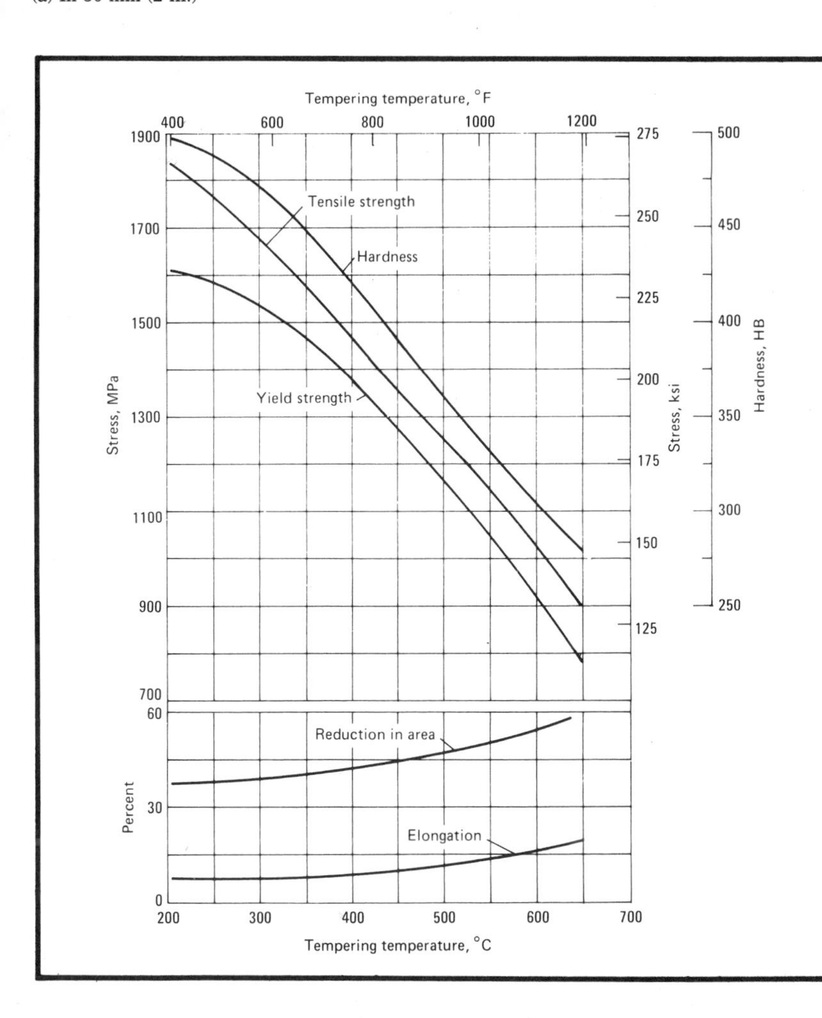

AISI 4140: Effect of Tempering Temperature on Tensile Properties.

Normalized at 870 °C (1600 °F); oil quenched from 845 °C (1550 °F); tempered at 56 °C (100 °F) intervals in 13.7-mm (0.540-in.) rounds; tested in 12.8-mm (0.505-in.) rounds. Tests were conducted using specimens machined to English units. Elongation was measured in 50 mm (2 in.). (Ref 2)

Mechanical Properties (continued)

AISI 4140, 4140H: Tensile Properties (Ref 5)

Condition or treatment	Tensile strength MPa	Tensile strength ksi	Yield strength MPa	Yield strength ksi	Elongation(a), %	Reduction in area, %	Hardness, HB	Izod impact energy J	Izod impact energy ft·lb
Normalized at 870 °C (1600 °F)	1020	148	655	95	17.7	47	302	23	17
Annealed at 815 °C (1500 °F)	655	95	420	61	25.7	57	197	54	40

(a) In 50 mm (2 in.)

AISI 4140: Effect of the Mass on Tensile Properties After Tempering at 540 °C (1000 °F). Normalized at 870 °C (1600 °F) in oversize rounds; oil quenched from 845 °C (1550 °F) in sizes given; tempered at 540 °C (1000 °F); tested in 12.8-mm (0.505-in.) rounds. Tests from bar 38.1-mm (1.50-in.) diam and larger were taken at half-radius position. Tests were conducted using specimens machined to English units. Elongation was measured in 50 mm (2 in.). (Ref 2)

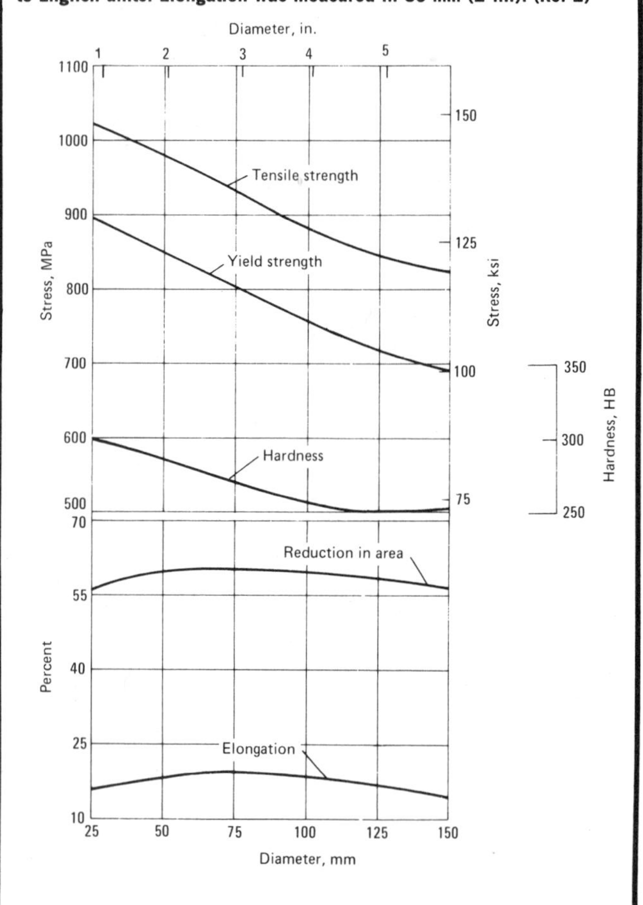

AISI 4140: Effect of the Mass on Tensile Properties After Tempering at 650 °C (1200 °F). Normalized at 870 °C (1600 °F) in oversize rounds; oil quenched from 845 °C (1550 °F) in sizes given; tempered at 650 °C (1200 °F); tested in 12.8-mm (0.505-in.) rounds. Tests from bar 38.1-mm (1.50-in.) diam and larger were taken at half-radius position. Tests were conducted using specimens machined to English units. Elongation was measured in 50 mm (2 in.). (Ref 2)

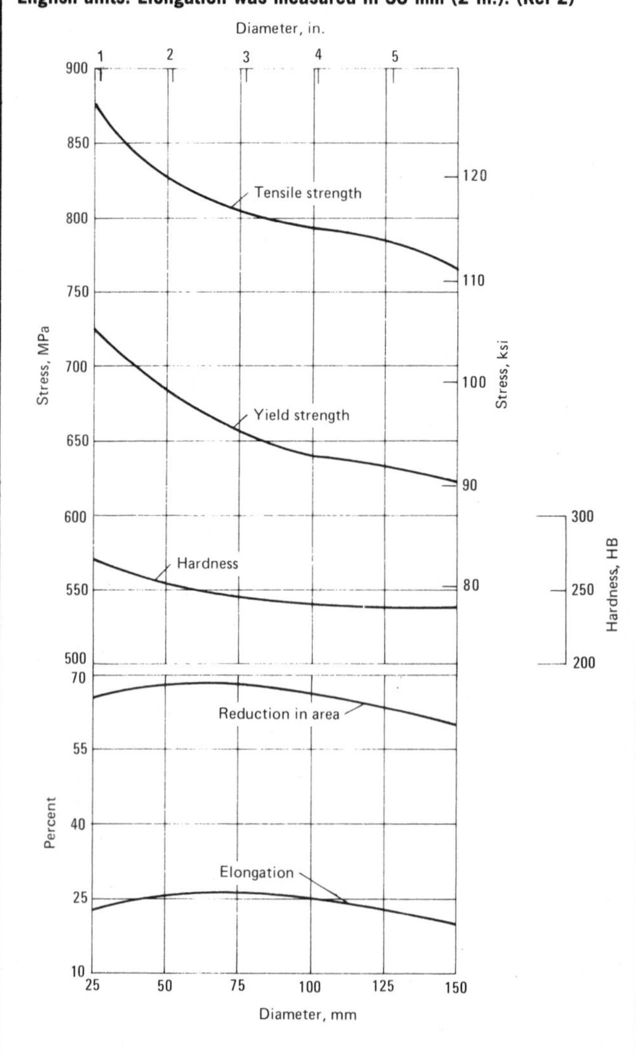

Machining Data

For machining data on AISI grades 4140 and 4140H, refer to the preceding machining tables for AISI grades 1340 and 1340H.

AISI 4142, 4142H

AISI 4142, 4142H: Chemical Composition

AISI grade	C	Mn	P max	S max	Si	Cr	Mo
	Chemical composition, %						
4142	0.40-0.45	0.75-1.10	0.035	0.040	0.15-0.30	0.80-1.10	0.15-0.25
4142H	0.39-0.46	0.65-1.10	0.035	0.040	0.15-0.30	0.75-1.20	0.15-0.25

Characteristics and Typical Uses. AISI grades 4142 and 4142H are available as hot rolled and cold rolled sheet and strip, hot rolled and cold finished bar, and cold heading quality wire. These steels have medium hardenability. Uses of grades 4142 and 4142H include small gears, pinions, high-strength bolts, cap screws, and socket- and recessed-head cap screws that can be heat treated.

AISI 4142, 4142H: Approximate Critical Points

Transformation point	Temperature(a) °C	°F	Transformation point	Temperature(a) °C	°F
Ac_1	730	1350	Ar_1	680	1255
Ac_3	805	1480	M_s	315	595
Ar_3	745	1370			

(a) On heating or cooling at 28 °C (50 °F) per hour

AISI 4142: Similar Steels (U.S. and/or Foreign). UNS G41420; ASTM A322, A331, A505, A519, A547; SAE J404, J412, J770

AISI 4142H: Similar Steels (U.S. and/or Foreign). UNS H41420; ASTM A304; SAE J1268; (W. Ger.) DIN 1.7223; (Ital.) UNI 38 CrMo 4

Physical Properties

AISI 4142, 4142H: Thermal Treatment Temperatures

Quenching medium: oil

Treatment	Temperature range °C	°F
Forging	1205 max	2200 max
Annealing	790-845	1450-1550
Hardening	845-870	1550-1600
Tempering	(a)	(a)

(a) To desired hardness

Mechanical Properties

AISI 4142H: End-Quench Hardenability Limits and Hardenability Band. Heat treating temperatures recommended by SAE: normalize at 870 °C (1600 °F), for forged or rolled specimens only; austenitize at 845 °C (1550 °F). (Ref 1)

Distance from quenched end, 1/16 in.	Hardness, HRC max	min
1	62	55
2	62	55
3	62	54
4	61	53
5	61	53
6	61	52
7	60	51
8	60	50
9	60	49
10	59	47
11	59	46
12	58	44
13	58	42
14	57	41
15	57	40
16	56	39
18	55	37
20	54	36
22	53	35
24	53	34
26	52	34
28	51	34
30	51	33
32	50	33

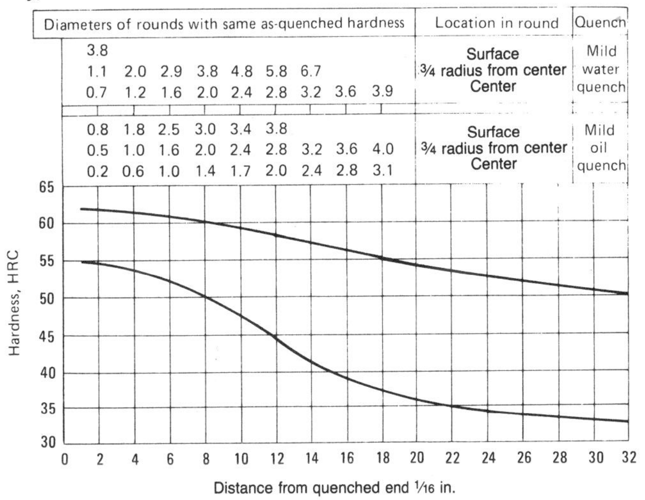

Mechanical Properties (continued)

AISI 4142, 4142H: Hardness and Machinability (Ref 7)

Condition	Hardness range, HB	Average machinability rating(a)
Annealed and cold drawn	187-229	65

(a) Based on AISI 1212 steel as 100% average machinability

Machining Data

For machining data on AISI grades 4142 and 4142H, refer to the preceding machining tables for AISI grades 1340 and 1340H.

AISI 4145, 4145H

AISI 4145, 4145H: Chemical Composition

AISI grade	C	Mn	P max	S max	Si	Cr	Mo
	Chemical composition, %						
4145	0.43-0.48	0.75-1.00	0.035	0.040	0.15-0.30	0.80-1.10	0.15-0.25
4145H	0.42-0.49	0.75-1.20	0.035	0.040	0.15-0.30	0.75-1.20	0.15-0.25

Characteristics and Typical Uses. AISI grades 4145 and 4145H are medium-hardenability, chromium-molybdenum steels. These grades are available as hot and cold rolled strip and sheet; hot rolled and cold finished bar; and seamless tube.

AISI 4145: Similar Steels (U.S. and/or Foreign). UNS G41450; ASTM A322, A331, A505, A519; MIL SPEC MIL-S-16974; SAE J404, J412, J770

AISI 4145H: Similar Steels (U.S. and/or Foreign). UNS H41450; ASTM A304; SAE J1268

AISI 4145, 4145H: Approximate Critical Points

Transformation point	Temperature(a) °C	Temperature(a) °F	Transformation point	Temperature(a) °C	Temperature(a) °F
Ac_1	725	1340	Ar_1	675	1250
Ac_3	800	1470	M_s	300	570
Ar_3	750	1380			

(a) On heating or cooling at 28 °C (50 °F) per hour

Mechanical Properties

AISI 4145H: End-Quench Hardenability Limits and Hardenability Band. Heat treating temperatures recommended by SAE: normalize at 870 °C (1600 °F), for forged or rolled specimens only; austenitize at 845 °C (1550 °F). (Ref 1)

Distance from quenched end, 1/16 in.	Hardness, HRC max	Hardness, HRC min
1	63	56
2	63	55
3	62	55
4	62	54
5	62	53
6	61	53
7	61	52
8	61	52
9	60	51
10	60	50
11	60	49
12	59	48
13	59	46
14	59	45
15	58	43
16	58	42
18	57	40
20	57	38
22	56	37
24	55	36
26	55	35
28	55	35
30	55	34
32	54	34

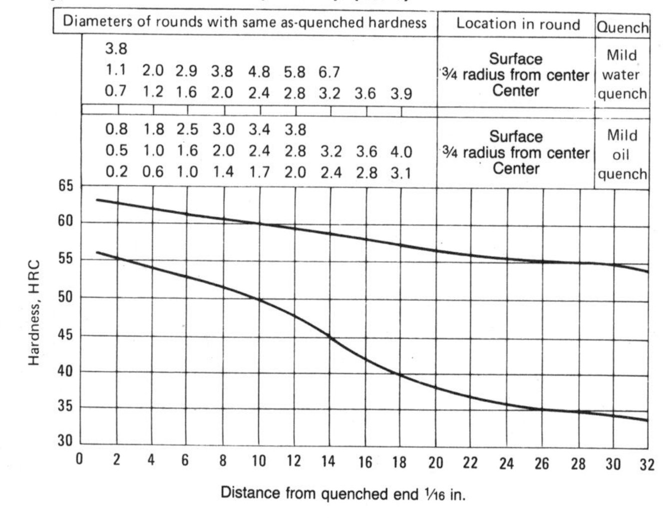

Mechanical Properties (continued)

AISI 4145, 4145H: Hardness and Machinability (Ref 7)

Condition	Hardness range, HB	Average machinability rating(a)
Annealed and cold drawn	187-229	60

(a) Based on AISI 1212 steel as 100% average machinability

Machining Data

For machining data on AISI grades 4145 and 4145H, refer to the preceding machining tables for AISI grades 1340 and 1340H.

Physical Properties

AISI 4145, 4145H: Thermal Treatment Temperatures

Quenching medium: oil

Treatment	Temperature range °C	°F
Annealing	790-845	1450-1550
Hardening	815-845	1500-1550
Tempering	(a)	(a)

(a) To desired hardness

AISI 4147, 4147H

AISI 4147, 4147H: Chemical Composition

AISI grade	C	Mn	P max	S max	Si	Cr	Mo
	Chemical composition, %						
4147	0.45-0.50	0.75-1.00	0.035	0.040	0.15-0.30	0.80-1.10	0.15-0.30
4147H	0.44-0.51	0.65-1.10	0.035	0.040	0.15-0.30	0.75-1.20	0.15-0.30

Characteristics and Typical Uses. AISI grades 4147 and 4147H are medium-hardenability, chromium-molybdenum, alloy steels, which are available as hot rolled and cold rolled sheet and strip, and hot rolled and cold finished bar and rod. These steels are used primarily for gears and other parts requiring fairly high hardness, as well as strength and toughness.

AISI 4147: Similar Steels (U.S. and/or Foreign). UNS G41470; ASTM A322, A331, A505, A519; SAE J404, J412, J770; (W. Ger.) DIN 1.7228; (Jap.) JIS SCM 5 H, SCM 5

AISI 4147H: Similar Steels (U.S. and/or Foreign). UNS H41470; ASTM A304; SAE J1268; (W. Ger.) DIN 1.7228; (Jap.) JIS SCM 5 H, SCM 5

AISI 4147, 4147H: Approximate Critical Points

Transformation point	Temperature(a) °C	°F	Transformation point	Temperature(a) °C	°F
Ac_1	735	1355	Ar_1	670	1240
Ac_3	790	1455	M_s	290	555
Ar_3	730	1350			

(a) On heating or cooling at 28 °C (50 °F) per hour

Machining Data

For machining data on AISI grades 4147 and 4147H, refer to the preceding machining tables for AISI grades 1340 and 1340H.

Mechanical Properties

AISI 4147, 4147H: Hardness and Machinability (Ref 7)

Condition	Hardness range, HB	Average machinability rating(a)
Annealed and cold drawn	187-235	60

(a) Based on AISI 1212 steel as 100% average machinability

Physical Properties

AISI 4147, 4147H: Thermal Treatment Temperatures

Quenching medium: oil

Treatment	Temperature range °C	°F
Annealing	790-845	1450-1550
Hardening	815-845	1500-1550
Tempering	(a)	(a)

(a) To desired hardness

Mechanical Properties (continued)

AISI 4147H: End-Quench Hardenability Limits and Hardenability Band. Heat treating temperatures recommended by SAE: normalize at 870 °C (1600 °F), for forged or rolled specimens only; austenitize at 845 °C (1550 °F). (Ref 1)

Distance from quenched end, 1/16 in.	Hardness, HRC max	Hardness, HRC min
1	64	57
2	64	57
3	64	56
4	64	56
5	63	55
6	63	55
7	63	55
8	63	54
9	63	54
10	62	53
11	62	52
12	62	51
13	61	49
14	61	48
15	60	46
16	60	45
18	59	42
20	59	40
22	58	39
24	57	38
26	57	37
28	57	37
30	56	37
32	56	36

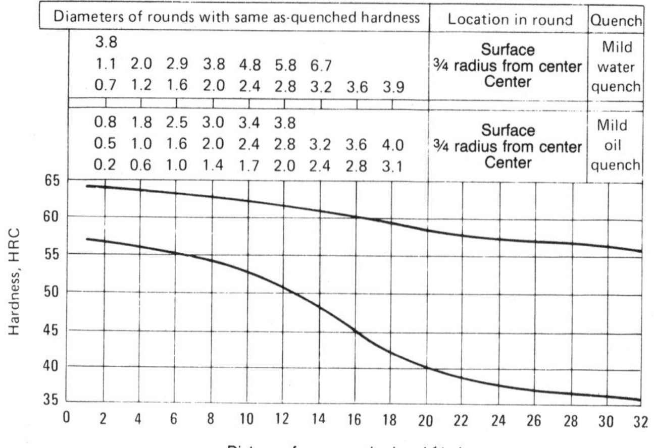

AISI 4150, 4150H

AISI 4150, 4150H: Chemical Composition

AISI grade	C	Mn	P max	S max	Si	Cr	Mo
4150	0.48-0.53	0.75-1.00	0.035	0.040	0.15-0.30	0.80-1.10	0.15-0.25
4150H	0.47-0.54	0.65-1.10	0.035	0.040	0.15-0.30	0.75-1.20	0.15-0.25

(Chemical composition, %)

Characteristics and Typical Uses. AISI grades 4150 and 4150H are medium-carbon, chromium-molybdenum steels. They provide deeper hardenability for uniform through hardening in larger sections, with small loss in machinability. In the preheat-treated condition, these steels provide optimum strength and hardness. Many producers match the carbon content to the bar size, giving uniform properties throughout each section and from bar to bar. Both annealed and heat treated bar are available resulfurized for increased machinability.

AISI 4150: Similar Steels (U.S. and/or Foreign). UNS G41500; ASTM A322, A331, A505, A519; MIL SPEC MIL-S-11595 (ORD4150); SAE J404, J412, J770; (W. Ger.) DIN 1.7228; (Jap.) JIS SCM 5 H, SCM 5

AISI 4150H: Similar Steels (U.S. and/or Foreign). UNS H41500; ASTM A304; SAE J1268; (W. Ger.) DIN 1.7228; (Jap.) JIS SCM 5 H, SCM 5

AISI 4150, 4150H: Approximate Critical Points

Transformation point	Temperature(a) °C	Temperature(a) °F	Transformation point	Temperature(a) °C	Temperature(a) °F
Ac_1	745	1370	Ar_1	670	1240
Ac_3	765	1410	M_s	275	530
Ar_3	730	1345			

(a) On heating or cooling at 28 °C (50 °F) per hour

Physical Properties

AISI 4150, 4150H: Thermal Treatment Temperatures
Quenching medium: oil

Treatment	Temperature range °C	Temperature range °F
Forging	1205 max	2200 max
Annealing(a)	790-845	1450-1550
Normalizing(b)	845-900	1550-1650
Hardening	830-855	1525-1575
Tempering	(c)	(c)

(a) Maximum hardness of 212 HB. (b) Hardness of approximately 387 HB. (c) To desired hardness

Mechanical Properties

AISI 4150, 4150H: Mass Effect on Mechanical Properties (Ref 4)

Condition or treatment	Size round mm	Size round in.	Tensile strength MPa	Tensile strength ksi	Yield strength MPa	Yield strength ksi	Elongation(a), %	Reduction in area, %	Hardness, HB
Annealed (heated to 830 °C or 1525 °F; furnace cooled 11 °C or 20 °F per hour to 645 °C or 1190 °F; cooled in air)	25	1	731	106	380	55	20.2	40.2	197
Normalized (heated to 870 °C or 1600 °F; cooled in air)	13	0.5	1340	194	896	130	10.0	24.8	375
	25	1	1160	168	731	106	11.7	30.8	321
	50	2	1095	159	717	104	13.5	40.6	311
	100	4	1005	146	635	92	19.5	56.5	293
Oil quenched from 830 °C (1525 °F); tempered at 540 °C (1000 °F)	13	0.5	1310	190	1215	176	13.5	47.2	375
	25	1	1205	175	1105	160	14.0	46.5	352
	50	2	1165	169	1040	151	15.5	51.0	341
	100	4	1095	159	883	128	15.0	46.7	311
Oil quenched from 830 °C (1525 °F); tempered at 595 °C (1100 °F)	13	0.5	1170	170	1076	156	14.6	45.5	341
	25	1	1145	166	1034	150	15.7	51.1	331
	50	2	1035	150	910	132	18.7	56.4	302
	100	4	910	132	675	98	20.0	57.5	269
Oil quenched from 830 °C (1525 °F); tempered at 650 °C (1200 °F)	13	0.5	1020	148	945	137	17.4	53.3	302
	25	1	972	141	883	128	18.7	55.7	285
	50	2	931	135	814	118	20.5	60.0	269
	100	4	855	124	625	91	21.5	61.4	255

(a) In 50 mm (2 in.)

AISI 4150, 4150H: Effect of the Mass on Hardness at Selected Points (Ref 4)

Size round mm	Size round in.	As-quenched hardness after quenching in oil at: Surface	½ radius	Center
13	0.5	64 HRC	64 HRC	63 HRC
25	1	62 HRC	62 HRC	62 HRC
50	2	58 HRC	57 HRC	56 HRC
100	4	47 HRC	43 HRC	42 HRC

AISI 4150, 4150H: Hardness and Machinability (Ref 7)

Condition	Hardness range, HB	Average machinability rating(a)
Annealed and cold drawn	187-241	55

(a) Based on AISI 1212 steel as 100% average machinability

AISI 4150H: End-Quench Hardenability Limits and Hardenability Band. Heat treating temperatures recommended by SAE: normalize at 870 °C (1600 °F), for forged or rolled specimens only; austenitize at 845 °C (1550 °F). (Ref 1)

Distance from quenched end, 1/16 in.	Hardness, HRC max	Hardness, HRC min
1	65	59
2	65	59
3	65	59
4	65	58
5	65	58
6	65	57
7	65	57
8	64	56
9	64	56
10	64	55
11	64	54
12	63	53
13	63	51
14	62	50
15	62	48
16	62	47
18	61	45
20	60	43
22	59	41
24	59	40
26	58	39
28	58	38
30	58	38
32	58	38

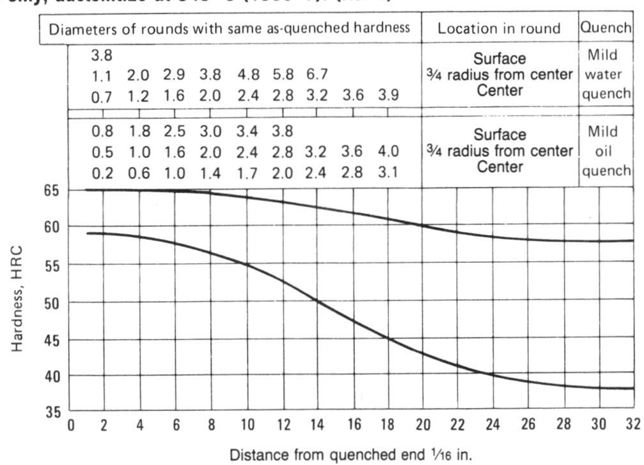

Mechanical Properties (continued)

AISI 4150, 4150H: Tensile Properties (Ref 5)

Condition or treatment	Tensile strength MPa	Tensile strength ksi	Yield strength MPa	Yield strength ksi	Elongation(a), %	Reduction in area, %	Hardness, HB	Izod impact energy J	Izod impact energy ft·lb
Normalized at 870 °C (1600 °F)	1160	168	740	107	11.7	31	321	12	9
Annealed at 815 °C (1500 °F)	731	106	380	55	20.2	40	197	24	18

(a) In 50 mm (2 in.)

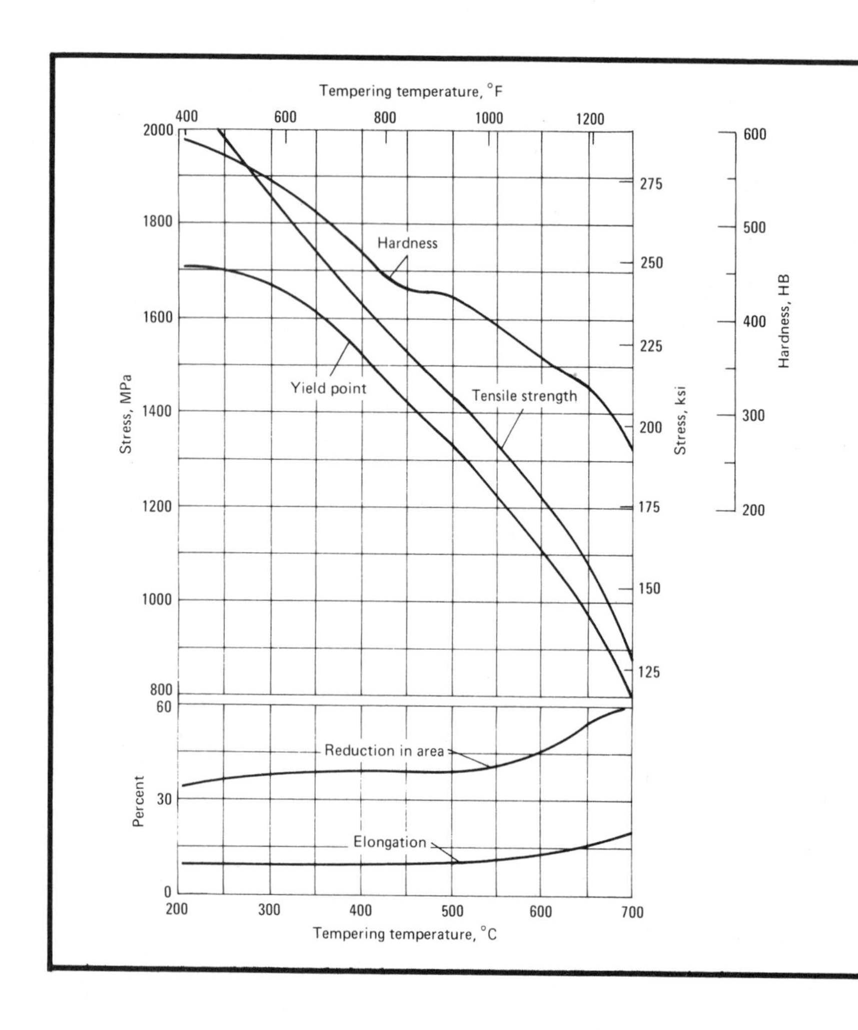

AISI 4150: Effect of Tempering Temperature on Tensile Properties. Normalized at 870 °C (1600 °F); reheated to 830 °C (1525 °F); quenched in agitated oil. Specimens were treated in 13.5-mm (0.530-in.) diam and machined to 12.8-mm (0.505-in.) diam for testing. Tests were conducted using specimens machined to English units. As-quenched hardness of 656 HB; elongation measured in 50 mm (2 in.) (Ref 4)

Machining Data (Ref 3)

AISI 4150, 4150H: Drilling

High speed steel tool material	Speed m/min	Speed ft/min	1.5 mm (0.062 in.) mm	1.5 mm (0.062 in.) in.	3 mm (0.125 in.) mm	3 mm (0.125 in.) in.	6 mm (0.25 in.) mm	6 mm (0.25 in.) in.	12 mm (0.5 in.) mm	12 mm (0.5 in.) in.	18 mm (0.75 in.) mm	18 mm (0.75 in.) in.	25 mm (1 in.) mm	25 mm (1 in.) in.
			Feed per revolution for nominal hole diameter of:											
Hardness, 175 to 225 HB														
M10, M7, and M1	18	60	0.025	0.001	0.075	0.003	0.15	0.006	0.25	0.010	0.36	0.014	0.40	0.016
Hardness, 225 to 275 HB														
M10, M7, and M1	15	50	0.025	0.001	0.075	0.003	0.102	0.004	0.18	0.007	0.25	0.010	0.30	0.012
Hardness, 275 to 325 HB														
M10, M7, and M1	12	40	...	...	0.050	0.002	0.102	0.004	0.15	0.006	0.20	0.008	0.23	0.009
Hardness, 325 to 375 HB														
M10, M7, and M1	11	35	...	...	0.050	0.002	0.075	0.003	0.13	0.005	0.20	0.008	0.23	0.009

Machining Data (Ref 3) (continued)

AISI 4150, 4150H: Turning (Single Point and Box Tools)

Depth of cut		High speed steel				Uncoated carbide						Coated carbide			
		Speed		Feed		Speed, brazed		Speed, inserted		Feed		Speed		Feed	
mm	in.	m/min	ft/min	mm/rev	in./rev	m/min	ft/min	m/min	ft/min	mm/rev	in./rev	m/min	ft/min	mm/rev	in./rev
Hardness, 175 to 225 HB															
1	0.040	40(a)	130(a)	0.18	0.007	115(b)	380(b)	150(b)	500(b)	0.18	0.007	200(c)	650(c)	0.18	0.007
4	0.150	30(a)	100(a)	0.40	0.015	90(d)	300(d)	120(d)	390(d)	0.50	0.020	150(e)	500(e)	0.40	0.015
8	0.300	23(a)	75(a)	0.50	0.020	72(d)	235(d)	95(d)	310(d)	0.75	0.030	120(e)	400(e)	0.50	0.020
16	0.625	18(a)	60(a)	0.75	0.030	56(d)	185(d)	73(d)	240(d)	1.00	0.040	...	...	...	...
Hardness, 225 to 275 HB															
1	0.040	34(a)	110(a)	0.18	0.007	105(b)	350(b)	130(b)	430(b)	0.18	0.007	175(c)	575(c)	0.18	0.007
4	0.150	26(a)	85(a)	0.40	0.015	85(d)	280(d)	105(d)	340(d)	0.50	0.020	135(e)	450(e)	0.40	0.015
8	0.300	20(a)	65(a)	0.50	0.020	67(d)	220(d)	82(d)	270(d)	0.75	0.030	105(e)	350(e)	0.50	0.020
16	0.625	15(a)	50(a)	0.75	0.030	53(d)	175(d)	64(d)	210(d)	1.00	0.040	...	...	...	...
Hardness, 275 to 325 HB															
1	0.040	26(f)	85(f)	0.18	0.007	100(b)	330(b)	125(b)	415(b)	0.18	0.007	170(c)	550(c)	0.18	0.007
4	0.150	20(f)	65(f)	0.40	0.015	79(d)	260(d)	100(d)	325(d)	0.40	0.015	130(e)	425(e)	0.40	0.015
8	0.300	15(f)	50(f)	0.50	0.020	60(d)	200(d)	76(d)	250(d)	0.75	0.020	100(e)	325(e)	0.50	0.020
Hardness, 325 to 375 HB															
1	0.040	18(f)	60(f)	0.18	0.007	84(b)	275(b)	105(b)	350(b)	0.18	0.007	135(c)	450(c)	0.18	0.007
4	0.150	15(f)	50(f)	0.40	0.015	66(d)	215(d)	84(d)	275(d)	0.40	0.015	105(e)	350(e)	0.40	0.015
8	0.300	11(f)	35(f)	0.50	0.020	53(d)	175(d)	66(d)	215(d)	0.75	0.020	84(e)	275(e)	0.50	0.020

(a) High speed steel tool material: M2 or M3. (b) Carbide tool material: C-7. (c) Carbide tool material: CC-7. (d) Carbide tool material: C-6. (e) Carbide tool material: CC-6. (f) Any premium high speed steel tool material (T15, M33, M41 to M47)

AISI 4150, 4150H: End Milling (Profiling)

Tool material	Depth of cut		Speed		Feed per tooth for cutter diameter of:							
					10 mm (0.375 in.)		12 mm (0.5 in.)		18 mm (0.75 in.)		25-50 mm (1-2 in.)	
	mm	in.	m/min	ft/min	mm	in.	mm	in.	mm	in.	mm	in.
Hardness, 175 to 225 HB												
M2, M3, and M7 high speed steel	0.5	0.020	37	120	0.025	0.001	0.050	0.002	0.075	0.003	0.102	0.004
	1.5	0.060	27	90	0.050	0.002	0.075	0.003	0.102	0.004	0.13	0.005
	diam/4	diam/4	24	80	0.038	0.0015	0.050	0.002	0.075	0.003	0.102	0.004
	diam/2	diam/2	21	70	0.025	0.001	0.038	0.0015	0.050	0.002	0.075	0.003
C-5 carbide	0.5	0.020	140	455	0.038	0.0015	0.075	0.003	0.13	0.005	0.15	0.006
	1.5	0.060	105	350	0.063	0.0025	0.102	0.004	0.15	0.006	0.18	0.007
	diam/4	diam/4	90	295	0.050	0.002	0.075	0.003	0.13	0.005	0.15	0.006
	diam/2	diam/2	84	275	0.038	0.0015	0.050	0.002	0.102	0.004	0.13	0.005
Hardness, 225 to 275 HB												
M2, M3, and M7 high speed steel	0.5	0.020	30	100	0.025	0.001	0.050	0.002	0.075	0.003	0.102	0.004
	1.5	0.060	23	75	0.050	0.002	0.075	0.003	0.102	0.004	0.13	0.005
	diam/4	diam/4	20	65	0.038	0.0015	0.050	0.002	0.075	0.003	0.102	0.004
	diam/2	diam/2	18	60	0.025	0.001	0.038	0.0015	0.050	0.002	0.075	0.003
C-5 carbide	0.5	0.020	120	390	0.025	0.001	0.050	0.002	0.102	0.004	0.13	0.005
	1.5	0.060	90	300	0.050	0.002	0.075	0.003	0.13	0.005	0.18	0.007
	diam/4	diam/4	78	255	0.038	0.0015	0.050	0.002	0.102	0.004	0.13	0.005
	diam/2	diam/2	72	235	0.025	0.001	0.038	0.0015	0.075	0.003	0.102	0.004
Hardness, 275 to 325 HB												
M2, M3, and M7 high speed steel	0.5	0.020	26	85	0.018	0.0007	0.038	0.0015	0.075	0.003	0.102	0.004
	1.5	0.060	20	65	0.025	0.001	0.050	0.002	0.102	0.004	0.13	0.005
	diam/4	diam/4	17	55	0.018	0.0007	0.038	0.0015	0.075	0.003	0.102	0.004
	diam/2	diam/2	15	50	0.013	0.0005	0.025	0.001	0.050	0.002	0.075	0.003
C-5 carbide	0.5	0.020	95	310	0.025	0.001	0.050	0.002	0.102	0.004	0.13	0.005
	1.5	0.060	72	235	0.050	0.002	0.075	0.003	0.13	0.005	0.15	0.006
	diam/4	diam/4	62	205	0.038	0.0015	0.050	0.002	0.102	0.004	0.13	0.005
	diam/2	diam/2	58	190	0.025	0.001	0.038	0.0015	0.075	0.003	0.102	0.004
Hardness, 325 to 375 HB												
M2, M3, and M7 high speed steel	0.5	0.020	20	65	0.013	0.0005	0.038	0.0015	0.075	0.003	0.102	0.004
	1.5	0.060	17	55	0.013	0.0005	0.038	0.0015	0.102	0.004	0.13	0.005
	diam/4	diam/4	15	50	0.013	0.0005	0.038	0.0015	0.075	0.003	0.102	0.004
	diam/2	diam/2	12	40	0.013	0.0005	0.025	0.001	0.050	0.002	0.075	0.003
C-5 carbide	0.5	0.020	79	260	0.025	0.001	0.038	0.0015	0.075	0.003	0.13	0.005
	1.5	0.060	60	200	0.038	0.0015	0.075	0.003	0.13	0.005	0.15	0.006
	diam/4	diam/4	52	170	0.038	0.0015	0.050	0.002	0.102	0.004	0.13	0.005
	diam/2	diam/2	49	160	0.025	0.001	0.050	0.002	0.075	0.003	0.102	0.004

Machining Data (Ref 3) (continued)

AISI 4150, 4150H: Face Milling

Depth of cut		High speed steel				Uncoated carbide						Coated carbide			
		Speed		Feed/tooth		Speed, brazed		Speed, inserted		Feed/tooth		Speed		Feed/tooth	
mm	in.	m/min	ft/min	mm	in.	m/min	ft/min	m/min	ft/min	mm	in.	m/min	ft/min	mm	in.
Hardness, 175 to 225 HB															
1	0.040	52(a)	170(a)	0.20	0.008	135(b)	450(b)	170(b)	550(b)	0.20	0.008	250(c)	825(c)	0.20	0.008
4	0.150	40(a)	130(a)	0.30	0.012	105(b)	350(b)	130(b)	450(b)	0.30	0.012	170(c)	560(c)	0.30	0.012
8	0.300	30(a)	100(a)	0.40	0.016	84(d)	275(d)	100(d)	335(d)	0.40	0.016	135(e)	435(e)	0.40	0.016
Hardness, 225 to 275 HB															
1	0.040	41(a)	135(a)	0.15	0.006	120(b)	390(b)	145(b)	475(b)	0.18	0.007	215(c)	700(c)	0.18	0.007
4	0.150	29(a)	95(a)	0.25	0.010	95(b)	310(b)	115(b)	375(b)	0.25	0.010	150(c)	485(c)	0.25	0.010
8	0.300	21(a)	70(a)	0.36	0.014	72(d)	235(d)	88(d)	290(d)	0.36	0.014	115(e)	375(e)	0.36	0.014
Hardness, 275 to 325 HB															
1	0.040	29(f)	95(f)	0.15	0.006	115(b)	375(b)	140(b)	455(b)	0.15	0.006	205(c)	680(c)	0.13	0.005
4	0.150	23(f)	75(f)	0.23	0.009	88(b)	290(b)	110(b)	355(b)	0.20	0.008	140(c)	460(c)	0.18	0.007
8	0.300	18(f)	60(f)	0.30	0.012	69(d)	225(d)	84(d)	275(d)	0.25	0.010	110(e)	355(e)	0.23	0.009
Hardness, 325 to 375 HB															
1	0.040	24(f)	80(f)	0.13	0.005	95(b)	315(b)	115(b)	385(b)	0.13	0.005	175(c)	575(c)	0.102	0.004
4	0.150	18(f)	60(f)	0.20	0.008	75(b)	245(b)	90(b)	300(b)	0.18	0.007	120(c)	390(c)	0.15	0.006
8	0.300	15(f)	50(f)	0.25	0.010	58(d)	190(d)	72(d)	235(d)	0.23	0.009	95(e)	305(e)	0.20	0.008

(a) High speed steel tool material: M2 or M3. (b) Carbide tool material: C-6. (c) Carbide tool material: CC-6. (d) Carbide tool material: C-5. (e) Carbide tool material: CC-5. (f) Any premium high speed steel tool material (T15, M33, M41 to M47)

AISI 4150, 4150H: Boring

Depth of cut		High speed steel				Uncoated carbide						Coated carbide			
		Speed		Feed		Speed, brazed		Speed, inserted		Feed		Speed		Feed	
mm	in.	m/min	ft/min	mm/rev	in./rev	m/min	ft/min	m/min	ft/min	mm/rev	in./rev	m/min	ft/min	mm/rev	in./rev
Hardness, 175 to 225 HB															
0.25	0.010	38(a)	125(a)	0.075	0.003	115(b)	370(b)	135(b)	435(b)	0.075	0.003	175(c)	570(c)	0.075	0.003
1.25	0.050	32(a)	105(a)	0.13	0.005	90(d)	300(d)	105(d)	350(d)	0.13	0.005	140(e)	455(e)	0.13	0.005
2.5	0.100	24(a)	80(a)	0.30	0.012	70(d)	230(d)	82(d)	270(d)	0.40	0.015	105(e)	350(e)	0.30	0.012
Hardness, 225 to 275 HB															
0.25	0.010	34(a)	110(a)	0.075	0.003	100(b)	320(b)	115(b)	375(b)	0.075	0.003	150(c)	500(c)	0.075	0.003
1.25	0.050	27(a)	90(a)	0.13	0.005	78(d)	255(d)	90(d)	300(d)	0.13	0.005	120(e)	400(e)	0.13	0.005
2.5	0.100	21(a)	70(a)	0.30	0.012	62(d)	205(d)	73(d)	240(d)	0.40	0.015	95(e)	315(e)	0.30	0.012
Hardness, 275 to 325 HB															
0.25	0.010	27(f)	90(f)	0.075	0.003	95(b)	305(b)	110(b)	360(b)	0.075	0.003	145(c)	480(c)	0.075	0.003
1.25	0.050	21(f)	70(f)	0.13	0.005	75(d)	245(d)	88(d)	290(d)	0.13	0.005	115(e)	385(e)	0.13	0.005
2.5	0.100	15(f)	50(f)	0.30	0.012	58(d)	190(d)	69(d)	225(d)	0.30	0.012	90(e)	300(e)	0.30	0.012
Hardness, 325 to 375 HB															
0.25	0.010	18(f)	60(f)	0.075	0.003	79(b)	260(b)	95(b)	305(b)	0.075	0.003	120(c)	395(c)	0.075	0.003
1.25	0.050	15(f)	50(f)	0.13	0.005	64(d)	210(d)	75(d)	245(d)	0.13	0.005	95(e)	315(e)	0.13	0.005
2.5	0.100	12(f)	40(f)	0.30	0.012	49(d)	160(d)	58(d)	190(d)	0.30	0.012	75(e)	245(e)	0.30	0.012

(a) High speed steel tool material: M2 or M3. (b) Carbide tool material: C-7. (c) Carbide tool material: CC-7. (d) Carbide tool material: C-6. (e) Carbide tool material: CC-6. (f) Any premium high speed steel tool material (T15, M33, M41 to M47)

AISI 4150, 4150H: Planing

Tool material	Depth of cut mm	Depth of cut in.	Speed m/min	Speed ft/min	Feed/stroke mm	Feed/stroke in.
Hardness, 175 to 225 HB						
M2 and M3 high speed steel	0.1	0.005	9	30	(a)	(a)
	2.5	0.100	14	45	1.25	0.050
	12	0.500	8	25	1.50	0.060
C-6 carbide	0.1	0.005	58	190	(a)	(a)
	2.5	0.100	70	230	2.05	0.080
	12	0.500	55	180	1.50	0.060
Hardness, 225 to 275 HB						
M2 and M3 high speed steel	0.1	0.005	8	25	(a)	(a)
	2.5	0.100	12	40	0.75	0.030
	12	0.500	6	20	1.15	0.045
C-6 carbide	0.1	0.005	49	160	(a)	(a)
	2.5	0.100	58	190	1.50	0.060
	12	0.500	46	150	1.25	0.050
Hardness, 275 to 325 HB						
M2 and M3 high speed steel	0.1	0.005	6	20	(a)	(a)
	2.5	0.100	11	35	0.75	0.030
	12	0.500	5	15	1.15	0.045
C-6 carbide	0.1	0.005	43	140	(a)	(a)
	2.5	0.100	52	170	1.50	0.060
	12	0.500	37	120	1.25	0.050
Hardness, 325 to 375 HB						
M2 and M3 high speed steel	0.1	0.005	6	20	(a)	(a)
	2.5	0.100	9	30	0.75	0.030
	12	0.500	5	15	1.15	0.045
C-6 carbide	0.1	0.005	37	120	(a)	(a)
	2.5	0.100	49	160	1.50	0.060
	12	0.500	34	110	1.25	0.050

(a) Feed is 75% the width of the square nose finishing tool

Machining Data (Ref 3) (continued)

AISI 4150, 4150H: Turning (Cutoff and Form Tools)

Tool material	Speed, m/min (ft/min)	Feed per revolution for cutoff tool width of: 1.5 mm (0.062 in.) mm	in.	3 mm (0.125 in.) mm	in.	6 mm (0.25 in.) mm	in.	Feed per revolution for form tool width of: 12 mm (0.5 in.) mm	in.	18 mm (0.75 in.) mm	in.	25 mm (1 in.) mm	in.	50 mm (2 in.) mm	in.
Hardness, 175 to 225 HB															
M2 and M3 high speed steel	27 (90)	0.038	0.0015	0.046	0.0018	0.056	0.0022	0.046	0.0018	0.041	0.0016	0.038	0.0015	0.028	0.0011
C-6 carbide	88 (290)	0.038	0.0015	0.046	0.0018	0.056	0.0022	0.046	0.0018	0.041	0.0016	0.038	0.0015	0.028	0.0011
Hardness, 225 to 275 HB															
M2 and M3 high speed steel	21 (70)	0.036	0.0014	0.043	0.0017	0.053	0.0021	0.043	0.0017	0.038	0.0015	0.036	0.0014	0.025	0.0010
C-6 carbide	69 (225)	0.036	0.0014	0.043	0.0017	0.053	0.0021	0.043	0.0017	0.038	0.0015	0.036	0.0014	0.025	0.0010
Hardness, 275 to 325 HB															
M2 and M3 high speed steel	15 (50)	0.033	0.0013	0.041	0.0016	0.050	0.0020	0.041	0.0016	0.036	0.0014	0.033	0.0013	0.023	0.0009
C-6 carbide	46 (150)	0.033	0.0013	0.041	0.0016	0.050	0.0020	0.041	0.0016	0.036	0.0014	0.033	0.0013	0.023	0.0009
Hardness, 325 to 375 HB															
Any premium high speed steel (T15, M33, M41-M47)	12 (40)	0.030	0.0012	0.038	0.0015	0.048	0.0019	0.038	0.0015	0.033	0.0013	0.030	0.0012	0.020	0.0008
C-6 carbide	38 (125)	0.030	0.0012	0.038	0.0015	0.048	0.0019	0.038	0.0015	0.033	0.0013	0.030	0.0012	0.020	0.0008

AISI 4150, 4150H: Reaming

Based on 4 flutes for 3- and 6-mm (0.125- and 0.25-in.) reamers, 6 flutes for 12-mm (0.5-in.) reamers, and 8 flutes for 25-mm (1-in.) and larger reamers

Tool material	Speed m/min	Speed ft/min	Feed per revolution for reamer diameter of: 3 mm (0.125 in.) mm	in.	6 mm (0.25 in.) mm	in.	12 mm (0.5 in.) mm	in.	25 mm (1 in.) mm	in.	35 mm (1.5 in.) mm	in.	50 mm (2 in.) mm	in.
Hardness, 175 to 225 HB														
Roughing														
M1, M2, and M7 high speed steel	21	70	0.102	0.004	0.18	0.007	0.30	0.012	0.50	0.020	0.65	0.025	0.75	0.030
C-2 carbide	26	85	0.102	0.004	0.18	0.007	0.30	0.012	0.50	0.020	0.65	0.025	0.75	0.030
Finishing														
M1, M2, and M7 high speed steel	11	35	0.13	0.005	0.20	0.008	0.30	0.012	0.50	0.020	0.65	0.025	0.75	0.030
C-2 carbide	14	45	0.13	0.005	0.20	0.008	0.30	0.012	0.50	0.020	0.65	0.025	0.75	0.030
Hardness, 225 to 275 HB														
Roughing														
M1, M2, and M7 high speed steel	18	60	0.075	0.003	0.15	0.006	0.25	0.010	0.38	0.015	0.50	0.020	0.65	0.025
C-2 carbide	23	75	0.102	0.004	0.15	0.006	0.25	0.010	0.38	0.015	0.50	0.020	0.65	0.025
Finishing														
M1, M2, and M7 high speed steel	9	30	0.102	0.004	0.18	0.007	0.25	0.010	0.38	0.015	0.50	0.020	0.65	0.025
C-2 carbide	12	40	0.102	0.004	0.18	0.007	0.25	0.010	0.38	0.015	0.50	0.020	0.65	0.025
Hardness, 275 to 325 HB														
Roughing														
M1, M2, and M7 high speed steel	15	50	0.075	0.003	0.13	0.005	0.20	0.008	0.30	0.012	0.38	0.015	0.50	0.020
C-2 carbide	20	65	0.102	0.004	0.15	0.006	0.20	0.008	0.30	0.012	0.38	0.015	0.50	0.020
Finishing														
M1, M2, and M7 high speed steel	8	25	0.075	0.003	0.15	0.006	0.20	0.008	0.30	0.012	0.38	0.015	0.50	0.020
C-2 carbide	11	35	0.102	0.004	0.15	0.006	0.20	0.008	0.30	0.012	0.38	0.015	0.50	0.020

(continued)

Machining Data (Ref 3) (continued)

AISI 4150, 4150H: Reaming (continued)

Tool material	Speed m/min	Speed ft/min	3 mm (0.125 in.) mm	3 mm (0.125 in.) in.	6 mm (0.25 in.) mm	6 mm (0.25 in.) in.	12 mm (0.5 in.) mm	12 mm (0.5 in.) in.	25 mm (1 in.) mm	25 mm (1 in.) in.	35 mm (1.5 in.) mm	35 mm (1.5 in.) in.	50 mm (2 in.) mm	50 mm (2 in.) in.
			Feed per revolution for reamer diameter of:											
			Hardness, 325 to 375 HB											
Roughing														
M1, M2, and M7 high speed steel	11	35	0.050	0.002	0.102	0.004	0.13	0.005	0.20	0.008	0.25	0.010	0.30	0.012
C-2 carbide	15	50	0.102	0.004	0.15	0.006	0.20	0.008	0.25	0.010	0.30	0.012	0.38	0.015
Finishing														
M1, M2, and M7 high speed steel	8	25	0.050	0.002	0.13	0.005	0.15	0.006	0.25	0.010	0.30	0.012	0.38	0.015
C-2 carbide	11	35	0.102	0.004	0.15	0.006	0.20	0.008	0.25	0.010	0.30	0.012	0.38	0.015

AISI 4161, 4161H

AISI 4161, 4161H: Chemical Composition

AISI grade	C	Mn	P max	S max	Si	Cr	Mo
	Chemical composition, %						
4161	0.55-0.65	0.75-1.10	0.035	0.040	0.15-0.30	0.70-0.90	0.25-0.35
4161H	0.55-0.65	0.65-1.10	0.035	0.040	0.15-0.30	0.65-0.95	0.25-0.35

Characteristics and Typical Uses. AISI grades 4161 and 4161H are chromium-molybdenum, alloy steels that have high hardenability. The hardenability required depends on the thickness of the material and the quenching process. These steels are available as hot rolled and cold finished bar, rod, and wire.

AISI 4161: Similar Steels (U.S. and/or Foreign). UNS G41610; ASTM A322, A331; SAE J404, J412, J770

AISI 4161H: Similar Steels (U.S. and/or Foreign). UNS H41610; ASTM A304; SAE J1268

Mechanical Properties

AISI 4161H: End-Quench Hardenability Limits and Hardenability Band. Heat treating temperatures recommended by SAE: normalize at 870 °C (1600 °F), for forged or rolled specimens only; austenitize at 845 °C (1550 °F). (Ref 1)

Distance from quenched end, 1/16 in.	Hardness, HRC max	Hardness, HRC min
1	65	60
2	65	60
3	65	60
4	65	60
5	65	60
6	65	60
7	65	60
8	65	60
9	65	59
10	65	59
11	65	59
12	64	59
13	64	58
14	64	58
15	64	57
16	64	56
18	64	55
20	63	53
22	63	50
24	63	48
26	63	45
28	63	43
30	63	42
32	63	41

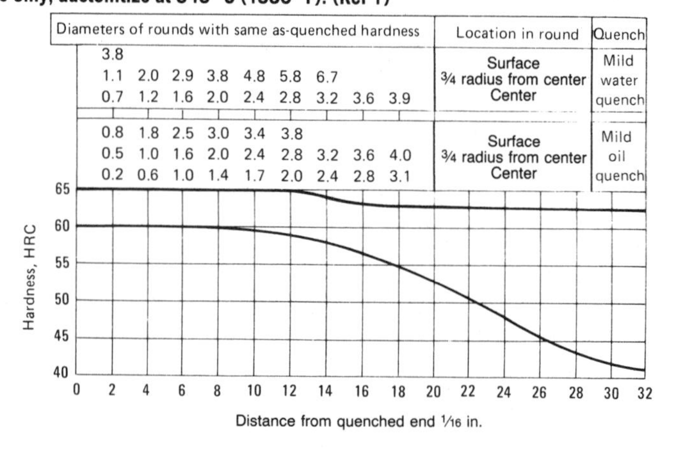

Mechanical Properties (continued)

AISI 4161, 4161H: Hardness and Machinability (Ref 7)

Condition	Hardness range, HB	Average machinability rating(a)
Spheroidized and cold drawn	187-241	50

(a) Based on AISI 1212 steel as 100% average machinability

Machining Data

For machining data on AISI grades 4161 and 4161H, refer to the preceding machining tables for AISI grades 4150 and 4150H.

Physical Properties

AISI 4161, 4161H: Thermal Treatment Temperatures

Quenching medium: oil

Treatment	Temperature range °C	°F
Annealing	790-845	1450-1550
Hardening	815-845	1500-1550
Tempering(a)	370 min	700 min

(a) To desired hardness

AISI 4320, 4320H

AISI 4320, 4320H: Chemical Composition

AISI grade	C	Mn	P max	S max	Si	Ni	Cr	Mo
	Chemical composition, %							
4320	0.17-0.22	0.45-0.65	0.035	0.040	0.15-0.30	1.65-2.00	0.40-0.60	0.20-0.30
4320H	0.17-0.23	0.40-0.70	0.035	0.040	0.15-0.30	1.55-2.00	0.35-0.65	0.20-0.30

Characteristics and Typical Uses. AISI 4320 and 4320H are carburizing grades of nickel-chromium-molybdenum steel with intermediate case hardenability. These steels are available as hot rolled and cold finished bar; special bearing quality, forging billets; hot and cold rolled strip and sheet; and seamless tube. Applications of AISI 4320 and 4320H include tractor and automotive gears, piston pins, ball studs, universal crosses, and roller bearings.

AISI 4320: Similar Steels (U.S. and/or Foreign). UNS G43200; ASTM A322, A331, A505, A519, A535; SAE J404, J412, J770

AISI 4320H: Similar Steels (U.S. and/or Foreign). UNS H43200; ASTM A304; SAE J1268

AISI 4320, 4320H: Approximate Critical Points

Transformation point	Temperature(a) °C	°F	Transformation point	Temperature(a) °C	°F
Ac_1	725	1335	Ar_1	630	1170
Ac_3	810	1490	M_s	380	720
Ar_3	740	1365			

(a) On heating or cooling at 28 °C (50 °F) per hour

Physical Properties

AISI 4320, 4320H: Thermal Treatment Temperatures

Quenching medium: oil

Treatment	Temperature range °C	°F
Forging	1230 max	2250 max
Cycle annealing	(a)	(a)
Normalizing	900-955	1650-1750
Carburizing	900-925	1650-1700
Reheating(b)	830-845	1525-1550
Tempering	120-175	250-350

(a) Heat at least as high as the carburizing temperature; hold for uniformity; cool rapidly to 540 to 675 °C (1000 to 1250 °F); hold for 1 to 3 h; air or furnace cool to obtain a structure suitable for machining and finish. (b) After slow cool

Machining Data

For machining data on AISI grades 4320 and 4320H, refer to the preceding machining tables for AISI grade A2317.

Mechanical Properties

AISI 4320, 4320H: Tensile Properties (Ref 5)

Condition or treatment	Tensile strength MPa	ksi	Yield strength MPa	ksi	Elongation(a), %	Reduction in area, %	Hardness, HB	Izod impact energy J	ft·lb
Normalized at 895 °C (1640 °F)	793	115	460	67	20.8	51	235	73	54
Annealed at 850 °C (1560 °F)	580	84	425	62	29.0	58	163	110	81

(a) In 50 mm (2 in.)

Mechanical Properties (continued)

AISI 4320, 4320H: Core Properties of Specimens Treated for Maximum Case Hardness and Core Toughness (Ref 4)

Treatment	Case depth, mm (in.)	Case hardness, HRC	Core properties: Tensile strength, MPa (ksi)	Core properties: Yield strength, MPa (ksi)	Core properties: Elongation(a), %	Core properties: Reduction in area, %	Core properties: Hardness, HB
Recommended practice for maximum case hardness							
Direct quenched from pot: carburized at 925 °C (1700 °F) for 8 h; quenched in agitated oil; tempered at 150 °C (300 °F)	1.5 (0.060)	60.5	1495 (217)	1105 (160)	13	50.1	429
Single quenched and tempered (for good case and core properties): carburized at 925 °C (1700 °F) for 8 h; pot cooled; reheated to 830 °C (1500 °F); quenched in agitated oil; tempered at 150 °C (300 °F)	1.9 (0.075)	62.5	1505 (218)	1225 (178)	13.5	48.2	429
Double quenched and tempered (for maximum refinement of case and core): carburized at 925 °C (1700 °F) for 8 h; pot cooled; reheated to 815 °C (1500 °F); quenched in agitated oil; reheated to 775 °C (1425 °F); quenched in agitated oil; tempered at 150 °C (300 °F)	1.9 (0.075)	62	1050 (152)	670 (97)	19.5	49.4	302
Recommended practice for maximum core toughness							
Direct quenched from pot: carburized at 925 °C (1700 °F) for 8 h; quenched in agitated oil; tempered at 230 °C (450 °F)	1.5 (0.060)	58.5	1490 (216)	1095 (159)	12.5	49.4	415
Single quenched and tempered (for good case and core properties): carburized at 925 °C (1700 °F) for 8 h; pot cooled; reheated to 815 °C (1500 °F); quenched in agitated oil; tempered at 230 °C (450 °F)	1.9 (0.075)	59	1460 (212)	1195 (173)	12.5	50.9	415
Double quenched and tempered (for maximum refinement of case and core): carburized at 925 °C (1700 °F) for 8 h; pot cooled; reheated to 815 °C (1500 °F); quenched in agitated oil; reheated to 775 °C (1425 °F); quenched in agitated oil; tempered at 230 °C (450 °F)	1.9 (0.075)	59	1005 (146)	650 (94)	21.8	56.3	293

(a) In 50 mm (2 in.)

AISI 4320, 4320H: Approximate Core Mechanical Properties (Ref 2)

Heat treatment of test specimens: (1) normalized at 925 °C (1700 °F) in 32-mm (1.25-in.) rounds; (2) machined to 25- or 13.7-mm (1- or 0.540-in.) rounds; (3) pseudocarburized at 925 °C (1700 °F) for 8 h; (4) box cooled to room temperature; (5) reheated to temperatures given in the table and oil quenched; (6) tempered at 150 °C (300 °F); (7) tested in 12.8-mm (0.505-in.) rounds; tests were conducted using test specimens machined to English units

Reheat temperature °C	Reheat temperature °F	Tensile strength MPa	Tensile strength ksi	Yield strength(a) MPa	Yield strength(a) ksi	Elongation(b), %	Reduction in area, %	Hardness, HB
Tested in 25-mm (1-in.) rounds								
775	1425	958	139	793	115	15	52	293
800	1475	1025	149	848	123	15.5	52	302
830	1525	1110	161	958	139	16	54	331
(c)	(c)	1150	167	972	141	14.5	50	341
Tested in 13.7-mm (0.540-in.) rounds								
775	1425	1040	151	869	126	15	49	321
800	1475	1080	157	896	130	16	50	331
830	1525	1180	171	1015	147	16	53	352
(c)	(c)	1240	180	1020	148	14.5	46	375

(a) 0.2% offset. (b) In 50 mm (2 in.). (c) Quenched from step 3

AISI 4320, 4320H: Permissible Compressive Stresses

Taken from gears with a case depth of 1.1 to 1.5 mm (0.045 to 0.060 in.) and minimum case hardness of 60 HRC

Data point	Compressive stress MPa	Compressive stress ksi
Mean	1590	230
Maximum	1790	260

Mechanical Properties (continued)

AISI 4320H: End-Quench Hardenability Limits and Hardenability Band. Heat treating temperatures recommended by SAE: normalize at 925 °C (1700 °F), for forged or rolled specimens only; austenitize at 925 °C (1700 °F). (Ref 1)

Distance from quenched end, 1/16 in.	Hardness, HRC max	Hardness, HRC min
1	48	41
2	47	38
3	45	35
4	43	32
5	41	29
6	38	27
7	36	25
8	34	23
9	33	22
10	31	21
11	30	20
12	29	20
13	28	...
14	27	...
15	27	...
16	26	...
18	25	...
20	25	...
22	24	...
24	24	...
26	24	...
28	24	...
30	24	...
32	24	...

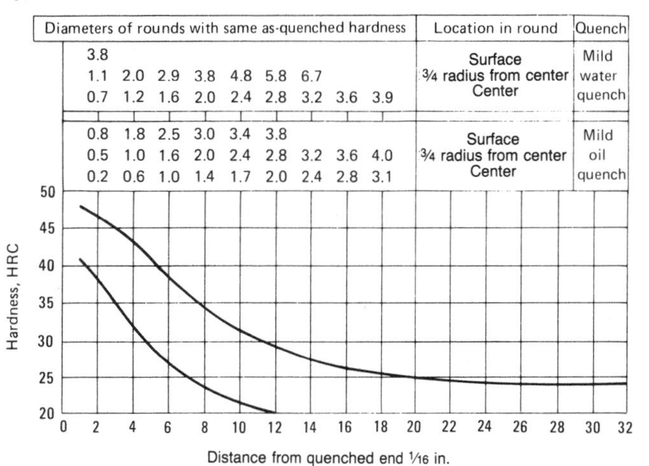

AISI 4320, 4320H: Mass Effect on Mechanical Properties (Ref 4)

Condition or treatment	Size round mm	Size round in.	Tensile strength MPa	Tensile strength ksi	Yield strength MPa	Yield strength ksi	Elongation(a), %	Reduction in area, %	Hardness, HB	Izod impact energy J	Izod impact energy ft·lb
Annealed (heated to 850 °C or 1560 °F; furnace cooled 15 °C or 30 °F per hour to 420 °C or 790 °F; cooled in air)	25	1	580	84	425	62	29.0	58.4	163	...	...
Normalized (heated to 895 °C or 1640 °F; cooled in air)	13	0.5	841	122	510	74	23.9	54.3	248	...	...
	25	1	793	115	460	67	20.8	50.7	235	...	...
	50	2	703	102	405	59	23.3	59.2	212	...	...
	100	4	703	102	395	57	22.3	54.7	201	...	...
Mock carburized at 925 °C (1700 °F) for 8 h; reheated to 815 °C (1500 °F); quenched in oil; tempered at 150 °C (300 °F)	13	0.5	1460	212	1125	163	11.8	45.5	415	38	28
	25	1	1050	152	738	107	17.0	51.0	302	60	44
	50	2	910	132	595	86	22.5	56.4	255	79	58
	100	4	827	120	515	75	24.0	57.1	248	72	53
Mock carburized at 925 °C (1700 °F) for 8 h; reheated to 815 °C (1500 °F); quenched in oil; tempered at 230 °C (450 °F)	13	0.5	1295	188	1035	150	13.9	52.8	388	...	...
	25	1	1025	149	724	105	17.8	55.2	285	...	...
	50	2	896	130	585	85	20.8	63.8	255	...	...
	100	4	814	118	515	75	22.5	51.9	241	...	...

(a) In 50 mm (2 in.)

AISI 4320, 4320H: Effect of the Mass on Hardness at Selected Points (Ref 4)

Size round mm	Size round in.	As-quenched hardness after quenching in oil at: Surface	½ radius	Center
13	0.5	44.5 HRC	44.5 HRC	44.5 HRC
25	1	39 HRC	37 HRC	36 HRC
50	2	35 HRC	30 HRC	27 HRC
100	4	25 HRC	24 HRC	24 HRC

AISI 4320, 4320H: Hardness and Machinability (Ref 7)

Condition	Hardness range, HB	Average machinability rating(a)
Annealed and cold drawn	187-229	60

(a) Based on AISI 1212 steel as 100% average machinability

Mechanical Properties (continued)

AISI 4320, 4320H: Low-Temperature Impact Properties (Ref 17)

Treatment	Hardness, HRC	Charpy impact (V-notch) energy at temperature of: −185 °C (−300 °F) J	ft·lb	−130 °C (−200 °F) J	ft·lb	−73 °C (−100 °F) J	ft·lb	−18 °C (0 °F) J	ft·lb	38 °C (100 °F) J	ft·lb	Transition temperature(a) °C	°F
Quenched from 900 °C (1650 °F); tempered at 205 °C (400 °F)	43	12	9	23	17	34	25	37	27	38	28	...	...
Quenched from 900 °C (1650 °F); tempered at 425 °C (800 °F)	37	7	5	14	10	22	16	34	25	41	30	−1	30
Quenched from 900 °C (1650 °F); tempered at 540 °C (1000 °F)	33	11	8	22	16	38	28	54	40	61	45	−54	−65
Quenched from 900 °C (1650 °F); tempered at 650 °C (1200 °F)	21	26	19	47	35	108	80	122	90	122	90	−115	−175

(a) 50% brittle

AISI 4340, 4340H, E4340, E4340H

AISI 4340, 4340H, E4340, E4340H: Chemical Composition

AISI grade	C	Mn	P max	S max	Si	Ni	Cr	Mo
4340	0.37-0.43	0.60-0.80	0.035	0.040	0.15-0.30	1.65-2.00	0.70-0.90	0.20-0.30
4340H	0.37-0.44	0.55-0.90	0.035	0.040	0.15-0.30	1.55-2.00	0.65-0.95	0.20-0.30
E4340 (a)	0.38-0.43	0.65-0.85	0.025	0.020	0.15-0.30	1.65-2.00	0.70-0.90	0.20-0.30
E4340H (a)	0.37-0.44	0.60-0.95	0.025	0.020	0.15-0.30	1.55-2.00	0.65-0.95	0.20-0.30

(a) The letter E as a prefix denotes the steel is made by electric furnace process

Characteristics. AISI grades 4340, 4340H, E4340, and E4340H are nickel-chromium-molybdenum, constructional, alloy steels. They are normally heat treated by quenching in oil and tempering to the desired hardness. These grades exhibit good response to heat treatment, especially in large sections, and possess a good combination of strength, ductility, and toughness in the quenched and tempered condition. The high depth in hardness is reflected in excellent torque properties.

Typical Uses. Because AISI 4340, 4340H, E4340, and E4340H are also carburizing steels, they are used where superior case hardness or core properties are desired. Their choice is determined primarily by the hardenability necessary to obtain the desired core properties under the given conditions of section size and heat treatment.

These grades are suitable for application in the production of piston pins, bearings, and similar parts for relatively severe service. Rods approximately 13-mm (0.5-in.) diam and rod mill rounds up to 28.6-mm (1.125-in.) diam are used for heavy sections, or for parts subject to particularly severe service conditions which require a very mild quench to prevent distortion.

AISI 4340, 4340H, E4340, and E4340H steels are found in aircraft and truck parts, and some ordnance materials. Additional uses include gears, fittings, forging dies, machine tool arbors, pressure vessels, and die blocks.

AISI 4340: Similar Steels (U.S. and/or Foreign). UNS G43400; AMS 5331, 6359, 6414, 6415; ASTM A322, A331, A505, A519, A547, A646; MIL SPEC MIL-S-16974; SAE J404, J412, J770; (W. Ger.) DIN 1.6565; (Jap.) JIS SNCM 8; (U.K.) B.S. 817 M 40, 3111 Type 6, 2 S 119, 3 S 95

AISI 4340H: Similar Steels (U.S. and/or Foreign). UNS H43400; ASTM A304; SAE J1268; (W. Ger.) DIN 1.6565; (Jap.) JIS SNCM 8; (U.K.) B.S. 817 M 40, 3111 Type 6, 2 S 119, 3 S 95

AISI E4340: Similar Steels (U.S. and/or Foreign). UNS G43406; ASTM A304, A331, A505, A519; MIL SPEC MIL-S-5000; SAE J404, J770; (W. Ger.) DIN 1.6562; (Ital.) UNI 40 NiCrMo 7, 40 NiCrMo 7 KB; (U.K.) B.S. Type 8, S 139

AISI E4340H: Similar Steels (U.S. and/or Foreign). UNS H43406; ASTM A304; SAE J1268; (W. Ger.) DIN 1.6562; (Ital.) UNI 40 NiCrMo 7, 40 NiCrMo 7 KB; (U.K.) B.S. Type 8, S 139

AISI 4340, 4340H, E4340, E4340H: Approximate Critical Points

Transformation point	Temperature(a) °C	°F	Transformation point	Temperature (a) °C	°F
Ac_1	725	1335	Ar_1	655	1210
Ac_3	775	1425	M_s	285	545
Ar_3	710	1310			

(a) On heating or cooling at 28 °C (50 °F) per hour

Physical Properties

AISI 4340, 4340H, E4340, E4340H: Mean Coefficients of Linear Thermal Expansion (Ref 9)

Austenitized at 870 °C (1600 °F), tempered at 545 °C (1010 °F) for 2 h

Temperature range °C	Temperature range °F	Material A(a) μm/m·K	Material A(a) μin./in.·°F	Material B(b) μm/m·K	Material B(b) μin./in.·°F
−195-21	−320-70	9.0	5.0	8.8	4.9
−130-21	−200-70	9.7	5.4	9.7	5.4
−73-20	−100-70	11.3	6.3	11.2	6.2
−73-95	−100-200	11.5	6.4	11.3	6.3
21-95	70-200	11.5	6.4	11.5	6.4
21-150	70-300	11.9	6.6	11.9	6.6
21-205	70-400	12.1	6.7	12.2	6.8
21-260	70-500	12.4	6.9	12.6	7.0
21-315	70-600	12.6	7.0	13.0	7.2
21-370	70-700	13.0	7.2	13.1	7.3
21-425	70-800	13.3	7.4	13.5	7.5
21-480	70-900	13.5	7.5	13.7	7.6
21-540	70-1000	13.7	7.6	13.9	7.7

(a) 1.88% nickel, normalized and tempered. (b) 1.90% nickel, quenched and tempered

AISI 4340, 4340H, E4340, E4340H: Thermal Treatment Temperatures

Quenching medium: oil

Treatment	Temperature range °C	Temperature range °F
Forging	1230 max	2250 max
Annealing (a)	595-665	1100-1225
Normalizing (b)	845-900	1550-1650
Hardening	830-855	1525-1575
Tempering	(c)	(c)

(a) Maximum hardness of 223 HB. (b) Hardness of approximately 415 HB. (c) To desired hardness

AISI 4340, 4340H, E4340, E4340H: Electrical Resistivity (Ref 6)

Temperature °C	Temperature °F	Electrical resistivity, μΩ·m
20	68	0.248
100	212	0.298
200	390	0.367
400	750	0.552
600	1110	0.797

Machining Data

For machining data on AISI grades 4340, 4340H, E4340, and E4340H, refer to the preceding machining tables for AISI grades 1340 and 1340H.

Mechanical Properties

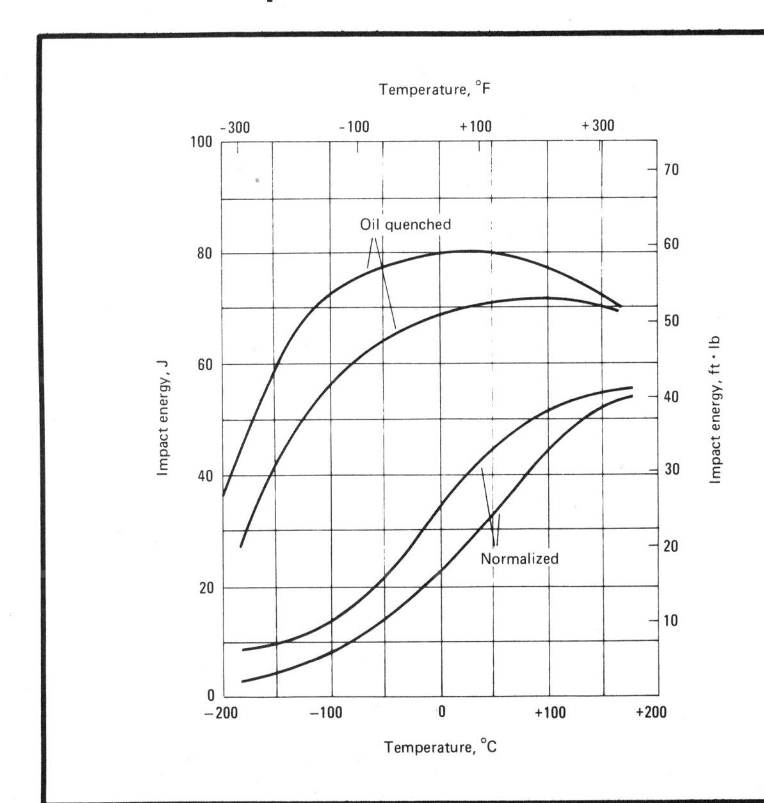

AISI 4340 (Modified): Effect of Heat Treatment on Impact Properties. Heat treatments: oil quenched from 830 °C (1525 °F) and tempered at 650 °C (1200 °F) to hardness of 29 to 30 HRC, or normalized (air cooled) from 830 °C (1525 °F) and tempered at 540 °C (1000 °F). Impact energy tests used Charpy keyhole notch specimen. (Ref 17)

Mechanical Properties (continued)

AISI 4340H: End-Quench Hardenability Limits and Hardenability Band. Heat treating temperatures recommended by SAE: normalize at 870 °C (1600 °F), for forged or rolled specimens only; austenitize at 845 °C (1550 °F). (Ref 1)

Distance from quenched end, 1/16 in.	Hardness, HRC max	Hardness, HRC min
1	60	53
2	60	53
3	60	53
4	60	53
5	60	53
6	60	53
7	60	53
8	60	52
9	60	52
10	60	52
11	59	51
12	59	51
13	59	50
14	58	49
15	58	49
16	58	48
18	58	47
20	57	46
22	57	45
24	57	44
26	57	43
28	56	42
30	56	41
32	56	40

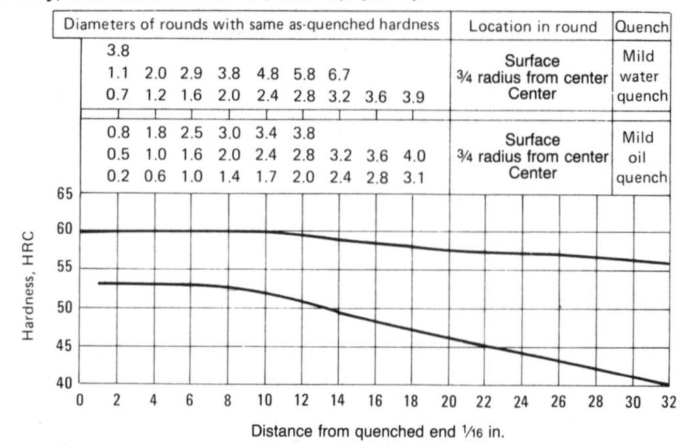

AISI 4340, 4340H, E4340, E4340H: Mass Effect on Mechanical Properties (Ref 4)

Condition or treatment	Size round mm	Size round in.	Tensile strength MPa	Tensile strength ksi	Yield strength MPa	Yield strength ksi	Elongation(a), %	Reduction in area, %	Hardness, HB
Annealed (heated to 810 °C or 1490 °F; furnace cooled 11 °C or 20 °F per hour to 355 °C or 670 °F; cooled in air)	25	1	745	108	470	68	22.0	50.0	217
Normalized (heated to 870 °C or 1600 °F; cooled in air)	13	0.5	1448	210	972	141	12.1	35.3	388
	25	1	1282	186	862	125	12.2	36.3	363
	50	2	1220	177	786	114	13.5	37.3	341
	100	4	1110	161	710	103	13.2	36.0	321
Oil quenched from 800 °C (1475 °F); tempered at 540 °C (1000 °F)	13	0.5	1255	182	1165	169	13.7	45.0	363
	25	1	1207	175	1145	166	14.2	45.9	352
	50	2	1172	170	1103	160	16.0	54.8	341
	100	4	1138	165	1000	145	15.5	53.4	331
Oil quenched from 800 °C (1475 °F); tempered at 595 °C (1100 °F)	13	0.5	1145	166	1117	162	17.1	57.0	331
	25	1	1138	165	1096	159	16.5	54.1	331
	50	2	1014	147	958	139	19.0	60.4	293
	100	4	924	134	786	114	19.7	60.7	269
Oil quenched from 800 °C (1475 °F); tempered at 650 °C (1200 °F)	13	0.5	1000	145	938	136	20.0	59.3	285
	25	1	958	139	883	128	20.0	59.7	277
	50	2	931	135	834	121	20.5	62.5	269
	100	4	855	124	731	106	21.7	63.0	255

(a) In 50 mm (2 in.)

AISI 4340, 4340H, E4340, E4340H: Effect of the Mass on Hardness at Selected Points (Ref 4)

Size round mm	Size round in.	As-quenched hardness after quenching in oil at: Surface	½ radius	Center
13	0.5	58 HRC	58 HRC	56 HRC
25	1	57 HRC	57 HRC	56 HRC
50	2	56 HRC	55 HRC	54 HRC
100	4	53 HRC	49 HRC	47 HRC

AISI 4340, 4340H, E4340, E4340H: Hardness and Machinability (Ref 7)

Condition	Hardness range, HB	Average machinability rating(a)
Annealed and cold drawn	187-241	50

(a) Based on AISI 1212 steel as 100% average machinability

Mechanical Properties (continued)

AISI E4340H: End-Quench Hardenability Limits and Hardenability Band. Heat treating temperatures recommended by SAE: normalize at 870 °C (1600 °F), for forged or rolled specimens only; austenitize at 845 °C (1550 °F). (Ref 1)

Distance from quenched end, 1/16 in.	Hardness, HRC max	Hardness, HRC min
1	60	53
2	60	53
3	60	53
4	60	53
5	60	53
6	60	53
7	60	53
8	60	53
9	60	53
10	60	53
11	60	53
12	60	53
13	60	52
14	59	52
15	59	52
16	59	51
18	58	51
20	58	50
22	58	49
24	57	48
26	57	47
28	57	46
30	57	45
32	57	44

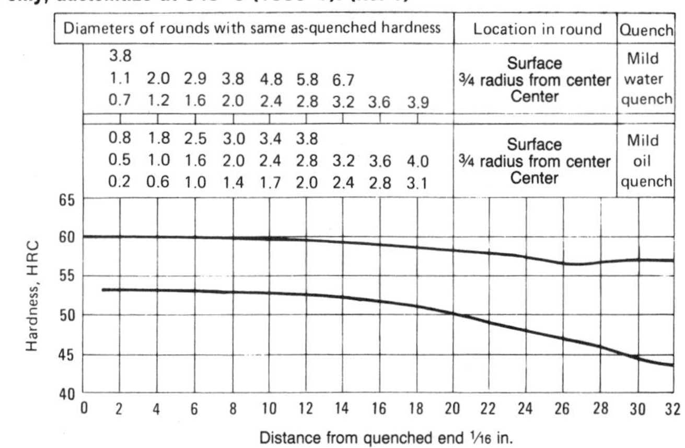

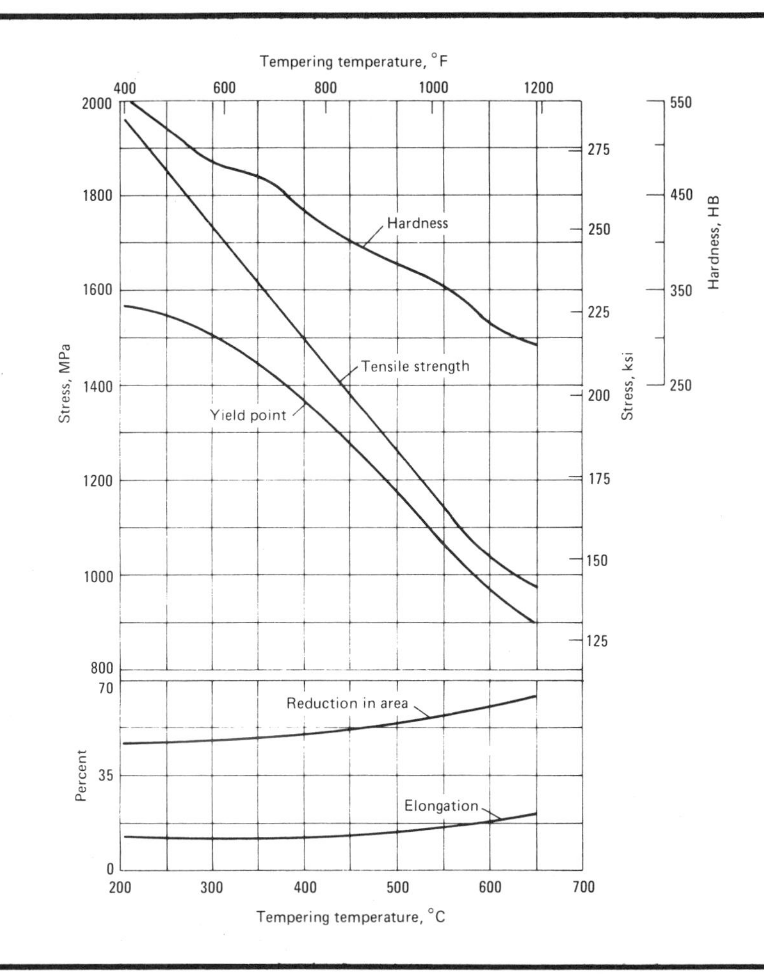

AISI 4340: Effect of Tempering Temperature on Tensile Properties. Normalized at 870 °C (1600 °F); reheated to 800 °C (1475 °F); quenched in agitated oil. Specimens were treated in 13.5-mm (0.530-in.) diam and machined to 12.8-mm (0.505-in.) diam for testing. Tests were conducted using specimens machined to English units. As-quenched hardness of 601 HB; elongation measured in 50 mm (2 in.). (Ref 4)

Mechanical Properties (continued)

AISI 4340: Effect of Mass and Tempering Temperature on Tensile Properties After Tempering at 540 °C (1000 °F).

Normalized at 870 °C (1600 °F) in oversize rounds; oil quenched from 845 °C (1550 °F) in sizes given; tempered at 540 °C (1000 °F); tested in 12.8-mm (0.505-in.) rounds. Tests from bar 38.1-mm (1.50-in.) diam and larger were taken at half-radius position. Tests were conducted using specimens machined to English units. Elongation was measured in 50 mm (2 in.). (Ref 2)

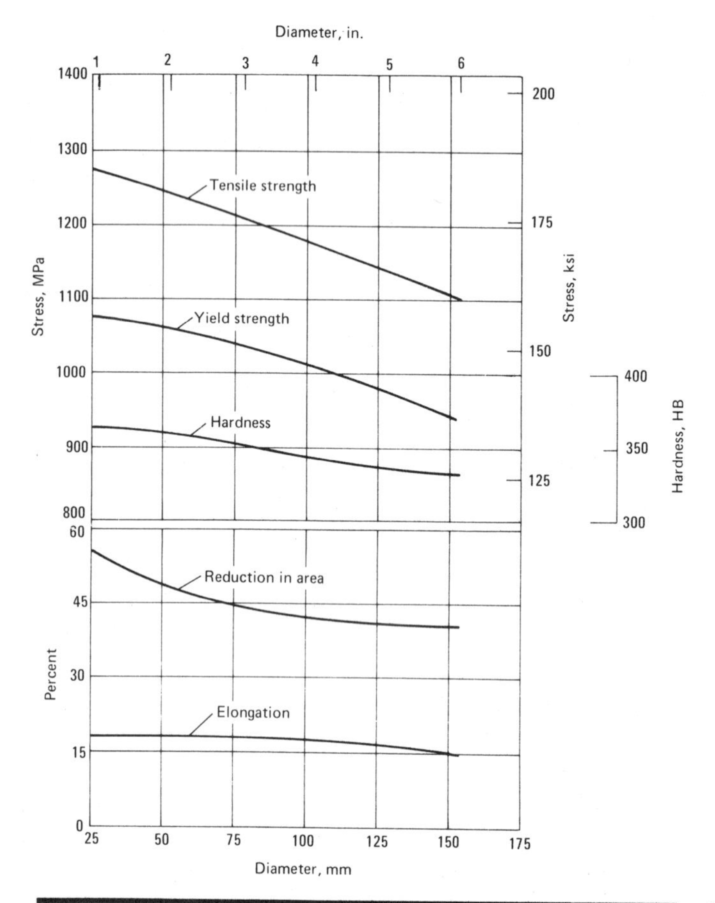

AISI 4340: Effect of Mass and Tempering Temperature on Tensile Properties After Tempering at 650 °C (1200 °F).

Normalized at 870 °C (1600 °F) in oversize rounds; oil quenched from 845 °C (1550 °F) in sizes given; tempered at 650 °C (1200 °F); tested in 12.8-mm (0.505-in.) rounds. Tests from bar 38.1-mm (1.50-in.) diam and larger were taken at half-radius position. Tests were conducted using specimens machined to English units. Elongation was measured in 50 mm (2 in.). (Ref 2)

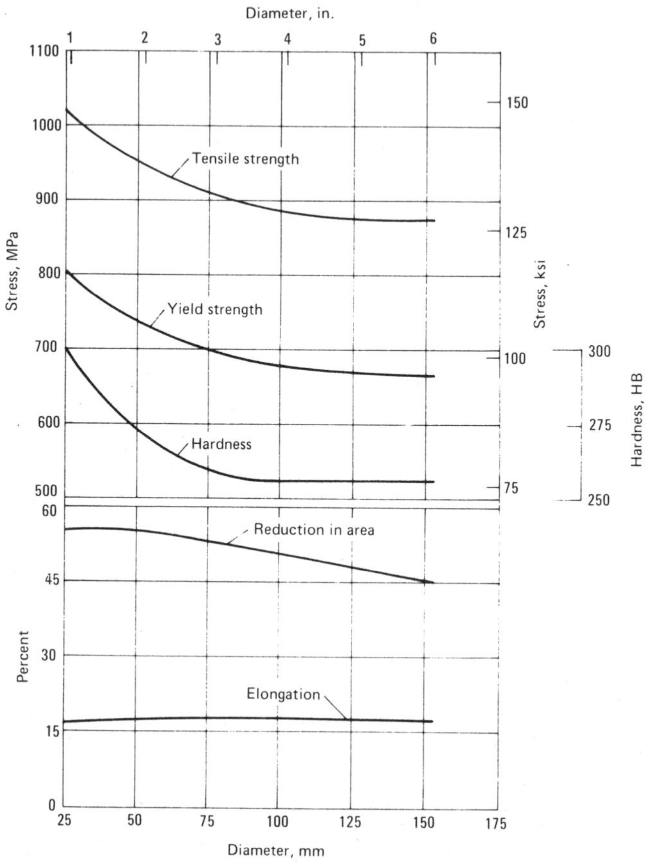

AISI 4340, 4340H, E4340, 4340H: Tensile and Fatigue Properties (Ref 17)

Condition or treatment	Temperature °C	Temperature °F	Tensile strength MPa	Tensile strength ksi	Yield strength(a) MPa	Yield strength(a) ksi	Elongation(b), %	Reduction in area, %	Hardness, HV	Modulus of elasticity GPa	Modulus of elasticity 10^6 psi	Endurance limit Unnotched MPa	Endurance limit Unnotched ksi	Endurance limit Notched MPa	Endurance limit Notched ksi
Oil quenched from 845 °C (1550 °F); tempered at 650 °C (1200 °F)	25	77	1005	146	938	136	20	60	...	213	30.9	510	74	138	20
	−195	−315	1435	208	1380	200	20	44	...	219	31.8	800	116	138	20
Oil quenched from 845 °C (1550 °F); tempered at 425 °C (800 °F)	25	77	1595	231	1475	214	12	46	...	212	30.7	614	89	338	49
	−195	−315	1985	288	1840	267	4	11	...	213	30.9	841	122	248	36
Oil quenched from 855 °C (1575 °F); tempered at 230 °C (450 °F) for 4 h	26	79	1855	269	1550	225	12	40	531	196	28.4	...	...	...	...
	−78	−108	1935	281	2020	293	12	40	548	211	30.6	...	...	...	...
	−195	−321	2200	319	1915	278	10	24	639	211	30.6	...	...	...	...
	−255	−423	2290	332	...	...	0.6	0.2	818	210	30.4	...	...	...	...

(a) 0.2% offset. (b) For wire and rod the gage length is four times the specimen diameter

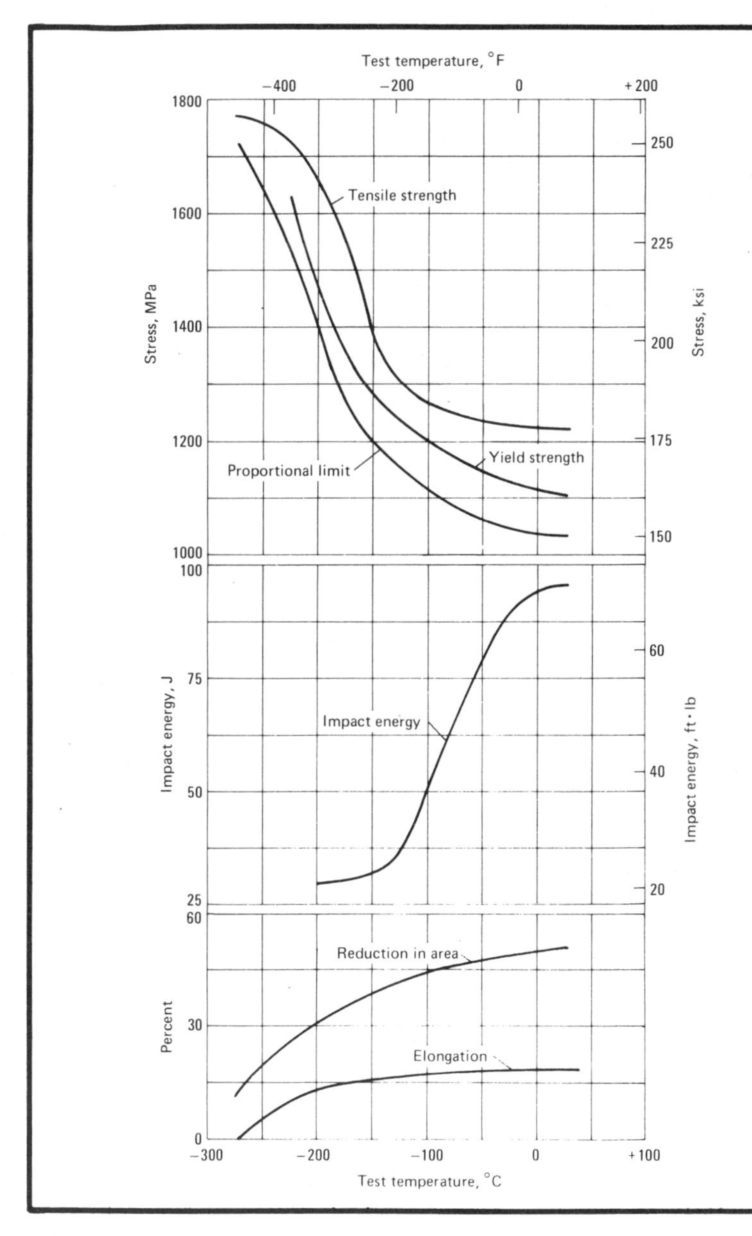

AISI 4340: Effect of Test Temperature on Tensile and Impact Properties. Oil quenched and tempered. When tested for tensile properties, material hardness was 37 to 39 HRC. Impact energy tests used Charpy V-notch specimens with a material hardness of 33 to 35 HRC. Proportional limit, 0.01% strain; yield strength, 0.2% offset; elongation in 50 mm (2 in.). (Ref 17)

AISI 4340, 4340H, E4340, E4340H: Effect of Nitriding on Fatigue Properties (Ref 16)

Chemical composition of test specimens: 0.39% carbon, 0.65% manganese, 0.25% silicon, 1.85% nickel, 0.82% chromium, 0.25% molybdenum; preliminary treatment: normalized at 940 to 955 °C (1725 to 1750 °F); R.R. Moore rotating beam fatigue specimens with a minimum diameter of 5.84 mm (0.230 in.)

Heat and nitriding treatments	Case hardness, HR15N	Core hardness, HRC	Fatigue properties — Smooth specimen — Fatigue limit MPa	Fatigue limit ksi	Number of cycles	Fatigue properties — Notched specimen(a) — Fatigue limit MPa	Fatigue limit ksi	Number of cycles
Specimens not nitrided — true fatigue limit								
Oil quenched 830 °C (1525 °F); tempered 540 °C (1000 °F)	...	37	505	73	$>10^7$	205	30	$>10^7$
Oil quenched 830 °C (1525 °F); tempered 540 °C (1000 °F); pseudonitrided 525 °C (975 °F) for 48 h	...	33	470	68	$>10^7$	180	26	$>10^7$
Nitrided specimens — no true fatigue limit(b)								
Oil quenched 830 °C (1525 °F); tempered 540 °C (1000 °F); nitrided 525 °C (975 °F) for 48 h	89	33	595	86	3.5×10^8(c)	350	51	$>1.6 \times 10^9$
Oil quenched 830 °C (1525 °F); tempered 540 °C (1000 °F); nitrided 525 °C (975 °F) for 10 h and 565 °C (1050 °F) for 40 h	86	32	620	90	$>4.6 \times 10^8$	360	52	$>4.6 \times 10^8$

(a) Specimens have 45°, 0.25-mm (0.010-in.) root-radius notch, 0.51 mm (0.020 in.) deep. Notching preceded nitriding. (b) Nitrided specimens do not show a true endurance or fatigue limit but give an "asymptotic" type of S-N curve; however, the notched specimens come closer to showing a true fatigue limit than the smooth ones. (c) Specimen failed at 595 MPa (86 ksi)

Mechanical Properties (continued)

AISI 4340: Transition Curves After Austempering.

—— austempered at 305 °C (585 °F) for 2 h to give 100% bainite; --- austempered at 285 °C (545 °F) for 2 h to give 90% bainite, 10% martensite. (Ref 17)

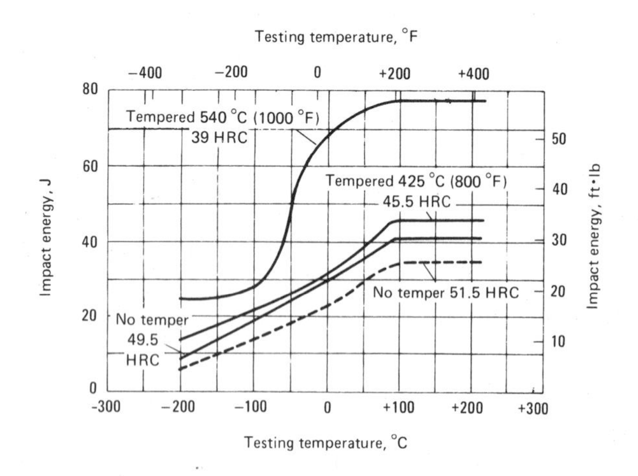

AISI 4340: Transition Curves After Quenching and Tempering.

Quenched in oil; tempered for 1 h. (Ref 17)

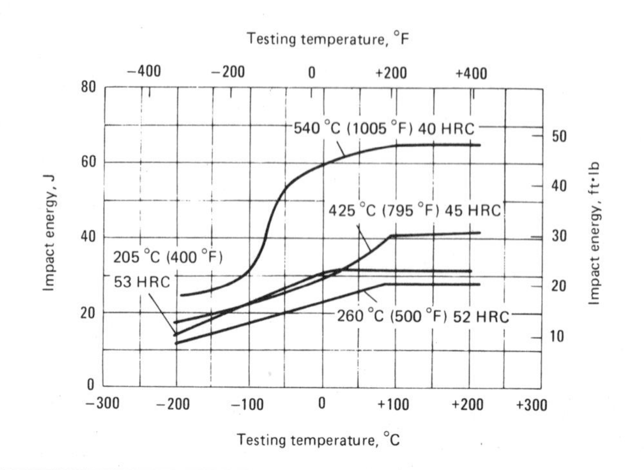

AISI 4340, 4340H, E4340, E4340H: Low-Temperature Impact Properties (Ref 17)

Treatment	Hardness, HRC	Charpy impact (V-notch) energy at temperature of: −185 °C (−300 °F) J	ft·lb	−130 °C (−200 °F) J	ft·lb	−73 °C (−100 °C) J	ft·lb	−18 °C (0 °F) J	ft·lb	38 °C (100 °F) J	ft·lb	Transition temperature(a) °C	°F
Quenched from 845 °C (1550 °F); tempered at 205 °C (400 °F)	52	15	11	20	15	27	20	28	21	28	21	...	...
Quenched from 845 °C (1550 °F); tempered at 315 °C (600 °F)	48	14	10	19	14	20	15	20	15	22	16	...	...
Quenched from 845 °C (1550 °F); tempered at 425 °C (800 °F)	44	12	9	18	13	22	16	28	21	34	25	...	...
Quenched from 845 °C (1550 °F); tempered at 540 °C (1000 °F)	38	20	15	24	18	38	28	47	35	49	36	−90	−130
Quenched from 845 °C (1550 °F); tempered at 650 °C (1200 °F)	30	20	15	38	28	75	55	75	55	75	55	−120	−185

(a) 50% brittle

AISI 4340, 4340H, E4340, E4340H: Tensile Properties (Ref 5)

Condition or treatment	Tensile strength MPa	ksi	Yield strength MPa	ksi	Elongation(a), %	Reduction in area, %	Hardness, HB	Izod impact energy J	ft·lb
Normalized at 870 °C (1600 °F)	1280	186	860	125	12.2	36	363	16	12
Annealed at 810 °C (1490 °F)	745	108	475	69	22.0	50	217	52	38

(a) In 50 mm (2 in.)

AISI 4419, 4419H

AISI 4419, 4419H: Chemical Composition

AISI grade	C	Mn	P max	S max	Si	Mo
			Chemical composition, %			
4419	0.18-0.23	0.45-0.65	0.035	0.040	0.15-0.35(a)	0.45-0.60
4419H	0.17-0.23	0.35-0.75	0.035	0.040	0.15-0.35(a)	0.45-0.60

(a) UNS composition: 0.15 to 0.30% silicon

Characteristics and Typical Uses. AISI grades 4419 and 4419H are carburizing steels with intermediate case hardenability and low core hardenability. Available as hot rolled and cold finished bar, rod, and wire, these steels are suitable for use in tractor and automotive gears, piston pins, and universal crosses. In most instances satisfactory case hardness can be produced by oil quenching.

AISI 4419: Similar Steels (U.S. and/or Foreign). UNS G44190; ASTM A322, A331; SAE J404, J412, J770; (W. Ger.) DIN 1.5419; (Ital.) UNI G 22 Mo 5; (Jap.) JIS SCPH 11

AISI 4419H: Similar Steels (U.S. and/or Foreign). UNS H44190; ASTM A304; J1268; (W. Ger.) DIN 1.5419; (Ital.) UNI G 22 Mo 5; (Jap.) JIS SCPH 11

AISI 4419, 4419H: Approximate Critical Points

Transformation point	Temperature(a) °C	°F	Transformation point	Temperature(a) °C	°F
Ac_1	695	1280	Ar_3	820	1510
Ac_3	870	1600	Ar_1	770	1420

(a) On heating or cooling at 28 °C (50 °F) per hour

Machining Data

For machining data on AISI grades 4119 and 4119H, refer to the preceding machining tables for AISI grade A2317.

Physical Properties

AISI 4419, 4419H: Thermal Treatment Temperatures

Quenching medium: oil

Treatment	Temperature range °C	°F
Forging	1230 max	2250 max
Carburizing	900-925	1650-1700
Tempering	120-175	250-350

Mechanical Properties

AISI 4419, 4419H: Mass Effect on Mechanical Properties (Ref 4)

Condition or treatment	Size round mm	in.	Tensile strength MPa	ksi	Yield strength MPa	ksi	Elongation(a), %	Reduction in area, %	Hardness, HB
Annealed (heated to 915 °C or 1675 °F; furnace cooled 11 °C or 20 °F per hour to 480 °C or 900 °F; cooled in air)	25	1	450	65	330	48	31.2	62.8	121
Normalized (heated to 955 °C or 1750 °F; cooled in air)	14	0.57	540	78	360	52	33.2	69.9	149
	25	1	515	75	350	51	32.5	69.4	143
	50	2	495	72	345	50	30.8	64.9	143
	100	4	505	73	330	48	30.0	60.8	143
Mock carburized at 925 °C (1700 °F) for 8 h; reheated to 845 °C (1550 °F); quenched in oil; tempered at 150 °C (300 °F)	14	0.57	710	103	450	65	24.3	60.3	217
	25	1	670	97	435	63	24.2	66.4	201
	50	2	660	96	415	60	25.3	64.7	201
	100	4	595	86	365	53	27.7	66.3	179
Mock carburized at 925 °C (1700 °F) for 8 h; reheated to 845 °C (1550 °F); quenched in oil; tempered at 230 °C (450 °F)	14	0.57	710	103	425	62	24.8	63.6	212
	25	1	650	94	405	59	25.0	68.6	197
	50	2	635	92	400	58	26.2	68.2	192
	100	4	580	84	330	48	27.0	67.1	170

(a) In 50 mm (2 in.)

Mechanical Properties (continued)

AISI 4419, 4419H: Effect of the Mass on Hardness at Selected Points (Ref 4)

Size round mm	in.	As-quenched hardness after quenching in oil at: Surface	½ radius	Center
14	0.57	96 HRB	95 HRB	93 HRB
25	1	94 HRB	93 HRB	89 HRB
50	2	94 HRB	92 HRB	88 HRB
100	4	93 HRB	90 HRB	82 HRB

AISI 4419, 4419H: Hardness and Machinability (Ref 7)

Condition	Hardness range, HB	Average machinability rating(a)
Annealed and cold drawn	170-212	65

(a) Based on AISI 1212 steel as 100% average machinability

AISI 4419, 4419H: Core Properties of Specimen Treated for Maximum Case Hardness and Core Toughness

Treatment	Case depth, mm (in.)	Case hardness, HRC	Core properties: Tensile strength, MPa (ksi)	Yield strength(a), MPa (ksi)	Elongation(b), %	Reduction in area, %	Hardness, HB
Recommended practice for maximum case hardness							
Direct quenched from pot: carburized at 925 °C (1700 °F) for 8 h; quenched in agitated oil; tempered at 150 °C (300 °F)	1.4 (0.054)	64	831 (120.50)	608 (88.25)	19.7	64.7	241
Single quenched and tempered (for good case and core properties): carburized at 925 °C (1700 °F) for 8 h; pot cooled; reheated to 845 °C (1550 °F); quenched in agitated oil; tempered at 150 °C (300 °F)	1.6 (0.062)	65	712 (103.25)	450 (65.25)	24.3	60.3	217
Double quenched and tempered (for maximum refinement of case and core): carburized at 925 °C (1700 °F) for 8 h; pot cooled; reheated to 855 °C (1575 °F); quenched in agitated oil; reheated to 830 °C (1525 °F); quenched in agitated oil; tempered at 150 °C (300 °F)	1.8 (0.070)	66	734 (106.50)	377 (54.75)	21.7	49.7	217
Recommended practice for maximum core toughness							
Direct quenched from pot: carburized at 925 °C (1700 °F) for 8 h; quenched in agitated oil; tempered at 230 °C (450 °F)	1.4 (0.054)	59	817 (118.50)	596 (86.50)	18.8	67.0	235
Single quenched and tempered (for good case and core properties): carburized at 925 °C (1700 °F) for 8 h; pot cooled; reheated to 845 °C (1550 °F); quenched in agitated oil; tempered at 230 °C (450 °F)	1.6 (0.062)	60.5	708 (102.75)	431 (62.50)	24.8	63.6	212
Double quenched and tempered (for maximum refinement of case and core): carburized at 925 °C (1700 °F) for 8 h; pot cooled; reheated to 855 °C (1575 °F); quenched in agitated oil; reheated to 830 °C (1525 °F); quenched in agitated oil; tempered at 230 °C (450 °F)	1.8 (0.070)	61	679 (98.50)	376 (54.50)	23.4	59.7	201

(a) In 50 mm (2 in.)

AISI 4422

AISI 4422: Chemical Composition

AISI grade	C	Mn	P max	S max	Si	Mo
4422	0.20-0.25	0.70-0.80	0.035	0.040	0.15-0.30	0.35-0.45

Characteristics and Typical Uses. AISI grade 4422 is a low-carbon, molybdenum steel with low case hardenability and intermediate core hardenability. This steel is available as hot rolled and cold finished bar and seamless mechanical tubing.

AISI 4422: Similar Steels (U.S. and/or Foreign). UNS G44220; ASTM A519; SAE J404, J412, J770; (W. Ger.) DIN 1.5419; (Ital.) UNI G 22 Mo 5; (Jap.) JIS SCPH 11

AISI 4422: Approximate Critical Points

Transformation point	Temperature(a) °C	°F	Transformation point	Temperature(a) °C	°F
Ac_1	730	1350	Ar_1	645	1195
Ac_3	845	1550	M_s	415	780
Ar_3	810	1490			

(a) On heating or cooling at 28 °C (50 °F) per hour

Mechanical Properties

AISI 4422: Approximate Core Mechanical Properties (Ref 2)

Heat treatment of test specimens: (1) normalized at 925 °C (1700 °F) in 32-mm (1.25-in.) rounds; (2) machined to 25- or 13.7-mm (1- or 0.540-in.) rounds; (3) pseudocarburized at 925 °C (1700 °F) for 8 h; (4) box cooled to room temperature; (5) reheated to temperatures given in the table and oil quenched; (6) tempered at 150 °C (300 °F); (7) tested in 12.8-mm (0.505-in.) rounds; tests were conducted using test specimens machined to English units

Reheat temperature °C	°F	Tensile strength MPa	ksi	Yield strength(a) MPa	ksi	Elongation(b), %	Reduction in area, %	Hardness, HB
Heat treated in 25-mm (1-in.) rounds								
775	1430	820	119	385	56	17.2	34.3	255
820	1505	758	110	400	58	18.3	44.4	235
870	1600	827	120	540	78	20.0	57.0	255
(c)	(c)	862	125	560	81	18.0	55.0	262
Heat treated in 13.7-mm (0.540-in.) rounds								
775	1430	827	120	370	54	14.0	28.8	255
820	1505	841	122	455	66	16.0	36.5	255
870	1600	896	130	640	93	17.0	54.0	277
(c)	(c)	965	140	670	97	14.3	50.4	293

(a) 0.2% offset. (b) In 50 mm (2 in.). (c) Quenched from step 3

AISI 4422: Hardness and Machinability (Ref 7)

Condition	Hardness range, HB	Average machinability rating(a)
Hot rolled and cold drawn	170-212	65

(a) Based on AISI 1212 steel as 100% average machinability

Physical Properties

AISI 4422: Thermal Treatment Temperatures

Quenching medium: oil

Treatment	Temperature range °C	°F
Forging	1230 max	2250 max
Cycle annealing	(a)	(a)
Normalizing	(b)	(b)
Carburizing	900-925	1650-1700
Tempering	120-175	250-350

(a) Heat at least as high as the carburizing temperature; hold for uniformity; cool rapidly to 540 to 675 °C (1000 to 1250 °F); air or furnace cool to obtain a structure suitable for machining and finish. (b) Temperature at least as high as the carburizing temperature, followed by air cooling

Machining Data

For machining data on AISI grade 4422, refer to the preceding machining tables for AISI grade A2317.

AISI 4427

AISI 4427: Chemical Composition

AISI grade	C	Mn	P max	S max	Si	Mo
	Chemical composition, %					
4427	0.24-0.29	0.70-0.90	0.035	0.040	0.15-0.30	0.35-0.45

Characteristics and Typical Uses. AISI grade 4427 is a medium-carbon, molybdenum, carburizing steel with intermediate case hardenability and medium core hardenability. This steel is available as hot rolled and cold finished bar and seamless mechanical tubing.

AISI 4427: Similar Steels (U.S. and/or Foreign). UNS G44270; ASTM A519; SAE J404, J412, J470

AISI 4427: Approximate Critical Points

Transformation point	Temperature(a) °C	°F
Ac_1	720	1330
Ac_3	840	1540
Ar_3	775	1425
Ar_1	650	1200
M_s	400	750

(a) On heating or cooling at 28 °C (50 °F) per hour

Machining Data

For machining data on AISI 4427, refer to the preceding machining tables for AISI grade A2317.

Mechanical Properties

AISI 4427: Approximate Core Mechanical Properties (Ref 2)

Heat treatment of test specimens: (1) normalized at 925 °C (1700 °F) in 32-mm (1.25-in.) rounds; (2) machined to 25- or 13.7-mm (1- or 0.540-in.) rounds; (3) pseudocarburized at 925 °C (1700 °F) for 8 h; (4) box cooled to room temperature; (5) reheated to temperatures given in the table and oil quenched; (6) tempered at 150 °C (300 °F); (7) tested in 12.8-mm (0.505-in.) rounds; tests were conducted using test specimens machined to English units

Reheat temperature °C	Reheat temperature °F	Tensile strength MPa	Tensile strength ksi	Yield strength(a) MPa	Yield strength(a) ksi	Elongation(b), %	Reduction in area, %	Hardness, HB
Heat treated in 25-mm (1-in.) rounds								
765	1410	951	138	560	81	17.3	45.7	293
805	1480	945	137	550	80	17.0	46.0	293
865	1590	1034	150	620	90	14.0	44.0	321
(c)	(c)	1076	156	670	97	14.6	44.6	321
Heat treated in 13.7-mm (0.540-in.) rounds								
765	1410	1096	159	540	78	10.0	16.0	331
805	1480	924	134	724	105	14.0	42.4	285
865	1590	1255	182	896	130	13.5	42.8	363
(c)	(c)	1434	208	903	131	11.3	37.4	415

(a) 0.2% offset. (b) In 50 mm (2 in.). (c) Quenched from step 3

AISI 4427: Hardness and Machinability (Ref 7)

Condition	Hardness range, HB	Average machinability rating(a)
Annealed and cold drawn	170-212	65

(a) Based on AISI 1212 steel as 100% average machinability

Physical Properties

AISI 4427: Thermal Treatment Temperatures

Quenching medium: oil

Treatment	Temperature range °C	Temperature range °F
Forging	1230 max	2250 max
Cycle annealing	(a)	(a)
Normalizing	(b)	(b)
Carburizing	900-925	1650-1700
Tempering	120-175	250-350

(a) Heat at least as high as the carburizing temperature; hold for uniformity; cool rapidly to 540 to 675 °C (1000 to 1250 °F); air or furnace cool to obtain a structure suitable for machining and finish. (b) Temperature should be at least as high as the carburizing temperature, followed by air cooling

AISI 4615

AISI 4615: Chemical Composition

AISI grade	C	Mn	P max	S max	Si	Ni	Mo
4615	0.13-0.18	0.45-0.65	0.035	0.040	0.15-0.30	1.65-2.00	0.20-0.30

Characteristics and Typical Uses. AISI grade 4615 is a nickel-molybdenum, carburizing, alloy steel with intermediate case hardenability and low core hardenability. This grade is available as hot rolled and cold rolled sheet and strip; hot rolled and cold finished bar, rod, wire, and tube; and aircraft quality steel.

AISI 4615: Similar Steels (U.S. and/or Foreign). UNS G46150; AMS 6290; ASTM A322, A331, A505; MIL SPEC MIL-S-7493 (A4615); SAE J404, J412, J770

AISI 4615: Approximate Critical Points

Transformation point	Temperature(a) °C	Temperature(a) °F
Ac_1	725	1340
Ac_3	810	1490
Ar_3	760	1400
Ar_1	650	1200
M_s	415	780

(a) On heating or cooling at 28 °C (50 °F) per hour

Mechanical Properties

AISI 4615: Hardness and Machinability (Ref 7)

Condition	Hardness range, HB	Average machinability rating(a)
Hot rolled and cold drawn	174-223	65

(a) Based on AISI 1212 steel as 100% average machinability

Machining Data

For machining data on AISI 4615, refer to the preceding machining tables for AISI grade A2317.

Physical Properties

AISI 4615: Mean Coefficients of Linear Thermal Expansion (Ref 9)

Temperature range °C	°F	Coefficient µm/m·K	µin./in.·°F
21-100	70-212	11	6.3
21-205	70-400	12	6.7
21-300	70-570	13	7.1
21-425	70-800	13	7.4
21-500	70-930	14	7.8
21-700	70-1290	15	8.1

AISI 4615: Electrical Resistivity

Temperature °C	°F	Electrical resistivity, µΩ·m
21	70	0.249
100	212	0.296
200	390	0.371

AISI 4615: Thermal Treatment Temperatures

Quenching medium: oil

Treatment	Temperature range °C	°F
Forging	1290 max	2350 max
Cycle annealing	(a)	(a)
Normalizing	(b)	(b)
Carburizing	900-925	1650-1700
Reheating	815-845	1500-1550
Tempering	120-175	250-350

(a) Heat at least as high as the carburizing temperature; hold for uniformity; cool rapidly to 540 to 675 °C (1000 to 1250 °F); air or furnace cool to obtain a structure suitable for machining and finish. (b) Temperature should be at least as high as the carburizing temperature, followed by air cooling

AISI 4620, 4620H

AISI 4620, 4620H: Chemical Composition

AISI grade	C	Mn	P max	S max	Si	Ni	Mo
4620	0.17-0.22	0.45-0.65	0.035	0.040	0.15-0.35(a)	1.65-2.00	0.20-0.30
4620H	0.17-0.23	0.35-0.75	0.035	0.040	0.15-0.35(a)	1.55-2.00	0.20-0.30

(a) UNS composition: 0.15 to 0.30% silicon

Characteristics and Typical Uses. AISI 4620 and 4620H are low-carbon, nickel-molybdenum, carburizing steels with intermediate case hardenability and medium core hardenability. These widely used, case hardening, alloy steels have good core toughness and hardness penetration with minimum distortion and growth characteristics. They are available as hot and cold rolled strip and sheet; hot rolled and cold finished bar, rod, and wire; and special quality, roller-bearing steel.

AISI 4620: Similar Steels (U.S. and/or Foreign). UNS G46200; AMS 6294; ASTM A322, A331, A505, A535; MIL SPEC MIL-S-7493 (A4620); SAE J404, J412, J770

AISI 4620H: Similar Steels (U.S. and/or Foreign). UNS H46200; ASTM A304; SAE J1268

Mechanical Properties

AISI 4620, 4620H: Hardness and Machinability (Ref 7)

Condition	Hardness range, HB	Average machinability rating(a)
Hot rolled and cold drawn	183-229	65

(a) Based on AISI 1212 steel as 100% average machinability

AISI 4620, 4620H: Approximate Critical Points

Transformation point	Temperature(a) °C	°F	Transformation point	Temperature(a) °C	°F
Ac_1	720	1330	Ar_1	645	1190
Ac_3	800	1475	M_s	400	755
Ar_3	750	1380			

(a) On heating or cooling at 28 °C (50 °F) per hour

Physical Properties

AISI 4620, 4620H: Thermal Treatment Temperatures

Quenching medium: oil

Treatment	Temperature range °C	°F	Treatment	Temperature range °C	°F
Forging	1260 max	2300 max	Carburizing	900-925	1650-1700
Cycle annealing	(a)	(a)	Reheating	815-845	1500-1550
Normalizing	(b)	(b)	Tempering	120-175	250-350

(a) Heat at least as high as the carburizing temperature; hold for uniformity; cool rapidly to 540 to 675 °C (1000 to 1250 °F); air or furnace cool to obtain a suitable structure for machining and finish. (b) Temperature should be at least as high as the carburizing temperature, followed by air cooling

Machining Data

For machining data on AISI grades 4620 and 4620H, refer to the preceding machining tables for AISI grade A2317.

Mechanical Properties (continued)

AISI 4620H: End-Quench Hardenability Limits and Hardenability Band. Heat treating temperatures recommended by SAE: normalize at 925 °C (1700 °F), for forged or rolled specimens only; austenitize at 925 °C (1700 °F). (Ref 1)

Distance from quenched end, 1/16 in.	Hardness, HRC max	Hardness, HRC min
1	48	41
2	45	35
3	42	27
4	39	24
5	34	21
6	31	...
7	29	...
8	27	...
9	26	...
10	25	...
11	24	...
12	23	...
13	22	...
14	22	...
15	22	...
16	21	...
18	21	...
20	20	...

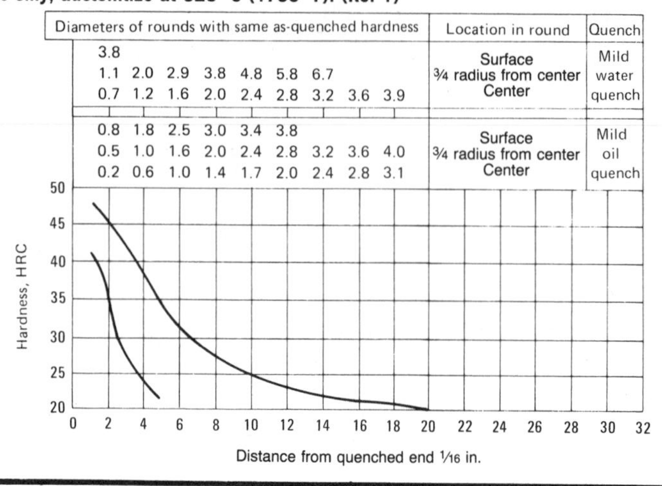

AISI 4620, 4620H: Mass Effect on Mechanical Properties (Ref 4)

Condition or treatment	Size round mm	Size round in.	Tensile strength MPa	Tensile strength ksi	Yield strength MPa	Yield strength ksi	Elongation(a), %	Reduction in area, %	Hardness, HB	Izod impact energy J	Izod impact energy ft·lb
Annealed (heated to 855 °C or 1575 °F; furnace cooled 17 °C or 30 °F per hour to 480 °C or 900 °F; cooled in air)	25	1	510	74	370	54	31.3	60.3	149	...	...
Normalized (heated to 900 °C or 1650 °F; cooled in air)	13	0.5	600	87	380	55	30.7	68.0	192	...	...
	25	1	570	83	365	53	29.0	66.7	174	...	...
	50	2	550	80	365	53	29.5	67.1	167	...	...
	100	4	530	77	360	52	30.5	65.2	163	...	...
Mock carburized at 925 °C (1700 °F) for 8 h; reheated to 815 °C (1500 °F); quenched in oil; tempered at 150 °C (300 °F)	13	0.5	876	127	620	90	20.0	59.8	255	58	43
	25	1	675	98	460	67	25.8	70.0	197	135	98
	50	2	660	96	450	65	27.0	69.7	192	138	102
	100	4	585	85	360	52	29.5	69.2	170	136	100
Mock carburized at 925 °C (1700 °F) for 8 h; reheated to 815 °C (1500 °F); quenched in oil; tempered at 230 °C (450 °F)	13	0.5	814	118	560	81	21.4	65.3	241	...	...
	25	1	675	98	455	66	27.5	68.9	192	...	...
	50	2	660	96	425	62	26.8	69.2	187	...	...
	100	4	580	84	365	53	29.8	70.3	170	...	...

(a) In 50 mm (2 in.)

AISI 4620, 4620H: Effect of the Mass on Hardness at Selected Points (Ref 4)

Size round mm	Size round in.	As-quenched hardness after quenching in oil at: Surface	½ radius	Center
13	0.5	40 HRC	32 HRC	31 HRC
25	1	27 HRC	99 HRB	97 HRB
50	2	24 HRC	94 HRB	91 HRB
100	4	96 HRB	91 HRB	88 HRB

AISI 4620, 4620H: Fatigue Limit Data (Ref 10)

Effective case depth mm	Effective case depth in.	Peak case hardness, HRC	Nominal fatigue limit MPa	Nominal fatigue limit ksi
0.94	0.037	64	903	131
0.38	0.015	63	862	125

Mechanical Properties (continued)

AISI 4620, 4620H: Bending Fatigue Behavior

Carburized to an effective case depth of 0.38 mm or 0.015 in. (○) and 0.94 mm or 0.037 in. (●). (Ref 10)

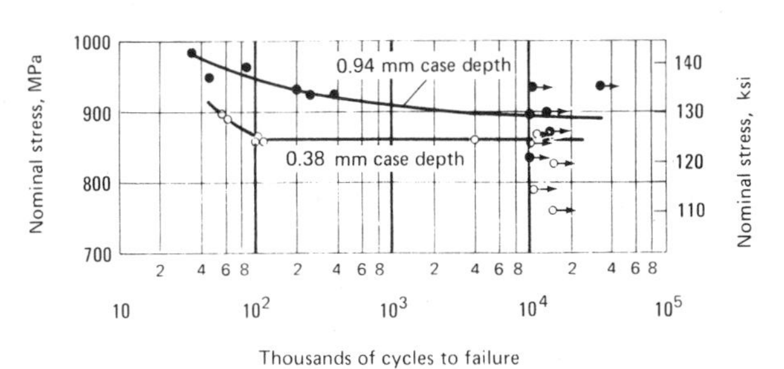

Carburized with effective case depth of 0.31 to 0.46 mm (0.012 to 0.018 in.). (Ref 10)

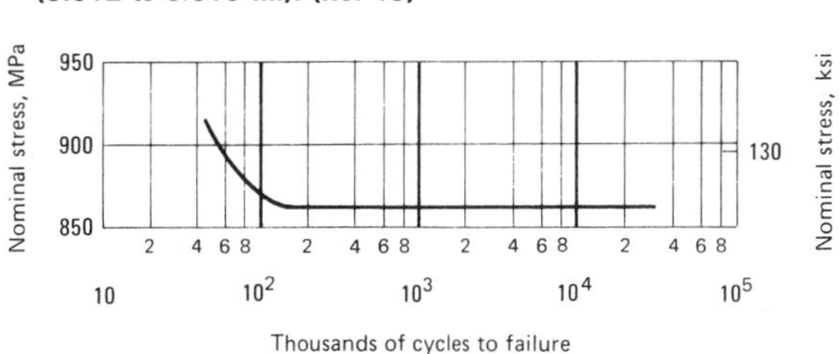

Carburized with effective case depth of 0.94 to 1.2 mm (0.037 to 0.046 in.). (Ref 10)

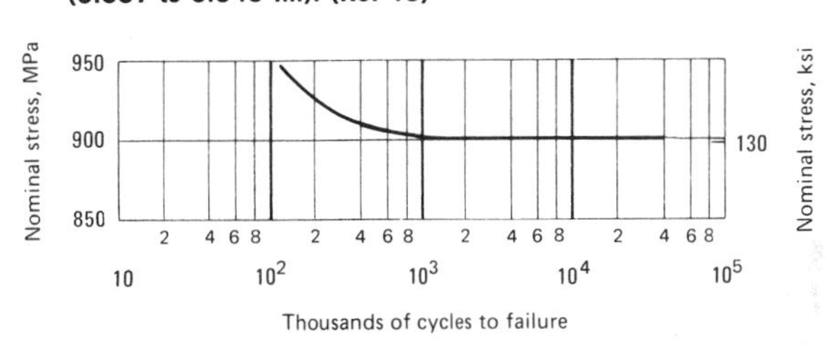

AISI 4620, 4620H: Low-Temperature Impact Properties (Ref 17)

Condition or treatment	Hardness, HRC	Charpy impact (V-notch) energy at temperature of: −185 °C (−300 °F) J	ft·lb	−130 °C (−200 °F) J	ft·lb	−73 °C (−100 °F) J	ft·lb	−18 °C (0 °F) J	ft·lb	38 °C (100 °F) J	ft·lb
Quenched from 900 °C (1650 °F); tempered at 150 °C (300 °F)	42	19	14	27	20	38	28	47	35	47	35
Quenched from 900 °C (1650 °F); tempered at 425 °C (800 °F)	34	15	11	22	16	45	33	75	55	75	55
Quenched from 900 °C (1650 °F); tempered at 540 °C (1000 °F)	29	22	16	46	34	75	55	106	78	106	78
Quenched from 900 °C (1650 °F); tempered at 650 °C (1200 °F)	19	23	17	65	48	140	103	156	115	159	117

AISI 4620, 4620H: Approximate Core Mechanical Properties (Ref 2)

Heat treatment of test specimens: (1) normalized at 925 °C (1700 °F) in 32-mm (1.25-in.) rounds; (2) machined to 25- or 13.7-mm (1- or 0.540-in.) rounds; (3) pseudocarburized at 925 °C (1700 °F) for 8 h; (4) box cooled to room temperature; (5) reheated to temperatures given in the table and oil quenched; (6) tempered at 150 °C (300 °F); (7) tested in 12.8-mm (0.505-in.) rounds; tests were conducted using test specimens machined to English units

Reheat temperature °C	°F	Tensile strength MPa	ksi	Yield strength(a) MPa	ksi	Elongation(b), %	Reduction in area, %	Hardness, HB
Heat treated in 25-mm (1-in.) rounds								
775	1425	883	128	758	110	16	54	277
800	1475	910	132	724	105	16.5	55	277
830	1525	945	137	752	109	17.5	55	285
(c)	(c)	965	140	772	112	15	52	293
Heat treated in 13.7-mm (0.540-in.) rounds								
775	1425	910	132	738	107	14	48	277
800	1475	958	139	786	114	16.5	52	293
830	1525	1000	145	807	117	17	55	302
(c)	(c)	1014	147	841	122	16	50	311

(a) 0.2% offset. (b) In 50 mm (2 in.). (c) Quenched from step 3

AISI 4620, 4620H: Permissible Compressive Stresses

Taken from gears with a case depth of 1.1 to 1.5 mm (0.045 to 0.060 in.) and minimum case hardness of 60 HRC

Data point	Compressive stress MPa	ksi
Mean	1450	210
Maximum	1650	240

Mechanical Properties (continued)

AISI 4620, 4620H: Core Properties of Specimen Treated for Maximum Case Hardness and Core Toughness (Ref 4)

Treatment	Case depth, mm (in.)	Case hardness, HRC	Core properties: Tensile strength, MPa (ksi)	Yield strength(a), MPa (ksi)	Elongation(b), %	Reduction in area, %	Hardness, HB
Recommended practice for maximum case hardness							
Direct quenched from pot: carburized at 925 °C (1700 °F) for 8 h; quenched in agitated oil; tempered at 150 °C (300 °F)	1.9 (0.075)	60.5	1022 (148.3)	802.2 (116.5)	17.0	55.7	311
Single quenched and tempered (for good case and core properties): carburized at 925 °C (1700 °F) for 8 h; pot cooled; reheated to 815 °C (1500 °F); quenched in agitated oil; tempered at 150 °C (300 °F)	1.9 (0.075)	62.5	822.5 (119.3)	576 (83.5)	19.5	59.4	277
Double quenched and tempered (for maximum refinement of case and core): carburized at 925 °C (1700 °F) for 8 h; pot cooled; reheated to 830 °C (1525 °F); quenched in agitated oil; reheated to 800 °C (1475 °F); quenched in agitated oil; tempered at 150 °C (300 °F)	1.5 (0.060)	62	841.2 (122.0)	533 (77.3)	22.0	55.7	248
Recommended practice for maximum core toughness							
Direct quenched from pot: carburized at 925 °C (1700 °F) for 8 h; quenched in agitated oil; tempered at 230 °C (450 °F)	1.5 (0.060)	58.5	1017 (147.5)	798.4 (115.8)	16.8	57.9	302
Single quenched and tempered (for good case and core properties): carburized at 925 °C (1700 °F) for 8 h; pot cooled; reheated to 815 °C (1500 °F); quenched in agitated oil; tempered at 230 °C (450 °F)	1.7 (0.065)	59	796.3 (115.5)	557 (80.8)	20.5	63.6	248
Double quenched and tempered (for maximum refinement of case and core): carburized at 925 °C (1700 °F) for 8 h; pot cooled; reheated to 815 °C (1525 °F); quenched in agitated oil; reheated to 800 °C (1475 °F); quenched in agitated oil; tempered at 230 °C (450 °F)	1.5 (0.060)	59	795.0 (115.3)	531 (77.0)	22.5	62.1	235

(a) In 50 mm (2 in.)

AISI 4626, 4626H

AISI 4626, 4626H: Chemical Composition

AISI grade	C	Mn	P max	S max	Si	Ni	Mo
4626	0.24-0.29	0.45-0.65	0.035	0.040	0.15-0.30	0.70-1.00	0.15-0.25
4626H	0.23-0.29	0.40-0.70	0.035	0.040	0.15-0.30	0.65-1.05	0.15-0.25

Characteristics and Typical Uses. AISI grades 4626 and 4626H are carburizing, nickel-molybdenum alloy steels. These steels have intermediate case hardenability, but low core hardenability. Grades 4626 and 4626H are available as hot rolled and cold finished bar, rod, and wire suitable for machining into components, heat treatment, or, in the as-finished condition, constructional applications.

AISI 4626: Similar Steels (U.S. and/or Foreign). UNS G46260; ASTM A322, A331; SAE J404, J412, J770

AISI 4626H: Similar Steels (U.S. and/or Foreign). UNS H46260; ASTM A304; SAE J1268

Mechanical Properties

AISI 4626H: End-Quench Hardenability Limits and Hardenability Band.

Heat treating temperatures recommended by SAE: normalize at 925 °C (1700 °F), for forged or rolled specimens only; austenitize at 925 °C (1700 °F). (Ref 1)

Distance from quenched end, 1/16 in.	Hardness, HRC max	Hardness, HRC min
1	51	45
2	48	36
3	41	29
4	33	24
5	29	21
6	27	...
7	25	...
8	24	...
9	23	...
10	22	...
11	22	...
12	21	...
13	21	...
14	20	...

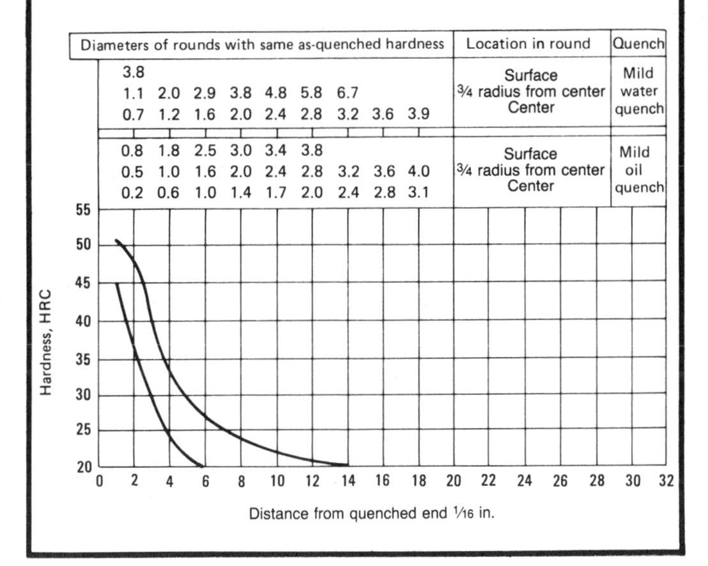

AISI 4626, 4626H: Bending Fatigue Behavior

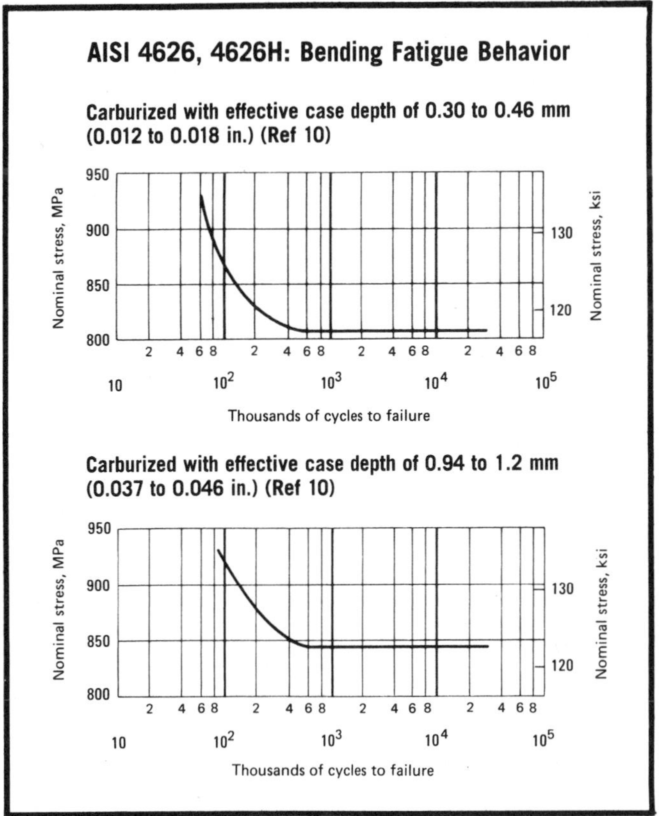

AISI 4626, 4626H: Hardness and Machinability (Ref 7)

Condition	Hardness range, HB	Average machinability rating(a)
Hot rolled and cold drawn	170-212	70

(a) Based on AISI 1212 steel as 100% average machinability

Physical Properties

AISI 4626, 4626H: Thermal Treatment Temperatures

Quenching medium: oil

Treatment	Temperature range °C	Temperature range °F
Cycle annealing	(a)	(a)
Normalizing	(b)	(b)
Carburizing	900-925	1650-1700
Reheating	815-845	1500-1550
Tempering	120-175	250-350

(a) Heat at least as high as the carburizing temperature; hold for uniformity; cool rapidly to 540 to 675 °C (1000 to 1250 °F); hold for 1 to 3 h, air or furnace cool to obtain a structure suitable for machining and finish. (b) Temperature should be at least as high as the carburizing temperature, followed by air cooling

Machining Data

For machining data on AISI grades 4626 and 4626H, refer to the preceding machining tables for AISI grades 1330 and 1330H.

AISI 4718, 4718H

AISI 4718, 4718H: Chemical Composition

AISI grade	Chemical composition, % C	Mn	P max	S max	Si	Ni	Cr	Mo
4718	0.16-0.21	0.70-0.90	0.040	0.040	0.20-0.35	0.90-1.20	0.35-0.55	0.30-0.40
4718H	0.15-0.21	0.60-0.95	0.040	0.040	0.15-0.30	0.85-1.25	0.30-0.60	0.30-0.40

Characteristics and Typical Uses. AISI grade 4718 and 4718H steels exhibit intermediate case hardenability and high core hardenability. They are available as hot rolled and cold rolled sheet and strip; hot rolled and cold finished bar, rod, and wire; and bar subject to end-quench hardenability requirements.

AISI 4718: Similar Steels (U.S. and/or Foreign). UNS G47180; ASTM A322, A331, A505; SAE J404, J412, J770; (W. Ger.) DIN 1.6755

AISI 4718H: Similar Steels (U.S. and/or Foreign). UNS H47180; ASTM A304; SAE J1268; (W. Ger.) DIN 1.6755

AISI 4718, 4718H: Approximate Critical Points

Transformation point	Temperature(a) °C	°F	Transformation point	Temperature(a) °C	°F
Ac_1	695	1285	Ar_1	635	1175
Ac_3	820	1510	M_s	395	740
Ar_3	755	1395			

(a) On heating or cooling at 28 °C (50 °F) per hour

Physical Properties

AISI 4718, 4718H: Thermal Treatment Temperatures

Quenching medium: oil

Treatment	Temperature range °C	°F
Forging	1230 max	2250 max
Cycle annealing	(a)	(a)
Normalizing	(b)	(b)
Carburizing	900-925	1650-1700
Tempering	120-175	250-350

(a) Heat at least as high as the carburizing temperature; hold for uniformity; cool rapidly to 540 to 675 °C (1000 to 1250 °F); hold for 1 to 3 h; air or furnace cool to obtain a structure suitable for machining and finish. (b) Temperature should be at least as high as the carburizing temperature, followed by air cooling

Machining Data

For machining data on AISI grades 4718 and 4718H, refer to the preceding machining tables for AISI grade A2317.

Mechanical Properties

AISI 4718, 4718H: Approximate Core Mechanical Properties (Ref 2)

Heat treatment of test specimens: (1) normalized at 925 °C (1700 °F) in 32-mm (1.25-in.) rounds; (2) machined to 25- or 13.7-mm (1- or 0.540-in.) rounds; (3) pseudocarburized at 925 °C (1700 °F) for 8 h; (4) box cooled to room temperature; (5) reheated to temperatures given in the table and oil quenched; (6) tempered at 150 °C (300 °F); (7) tested in 12.8-mm (0.505-in.) rounds; tests were conducted using test specimens machined to English units

Reheat temperature °C	°F	Tensile strength MPa	ksi	Yield strength(a) MPa	ksi	Elongation(b), %	Reduction in area, %	Hardness, HB
Heat treated in 25-mm (1-in.) rounds								
765	1405	1096	158.9	601	87.2	14.5	36.8	331
800	1475	1068	154.9	714	103.6	17.8	56.5	321
850	1560	1096	159.0	724	105.0	16.5	57.0	331
(c)	(c)	1127	163.5	812	117.8	15.3	51.0	341
Heat treated in 13.7-mm (0.540-in.) rounds								
765	1405	1131	164.0	735	106.6	9.6	32.4	341
800	1475	1159	168.1	834	121.0	13.0	48.7	352
850	1560	1172	170.0	841	122.0	13.0	52.0	352
(c)	(c)	1324	192.0	1010	146.5	14.5	50.5	388

(a) 0.2% offset. (b) In 50 mm (2 in.). (c) Quenched from step 3

AISI 4718, 4718H: Hardness and Machinability (Ref 7)

Condition	Hardness range, HB	Average machinability rating(a)
Hot rolled and cold drawn	187-229	60

(a) Based on AISI 1212 steel as 100% average machinability

AISI 4718, 4718H: Permissible Compressive Stresses

Taken from gears with a case depth of 1.1 to 1.5 mm (0.045 to 0.060 in.) and a minimum case hardness of 60 HRC

Data point	Compressive stress MPa	ksi
Mean	1450	210
Maximum	1655	240

Mechanical Properties (continued)

AISI 4718H: End-Quench Hardenability Limits and Hardenability Band. Heat treating temperatures recommended by SAE: normalize at 925 °C (1700 °F), for forged or rolled specimens only; austenitize at 925 °C (1700 °F). (Ref 1)

Distance from quenched end, 1/16 in.	Hardness, HRC max	Hardness, HRC min
1	47	40
2	47	40
3	45	38
4	43	33
5	40	29
6	37	27
7	35	25
8	33	24
9	32	23
10	31	22
11	30	22
12	29	21
13	29	21
14	28	21
15	27	20
16	27	20
18	27	...
20	26	...
22	26	...
24	25	...
26	25	...
28	24	...
30	24	...
32	24	...

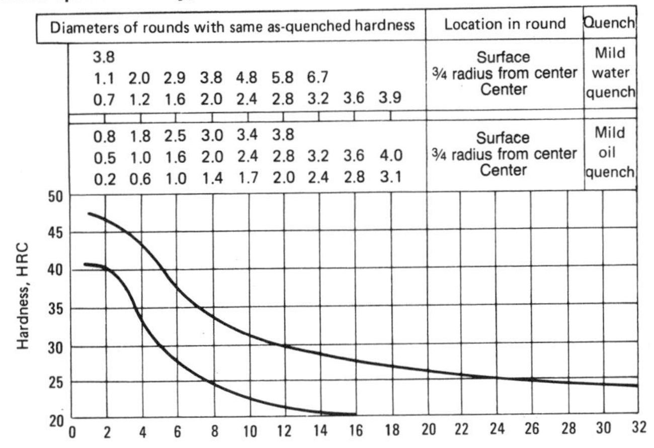

AISI 4720, 4720H

AISI 4720, 4720H: Chemical Composition

AISI grade	C	Mn	P max	S max	Si	Ni	Cr	Mo
	Chemical composition, %							
4720	0.17-0.22	0.50-0.70	0.035	0.040	0.15-0.30	0.90-1.20	0.35-0.55	0.15-0.25
4720H	0.17-0.23	0.45-0.75	0.035	0.040	0.15-0.30	0.85-1.25	0.30-0.60	0.15-0.25

Characteristics and Typical Uses. AISI grades 4720 and 4720H are carburizing steels with intermediate case hardenability and medium core hardenability. These steels are available as forging billets; hot rolled and cold finished bar, rod, and wire; seamless mechanical tubing; and special quality steel for roller bearings. Applications include tractor and automotive gears, piston pins, universal crosses, and roller bearings.

AISI 4720: Similar Steels (U.S. and/or Foreign). UNS G47200; ASTM A322, A331, A519, A535, A711; SAE J404, J412, J770

AISI 4720H: Similar Steels (U.S. and/or Foreign). UNS H47200; ASTM A304; SAE J1268

Machining Data

For machining data on AISI grades 4720 and 4720H, refer to the preceding machining tables for AISI grade A2317.

Physical Properties

AISI 4720, 4720H: Thermal Treatment Temperatures

Quenching medium: oil

Treatment	Temperature range °C	Temperature range °F
Forging	1230 max	2250 max
Cycle annealing	(a)	(a)
Normalizing	(b)	(b)
Carburizing	900-925	1650-1700
Reheating	815-845	1500-1550
Tempering	120-175	250-350

(a) Heat to at least the carburizing temperature; hold for uniformity; cool rapidly to 540 to 675 °C (1000 to 1250 °F); hold for 1 to 3 h; air or furnace cool to obtain a structure suitable for machining and finish. (b) Temperature should be at least as high as the carburizing temperature, followed by air cooling

Mechanical Properties

AISI 4720H: End-Quench Hardenability Limits and Hardenability Band. Heat treating temperatures recommended by SAE: normalize at 925 °C (1700 °F), for forged or rolled specimens only; austenitize at 925 °C (1700 °F). (Ref 1)

Distance from quenched end, 1/16 in.	Hardness, HRC max	Hardness, HRC min
1	48	41
2	47	39
3	43	31
4	39	27
5	35	23
6	32	21
7	29	...
8	28	...
9	27	...
10	26	...
11	25	...
12	24	...
13	24	...
14	23	...
15	23	...
16	22	...
18	21	...
20	21	...
22	21	...
24	20	...

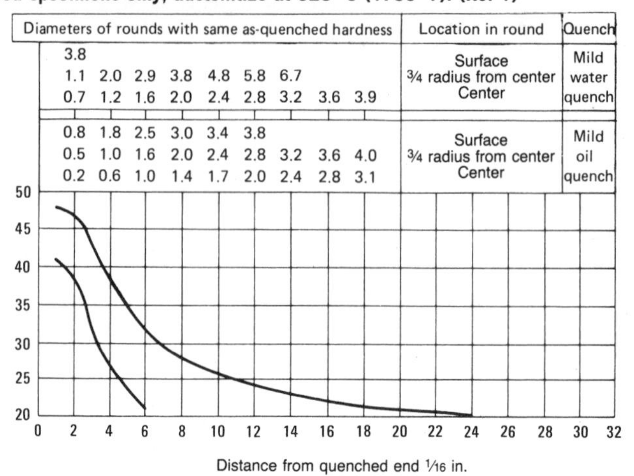

AISI 4720, 4720H: Hardness and Machinability (Ref 7)

Condition	Hardness range, HB	Average machinability rating(a)
Hot rolled and cold drawn	187-229	65

(a) Based on AISI 1212 steel as 100% average machinability

AISI 4815, 4815H

AISI 4815, 4815H: Chemical Composition

AISI grade	C	Mn	P max	S max	Si	Ni	Mo
4815	0.13-0.18	0.40-0.60	0.035	0.040	0.15-0.30	3.25-3.75	0.20-0.30
4815H	0.12-0.18	0.30-0.70	0.035	0.040	0.15-0.30	3.20-3.80	0.20-0.30

(Chemical composition, %)

Characteristics and Typical Uses. AISI grades 4815 and 4815H are carburizing, nickel-molybdenum steels with high case hardenability and medium to high core hardenability.

These steels are available as hot rolled and cold finished bar, rod, and wire; hot rolled and cold rolled sheet and strip; and bar subject to end-quench hardenability requirements. Applications include tractor and automotive gears, rock bit cutters, pump parts to resist wear, and spline shafts.

AISI 4815: Similar Steels (U.S. and/or Foreign). UNS G48150; ASTM A322, A331, A505; SAE J404, J412, J770

AISI 4815H: Similar Steels (U.S. and/or Foreign). UNS H48150; ASTM A304; SAE J1268

AISI 4815, 4815H: Approximate Critical Points

Transformation point	Temperature(a) °C	Temperature(a) °F
Ac_1	690	1275
Ac_3	790	1450
Ar_3	710	1310
Ar_1	425	800
M_s	385	725

(a) On heating or cooling at 28 °C (50 °F) per hour

Physical Properties

AISI 4815, 4815H: Thermal Treatment Temperatures

Quenching medium: oil

Treatment	Temperature range °C	Temperature range °F
Forging	1230 max	2250 max
Annealing (a)	815-855	1500-1575
Normalizing (b)	870-925	1600-1700
Carburizing	900-955	1650-1750
Reheating	800-830	1475-1525
Tempering	120-175	250-350

(a) Heat at least as high as the carburizing temperature; hold for uniformity; cool rapidly to 540 to 675 °C (1000 to 1250 °F); hold for 1 to 3 h; air or furnace cool to obtain a structure suitable for machining and finish. (b) Normalize and temper. After normalizing, reheat to 595 to 650 °C (1100 to 1200 °F) and hold at temperature approximately 1 h/25 mm (1 h/in.) of maximum section, or 4 h minimum time

Mechanical Properties

AISI 4815H: End-Quench Hardenability Limits and Hardenability Band.

Heat treating temperatures recommended by SAE: normalize at 925 °C (1700 °F), for forged or rolled specimens only; austenitize at 845 °C (1550 °F). (Ref 1)

Distance from quenched end, 1/16 in.	Hardness, HRC max	Hardness, HRC min
1	45	38
2	44	37
3	44	34
4	42	30
5	41	27
6	39	24
7	37	22
8	35	21
9	33	20
10	31	...
11	30	...
12	29	...
13	28	...
14	28	...
15	27	...
16	27	...
18	26	...
20	25	...
22	24	...
24	24	...
26	24	...
28	23	...
30	23	...
32	23	...

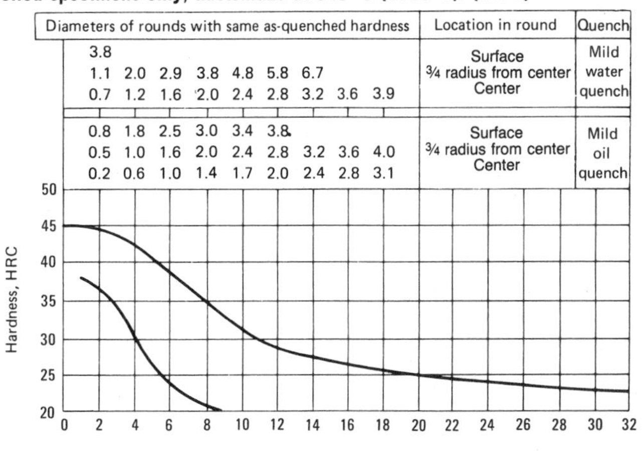

AISI 4815, 4815H: Approximate Core Mechanical Properties (Ref 2)

Heat treatment of test specimens: (1) normalized at 925 °C (1700 °F) in 32-mm (1.25-in.) rounds; (2) machined to 25- or 13.7-mm (1- or 0.540-in.) rounds; (3) pseudocarburized at 925 °C (1700 °F) for 8 h; (4) box cooled to room temperature; (5) reheated to temperatures given in the table and oil quenched; (6) tempered at 150 °C (300 °F); (7) tested in 12.8-mm (0.505-in.) rounds; tests were conducted using test specimens machined to English units

Reheat temperature °C	Reheat temperature °F	Tensile strength MPa	Tensile strength ksi	Yield strength(a) MPa	Yield strength(a) ksi	Elongation(b), %	Reduction in area, %	Hardness, HB
Heat treated in 25-mm (1-in.) rounds								
775	1425	1025	149	855	124	15.0	55	311
800	1475	1055	153	889	129	14.5	56	321
830	1525	1075	156	910	132	15.0	55	331
(c)	(c)	1095	159	917	133	16.0	50	331
Heat treated in 13.7-mm (0.540-in.) rounds								
775	1425	1115	162	951	138	15.0	55	331
800	1475	1160	168	993	144	15.5	56	352
830	1525	1165	169	1005	146	15.5	56	352
(c)	(c)	1170	170	1070	155	14.5	50	352

(a) 0.2% offset. (b) In 50 mm (2 in.). (c) Quenched from step 3

AISI 4815, 4815H: Hardness and Machinability (Ref 7)

Condition	Hardness range, HB	Average machinability rating(a)
Annealed and cold drawn	187-229	50

(a) Based on AISI 1212 steel as 100% average machinability

Machining Data

For machining data on AISI grades 4815 and 4815H, refer to the preceding machining tables for AISI grade A2317.

AISI 4817, 4817H

AISI 4817, 4817H: Chemical Composition

AISI grade	C	Mn	P max	S max	Si	Ni	Mo
	Chemical composition, %						
4817	0.15-0.20	0.40-0.60	0.035	0.040	0.15-0.30	3.25-3.75	0.20-0.30
4817H	0.14-0.20	0.30-0.70	0.035	0.040	0.15-0.30	3.20-3.80	0.20-0.30

Characteristics and Typical Uses. AISI grades 4817 and 4817H, which have characteristics similar to AISI grades 4815 and 4815H, have slightly higher hardness and strength. AISI 4817 and 4817H also have high case and core hardenability.

These steels are available as hot rolled and cold finished bar, rod, and wire; seamless mechanical pipe; and bar subject to end-quench hardenability requirements.

AISI 4817: Similar Steels (U.S. and/or Foreign). UNS G48170; ASTM A304, A322, A331, A519; SAE J404, J412, J770

AISI 4817H: Similar Steels (U.S. and/or Foreign). UNS H48170; ASTM A304; SAE J1268

Machining Data

For machining data on AISI grades 4817 and 4817H, refer to the preceding machining tables for AISI grade A2317.

Physical Properties

AISI 4817, 4817H: Thermal Treatment Temperatures

Quenching medium: oil

Treatment	Temperature range °C	°F
Cycle annealing	(a)	(a)
Normalizing	(b)	(b)
Carburizing	900-925	1650-1700
Reheating	800-830	1475-1525
Tempering	120-165	250-325

(a) Heat at least as high as the carburizing temperature; hold for uniformity; cool rapidly to 540 to 675 °C (1000 to 1250 °F); holds for 1 to 3 h; air or furnace cool to obtain a structure suitable for machining and finish. (b) Normalize and temper. After normalizing, reheat to 595 to 650 °C (1100 to 1200 °F) and hold at temperature for approximately 1 h/25 mm (1 h/in.) of maximum section, or 4 h minimum time

Mechanical Properties

AISI 4817H: End-Quench Hardenability Limits and Hardenability Band.

Heat treating temperatures recommended by SAE: normalize at 925 °C (1700 °F), for forged or rolled specimens only; austenitize at 845 °C (1550 °F). (Ref 1)

Distance from quenched end, 1/16 in.	Hardness, HRC max	min
1	46	39
2	46	38
3	45	35
4	44	32
5	42	29
6	41	27
7	39	25
8	37	23
9	35	22
10	33	21
11	32	20
12	31	20
13	30	...
14	29	...
15	28	...
16	28	...
18	27	...
20	26	...
22	25	...
24	25	...
26	25	...
28	25	...
30	24	...
32	24	...

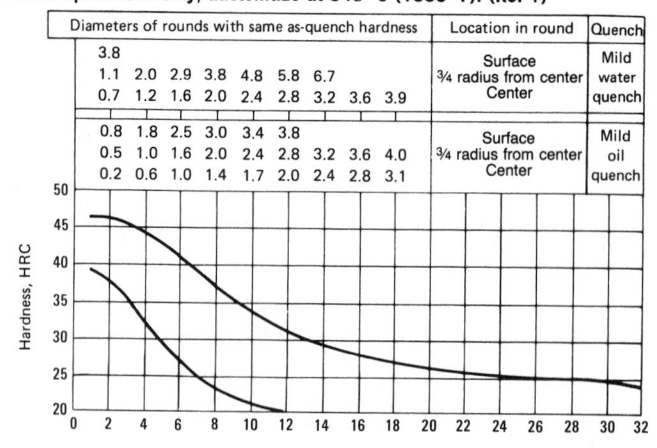

Mechanical Properties (continued)

AISI 4817, 4817H: Bending Fatigue Behavior

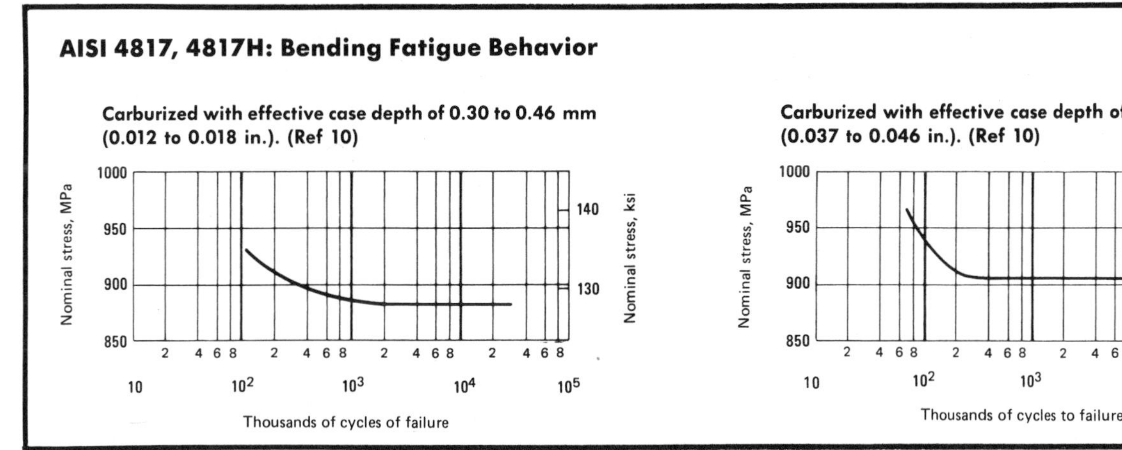

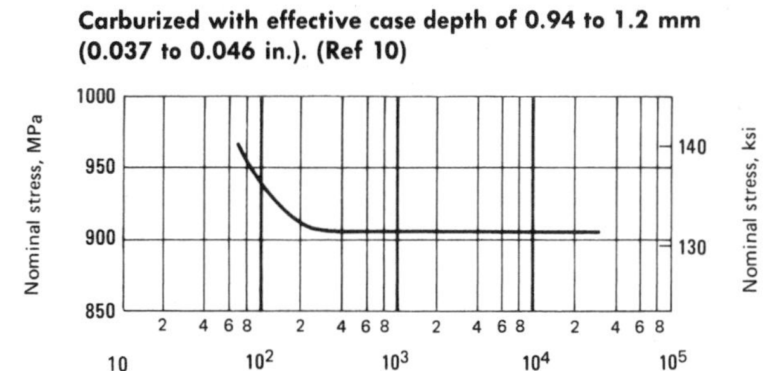

AISI 4817, 4817H: Hardness and Machinability (Ref 7)

Condition	Hardness range, HB	Average machinability rating(a)
Hot rolled and cold drawn	187-229	60

(a) Based on AISI 1212 steel as 100% average machinability

AISI 4817, 4817H: Permissible Compressive Stresses

Taken from gears having case depth of 1.1 to 1.5 mm (0.045 to 0.060 in.) and minimum case hardness of 60 HRC

Data point	Compressible stress MPa	ksi
Mean	1585	230
Maximum	1795	260

AISI 4820, 4820H

AISI 4820, 4820H: Chemical Composition

AISI grade	C	Mn	P max	S max	Si	Ni	Mo
4820	0.18-0.23	0.50-0.70	0.035	0.040	0.15-0.30	3.25-3.75	0.20-0.30
4820H	0.17-0.23	0.40-0.80	0.035	0.040	0.15-0.30	3.20-3.80	0.20-0.30

Chemical composition, %

Characteristics and Typical Uses. AISI grades 4820 and 4820H are medium-carbon, nickel-molybdenum, carburizing steels with high hardenability in both the case and core. These grades are used for case hardened gears, as well as bearings and races.

AISI 4020 and 4020H steels are available in hot rolled and cold rolled sheet and strip; hot rolled and cold finished bar, rod, and wire; seamless mechanical pipe; special quality ball- and roller-bearing steel; and bar subject to end-quench hardenability requirements.

AISI 4820: Similar Steels (U.S. and/or Foreign). UNS G48200; ASTM A322, A331, A505, A519, A535; SAE J404, J412, J770

AISI 4820H: Similar Steels (U.S. and/or Foreign). UNS H48200; ASTM A304; SAE J1268

AISI 4820, 4820H: Approximate Critical Points

Transformation point	Temperature(a) °C	°F
Ac_1	690	1270
Ac_3	780	1440
Ar_3	675	1245
Ar_1	415	780
M_s	370	695

(a) On heating or cooling at 28 °C (50 °F) per hour

Physical Properties

AISI 4820, 4820H: Thermal Treatment Temperatures

Quenching medium: oil

Treatment	Temperature range °C	°F
Cycle annealing	(a)	(a)
Normalizing	(b)	(b)
Carburizing	900-925	1650-1700
Reheating	800-830	1475-1525
Tempering	120-165	250-325

(a) Heat at least as high as the carburizing temperature; hold for uniformity; cool rapidly to 540 to 675 °C (1000 to 1250 °F); hold for 1 to 3 h; air or furnace cool to obtain a structure suitable for machining and finish. (b) Normalize and temper. After normalizing at carburizing temperature or above, reheat to 595 to 650 °C (1100 to 1200 °F) and hold at temperature for approximately 1 h/25 mm (1 h/in.) of maximum section, or 4 h minimum time

Mechanical Properties

AISI 4820, 4820H: Core Properties of Specimen Treated for Maximum Case Hardness and Core Toughness (Ref 4)

Treatment	Case depth, mm (in.)	Core properties: Case hardness, HRC	Tensile strength, MPa (ksi)	Yield strength, MPa (ksi)	Elongation(a), %	Reduction in area, %	Hardness HB
Recommended practice for maximum case hardness							
Direct quenched from pot: carburized at 925 °C (1700 °F) for 8 h; quenched in agitated oil; tempered at 150 °C (300 °F)	0.99 (0.039)	60	1415 (205)	1141 (165.5)	13.3	53.3	415
Single quenched and tempered (for good case and core properties): carburized at 925 °C (1700 °F) for 8 h; pot cooled; reheated to 800 °C (1475 °F); quenched in agitated oil; tempered at 150 °C (300 °F)	1.2 (0.047)	61	1431 (207.5)	1150 (167)	13.8	52.2	415
Double quenched and tempered (for maximum refinement of case and core): carburized at 925 °C (1700 °F) for 8 h; pot cooled; reheated to 815 °C (1500 °F); quenched in agitated oil; reheated to 790 °C (1450 °F); quenched in agitated oil; tempered at 150 °C (300 °F)	1.2 (0.047)	60	1410 (204.5)	1141 (165.5)	13.8	52.4	415
Recommended practice for maximum core toughness							
Direct quenched from pot: carburized at 925 °C (1700 °F) for 8 h; quenched in agitated oil; tempered at 230 °C (450 °F)	0.99 (0.039)	56	1382 (200.5)	1170 (170)	12.8	53	401
Single quenched and tempered (for good case and core properties): carburized at 925 °C (1700 °F) for 8 h; pot cooled; reheated to 800 °C (1475 °F); quenched in agitated oil; tempered at 230 °C (450 °F)	1.2 (0.047)	57.5	1415 (205)	1272 (184.5)	13	53.3	415
Double quenched and tempered (for maximum refinement of case and core): carburized at 925 °C (1700 °F) for 8 h; pot cooled; reheated to 815 °C (1500 °F); quenched in agitated oil; reheated to 790 °C (1450 °F); quenched in agitated oil; tempered at 230 °C (450 °F)	1.2 (0.047)	56.5	1355 (196.5)	1182 (171.5)	13.0	53.4	401

(a) In 50 mm (2 in.)

AISI 4820, 4820H: Approximate Core Mechanical Properties (Ref 2)

Heat treatment of test specimens: (1) normalized at 925 °C (1700 °F) in 32-mm (1.25-in.) rounds; (2) machined to 25- or 13.7-mm (1- or 0.540-in.) rounds; (3) pseudocarburized at 925 °C (1700 °F) for 8 h; (4) box cooled to room temperature; (5) reheated to temperatures given in the table and oil quenched; (6) tempered at 150 °C (300 °F); (7) tested in 12.8-mm (0.505-in.) rounds; tests were conducted using test specimens machined to English units

Reheat temperatures °C	°F	Tensile strength MPa	ksi	Yield strength(a) MPa	ksi	Elongation(b), %	Reduction in area, %	Hardness, HB
Heat treated in 25-mm (1-in.) rounds								
730	1345	1230	178.4	806	116.9	13.7	38.0	363
755	1395	1234	179.0	820	119.0	14.8	43.0	363
835	1535	1322	191.8	1021	148.1	12.5	43.1	388
(c)	(c)	1413	205.0	1085	157.3	11.3	36.7	401
Heat treated in 13.7-mm (0.540-in.) rounds								
730	1345	1363	197.7	945	137.1	13.1	42.2	401
755	1395	1328	192.6	920	133.4	13.8	43.8	388
835	1535	1422	206.3	1051	152.5	13.3	46.0	401
(c)	(c)	1448	210.0	1138	165.0	12.3	43.0	415

(a) 0.2% offset. (b) In 50 mm (2 in.). (c) Quenched from step 3

Mechanical Properties (continued)

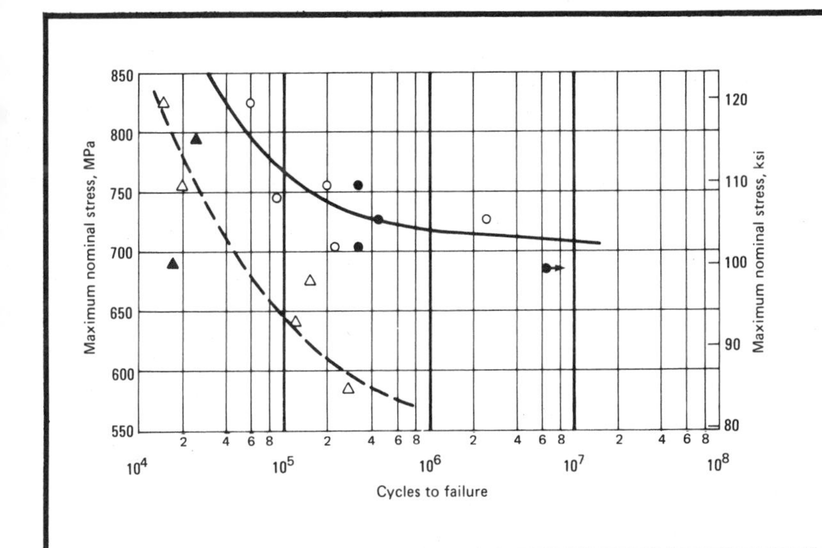

AISI 4820: Bending Fatigue Properties. —— standard treatment: as-heat treated (○), 437 impacts at 5.62 J or 4.14 ft·lb (●); – – – refrigerated at −73 °C (−100 °F): as-heat treated (△), 115 impacts at 5.62 J or 4.14 ft·lb (▲). The standard treatment used direct quenching from carburizing temperature or cooling under atmosphere from carburizing temperature to a lower temperature (around 845 °C or 1550 °F) before quenching. The refrigerated treatment used control cooling from the carburizing temperature followed by austenitizing at a temperature above the initial temperatures of both case and core; then refrigeration at −73 °C (−100 °F). (Ref 10)

AISI 4820, 4820H: Mass Effect on Mechanical Properties (Ref 4)

Condition or treatment	Size round mm	Size round in.	Tensile strength MPa	Tensile strength ksi	Yield strength MPa	Yield strength ksi	Elongation(a), %	Reduction in area, %	Hardness, HB
Annealed (heated to 815 °C or 1500 °F; furnace cooled 17 °C or 30 °F per hour to 260 °C or 500 °F; cooled in air)	25	1	685	99	460	67	22.3	58.8	197
Normalized (heated to 860 °C or 1580 °F; cooled in air)	13	0.5	772	112	495	72	26.0	57.8	235
	25	1	758	110	485	70	24.0	59.2	229
	50	2	738	107	475	69	23.0	59.8	223
	100	4	717	104	470	68	22.0	58.4	212
Mock carburized at 925 °C (1700 °F) for 8 h; reheated to 800 °C (1475 °F); quenched in oil; tempered at 150 °C (300 °F)	13	0.5	1440	209	1195	173	14.2	54.3	401
	25	1	1170	170	869	126	15.0	51.0	352
	50	2	938	136	640	93	19.8	56.3	277
	100	4	820	119	560	81	23.0	59.4	241
Mock carburized at 925 °C (1700 °F) for 8 h; reheated to 1475 °F; quenched in oil; tempered at 230 °C (450 °F)	13	0.5	1415	205	1170	170	13.2	52.3	388
	25	1	1125	163	827	120	15.5	53.1	331
	50	2	896	130	635	92	19.0	62.7	269
	100	4	807	117	550	80	21.0	63.8	235

(a) In 50 mm (2 in.)

AISI 4820, 4820H: Effect of the Mass on Hardness at Selected Points (Ref 4)

Size round mm	Size round in.	As-quenched hardness after quenching in oil at: Surface	½ radius	Center
13	0.5	45 HRC	45 HRC	44 HRC
25	1	43 HRC	39 HRC	37 HRC
50	2	36 HRC	31 HRC	27 HRC
100	4	27 HRC	24 HRC	24 HRC

AISI 4820, 4820H: Hardness and Machinability (Ref 7)

Condition	Hardness range, HB	Average machinability rating(a)
Annealed and cold drawn	187-229	50

(a) Based on AISI 1212 steel as 100% average machinability

AISI 4820, 4820H: Izod Impact Energy Values of Core for Carburized Steel (Ref 10)

Bar diameter mm	Bar diameter in.	Yield strength MPa	Yield strength ksi	Energy J	Energy ft·lb
13	0.5	1205	173	47.5	35
25	1	875	126	40.7	30
50	2	645	93	69.2	51
100	4	562	81	97.6	72
150	6	520	75	65.1(a)	48(a)

(a) Charpy impact (V-notch)

Mechanical Properties (continued)

AISI 4820H: End-Quench Hardenability Limits and Hardenability Band. Heat treating temperatures recommended by SAE: normalize at 925 °C (1700 °F), for forged or rolled specimens only; austenitize at 845 °C (1550 °F). (Ref 1)

Distance from quenched end, 1/16 in.	Hardness, HRC max	Hardness, HRC min
1	48	41
2	48	40
3	47	39
4	46	38
5	45	34
6	43	31
7	42	29
8	40	27
9	39	26
10	37	25
11	36	24
12	35	23
13	34	22
14	33	22
15	32	21
16	31	21
18	29	20
20	28	20
22	28	...
24	27	...
26	27	...
28	26	...
30	26	...
32	25	...

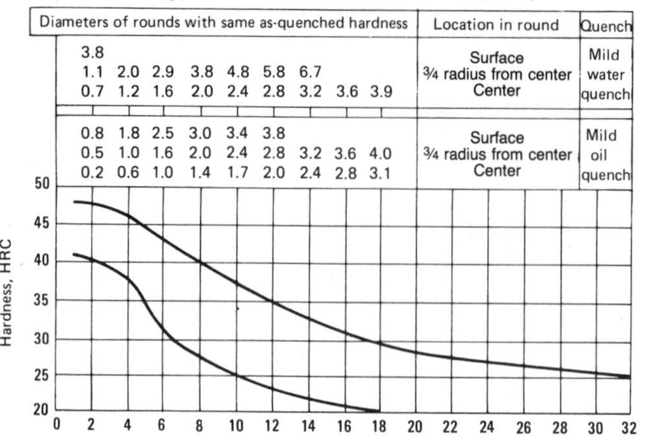

AISI 4820, 4820H: Tensile Properties (Ref 5)

Condition or treatment	Tensile strength MPa	Tensile strength ksi	Yield strength MPa	Yield strength ksi	Elongation(a), %	Reduction in area, %	Hardness, HB	Izod impact energy J	Izod impact energy ft·lb
Normalized at 860 °C (1580 °F)	760	110	485	70	24.0	59	229	110	81
Annealed at 815 °C (1500 °F)	685	99	460	67	22.3	59	197	94	69

(a) In 50 mm (2 in.)

Machining Data

For machining data on AISI grades 4820 and 4820H, refer to the preceding machining tables for AISI grade A2317.

AISI E50100, E51100, E52100

AISI E50100, E51100, E52100: Chemical Composition

AISI grade	C	Mn	P max	S max	Si	Cr
E50100 (a)	0.98-1.10	0.25-0.45	0.025	0.025	0.15-0.30	0.40-0.60
E51100	0.98-1.10	0.25-0.45	0.025	0.025	0.15-0.30	0.90-1.15
E52100	0.98-1.10	0.25-0.45	0.025	0.025	0.15-0.30	1.30-1.60

(a) Nonstandard for AISI

Characteristics. AISI grades E50100, E51100, and E52100 are high-carbon, chromium, electric-furnace, alloy steels. The compositions of the three grades are identical except for a variation in chromium content with a corresponding variation in hardenability.

Typical Uses. AISI grades E50100, E51100, and E52100 are available as hot rolled and cold finished bar, rod, and wire; tubes for the manufacture of ball and roller bearings; blooms, billets, and slabs for forging; and aircraft quality steels.

These steels are used primarily for the races and balls or rollers of rolling-element (antifriction) bearings. They are also suitable for other parts requiring high hardness and wear resistance.

AISI E50100: Similar Steels (U.S. and/or Foreign). UNS G50986; AMS 6442 B; ASTM A295, A519, A711; SAE J404, J412, J770; (W. Ger.) DIN 1.3501

AISI E51100: Similar Steels (U.S. and/or Foreign). UNS G51986; AMS 6443, 6446, 6449; ASTM A295, A322, A505, A519, A711; SAE J404, J412, J770; (W. Ger.) DIN 1.3503

AISI E52100: Similar Steels (U.S. and/or Foreign). UNS G52986; AMS 6440, 6441, 6444, 6447; ASTM A322, A331, A505, A519, A535, A646, A711; MIL SPEC MIL-S-980, MIL-S-7420, MIL-S-22141; SAE J404, J412, J770; (W. Ger.) DIN 1.3505; (Fr.) AFNOR 100 C 6; (Ital.) UNI 100 Cr 6; (U.K.) B.S. 534 A99, 535 A 99

AISI E50100, E51100, E52100: Approximate Critical Points

Transformation point	Temperature(a) E51100 °C	°F	E52100 °C	°F
Ac_1	750	1385	725	1340
Ac_3	770	1415	770	1415
Ar_3	715	1320	715	1320
Ar_1	705	1300	690	1270
M_s	...	...	250	485

(a) On heating or cooling at 28 °C (50 °F) per hour

Physical Properties

AISI E50100, E51100, E52100: Thermal Treatment Temperatures

Quenching medium: water or oil

Treatment	Temperature range °C	°F
Forging(a)	1150 max	2100 max
Forging(b)	1120 max	2050 max
Annealing	730-790	1350-1450
Austenitizing(c)	775-800	1425-1475
Austenitizing(d)	815-870	1500-1600
Tempering	(e)	(e)

(a) AISI E51100. (b) AISI E52100. (c) Water quench. (d) Oil quench. (e) To desired hardness

AISI E52100: Average Coefficients of Linear Thermal Expansion (Ref 6)

Condition	Temperature range °C	°F	Coefficient µm/m·K	µin./in.·°F
Annealed	23-280	74-540	11.9	6.60
Hardened	23-280	74-540	12.5	6.96

AISI E52100: Density (Ref 6)

Condition	Density g/cm^3	$lb/in.^3$
Annealed	7.81	0.282

Mechanical Properties

TBS-9: Effects of Tempering Temperatures on Hardness (Ref 12)

The alloy TBS-9 has metallurgical and processing characteristics similar to AISI E52100; chemical composition: 0.84 to 0.95% carbon, 0.50 to 0.80% manganese, 0.20 to 0.35% silicon, 0.25 to 0.40% chromium, 0.05 to 0.10% molybdenum; test specimens were austenitized at 845 °C (1550 °F)

Tempering temperature °C	°F	Hardness, HRC After 1 h tempering	After 2 h tempering
93	200	65.0	65.0
150	300	64.0	63.5
205	400	60.5	60.0
260	500	58.5	58.0
315	600	55.5	55.5
370	700	52.5	52.0
425	800	49.0	49.0
480	900	45.0	45.0
540	1000	40.0	39.0
595	1100	36.0	35.0
650	1200	32.0	30.0
705	1300	26.0	24.0

AISI E50100, E51100, E52100: Hardness and Machinability (Ref 7)

Condition	Hardness range, HB	Average machinability rating(a)
Spheroidized annealed and cold drawn	183-241	40

(a) Based on AISI 1212 steel as 100% average machinability

TBS-9: Effects of Austenitizing Temperatures on Hardness (Ref 12)

The alloy TBS-9 has metallurgical and processing characteristics similar to AISI E52100; chemical composition: 0.84 to 0.95% carbon, 0.50 to 0.80% manganese, 0.20 to 0.35% silicon, 0.25 to 0.40% chromium, 0.05 to 0.10% molybdenum

Austenitizing temperature °C	°F	Hardness, HRC As quenched	After tempering(a)
800	1475	66	63.5
815	1500	65	64.0
830	1525	66	63.5
845	1550	66	63.5
855	1575	65	63.0

(a) At 150 °C (300 °F)

AISI E52100: Rockwell C Hardness Values (Ref 11)

After tempering for 1 h at various temperatures

Tempering temperature °C	°F	Hardness, HRC Water quenched	Oil quenched
(a)	(a)	66	64
150	300	64	62
205	400	61	60
260	500	60	58
315	600	57	57
370	700	54	54
425	800	...	51
480	900	...	48

(a) As hardened

Machining Data (Ref 3)

AISI E50100, E51100, E52100: Turning (Single Point and Box Tools)

Depth of cut		High speed steel				Uncoated carbide						Coated carbide			
		Speed		Feed		Speed, brazed		Speed, inserted		Feed		Speed		Feed	
mm	in.	m/min	ft/min	mm/rev	in./rev	m/min	ft/min	m/min	ft/min	mm/rev	in./rev	m/min	ft/min	mm/rev	in./rev
Hardness, 175 to 225 HB															
1	0.040	37(a)	120(a)	0.18	0.007	115(b)	380(b)	145(b)	475(b)	0.18	0.007	190(c)	625(c)	0.18	0.007
4	0.150	27(a)	90(a)	0.40	0.015	90(d)	300(d)	115(d)	375(d)	0.50	0.020	150(e)	500(e)	0.40	0.015
8	0.300	21(a)	70(a)	0.50	0.020	72(d)	235(d)	90(d)	300(d)	0.75	0.030	120(e)	400(e)	0.50	0.020
16	0.625	17(a)	55(a)	0.75	0.030	56(d)	185(d)	70(d)	230(d)	1.00	0.040	...	...	...	...
Hardness, 225 to 275 HB															
1	0.040	30(a)	100(a)	0.18	0.007	105(b)	350(b)	135(b)	435(b)	0.18	0.007	175(c)	575(c)	0.18	0.007
4	0.150	24(a)	80(a)	0.40	0.015	84(d)	275(d)	105(d)	340(d)	0.50	0.020	135(e)	450(e)	0.40	0.015
8	0.300	18(a)	60(a)	0.50	0.020	66(d)	215(d)	82(d)	270(d)	0.75	0.030	105(e)	350(e)	0.50	0.020
16	0.625	14(a)	45(a)	0.75	0.030	52(d)	170(d)	64(d)	210(d)	1.00	0.040	...	...	...	...
Hardness, 275 to 325 HB															
1	0.040	24(f)	80(f)	0.18	0.007	105(b)	340(b)	120(b)	400(b)	0.18	0.007	160(c)	525(c)	0.18	0.007
4	0.150	18(f)	60(f)	0.40	0.015	81(d)	265(d)	95(d)	315(d)	0.40	0.015	120(e)	400(e)	0.40	0.015
8	0.300	14(f)	45(f)	0.50	0.020	64(d)	210(d)	76(d)	250(d)	0.50	0.020	100(e)	325(e)	0.50	0.020
Hardness, 325 to 375 HB															
1	0.040	17(f)	55(f)	0.18	0.007	87(b)	285(b)	105(b)	350(b)	0.18	0.007	135(c)	450(c)	0.18	0.007
4	0.150	14(f)	45(f)	0.40	0.015	69(d)	225(d)	84(d)	275(d)	0.40	0.015	105(e)	350(e)	0.40	0.015
8	0.300	9(f)	30(f)	0.50	0.020	53(d)	175(d)	66(d)	215(d)	0.50	0.020	84(e)	275(e)	0.50	0.020
Hardness, 375 to 425 HB															
1	0.040	14(f)	45(f)	0.18	0.007	70(b)	230(b)	85(b)	280(b)	0.18	0.007	115(c)	375(c)	0.18	0.007
4	0.150	11(f)	35(f)	0.40	0.015	55(d)	180(d)	67(d)	220(d)	0.40	0.015	90(e)	300(e)	0.40	0.015
8	0.300	8(f)	25(f)	0.50	0.020	43(d)	140(d)	52(d)	170(d)	0.50	0.020	69(e)	225(e)	0.50	0.020

(a) High speed steel tool material: M2 or M3. (b) Carbide tool material: C-7. (c) Carbide tool material: CC-7. (d) Carbide tool material: C-6. (e) Carbide tool material: CC-6. (f) Any premium high speed steel tool material (T15, M33, M41 to M47)

AISI E50100, E51100, E52100: Turning (Cutoff and Form Tools)

Tool material	Speed, m/min (ft/min)	Feed per revolution for cutoff tool width of:						Feed per revolution for form tool width of:							
		1.5 mm (0.062 in.)		3 mm (0.125 in.)		6 mm (0.25 in.)		12 mm (0.5 in.)		18 mm (0.75 in.)		25 mm (1 in.)		50 mm (2 in.)	
		mm	in.	mm	in.	mm	in.	mm	in.	mm	in.	mm	in.	mm	in.
Hardness, 175 to 225 HB															
M2 and M3 high speed steel	26 (85)	0.038	0.0015	0.046	0.0018	0.056	0.0022	0.046	0.0018	0.041	0.0016	0.038	0.0015	0.028	0.0011
C-6 carbide	84 (275)	0.038	0.0015	0.046	0.0018	0.056	0.0022	0.046	0.0018	0.041	0.0016	0.038	0.0015	0.028	0.0011
Hardness, 225 to 275 HB															
M2 and M3 high speed steel	20 (65)	0.036	0.0014	0.043	0.0017	0.053	0.0021	0.043	0.0017	0.038	0.0015	0.036	0.0014	0.025	0.0010
C-6 carbide	62 (205)	0.036	0.0014	0.043	0.0017	0.053	0.0021	0.043	0.0017	0.038	0.0015	0.036	0.0014	0.025	0.0010
Hardness, 275 to 325 HB															
M2 and M3 high speed steel	14 (45)	0.033	0.0013	0.041	0.0016	0.050	0.0020	0.041	0.0016	0.036	0.0014	0.033	0.0013	0.023	0.0009
C-6 carbide	44 (145)	0.033	0.0013	0.041	0.0016	0.050	0.0020	0.041	0.0016	0.036	0.0014	0.033	0.0013	0.023	0.0009
Hardness, 325 to 375 HB															
Any premium high speed steel (T15, M33, M41-M47)	11 (35)	0.030	0.0012	0.038	0.0015	0.048	0.0019	0.038	0.0015	0.033	0.0013	0.030	0.0012	0.020	0.0008
C-6 carbide	35 (115)	0.030	0.0012	0.038	0.0015	0.048	0.0019	0.038	0.0015	0.033	0.0013	0.030	0.0012	0.020	0.0008
Hardness, 375 to 425 HB															
Any premium high speed steel (T15, M33, M41-M47)	9 (95)	0.028	0.0011	0.036	0.0014	0.046	0.0018	0.036	0.0014	0.030	0.0012	0.028	0.0011	0.018	0.0007
C-6 carbide	29 (95)	0.028	0.0011	0.036	0.0014	0.046	0.0018	0.036	0.0014	0.030	0.0012	0.028	0.0011	0.018	0.0007

AISI E50100, E51100, E52100: Boring

Depth of cut		High speed steel				Uncoated carbide						Coated carbide			
		Speed		Feed		Speed, brazed		Speed, inserted		Feed		Speed		Feed	
mm	in.	m/min	ft/min	mm/rev	in./rev	m/min	ft/min	m/min	ft/min	mm/rev	in./rev	m/min	ft/min	mm/rev	in./rev
Hardness, 175 to 225 HB															
0.25	0.010	37(a)	120(a)	0.075	0.003	105(b)	350(b)	125(b)	410(b)	0.075	0.003	165(c)	545(c)	0.075	0.003
1.25	0.050	29(a)	95(a)	0.13	0.005	85(d)	280(d)	100(d)	330(d)	0.13	0.005	135(e)	435(e)	0.13	0.005
2.5	0.100	21(a)	70(a)	0.30	0.012	67(d)	220(d)	79(d)	260(d)	0.40	0.015	105(e)	350(e)	0.30	0.012
Hardness, 225 to 275 HB															
0.25	0.010	30(a)	100(a)	0.075	0.003	100(b)	325(b)	115(b)	380(b)	0.075	0.003	150(c)	500(c)	0.075	0.003
1.25	0.050	24(a)	80(a)	0.13	0.005	79(d)	260(d)	95(d)	305(d)	0.13	0.005	125(e)	405(e)	0.13	0.005
2.5	0.100	20(a)	65(a)	0.30	0.012	62(d)	205(d)	73(d)	240(d)	0.40	0.015	85(e)	280(e)	0.30	0.012
Hardness, 275 to 325 HB															
0.25	0.010	24(f)	80(f)	0.075	0.003	90(b)	300(b)	105(b)	350(b)	0.075	0.003	140(c)	460(c)	0.075	0.003
1.25	0.050	20(f)	65(f)	0.13	0.005	73(d)	240(d)	85(d)	280(d)	0.13	0.005	115(e)	370(e)	0.13	0.005
2.5	0.100	15(f)	50(f)	0.30	0.012	32(d)	105(d)	38(d)	125(d)	0.30	0.012	85(e)	280(e)	0.30	0.012
Hardness, 325 to 375 HB															
0.25	0.010	21(f)	70(f)	0.075	0.003	79(b)	260(b)	95(b)	305(b)	0.075	0.003	120(c)	395(c)	0.075	0.003
1.25	0.050	17(f)	55(f)	0.13	0.005	64(d)	210(d)	75(d)	245(d)	0.13	0.005	95(e)	315(e)	0.13	0.005
2.5	0.100	14(f)	45(f)	0.30	0.012	49(d)	160(d)	58(d)	190(d)	0.30	0.012	75(e)	245(e)	0.30	0.012
Hardness, 375 to 425 HB															
0.25	0.010	18(f)	60(f)	0.075	0.003	64(b)	210(b)	75(b)	245(b)	0.075	0.003	100(c)	325(c)	0.075	0.003
1.25	0.050	15(f)	50(f)	0.13	0.005	50(d)	165(d)	59(d)	195(d)	0.13	0.005	79(e)	260(e)	0.13	0.005
2.5	0.100	11(f)	35(f)	0.30	0.012	40(d)	130(d)	47(d)	155(d)	0.30	0.012	64(e)	210(e)	0.30	0.012

(a) High speed steel tool material: M2 or M3. (b) Carbide tool material: C-7. (c) Carbide tool material: CC-7. (d) Carbide tool material: C-6. (e) Carbide tool material: CC-6. (f) Any premium high speed steel tool material (T15, M33, M41 to M47)

AISI E50100, E51100, E52100: Drilling

Tool material	Speed		Feed per revolution for nominal hole diameter of:											
			1.5 mm (0.062 in.)		3 mm (0.125 in.)		6 mm (0.25 in.)		12 mm (0.5 in.)		18 mm (0.75 in.)		25 mm (1 in.)	
	m/min	ft/min	mm	in.	mm	in.	mm	in.	mm	in.	mm	in.	mm	in.
Hardness, 175 to 225 HB														
M10, M7, and M1 high speed steel	17	55	0.025	0.001	0.075	0.003	0.15	0.006	0.28	0.011	0.36	0.014	0.40	0.016
Hardness, 225 to 275 HB														
M10, M7, and M1 high speed steel	15	50	0.025	0.001	0.075	0.003	0.102	0.004	0.18	0.007	0.25	0.010	0.30	0.012
Hardness, 275 to 325 HB														
M10, M7, and M1 high speed steel	11	35	...	...	0.050	0.002	0.102	0.004	0.15	0.006	0.20	0.008	0.23	0.009
Hardness, 325 to 375 HB														
M10, M7, and M1 high speed steel	9	30	...	...	0.050	0.002	0.075	0.003	0.13	0.005	0.20	0.008	0.23	0.009
Hardness, 375 to 425 HB														
Any premium high speed steel (T15, M33, M41-M47)	8	25	...	...	0.050	0.002	0.075	0.003	0.102	0.004	0.15	0.006	0.20	0.008

AISI E50100, E51100, E52100: Planing

Tool material	Depth of cut		Speed		Feed/stroke	
	mm	in.	m/min	ft/min	mm	in.
Hardness, 175 to 225 HB						
M2 and M3 high speed steel	0.1	0.005	8	25	(a)	(a)
	2.5	0.100	12	40	0.75	0.030
	12	0.500	6	20	1.15	0.045
C-6 carbide	0.1	0.005	55	180	(a)	(a)
	2.5	0.100	67	220	1.50	0.060
	12	0.500	53	175	1.25	0.050
Hardness, 225 to 275 HB						
M2 and M3 high speed steel	0.1	0.005	8	25	(a)	(a)
	2.5	0.100	11	35	0.75	0.030
	12	0.500	6	20	1.15	0.045
C-6 carbide	0.1	0.005	43	140	(a)	(a)
	2.5	0.100	55	180	1.50	0.060
	12	0.500	43	140	1.25	0.050
Hardness, 275 to 325 HB						
M2 and M3 high speed steel	0.1	0.005	6	20	(a)	(a)
	2.5	0.100	9	30	0.50	0.020
	12	0.500	5	15	0.75	0.030
C-6 carbide	0.1	0.005	40	130	(a)	(a)
	2.5	0.100	49	160	1.50	0.060
	12	0.500	34	110	1.25	0.050
Hardness, 325 to 375 HB						
M2 and M3 high speed steel	0.1	0.005	5	15	(a)	(a)
	2.5	0.100	8	25	0.50	0.020
	12	0.500	5	15	0.75	0.030
C-6 carbide	0.1	0.005	34	110	(a)	(a)
	2.5	0.100	46	150	1.50	0.060
	12	0.500	30	100	1.25	0.050

(a) Feed is 75% the width of the square nose finishing tool

Machining Data (Ref 3) (continued)

AISI E50100, E51100, E52100: Reaming

Based on 4 flutes for 3- and 6-mm (0.125- and 0.25-in.) reamers, 6 flutes for 12-mm (0.5-in.) reamers, and 8 flutes for 25-mm (1-in.) and larger reamers

Tool material	Speed		Feed per revolution for reamer diameter of: 3 mm (0.125 in.)		6 mm (0.25 in.)		12 mm (0.5 in.)		25 mm (1 in.)		35 mm (1.5 in.)		50 mm (2 in.)	
	m/min	ft/min	mm	in.	mm	in.	mm	in.	mm	in.	mm	in.	mm	in.
Hardness, 175 to 225 HB														
Roughing														
M1, M2, and M7 high speed steel	18	60	0.075	0.003	0.15	0.006	0.20	0.008	0.30	0.012	0.40	0.015	0.50	0.020
C-2 carbide	21	70	0.102	0.004	0.15	0.006	0.20	0.008	0.30	0.012	0.40	0.015	0.50	0.020
Finishing														
M1, M2, and M7 high speed steel	11	35	0.13	0.005	0.20	0.008	0.30	0.012	0.50	0.020	0.65	0.025	0.75	0.030
C-2 carbide	14	45	0.13	0.005	0.20	0.008	0.30	0.012	0.50	0.020	0.65	0.025	0.75	0.030
Hardness, 225 to 275 HB														
Roughing														
M1, M2, and M7 high speed steel	15	50	0.075	0.003	0.15	0.006	0.20	0.008	0.30	0.012	0.40	0.015	0.50	0.020
C-2 carbide	20	65	0.102	0.004	0.15	0.006	0.20	0.008	0.30	0.012	0.40	0.015	0.50	0.020
Finishing														
M1, M2, and M7 high speed steel	9	30	0.102	0.004	0.18	0.007	0.25	0.010	0.40	0.015	0.50	0.020	0.65	0.025
C-2 carbide	12	40	0.102	0.004	0.18	0.007	0.25	0.010	0.40	0.015	0.50	0.020	0.65	0.025
Hardness, 275 to 325 HB														
Roughing														
M1, M2, and M7 high speed steel	12	40	0.075	0.003	0.102	0.004	0.15	0.006	0.25	0.010	0.30	0.012	0.40	0.015
C-2 carbide	17	55	0.102	0.004	0.13	0.005	0.15	0.006	0.25	0.010	0.30	0.012	0.40	0.015
Finishing														
M1, M2, and M7 high speed steel	8	25	0.075	0.003	0.15	0.006	0.20	0.008	0.30	0.012	0.40	0.015	0.50	0.020
C-2 carbide	11	35	0.102	0.004	0.15	0.006	0.20	0.008	0.30	0.012	0.40	0.015	0.50	0.020
Hardness, 325 to 375 HB														
Roughing														
M1, M2, and M7 high speed steel	9	30	0.050	0.002	0.102	0.004	0.15	0.006	0.20	0.008	0.25	0.010	0.30	0.012
C-2 carbide	14	45	0.102	0.004	0.15	0.006	0.20	0.008	0.25	0.010	0.30	0.012	0.40	0.015
Finishing														
M1, M2, and M7 high speed steel	8	25	0.075	0.003	0.13	0.005	0.15	0.006	0.25	0.010	0.30	0.012	0.40	0.015
C-2 carbide	11	35	0.102	0.004	0.15	0.006	0.20	0.008	0.25	0.010	0.30	0.012	0.40	0.015
Hardness, 375 to 425 HB														
Roughing														
Any premium high speed steel (T15, M33, M41-M47)	8	25	0.050	0.002	0.102	0.004	0.13	0.005	0.15	0.006	0.18	0.007	0.20	0.008
C-2 carbide	12	40	0.102	0.004	0.15	0.006	0.20	0.008	0.25	0.010	0.28	0.011	0.30	0.012
Finishing														
Any premium high speed steel (T15, M33, M41-M47)	6	20	0.050	0.002	0.102	0.004	0.13	0.005	0.15	0.006	0.18	0.007	0.20	0.008
C-2 carbide	9	30	0.102	0.004	0.15	0.006	0.20	0.008	0.25	0.010	0.28	0.011	0.30	0.012

AISI E50100, E51100, E52100: End Milling (Profiling)

Tool material	Depth of cut		Speed		Feed per tooth for cutter diameter of: 10 mm (0.375 in.)		12 mm (0.5 in.)		18 mm (0.75 in.)		25-50 mm (1-2 in.)	
	mm	in.	m/min	ft/min	mm	in.	mm	in.	mm	in.	mm	in.
Hardness, 175 to 225 HB												
M2, M3, and M7 high speed steel	0.5	0.020	34	110	0.025	0.001	0.050	0.002	0.075	0.003	0.102	0.004
	1.5	0.060	26	85	0.050	0.002	0.075	0.003	0.102	0.004	0.13	0.005
	diam/4	diam/4	21	70	0.025	0.001	0.050	0.002	0.075	0.003	0.102	0.004
	diam/2	diam/2	18	60	0.018	0.0007	0.025	0.001	0.050	0.002	0.075	0.003
C-5 carbide	0.5	0.020	140	455	0.038	0.0015	0.075	0.003	0.13	0.005	0.15	0.006
	1.5	0.060	105	350	0.063	0.0025	0.102	0.004	0.15	0.006	0.18	0.007
	diam/4	diam/4	90	295	0.050	0.002	0.075	0.003	0.13	0.005	0.15	0.006
	diam/2	diam/2	84	275	0.038	0.0015	0.050	0.002	0.102	0.004	0.13	0.005

(continued)

AISI E50100, E51100, E52100: End Milling (Profiling) (continued)

Tool material	Depth of cut mm	Depth of cut in.	Speed m/min	Speed ft/min	Feed per tooth for cutter diameter of: 10 mm (0.375 in.) mm	10 mm (0.375 in.) in.	12 mm (0.5 in.) mm	12 mm (0.5 in.) in.	18 mm (0.75 in.) mm	18 mm (0.75 in.) in.	25-50 mm (1-2 in.) mm	25-50 mm (1-2 in.) in.
Hardness, 225 to 275 HB												
M2, M3, and M7 high speed steel	0.5	0.020	30	100	0.025	0.001	0.050	0.002	0.075	0.003	0.102	0.004
	1.5	0.060	23	75	0.050	0.002	0.075	0.003	0.102	0.004	0.13	0.005
	diam/4	diam/4	20	65	0.025	0.001	0.050	0.002	0.075	0.003	0.102	0.004
	diam/2	diam/2	18	60	0.018	0.0007	0.025	0.001	0.050	0.002	0.075	0.003
C-5 carbide	0.5	0.020	120	390	0.025	0.001	0.050	0.002	0.102	0.004	0.13	0.005
	1.5	0.060	90	300	0.050	0.002	0.075	0.003	0.13	0.005	0.15	0.006
	diam/4	diam/4	78	255	0.038	0.0015	0.050	0.002	0.102	0.004	0.13	0.005
	diam/2	diam/2	72	235	0.025	0.001	0.038	0.0015	0.075	0.003	0.102	0.004
Hardness, 275 to 325 HB												
M2, M3, and M7 high speed steel	0.5	0.020	26	85	0.018	0.0007	0.038	0.0015	0.075	0.003	0.102	0.004
	1.5	0.060	20	65	0.025	0.001	0.050	0.002	0.102	0.004	0.13	0.005
	diam/4	diam/4	17	55	0.018	0.0007	0.038	0.0015	0.075	0.003	0.102	0.004
	diam/2	diam/2	15	50	0.013	0.0005	0.025	0.001	0.050	0.002	0.075	0.003
C-5 carbide	0.5	0.020	95	310	0.025	0.001	0.050	0.002	0.075	0.003	0.102	0.004
	1.5	0.060	72	235	0.050	0.002	0.075	0.003	0.102	0.004	0.13	0.005
	diam/4	diam/4	62	205	0.038	0.0015	0.050	0.002	0.075	0.003	0.102	0.004
	diam/2	diam/2	58	190	0.025	0.001	0.038	0.0015	0.050	0.002	0.075	0.003
Hardness, 325 to 375 HB												
M2, M3, and M7 high speed steel	0.5	0.020	20	65	0.013	0.0005	0.013	0.0015	0.075	0.003	0.102	0.004
	1.5	0.060	17	55	0.013	0.0005	0.013	0.0015	0.102	0.004	0.13	0.005
	diam/4	diam/4	15	50	0.013	0.0005	0.013	0.0015	0.075	0.003	0.102	0.004
	diam/2	diam/2	12	40	0.013	0.0005	0.025	0.001	0.050	0.002	0.075	0.003
C-5 carbide	0.5	0.020	79	260	0.025	0.001	0.038	0.0015	0.075	0.003	0.102	0.004
	1.5	0.060	60	200	0.038	0.0015	0.075	0.003	0.102	0.004	0.13	0.005
	diam/4	diam/4	52	170	0.038	0.0015	0.050	0.002	0.075	0.003	0.102	0.004
	diam/2	diam/2	49	160	0.025	0.001	0.038	0.0015	0.050	0.002	0.075	0.003
Hardness, 375 to 425 HB												
Any premium high speed steel (T15, M33, M41-M47)	0.5	0.020	17	55	0.013	0.0005	0.018	0.0007	0.025	0.001	0.050	0.002
	1.5	0.060	14	45	0.013	0.0005	0.025	0.001	0.050	0.002	0.075	0.003
	diam/4	diam/4	12	40	0.013	0.0005	0.013	0.0005	0.050	0.002	0.063	0.0025
	diam/2	diam/2	11	35	...	...	...	...	0.038	0.0015	0.050	0.002
C-5 carbide	0.5	0.020	60	200	0.025	0.001	0.025	0.001	0.050	0.002	0.075	0.003
	1.5	0.060	46	150	0.038	0.0015	0.050	0.002	0.075	0.003	0.102	0.004
	diam/4	diam/4	40	130	0.038	0.0015	0.038	0.0015	0.063	0.0025	0.089	0.0035
	diam/2	diam/2	35	115	0.025	0.001	0.025	0.001	0.050	0.002	0.075	0.003

AISI E50100, E51100, E52100: Face Milling

Depth of cut mm	Depth of cut in.	High speed steel: Speed m/min	Speed ft/min	Feed/tooth mm	Feed/tooth in.	Uncoated carbide: Speed, brazed m/min	Speed, brazed ft/min	Speed, inserted m/min	Speed, inserted ft/min	Feed/tooth mm	Feed/tooth in.	Coated carbide: Speed m/min	Speed ft/min	Feed/tooth mm	Feed/tooth in.
Hardness, 175 to 225 HB															
1	0.040	43(a)	140(a)	0.20	0.008	145(b)	480(b)	155(b)	510(b)	0.20	0.008	235(c)	765(c)	0.20	0.008
4	0.150	30(a)	100(a)	0.30	0.012	110(b)	360(b)	120(b)	400(b)	0.30	0.012	160(c)	520(c)	0.30	0.012
8	0.300	24(a)	80(a)	0.40	0.016	78(d)	255(d)	95(d)	310(d)	0.40	0.016	120(e)	400(e)	0.40	0.016
Hardness, 225 to 275 HB															
1	0.040	37(a)	120(a)	0.15	0.006	130(b)	420(b)	135(b)	450(b)	0.18	0.007	205(c)	675(c)	0.18	0.007
4	0.150	27(a)	90(a)	0.25	0.010	100(b)	320(b)	105(b)	350(b)	0.25	0.010	140(c)	455(c)	0.25	0.010
8	0.300	21(a)	70(a)	0.36	0.014	69(d)	225(d)	84(d)	275(d)	0.36	0.014	110(e)	355(e)	0.36	0.014
Hardness, 275 to 325 HB															
1	0.040	29(f)	95(f)	0.15	0.006	110(b)	360(b)	120(b)	400(b)	0.15	0.006	185(c)	600(c)	0.13	0.005
4	0.150	23(f)	75(f)	0.23	0.009	84(b)	275(b)	90(b)	300(b)	0.20	0.008	120(c)	390(c)	0.18	0.007
8	0.300	18(f)	60(f)	0.30	0.012	58(d)	190(d)	72(d)	235(d)	0.25	0.010	95(e)	305(e)	0.23	0.009
Hardness, 325 to 375 HB															
1	0.040	23(f)	75(f)	0.13	0.005	88(b)	290(b)	90(b)	300(b)	0.13	0.005	135(c)	450(c)	0.102	0.004
4	0.150	17(f)	55(f)	0.20	0.008	69(b)	225(b)	76(b)	250(b)	0.18	0.007	100(c)	325(c)	0.15	0.006
8	0.300	12(f)	40(f)	0.25	0.010	49(d)	160(d)	59(d)	195(d)	0.23	0.009	76(e)	250(e)	0.20	0.008
Hardness, 375 to 425 HB															
1	0.040	18(f)	60(f)	0.102	0.004	72(b)	235(b)	79(b)	260(b)	0.102	0.004	120(c)	390(c)	0.075	0.003
4	0.150	12(f)	40(f)	0.15	0.006	53(b)	175(b)	59(b)	195(b)	0.15	0.006	76(c)	250(c)	0.13	0.005
8	0.300	9(f)	30(f)	0.20	0.008	37(b)	120(b)	46(b)	150(b)	0.20	0.008	59(c)	195(c)	0.18	0.007

(a) High speed steel tool material: M2 or M7. (b) Carbide tool material: C-6. (c) Carbide tool material: CC-6. (d) Carbide tool material: C-5. (e) Carbide tool material: CC-5. (f) Any premium high speed steel tool material (T15, M33, M41 to M47)

AISI 50B40, 50B40H

AISI 50B40, 50B40H: Chemical Composition

AISI grade	Chemical composition, % C	Mn	P max	S max	Si	Cr	B
50B40	0.38-0.43	0.75-1.00	0.035	0.040	0.15-0.30	0.40-0.60	0.0005-0.003
50B40H	0.37-0.43	0.65-1.10	0.035	0.040	0.15-0.30	0.30-0.70	0.0005-0.003

Characteristics and Typical Uses. AISI grades 50B40 and 50B40H are directly hardenable, medium-carbon, chromium steels with medium hardenability characteristics. Hardenability has been improved by the addition of boron. These grades are available as hot rolled and cold finished bar, seamless mechanical tubing, and hot rolled bar subject to end-quench hardenability requirements.

AISI 50B40: Similar Steels (U.S. and/or Foreign). UNS G50401; ASTM A322, A519; SAE J404, J412, J770; (W. Ger.) DIN 1.7007; (Ital.) UNI 38 CrB 1 KB

AISI 50B40H: Similar Steels (U.S. and/or Foreign). UNS H50401; ASTM A304; SAE J1268; (W. Ger.) DIN 1.7007; (Ital.) UNI 38 CrB 1 KB

Mechanical Properties

AISI 50B40H: End-Quench Hardenability Limits and Hardenability Band. Heat treating temperatures recommended by SAE: normalize at 870 °C (1600 °F), for forged or rolled specimens only; austenitize at 845 °C (1550 °F). (Ref 1)

Distance from quenched end, 1/16 in.	Hardness, HRC max	min
1	60	53
2	60	53
3	59	52
4	59	51
5	58	50
6	58	48
7	57	44
8	57	39
9	56	34
10	55	31
11	53	29
12	51	28
13	49	27
14	47	26
15	44	25
16	41	25
18	38	23
20	36	21
22	35	...
24	34	...
26	33	...
28	32	...
30	30	...
32	29	...

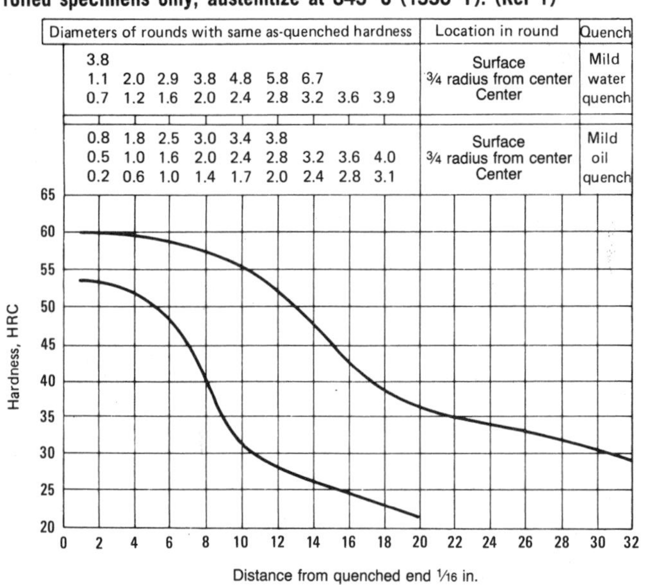

AISI 50B40, 50B40H: Hardness and Machinability (Ref 7)

Condition	Hardness range, HB	Average machinability rating(a)
Annealed and cold drawn	174-223	65

(a) Based on AISI 1212 steel as 100% average machinability

Machining Data

For machining data on AISI grades 50B40 and 50B40H, refer to the preceding machining tables for AISI grades 1340 and 1340H.

Physical Properties

AISI 50B40, 50B40H: Thermal Treatment Temperatures

Anneal or normalize for optimum machinability; quenching medium: oil

Treatment	Temperature range °C	°F
Annealing(a)	815-870	1500-1600
Normalizing	870-925	1600-1700
Austenitizing	815-845	1500-1550
Tempering	(b)	(b)

(a) Maximum hardness of 187 HB. (b) To desired hardness

AISI 50B44, 50B44H

AISI 50B44, 50B44H: Chemical Composition

AISI grade	Chemical composition, % C	Mn	P max	S max	Si	Cr	B
50B44	0.43-0.48	0.75-1.00	0.035	0.040	0.15-0.30	0.40-0.60	0.0005-0.003
50B44H	0.42-0.49	0.65-1.10	0.035	0.040	0.15-0.30	0.30-0.70	0.0005-0.003

Characteristics. AISI grades 50B44 and 50B44H are directly hardenable, chromium alloy steels with boron added to improve hardenability. These grades have a low hardenability rating within the 0.45 to 0.50% mean classification of carbon content.

Typical Uses. AISI 50B44 and 50B44H steels are used for average size automotive parts. They are available as hot rolled and cold finished bar, seamless mechanical tubing, and bar subject to end-quench hardenability requirements.

AISI 50B44: Similar Steels (U.S. and/or Foreign). UNS G50441; ASTM A322, A519; SAE J404, J412, J770

AISI 50B44H: Similar Steels (U.S. and/or Foreign). UNS H50441; ASTM A304; SAE J1268

Mechanical Properties

AISI 50B44H: End-Quench Hardenability Limits and Hardenability Band. Heat treating temperatures recommended by SAE: normalize at 870 °C (1600 °F), for forged or rolled specimens only; austenitize at 845 °C (1550 °F). (Ref 1)

Distance from quenched end, 1/16 in.	Hardness, HRC max	min
1	63	56
2	63	56
3	62	55
4	62	55
5	61	54
6	61	52
7	60	48
8	60	43
9	59	38
10	58	34
11	57	31
12	56	30
13	54	29
14	52	29
15	50	28
16	48	27
18	44	26
20	40	24
22	38	23
24	37	21
26	36	20
28	35	...
30	34	...
32	33	...

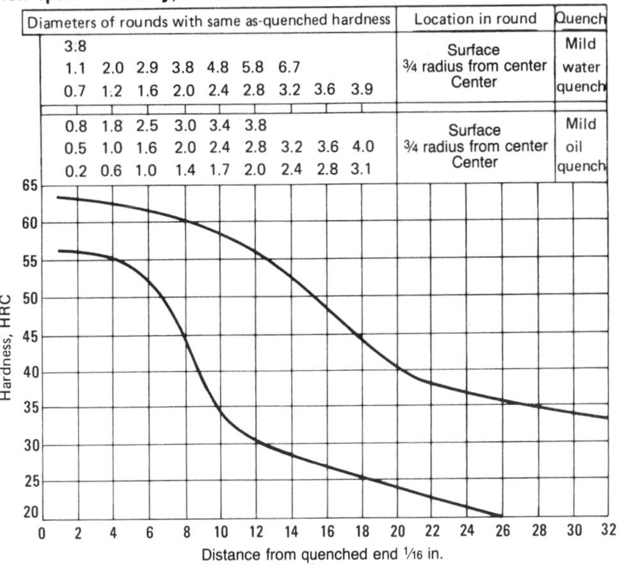

AISI 50B44, 50B44H: Hardness and Machinability (Ref 7)

Condition	Hardness range, HB	Average machinability rating(a)
Annealed and cold drawn	174-223	65

(a) Based on AISI 1212 steel as 100% average machinability

Machining Data

For machining data on AISI grades 50B44 and 50B44H, refer to the preceding machining tables for AISI grades 1340 and 1340H.

Physical Properties

AISI 50B44, 50B44H: Thermal Treatment Temperatures

Anneal or normalize for optimum machinability; quenching medium: oil

Treatment	Temperature range °C	°F
Annealing(a)	815-870	1500-1600
Normalizing	870-925	1600-1700
Austenitizing	815-845	1500-1550
Tempering	(b)	(b)

(a) Maximum hardness of 197 HB. (b) To desired hardness

AISI 5046, 5046H, 50B46, 50B46H

AISI 5046, 5046H, 50B46, 50B46H: Chemical Composition

AISI grade	Chemical composition, % C	Mn	P max	S max	Si	Cr	B
5046	0.43-0.50	0.75-1.00	0.035	0.040	0.15-0.30	0.20-0.35	...
5046H	0.44-0.50	0.65-1.10	0.035	0.040	0.15-0.30	0.13-0.43	...
50B46	0.44-0.49	0.75-1.00	0.035	0.040	0.15-0.30	0.20-0.35	0.0005-0.003
50B46H	0.43-0.49	0.65-1.10	0.035	0.040	0.15-0.30	0.13-0.43	0.0005-0.003

Characteristics and Typical Uses. AISI 5046, 5046H, 50B46, and 50B46H are directly hardenable, chromium steels with a low hardenability rating within the 0.45 to 0.50% mean classification of carbon content. AISI 50B46 and 50B46H contain boron to improve hardenability. Characteristics and uses are similar to AISI grades 50B44 and 50B44H.

AISI 5046, 5046H, 50B46, and 50B46H steels are available as hot rolled and cold finished bar, seamless mechanical tubing, and bar subject to end-quench hardenability requirements.

AISI 5046: Similar Steels (U.S. and/or Foreign). UNS G50460; ASTM A519; SAE J404, J412, J770

AISI 5046H: Similar Steels (U.S. and/or Foreign). UNS H50460; ASTM A304; SAE J1268

AISI 50B46: Similar Steels (U.S. and/or Foreign). UNS G50461; ASTM A322, A519; SAE J404, J412, J770

AISI 50B46H: Similar Steels (U.S. and/or Foreign). UNS H50461; ASTM A304; SAE J1268

AISI 5046, 50B46: Approximate Critical Points

Transformation point	Temperature(a) AISI 5046 °C	°F	AISI 50B46 °C	°F
Ac_1	715	1320	720	1330
Ac_3	770	1420	780	1440
Ar_3	730	1350	725	1340
Ar_1	680	1260	655	1210
M_s	325	620	325	620

(a) On heating or cooling at 28 °C (50 °F) per hour

Physical Properties

AISI 5046, 50B46: Thermal Treatment Temperatures

Quenching medium: oil

Treatment	Temperature range AISI 5046 °C	°F	AISI 50B46 °C	°F
Forging	1205 max	2200 max	1205 max	2200 max
Annealing(a)	790-845	1450-1550	815-870	1500-1600
Normalizing	845-900(b)	1550-1650(b)	845-900(c)	1550-1650(c)
Austenitizing	830-855	1525-1575	825-855	1525-1575
Tempering	(d)	(d)	(d)	(d)

(a) Maximum hardness of 192 HB. (b) Hardness of approximately 197 HB. (c) Hardness of approximately 223 HB. (d) To desired hardness

Machining Data

For machining data on AISI grades 5046, 5046H, 50B46, and 50B46H, refer to the preceding machining tables for AISI grades 1340 and 1340H.

Mechanical Properties

AISI 5046: Effect of Tempering Temperature on Tensile Properties.

Normalized at 870 °C (1600 °F); oil quenched from 845 °C (1550 °F); tempered at 56 °C (100 °F) intervals. Specimens were treated in 13.7-mm (0.540-in.) diam and machined to 12.8-mm (0.505-in.) diam for testing. Tests were conducted using specimens machined to English units. Elongation was measured in 50 mm (2 in.). (Ref 2)

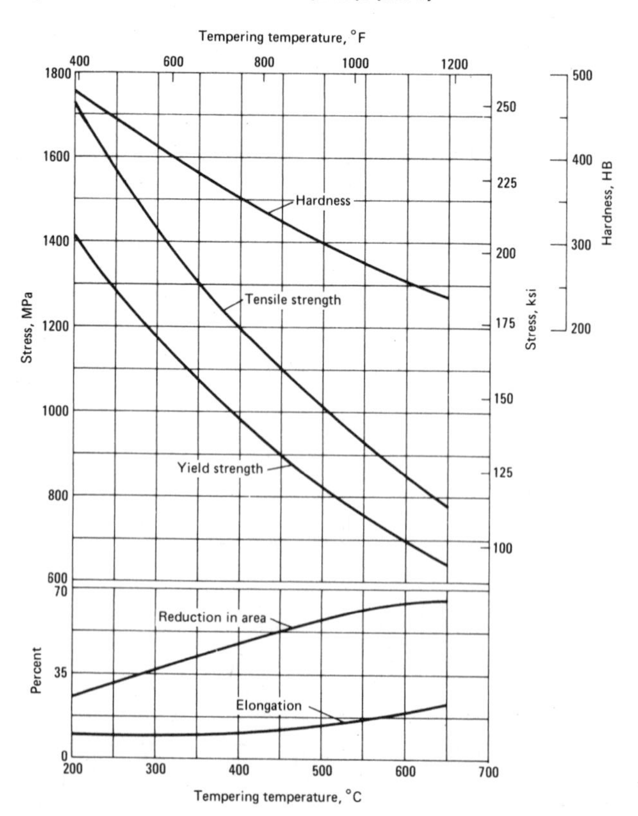

AISI 5046H: End-Quench Hardenability Limits and Hardenability Band. Heat treating temperatures recommended by SAE: normalize at 870 °C (1600 °F), for forged or rolled specimens only; austenitize at 845 °C (1550 °F). (Ref 1)

Distance from quenched end, 1/16 in.	Hardness, HRC max	Hardness, HRC min
1	63	56
2	62	55
3	60	45
4	56	32
5	52	28
6	46	27
7	39	26
8	35	25
9	34	24
10	33	24
11	33	23
12	32	23
13	32	22
14	31	22
15	31	21
16	30	21
18	29	20
20	28	...
22	27	...
24	26	...
26	25	...
28	24	...
30	23	...
32	23	...

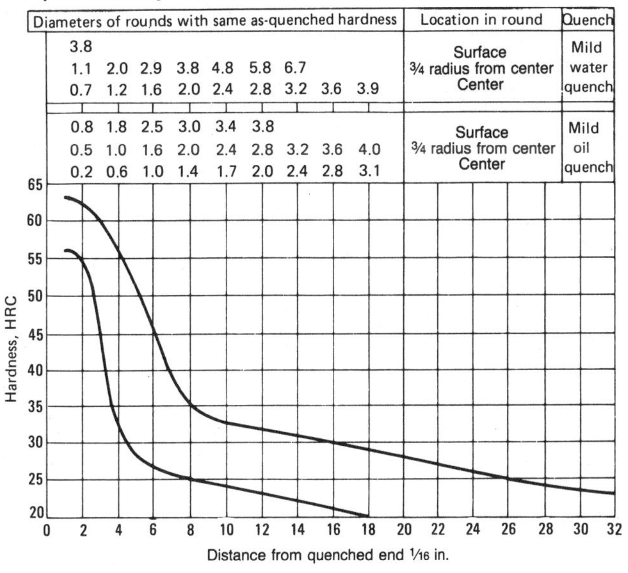

AISI 50B46H: End-Quench Hardenability Limits and Hardenability Band. Heat treating temperatures recommended by SAE: normalize at 870 °C (1600 °F), for forged or rolled specimens only; austenitize at 845 °C (1550 °F). (Ref 1)

Distance from quenched end, 1/16 in.	Hardness, HRC max	Hardness, HRC min
1	63	56
2	62	54
3	61	52
4	60	50
5	59	41
6	58	32
7	57	31
8	56	30
9	54	29
10	51	28
11	47	27
12	43	26
13	40	26
14	38	25
15	37	25
16	36	24
18	35	23
20	34	22
22	33	21
24	32	20
26	31	...
28	30	...
30	29	...
32	28	...

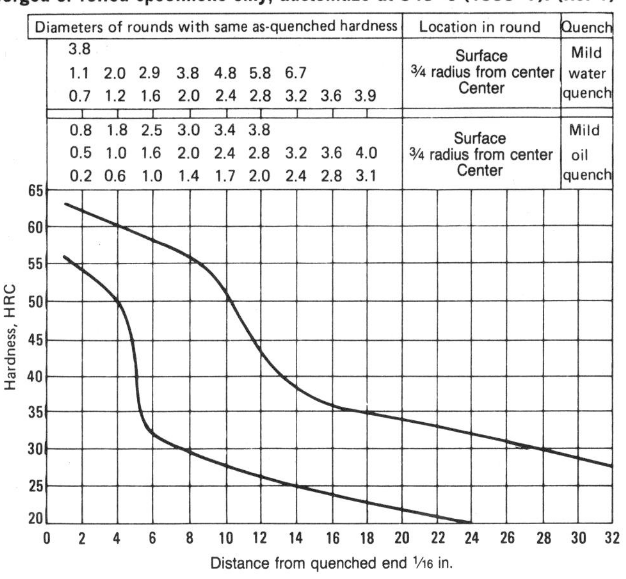

AISI 5046, 5046H, 50B46, 50B46H: Hardness and Machinability (Ref 7)

Condition	Hardness range, HB	Average machinability rating(a)
Annealed and cold drawn	174-223	60

(a) Based on AISI 1212 steel as 100% average machinability

Mechanical Properties (continued)

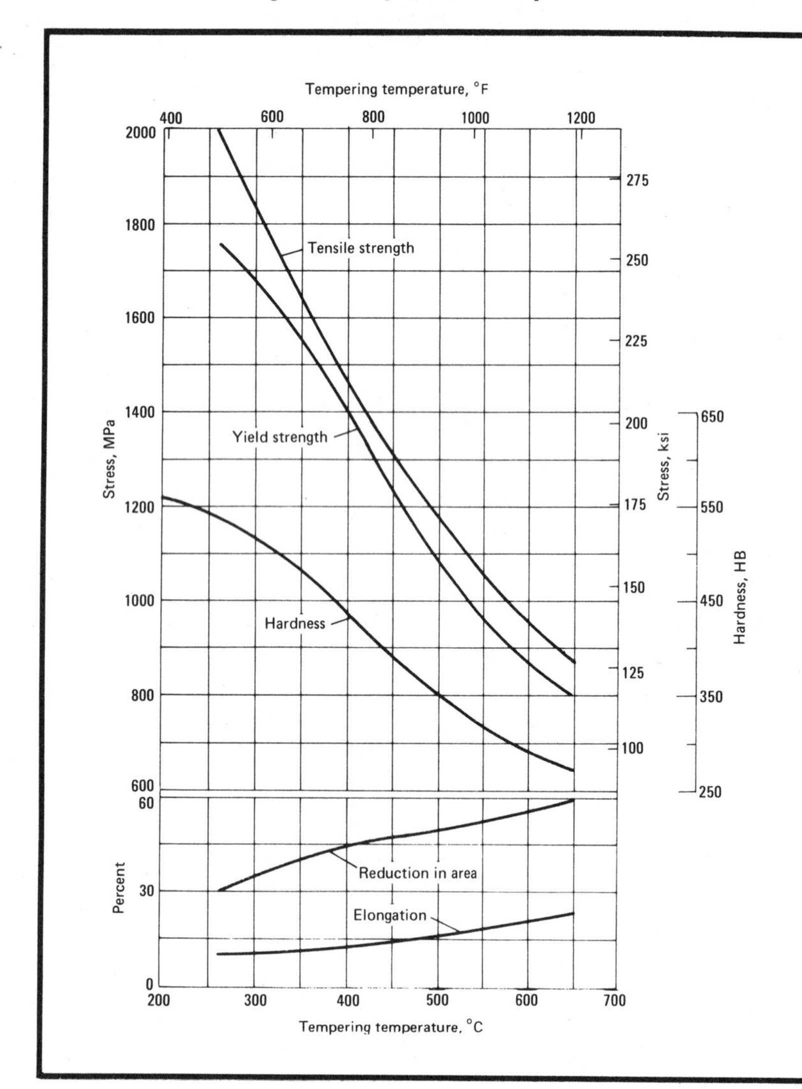

AISI 50B46: Effect of Tempering Temperature on Tensile Properties. Normalized at 870 °C (1600 °F); oil quenched from 845 °C (1550 °F); tempered at 56 °C (100 °F) intervals. Specimens were treated in 13.7-mm (0.540-in.) diam and machined to 12.8-mm (0.505-in.) diam for testing. Tests were conducted using specimens machined to English units. Elongation was measured in 50 mm (2 in.). (Ref 2)

AISI 50B50, 50B50H

AISI 50B50, 50B50H: Chemical Composition

AISI grade	Chemical composition, % C	Mn	P max	S max	Si	Cr	B
50B50	0.48-0.53	0.75-1.00	0.035	0.040	0.15-0.30	0.40-0.60	0.0005-0.003
50B50H	0.47-0.54	0.65-1.10	0.035	0.040	0.15-0.30	0.35-0.70	0.0005-0.003

Characteristics and Typical Uses. AISI grades 50B50 and 50B50H are directly hardenable, chromium alloy steels with a medium hardenability rating in the 0.50 to 0.60% mean classification of carbon content. These steels are used primarily for springs and hand tools. They are available as hot rolled and cold finished bar, seamless mechanical tubing, and bar subject to end-quench hardenability requirements.

AISI 50B50: Similar Steels (U.S. and/or Foreign). UNS G50501; ASTM A322, A519; SAE J404, J412, J770; (W. Ger.) DIN 1.7138; (Jap.) JIS SUP 11

AISI 50B50H: Similar Steels (U.S. and/or Foreign). UNS H50501; ASTM A304; SAE J1268; (W. Ger.) DIN 1.7138; (Jap.) JIS SUP 11

Mechanical Properties

AISI 50B50H: End-Quench Hardenability Limits and Hardenability Band. Heat treating temperatures recommended by SAE: normalize at 870 °C (1600 °F), for forged or rolled specimens only; austenitize at 845 °C (1550 °F). (Ref 1)

Distance from quenched end, 1/16 in.	Hardness, HRC max	Hardness, HRC min
1	65	59
2	65	59
3	64	58
4	64	57
5	63	56
6	63	55
7	62	52
8	62	47
9	61	42
10	60	37
11	60	35
12	59	33
13	58	32
14	57	31
15	56	30
16	54	29
18	50	28
20	47	27
22	44	26
24	41	25
26	39	24
28	38	22
30	37	21
32	36	20

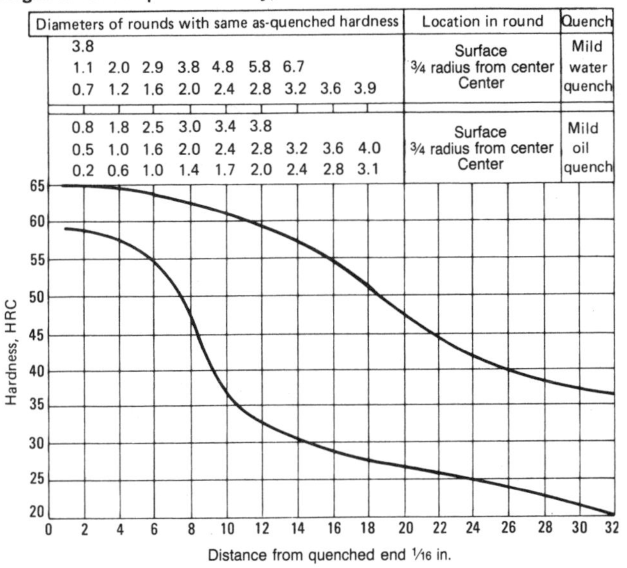

AISI 50B50, 50B50H: Hardness and Machinability (Ref 7)

Condition	Hardness range, HB	Average machinability rating(a)
Annealed and cold drawn	183-235	55

(a) Based on AISI 1212 steel as 100% average machinability

Machining Data

For machining data on AISI grades 50B50 and 50B50H, refer to the preceding machining tables on AISI grades 4150 and 4150H.

Physical Properties

AISI 50B50, 50B50H: Thermal Treatment Temperatures

Anneal or normalize for optimum machinability; quenching medium: oil

Treatment	Temperature range °C	Temperature range °F
Annealing(a)	815-870	1500-1600
Normalizing	870-925	1600-1700
Austenitizing	800-845	1475-1550
Tempering	(b)	(b)

(a) Maximum hardness of 201 HB. (b) To desired hardness

AISI 5060, 50B60, 50B60H

AISI 5060, 50B60, 50B60H: Chemical Composition

AISI grade	C	Mn	P max	S max	Si	Cr	B
5060	0.56-0.64	0.75-1.00	0.035	0.040	0.15-0.30	0.40-0.60	...
50B60	0.56-0.64	0.75-1.00	0.035	0.040	0.15-0.30	0.40-0.60	0.0005-0.003
50B60H	0.55-0.65	0.70-1.10	0.035	0.040	0.15-0.30	0.35-0.70	0.0005-0.003

(Chemical composition, %)

Characteristics. AISI grades 5060, 50B60, and 50B60H are directly hardenable, chromium alloy steels with medium hardenability. The addition of boron to AISI 50B60 and 50B60H improves hardenability.

Typical Uses. AISI 5060, 50B60, and 50B60H steels are used primarily for springs and hand tools. They are available as hot rolled and cold finished bar, rod, and wire; seamless mechanical tubing; and bar subject to end-quench hardenability requirements.

AISI 5060: Similar Steels (U.S. and/or Foreign). UNS G50600; SAE J404, J770

AISI 50B60: Similar Steels (U.S. and/or Foreign). UNS G50601; ASTM A322, A331, A519; SAE J404, J412, J770

AISI 50B60H: Similar Steels (U.S. and/or Foreign). UNS H50601; ASTM A304; SAE J1268

AISI 5060, 50B60, 50B60H: Approximate Critical Points

Transformation point	Temperature(a) °C	Temperature(a) °F	Transformation point	Temperature(a) °C	Temperature(a) °F
Ac_1	730	1345	Ar_1	675	1250
Ac_3	770	1420	M_s	270	515
Ar_3	730	1345			

(a) On heating or cooling at 28 °C (50 °F) per hour

Physical Properties

AISI 50B60: Thermal Treatment Temperatures

Anneal or normalize for optimum machinability; quenching medium: oil

Treatment	Temperature range °C	Temperature range °F
Forging	1205 max	2200 max
Annealing(a)	790-845	1450-1550
Normalizing	845-900	1550-1650
Austenitizing(b)	830-855	1525-1575
Tempering	(c)	(c)

(a) Maximum hardness of 217 HB. (b) Hardness of approximately 255 HB. (c) To desired hardness

Mechanical Properties

AISI 50B60H: End-Quench Hardenability Limits and Hardenability Band.

Heat treating temperatures recommended by SAE: normalize at 870 °C (1600 °F), for forged or rolled specimens only; austenitize at 845 °C (1550 °F). (Ref 1)

Distance from quenched end, 1/16 in.	Hardness, HRC max	Hardness, HRC min
1	...	60
2	...	60
3	...	60
4	...	60
5	...	60
6	...	59
7	...	57
8	65	53
9	65	47
10	64	42
11	64	39
12	64	37
13	63	36
14	63	35
15	63	34
16	62	34
18	60	33
20	58	31
22	55	30
24	53	29
26	51	28
28	49	27
30	47	26
32	44	25

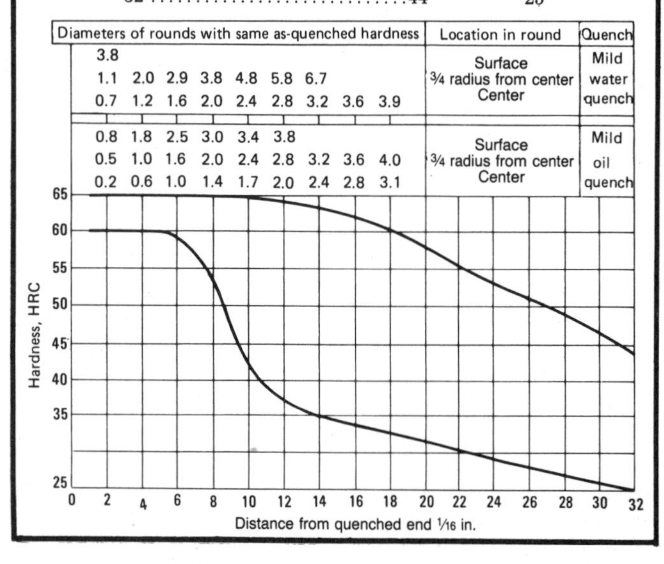

AISI 50B60: Effect of Tempering Temperature on Tensile Properties.

Normalized at 870 °C (1600 °F); oil quenched from 845 °C (1550 °F); tempered at 56 °C (100 °F) intervals. Specimens were treated in 13.7-mm (0.540-in.) diam and machined to 12.8-mm (0.505-in.) diam for testing. Tests were conducted using specimens machined to English units. Elongation was measured in 50 mm (2 in.). (Ref 2)

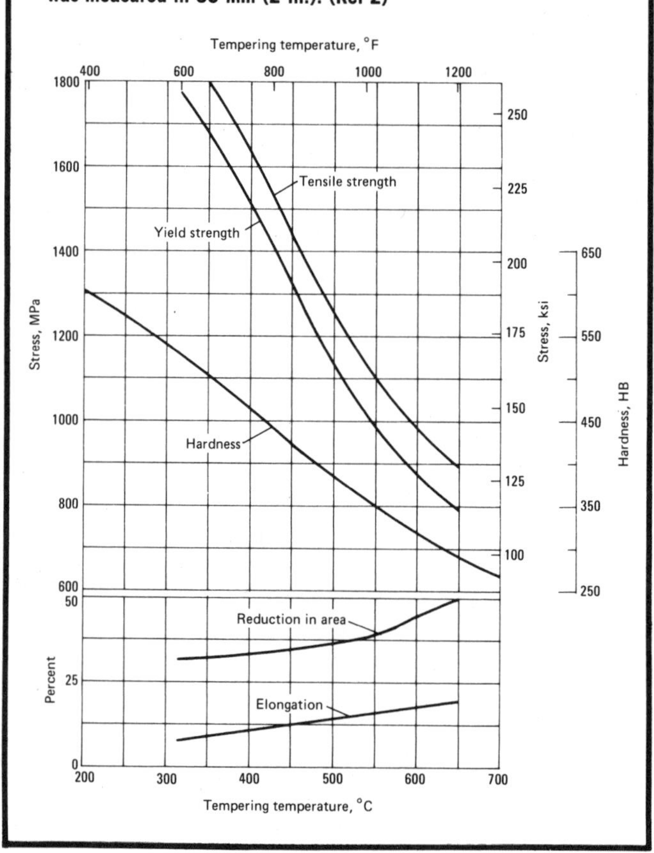

Mechanical Properties (continued)

AISI 5060, 50B60, 50B60H: Hardness and Machinability (Ref 7)

Condition	Hardness range, HB	Average machinability rating(a)
Spheroidized annealed and cold drawn	170-212	55

(a) Based on AISI 1212 steel as 100% average machinability

Machining Data

For machining data on AISI grades 5060, 50B60, and 50B60H, refer to the preceding machining tables for AISI grades 4150 and 4150H.

AISI 5120, 5120H

AISI 5120, 5120H: Chemical Composition

AISI grade	C	Mn	P max	S max	Si	Cr
5120	0.17-0.22	0.70-0.90	0.035	0.040	0.15-0.30	0.70-0.90
5120H	0.17-0.23	0.60-1.00	0.035	0.040	0.15-0.30	0.60-1.00

Chemical composition, %

Characteristics and Typical Uses. AISI grades 5120 and 5120H are low-carbon, chromium, carburizing steels with low case hardenability and low-to-medium core hardenability. These steels are available as hot rolled and cold finished bar, rod, and wire; seamless mechanical tubing; and bar subject to end-quench hardenability requirements.

AISI 5120: Similar Steels (U.S. and/or Foreign). UNS G51200; ASTM A322, A331, A519; SAE J404, J770; (W. Ger.) DIN 1.7147; (Fr.) AFNOR 20 MC 5

AISI 5120H: Similar Steels (U.S. and/or Foreign). UNS H51200; ASTM A304; SAE J1268; (W. Ger.) DIN 1.7147; (Fr.) AFNOR 20 MC 5

AISI 5120, 5120H: Approximate Critical Points

Transformation point	Temperature(a) °C	°F
Ac_1	765	1410
Ac_3	840	1540
Ar_3	800	1470
Ar_1	700	1290
M_s	405	760

(a) On heating or cooling at 28 °C (50 °F) per hour

Physical Properties

AISI 5120, 5120H: Thermal Treatment Temperatures

Quenching medium: oil

Treatment	Temperature range °C	°F
Forging	1230 max	2250 max
Normalizing	(a)	(a)
Carburizing	900-925	1650-1700
Tempering	120-175	250-350

(a) Temperature should be as high as carburizing temperature, followed by air cooling

Machining Data

For machining data on AISI grades 5120 and 5120H, refer to the preceding machining tables for AISI grade A2317.

Mechanical Properties

AISI 5120, 5120H: Approximate Core Mechanical Properties (Ref 2)

Heat treatment of test specimens: (1) normalized at 925 °C (1700 °F) in 32-mm (1.25-in.) rounds; (2) machined to 25- or 13.7-mm (1- or 0.540-in.) rounds; (3) pseudocarburized at 925 °C (1700 °F) for 8 h; (4) box cooled to room temperature; (5) reheated to temperatures given in the table and oil quenched; (6) tempered at 150 °C (300 °F); (7) tested in 12.8-mm (0.505-in.) rounds; tests were conducted using test specimens machined to English units

Reheat temperature °C	°F	Tensile strength MPa	ksi	Yield strength(a) MPa	ksi	Elongation(b), %	Reduction in area, %	Hardness, HB
Heat treated in 25-mm (1-in.) rounds								
775	1425	834	121	635	92	14.0	41	262
800	1475	883	128	696	101	15.0	42	269
845	1550	938	136	758	110	16.0	45	285
(c)	(c)	986	143	786	114	13.5	45	302
Heat treated in 13.7-mm (0.540-in.) rounds								
775	1425	848	123	650	94	14.5	40	269
800	1475	910	132	717	104	15.0	43	277
845	1550	979	142	786	114	16.0	50	293
(c)	(c)	1020	148	848	123	14.0	40	311

(a) 0.2% offset. (b) In 50 mm (2 in.). (c) Quenched from step 3

AISI 5120, 5120H: Hardness and Machinability (Ref 7)

Condition	Hardness range, HB	Average machinability rating(a)
Hot rolled and cold drawn	163-201	70

(a) Based on AISI 1212 steel as 100% average machinability

Mechanical Properties (continued)

AISI 5120H: End-Quench Hardenability Limits and Hardenability Band. Heat treating temperatures recommended by SAE: normalize at 925 °C (1700 °F), for forged or rolled specimens only; austenitize at 925 °C (1700 °F). (Ref 1)

Distance from quenched end, 1/16 in.	Hardness, HRC max	Hardness, HRC min
1	48	40
2	46	34
3	41	28
4	36	23
5	33	20
6	30	...
7	28	...
8	27	...
9	25	...
10	24	...
11	23	...
12	22	...
13	21	...
14	21	...
15	20	...

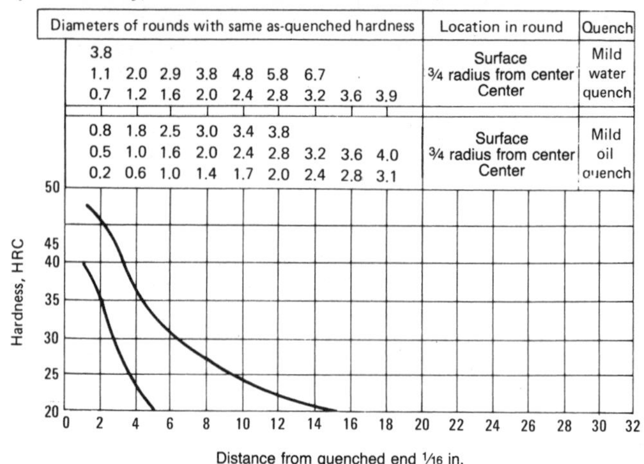

AISI 5130, 5130H

AISI 5130, 5130H: Chemical Composition

AISI grade	C	Mn	P max	S max	Si	Cr
5130	0.28-0.33	0.70-0.90	0.035	0.040	0.15-0.30	0.80-1.10
5130H	0.27-0.33	0.60-1.00	0.035	0.045	0.15-0.30	0.75-1.20

Characteristics and Typical Uses. AISI grades 5130 and 5130H are medium-carbon, chromium, alloy steels with low hardenability for water hardening. Parts of small section size from these steels can be oil quenched where deformation may be a problem.

Grades 5130 and 5130H are available as hot rolled and cold finished bar; seamless mechanical tubing; and, on a limited basis, hot rolled and cold rolled sheet and strip.

AISI 1530: Similar Steels (U.S. and/or Foreign). UNS G51300; ASTM A322, A331, A505, A519; SAE J404, J412, J770; (W. Ger.) DIN 1.7030; (U.K.) B.S. 530 A 30, 530 H 30

AISI 5130H: Similar Steels (U.S. and/or Foreign). UNS H51300; ASTM A304; SAE J1268; (W. Ger.) DIN 1.7033; (Fr.) AFNOR 32 C 4; (Ital.) UNI 34 Cr 4 KB; (Jap.) JIS SCr 2 H, SCr 2; (U.K.) B.S. 530 A 32, 530 H 32

Machining Data

For machining data on AISI grades 5130 and 5130H, refer to the preceding machining tables for AISI grades 1330 and 1330H.

AISI 5130, 5130H: Approximate Critical Points

Transformation point	Temperature(a) °C	Temperature(a) °F	Transformation point	Temperature(a) °C	Temperature(a) °F
Ac_1	745	1370	Ar_1	695	1280
Ac_3	810	1490	M_s	360	680
Ar_3	745	1370			

(a) On heating or cooling at 28 °C (50 °F) per hour

Physical Properties

AISI 5130, 5130H: Thermal Treatment Temperatures

Quenching mediums: water, caustic, oil

Treatment	Temperature range °C	Temperature range °F
Forging	1230 max	2250 max
Annealing(a)	815-870	1500-1600
Normalizing(b)	870-925	1600-1700
Austenitizing	855-885	1575-1625

(a) Maximum hardness of 170 HB. (b) Hardness of approximately 241 HB

AISI 5130, 5130H: Density (Ref 6)

Condition	Density g/cm³	Density lb/in.³
Hardened and tempered	7.83	0.283

Mechanical Properties

AISI 5130: Effect of Tempering Temperature on Tensile Properties.

Normalized at 900 °C (1650 °F); water quenched from 870 °C (1600 °F); tempered at 56 °C (100 °F) intervals. Specimens were treated in 13.7-mm (0.540-in.) diam and machined to 12.8-mm (0.505-in.) diam for testing. Tests were conducted using specimens machined to English units. Elongation was measured in 50 mm (2 in.). (Ref 2)

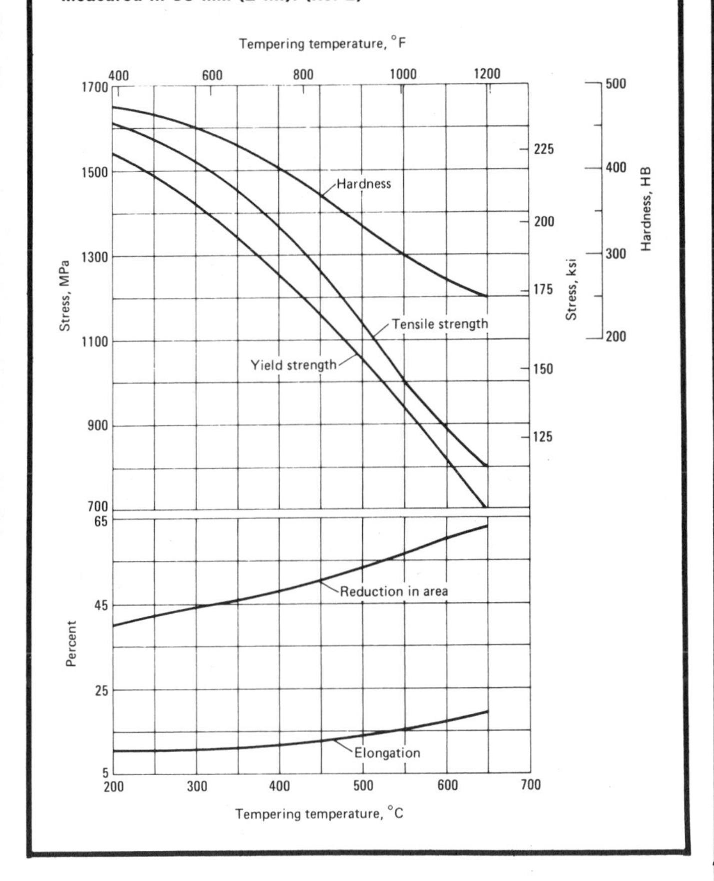

AISI 5130, 5130H: Hardness and Machinability (Ref 7)

Condition	Hardness range, HB	Average machinability rating(a)
Annealed and cold drawn	174-212	70

(a) Based on AISI 1212 steel as 100% average machinability

AISI 5130H: End-Quench Hardenability Limits and Hardenability Band.

Heat treating temperatures recommended by SAE: normalize at 900 °C (1650 °F), for forged or rolled specimens only; austenitize at 870 °C (1600 °F). (Ref 1)

Distance from quenched end, 1/16 in.	Hardness, HRC max	Hardness, HRC min
1	56	49
2	55	46
3	53	42
4	51	39
5	49	35
6	47	32
7	45	30
8	42	28
9	40	26
10	38	25
11	37	23
12	36	22
13	35	21
14	34	20
15	34	...
16	33	...
18	32	...
20	31	...
22	30	...
24	29	...
26	27	...
28	26	...
30	25	...
32	24	...

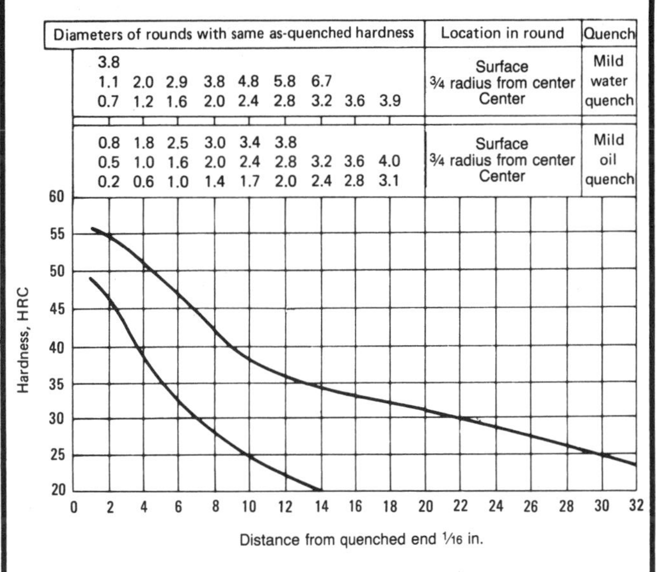

AISI 5132, 5132H

AISI 5132, 5132H: Chemical Composition

AISI grade	C	Mn	P max	S max	Si	Cr
5132	0.30-0.35	0.60-0.80	0.035	0.040	0.15-0.30	0.75-1.00
5132H	0.29-0.35	0.50-0.90	0.035	0.040	0.15-0.30	0.65-1.10

Chemical composition, %

Characteristics and Typical Uses. AISI grades 5132 and 5132H are medium-carbon, chromium, alloy steels with low hardenability for water or caustic solution quenching. Parts of small section size made from these steels can be oil quenched where stability is required.

Grades 5132 and 5132H are available as hot and cold rolled sheet and strip; seamless mechanical tubing; hot rolled and cold finished bar, rod, and wire; and bar subject to end-quench hardenability requirements.

AISI 5132: Similar Steels (U.S. and/or Foreign). UNS G51320; ASTM A322, A331, A505, A519; SAE J404, J412, J770; (W. Ger.) DIN 1.7033; (Fr.) AFNOR 32 C 4; (Ital.) UNI 34 Cr 4 KB; (Jap.) JIS SCr 2 H, SCr 2; (U.K.) B.S. 530 A 32, 530 H 32

AISI 5132H: Similar Steels (U.S. and/or Foreign). UNS H51320; ASTM A304; SAE J1268; (W. Ger.) DIN 1.7034; (Fr.) AFNOR 38 C 4; (Ital.) UNI 38 Cr 4 KB; (Jap.) JIS SCr 3 H; (U.K.) B.S. 530 A 36, 530 H 36, Type 3

Mechanical Properties

AISI 5132H: End-Quench Hardenability Limits and Hardenability Band. Heat treating temperatures recommended by SAE: normalize at 900 °C (1650 °F), for forged or rolled specimens only; austenitize at 870 °C (1600 °F). (Ref 1)

Distance from quenched end, 1/16 in.	Hardness, HRC max	Hardness, HRC min
1	57	50
2	56	47
3	54	43
4	52	40
5	50	35
6	48	32
7	45	29
8	42	27
9	40	25
10	38	24
11	37	23
12	36	22
13	35	21
14	34	20
15	34	...
16	33	...
18	32	...
20	31	...
22	30	...
24	29	...
26	28	...
28	27	...
30	26	...
32	25	...

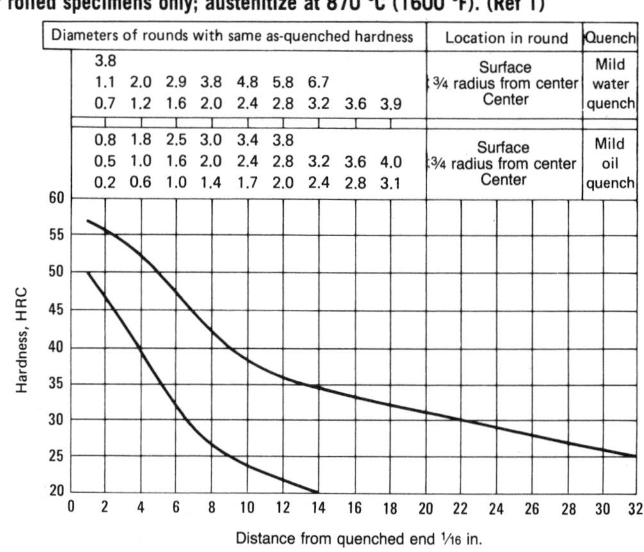

AISI 5132, 5132H: Hardness and Machinability (Ref 7)

Condition	Hardness range, HB	Average machinability rating(a)
Annealed and cold drawn	174-212	70

(a) Based on AISI 1212 steel as 100% average machinability

Machining Data

For machining data on AISI grades 5132 and 5132H, refer to the preceding machining tables for AISI grades 1330 and 1330H.

Physical Properties

AISI 5132, 5132H: Mean Apparent Specific Heat (Ref 6)

Test specimens were in the annealed condition

Temperature range °C	°F	Specific heat J/kg·K	Btu/lb·°F
50-100	10-38	494	0.118
150-200	66-93	502	0.120
200-250	93-120	536	0.128
250-300	120-150	553	0.132
300-350	150-175	578	0.138
350-400	175-205	599	0.143
450-500	230-260	650	0.155
550-600	290-315	733	0.175
650-700	345-370	842	0.201
700-750	370-400	1499	0.358
750-800	400-425	938	0.224
850-900	455-480	578	0.138

AISI 5132, 5132H: Thermal Conductivity (Ref 6)

Test specimens were in the annealed condition

Temperature °C	°F	Thermal conductivity W/m·K	Btu/ft·h·°F
0	32	48.6	28.1
100	212	46.5	26.9
200	390	44.3	25.6
300	570	42.2	24.4
400	750	38.4	22.2
500	930	35.6	20.6
600	1110	31.8	18.4
700	1290	28.9	16.7
800	1470	31.1	18.0
1000	1830	28.0	16.2
1200	2190	30.6	17.7

AISI 5132, 5132H: Electrical Resistivity (Ref 6)

Test specimens were in the hardened and tempered condition

Temperature °C	°F	Electrical resistivity, μΩ·m	Temperature °C	°F	Electrical resistivity, μΩ·m
20	68	0.210	900	1650	1.145
100	212	0.259	1000	1830	1.177
200	390	0.330	1100	2010	1.205
400	750	0.517	1200	2190	1.230
600	1110	0.934	1300	2370	1.251
800	1470	1.106			

AISI 5132, 5132H: Thermal Treatment Temperatures

Quenching mediums: water, caustic, oil

Treatment	Temperature range °C	°F
Forging	1230 max	2250 max
Annealing	815-870	1500-1600
Normalizing	870-925	1600-1700
Austenitizing	830-855	1525-1575
Tempering	(a)	(a)

(a) To desired hardness

AISI 5135, 5135H

AISI 5135, 5135H: Chemical Composition

AISI grade	C	Mn	P max	S max	Si	Cr
5135	0.33-0.38	0.60-0.80	0.035	0.040	0.15-0.30	0.75-1.00
5135H	0.32-0.38	0.50-0.60	0.035	0.040	0.15-0.30	0.70-1.15

Characteristics and Typical Uses. AISI grades 5135 and 5135H are low-hardenability, medium-carbon, chromium, alloy steels. These steels are frequently used for water-quenched parts of moderate section size and oil-quenched parts of small section size. They are suitable, too, for parts requiring strength and toughness which have section size small enough to permit use of the customary heat treatment to obtain desired mechanical properties.

AISI 5135 and 5135H are available as hot rolled and cold finished bar, rod, and wire; seamless mechanical tubing; and bar subject to end-quench hardenability requirements.

AISI 5135: Similar Steels (U.S. and/or Foreign). UNS G51350; ASTM A322, A331, A519; SAE J404, J412, J770; (W. Ger.) DIN 1.7034; (Fr.) AFNOR 38 C 4; (Ital.) UNI 38 Cr 4 KB; (Jap.) JIS SCr 3 H; (U.K.) B.S. 530 A 36, 530 H 36, Type 3

AISI 5135H: Similar Steels (U.S. and/or Foreign). UNS H51350; ASTM A304; SAE J1268; (W. Ger.) DIN 1.7035; (Fr.) AFNOR 42 C 4; (Ital.) UNI 41 Cr 4 KB, 40 Cr 4; (Jap.) JIS SCr 4 H; (U.K.) B.S. 530 A 40, 530 H 40, 530 M 40, 2 S. 117

Physical Properties

AISI 5135, 5135H: Thermal Treatment Temperatures

Quenching medium: oil or water

Treatment	Temperature range °C	°F
Forging	1205 max	2200 max
Annealing	815-870	1500-1600
Normalizing	870-925	1600-1700
Austenitizing	815-845	1500-1550
Tempering	(a)	(a)

(a) To desired hardness

Machining Data

For machining data on AISI grades 5135 and 5135H, refer to the preceding machining tables for AISI grades 1330 and 1330H.

Mechanical Properties

AISI 5135H: End-quench Hardenability Limits and Hardenability Band. Heat treating temperatures recommended by SAE: normalize at 870 °C (1600 °F), for forged or rolled specimens only; austenitize at 845 °C (1550 °F). (Ref 1)

Distance from quenched end, 1/16 in.	Hardness, HRC max	Hardness, HRC min
1	58	51
2	57	49
3	56	47
4	55	43
5	54	38
6	52	35
7	50	32
8	47	30
9	45	28
10	43	27
11	41	25
12	40	24
13	39	23
14	38	22
15	37	21
16	37	21
18	36	20
20	35	...
22	34	...
24	33	...
26	32	...
28	32	...
30	31	...
32	30	...

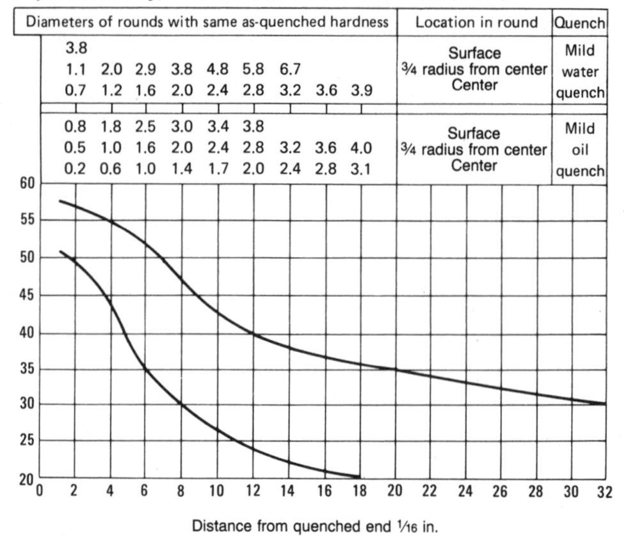

AISI 5135, 5135H: Hardness and Machinability (Ref 7)

Condition	Hardness range, HB	Average machinability rating(a)
Annealed and cold drawn	179-217	70

(a) Based on AISI 1212 steel as 100% average machinability

AISI 5140, 5140H

AISI 5140, 5140H: Chemical Composition

AISI grade	C	Mn	P max	S max	Si	Cr
5140	0.38-0.43	0.70-0.90	0.035	0.040	0.15-0.30	0.70-0.90
5140H	0.37-0.44	0.60-1.00	0.035	0.040	0.15-0.30	0.60-1.00

Characteristics and Typical Uses. AISI grades 5140 and 5140H are medium-carbon, chromium steels with low, direct hardenability. These steels are used when superior properties are desired. Applications include fasteners—high-strength screws, bolts, cap screws, and recessed-head screws—and severely cold headed, upset, and extruded parts.

Grades 5140 and 5140H are available as hot rolled and cold finished bar, rod, and wire; hot and cold rolled sheet and strip; seamless mechanical tubing; and bar subject to end-quench hardenability requirements.

AISI 5140: Similar Steels (U.S. and/or Foreign). UNS G51400; ASTM A322, A331, A505, A519; SAE J404, J412, J770; (W. Ger.) DIN 1.7035; (Fr.) AFNOR 42 C 4; (Ital.) UNI 40 Cr 4, 41 Cr 4 KB; (Jap.) JIS SCr 4 H; (U.K.) B.S. 530 A 40, 530 H 40, 530 M 40, 2 S 117

AISI 5140H: Similar Steels (U.S. and/or Foreign). UNS H51400; ASTM A304; SAE J1268; (W. Ger.) DIN 1.7006; (Fr.) AFNOR 42 C 2, 45 C 2

AISI 5140, 5140H: Approximate Critical Points

Transformation point	Temperature(a) °C	Temperature(a) °F	Transformation point	Temperature(a) °C	Temperature(a) °F
Ac_1	745	1370	Ar_1	695	1280
Ac_3	810	1490	M_s	360	680
Ar_3	745	1370			

(a) On heating or cooling at 28 °C (50 °F) per hour

Machining Data

For machining data on AISI grades 5140 and 5140H, refer to the preceding machining tables for AISI grades 1340 and 1340H.

Physical Properties

AISI 5140, 5140H: Electrical Resistivity and Thermal Conductivity (Ref 6)

Test specimens were in the hardened and tempered condition

Temperature °C	Temperature °F	Electrical resistivity, μΩ·m	Thermal conductivity W/m·K	Thermal conductivity Btu/ft·h·°F
20	68	0.228	...	...
100	212	0.281	44.6	25.8
200	390	0.352	43.4	25.1
400	750	0.530	37.7	21.8
600	1110	0.785	31.3	18.1

AISI 5140, 5140H: Average Coefficients of Linear Thermal Expansion (Ref 6)

Temperature range °C	Temperature range °F	Coefficient μm/m·K	Coefficient μin./in.·°F
20-200	68-390	12.6	7.0
20-300	68-570	13.3	7.4
20-400	68-750	13.9	7.7
20-500	68-930	14.0	7.8
20-600	68-1110	14.6	8.1
20-700	68-1290	14.9	8.3

AISI 5140, 5140H: Mean Apparent Specific Heat (Ref 6)

Test specimens were in the hardened and tempered condition

Temperature range °C	Temperature range °F	Specific heat J/kg·K	Specific heat Btu/lb·°F
20-100	68-212	452	0.108
20-200	68-390	473	0.113
20-400	68-750	519	0.124
20-600	68-1110	565	0.135

AISI 5140, 5140H: Thermal Treatment Temperatures

Quenching medium: oil or water

Treatment	Temperature range °C	Temperature range °F
Forging	1230 max	2250 max
Annealing(a)	815-870	1500-1600
Normalizing(b)	870-925	1600-1700
Austenitizing	855-885	1575-1625
Tempering	(c)	(c)

(a) Maximum hardness of 170 HB. (b) Hardness of approximately 241 HB. (c) To desired hardness

Mechanical Properties

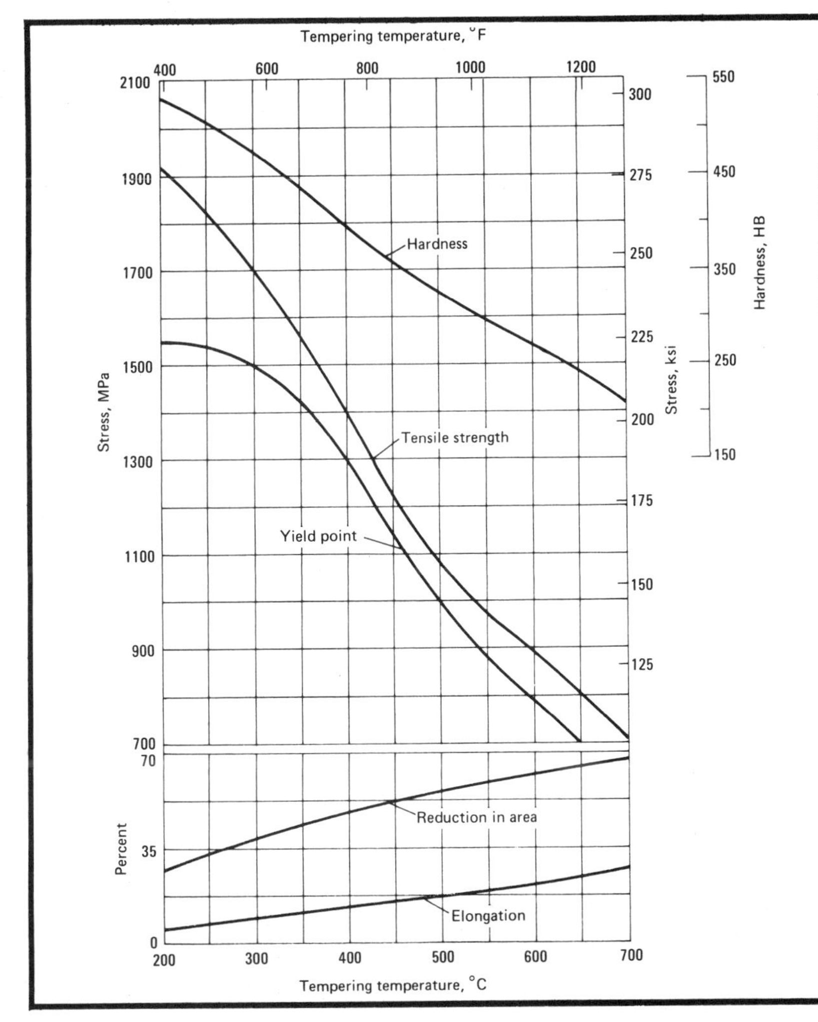

AISI 5140: Effect of Tempering Temperature on Tensile Properties. Normalized at 870 °C (1600 °F); reheated to 845 °C (1550 °F); quenched in agitated oil. Specimens were treated in 13.5-mm (0.530-in.) diam and machined to 12.8-mm (0.505-in.) diam for testing. Tests were conducted using specimens machined to English units. Elongation was measured in 50 mm (2 in.). (Ref 4)

Mechanical Properties (continued)

AISI 5140, 5140H: End-Quench Hardenability Limits and Hardenability Band.

Heat treating temperatures recommended by SAE: normalize at 870 °C (1600 °F), for forged r rolled specimens only; austenitize at 845 °C (1550 °F). (Ref 1)

Distance from quenched end, 1/16 in.	Hardness, HRC max	Hardness, HRC min
1	60	53
2	59	52
3	58	50
4	57	48
5	56	43
6	54	38
7	52	35
8	50	33
9	48	31
10	46	30
11	45	29
12	43	28
13	42	27
14	40	27
15	39	26
16	38	25
18	37	24
20	36	23
22	35	21
24	34	20
26	34	...
28	33	...
30	33	...
32	32	...

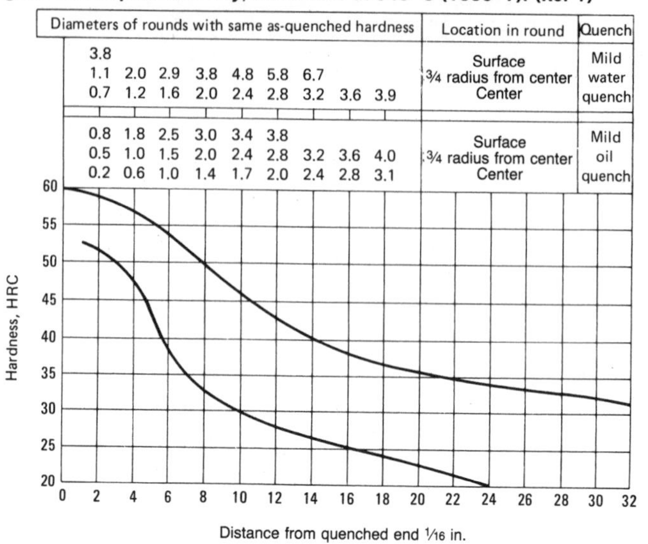

AISI 5140, 5140H: Mass Effect on Mechanical Properties (Ref 4)

Condition or treatment	Size round mm	Size round in.	Tensile strength MPa	Tensile strength ksi	Yield strength MPa	Yield strength ksi	Elongation(a), %	Reduction in area, %	Hardness, HB
Annealed (heated to 830 °C or 1525 °F; furnace cooled 11 °C or 20 °F per hour to 650 °C or 1200 °F; cooled in air)	25	1	570	83	290	42	28.6	57.3	167
Normalized (heated to 870 °C or 1600 °F; cooled in air	13	0.5	827	120	525	76	22.0	62.3	235
	25	1	793	115	470	68	22.7	59.2	229
	50	2	779	113	455	66	21.8	55.8	223
	100	4	765	111	415	60	21.6	52.3	217
Oil quenched from 845 °C (1550 °F); tempered at 540 °C (1000 °F)	13	0.5	1015	147	910	132	17.8	57.1	302
	25	1	972	141	841	122	18.5	58.9	293
	50	2	883	128	689	100	19.7	59.1	255
	100	4	862	125	565	82	20.2	55.4	248
Oil quenched from 845 °C (1550 °F); tempered at 595 °C (1100 °F)	13	0.5	896	130	779	113	20.2	61.4	269
	25	1	876	127	724	105	20.5	61.7	262
	50	2	814	118	615	89	22.0	63.2	241
	100	4	800	116	510	74	22.1	59.0	235
Oil quenched from 845 °C (1550 °F); tempered at 650 °C (1200 °F)	13	0.5	827	120	703	102	22.2	63.4	241
	25	1	807	117	650	94	22.5	63.5	235
	50	2	758	110	565	82	24.5	67.1	223
	100	4	731	106	470	68	24.6	63.1	217

(a) In 50 mm (2 in.)

AISI 5140, 5140H: Effect of the Mass on Hardness at Selected Points (Ref 4)

Size round mm	Size round in.	As-quenched hardness after quenching in oil at: Surface	½ radius	Center
13	0.5	57 HRC	57 HRC	56 HRC
25	1	53 HRC	48 HRC	45 HRC
50	2	46 HRC	38 HRC	35 HRC
100	4	35 HRC	29 HRC	20 HRC

AISI 5140, 5140H: Hardness and Machinability (Ref 7)

Condition	Hardness range, HB	Average machinability rating(a)
Annealed and cold drawn	179-217	65

(a) Based on AISI 1212 steel as 100% average machinability

Mechanical Properties (continued)

AISI 5140, 5140H: Tensile Properties (Ref 5)

Condition or treatment	Tensile strength MPa	Tensile strength ksi	Yield strength MPa	Yield strength ksi	Elongation(a), %	Reduction in area, %	Hardness, HB	Izod impact energy J	Izod impact energy ft·lb
Normalized at 870 °C (1600 °F)	793	115	475	69	22.7	59	229	38	28
Annealed at 830 °C (1525 °F)	570	83	295	43	28.6	57	167	41	30

(a) In 50 mm (2 in.)

AISI 5150, 5150H

AISI 5150, 5150H: Chemical Composition

AISI grade	C	Mn	P max	S max	Si	Cr
5150	0.48-0.53	0.70-0.90	0.035	0.040	0.15-0.30	0.70-0.90
5150H	0.47-0.54	0.60-1.00	0.035	0.040	0.15-0.30	0.60-1.00

Characteristics. AISI grades 5150 and 5150H are relatively inexpensive, carbon-chromium alloy steels with higher hardness, strength, and elasticity than AISI 5140. They are directly hardenable with a medium hardenability rating.

These grades are available as hot rolled and cold finished bar, rod, and wire; hot and cold rolled sheet and strip; seamless mechanical tubing; and bar subject to end-quench hardenability requirements.

Typical Uses. Applications of AISI grades 5150 and 5150H include coil and flat springs 3.2-mm (0.125-in.) thick and heavier, and hand tools. These parts are usually oil-quenched. A minimum of 90% martensite is desirable in the as-quenched condition.

AISI 5150: Similar Steels (U.S. and/or Foreign). UNS G51500; ASTM A322, A331, A505, A519; SAE J404, J412, J770; (W. Ger.) DIN 1.7006; (Fr.) AFNOR 42 C 2, 45 C 2

AISI 5150H: Similar Steels (U.S. and/or Foreign). UNS H51500; ASTM A304; SAE J1268

AISI 5150, 5150H: Approximate Critical Points

Transformation point	Temperature(a) °C	Temperature(a) °F	Transformation point	Temperature(a) °C	Temperature(a) °F
Ac_1	720	1330	Ar_1	700	1290
Ac_3	770	1420	M_s	290	555
Ar_3	720	1330			

(a) On heating or cooling at 28 °C (50 °F) per hour

Physical Properties

AISI 5150, 5150H: Thermal Treatment Temperatures

Quenching medium: oil

Treatment	Temperature range °C	Temperature range °F
Forging	1205 max	2200 max
Annealing(a)	815-870	1500-1600
Normalizing(b)	845-900	1550-1650
Austenitizing	830-855	1525-1575
Tempering	(c)	(c)

(a) Maximum hardness of 202 HB. (b) Hardness of approximately 293 HB. (c) To desired hardness

Machining Data

For machining data on AISI grades 5150 and 5150H, refer to the preceding machining tables for AISI grades 4150 and 4150H.

Mechanical Properties

AISI 5150, 5150H: Tensile Properties (Ref 5)

Condition or treatment	Tensile strength MPa	Tensile strength ksi	Yield strength MPa	Yield strength ksi	Elongation(a), %	Reduction in area, %	Hardness, HB	Izod impact energy J	Izod impact energy ft·lb
Normalized at 870 °C (1600 °F)	869	126	530	77	20.7	59	255	31	23
Annealed at 825 °C (1520 °F)	675	98	360	52	22.0	44	197	26	19

(a) In 50 mm (2 in.)

Mechanical Properties (continued)

AISI 5150H: End-Quench Hardenability Limits and Hardenability Band.
Heat treating temperatures recommended by SAE: normalize at 870 °C (1600 °F), for forged or rolled specimens only; austenitize at 845 °C (1550 °F). (Ref 1)

Distance from quenched end, 1/16 in.	Hardness, HRC max	Hardness, HRC min
1	65	59
2	65	58
3	64	57
4	63	56
5	62	53
6	61	49
7	60	42
8	59	38
9	58	36
10	56	34
11	55	33
12	53	32
13	51	31
14	50	31
15	48	30
16	47	30
18	45	29
20	43	28
22	42	27
24	41	26
26	40	25
28	39	24
30	39	23
32	38	22

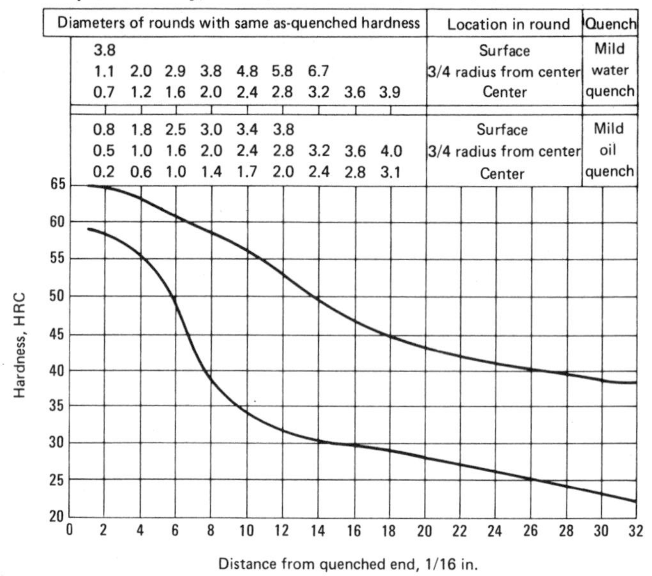

AISI 5150, 5150H: Mass Effect on Mechanical Properties (Ref 4)

Condition or treatment	Size round mm	Size round in.	Tensile strength MPa	Tensile strength ksi	Yield strength MPa	Yield strength ksi	Elongation(a), %	Reduction in area, %	Hardness, HB
Annealed (heated to 825 °C or 1520 °F; furnace cooled 11 °C or 20 °F per hour to 645 °C or 1190 °F; cooled in air)	25	1	675	98	360	52	22.0	43.7	197
Normalized (heated to 870 °C or 1600 °F; cooled in air)	13	0.5	903	131	565	82	21.0	60.6	262
	25	1	869	126	530	77	20.7	58.7	255
	50	2	848	123	495	72	20.0	53.3	248
	100	4	841	122	435	63	18.2	48.2	241
Oil quenched from 830 °C (1525 °F); tempered at 540 °C (1000 °F)	13	0.5	1095	159	1000	145	16.4	52.9	311
	25	1	1055	153	910	132	17.0	54.1	302
	50	2	910	132	670	97	18.5	55.5	255
	100	4	862	125	595	86	20.0	57.5	248
Oil quenched from 830 °C (1525 °F); tempered at 595 °C (1100 °F)	13	0.5	993	144	903	131	19.2	55.2	285
	25	1	945	137	793	115	20.2	59.5	277
	50	2	876	127	600	87	20.0	58.8	255
	100	4	827	120	550	80	19.7	56.4	241
Oil quenched from 830 °C (1525 °F); tempered at 650 °C (1200 °F)	13	0.5	938	136	834	121	21.7	59.7	269
	25	1	883	128	745	108	21.2	61.9	255
	50	2	821	119	605	88	22.7	63.0	241
	100	4	793	115	525	76	21.5	60.8	235

(a) In 50 mm (2 in.)

AISI 5150, 5150H: Effect of the Mass on Hardness at Selected Points (Ref 4)

Size round mm	Size round in.	As-quenched hardness after quenching in oil at: Surface	½ radius	Center
13	0.5	60 HRC	60 HRC	59 HRC
25	1	59 HRC	52 HRC	50 HRC
50	2	55 HRC	44 HRC	40 HRC
100	4	37 HRC	31 HRC	29 HRC

AISI 5150, 5150H: Hardness and Machinability (Ref 7)

Condition	Hardness range, HB	Average machinability rating(a)
Annealed and cold drawn	183-235	60

(a) Based on AISI 1212 steel as 100% average machinability

Mechanical Properties (continued)

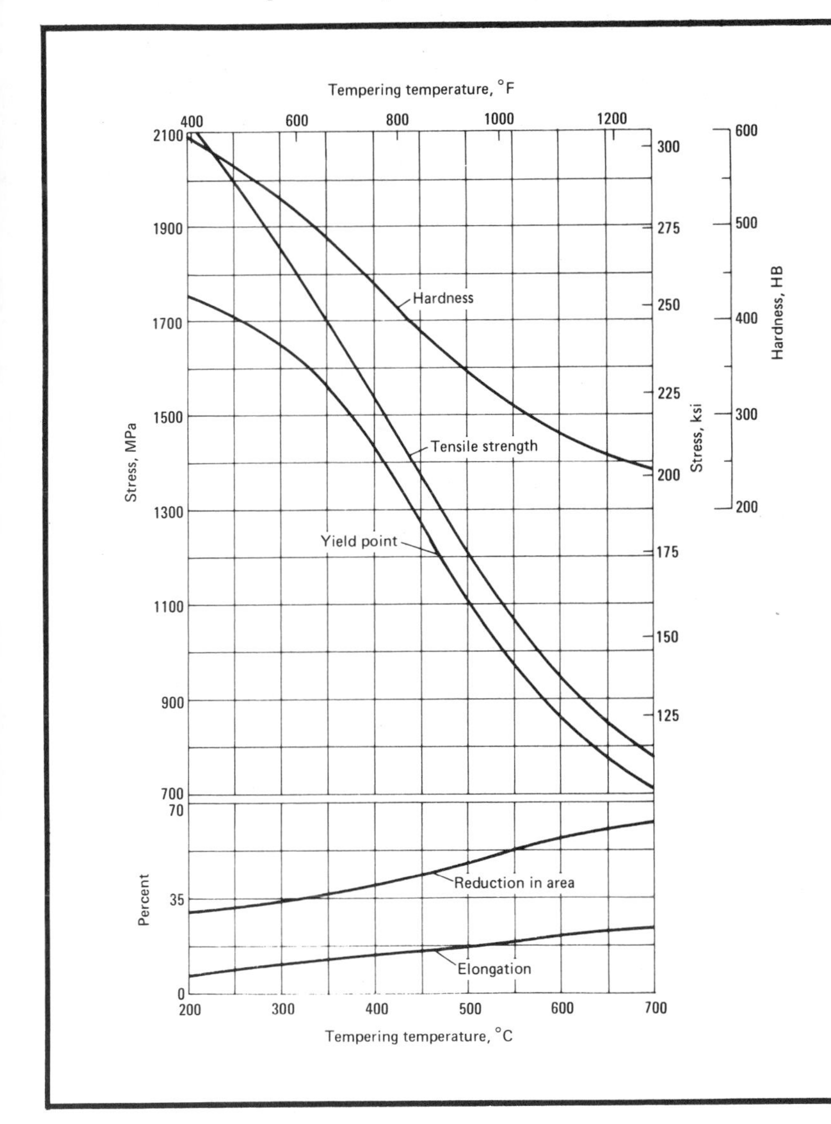

AISI 5150: Effect of Tempering Temperature on Tensile Properties. Normalized at 900 °C (1650 °F); water quenched from 870 °C (1600 °F); tempered at 56 °C (100 °F) intervals. Specimens were treated in 13.7-mm (0.540-in.) diam and machined to 12.8-mm (0.505-in.) diam for testing. Tests were conducted using specimens machined to English units. Elongation was measured in 50 mm (2 in.). (Ref 2)

AISI 5155, 5155H

AISI 5155, 5155H: Chemical Composition

AISI grade	Chemical composition, % C	Mn	P max	S max	Si	Cr
5155	0.51-0.59	0.70-0.90	0.035	0.040	0.15-0.30	0.70-0.90
5155H	0.50-0.60	0.60-1.00	0.035	0.040	0.15-0.30	0.60-1.00

Characteristics. AISI grades 5155 and 5155H are directly hardenable, medium-carbon, chromium, alloy steels with medium hardenability. These steels are available as hot rolled and cold finished bar, rod, and wire; seamless mechanical tubing; and bar subject to end-quench hardenability requirements.

Typical Uses. AISI 5155 and 5155H steels are used for springs and hand tools. These grades are also suitable for services involving relatively small cross-sectional areas that are subject to severe service conditions, such as rod and wire with broad applications as fasteners.

AISI 5155: Similar Steels (U.S. and/or Foreign). UNS G51550; ASTM A322, A331, A519; SAE J404, J412, J770; (W. Ger.) DIN 1.7176; (Fr.) AFNOR 55 C 3

AISI 5155H: Similar Steels (U.S. and/or Foreign). UNS H51550; ASTM A304; SAE J1268; (W. Ger.) DIN 1.7176; (Fr.) AFNOR 55 C 3

Physical Properties

AISI 5155, 5155H: Thermal Treatment Temperatures

Quenching medium: oil

Treatment	Temperature range °C	°F
Annealing	815-870	1500-1600
Normalizing	870-925	1600-1700
Austenitizing	800-845	1475-1550
Tempering	(a)	(a)

(a) To desired hardness

Machining Data

For machining data on AISI grades 5155 and 5155H, refer to the preceding machining tables for AISI grades 4150 and 4150H.

Mechanical Properties

AISI 5155, 5155H: Hardness and Machinability (Ref 7)

Condition	Hardness range, HB	Average machinability rating(a)
Annealed and cold drawn	183-235	55

(a) Based on AISI 1212 steel as 100% average machinability

AISI 5160, 5160H, 51B60, 51B60H

AISI 5160, 5160H, 51B60, 51B60H: Chemical Composition

AISI grade	C	Mn	P max	S max	Si	Cr	B
5160	0.56-0.64	0.75-1.00	0.035	0.040	0.15-0.30	0.70-0.90	...
5160H	0.55-0.65	0.65-1.10	0.035	0.040	0.15-0.30	0.60-1.00	...
51B60	0.56-0.64	0.75-1.00	0.035	0.040	0.15-0.30	0.70-0.90	0.0005-0.0030
51B60H	0.55-0.65	0.65-1.10	0.035	0.040	0.15-0.30	0.60-1.00	0.0005-0.0030

Characteristics. AISI grades 5160, 5160H, 51B60, and 51B60H are medium-high carbon, chromium, directly hardenable, alloy steels with medium hardenability. The addition of boron to 51B60 and 51B60H improves hardenability by increasing the depth of hardening during quenching.

The four grades are available as hot rolled and cold finished bar, rod, and wire; hot and cold rolled sheet and strip; seamless mechanical tubing; and bar subject to end-quench hardenability requirements.

Typical Uses. AISI grades 5160, 5160H, 51B60, and 51B60H are used for applications involving relatively small cross-sectional areas that are subject to severe service conditions, particularly rod and wire for the fastener and spring fields.

AISI 5160: Similar Steels (U.S. and/or Foreign). UNS G51600; ASTM A322, A331, A505, A519; SAE J404, J412, J770

AISI 5160H: Similar Steels (U.S. and/or Foreign). UNS H51600; ASTM A304; SAE J1268

AISI 51B60: Similar Steels (U.S. and/or Foreign). UNS G51601; ASTM A322, A331, A519; SAE J404, J412, J770

AISI 51B60H: Similar Steels (U.S. and/or Foreign). UNS H51601; ASTM A304; SAE J1268

Machining Data

For machining data on AISI grades 5160, 5160H, 51B60, and 51B60H, refer to the preceding machining tables for AISI grades 4150 and 4150H.

AISI 5160, 51B60: Approximate Critical Points

Transformation point	AISI 5160 °C	AISI 5160 °F	AISI 51B60 °C	AISI 51B60 °F
Ac_1	710	1310	725	1335
Ac_3	765	1410	770	1420
Ar_3	715	1320	730	1345
Ar_1	675	1250	675	1250
M_s	255	490	255	490

(a) On heating or cooling at 28 °C (50 °F) per hour

Physical Properties

AISI 5160, 5160H, 51B60, 51B60H: Thermal Treatment Temperatures

Quenching medium: oil

Treatment	Temperature range °C	°F
Forging	1205 max	2200 max
Annealing(a)	790-845	1450-1550
Normalizing(b)	845-900	1550-1650
Austenitizing	830-855	1525-1575
Tempering	(c)	(c)

(a) Maximum hardness of 223 HB. (b) Hardness of approximately 286 HB. (c) To desired hardness

Mechanical Properties

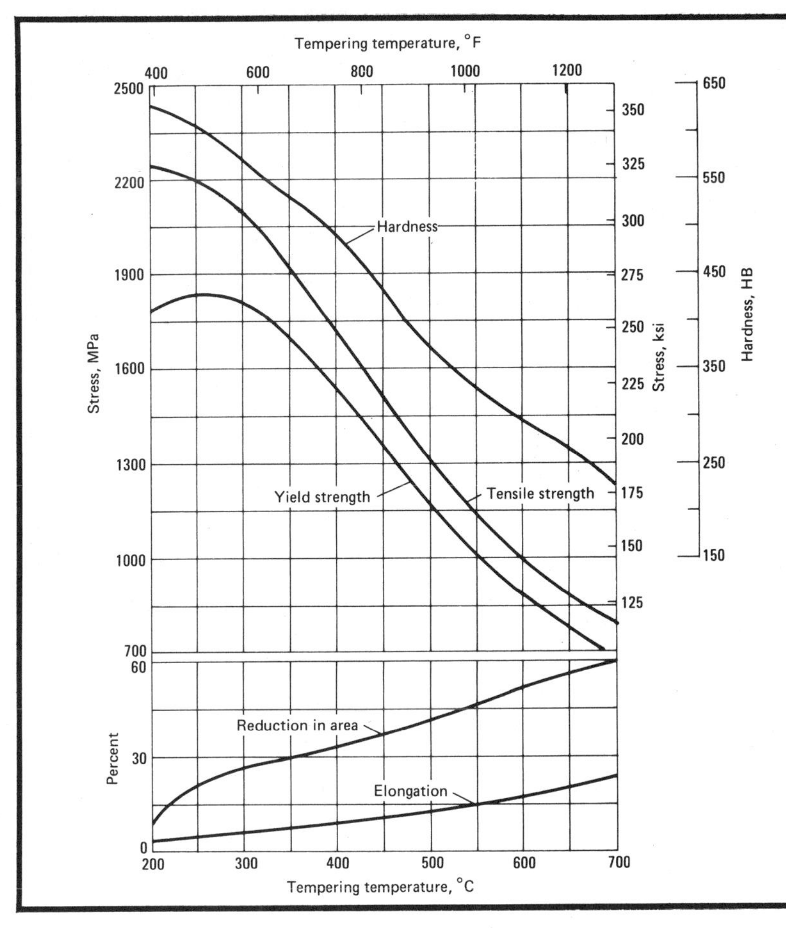

AISI 5160: Effect of Tempering Temperature on Tensile Properties. Normalized at 855 °C (1575 °F); reheated to 830 °C (1525 °F); quenched in oil. Specimens were treated in 13.5-mm (0.530-in.) diam and machined to 12.8-mm (0.505-in.) diam for testing. Tests were conducted using specimens machined to English units. As-quenched hardness of 682 HB; elongation measured in 50 mm (2 in.). (Ref 4)

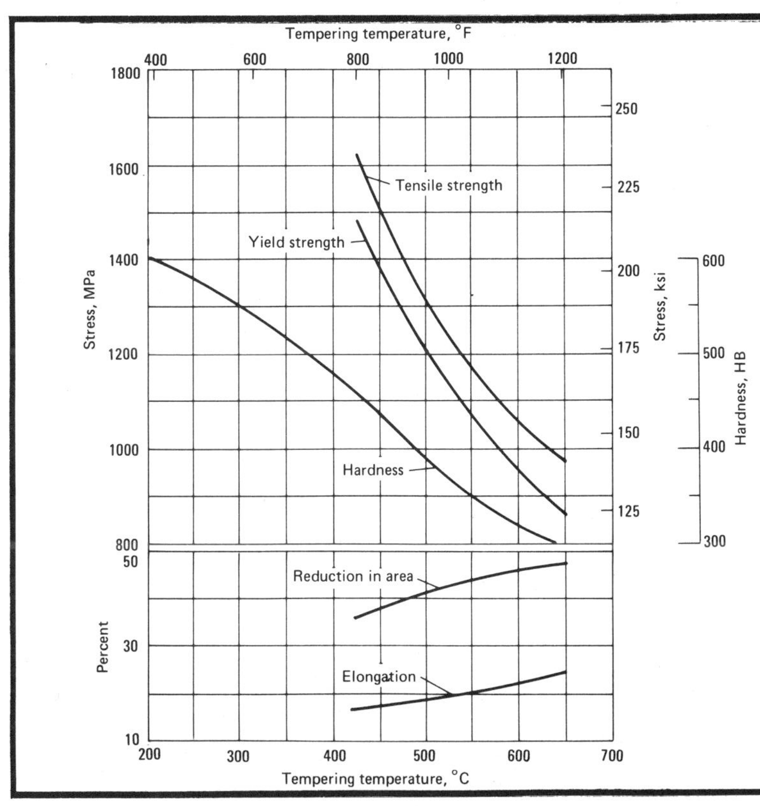

AISI 51B60: Effect of Tempering Temperature on Tensile Properties. Normalized at 870 °C (1600 °F); oil quenched from 845 °C (1550 °F); tempered at 56 °C (100 °F) intervals. Specimens were treated in 13.7-mm (0.540-in.) diam and machined to 12.8-mm (0.505-in.) diam for testing. Tests were conducted using specimens machined to English units. Elongation was measured in 50 mm (2 in.). (Ref 2)

Mechanical Properties (continued)

AISI 5160H: End-Quench Hardenability Limits and Hardenability Band.

Heat treating temperatures recommended by SAE: normalize at 870 °C (1600 °F), for forged or rolled specimens only; austenitize at 845 °C (1550 °F). (Ref 1)

Distance from quenched end, 1/16 in.	Hardness, HRC max	Hardness, HRC min
1	...	60
2	...	60
3	...	60
4	65	59
5	65	58
6	64	58
7	64	52
8	63	47
9	62	42
10	61	39
11	60	37
12	59	36
13	58	35
14	56	35
15	54	34
16	52	34
18	48	33
20	47	32
22	46	31
24	45	30
26	44	29
28	43	28
30	43	28
32	42	27

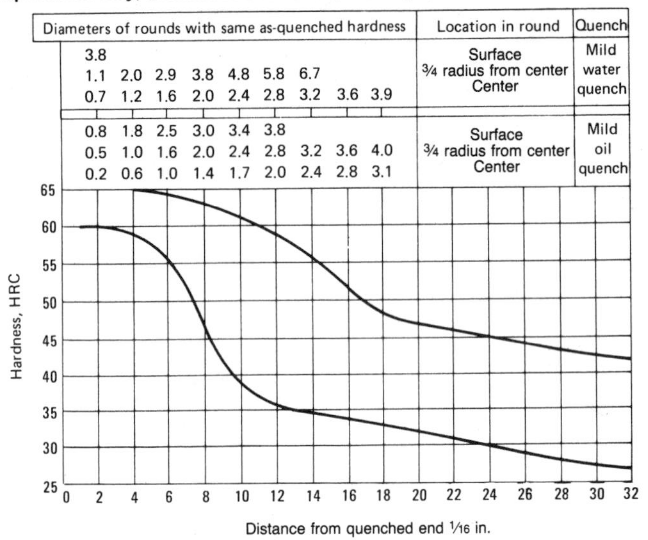

AISI 5160, 5160H, 51B60, 51B60H: Mass Effect on Mechanical Properties (Ref 4)

Condition or treatment	Size round mm	Size round in.	Tensile strength MPa	Tensile strength ksi	Yield strength MPa	Yield strength ksi	Elongation(a), %	Reduction in area, %	Hardness, HB
Annealed (heated to 815 °C or 1495 °F; furnace cooled 11 °C or 20 °F per hour to 480 °C or 900 °F; cooled in air)	25	1	724	105	275	40	17.2	30.6	197
Normalized (heated to 855 °C or 1575 °F; cooled in air)	13	0.5	1025	149	650	94	18.2	50.7	285
	25	1	958	139	530	77	17.5	44.8	269
	50	2	924	134	510	74	16.0	39.0	262
	100	4	924	134	485	70	14.8	34.2	255
Oil quenched from 830 °C (1525 °F); tempered at 540 °C (1000 °F)	13	0.5	1170	170	1070	155	14.2	45.1	341
	25	1	1145	166	1005	146	14.5	45.7	341
	50	2	1060	154	703	102	17.8	51.2	293
	100	4	965	140	703	102	18.5	52.0	285
Oil quenched from 830 °C (1525 °F); tempered at 595 °C (1100 °F)	13	0.5	1050	152	924	134	16.6	50.6	302
	25	1	1000	145	869	126	18.0	53.6	302
	50	2	931	135	635	92	20.0	54.6	277
	100	4	889	129	615	89	21.2	57.0	262
Oil quenched from 830 °C (1525 °F); tempered at 650 °C (1200 °F)	13	0.5	917	133	793	115	19.8	55.5	269
	25	1	889	129	765	111	20.7	55.6	262
	50	2	779	113	580	84	21.8	57.5	248
	100	4	827	120	540	78	22.8	60.8	241

(a) In 50 mm (2 in.)

AISI 5160, 5160H, 51B60, 51B60H: Effect of the Mass on Hardness at Selected Points (Ref 4)

Size round mm	Size round in.	As-quenched hardness after quenching in oil at: Surface	½ radius	Center
13	0.5	63 HRC	62 HRC	62 HRC
25	1	62 HRC	61 HRC	60 HRC
50	2	53 HRC	46 HRC	43 HRC
100	4	40 HRC	32 HRC	29 HRC

AISI 5160, 5160H, 51B60, 51B60H: Hardness and Machinability (Ref 7)

Condition	Hardness range, HB	Average machinability rating(a)
Spheroidized annealed and cold drawn	179-217	55

(a) Based on AISI 1212 steel as 100% average machinability

Mechanical Properties (continued)

AISI 51B60H: End-Quench Hardenability Limits and Hardenability Band. Heat treating temperatures recommended by SAE: normalize at 870 °C (1600 °F), for forged or rolled specimens only; austenitize at 845 °C (1550 °F). (Ref 1)

Distance from quenched end, 1/16 in.	Hardness, HRC max	Hardness, HRC min
1	...	60
2	...	60
3	...	60
4	...	60
5	...	60
6	...	59
7	...	58
8	...	57
9	...	54
10	...	50
11	...	44
12	65	41
13	65	40
14	64	39
15	64	38
16	63	37
18	61	36
20	59	34
22	57	33
24	55	31
26	53	30
28	51	28
30	49	27
32	47	25

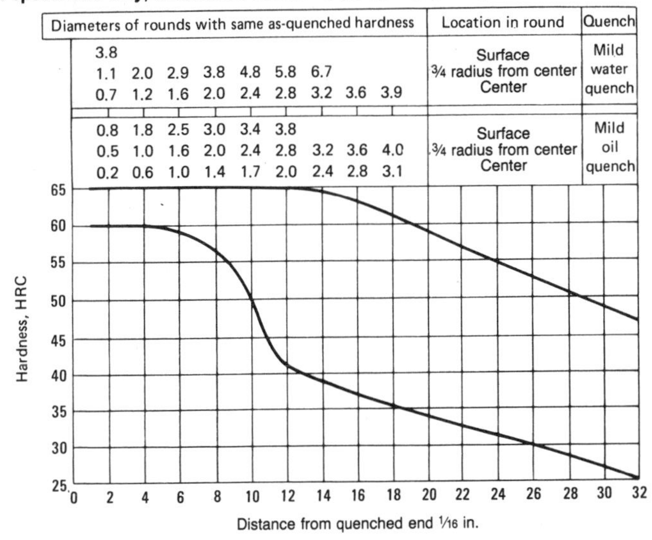

AISI 5160, 5160H, 51B60, 51B60H: Tensile Properties (Ref 5)

Condition or treatment	Tensile strength MPa	Tensile strength ksi	Yield strength MPa	Yield strength ksi	Elongation(a), %	Reduction in area, %	Hardness, HB	Izod impact energy J	Izod impact energy ft·lb
Normalized at 855 °C (1575 °F)	958	139	530	77	17.5	45	269	11	8
Annealed at 815 °C (1495 °F)	724	105	275	40	17.2	31	197	10	7

(a) In 50 mm (2 in.)

AISI 6118, 6118H

AISI 6118, 6118H: Chemical Composition

AISI grade	C	Mn	P max	S max	Si	Cr	V
6118	0.16-0.21	0.50-0.70	0.035	0.040	0.15-0.30	0.50-0.70	0.10-0.15
6118H	0.15-0.21	0.40-0.80	0.035	0.040	0.15-0.30	0.40-0.80	0.10-0.15

Characteristics and Typical Uses. AISI 6118 and 6118H are carburizing grades of chromium-vanadium, alloy steels with a low case hardenability and low-to-medium core hardenability. They are used for small parts that are not subject to heavy loading. AISI 6118 and 6118H are available as hot rolled and cold finished bar, rod, and wire, and as bar subject to end-quench hardenability requirements.

AISI 6118: Similar Steels (U.S. and/or Foreign). UNS G61180; ASTM A322; SAE J404, J770; (W. Ger.) DIN 1.7511

AISI 6118H: Similar Steels (U.S. and/or Foreign). UNS H61180; ASTM A304; SAE J1268; (W. Ger.) DIN 1.7511

AISI 6118, 6118H: Approximate Critical Points

Transformation point	Temperature(a) °C	Temperature(a) °F
Ac_1	760	1400
Ac_3	850	1560
Ar_3	775	1430
Ar_1	690	1270

(a) On heating or cooling at 28 °C (50 °F) per hour

Physical Properties

AISI 6118, 6118H: Thermal Treatment Temperatures

Quenching medium: oil

Treatment	Temperature °C	Temperature °F
Forging	1230 max	2250 max
Normalizing	(a)	(a)
Carburizing	900	1650
Tempering	165	325

(a) At least as high as the carburizing temperature, followed by air cooling

Mechanical Properties

AISI 6118, 6118H: End-Quench Hardenability Limits and Hardenability Band. Heat treating temperatures recommended by SAE: normalize at 925 °C (1700 °F), for forged or rolled specimens only; austenitize at 925 °C (1700 °F). (Ref 1)

Distance from quenched end, 1/16 in.	Hardness, HRC max	Hardness, HRC min
1	46	39
2	44	36
3	38	28
4	33	24
5	30	22
6	28	20
7	27	...
8	26	...
9	26	...
10	25	...
11	25	...
12	24	...
13	24	...
14	23	...
15	23	...
16	22	...
18	22	...
20	21	...
22	21	...
24	20	...

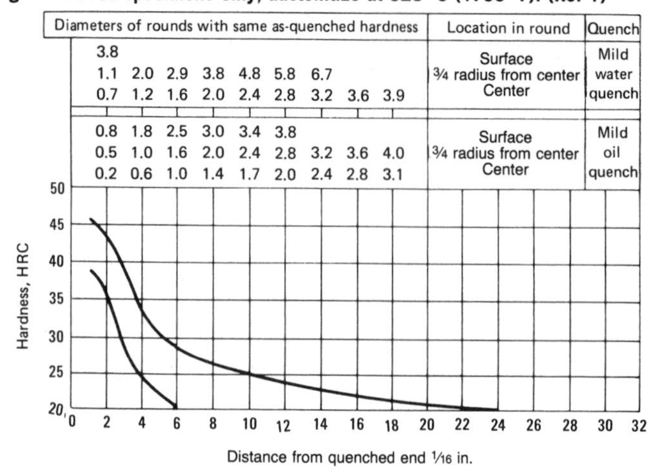

AISI 6118, 6118H: Hardness and Machinability (Ref 7)

Condition	Hardness range, HB	Average machinability rating(a)
Hot rolled and cold drawn	179-217	60

(a) Based on AISI 1212 steel as 100% average machinability

Machining Data

For machining data on AISI grades 6118 and 6118H, refer to the preceding machining tables for AISI grade A2317.

AISI 6150, 6150H

AISI 6150, 6150H: Chemical Composition

AISI grade	C	Mn	P max	S max	Si	Cr	V min
6150	0.48-0.53	0.70-0.90	0.035	0.040	0.15-0.30	0.80-1.10	0.15
6150H	0.47-0.54	0.60-1.00	0.035	0.040	0.15-0.30	0.75-1.20	0.15

(Chemical composition, %)

Characteristics. AISI grades 6150 and 6150H are directly hardenable, chromium-vanadium alloy steels with medium hardenability. These steels are available as hot rolled and cold finished bar, rod, and wire; bar subject to end-quench hardenability requirements; and bar that meet aircraft quality requirements.

Typical Uses. AISI 6150 and 6150H steels are used for: cold formed flat springs up to 3.2-mm (0.125-in.) thick; flat springs 3.2 mm (0.125 in.) and over that are hot formed, oil quenched, and tempered at 390 to 480 °C (725 to 900 °F) to a hardness of 44 to 48 or 48 to 52 HRC; and coil springs over 12.7-mm (0.500-in.) diam with the same treatment. Other applications include valve springs, piston rods, pump parts, and spline shafts.

AISI 6150: Similar Steels (U.S. and/or Foreign). UNS G61500; AMS 6448, 6450, 6455, 7301; ASTM A231, A322, A331, A505; MIL SPEC MIL-S-8503; SAE J404, J412, J770; (W. Ger.) DIN 1.8159; (Fr.) AFNOR 50 CV 4; (Ital.) UNI 50 CrV 4; (Jap.) JIS SUP 10; (Swed.) SS_{14} 2230; (U.K.) B.S. 735 A 50, En. 47

AISI 6150H: Similar Steels (U.S. and/or Foreign). UNS H61500; ASTM A304; SAE J1268; (W. Ger.) DIN 1.8159; (Fr.) AFNOR 50 CV 4; (Ital.) UNI 50 CrV 4; (Jap.) JIS SUP 10; (Swed.) SS_{14} 2230; (U.K.) B.S. En. 47, 735 A 50

AISI 6150, 6150H: Approximate Critical Points

Transformation point	Temperature(a) °C	Temperature(a) °F	Transformation point	Temperature(a) °C	Temperature(a) °F
Ac_1	750	1380	Ar_1	695	1280
Ac_3	790	1450	M_s	285	545
Ar_3	745	1370			

(a) On heating or cooling at 28 °C (50 °F) per hour

Physical Properties

AISI 6150, 6150H: Average Coefficients of Linear Thermal Expansion (Ref 6)

Condition or treatment	Coefficient for temperatures from 20 °C (68 °F)(a) to: 100 °C (212 °F) μm/m·K (μin./in.·°F)	200 °C (390 °F) μm/m·K (μin./in.·°F)	300 °C (570 °F) μm/m·K (μin./in.·°F)	400 °C (750 °F) μm/m·K (μin./in.·°F)	500 °C (930 °F) μm/m·K (μin./in.·°F)	600 °C (1110 °F) μm/m·K (μin./in.·°F)
Material A(b)						
Annealed	12.2 (6.78)	12.7 (7.06)	13.3 (7.39)	13.7 (7.62)	14.1 (7.84)	14.4 (8.01)
Hardened, tempered at 205 °C (400 °F)	12.0 (6.67)	12.5 (6.95)	12.9 (7.17)	13.0 (7.23)	13.3 (7.39)	13.7 (7.62)
Material B(c)						
Annealed	12.4 (6.89)	12.6 (7.01)	13.3 (7.39)	13.8 (7.67)	14.2 (7.90)	14.5 (8.06)
Material C(d)						
Annealed	12.4 (6.89)	12.8 (7.12)	13.4 (7.45)	13.9 (7.73)	14.2 (7.90)	14.5 (8.06)
Hardened, tempered at 425 °C (800 °F)	11.8 (6.56)	12.4 (6.89)	13.1 (7.28)	13.6 (7.56)	13.9 (7.73)	14.1 (7.84)
Hardened, tempered at 650 °C (1200 °F)	12.3 (6.84)	12.7 (7.06)	13.4 (7.45)	13.9 (7.73)	14.3 (7.95)	14.7 (8.17)

AISI 6150, 6150H: Thermal Treatment Temperatures

Quenching medium: oil

Treatment	Temperature range °C	°F
Forging	1205 max	2200 max
Annealing(a)	845-900	1550-1650
Normalizing(b)	870-925	1600-1700
Austenitizing	855-885	1575-1625
Tempering	(c)	(c)

(a) Maximum hardness of 202 HB. (b) Hardness of approximately 302 HB. (c) To desired hardness

Machining Data

For machining data on AISI grades 6150 and 6150H, refer to the preceding machining tables for AISI grades 4150 and 4150H.

Mechanical Properties

AISI 6150H: End-Quench Hardenability Limits and Hardenability Band. Heat treating temperatures recommended by SAE: normalize at 900 °C (1650 °F), for forged or rolled specimens only; austenitize at 870 °C (1600 °F). (Ref 1)

Distance from quenched end, 1/16 in.	Hardness, HRC max	min
1	65	59
2	65	58
3	64	57
4	64	56
5	63	55
6	63	53
7	62	50
8	61	47
9	61	43
10	60	41
11	59	39
12	58	38
13	57	37
14	55	36
15	54	35
16	52	35
18	50	34
20	48	32
22	47	31
24	46	30
26	45	29
28	44	27
30	43	26

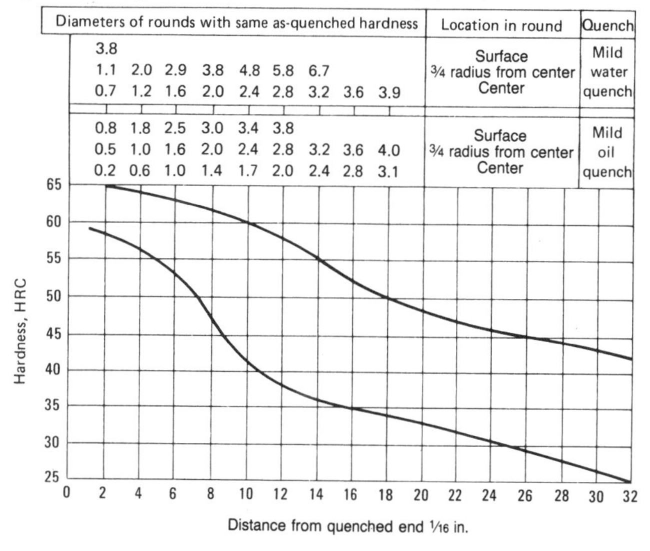

Mechanical Properties (continued)

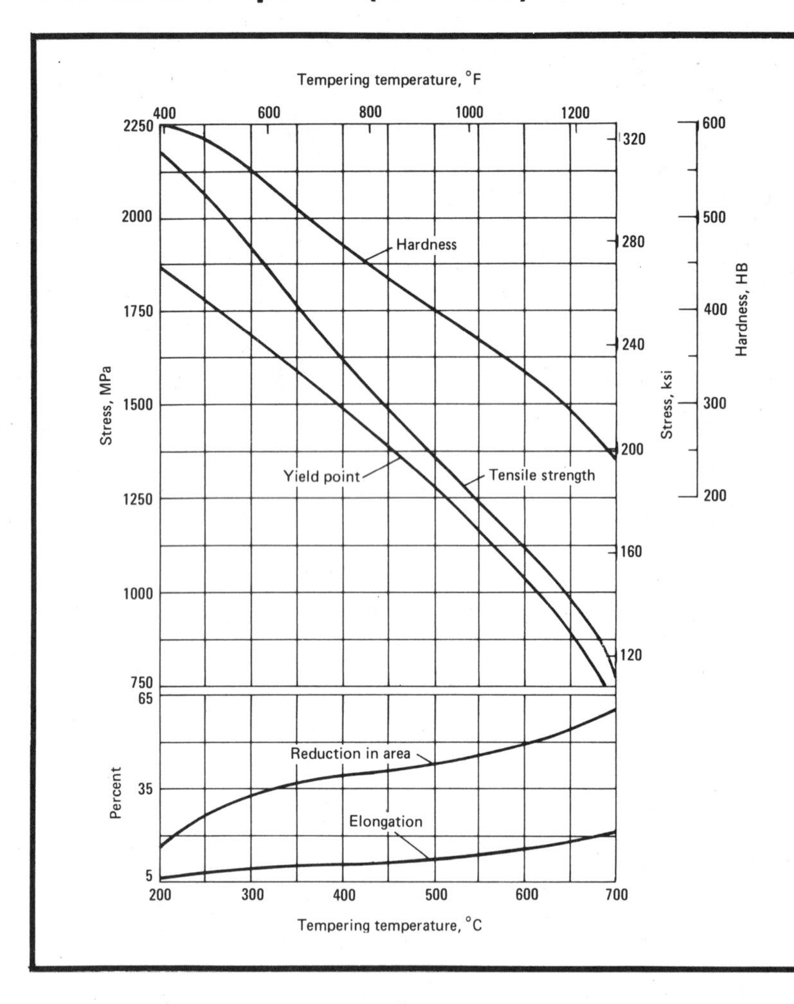

AISI 6150: Effect of Tempering Temperature on Tensile Properties. Normalized at 870 °C (1600 °F); reheated to 845 °C (1550 °F); quenched in agitated oil. Specimens were treated in 14.4-mm (0.565-in.) diam and machined to 12.8-mm (0.505-in.) diam for testing. Tests were conducted using specimens machined to English units. As-quenched hardness of 627 HB; elongation measured in 50 mm (2 in.). (Ref 4)

AISI 6150, 6150H: Mass Effect on Mechanical Properties (Ref 4)

Condition or treatment	Size round, mm	Size round, in.	Tensile strength, MPa	Tensile strength, ksi	Yield strength, MPa	Yield strength, ksi	Elongation(a), %	Reduction in area, %	Hardness, HB
Annealed (heated to 815 °C or 1500 °F; furnace cooled 11 °C or 20 °F per hour to 670 °C or 1240 °F; cooled in air)	25	1	670	97	415	60	23.0	48.4	197
Normalized (heated to 870 °C or 1600 °F; cooled in air)	13	0.5	972	141	640	93	20.6	63.0	285
	25	1	938	136	615	89	21.8	61.0	269
	50	2	896	130	515	75	20.7	56.5	262
	100	4	883	128	460	67	18.2	49.6	255
Oil quenched from 845 °C (1550 °F); tempered at 540 °C (1000 °F)	13	0.5	1240	180	1225	178	14.6	49.4	363
	25	1	1200	174	1160	168	14.5	48.2	352
	50	2	1145	166	1000	145	14.5	46.7	331
	100	4	1050	152	876	127	16.0	48.7	302
Oil quenched from 845 °C (1550 °F); tempered at 595 °C (1100 °F)	13	0.5	1105	160	1090	158	16.4	52.3	321
	25	1	1090	158	1035	150	16.0	53.2	311
	50	2	1020	148	910	132	17.7	55.2	293
	102	4	896	130	745	108	19.0	55.4	262
Oil quenched from 845 °C (1550 °F); tempered at 650 °C (1200 °F)	13	0.5	1015	147	979	142	17.8	53.9	293
	25	1	972	141	896	130	18.7	56.3	293
	50	2	924	134	800	116	19.5	57.4	269
	100	4	841	122	650	94	21.0	59.7	241

(a) In 50 mm (2 in.)

Mechanical Properties (continued)

AISI 6150, 6150H: Effect of the Mass on Hardness at Selected Points (Ref 4)

Size round mm	in.	As-quenched hardness after quenching in oil at: Surface	½ radius	Center
13	0.5	61 HRC	60 HRC	60 HRC
25	1	60 HRC	58 HRC	57 HRC
50	2	54 HRC	47 HRC	44 HRC
100	4	42 HRC	36 HRC	35 HRC

AISI 6150, 6150H: Hardness and Machinability (Ref 7)

Condition	Hardness range, HB	Average machinability rating(a)
Annealed and cold drawn	183-241	55

(a) Based on AISI 1212 steel as 100% average machinability

AISI 6150, 6150H: Tensile Properties (Ref 5)

Condition or treatment	Tensile strength MPa	ksi	Yield strength MPa	ksi	Elongation(a), %	Reduction in area, %	Hardness, HB	Izod impact energy J	ft·lb
Normalized at 870 °C (1600 °F)	938	136	615	89	21.8	61	269	35	26
Annealed at 815 °C (1500 °F)	670	97	415	60	23.0	48	197	27	20

(a) In 50 mm (2 in.)

AISI 81B45, 81B45H

AISI 81B45, 81B45H: Chemical Composition

AISI grade	C	Mn	P max	S max	Si	Ni	Cr	Mo	B
81B45	0.43-0.48	0.75-1.00	0.035	0.040	0.15-0.30	0.20-0.40	0.35-0.55	0.08-0.15	0.0005-0.003
81B45H	0.42-0.49	0.70-1.05	0.035	0.040	0.15-0.30	0.15-0.45	0.30-0.60	0.08-0.15	0.0005-0.003

Chemical composition, %

Characteristics and Typical Uses. AISI grades 81B45 and 81B45H are directly hardneable, nickel-chromium-molybdenum alloy steels, which contain boron to improve hardenability. These steels are used primarily for gears and other parts requiring fairly high hardness, as well as strength and toughness. They are available as hot rolled and cold finished bar, seamless mechanical tubing, bar subject to end-quench hardenability requirements, and aircraft quality stock.

AISI 81B45: Similar Steels (U.S. and/or Foreign). UNS G81451; ASTM A322, A519; SAE J404, J412, J770

AISI 81B45H: Similar Steels (U.S. and/or Foreign). UNS H81451; ASTM A304; SAE J1268

AISI 81B45, 81B45H: Approximate Critical Points

Transformation point	Temperature(a) °C	°F	Transformation point	Temperature(a) °C	°F
Ac_1	710	1310	Ar_1	655	1215
Ac_3	790	1450	M_s	315	600
Ar_3	720	1325			

(a) On heating or cooling at 28 °C (50 °F) per hour

Physical Properties

AISI 81B45, 81B45H: Thermal Treatment Temperatures

Anneal or normalize for optimum machinability; quenching medium: oil or water

Treatment	Temperature range °C	°F
Forging	1205 max	2200 max
Annealing(a)	815-870	1500-1600
Normalizing(b)	845-900	1550-1650
Austenitizing	830-855	1525-1575
Tempering	(c)	(c)

(a) Maximum hardness of 192 HB. (b) Hardness of approximately 285 HB. (c) To desired hardness

Machining Data

For machining data on AISI grades 81B45 and 81B45H, refer to the preceding machining tables for AISI grades 1340 and 1340H.

Mechanical Properties

AISI 81B45, 81B45H: Hardness and Machinability (Ref 7)

Condition	Hardness range, HB	Average machinability rating(a)
Annealed and cold drawn	179-223	65

(a) Based on AISI 1212 steel as 100% average machinability

Mechanical Properties (continued)

AISI 81B45H: End-Quench Hardenability Limits and Hardenability Band. Heat treating temperatures recommended by SAE: normalize at 870 °C (1600 °F), for forged or rolled specimens only; austenitize at 845 °C (1550 °F). (Ref 1)

Distance from quenched end, 1/16 in.	Hardness, HRC max	Hardness, HRC min
1	63	56
2	63	56
3	63	56
4	63	56
5	63	55
6	63	54
7	62	53
8	62	51
9	61	48
10	60	44
11	60	41
12	59	39
13	58	38
14	57	37
15	57	36
16	56	35
18	55	34
20	53	32
22	52	31
24	50	30
26	49	29
28	47	28
30	45	28
32	43	27

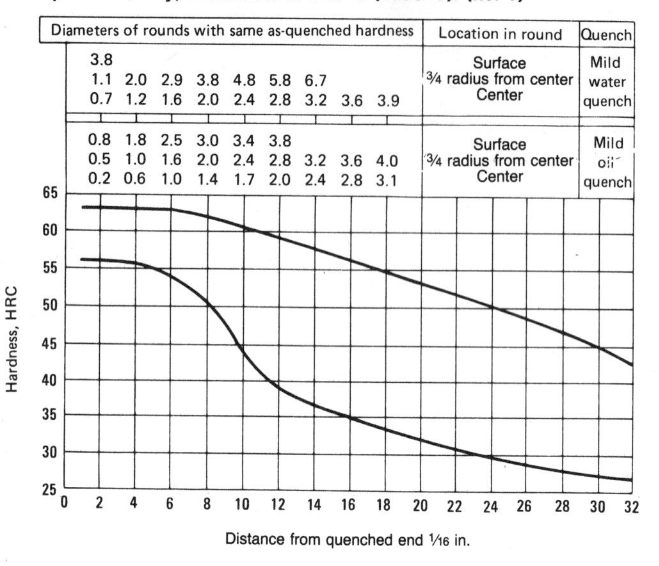

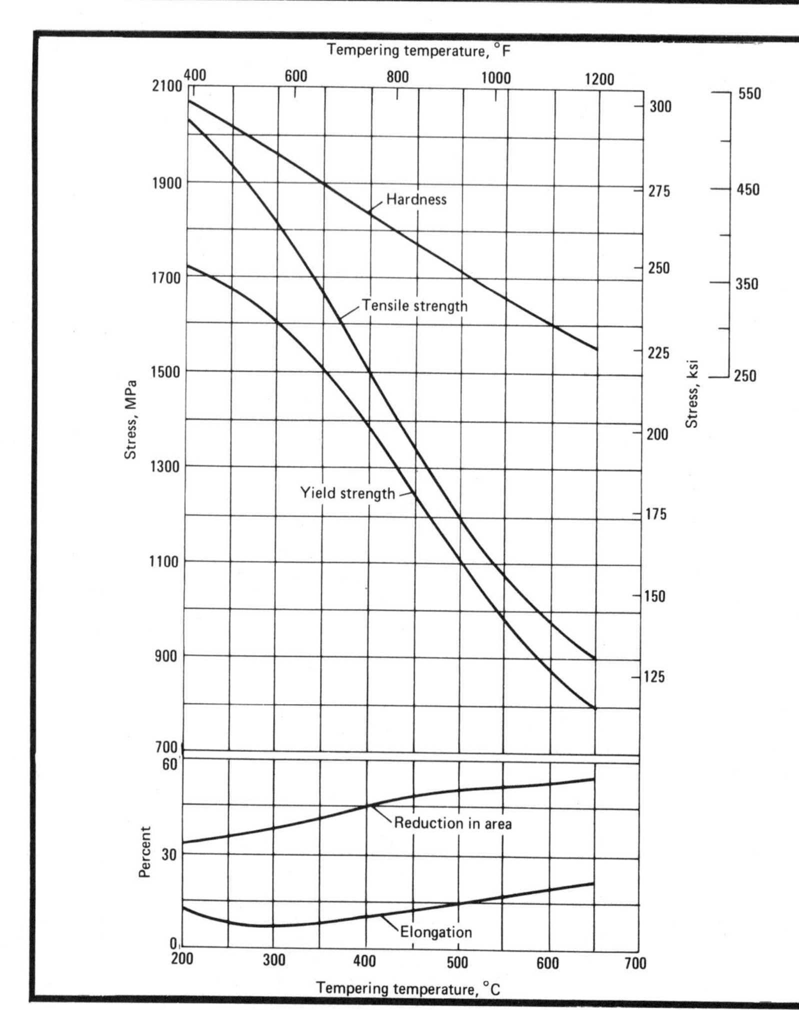

AISI 81B45: Effect of Tempering Temperature on Tensile Properties. Normalized at 870 °C (1600 °F); oil quenched from 845 °C (1550 °F); tempered at 56 °C (100 °F) intervals. Specimens were treated in 13.7-mm (0.540-in.) diam and machined to 12.8-mm (0.505-in.) diam for testing. Tests were conducted using specimens machined to English units. Elongation was measured in 50 mm (2 in.). (Ref 2)

AISI 8615

AISI 8615: Chemical Composition

AISI grade	C	Mn	P max	S max	Si	Ni	Cr	Mo
	Chemical composition, %							
8615	0.13-0.18	0.70-0.90	0.035	0.040	0.15-0.30	0.40-0.70	0.40-0.60	0.15-0.25

Characteristics and Typical Uses. AISI grade 8615 is a nickel-chromium-molybdenum, carburizing, alloy steel with medium case and core hardenability. This steel is available as hot rolled bar and aircraft quality stock. Uses of AISI 8615 include automotive gears and shafts, knuckle pins, piston pins, pump shafts, and spline shafts.

AISI 8615: Similar Steels (U.S. and/or Foreign). UNS G86150; AMS 5333; ASTM A322, A505; MIL SPEC MIL-S-866; SAE J404, J770

AISI 8615: Approximate Critical Points

Transformation point	Temperature(a) °C	°F
Ac_1	740	1360
Ac_3	845	1550
Ar_3	790	1455
Ar_1	685	1265

(a) On heating or cooling at 28 °C (50 °F) per hour

Physical Properties

AISI 8615: Thermal Treatment Temperatures

Quenching medium: oil

Treatment	Temperature range °C	°F
Forging	1230 max	2250 max
Normalizing	(a)	(a)
Carburizing	900-925	1650-1700
Reheating	845-870	1550-1600
Tempering	120-175	250-350

(a) At least as high as the carburizing temperature, followed by air cooling

Mechanical Properties

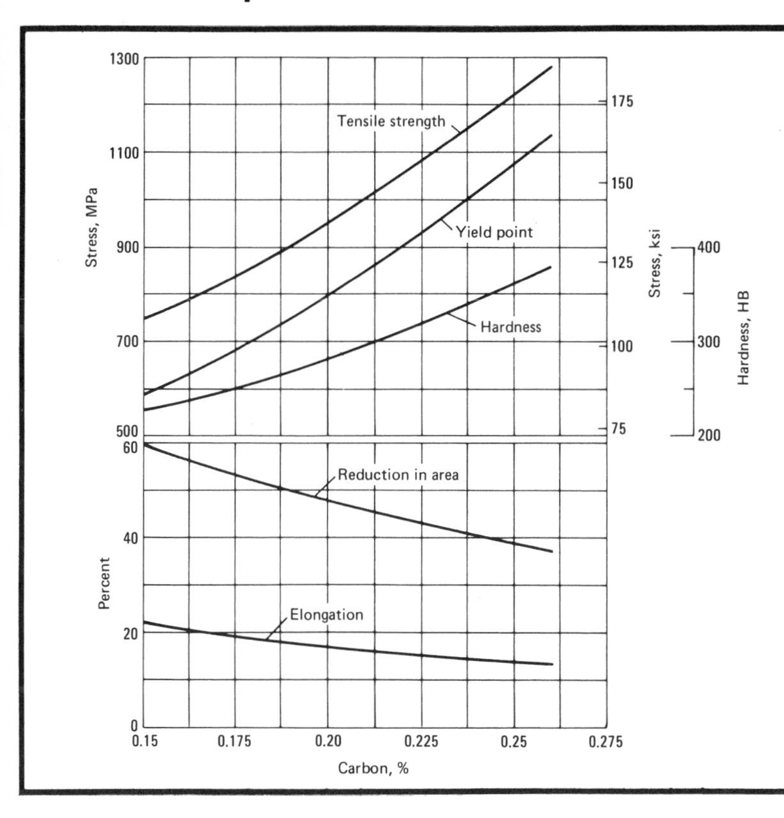

AISI 8600 and 8700 Series Steels: Average Core Properties. Test specimens were 25-mm (1-in.) bar pseudocarburized at 900 to 925 °C (1650 to 1700 °F); furnace cooled; reheated at 845 °C (1550 °F); oil quenched; tempered at 150 °C (300 °F). Elongation was measured in 50 mm (2 in.). In small sizes, the 8700 steels show properties similar to those of the 8600 steels of equivalent carbon content; however, because of increased molybdenum content the 8700 series will maintain these values to somewhat heavier section sizes. (Ref 10)

AISI 8615: Tensile Properties

Test specimens were 25-mm (1-in.) bar; tests were conducted using specimens machined to English units

Treatment	Tensile strength MPa	Tensile strength ksi	Yield strength MPa	Yield strength ksi	Elongation(a), %	Reduction in area, %
Pseudocarburized at 900-925 °C (1650-1700 °F); box cooled; reheated at 840 °C (1550 °F); oil quenched; tempered at 150 °C (300 °F)	690	100	500	72.5	20	58

(a) In 50 mm (2 in.)

Mechanical Properties (continued)

AISI 8615: Hardness and Machinability (Ref 7)

Condition	Hardness range, HB	Average machinability rating(a)
Hot rolled and cold drawn	179-235	70

(a) Based on AISI 1212 steel as 100% average machinability

Machining Data

For machining data on AISI grade 8615, refer to the preceding machining tables for AISI grade A2317.

AISI 8617, 8617H

AISI 8617, 8617H: Chemical Composition

AISI grade	Chemical composition, % C	Mn	P max	S max	Si	Ni	Cr	Mo
8617	0.15-0.20	0.70-0.90	0.035	0.040	0.15-0.30	0.40-0.70	0.40-0.60	0.15-0.25
8617H	0.14-0.20	0.65-0.95	0.035	0.040	0.15-0.30	0.35-0.75	0.35-0.65	0.15-0.25

Characteristics and Typical Uses. AISI grades 8617 and 8617H are nickel-chromium-molybdenum alloy steels with intermediate case hardenability and medium core hardenability.

These steels are available as hot rolled and cold finished bar, rod, and wire; bar subject to end-quench hardenability requirements; and aircraft quality stock. Uses of AISI 8617 and 8617H include automotive and tractor gears, piston pins, ball studs, universal crosses, and roller bearings.

AISI 8617: Similar Steels (U.S. and/or Foreign). UNS G86170; ASTM A322, A331, A505; AMS 6272; SAE J404, J770; (W. Ger.) DIN 1.6523; (Fr.) AFNOR 20 NCD 2, 22 NCD 2; (Ital.) UNI 20 NiCrMo 2; (Jap.) JIS SNCM 21 H, SNCM 21; (U.K.) B.S. 805 H 20, 805 M 20

AISI 8617H: Similar Steels (U.S. and/or Foreign). UNS H86170; ASTM A304; SAE J1268; (W. Ger.) DIN 1.6523; (Fr.) AFNOR 20 NCD 2, 22 NCD 2; (Ital.) UNI 20 NiCrMo; (Jap.) JIS SNCM 21 H, SNCM 21; (U.K.) B.S. 805 H 20, 805 M 20

AISI 8617, 8617H: Approximate Critical Points

Transformation point	Temperature(a) °C	°F
Ac_1	745	1370
Ac_3	845	1550
Ar_3	775	1430
Ar_1	680	1260

(a) On heating or cooling at 28 °C (50 °F) per hour

Physical Properties

AISI 8617, 8617H: Thermal Treatment Temperatures

Quenching medium: oil

Treatment	Temperature range °C	°F
Forging	1230 max	2250 max
Annealing	855-900	1575-1650
Normalizing	900-940	1650-1725
Carburizing	925	1700
Hardening	800-855	1475-1575

Machining Data

For machining data on AISI grades 8617 and 8617H, refer to the preceding machining tables for AISI grade A2317.

Mechanical Properties

AISI 8617, 8617H: Approximate Core Mechanical Properties

Heat treatment of test specimens: (1) normalized at 925 °C (1700 °F) in 32-mm (1.25-in.) rounds; (2) machined to 25- or 13.7-mm (1- or 0.540-in.) rounds; (3) pseudocarburized at 925 °C (1700 °F) for 8 h; (4) box cooled to room temperature; (5) reheated to temperatures given in the table and oil quenched; (6) tempered at 150 °C (300 °F); (7) tested in 12.8-mm (0.505-in.) rounds; tests were conducted using test specimens machined to English units

Reheat temperature °C	°F	Tensile strength MPa	ksi	Yield strength(a) MPa	ksi	Elongation(b), %	Reduction in area, %	Hardness, HB
(c)	(c)	883	128	650	94	19	54	255
855	1575	834	121	625	91	19	52	248
800	1475	786	114	530	77	22	54	233
855, 800(d)	1575, 1475(d)	820	119	550	80	23	56	243

(a) 0.2% offset. (b) In 50 mm (2 in.). (c) Quenched from step 3. (d) Double oil quench

AISI 8617, 8617H: Permissible Compressive Stresses

Taken from gears having a case depth of 1.1 to 1.5 mm (0.045 to 0.060 in.) and a maximum case hardness of 60 HRC

Data point	Compressive stress MPa	ksi
Mean	1379	200
Maximum	1517	220

AISI 8617, 8617H: As-Quenched Core Hardness

After quenching in oil at 925 °C (1700 °F)

Size round mm	in.	As-quenched core hardness, HB, at: 7/8 radius	1/2 radius	Center
25	1	302	265	258
50	2	235	215	207
75	3	212	195	192

Mechanical Properties (continued)

AISI 8617H: End-Quench Hardenability Limits and Hardenability Band. Heat treating temperatures recommended by SAE: normalize at 925 °C (1700 °F), for forged or rolled specimens only; austenitize at 925 °C (1700 °F). (Ref 1)

Distance from quenched end, 1/16 in.	Hardness, HRC max	Hardness, HRC min
1	46	39
2	44	33
3	41	27
4	38	24
5	34	20
6	31	...
7	28	...
8	27	...
9	26	...
10	25	...
11	24	...
12	23	...
13	23	...
14	22	...
15	22	...
16	21	...
18	21	...
20	20	...

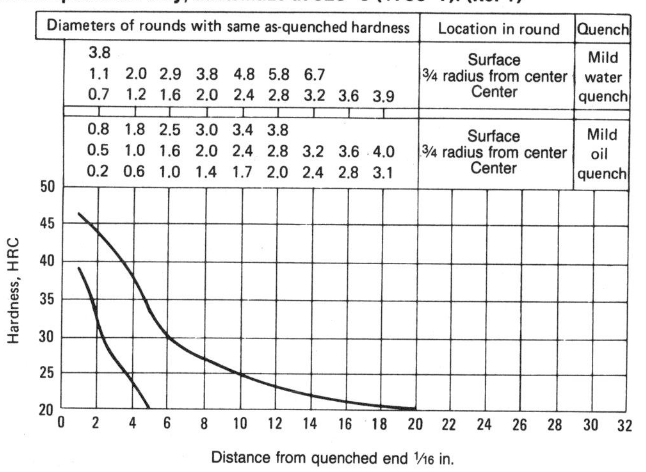

AISI 8617, 8617H: Hardness and Machinability (Ref 7)

Condition	Hardness range, HB	Average machinability rating(a)
Hot rolled and cold drawn	179-235	70

(a) Based on AISI 1212 steel as 100% average machinability

For additional data on mechanical properties of AISI grades 8617 and 8617H, refer to the figure on AISI 8600 and 8700 series steels: average core properties, which is included in AISI 8615.

AISI 8620, 8620H

AISI 8620, 8620H: Chemical Composition

AISI grade	C	Mn	P max	S max	Si	Ni	Cr	Mo
8620	0.18-0.23	0.70-0.90	0.035	0.040	0.15-0.30	0.40-0.70	0.40-0.60	0.15-0.25
8620H	0.19-0.25	0.60-0.95	0.035	0.040	0.15-0.30	0.35-0.75	0.35-0.65	0.15-0.25

(Chemical composition, %)

Characteristics and Typical Uses. AISI grades 8620 and 8620H are carburizing, nickel-chromium-molybdenum, alloy steels, characteristically used in moderate section sizes which require medium hardenability, strength, and shock resistance. The hardening treatment is selected through evaluation of the properties required, in both the case and core, for particular applications of parts.

AISI 8620 and 8620H are used for differential ring gears, camshafts, piston pins, transmission gears, steering worm gears, spline shafts, high-strength fasteners, carburized chain, chain pins, and applications requiring medium strength, tough core properties, and hard, wear-resistant case properties. AISI grade 8620 may be satisfactorily welded by any commercial process; however, the use of shielded-arc, carbon-molybdenum electrodes is recommended.

Grades 8620 and 8620H are available as hot rolled and cold finished bar, rod, and wire; bar subject to end-quench hardenability requirements; and aircraft quality stock.

AISI 8620: Similar Steels (U.S. and/or Foreign). UNS G86200; AMS 6274, 6276, 6277; ASTM A322, A331, A505, A513; MIL SPEC MIL-S-16974; SAE J404, J770; (W. Ger.) DIN 1.6523; (Fr.) AFNOR 20 NCD 2, 22 NCD 2; (Ital.) UNI 20 NiCrMo 2; (Jap.) JIS SNCM 21 H, SNCM 21; (U.K.) B.S. 805 H 20, 805 M 20

AISI 8620H: Similar Steels (U.S. and/or Foreign). UNS H86200; ASTM A304; SAE J1268; (W. Ger.) DIN 1.6523; (Fr.) AFNOR 20 NCD 2, 22 NCD 2; (Ital.) UNI 20 NiCrMo 2; (Jap.) JIS SNCM 21 H, SNCM 21; (U.K.) B.S. 805 H 20, 805 M 20

AISI 8620, 8620H: Approximate Critical Points

Transformation point	Temperature(a) °C	Temperature(a) °F	Transformation point	Temperature(a) °C	Temperature(a) °F
Ac_1	730	1350	Ar_1	660	1220
Ac_3	830	1525	M_s	395	745
Ar_3	770	1415			

(a) On heating or cooling at 28 °C (50 °F) per hour

Physical Properties

AISI 8620, 8620H: Thermal Treatment Temperatures

Quenching medium: oil

Treatment	Temperature range °C	°F
Forging	1230 max	2250 max
Annealing	855-885	1575-1625
Normalizing	870-955	1600-1750
Carburizing	900-925	1650-1700
Reheating	800-835	1475-1535
Tempering	(a)	(a)

(a) To desired hardness

Machining Data

For machining tables on AISI grades 8620 and 8620H, refer to the preceding machining tables for AISI grade A2317.

Mechanical Properties

AISI 8620, 8620H: Effect of the Mass on Core Properties in Response to Heat Treatment (Ref 13)

Test locations: 25-mm (1-in.) rounds at center, 50-mm (2-in.) rounds at ½ radius; tests were conducted using test specimens machined to English units

Treatment or condition	Size round mm	in.	Tensile strength MPa	ksi	Yield strength MPa	ksi	Elongation(a), %	Reduction in area, %	Hardness, HB	Izod impact energy J	ft·lb
Annealed (heated to 870 °C or 1600 °F; furnace cooled at 17 °C or 30 °F per hour to 620 °C or 1150 °F; air cooled)	13	0.5	530	77	385	56	31	62	149	115	83
Normalized (heated to 915 °C or 1675 °F; air cooled)	13	0.5	635	92	360	52	26	60	183	98	73
Quenched from carburizing box (carburized at 925 °C or 1700 °F for 8 h; oil quenched; tempered at 150 °C or 300 °F)	13	0.5	1325	192	1035	150	12	49	388	37	27
	25	1	931	135	758	110	16	40	321	...	...
	50	2	745	108	560	81	25	58	277	...	...
Cooled in carburizing box, reheated and quenched (carburized at 925 °C or 1700 °F for 8 h; cooled to room temperature in box; reheated to 845 °C or 1550 °F; oil quenched; tempered at 150 °C or 300 °F)	13	0.5	1296	188	1035	150	11	52	388	35	26
	25	1	958	139	814	118	19	48	321	...	...
	50	2	675	98	490	71	21	54	269	...	...
Cooled in carburizing box, reheated and quenched twice (carburized at 925 °C or 1700 °F for 8 h; cooled to room temperature in box; reheated to 845 °C or 1550 °F; oil quenched; reheated again to 800 °C or 1475 °F; oil quenched; tempered at 150 °C or 300 °F)	13	0.5	917	133	570	83	20	57	269	76	56

(a) In 50 mm (2 in.)

AISI 8620, 8620H: Low-Temperature Impact Properties (Ref 17)

Quenched from 900 °C (1650 °F)

Tempering temperature °C	°F	Hardness, HRC	Charpy impact (V-notch) energy at temperature of: −185 °C (−300 °F) J	ft·lb	−130 °C (−200 °F) J	ft·lb	−73 °C (−100 °F) J	ft·lb	−18 °C (0 °F) J	ft·lb	38 °C (100 °F) J	ft·lb	Transition temperature(a) °C	°F
150	300	43	15	11	22	16	31	23	47	35	47	35	...	...
425	800	36	11	8	18	13	27	20	47	35	61	45	−29	−20
540	1000	29	34	25	45	33	88	65	103	76	103	76	−101	−150
650	1200	21	14	10	115	85	145	107	156	115	159	117	−126	−195

(a) 50% brittle

Mechanical Properties (continued)

AISI 8620, 8620H: Core Properties of Specimens Treated for Maximum Case Hardness and Core Toughness (Ref 4)

Specimens were treated in 14.4-mm (0.565-in.) rounds and machined to 12.8-mm (0.505-in.) rounds for testing; tests were conducted using test specimens machined to English units

Treatment	Case depth, mm (in.)	Case hardness, HRC	Core properties: Tensile strength, MPa (ksi)	Yield strength, MPa (ksi)	Elongation(a), %	Reduction in area, %	Hardness, HB
Recommended practice for maximum case hardness							
Direct quenched from pot: carburized at 925 °C (1700 °F) for 8 h; quenched in agitated oil; tempered at 150 °C (300 °F)	1.4 (0.056)	63	1324 (192.0)	1036 (150.3)	12.5	49.4	388
Single quenched and tempered (for good case and core properties): carburized at 925 °C (1700 °F) for 8 h; pot cooled; reheated to 845 °C (1550 °F); quenched in agitated oil; tempered at 150 °C (300 °F)	1.9 (0.075)	64	1300 (188.5)	1033 (149.8)	11.5	51.6	388
Double quenched and tempered (for maximum refinement of case and core): carburized at 925 °C (1700 °F) for 8 h; pot cooled; reheated to 845 °C (1550 °F); quenched in agitated oil; reheated to 800 °C (1475 °F); quenched in agitated oil; tempered at 150 °C (300 °F)	1.8 (0.070)	64	917 (133.0)	572 (83.0)	20.0	56.8	269
Recommended practice for maximum core toughness							
Direct quenched from pot: carburized at 925 °C (1700 °F) for 8 h; quenched in agitated oil; tempered at 230 °C (450 °F)	1.3 (0.050)	58	1250 (181.3)	926 (134.3)	12.8	50.6	52
Single quenched and tempered (for good case and core properties): carburized at 925 °C (1700 °F) for 8 h; pot cooled; reheated to 845 °C (1550 °F); quenched in agitated oil; tempered at 230 °C (450 °F)	1.9 (0.076)	61	1157 (167.8)	833 (120.8)	14.3	53.2	341
Double quenched and tempered (for maximum refinement of case and core): carburized at 925 °C (1700 °F) for 8 h; pot cooled; reheated to 845 °C (1550 °F); quenched in agitated oil; reheated to 800 °C (1475 °F); quenched in agitated oil; tempered at 230 °C (450 °F)	1.8 (0.070)	61	898 (130.3)	533 (77.3)	22.5	51.7	262

(a) In 50 mm (2 in.)

AISI 8620, 8620H: Approximate Core Mechanical Properties (Ref 2)

Heat treatment of test specimens: (1) normalized at 925 °C (1700 °F) in 32-mm (1.25-in.) rounds; (2) machined to 25- or 13.7-mm (1- or 0.540-in.) rounds; (3) pseudocarburized at 925 °C (1700 °F) for 8 h; (4) box cooled to room temperature; (5) reheated to temperatures given in the table and oil quenched; (6) tempered at 150 °C (300 °F); (7) tested in 12.8-mm (0.505-in.) rounds; tests were conducted using test specimens machined to English units

Reheat temperature, °C	°F	Tensile strength, MPa	ksi	Yield strength(a), MPa	ksi	Elongation(b), %	Reduction in area, %	Hardness, HB
Heat treated in 25-mm (1-in.) rounds								
775	1425	903	131	710	103	15.0	52	277
800	1475	958	139	765	111	16.0	54	293
845	1550	1050	152	855	124	17.0	55	321
(c)	(c)	1075	156	876	127	14.0	50	321
Heat treated in 13.7-mm (0.540-in.) rounds								
775	1425	938	136	772	112	15.0	49	285
800	1475	1040	151	848	123	15.5	50	321
845	1550	1095	159	910	132	16.5	56	331
(c)	(c)	1110	161	924	134	15.0	53	331

(a) 0.2% offset. (b) In 50 mm (2 in.). (c) Quenched from step 3

Mechanical Properties (continued)

AISI 8620H: End-Quench Hardenability Limits and Hardenability Band. Heat treating temperatures recommended by SAE: normalize at 925 °C (1700 °F), for forged or rolled specimens only; austenitize at 925 °C (1700 °F). (Ref 1)

Distance from quenched end, 1/16 in.	Hardness, HRC max	Hardness, HRC min
1	48	41
2	47	37
3	44	32
4	41	27
5	37	23
6	34	21
7	32	...
8	30	...
9	29	...
10	28	...
11	27	...
12	26	...
13	25	...
14	25	...
15	24	...
16	24	...
18	23	...
20	23	...
22	23	...
24	23	...
26	23	...
28	22	...
30	22	...
32	22	...

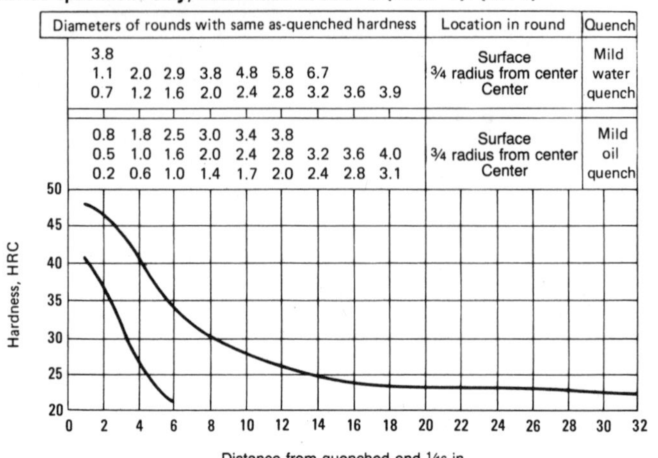

AISI 8620, 8620H: Mass Effect on Mechanical Properties (Ref 4)

Condition or treatment	Size round mm	Size round in.	Tensile strength MPa	Tensile strength ksi	Yield strength MPa	Yield strength ksi	Elongation(a), %	Reduction in area, %	Hardness, HB
Annealed (heated to 870 °C or 1600 °F; furnace cooled 17 °C or 30 °F per hour to 620 °C or 1150 °F; cooled in air)	25	1	540	78	385	56	31.3	62.1	149
Normalized (heated to 915 °C or 1675 °F; cooled in air)	13	0.5	660	96	370	54	26.3	62.5	197
	25	1	635	92	360	52	26.3	59.7	183
	50	2	600	87	360	52	27.8	62.1	179
	100	4	565	82	360	52	28.5	62.3	163
Mock carburized at 925 °C (1700 °F) for 8 h; reheated to 845 °C (1550 °F); quenched in oil; tempered at 150 °C (300 °F)	13	0.5	1380	200	1085	157	13.2	49.4	388
	25	1	876	127	580	84	20.8	52.7	255
	50	2	807	117	505	73	23.0	57.8	235
	100	4	675	98	400	58	24.3	57.6	207
Mock carburized at 925 °C (1700 °F) for 8 h; reheated to 845 °C (1550 °F); quenched in oil; tempered at 230 °C (450 °F)	13	0.5	1225	178	965	140	14.6	53.9	352
	25	1	855	124	560	81	19.5	54.2	248
	50	2	786	114	495	72	22.0	59.0	229
	100	4	675	98	385	56	25.5	57.8	201

(a) In 50 mm (2 in.)

AISI 8620, 8620H: Effect of the Mass on Hardness at Selected Points (Ref 4)

Size round mm	Size round in.	As-quenched hardness after quenching in oil at: Surface	½ radius	Center
13	0.5	43 HRC	43 HRC	43 HRC
25	1	29 HRC	27 HRC	25 HRC
50	2	23 HRC	22 HRC	97 HRB
100	4	22 HRC	95 HRB	93 HRB

AISI 8620, 8620H: Hardness and Machinability (Ref 7)

Condition	Hardness range, HB	Average machinability rating(a)
Hot rolled and cold drawn	179-235	65

(a) Based on AISI 1212 steel as 100% average machinability

AISI 8620, 8620H: Tensile Properties (Ref 5)

Condition or treatment	Tensile strength MPa	Tensile strength ksi	Yield strength MPa	Yield strength ksi	Elongation(a), %	Reduction in area, %	Hardness, HB	Izod impact energy J	Izod impact energy ft·lb
Normalized at 915 °C (1675 °F)	635	92	360	52	26.3	60	183	100	74
Annealed at 870 °C (1600 °F)	650	94	385	56	31.3	62	149	115	83

(a) In 50 mm (2 in.)

For additional data on mechanical properties of AISI grades 8620 and 8620H, refer to the figure on AISI 8600 and 8700 series steels: average core properties, which is included in AISI 8615.

AISI 8622, 8622H

AISI 8622, 8622H: Chemical Composition

AISI grade	C	Mn	P max	S max	Si	Ni	Cr	Mo
8622	0.20-0.25	0.70-0.90	0.035	0.040	0.15-0.30	0.40-0.70	0.40-0.60	0.15-0.25
8622H	0.19-0.25	0.60-0.95	0.035	0.040	0.15-0.30	0.35-0.75	0.35-0.65	0.15-0.25

Characteristics. AISI grades 8622 and 8622H are triple-alloy, carburizing steels with intermediate case and core hardenability. They are used where medium case hardness and good core properties are desired. In the AISI 8600 carburizing grades, minimum hardness increases with greater carbon content—a factor that must be considered when making a grade selection.

Typical Uses. High-strength fasteners, carburized chain and chain pins, and other severe-duty parts are the primary applications of rods and wires produced from these steels. Grades 8622 and 8622H are available as hot rolled and cold finished bar, rod, and wire, and bar subject to end-quench hardenability requirements.

AISI 8622: Similar Steels (U.S. and/or Foreign). UNS G86220; ASTM A322, A331; SAE J404, J770; (W. Ger.) DIN 1.6543; (U.K.) B.S. 805 A 20

AISI 8622H: Similar Steels (U.S. and/or Foreign). UNS H86220; ASTM A304; SAE J1268; (W. Ger.) DIN 1.6543; (U.K.) B.S. 805 A 20

Physical Properties

AISI 8622, 8622H: Thermal Treatment Temperatures

Quenching medium: oil

Treatment	Temperature range °C	Temperature range °F
Forging	1230 max	2250 max
Cycle annealing	(a)	(a)
Normalizing	(b)	(b)
Carburizing	900-925	1650-1700
Reheating	845-870	1550-1600
Tempering	120-150	250-300

(a) Heat at least as high as the carburizing temperature; hold for uniformity; cool rapidly to 540 to 675 °C (1000 to 1250 °F); hold for 1 to 3 h; air cool to obtain a structure suitable for machining and finish. (b) At least as high as the carburizing temperature, followed by air cooling

Mechanical Properties

AISI 8622, 8622H: Approximate Core Mechanical Properties (Ref 2)

Heat treatment of test specimens: (1) normalized at 925 °C (1700 °F) in 32-mm (1.25-in.) rounds; (2) machined to 25- or 13.7-mm (1- or 0.540-in.) rounds; (3) pseudocarburized at 925 °C (1700 °F) for 8 h; (4) box cooled to room temperature; (5) reheated to temperatures given in the table and oil quenched; (6) tempered at 150 °C (300 °F); (7) tested in 12.8-mm (0.505-in.) rounds; tests were conducted using test specimens machined to English units

Reheat temperature °C	Reheat temperature °F	Tensile strength MPa	Tensile strength ksi	Yield strength(a) MPa	Yield strength(a) ksi	Elongation(b), %	Reduction in area, %	Hardness, HB
Heat treated in 25-mm (1-in.) rounds								
775	1425	1176	170.5	654	94.8	12.1	24.9	352
820	1510	1204	174.6	805	116.8	14.8	44.5	363
865	1590	1250	181.3	900	130.6	13.6	46.0	388
(c)	(c)	1314	190.6	928	134.6	12.8	42.5	388
Heat treated in 13.7-mm (0.540-in.) rounds								
775	1425	1198	173.7	693	100.5	11.3	24.6	352
820	1510	1451	210.5	1041	151.0	13.0	41.9	415
865	1590	1510	219.0	1093	158.5	13.3	47.3	429
(c)	(c)	1524	221.1	1149	166.6	13.5	47.7	429

(a) 0.2% offset. (b) In 50 mm (2 in.). (c) Quenched from step 3

AISI 8622, 8622H: Permissible Compressive Stresses

Taken from gears having a case depth of 1.1 to 1.5 mm (0.045 to 0.060 in.) and minimum case hardness of 60 HRC

Data point	Compressive stress MPa	Compressive stress ksi
Mean	1379	200
Maximum	1517	220

Mechanical Properties (continued)

AISI 8622H: End-Quench Hardenability Limits and Hardenability Band. Heat treating temperatures recommended by SAE: normalize at 925 °C (1700 °F), for forged or rolled specimens only; austenitize at 925 °C (1700 °F). (Ref 1)

Distance from quenched end, 1/16 in.	Hardness, HRC max	Hardness, HRC min
1	50	43
2	49	39
3	47	34
4	44	30
5	40	26
6	37	24
7	34	22
8	32	20
9	31	...
10	30	...
11	29	...
12	28	...
13	27	...
14	26	...
15	26	...
16	25	...
18	25	...
20	24	...
22	24	...
24	24	...
26	24	...
28	24	...
30	24	...
32	24	...

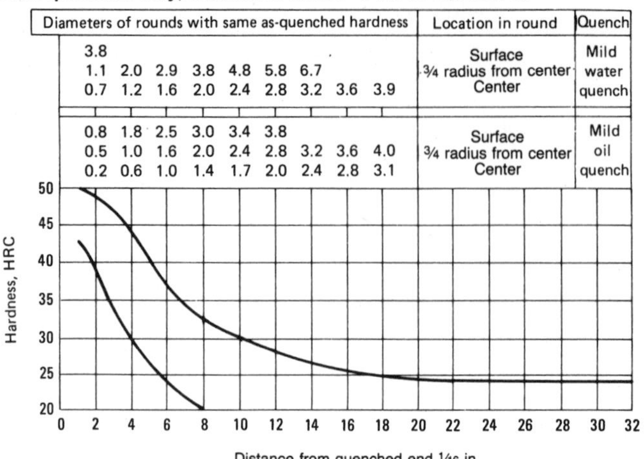

AISI 8622, 8622H: Hardness and Machinability (Ref 7)

Condition	Hardness range, HB	Average machinability rating(a)
Hot rolled and cold drawn	179-235	65

(a) Based on AISI 1212 steel as 100% average machinability

For additional data on mechanical properties of AISI grades 8622 and 8622H, refer to the figure on AISI 8600 and 8700 series steels: average core properties, which is included in AISI 8615.

Machining Data

For machining data on AISI grades 8622 and 8622H, refer to the preceding machining tables for AISI grade A2317.

AISI 8625, 8625H

AISI 8625, 8625H: Chemical Composition

AISI grade	C	Mn	P max	S max	Si	Ni	Cr	Mo
8625	0.23-0.28	0.70-0.90	0.035	0.040	0.15-0.30	0.40-0.70	0.40-0.60	0.15-0.25
8625H	0.22-0.28	0.65-0.95	0.035	0.040	0.15-0.30	0.35-0.75	0.35-0.65	0.15-0.25

(Column headers C through Mo: Chemical composition, %)

Characteristics and Typical Uses. AISI grades 8625 and 8625H are triple-alloy, carburizing steels with intermediate case hardenability and high core hardenability. Uses of these steels are similar to those described for AISI grades 8620 and 8622, and include applications where higher case and core hardenability are desirable.

AISI 8625 and 8625H are available as hot rolled and cold finished bar, rod, and wire; bar subject to end-quench hardenability requirements; and aircraft quality bar.

AISI 8625: Similar Steels (U.S. and/or Foreign). UNS G86250; ASTM A322, A331; MIL SPEC MIL-S-16974; SAE J404, J770

AISI 8625H: Similar Steels (U.S. and/or Foreign). UNS H86250; ASTM A304; SAE J1268

AISI 8625, 8625H: Approximate Critical Points

Transformation point	Temperature(a) °C	Temperature(a) °F
Ac_1	730	1350
Ac_3	805	1485
Ar_3	755	1390
Ar_1	660	1220
M_s	375	710

(a) On heating or cooling at 28 °C (50 °F) per hour

Mechanical Properties

AISI 8625H: End-Quench Hardenability Limits and Hardenability Band. Heat treating temperatures recommended by SAE: normalize at 900 °C (1650 °F), for forged or rolled specimens only; austenitize at 870 °C (1600 °F). (Ref 1)

Distance from quenched end, 1/16 in.	Hardness, HRC max	min
1	52	45
2	51	41
3	48	36
4	46	32
5	43	29
6	40	27
7	37	25
8	35	23
9	33	22
10	32	21
11	31	20
12	30	...
13	29	...
14	28	...
15	28	...
16	27	...
18	27	...
20	26	...
22	26	...
24	26	...
26	26	...
28	25	...
30	25	...
32	25	...

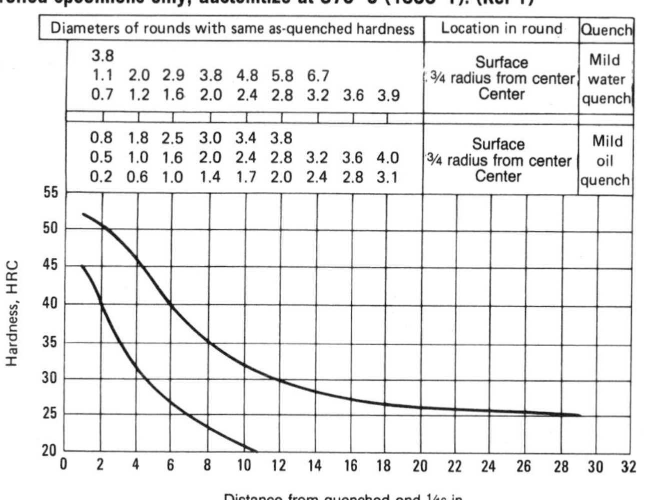

AISI 8625, 8625H: Hardness and Machinability (Ref 7)

Condition	Hardness range, HB	Average machinability rating(a)
Annealed and cold drawn	179-223	60

(a) Based on AISI 1212 steel as 100% average machinability

For additional data on mechanical properties of AISI grades 8625 and 8625H, refer to the figure on AISI 8600 and 8700 series steels: average core properties, which is included in AISI 8615.

Physical Properties

AISI 8625, 8625H: Thermal Treatment Temperatures

Quenching medium: oil

Treatment	Temperature range °C	°F
Forging	1230 max	2250 max
Cycle annealing	(a)	(a)
Normalizing	(b)	(b)
Carburizing	900-925	1650-1700
Reheating	845-870	1550-1600
Tempering	120-175	250-350

(a) Heat at least as high as the carburizing temperature; hold for uniformity; cool rapidly to 540 to 675 °C (1000 to 1250 °F); hold for 1 to 3 h; air cool to obtain a structure suitable for machining and finish. (b) At least as high as the carburizing temperature, followed by air cooling

Machining Data

For machining data on AISI grades 8625 and 8625H, refer to the preceding machining tables for AISI grade A2317.

AISI 8627, 8627H

AISI 8627, 8627H: Chemical Composition

AISI grade	C	Mn	P max	S max	Si	Ni	Cr	Mo
				Chemical composition, %				
8627	0.25-0.30	0.70-0.90	0.035	0.045	0.15-0.30	0.40-0.70	0.40-0.60	0.15-0.25
8627H	0.24-0.30	0.60-0.95	0.035	0.045	0.15-0.30	0.35-0.75	0.35-0.65	0.15-0.25

Characteristics and Typical Uses. AISI grades 8627 and 8627H are triple-alloy, carburizing steels with intermediate case hardenability and high core hardenability. These steels are used for parts requiring good core properties after heat treatment.

Grades 8627 and 8627H are available as hot rolled and cold finished bar, rod, and wire, and bar subject to end-quench hardenability requirements.

AISI 8627: Similar Steels (U.S. and/or Foreign). UNS G86270; ASTM A322, A331; SAE J404, J770

AISI 8627H: Similar Steels (U.S. and/or Foreign). UNS H86270; ASTM A304, SAE J1268

Machining Data

For machining data on AISI grades 8627 and 8627H, refer to the preceding machining tables for AISI grade A2317.

Mechanical Properties

AISI 8627H: End-Quench Hardenability Limits and Hardenability Band. Heat treating temperatures recommended by SAE: normalize at 900 °C (1650 °F), for forged or rolled specimens only; austenitize at 870 °C (1600 °F). (Ref 1)

Distance from quenched end, 1/16 in.	Hardness, HRC max	Hardness, HRC min
1	54	47
2	52	43
3	50	38
4	48	35
5	45	32
6	43	29
7	40	27
8	38	26
9	36	24
10	34	24
11	33	23
12	32	22
13	31	21
14	30	21
15	30	20
16	29	20
18	28	...
20	28	...
22	28	...
24	27	...
26	27	...
28	27	...
30	27	...
32	27	...

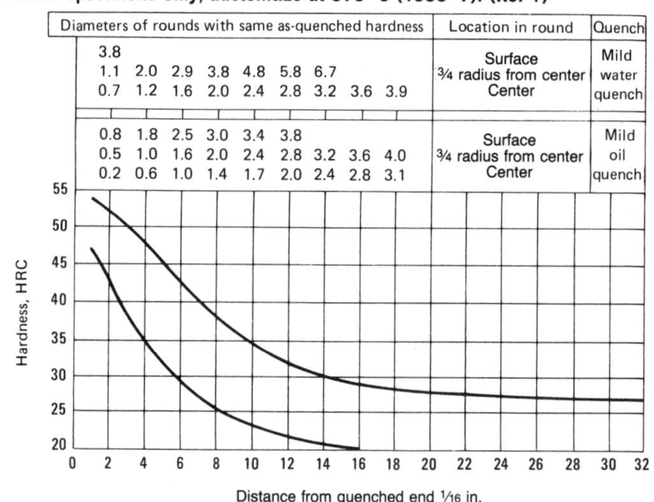

AISI 8627, 8627H: Hardness and Machinability (Ref 7)

Condition	Hardness range, HB	Average machinability rating(a)
Annealed and cold drawn	179-223	60

(a) Based on AISI 1212 steel as 100% average machinability

For additional data on mechanical properties of AISI grades 8627 and 8627H, refer to the figure on AISI 8600 and 8700 series steels: average core properties, which is included in AISI 8615.

AISI 8630, 8630H, 86B30H

AISI 8630, 8630H, 86B30H: Chemical Composition

AISI grade	C	Mn	P max	S max	Si	Ni	Cr	Mo
	Chemical composition, %							
8630	0.28-0.33	0.70-0.90	0.035	0.040	0.15-0.30	0.40-0.70	0.40-0.60	0.15-0.25
8630H	0.27-0.33	0.65-0.95	0.035	0.040	0.15-0.30	0.35-0.75	0.35-0.65	0.15-0.25
86B30H (a)	0.27-0.33	0.65-0.95	0.035	0.040	0.15-0.30	0.35-0.75	0.35-0.65	0.15-0.25

(a) Contains 0.0005 to 0.003% boron

Characteristics and Typical Uses. AISI grades 8630, 8630H, and 86B30H are directly hardenable, triple-alloy steels with low hardenability; although the addition of boron improves hardenability of grade 86B30H. These steels have good mechanical properties when normalized in light sections. They are air hardening after welding.

AISI grades 8630, 8630H, and 86B30H are available as hot rolled and cold finished bar, rod, and wire; mechanical seamless tubing; and aircraft quality stock. Applications include connecting rods, engine bolts and studs, shapes, and tubing.

AISI 8630: Similar Steels (U.S. and/or Foreign). UNS G86300; AMS 6280, 6281, 6355, 6530, 6550; ASTM A322, A331, A505, A519; MIL SPEC MIL-S-16974; SAE J404, J412, J770; (W. Ger.) DIN 1.6545; (Ital.) UNI 30 NiCrMo 2 KB

AISI 8630H: Similar Steels (U.S. and/or Foreign). UNS H86300; ASTM A304; SAE J1268; (W. Ger.) DIN 1.6545; (Ital.) UNI 30 NiCrMo 2 KB

AISI 86B30H: Similar Steels (U.S. and/or Foreign). UNS H86301; ASTM A304; SAE J1268

AISI 8630, 8630H, 86B30H: Approximate Critical Points

Transformation point	Temperature(a) °C	°F	Transformation point	Temperature(a) °C	°F
Ac_1	735	1355	Ar_1	660	1220
Ac_3	795	1460	M_s	360	680
Ar_3	745	1370			

(a) On heating or cooling at 28 °C (50 °F) per hour

Physical Properties

AISI 8630, 8630H, 86B30H: Thermal Treatment Temperatures

Anneal or normalize for optimum machinability; quenching medium: oil or water

Treatment	Temperature range °C	°F
Forging	1230 max	2250 max
Annealing (a)	815-870	1500-1600
Normalizing (b)	870-925	1600-1700
Austenitizing	855-885	1575-1625
Tempering	(c)	(c)

(a) Maximum hardness of 179 HB. (b) Hardness of approximately 225 HB. (c) To desired hardness

AISI 8630, 8630H, 86B30H: Average Coefficients of Linear Thermal Expansion (Ref 9)

Temperature range °C	°F	Coefficient μm/m·K	μin./in.·°F
21-100	70-212	11.2	6.2
21-200	70-390	12.2	6.8
21-300	70-570	13.0	7.2
21-400	70-750	13.5	7.5
21-500	70-930	14.0	7.8
21-600	70-1110	14.4	8.0
21-700	70-1290	14.9	8.3

Mechanical Properties

AISI 8630, 8630H, 86B30H: Tensile and Fatigue Properties (Ref 17)

Temperature °C	°F	Tensile strength MPa	ksi	Yield strength(a) MPa	ksi	Elongation(b), %	Reduction in area, %	Hardness, HV	Modulus of elasticity GPa	10^6 psi	Endurance limit Unnotched MPa	ksi	Notched MPa	ksi
Normalized														
25	77	620	90	550	80	16	58	217	187	27.1	315	46	...	...
−79	−110	717	104	635	92	18	57	252	199	28.9	400	58	...	...
−195	−315	1035	150	924	134	12	28	289	220	31.9	640	93	...	...
Oil quenched from 845 °C (1550 °F); tempered at 455 °C (850 °F)														
25	77	979	142	841	122	16	60	302	205	29.8	485	70	205	30
−79	−110	1060	154	876	127	17	57	332	214	31.1	550	80	...	...
−195	−315	1325	192	1145	166	12	50	376	223	32.3	862	125	240	35

(a) 0.2% offset. (b) In 50 mm (2 in.)

Mechanical Properties (continued)

AISI 8630H: End-Quench Hardenability Limits and Hardenability Band. Heat treating temperatures recommended by SAE: normalize at 900 °C (1650 °F), for forged or rolled specimens only; austenitize at 870 °C (1600 °F). (Ref 1).

Distance from quenched end, 1/16 in.	Hardness, HRC max	Hardness, HRC min
1	56	49
2	55	46
3	54	43
4	52	39
5	50	35
6	47	32
7	44	29
8	41	28
9	39	27
10	37	26
11	35	25
12	34	24
13	33	23
14	33	22
15	32	22
16	31	21
18	30	21
20	30	20
22	29	20
24	29	...
26	29	...
28	29	...
30	29	...
32	29	...

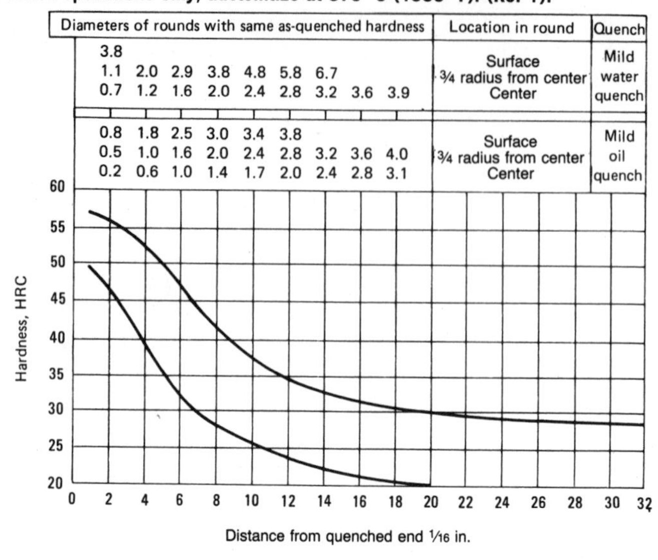

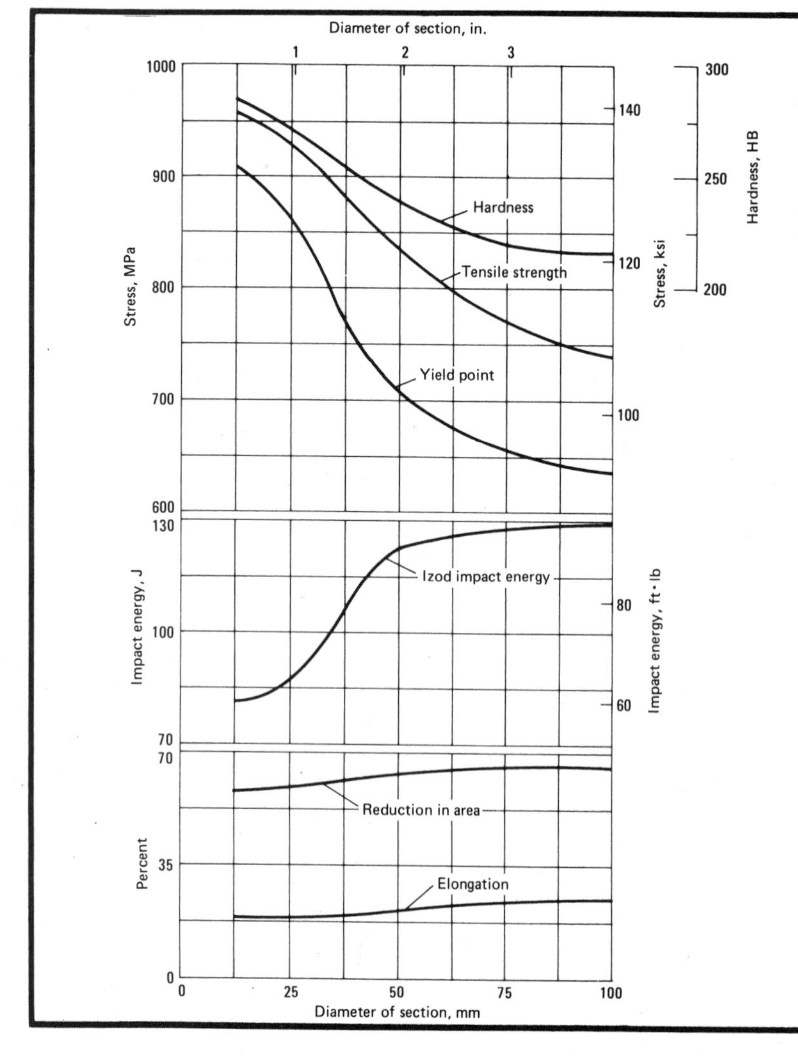

AISI 8630: Effect of Mass on Tensile Properties. Water quenched from 845 °C (1550 °F); tempered at 540 °C (1000 °F). On sizes 25 mm (1 in.) and larger, properties are those at half-radius on samples cut longitudinally. Elongation was measured in 50 mm (2 in.). (Ref 14)

Mechanical Properties (continued)

AISI 8630H: End-Quench Hardenability Limits and Hardenability Band.

Heat treating temperatures recommended by SAE: normalize at 900 °C (1650 °F), for forged or rolled specimens only; austenitize at 870 °C (1600 °F). (Ref 1).

Distance from quenched end, 1/16 in.	Hardness, HRC max	Hardness, HRC min
1	56	49
2	55	46
3	54	43
4	52	39
5	50	35
6	47	32
7	44	29
8	41	28
9	39	27
10	37	26
11	35	25
12	34	24
13	33	23
14	33	22
15	32	22
16	31	21
18	30	21
20	30	20
22	29	20
24	29	...
26	29	...
28	29	...
30	29	...
32	29	...

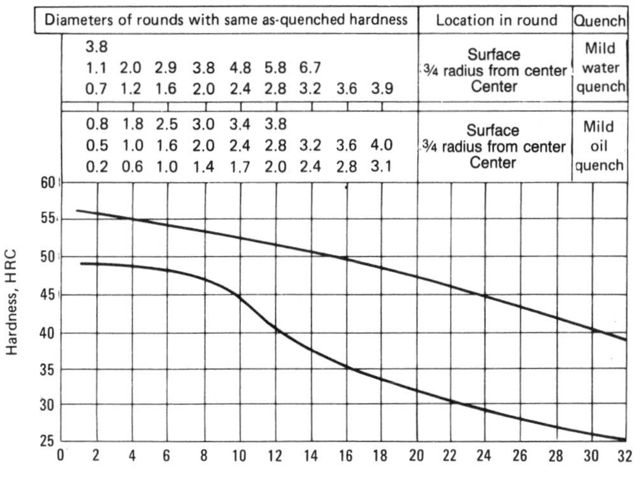

AISI 8630, 8630H, 86B30H: Mass Effect on Mechanical Properties

Condition or treatment	Size round mm	Size round in.	Tensile strength MPa	Tensile strength ksi	Yield strength MPa	Yield strength ksi	Elongation(a), %	Reduction in area, %	Hardness, HB
Annealed (heated to 845 °C or 1550 °F; furnace cooled 11 °C or 20 °F per hour to 625 °C or 1155 °F; cooled in air)	25	1	565	82	370	54	29.0	58.9	156
Normalized (heated to 870 °C or 1600 °F; cooled in air)	13	0.5	655	95	425	62	25.2	60.2	201
	25	1	650	94	425	62	23.5	53.5	187
	50	2	640	93	425	62	26.2	59.2	187
	100	4	635	92	385	56	24.5	57.3	187
Water quenched from 845 °C (1550 °F); tempered at 480 °C (900 °F)	13	0.5	1050	152	1035	150	16.4	59.4	302
	25	1	1015	147	910	132	16.2	56.5	293
	50	2	895	130	738	107	19.2	63.7	269
	100	4	780	113	595	86	21.2	64.7	235
Water quenched from 845 °C (1550 °F); tempered at 540 °C (1000 °F)	13	0.5	958	139	910	132	18.9	58.1	285
	25	1	931	135	850	123	18.7	59.6	269
	50	2	827	120	689	100	21.2	65.6	235
	100	4	738	107	565	82	23.0	63.0	217
Water quenched from 845 °C (1550 °F); tempered at 595 °C (1100 °F)	13	0.5	924	134	910	132	19.2	61.0	269
	25	1	814	118	695	101	18.7	58.2	241
	50	2	765	111	615	89	22.5	68.6	223
	100	4	660	96	495	72	25.5	68.1	197

(a) In 50 mm (2 in.)

AISI 8630, 8630H, 86B30H: Effect of the Mass on Hardness at Selected Points

Size round mm	Size round in.	As-quenched hardness after quenching in water at: Surface	½ radius	Center
13	0.5	52 HRC	49 HRC	47 HRC
25	1	52 HRC	48 HRC	43 HRC
50	2	51 HRC	31 HRC	30 HRC
100	4	47 HRC	25 HRC	22 HRC

AISI 8630, 8630H, 86B30H: Hardness and Machinability (Ref 7)

Condition	Hardness range, HB	Average machinability rating(a)
Annealed and cold drawn	179-229	70

(a) Based on AISI 1212 steel as 100% average machinability

Mechanical Properties (continued)

AISI 8630: Effect of Tempering Temperature on Tensile Properties After Quenching in Oil.

Normalized at 870 °C (1600 °F); reheated to 845 °C (1550 °F); quenched in oil. Test specimens were 25-mm (1-in.) sections, 13.5-mm (0.530-in.) diam when treated and machined to 12.8-mm (0.505-in.) for testing. Tests were conducted using specimens machined to English units. As-quenched hardness of 534 HB; elongation measured in 50 mm (2 in.). (Ref 14)

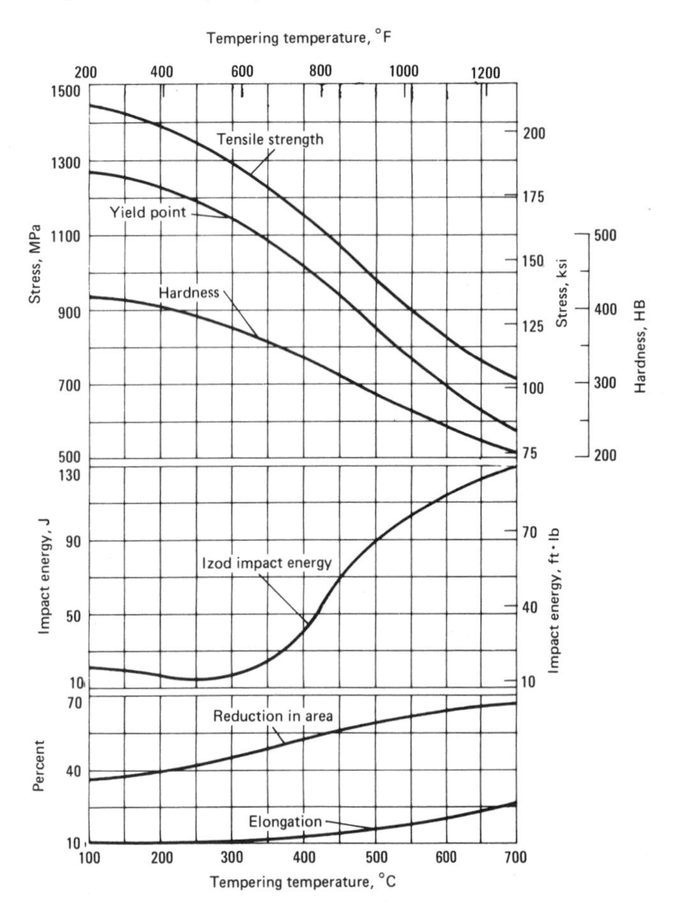

AISI 8630: Effect of Tempering Temperature on Tensile Properties After Quenching in Water.

Normalized at 870 °C (1600 °F); reheated to 845 °C (1550 °F); quenched in water. Test specimens were 13-mm (0.5-in.) sections, 13.5-mm (0.530-in.) diam when treated and machined to 12.8-mm (0.505-in.) for testing. Tests were conducted using specimens machined to English units. As-quenched hardness of 534 HB; elongation measured in 50 mm (2 in.). (Ref 14)

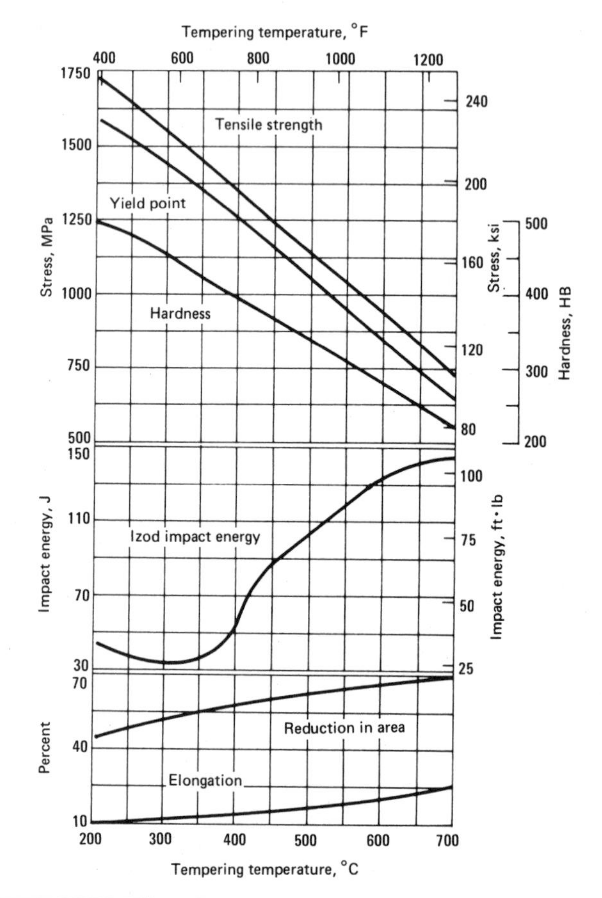

AISI 8630, 8630H, 86B30H: Low-Temperature Impact Properties (Ref 17)

Quenched from 855 °C (1575 °F)

Tempering temperature		Hardness,	Charpy impact (V-notch) energy at temperature of:										Transition temperature(a)	
			−185 °C (−300 °F)		−130 °C (−200 °F)		−73 °C (−100 °F)		−18 °C (0 °F)		38 °C (100 °F)			
°C	°F	HRC	J	ft·lb	J	ft·lb	J	ft·lb	J	ft·lb	J	ft·lb	°C	°F
425	800	41	9	7	16	12	23	17	34	25	42	31	−18	0
540	1000	34	15	11	27	20	58	43	72	53	73	54	−105	−155
650	1200	27	24	18	38	28	100	74	108	80	111	82	−110	−165

(a) 50% brittle

AISI 8630, 8630H, 86B30H: Tensile Properties (Ref 5)

Condition or treatment	Tensile strength MPa	Tensile strength ksi	Yield strength MPa	Yield strength ksi	Elongation(a), %	Reduction in area, %	Hardness, HB	Izod impact energy J	Izod impact energy ft·lb
Normalized at 870 °C (1600 °F)	650	94	425	62	23.5	54	187	95	70
Annealed at 845 °C (1550 °F)	565	82	370	54	29.0	59	156	95	70

(a) In 50 mm (2 in.)

Machining Data

For machining data on AISI grades 8630, 8630H, and 86B30H, refer to the preceding machining tables for AISI grades 1330 and 1330H.

AISI 8637, 8637H

AISI 8637, 8637H: Chemical Composition

AISI grade	Chemical composition, % C	Mn	P max	S max	Si	Ni	Cr	Mo
8637	0.35-0.40	0.75-1.00	0.035	0.040	0.15-0.30	0.40-0.70	0.40-0.60	0.15-0.25
8637H	0.34-0.41	0.70-1.05	0.035	0.040	0.15-0.30	0.35-0.75	0.35-0.65	0.15-0.25

Characteristics and Typical Uses. AISI grades 8637 and 8637H are triple-alloy, directly hardenable steels with medium hardenability. High-strength fasteners and similar severe-duty, cold formed parts are the primary uses of rod and wire produced from these steels. Electric-furnace grade E8637 is produced as aircraft quality, killed, fine-grain steel.

AISI 8637 and 8637H steels are available as hot rolled and cold finished bar, rod, and wire; seamless mechanical tubing; and bar subject to end-quench hardenability requirements.

AISI 8637: Similar Steels (U.S. and/or Foreign). UNS G86370; ASTM A322, A331, A519; SAE J404, J412, J770

AISI 8637H: Similar Steels (U.S. and/or Foreign). UNS H86370; ASTM A304; SAE J1268

AISI 8637, 8637H: Approximate Critical Points

Transformation point	Temperature(a) °C	°F
Ac_1	730	1350
Ac_3	790	1450
Ar_3	730	1345
Ar_1	665	1225
M_s	340	640

(a) On heating or cooling at 28 °C (50 °F) per hour

Physical Properties

AISI 8637, 8637H: Thermal Treatment Temperatures

Quenching medium: oil

Treatment	Temperature range °C	°F
Forging	1230 max	2250 max
Annealing	815-870	1500-1600
Normalizing	870	1600
Austenitizing	800-845	1475-1550
Tempering	(a)	(a)

(a) To desired hardness

Mechanical Properties

AISI 8637H: End-Quench Hardenability Limits and Hardenability Band. Heat treating temperatures recommended by SAE: normalize at 870 °C (1600 °F), for forged or rolled specimens only; austenitize at 845 °C (1550 °F). (Ref 1)

Distance from quenched end, 1/16 in.	Hardness, HRC max	min
1	59	52
2	58	51
3	58	50
4	57	48
5	56	45
6	55	42
7	54	39
8	53	36
9	51	34
10	49	32
11	47	31
12	46	30
13	44	29
14	43	28
15	41	27
16	40	26
18	39	25
20	37	25
22	36	24
24	36	24
26	35	24
28	35	24
30	35	23
32	35	23

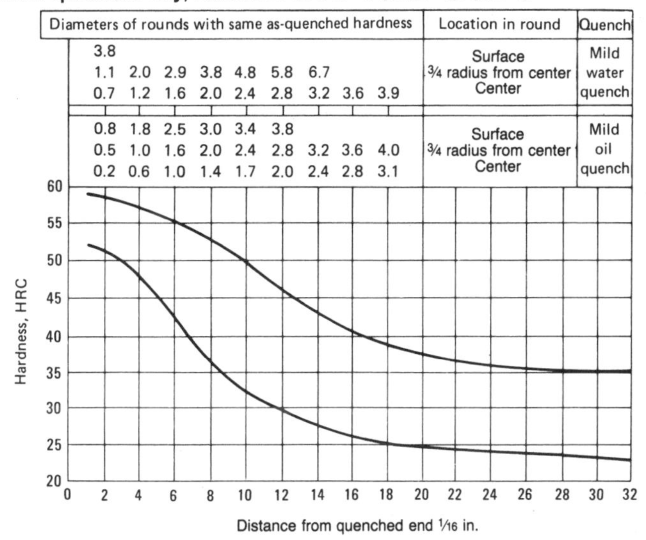

Mechanical Properties (continued)

AISI 8637, 8637H: Hardness and Machinability (Ref 7)

Condition	Hardness range, HB	Average machinability rating(a)
Annealed and cold drawn	179-229	65

(a) Based on AISI 1212 steel as 100% average machinability

Machining Data

For machining data on AISI grades 8637 and 8637H, refer to the preceding machining tables for AISI grades 1330 and 1330H.

AISI 8640, 8640H

AISI 8640, 8640H: Chemical Composition

AISI grade	C	Mn	P max	S max	Si	Ni	Cr	Mo
	Chemical composition, %							
8640	0.38-0.43	0.75-1.00	0.035	0.040	0.15-0.30	0.40-0.70	0.40-0.60	0.15-0.25
8640H	0.37-0.44	0.70-1.05	0.035	0.040	0.15-0.30	0.35-0.75	0.35-0.65	0.15-0.25

Characteristics. AISI 8640 and 8640H are nickel-chromium-molybdenum alloy steels with medium hardenability characteristics. These grades are available as hot rolled and cold finished bar, rod, and wire; seamless mechanical tubing; hot and cold rolled sheet and strip; bar subject to end-quench hardenability requirements; and aircraft quality stock.

Typical Uses. AISI 8640 and 8640H steels are used for medium and large parts requiring a high degree of strength and toughness; for average size automotive parts, such as steering knuckles, axle shafts, and propeller shafts; and for aircraft parts. Electric furnace grade E8640 is produced as aircraft quality, killed, fine-grain steels, which are utilized in manufacturing cap, socket, and recessed-head aircraft screws, and similar severe-duty cold formed parts.

AISI 8640: Similar Steels (U.S. and/or Foreign). UNS G86400; ASTM A322, A331, A505, A519; MIL SPEC MIL-S-16974; SAE J404, J412, J770; (W. Ger.) DIN 1.6546; (Ital.) UNI 40 NiCrMo 2 KB; (U.K.) B.S. Type 7

AISI 8640H: Similar Steels (U.S. and/or Foreign). UNS H86400; ASTM A304; SAE J1268; (W. Ger.) DIN 1.6546; (Ital.) UNI 40 NiCrMo 2 KB; (U.K.) B.S. Type 7

Physical Properties

AISI 8640, 8640H: Thermal Treatment Temperatures

Quenching medium: oil

Treatment	Temperature range °C	°F
Forging	1205 max	2200 max
Annealing(a)	790-845	1450-1550
Normalizing(b)	845-900	1550-1650
Austenitizing	830-855	1525-1575
Tempering	(c)	(c)

(a) Maximum hardness of 197 HB. (b) Hardness of approximately 302 HB. (c) To desired hardness

AISI 8640, 8640H: Approximate Critical Points

Transformation point	Temperature(a) °C	°F	Transformation point	Temperature(a) °C	°F
Ac_1	730	1350	Ar_1	665	1230
Ac_3	780	1435	M_s	320	610
Ar_3	725	1340			

(a) On heating or cooling at 28 °C (50 °F) per hour

Machining Data

For machining data on AISI grades 8640 and 8640H, refer to the preceding machining tables for AISI grades 1340 and 1340H.

Mechanical Properties

AISI 8640, 8640H: Low-Temperature Impact Properties (Ref 17)

Quenched from 845 °C (1550 °F)

Tempering temperature °C	°F	Hardness, HRC	Charpy impact (V-notch) energy at temperature of: −185 °C (−300 °F) J	ft·lb	−130 °C (−200 °F) J	ft·lb	−73 °C (−100 °F) J	ft·lb	−18 °C (0 °F) J	ft·lb	38 °C (100 °F) J	ft·lb	Transition temperature(a) °C	°F
425	800	46	7	5	14	10	19	14	27	20	31	23	...	...
540	1000	38	15	11	20	15	33	24	54	40	54	40	−79	−110
650	1200	30	24	18	30	22	66	49	85	63	89	66	−96	−140

(a) 50% brittle

AISI 8640: Effect of the Mass on Tensile Properties After Tempering at 540 °C (1000 °F). Normalized at 870 °C (1600 °F) in oversize rounds; oil quenched from 845 °C (1550 °F) in sizes indicated; tempered at 540 °C (1000 °F). Specimens were tested in 12.8-mm (0.505-in.) rounds. Tests from bar 38 mm (1.5 in.) and larger were taken at half-radius position. Tests were conducted using specimens machined to English units. Elongation was measured in 50 mm (2 in.). (Ref 2)

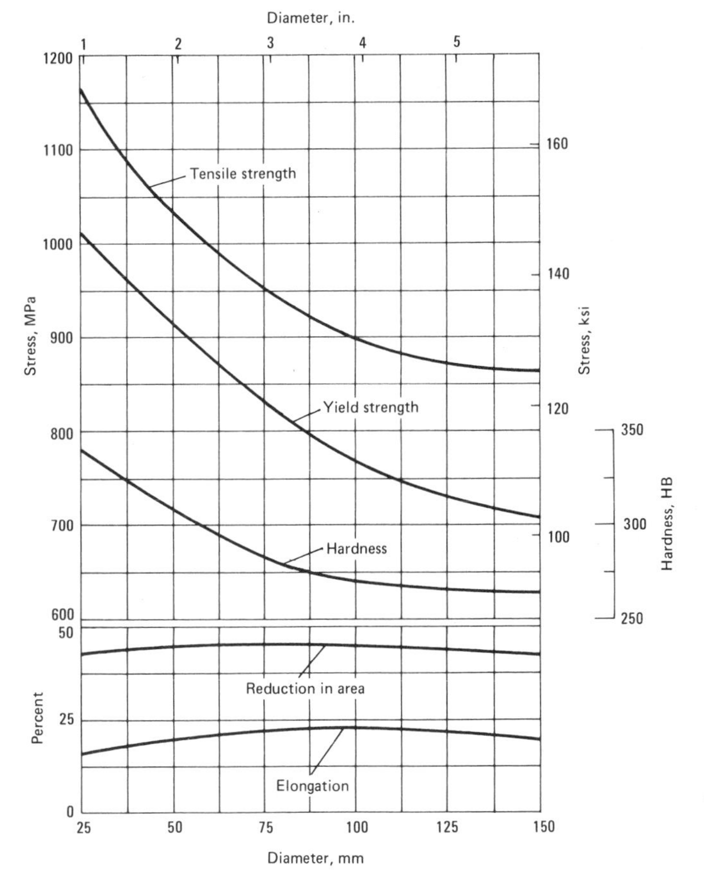

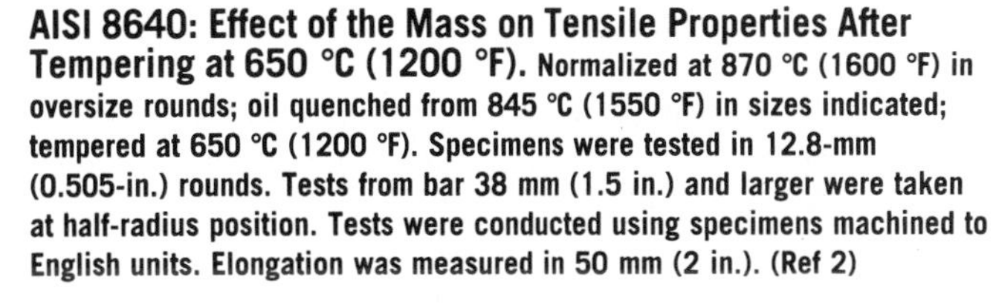

AISI 8640: Effect of the Mass on Tensile Properties After Tempering at 650 °C (1200 °F). Normalized at 870 °C (1600 °F) in oversize rounds; oil quenched from 845 °C (1550 °F) in sizes indicated; tempered at 650 °C (1200 °F). Specimens were tested in 12.8-mm (0.505-in.) rounds. Tests from bar 38 mm (1.5 in.) and larger were taken at half-radius position. Tests were conducted using specimens machined to English units. Elongation was measured in 50 mm (2 in.). (Ref 2)

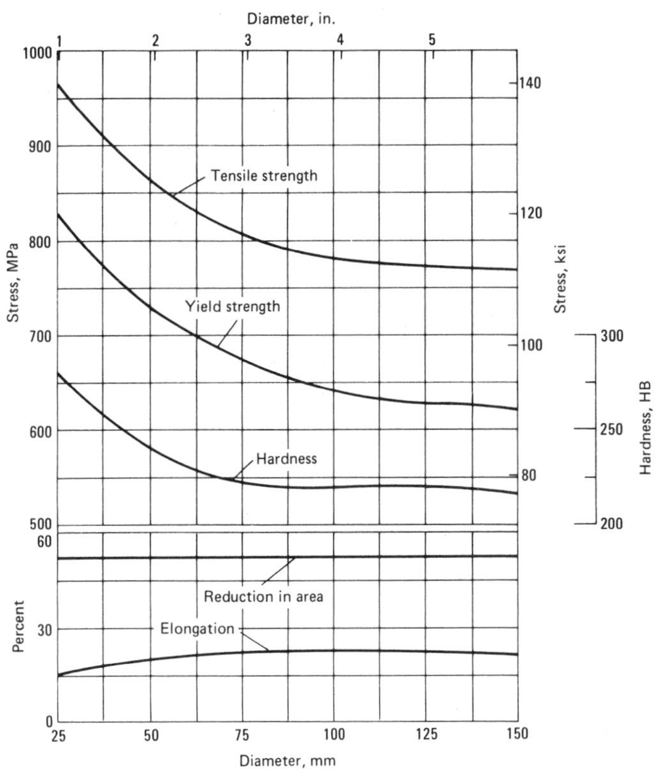

AISI 8640, 8640H: Tensile Properties (Ref 15)

Condition or treatment	Size round mm	Size round in.	Tensile strength MPa	Tensile strength ksi	Yield strength MPa	Yield strength ksi	Elongation(a), %	Reduction in area, %	Hardness, HB
Cold drawn	25	1	945	137	931	135	9	43	277
Oil quenched from 845 °C (1550 °F); tempered at 540 °C (1000 °F)	25	1	1165	169	1000	145	17	44	342
	50	2	1035	150	910	132	20	45	310
	75	3	945	137	814	118	22	45	283
	100	4	889	129	752	109	23	45	270
Oil quenched from 845 °C (1550 °F); tempered at 650 °C (1200 °F)	25	1	972	141	827	120	16	52	280
	50	2	848	123	717	104	19	52	221
	75	3	793	115	655	95	21	52	211
	100	4	772	112	635	92	22	53	219

(a) In 50 mm (2 in.)

AISI 8640, 8640H: Hardness and Machinability (Ref 7)

Condition	Hardness range, HB	Average machinability rating(a)
Annealed and cold drawn	184-229	65

(a) Based on AISI 1212 steel as 100% average machinability

Mechanical Properties (continued)

AISI 8640H: End-Quench Hardenability Limits and Hardenability Band. Heat treating temperatures recommended by SAE: normalize at 870 °C (1600 °F), for forged or rolled specimens only; austenitize at 845 °C (1550 °F). (Ref 1)

Distance from quenched end, 1/16 in.	Hardness, HRC max	Hardness, HRC min
1	60	53
2	60	53
3	60	52
4	59	51
5	59	49
6	58	46
7	57	42
8	55	39
9	54	36
10	52	34
11	50	32
12	49	31
13	47	30
14	45	29
15	44	28
16	42	28
18	41	26
20	39	26
22	38	25
24	38	25
26	37	24
28	37	24
30	37	24
32	37	24

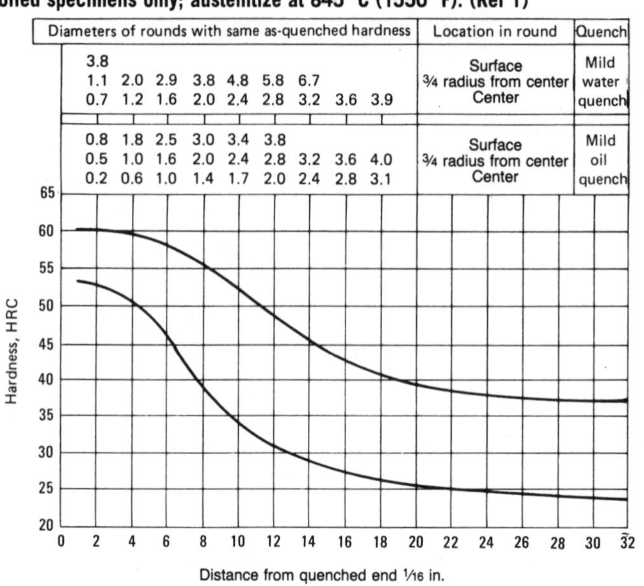

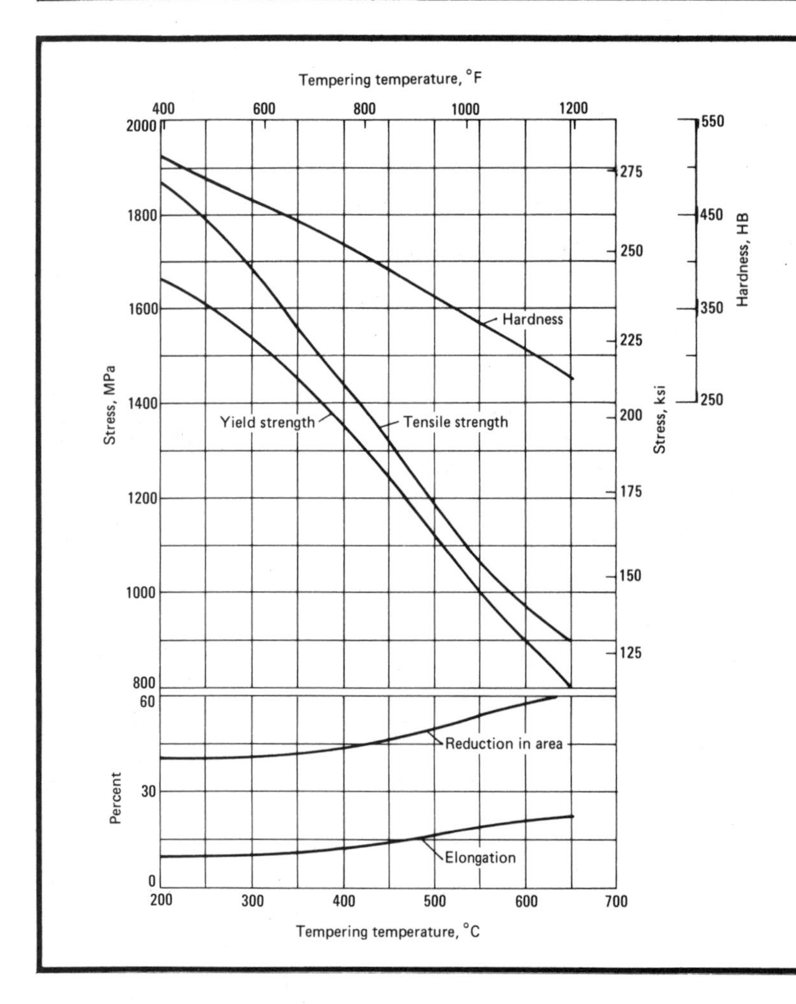

AISI 8640: Effect of Tempering Temperature on Tensile Properties. Oil quenched at 875 °C (1550 °F) in 13-mm (0.5-in.) sections. Elongation was measured in 50 mm (2 in.); yield strength at 0.2% offset. (Ref 14)

AISI 8642, 8642H

AISI 8642, 8642H: Chemical Composition

AISI grade	C	Mn	P max	S max	Si	Ni	Cr	Mo
	Chemical composition, %							
8642	0.40-0.45	0.75-1.00	0.035	0.040	0.15-0.30	0.40-0.70	0.40-0.60	0.15-0.25
8642H	0.39-0.46	0.70-1.05	0.035	0.040	0.15-0.30	0.35-0.75	0.35-0.65	0.15-0.25

Characteristics and Typical Uses. AISI grades 8642 and 8642H are directly hardenable, triple-alloy steels with medium hardenability. Other characteristics and uses of these steels are similar to those described for AISI grades 8640 and 8640H.

AISI 8642 and 8642H steels are available as hot rolled and cold finished bar, rod, and wire; seamless mechanical tubing; bar subject to end-quench hardenability requirements; and, on a limited basis, hot and cold rolled sheet and strip.

AISI 8642: Similar Steels (U.S. and/or Foreign). UNS G86420; ASTM A322, A331, A505, A519; SAE J404, J412, J770

AISI 8642H: Similar Steels (U.S. and/or Foreign). UNS H86420; ASTM A304; SAE J1268

AISI 8642, 8642H: Approximate Critical Points

Transformation point	Temperature(a) °C	Temperature(a) °F
Ac_1	730	1350
Ac_3	780	1435
Ar_3	715	1320
Ar_1	665	1230

(a) On heating or cooling at 28 °C (50 °F) per hour

Physical Properties

AISI 8642, 8642H: Thermal Treatment Temperatures

Quenching medium: oil

Treatment	Temperature range °C	Temperature range °F
Forging	1205 max	2200 max
Annealing	845-870	1550-1600
Austenitizing	815-855	1500-1575
Tempering	(a)	(a)

(a) To desired hardness

Machining Data

For machining data on AISI grades 8642 and 8642H, refer to the preceding machining tables for AISI grades 1340 and 1340H.

Mechanical Properties

AISI 8642H: End-Quench Hardenability Limits and Hardenability Band.

Heat treating temperatures recommended by SAE: normalize at 870 °C (1600 °F), for forged or rolled specimens only; austenitize at 845 °C (1550 °F). (Ref 1)

Distance from quenched end, 1/16 in.	Hardness, HRC max	Hardness, HRC min
1	62	55
2	62	54
3	62	53
4	61	52
5	61	50
6	60	48
7	59	45
8	58	42
9	57	39
10	55	37
11	54	34
12	52	33
13	50	32
14	49	31
15	48	30
16	46	29
18	44	28
20	42	28
22	41	27
24	40	27
26	40	26
28	39	26
30	39	26
32	39	26

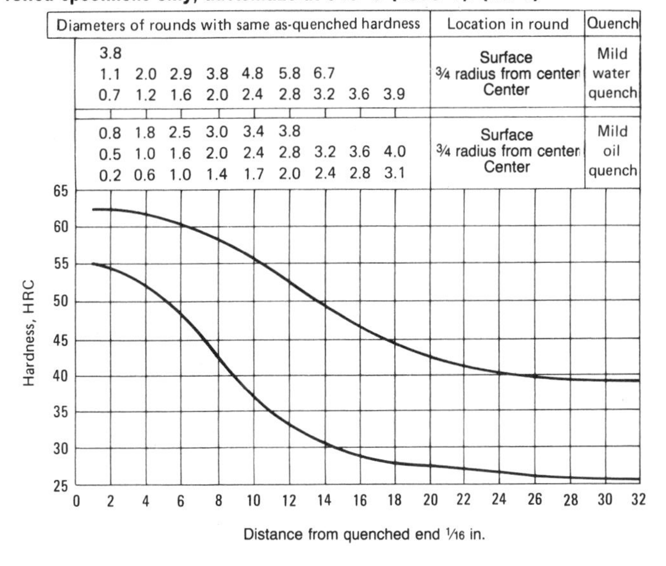

AISI 8642, 8642H: Hardness and Machinability (Ref 7)

Condition	Hardness range, HB	Average machinability rating(a)
Annealed and cold drawn	184-229	65

(a) Based on AISI 1212 steel as 100% average machinability

Mechanical Properties (continued)

AISI 8642, 8642H: Mass Effect on Mechanical Properties

Condition or treatment	Size round mm	in.	Tensile strength MPa	ksi	Yield strength MPa	ksi	Elongation(a), %	Reduction in area, %	Hardness, HB
Oil quenched from 845 °C (1550 °F); tempered at 540 °C (1000 °F)	25	1	1240	180	1110	161	15.0	46.1	363
	50	2	1125	163	986	143	16.8	49.5	331
	75	3	1040	151	889	129	18.2	52.0	306
	100	4	972	141	814	118	19.2	54.0	285
Oil quenched from 845 °C (1550 °F); tempered at 595 °C (1100 °F)	25	1	1095	159	945	137	17.2	50.5	321
	50	2	1040	151	834	121	18.2	52.0	306
	75	3	951	138	786	114	19.7	54.6	277
	100	4	896	130	731	106	20.6	56.3	262
Oil quenched from 845 °C (1550 °F); tempered at 650 °C (1200 °F)	25	1	972	141	814	118	19.2	54.6	385
	50	2	903	131	738	107	20.4	56.0	269
	75	3	855	124	685	99	21.5	57.5	248
	100	4	807	117	625	91	22.0	59.0	235

(a) In 50 mm (2 in.)

AISI 8645, 8645H, 86B45, 86B45H

AISI 8645, 8645H, 86B45, 86B45H: Chemical Composition

AISI grade	C	Mn	P max	S max	Si	Ni	Cr	Mo
8645	0.43-0.48	0.75-1.00	0.035	0.040	0.15-0.30	0.40-0.70	0.40-0.60	0.15-0.25
8645H	0.42-0.49	0.70-1.05	0.035	0.040	0.15-0.30	0.35-0.75	0.35-0.65	0.15-0.25
86B45 (a)	0.43-0.48	0.75-1.00	0.035	0.040	0.15-0.30	0.40-0.70	0.40-0.60	0.15-0.25
86B45H (a)	0.42-0.49	0.70-1.05	0.035	0.040	0.15-0.30	0.35-0.75	0.35-0.65	0.15-0.25

(a) Can be expected to contain 0.0005 to 0.003% boron

Characteristics. AISI grades 8645, 8645H, 86B45, and 86B45H are directly hardenable, nickel-chromium-molybdenum alloy steels with medium hardenability. The addition of boron to AISI 86B45 and 86B45H increases maximum hardenability.

Typical Uses. AISI 8645, 8645H, 86B45, and 86B45H steels are used for springs and parts such as gears, which require fairly high hardness, as well as strength and toughness. These steels are available as hot rolled and cold finished bar, rod, and wire; seamless mechanical tubing; bar subject to end-quench hardenability requirements; and, on a limited basis, hot and cold rolled sheet and strip.

AISI 8645: Similar Steels (U.S. and/or Foreign). UNS G86450; ASTM A322, A331, A505, A519; MIL SPEC MIL-S-16974; SAE J404, J412, J770

AISI 8645H: Similar Steels (U.S. and/or Foreign). UNS H86450; ASTM A304; SAE J1268

AISI 86B45: Similar Steels (U.S. and/or Foreign). UNS G86451; ASTM A519; SAE J404, J412, J770

AISI 86B45H: Similar Steels (U.S. and/or Foreign). UNS H86451; ASTM A304; SAE J1268

AISI 8645, 8645H, 86B45, 86B45H: Approximate Critical Points

Transformation point	Temperature(a) AISI 8645, 8645H °C	°F	AISI 86B45, 86B45H °C	°F
Ac_1	730	1350	720	1330
Ac_3	775	1430	770	1420
Ar_3	710	1310	695	1280
Ar_1	665	1230	650	1200
M_s	300	575	...	...

(a) On heating or cooling at 28 °C (50 °F) per hour

Physical Properties

AISI 8645, 8645H, 86B45, 86B45H: Thermal Treatment Temperatures

Quenching medium: oil

Treatment	Temperature range °C	°F
Forging	1205 max	2200 max
Annealing(a)	790-845	1450-1550
Normalizing(b)	845-900	1550-1650
Austenitizing	830-855	1525-1575
Tempering	(c)	(c)

(a) Maximum hardness of 207 HB. (b) Hardness of approximately 277 HB. (c) To desired hardness

Mechanical Properties

AISI 8645H: End-Quench Hardenability Limits and Hardenability Band. Heat treating temperatures recommended by SAE: normalize at 870 °C (1600 °F), for forged or rolled specimens only; austenitize at 845 °C (1550 °F). (Ref 1)

Distance from quenched end, 1/16 in.	Hardness, HRC max	Hardness, HRC min
1	63	56
2	63	56
3	63	55
4	63	54
5	62	52
6	61	50
7	61	48
8	60	45
9	59	41
10	58	39
11	56	37
12	55	35
13	54	34
14	52	33
15	51	32
16	49	31
18	47	30
20	45	29
22	43	28
24	42	28
26	42	27
28	41	27
30	41	27
32	41	27

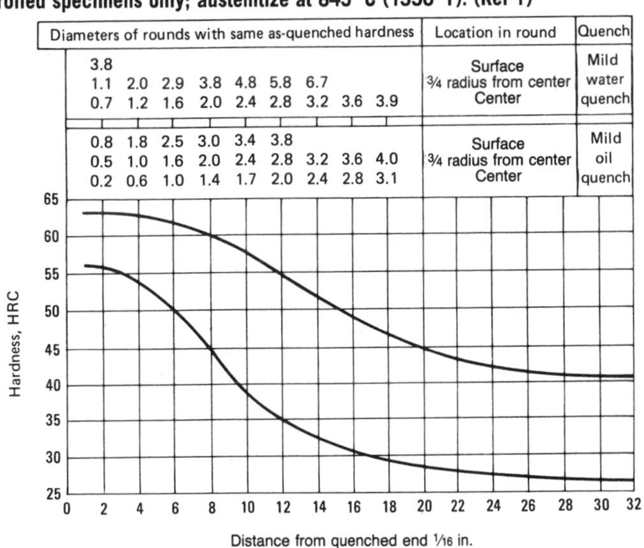

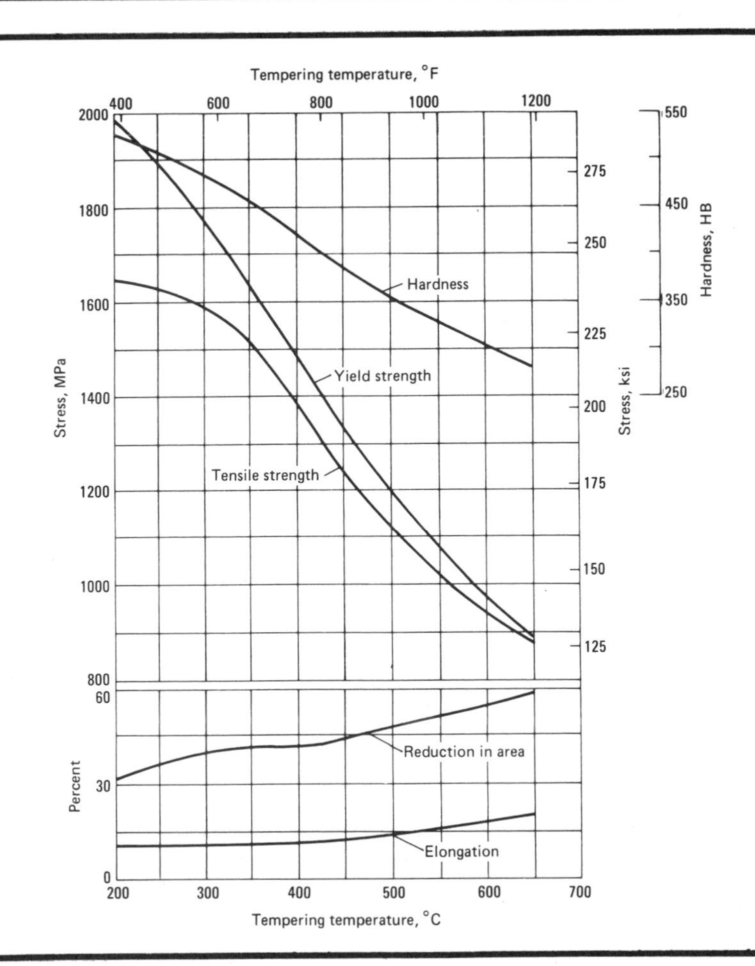

AISI 86B45: Effect of Tempering Temperature on Tensile Properties. Normalized at 870 °C (1600 °F); oil quenched from 845 °C (1550 °F); tempered at 56 °C (100 °F) intervals. Specimens were treated in 13.7-mm (0.540-in.) diam and machined to 12.8-mm (0.505-in.) diam for testing. Tests were conducted using specimens machined to English units. Elongation was measured in 50 mm (2 in.). (Ref 2)

AISI 8650H: End-Quench Hardenability Limits and Hardenability Band.

Heat treating temperatures recommended by SAE: normalize at 870 °C (1600 °F), for forged or rolled specimens only; austenitize at 845 °C (1550 °F). (Ref 1)

Distance from quenched end, 1/16 in.	Hardness, HRC max	min
1	65	59
2	65	58
3	65	57
4	64	57
5	64	56
6	63	54
7	63	53
8	62	50
9	61	47
10	60	44
11	60	41
12	59	39
13	58	37
14	58	36
15	57	35
16	56	34
18	55	33
20	53	32
22	52	31
24	50	31
26	49	30
28	47	30
30	46	29
32	45	29

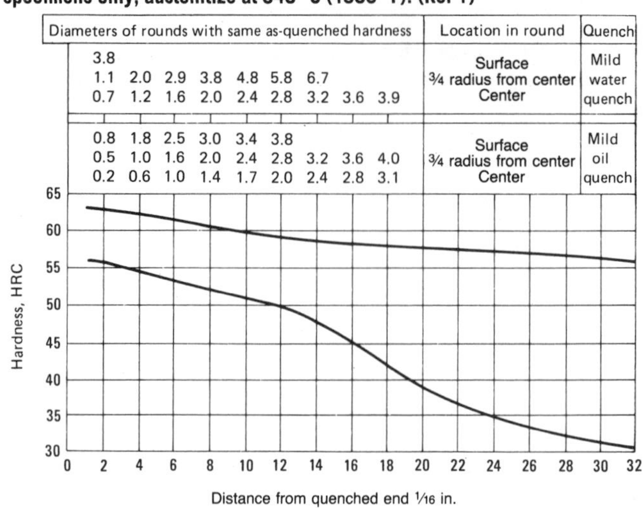

AISI 86B45: Effect of Mass on Tensile Properties After Tempering at 540 °C (1000 °F).

Normalized at 870 °C (1600 °F) in oversize rounds; oil quenched from 845 °C (1550 °F) in sizes indicated; tempered at 540 °C (1000 °F). Test specimens were 12.8-mm (0.505-in.) rounds. Tests from bar 38 mm (1.5 in.) and larger were taken at half-radius position. Tests were conducted using specimens machined to English units. Elongation was measured in 50 mm (2 in.). (Ref 2)

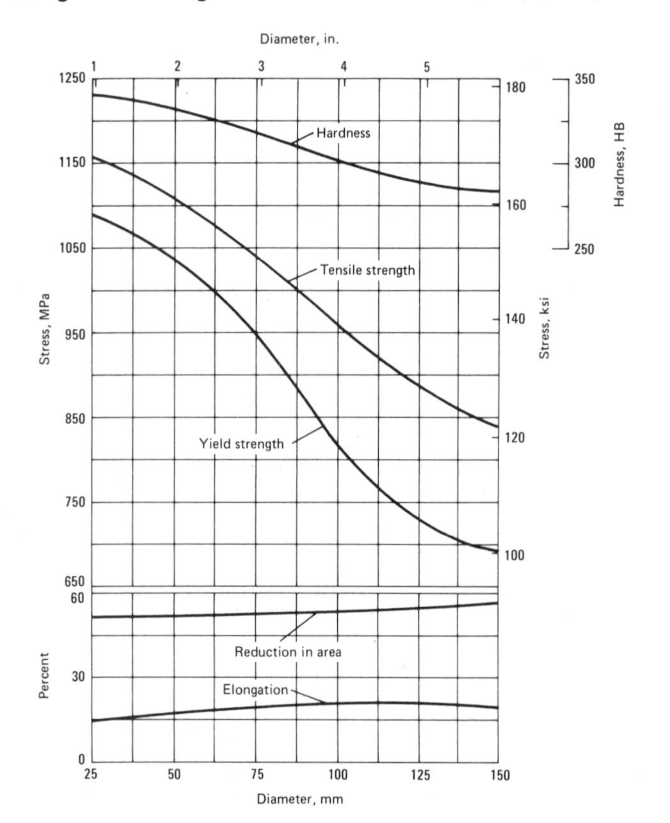

AISI 86B45: Effect of Mass on Tensile Properties After Tempering at 650 °C (1200 °F).

Normalized at 870 °C (1600 °F) in oversize rounds; oil quenched from 845 °C (1550 °F) in sizes indicated; tempered at 650 °C (1200 °F). Test specimens were 12.8-mm (0.505-in.) rounds. Tests from bar 38 mm (1.5 in.) and larger were taken at half-radius position. Tests were conducted using specimens machined to English units. Elongation was measured in 50 mm (2 in.). (Ref 2)

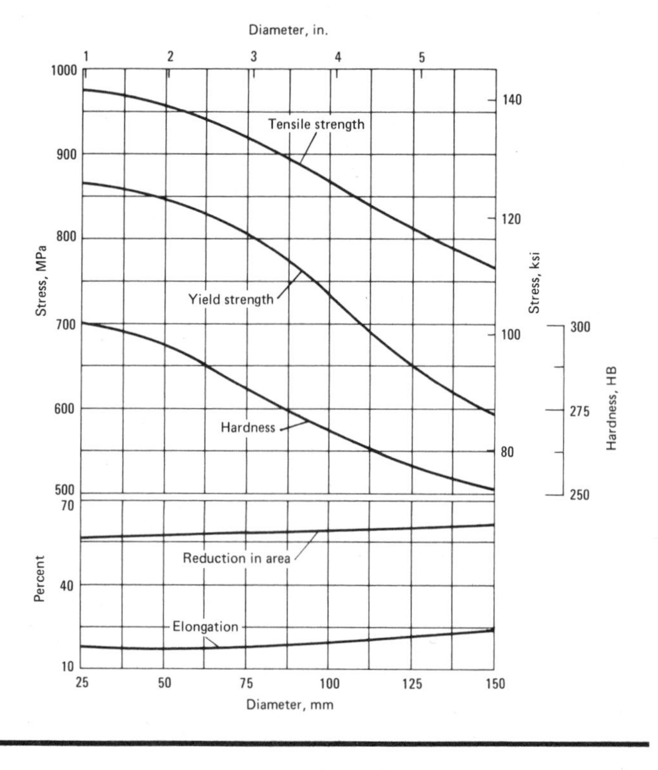

Mechanical Properties (continued)

AISI 8645, 8645H, 86B45, 86B45H: Hardness and Machinability (Ref 7)

Condition	Hardness range, HB	Average machinability rating(a)
Annealed and cold drawn	184-235	65

(a) Based on AISI 1212 steel as 100% average machinability

Machining Data

For machining data on AISI grades 8645, 8645H, 86B45, and 86B45H, refer to the preceding machining tables for AISI grades 1340 and 1340H.

AISI 8650, 8650H, 86B50

AISI 8650, 8650H, 86B50: Chemical Composition

AISI grade	Chemical composition, % C	Mn	P max	S max	Si	Ni	Cr	Mo
8650	0.48-0.53	0.75-1.00	0.035	0.045	0.15-0.30	0.40-0.70	0.40-0.60	0.15-0.25
8650H	0.47-0.54	0.70-1.05	0.035	0.045	0.15-0.35	0.35-0.75	0.35-0.65	0.15-0.25
86B50	0.47-0.54	0.70-1.05	0.035	0.045	0.15-0.35	0.35-0.75	0.35-0.65	0.15-0.25

(a) Can be expected to contain 0.0005 to 0.003% boron

Characteristics. AISI grades 8650, 8650H, and 86B50 are directly hardenable, triple-alloy steels with medium hardenability. The addition of boron to AISI 86B50 enhances hardenability of that grade.

Typical Uses. AISI 8650, 8650H, and 86B50 steels are available on a limited basis as hot and cold rolled strip and sheet; hot rolled and cold finished bar, rod, and wire; and bar subject to end-quench hardenability requirements. Primary uses include springs, hand tools, and automotive axle shafts.

AISI 8650: Similar Steels (U.S. and/or Foreign). UNS G86500; ASTM A322, A505, A519; SAE J404, J412, J770

AISI 8650H: Similar Steels (U.S. and/or Foreign). UNS H86500; ASTM A304; SAE J1268

AISI 86B50: Similar Steels (U.S. and/or Foreign). UNS G86501

AISI 8650, 8650H, 86B50: Approximate Critical Points

Transformation point	Temperature(a) °C	°F	Transformation point	Temperature(a) °C	°F
Ac_1	730	1350	Ar_1	655	1210
Ac_3	770	1420	M_s	285	545
Ar_3	700	1295			

(a) On heating or cooling at 28 °C (50 °F) per hour

Machining Data

For machining data on AISI grades 8650, 8650H, and 86B50, refer to the preceding machining tables for AISI grades 4150 and 4150H.

Physical Properties

AISI 8650, 8650H, 86B50: Thermal Treatment Temperatures

Quenching medium: oil

Treatment	Temperature range °C	°F
Forging	1175 max	2150 max
Annealing(a)	790-845	1450-1550
Normalizing(b)	845-900	1550-1650
Austenitizing	830-855	1525-1575
Tempering	(c)	(c)

(a) Maximum hardness of 212 HB. (b) Hardness of approximately 355 HB. (c) To desired hardness

Mechanical Properties

AISI 8650, 8650H, 86B50: Tensile Properties (Ref 5)

Condition or treatment	Tensile strength MPa	ksi	Yield strength MPa	ksi	Elongation(a), %	Reduction in area, %	Hardness, HB	Izod impact energy J	ft·lb
Normalized at 870 °C (1600 °F)	1025	149	690	100	14.0	45	302	14	10
Annealed at 795 °C (1465 °F)	715	104	385	56	22.5	46	212	29	22

(a) In 50 mm (2 in.)

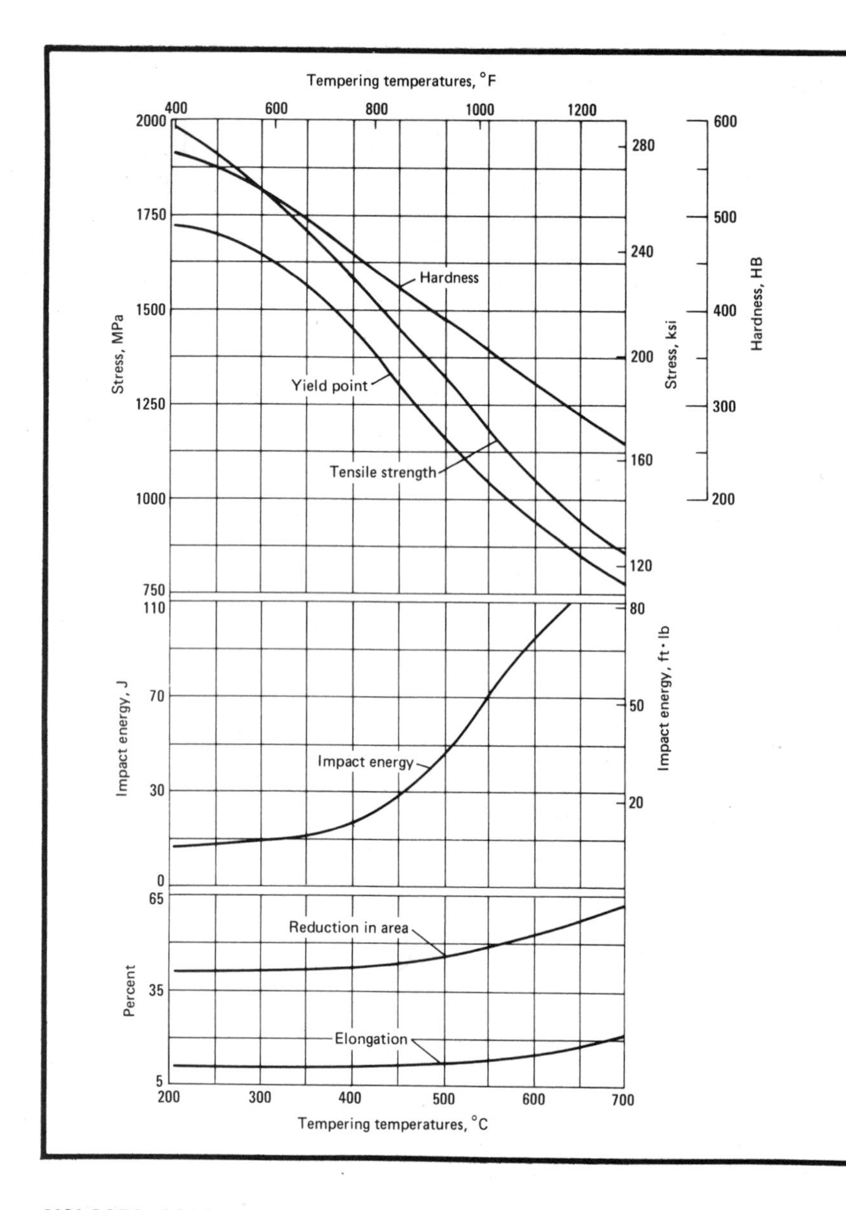

AISI 8650: Effect of Tempering Temperature on Tensile Properties. Normalized at 870 °C (1600 °F); reheated to 800 °C (1475 °F); quenched in agitated oil. Specimens were 13-mm (0.5-in.) sections treated in 13.5-mm (0.530-in.) diam and machined to 12.8-mm (0.505-in.) diam for testing. Tests were conducted using specimens machined to English units. As-quenched hardness of 638 HB; elongation measured in 50 mm (2 in.). (Ref 14)

AISI 8650, 8650H, 86B50: Mass Effect on Mechanical Properties (Ref 4)

Condition or treatment	Size round mm	Size round in.	Tensile strength MPa	Tensile strength ksi	Yield strength MPa	Yield strength ksi	Elongation(a), %	Reduction in area, %	Hardness, HB
Annealed (heated to 795 °C or 1465 °F; furnace cooled 11 °C or 20 °F per hour to 460 °C or 860 °F; cooled in air)	25	1	1255	182	1076	156	22.5	46.4	212
Normalized (heated to 870 °C or 1600 °F; cooled in air)	13	0.5	1255	182	903	131	10.3	25.3	363
	25	1	1020	148	689	100	14.0	40.4	302
	50	2	993	144	660	96	15.5	44.8	293
	100	4	958	139	640	93	15.0	40.5	285
Oil quenched from 800 °C (1475 °F); tempered at 540 °C (1000 °F)	13	0.5	1225	178	1165	169	14.6	48.2	363
	25	1	1185	172	1105	160	14.5	49.1	352
	50	2	1140	165	1020	148	17.0	55.6	331
	100	4	986	143	779	113	18.7	54.9	285
Oil quenched from 800 °C (1475 °F); tempered at 595 °C (1100 °F)	13	0.5	1060	154	1040	151	17.8	54.9	321
	25	1	1060	154	986	143	17.7	57.3	311
	50	2	1000	145	903	131	20.0	61.0	293
	100	4	869	126	675	98	22.0	61.2	255
Oil quenched from 800 °C (1475 °F); tempered at 650 °C (1200 °F)	13	0.5	1020	148	945	137	18.5	54.8	293
	25	1	972	141	910	132	19.5	59.8	285
	50	2	931	135	834	121	21.2	62.3	277
	100	4	841	122	650	94	22.5	59.8	241

(a) In 50 mm (2 in.)

Mechanical Properties (continued)

AISI 8650, 8650H, 86B50: Effect of the Mass on Hardness at Selected Points (Ref 4)

Size round mm	in.	As-quenched hardness after quenching in oil at: Surface	½ radius	Center
13	0.5	61 HRC	61 HRC	61 HRC
25	1	58 HRC	58 HRC	57 HRC
50	2	53 HRC	53 HRC	52 HRC
100	4	42 HRC	39 HRC	38 HRC

AISI 8650, 8650H, 86B50: Hardness and Machinability (Ref 7)

Condition	Hardness range, HB	Average machinability rating(a)
Annealed and cold drawn	187-248	60

(a) Based on AISI 1212 steel as 100% average machinability

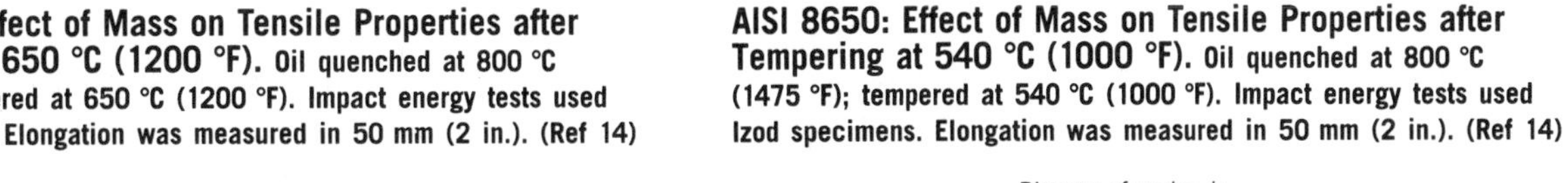

AISI 8650: Effect of Mass on Tensile Properties after Tempering at 650 °C (1200 °F). Oil quenched at 800 °C (1475 °F); tempered at 650 °C (1200 °F). Impact energy tests used Izod specimens. Elongation was measured in 50 mm (2 in.). (Ref 14)

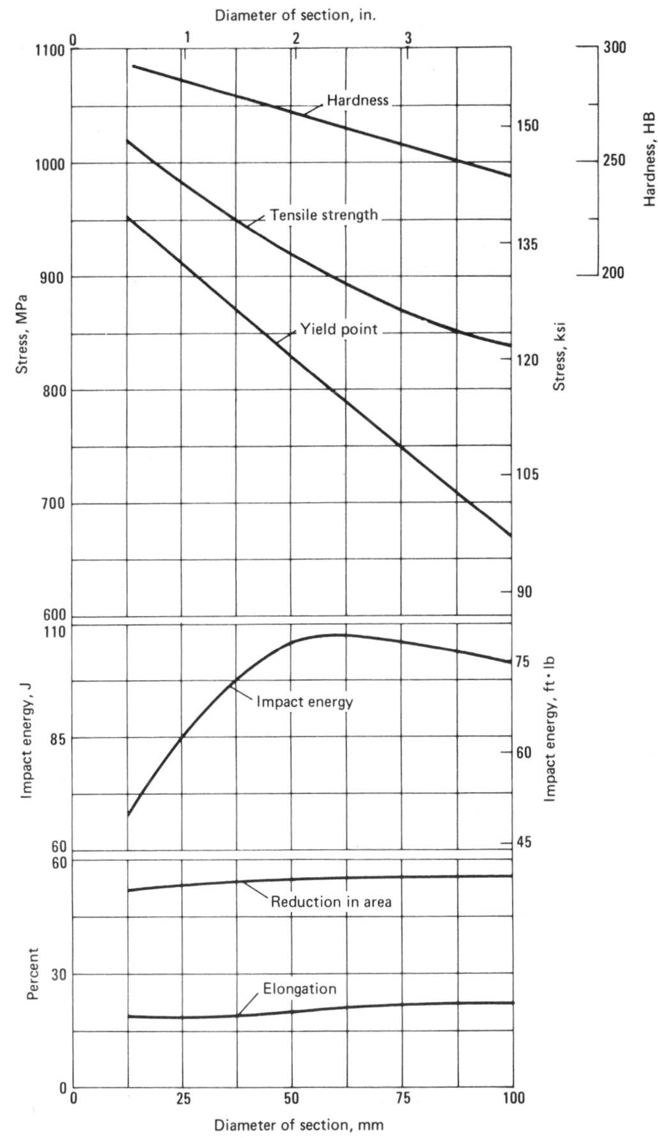

AISI 8650: Effect of Mass on Tensile Properties after Tempering at 540 °C (1000 °F). Oil quenched at 800 °C (1475 °F); tempered at 540 °C (1000 °F). Impact energy tests used Izod specimens. Elongation was measured in 50 mm (2 in.). (Ref 14)

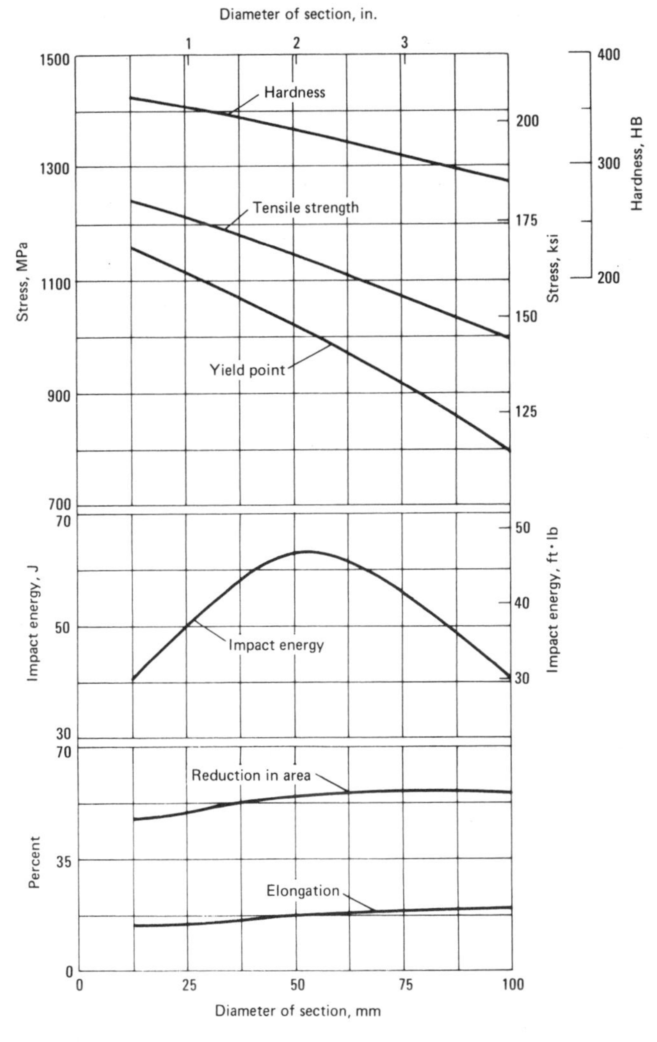

Mechanical Properties (continued)

AISI 8650H: End-Quench Hardenability Limits and Hardenability Band. Heat treating temperatures recommended by SAE: normalize at 870 °C (1600 °F), for forged or rolled specimens only; austenitize at 845 °C (1550 °F). (Ref 1)

Distance from quenched end, 1/16 in.	Hardness, HRC max	Hardness, HRC min
1	65	59
2	65	58
3	65	57
4	64	57
5	64	56
6	63	54
7	63	53
8	62	50
9	61	47
10	60	44
11	60	41
12	59	39
13	58	37
14	58	36
15	57	35
16	56	34
18	55	33
20	53	32
22	52	31
24	50	31
26	49	30
28	47	30
30	46	29
32	45	29

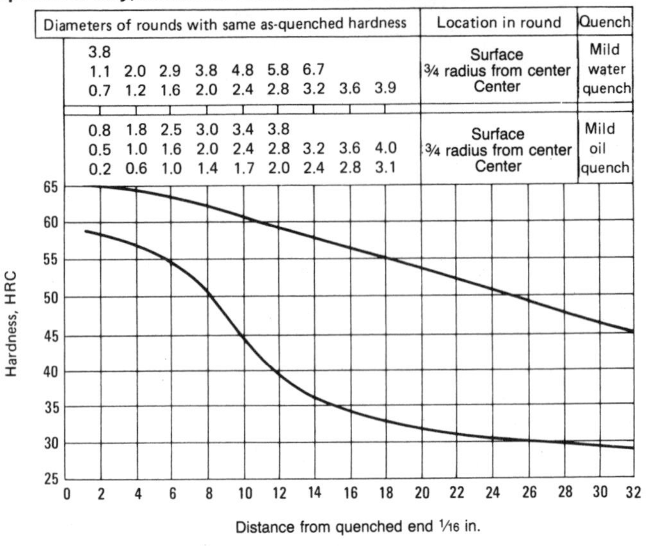

AISI 8655, 8655H

AISI 8655, 8655H: Chemical Composition

AISI grade	C	Mn	P max	S max	Si	Ni	Cr	Mo
8655	0.51-0.59	0.75-1.00	0.035	0.050	0.15-0.30	0.40-0.70	0.40-0.60	0.15-0.25
8655H	0.50-0.60	0.70-1.05	0.035	0.050	0.15-0.30	0.35-0.75	0.35-0.65	0.15-0.25

(Chemical composition, %)

Characteristics and Typical Uses. AISI 8655 and 8655H are directly hardenable, triple-alloy steels with high hardenability. Primary uses of these grades are springs and hand tools.

AISI 8655 and 8655H are available as hot and cold rolled sheet and strip; hot rolled and cold finished bar; seamless mechanical tubing; and bar subject to end-quench hardenability requirements.

AISI 8655: Similar Steels (U.S. and/or Foreign). UNS G86550; ASTM A322, A331, A505; SAE J404, J412, J770

AISI 8655H: Similar Steels (U.S. and/or Foreign). UNS H86550; ASTM A304; SAE J1268

Machining Data

For machining data on AISI grades 8655 and 8655H, refer to the preceding machining tables for AISI grades 4150 and 4150H.

AISI 8655, 8655H: Approximate Critical Points

Transformation point	Temperature(a) °C	Temperature(a) °F
Ac_1	730	1345
Ac_3	765	1410
Ar_3	690	1270
Ar_1	660	1220
M_s	270	515

(a) On heating or cooling at 28 °C (50 °F) per hour

Mechanical Properties

AISI 8655, 8655H: Hardness and Machinability (Ref 7)

Condition	Hardness range, HB	Average machinability rating(a)
Annealed and cold drawn	187-248	55

(a) Based on AISI 1212 steel as 100% average machinability

Physical Properties

AISI 8655, 8655H: Thermal Treatment Temperatures

Quenching medium: oil

Treatment	Temperature range °C	Temperature range °F
Forging	1175 max	2150 max
Annealing	815-870	1500-1600
Austenitizing	800-845	1475-1550
Tempering	(a)	(a)

(a) To desired hardness

Mechanical Properties (continued)

AISI 8655H: End-Quench Hardenability Limits and Hardenability Band. Heat treating temperatures recommended by SAE: normalize at 870 °C (1600 °F), for forged or rolled specimens only; austenitize at 845 °C (1550 °F). (Ref 1)

Distance from quenched end, 1/16 in.	Hardness, HRC max	min
1	...	60
2	...	59
3	...	59
4	...	58
5	...	57
6	...	56
7	...	55
8	...	54
9	...	52
10	65	49
11	65	46
12	64	43
13	64	41
14	63	40
15	63	39
16	62	38
18	61	37
20	60	35
22	59	34
24	58	34
26	57	33
28	56	33
30	55	32
32	53	32

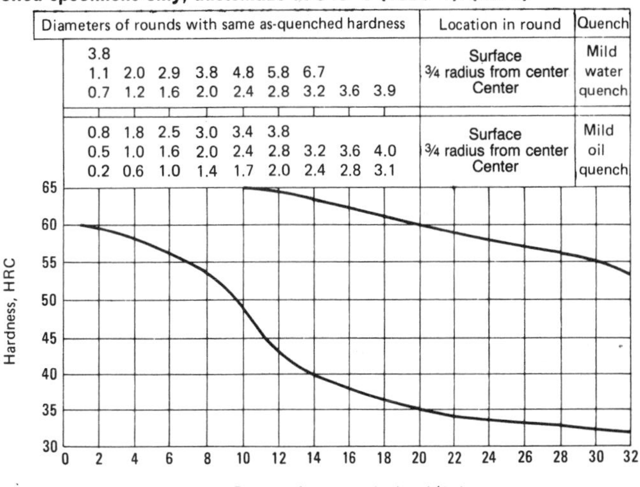

AISI 8660, 8660H

AISI 8660, 8660H: Chemical Composition

AISI grade	C	Mn	P max	S max	Si	Ni	Cr	Mo
8660	0.55-0.65	0.75-1.00	0.035	0.040	0.15-0.30	0.40-0.70	0.40-0.60	0.15-0.25
8660H	0.55-0.65	0.70-1.05	0.035	0.040	0.15-0.30	0.35-0.75	0.35-0.65	0.15-0.25

(Chemical composition, %)

Characteristics and Typical Uses. AISI 8660 and 8660H are directly hardenable, nickel-chromium-molybdenum alloy steels with good hardenability and surface qualities. These grades are used in a variety of parts, including springs and hand tools.

AISI 8660 and 8660H steels are available as hot rolled and cold finished bar, rod, and wire; seamless mechanical tubing; forging stock; bar subject to end-quench hardenability requirements; and hot and cold rolled sheet and strip.

AISI 8660: Similar Steels (U.S. and/or Foreign). UNS G86600; ASTM A322, A332, A505, A519, A711; SAE J404, J412, J770

AISI 8660H: Similar Steels (U.S. and/or Foreign). UNS H86600; ASTM A304; SAE J1268

Machining Data

For machining data on AISI grades 8660 and 8660H, refer to the preceding machining tables for AISI grades 4150 and 4150H.

AISI 8660, 8660H: Approximate Critical Points

Transformation point	Temperature(a) °C	°F
Ac_1	730	1345
Ac_3	765	1410
Ar_3	690	1270
Ar_1	665	1230
M_s	250	485

(a) On heating or cooling at 28 °C (50 °F) per hour

Mechanical Properties

AISI 8660, 8660H: Hardness and Machinability (Ref 7)

Condition	Hardness range, HB	Average machinability rating(a)
Spheroidized annealed and cold drawn	179-217	55

(a) Based on AISI 1212 steel as 100% average machinability

Physical Properties

AISI 8660, 8660H: Thermal Treatment Temperatures

Quenching medium: oil

Treatment	Temperature range °C	°F
Forging	1175 max	2150 max
Annealing (a)	790-845	1450-1550
Normalizing (b)	845-900	1550-1650
Austenitizing	830-855	1525-1575
Tempering	(c)	(c)

(a) Maximum hardness of 229 HB. (b) Hardness of approximately 321 HB. (c) To desired hardness

Mechanical Properties (continued)

AISI 8660H: End-Quench Hardenability Limits and Hardenability Band. Heat treating temperatures recommended by SAE: normalize at 870 °C (1600 °F), for forged or rolled specimens only; austenitize at 845 °C (1550 °F). (Ref 1)

Distance from quenched end, 1/16 in.	Hardness, HRC max	Hardness, HRC min
1	...	60
2	...	60
3	...	60
4	...	60
5	...	60
6	...	59
7	...	58
8	...	57
9	...	55
10	...	53
11	...	50
12	...	47
13	...	45
14	...	44
15	...	43
16	65	42
18	64	40
20	64	39
22	63	38
24	62	37
26	62	36
28	61	36
30	60	35
32	60	35

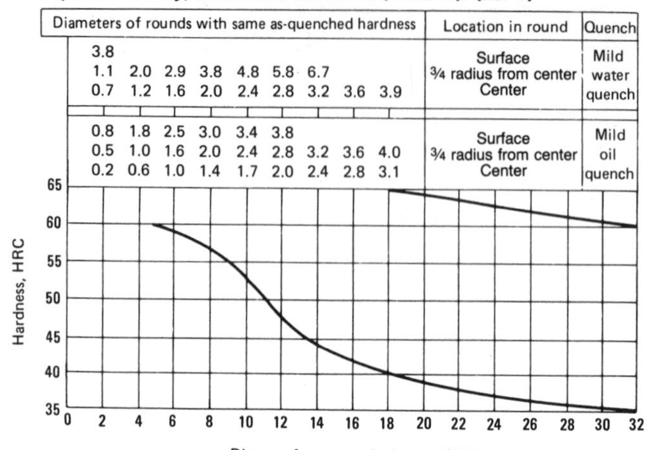

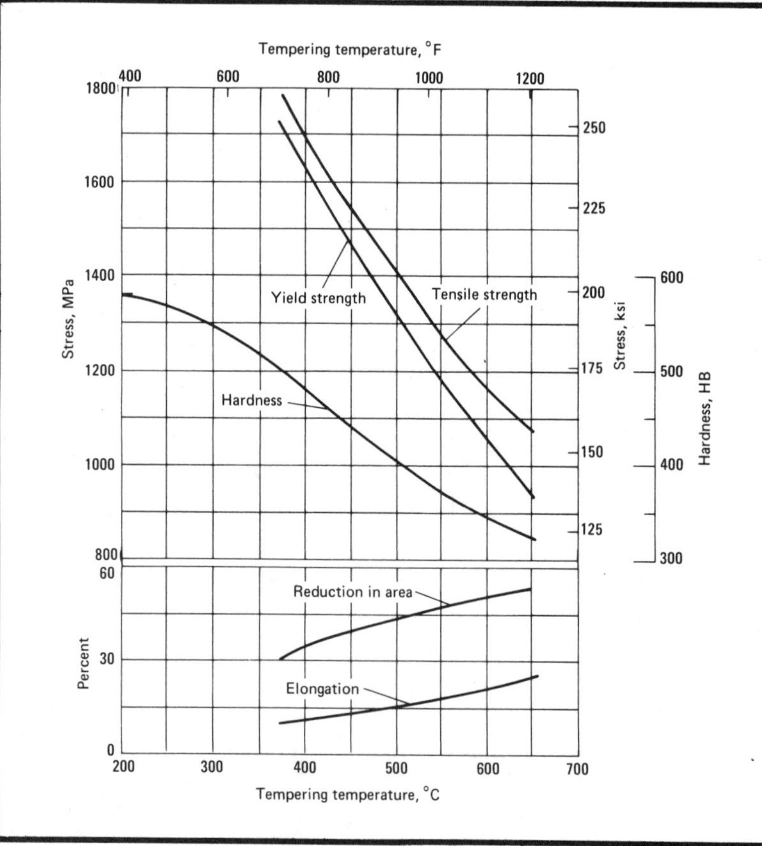

AISI 8660: Effect of Tempering Temperature on Tensile Properties. Normalized at 870 °C (1600 °F); oil quenched from 845 °C (1550 °F); tempered at 56 °C (100 °F) intervals. Specimens were treated in 13.7-mm (0.540-in.) diam and machined to 12.8-mm (0.505-in.) diam for testing. Tests were conducted using specimens machined to English units. Elongation was measured in 50 mm (2 in.). (Ref 2)

AISI 8660, 8660H: Low-Temperature Impact Properties (Ref 17)

Quenched from 800 °C (1475 °F)

Tempering temperature °C	°F	Hardness, HRC	−185 °C (−300 °F) J	ft·lb	−130 °C (−200 °F) J	ft·lb	−73 °C (−100 °F) J	ft·lb	−18 °C (0 °F) J	ft·lb	38 °C (100 °F) J	ft·lb	Transition temperature(a) °C	°F
			Charpy impact (V-notch) energy at temperature of:											
425	800	47	5	4	8	6	14	10	18	13	22	16	...	...
540	1000	41	14	10	16	12	20	15	27	20	41	30	−25	−10
650	1200	30	22	16	24	18	34	25	73	54	81	60	−70	−90

(a) 50% brittle

AISI 8720, 8720H

AISI 8720, 8720H: Chemical Composition

AISI grade	C	Mn	P max	S max	Si	Ni	Cr	Mo
	Chemical composition, %							
8720	0.18-0.23	0.70-0.90	0.035	0.040	0.15-0.30	0.40-0.70	0.40-0.60	0.20-0.30
8720H	0.17-0.23	0.60-0.95	0.035	0.040	0.15-0.30	0.35-0.75	0.35-0.65	0.20-0.30

Characteristics. AISI grades 8720 and 8720H are carburizing, triple-alloy steels with medium to high case hardenability and medium core hardenability. These steels have characteristics similar to AISI grade 8620, but are suitable for larger parts because of greater hardenability.

Typical Uses. AISI grades 8720 and 8720H are available as hot rolled and cold finished bar, rod, and wire; seamless mechanical tubing; bar subject to end-quench hardenability requirements; and hot and cold rolled sheet and strip. Typical uses include differential pinions, gears, steering worms, wear-resistant pump parts, and spline shafts.

AISI 8720: Similar Steels (U.S. and/or Foreign). UNS G87200; ASTM A322, A331, A505, A519; SAE J404, J770; (W. Ger.) DIN 1.6543; (U.K.) B.S. 805 A 20

AISI 8720H: Similar Steels (U.S. and/or Foreign). UNS H87200; ASTM A304; SAE J1268; (W. Ger.) DIN 1.6543; (U.K.) B.S. 805 A 20

Physical Properties

AISI 8720, 8720H: Thermal Treatment Temperatures

Quenching medium: oil

Treatment	Temperature range °C	°F
Forging	1230 max	2250 max
Cycle annealing(a)	855-885	1575-1625
Normalizing(b)	900-955	1650-1750
Carburizing	900-925	1650-1700
Reheating	845-870	1550-1600
Tempering	120-175	250-350

(a) Heat at least as high as the carburizing temperature; hold for uniformity; cool rapidly to 540 to 675 °C (1000 to 1250 °F); air or furnace cool to obtain a structure suitable for machining and finish. (b) Temperature should be at least as high as the carburizing temperature, followed by air cooling

AISI 8720, 8720H: Approximate Critical Points

Transformation point	Temperature(a) °C	°F	Transformation point	Temperature(a) °C	°F
Ac_1	730	1350	Ar_1	660	1220
Ac_3	830	1530	M_s	395	740
Ar_3	770	1420			

(a) On heating or cooling at 28 °C (50 °F) per hour

Machining Data

For machining data on AISI grades 8720 and 8720H, refer to the preceding machining tables for AISI grade A2317.

Mechanical Properties

AISI 8720, 8720H: Hardness and Machinability (Ref 7)

Condition	Hardness range, HB	Average machinability rating(a)
Hot rolled and cold drawn	179-235	65

(a) Based on AISI 1212 steel as 100% average machinability

For additional data on mechanical properties of AISI grades 8720 and 8720H, refer to the figure on AISI 8600 and 8700 series steels: average core properties, which is included in AISI 8615.

Mechanical Properties (continued)

AISI 8720H: End-Quench Hardenability Limits and Hardenability Band. Heat treating temperatures recommended by SAE: normalize at 925 °C (1700 °F), for forged or rolled specimens only; austenitize at 925 °C (1700 °F). (Ref 1)

Distance from quenched end, 1/16 in.	Hardness, HRC max	min
1	48	41
2	47	38
3	45	35
4	42	30
5	38	26
6	35	24
7	33	22
8	31	21
9	30	20
10	29	...
11	28	...
12	27	...
13	26	...
14	26	...
15	25	...
16	25	...
18	24	...
20	24	...
22	23	...
24	23	...
26	23	...
28	23	...
30	22	...
32	22	...

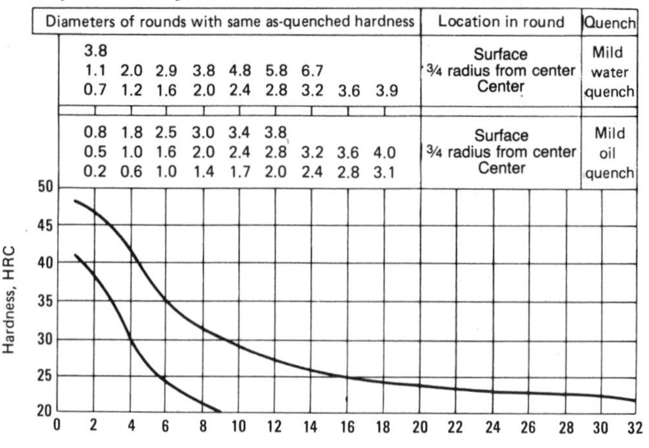

AISI 8740, 8740H

AISI 8740, 8740H: Chemical Composition

AISI grade	C	Mn	P max	S max	Si	Ni	Cr	Mo
8740	0.38-0.43	0.75-1.00	0.035	0.040	0.15-0.30	0.40-0.70	0.40-0.60	0.20-0.30
8740H	0.37-0.44	0.70-1.05	0.035	0.040	0.15-0.30	0.35-0.75	0.35-0.65	0.20-0.30

Chemical composition, %

Characteristics. AISI grades 8740 and 8740H are directly hardenable, nickel-chromium-molybdenum alloy steels with medium hardenability. These steels are moderately machinable. They can be hardened in oil in moderate sections, and thus can be heat treated with less distortion than types requiring water quenching.

AISI 8740 and 8740H steels are available as hot rolled and cold finished bar, rod, and wire; bar subject to end-quench hardenability requirements; seamless mechanical tubing; hot and cold rolled sheet and strip; and aircraft quality stock.

Typical Uses. AISI 8740 and 8740H steels are used in cam shafts, wrist pins, clutch fingers, and other automotive parts where high-strength and core property requirements are not as severe as in the higher alloy steels. Additional applications include torsion bar, springs, high-strength fasteners, and other cold headed parts.

AISI 8740: Similar Steels (U.S. and/or Foreign). UNS G87400; AMS 6322, 6323, 6325, 6327, 6358; ASTM A322, A331, A505, A519; MIL SPEC MIL-S-6049; SAE J404, J412, J770; (W. Ger.) DIN 1.6546; (Ital.) UNI 40 NiCrMo 2 KB; (U.K.) B.S. Type 7

AISI 8740H: Similar Steels (U.S. and/or Foreign). UNS H87400; ASTM A304; SAE J1268; (W. Ger.) DIN 1.6546; (Ital.) UNI 40 NiCrMo 2 KB; (U.K.) B.S. Type 7

Machining Data

For machining data on AISI grades 8740 and 8740H, refer to the preceding machining tables for AISI grades 1340 and 1340H.

AISI 8740, 8740H: Approximate Critical Points

Transformation point	Temperature(a) °C	°F
Ac_1	730	1345
Ac_3	780	1435
Ar_3	720	1330
Ar_1	660	1220
M_s	320	605

(a) On heating or cooling at 28 °C (50 °F) per hour

Physical Properties

AISI 8740, 8740H: Thermal Treatment Temperatures

Quenching medium: oil

Treatment	Temperature range °C	°F
Forging	1205 max	2200 max
Annealing	815-870	1500-1600
Austenitizing	830-855	1525-1575
Tempering	(a)	(a)

(a) To desired hardness

Mechanical Properties

AISI 8740H: End-Quench Hardenability Limits and Hardenability Band. Heat treating temperatures recommended by SAE: normalize at 870 °C (1600 °F), for forged or rolled specimens only; austenitize at 845 °C (1550 °F). (Ref 1)

Distance from quenched end, 1/16 in.	Hardness, HRC max	Hardness, HRC min
1	60	53
2	60	53
3	60	52
4	60	51
5	59	49
6	58	46
7	57	43
8	56	40
9	55	37
10	53	35
11	52	34
12	50	32
13	49	31
14	48	31
15	46	30
16	45	29
18	43	28
20	42	28
22	41	27
24	40	27
26	39	27
28	39	27
30	38	26
32	38	26

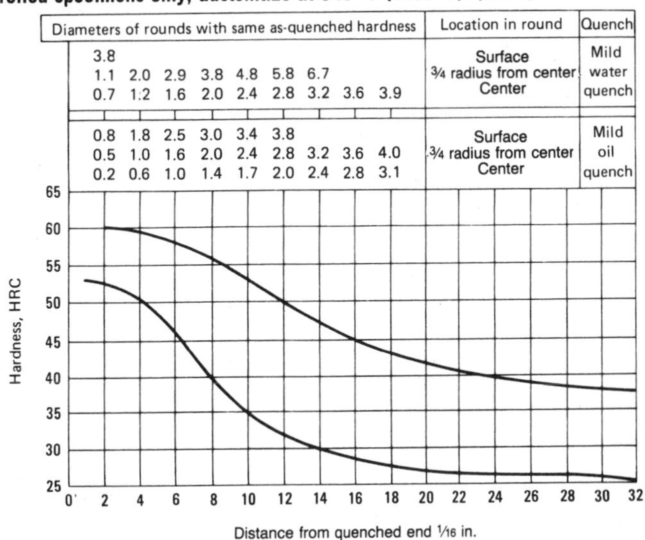

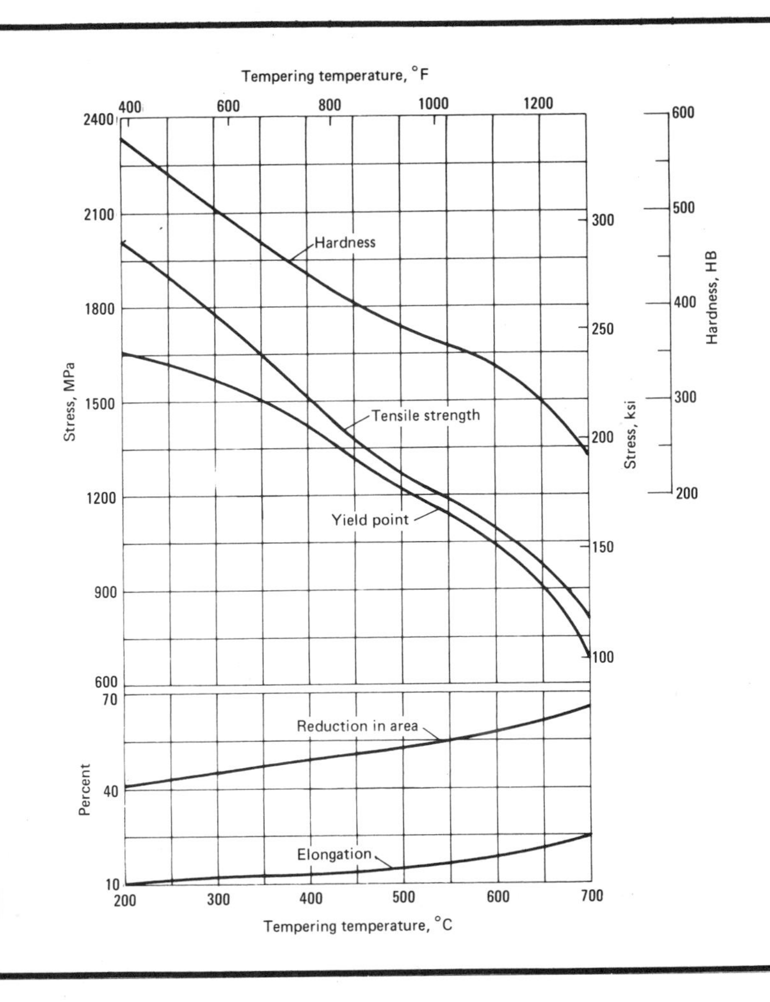

AISI 8740: Effect of Tempering Temperature on Tensile Properties. Normalized at 870 °C (1600 °F); reheated to 830 °C (1525 °F); quenched in agitated oil. Specimens were treated in 14.4-mm (0.565-in.) diam and machined to 12.8-mm (0.505-in.) diam for testing. Tests were conducted using specimens machined to English units. As-quenched hardness of 601 HB; elongation measured in 50 mm (2 in.). (Ref 4)

Mechanical Properties (continued)

AISI 8740, 8740H: Tensile Properties (Ref 5)

Condition or treatment	Tensile strength MPa	Tensile strength ksi	Yield strength MPa	Yield strength ksi	Elongation(a), %	Reduction in area, %	Hardness, HB	Izod impact energy J	Izod impact energy ft·lb
Normalized at 870 °C (1600 °F)	930	135	550	80	16.0	48	269	18	13
Annealed at 815 °C (1500 °F)	695	101	415	60	22.2	46	201	41	30

(a) In 50 mm (2 in.)

AISI 8740, 8740H: Mass Effect on Mechanical Properties (Ref 4)

Condition or treatment	Size round mm	Size round in.	Tensile strength MPa	Tensile strength ksi	Yield strength MPa	Yield strength ksi	Elongation(a), %	Reduction in area, %	Hardness, HB
Annealed (heated to 815 °C or 1500 °F; furnace cooled 11 °C or 20 °F per hour to 595 °C or 1100 °F; cooled in air)	25	1	696	101	415	60	22.2	46.4	201
Normalized (heated to 870 °C or 1600 °F; cooled in air)	13	0.5	938	136	620	90	16.0	47.1	269
	25	1	931	135	605	88	16.0	47.9	269
	50	2	910	132	605	88	16.7	50.1	262
	100	4	910	132	600	87	15.5	46.1	255
Oil quenched from 830 °C (1525 °F); tempered at 540 °C (1000 °F)	13	0.5	1235	179	1140	165	13.5	47.4	352
	25	1	1225	178	1130	164	16.0	53.0	352
	50	2	1180	171	1060	154	15.7	52.8	331
	100	4	958	139	745	108	18.0	55.6	277
Oil quenched from 830 °C (1525 °F); tempered at 595 °C (1100 °F)	13	0.5	1060	154	965	140	17.4	55.1	311
	25	1	1025	149	924	134	18.2	59.9	302
	50	2	979	142	841	122	18.5	62.0	277
	100	4	855	124	670	97	20.5	59.8	248
Oil quenched from 830 °C (1525 °F); tempered at 650 °C (1200 °F)	13	0.5	965	140	876	127	19.9	60.7	285
	25	1	952	138	848	123	20.0	60.7	285
	50	2	876	127	731	106	21.5	65.4	255
	100	4	800	116	605	88	22.7	62.9	229

(a) In 50 mm (2 in.)

AISI 8740, 8740H: Effect of the Mass on Hardness at Selected Points (Ref 4)

Size round mm	Size round in.	As-quenched hardness after quenching in oil at: Surface	½ radius	Center
13	0.5	57 HRC	56 HRC	55 HRC
25	1	56 HRC	55 HRC	54 HRC
50	2	52 HRC	49 HRC	45 HRC
100	4	42 HRC	37 HRC	36 HRC

AISI 8740, 8740H: Hardness and Machinability (Ref 7)

Condition	Hardness range, HB	Average machinability rating(a)
Annealed and cold drawn	184-235	65

(a) Based on AISI 1212 steel as 100% average machinability

AISI 8822, 8822H

AISI 8822, 8822H: Chemical Composition

AISI grade	C	Mn	P max	S max	Si	Ni	Cr	Mo
	Chemical composition, %							
8822	0.20-0.25	0.75-1.00	0.035	0.040	0.15-0.30	0.40-0.70	0.40-0.60	0.30-0.40
8822H	0.19-0.25	0.70-1.05	0.035	0.040	0.15-0.30	0.35-0.75	0.35-0.65	0.30-0.40

Characteristics. AISI grades 8822 and 8822H are carburizing, nickel-chromium-molybdenum alloy steels with medium to high case hardenability and high core hardenability. Core hardenability depends on the carbon content of these basic steels, as well as the alloy content. Good case properties can be obtained by oil quenching.

Typical Uses. AISI 8822 and 8822H steels are used specifically for heavy-duty, bevel-drive pinions and gears, and large roller bearings. These grades are available as hot rolled and cold finished bar, rod, and wire; seamless mechanical tubing; and bar subject to end-quench hardenability requirements.

AISI 8822: Similar Steels (U.S. and/or Foreign). UNS G88220; ASTM A322, A331, A519; SAE J404, J770; (W. Ger.) DIN 1.6543; (U.K.) B.S. 805 A 20

AISI 8822H: Similar Steels (U.S. and/or Foreign). UNS H88220; ASTM A304; SAE J1268; (W. Ger.) DIN 1.6543; (U.K.) B.S. 805 A 20

AISI 8822, 8822H: Approximate Critical Points

Transformation point	Temperature(a) °C	°F	Transformation point	Temperature(a) °C	°F
Ac_1	720	1330	Ar_1	645	1195
Ac_3	840	1540	M_s	385	725
Ar_3	785	1445			

(a) On heating or cooling at 28 °C (50 °F) per hour

Physical Properties

AISI 8822, 8822H: Thermal Treatment Temperatures

Quenching medium: oil

Treatment	Temperature range °C	°F
Forging	1230 max	2250 max
Cycle annealing	(a)	(a)
Normalizing	(b)	(b)
Carburizing	900-925	1650-1700
Reheating	845-870	1550-1600
Tempering	120-175	250-350

(a) Heat at least as high as the carburizing temperature; hold for uniformity; cool rapidly to 540 to 675 °C (1000 to 1250 °F); air or furnace cool to obtain a structure suitable for machining and finish. (b) At least as high as the carburizing temperature, followed by air cooling

Machining Data

For machining data on AISI grades 8822 and 8822H, refer to the preceding machining tables for AISI grade A2317.

Mechanical Properties

AISI 8822H: End-Quench Hardenability Limits and Hardenability Band.

Heat treating temperatures recommended by SAE: normalize at 925 °C (1700 °F), for forged or rolled specimens only; austenitize at 925 °C (1700 °F). (Ref 1)

Distance from quenched end, 1/16 in.	Hardness, HRC max	min
1	50	43
2	49	42
3	48	39
4	46	33
5	43	29
6	40	27
7	37	25
8	35	24
9	34	24
10	33	23
11	32	23
12	31	22
13	31	22
14	30	22
15	30	21
16	29	21
18	29	20
20	28	...
22	27	...
24	27	...
26	27	...
28	27	...
30	27	...
32	27	...

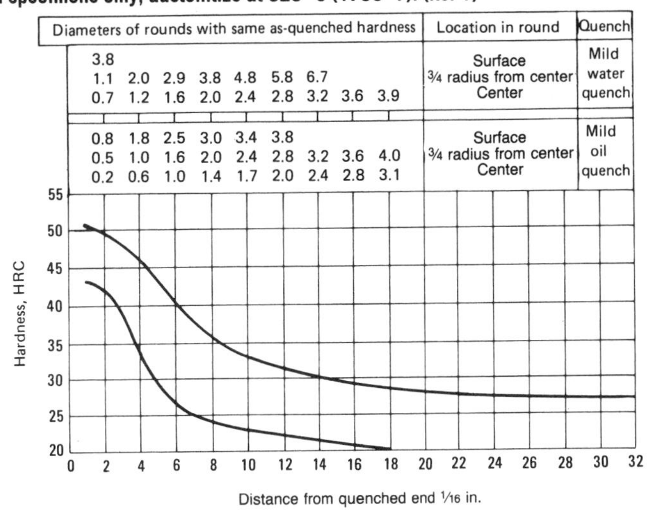

AISI 8822, 8822H: Hardness and Machinability (Ref 7)

Condition	Hardness range, HB	Average machinability rating(a)
Hot rolled and cold drawn	179-223	55

(a) Based on AISI 1212 steel as 100% average machinability

Mechanical Properties (continued)

AISI 8822, 8822H: Approximate Core Mechanical Properties (Ref 2)

Heat treatment of test specimens: (1) normalized at 925 °C (1700 °F) in 32-mm (1.25-in.) rounds; (2) machined to 25- or 13.7-mm (1- or 0.540-in.) rounds; (3) pseudocarburized at 925 °C (1700 °F) for 8 h; (4) box cooled to room temperature; (5) reheated to temperatures given in the table and oil quenched; (6) tempered at 150 °C (300 °F); (7) tested in 12.8-mm (0.505-in.) rounds; tests were conducted using test specimens machined to English units

Reheat temperature		Tensile strength		Yield strength(a)		Elongation(b),	Reduction	Hardness,
°C	°F	MPa	ksi	MPa	ksi	%	in area, %	HB
Heat treated in 25-mm (1-in.) rounds								
775	1425	1176	170.5	654	94.8	12.1	24.9	352
820	1510	1204	174.6	805	116.8	14.8	44.5	363
865	1590	1250	181.3	900	130.6	13.6	46.0	388
(c)	(c)	1314	190.6	928	134.6	12.8	42.5	388
Heat treated in 13.7-mm (0.540-in.) rounds								
775	1425	1198	173.7	693	100.5	11.3	24.6	352
820	1510	1451	210.5	1041	151.0	13.0	41.9	415
865	1590	1510	219.0	1093	158.5	13.3	47.3	429
(c)	(c)	1524	221.1	1149	166.6	13.5	47.7	429

(a) 0.2% offset. (b) In 50 mm (2 in.). (c) Quenched from step 3

AISI 8822, 8822H: Permissible Compressive Stresses

Taken from gears with a case depth of 1.1 to 1.5 mm (0.045 to 0.060 in.) and a minimum case hardness of 60 HRC

Data	Compressive stress	
point	MPa	ksi
Mean	1448	210
Maximum	1655	240

AISI 9255

AISI 9255: Chemical Composition

AISI	Chemical composition, %				
grade	C	Mn	P max	S max	Si
9255	0.51-0.59	0.70-0.95	0.035	0.040	1.80-2.20

Characteristics and Typical Uses. AISI grade 9255 is a directly hardenable, silicon alloy steel with the capacity for medium hardenability. This steel is widely used for springs subject to shock loads and moderately elevated temperature. AISI 9255 steel is available as hot rolled and cold finished bar, rod, and wire, and seamless mechanical tubing.

AISI 9255: Similar Steels (U.S. and/or Foreign). UNS G92550; ASTM A322, A519; SAE J404, J412, J770; (W. Ger.) DIN 1.0904; (Fr.) AFNOR 55 S 7; (Ital.) UNI 55 Si 8; (Swed.) SS_{14} 2085**; (U.K.) B.S. 250 A 53

Machining Data

For machining data on AISI grade 9255, refer to the preceding machining tables for AISI grades 4150 and 4150H.

AISI 9255: Approximate Critical Points

Transformation point	Temperature(a) °C	°F	Transformation point	Temperature(a) °C	°F
Ac_1	760	1400	Ar_1	715	1320
Ac_3	815	1500	M_s	305	585
Ar_3	750	1380			

(a) On heating or cooling at 28 °C (50 °F) per hour

Physical Properties

AISI 9255: Thermal Treatment Temperature

Treatment	Temperature range °C	°F
Forging	1205 max	2200 max
Normalizing	870-925	1600-1700
Austenitizing	815-900	1500-1650
Tempering	(a)	(a)

(a) To desired hardness

Mechanical Properties

AISI 9255: Tensile Properties (Ref 5)

Condition or treatment	Tensile strength		Yield strength		Elongation(a),	Reduction	Hardness,	Izod impact energy	
	MPa	ksi	MPa	ksi	%	in area, %	HB	J	ft·lb
Normalized at 900 °C (1650 °F)	930	135	580	84	19.7	43	69	14	10
Annealed at 845 °C (1550 °F)	770	112	490	71	21.7	41	229	9	7

(a) In 50 mm (2 in.)

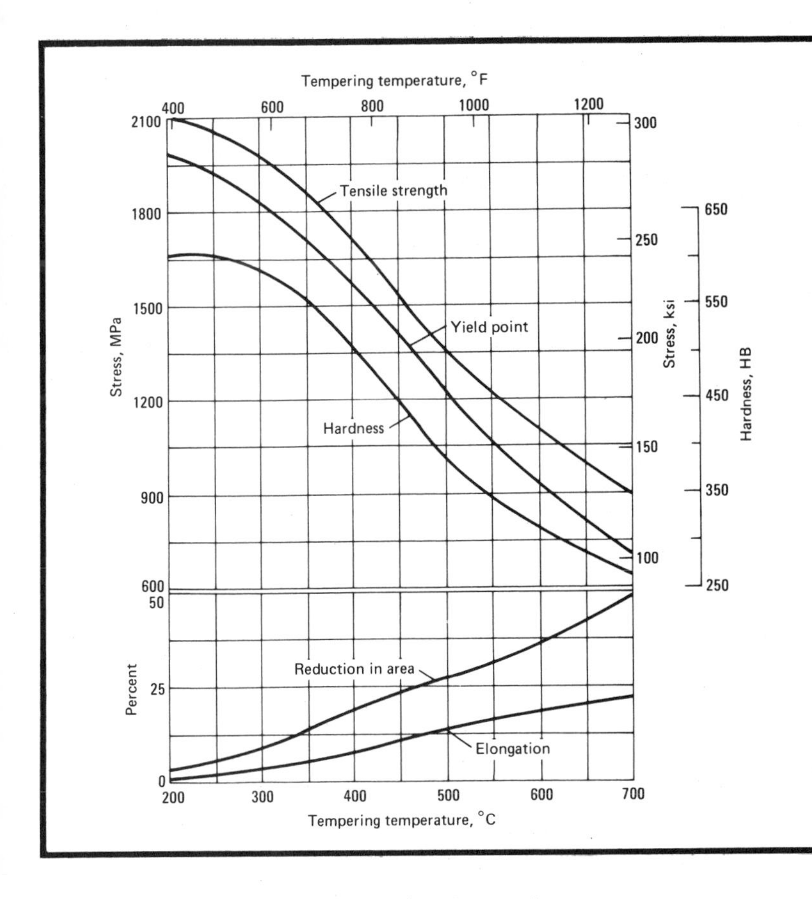

AISI 9255: Effect of Tempering Temperature on Tensile Properties. Normalized at 900 °C (1650 °F); reheated to 885 °C (1625 °F); quenched in agitated oil. Specimens were treated in 25-mm (1-in.) diam and machined to 12.8-mm (0.505-in.) diam for testing. Tests were conducted using specimens machined to English units. As-quenched hardness of 653 HB; elongation measured in 50 mm (2 in.). (Ref 4)

AISI 9255: Mass Effect on Mechanical Properties

Condition or treatment	Size round mm	Size round in.	Tensile strength MPa	Tensile strength ksi	Yield strength MPa	Yield strength ksi	Elongation(a), %	Reduction in area, %	Hardness, HB
Annealed (heated to 845 °C or 1550 °F; furnace cooled 11 °C or 20 °F per hour to 660 °C or 1220 °F; cooled in air)	25	1	779	113	485	70	21.7	41.1	229
Normalized (heated to 900 °C or 1650 °F; cooled in air)	13	0.5	951	138	585	85	20.0	45.5	277
	25	1	931	135	580	84	19.7	43.4	269
	50	2	931	135	565	82	19.5	39.5	269
	100	4	917	133	550	80	18.7	36.1	269
Oil quenched from 885 °C (1625 °F); tempered at 540 °C (1000 °F)	13	0.5	1170	170	1007	146	14.9	40.0	331
	25	1	1130	164	924	134	16.7	38.3	321
	50	2	1070	155	703	102	18.0	45.6	302
	100	4	1025	149	650	94	19.2	43.7	293
Oil quenched from 885 °C (1625 °F); tempered at 595 °C (1100 °F)	13	0.5	1070	155	910	132	18.1	45.3	302
	25	1	1035	150	814	118	19.2	44.8	293
	50	2	1005	146	635	92	20.0	48.7	293
	100	4	945	137	570	83	21	46.0	277
Oil quenched from 885 °C (1625 °F); tempered at 650 °C (1200 °F)	13	0.5	1000	145	848	123	21	50.4	285
	25	1	951	138	731	106	21.2	48.2	277
	50	2	951	138	600	87	21	50.7	277
	100	4	910	132	565	82	21.7	48.3	262

(a) In 50 mm (2 in.)

AISI 9255: Effect of the Mass on Hardness at Selected Points

Size round mm	Size round in.	As-quenched hardness after quenching in oil at: Surface	½ radius	Center
13	0.5	61 HRC	59 HRC	58 HRC
25	1	57 HRC	55 HRC	48 HRC
50	2	52 HRC	37 HRC	33 HRC
100	4	35.5 HRC	31.5 HRC	27.5 HRC

AISI 9255: Hardness and Machinability (Ref 7)

Condition	Hardness range, HB	Average machinability rating(a)
Spheroidized annealed and cold drawn	179-229	40

(a) Based on AISI 1212 steel as 100% average machinability

AISI 9260, 9260H

AISI 9260, 9260H: Chemical Composition

AISI grade	C	Mn	P max	S max	Si
9260	0.56-0.64	0.75-1.00	0.035	0.040	1.80-2.20
9260H	0.55-0.65	0.65-1.10	0.035	0.040	1.70-2.20

Chemical composition, %

Characteristics and Typical Uses. AISI grades 9260 and 9260H are directly hardenable, silicon alloy steels with medium hardenability. These steels are used in springs for heavy duty service and are available as hot rolled and cold finished bar, rod, and wire; seamless mechanical tubing; bar subject to end-quench hardenability requirements; and hot and cold rolled strip and sheet.

AISI 9260: Similar Steels (U.S. and/or Foreign). UNS G92600; ASTM A29, A59, A322, A331, A505, A519; SAE J404, J412, J770; (W. Ger.) DIN 1.0909; (Fr.) AFNOR 60 SC 7, 61 SC 7; (U.K.) B.S. 250 A 58

AISI 9260H: Similar Steels (U.S. and/or Foreign). UNS H92600; ASTM A304; SAE J1268; (W. Ger.) DIN 1.0909; (Fr.) AFNOR 60 S 7, 61 Sc 7; (U.K.) B.S. 250 A 58

AISI 9260, 9260H: Approximate Critical Points

Transformation point	Temperature(a) °C	°F	Transformation point	Temperature(a) °C	°F
Ac_1	745	1370	Ar_1	715	1315
Ac_3	815	1500	M_s	270	515
Ar_3	750	1380			

(a) On heating or cooling at 28 °C (50 °F) per hour

Physical Properties

AISI 9260, 9260H: Thermal Treatment Temperatures

Quenching medium: oil

Treatment	Temperature range °C	°F
Forging	1205 max	2200 max
Annealing (a)	815-870	1500-1600
Normalizing (b)	870-925	1600-1700
Austenitizing	855-885	1575-1625
Tempering	(c)	(a)

(a) Maximum hardness of 229 HB. (b) Hardness of approximately 302 HB. (c) To desired hardness

Mechanical Properties

AISI 9260H: End-Quench Hardenability Limits and Hardenability Band.

Heat treating temperatures recommended by SAE: by SAE: normalize at 900 °C (1650 °F), for forged or rolled specimens only; austenitize at 870 °C (1600 °F). (Ref 1)

Distance from quenched end, 1/16 in.	Hardness, HRC max	min
1	...	60
2	...	60
3	65	57
4	64	53
5	63	46
6	62	41
7	60	38
8	58	36
9	55	36
10	52	35
11	49	34
12	47	34
13	45	33
14	43	33
15	42	32
16	40	32
18	38	31
20	37	31
22	36	30
24	36	30
26	35	29
28	35	29
30	35	28
32	34	28

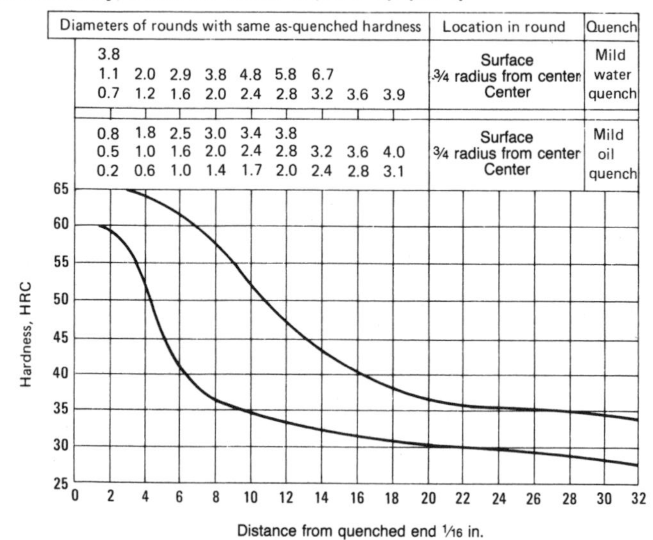

Mechanical Properties (continued)

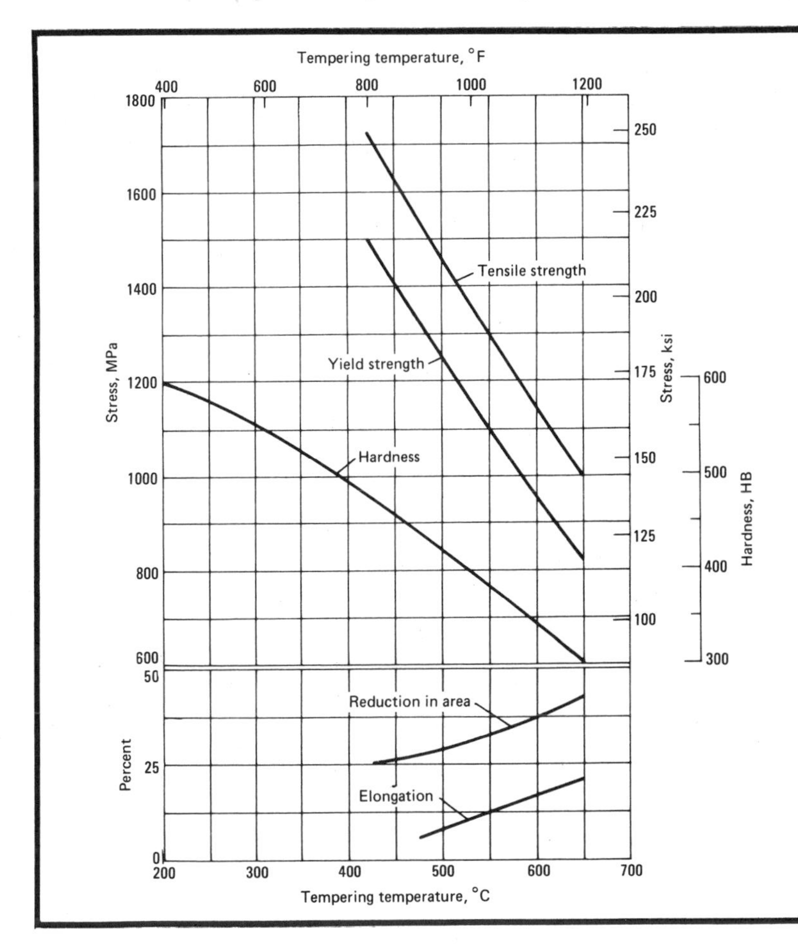

AISI 9260: Effect of Tempering Temperature on Tensile Properties. Normalized at 900 °C (1650 °F); oil quenched from 870 °C (1600 °F); tempered at 56 °C (100 °F) intervals. Specimens were treated in 13.7-mm (0.540-in.) diam and machined to 12.8-mm (0.505-in.) diam for testing. Tests were conducted using specimens machined to English units. Elongation was measured in 50 mm (2 in.). (Ref 2)

AISI 9260, 9260H: Hardness and Machinability

Condition	Hardness range, HB	Average machinability rating(a)
Spheroidized annealed and cold drawn	184-235	40

(a) Based on AISI 1212 steel as 100% average machinability

Machining Data

For machining data on AISI grades 9260 and 9260H, refer to the preceding machining tables for AISI 4150 and 4150H.

AISI E9310, E9310H

AISI E9310, E9310H: Chemical Composition

AISI grade	C	Mn	P max	S max	Si	Ni	Cr	Mo
	Chemical composition, %							
E9310	0.08-0.13	0.45-0.65	0.025	0.025	0.20-0.35	3.00-3.50	1.00-1.40	0.08-0.15
E9310H	0.07-0.13	0.40-0.70	...	0.050	0.15-0.30	2.95-3.55	1.00-1.45	0.08-0.15

Characteristics and Typical Uses. AISI grades E9310 and E9310H are electric-furnace, carburizing, nickel-chromium-molybdenum, alloy steels with high case and core hardenability. They are typically used for aircraft and heavy-duty truck gears. AISI E9310 and E9310H steels are available as hot rolled and cold finished bar, rod, and wire; seamless mechanical tubing; bar subject to end-quench hardenability requirements; forging stock; and aircraft quality bar.

AISI E9310: Similar Steels (U.S. and/or Foreign). UNS G93106; AMS 6260 F, 6265 B; ASTM A322, A519, A711; SAE J404, J770

AISI E9310H: Similar Steels (U.S. and/or Foreign). UNS H93100; ASTM A304; SAE J1268

AISI E9310, E9310H: Approximate Critical Points

Transformation point	Temperature(a) °C	°F	Transformation point	Temperature(a) °C	°F
Ac_1	715	1320	Ar_1	580	1080
Ac_3	820	1510	M_s	365	685
Ar_3	665	1230			

(a) On heating or cooling at 28 °C (50 °F) per hour

Physical Properties

AISI E9310, E9310H: Thermal Treatment Temperatures

Quenching medium: oil

Treatment	Temperature range °C	°F
Forging	1230 max	2250 max
Normalizing	900-955	1650-1750
Carburizing	870-925	1600-1700
Reheating	790-830	1450-1525
Tempering	120-165	250-325

Machining Data

For machining data on AISI grades E9310 and E9310H, refer to the preceding machining tables for AISI A2317.

Mechanical Properties

AISI E9310, E9310H: Core Properties of Specimens Treated for Maximum Case Hardness and Core Toughness (Ref 4)

Specimens were treated in 14.4-mm (0.565-in.) rounds and machined to 12.8-mm (0.505-in.) rounds for testing; tests were conducted using test specimens machined to English units

Treatment	Case depth, mm (in.)	Case hardness, HRC	Core properties: Tensile strength, MPa (ksi)	Yield strength, MPa (ksi)	Elongation(a), %	Reduction in area, %	Hardness HB
Recommended practice for maximum case hardness							
Direct quenched from pot: carburized at 925 °C (1700 °F) for 8 h; quenched in agitated oil; tempered at 150 °C (300 °F)	0.99 (0.039)	59.5	1238 (179.5)	993 (144.0)	15.3	59.1	375
Single quenched and tempered (for good case and core properties): carburized at 925 °C (1700 °F) for 8 h; pot cooled; reheated to 790 °C (1450 °F); quenched in agitated oil; tempered at 150 °C (300 °F)	1.2 (0.047)	62.0	1193 (173.0)	931 (135.0)	15.5	60.0	363
Double quenched and tempered (for maximum refinement of case and core): carburized at 925 °C (1700 °F) for 8 h; pot cooled; reheated to 800 °C (1475 °F); quenched in agitated oil; reheated to 775 °C (1425 °F); quenched in agitated oil; tempered at 150 °C (300 °F)	1.4 (0.055)	60.5	1203 (174.5)	958 (139.0)	15.3	62.1	363
Recommended practice for maximum core toughness							
Direct quenched from pot: carburized at 925 °C (1700 °F) for 8 h; quenched in agitated oil; tempered at 230 °C (450 °F)	0.99 (0.039)	54.5	1227 (178.0)	1010 (146.5)	15.0	59.7	363
Single quenched and tempered (for good case and core properties): carburized at 925 °C (1700 °F) for 8 h; pot cooled; reheated to 790 °C (1450 °F); quenched in agitated oil; tempered at 230 °C (450 °F)	1.2 (0.047)	59.5	1158 (168.0)	948 (137.5)	15.5	60.0	341
Double quenched and tempered (for maximum refinement of case and core): carburized at 925 °C (1700 °F) for 8 h; pot cooled; reheated to 800 °C (1475 °F); quenched in agitated oil; reheated to 775 °C (1425 °F); quenched in agitated oil; tempered at 230 °C (450 °F)	1.4 (0.055)	58.0	1169 (169.5)	951 (138.0)	14.8	61.8	352

(a) In 50 mm (2 in.)

AISI E9310H: End-Quench Hardenability Limits and Hardenability Band. Heat treating temperatures recommended by SAE: normalize at 925 °C (1700 °F), for forged or rolled specimens only; austenitize at 845 °C (1550 °F). (Ref 1)

Distance from quenched end, 1/16 in.	Hardness, HRC max	min
1	43	36
2	43	35
3	43	35
4	42	34
5	42	32
6	42	31
7	42	30
8	41	29
9	40	28
10	40	27
11	39	27
12	38	26
13	37	26
14	36	26
15	36	26
16	35	26
18	35	26
20	35	25
22	34	25
24	34	25
26	34	25
28	34	25
30	33	24
32	33	24

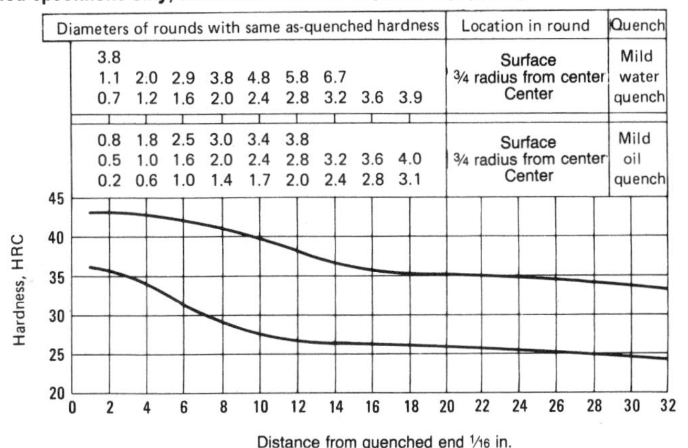

AISI E9310, E9310H: Mass Effect on Mechanical Properties (Ref 4)

Condition or treatment	Size round mm	in.	Tensile strength MPa	ksi	Yield strength MPa	ksi	Elongation(a), %	Reduction in area, %	Hardness, HB
Annealed (heated to 845 °C or 1550 °F; furnace cooled 17 °C or 30 °F per hour to 405 °C or 760 °F; cooled in air)	25	1	820	119	450	65	17.3	42.1	241
Normalized (heated to 890 °C or 1630 °F; cooled in air)	13	0.5	917	133	605	88	20.0	63.7	285
	25	1	910	132	570	83	18.8	58.1	269
	50	2	903	131	565	82	19.5	60.5	262
	100	4	862	125	565	82	19.5	61.7	255
Mock carburized at 925 °C (1700 °F) for 8 h; reheated to 790 °C (1450 °F); quenched in oil; tempered at 150 °C (300 °F)	13	0.5	1234	179	986	143	15.7	58.9	363
	25	1	1096	159	848	123	15.5	57.5	321
	50	2	1000	145	745	108	18.5	66.7	293
	100	4	938	136	655	95	19.0	62.3	277
Mock carburized at 925 °C (1700 °F) for 8 h; reheated to 790 °C (1450 °F); quenched in oil; tempered at 230 °C (450 °F)	13	0.5	1227	178	979	142	15.0	60.3	363
	25	1	1089	158	848	123	16.0	61.7	321
	50	2	993	144	731	106	17.8	68.1	293
	100	4	910	132	660	96	20.5	67.0	269

(a) In 50 mm (2 in.)

AISI E9310, E9310H: Effect of the Mass on Hardness at Selected Points (Ref 4)

Size round mm	in.	As-quenched hardness after quenching in oil at: Surface	1/2 radius	Center
13	0.5	40 HRC	40 HRC	38 HRC
25	1	40 HRC	38 HRC	37 HRC
50	2	38 HRC	35 HRC	32 HRC
100	4	31 HRC	30 HRC	29 HRC

AISI E9310H: Hardness and Machinability

Condition	Hardness range, HB	Average machinability rating(a)
Annealed and cold drawn	184-229	50

(a) Based on AISI 1212 steel as 100% average machinability

Mechanical Properties (continued)

AISI E9310, E9310H: Tensile Properties (Ref 5)

Condition or treatment	Tensile strength MPa	Tensile strength ksi	Yield strength MPa	Yield strength ksi	Elongation(a), %	Reduction in area, %	Hardness, HB	Izod impact energy J	Izod impact energy ft·lb
Normalized at 890 °C (1630 °F)	910	132	570	83	18.8	58	269	119	88
Annealed at 845 °C (1550 °F)	820	119	440	64	17.3	42	241	79	58

(a) In 50 mm (2 in.)

AISI E9310, E9310H: Approximate Core Mechanical Properties (Ref 2)

Heat treatment of test specimens: (1) normalized at 925 °C (1700 °F) in 32-mm (1.25-in.) rounds; (2) machined to 25- or 13.7-mm (1- or 0.540-in.) rounds; (3) pseudocarburized at 925 °C (1700 °F) for 8 h; (4) box cooled to room temperature; (5) reheated to temperatures given in the table and oil quenched; (6) tempered at 150 °C (300 °F); (7) tested in 12.8-mm (0.505-in.) rounds; tests were conducted using test specimens machined to English units

Reheat temperature °C	Reheat temperature °F	Tensile strength MPa	Tensile strength ksi	Yield strength(a) MPa	Yield strength(a) ksi	Elongation(b), %	Reduction in area, %	Hardness, HB
Heat treated in 25-mm (1-in.) rounds								
775	1425	1000	145	814	118	15.5	54	302
800	1475	1076	156	917	133	15.5	54	331
830	1525	1179	171	1020	148	15.5	54	352
(c)	(c)	1200	174	1034	150	15.0	53	363
Heat treated in 13.7-mm (0.540-in.) rounds								
775	1425	1069	155	896	130	15.5	52	331
800	1475	1131	164	965	140	16.0	53	341
830	1525	1200	174	1055	153	16.0	53	363
(c)	(c)	1289	187	1117	162	15.0	51	375

(a) 0.2% offset. (b) In 50 mm (2 in.). (c) Quenched from step 3

AISI E9310, E9310H: Permissible Compressive Stresses

Taken from gears with case depth of 1.1 to 1.5 mm (0.045 to 0.060 in.) and a minimum case hardness of 60 HRC

Data point	Compressive stress MPa	Compressive stress ksi
Mean	1725	250
Maximum	2000	290

AISI 94B15, 94B15H:

AISI 94B15, 94B15H: Chemical Composition

AISI grade	C	Mn	P max	S max	Si	Ni	Cr	Mo	B
94B15	0.13-0.18	0.75-1.00	0.035	0.040	0.15-0.30	0.30-0.60	0.30-0.50	0.08-0.15	0.0005-0.003
94B15H	0.12-0.18	0.70-1.05	0.035	0.040	0.15-0.30	0.25-0.65	0.25-0.55	0.08-0.15	0.0005-0.003

Characteristics and Typical Uses. AISI grades 945B15 and 945B15H are carburizing, nickel-chromium-molybdenum alloy steels with intermediate case and high core hardenability. These steels are available as aircraft quality stock, seamless mechanical tubing, and bar subject to end-quench hardenability requirements.

AISI 94B15: Similar Steels (U.S. and/or Foreign). UNS G94151; AMS 6275 A; ASTM A519; SAE J404, J770

AISI 94B15H: Similar Steels (U.S. and/or Foreign). UNS H94151; ASTM A304; SAE J1268

Physical Properties

AISI 94B15, 94B15H: Thermal Treatment Temperatures

Quenching medium: oil

Treatment	Temperature range °C	Temperature range °F
Forging	1230 max	2250 max
Normalizing	(a)	(a)
Carburizing	900-925	1650-1700
Tempering	120-150	250-300

(a) At least as high as the carburizing temperature, followed by air cooling

Machining Data

For machining data on AISI grades 94B15 and 94B15H, refer to the preceding machining tables for AISI grade A2317.

Mechanical Properties

AISI 94B15H: End-Quench Hardenability Limits and Hardenability Band. Heat treating temperatures recommended by SAE: normalize at 925 °C (1700 °F), for forged or rolled specimens only; austenitize at 925 °C (1700 °F). (Ref 1)

Distance from quenched end, 1/16 in.	Hardness, HRC max	Hardness, HRC min
1	45	38
2	45	38
3	44	37
4	44	36
5	43	32
6	42	28
7	40	25
8	38	23
9	36	21
10	34	20
11	33	...
12	31	...
13	30	...
14	29	...
15	28	...
16	27	...
18	26	...
20	25	...
22	24	...
24	23	...
26	23	...
28	22	...
30	22	...
32	22	...

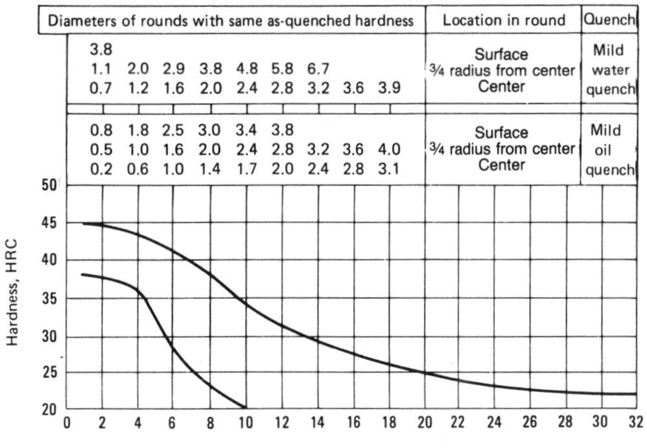

AISI 94B15, 94B15H: Hardness and Machinability (Ref 7)

Condition	Hardness range, HB	Average machinability rating(a)
Hot rolled and cold drawn	163-202	70

(a) Based on AISI 1212 steel as 100% average machinability

AISI 94B17, 94B17H

AISI 94B17, 94B17H: Chemical Composition

AISI grade	C	Mn	P max	S max	Si	Ni	Cr	Mo	B
94B17	0.15-0.20	0.75-1.00	0.035	0.040	0.15-0.30	0.30-0.60	0.30-0.50	0.08-0.15	0.0005-0.003
94B17H	0.14-0.20	0.70-1.05	0.035	0.040	0.15-0.30	0.25-0.65	0.25-0.55	0.08-0.15	0.0005-0.003

(Column group header: Chemical composition, %)

Characteristics and Typical Uses. AISI grades 94B17 and 94B17H are carburizing, triple-alloy steels with intermediate case and high core hardenability. The addition of boron improves machinability. These relatively low-carbon steels were designed as substitutes for the higher carbon, higher alloy steels in manufacturing fasteners and cold headed parts that are quenched and tempered after forming. Parts made from these steels are used in moderate temperature environments.

AISI 94B17 and 94B17H steels are available as hot rolled and cold finished bar, rod, and wire; seamless mechanical tubing; aircraft quality stock; and bar subject to end-quench hardenability requirements.

AISI 94B17: Similar Steels (U.S. and/or Foreign). UNS G94171; AMS 6275; ASTM A322, A331, A519; SAE J404, J770

AISI 94B17H: Similar Steels (U.S. and/or Foreign). UNS H94171; ASTM A304; SAE J1268

AISI 94B17, 94B17H: Approximate Critical Points

Transformation point	Temperature(a) °C	Temperature(a) °F
Ac_1	705	1300
Ac_3	840	1540
Ar_3	770	1420
Ar_1	640	1180
M_s	415	780

(a) On heating or cooling at 28 °C (50 °F) per hour

Mechanical Properties

AISI 94B17H: End-Quench Hardenability Limits and Hardenability Band. Heat treating temperatures recommended by SAE: normalize at 925 °C (1700 °F), for forged or rolled specimens only; austenitize at 925 °C (1700 °F). (Ref 1)

Distance from quenched end, 1/16 in.	Hardness, HRC max	Hardness, HRC min
1	46	39
2	46	39
3	45	38
4	45	37
5	44	34
6	43	29
7	42	26
8	41	24
9	40	23
10	38	21
11	36	20
12	34	...
13	33	...
14	32	...
15	31	...
16	30	...
18	28	...
20	27	...
22	26	...
24	25	...
26	24	...
28	24	...
30	23	...
32	23	...

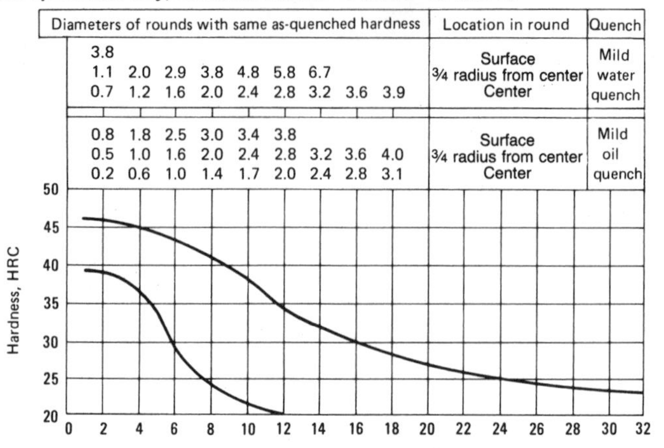

AISI 94B17, 94B17H: Approximate Core Mechanical Properties (Ref 2)

Heat treatment of test specimens: (1) normalized at 925 °C (1700 °F) in 32-mm (1.25-in.) rounds; (2) machined to 25- or 13.7-mm (1- or 0.540-in.) rounds; (3) pseudocarburized at 925 °C (1700 °F) for 8 h; (4) box cooled to room temperature; (5) reheated to temperatures given in the table and oil quenched; (6) tempered at 150 °C (300 °F); (7) tested in 12.8-mm (0.505-in.) rounds; tests were conducted using test specimens machined to English units

Reheat temperature °C	Reheat temperature °F	Tensile strength MPa	Tensile strength ksi	Yield strength(a) MPa	Yield strength(a) ksi	Elongation(b), %	Reduction in area, %	Hardness, HB
Heat treated in 25-mm (1-in.) rounds								
765	1410	959	139.1	481	69.8	14.0	27.0	293
810	1490	1048	152.0	695	100.8	14.3	47.7	321
865	1590	1235	179.1	889	129.0	15.0	55.3	363
(c)	(c)	1301	188.7	987	143.1	15.2	56.0	388
Heat treated in 13.7-mm (0.540-in.) rounds								
765	1410	936	135.8	509	73.8	13.3	28.1	285
810	1490	1131	164.0	776	112.5	13.0	44.1	341
865	1590	1253	181.8	920	133.5	14.0	57.7	375
(c)	(c)	1331	193.1	997	144.6	15.0	55.1	388

(a) 0.2% offset. (b) In 50 mm (2 in.). (c) Quenched from step 3

AISI 94B17, 94B17H: Hardness and Machinability (Ref 7)

Condition	Hardness range, HB	Average machinability rating(a)
Hot rolled and cold drawn	163-202	70

(a) Based on AISI 1212 steel as 100% average machinability

Physical Properties

AISI 94B17, 94B17H: Thermal Treatment Temperatures

Quenching medium: oil

Treatment	Temperature range °C	Temperature range °F
Forging	1230 max	2250 max
Normalizing	(a)	(a)
Carburizing	900-925	1650-1700
Tempering	120-175	250-350

(a) At least as high as the carburizing temperature, followed by air cooling

Machining Data

For machining data on AISI grades 94B17 and 94B17H, refer to the preceding machining tables for AISI grade A2317.

AISI 94B30, 94B30H

AISI 94B30, 94B30H: Chemical Composition

AISI grade	C	Mn	P max	S max	Si	Ni	Cr	Mo	B
94B30	0.28-0.33	0.75-1.00	0.035	0.040	0.15-0.30	0.30-0.60	0.30-0.50	0.08-0.15	0.0005-0.003
94B30H	0.27-0.33	0.70-1.05	0.035	0.040	0.15-0.30	0.25-0.65	0.25-0.55	0.08-0.15	0.0005-0.003

Chemical composition, %

Characteristics. AISI grades 94B30 and 94B30H are directly hardenable, triple-alloy steels with medium hardenability in the 0.30 to 0.37% mean classification of carbon content. The boron addition improves depth of hardening.

Typical Uses. AISI 94B30 and 94B30H steels can be substituted for higher carbon, higher alloy steels in manufacturing connecting rods, axle shafts, bolts, studs, screws, and other cold headed parts that are quenched and tempered after forming. They are not recommended, however, for parts in high-temperature environments. Grade 94B30 and 94B30H steels are available as hot rolled and cold finished bar, rod, and wire; seamless mechanical tubing; and bar subject to end-quench hardenability requirements.

AISI 94B30: Similar Steels (U.S. and/or Foreign). UNS G94301; ASTM A322, A331, A519; SAE J404, J412, J770

AISI 94B30H: Similar Steels (U.S. and/or Foreign). UNS H94301; ASTM A304; SAE J1268

AISI 94B30, 94B30H: Approximate Critical Points

Transformation point	Temperature(a) °C	°F	Transformation point	Temperature(a) °C	°F
Ac_1	720	1330	Ar_1	655	1210
Ac_3	805	1485	M_s	370	695
Ar_3	750	1380			

(a) On heating or cooling at 28 °C (50 °F) per hour

Physical Properties

AISI 94B30, 94B30H: Thermal Treatment Temperatures

Quenching medium: oil

Treatment	Temperature range °C	°F
Forging	1230 max	2250 max
Annealing(a)	815-870	1500-1600
Normalizing(b)	870-925	1600-1700
Austenitizing	855-885	1575-1625
Tempering	(c)	(c)

(a) Maximum hardness of 174 HB. (b) Hardness of approximately 217 HB. (c) To desired hardness

Machining Data

For machining data on AISI grades 94B30 and 94B30H, refer to the preceding machining tables for AISI grades 1330 and 1330H.

Mechanical Properties

AISI 94B30, 94B30H: Hardness and Machinability (Ref 7)

Condition	Hardness range, HB	Average machinability rating(a)
Annealed and cold drawn	170-223	70

(a) Based on AISI 1212 steel as 100% average machinability

AISI 94B30: Effect of Tempering Temperature on Tensile Properties. Normalized at 900 °C (1650 °F); oil quenched from 870 °C (1600 °F); tempered at 56 °C (100 °F) intervals. Specimens were treated in 13.7-mm (0.540-in.) diam and machined to 12.8-mm (0.505-in.) diam for testing. Tests were conducted using specimens machined to English units. Elongation was measured in 50 mm (2 in.). (Ref 2)

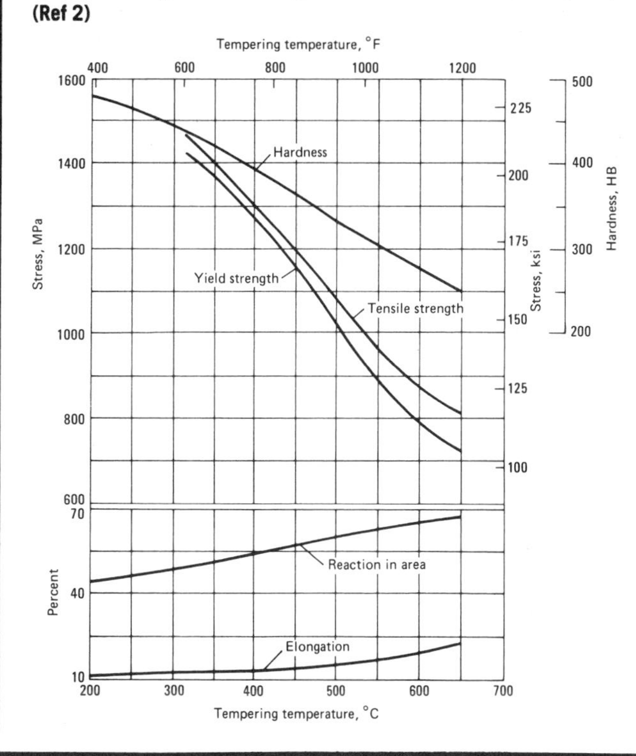

Mechanical Properties (continued)

AISI 94B30H: End-Quench Hardenability Limits and Hardenability Band. Heat treating temperatures recommended by SAE: normalize at 900 °C (1650 °F), for forged or rolled specimens only; austenitize at 870 °C (1600 °F). (Ref 1)

Distance from quenched end, 1/16 in.	Hardness, HRC max	Hardness, HRC min
1	56	49
2	56	49
3	55	48
4	55	48
5	54	47
6	54	46
7	53	44
8	53	42
9	52	39
10	52	37
11	51	34
12	51	32
13	50	30
14	49	29
15	48	28
16	46	27
18	44	25
20	42	24
22	40	23
24	38	23
26	37	22
28	35	21
30	34	21
32	34	20

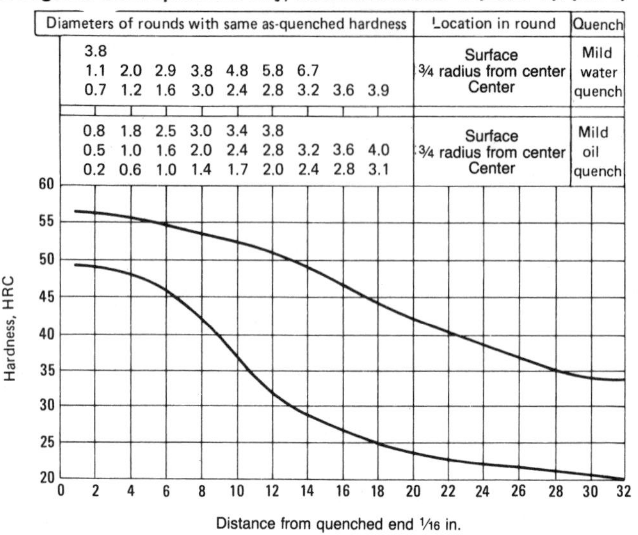

AISI 94B40

AISI 94B40: Chemical Composition

AISI grade	C	Mn	P max	S max	Si	Ni	Cr	Mo	B
94B40	0.38-0.43	0.75-1.00	0.035	0.040	0.15-0.30	0.30-0.60	0.30-0.50	0.08-0.15	0.0005-0.003

(Chemical composition, %)

Characteristics. AISI grade 94B40 is a directly hardenable, nickel-chromium-molybdenum alloy steel with medium hardenability in the 0.40 to 0.42% mean classification of carbon content. The boron addition improves depth of hardening for this grade.

Typical Uses. AISI 94B40 steel is substituted for higher carbon, higher alloy steels in manufacturing parts that are quenched and tempered after forming. It is not recommended, however, for applications in high-temperature environments. This grade is available as hot rolled and cold finished bar and seamless mechanical tubing.

AISI 94B40: Similar Steels (U.S. and/or Foreign). UNS G94401; ASTM A322, A519; SAE J778

AISI 94B40: Approximate Critical Points

Transformation point	Temperature(a) °C	Temperature(a) °F
Ac_1	725	1335
Ac_3	790	1455
Ar_3	730	1350
Ar_1	660	1220
M_s	265	510

(a) On heating or cooling at 28 °C (50 °F) per hour

Machining Data

For machining data on AISI 94B40, refer to the preceding machining tables for AISI grades 1340 and 1340H.

Physical Properties

AISI 94B40: Thermal Treatment Temperatures

Quenching medium: oil

Treatment	Temperature range °C	Temperature range °F
Forging	1230 max	2250 max
Annealing(a)	815-870	1500-1600
Normalizing(b)	845-900	1550-1650
Austenitizing	830-855	1525-1575
Tempering	(c)	(c)

(a) Maximum hardness of 192 HB. (b) Hardness of approximately 255 HB. (c) To desired hardness

Mechanical Properties

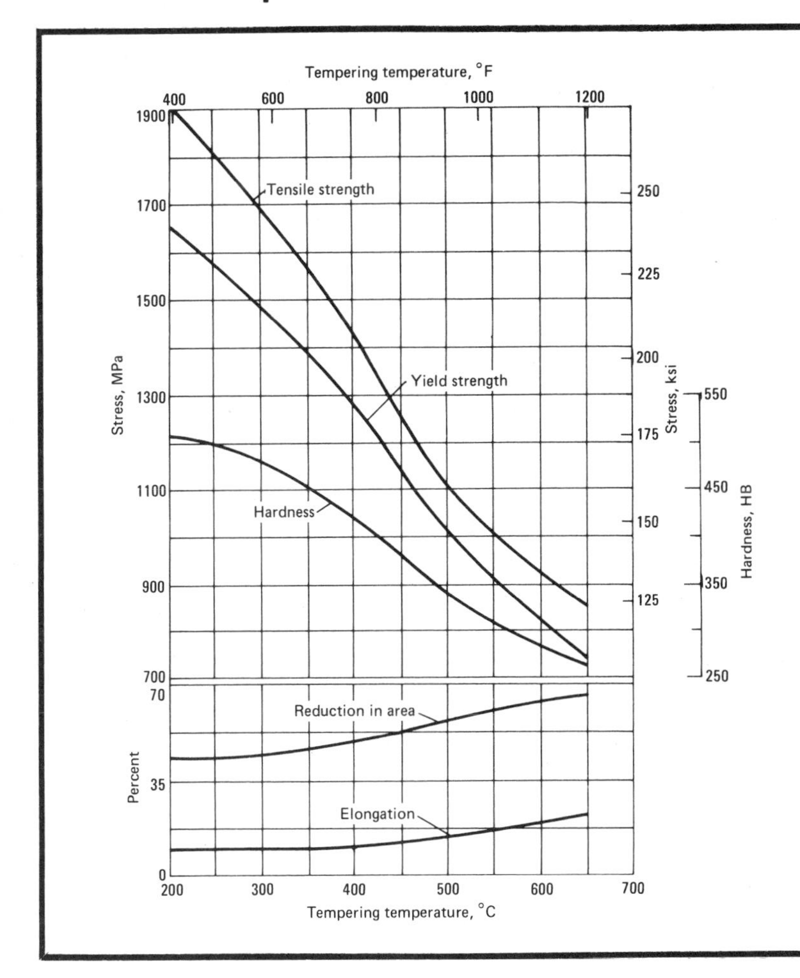

AISI 94B40: Effect of Tempering Temperature on Tensile Properties. Normalized at 870 °C (1600 °F); oil quenched from 845 °C (1550 °F); tempered at 56 °C (100 °F) intervals. Specimens were treated in 13.7-mm (0.540-in.) diam and machined to 12.8-mm (0.505-in.) for testing. Tests were conducted using specimens machined to English units. Elongation was measured in 50 mm (2 in.). (Ref 2)

AISI 9840

AISI 9840: Chemical Composition

AISI grade	C	Mn	P max	S max	Si	Ni	Cr	Mo
	Chemical composition, %							
9840	0.38-0.43	0.70-0.90	0.035	0.040	0.15-0.30	0.85-1.15	0.70-0.90	0.20-0.30

Characteristics and Typical Uses. AISI grade 9840 is a medium-carbon, constructional, alloy steel. It has characteristics similar to AISI grade 4340 and can be nitrided to improve wear and galling resistance or fatigue properties. This steel does not age harden. Core properties of AISI 9840 are limited to those which, preceding nitriding, are produced by conventional quenching and tempering. Case hardness lower than the nickel-aluminum steels may be developed, depending on prior heat treatment and nitriding cycle.

AISI 9840 steel is available as aircraft quality stock; forging stock; hot rolled and cold finished bar, rod, and wire; and seamless mechanical tubing.

AISI 9840: Similar Steels (U.S. and/or Foreign). UNS G98400; AMS 6342 C; ASTM A274, A322, A519; SAE J778; (W. Ger.) DIN 1.6511; (Ital.) UNI 38 NiCrMo 4; (U.K.) B.S. 816 M 40

AISI 9840: Approximate Critical Points

Transformation point	Temperature(a) °C	°F	Transformation point	Temperature(a) °C	°F
Ac_1	725	1340	Ar_1	675	1250
Ac_3	780	1435	M_s	300	575
Ar_3	700	1290			

(a) On heating or cooling at 28 °C (50 °F) per hour

Physical Properties

AISI 9840: Thermal Treatment Temperatures

Quenching medium: oil

Treatment	Temperature range °C	°F
Forging	1205 max	2200 max
Annealing(a)	790-845	1450-1550
Normalizing(b)	845-900	1550-1650
Austenitizing	830-855	1525-1575
Tempering	(c)	(c)

(a) Maximum hardness of 207 HB. (b) Hardness of approximately 341 HB. (c) To desired hardness

Machining Data

For machining data on AISI grade 9840, refer to the preceding machining tables for AISI grades 4340 and 4340H.

Mechanical Properties

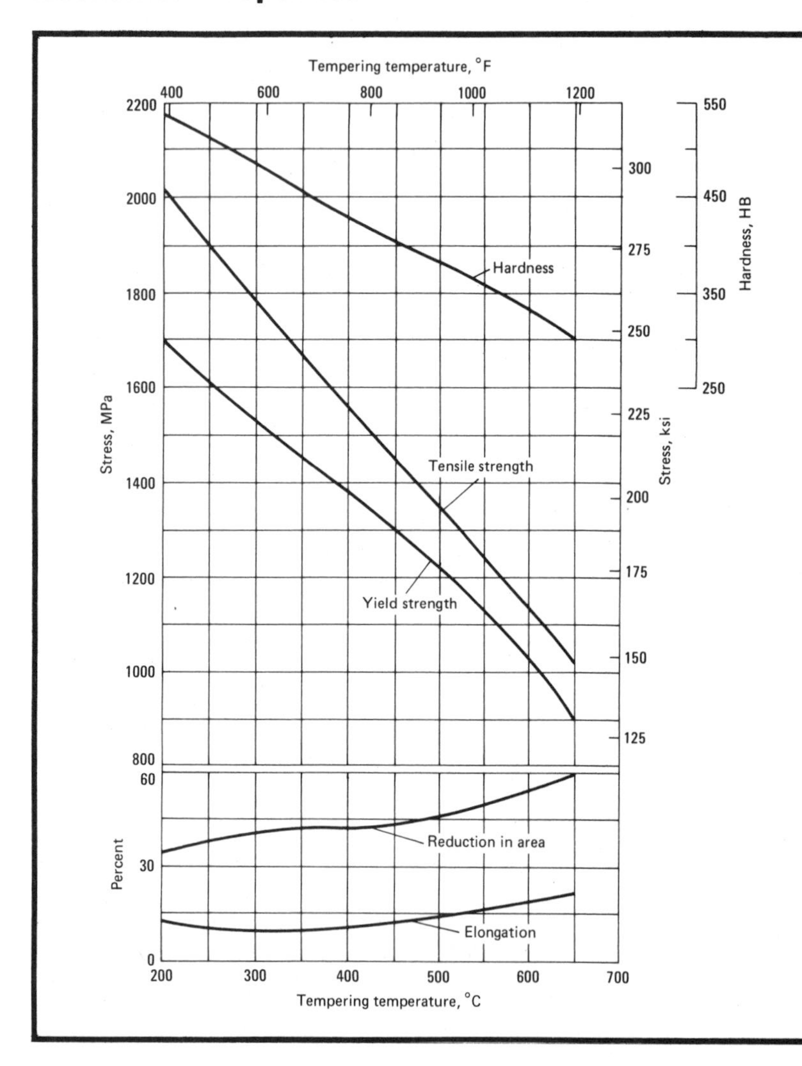

AISI 9840: Effect of Tempering Temperature on Tensile Properties. Normalized at 870 °C (1600 °F); oil quenched from 845 °C (1550 °F); tempered at 56 °C (100 °F) intervals. Specimens were treated in 13.7-mm (0.540-in.) diam and machined to 12.8-mm (0.505-in.) for testing. Tests were conducted using specimens machined to English units. Elongation was measured in 50 mm (2 in.). (Ref 2)

Mechanical Properties (continued)

AISI 9840: Effect of Mass on Tensile Properties After Tempering at 540 °C (1000 °F).

Normalized at 870 °C (1600 °F) in oversize rounds; oil quenched from 845 °C (1550 °F) in sizes indicated; tempered at 540 °C (1000 °F). Specimens were tested in 12.8-mm (0.505-in.) rounds. Tests from bar 38-mm (1.5-in.) diam and larger were taken at half-radius position. Tests were conducted using specimens machined to English units. Elongation was measured in 50 mm (2 in.). (Ref 2)

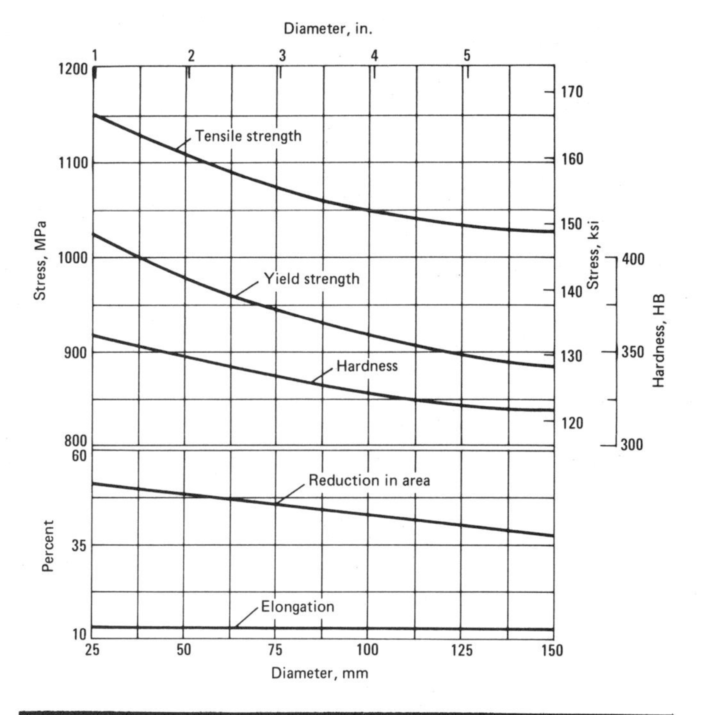

AISI 9840: Effect of Mass on Tensile Properties After Tempering at 650 °C (1200 °F).

Normalized at 870 °C (1600 °F) in oversize rounds; oil quenched from 845 °C (1550 °F) in sizes indicated; tempered at 650 °C (1200 °F). Specimens were tested in 12.8-mm (0.505-in.) rounds. Tests from bar 38-mm (1.5-in.) diam and larger were taken at half-radius position. Tests were conducted using specimens machined to English units. Elongation was measured in 50 mm (2 in.). (Ref 2)

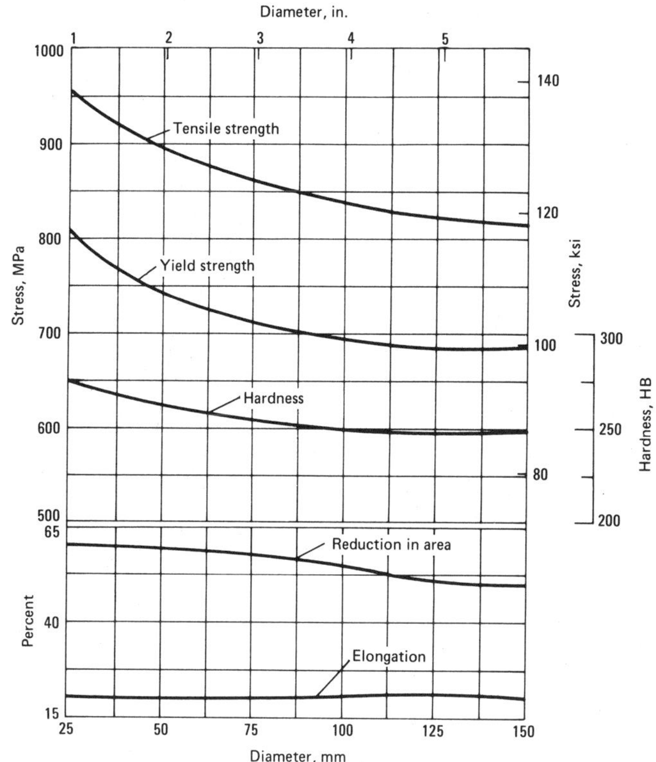

AISI 9840: Case Depth and Hardness After Nitriding (Ref 16)

Specimens which had been quenched and tempered at 540 °C (1000 °F) were nitrided at 525 °C (975 °F) for 40 h with 20 to 30% ammonia dissociation

Nitrided case depth, mm	Nitrided case depth, in.	Nitrided case hardness, HR15N	Core hardness, HRC
0.635-0.762	0.025-0.030	88	35

AISI 9840: Tensile Properties After Quenching in Oil (Ref 16)

Specimens were 25-mm (1-in.) round; tests were conducted using test specimens machined to English units

Tempering temperature		Tensile strength		Yield strength		Elongation(a),	Reduction	Hardness,
°C	°F	MPa	ksi	MPa	ksi	%	in area, %	HB
540	1000	1240	180	1105	160	15	54	361
595	1100	1105	160	965	140	16	56	321
650	1200	965	140	825	120	19	60	280

(a) In 50 mm (2 in.)

High-Strength Steels: Carbon and Low-Alloy

High-strength carbon and high-strength, low-alloy (HSLA) steels provide higher as-rolled tensile properties with higher minimum yield strengths (240 to 825 MPa or 35 to 120 ksi) than the regular carbon steels. Another advantage over regular carbon and structural steels is higher fatigue strength.

With appropriate design considerations, high-strength steels may be used in three ways to achieve durable and economical end products:

- *In reduced thicknesses* to lower weight without impairing strength
- *In the same thickness* to give greater strength and/or fatigue resistance at no increase in weight
- *Combinations of these* to give a stronger yet lighter weight part or product

High-strength steels can be welded using conventional welding processes. However, minor adjustments to standard welding practices may be necessary. Formability and machinability of these steels is less than that of comparable low-carbon and structural steels. Up to 25% more power and up to 30% lower machining speeds may be needed to form these steels. When considering applications of high-strength steels, it is best to consult a steel producer. High-strength steels are available in hot and cold rolled strip and sheet, galvanized strip, and sheet, plate, and bar.

Applications of high-strength steels include heavy-duty highway and off-road vehicles, construction and farm machinery, industrial equipment, storage tanks, mine and railroad cars, barges and dredges, snowmobiles, lawnmowers, and passenger car components. Bridges, power transmission towers, light poles, and building beams and panels are additional uses of these steels.

The choice of a specific high-strength steel depends on a number of application requirements including thickness reduction, corrosion resistance, formability, and weldability. For many applications, the most important factor in the steel selection process is favorable strength-to-weight ratio.

AISI Designation System

The American Iron and Steel Institute (AISI), the American Society for Testing and Materials (ASTM), and the Society of Automotive Engineers (SAE) have designation systems for the high-strength steels. The AISI system governs sheet and strip; ASTM and SAE designations cover sheet, strip, plate, shapes, and bar.

The AISI designation system (Ref 1) for high-strength sheet steel has three basic components: (*a*) the minimum yield strength in ksi, (*b*) a chemical composition classification, and (*c*) a classification for the deoxidation practice.

Yield strength is categorized in 5-ksi increments from 35 to 60 ksi, 10-ksi increments from 60 to 80 ksi, and 20-ksi increments from 80 to 140 ksi. In many of the ASTM specifications the grade number indicates the yield strength.

Chemical composition of steels is designated by the following letter classifications: S, for structural quality; X, for low-alloy; W, for weathering; and D, for dual phase.

S indicates that the steel is structural quality. These steels have compositions containing carbon and manganese; carbon, manganese, and phosphorus; carbon, manganese, and nitrogen; or carbon, manganese, nitrogen, and phosphorus. Included in this category are stress-relief annealed (recovery-annealed) steels, except those of low-alloy (X) composition.

X describes the low-alloy steel grades. Compositions of these steels include niobium, chromium, copper, molybdenum, nickel, silicon, titanium, vanadium, and zirconium, singly or in combination.

W describes a weathering-steel composition containing silicon, phosphorus, copper, nickel, and chromium in various combinations. The weathering-steels have an atmospheric corrosion resistance that is four times greater than that of conventional low-carbon steel (copper content not exceeding 0.02%). These low-alloy steels are grouped separately to make them more accessible to users interested in weathering characteristics.

D designates dual phase grades which have martensite or transformation products dispersed in a ferrite matrix. These specially processed steels develop unusually high work-hardening characteristics. For example, an 80D grade attains 550-MPa (80-ksi) yield strength after a 5% strain in forming. Thus, in processing the steel, the yield strength has been increased by approximately 205 MPa (30 ksi).

Deoxidation practice is also designated by a letter classification: F, for killed-plus-sulfide inclusion control; K, for killed; and O, for nonkilled or semikilled. In the killed-plus-sulfide inclusion control practice (F), special mill practices reduce or modify the inclusions for improved formability.

As a general rule, deoxidation practice is related to formability. The formability ranking (excellent to good to fair) for a steel of any given strength level and any chemical composition level is F > K > O.

Product Descriptors

Structural quality grade (S) steels are typically strengthened by various additions of carbon, manganese, nitrogen, and/or phosphorus, usually without the addition of other alloying elements. The strength levels above 380 MPa (55 ksi) are obtained with variations in the hot rolling, cold rolling, and/or annealing cycles. Recovery annealing is often used for structural quality cold rolled sheet.

Structural quality grades containing nitrogen have a pronounced strain-aging effect and are frequently specified by users for this characteristic. For instance, sheets containing nitrogen typically gain about 105 to 140 MPa (15 to 20 ksi) in yield strength after being strained 10% and aged at room temperature (20 °C or 68 °F) for a week. The addition of heat, as in paint-bake cycles, accelerates this mechanism. However, this hardening effect occurs only where sufficient strain is induced during forming.

Steel grades of structural quality, high-strength sheet are readily weldable with conventional equipment used in joining low-carbon sheet steel. For certain grades, some welding practice modifications are required. Although the formability of these steels is reasonably good, they generally do not form as easily as most low-carbon steels; nor do they form as easily as most microalloyed sheet steels of the same strength level.

Low-alloy grade (X) steels include steels microalloyed with various combinations of niobium, chromium, copper, molybdenum, nickel, silicon, titanium, vanadium, and/or zirconium. Their strength is obtained through the addition of small quantities of these alloying elements. Therefore, the low-alloy grade steels usually provide higher production and service performance.

The inclusion-control, low-alloy grades (XF) are frequently referred to as "improved or better forming" steels, because the sulfide present is reduced in volume to allow more severe forming. The low-alloy grades exhibit good weldability using conventional equipment, but some welding practice adjustments may be required.

Weathering grade (W) steels contain phosphorus, copper, nickel, and chromium in various combinations. They provide atmospheric corrosion resistance that is at least four times that of conventional low-carbon steels without copper (copper not exceeding 0.02%).

Weathering sheet steels and structural shapes have been used in structural applications which allow the development of stable oxide surface layers. Their use is not appropriate, however, in conditions subject to continuous wetting, which prevents stable oxide formation, or where strong electrolytes can form, such as in the presence of chlorides or sulfates. Because of these limitations, weathering steels offer no advantage in highway vehicle use.

In general, surface appearance of these steels after weathering is pleasing and maintenance cost is low. Frequently, painting or other surface protection is not required.

The weathering grades are readily weldable, but may require special welding practices. Changes in welding materials and procedures may be necessary to match base metal properties and help retain weathering characteristics in the welding zone.

Dual phase grade (D) steels have martensite or other hard transformation products dispersed in a ferrite matrix. High work-hardening characteristics and very good ductility are the basis for defining minimum yield strength levels after 5% strain. As-received yield strength levels will typically be 140 to 205 MPa (20 to 30 ksi) less than the yield strength designation.

In certain applications, the use of dual phase steels may permit production of more intricately shaped parts than can be made satisfactorily with other high-strength grades. The work-hardening effect in dual phase steels occurs only where sufficient strain is induced during manufacture of the part. Weldability of these products is similar to that of low-alloy grades.

Applicable Public Specifications. Typical ASTM and SAE designations for high-strength steels in the various product descriptors are given in the table on High-Strength Steels: ASTM and SAE Specifications.The Canadian Standards Association uses the G40.21 classification on the high-strength, low-alloy steels.

Fabricating and Processing

Fabricating and processing procedures for HSLA and high-strength carbon steels are similar to those for the structural and low-carbon steels. The following discusses the slight procedural differences required for HSLA and high-strength carbon steels.

Cold Forming. Any conventional equipment used for cold forming operations, such as a press brake, can be used with HSLA and high-strength carbon steels. In general, these steels are stiffer than carbon steels and require greater force to produce a permanent set. However, thinner sections are often used; therefore forming pressures are slightly greater. Liberal bend radii and increased die clearances are required, along with provision for more springback.

Shearing and Punching. Greater force is required for punching HSLA and high-strength carbon steels of the same thickness as carbon steel, because of the difference in strength. For example, a 345-MPa (50-ksi) minimum yield point HSLA steel would require about 25% greater force than is needed for the ASTM A36 steel or a low-carbon steel with copper. Also, a greater hold-down pressure may be needed in shearing HSLA and high-carbon steels because they have greater shear strengths than the structural carbon grades. It is often necessary to provide larger clearances when shearing HSLA and high-carbon steels.

Drilling and Reaming. Speeds slightly slower than those used for carbon steels produce the best results when drilling and reaming HSLA steels. However, the same feeds should be used.

Sawing and Milling. Speeds about 30% less than those used for structural carbon steel produce the best tool life for high-strength steels. Further benefits can be gained from the use of a coolant.

Gas Cutting. The same practices used with structural carbon steel should be followed; however, when HSLA steels are being cut, the gas cut edge may harden to some degree. The amount of hardening will depend on the thickness and grade of the steel. It is not usually necessary to postheat the cut edge.

Stress Relieving. When necessary to stress relieve HSLA and high-strength carbon steels after cold forming or welding to relieve residual stresses, the temperature should be about 620 °C (1150 °F). Allow about 1 h for each 25 mm (1 in.) of maximum sheet or plate thickness.

Weldability. The HSLA and high-strength carbon steels are readily weldable by all conventional fusion welding processes, as well as by all resistance welding processes, in sheet and strip thicknesses. The principle concern in developing welding procedures for these grades is alloy content or section size. As alloy content or section size of iron-base materials increases, these grades become more susceptible to hydrogen embrittlement or underbead cracking, necessitating the use of low-hydrogen processes or adequate preheats. Sources of hydrogen that must be avoided, other than welding electrodes, include all organic materials, such as oil or grease. Rust scale, which holds moisture, and adverse atmospheric conditions should be avoided.

Toughness. HSLA steels have better notch toughness and less susceptibility to brittle fracture than high-strength carbon and structural low-carbon steels. Generally, all materials approach brittleness as the temperature is lowered.

Fatigue Limit. HSLA steels have a higher endurance limit than that of carbon steels of the same size and shape. The axial tension-compression endurance limit of samples with as-rolled mill surfaces is approximately 45% of the ultimate strength.

Modulus of elasticity in tension is 195 to 205 GPa (28 000 to 30 000 ksi). The modulus of elasticity of HSLA and high-strength carbon steels is similar to that of other carbon and alloy steels.

Shear modulus or modulus of rigidity is about 83 GPa (12 000 ksi). Shear modulus, too, is similar for the HSLA and high-strength carbon steels and for other carbon and alloy steels.

SAE Recommended Practice – SAE J410c (Ref 6)

The SAE numbering system for the high-strength, low-alloy steels uses a 900 number. The last two digits of the number represent the minimum yield point. Steels containing strengthening elements, such as niobium, vanadium, or nitrogen, added singly or in combination, are designated by the suffix X. These steels are usually made semikilled. Fully killed steel made to fine-grain practice may be specified with the additional suffix K. An example is SAE 950XK. Grades 945 and 950, which have more than one composition, are designated by the suffix A or other letters.

Description of Grades

Grade 942X is a niobium- or vanadium-treated, carbon-manganese, high-strength steel similar to 945X and 945C, except for somewhat improved welding and forming properties.

Grade 945A is a high-strength, low-alloy steel with excellent welding characteristics, both arc and resistance, and the best formability, weldability, and low-temperature notch toughness of the high-strength steels. It is generally used in sheet, strip, and light plate thickness.

Grade 945C is a carbon-manganese high strength steel with satisfactory arc welding properties, if adequate precautions are observed. It is similar to grade 950C, except that lower carbon and manganese improve arc welding characteristics, formability, and low-temperature notch toughness at some sacrifice in strength.

Grade 945X is a niombium- or vanadium-treated carbon-manganese, high-strength steel similar to 945C, except for somewhat improved welding and forming properties.

Grade 950A is a high-strength, low-alloy steel with good weldability, both arc and resistance, good low-temperature notch toughness, and good formability, It is generally used in sheet, strip, and light plate thickness.

Grade 950B is a high-strength, low-alloy steel with satisfactory arc welding properties and fairly good low-temperature notch toughness and formability.

Grade 950C is a carbon-manganese, high-strength steel which can be arc welded with special precautions, but is unsuitable for resistance welding. The formability and toughness are fair.

Grade 950D is a high-strength, low-alloy steel with good weldability, both arc and resistance, and fairly good formability. Where low-temperature properties are important, the effect of phosphorus in conjunction with other elements present should be considered.

Grade 950X is a niobium- or vanadium-treated, carbon-manganese, high-strength steel similar to 950C, except for somewhat improved welding and forming properties.

Grades 955X, 960X, 965X, 970X, and 980X are steels similar to 945X and 950X, but with higher strength obtained by increased amounts of strengthening elements such as carbon or manganese, or by the addition of nitrogen up to about 0.015%. This increased strength is accompanied by reduced formability and, usually, decreased weldability. Toughness will vary considerably with composition and mill practice.

REFERENCES

1. High Strength Steel Source Guide, AISI SG603D, Committee of Sheet Steel Producers, American Iron and Steel Institute, Washington, D.C., Feb 1981
2. DOFASCO High Strength Low Alloy Steels, Dominion Foundries and Steel, Ltd., Hamilton, Ontario, Canada, June 1978
3. High Strength Alloy Steels and High Strength Carbon Steels, ADV 2577, Republic Steel Corporation, Cleveland, Ohio
4. Van Steels Data Sheet, Jones & Laughlin Steel Corporation, Pittsburgh, PA
5. Mayari R Weathering Steel, Bethlehem Steel Corporation, Bethlehem, PA, Dec 1974
6. SAE Recommended Practice for High Strength Low Alloy Steel—SAE J410c, *SAE Handbook,* Part 1, Society of Automotive Engineers, Warrendale, PA, 1981

Structural Quality (S) Steels: ASTM Recommended Practice for Mechanical Properties

ASTM designation	Grade	Tensile strength MPa	Tensile strength ksi	Yield strength MPa	Yield strength ksi	Elongation(a), %	Formability rating(b)
Sheet and strip							
A570	A	310	45	170	25	27(c)	...
	B	340	49	205	30	25(c)	...
	C	360	52	230	33	23(c)	9
	D	380	55	275	40	21(c)	7
	E	400	58	290	42	19(c)	...
A611	A	290	42	170	25	26	...
	B	310	45	205	30	24	...
	C	330	48	230	33	22	9
	D	360	52	275	40	20	8
	E	565	82	550	80(d)	...	1
A446	A	310	45	230	33	20	9
	B	360	52	255	37	18	7
	C	380	55	275	40	16	7
	D	450	65	345	50	12	4
	E	565	82	550	80(d)	...	1
	F	480	70	345	50	12	...
Plate, bar, and shapes							
A440	(e)	485	70	345	50	18(f)	...
	(g)	460	67	315	46	21	...

(a) In 50 mm (2 in.). (b) Formability is rated from 10 to 1 for best to worst, respectively. A rating of 10 represents AISI 1010 low-carbon, commercial quality, sheet steel (Ref 6). (c) Thickness of 5.84 to 2.47 mm (0.2299 to 0.0972 in.). (d) Because there is no halt in the gage or drop in the beam, the yield strength, which approaches the tensile strength, should be taken as the stress at 0.5% elongation. (e) Thickness of 19 mm (0.75 in.) and under. (f) Elongation in 200 mm (8 in.). (g) Thickness of 19 to 38 mm (0.75 to 1.5 in.)

High-Strength Steels: ASTM and SAE Specifications

Quality descriptor and abbreviation	Form	Specification
Structural quality (S)	Sheet and strip	ASTM A570, A611, A446
	Plate, bar, shapes	ASTM A440
Low-alloy (X)	Sheet and strip	ASTM A607, A715; SAE J410c
	Plate, bar, shapes	ASTM A242, A441, A572, A588, A656, A709 grade 50; SAE J410c
Weathering (W)	Sheet and strip	ASTM A606 type 4
	Plate, bar, shapes	ASTM A242 type 2, A588 grade D, A709 grades 50W and 100W
Dual phase (D)	...	None

Structural Quality (S) Steels: ASTM Recommended Practice for Chemical Composition

ASTM designation	Grade	C max	Mn	P max	S max	Cu(a)
A570	A, B, C	0.25	0.25-0.60	0.04	0.04	0.20
	D, E	0.25	0.60-0.90	0.04	0.04	0.20
A611	A, B, C, E	0.20	0.60 max	0.04	0.04	0.20
	D	0.20	0.90 max	0.04	0.04	0.20
A446	A, E	0.20	...	0.04	0.04	0.20
	B	0.20	...	0.10	0.04	0.20
	C	0.25	...	0.10	0.04	0.20
	D	0.40	...	0.20	0.04	0.20
	F	0.50	...	0.04	0.04	0.20
A440 (b)	...	0.28	1.1-1.6	0.04	0.05	0.20

Chemical composition, %

(a) Minimum content when a copper steel is specified. (b) Contains a maximum of 0.30% silicon

Low-Alloy Sheet, Strip, and Plate: Recommended Inside Bend Radius (Ref 3)

***t* is material thickness**

Grade	Up to 4.57 mm (0.180 in.)	4.57-5.82 mm (0.180-0.229 in.)
Low-alloy, killed or semikilled (XK) steels		
45	1.5*t*	2*t*
50	2*t*	2.5*t*
55	2*t*	2.5*t*
60	2.5*t*	3*t*
65	3*t*	3.5*t*
70	3.5*t*	4*t*
Low-alloy, improved formability (XF) steels		
50	0.5*t*	1*t*
60	1*t*	1.5*t*
70	1*t*	1.5*t*
80	1*t*	1.5*t*

Recommended inside bend radius for material thickness of:

Low-Alloy Sheet, Strip, and Plate: Recommended Inside Bend Radius (Ref 2)

***t* is material thickness**

Grade	Up to 3.18 mm (0.125 in.)	3.18-6.4 mm (0.125-0.25 in.)	6.4-13 mm (0.25-0.50 in.)
Low-alloy, nonkilled (XO) steels			
45	1*t*	1.5*t*	2*t*
50	1.5*t*	2*t*	2.5*t*
55	2*t*	2.5*t*	3*t*
60	2.5*t*	3*t*	3.5*t*
65	3*t*	3.5*t*	4*t*
70	3.5*t*	4*t*	4.5*t*
80	4*t*	4.5*t*	5*t*
Low-alloy, improved formability (XF) steels			
45	1*t*	1*t*	1.5*t*
50	1*t*	1*t*	2*t*
60	1.5*t*	2*t*	2.5*t*
70	2*t*	2.5*t*	3*t*
80	2.5*t*	3*t*	3.5*t*

Recommended inside bend radius for material thickness of:

Low-Alloy (X) Steels: ASTM Recommended Practice for Chemical Composition

ASTM designation	Grade or type	C	Mn	P max	S max	Si max	Nb min	V min	Other
		Chemical composition, %							
A607	Grade 45	0.22	1.35	0.04	0.05	...	(a)	(a)	...
	Grade 50	0.23	1.35	0.04	0.05	...	(a)	(a)	...
	Grade 55	0.25	1.35	0.04	0.05	...	(a)	(a)	...
	Grade 60	0.26	1.50	0.04	0.05	...	(a)	(a)	...
	Grade 65	0.26	1.50	0.04	0.05	...	(a)	(a)	0.012 N
	Grade 70	0.26	1.65	0.04	0.05	...	(a)	(a)	0.012 N
A715	Type 1(b)	0.15	1.65	0.025	0.035	0.01	...	...	0.05 Ti
	Type 2(b)	0.15	1.65	0.025	0.035	0.60	...	0.02	0.005 N
	Type 3(b)	0.15	1.65	0.025	0.035	0.60	0.005	0.08	0.02 N
	Type 4(b)	0.15	1.65	0.025	0.035	0.90	0.005-0.06	...	0.05 Zr, 0.08 Cr, 0.01 Ti, 0.0025 B
	Type 5(c)	0.15	1.65	0.025	0.035	0.30	0.03	...	0.20 Mo
	Type 6(d)	0.15	1.65	0.025	0.035	0.90	0.005-0.10	...	...
	Type 7(d)	0.15	1.65	0.025	0.035	...	(e)	(e)	0.02 N
A242	Type 1	0.15	1.00	0.15	0.05	...	...	...	0.20 Cu
	Type 2	0.20	1.35	0.04	0.05	...	...	...	0.20 Cu
A441	...	0.22	0.85-1.25	0.04	0.05	0.30	...	0.02	0.20 Cu
A572	Grade 42	0.21	1.35	0.04	0.05	0.30	...	...	...
	Grade 45	0.22	1.35	0.04	0.05	0.30	...	...	...
	Grade 50	0.23	1.35	0.04	0.05	0.30	...	...	...
	Grade 55	0.25	1.35	0.04	0.05	0.30	...	...	...
	Grade 60	0.26	1.35	0.04	0.05	0.30	...	...	...
	Grade 65(f)	0.23	1.65	0.04	0.05	0.30	...	...	...
	Grade 65(g)	0.26	1.35	0.04	0.05	0.30	...	...	...
A588	Grade A	0.10-0.19	0.90-1.25	0.04	0.05	0.15-0.30	...	0.02-0.10	0.04-0.65 Cr, 0.25-0.40 Cu
	Grade B	0.20	0.75-1.25	0.04	0.05	0.15-0.30	...	0.01-0.10	0.25-0.50 Ni, 0.40-0.70 Cr, 0.20-0.40 Cu
	Grade C	0.15	0.80-1.35	0.04	0.05	0.15-0.30	...	0.01-0.10	0.25-0.50 Ni, 0.30-0.50 Cr, 0.20-0.50 Cu
	Grade D	0.10-0.20	0.75-1.25	0.04	0.05	0.50-0.90	0.04	...	0.50-0.90 Cr, 0.30 max Cu, 0.05-0.15 Zr
	Grade E	0.15	1.20	0.04	0.05	0.15-0.30	0.04	...	0.75-1.25 Ni, 0.10-0.25 Mo, 0.50-0.80 Cu, 0.05-0.15 Zr
	Grade F	0.10-0.20	0.50-1.00	0.04	0.05	0.30	...	0.05	0.40-1.10 Ni, 0.30 max Cr, 0.10-0.20 Mo, 0.30-100 Cu
	Grade G	0.20	1.20	0.04	0.05	0.25-0.70	...	0.02-0.10	0.80 max Ni, 0.50-1.00 Cr, 0.10 max Mo, 0.30-0.50 Cu, 0.07 max Ti
	Grade H	0.20	1.25	0.035	0.04	0.25-0.75	...	...	0.30-0.60 Ni, 1.00-0.25 Cr, 0.15 max Mo, 0.20-0.35 Cu, 0.005-0.03 Ti
	Grade J	0.30	0.60-1.00	0.04	0.05	0.30-0.50	...	...	0.50-0.70 Ni, 0.30 min Cu, 0.03-0.05 Ti
A656	Grade 1	0.18	1.60	0.04	0.05	0.60	0.05-0.15	...	0.02 Al, 0.005-0.03 Ni
	Grade 2	0.15	0.90	0.04	0.05	0.10	...	...	0.01 Al, 0.05-0.50 Ti
A709	Grade 50	0.23	1.35	0.04	0.05	...	...	...	...

(a) Chemical composition includes either 0.005% niobium or 0.01% vanadium. (b) AISI grades 70 and 80. (c) AISI grade 80. (d) AISI grades 50, 60, 70, and 80. (e) Chemical composition includes 0.005% of either niobium or vanadium. (f) Thickness of 13 to 32 mm (0.50 to 1.25 in.). (g) Thickness of 13 mm (0.5 in.)

Weathering, Low-Alloy Steels: ASTM Recommended Practice for Chemical Composition

ASTM designation	Grade or type	C	Mn	P max	S max	Cu
		Chemical composition, %				
A606	Type 4	0.22	1.25	...	0.05	...
A242	Type 2	0.20	1.35	0.04	0.05	0.20 min
A588(a)	Grade D	0.10-0.20	0.75-1.25	0.04	0.05	0.20-0.50
A709	Grade 50W	0.20	1.35	0.04	0.05	0.20 min
	Grade 100W	0.10-0.21	0.40-1.50	0.035	0.05	0.20 min

(a) Chemical composition also contains 0.50 to 0.90% silicon, 0.50 to 0.90% chromium, 0.05 to 0.15% zirconium, and 0.04% niobium

Low-Alloy (X) Steels: ASTM Recommended Practice for Mechanical Properties

ASTM designation	Grade or type	Tensile strength MPa	Tensile strength ksi	Yield strength MPa	Yield strength ksi	Elongation, min %, In 50 mm (2 in.)	Elongation, min %, In 200 mm (8 in.)	Formability rating(a)
Sheet and strip								
A607	Grade 45	415	60	310	45	25 (hot rolled) 22 (cold rolled)	...	7(XO), 8(XK)
	Grade 50	450	65	345	50	22 (hot rolled) 20 (cold rolled)	...	6(XO), 7(XK)
	Grade 55	485	70	380	55	20 (hot rolled) 18 (cold rolled)	...	...
	Grade 60	515	75	415	60	18 (hot rolled) 16 (cold rolled)	...	5(XK)
	Grade 65	550	80	450	65	16 (hot rolled) 15 (cold rolled)	...	...
	Grade 70	585	85	485	70	14 (hot rolled) 14 (cold rolled)	...	...
A715	Grade 50	415	60	345	50	24	19	7(XK), 8(XF)
	Grade 60	485	70	415	60	22	18	6(XF)
	Grade 70	550	80	485	70	20	17	5(XF)
	Grade 80	620	90	550	80	18	16	5(XF)
Plate, bar, and shapes								
A242	(b)	485	70	345	50	18	21	...
	(c)	460	67	315	46	18	21	...
A441	(b)	485	70	345	50	18	...	...
	(c)	460	67	315	46	18	21	...
A572	Grade 42	415	60	290	42	24	20	...
	Grade 45	415	60	310	45	22	19	...
	Grade 50	450	65	345	50	21	18	...
	Grade 55	485	70	380	55	20	17	...
	Grade 60	515	75	415	60	18	16	...
	Grade 65	550	80	450	65	17	15	...
A588	(d)	485	70	345	50	21	18	...
A656	...	655	95	550	80	...	12	...
A709	Grade 50	450	65	345	50	21	18	...

(a) Formability is rated from 10 to 1 for best to worst, respectively. A rating of 10 represents AISI 1010 low-carbon, commercial quality, sheet steel (Ref 6). The letters XO indicate low-alloy nonkilled, XK indicates low-alloy killed, XF indicates low-alloy improved formability. (b) Thickness of 19 mm (0.75 in.) (c) Thickness of 19 to 38 mm (0.75 to 1.5 in.) (d) Thickness of up to 100 mm (4 in.)

Weathering, Low-Alloy Steels: ASTM Recommended Practice for Mechanical Properties

ASTM designation	Grade or condition	Tensile strength MPa	Tensile strength ksi	Yield strength MPa	Yield strength ksi	Elongation, min %, In 50 mm (2 in.)	Elongation, min %, In 200 mm (8 in.)	Formability rating(a)
Sheet and strip								
A606	Hot rolled, annealed	450	65	310	45	22	...	5
	Cold rolled	450	65	310	45	22	...	5
Plate, bar, and shapes								
A242	(b)	485	70	345	50	21	18	...
A709	Grade 50W	485	70	345	50	22	18	...
	Grade 100W	690	100	620	90	17	...	...

(a) Formability is rated from 10 to 1 for best to worst, respectively. A rating of 10 represents AISI 1010 low-carbon, commercial quality, sheet steel (Ref 6). (b) Thickness of up to 19 mm (0.75 in.)

Low-Alloy, Improved Formability (XF) Sheet, Strip, and Plate: Recommended Inside Bend Radius (Ref 4)

***t* is material thickness**

Grade	Up to 6.4 mm (0.25 in.) Longitudinal	Up to 6.4 mm (0.25 in.) Transverse	6.4-13 mm (0.25-0.50 in.) Longitudinal	6.4-13 mm (0.25-0.50 in.) Transverse
50	0.5*t*	1*t*	1*t*	1.5*t*
60	1*t*	1.5*t*	1.5*t*	2*t*
70	1.5*t*	2*t*	2*t*	2.5*t*
80	1.5*t*	2*t*	2*t*	2.5*t*

Weathering Killed (WK), Low-Alloy Sheet, Strip, and Plate: Recommended Inside Bend Radius (Ref 5)

***t* is material thickness**

Grade	Up to 1.6 mm (0.063 in.)	1.6-6.4 mm (0.063-0.25 in.)	6.4-13 mm (0.25-0.50 in.)
50	1*t*	2*t*	3*t*

SAE Designations: Chemical Composition

Chemical compositions include 0.05% maximum sulfur and 0.90% maximum silicon

SAE grade	C	Mn	P	SAE grade	C	Mn	P
	Chemical composition, max %				Chemical composition, max %		
942X	0.21	1.35	0.04	950D	0.15	1.00	0.15
945A	0.15	1.00	0.04	950X	0.23	1.35	0.04
945C	0.23	1.40	0.04	955X	0.25	1.35	0.04
945X	0.22	1.35	0.04	960X	0.26	1.45	0.04
950A	0.15	1.30	0.04	965X	0.26	1.45	0.04
950B	0.22	1.30	0.04	970X	0.26	1.65	0.04
950C	0.25	1.60	0.04	980X	0.26	1.65	0.04

SAE Designations: Tensile Properties

Determined in accordance with ASTM A370; all values given are minimums

SAE grade	Product form	Tensile strength MPa	Tensile strength ksi	Yield strength(a) MPa	Yield strength(a) ksi	Elongation, min %: In 50 mm (2 in.)	Elongation, min %: In 200 mm (8 in.)
942X	Plate, shapes, and bar to 100 mm (4 in.) incl	415	60	290	42	24	20
945A, C	Sheet and strip	415	60	310	45	22	...
	Plate, shapes, and bar to 13 mm (0.5 in.) incl	450	65	310	45	22	18
	13-38 mm (0.5-1.5 in.) incl	425	62	290	42	24	19
	38-76 mm (1.5-3.0 in.) incl	425	62	276	40	24	19
945X	Sheet and strip	415	60	310	45	25	...
	Plate, shapes, and bar to 38 mm (1.5 in.) incl	415	60	310	45	22	19
950A, B, C, D	Sheet and strip	485	70	345	50	22	...
	Plate, shapes, and bar to 13 mm (0.5 in.)	485	70	345	50	22	18
	13-38 mm (0.5-1.5 in.) incl	460	67	310	45	24	19
	38-76 mm (1.5-3.0 in.) incl	435	63	290	42	24	19
950X	Sheet and strip	450	65	345	50	22	...
	Plate, shapes, and bar to 38 mm (1.5 in.) incl	450	65	345	50	...	18
955X	Sheet and strip	485	70	380	55	20	...
	Plate, shapes, and bar to 38 mm (1.5 in.) incl	485	70	380	55	...	17
960X	Sheet and strip	515	75	415	60	18	...
	Plate, shapes, and bar to 38 mm (1.5 in.) incl	515	75	415	60	...	16
965X	Sheet and strip	550	80	450	65	16	...
	Plate, shapes, and bar to 19 mm (0.75 in.) incl	550	80	450	65	...	15
970X	Sheet and strip	585	85	485	70	14	...
	Plate, shapes, and bar to 19 mm (0.75 in.) incl	585	85	485	70	...	14
980X	Sheet and strip	655	95	550	80	12	...
	Plate to 10 mm (0.38 in.) incl	655	95	550	80	...	10

(a) 0.2% offset

SAE High-Strength, Low-Alloy Steel Plate, Shapes, and Bar: 180° Bend Test Requirements

SAE grade	Ratio of bend diameter to material thickness of: To 10 mm (0.38 in.) incl	10-13 mm (0.38-0.50 in.) incl	13-19 mm (0.50-0.75 in.) incl	19-25 mm (0.75-1.0 in.) incl	25-38 mm (1.0-1.5 in.) incl	38-50 mm (1.5-2 in.) incl	50-100 mm (2-4 in.) incl
942X	1	1	1	1.5	2	2.5	3
945A	1	1	2	2	2	3	3
945C	1	1	2	2	2	3	3
945X	1	1	1	1.5	2	2.5	...
950A	1	1	2	2	2	3	3(a)
950B	1	1	2	2	2	3	3(a)
950C	1	1	2	2	2	3	3(a)
950D	1	1	2	2	2	3	3(a)
950X	1	1	1	1.5	2.5	3	...
955X	1	1.5	1.5	2	3	3.5(b)	...
960X	1.5	2	2	2.5	3	...	...
965X	2	2.5	3	...	...	...	...
970X	3	...	...	...	...	...	...
980X	3	...	...	...	...	...	...

(a) Thickness of 50-75 mm (2-3 in.) inclusive. (b) Applicable to webs of structural shapes

SAE High-Strength, Low-Alloy Steels: Approximate Order of Increasing Excellence

Weldability	Formability	Toughness
980X	980X	980X
970X	970X	970X
965X	965X	965X
960X	960X	960X
955X, 950C, 942X	955X	955X
945C	950C	945C, 950C, 942X
950B, 950X	950D	945X, 950X
945X	950B, 950X, 942X	950D
950D	945C, 945X	950B
950A	950A	950A
945A	945A	945A

SAE High-Strength, Low-Alloy Steel Sheet, Strip, and Plate: Suggested Minimum Inside Radii for Cold Bending

SAE grade	Ratio of bend radius to material thickness of: To 4.57 mm (0.180 in.)	4.60-6.35 mm (0.181-0.250 in.)	6.38-12.7 mm (0.251-0.500 in.)
942X	...	1	2
945A, C	1	2	2.5
945X	1	1	2
950A, B, C, D	1	2	3
950X	1.5	2.5	2.5
955X	2	3	3
960X	2.5	3.5	3.5
965X	3	4	4
970X	3.5	4.5	4.5
980X	3.5	4.5	4.5(a)

(a) Available to 9.53 mm (0.375 in.) inclusive

Stainless and Heat-Resisting Steels

Stainless and heat-resisting alloys possess unusual resistance to attack from corrosive media at atmospheric and elevated temperatures. These steels have widely differing characteristics, and frequently, special qualifications which make them particularly well suited for specific uses. Thus, stainless and heat-resisting alloys are produced to cover a wide range of mechanical and physical properties for particular applications. The most familiar standard types are listed in the table on chemical compositions of stainless and heat-resisting steels.

Heat-resisting steels contain at least 4% chromium; stainless steels are basically low-carbon alloy steels containing at least 11% chromium. The chromium addition in these steels provides resistance to corrosion and scaling at elevated temperatures. Carbon contents are normally held at 0.20% or less, except in some steel types which require higher hardenability for applications such as bearings and fine cutlery. To maintain stainless qualities in these steels, however, the increased carbon content requires increased chromium. Other elements, such as nickel, molybdenum, niobium (columbium), titanium, aluminum, sulfur, copper, and selenium are added to produce types with special properties.

The "stainlessness" of all stainless steels is attributable to the chromium addition. Chromium contributes to the formation of a thin, passive, surface film, probably oxide, that protects against corrosive attack. If the passive film is penetrated in an oxidizing atmosphere, it will re-form and continue to protect against corrosion. Under certain reducing atmospheres, when the passive surface deteriorates and cannot re-form, the stainless steel exhibits poor corrosion resistance.

Stainless steels are selected for the qualities of high strength, excellent workability, abrasion and erosion resistance, magnetic properties, attractive appearance, and ease of cleaning and sterilizing the smooth, dense surfaces.

Four general classes of stainless steels have been developed to provide such varied properties. The four classes are: the austenitic types of the chromium-nickel-manganese 200 series and the chromium-nickel 300 series; the straight-chromium, hardenable 400-series, martensitic types; the straight-chromium, nonhardenable, 400-series ferritic types; and the precipitation-hardening, chromium-nickel alloys with additional elements that are hardenable by solution treating and aging.

The austenitic types have high ductility, low yield strength, and high tensile strength which permit deep drawing and forming. These steels are readily welded by most of the conventional fusion and resistance welding processes, and by brazing. Through cold working they may be work hardened to high levels, but not as high as can be obtained by heat treating the hardenable types of the 400 series. The austenitic types are nonmagnetic as annealed, and depending on the composition, may become slightly magnetic when cold worked. They have desirable properties at cryogenic temperatures, and surpass the strength of any 400 series type at temperatures above 540 °C (1000 °F).

The austenitic stainless steels provide the best corrosion resistance of all the stainless steels. However, they are subject to chromium-carbide formation under certain circumstances, which usually results in intergranular corrosion. The fully annealed steel is free from this weakness, but if, in fabrication or use, the alloy is heated within the range of 425 to 815 °C (800 to 1500 °F), or cooled slowly through this range after heating to a higher temperature, the carbides in the steel tend to accumulate at the grain boundaries, and form continuous paths for corrosion. When such heating is done, it should be followed by a corrective anneal.

With extra low carbon types, or types 321 or 347 which are stabilized by titanium or niobium (columbium) respectively, annealing is not usually required for corrosion resistance, but may be necessary to relieve strains induced by severe forming operations.

The martensitic steels of the 400 series can be heat treated to a wide range of useful hardness and strength levels. These steels contain chromium as the major alloying element. Chromium and carbon contents are balanced so that the soft austenitic phase which develops at high temperatures transforms to the hard martensitic phase during cooling to room temperature.

The martensitic steels are magnetic at all temperatures and are selected for resistance to wear and abrasion. These types are generally not as corrosion resistant as the ferritic and austenitic types. They have fair cold forming characteristics and can be welded, but usually require annealing after forming or welding to prevent cracking, then hardening to develop high strength and optimum corrosion resistance.

During heating at the hardening temperature of the martensitic steels, the chromium carbides are dissolved and the carbon is uniformly distributed throughout the metallic structure. If the quench is sufficiently rapid, the carbon does not recombine with chromium, which is then present in sufficient amounts to provide optimum corrosion resistance.

The ferritic types of the 400 series are normally used for their corrosion resistance and resistance to scaling at elevated temperatures, rather than for high-strength purposes. They are characterized by a chromium-to-carbon balance which suppresses the development of austenite at high temperatures. Because little or no austenite is present, these grades do not transform to martensite upon cooling, but remain ferritic throughout their normal operating range. For all practical purposes, the ferritic steels are nonhardenable by the usual method of heating and quenching.

The nonhardenable grades can be formed and welded, and can be buffed to a high luster resembling chromium plate. They work harden to a moderate extent, normally to a maximum hardness level of 25 HRC and are magnetic in all conditions.

Precipitation-hardening stainless steels are divided into three types: austenitic, semiaustenitic, and martensitic or maraging. The difference in the types is the metallurgical microstructure that exists after heat treatment. There is no major distinction based on the precipitate formed.

The austenitic precipitation-hardening grades retain the austenitic structure at room temperature and at all times during and after heat treatment. Because there is no martensitic transformation, the austenitic grades have the lowest strength of the precipitation-hardening stainless steels at room temperature.

The semiaustenitic grades have compositions lower in nickel and higher in carbon than the straight austenitic grades. The initial high-temperature anneal or solution treatment produces an austenitic solid-solution microstructure similar to that of the austenitic grades. The semiaustenitic grades then receive a conditioning treatment at 925 to 1035 °C (1700 to 1900 °F) which precipitates a portion of the carbon as a complex high-alloy carbide, alters the matrix composition, and raises the M_s temperature. Transformation to martensite occurs when cooled to room temperature or by subzero cooling to −75 °C (−100 °F). The transformed martensite is then further tempered or aged to develop its final properties.

The martensitic precipitation-hardening stainless steels have their alloy content reduced or adjusted so that the M_s temperature is high enough to transform the elevated-temperature austenite phase to martensite on cooling to room temperature. The final properties are then developed by one final low-temperature heat treatment.

The martensitic structure will not stretch form as well as austenite, and the lower chromium content may somewhat reduce corrosion resistance. The martensitic grades may not be as useful at elevated temperatures as the semiaustenitic grades because they have a tendency to overage or soften by the formation of reverted or retained austenite, or overage simply by particle growth.

REFERENCES

1. *Metals Handbook,* 9th ed., Vol 3, American Society for Metals, 1980
2. *Machining Data Book,* 3rd ed., Vol 1 and 2, Metcut Research Associates, Inc., Cincinnati, 1980
3. Mechanical and Physical Properties of the Austenitic Chromium-Nickel Stainless Steels at Subzero Temperatures, The International Nickel Company, Inc., New York, 1970
4. Mechanical and Physical Properties of the Austenitic Chromium-Nickel Stainless Steels at Ambient Temperatures, The International Nickel Company, Inc., New York, 1963
5. Mechanical and Physical Properties of the Austenitic Chromium-Nickel Stainless Steels at Elevated Temperatures, The International Nickel Company, Inc., New York, 1963
6. *Steel Products Manual: Stainless and Heat Resisting Steels,* American Iron and Steel Institute, Washington, D.C., March 1979
7. Universal-Cyclops Specialty Steel Div., Cyclops Corp., 650 Washington Rd., Pittsburgh, PA 15228
8. Allegheny Ludlum Industries, Inc., 1800 Two Oliver Plaza, Pittsburgh, PA 15222
9. Carpenter Technology Corp., 101 W. Bern St., Reading, PA 19603
10. Crucible Specialty Metals Div., P.O. Box 977, Syracuse, NY 13201
11. Eastern Stainless Steel Co., Div. of Eastmet Corp., P.O. Box 1975, Baltimore, MD 21203
12. Armco Steel Corp., Stainless Steel Div., Middletown, OH 45043
13. Lewis Laboratories of the National Aeronautics and Space Administration, 21000 Brookpark Rd., Cleveland, OH 44135
14. Technical Bulletin: Ferralium® Alloy 255, Cabot Corporation, High Technology Materials Div., 1020 W. Park Ave., Kokomo, IN 46901

Chemical Compositions of Standard Stainless and Heat Resisting Steels

Compositions are maximum unless otherwise stated

Type No.	UNS No.	Chemical composition, % C	Mn	P	S	Si	Cr	Ni	Mo	Other elements
201	S20100	0.15	5.50-7.50	0.060	0.030	1.00	16.00-18.00	3.50-5.50		N 0.25
202	S20200	0.15	7.50-10.00	0.060	0.030	1.00	17.00-19.00	4.00-6.00		N 0.25
205	S20500	0.12-0.25	14.00-15.50	0.060	0.030	1.00	16.50-18.00	1.00-1.75		N 0.32-0.40
301	S30100	0.15	2.00	0.045	0.030	1.00	16.00-18.00	6.00-8.00		
302	S30200	0.15	2.00	0.045	0.030	1.00	17.00-19.00	8.00-10.00		
302B	S30215	0.15	2.00	0.045	0.030	2.00-3.00	17.00-19.00	8.00-10.00		
303	S30300	0.15	2.00	0.20	0.15 min	1.00	17.00-19.00	8.00-10.00	0.60(a)	
303Se	S30323	0.15	2.00	0.20	0.060	1.00	17.00-19.00	8.00-10.00		Se 0.15 min
304	S30400	0.08	2.00	0.045	0.030	1.00	18.00-20.00	8.00-10.50		
304L	S30403	0.03	2.00	0.045	0.030	1.00	18.00-20.00	8.00-12.00		
	S30430	0.08	2.00	0.045	0.030	1.00	17.00-19.00	8.00-10.00		Cu 3.00-4.00
304N	S30451	0.08	2.00	0.045	0.030	1.00	18.00-20.00	8.00-10.50		N 0.10-0.16
305	S30500	0.12	2.00	0.045	0.030	1.00	17.00-19.00	10.50-13.00		
308	S30800	0.08	2.00	0.045	0.030	1.00	19.00-21.00	10.00-12.00		
309	S30900	0.20	2.00	0.045	0.030	1.00	22.00-24.00	12.00-15.00		
309S	S30908	0.08	2.00	0.045	0.030	1.00	22.00-24.00	12.00-15.00		
310	S31000	0.25	2.00	0.045	0.030	1.50	24.00-26.00	19.00-22.00		
310S	S31008	0.08	2.00	0.045	0.030	1.50	24.00-26.00	19.00-22.00		
314	S31400	0.25	2.00	0.045	0.030	1.50-3.00	23.00-26.00	19.00-22.00		
316	S31600	0.08	2.00	0.045	0.030	1.00	16.00-18.00	10.00-14.00	2.00-3.00	
316F	S31620	0.08	2.00	0.20	0.10 min	1.00	16.00-18.00	10.00-14.00	1.75-2.50	
316L	S31603	0.03	2.00	0.045	0.030	1.00	16.00-18.00	10.00-14.00	2.00-3.00	
316N	S31651	0.08	2.00	0.045	0.030	1.00	16.00-18.00	10.00-14.00	2.00-3.00	N 0.10-0.16
317	S31700	0.08	2.00	0.045	0.030	1.00	18.00-20.00	11.00-15.00	3.00-4.00	
317L	S31703	0.03	2.00	0.045	0.030	1.00	18.00-20.00	11.00-15.00	3.00-4.00	
321	S32100	0.08	2.00	0.045	0.030	1.00	17.00-19.00	9.00-12.00		Ti 5 × C min
329	S32900	0.10	2.00	0.040	0.030	1.00	25.00-30.00	3.00-6.00	1.00-2.00	
330	N08330	0.08	2.00	0.040	0.030	0.75-1.50	17.00-20.00	34.00-37.00		
347	S34700	0.08	2.00	0.045	0.030	1.00	17.00-19.00	9.00-13.00		Cb + Ta 10 × C min
348	S34800	0.08	2.00	0.045	0.030	1.00	17.00-19.00	9.00-13.00		Cb + Ta 10 × C min; Ta 0.10 max; Co 0.20 max
384	S38400	0.08	2.00	0.045	0.030	1.00	15.00-17.00	17.00-19.00		
403	S40300	0.15	1.00	0.040	0.030	0.50	11.50-13.00			
405	S40500	0.08	1.00	0.040	0.030	1.00	11.50-14.50			Al 0.10-0.30
409	S40900	0.08	1.00	0.045	0.045	1.00	10.50-11.75			Ti 6 × C min; 0.75 max
410	S41000	0.15	1.00	0.040	0.030	1.00	11.50-13.50			
414	S41400	0.15	1.00	0.040	0.030	1.00	11.50-13.50	1.25-2.50		
416	S41600	0.15	1.25	0.060	0.15 min	1.00	12.00-14.00		0.60(a)	
416Se	S41623	0.15	1.25	0.060	0.060	1.00	12.00-14.00			Se 0.15 min
420	S42000	>0.15	1.00	0.040	0.030	1.00	12.00-14.00			
420F	S42020	>0.15	1.25	0.060	0.15 min	1.00	12.00-14.00		0.60(a)	
422	S42200	0.20-0.25	1.00	0.025	0.025	0.75	11.00-13.00	0.50-1.00	0.75-1.25	V 0.15-0.30 W 0.75-1.25
429	S42900	0.12	1.00	0.040	0.030	1.00	14.00-16.00			
430	S43000	0.12	1.00	0.040	0.030	1.00	16.00-18.00			
430F	S43020	0.12	1.25	0.060	0.15 min	1.00	16.00-18.00		0.60(a)	
430FSe	S43023	0.12	1.25	0.060	0.060	1.00	16.00-18.00			Se 0.15 min
431	S43100	0.20	1.00	0.040	0.030	1.00	15.00-17.00	1.25-2.50		
434	S43400	0.12	1.00	0.040	0.030	1.00	16.00-18.00		0.75-1.25	
436	S43600	0.12	1.00	0.040	0.030	1.00	16.00-18.00		0.75-1.25	Cb + Ta 5 × C min, 0.70 max
440A	S44002	0.60-0.75	1.00	0.040	0.030	1.00	16.00-18.00		0.75	
440B	S44003	0.75-0.95	1.00	0.040	0.030	1.00	16.00-18.00		0.75	
440C	S44004	0.95-1.20	1.00	0.040	0.030	1.00	16.00-18.00		0.75	
442	S44200	0.20	1.00	0.040	0.030	1.00	18.00-23.00			
446	S44600	0.20	1.50	0.040	0.030	1.00	23.00-27.00			N 0.25
501	S50100	>0.10	1.00	0.040	0.030	1.00	4.00-6.00		0.40-0.65	
502	S50200	0.10	1.00	0.040	0.030	1.00	4.00-6.00		0.40-0.65	
503	S50300	0.15	1.00	0.040	0.040	1.00	6.00-8.00		0.45-0.65	
504	S50400	0.15	1.00	0.040	0.040	1.00	8.00-10.00		0.90-1.10	
	S13800	0.05	0.10	0.01	0.008	0.10	12.25-13.25	7.50-8.50	2.00-2.50	Al 0.90-1.35 N 0.010
	S15500	0.07	1.00	0.040	0.030	1.00	14.00-15.50	3.50-5.50		Cu 2.50-4.50 Cb + Ta 0.15-0.45
	S17400	0.07	1.00	0.040	0.030	1.00	15.50-17.50	3.00-5.00		Cu 3.00-5.00 Cb + Ta 0.15-0.45
	S17700	0.09	1.00	0.040	0.040	1.00	16.00-18.00	6.50-7.75		Al 0.75-1.50

(a) May be added at manufacturer's option

AISI Type 201

AISI Type 201: Chemical Composition

Element	Composition, %
Carbon	0.150 max
Manganese	5.5-7.5
Phosphorus	0.060 max
Sulfur	0.030 max
Silicon	1.0 max
Chromium	16-18
Nickel	3.5-5.5
Nitrogen	0.250 max

Characteristics. AISI type 201 is an austenitic chromium-nickel-manganese stainless steel that was developed to conserve nickel. This steel serves as a satisfactory alternate for AISI type 301. With few exceptions, fabrication and corrosion resistance of the two grades are similar (see AISI type 301).

Type 201 stainless steel resists corrosion attack over relatively long periods of exposure in both industrial and marine atmospheres. Depending upon the severity of the corrodant, annealing may be required in order to avoid intergranular corrosion after exposure to temperatures in the 480 to 815 °C (900 to 1500 °F) range.

Forming methods for type 201 stainless steel include drawing, bending, stretch bending, roll forming, end forming, and press-brake forming. Severe deep drawing is not recommended, particularly when two or more successive draws are required. Because of the work-hardening rate, type 201 sometimes shows severe springback or causes excessive tool wear.

When machining this stainless steel, heavy feeds, low surface speeds, rigid machine setups, and sharp tools are needed. The machinability rating of type 201 is about 45% of AISI 1212 steel. Type 201 steel may be welded by the conventional fusion and resistance welding processes. Annealing after welding may be necessary, depending on service conditions. To minimize oxidation, the temperature limit for continuous service of this stainless steel is about 845 °C (1550 °F); for intermittent service the limit is about 790 °C (1450 °F).

Typical Uses. Applications of AISI type 201 include railroad, rapid transit, and subway cars; truck trailers; automotive trim; architectural applications such as window frames and doors; and shallow drawn cooking ware, such as pans and lids.

AISI Type 201: Similar Steels (U.S. and/or Foreign). UNS S20100; ASME SA412; ASTM A412, A429, A666; FED QQ-S-766; SAE J405 (30201)

Physical Properties

AISI Type 201: Selected Physical Properties (Ref 1)

Melting range	1400-1455 °C (2550-2650 °F)
Density	7.83 g/cm^3 (0.283 lb/in.^{3})
Specific heat at 0-100 °C (32-212 °F)	500 J/kg·K (0.12 Btu/lb·°F)
Electrical resistivity at 20 °C (68 °F)	68.5 nΩ·m
Magnetic permeability, when H is 200 Oe, in the annealed condition	1.02 max

AISI Type 201: Average Coefficients of Linear Thermal Expansion (Ref 1)

Temperature range		Coefficient	
°C	°F	μm/m·K	μin./in.·°F
0-100	32-212	15.7	8.7
0-315	32-600	17.5	9.7
0-540	32-1000	18.4	10.2
0-650	32-1200	18.9	10.5
0-870	32-1600	20.3	11.3

AISI Type 201: Thermal Conductivity (Ref 1)

Temperature		Thermal conductivity	
°C	°F	W/m·K	Btu/ft·h·°F
100	212	16.3	9.4
500	930	21.5	12.4

AISI Type 201: Thermal Treatment Temperatures

Treatment	Temperature range °C	°F
Forging, start	1150-1230	2100-2250
Forging, finish	925 min	1700 min
Annealing (a)	1010-1120	1850-2050
Hardening (b)	...	...

(a) Water quench or rapidly air cool.
(b) Hardenable by cold work only

Mechanical Properties

AISI Type 201: Modulus of Elasticity (Ref 1)

197 GPa	28.6 × 10^6 psi

Mechanical Properties (continued)

AISI Type 201: Tensile Properties (Ref 7)

Condition	Tensile strength MPa	Tensile strength ksi	Yield strength(a) MPa	Yield strength(a) ksi	Elongation(b), %	Hardness
Annealed(c)	760	110	380	55	52	87 HRB
25% hard	860	125	515	75	25	25 HRC
50% hard	1035	150	760	110	12	32 HRC
75% hard	1205	175	930	135	12	37 HRC
Fully hard	1275	185	965	140	8	41 HRC
Extra hard	1550	225	1480	215	1	43 HRC

(a) 0.2% offset. (b) In 50 mm (2 in.). (c) Izod impact energy for this condition is 163 J (120 ft·lb)

AISI Type 201: Izod Impact Energy (Ref 8)

Test specimens were in the annealed condition

Test temperature °C	Test temperature °F	Energy absorbed J	Energy absorbed ft·lb
21	70	165	120
−185	−300	100	75

AISI Type 201: Short-Time Elevated Temperature Properties (Ref 8)

Test temperature °C	Test temperature °F	Tensile strength MPa	Tensile strength ksi	Yield strength(a) MPa	Yield strength(a) ksi	Elongation(b), %
21	70	809	117.3	365	53.0	56.5
95	200	673	97.6	268	38.9	62.5
205	400	561	81.4	208	30.1	46.5
315	600	545	79.0	189	27.4	44.0
425	800	527	76.5	181	26.2	45.5
540	1000	479	69.5	161	23.4	33.0
650	1200	328	47.6	141	20.5	28.5
760	1400	188	27.2	125	18.2	27.5
870	1600	130	18.9	97	14.0	55.0

(a) 0.2% offset. (b) In 50 mm (2 in.)

AISI Type 201: Stress-Rupture Properties (Ref 8)

Test temperature °C	Test temperature °F	Stress for rupture in: 100 h MPa	100 h ksi	1000 h MPa	1000 h ksi	10 000 h(a) MPa	10 000 h(a) ksi	Elongation(b), % 100 h	Elongation(b), % 1000 h
650	1200	200	29.0	150	22	114	16.5	14	14
730	1350	103	15.0	689	100	46	6.6	16	22
815	1500	51	7.4	30	4	...	...	35	42

(a) Extrapolated data. (b) In 50 mm (2 in.)

AISI Type 201: Stress-Strain Diagram (Ref 7)

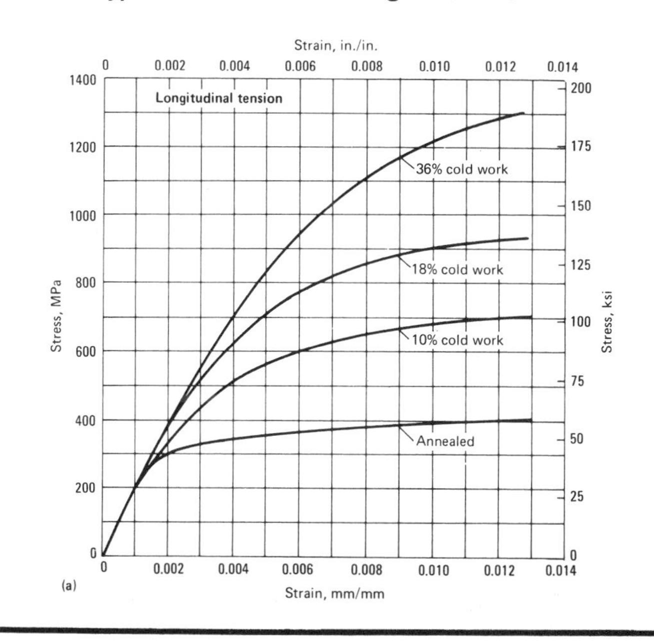

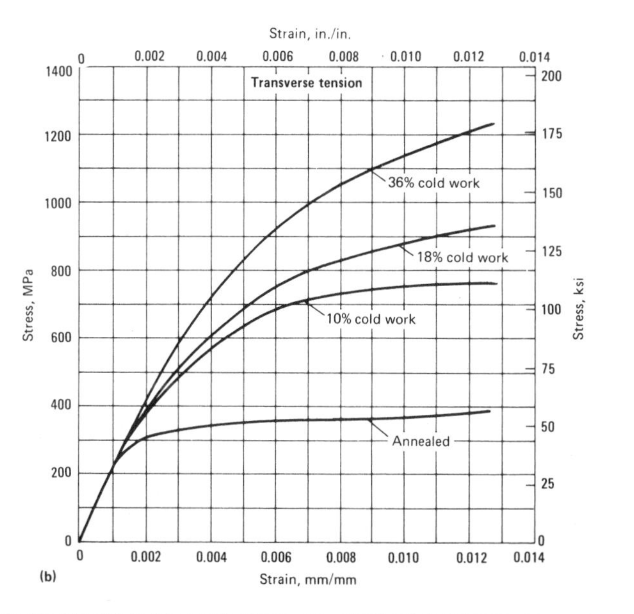

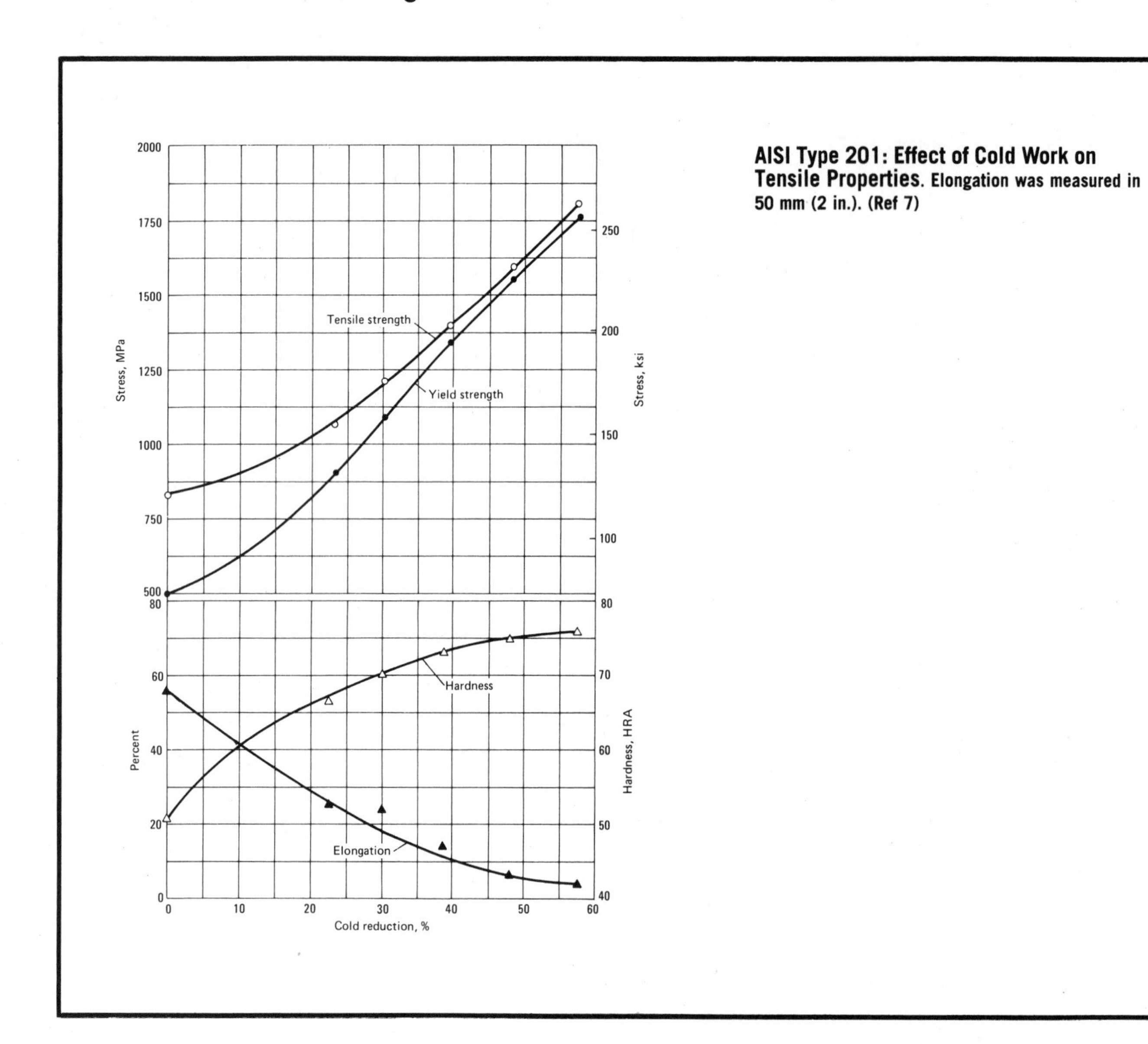

AISI Type 201: Effect of Cold Work on Tensile Properties. Elongation was measured in 50 mm (2 in.). (Ref 7)

Machining Data (Ref 2)

AISI Type 201: Turning (Cutoff and Form Tools)

Tool material	Speed, m/min (ft/min)	Feed per revolution for cutoff tool width of: 1.5 mm (0.062 in.) mm (in.)	3 mm (0.125 in.) mm (in.)	6 mm (0.25 in.) mm (in.)	Feed per revolution for form tool width of: 12 mm (0.5 in.) mm (in.)	18 mm (0.75 in.) mm (in.)	25 mm (1 in.) mm (in.)	35 mm (1.5 in.) mm (in.)	50 mm (2 in.) mm (in.)
Annealed, with hardness of 135 to 185 HB									
M2 and M3 high speed steel	26	0.033	0.041	0.050	0.050	0.041	0.036	0.030	0.025
	(85)	(0.0013)	(0.0016)	(0.002)	(0.002)	(0.0016)	(0.0014)	(0.0012)	(0.001)
C-2 carbide	84	0.033	0.041	0.050	0.050	0.041	0.036	0.030	0.025
	(275)	(0.0013)	(0.0016)	(0.002)	(0.002)	(0.0016)	(0.0014)	(0.0012)	(0.001)
Cold drawn, with hardness of 225 to 275 HB									
M2 and M3 high speed steel	20	0.025	0.038	0.050	0.050	0.046	0.038	0.030	0.025
	(65)	(0.0010)	(0.0015)	(0.002)	(0.002)	(0.0018)	(0.0015)	(0.0012)	(0.001)
C-2 carbide	69	0.025	0.038	0.050	0.050	0.046	0.038	0.030	0.025
	(225)	(0.0010)	(0.0015)	(0.002)	(0.002)	(0.0018)	(0.0015)	(0.0012)	(0.001)

Machining Data (Ref 2) (continued)

AISI Type 201: Turning (Single Point and Box Tools)

Depth of cut mm	Depth of cut in.	High speed steel Speed m/min	High speed steel Speed ft/min	High speed steel Feed mm/rev	High speed steel Feed in./rev	Uncoated carbide Speed, brazed m/min	Uncoated carbide Speed, brazed ft/min	Uncoated carbide Speed, inserted m/min	Uncoated carbide Speed, inserted ft/min	Uncoated carbide Feed mm/rev	Uncoated carbide Feed in./rev	Coated carbide Speed m/min	Coated carbide Speed ft/min	Coated carbide Feed mm/rev	Coated carbide Feed in./rev
Annealed, with hardness of 135 to 185 HB															
1	0.040	34(a)	110(a)	0.18	0.007	105(b)	350(b)	120(b)	400(b)	0.18	0.007	160(c)	525(c)	0.18	0.007
4	0.150	27(a)	90(a)	0.40	0.015	100(b)	325(b)	105(b)	350(b)	0.40	0.015	135(c)	450(c)	0.40	0.015
8	0.300	21(a)	70(a)	0.50	0.020	76(d)	250(d)	84(d)	275(d)	0.50	0.020	105(e)	350(e)	0.50	0.020
16	0.625	17(a)	55(a)	0.75	0.030	60(d)	200(d)	64(d)	210(d)	0.75	0.030	...	...	...	...
Cold drawn, with hardness of 225 to 275 HB															
1	0.040	30(f)	100(f)	0.18	0.007	100(b)	325(b)	105(b)	350(b)	0.18	0.007	135(c)	450(c)	0.18	0.007
4	0.150	24(f)	80(f)	0.40	0.015	84(b)	275(b)	90(b)	300(b)	0.40	0.015	120(c)	400(c)	0.40	0.015
8	0.300	20(f)	65(f)	0.50	0.020	64(d)	210(d)	72(d)	235(d)	0.50	0.020	90(e)	300(e)	0.50	0.020
16	0.625	15(f)	50(f)	0.75	0.030	49(d)	160(d)	55(d)	180(d)	0.75	0.030	...	...	...	...

(a) High speed steel tool material: M2 or M3. (b) Carbide tool material: C-3. (c) Carbide tool material: CC-3. (d) Carbide tool material: C-2. (e) Carbide tool material: CC-2. (f) Any premium high speed steel tool material (T15, M33, M41 to M47)

AISI Type 201: Drilling

Tool material	Speed, m/min (ft/min)	Feed per revolution for nominal hole diameter of: 1.5 mm (0.062 in.) mm (in.)	3 mm (0.125 in.) mm (in.)	6 mm (0.25 in.) mm (in.)	12 mm (0.5 in.) mm (in.)	18 mm (0.75 in.) mm (in.)	25 mm (1 in.) mm (in.)	35 mm (1.5 in.) mm (in.)	50 mm (2 in.) mm (in.)
Annealed, with hardness of 135 to 185 HB									
M10, M7, and M1 high speed steel	18 (60)	0.025 (0.001)	0.050 (0.002)	0.102 (0.004)	0.18 (0.007)	0.25 (0.010)	0.30 (0.012)	0.40 (0.015)	0.45 (0.018)
Cold drawn, with hardness of 225 to 275 HB									
M10, M7, and M1 high speed steel	17 (55)	0.025 (0.001)	0.050 (0.002)	0.102 (0.004)	0.18 (0.007)	0.25 (0.010)	0.30 (0.012)	0.40 (0.015)	0.45 (0.018)

AISI Type 201: Reaming

Based on 4 flutes for 3- and 6-mm (0.125- and 0.25-in.) reamers, 6 flutes for 12-mm (0.5-in.) reamers, and 8 flutes for 25-mm (1-in.) and larger reamers

Tool material	Speed m/min	Speed ft/min	3 mm (0.125 in.) mm	3 mm (0.125 in.) in.	6 mm (0.25 in.) mm	6 mm (0.25 in.) in.	12 mm (0.5 in.) mm	12 mm (0.5 in.) in.	25 mm (1 in.) mm	25 mm (1 in.) in.	35 mm (1.5 in.) mm	35 mm (1.5 in.) in.	50 mm (2 in.) mm	50 mm (2 in.) in.
			Feed per revolution for reamer diameter of:											
Annealed, with hardness of 135 to 185 HB														
Roughing														
M1, M2, and M7 high speed steel	21	70	0.075	0.003	0.13	0.005	0.20	0.008	0.25	0.010	0.30	0.012	0.40	0.015
C-2 carbide	26	85	0.102	0.004	0.20	0.008	0.30	0.012	0.40	0.016	0.50	0.020	0.60	0.024
Finishing														
M1, M2, and M7 high speed steel	11	35	0.075	0.003	0.075	0.003	0.102	0.004	0.15	0.006	0.18	0.007	0.20	0.008
C-2 carbide	15	50	0.075	0.003	0.102	0.004	0.150	0.006	0.20	0.008	0.23	0.009	0.25	0.010
Cold drawn, with hardness of 225 to 275 HB														
Roughing														
M1, M2, and M7 high speed steel	18	60	0.075	0.003	0.13	0.005	0.20	0.008	0.25	0.010	0.30	0.012	0.40	0.015
C-2 carbide	23	75	0.102	0.004	0.20	0.008	0.30	0.012	0.40	0.016	0.50	0.020	0.60	0.024
Finishing														
M1, M2, and M7 high speed steel	11	35	0.075	0.003	0.075	0.003	0.102	0.004	0.15	0.006	0.18	0.007	0.20	0.008
C-2 carbide	15	50	0.075	0.003	0.102	0.004	0.150	0.006	0.20	0.008	0.23	0.009	0.25	0.010

Machining Data (Ref 2) (continued)

AISI Type 201: Surface Grinding With a Horizontal Spindle and Reciprocating Table

Test specimens were annealed or cold drawn, with hardness of 135 to 275 HB

Operation	Wheel speed m/s	Wheel speed ft/min	Table speed m/min	Table speed ft/min	Downfeed mm/pass	Downfeed in./pass	Crossfeed mm/pass	Crossfeed in./pass	Wheel identification
Roughing	28-33	5500-6500	15-30	50-100	0.050	0.002	1.25-12.5	0.050-0.500	C46JV
Finishing	28-33	5500-6500	15-30	50-100	0.013 max	0.0005 max	(a)	(a)	C46JV

(a) Maximum ¼ of wheel width

AISI Type 201: Cylindrical and Internal Grinding

Test specimens were annealed or cold drawn, with hardness of 135 to 275 HB

Operation	Wheel speed m/s	Wheel speed ft/min	Work speed m/min	Work speed ft/min	Infeed on diameter mm/pass	Infeed on diameter in./pass	Traverse(a)	Wheel identification
Cylindrical grinding								
Roughing	28-33	5500-6500	15-30	50-100	0.050	0.002	½	C54JV
Finishing	28-33	5500-6500	15-30	50-100	0.013 max	0.0005 max	⅙	C54JV
Internal grinding								
Roughing	25-33	5000-6500	23-60	75-200	0.013	0.0005	⅓	C60KV
Finishing	25-33	5000-6500	23-60	75-200	0.005 max	0.0002 max	⅙	C60KV

(a) Wheel width per revolution of work

AISI Type 201: Face Milling

Depth of cut mm	Depth of cut in.	M2 and M7 high speed steel Speed m/min	Speed ft/min	Feed/tooth mm	Feed/tooth in.	Uncoated C-2 carbide Speed, brazed m/min	Speed, brazed ft/min	Speed, inserted m/min	Speed, inserted ft/min	Feed/tooth mm	Feed/tooth in.	Coated CC-2 carbide Speed m/min	Speed ft/min	Feed/tooth mm	Feed/tooth in.
Annealed, with hardness of 135 to 185 HB															
1	0.040	40	130	0.20	0.008	130	430	145	475	0.20	0.008	215	700	0.20	0.008
4	0.150	30	100	0.30	0.012	100	325	110	360	0.30	0.012	145	475	0.30	0.012
8	0.300	24	80	0.40	0.016	70	230	85	280	0.40	0.016	105	350	0.40	0.016
Cold drawn, with hardness of 225 to 275 HB															
1	0.040	35	115	0.15	0.006	120	400	135	440	0.18	0.007	200	650	0.18	0.007
4	0.150	27	90	0.25	0.010	90	300	100	330	0.25	0.010	130	425	0.25	0.010
8	0.300	21	70	0.36	0.014	60	200	76	250	0.36	0.014	100	325	0.36	0.014

AISI Type 201: Boring

Depth of cut mm	Depth of cut in.	High speed steel Speed m/min	Speed ft/min	Feed mm/rev	Feed in./rev	Uncoated carbide Speed, brazed m/min	Speed, brazed ft/min	Speed, inserted m/min	Speed, inserted ft/min	Feed mm/rev	Feed in./rev	Coated carbide Speed m/min	Speed ft/min	Feed mm/rev	Feed in./rev
Annealed, with hardness of 135 to 185 HB															
0.25	0.010	35(a)	115(a)	0.075	0.003	90(b)	300(b)	105(b)	350(b)	0.075	0.003	145(c)	475(c)	0.075	0.003
1.25	0.050	27(a)	90(a)	0.13	0.005	72(d)	235(d)	84(d)	275(d)	0.13	0.005	115(e)	375(e)	0.13	0.005
2.5	0.100	21(a)	70(a)	0.30	0.012	66(d)	215(d)	76(d)	250(d)	0.30	0.012	95(e)	315(e)	0.30	0.012
Cold drawn, with hardness of 225 to 275 HB															
0.25	0.010	30(f)	100(f)	0.075	0.003	84(b)	275(b)	95(b)	315(b)	0.075	0.003	120(c)	400(c)	0.075	0.003
1.25	0.050	24(f)	80(f)	0.13	0.005	66(d)	215(d)	76(d)	250(d)	0.13	0.005	95(e)	315(e)	0.13	0.005
2.5	0.100	20(f)	65(f)	0.30	0.012	53(d)	175(d)	64(d)	210(d)	0.30	0.012	84(e)	275(e)	0.30	0.012

(a) High speed steel tool material: M2 or M3. (b) Carbide tool material: C-3. (c) Carbide tool material: CC-3. (d) Carbide tool material: C-2. (e) Carbide tool material: CC-2. (f) Any premium high speed steel tool material (T15, M33, M41 to M47)

AISI Type 201: Planing

Tool material	Depth of cut mm	Depth of cut in.	Speed m/min	Speed ft/min	Feed/stroke mm	Feed/stroke in.
Annealed, with hardness of 135 to 185 HB						
M2 and M3 high speed steel	0.1	0.005	9	30	(a)	(a)
	2.5	0.100	12	40	1.25	0.050
	12	0.500	8	25	1.50	0.060
C-2 carbide	0.1	0.005	38	125	(a)	(a)
	2.5	0.100	43	140	2.05	0.080
Cold drawn, with hardness of 225 to 275 HB						
M2 and M3 high speed steel	0.1	0.005	8	25	(a)	(a)
	2.5	0.100	11	35	0.75	0.030
	12	0.500	8	25	1.15	0.045
C-2 carbide	0.1	0.005	38	125	(a)	(a)
	2.5	0.100	43	140	1.50	0.060

(a) Feed is 75% the width of the square nose finishing tool

Machining Data (Ref 2) (continued)

AISI Type 201: End Milling (Profiling)

Tool material	Depth of cut mm	Depth of cut in.	Speed m/min	Speed ft/min	Feed per tooth for cutter diameter of: 10 mm (0.375 in.) mm	10 mm (0.375 in.) in.	12 mm (0.5 in.) mm	12 mm (0.5 in.) in.	18 mm (0.75 in.) mm	18 mm (0.75 in.) in.	25-50 mm (1-2 in.) mm	25-50 mm (1-2 in.) in.
Annealed, with hardness of 135 to 185 HB												
M2, M3, and M7 high speed steel	0.5	0.020	34	110	0.025	0.001	0.050	0.002	0.102	0.004	0.13	0.005
	1.5	0.060	24	80	0.050	0.002	0.075	0.003	0.13	0.005	0.15	0.006
	diam/4	diam/4	21	70	0.025	0.001	0.050	0.002	0.102	0.004	0.13	0.005
	diam/2	diam/2	18	60	0.025	0.001	0.038	0.0015	0.075	0.003	0.102	0.004
C-2 carbide	0.5	0.020	110	360	0.013	0.0005	0.025	0.001	0.050	0.002	0.102	0.004
	1.5	0.060	82	270	0.025	0.001	0.050	0.002	0.075	0.003	0.13	0.005
	diam/4	diam/4	72	235	0.025	0.001	0.038	0.0015	0.063	0.0025	0.102	0.004
	diam/2	diam/2	67	220	0.013	0.0005	0.025	0.001	0.050	0.002	0.075	0.003
Cold drawn, with hardness of 225 to 275 HB												
M2, M3, and M7 high speed steel	0.5	0.020	29	95	0.025	0.001	0.050	0.002	0.102	0.004	0.13	0.005
	1.5	0.060	21	70	0.050	0.002	0.075	0.003	0.13	0.005	0.15	0.006
	diam/4	diam/4	18	60	0.025	0.001	0.050	0.002	0.102	0.004	0.13	0.005
	diam/2	diam/2	15	50	0.025	0.001	0.038	0.0015	0.075	0.003	0.102	0.004
C-2 carbide	0.5	0.020	100	325	0.013	0.0005	0.025	0.001	0.050	0.002	0.075	0.003
	1.5	0.060	75	245	0.025	0.001	0.050	0.002	0.075	0.003	0.102	0.004
	diam/4	diam/4	64	210	0.025	0.001	0.038	0.0015	0.063	0.0025	0.075	0.003
	diam/2	diam/2	60	200	0.013	0.0005	0.025	0.001	0.050	0.002	0.063	0.0025

AISI Type 202

AISI Type 202: Chemical Composition

Element	Composition, %
Carbon	0.150 max
Manganese	7.5-10.0
Phosphorus	0.060 max
Sulfur	0.030 max
Silicon	1.0 max
Chromium	17.0-19.0
Nickel	4-6
Nitrogen	0.250 max

Characteristics. AISI type 202 is an austenitic, chromium-nickel-manganese steel which was developed to conserve nickel. This steel, an alternate for type 302, was designed for applications where good forming characteristics are important. Type 202 is essentially nonmagnetic when annealed and becomes slightly magnetic when cold worked.

This grade work hardens rapidly. It can be cold worked by drawing, bending, roll forming, and stamping, and has good deep drawing properties as well. Type 202 may be welded by the conventional fusion and resistance welding processes. Annealing after welding may be necessary to avoid intergranular corrosion.

For machining, rigid machine setups and sharp tools are required; dull tools ride over the surface causing work hardening and difficult cutting. The machinability rating of AISI type 202 is 45% compared to 100% for AISI 1212.

This stainless steel resists corrosion in moderate corrosive environments; most organic and strongly oxidizing acids; industrial and marine atmospheres; and solutions used in chemical, textile, petroleum, dairy, and foodstuff industries. Type 202 resists scaling in continuous exposure to 845 °C (1550 °F) maximum and intermittent exposure to 815 °C (1500 °F) maximum.

Typical Uses. Applications of AISI type 202 include kitchen utensils such as pots and pans, washing machine tubs, wash-fountain bowls, door hardware, dairy equipment, sink drains, bowls, countertops, and architectural trim.

AISI Type 202: Similar Steels (U.S. and/or Foreign). UNS S20200; ASTM A314, A412, A429, A473, A666; FED QQ-S-763, QQ-S-766, STD-66; SAE J405 (30202)

Machining Data

For machining data on AISI type 202, refer to the preceding machining tables for AISI type 201.

Physical Properties

AISI Type 202: Selected Physical Properties (Ref 1)

Melting range	1400-1455 °C (2550-2650 °F)
Density	7.81 g/cm³ (0.282 lb/in.³)
Specific heat at 0-100 °C (32-212 °F)	500 J/kg·K (0.12 Btu/lb·°F)
Electrical resistivity at 20 °C (68 °F)	72 nΩ·m
Magnetic permeability, when H is 200 Oe, in the annealed condition	1.02 max

AISI Type 202: Thermal Treatment Temperatures

Treatment	Temperature range °C	°F
Forging, start	1150-1230	2100-2250
Forging, finish	925 min	1700 min
Annealing(a)	1010-1120	1850-2050
Hardening(b)	...	...

(a) Water quench or rapidly air cool. (b) Hardenable by cold work only

AISI Type 202: Average Coefficients of Linear Thermal Expansion (Ref 1)

Temperature range °C	°F	Coefficient μm/m·K	μin./in.·°F
0-100	32-212	17.5	9.7
0-315	32-600	18.4	10.2
0-540	32-1000	19.3	10.7
0-650	32-1200	19.8	11.0
0-760	32-1400	20.3	11.3

AISI Type 202: Thermal Conductivity (Ref 1)

Temperature °C	°F	Thermal conductivity W/m·K	Btu/ft·h·°F
100	212	16.3	9.4
500	930	21.6	12.5

Mechanical Properties

AISI Type 202: Stress-Rupture Properties (Ref 7)

Test temperature °C	°F	Stress for rupture in: 100 h MPa	100 h ksi	1000 h MPa	1000 h ksi	10 000 h(a) MPa	10 000 h(a) ksi
540	1000	340	49	285	41	...	...
595	1100	270	39	220	32	...	...
650	1200	200	29	160	24	120	17. 5
705	1300	145	21	97	14	69	10
760	1400	97	14	60	8.5	...	...
815	1500	55	8	30	4.5	...	...

(a) Extrapolated data

AISI Type 202: Modulus of Elasticity (Ref 1)

197 GPa....28.6 × 10⁶ psi

AISI Type 202: Tensile Properties (Ref 1)

Product form	Condition	Tensile strength MPa	Tensile strength ksi	Yield strength MPa	Yield strength ksi	Elongation(a), %
Bar	Annealed	515	75	275	40	40
Wire, bar, sheet, strip	Annealed	655	95	310	45	40
Sheet, strip	25% hard	860	125	515	75	12
Sheet, strip	50% hard	1030	150	760	110	10

(a) In 50 mm (2 in.)

AISI Type 202: Elevated-Temperature Tensile Properties (Ref 7)

Test temperature °C	°F	Tensile strength MPa	Tensile strength ksi	Yield strength(a) MPa	Yield strength(a) ksi	Elongation(b), %
205	400	595	86	235	34	45
315	600	560	81	205	30	45
425	800	540	78	195	28	45
540	1000	495	72	170	25	36
650	1200	370	54	150	22	28
760	1400	215	31	130	19	31
870	1600	125	18	105	15	48

(a) 0.2% offset. (b) In 50 mm (2 in.)

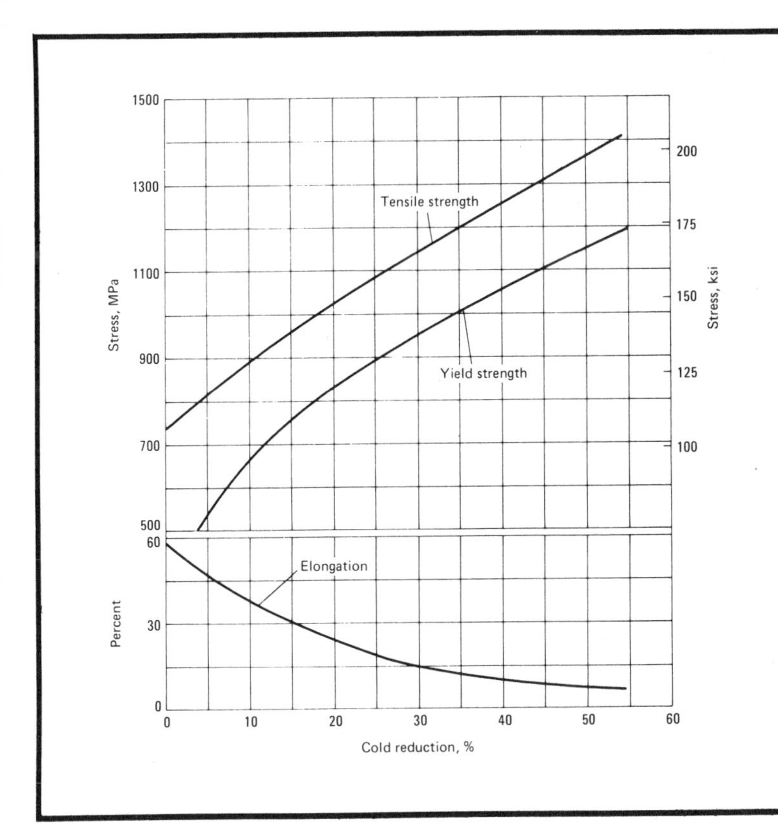

AISI Type 202: Effect of Cold Reduction on Tensile Properties. Elongation was measured in 50 mm (2 in.); yield strength at 0.2% offset. (Ref 7)

AISI Type 301

AISI Type 301: Chemical Composition

Element	Composition, %
Carbon	0.150 max
Manganese	2.0 max
Phosphorus	0.045 max
Sulfur	0.030 max
Silicon	1.0 max
Chromium	16-18
Nickel	6-8

Characteristics. AISI type 301 is an austenitic, chromium-nickel steel which can attain high tensile strength and ductility through moderate or severe cold working. This grade is most often used in the cold rolled or cold drawn condition in the form of sheet, strip, and wire. It is essentially nonmagnetic when annealed, and becomes magnetic when cold worked.

The high strengths of type 301 in the six available conditions or tempers, excellent resistance to atmospheric corrosion, and bright, attractive surface make this grade a good choice for decorative structural applications.

AISI type 301 also exhibits resistance to atmospheric corrosion, various chemicals used to deice roads, and foodstuffs and juices. However, the corrosion properties are not as good as those of the 18-8 chromium-nickel steels. Best resistance to corrosion is obtained in the annealed or cold worked condition. Intergranular corrosion may occur if, during fabrication or service, the material is heated in the range of 480 to 870 °C (900 to 1600 °F) and not subsequently annealed. Type 301 steel resists scaling in continuous exposure to 900 °C (1650 °F) maximum, and intermittent exposure to 790 °C (1450 °F) maximum.

Because of a high rate of work hardening, type 301 requires heavy feeds, low surface speeds, rigid machine setups, and sharp tools. Speeds used for machining this stainless steel are about 45% of those used for AISI grade 1212.

Type 301 may be welded using most common welding methods. For methods requiring filler metal, type 308 electrodes are recommended. Depending on welding techniques, carbide precipitation can occur; therefore a post welding treatment at 980 °C (1800 °F), or above, is recommended.

Typical Uses. Applications of AISI type 301 include automobile wheel covers, transportation equipment, aircraft and missile structures, roof drainage products, kitchen utensils, storm door frames, and tableware.

AISI Type 301: Similar Steels (U.S. and/or Foreign). UNS S30100; AMS 5517, 5518, 5519; ASTM A167, A177, A554, A666; FED QQ-S-766; MIL SPEC MIL-S-5059; SAE J405 (30301); (W. Ger.) DIN 1.4310; (Fr.) AFNOR Z 12 CN 17.08; (Ital.) UNI X 12 CrNi 17 07; (Jap.) JIS SUS 301

Physical Properties

AISI Type 301: Thermal Treatment Temperatures

Treatment	Temperature range °C	°F
Forging, start	1150-1260	2100-2300
Forging, finish	925 min	1700 min
Annealing (a)	1010-1120	1850-2050
Hardening (b)	...	...

(a) Water quenched. (b) Hardenable by cold work only

AISI Type 301: Electrical Resistivity (Ref 8)

Temperature °C	°F	Specific electrical resistance, nΩ·m
20	68	72
100	212	78
200	390	86
400	750	100
600	1110	111
800	1470	121
900	1650	126

AISI Type 301: Selected Physical Properties (Ref 8)

Melting range	1400-1420 °C (2550-2590 °F)
Density	7.92-8.08 g/cm³ (0.286-0.292 lb/in.³)
Specific heat at 0-100 °C (32-212 °F)	500 J/kg·K (0.12 Btu/lb·°F)
Magnetic permeability, when H is 200 Oe, in the annealed condition	1.02 max

AISI Type 301: Average Coefficients of Linear Thermal Expansion (Ref 8)

Temperature range °C	°F	Coefficient µm/m·K	µin./in.·°F
20-100	68-212	16.6	9.2
20-300	68-570	17.6	9.8
20-500	68-930	18.5	10.3
20-700	68-1290	19.4	10.8
20-870	68-1600	19.8	11.0

AISI Type 301: Thermal Conductivity (Ref 8)

Temperature °C	°F	Thermal conductivity W/m·K	Btu/ft·h·°F
100	212	16.3	9.4
500	930	21.5	12.4

Machining Data

For machining data on AISI type 301, refer to the preceding machining tables for AISI type 201.

Mechanical Properties

AISI Type 301: Tensile Properties at Low Temperatures (Ref 8)

Test temperature °C	°F	Tensile strength MPa	ksi	Yield strength(a) MPa	ksi	Elongation(b), %	Reduction in area, %
Annealed							
26	78	725	105	275	40	60	70
0	32	1070	155	295	43	53	64
−40	−40	1240	180	330	48	42	63
−62	−80	1345	195	345	50	40	62
−195	−320	1895	275	515	75	30	57
50% hard							
26	78	1035	150	655	95	54	68
0	32	1170	170	675	98	46	65
−40	−40	1295	188	696	101	38	60
−62	−80	1415	205	724	105	37	62
−195	−320	2000	290	800	116	25	50
75% hard							
26	78	1310	190	1180	171	17	...
−73	−100	1545	224	1060	154	19	...
−195	−320	2000	290	1330	193	20	...
−255	−425	2185	317	...	...	14	...
Full hard							
26	78	1415	205	1330	193	6	...
−195	−320	2080	302	1480	215	20	...
−255	−425	2345	340	1725	250	15	...

(a) 0.2% offset. (b) In 50 mm (2 in.)

AISI Type 301: Modulus of Elasticity (Ref 7)

Stressed in:	Modulus of elasticity GPa	10^6 psi
Tension	193	28.0
Torsion	77.2	11.2

AISI Type 301: Endurance Limit (Ref 4)

Condition	Stress MPa	ksi
Full hard temper	550	80

AISI Type 301: Tensile and Compressive Properties (Ref 8)

Temper	Condition	Longitudinal Yield strength(a) MPa	Longitudinal Yield strength(a) ksi	Longitudinal Modulus of elasticity GPa	Longitudinal Modulus of elasticity 10^6 psi	Transverse Yield strength(a) MPa	Transverse Yield strength(a) ksi	Transverse Modulus of elasticity GPa	Transverse Modulus of elasticity 10^6 psi
Tensile properties									
Annealed	As rolled	250	36	214	31.0	250	36	211	30.6
25% hard	As rolled	550	80	193	28.0	580	84	197	28.6
	Stress relieved	530	77	198	28.7	545	79	186	27.0
50% hard	As rolled	841	122	185	26.8	848	123	194	28.1
	Stress relieved	883	128	192	27.9	896	130	197	28.6
75% hard	As rolled	979	142	178	25.8	1000	145	190	27.5
	Stress relieved	1070	155	188	27.3	1070	155	199	28.8
Full hard	As rolled	1105	160	174	25.2	1125	163	196	28.4
	Stress relieved	1205	175	196	28.4	1250	181	210	30.5
Compressive properties									
Annealed	As rolled	260	38	211	30.6	260	38	209	30.3
25% hard	As rolled	345	50	194	28.2	625	91	194	28.2
	Stress relieved	505	73	199	28.8	580	84	210	30.5
50% hard	As rolled	620	90	190	27.5	979	142	190	27.5
	Stress relieved	765	111	201	29.2	993	144	205	29.8
75% hard	As rolled	689	100	183	26.5	1170	170	192	27.9
	Stress relieved	917	133	190	27.5	1215	176	203	29.5
Full hard	As rolled	793	115	170	24.6	1315	191	203	29.4
	Stress relieved	1165	169	191	27.7	1440	209	204	29.6

(a) 0.2% offset

AISI Type 301: Tensile Properties at Elevated Temperatures (Ref 8)

Temperature °C	Temperature °F	Condition	Tensile strength MPa	Tensile strength ksi	Yield strength MPa	Yield strength ksi	Elongation(a), %
Room temperature		Annealed	724	105.0	276	40.0	55
		25% hard	889	129.0	503	73.0	43.5
		50% hard	1138	165.0	772	112.0	28.5
205	400	Annealed	552	80.0	152	22.0	46
		25% hard	625	90.6	424	61.5	23
		50% hard	876	127.0	731	106.0	9
315	600	Annealed	485	70.4	134	19.4	40
		25% hard	594	86.2	412	59.8	20
		50% hard	846	122.7	656	95.2	6.5
425	800	Annealed	463	67.2	134	19.5	39
		25% hard	806	116.9	377	54.7	17.5
		50% hard	134	19.5	590	85.5	7
540	1000	Annealed	401	58.2	126	18.3	34
		25% hard	447	64.9	353	51.2	16.5
		50% hard	538	78.0	464	67.3	7
650	1200	Annealed	282	40.9	106	15.4	36
		25% hard	352	51.0	276	40.0	20
		50% hard	396	57.5	331	48.0	10
760	1400	Annealed	204	29.6	99	14.4	30
		25% hard	248	36.0	186	27.0	17
		50% hard	241	35.0	214	31.0	10
870	1600	Annealed	109	15.8	66	9.5	29
		25% hard	134	19.4	106	15.4	15
		50% hard	113	16.4	96	13.9	12.5

(a) In 50 mm (2 in.)

AISI Type 301: Tensile Properties at Room Temperature (Ref 8)

Condition	Tensile strength MPa	Tensile strength ksi	Yield strength(a) MPa	Yield strength(a) ksi	Elongation(b), %	Hardness
Annealed	725	105	275	40	55	85 HRB
25% hard	860	125	515	75	25	25 HRC
50% hard	1035	150	760	110	15	30 HRC
75% hard	1205	175	930	135	10	35 HRC
Full hard	1275	185	965	140	8	40 HRC
Extra hard	1550	225	1480	215	1	43 HRC

(a) 0.2% offset. (b) In 50 mm (2 in.)

Mechanical Properties (continued)

AISI Type 301: Tensile Strength and Modulus of Elasticity after Cold Work (Ref 3)

Tensile strength(a) MPa	Tensile strength(a) ksi	Modulus of elasticity at: Room temperature GPa	Room temperature 10^6 psi	−195 °C (−320 °F) GPa	−195 °C (−320 °F) 10^6 psi
1303	189	175	25.4	197	28.6
1593	231	178	25.8	192	27.8

(a) At room temperature

AISI Type 301: Effect of Cold Work on Tensile Properties. Elongation was measured in 50 mm (2 in.). (Ref 7)

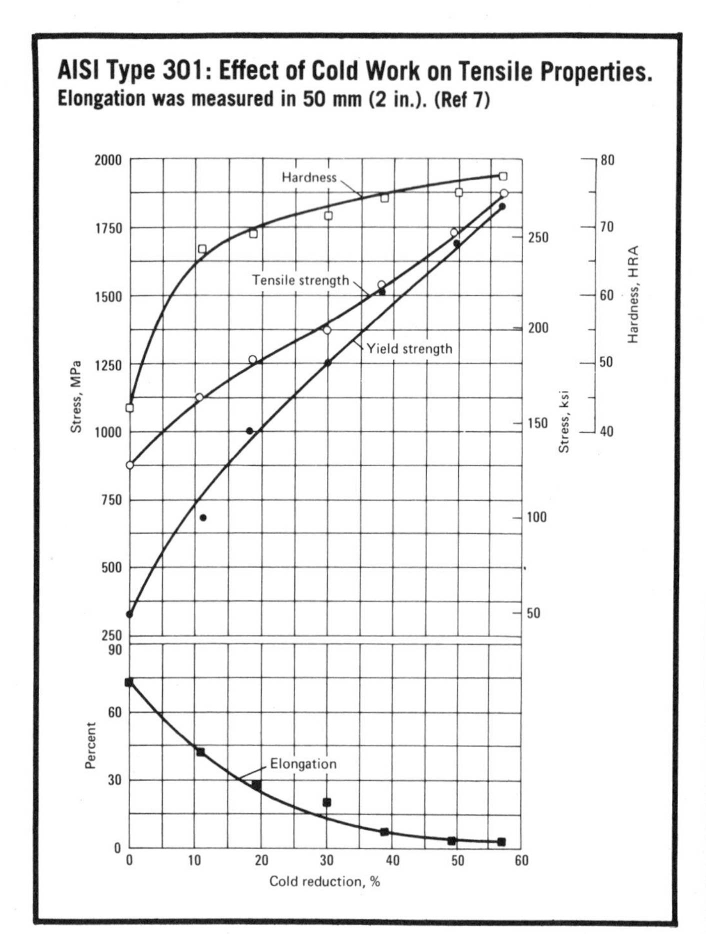

Machining Data

For machining data on AISI type 301, refer to the preceding machining tables for AISI type 201.

AISI Type 301: Effect of Test Temperature on Tensile Properties of Annealed Material. Elongation was measured in 50 mm (2 in.). (Ref 3)

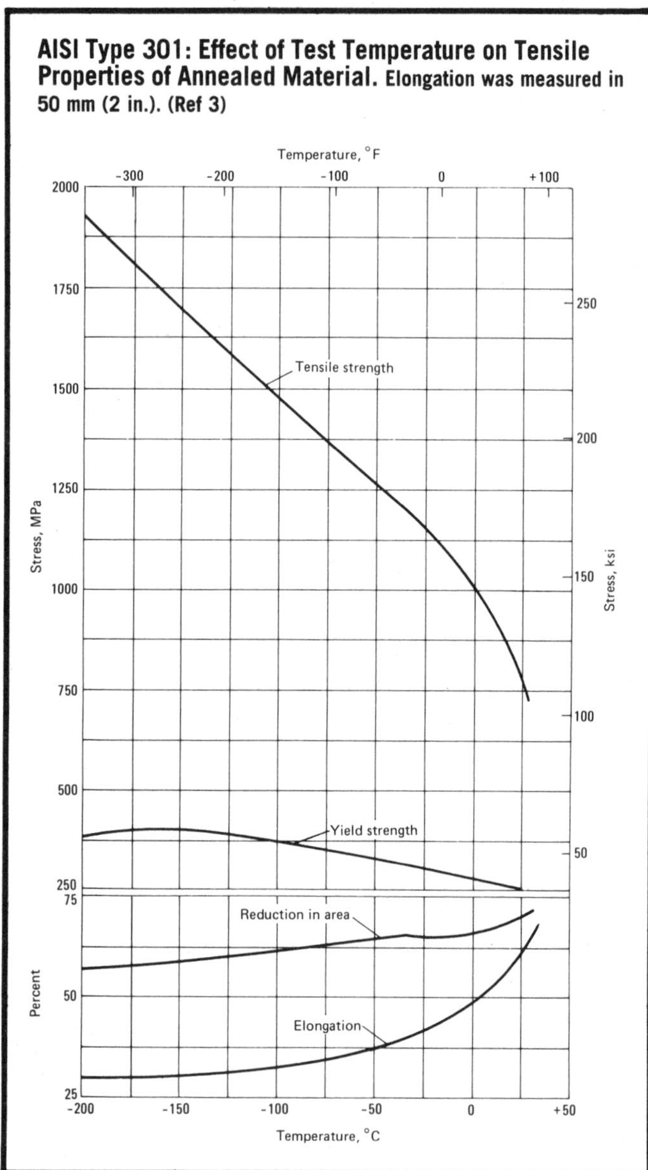

AISI Type 301: Effect of Test Temperature on Tensile and Yield Strengths of Cold Worked Material. — tensile strength; ---- yield strength (Ref 3)

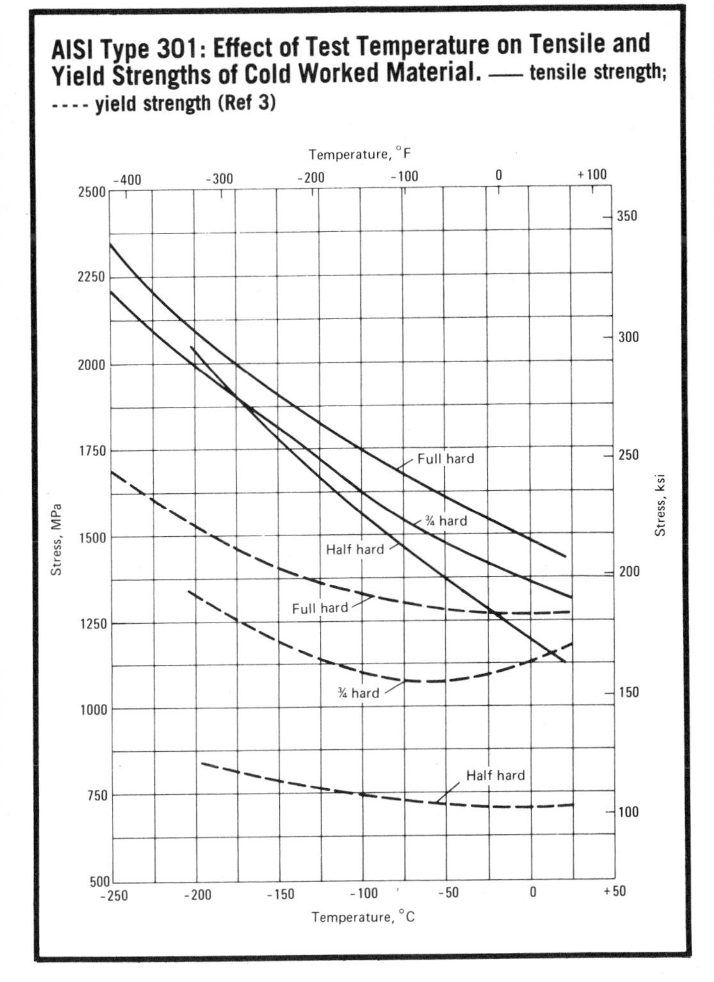

AISI Type 301: Effect of Test Temperature on Elongation and Reduction in Area of Cold Worked Material. Elongation was measured in 50 mm (2 in.). (Ref 3)

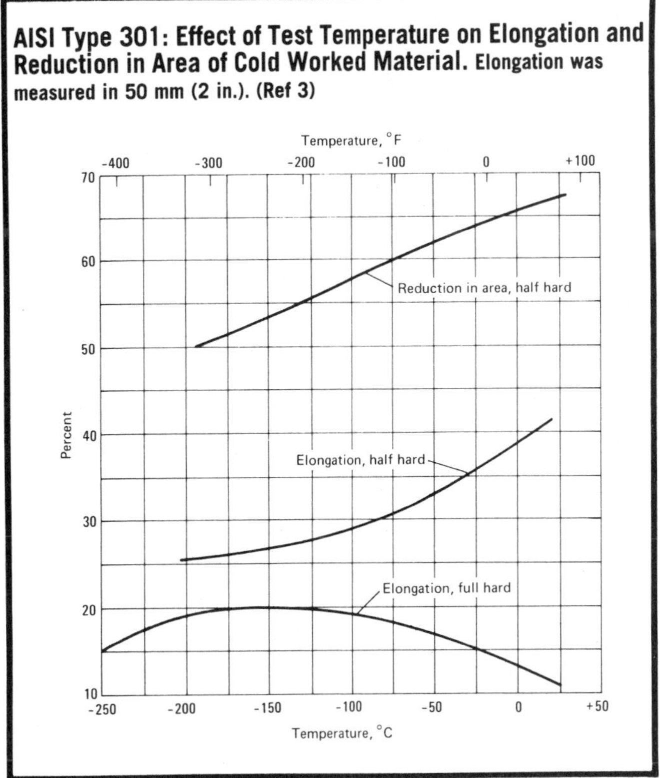

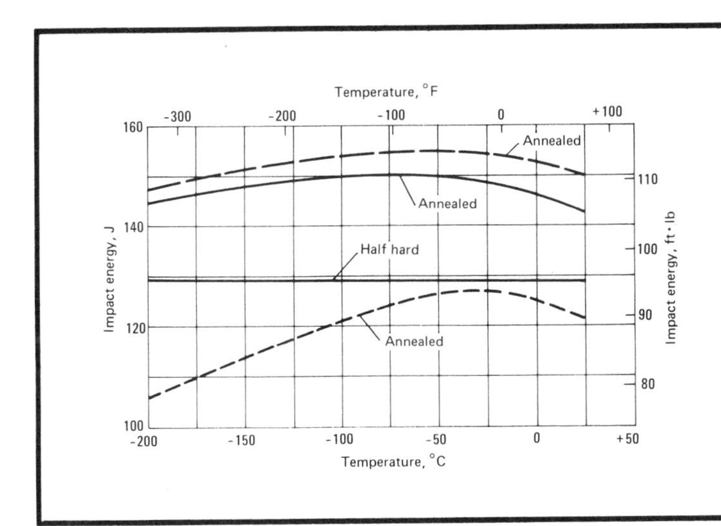

AISI Type 301: Impact Properties. Impact tests: — Izod, ——— V-notch, --- keyhole notch (Ref 3)

AISI Type 302

AISI Type 302: Chemical Composition

Element	Composition, %
Carbon	0.150 max
Manganese	2.0 max
Phosphorus	0.045 max
Sulfur	0.030 max
Silicon	1.0 max
Chromium	17-19
Nickel	8-10

Characteristics. AISI type 302 is the original, general purpose, austenitic, chromium-nickel stainless steel. The corrosion resistance of this grade is superior to that of type 301. Type 302 is used most often in the annealed condition. It can be cold worked to high tensile strengths, but with slightly lower ductility than type 301. In the annealed condition, type 302 is austenitic, nonmagnetic, extremely tough and ductile, and has excellent corrosion resistance. The steel becomes slightly magnetic when cold worked.

Type 302 is readily fabricated by cold working and responds well to drawing, bending, forming, spinning, roll forming, stamping, and upsetting. Because of toughness and high work-hardening characteristics, type 302 requires rigid machine setups and sharp tools. Better chip action and finish are obtained when the material is in a moderately cold worked, rather than in an annealed condition. Heavy feeds and an adequate supply of coolant are advised. The machinability rating of type 302 is about 45% of that for AISI 1212 steel.

Electrodes recommended for gas or arc welding are type 308 or type 347. Welds are usually as tough and ductile as the base metal. Because areas adjacent to the weld are subject to temperatures within the carbide precipitation range, annealing followed by rapid cooling is advised after welding.

In the annealed condition, type 302 is resistant to atmospheric corrosion, all foodstuffs, sterilizing solutions, most organic chemicals and dyestuffs, and a wide variety of chemicals. This stainless steel has good resistance to many acidic environments, especially nitric acid, but has poor resistance to halogen acids. Type 302 resists scaling in continuous exposure to 900 °C (1650 °F), and, for intermittent exposure, to 815 °C (1500 °F).

Typical Uses. Applications of AISI type 302 include food serving and handling equipment, food processing machinery, dairy equipment, hospital equipment, household appliances and many types of kitchen ware, exterior architectural trim, fishing tackle, camera parts, jewelry, aircraft cowling, furnace parts, steam and oil piping, nitric acid vessels, springs, screens, and wire rope.

AISI Type 302: Similar Steels (U.S. and/or Foreign). UNS S30200; AMS 5515, 5516, 5636, 5637, 5688; ASME SA240, SA479; ASTM A167, A240, A276, A313, A314, A368, A473, A478, A479, A492, A493, A511, A554, A666; FED QQ-S-763, QQ-S-766, QQ-W-423; MIL SPEC MIL-S-862; SAE J230, J405 (30302)

Physical Properties

AISI Type 302: Selected Physical Properties

Melting range	1400-1420 °C (2550-2590 °F)
Density	7.86 g/cm³ (0.284 lb/in.³)
Specific heat at 0-100 °C (32-212 °F)	500 J/kg·K (0.12 Btu/lb·°F)
Magnetic permeability, when H is 200 Oe, in the annealed condition	1.02 max

AISI Type 302: Average Coefficients of Linear Thermal Expansion

Temperature range °C	°F	Coefficient μm/m·K	μin./in.·°F
0-100	32-212	17.3	9.6
0-315	32-600	17.8	9.9
0-540	32-1000	18.4	10.2
0-650	32-1200	18.7	10.4

AISI Type 302: Thermal Conductivity

Temperature °C	°F	Thermal conductivity W/m·K	Btu/ft·h·°F
100	212	16.3	9.4
500	930	21.5	12.4

AISI Type 302: Thermal Treatment Temperatures

Treatment	Temperature range °C	°F
Forging, start	1150-1260	2100-2300
Forging, finish	925 min	1700 min
Annealing (a)	1010-1120	1850-2050
Hardening (b)	...	...

(a) Water quenched. (b) Hardenable by cold work only

AISI Type 302: Electrical Resistivity

Temperature °C	°F	Electrical resistivity, nΩ·m
20	68	72
100	212	78
200	390	86
400	750	100
600	1110	111
800	1470	121
900	1650	126

Machining Data

For machining data on AISI type 302, refer to the preceding machining tables for AISI type 201.

Mechanical Properties

AISI Type 302: Stress-Rupture and Creep Properties (Ref 7)

Test temperature °C	°F	Stress for rupture in: 1000 h MPa	ksi	10 000 h(a) MPa	ksi	100 000 h(a) MPa	ksi	Stress for secondary creep rate of(a): 1% in 10 000 h MPa	ksi	1% in 100 000 h MPa	ksi
480	900	338	49.0	262	38.0	...	...	...	...	...	...
540	1000	241	35.0	183	26.5	145	21.0	138	20.0	83	12.0
595	1100	152	22.0	117	17.0	90	13.0	90	13.0	52	7.5
650	1200	97	14.0	69	10.0	48	7.0	57	8.3	30	4.3
705	1300	59	8.5	41	6.0	28	4.0	34	5.0	17	2.5
760	1400	41	6.0	26	3.7	17	2.5	21	3.0	10	1.5
815	1500	26	3.7	19	2.7	10	1.5	17	2.4	8	1.2
870	1600	19	2.7	...	...	...	...	...	...	...	...

(a) Extrapolated data

AISI Type 302: Tensile Properties at Room Temperature (Ref 7)

Condition and product form	Tensile strength MPa	ksi	Yield strength(a) MPa	ksi	Elongation(b), %	Reduction in area, %	Hardness
Annealed strip	620	90	275	40	55	...	80 HRB
25% hard strip(c)	860	125	515	75	12	...	25 HRC
Annealed bar	585	85	240	35	60	70	80 HRB

(a) 0.2% offset. (b) In 50 mm (2 in.). (c) Minimum properties. Standard practice is to produce to either minimum tensile strength, minimum yield strength, or minimum hardness, but not to combinations of these properties

AISI Type 302: Modulus of Elasticity (Ref 7)

Stressed in:	Modulus GPa	10^6 psi
Tension	193	28.0
Torsion	77.2	11.2

AISI Type 302: Impact Properties (Ref 4)

Cold rolled to tensile strength of: MPa	ksi	Izod impact energy J	ft·lb
703	102	163	120
986	143	110	80
1140	165	68	50
1345	195	34	25
1550	225	23	17

AISI Type 302: Tensile Properties at Subzero and Elevated Temperatures (Ref 7)

Test temperature °C	°F	Tensile strength MPa	ksi	Yield strength(a) MPa	ksi	Elongation(b), %	Reduction in area, %
−195	−320	1620	235	385	56	40	59
−60	−80	1110	161	345	50	57	70
−40	−40	1000	145	330	48	60	73
0	32	840	122	275	40	65	76
21	70	585	85	255	37	57	70
205	400	495	72	160	23	51	73
315	600	470	68	135	19.5	45	71
425	800	440	64	115	16.5	40	69
540	1000	385	56	97	14	36	66
650	1200	305	44	86	12.5	34	58
760	1400	200	29	76	11	36	46
870	1600	110	16	...	...	40	40

(a) 0.2% offset. (b) In 50 mm (2 in.)

AISI Type 302: Endurance Limit (Ref 4)

Treatment	Endurance limit MPa	ksi
Full hard temper	485-550	70-80

AISI Type 302: Tensile Strength and Modulus of Elasticity After Cold Work (Ref 3)

Tensile strength MPa	ksi	Modulus of elasticity: Room temperature GPa	10^6 psi	−195 °C (−320 °F) GPa	ksi
1115	162	181	26.2	188	27.2

AISI Type 302: Impact Properties of Weld Metal (Ref 3)

Test temperature °C	°F	Charpy keyhole notch impact energy for weld metal condition of: As welded J	ft·lb	Annealed J	ft·lb
Room temperature		43-46	32-34	50-54	37-40
−76	−105	35-38	26-28	42-47	31-35
−195	−320	12-14	9-10	34-38	25-28

AISI Type 302: Effect of Cold Reduction on Tensile Properties (Ref 8)

Cold reduction, %	Tensile strength MPa	ksi	Yield strength(a) MPa	ksi	Elongation(b), %
0	640	93	250	36	67.0
10	807	117	635	92	43.3
20	848	123	731	106	31.0
30	945	137	827	120	19.5
40	1050	152	917	133	14.0
50	1165	169	1005	146	10.5
60	1250	181	1110	161	7.0

(a) 0.2% offset. (b) In 50 mm (2 in.)

Mechanical Properties (continued)

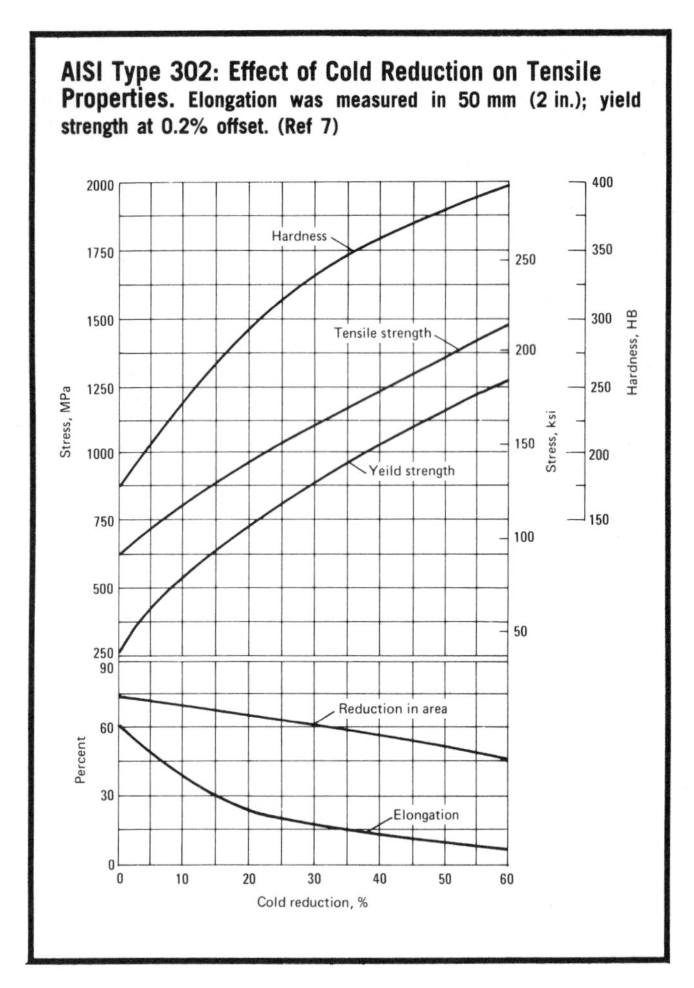

AISI Type 302: Effect of Cold Reduction on Tensile Properties. Elongation was measured in 50 mm (2 in.); yield strength at 0.2% offset. (Ref 7)

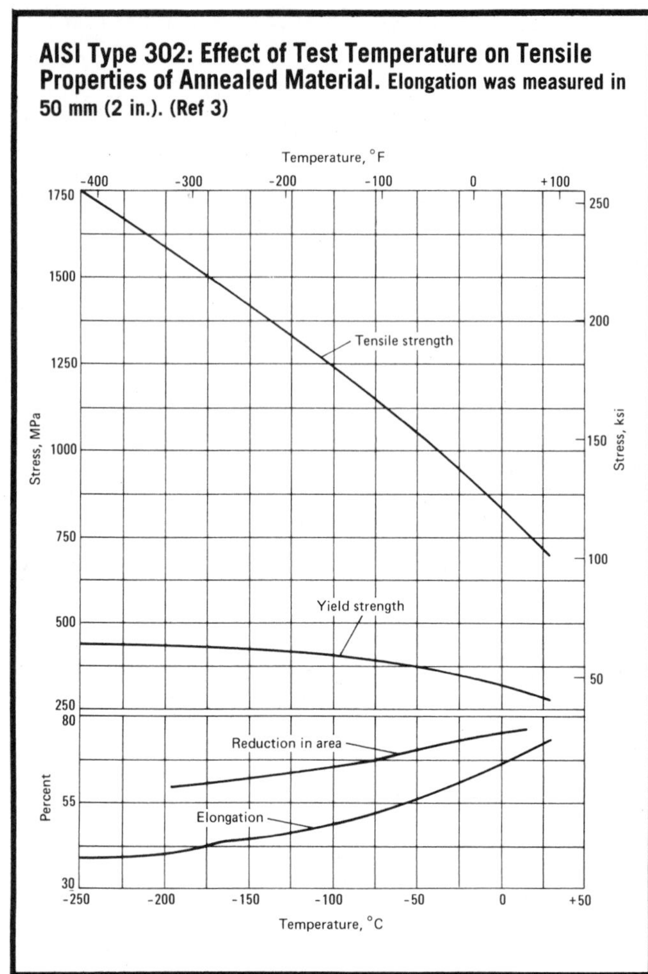

AISI Type 302: Effect of Test Temperature on Tensile Properties of Annealed Material. Elongation was measured in 50 mm (2 in.). (Ref 3)

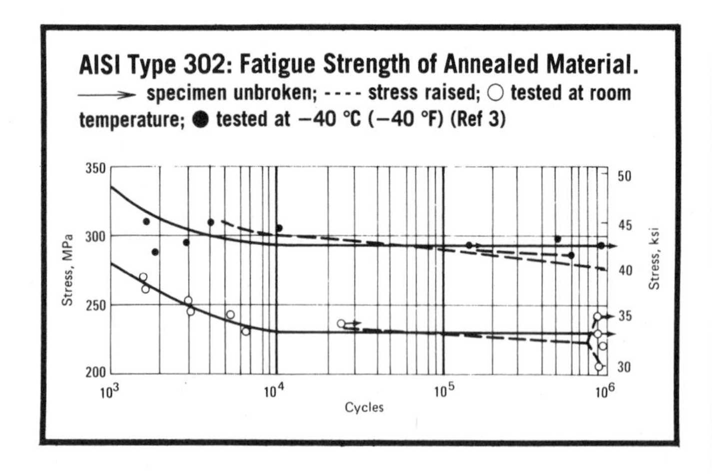

AISI Type 302: Fatigue Strength of Annealed Material. ——> specimen unbroken; - - - - stress raised; ○ tested at room temperature; ● tested at −40 °C (−40 °F) (Ref 3)

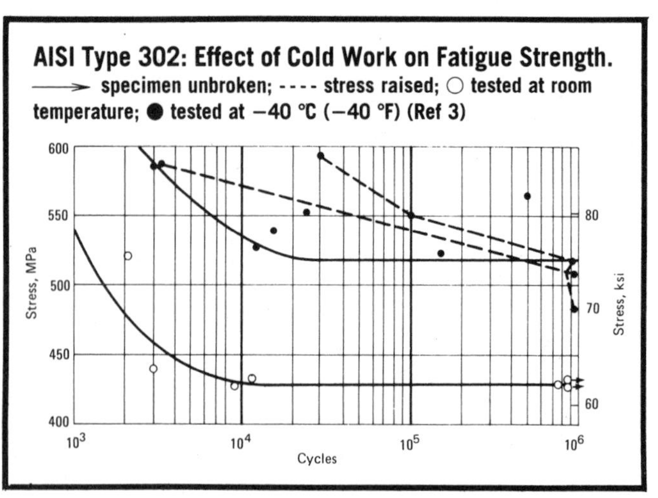

AISI Type 302: Effect of Cold Work on Fatigue Strength. ——> specimen unbroken; - - - - stress raised; ○ tested at room temperature; ● tested at −40 °C (−40 °F) (Ref 3)

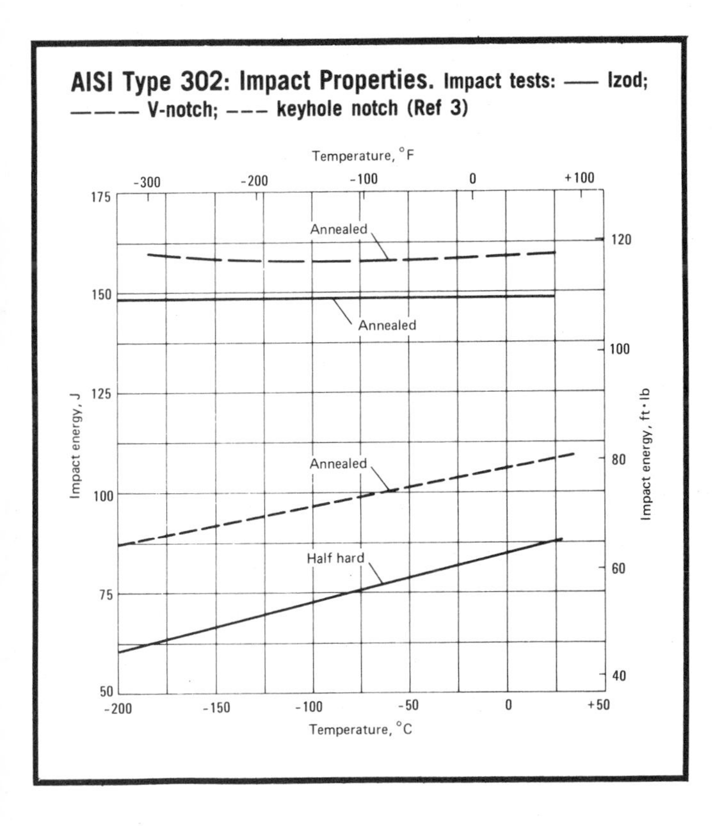

AISI Type 302: Impact Properties. Impact tests: —— Izod; – – – – V-notch; - - - keyhole notch (Ref 3)

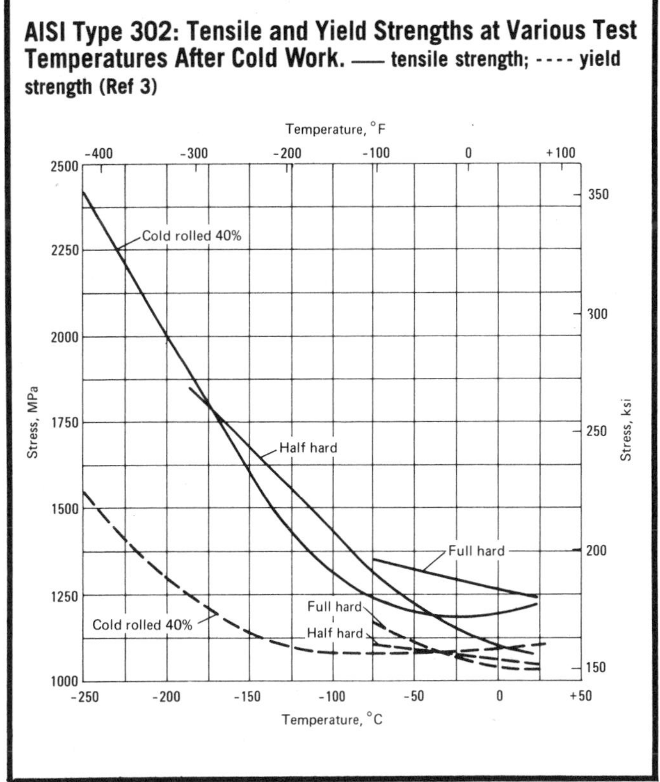

AISI Type 302: Tensile and Yield Strengths at Various Test Temperatures After Cold Work. —— tensile strength; - - - - yield strength (Ref 3)

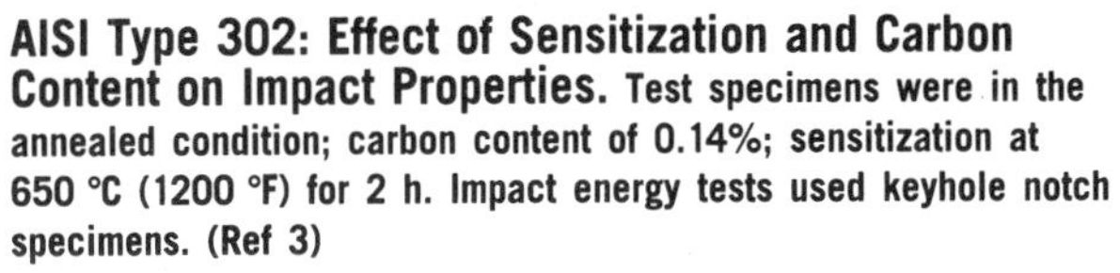

AISI Type 302: Effect of Sensitization and Carbon Content on Impact Properties. Test specimens were in the annealed condition; carbon content of 0.14%; sensitization at 650 °C (1200 °F) for 2 h. Impact energy tests used keyhole notch specimens. (Ref 3)

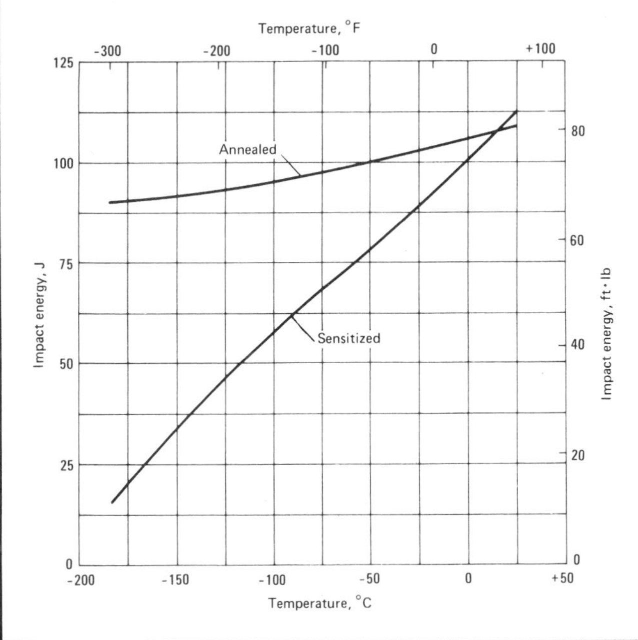

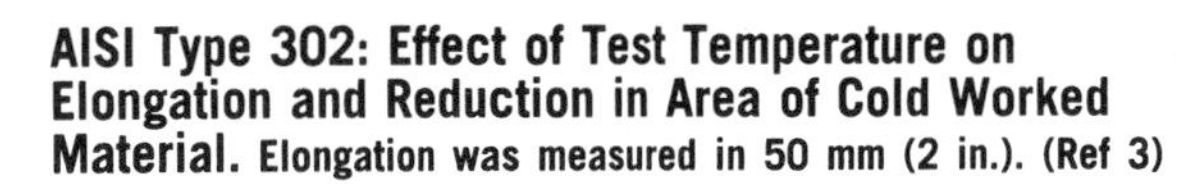

AISI Type 302: Effect of Test Temperature on Elongation and Reduction in Area of Cold Worked Material. Elongation was measured in 50 mm (2 in.). (Ref 3)

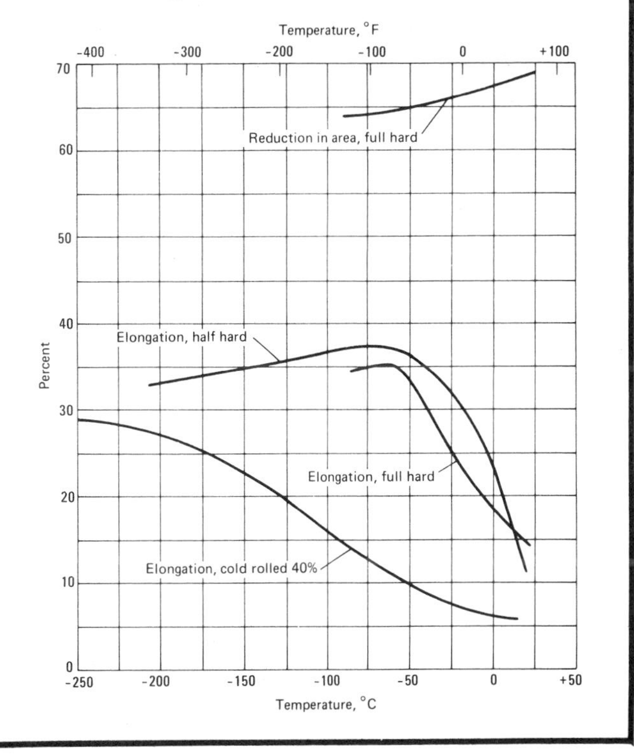

AISI Type 302B

AISI Type 302B: Chemical Composition

Element	Composition, %
Carbon	0.150 max
Manganese	2.0 max
Phosphorus	0.045 max
Sulfur	0.030 max
Silicon	2-3
Chromium	17-19
Nickel	8-10

Characteristics. AISI type 302B is an austenitic, chromium-nickel steel. The resistance to scaling of this stainless steel is superior to that of type 302. AISI type 302B is essentially nonmagnetic when annealed, and becomes slightly magnetic when cold worked.

This steel can be deep drawn, stamped, headed, and upset without difficulty. Because the steel work hardens, severe forming operations should be followed by an anneal. Type 302B requires rigid machine setups and sharp tools. Better chip action and finish are obtained with material in a moderately cold worked condition than in an annealed condition. This grade is machined at speeds about 40% of those used for free-machining AISI grade 1212.

Type 302B can be satisfactorily brazed and welded by all methods. Electrodes recommended for gas or arc welding are type 308 or 347.

Typical Uses. AISI type 302B is used exclusively for parts which are exposed to high temperatures.

AISI Type 302B: Similar Steels (U.S. and/or Foreign). UNS S30215; ASTM A167, A276, A314, A473, A580; SAE J405 (30302B)

Physical Properties

AISI Type 302B: Selected Physical Properties (Ref 6)

Melting range	1370-1400 °C (2500-2550 °F)
Density	8.0 g/cm³ (0.29 lb/in.³)
Specific heat at 0-100 °C (32-212 °F)	500 J/kg·K (0.12 Btu/lb·°F)
Electrical resistivity at 20 °C (68 °F)	72 nΩ·m

AISI Type 302B: Average Coefficients of Linear Thermal Expansion (Ref 6)

Temperature range °C	°F	Coefficient µm/m·K	µin./in.·°F
0-100	32-212	16.2	9.0
0-315	32-600	18.0	10.0
0-540	32-1000	19.4	10.8
0-650	32-1200	20.2	11.2

AISI Type 302B: Thermal Conductivity (Ref 6)

Temperature °C	°F	Thermal conductivity W/m·K	Btu/ft·h·°F
100	212	15.9	9.2
500	930	21.6	12.5

AISI Type 302B: Thermal Treatment Temperatures

Treatment	Temperature range °C	°F
Forging, start	1120-1230	2050-2250
Forging, finish	925 min	1700 min
Annealing(a)	1010-1120	1850-2050
Hardening(b)	...	...

(a) Water quenched. (b) Hardenable by cold work only

Mechanical Properties

AISI Type 302B: Representative Tensile Properties (Ref 6)

Test specimens were in the annealed condition

Product form	Tensile strength MPa	ksi	Yield strength(a) MPa	ksi	Elongation(b), %	Reduction in area, %	Hardness, HRB
Sheet	655	95	275	40	55	...	85
Strip	655	95	275	40	55	...	85
Plate	620	90	275	40	50	65	85
Bar	620	90	275	40	50	65	85

(a) 0.2% offset. (b) In 50 mm (2 in.)

AISI 302B: Selected Mechanical Properties (Ref 6)

Modulus of elasticity, in tension (Ref 6)	195 GPa (28 × 10⁶ psi)
Izod impact properties, in the annealed condition (Ref 6)	95-120 J (70-90 ft·lb)

Machining Data (Ref 2)

AISI Type 302B: Turning (Single Point and Box Tools)

Depth of cut		High speed steel				Uncoated carbide						Coated carbide			
		Speed		Feed		Speed, brazed		Speed, inserted		Feed		Speed		Feed	
mm	in.	m/min	ft/min	mm/rev	in./rev	m/min	ft/min	m/min	ft/min	mm/rev	in./rev	m/min	ft/min	mm/rev	in./rev
Annealed, with hardness of 135 to 185 HB															
1	0.040	29(a)	95(a)	0.18	0.007	100(b)	325(b)	115(b)	375(b)	0.18	0.007	150(c)	500(c)	0.18	0.007
4	0.150	23(a)	75(a)	0.40	0.015	90(b)	300(b)	100(b)	325(b)	0.40	0.015	130(c)	425(c)	0.40	0.015
8	0.300	18(a)	60(a)	0.50	0.020	69(d)	250(d)	76(d)	250(d)	0.50	0.020	100(e)	325(e)	0.50	0.020
16	0.625	14(a)	45(a)	0.75	0.030	53(d)	200(d)	60(d)	200(d)	0.75	0.030	...	...	...	...
Cold drawn, with hardness of 225 to 275 HB															
1	0.040	24(f)	80(f)	0.18	0.007	90(b)	300(b)	100(b)	325(b)	0.18	0.007	130(c)	425(c)	0.18	0.007
4	0.150	20(f)	65(f)	0.40	0.015	76(b)	250(b)	84(b)	275(b)	0.40	0.015	105(c)	360(c)	0.40	0.015
8	0.300	15(f)	50(f)	0.50	0.020	58(d)	190(d)	66(d)	215(d)	0.50	0.020	84(e)	275(e)	0.50	0.020
16	0.625	12(f)	40(f)	0.75	0.030	43(d)	140(d)	50(d)	165(d)	0.75	0.030	...	...	...	...

(a) High speed steel tool material: M2 or M3. (b) Carbide tool material: C-3. (c) Carbide tool material: CC-3. (d) Carbide tool material: C-2. (e) Any premium high speed steel tool material (T15, M33, M41 to M47)

AISI Type 302B: Turning (Cutoff and Form Tools)

Tool material	Speed, m/min (ft/min)	Feed per revolution for cutoff tool width of: 1.5 mm (0.062 in.) mm (in.)	3 mm (0.125 in.) mm (in.)	6 mm (0.25 in.) mm (in.)	Feed per revolution for form tool width of: 12 mm (0.5 in.) mm (in.)	18 mm (0.75 in.) mm (in.)	25 mm (1 in.) mm (in.)	35 mm (1.5 in.) mm (in.)	50 mm (2 in.) mm (in.)
Annealed, with hardness of 135 to 185 HB									
M2 and M3 high speed steel	23	0.038	0.046	0.053	0.050	0.046	0.038	0.033	0.028
	(75)	(0.0015)	(0.0018)	(0.0021)	(0.002)	(0.0018)	(0.0015)	(0.0013)	(0.0011)
C-2 carbide	76	0.038	0.046	0.053	0.050	0.046	0.038	0.033	0.028
	(250)	(0.0015)	(0.0018)	(0.0021)	(0.002)	(0.0018)	(0.0015)	(0.0013)	(0.0011)
Cold drawn, with hardness of 225 to 275 HB									
M2 and M3 high speed steel	17	0.025	0.038	0.050	0.050	0.046	0.038	0.030	0.025
	(55)	(0.001)	(0.0015)	(0.0020)	(0.002)	(0.0018)	(0.0015)	(0.0012)	(0.001)
C-2 carbide	53	0.025	0.038	0.050	0.050	0.046	0.038	0.030	(0.001)
	(175)	(0.001)	(0.0015)	(0.0020)	(0.002)	(0.0018)	(0.0015)	(0.0012)	(0.001)

AISI Type 302B: Face Milling

Depth of cut		High speed steel(a)				Uncoated C-2 carbide						Coated CC-2 carbide			
		Speed		Feed/tooth		Speed, brazed		Speed, inserted		Feed/tooth		Speed		Feed/tooth	
mm	in.	m/min	ft/min	mm	in.	m/min	ft/min	m/min	ft/min	mm	in.	m/min	ft/min	mm	in.
Annealed, with hardness of 135 to 185 HB															
1	0.040	35	115	0.20	0.008	120	400	135	440	0.20	0.008	200	650	0.20	0.008
4	0.150	27	90	0.30	0.012	90	300	100	330	0.30	0.012	130	425	0.30	0.012
8	0.300	21	70	0.40	0.016	60	200	76	250	0.40	0.016	100	325	0.40	0.016
Cold drawn, with hardness of 225 to 275 HB															
1	0.040	30	100	0.15	0.006	88	290	105	350	0.15	0.007	160	525	0.15	0.007
4	0.150	23	75	0.25	0.010	75	245	90	300	0.25	0.010	120	400	0.25	0.010
8	0.300	17	55	0.36	0.014	58	190	72	235	0.36	0.014	90	300	0.36	0.014

(a) High speed steel tool material: for hardness of 135 to 185 HB, M2 or M7; for hardness of 225 to 275 HB, use any premium high speed steel (T15, M33, M41 to M47)

AISI Type 302B: Boring

Depth of cut		High speed steel				Uncoated carbide						Coated carbide			
		Speed		Feed		Speed, brazed		Speed, inserted		Feed		Speed		Feed	
mm	in.	m/min	ft/min	mm/rev	in./rev	m/min	ft/min	m/min	ft/min	mm/rev	in./rev	m/min	ft/min	mm/rev	in./rev
Annealed, with hardness of 135 to 185 HB															
0.25	0.01	29(a)	95(a)	0.075	0.003	84(b)	275(b)	100(b)	325(b)	0.075	0.003	135(c)	450(c)	0.075	0.003
1.25	0.05	23(a)	75(a)	0.13	0.005	69(d)	225(d)	79(d)	260(d)	0.13	0.005	105(e)	350(e)	0.13	0.005
2.50	0.10	18(a)	60(a)	0.30	0.012	58(d)	190(d)	69(d)	225(d)	0.30	0.012	90(e)	300(e)	0.30	0.012
Cold drawn, with hardness of 225 to 275 HB															
0.25	0.01	24(f)	80(f)	0.075	0.003	75(b)	245(b)	85(b)	280(b)	0.075	0.003	115(c)	375(c)	0.075	0.003
1.25	0.05	20(f)	65(f)	0.13	0.005	58(d)	190(d)	69(d)	225(d)	0.13	0.005	90(e)	300(e)	0.13	0.005
2.50	0.10	15(f)	50(f)	0.30	0.012	53(d)	175(d)	60(d)	200(d)	0.30	0.012	76(e)	250(e)	0.30	0.012

(a) High speed steel tool material: M2 or M3. (b) Carbide tool material: C-3. (c) Carbide tool material: CC-3. (d) Carbide tool material: C-2. (e) Carbide tool material: CC-2. (f) Any premium high speed steel tool material (T15, M33, M41 to M47)

Machining Data (Ref 2) (continued)

AISI Type 302B: End Milling (Profiling)

Tool material	Depth of cut mm	Depth of cut in.	Speed m/min	Speed ft/min	Feed per tooth for cutter diameter of: 10 mm (0.375 in.) mm	10 mm in.	12 mm (0.5 in.) mm	12 mm in.	18 mm (0.75 in.) mm	18 mm in.	25-50 mm (1-2 in.) mm	25-50 mm in.
Annealed, with hardness of 135 to 185 HB												
M2, M3, and M7 high speed steel	0.5	0.020	30	100	0.025	0.001	0.050	0.002	0.102	0.004	0.13	0.0050
	1.5	0.060	23	75	0.050	0.002	0.075	0.003	0.130	0.005	0.15	0.0060
	diam/4	diam/4	20	65	0.025	0.001	0.050	0.002	0.102	0.004	0.13	0.0050
	diam/2	diam/2	17	55	0.025	0.001	0.025	0.001	0.075	0.003	0.102	0.0040
C-2 carbide	0.5	0.020	105	340	0.013	0.0005	0.025	0.001	0.050	0.002	0.102	0.0040
	1.5	0.060	79	260	0.025	0.001	0.050	0.002	0.075	0.003	0.13	0.0050
	diam/4	diam/4	69	225	0.025	0.001	0.038	0.0015	0.063	0.0025	0.102	0.0040
	diam/2	diam/2	64	210	0.013	0.0005	0.025	0.001	0.050	0.002	0.075	0.0030
Cold drawn, with hardness of 225 to 275 HB												
M2, M3, and M7 high speed steel	0.5	0.020	27	90	0.025	0.001	0.050	0.002	0.102	0.004	0.13	0.0050
	1.5	0.060	20	65	0.050	0.002	0.075	0.003	0.13	0.005	0.15	0.0060
	diam/4	diam/4	17	55	0.025	0.001	0.050	0.002	0.102	0.004	0.13	0.0050
	diam/2	diam/2	15	50	0.025	0.001	0.025	0.001	0.075	0.003	0.102	0.0040
C-2 carbide	0.5	0.020	90	300	0.013	0.0005	0.025	0.001	0.050	0.002	0.075	0.0030
	1.5	0.060	70	230	0.025	0.001	0.050	0.002	0.075	0.003	0.102	0.0040
	diam/4	diam/4	60	200	0.025	0.001	0.038	0.0015	0.063	0.0025	0.075	0.0030
	diam/2	diam/2	55	180	0.013	0.0005	0.025	0.001	0.050	0.002	0.063	0.0025

AISI Type 302B: Drilling

Tool material	Speed, m/min (ft/min)	Feed per revolution for nominal hole diameter of: 1.5 mm (0.062 in.) mm (in.)	3 mm (0.125 in.) mm (in.)	6 mm (0.25 in.) mm (in.)	12 mm (0.5 in.) mm (in.)	18 mm (0.75 in.) mm (in.)	25 mm (1 in.) mm (in.)	35 mm (1.5 in.) mm (in.)	50 mm (2 in.) mm (in.)
Annealed, with hardness of 135 to 185 HB									
M10, M7 and M1 high speed steel	17 (55)	0.025 (0.001)	0.05 (0.002)	0.102 (0.004)	0.18 (0.007)	0.25 (0.010)	0.30 (0.012)	0.40 (0.015)	0.45 (0.018)
Cold drawn, with hardness of 225 to 275 HB									
M10, M7 and M1 high speed steel	15 (50)	0.025 (0.001)	0.05 (0.002)	0.102 (0.004)	0.18 (0.007)	0.25 (0.010)	0.30 (0.012)	0.40 (0.015)	0.45 (0.018)

AISI Type 302B: Cylindrical and Internal Grinding

Test specimens were annealed or cold drawn, with hardness of 135 to 275 HB

Operation	Wheel speed m/s	Wheel speed ft/min	Work speed m/min	Work speed ft/min	Infeed on diameter mm/pass	Infeed on diameter in./pass	Traverse(a)	Wheel identification
Cylindrical grinding								
Roughing	28-33	5500-6500	15-30	50-100	0.050	0.002	½	C54JV
Finishing	28-33	5500-6500	15-30	50-100	0.013 max	0.0005 max	1/16	C54JV
Internal grinding								
Roughing	25-33	5000-6500	23-60	75-200	0.013	0.0005	⅓	C60KV
Finishing	25-33	5000-6500	23-60	75-200	0.005 max	0.0002 max	⅙	C60KV

(a) Wheel width per revolution of work

AISI Type 302B: Planing

Tool material	Depth of cut mm	Depth of cut in.	Speed m/min	Speed ft/min	Feed/stroke mm	Feed/stroke in.
Annealed, with hardness of 135 to 185 HB						
M2 and M3 high speed steel	0.1	0.005	8	25	(a)	(a)
	2.5	0.100	11	35	1.25	0.050
	12	0.500	8	25	1.50	0.060
C-2 carbide	0.1	0.005	38	125	(a)	(a)
	2.5	0.100	43	140	1.50	0.060
Cold drawn, with hardness of 225 to 275 HB						
M2 and M3 high speed steel	0.1	0.005	6	20	(a)	(a)
	2.5	0.100	9	30	0.75	0.030
	12	0.500	6	20	1.15	0.045
C-2 carbide	0.1	0.005	34	110	(a)	(a)
	2.5	0.100	37	120	1.50	0.060

(a) Feed is 75% the width of the square nose finishing tool

AISI Type 302B: Reaming

Based on 4 flutes for 3- and 6-mm (0.125- and 0.25-in.) reamers, 6 flutes for 12-mm (0.5-in.) reamers, and 8 flutes for 25-mm (1-in.) and larger reamers

Tool material	Speed m/min	Speed ft/min	Feed per revolution for reamer diameter of: 3 mm (0.125 in.) mm	3 mm (0.125 in.) in.	6 mm (0.25 in.) mm	6 mm (0.25 in.) in.	12 mm (0.5 in.) mm	12 mm (0.5 in.) in.	25 mm (1 in.) mm	25 mm (1 in.) in.	35 mm (1.5 in.) mm	35 mm (1.5 in.) in.	50 mm (2 in.) mm	50 mm (2 in.) in.
Annealed, with hardness of 135 to 185 HB														
Roughing														
M1, M2, and M7 high speed steel	21	70	0.075	0.003	0.13	0.005	0.20	0.008	0.25	0.010	0.30	0.012	0.40	0.015
C-2 carbide	26	85	0.102	0.004	0.20	0.008	0.30	0.012	0.40	0.016	0.50	0.020	0.60	0.024
Finishing														
M1, M2, and M7 high speed steel	11	35	0.075	0.003	0.075	0.003	0.102	0.004	0.15	0.006	0.18	0.007	0.20	0.008
C-2 carbide	15	50	0.075	0.003	0.102	0.004	0.15	0.006	0.20	0.008	0.23	0.009	0.25	0.010
Cold drawn, with hardness of 225 to 275 HB														
Roughing														
M1, M2, and M7 high speed steel	18	60	0.075	0.003	0.13	0.005	0.20	0.008	0.25	0.010	0.30	0.012	0.40	0.015
C-2 carbide	23	75	0.102	0.004	0.20	0.008	0.30	0.012	0.40	0.016	0.50	0.020	0.60	0.024
Finishing														
M1, M2, and M7 high speed steel	11	35	0.075	0.003	0.075	0.003	0.102	0.004	0.15	0.006	0.18	0.007	0.20	0.008
C-2 carbide	15	50	0.075	0.003	0.102	0.004	0.15	0.006	0.20	0.008	0.23	0.009	0.25	0.010

AISI Type 302B: Surface Grinding with a Horizontal Spindle and Reciprocating Table

Test specimens were annealed or cold drawn, with hardness of 135 to 275 HB

Operation	Wheel speed m/s	Wheel speed ft/min	Table speed m/min	Table speed ft/min	Downfeed mm/pass	Downfeed in./pass	Crossfeed mm/pass	Crossfeed in./pass	Wheel identification
Roughing	28-33	5500-6500	15-30	50-100	0.050	0.002	1.25-12.5	0.050-0.500	C46JV
Finishing	28-33	5500-6500	15-30	50-100	0.013 max	0.0005 max	(a)	(a)	C46JV

(a) Maximum ¼ of wheel width

AISI Types 303, 303Se

AISI Types 303, 303Se: Chemical Composition

Element	Composition, % Type 303	Type 303Se
Carbon	0.15	0.15
Manganese	2.00	2.00
Phosphorus	0.20	0.20
Sulfur	0.15	0.06
Silicon	1.00	1.00
Chromium	17-19	17-19
Nickel	8-10	8-10
Molybdenum(a)	0.60	...
Selenium	...	0.15

(a) Optional

Characteristics. AISI types 303 and 303Se are austenitic, chromium-nickel steels with elements added to improve machining and nonseizing characteristics. These grades are the most readily machinable of all the chromium-nickel grades and are suitable for use in automatic screw machines. Both types are also widely used to minimize seizing or galling.

If types 303 or 303Se are exposed in the temperature range 425 to 815 °C (800 to 1500 °F) during fabrication or use, a corrective treatment, consisting of heating to at least 1040 °C (1900 °F) and water quenching, should be used. The safe scaling temperature for continuous exposure is 900 °C (1650 °F), and for intermittent exposure, about 815 °C (1500 °F).

Type 303 stainless steel will withstand only moderate cold working; it is not adaptable to severe cold-finishing operations. However, type 303Se is preferred where free-machining grades are required for parts that involve some cold forming operations. Welding of either grade is not recommended because of the high sulfur or selenium content.

In machining, these grades require a rigid machine setup and positive cutting action to prevent work hardening of the surface. They machine up to about 80% as well as AISI grade 1212 screw stock.

The corrosion resistance of these grades has been sacrificed somewhat in order to obtain ease of machining. Although satisfactory for many of the milder corrodants, atmospheric service, most foodstuffs, and most organic chemicals, types 303 and 303Se are inferior to their non-free-machining counterparts in resisting severe corrodants such as reducing acids and chlorides. Finished machined parts should be passivated for optimum corrosion resistance. The correct procedures must be followed when passivation is specified.

Typical Uses. Applications of AISI types 303 and 303Se include screw machine products, bar and fountain accessories, machined shafts, fishline guides, homogenizers, and valves and accessories for chemical handling equipment.

AISI Type 303: Similar Steels (U.S. and/or Foreign). UNS S30300; AMS 5640 (1); ASME SA194, SA320; ASTM A194, A314, A320, A473, A581, A582; MIL SPEC MIL-S-862; SAE J405 (30303); (W. Ger.) DIN 1.4305; (Fr.) AFNOR Z 10 CNF 18.09; (Ital.) UNI X 10 CrNiS 18 09; (Jap.) SUS 303; (Swed.) SS_{14} 2346; (U.K.) B.S. 303 S 21

AISI Type 303Se: Similar Steels (U.S. and/or Foreign). UNS S30323; AMS 5640 (type 2), 5641, 5738; ASME SA194, SA320; ASTM A194, A194 (8 F), A314, A320, A473, A581, A582; MIL SPEC MIL-S-862; SAE J405 (30303Se)

Physical Properties

AISI Types 303, 303Se: Selected Physical Properties (Ref 6)

Melting range	1400-1420 °C (2550-2590 °F)
Density	8.0 g/cm³ (0.29 lb/in.³)
Specific heat at 0-100 °C (32-212 °F)	500 J/kg·K (0.12 Btu/lb·°F)
Electrical resistivity at 20 °C (68 °F)	72 nΩ·m
Magnetic permeability, when H is 200 Oe, in the annealed condition	1.02 max

AISI Types 303, 303Se: Thermal Treatment Temperatures

Treatment	Temperature range °C	°F
Forging, start	1150-1290	2100-2350
Forging, finish	925 min	1700 min
Annealing(a)	1065	1950
Full annealing(b)	1010-1120	1850-2050

(a) Water quench for maximum resistance to corrosion. (b) Cool rapidly to room temperature

AISI Types 303, 303Se: Average Coefficients of Linear Thermal Expansion (Ref 6)

Temperature range °C	°F	Coefficient µm/m·K	µin./in.·°F
0-100	32-212	17.3	9.6
0-315	32-600	17.8	9.9
0-540	32-1000	18.4	10.2
0-650	32-1200	18.7	10.4

AISI Types 303, 303Se: Thermal Conductivity (Ref 6)

Temperature range °C	°F	Thermal conductivity W/m·K	Btu/ft·h·°F
100	212	16.3	9.4
500	930	21.5	12.4

Mechanical Properties

AISI Types 303, 303Se: Tensile Properties at Elevated Temperature (Ref 9)

Test specimens were annealed and tested in short-time tensile tests

Test temperature °C	°F	Tensile strength MPa	ksi	Yield strength(a) MPa	ksi	Elongation(b), %	Reduction in area, %	Stress(c) MPa	ksi
21	70	620	90	240	35	50	55	...	...
425	800	420	61	240	35	35	51	...	...
540	1000	380	55	235	34	34	55	115	17
650	1200	310	45	205	30	30	54	50	7
760	1400	205	30	145	21	31	45	5	2
870	1600	140	20	70	10	34	43	...	...

(a) 0.2% offset. (b) In 50 mm (2 in.). (c) For 1% creep in 10 000 h

AISI Types 303, 303Se: Tensile Properties at Room Temperature (Ref 7)

Test specimens were bar

Condition	Tensile strength MPa	ksi	Yield strength(a) MPa	ksi	Elongation(b), %	Reduction in area, %	Hardness, HB
Annealed	620	90	240	35	50	55	160
Cold drawn	690	100	415	60	40	53	228

(a) 0.2% offset. (b) In 50 mm (2 in.)

AISI Types 303, 303Se: Modulus of Elasticity (Ref 7)

Stressed in:	Modulus GPa	10^6 psi
Tension	193	28.0
Torsion	77.2	11.2

AISI Type 303: Endurance Limit (Ref 4)

Condition	Endurance limit MPa	ksi
25% hard	330	48
Annealed	240	35

Mechanical Properties (continued)

AISI Type 303: Effect of Test Temperature on Tensile Properties of Cold Worked Material. Test specimens were reduced by cold drawing 10%. Elongation was measured in 50 mm (2 in.). (Ref 3)

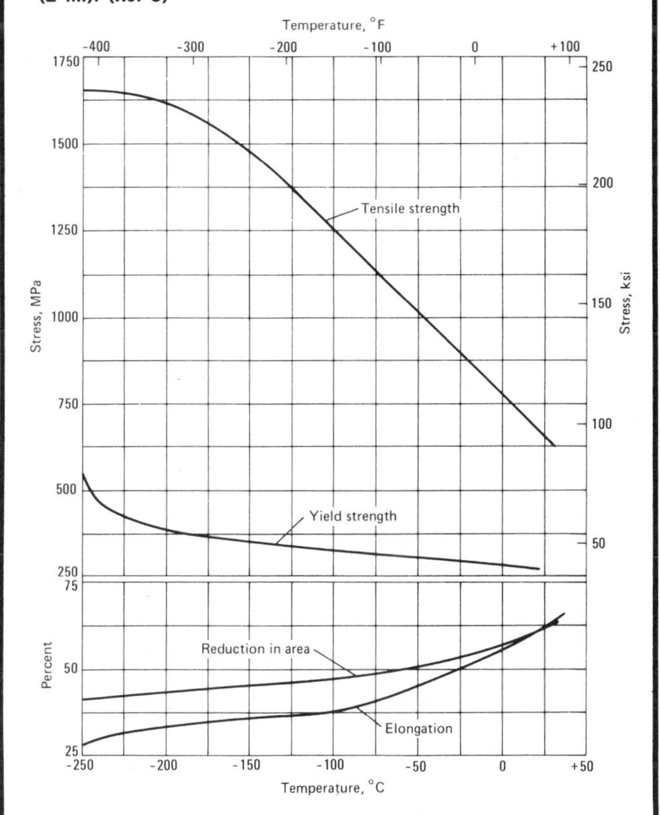

AISI Type 303: Izod Impact Properties. Test specimens were in the annealed condition. (Ref 3)

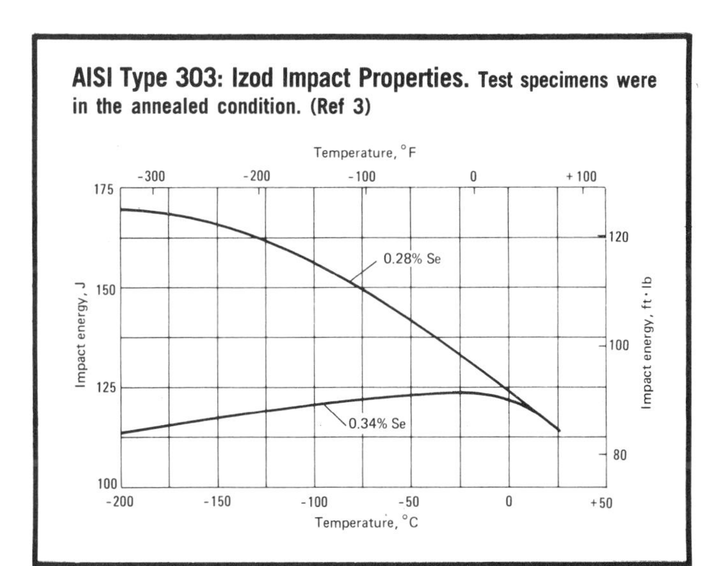

AISI Type 303: Effect of Test Temperature on Tensile Properties of Annealed Material. Elongation was measured in 50 mm (2 in.). (Ref 3)

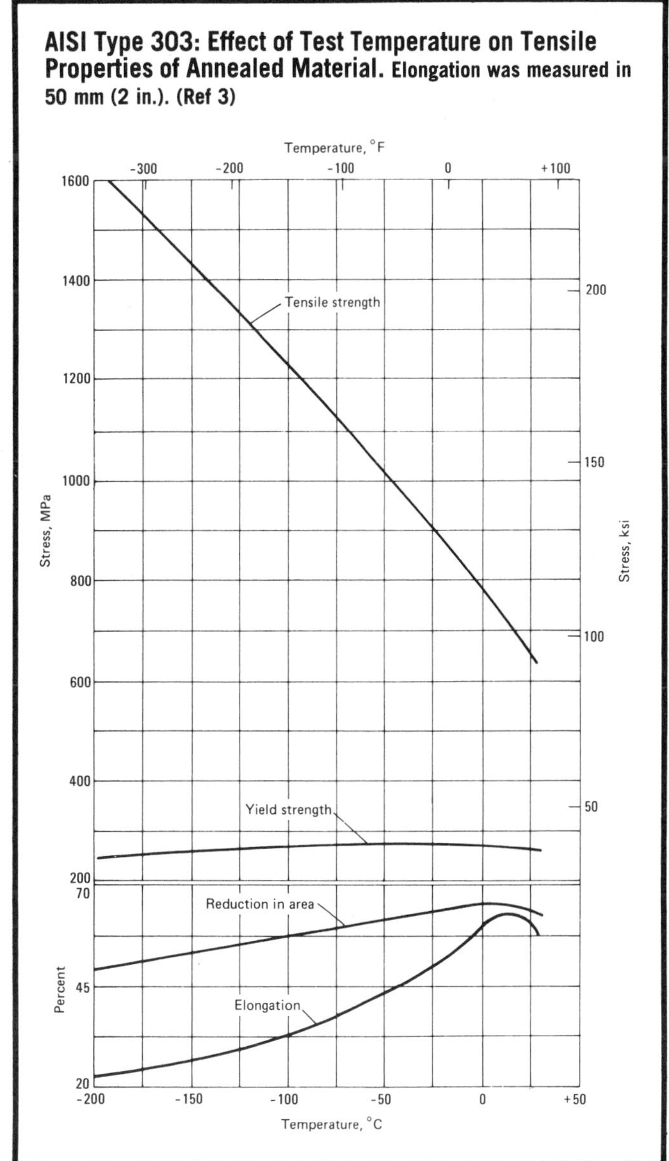

Mechanical Properties (continued)

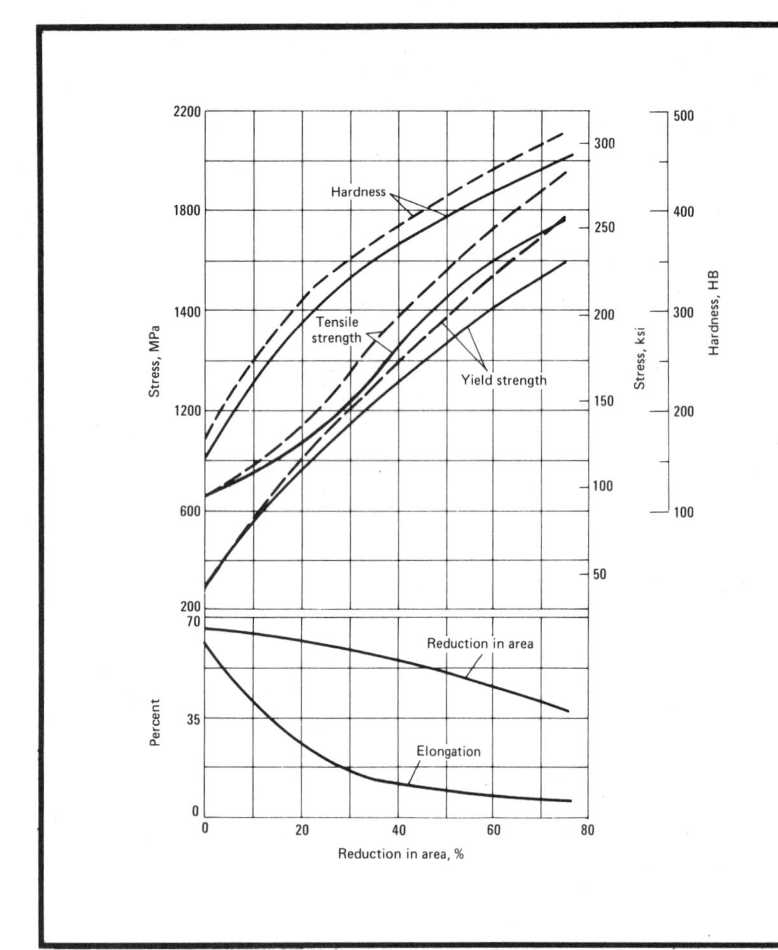

AISI Types 303, 303Se: Effect of Cold Reduction on Tensile Properties. —— as cold drawn; ---- as cold drawn and stress relieved for 4 h at 400 °C (750 °F). Elongation was measured in 4 × diam of test specimen; yield strength at 0.2%. (Ref 7)

Machining Data (Ref 2)

AISI Types 303, 303Se: Face Milling

Depth of cut		M2 and M7 high speed steel: Speed		Feed/tooth		Uncoated C-2 carbide: Speed, brazed		Speed, inserted		Feed/tooth		Coated CC-2 carbide: Speed		Feed/tooth	
mm	in.	m/min	ft/min	mm	in.	m/min	ft/min	m/min	ft/min	mm	in.	m/min	ft/min	mm	in.
Annealed, with hardness of 135 to 185 HB															
1	0.040	49	160	0.20	0.008	150	500	170	550	0.20	0.008	250	825	0.20	0.008
4	0.150	40	130	0.30	0.012	115	375	125	415	0.30	0.012	170	550	0.30	0.012
8	0.300	30	100	0.40	0.016	81	265	100	325	0.40	0.016	130	425	0.40	0.016
Cold drawn, with hardness of 225 to 275 HB															
1	0.040	46	150	0.15	0.006	130	430	145	475	0.18	0.007	215	700	0.18	0.007
4	0.150	37	120	0.25	0.010	100	325	110	360	0.25	0.010	145	475	0.25	0.010
8	0.300	27	90	0.36	0.014	70	230	85	280	0.36	0.014	115	375	0.36	0.014

AISI Types 303, 303Se: Planing

Tool material	Depth of cut mm	in.	Speed m/min	ft/min	Feed/stroke mm	in.
Annealed, with hardness of 135 to 185 HB						
M2 and M3 high speed steel	0.1	0.005	11	35	(a)	(a)
	2.5	0.100	14	45	1.25	0.050
	12	0.500	9	30	1.50	0.060
C-2 carbide	0.1	0.005	46	150	(a)	(a)
	2.5	0.100	49	160	2.05	0.080

Tool material	Depth of cut mm	in.	Speed m/min	ft/min	Feed/stroke mm	in.
Cold drawn, with hardness of 225 to 275 HB						
M2 and M3 high speed steel	0.1	0.005	9	30	(a)	(a)
	2.5	0.100	12	40	.75	0.030
	12	0.500	8	25	1.15	0.045
C-2 carbide	0.1	0.005	38	125	(a)	(a)
	2.5	0.100	43	140	1.50	0.060

(a) Feed is 75% the width of the square nose finishing tool

Machining Data (Ref 2) (continued)

AISI Types 303, 303Se: Turning (Single Point and Box Tools)

Depth of cut		High speed steel(a)				Uncoated C-2 carbide tools						Coated CC-2 carbide			
		Speed		Feed		Speed, brazed		Speed, inserted		Feed		Speed		Feed	
mm	in.	m/min	ft/min	mm/rev	in./rev	m/min	ft/min	m/min	ft/min	mm/rev	in./rev	m/min	ft/min	mm/rev	in./rev
Annealed, with hardness of 135 to 185 HB															
1	0.040	40	130	0.18	0.007	150	500	170	550	0.18	0.007	215	700	0.18	0.007
4	0.150	32	105	0.40	0.015	130	425	145	475	0.40	0.015	185	600	0.40	0.015
8	0.300	24	80	0.50	0.020	100	325	115	370	0.75	0.030	145	475	0.50	0.020
16	0.625	20	65	0.75	0.030	176	250	85	280	1.00	0.040	...	...	...	...
Cold drawn, with hardness of 225 to 275 HB															
1	0.040	37	120	0.18	0.007	120	400	145	475	0.18	0.007	185	600	0.18	0.007
4	0.150	30	100	0.40	0.015	110	365	125	410	0.40	0.015	160	525	0.40	0.015
8	0.300	24	80	0.50	0.020	84	275	100	320	0.75	0.030	90	300	0.50	0.020
16	0.625	18	60	0.75	0.030	69	225	73	240	1.00	0.040	...	...	...	...

(a) High speed steel tool material: for hardness of 135 to 185 HB, use M2 or M3; for hardness of 225 to 275 HB, use any premium high speed steel (T15, M33, M41 to M47)

AISI Types 303, 303Se: Turning (Cutoff and Form Tools)

Tool material	Speed, m/min	Feed per revolution for cutoff tool width of: 1.5 mm (0.062 in.) mm (in.)	3 mm (0.125 in.) mm (in.)	6 mm (0.25 in.) mm (in.)	Feed per revolution for form tool width of: 12 mm (0.5 in.) mm (in.)	18 mm (0.75 in.) mm (in.)	25 mm (1 in.) mm (in.)	35 mm (1.5 in.) mm (in.)	50 mm (2 in.) mm (in.)
Annealed, with hardness of 135 to 185 HB									
M2 and M3 high speed steel	30 (100)	0.038 (0.0015)	0.050 (0.002)	0.061 (0.0024)	0.063 (0.0025)	0.050 (0.002)	0.046 (0.0018)	0.038 (0.0015)	0.025 (0.001)
C-6 carbide	100 (325)	0.038 (0.0015)	0.050 (0.002)	0.061 (0.0024)	0.063 (0.0025)	0.050 (0.002)	0.046 (0.0018)	0.038 (0.0015)	0.025 (0.001)
Cold drawn, with hardness of 225 to 275 HB									
M2 and M3 high speed steel	43 (140)	0.050 (0.002)	0.063 (0.0025)	0.075 (0.003)	0.063 (0.0025)	0.056 (0.0022)	0.050 (0.002)	0.046 (0.0018)	0.038 (0.0015)
C-6 carbide	115 (375)	0.050 (0.002)	0.063 (0.0025)	0.075 (0.003)	0.063 (0.0025)	0.056 (0.0022)	0.050 (0.002)	0.046 (0.0018)	0.038 (0.0015)

AISI Types 303, 303Se: Boring

Depth of cut		High speed steel(a)				Uncoated C-2 carbide						Coated CC-2 carbide			
		Speed		Feed		Speed, brazed		Speed, inserted		Feed		Speed		Feed	
mm	in.	m/min	ft/min	mm/rev	in./rev	m/min	ft/min	m/min	ft/min	mm/rev	in./rev	m/min	ft/min	mm/rev	in./rev
Annealed, with hardness of 135 to 185 HB															
0.25	0.010	41	135	0.075	0.003	120	400	145	475	0.075	0.003	185	600	0.075	0.003
1.25	0.050	32	105	0.13	0.005	100	325	115	375	0.13	0.005	150	490	0.13	0.005
2.5	0.100	26	85	0.30	0.012	84	275	100	335	0.30	0.012	130	425	0.30	0.012
Cold drawn, with hardness of 225 to 275 HB															
0.25	0.010	37	120	0.075	0.003	115	375	130	425	0.075	0.003	160	525	0.075	0.003
1.25	0.050	29	95	0.13	0.005	84	275	100	335	0.13	0.005	130	420	0.13	0.005
2.5	0.100	24	80	0.30	0.012	76	250	87	285	0.30	0.012	115	370	0.30	0.012

(a) High speed steel tool material: for hardness of 135 to 185 HB, M2 or M3 high speed steel; for hardness of 225 to 275 HB, use any premium high speed steel (T15, M33, M41 to M47)

AISI Types 303, 303Se: Drilling

Tool material	Speed, m/min (ft/min)	Feed per revolution for nominal hole diameter of: 1.5 mm (0.062 in.) mm (in.)	3 mm (0.125 in.) mm (in.)	6 mm (0.25 in.) mm (in.)	12 mm (0.5 in.) mm (in.)	18 mm (0.75 in.) mm (in.)	25 mm (1 in.) mm (in.)	35 mm (1.5 in.) mm (in.)	50 mm (2 in.) mm (in.)
Annealed, with hardness of 135 to 185 HB									
M10, M7 and M1 high speed steel	37 (120)	0.025 (0.001)	0.075 (0.003)	0.15 (0.006)	0.25 (0.010)	0.36 (0.014)	0.45 (0.017)	0.55 (0.021)	0.65 (0.025)
Cold drawn, with hardness of 225 to 275 HB									
M10, M7 and M1 high speed steel	30 (100)	0.025 (0.001)	0.075 (0.003)	0.15 (0.006)	0.25 (0.010)	0.36 (0.014)	0.45 (0.017)	0.55 (0.021)	0.65 (0.025)

Machining Data (Ref 2) (continued)

AISI Types 303, 303Se: Reaming (Ref 2)

Based on 4 flutes for 3- and 6-mm (0.125- and 0.25-in.) reamers, 6 flutes for 12-mm (0.5-in.) reamers, and 8 flutes for 25-mm (1-in.) and larger reamers

Tool material	Speed		Feed per revolution for reamer diameter of:											
			3 mm (0.125 in.)		6 mm (0.25 in.)		12 mm (0.5 in.)		25 mm (1 in.)		35 mm (1.5 in.)		50 mm (2 in.)	
	m/min	ft/min	mm	in.	mm	in.	mm	in.	mm	in.	mm	in.	mm	in.
Annealed, with hardness of 135 to 185 HB														
Roughing														
M1, M2, and M7 high speed steel	27	90	0.075	0.003	0.15	0.006	0.23	0.009	0.28	0.011	0.40	0.015	0.45	0.018
C-2 carbide	34	110	0.102	0.004	0.20	0.008	0.30	0.012	0.40	0.016	0.50	0.020	0.60	0.024
Finishing														
M1, M2, and M7 high speed steel	15	50	0.075	0.003	0.102	0.004	0.15	0.006	0.20	0.008	0.23	0.009	0.25	0.010
C-2 carbide	20	65	0.102	0.004	0.15	0.006	0.20	0.008	0.25	0.010	0.28	0.011	0.30	0.012
Cold drawn, with hardness of 225 to 275 HB														
Roughing														
M1, M2, and M7 high speed steel	26	85	0.075	0.003	0.15	0.006	0.23	0.009	0.28	0.011	0.40	0.015	0.45	0.018
C-2 carbide	30	100	0.102	0.004	0.20	0.008	0.30	0.012	0.40	0.016	0.50	0.020	0.60	0.024
Finishing														
M1, M2, and M7 high speed steel	12	40	0.075	0.003	0.102	0.004	0.15	0.006	0.20	0.008	0.23	0.009	0.25	0.010
C-2 carbide	18	60	0.102	0.004	0.15	0.006	0.20	0.008	0.25	0.010	0.28	0.011	0.30	0.012

AISI Types 303, 303Se: End Milling (Profiling)

Tool material	Depth of cut		Speed		Feed per tooth for cutter diameter of:							
					10 mm (0.375 in.)		12 mm (0.5 in.)		18 mm (0.75 in.)		25-50 mm (1-2 in.)	
	mm	in.	m/min	ft/min	mm	in.	mm	in.	mm	in.	mm	in.
Annealed, with hardness of 135 to 185 HB												
M2, M3, and M7 high speed steel	0.5	0.020	53	175	0.025	0.001	0.050	0.002	0.102	0.004	0.13	0.005
	1.5	0.060	41	135	0.050	0.002	0.075	0.003	0.13	0.005	0.15	0.006
	diam/4	diam/4	37	120	0.025	0.001	0.050	0.002	0.102	0.004	0.13	0.005
	diam/2	diam/2	30	100	0.025	0.001	0.038	0.0015	0.075	0.003	0.102	0.004
C-2 carbide	0.5	0.020	140	455	0.025	0.001	0.038	0.0015	0.102	0.004	0.13	0.005
	1.5	0.060	105	350	0.038	0.0015	0.050	0.002	0.13	0.005	0.15	0.006
	diam/4	diam/4	90	295	0.038	0.0015	0.050	0.002	0.102	0.004	0.13	0.005
	diam/2	diam/2	84	275	0.025	0.001	0.038	0.0015	0.075	0.003	0.102	0.004
Cold drawn, with hardness of 225 to 275 HB												
M2, M3, and M7 high speed steel	0.5	0.020	44	145	0.025	0.001	0.050	0.002	0.102	0.004	0.13	0.005
	1.5	0.060	34	110	0.050	0.002	0.076	0.003	0.13	0.005	0.15	0.006
	diam/4	diam/4	29	95	0.025	0.001	0.050	0.002	0.102	0.004	0.13	0.005
	diam/2	diam/2	26	85	0.025	0.001	0.025	0.001	0.075	0.003	0.102	0.004
C-2 carbide	0.5	0.020	130	420	0.025	0.001	0.038	0.0015	0.102	0.004	0.13	0.005
	1.5	0.060	100	325	0.038	0.0015	0.050	0.002	0.13	0.005	0.15	0.006
	diam/4	diam/4	85	280	0.038	0.0015	0.050	0.002	0.102	0.004	0.13	0.005
	diam/2	diam/2	79	260	0.025	0.001	0.038	0.0015	0.13	0.003	0.102	0.004

AISI Types 303, 303Se: Cylindrical and Internal Grinding

Test specimens were annealed or cold drawn, with hardness of 135 to 275 HB

Operation	Wheel speed		Work speed		Infeed on diameter		Traverse(a)	Wheel identification
	m/s	ft/min	m/min	ft/min	mm/pass	in./pass		
Cylindrical grinding								
Roughing	28-33	5500-6500	15-30	50-100	0.050	0.002	1/2	C54JV
Finishing	28-33	5500-6500	15-30	50-100	0.013 max	0.0005 max	1/6	C54JV
Internal grinding								
Roughing	25-33	5000-6500	23-60	75-200	0.013	0.0005	1/3	C60KV
Finishing	25-33	5000-6500	23-60	75-200	0.005 max	0.0002 max	1/6	C60KV

(a) Wheel width per revolution of work

Machining Data (Ref 2) (continued)

AISI Types 303, 303Se: Surface Grinding with a Horizontal Spindle and Reciprocating Table

Test specimens were annealed or cold drawn, with hardness of 135 to 275 HB

Operation	Wheel speed m/s	Wheel speed ft/min	Table speed m/min	Table speed ft/min	Downfeed mm/pass	Downfeed in./pass	Crossfeed mm/pass	Crossfeed in./pass	Wheel identification
Roughing	28-33	5500-6500	15-30	50-100	0.050	0.002	1.25-12.5	0.050-0.500	C46JV
Finishing	28-33	5500-6500	15-30	50-100	0.013 max	0.0005 max	(a)	(a)	C46JV

(a) Maximum 1/4 of wheel width

AISI Type 303MA

AISI Type 303MA: Chemical Composition

Element	Composition, %
Carbon	0.15
Manganese	2.0
Phosphorus	0.04
Sulfur	0.11-0.16
Silicon	1.0
Chromium	17-19
Nickel	8-10
Molybdenum	0.40-0.60
Aluminum	0.60-1.00

Characteristics. AISI type 303MA (U.S. Patent No. 2 900 250) is a free-machining, chromium-nickel steel which offers several advantages over AISI type 303 and 303Se steels. Type 303MA stainless steel machines faster with better tool life and improved finish. A lower sulfur content and the addition of aluminum overcome many problems of corrosion resistance and longitudinal splitting normally associated with free-machining steels. This alloy is better than types 303 and 303Se in food handling, soft-drink dispensing, and chemical industry applications which require added corrosion resistance.

AISI grade 303MA has good deforming properties for applications such as swaging, staking, extruding, and thread rolling. This steel is not recommended for severe deep drawing or severe cold heading operations.

Maximum temperature for resistance to scaling of type 303MA is about 900 °C (1650 °F) in continuous service, and 815 °C (1500 °F) in intermittent service. Type 303MA can be welded using all the common welding processes. Type 308 welding wire is recommended as filler metal.

Because type 303MA work hardens rapidly, rigid machine setups and positive cutting action are needed to avoid hardening of the surface. Machinability is about 85% of that for AISI 1212.

This alloy is satisfactory for atmospheric service, foodstuffs, and most organic chemicals. Service trials are recommended before using under the most severe corrosive conditions.

Typical Uses. Applications of AISI type 303MA include screw machine products, fasteners, bushings, hose and cable connectors, valve parts, gears, and pulleys.

AISI 303MA: Similar Steels (U.S. and/or Foreign). UNS S30345; ASTM A581 (XM-2), A582 (XM-2)

Physical Properties

AISI Type 303MA: Selected Physical Properties (Ref 7)

Density	7.81 g/cm³ (0.282 lb/in.³)
Specific heat at 0-100 °C (32-212 °F)	500 J/kg·K (0.12 Btu/lb·°F)
Electrical resistivity, in the annealed condition	80.5 nΩ·m
Electrical resistivity, in the cold drawn condition	83.0 nΩ·m
Magnetic permeability, when H is 200 Oe, in the annealed condition	1.02 max

AISI Type 303MA: Thermal Treatment Temperatures

Treatment	Temperature range °C	Temperature range °F
Forging, start	1150-1260	2100-2300
Forging, finish	925	1700 min
Annealing(a)	1040-1065	1900-1950
Hardening(b)	...	...

(a) If product is to be used in a severely corrosive service, air or water quench. Quench rapidly through 480 to 870 °C (900 to 1600 °F). (b) Hardenable by cold work only

AISI Type 303MA: Average Coefficients of Linear Thermal Expansion (Ref 7)

Temperature range °C	Temperature range °F	Coefficient µm/m·K	Coefficient µin./in.·°F
21-315	70-600	16.4	9.1
21-540	70-1000	18.5	10.3
21-650	70-1200	19.1	10.6

AISI Type 303MA: Thermal Conductivity (Ref 7)

Temperature °C	Temperature °F	Thermal conductivity W/m·K	Thermal conductivity Btu/ft·h·°F
100	212	16.3	9.4
500	930	21.5	12.4

Mechanical Properties

AISI Type 303MA: Modulus of Elasticity (Ref 7)

Stressed in:	Modulus GPa	10^6 psi
Tension	190	27.6

AISI Type 303MA: Izod V-Notch Impact Properties

Test temperature °C	°F	Energy J	ft·lb
−195	−320	150	110
21	70	115	85

AISI Type 303MA: Tensile Properties at Room Temperature (Ref 7)

Test specimens were bar

Condition	Tensile strength MPa	ksi	Yield strength(a) MPa	ksi	Elongation(b), %	Reduction in area, %	Hardness, HB
Annealed	620	90	275	40	50	60	160
Cold drawn	690	100	415	60	40	55	228

(a) 0.2% offset. (b) In 50 mm (2 in.)

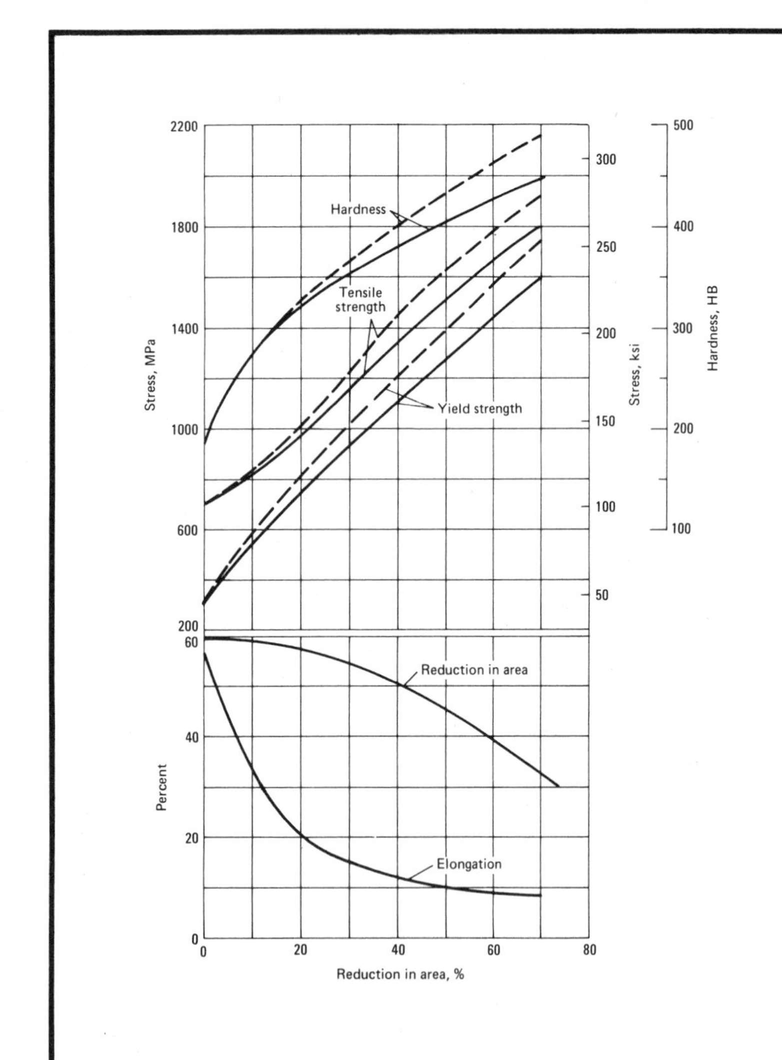

AISI Type 303MA: Effect of Cold Reduction on Tensile Properties. —— as cold drawn; - - - - as cold drawn and stress relieved at 400 °C (750 °F) for 4 h. Ductility curves are averages for cold drawn and cold drawn plus stress relieved conditions. Elongation was measured in 4 × diam of test specimen; yield strength at 0.2% offset. (Ref 7)

Machining Data

For machining data on AISI type 303MA, refer to the preceding machining tables for AISI type 303 and 303Se.

AISI Type 304

AISI Type 304: Chemical Composition

Element	Composition, max %
Carbon	0.08
Manganese	2.0
Phosphorus	0.045
Sulfur	0.030
Silicon	1.0
Chromium	18-20
Nickel	8.0-10.5

Characteristics. AISI type 304 is an austenitic, chromium-nickel, stainless steel characterized by a comparatively low carbon content and higher chromium and nickel contents than those of AISI types 301 and 302. Therefore, it is particularly well suited where welded construction is required and where the finished product must resist most severe forms of corrosion.

In light sections, type 304 can be welded with a small amount of carbide precipitation or loss of corrosion resistance resulting. Postweld annealing is not necessary in most cases. The homogeneous structure, high ductility, and excellent strength of this stainless steel provide good performance in deep drawing, cold forming, spinning, roll forming, and bending. It is nonmagnetic in the annealed condition, and becomes slightly magnetic when cold worked.

In the fully annealed condition, type 304 is highly resistant to ordinary rusting and corrosive action, and is immune to foodstuffs, sterilizing solutions, most organic chemicals and dyestuffs, and a variety of inorganic acids. It resists nitric acid well, sulfuric acids moderately, and halogen acids poorly. This grade also has good scaling resistance up to 870 °C (1600 °F) in continuous service, and up to 790 °C (1450 °F) in intermittent service.

In machining, rigid machine setups, heavy feeds, and slow spindle speeds are most suitable. The machinability rating of type 304 is about 60% of that of AISI grade 1212.

Typical Uses. Applications of AISI type 304 stainless steel include screw machine products, bar and fountain accessories, machined shafts, dairy equipment, homogenizers, and valves and accessories for chemical handling equipment.

AISI Type 304: Similar Steels (U.S. and/or Foreign). UNS S30400; AMS 5501, 5513, 5560, 5565, 5566, 5567, 5639, 5697; ASME SA182, SA194 (8), SA213, SA240, SA249, SA312, SA320 (B8), SA358, SA376, SA403, SA409, SA430, SA479, SA688; ASTM A167, A182, A193, A194, A213, A240, A249, A269, A270, A271, A276, A312, A313, A314, A320, A368, A376, A409, A430, A473, A478, A479, A492, A493, A511, A554, A580, A632, A651, A666, A688; FED QQ-W-423, QQ-S-763, QQ-S-766, STD-66; MIL SPEC MIL-F-20138, MIL-S-862, MIL-S-5059, MIL-S-23195, MIL-S-23196, MIL-T-6845, MIL-T-8504, MIL-T-8506; SAE J405 (30304); (W. Ger.) DIN 1.4301; (Fr.) AFNOR Z 6 CN 18.09; (Ital.) UNI X 5 CrNi 18 10; (Jap.) SUS 304; (Swed.) SS_{14} 2332; (U.K.) B.S. 304 S 15, 302 S 17, 304 S 16, 304 S 18, 304 S 25, 304 S 40, En. 58 E

Physical Properties

AISI Type 304: Selected Physical Properties (Ref 1, 6)

Melting range	1400-1455 °C (2550-2650 °F)
Density	8.0 g/cm³ (0.29 lb/in.³)
Specific heat at 0-100 °C (32-212 °F)	502 J/kg·K (0.12 Btu/lb·°F)
Electrical resistivity at 20 °C (68 °F)	72 nΩ·m
Electrical resistivity at 650 °C (1200 °F)	116 nΩ·m
Magnetic permeability, when H is 200 Oe, in the annealed condition	1.02 max

AISI Type 304: Average Coefficients of Linear Thermal Expansion (Ref 1, 6)

Temperature range °C	°F	Coefficient μm/m·K	μin./in.·°F
0-100	32-212	17.3	9.6
0-315	32-600	17.8	9.9
0-540	32-1000	18.4	10.2
0-650	32-1200	18.7	10.4

AISI Type 304: Thermal Treatment Temperatures

Treatment	Temperature range °C	°F
Forging, start	1150-1260	2100-2300
Forging, finish	925 min	1700 min
Annealing(a)	1010-1120	1850-2050
Stress relieving	205-425	400-800
Hardening(b)	...	...

(a) Cool rapidly in air or water quench. (b) Hardenable by cold work only

AISI Type 304: Thermal Conductivity (Ref 1, 6)

Temperature °C	°F	Thermal conductivity W/m·K	Btu/ft·h·°F
100	212	16.3	9.4
500	930	21.5	12.4

Machining Data

For machining data on AISI type 304, refer to the preceding machining tables for AISI type 301.

Mechanical Properties

For data on sheet formability and corrosion rates in boiling organic acids, refer to the mechanical properties section of ASTM XM-27.

AISI Type 304: Tensile Properties at Room Temperature (Ref 6)

Condition	Size round mm	Size round in.	Tensile strength MPa	Tensile strength ksi	Yield strength(a) MPa	Yield strength(a) ksi	Elongation(b), %	Reduction in area, %	Hardness
Sheet and strip									
Annealed	...	...	580	84	290	42	55	...	80 HRB
Wire(c)									
Annealed	1.6	0.062	725	105	240	35	55	...	...
Annealed	13	0.50	620	90	240	35	60	55	83 HRB
Soft tempered	1.6	0.062	860	125	620	90	40	...	...
Soft tempered	13	0.50	690	100	415	60	45	65	95 HRB
Hard tempered	1.6	0.062	1105	160	860	125	20	...	...
Hard tempered	13	0.50	965	140	725	105	25	55	33 HRC
Spring tempered	1.6	0.062	1795	260	...	...	...	...	...
Spring tempered	6.4	0.25	1240	180	...	...	...	...	...
Spring tempered	7.80	0.307	1170	170	...	...	...	...	...
Plate									
Annealed	...	...	565	82	240	35	60	70	149 HB
Bar									
Annealed	...	...	585	85	235	34	60	70	149 HB
Annealed and cold drawn	25	1.0	690	100	415	60	45	...	212 HB
Cold drawn, high tensile	22.2	0.875	860	125	655	95	25	...	277 HB
Cold drawn, high tensile	38	1.5	760	110	515	75	60	...	240 HB

(a) 0.2% offset. (b) In 50 mm (2 in.), except for wire. For wire 3.18-mm (0.125-in.) diam and larger, gage length is 4 × diam of test specimen; for smaller diameters, the gage length is 254 mm (10.0 in.). (c) Tensile and yield strengths are slightly higher when in straightened form, spring temper up to 10% less

AISI Type 304: Tensile Properties at Elevated Temperatures (Ref 7)

Test specimens were 16-mm (0.625-in.) diam bar annealed and cold worked to hardness of 95 HRB

Test temperature °C	Test temperature °F	Tensile strength MPa	Tensile strength ksi	Yield strength(a) MPa	Yield strength(a) ksi	Elongation(b), %	Reduction in area, %
21	70	669	97.0	410	59.5	66.5	81.5
205	400	483	70.0	331	48.0	36.5	72.5
425	800	472	68.5	290	42.0	35.5	62.0
540	1000	427	62.0	265	38.5	34.5	68.0
650	1200	324	47.0	214	31.0	35.0	54.5
760	1400	207	30.0	162	23.5	44.5	63.0
870	1600	119	17.2	112	16.3	58.5	61.5
980	1800	68	9.9	66	9.6	75.5	89.5

(a) 0.2% offset. (b) In 4 × diam of test specimen

AISI Type 304: Tensile Properties at Subzero and Elevated Temperatures (Ref 7)

Test temperature °C	Test temperature °F	Tensile strength MPa	Tensile strength ksi	Yield strength(a) MPa	Yield strength(a) ksi	Elongation(b), %	Reduction in area, %
−195	−320	1620	235	386	56	40	59
−62	−80	1110	161	345	50	57	70
−40	−40	1000	145	385	48	60	73
0	32	840	122	345	40	65	76
21	70	585	85	330	37	57	70
205	400	495	72	275	23	51	73
315	600	470	68	255	19.5	45	71
425	800	440	64	160	16.5	40	69
540	1000	385	56	135	14	36	66
650	1200	305	44	115	12.5	34	58
760	1400	200	29	76	11	36	46
870	1600	110	16	...	...	40	40

(a) 0.2% offset. (b) In 50 mm (2 in.)

AISI Type 304: Modulus of Elasticity

Stressed in:	Modulus GPa	Modulus 10^6 psi
Tension	193	28.0
Torsion	77	11.2

AISI Type 304: Izod V-Notch Impact Energy (Ref 3)

Test temperature °C	Test temperature °F	Energy J	Energy ft·lb
−195	−320	150	110
21	70	150	110

AISI Type 304: Endurance Limits After Cold Work (Ref 4)

Condition	Stress MPa	Stress ksi
50% hard temper	483	70.0
75% hard temper	638	92.5

AISI Type 304: Stress-Rupture and Creep Properties (Ref 7)

Test temperature °C	°F	Stress for rupture in: 1000 h MPa	ksi	10 000 h(a) MPa	ksi	100 000 h(a) MPa	ksi	Stress for secondary creep rate of(a): 1% in 10 000 h MPa	ksi	1% in 100 000 h MPa	ksi
480	900	338	49.0	262	38.0	...	...	...	...	...	...
540	1000	241	35.0	183	26.5	145	21.0	138	20.0	83	12.0
595	1100	152	22.0	117	17.0	90	13.0	90	13.0	52	7.5
650	1200	97	14.0	69	10.0	48	7.0	57	8.3	30	4.3
705	1300	59	8.5	41	6.0	28	4.0	34	5.0	17	2.5
760	1400	41	6.0	26	3.7	17	2.5	21	3.0	10	1.5
815	1500	26	3.7	19	2.7	10	1.5	17	2.4	8	1.2
870	1600	19	2.7	...	...	...	...	...	...	...	...

(a) Extrapolated data

AISI Type 304: Charpy Impact Properties for Weld Metals (Ref 3)

Condition	Charpy keyhole notch impact energy at test temperature of: Room temperature J	ft·lb	−76 °C (−105 °F) J	ft·lb	−195 °C (−320 °F) J	ft·lb
Type 304 electrode						
As welded	43-50	32-37	33-38	24-28	24-31	18-23
Annealed	56-66	41-49	53-57	39-42	43-57	32-42
Type 308 electrode						
As welded	42-45	31-33	28-34	21-25	19-27	14-20
Annealed	49-52	36-38	41	30	41-42	30-31

AISI Type 304: Effect of Cold Reduction on Tensile Properties (Ref 8)

Cold reduction, %	Tensile strength MPa	ksi	Yield strength(a) MPa	ksi	Elongation(b), %
0	595	86.3	225	32.7	54.5
10	675	97.9	483	70.0	36.5
20	780	113.2	663	96.2	24.0
30	904	131.1	817	118.5	15.5
40	1005	145.7	931	135.0	12.0
50	1091	158.3	1000	145.0	9.0
60	1158	168.0	1047	151.8	6.0

(a) 0.2% offset. (b) In 50 mm (2 in.)

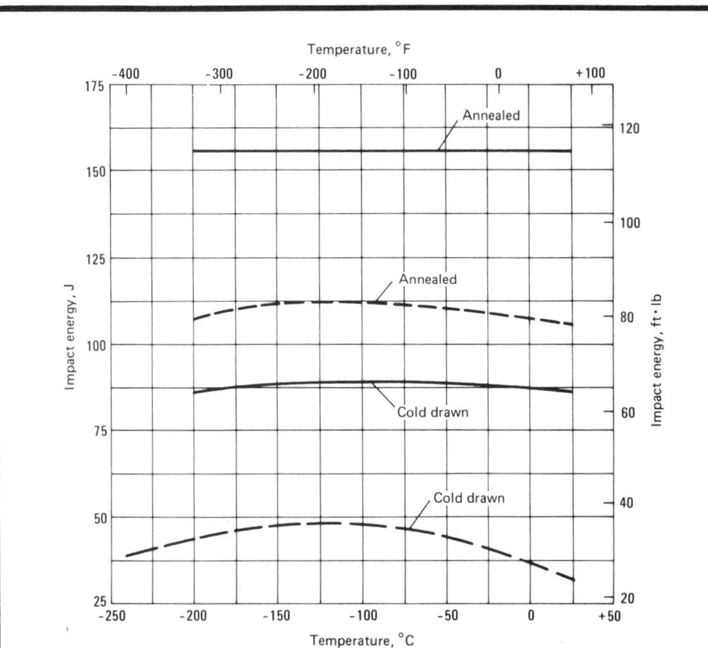

AISI Type 304: Impact Properties. —— Izod impact energy: specimens were annealed or cold drawn to tensile strength of 1450 MPa (210 ksi); ——— Charpy keyhole impact energy: specimens were annealed or cold drawn to 937 MPa (135.9 ksi). (Ref 3)

Mechanical Properties (continued)

AISI Type 304: Effect of Test Temperature on Tensile Properties of Annealed Material. Elongation was measured in 50 mm (2 in.). (Ref 3)

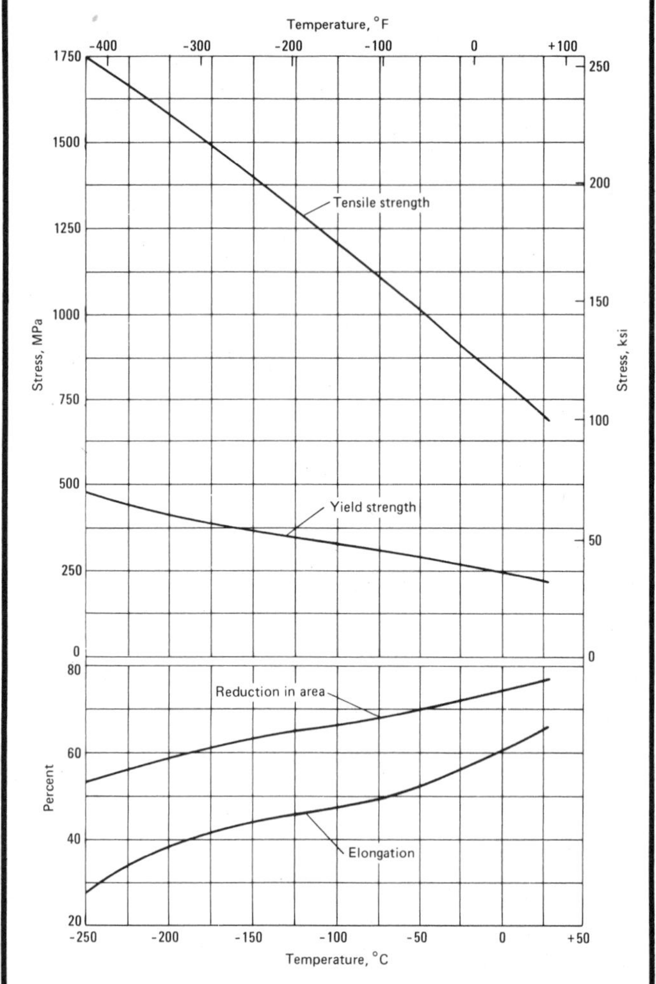

AISI Type 304: Effect of Sensitization and Carbon Content on Impact Properties. Test specimens were in the annealed condition; carbon content of 0.07%; sensitization at 650 °C (1200 °F) for 2 h. Impact energy tests used keyhole notch specimens. (Ref 3)

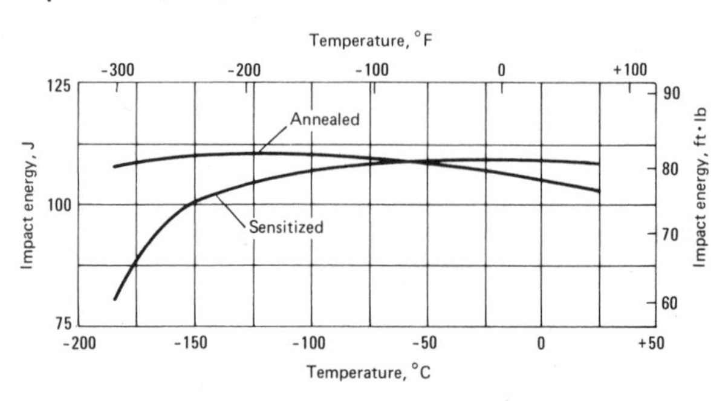

AISI Type 304: Effect of Cold Reduction on Tensile Properties. Annealed at 1065 °C (1950 °F); water quenched. Specimens were 9.53-mm (0.375-in.) rounds. unstraightened and untempered. Tests were conducted using specimens machined to English units. Elongation was measured in 50 mm (2 in.); yield strength at 0.2% offset. (Ref 10)

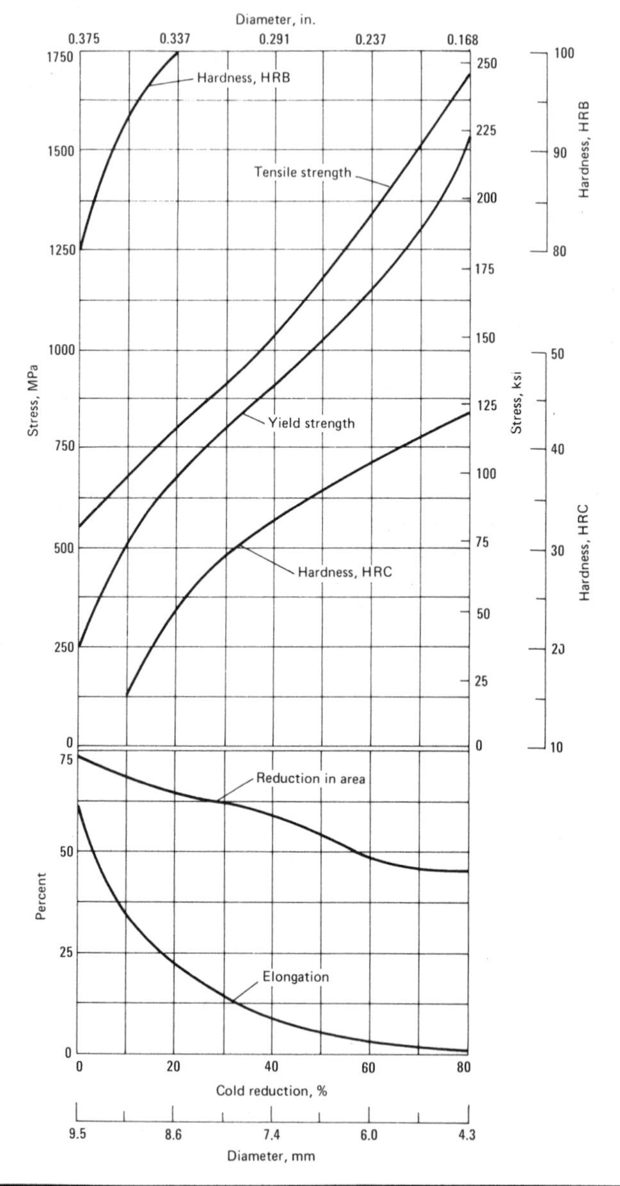

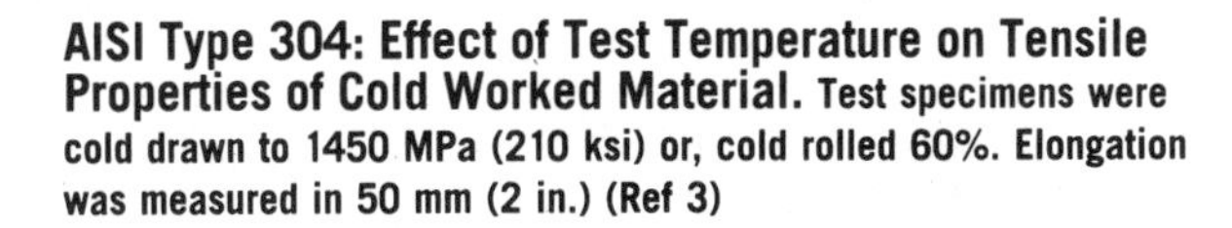
AISI Type 304: Effect of Test Temperature on Tensile Properties of Cold Worked Material. Test specimens were cold drawn to 1450 MPa (210 ksi) or, cold rolled 60%. Elongation was measured in 50 mm (2 in.) (Ref 3)

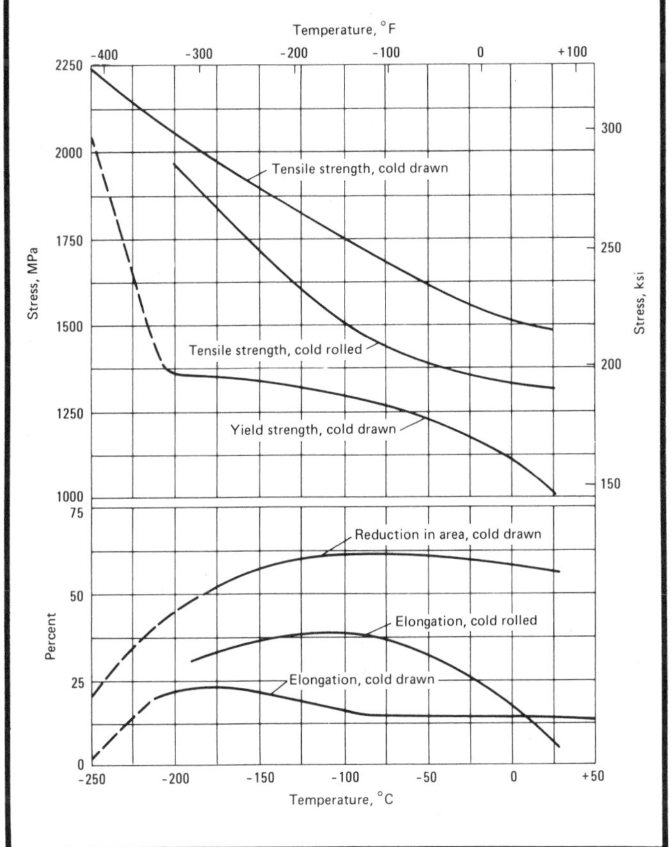

AISI Type 304: Reciprocating Beam Fatigue Strength of Cold Worked Material. Material tested was 18-8 stainless steel, cold worked to 1450 MPa (210 ksi). Unnotched specimens were tested at 25 °C or 77 °F (○) and at −195 °C or −320 °F (△). Notched specimens were tested at 25 °C or 77 °F (●); at −195 °C or −320 °F (▲); and at −255 °C or −425 °F (■). (Ref 3)

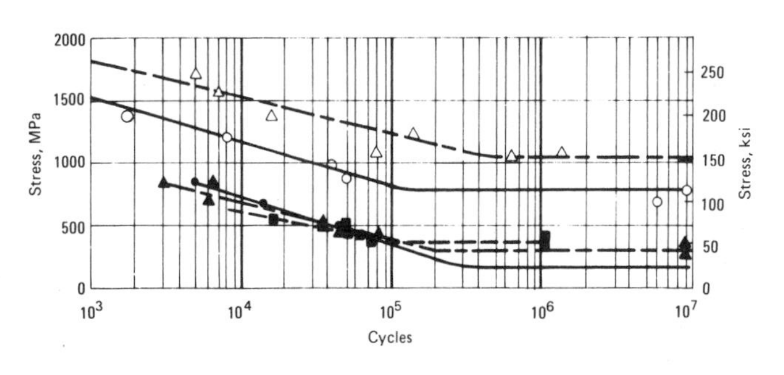

AISI Type 304L

AISI Type 304L: Chemical Composition

Element	Composition, max %
Carbon	0.030
Manganese	2.0
Phosphorus	0.045
Sulfur	0.030
Silicon	1.0
Chromium	18-20
Nickel	8-12

Characteristics. AISI type 304L is a very low-carbon, austenitic, chromium-nickel steel. This grade has general corrosion resistance similar to that of type 304 but superior resistance to intergranular corrosion after welding or stress relieving. Type 304L is recommended for use in parts that are fabricated by brazing or welding and cannot be subsequently annealed. When filler metal is required, type 308L is suggested.

Type 304L stainless steel is essentially nonmagnetic when annealed, and becomes slightly magnetic when cold worked. This grade is very ductile and can be easily cold worked by the common cold forming methods. Because it work hardens rapidly, in-process annealing may be necessary to restore ductility and lower hardness.

In most severe corrodants or when operating temperatures exceed about 425 °C (800 °F), type 304L may not be satisfactory; one of the stabilized types 321 or 347 should be considered. The maximum temperature for scaling resistance in continuous exposure is about 900 °C (1650 °F), and for intermittent exposure about 815 °C (1500 °F).

In machining, better results are obtained if the type 304 material is in a moderately cold worked condition than in an annealed condition. Machinability of type 304L is about 60% of that for free-machining AISI grade 1212.

AISI Type 304L: Similar Steels (U.S. and/or Foreign). UNS S30403; AMS 5511, 5647; ASME SA182, SA213, SA240, SA249, SA312, SA403, SA479, SA688; ASTM A167, A182, A213, A240, A249, A276, A312, A314, A403, A473, A478, A479, A511, A554, A580, A632, A688; FED QQ-S-763, QQ-S-766; MIL SPEC MIL-S-862, MIL-S-23195, MIL-S-23196; SAE J405 (30304L); (W. Ger.) DIN 1.4306; (Ital.) UNI X 2 CrNi 18 11, X 3 CrNi 18 11, X 2 CrNi 18 11 KG, X 2 CrNi 18 11 KW; (Jap.) JIS SUS 304 L, SCS 19; (Swed.) SS_{14} 2352; (U.K.) B.S. 304 S 12, 304 S 14, 304 S 22, S 536

Typical Uses. Applications of AISI type 304L include architectural trim, beer barrels, brewing equipment, chemical handling equipment, household appliances, textile drying equipment, oil refinery tubes, steam piping, cookware, nuclear reactor component vessels, vacuum pump parts, dairy equipment, and winery tanks.

Physical Properties

AISI Type 304L: Thermal Treatment Temperatures

Treatment	Temperature range °C	°F
Forging, start	1150-1260	2100-2300
Forging, finish	925 min	1700 min
Annealing(a)	1010-1120	1850-2050
Hardening(b)	...	...

(a) Thin sections such as strip may be air cooled; however, large forgings should be water quenched to minimize exposure in the carbide precipitation region. (b) Hardenable by cold work only

For additional data on physical properties of AISI type 304L, refer to the preceding tables for AISI type 304.

Machining Data

For machining data on AISI type 304L, refer to the preceding machining tables for AISI type 201.

Mechanical Properties

AISI Type 304L: Effect of Cold Reduction on Tensile Properties (Ref 8)

Cold reduction, %	Tensile strength MPa	ksi	Yield strength(a) MPa	ksi	Elongation(b), %
0	210	30.5	565	82.0	58
5	441	63.9	674	97.8	43
10	503	73.0	727	105.5	35
20	669	97.0	840	121.9	24
30	836	121.3	951	137.9	15
50	1058	153.5	1172	170.0	4.8

(a) 0.2% offset. (b) In 50 mm (2 in.)

The following tables given for AISI type 304 are also applicable for AISI grade 304L: Tensile Properties at Subzero and Elevated Temperatures, Tensile Properties at Elevated Temperatures, and Stress-Rupture and Creep Properties.

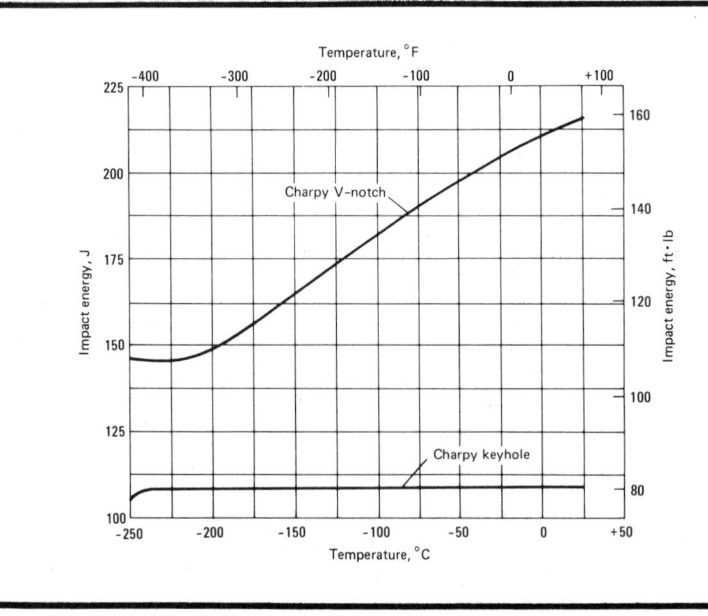

AISI Type 304L: Charpy Impact Properties (Ref 3)

Mechanical Properties (continued)

AISI Type 304L: Effect of Test Temperature on Tensile Properties of Annealed Material. Elongation was measured in 50 mm (2 in.). (Ref 3)

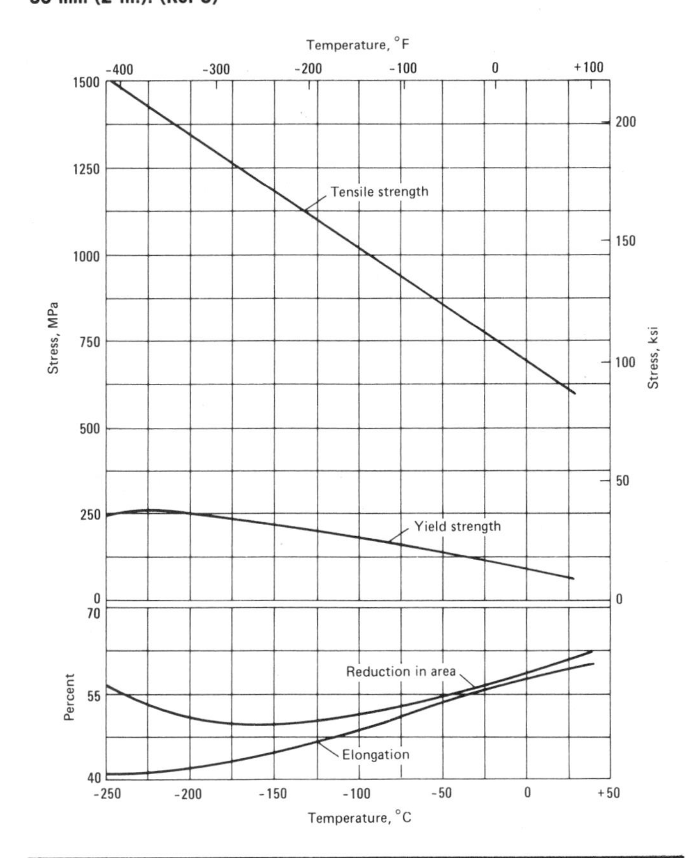

AISI Type 304L: Effect of Sensitization and Carbon Content on Impact Properties. Test specimens were in the annealed condition; carbon content of 0.03% maximum; sensitization at 650 °C (1200 °F) for 2 h. Impact energy tests used keyhole notch specimens. (Ref 3)

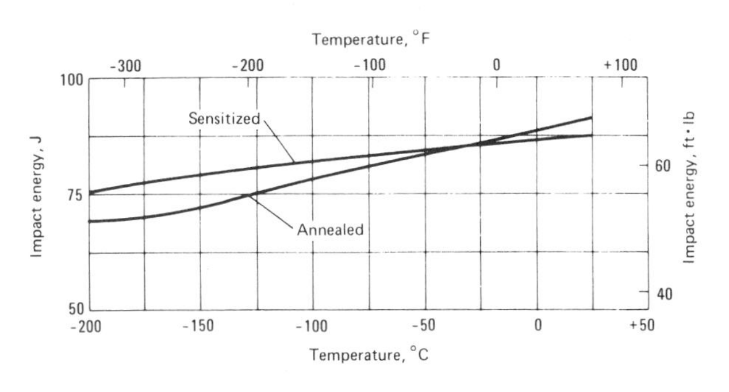

AISI Type 304L: Effect of Cold Reduction on Tensile Properties. Annealed to 1065 °C (1950 °F); water quenched. Specimens were 9.53-mm (0.375-in.) rounds, unstraightened and untempered. Tests were conducted using specimens machined to English units. Elongation was measured in 50 mm (2 in.); yield strength at 0.2% offset. (Ref 10)

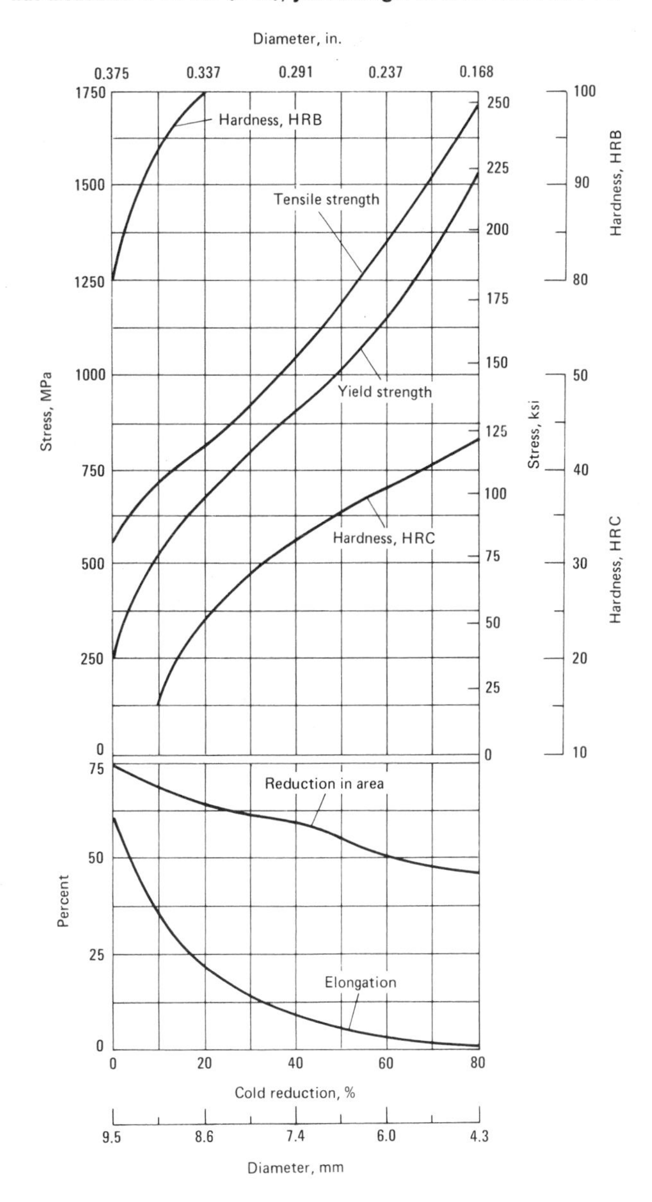

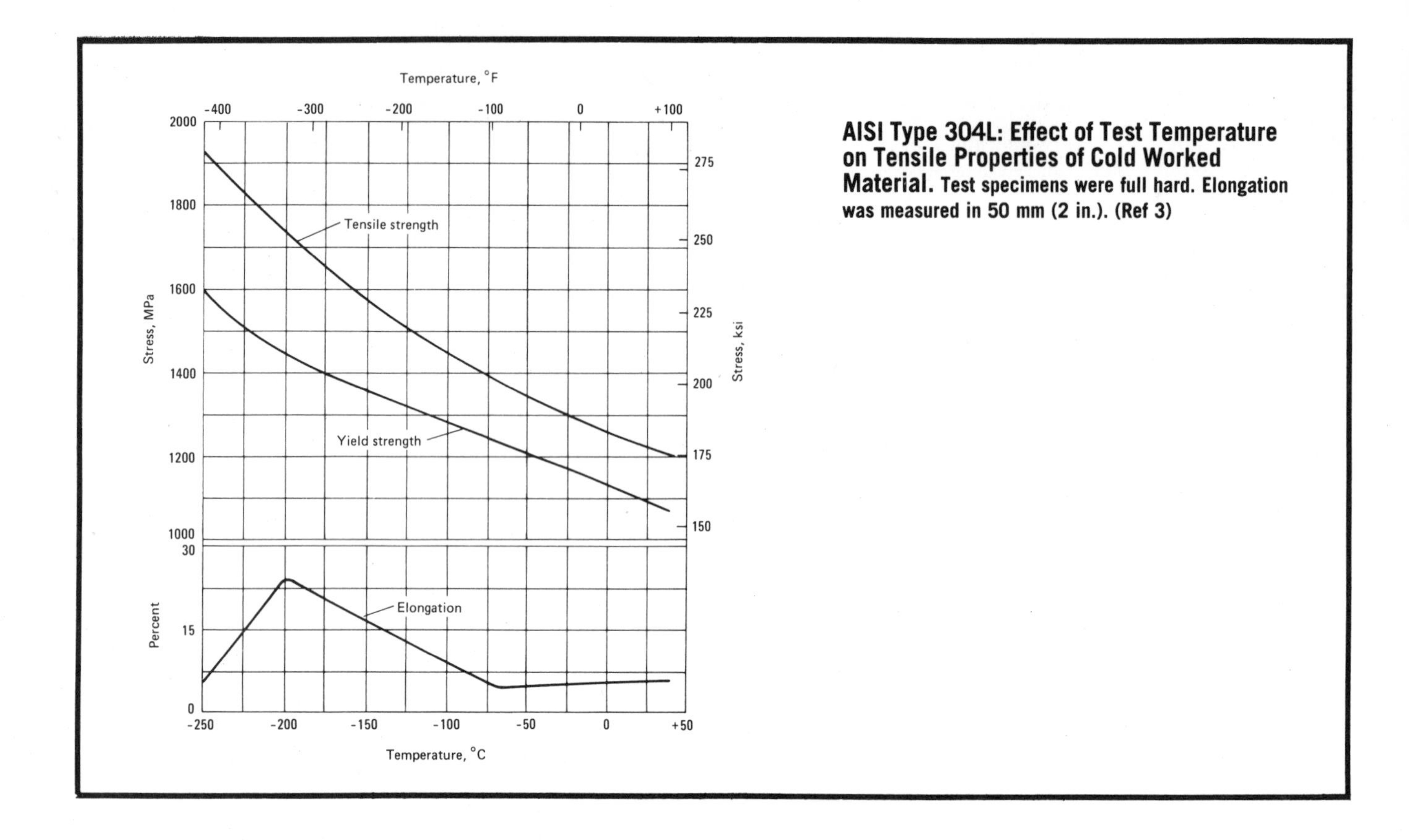

AISI Type 304L: Effect of Test Temperature on Tensile Properties of Cold Worked Material. Test specimens were full hard. Elongation was measured in 50 mm (2 in.). (Ref 3)

AISI Types 304N, 304HN

AISI Types 304N, 304HN: Chemical Composition

Element	Composition, % Type 304N	Type 304HN
Carbon	0.08	0.08
Manganese	2.00	2.00
Phosphorus	0.045	0.045
Sulfur	0.030	0.030
Silicon	1.00	1.00
Chromium	18-20	18-20
Nickel	8-10.5	8-10.5
Nitrogen	0.10-0.16	0.16-0.30

Characteristics. AISI types 304N and 304HN are essentially high-nitrogen versions of type 304. Through solid-solution strengthening, the nitrogen provides higher yield strength and tensile strength in annealed types 304N and 304HN than in conventional type 304, yet does not adversely affect ductility, corrosion resistance, or nonmagnetic properties. In some cases, solid-solution strengthening eliminates the need for cold working to increase strength.

The characteristics of corrosion resistance and susceptibility to carbide precipitation for these stainless steels are similar to those for conventional type 304. Scale resistance in continuous exposure is 870 °C (1600 °F) maximum.

Types 304N and 304HN are readily fabricated by cold working, and can be heavily cold worked without intermediate annealing. Because of the higher initial strength of these stainless steels, more power input is required than for other stainless steels.

Machinability of types 304N and 304HN is about the same as for types 302 and 304; however, because of their higher strength, more power input is required. Heavy positive feeds should be used to avoid work hardening of the surface.

These steels can be satisfactorily brazed and welded using common methods. Grain-boundary carbide precipitation may occur in the heat-affected zone of the steels. The practice of annealing will restore corrosion resistance to the weldment.

Typical Uses. Applications of AISI types 304N and 304HN include: aircraft and aerospace components; marine shafting and hardware; pump parts, unfired boilers, and pressure vessels; seamless and welded tubing for boilers, superheaters, and heat exchangers; forged fittings, pipe flanges, and valves; and parts for high-temperature service.

AISI Type 304N: Similar Steels (U.S. and/or Foreign). UNS S30450; ASME SA182, SA213, SA240, SA249, SA312, SA358, SA376, SA430, SA479; ASTM A182, A213, A240, A249, A276, A312, A358, A376, A403, A430, A479

AISI Type 304HN: Similar Steels (U.S. and/or Foreign). UNS S30452; ASME S240 (XM-21); ASTM A240 (XM-21), A276 (XM-21)

Physical Properties

AISI Types 304N, 304HN: Selected Physical Properties (Ref 9)

Melting range	1400-1455 °C (2550-2650 °F)
Density	8.0 g/cm³ (0.29 lb/in.³)
Specific heat at 0-100 °C (32-212 °F)	500 J/kg·K (0.12 Btu/lb·°F)
Average coefficient of linear thermal expansion at 0-650 °C (32-1200 °F)	18.9 μm/m·K (10.5 μin./in.·°F)
Electrical resistivity at 20 °C (68 °F)	72 nΩ·m

AISI Types 304N, 304HN: Thermal Treatment Temperatures

Treatment	Temperature range °C	°F
Forging, start	1150-1260	2100-2300
Forging, finish	925 min	1700 min
Annealing(a)	1010-1120	1850-2050
Hardening(b)	...	...

(a) Water quench. (b) Hardenable by cold work only

Mechanical Properties

AISI Types 304N, 304HN: Effect of Cold Work on Tensile Properties (Ref 9)

Cold reduction, %	Tensile strength MPa	ksi	Yield strength(a) MPa	ksi	Elongation(b), %
(c)	760	110	450	65	48
15	1035	150	825	120	25
20	1105	160	930	135	18
30	1240	180	1105	160	13
40	1345	195	1205	175	11
45	1380	200	1240	180	9
60	1480	215	1275	185	7
75	1515	220	1310	190	6

(a) 0.2% offset. (b) In 50 mm (2 in.). (c) As annealed

AISI Types 304N, 304HN: Tensile Properties at Elevated Temperatures (Ref 9)

Test temperature °C	°F	Tensile strength MPa	ksi	Yield strength(a) MPa	ksi
150	300	285	41	615	89
370	700	215	31	570	83
595	1100	185	27	475	69
705	1300	165	24	365	53

(a) 0.2% offset

AISI Types 304N, 304HN: Tensile Properties at Room Temperature

Test specimens were in the annealed condition

Form	Tensile strength MPa	ksi	Yield strength(a) MPa	ksi	Elongation(b), %	Reduction in area, %	Hardness
Type 304N (Ref 6)							
Sheet	620	90	330	48	50	...	85 HRB
Bar	620	90	290	42	55	65	180 HB
Type 304HN (Ref 9)							
Bar	725	105	415	60	50	75	94 HRB

(a) 0.2% offset. (b) In 50 mm (2 in.)

Machining Data

For machining data on AISI types 304N and 304HN, refer to the preceding machining tables for AISI type 201.

AISI Type 305

AISI Type 305: Chemical Composition

Element	Composition, max %
Carbon	0.12
Manganese	2.0
Phosphorus	0.045
Sulfur	0.030
Silicon	1.0
Chromium	17-19
Nickel	10.5-13.0

Characteristics. AISI type 305 is an austenitic, chromium-nickel steel with higher nickel content than types 302 and 304. This higher nickel content lowers the tendency to work harden, so that a greater amount of deformation is possible before process annealing is necessary. Type 305 is used for free-spinning, deep-drawing, and severe cold heading operations where the ability to take large initial reductions, and larger reductions between annealing, is desirable.

This stainless steel may not be suitable for recessed-head and other extremely difficult cold heading jobs, unless the wire is heated as it is fed into the machine. Heating to temperatures between 95 and 260 °C (200 and 500 °F) lowers the strength of the wire and results in better heading characteristics and improved tool life.

Type 305 produces a very ductile, stringy chip and usually requires chip curlers. The machinability rating of this grade is about 45% of that for free-machining AISI 1212. Type 305 is readily brazed and can be welded using the common fusion and resistance welding processes.

The corrosion and scaling resistance of type 305 is similar to that of types 302 and 304. This steel is susceptible to carbide precipitation when heated or cooled in the range of 425 to 900 °C (800 to 1650 °F); therefore, a corrective anneal, water quenching from at least 1040 °C (1900 °F), should be used.

Typical Uses. Applications of AISI type 305 include all types of cold headed fasteners; spun parts, such as tank covers, coffee urn tops, and lids; and deep drawn kitchen utensils and appliances.

AISI Type 305: Similar Steels (U.S. and/or Foreign). UNS S30500; AMS 5514, 5685, 5686; ASME SA193, SA194, SA240; ASTM A167, A240, A249, A276, A313, A314, A368, A473, A478, A492, A493, A511, A554, A580; FED QQ-S-763, QQ-W-423; SAE J405 (30305); (W. Ger.) DIN 1.4303; (Ital.) UNI X 8 CrNi 19 10; (Jap.) JIS SUS 305, SUS 305 J1

Physical Properties

AISI Type 305: Thermal Treatment Temperatures

Treatment	Temperature range °C	Temperature range °F
Forging, start	1150-1260	2100-2300
Forging, finish	925 min	1700 min
Annealing(a)	1010-1120	1850-2050
Hardening(b)	...	...

(a) Thin sections, such as strip, may be air cooled, but larger sections should be water quenched to minimize exposure in the carbide precipitation region. (b) Hardenable by cold work only

AISI Type 305: Selected Physical Properties (Ref 6)

Melting range	1400-1455 °C (2550-2650 °F)
Density	8.0 g/cm^3 (0.29 lb/in.^{3})
Specific heat at 0-100 °C (32-212 °F)	500 J/kg·K (0.12 Btu/lb·°F)
Electrical resistivity at 20 °C (68 °F)	72 nΩ·m

AISI Type 305: Average Coefficients of Linear Thermal Expansion (Ref 6)

Temperature range °C	Temperature range °F	Coefficient μm/m·K	Coefficient μin./in.·°F
0-100	32-212	17.3	9.6
0-315	32-600	17.8	9.9
0-540	32-1000	18.4	10.2
0-650	32-1200	18.7	10.4

AISI Type 305: Thermal Conductivity (Ref 6)

Temperature °C	Temperature °F	Thermal conductivity W/m·K	Thermal conductivity Btu/ft·h·°F
100	212	16.3	9.4
500	930	21.5	12.4

Machining Data

For machining data on AISI type 305, refer to the preceding machining tables for AISI type 201.

Mechanical Properties

AISI Type 305: Tensile Properties (Ref 6)

Product form	Condition	Tensile strength MPa	Tensile strength ksi	Yield strength(a) MPa	Yield strength(a) ksi	Elongation(b), %	Reduction in area, %	Hardness, HRB
Sheet	Annealed	585	85	260	38	50	...	80
Strip	Annealed	585	85	260	38	50	...	80
Wire	Annealed	585	85	325	47	60	...	78
	Soft tempered	690	100	370	54	58	...	82
Plate	Annealed	585	85	240	35	55	...	...

(a) 0.2% offset. (b) In 50 mm (2 in.), except for wire. For wire 3.18-mm (0.125-in.) diam and larger, gage length is 4 × diam; for smaller diameters, the gage length is 254 mm (10.0 in.)

AISI Type 305: Modulus of Elasticity (Ref 6)

Stressed in:	Modulus GPa	Modulus 10^6 psi
Tension	193	28.0
Torsion	86	12.5

AISI Type 305: Effect of Cold Reduction on Tensile Properties (Ref 8)

Cold reduction, %	Tensile strength MPa	Tensile strength ksi	Yield strength(a) MPa	Yield strength(a) ksi	Elongation(b), %
0	588	85.3	233	33.8	62.3
10	643	93.3	480	69.6	45.5
20	734	106.4	618	89.6	30.3
30	876	127.0	777	112.7	17.5
40	998	144.7	896	129.9	11.0
50	1062	154.0	944	136.9	7.5
60	1131	164.0	1008	146.2	7.3

(a) 0.2% offset. (b) In 50 mm (2 in.)

Mechanical Properties (continued)

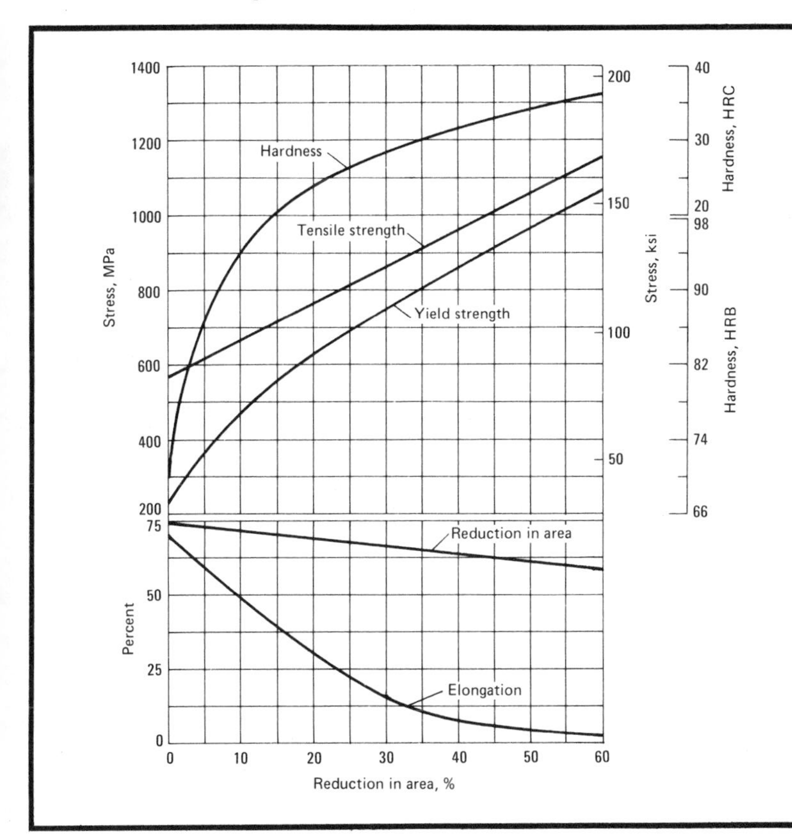

AISI Type 305: Effect of Cold Reduction on Tensile Properties at Room Temperature. Reduction in area by cold drawing. Elongation was measured in 50 mm (2 in.); yield strength at 0.2% offset. (Ref 7)

AISI Type 308

AISI Type 308: Chemical Composition

Element	Composition, max %
Carbon	0.08
Manganese	2.0
Phosphorus	0.045
Sulfur	0.030
Silicon	1.0
Chromium	19-21
Nickel	10-12

Characteristics. AISI type 308 is a high-chromium, high-nickel modification of the standard 18-8 austenitic stainless steel. The higher alloy content of this grade provides increased corrosion and heat resistance, and compensates for loss of alloying elements through oxidation during welding. Consequently, type 308 is used extensively in the form of welding rod.

Because this stainless steel produces tough, ductile, corrosion-resistant welds, it is used to join the austenitic chromium-nickel and nonhardenable straight chromium stainless steels. However, because of higher thermal expansion, type 308 is not recommended for joining the straight chromium grades if the weldment will be subject to extreme temperature cycling. A low-carbon modification of this grade is sometimes used to weld type 304L.

Typical Uses. Applications of AISI type 308 include welding rod, industrial furnaces, and equipment for handling sulfite liquor at elevated temperatures.

AISI Type 308: Similar Steels (U.S. and/or Foreign). UNS S30800; ASTM A167, A276, A314, A473, A580; SAE J405 (30308); (W. Ger.) DIN 1.4303; (Ital.) UNI X 8 CrNi 19 10; (Jap.) JIS SUS 305, SUS 305 J1

Physical Properties

AISI Type 308: Thermal Treatment Temperatures

Treatment	Temperature range °C	°F
Forging, start	1095-1205	2000-2200
Forging, finish	925 min	1700 min
Annealing	1010-1120	1850-2050
Hardening(a)	...	...

(a) Hardenable by cold work only

Physical Properties (continued)

AISI Type 308: Average Coefficients of Linear Thermal Expansion (Ref 6)

Temperature range °C	°F	Coefficient µm/m·K	µin./in.·°F
0-100	32-212	17.3	9.6
0-315	32-600	17.8	9.9
0-540	32-1000	18.4	10.2
0-650	32-1200	18.7	10.4

AISI Type 308: Electrical Resistivity (Ref 6)

Temperature °C	°F	Electrical resistivity, nΩ·m
20	68	72
650	1200	116

AISI Type 308: Selected Physical Properties (Ref 6)

Melting range ... 1400-1455 °C (2550-2650 °F)
Density ... 8.0 g/cm^3 (0.29 lb/in.^{3})
Specific heat at 0-100 °C (32-212 °F) ... 500 J/kg·K (0.12 Btu/lb·°F)

AISI Type 308: Thermal Conductivity (Ref 6)

Temperature °C	°F	Thermal conductivity W/m·K	Btu/ft·h·°F
100	212	15.2	8.8
500	930	21.6	12.5

Mechanical Properties

AISI Type 308: Tensile Properties of Annealed Material (Ref 4)

Form	Tensile strength MPa	ksi	Yield strength(a) MPa	ksi	Elongation(b), %	Reduction in area, %	Hardness
Sheet and strip	585	85	240	35	50	...	80 HRB
Plate	585	85	205	30	55	60	150 HB
Bar	585	85	205	30	55	65	150 HB
Wire, 0.051-0.51 mm (0.002-0.020 in.) diam.	655-895	95-130	240-485	35-70	30-55	...	...
Wire, 0.53-3.18 mm (0.021-0.125 in.) diam.	620-795	90-115	275-450	40-65	25-55	...	...
Wire, 3.20-9.53 mm (0.126-0.375 in.) diam.	585-725	85-105	345-485	50-70	25-55	...	83-93 HRB

(a) 0.2% offset. (b) In 50 mm (2 in.)

AISI Type 308: Tensile Properties of Cold Drawn Wire (Ref 4)

Diameter mm	in.	Tensile strength MPa	ksi	Yield strength(a) MPa	ksi	Elongation(b), %	Hardness, HRC
25% hard							
0.051-0.51	0.002-0.020	895-1140	130-165	655-895	95-130	15-20	...
0.53-3.18	0.021-0.125	760-895	110-130	550-825	80-120	15-25	...
3.20-9.53	0.126-0.375	760-895	110-130	515-655	75-95	15-25	23-28
50% hard							
0.051-0.51	0.002-0.020	1140-1380	165-200	895-1105	130-160	11-18	...
0.53-3.18	0.021-0.125	895-1105	130-160	760-1035	110-150	11-18	...
3.20-9.53	0.126-0.375	895-1035	130-150	725-860	105-125	11-18	29-32
75% hard							
0.051-0.51	0.002-0.020	1380-1550	200-225	1105-1380	160-200	5-10	...
0.53-3.18	0.021-0.125	1105-1310	160-190	1035-1240	150-180	6-12	...
3.20-9.53	0.126-0.375	1035-1170	150-170	860-1070	125-155	6-12	33-37
Full hard							
0.051-0.51	0.002-0.020	1550-1725	225-250	1450-1655	210-240	1-2	...
0.53-3.18	0.021-0.125	1310-1655	190-240	1240-1550	180-225	2-5	...
3.20-9.53	0.126-0.375	1310-1515	190-220	1070-1275	155-185	2-5	38-43

(a) 0.2% offset. (b) In 50 mm (2 in.)

AISI Type 308: Selected Mechanical Properties

Modulus of elasticity(a) ... 195 GPa (28 × 10^6 psi)

Izod impact energy ... 135-165 J (100-120 ft·lb)

Charpy V-notch impact energy, min ... 135 J (100 ft·lb)

(a) Stressed in tension

Machining Data

For machining data on AISI type 308, refer to the preceding machining tables for AISI type 201.

Mechanical Properties (continued)

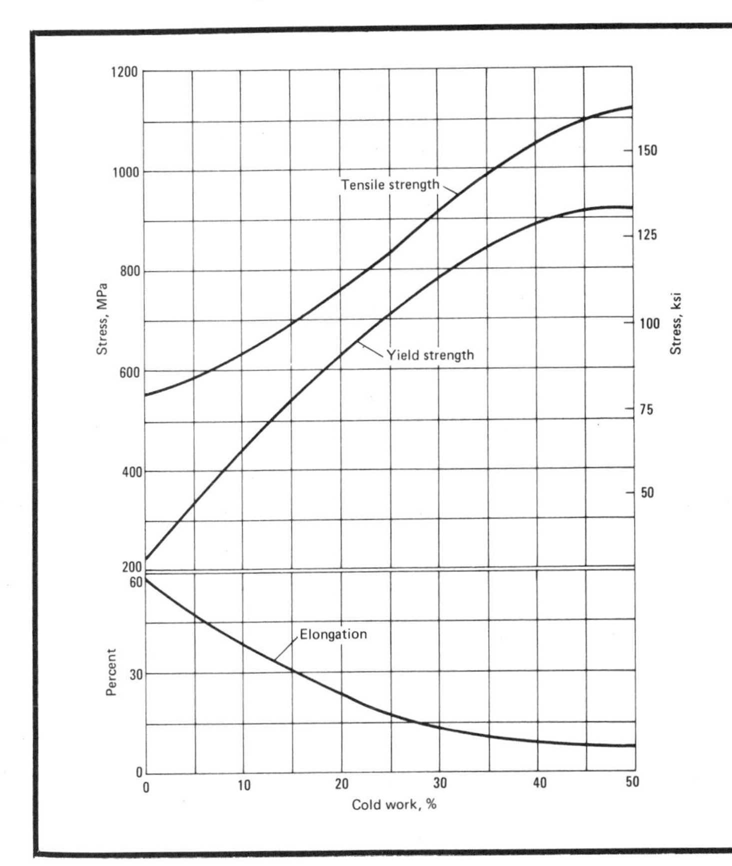

AISI Type 308: Effect of Cold Work on Tensile Properties. Elongation measured in 50 mm (2 in.); yield strength at 0.2% offset (Ref 4)

AISI Types 309, 309S, 309S(Cb)

AISI Types 309, 309S, 309S(Cb): Chemical Composition

Element	Composition, max % Type 309	Types 309S, 309S(Cb)
Carbon	0.20	0.08
Manganese	2.0	2.0
Phosphorus	0.045	0.045
Sulfur	0.030	0.030
Silicon	1.0	1.0
Chromium	22-24	22-24
Nickel	12-15	12-15
Colombium and tantalum	...	(a)

(a) For type 309S(Cb) only, the minimum content is ten times the carbon content

Characteristics. AISI types 309, 309S, and 309S(Cb) are austenitic, chromium-nickel steels with high heat-resisting characteristics. They are essentially nonmagnetic when annealed and become slightly magnetic when cold worked.

These alloys have high resistance to hot petroleum products and sulfite liquors in paper and paper pulp mills. They are also useful in handling nitric acid, nitric-sulfuric acid mixtures, and acetic, citric, and lactic acids. For improved service in boiling nitric acid, type 309S(Cb) is preferred.

Type 309, 309S, and 309S(Cb) steels can be deep drawn, stamped, headed, and upset without difficulty. Because they work harden, severe forming operations should be followed by an anneal. The scaling resistance for type 309 in continuous service is up to 1095 °C (2000 °F), and up to 1010 °C (1850 °F) where intermittent heating and cooling are encountered.

The type 309 alloys require rigid machine setups and sharp tools. Better chip action and finish are obtained with material in a moderately cold worked condition than in the annealed state. These grades are machined at speeds about 40% of those for free-machining AISI 1212 steel.

Type 309 can be satisfactorily welded or brazed by all methods. Where the process uses filler metal, type 309 or 310 may be used. The use of type 309S or 309S(Cb) stainless steels allows welding without subsequent annealing to avoid carbide precipitation.

Typical Uses. Applications of AISI types 309, 309S, and 309S(Cb) include various furnace parts, welding rod, aircraft and jet engine parts, heat exchangers, and chemical processing equipment.

AISI Type 309: Similar Steels (U.S. and/or Foreign). UNS S30900; ASME SA249, SA312, SA358, SA403, SA409; ASTM A167, A249, A276, A312, A314, A358, A403, A409, A473, A511, A554, A580; FED QQ-S-763; QQ-S-766; MIL SPEC MIL-S-862; SAE J405 (30309); (W. Ger.) DIN 1.4828; (Ital.) UNI X 16 CrNi 23 14

AISI Type 309S: Similar Steels (U.S. and/or Foreign). UNS S30908; AMS 5523, 5574, 5650; ASME SA240; ASTM A167, A240, A276, A314, A473, A511, A554, A580; SAE J405 (30309S); (W. Ger.) DIN 1.4833; (Fr.) AFNOR Z 15 CN 24.13; (Ital.) UNI X 6 CrNi 23 14

AISI Type 309S(Cb): Similar Steels (U.S. and/or Foreign). UNS S30940; ASTM A554

Physical Properties

AISI Types 309, 309S, 309S(Cb): Thermal Treatment Temperatures

Treatment	Temperature range °C	°F
Forging, start	1120-1230	2050-2250
Forging, finish	980 min	1800 min
Annealing(a)	1040-1120	1900-2050
Hardening(b)	···	···

(a) Water quench or cool rapidly to prevent carbide precipitation. (b) Hardenable by cold work only

AISI Types 309, 309S, 309S(Cb): Thermal Conductivity

Temperature °C	°F	Thermal conductivity W/m·K	Btu/ft·h·°F
100	212	15.6	9.0
500	930	18.7	10.8

AISI Types 309, 309S, 309S(Cb): Selected Physical Properties

Melting range	1400-1455 °C (2550-2650 °F)
Density	8.0 g/cm^3 (0.29 lb/in.^{3})
Specific heat at 0-100 °C (32-212 °F)	500 J/kg·K (0.12 Btu/lb·°F)
Electrical resistivity at 20 °C (68 °F)	78 nΩ·m

AISI Types 309, 309S, 309S(Cb): Average Coefficients of Linear Thermal Expansion

Temperature range °C	°F	Coefficient μm/m·K	μin./in.·°F
0-100	32-212	14.9	8.3
0-315	32-600	16.7	9.3
0-540	32-1000	17.3	9.6
0-650	32-1200	18.0	10.0
0-980	32-1800	20.7	11.5

Mechanical Properties

AISI Types 309, 309S, 309S(Cb): Representative Tensile Properties (Ref 6)

Product form	Tensile strength MPa	ksi	Yield strength(a) MPa	ksi	Elongation(b), %	Reduction in area, %	Hardness
Annealed							
Sheet	620	90	310	45	45	···	85 HRB
Strip	620	90	310	45	45	···	85 HRB
Plate	655	95	275	40	45	···	170 HB
Bar	655	95	275	40	45	65	83 HRB
Soft temper							
Wire, 1.57-mm (0.062-in.) diam	860	125	620	90	30	···	···
Wire, 12.7-mm (0.500-in.) diam	725	105	620	90	35	60	85 HRB

(a) 0.2% offset. (b) In 50 mm (2 in.), except for wire. For wire 3.18-mm (0.125-in.) diam and larger, gage length is 4 × diam; for smaller diameters, the gage length is 254 mm (10.0 in.)

AISI Types 309, 309S, 309S(Cb): Modulus of Elasticity (Ref 7)

Stressed in:	Modulus GPa	10^6 psi
Tension	200	29.0
Torsion	77	11.2

AISI Type 309: Impact Properties (Ref 4)

Impact test	Energy J	ft·lb
Izod	120-165	90-120

AISI Types 309, 309S, 309S(Cb): Stress-Rupture and Creep Properties (Ref 9)

Test temperature °C	°F	Stress for rupture in: 1000 h MPa	ksi	10 000 h MPa	ksi	100 000 h MPa	ksi	Stress causing 1% elongation in: 10 000 h MPa	ksi	100 000 h MPa	ksi
540	1000	···	···	···	···	···	···	110	16.0	···	···
595	1100	···	···	···	···	···	···	83	12.0	48	7.0
650	1200	141	20.5	103	15.0	78	11.3	48	7.0	31	4.5
705	1300	81	11.7	48	7.0	26	3.7	31	4.5	17	2.5
760	1400	52	7.5	31	4.5	21	3.0	15	2.2	10	1.5
815	1500	34	5.0	28	4.0	10	1.5	7	1.0	3	0.5
870	1600	22	3.2	10	1.5	6	0.9	···	···	···	···
980	1800	7	1.0	6	0.8	3	0.5	···	···	···	···

Mechanical Properties (continued)

AISI Types 309, 309S, 309S(Cb): Tensile Properties at Elevated Temperatures (Ref 9)

Test temperature °C	°F	Tensile strength MPa	ksi	Yield strength(a) MPa	ksi	Elongation(b), %	Reduction in area, %
21	70	620	90	290	42	50	77
150	300	560	81	255	37	47	77
205	400	550	80	240	35	46	75
260	500	530	77	220	32	45	75
315	600	515	75	205	30	44	73
370	700	510	74	195	28	43	72
425	800	495	72	185	27	40	67
480	900	475	69	170	25	39	66
540	1000	455	66	165	24	36	58
595	1100	405	59	160	23	35	54
650	1200	380	55	150	22	35	50
705	1300	305	44	145	21	37	41
760	1400	250	36	140	20	40	40
815	1500	185	27	130	19	46	37
870	1600	145	21	125	18	50	48
925	1700	105	15	...	...	59	47
980	1800	76	11	...	...	65	66

(a) 0.2% offset. (b) In 50 mm (2 in.)

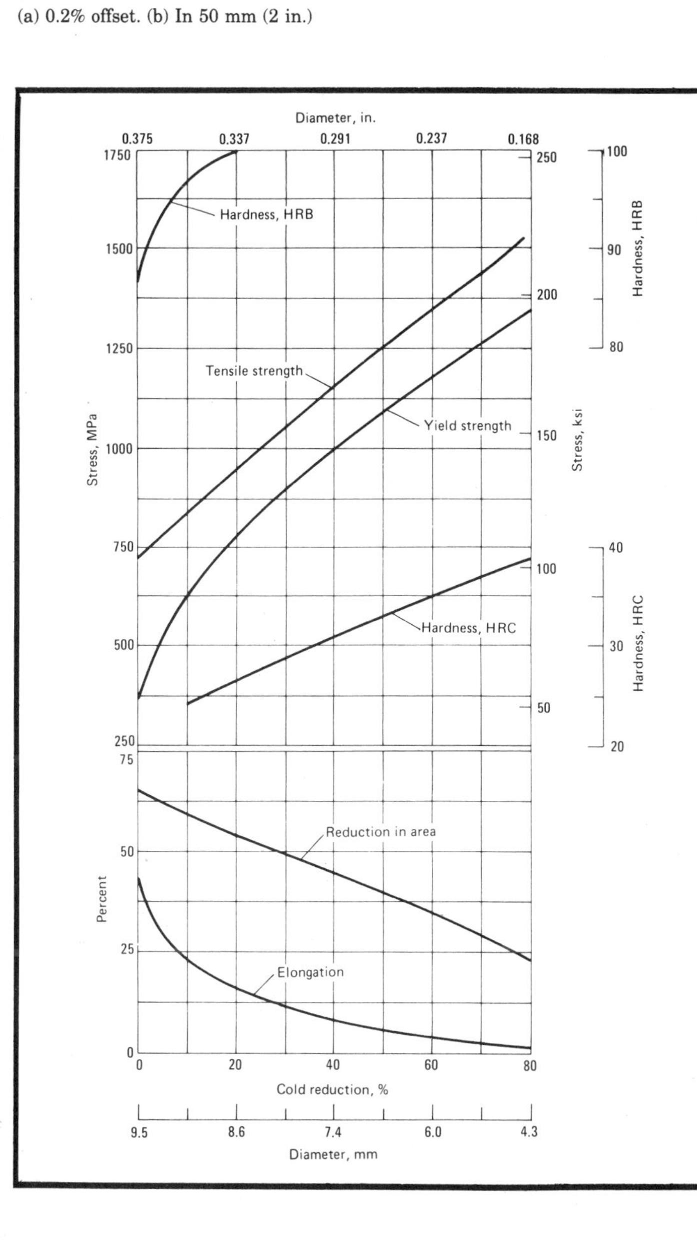

AISI Types 309, 309S, 309S(Cb): Effect of Cold Reduction on Tensile Properties. Annealed at 1065 °C (1950 °F); water quenched. Specimens were 9.53-mm (0.375-in.) round, unstraightened and untempered. Elongation was measured in 50 mm (2 in.); yield strength at 0.2% offset. (Ref 10)

Machining Data

For machining data on AISI types 309, 309S, and 309S (Cb), refer to the preceding machining tables for AISI type 302B.

AISI Types 310, 310S

AISI Types 310, 310S: Chemical Composition

Element	Composition, max % Type 310	Type 310S
Carbon	0.25	0.08
Manganese	2.0	2.0
Phosphorus	0.045	0.045
Sulfur	0.030	0.030
Silicon	1.5	1.5
Chromium	24-26	24-26
Nickel	19-22	19-22

Characteristics. AISI types 310 and 310S are austenitic, chromium-nickel steels with high strength and excellent corrosion resistance. They are used in applications which involve sulfur-bearing gases at elevated temperature. Marine exposures produce little or no evidence of stress corrosion for annealed or cold rolled types 310 and 310S up to 90% of their yield strength.

These alloys have excellent resistance to oxidizing acids and most common corrosive agents. They also have good resistance to molten salts such as tempering, neutral, cyaniding, and high-speed salts. Although not resistant to molten aluminum or molten magnesium, types 310 and 310S are somewhat resistant to molten lead.

Type 310 stainless steel has excellent resistance to scaling in continuous service up to 1150 °C (2100 °F) and up to 1095 °C (2000 °F) if used in intermittent heating and cooling.

Types 310 and 310S can be deep drawn, stamped, headed, and upset without difficulty. Because these steels work harden, severe forming operations should be followed by an anneal. Close clearances in blanking and punching are necessary to minimize burring. Higher press power and lower speeds are required for forming these steels than for ordinary carbon steels.

These stainless steels work harden rapidly and are tough to machine. The machinability rating for types 310 and 310S are about 55% of that for free-machining AISI 1212, and can be somewhat improved by using moderately cold drawn bars.

Type 310 can be satisfactorily welded and brazed by the common methods. It does not air harden on cooling and the weld is very tough. If carbide precipitation will be a problem, type 310S is suggested when the welded assembly cannot be heated to 1040 °C (1900 °F) and water quenched after welding.

Typical Uses. Applications of AISI types 310 and 310S include welding rod, furnace parts, oil refining equipment, heat exchangers, oil burner parts, chemical plant equipment, annealing boxes, aircraft and jet engine parts, tube hangers, and soot-blowing tubing.

AISI Type 310: Similar Steels (U.S. and/or Foreign). UNS S31000; AMS 5694, 5695; ASME SA182, SA213, SA249, SA312, SA358, SA403, SA409; ASTM A167, A182, A213, A249, A276, A312, A314, A358, A403, A409, A473, A511, A632; FED QQ-S-763, QQ-S-766, QQ-W-423, STD-66; MIL SPEC MIL-S-862; SAE J405 (30310); (W. Ger.) DIN 1.4841; (Fr.) AFNOR Z 12 CNS 25.20; (Ital.) UNI X 16 CrNiSi 25 20, X 22 CrNi 25 20; (Jap.) JIS SUS Y 310; (U.K.) B.S. 310 S 24

AISI Type 310S: Similar Steels (U.S. and/or Foreign). UNS S31008; AMS 5521, 5572, 5577, 5651; ASME SA240, SA479; ASTM A167, A240, A276, A314, A473, A479, A511, A554, A580; SAE J405 (30310S)

Physical Properties

AISI Types 310, 310S: Selected Physical Properties (Ref 6)

Melting range	1400-1455 °C (2550-2650 °F)
Density	8.0 g/cm^3 (0.29 lb/in.^{3})
Specific heat at 0-100 °C (32-212 °F)	500 J/kg·K (0.12 Btu/lb·°F)
Electrical resistivity at 20 °C (68 °F)	78 nΩ·m

AISI Types 310, 310S: Average Coefficients of Linear Thermal Expansion (Ref 6)

Temperature range °C	°F	Coefficient μm/m·K	μin./in.·°F
0-100	32-212	15.8	8.8
0-315	32-600	16.2	9.0
0-540	32-1000	16.9	9.4
0-650	32-1200	17.5	9.7
0-980	32-1800	19.1	10.6

AISI Types 310, 310S: Thermal Conductivity (Ref 6)

Temperature °C	°F	Thermal conductivity W/m·K	Btu/ft·h·°F
100	212	14.2	8.2
500	930	18.7	10.8

AISI Types 310, 310S: Thermal Treatment Temperatures

Treatment	Temperature range °C	°F
Forging, start	1095-1230	2000-2250
Forging, finish	955 min	1750 min
Annealing(a)	1040-1150	1900-2100
Hardening(b)	...	...

(a) To prevent carbide precipitation, water quench or cool rapidly. (b) Hardenable by cold work only

Machining Data

For machining data on AISI types 310 and 310S, refer to the preceding machining tables for AISI type 302B.

Mechanical Properties

AISI Types 310, 310S: Representative Tensile Properties (Ref 6)

Product form	Tensile strength MPa	Tensile strength ksi	Yield strength(a) MPa	Yield strength(a) ksi	Elongation(b), %	Reduction in area, %	Hardness
Annealed							
Sheet	620	90	310	45	45	...	85 HRB
Strip	620	90	310	45	45	...	85 HRB
Plate	655	95	275	40	45	...	170 HB
Bar	655	95	275	40	45	65	160 HB
Soft tempered							
Wire, 1.6-mm (0.062-in.) diam	860	125	620	90	30	...	...
Wire, 12.7-mm (0.500-in.) diam	725	105	485	70	35	60	85 HRB

(a) 0.2% offset. (b) In 50 mm (2 in.), except for wire. For wire 3.18-mm (0.125-in.) diam and larger, gage length is 4 × diam; for smaller diameters, the gage length is 254 mm (10.0 in.)

AISI Types 310, 310S: Modulus of Elasticity (Ref 7)

Stressed in:	Modulus GPa	Modulus 10^6 psi
Tension	200	29.0
Torsion	77	11.2

AISI Types 310, 310S: Izod Impact Energy

Temperature	Energy J	Energy ft·lb
Room temperature	120	90

AISI Types 310, 310S: Tensile Properties at Elevated Temperature (Ref 9)

Test temperature °C	Test temperature °F	Tensile strength MPa	Tensile strength ksi	Yield strength(a) MPa	Yield strength(a) ksi	Elongation(b), %	Reduction in area, %
21	70	625	91	290	42	47	70
150	300	600	87	240	35	39	70
205	400	585	85	230	33	38	69
260	500	570	83	220	32	37	69
315	600	565	82	205	30	37	69
370	700	560	81	200	29	37	67
425	800	525	76	185	27	36	63
480	900	515	75	180	26	34	61
540	1000	495	72	165	24	33	55
595	1100	450	65	150	22	33	50
650	1200	385	56	145	21	35	45
705	1300	340	49	140	20	36	40
760	1400	285	41	130	19	37	39
815	1500	235	34	125	18	40	38
870	1600	185	27	110	16	45	37
925	1700	215	31	...	...	50	40
980	1800	90	13	...	...	57	40

(a) 0.2% offset. (b) In 50 mm (2 in.)

AISI Types 310, 310S: Stress-Rupture Properties and Creep Strength for Annealed Bar (Ref 9)

Test temperature °C	Test temperature °F	Stress for rupture in: 1000 h MPa	1000 h ksi	10 000 h MPa	10 000 h ksi	100 000 h MPa	100 000 h ksi	Stress for 1% elongation in: 10 000 h MPa	10 000 h ksi	100 000 h MPa	100 000 h ksi
540	1000	220	32	170	25	125	18	140	20	90	13
595	1100	160	23	110	16	83	12	105	15	55	8
650	1200	110	16	69	10	55	8	55	8	40	6
705	1300	69	10	40	6	35	5	35	5	30	4
760	1400	50	7	30	4	20	3	15	2	15	2
815	1500	35	5	15	2	7	1	7	1	...	...
870	1600	20	3	7	1	...	...	...	...	...	...
980	1800	7	1	...	...	...	...	...	...	...	...

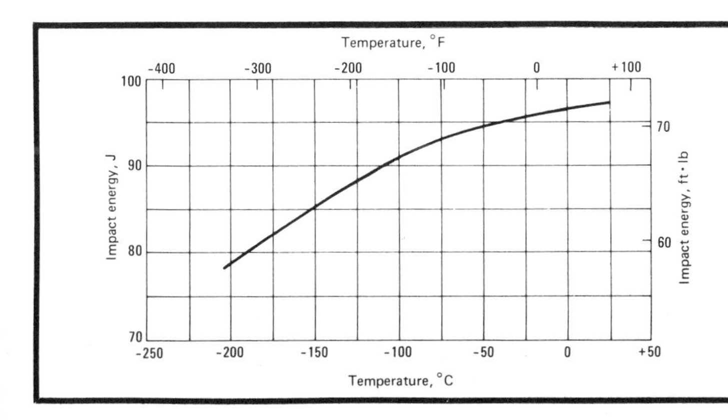

AISI Type 310: Charpy K Impact Properties. Specimens were in the annealed condition. (Ref 3)

AISI Types 310, 310S: Effect of Cold Reduction on Tensile Properties. Elongation was measured in 50 mm (2 in.); yield strength at 0.2% (Ref 7)

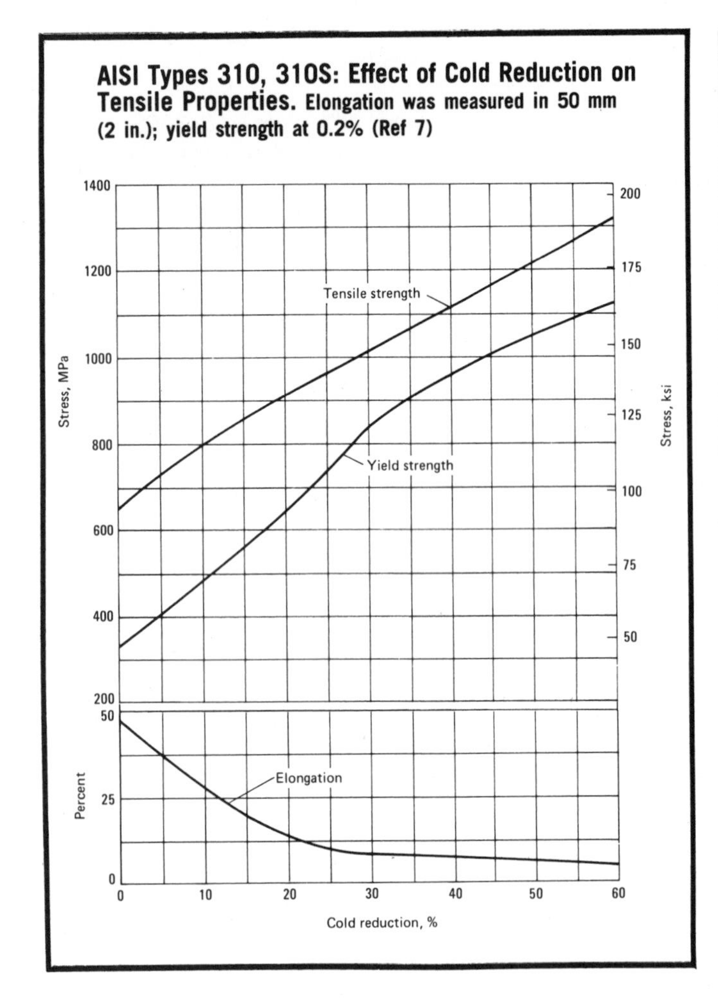

AISI Type 310: Effect of Test Temperature on Tensile Properties of Annealed Material. Elongation was measured in 50 mm (2 in.). (Ref 3)

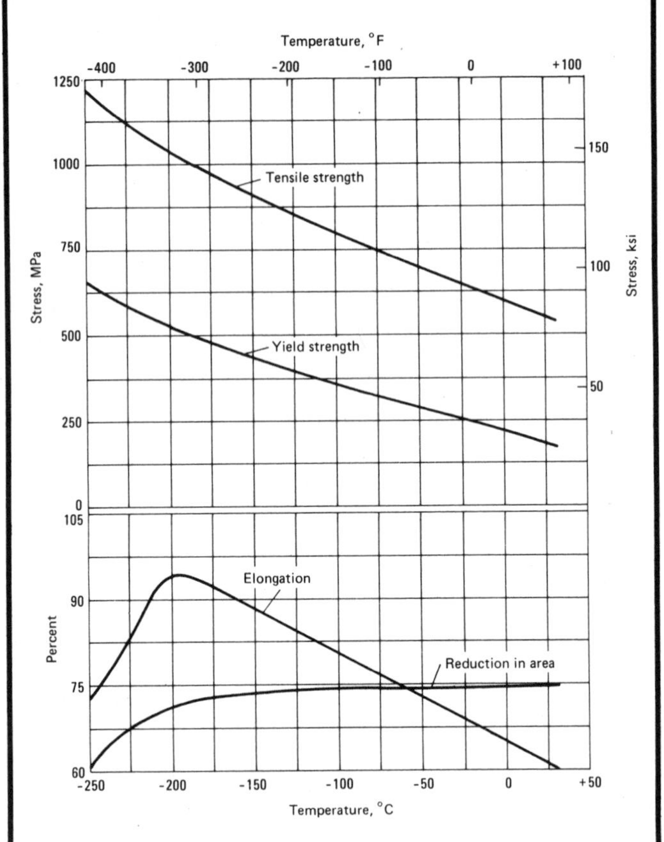

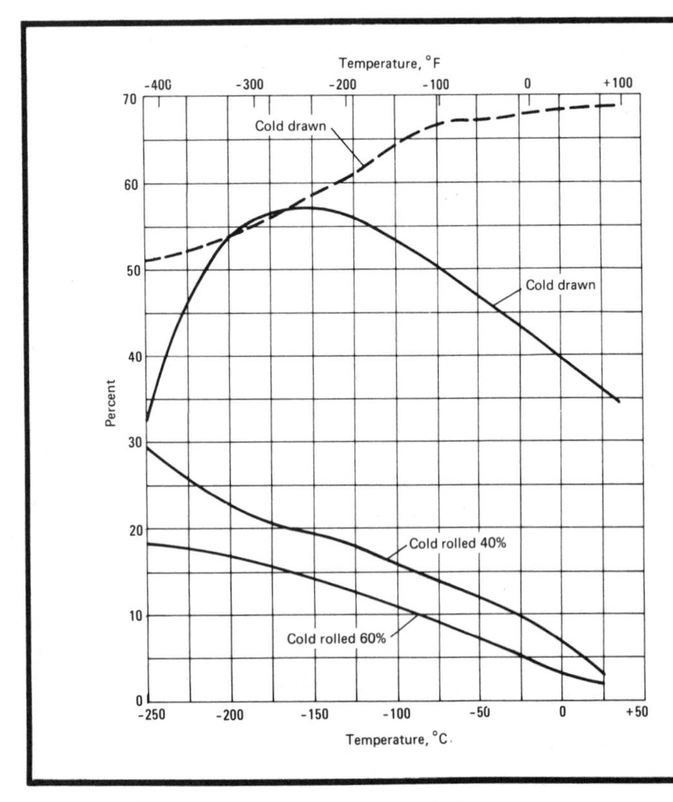

AISI Type 310: Effect of Test Temperature on Elongation and Reduction in Area of Cold Worked Material. —— elongation; - - - - reduction in area. Specimens were cold rolled 40 or 60%, or cold drawn to 655 MPa (95 ksi). Elongation was measured in 50 mm (2 in.). (Ref 3)

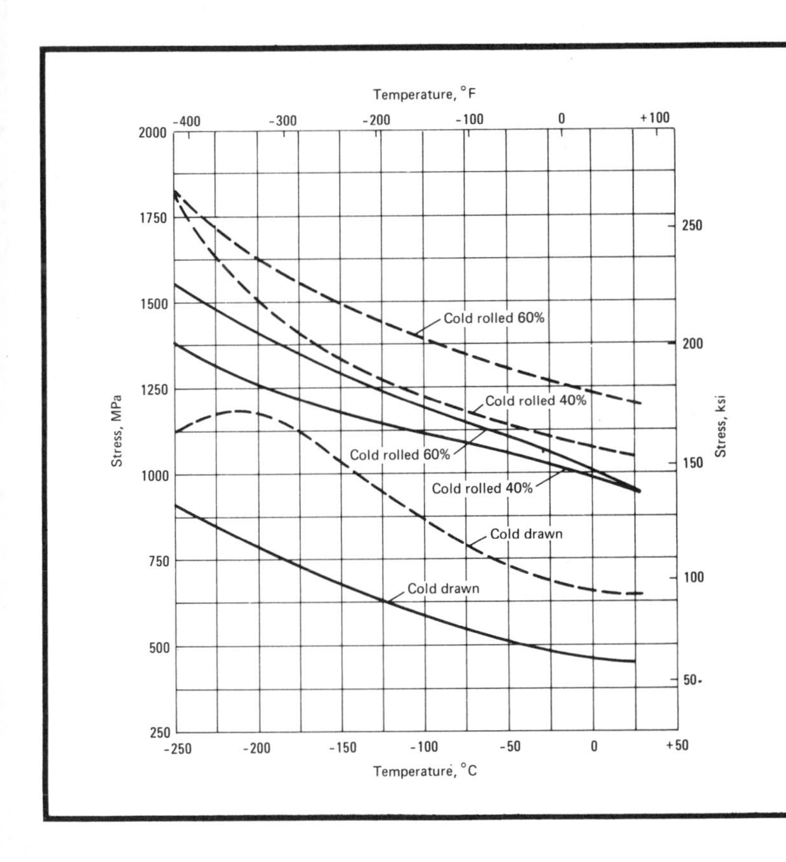

AISI Type 310: Effect of Test Temperature on Yield and Tensile Strengths of Cold Worked Material. —— yield strength; ---- tensile strength. Specimens were cold rolled 40 or 60%, or cold drawn to 655 MPa (95 ksi). (Ref 3)

AISI Type 314

AISI Type 314: Chemical Composition

Element	Composition, max %
Carbon	0.25
Manganese	2.0
Phosphorus	0.045
Sulfur	0.030
Silicon	1.5-3.0
Chromium	23-26
Nickel	19-22

Characteristics. AISI type 314 is an austenitic, chromium-nickel steel with the highest heat-resisting properties of any of the chromium-nickel steels. This steel is used when maximum resistance to carburization is desired. Type 314 is essentially nonmagnetic when annealed or cold worked.

Because of toughness and high work-hardening characteristics, type 314 requires rigid machine setups and sharp tools. Better chip action and finish are obtained with material in a moderately cold worked condition than in the annealed condition. In general, this steel is machined at speeds about 40% of those used for free-machining AISI 1212 steel.

Typical Uses. Applications of AISI type 314 include furnace parts, aircraft and jet engine parts, oil burner parts, and chemical processing equipment.

AISI Type 314: Similar Steels (U.S. and/or Foreign). UNS S31400; AMS 5522, 5652; ASTM A276, A314, A473, A580; SAE J405 (30314); (W. Ger.) DIN 1.4841; (Fr.) AFNOR Z 12 CNS 25.20; (Ital.) UNI X 16 CrNiSi 25 20, X 22 CrNi 25 20; (Jap.) JIS SUS Y 310; (U.K.) B.S. 310 S 24

Physical Properties

AISI Type 314: Selected Physical Properties (Ref 6)

Density	7.8 g/cm³ (0.28 lb/in.³)
Specific heat at 0-100 °C (32-212 °F)	500 J/kg·K (0.12 Btu/lb·°F)
Electrical resistivity at 20 °C (68 °F)	77 nΩ·m

AISI Type 314: Thermal Conductivity (Ref 6)

Temperature		Thermal conductivity	
°C	°F	W/m·K	Btu/ft·h·°F
100	212	17.5	10.1
500	930	21.1	12.2

Physical Properties (continued)

AISI Type 314: Average Coefficients of Linear Thermal Expansion (Ref 6)

Temperature range °C	°F	Coefficient μm/m·K	μin./in.·°F
0-315	32-600	15.1	8.4
0-815	32-1500	17.6	9.8

AISI Type 314: Thermal Treatment Temperatures

Treatment	Temperature range °C	°F
Forging, start	1040-1120	1900-2050
Forging, finish	925 min	1700 min
Annealing(a)	1150	2100
Hardening(b)	...	...

(a) Cool rapidly. (b) Hardenable by cold work only

Mechanical Properties

AISI Type 314: Representative Tensile Properties (Ref 4, 6)

Test specimens were in the annealed condition

Product form	Diameter mm	in.	Tensile strength MPa	ksi	Yield strength(a) MPa	ksi	Elongation(b), %	Reduction in area, %	Hardness
Sheet	...	...	689	100	345	50	40	...	85 HRB
Plate	...	...	689	100	345	50	45	60	180 HB
Bar	...	...	689	100	345	50	45	60	180 HB
Wire	0.051-0.508	0.002-0.020	655-896	95-130	275-515	40-75	35-55	...	...
	0.533-3.175	0.021-0.125	620-758	90-110	240-415	35-60	25-55	...	...
	3.200-9.525	0.126-0.375	585-724	85-105	240-275	35-40	25-55	...	...

(a) 0.2% offset. (b) In 50 mm (2 in.)

Machining Data

For machining data on AISI type 314, refer to the preceding machining tables for AISI type 302B.

AISI Type 316

AISI Type 316: Chemical Composition

Element	Composition, max %
Carbon	0.080
Manganese	2.0
Phosphorus	0.045
Sulfur	0.030
Silicon	1.0
Chromium	16-18
Nickel	10-14
Molybdenum	2.0-3.0

Characteristics. AISI type 316 is a molybdenum-bearing, chromium-nickel, stainless and heat-resistant steel. This grade has superior corrosion resistance over other chromium-nickel steels when exposed to many types of chemical corrodents, as well as marine atmospheres. In addition, type 316 offers higher creep, stress-to-rupture, and tensile strengths than any other stainless steel. However, after welding it is susceptible to intergranular corrosion unless the intergranular carbides are dissolved by annealing at 1040 to 1150 °C (1900 to 2100 °F), followed by a water quenching.

Type 316 is less susceptible to pitting or pinhole corrosion than other grades of stainless steel when exposed to hot and cold dilute solutions of sulfuric and hydrochloric acids; acetic, formic, and tartaric acids; acid sulfates; and alkaline chlorides.

In the annealed condition, type 316 exhibits excellent ductility and may be readily cold worked using methods such as roll forming, deep drawing, bending, and upsetting. Annealing is necessary to restore ductility and to lower hardness for subsequent forming operations. Severely formed parts should be annealed to remove stress.

For machining, type 316 requires rigid setups and sharp tools. Better chip action and finish are obtained with material in a moderately cold worked condition rather than in the annealed condition. Heavy feeds and an adequate supply of coolant are also recommended. The machinability of this grade is about 55% of that for AISI 1212.

Type 316 steel can be welded by most of the common fusion and resistance welding methods. The suitable filler wire is type 316.

Typical Uses. Applications of AISI type 316 include chemical storage and transportation tanks; food processing equipment and cooking kettles; digesters and evaporators for the paper industry; textile dyeing equipment; nuclear fuel cladding and heat exchangers; and oil refining equipment.

AISI Type 316: Similar Steels (U. S. and/or Foreign): UNS S31600; AMS 5524, 5573, 5648, 5690, 5691; ASME SA182, SA193, SA194, SA213, SA240, SA249, SA312, SA320, SA358, SA376, SA403, SA409, SA430, SA479, SA688; ASTM A167, A182, A193, A194, A213, A240, A249, A269, A276, A312, A313, A314, A320, A358, A368, A376, A403, A409, A430, A473, A478, A479; FED QQ-S-763, QQ-S-766, QQ-W-423; MIL SPEC MIL-S-862, MIL-S-5059; SAE J405 (30316); (W. Ger.) DIN 1.4401; (Fr.) AFNOR Z 6 CND 17.11; (Ital.) UNI X 5 CrNiMo 17 12; (Jap.) JIS SUS 316, SUH 309, SUS Y 316; (Swed.) SS_{14} 2347; (U.K.) B.S. 316 S 16, 316 S 18, 316 S 25, 316 S 26, 316 S 30, 316 S 40, 316 S 41, En. 58 H

Physical Properties

AISI Type 316: Thermal Treatment Temperatures

Treatment	Temperature range °C	°F
Forging, start	1150-1260	2100-2300
Forging, finish	925 min	1700 min
Annealing(a)	1010-1120	1850-2050
Hardening(b)	···	···

(a) Follow with rapid quenching. Thin sections, such as sheet or strip may be air cooled. Forgings should be water quenched. (b) Hardenable by cold work only

AISI Type 316: Thermal Conductivity (Ref 6)

Temperature °C	°F	Thermal conductivity W/m·K	Btu/ft·h·°F
100	212	16.3	9.4
500	930	21.5	12.4

AISI Type 316: Selected Physical Properties (Ref 6)

Melting range	1370-1400 °C (2500-2550 °F)
Density	8.0 g/cm³ (0.29 lb/in.³)
Specific heat at 0-100 °C (32-212 °F)	500 J/kg·K (0.12 Btu/lb·°F)
Electrical resistivity at 20 °C (68 °F)	74 nΩ·m

AISI Type 316: Average Coefficients of Linear Thermal Expansion (Ref 6)

Temperature range °C	°F	Coefficient µm/m·K	µin./in.·°F
0-100	32-212	16.0	8.9
0-315	32-600	16.2	9.0
0-540	32-1000	17.5	9.7
0-650	32-1200	18.5	10.3
0-815	32-1500	20.0	11.1

Mechanical Properties

AISI Type 316: Representative Tensile Properties (Ref 6)

Condition	Tensile strength MPa	ksi	Yield strength(a) MPa	ksi	Elongation(b), %	Reduction in area, %	Hardness	Endurance limit MPa	ksi
Sheet									
Annealed	580	84	290	42	50	···	79 HRB	270	39
Strip									
Annealed	580	84	290	42	50	···	79 HRB	270	39
Wire, 1.6-mm (0.062-in.) diam									
Annealed	620	90	240	35	55	···	···	···	···
Soft tempered	689	100	515	75	50	···	···	···	···
Spring tempered	1585	230	···	···	···	···	···	···	···
Wire, 13-mm (0.50-in.) diam									
Annealed	550	80	240	35	60	70	78 HRB	···	···
Soft tempered	620	90	415	60	50	65	83 HRB	···	···
Spring tempered	1105	160	···	···	···	···	···	···	···
Wire									
Spring tempered	1034	150	···	···	···	···	···	···	···
Plate									
Annealed	565	82	250	36	55	···	149 HB	270	39
Bar									
Annealed	550	80	240	35	60	70	149 HB	260	38
Annealed and cold drawn	620	90	415	60	45	65	190 HB	275	40

(a) 0.2% offset. (b) In 50 mm (2 in.), except for wire. For wire, gage length is 4 × diam for 3.18-mm (0.125-in.) diam and larger; for smaller diameters, the gage length is 254 mm (10 in.)

AISI Type 316: Modulus of Elasticity (Ref 7)

Stressed in:	Modulus GPa	10^6 psi
Tension	193	28.0
Torsion	77	11.2

AISI Type 316: Endurance Limit

Condition	Stress MPa	ksi
Annealed	270	39

AISI Type 316: Impact Properties (Ref 4)

Impact test	Energy J	ft·lb
Izod	95-163	70-120
Charpy keyhole notch	110	80
Charpy V-notch	105	78
	176	130

Mechanical Properties (continued)

AISI Type 316: Tensile Properties at Elevated Temperatures (Ref 7)

Test temperature °C	°F	Tensile strength MPa	ksi	Yield strength(a) MPa	ksi	Elongation(b), %	Reduction in area, %
205	400	560	81	240	35	51	76
315	600	540	78	215	31	48	72
425	800	525	76	195	28	47	66
540	1000	485	70	165	24	44	60
650	1200	395	57	145	21	40	53
760	1400	240	35	125	18	37	46
870	1600	165	24	110	16	39	44

(a) 0.2% offset. (b) In 50 mm (2 in.)

For data on corrosion rates in boiling organic acids or boiling, dilute sulfuric acid solutions, refer to the mechanical properties section of ASTM XM-27.

AISI Type 316: Stress-Rupture and Creep Properties (Ref 7)

Test temperature		Stress for rupture in:						Stress for secondary creep rate of(a):			
		1000 h		10 000 h(a)		100 000 h(a)		1% in 10 000 h		1% in 100 000 h	
°C	°F	MPa	ksi	MPa	ksi	MPa	ksi	MPa	ksi	MPa	ksi
595	1100	245	36	195	28	170	25	124	18.0	79	11.5
650	1200	165	24	115	17	97	14	83	12.0	48	7.0
705	1300	110	16	69	10	50	7	52	7.5	28	4.0
760	1400	69	10	40	6	30	4	31	4.5	15	2.2
815	1500	40	6	30	4	15	2	21	3.0	...	...
870	1600	30	4	...	...	...	...	...	...	...	...

(a) Extrapolated data

Machining Data

For machining data on AISI type 316, refer to the preceding machining tables for AISI type 302B.

AISI Type 316: Effect of Test Temperature on Tensile Properties of Annealed Material. Elongation was measured in 50 mm (2 in.). (Ref 4)

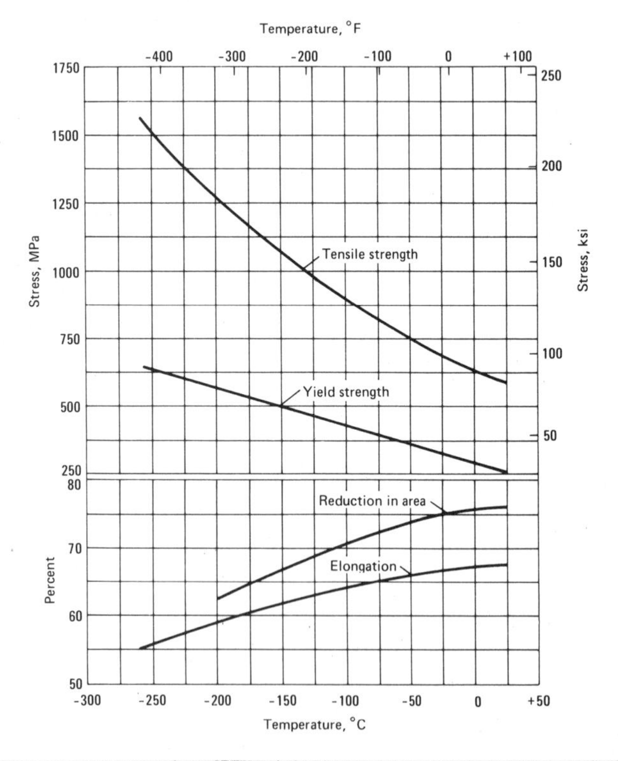

AISI Type 316: Effect of Test Temperature on Tensile Properties of Cold Worked Material. Specimens were cold drawn to 25% reduction; elongation was measured in 50 mm (2 in.). (Ref 4)

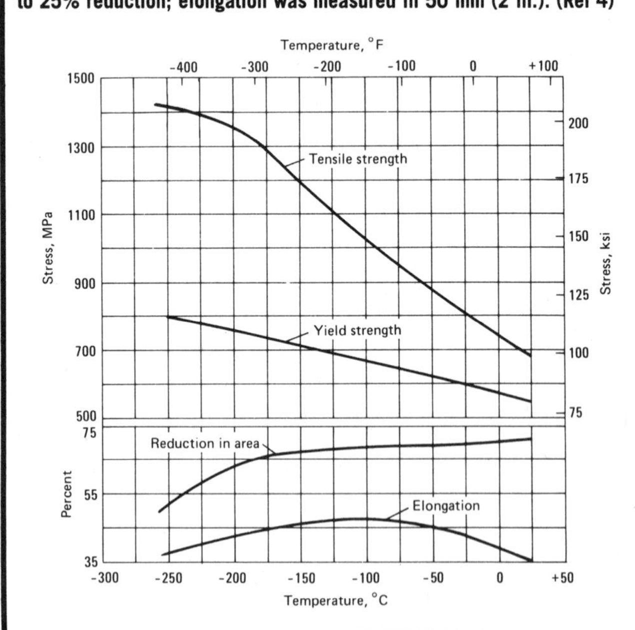

AISI Type 316: Effect of Cold Work on Tensile Properties.

Annealed at 1065 °C (1950 °F); water quenched. Specimens were 9.53-mm (0.375-in.) round, unstraightened and untempered. Tests were conducted using specimens machined to English units. Elongation was measured in 50 mm (2 in.); yield strength at 0.2% offset. (Ref 10)

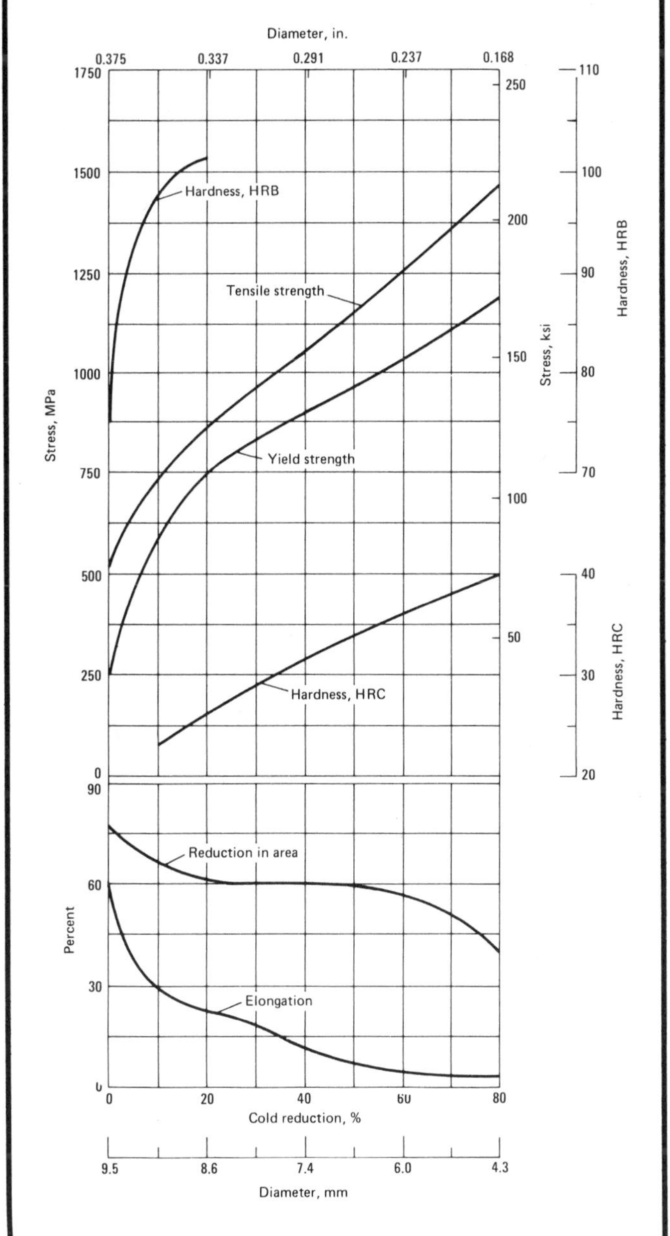

AISI Type 316: Rupture Strength of Annealed Steel (Ref 8)

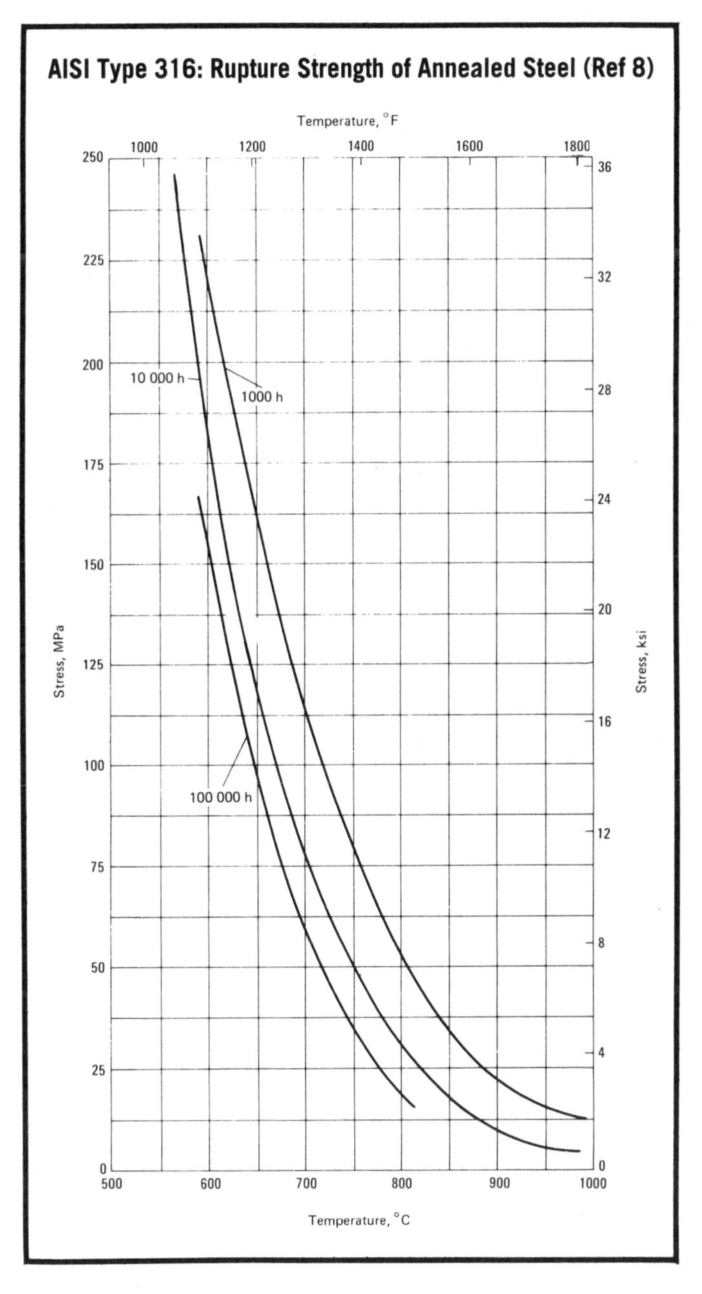

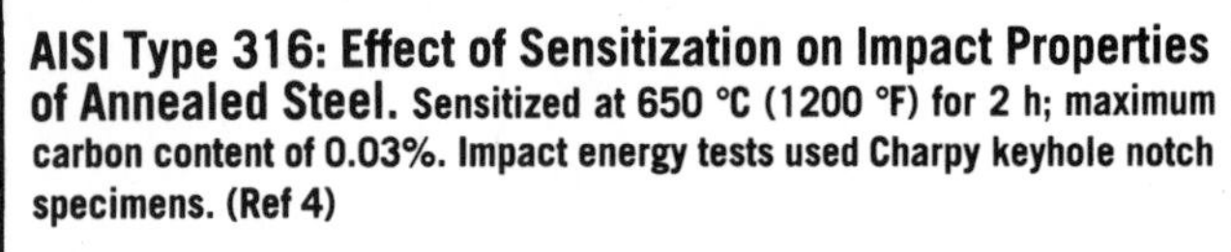

AISI Type 316: Effect of Sensitization on Impact Properties of Annealed Steel. Sensitized at 650 °C (1200 °F) for 2 h; maximum carbon content of 0.03%. Impact energy tests used Charpy keyhole notch specimens. (Ref 4)

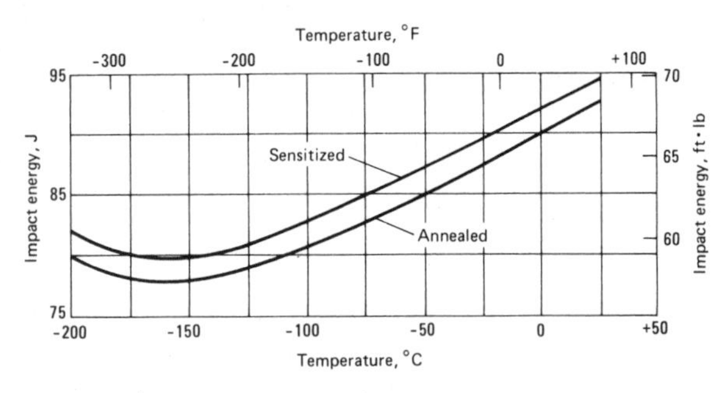

AISI Type 316: Effect of Sigmatizing on Impact Properties. Sigmatized at 730 to 900 °C (1350 to 1650 °F); carbon content of 0.02%. Impact energy test used Charpy keyhole notch specimens. (Ref 4)

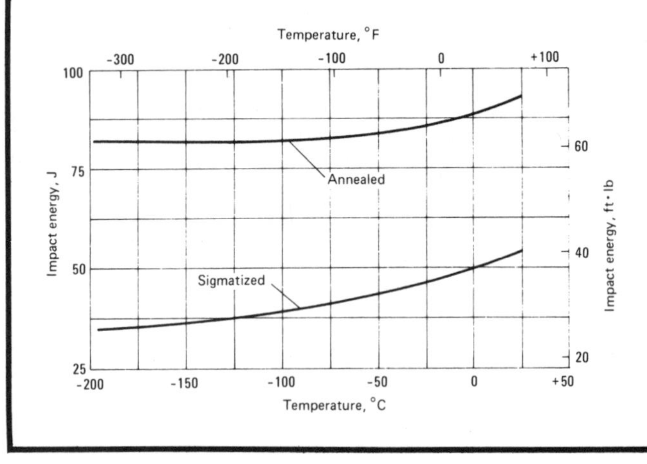

AISI Type 316: Creep Strength of Annealed Steel (Ref 8)

AISI Type 316L

AISI Type 316L: Chemical Composition

Element	Composition, max %
Carbon	0.030
Manganese	2.0
Phosphorus	0.045
Sulfur	0.030
Silicon	1.0
Chromium	16-18
Nickel	10-14
Molybdenum	2-3

Characteristics. AISI type 316L is a very low-carbon, austenitic, chromium-nickel, stainless steel. This grade has general corrosion resistance similar to type 316, but superior resistance to intergranular corrosion following welding or stress relieving. It is recommended for parts that are fabricated by welding and cannot be subsequently annealed. Parts made from type 316L are generally limited to service at temperatures up to 425 °C (800 °F).

Type 316L is essentially nonmagnetic when annealed, and becomes slightly magnetic when cold worked. It can be deep drawn, stamped, headed, and upset without difficulty. Because this steel work hardens, severe forming operations should be followed by an anneal.

The type 316L alloy machines with a tough and stringy chip. Rigidly supported tools with as heavy a cut as possible should be used to prevent glazing. Machinability for this grade is about 55% of that for free-machining AISI 1212, but can be improved by using a moderately cold worked material.

Type 316L can be satisfactorily brazed and welded by all methods, except oxyacetylene gas. When a filler metal is required, types 316L and 317L are satisfactory.

Typical Uses. Applications of AISI type 316L include pulp handling equipment; processing equipment for producing photographic chemicals, inks, rayon, rubber, textile bleaches, and dyestuffs; and high-temperature equipment.

AISI Type 316L: Similar Steels (U. S. and/or Foreign). UNS S31603; AMS 5507, 5653; ASME SA182, SA213, SA240, SA249, SA312, SA403, SA479, SA688; ASTM A167, A182, A213, A240, A249, A269, A276, A312, A314, A403, A473, A478, A479, A511, A554, A580, A632, A688; FED QQ-S-763, QQ-S-766; MIL SPEC MIL-S-862; SAE J405 (30316L); (W. Ger.) DIN 1.4404; (Fr.) AFNOR Z 2 CND 17.12; (Ital.) UNI X 2 CrNiMo 17 12; (Jap.) JIS SUS 316 L, SUH 310; (Swed.) SS_{14} 2348; (U.K.) B.S. 316 S 12, 316 S 14, 316 S 22, 316 S 24, 316 S 29, 316 S 30, 316 S 31, 316 S 37, 316 S 82, S. 537

Mechanical Properties

AISI Type 316L: Charpy Impact Properties of Weld Joints (Ref 3)

Test specimens were in the as-welded condition

Test temperature		Impact energy at keyhole notch location of: In weld		Heat-affected zone		Unaffected zone	
°C	°F	J	ft·lb	J	ft·lb	J	ft·lb
30	85	43	32	95	70	103	76
−195	−320	22	16	88	65	96	71
−260	−440	26	19	81	60	95	70
−260(a)	−440(a)	34	25	94	69	148	109

(a) Data for Charpy V-notch

AISI Type 316L: Representative Tensile Properties (Ref 6)

Product form	Tensile strength MPa	Tensile strength ksi	Yield strength(a) MPa	Yield strength(a) ksi	Elongation(b), %	Hardness
Annealed						
Sheet	560	81	290	42	50	79 HRB
Strip	560	81	290	42	50	79 HRB
Plate	560	81	235	34	55	146 HB
Bar	515	75	205	30	60	149 HB
Annealed and cold drawn						
Bar	585	85	380	55	45	190 HB

(a) 0.2% offset. (b) Elongation in 50 mm (2 in.)

AISI Type 316L: Modulus of Elasticity (Ref 7)

Stressed in:	Modulus of elasticity GPa	10^6 psi
Tension	193	28.0
Torsion	77	11.2

AISI Type 316L: Izod Impact Properties

Test temperature °C	°F	Impact energy J	ft·lb
−195	−320	150	110
21	70	150	110

Physical Properties

AISI Type 316L: Thermal Treatment Temperatures

Treatment	Temperature range °C	°F
Forging, start	1150-1260	2100-2300
Forging, finish	925 min	1700 min
Annealing (a)	1010-1120	1850-2050
Hardening (b)	...	...

(a) Cool rapidly in air or water quench. (b) Hardenable by cold work only

For additional data on physical properties of AISI type 316L, refer to the preceding tables for AISI type 316.

Machining Data

For machining data on AISI type 316L, refer to the preceding machining tables for AISI type 302B.

Mechanical Properties (continued)

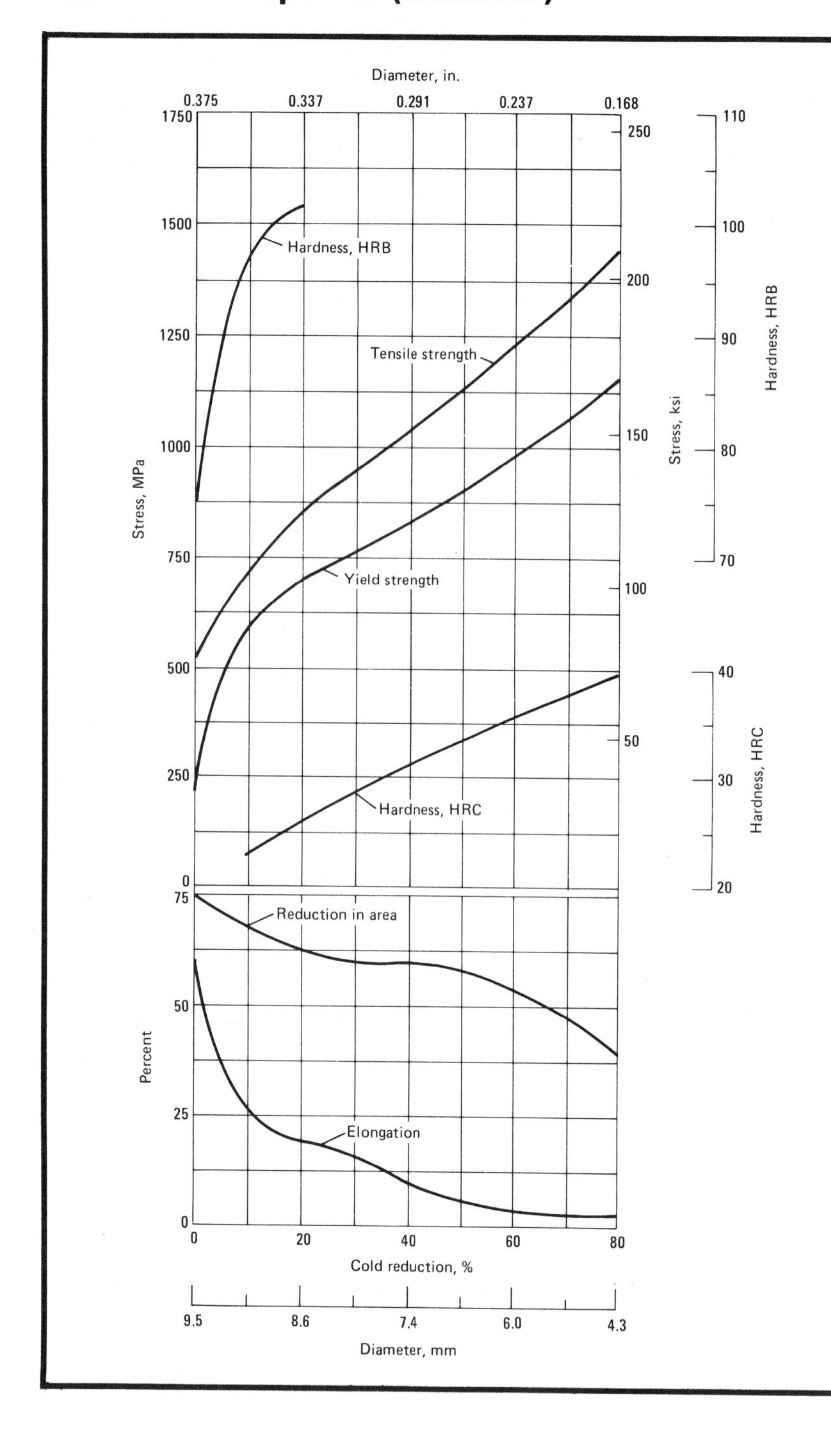

AISI Type 316L: Effect of Cold Reduction on Tensile Properties. Annealed at 1065 °C (1950 °F); water quenched. Specimens were 9.53-mm (0.375-in.) rounds, unstraightened and untempered. Tests were conducted using specimens machined to English units. Elongation was measured in 50 mm (2 in.); yield strength at 0.2% offset. (Ref 10)

AISI Type 316F

AISI Type 316F: Chemical Composition

Element	Composition, max %
Carbon	0.08
Manganese	2.00
Phosphorus	0.20
Sulfur	0.10
Silicon	1.00
Chromium	16-18
Nickel	10-14
Molybdenum	1.75-2.00

Characteristics. AISI type 316F is an austenitic, chromium-nickel, stainless and heat-resisting steel. It is similar to type 316 with elements added to improve machining and nonseizure characteristics.

Type 316F is nonmagnetic when annealed, but slightly magnetic when cold worked. Welding, cold working, and machining characteristics are similar to those of type 316. One exception is that type 316F has a machinability rating approaching 65% of that for free-machining AISI 1212; type 316 has a 55% machinability rating.

Typical Uses. AISI type 316F is suitable for use in automatic screw machines. Other applications are similar to those given for AISI types 316 and 316L.

AISI Type 316F: Similar Steels (U.S. and/or Foreign). UNS S31620; AMS 5649

Physical Properties

AISI Type 316F: Thermal Treatment Temperatures

Treatment	Temperature range °C	Temperature range °F
Forging, start	1205	2200
Forging, finish	925 min	1700 min
Annealing(a)	1095	2000
Hardening(b)	···	···

(a) Air cool rapidly or water quench. (b) Hardenable by cold work only

AISI Type 316F: Average Coefficients of Linear Thermal Expansion (Ref 6)

Temperature range °C	Temperature range °F	Coefficient μm/m·K	Coefficient μin./in.·°F
0-100	32-212	16.6	9.2
0-315	32-600	17.5	9.7
0-540	32-1000	18.2	10.1

AISI Type 316F: Selected Physical Properties (Ref 6)

Melting range	1370-1400 °C (2500-2550 °F)
Density	8.0 g/cm³ (0.29 lb/in.³)
Specific heat at 0-100 °C (32-212 °F)	485 J/kg·K (0.116 Btu/lb·°F)
Thermal conductivity at 100 °C (212 °F)	14.4 W/m·K (8.3 Btu/ft·h·°F)
Electrical resistivity at 20 °C (68 °F)	74 nΩ·m

Machining Data

For machining data on AISI type 316F, refer to the preceding machining tables for AISI type 302B.

Mechanical Properties

AISI Type 316F: Representative Tensile Properties (Ref 6)

Test specimens were in the annealed condition

Product form	Tensile strength MPa	Tensile strength ksi	Yield strength(a) MPa	Yield strength(a) ksi	Elongation(b), %	Hardness
Sheet	585	85	260	38	60	85 HRB
Bar	565	82	240	35	57	143 HB

(a) 0.2% offset. (b) In 50 mm (2 in.)

AISI Type 316N

AISI Type 316N: Chemical Composition

Element	Composition, max %
Carbon	0.08
Manganese	2.00
Phosphorus	0.045
Sulfur	0.030
Silicon	1.00
Chromium	16-18
Nickel	10-14
Molybdenum	2-3
Ntrogen	0.10-0.16

Characteristics. AISI type 316N is a nonhardenable, austenitic, chromium-nickel-molybdenum-nitrogen stainless steel. This grade has higher strength than conventional type 316 yet retains ductility, as well as good corrosion and heat resistance. It is nonmagnetic in the annealed condition, but may become slightly magnetic when cold worked.

The higher nitrogen content of type 316N provides improved resistance in environments that promote pitting and crevice corrosion in stainless steels. However, nitrogen has no apparent effect in nonpitting acid environments.

Type 316N can be readily cold formed by heading, drawing, bending, and upsetting. Type 316 or 316N electrodes should be used when welding this steel. For maximum corrosion resistance, assemblies should be annealed after welding.

Typical Uses. Applications of AISI type 316N include screw machine products, food processing equipment, photographic developing equipment, pulp handling equipment, and accessories for chemical handling equipment.

AISI Type 316N: Similar Steels (U.S. and/or Foreign). UNS S31651; ASME SA182, SA213, SA240, SA249, SA312, SA358, SA376, SA403, SA430, SA479; ASTM A182, A213, A240, A249, A276, A312, A358, A376, A403, A430, A479

Physical Properties

AISI Type 316N: Thermal Treatment Temperatures

Treatment	Temperature range °C	°F
Forging, start	1040-1150	1900-2100
Forging, finish	925 min	1700 min
Annealing(a)	1010-1120	1850-2050
Annealing(b)	1095	2000

(a) For full annealing; follow with water quench. (b) For maximum corrosion resistance; follow with water quench

AISI Type 316N: Thermal Conductivity (Ref 10)

Temperature °C	°F	Thermal conductivity W/m·K	Btu/ft·h·°F
95	200	16.1	9.3
540	1000	21.5	12.4

AISI Type 316N: Selected Physical Properties (Ref 10)

Melting range	1370-1400 °C (2500-2550 °F)
Density	7.92 g/cm³ (0.286 lb/in.³)
Specific heat at 0-100 °C (32-212 °F)	500 J/kg·K (0.12 Btu/lb·°F)
Electrical resistivity at 20 °C (68 °F)	72 nΩ·m

AISI Type 316N: Average Coefficients of Linear Thermal Expansion (Ref 10)

Temperature range °C	°F	Coefficient μm/m·K	μin./in.·°F
0-100	32-212	16.0	8.9
0-315	32-600	16.2	9.0
0-540	32-1000	17.5	9.7
0-650	32-1200	18.5	10.3

Mechanical Properties

AISI Type 316N: Modulus of Elasticity (Ref 10)

Stressed in:	Modulus GPa	10^6 psi
Tension	196	28.5

AISI Type 316N: Charpy Impact Energy (Ref 10)

Specimen diameter mm	in.	Energy J	ft·lb
13	0.5	80	59

AISI Type 316N: Representative Tensile Properties (Ref 6)

Test specimens were in the annealed condition

Product form	Tensile strength MPa	ksi	Yield strength(a) MPa	ksi	Elongation(b), %	Hardness
Sheet	620	90	330	48	48	85 HRB
Bar	620	90	290	42	55	180 HB

(a) 0.2% offset. (b) In 50 mm (2 in.)

Machining Data

For machining data on AISI type 316N, refer to the preceding machining tables for AISI type 302B.

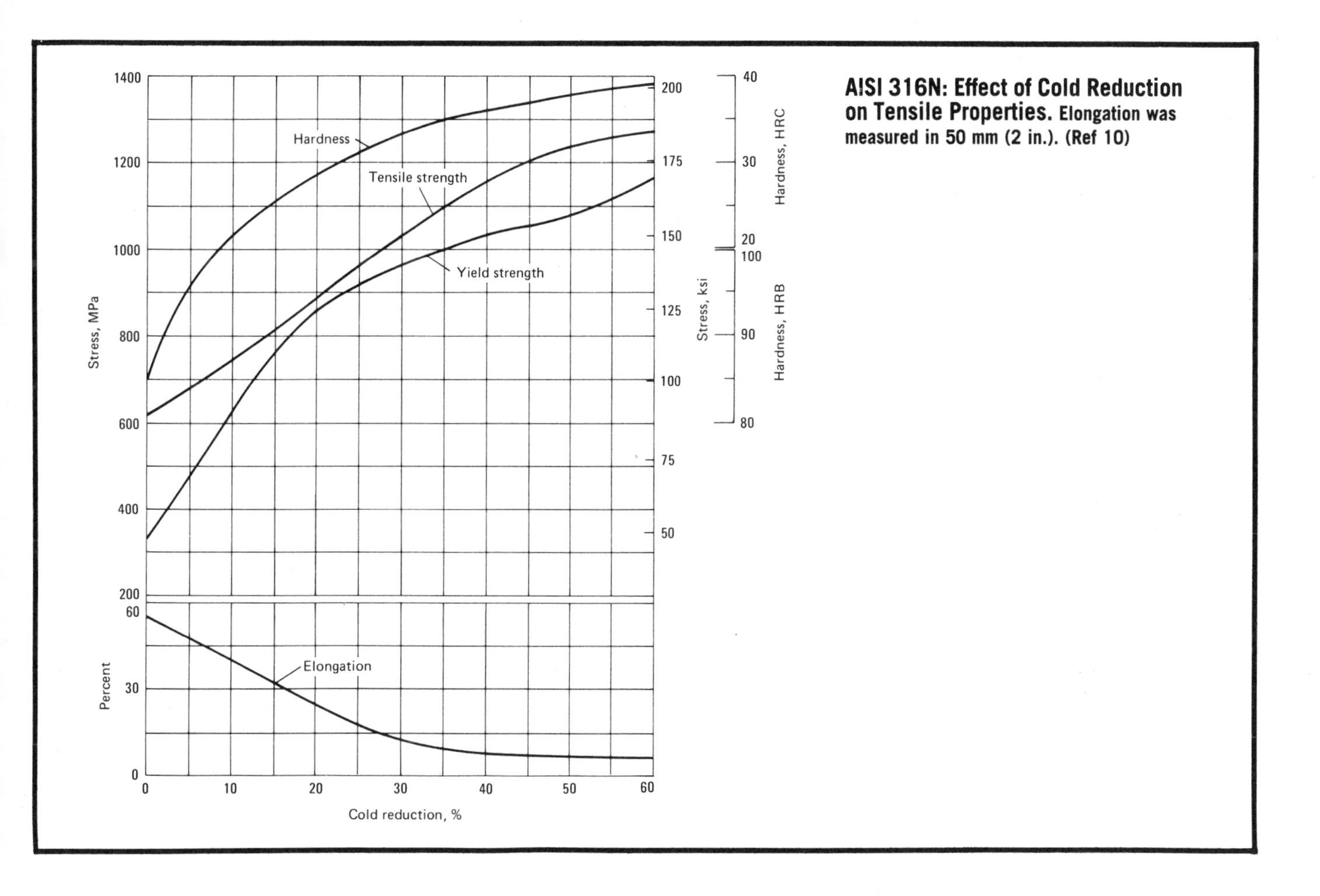

AISI 316N: Effect of Cold Reduction on Tensile Properties. Elongation was measured in 50 mm (2 in.). (Ref 10)

AISI Type 317

AISI Type 317: Chemical Composition

Element	Composition, max %
Carbon	0.080
Manganese	2.0
Phosphorus	0.045
Sulfur	0.030
Silicon	1.0
Chromium	18-20
Nickel	11-15
Molybdenum	3.0-4.0

Characteristics. AISI type 317 is a molybdenum-bearing, austenitic, chromium-nickel steel which exhibits excellent corrosion resistance. The steel was developed for use in sulfite pulp mills to resist the corrosion of sulfurous acid compounds more effectively. Type 317 also resists pitting in phosphoric and acetic acids.

In the annealed condition, this stainless steel has excellent intergranular corrosion resistance. However, if exposed to the temperature range 425 to 900 °C (800 to 1650 °F), a subsequent anneal at 1040 °C (1900 °F) minimum, followed by water quenching, is suggested. Type 317 has a safe scaling temperature for continuous service of 870 °C or 1600 °F.

This grade can be cold worked by deep drawing, stamping, heading, and upsetting without difficulty. Severe cold forming operations should be followed by an anneal. Type 317 steel machines with a tough and stringy chip. As heavy a cut as possible should be used to avoid glazing. The machinability rating of this grade is about 55% of that for free-machining AISI 1212. Moderate cold working usually improves machinability.

Typical Uses. Applications of AISI type 317 include paper pulp handling equipment; process equipment for producing photographic chemicals, inks, rayon, rubber, textile bleaches, and dyestuffs; and high-temperature equipment.

AISI Type 317: Similar Steels (U.S. and/or Foreign). UNS S31700; ASME SA240, SA249, SA312, SA403, SA409; ASTM A167, A240, A249, A269, A276, A312, A314, A403, A409, A473, A478, A511, A554, A580, A632; FED QQ-S-763; MIL SPEC MIL-S-862; SAE J405 (30317); (W. Ger.) DIN 1.4449; (Ital.) UNI X 5 CrNiMo 18 15; (Jap.) JIS SUS 317

Physical Properties

AISI Type 317: Thermal Treatment Temperatures

Treatment	Temperature range °C	°F
Forging, start	1150-1260	2100-2300
Forging, finish	925 min	1700 min
Annealing(a)	1010-1120	1850-2050
Hardening(b)	...	...

(a) Water quench. (b) Hardenable by cold work only

AISI Type 317: Average Coefficients of Linear Thermal Expansion (Ref 6)

Temperature range °C	°F	Coefficient μm/m·K	μin./in.·°F
0-100	32-212	16.0	8.9
0-315	32-600	16.2	9.0
0-540	32-1000	17.5	9.7
0-650	32-1200	18.5	10.3
0-815	32-1500	20.0	11.1

AISI Type 317: Selected Physical Properties (Ref 6)

Melting range	1370-1400 °C (2500-2550 °F)
Density	8.0 g/cm^3 (0.29 lb/in.^{3})
Specific heat at 0-100 °C (32-212 °F)	502 J/kg·K (0.12 Btu/lb·°F)
Electrical resistivity at 20 °C (68 °F)	74 nΩ·m

AISI Type 317: Thermal Conductivity (Ref 6)

Temperature °C	°F	Thermal conductivity W/m·K	Btu/ft·h·°F
100	212	16.3	9.4
500	930	21.5	12.4

Mechanical Properties

AISI Type 317: Tensile and Creep Properties at Elevated Temperatures

Test specimens were 25-mm (1-in.) diam bar, water quenched from 1120 °C (2050 °F)

Test temperature °C	°F	Stress for short-time tests MPa	ksi	Stress for 1% elongation in 10 000 h MPa	ksi
21	70	585	85	...	...
540	1000	505	73	172	25.0
595	1100	485	70	125	18.2
650	1200	460	67	88	12.7
705	1300	450	65	54	7.9
760	1400	350	51	31	4.5
815	1500	275	40	19	2.8

AISI Type 317: Charpy Impact Properties of Weld Metal (Ref 3)

Condition	21 °C (70 °F) J	ft·lb	−76 °C (−105 °F) J	ft·lb	−195 °C (−320 °F) J	ft·lb
2.0% ferrite						
As welded	28-31	21-23	22-23	16-17	12-19	9-14
Annealed	28	21	22-23	16-17	15-23	11-17
5.5% ferrite						
As welded	14-20	10-15	9-11	7-8	5	4
Annealed	14-16	10-12	7-9	5-7	4	3

(Charpy keyhole notch impact energy for test temperature of:)

AISI Type 317: Representative Tensile Properties (Ref 6)

Test specimens were in the annealed condition

Product form	Tensile strength MPa	ksi	Yield strength(a) MPa	ksi	Elongation(b), %	Hardness
Sheet	620	90	275	40	45	85 HRB
Strip	620	90	275	40	45	85 HRB
Plate	585	85	275	40	50	160 HB
Bar	585	85	275	40	50	160 HB

(a) 0.2% offset. (b) In 50 mm (2 in.)

AISI Type 317: Modulus of Elasticity (Ref 6)

Stressed in:	Modulus GPa	10^6 psi
Tension	195	28

Machining Data

For machining data on AISI type 317, refer to the preceding machining tables for AISI type 302B.

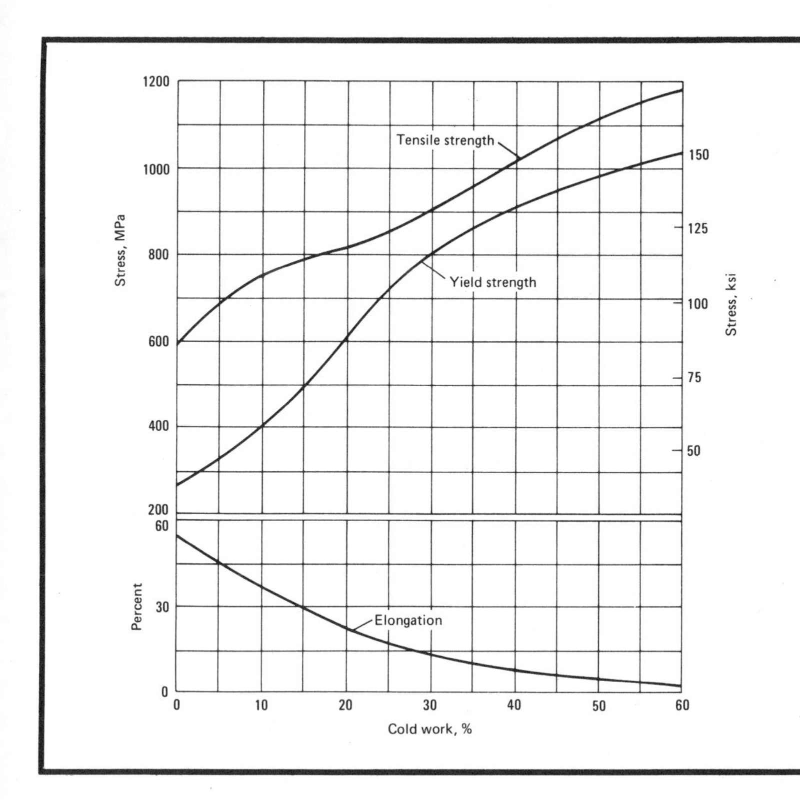

AISI Type 317: Effect of Cold Work on Mechanical Properties. Elongation was measured in 50 mm (2 in.); yield strength at 0.2% offset. (Ref 4)

AISI Type 317L

AISI Type 317L: Chemical Composition

Element	Composition, max %
Carbon	0.030
Manganese	2.0
Phosphorus	0.045
Sulfur	0.030
Silicon	1.0
Chromium	18-20
Nickel	11-15
Molybdenum	3.0-4.0

Characteristics. AISI type 317L is a molybdenum-bearing, austenitic, chromium-nickel steel similar to type 316L, but containing a higher alloy content. This stainless steel exhibits superior corrosion resistance in special applications where a minimum amount of contamination resulting from reaction between the material and the vessel is desired.

Type 317L was developed primarily to resist the attack of sulfurous acid compounds. Because it withstands corrosion, uses of this steel have expanded; it is now found in many other industrial applications, as well. The low carbon content of this grade provides immunity to intergranular corrosion in applications where parts with heavy cross sections cannot be annealed after welding, or where low-temperature stress relieving treatments are desired.

The safe scaling temperature for type 317L in continuous service is 925 °C (1700 °F). The maximum temperature for intermittent service is 815 °C (1500 °F).

Typical Uses. Applications of AISI type 317L include handling of sulfur, pulp liquor, acid dyestuffs, and acetylating and nitrating mixtures; bleaching solutions; flue gasses from fossil fuel plants; and sulfurous, acetic, formic, citric, and tartaric acids.

AISI Type 317L: Similar Steels (U. S. and/or Foreign). UNS S31703; ASME SA240; ASTM A167, A240; (W. Ger.) DIN 1.4438; (Fr.) AFNOR Z 2 CND 19.15; (Swed.) SS_{14} 2367; (U.K.) B.S. 317 S 12

Machining Data

For machining data on AISI type 317L, refer to the preceding machining tables for AISI type 302B.

Physical Properties

AISI Type 317L: Thermal Treatment Temperatures

Treatment	Temperature range °C	°F
Forging, start	1150-1260	2100-2300
Forging, finish	925 min	1700 min
Annealing(a)	1040-1095	1900-2000
Hardening(b)	...	...

(a) Water quench. (b) Hardenable by cold work only

AISI Type 317L: Thermal Conductivity (Ref 11)

Temperature °C	°F	Thermal conductivity W/m·K	Btu/ft·h·°F
100	212	16.3	9.4
500	930	21.5	12.4

AISI Type 317L: Average Coefficients of Linear Thermal Expansion (Ref 11)

Temperature range °C	°F	Coefficient μm/m·K	μin./in.·°F
0-100	32-212	14.4	8.0
0-315	32-600	16.2	9.0
0-540	32-1000	17.5	9.7
0-650	32-1200	18.5	10.3
0-815	32-1500	20.0	11.1

AISI Type 317L: Selected Physical Properties (Ref 11)

Melting point	1425-1430 °C (2595-2605 °F)
Density	8.0 g/cm^3 (0.29 lb/in.^{3})
Specific heat at 0-100 °C (32-212 °F)	486 J/kg·K (0.116 Btu/lb·°F)
Electrical resistivity at 20 °C (68 °F)	74 nΩ·m

Mechanical Properties

AISI Type 317L: Representative Tensile Properties (Ref 6)

Test specimens were in the annealed condition

Product form	Tensile strength MPa	ksi	Yield strength(a) MPa	ksi	Elongation(b), %	Hardness, HRB
Sheet	595	86	260	38	55	85
Plate	585	85	240	35	55	80
Tubing	595	86	345	50	55	...

(a) 0.2% offset. (b) In 50 mm (2 in.)

AISI Type 317L: Tensile Strength at Elevated Temperature and Creep Properties (Ref 11)

Test temperature °C	°F	Strength for short-time elevated temperature MPa	ksi	Stress for creep rate of 1% in 10 000 h MPa	ksi
540	1000	505	73	172	25.0
595	1100	485	70	125	18.2
650	1200	460	67	88	12.7
705	1300	450	65	54	7.8
760	1400	350	51	31	4.5
815	1500	275	40	19	2.8

AISI Type 321

AISI Type 321: Chemical Composition

Element	Composition, max %
Carbon	0.080
Manganese	2.0
Phosphorus	0.045
Sulfur	0.030
Silicon	1.0
Chromium	17-19
Nickel	9-12
Titanium	(a)

(a) Five times the minimum carbon content

Characteristics. AISI type 321 is a titanium-bearing, austenitic, chromium-nickel, stainless steel. This grade is stabilized against carbide precipitation and designed for operation within the temperature range where carbide precipitation develops. In this type of steel, the carbon combines preferentially with titanium to form a harmless titanium carbide, and leaves the chromium in solution to maintain full corrosion resistance.

Type 321 withstands all ordinary rusting and resists all foodstuffs, sterilizing solutions, and most of the organic and inorganic chemicals and dyestuffs. It has good resistance to nitric acid, but poor resistance to halogen acids. This stainless steel has excellent scale resistance up to 900 °C (1650 °F). For intermittent service, the temperature should not exceed 815 °C (1500 °F).

Type 321 is extremely tough and ductile, and responds to deep drawing, bending, forming, and upsetting. After cold working, it is slightly magnetic. In machining, the steel requires a powerful machine, heavy feeds, and slow spindle speeds. The machinability rating of type 321 is about 45% of that for free-machining AISI 1212.

This steel can be welded with all electric welding processes. If a filler metal is required, type 347 is suggested. Type 321 should not be used because most, if not all, titanium is lost in welding.

Typical Uses. Applications of AISI type 321 include furnace parts; aircraft collector rings and exhaust manifolds; expansion joints; and high-temperature chemical handling equipment.

AISI Type 321: Similar Steels (U.S. and/or Foreign). UNS S32100; AMS 5510, 5557, 5559, 5570, 5576, 5645, 5689; ASME SA182, SA193, SA194, SA213, SA240, SA249, SA312, SA320, SA358, SA376, SA403, SA409, SA430, SA479; ASTM A167, A182, A193, A194, A213, A240, A249, A269, A271, A276, A312, A314, A320, A358, A376, A403, A409, A430, A473, A479, A493, A511; FED QQ-S-763, QQ-S-766, QQ-W-423; MIL SPEC MIL-S-862; SAE J405 (30321); (W. Ger.) DIN 1.4541; (Fr.) AFNOR Z 6 CNT 18.10; (Ital.) UNI X 6 CrNiTi 18 11, X 6 CrNiTi 18 11 KG, X 6 CrNiTi 18 11 KW, X 6 CrNiTi 18 11 KT, ICL 472 T; (Jap.) JIS SUS 321; (Swed.) SS_{14} 2337; (U.K.) B.S. CDS-20, 321 S 12, 321 S 18, 321 S 22, 321 S 27, 321 S 40, 321 S 49, 321 S 50, 321 S 59, 321 S 87, En. 58 B, En. 58 C

Physical Properties

AISI Type 321: Thermal Treatment Temperatures

Treatment	Temperature range °C	°F
Forging, start	1150-1260	2100-2300
Forging, finish	925 min	1700 min
Annealing(a)	955-1120	1750-2050
Stabilizing(b)	845-900	1550-1650
Hardening(c)	...	...

(a) Follow with rapid air cooling or water quenching. (b) Recommended for added protection against intergranular corrosion for severe service conditions, particularly above 425 °C (800 °F). Follow with rapid air cooling or water quenching. (c) Hardenable by cold work only

AISI Type 321: Average Coefficients of Linear Thermal Expansion (Ref 6)

Temperature range °C	°F	Coefficient µm/m·K	µin./in.·°F
0-100	32-212	16.7	9.3
0-315	32-600	17.1	9.5
0-540	32-1000	18.5	10.3
0-650	32-1200	19.3	10.7
0-815	32-1500	20.2	11.2
20-925	68-1700	20.5	11.4

AISI Type 321: Thermal Conductivity (Ref 6)

Temperature °C	°F	Thermal conductivity W/m·K	Btu/ft·h·°F
100	212	16.1	9.3
500	930	22.1	12.8

AISI Type 321: Selected Physical Properties (Ref 6)

Melting range	1400-1425 °C (2550-2600 °F)
Density	8.0 g/cm³ (0.29 lb/in.³)
Specific heat at 0-100 °C (32-212 °F)	500 J/kg·K (0.12 Btu/lb·°F)
Electrical resistivity at 20 °C (68 °F)	72 nΩ·m

Mechanical Properties

AISI Type 321: Representative Tensile Properties (Ref 6)

Product form	Tensile strength MPa	ksi	Yield strength(a) MPa	ksi	Elongation(b), %	Reduction in area, %	Hardness
Annealed							
Sheet	620	90	240	35	45	...	80 HRB
Strip	620	90	240	35	45	...	80 HRB
Plate	585	85	240	35	35	...	160 HB
Bar	585	85	240	35	55	65	150 HB
Annealed and cold drawn							
Bar, 25-mm (1.0-in.) diam	655	95	415	60	40	60	185 HB
Soft tempered							
Wire, 1.6-mm (0.062-in.) diam	795	115	585	85	30	...	...
Wire, 13-mm (0.50-in.) diam	655	95	450	65	40	60	89 HRB

(a) 0.2% offset. (b) In 50 mm (2 in.), except for wire. For wire 3.18-mm (0.125-in.) diam and larger, gage length is 4 × diam; for smaller diameters, the gage length is 254 mm (10.0 in.)

AISI Type 321: Modulus of Elasticity (Ref 7)

Stressed in:	Modulus GPa	10^6 psi
Tension	193	28.0
Torsion	77	11.2

AISI Type 321: Impact Properties of Annealed Material (Ref 4)

Impact energy test	Energy J	ft·lb
Izod	120-150	90-110
Charpy keyhole notch	95	70
Charpy V-notch	165	120

AISI Type 321: Tensile Properties at Elevated Temperatures (Ref 7)

Test specimens were 16.0-mm (0.625-in.) diam rod, annealed at 1040 °C (1900 °F) for ½ h; water quenched; hardness of 75 HRB

Test temperature °C	°F	Tensile strength MPa	ksi	Yield strength(a) MPa	ksi	Elongation(b), %	Reduction in area, %
21	70	569	82.5	207	30.0	71.0	79.5
205	400	421	61.0	145	21.0	48.0	76.0
425	800	400	58.0	121	17.5	42.5	71.5
540	1000	379	55.0	124	18.0	42.0	70.5
650	1200	303	44.0	117	17.0	37.0	57.5
760	1400	210	30.5	114	16.5	44.5	53.0
870	1600	117	17.0	93	13.5	57.0	82.0
980	1800	63	9.1	50	7.3	78.0	96.0

(a) 0.2% offset. (b) Gage length is 4 × diam of test specimen

AISI Type 321: Endurance Limit

Condition	Stress MPa	ksi
Annealed	270	39

Mechanical Properties (continued)

AISI Type 321: Stress-Rupture and Creep Properties (Ref 7)

Test temperature		Stress for rupture in: 1000 h		10 000 h(a)		100 000 h(a)		Stress for secondary creep rate of(a): 1% in 10 000 h		1% in 100 000 h	
°C	°F	MPa	ksi	MPa	ksi	MPa	ksi	MPa	ksi	MPa	ksi
480	900	340	49.0	295	43.0	...	...	...	...	...	...
540	1000	290	42.0	235	34.0	...	...	...	...	...	...
595	1100	200	29.0	145	21.0	105	15.0	115	16.5	80	11.5
650	1200	130	18.5	85	12.0	50	7.5	65	9.5	45	6.5
705	1300	75	11.0	40	6.0	30	4.0	35	5.0	25	3.5
760	1400	45	6.5	25	3.5	17	2.5	17	2.5	15	2.0
815	1500	30	4.0	15	2.0	10	1.5	10	1.5	...	...

(a) Extrapolated data

AISI Type 321: Tensile Properties of Cold Drawn Wire (Ref 4)

Specimen diameter, mm	in.	Tensile strength, MPa	ksi	Yield strength, MPa	ksi	Elongation(a), %
25% hard						
0.05-0.51	0.002-0.020	1105-1380	160-200	825-1170	120-170	15-20
0.53-3.18	0.021-0.125	895-1105	130-160	690-895	100-130	15-25
3.20-9.53	0.126-0.375	825-965	120-140	620-760	90-110	20-30
50% hard						
0.05-0.51	0.002-0.020	1380-1585	200-230	1205-1480	175-215	11-18
0.53-3.18	0.021-0.125	1105-1380	160-200	895-1170	130-170	11-18
3.20-9.53	0.126-0.375	1035-1170	150-170	825-965	120-140	10-18
75% hard						
0.05-0.51	0.002-0.020	1585-1860	230-270	1310-1515	190-220	5-10
0.53-3.18	0.021-0.125	1380-1585	200-230	1170-1380	170-200	6-12
3.20-9.53	0.126-0.375	1170-1380	170-200	965-1170	140-170	5-10
Full hard						
0.05-0.51	0.002-0.020	1860-2345	270-340	1585-1795	230-260	1-2
0.53-3.18	0.021-0.125	1620-1895	235-275	1450-1690	210-245	2-5
3.20-9.53	0.126-0.375	1380-1655	200-240	1170-1450	170-210	2-5

(a) In 50 mm (2 in.)

AISI Type 321: Tensile Properties at Elevated Temperatures After Annealing and Cold Work (Ref 7)

Test specimens were 16-mm (0.625-in.) diam rod; annealed and cold worked to hardness of 96 HRB

Test temperature, °C	°F	Tensile strength, MPa	ksi	Yield strength(a), MPa	ksi	Elongation(b), %	Reduction in area, %
21	70	655	95.0	465	67.5	48.0	73.0
205	400	493	71.5	390	56.5	27.5	74.0
425	800	476	69.0	390	56.5	23.5	63.0
540	1000	448	65.0	365	53.0	22.0	66.0
650	1200	390	56.5	331	48.0	23.0	55.0
760	1400	296	43.0	279	40.5	18.0	32.5
870	1600	169	24.5	162	23.5	50.5	75.5
980	1800	69	10.0	63	9.2	102.0	93.5

(a) 0.2% offset. (b) Gage length is 4 × diam

AISI Type 321: Impact Properties of Weld Metal

Base metal type 321; filler metal type 347

Condition	Charpy keyhole notch impact energy at temperature of: 20 °C (68 °F), J	ft·lb	−76 °C (−105 °F), J	ft·lb	−195 °C (−320 °F), J	ft·lb
As welded	43	32	33-37	24-27	24-35	18-26
Annealed	39-43	29-32	38-49	28-36	31-37	23-27

Machining Data

For machining data on AISI type 321, refer to the preceding machining tables for AISI type 201.

AISI Type 321: Effect of Cold Reduction on Tensile Properties. Elongation was measured in 50 mm (2 in.); yield strength at 0.2% offset. (Ref 10)

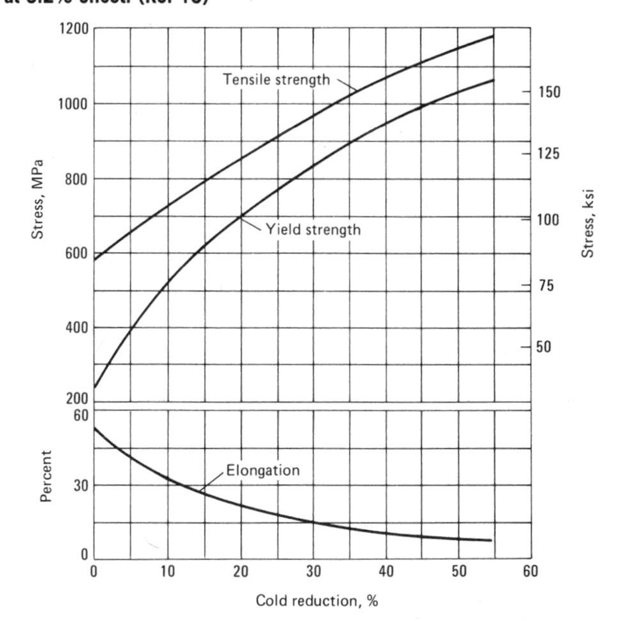

AISI Type 321: Effect of Test Temperature on Tensile Properties of Annealed Material. Elongation was measured in 50 mm (2 in.). (Ref 3)

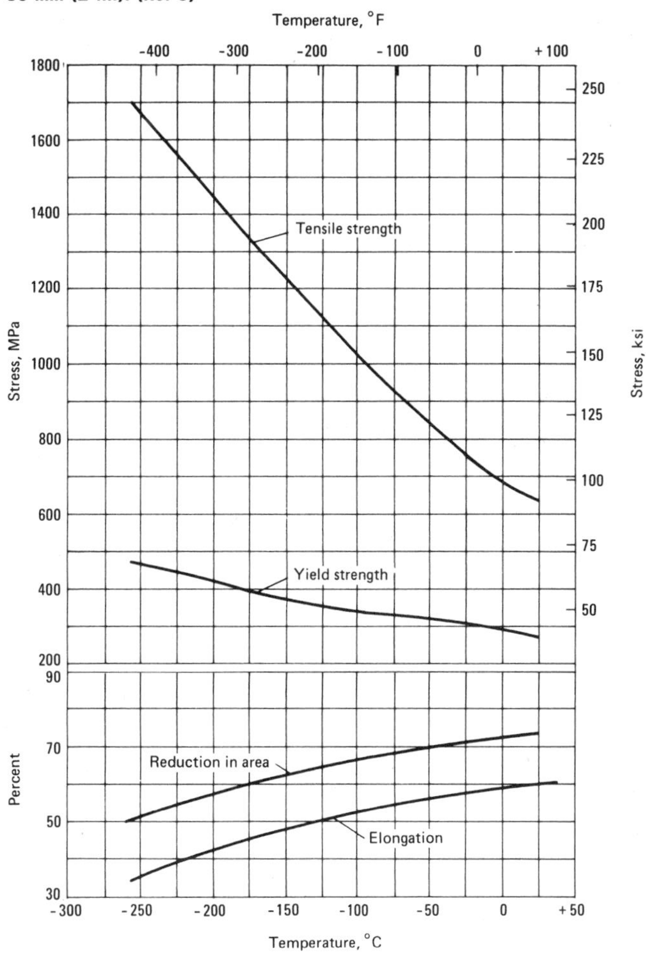

AISI Type 321: Impact Properties. Impact energy tests: —— Izod; ---- Charpy V-notch (Ref 4)

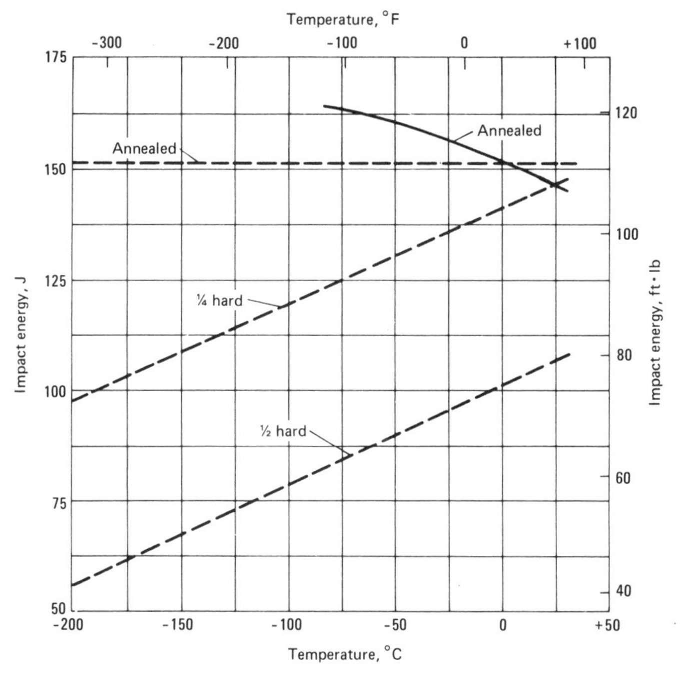

AISI Type 329

AISI Type 329: Chemical Composition

Element	Composition, max %
Carbon	0.10
Manganese	2.00
Phosphorus	0.04
Sulfur	0.03
Silicon	1.00
Chromium	25-30
Nickel	3.0-6.0
Molybdenum	1.0-2.0

Characteristics. AISI type 329 is a molybdenum-bearing, chromium-nickel, stainless steel. It is a two-phase alloy consisting of a ferrite matrix with pools of austenite. This steel has an annealed yield strength more than twice that available from the conventional austenitic stainless steels. It also has high resistance to stress corrosion cracking and pitting, as well as good general corrosion resistance to many environments.

Type 329 can be age hardened, but because it becomes brittle when hardened, should be restricted to compressive forces in this condition. This stainless steel is essentially nonmagnetic as age hardened. The annealed material has excellent resistance in nitric acid, mixtures of nitric acid and hydrofluoric acids, phosphoric acid, and the stronger organic acids.

The safe scaling temperature for continuous service of type 329 is 870 °C (1600 °F). The maximum temperature for continuous service is 1040 °C (1900 °F). Machinability of type 329 is about 40% of that for free-machining type AISI 1212. Welding is not recommended for this grade.

Typical Uses. Applications of AISI type 329 include heat exchangers in pulp mills, nitric acid plants, and various uses in the food-processing and power-generating industries, as well as valves, valve fittings, and pump parts where there are extremely corrosive conditions, or where resistance to wear and galling is desired.

AISI Type 329: Similar Steels (U.S. and/or Foreign). UNS S32900; ASME SA268; ASTM A268, A511

Physical Properties

AISI Type 329: Thermal Treatment Temperatures

Treatment	Temperature range °C	°F
Forging, start	1040-1095	1900-2000
Forging, finish	925 min	1700 min
Annealing(a)	955-980	1750-1800
Hardening(b)	730	1350
Tempering(c)	···	···

(a) For optimum ductility and impact strength, water quench sections 150 mm (6 in.) and smaller; because of the danger of cracking, larger sizes should be air cooled. (b) Hold for 12 h; slow cool in air or furnace. (c) Not recommended

AISI Type 329: Selected Physical Properties (Ref 6)

Density	7.8 g/cm^3 (0.28 lb/in.^{3})
Specific heat at 0-100 °C (32-212 °F)	460 J/kg·K (0.11 Btu/lb·°F)
Average coefficient of linear thermal expansion at 20-815 °C (68-1500 °F)	14.4 μm/m·K (8.0 μin./in.·°F)
Electrical resistivity at 20 °C (68 °F)	75 nΩ·m

Machining Data

For machining data on AISI type 329, refer to the preceding machining tables for AISI type 302B.

Mechanical Properties

AISI Type 329: Representative Tensile Properties (Ref 6)

Test specimens were in the annealed condition

Product form	Tensile strength MPa	Tensile strength ksi	Yield strength(a) MPa	Yield strength(a) ksi	Elongation(b), %	Reduction in area, %	Hardness, HB	Charpy impact energy J	Charpy impact energy ft·lb
Strip	725	105	550	80	25	50	230	54	40
Bar(c)	725	105	550	80	25	50	230	54	40

(a) 0.2% offset. (b) In 50 mm (2 in.). (c) Diameter of 25 to 75 mm (1 to 3 in.)

AISI Type 330

AISI Type 330: Chemical Composition

Element	Composition, %
Carbon	0.08 max
Manganese	2.00 max
Phosphorus	0.04 max
Sulfur	0.03 max
Silicon	0.75-1.50
Chromium	17-20
Nickel	34-37

Characteristics. AISI type 330 is an austenitic, chromium-nickel, stainless steel especially suited for application in temperatures from 815 °C (1500 °F) to about 1175 °C (2150 °F). This alloy will withstand prolonged heating in air up to 1140 °C (1900 °F) without excessive scaling. In reducing or neutral atmospheres, type 330 has been used at temperatures up to 1175 °C (2150 °F).

This steel is resistant to carburizing and carbonitriding atmospheres in the 870 to 955 °C (1600 to 1750 °F) range. Although it resists sensitization to a greater degree than standard 18-8 types, some degree of carbide precipitation may occur in the 480 to 815 °C (900 to 1500 °F) range.

In the annealed condition, type 330 may be cold formed. No special precautions are necessary, except that, because the alloy work hardens, it may require in-process annealing.

Type 330 may be fusion welded by the usual methods. For shielded-metal-arc welding, special electrodes containing higher carbon than the base metal are recommended to produce welds free from cracks. The machinability of this steel is about 40% of that for AISI 1212. Rigid machine setups and sharp tools are required for machining type 330 stainless steel.

Typical Uses. Applications of AISI type 330 include annealing fixtures, carburizing boxes, and furnace parts.

AISI Type 330: Similar Steels (U.S. and/or Foreign). UNS N08330; AMS 5592, 5716; ASTM B511, B512, B535, B536, B546; SAE J405 (30330), J412 (30330)

Physical Properties

AISI Type 330: Thermal Treatment Temperatures

Treatment	Temperature range °C	Temperature range °F
Forging, start	1150-1205	2100-2200
Forging, finish	980 min	1800 min
Annealing(a)	1120-1230	2050-2250
Hardening(b)	...	...

(a) Water or air quench. (b) Hardenable by cold work only

AISI Type 330: Thermal Conductivity (Ref 7)

Temperature °C	Temperature °F	Thermal conductivity W/m·K	Thermal conductivity Btu/ft·h·°F
24	75	12.5	7.2
100	212	16.3	9.4
425	800	19.4	11.2
500	930	21.6	12.5
870	1600	28.5	16.5

AISI Type 330: Average Coefficients of Linear Thermal Expansion (Ref 3, 7)

Temperature range °C	Temperature range °F	Coefficient μm/m·K	Coefficient μin./in.·°F
−185-21	−300-70	10.4	5.8
−130-21	−200-70	11.7	6.5
−73-21	−100-70	13.0	7.2
−18-21	0-70	13.7	7.6
0-100	32-212	14.4	8.0
0-315	32-600	16.0	8.9
0-540	32-1000	16.7	9.3
0-650	32-1200	17.3	9.6
0-870	32-1600	18.0	10.0

AISI Type 330: Selected Physical Properties (Ref 7)

Melting range	1400-1425 °C (2550-2600 °F)
Density	8.00 g/cm³ (0.289 lb/in.³)
Specific heat at 0-100 °C (32-212 °F)	460 J/kg·K (0.11 Btu/lb·°F)
Electrical resistivity at 20 °C (68 °F)	102 nΩ·m
Electrical resistivity at 425 °C (800 °F)	114 nΩ·m

Mechanical Properties

AISI Type 330: Modulus of Elasticity (Ref 7)

Test temperature °C	Test temperature °F	Modulus GPa	Modulus 10⁶ psi
21	70	197	28.5
315	600	179	26.0
540	1000	164	23.8
760	1400	145	21.0
870	1600	134	19.5
980	1800	124	18.0

AISI Type 330: Representative Tensile Properties (Ref 6)

Test specimens were in the annealed condition

Product form	Tensile strength MPa	Tensile strength ksi	Yield strength(a) MPa	Yield strength(a) ksi	Elongation(b), %	Hardness, HRB
Sheet	550	80	260	38	40	...
Strip	550	80	260	38	40	...
Plate	620	90	260	38	45	80
Bar(c)	585	85	290	42	45	80

(a) 0.2% offset. (b) In 50 mm (2 in.). (c) Specimens were 25- to 75-mm (1- to 3-in.) diam with reduction in area of 65% and Charpy impact energy of 325 J (240 ft·lb)

Mechanical Properties (continued)

AISI Type 330: Representative Tensile Properties at Elevated Temperatures (Ref 7)

Test temperature °C	°F	Tensile strength MPa	ksi	Yield strength(a) MPa	ksi	Elongation(b), %	Reduction in area, %
21	70	586	85.0	290	42.0	43	68
205	400	552	80.0	241	35.0	40	65
315	600	524	76.0	217	31.5	...	...
425	800	490	71.0	193	28.0	...	...
540	1000	448	65.0	172	25.0	...	...
650	1200	393	57.0	155	22.5	28	32
760	1400	276	40.0	138	20.0	22	27
870	1600	152	22.0	90	13.0	30	34
980	1800	76	11.0	50	7.2	53	48
1095	2000	35	5.1	27	3.9	72	50
1150	2100	30	4.4	23	3.4	93	75

(a) 0.2% offset. (b) In 50 mm (2 in.)

AISI Type 330: Stress-Rupture and Creep Properties (Ref 7)

Test temperature °C	°F	Stress for rupture in: 1000 h MPa	ksi	10 000 h MPa	ksi	Stress for secondary creep rate of 1% in 10 000 h MPa	ksi
815	1500	39	5.70	24	3.50	19	2.80
870	1600	26	3.80	17	2.40	14	2.00
925	1700	19	2.70	11	1.60	9	1.30
980	1800	12	1.70	7	1.00	6	0.80
1040	1900	8	1.20	5	0.68	3	0.42
1095	2000	5	0.78	3	0.45	2	0.27

Machining Data

For machining data on AISI type 330, refer to the preceding machining tables for AISI type 302B.

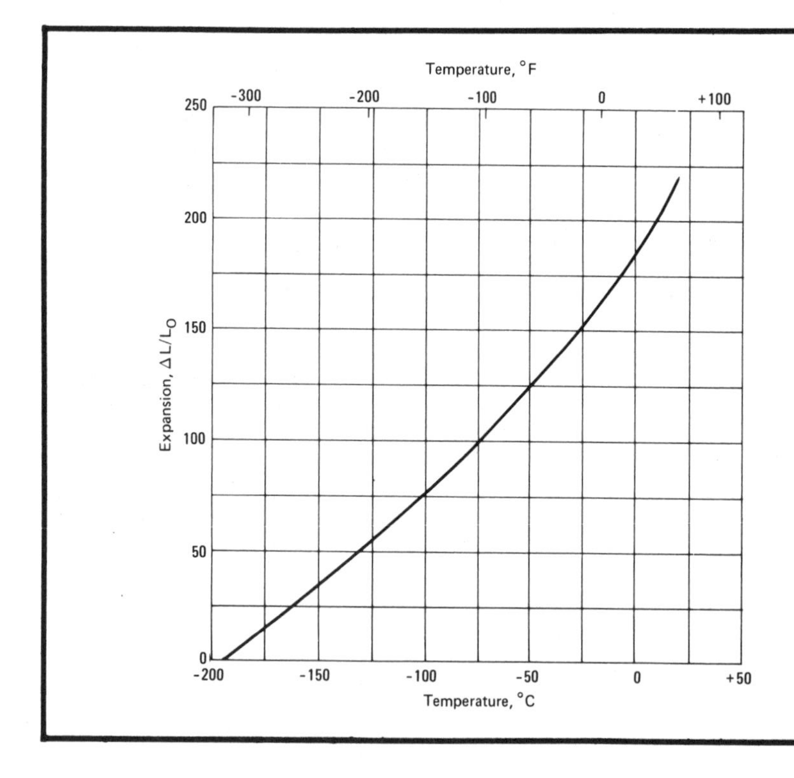

AISI Type 330: Expansion Characteristics (Ref 3)

AISI Types 347, 348

AISI Types 347, 348; Chemical Composition

Element	Composition, max % Type 347	Type 348
Carbon	0.080	0.080
Manganese	2.000	2.000
Phosphorus	0.045	0.045
Sulfur	0.030	0.030
Silicon	1.0	1.0
Chromium	17	19
Nickel	9-13	9-13
Niobium plus tantalum	(a)	(a)
Tantalum	···	0.1
Cobalt	···	0.2

(a) Ten times the minimum carbon content

Characteristics. AISI types 347 and 348 are austenitic, chromium-nickel steels containing niobium and tantalum. They are recommended for parts fabricated by welding that cannot be subsequently annealed. These alloys are also used for parts which are intermittently heated and cooled to temperatures between 425 and 870 °C (800 and 1600 °F).

The additions of niobium and tantalum produce stabilized types of stainless steel which eliminate carbide precipitation, and consequently, intergranular corrosion. Type 348 is recommended for use in applications that require a restricted tantalum content, such as those encountered in radioactive service.

Both types 347 and 348 are nonmagnetic when annealed, but slightly magnetic when cold worked. In the annealed condition, types 347 and 348 are sufficiently ductile to permit cold forming by most commercial methods. Machinability of these grades is about 45% of that for AISI 1212. A free-machining grade of type 347 containing selenium has a machinability rating of about 65%.

Types 347 and 348 are weldable by any of the electric welding processes. Because they are stabilized, these stainless steels are resistant to carbide precipitation and can be used in the as-welded condition. If a filler metal is required, type 347 is suggested.

The general corrosion resistance of these steels is similar to that of type 304. Type 347 may be used at temperatures up to about 900 °C (1650 °F) without excessive scaling. For intermittent service, a temperature of about 815 °C (1500 °F) should not be exceeded.

Typical Uses. Applications of AISI types 347 and 348 include aircraft exhaust manifolds, fire walls, pressure vessels, heavy welded equipment, and elevated-temperature, chemical handling equipment.

AISI Type 347: Similar Steels (U.S. and/or Foreign). UNS S34700; AMS 5512, 5556, 5558, 5571, 5575, 5646, 5654, 5674, 5680, 5681; ASME SA182, SA193, SA194, SA213, SA240, SA249, SA312, SA320 (B8C), SA358, SA376, SA403, SA409, SA430, SA479; ASTM A167, A182, A193, A194, A213, A249, A269, A271, A276, A312, A314, A320, A358, A376, A403, A409, A430, A473, A479, A493, A511, A554, A580, A633; FED QQ-S-763, QQ-S-766, QQ-W-423; MIL SPEC MIL-S-862, MIL-S-23195, MIL-S-23196; SAE J405 (30347); (W. Ger.) DIN 1.4550; (Fr.) AFNOR Z 6 CNNb 18.10; (Ital.) UNI X 8 CrNiNb 18 11; (Jap.) JIS SUS 347; (Swed.) SS_{14} 2338; (U.K.) B.S. 347 S 17, En. 58 F, En. 58 G, ANC 3 Grade B

AISI Type 348: Similar Steels (U. S. and/or Foreign). UNS S34800; ASME SA182, SA213, SA240, SA249, SA312, SA358, SA376, SA403, SA409, SA479; ASTM A167, A182, A213, A240, A249, A269, A276, A312, A314, A358, A376, A403, A479, A580, A632; FED QQ-S-766; MIL SPEC MIL-S-23195, MIL-S-23196; SAE J405 (30348); (W. Ger.) DIN 1.4546; (U.K.) B.S. 347 S 17, 347 S 18, 347 S 40, 2 S. 130, S. 525, S. 527

Physical Properties

AISI Types 347, 348: Thermal Treatment Temperatures

Treatment	Temperature range °C	°F
Forging, start	1150-1260	2100-2300
Forging, finish	925 min	1700 min
Annealing(a)	1010-1120	1850-2050
Stabilizing(b)	845-900	1550-1650
Hardening(c)	···	···

(a) Water quench or air cool. (b) A second treatment sometimes used when service temperatures will be in the 425 to 870 °C (800 to 1600 °F) range. (c) Hardenable by cold work only

AISI Types 347, 348: Selected Physical Properties (Ref 6)

Melting range	1400-1425 °C (2550-2600 °F)
Density	8.0 g/cm³ (0.29 lb/in.³)
Specific heat at 0-100 °C (32-212 °F)	500 J/kg·K (0.12 Btu/lb·°F)
Electrical resistivity at 20 °C (68 °F)	79.1 nΩ·m

AISI Types 347, 348: Average Coefficients of Linear Thermal Expansion (Ref 6)

Temperature range °C	°F	Coefficient µm/m·K	µin./in.·°F
0-100	32-212	17.3	9.6
0-260	32-500	17.8	9.9
0-540	32-1000	18.4	10.2
0-650	32-1200	18.7	10.4

AISI Types 347, 348: Thermal Conductivity (Ref 6)

Temperature °C	°F	Thermal conductivity W/m·K	Btu/ft·h·°F
100	212	16.3	9.4
500	930	21.5	12.4

Mechanical Properties

AISI Types 347, 348: Charpy Keyhole Notch Impact Properties of Weld and Base Metal (Ref 3)

Condition	Room temperature J	Room temperature ft·lb	−76 °C (−105 °F) J	−76 °C (−105 °F) ft·lb	−195 °C (−320 °F) J	−195 °C (−320 °F) ft·lb
As welded(a)	37-45	27-33	28-39	21-29	22-30	16-22
Stress relieved	33-37	24-27	20-22	15-16	12-18	9-13
Stabilized	26-30	19-22	20-24	15-18	9-27	7-20
Annealed	34-37	25-27	27-41	20-30	31-35	23-26
As welded(b)	34-38	25-28	24-31	18-23	18-37	13-27
Annealed(b)	38-42	28-31	27-37	20-27	28-31	21-23

(a) 3.5% ferrite. (b) Titanic-type coating; all other electrodes had lime-type coating

AISI Types 347, 348: Impact Properties of Annealed Material (Ref 4)

Impact energy test	Energy J	Energy ft·lb
Izod	120-160	90-120
Charpy keyhole notch	50	37
	80	59
Charpy V-notch	160	120

AISI Types 347, 348: Tensile Properties of Cold Drawn Wire (Ref 4)

Specimen diameter mm	Specimen diameter in.	Tensile strength MPa	Tensile strength ksi	Yield strength(a) MPa	Yield strength(a) ksi	Elongation(b), %
25% hard						
0.05-0.51	0.002-0.020	895-1170	130-170	690-965	100-140	15-20
0.53-3.18	0.021-0.125	895-1105	130-160	690-895	100-130	15-25
3.20-9.53	0.126-0.375	795-930	115-135	585-725	85-105	15-25
50% hard						
0.05-0.51	0.002-0.020	1170-1415	170-205	930-1170	135-170	11-18
0.53-3.18	0.021-0.125	1105-1380	160-200	895-1170	130-170	11-18
3.20-9.53	0.126-0.375	965-1105	140-160	760-895	110-130	10-13
75% hard						
0.05-0.51	0.002-0.020	1415-1585	205-230	1275-1450	185-210	5-10
0.53-3.18	0.021-0.125	1380-1585	200-230	1170-1380	170-200	6-12
3.20-9.53	0.126-0.375	1170-1380	170-200	965-1170	140-170	5-10
Full hard						
0.05-0.51	0.002-0.020	1860-2135	270-310	1585-1795	230-260	1-2
0.53-3.18	0.021-0.125	1585-1895	230-275	1450-1690	210-245	2-5
3.20-9.53	0.126-0.375	1310-1515	190-220	1105-1310	160-190	3-6

(a) 0.2% offset. (b) In 50 mm (2 in.)

AISI Types 347, 348: Modulus of Elasticity (Ref 7)

Stressed in:	Modulus GPa	Modulus 10^6 psi
Tension	195	28
Torsion	77	11.2

AISI Types 347, 348: Endurance Limit (Ref 4)

Condition	Stress MPa	Stress ksi
75% hard	605	88

AISI Types 347, 348: Representative Tensile Properties (Ref 6)

Product form	Tensile strength MPa	Tensile strength ksi	Yield strength(a) MPa	Yield strength(a) ksi	Elongation(b), %	Reduction in area, %	Hardness
Annealed							
Sheet	655	95	275	40	45	...	85 HRB
Strip	655	95	275	40	45	...	85 HRB
Plate	620	90	240	35	50	...	160 HB
Bar	620	90	240	35	50	65	160 HB
Annealed and cold drawn							
Bar(c)	690	100	450	65	40	60	212 HB
Soft tempered							
Wire, 1.6-mm (0.062-in.) diam	825	120	620	90	30	...	...
Wire, 13-mm (0.50-in.) diam	690	100	485	70	40	60	95 HRB

(a) 0.2% offset. (b) In 50 mm (2 in.), except for wire. For wire 3.18-mm (0.125-in.) diam and larger, gage length is 4 × diam; for smaller diameters, the gage length is 254 mm (10.0 in.). (c) Diameter of 25 mm (1 in.)

AISI Types 347, 348: Stress-Rupture and Creep Properties (Ref 7)

Test temperature °C	Test temperature °F	Stress for rupture in: 1000 h MPa	1000 h ksi	10 000 h MPa	10 000 h ksi	100 000 h(a) MPa	100 000 h(a) ksi	Stress for secondary creep rate of(a): 1% in 10 000 h MPa	1% in 10 000 h ksi	1% in 100 000 h MPa	1% in 100 000 h ksi
540	1000	340	49.0	260	38.0	230	33	22	32	185	27
595	1100	240	34.5	185	27.0	150	21.5	160	23	115	16.5
650	1200	155	22.5	120	17.5	90	13	110	16	65	9.5
705	1300	97	14.0	72	10.5	50	7.5	60	9.5	35	5
760	1400	55	8.0	40	5.5	25	3.5	35	5	15	2
815	1500	30	4.5	15	2.5	10	1.5	12	1.8	...	...

(a) Extrapolated data

Mechanical Properties (continued)

AISI Types 347, 348: Tensile Properties at Elevated Temperatures (Ref 9)

Test specimens were in the annealed condition

Test temperature °C	Test temperature °F	Short-time tensile tests: Tensile strength MPa	Tensile strength ksi	Yield strength(a) MPa	Yield strength(a) ksi	Elongation(b), %	Reduction in area, %	Stress for creep rate of 1% in 10 000 h MPa	ksi
21	70	620	90	255	37	50	65	...	...
95	200	580	84	250	36	45	70	...	...
150	300	540	78	240	35	42	72	...	...
205	400	525	76	205	30	38	74	...	...
260	500	510	74	195	28	36	71	...	...
315	600	495	72	170	25	35	72	...	...
370	700	470	68	160	23	35	72	...	...
425	800	455	66	150	22	35	70	...	...
480	900	435	63	145	21	34	69	...	...
540	1000	405	59	145	21	34	65	115	17
595	1100	380	55	140	20	35	67	90	13
650	1200	330	48	140	20	36	67	50	7
705	1300	275	40	140	20	36	70	30	4.5
760	1400	215	31	130	19	37	70	15	2
815	1500	170	25	115	17	40	70	10	1.5
870	1600	140	20	105	15	40	70	7	1
925	1700	105	15	83	12	45	72	...	...
980	1800	83	12	...	...	50	75	...	...
1040	1900	69	10	...	...	60	77	...	...

(a) 0.2% offset. (b) In 50 mm (2 in.)

AISI Type 347: Effect of Test Temperature on Tensile Properties of Annealed Material. Elongation was measured in 50 mm (2 in.). (Ref 3)

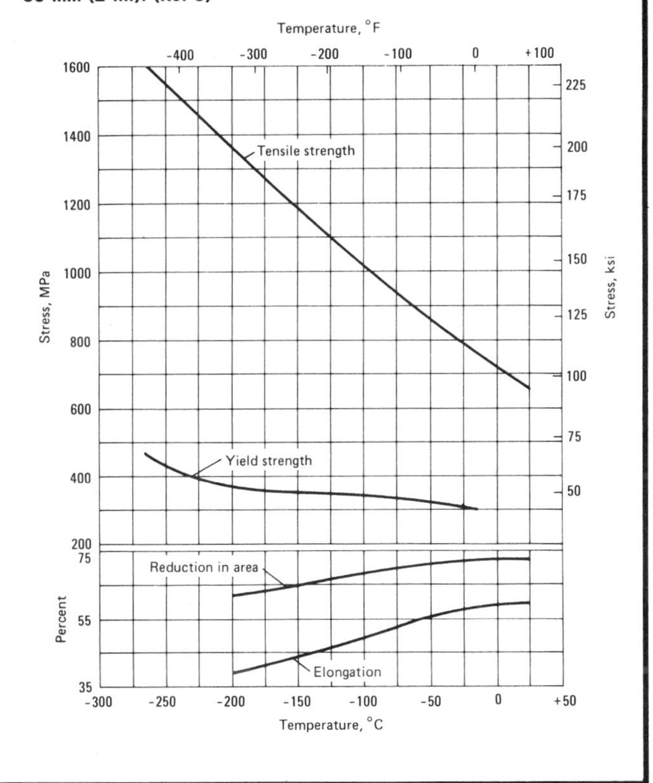

AISI Type 347: Impact Properties. Impact energy tests used Izod, Charpy K, or Charpy V specimens as indicated. (Ref 4)

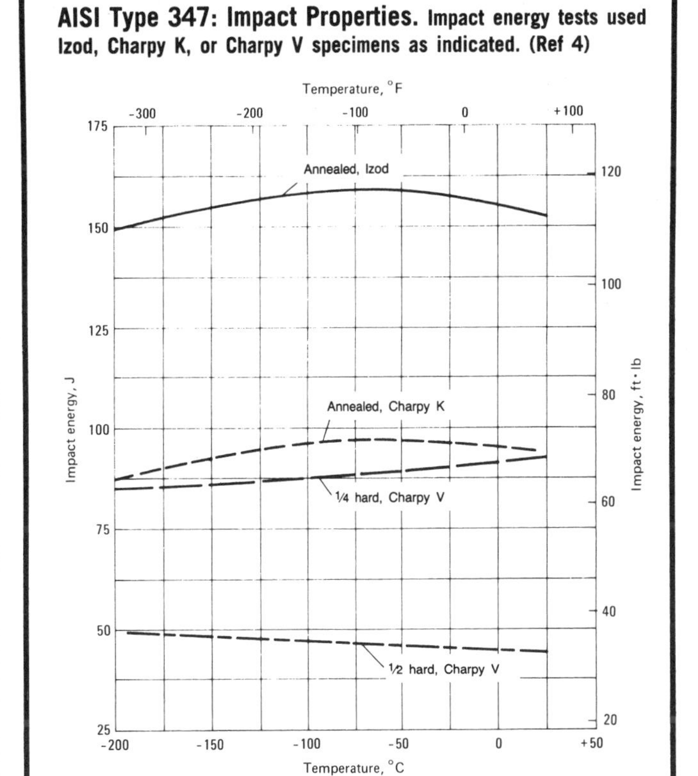

Machining Data

For machining data on AISI types 347 and 348, refer to the preceding machining tables for AISI type 201.

AISI Type 347: Effect of Cold Working on Tensile Properties.

Annealed at 1065 °C (1950 °F); water quenched. Specimens were 9.53-mm (0.375-in.) round, unstraightened and untempered. Elongation in 50 mm (2 in.); yield strength at 0.2% offset. (Ref 10)

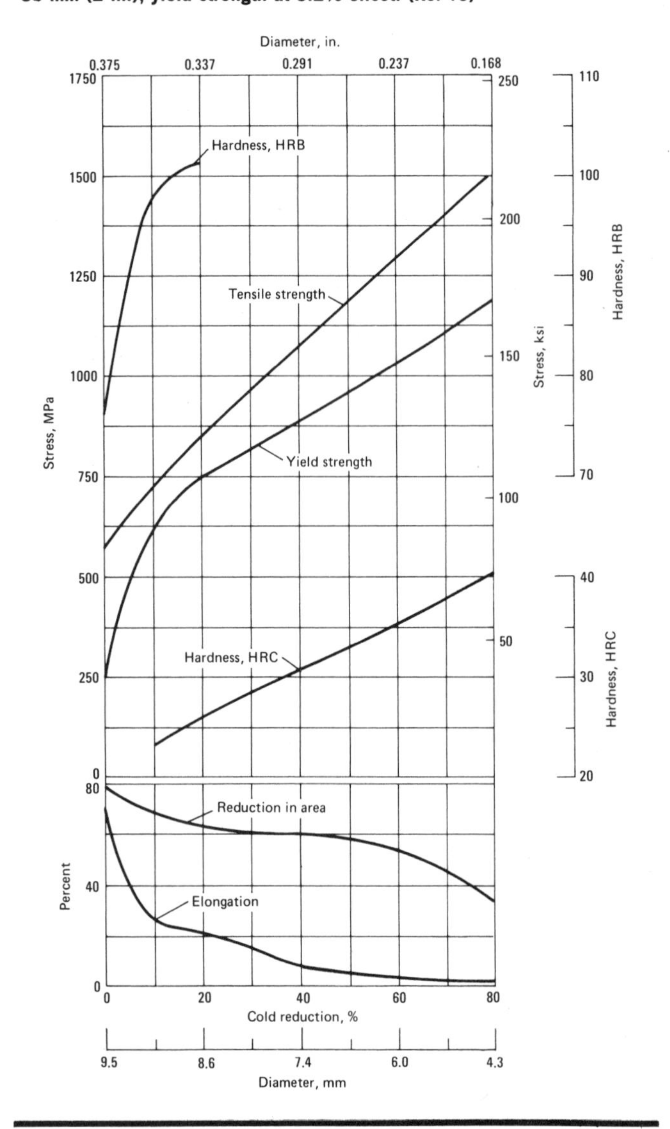

AISI Type 347: Effect of Sensitization on Impact Properties.

Annealed steel sensitized at 650 °C (1200 °F) for 2 h; 0.03% max carbon; Charpy keyhole notch specimens. (Ref 4)

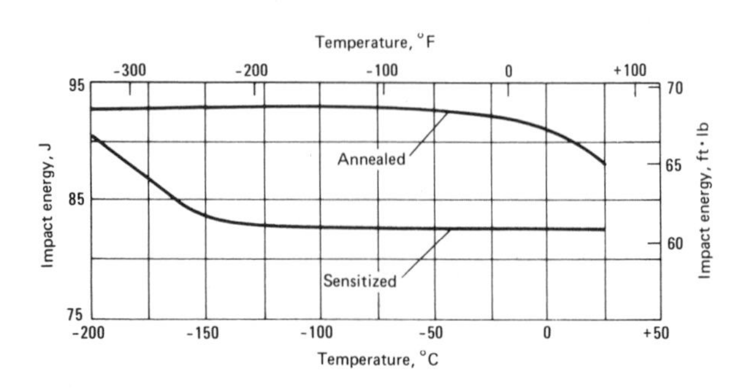

AISI Type 347: Rupture Strength of Annealed Steel (Ref 8)

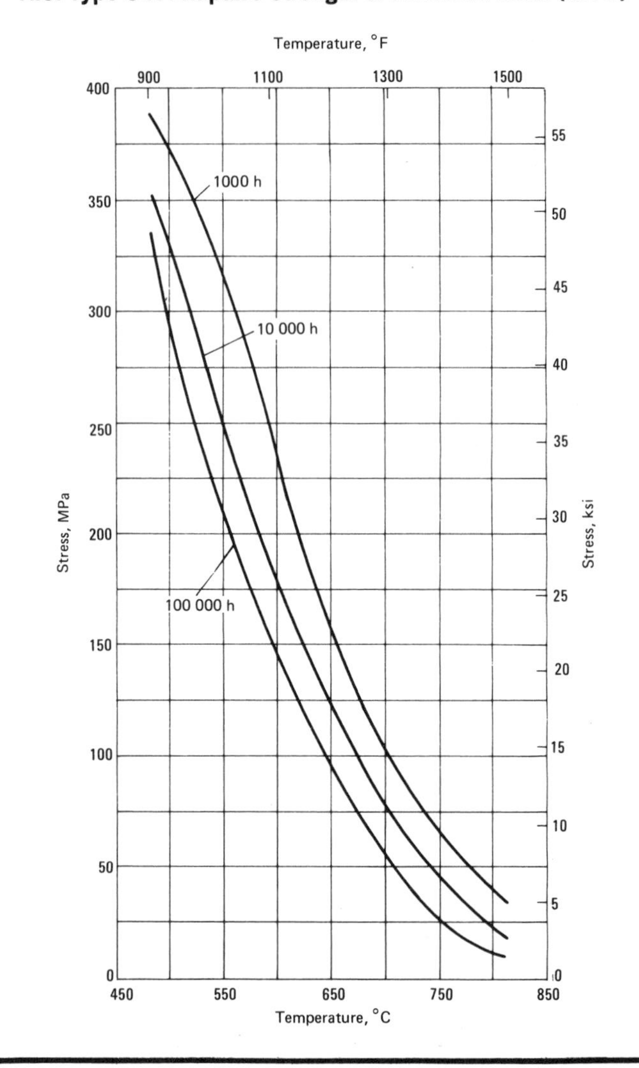

AISI Type 347: Effect of Sigmatizing on Impact Properties.

Sigmatized at 730 to 900 °C (1350 to 1650 °F); carbon content of 0.02%. Impact energy tests used Charpy keyhole notch specimens. (Ref 4)

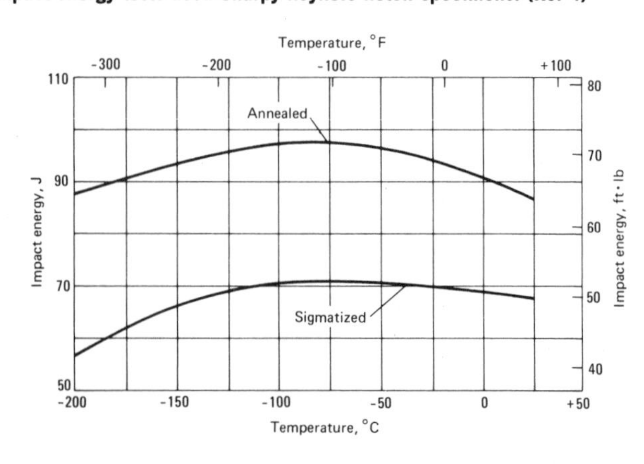

AISI Type 347: Effect of Test Temperature on Tensile Properties of Cold Worked Material. Specimens were cold drawn to 10% reduction. Elongation was measured in 50 mm (2 in.). (Ref 3)

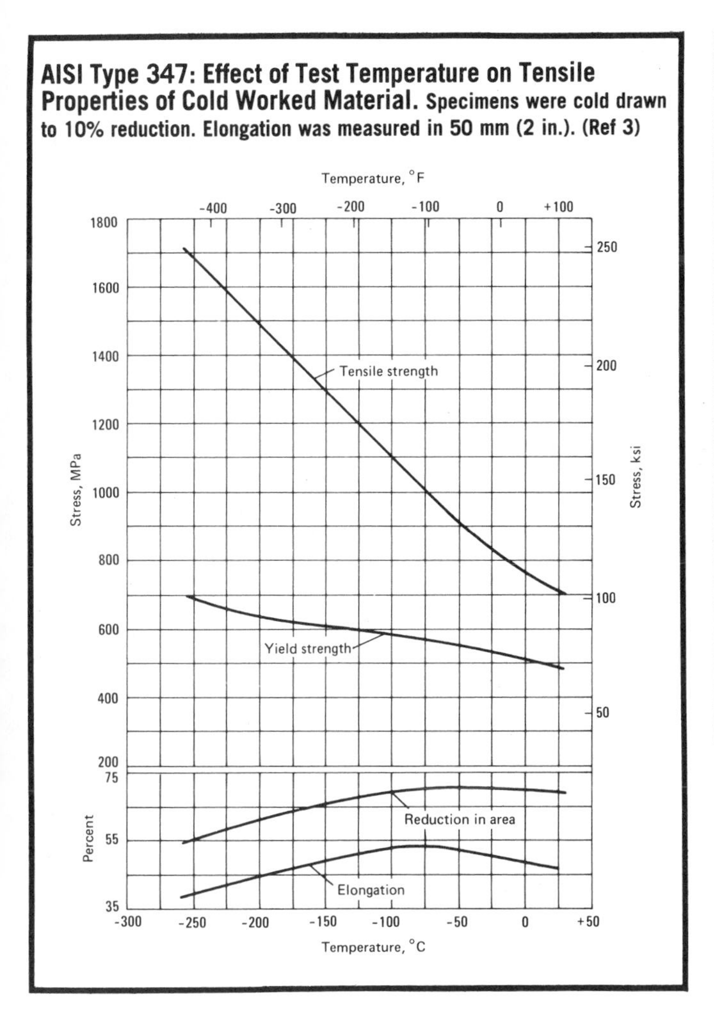

AISI Type 347: Creep Strength of Annealed Steel (Ref 8)

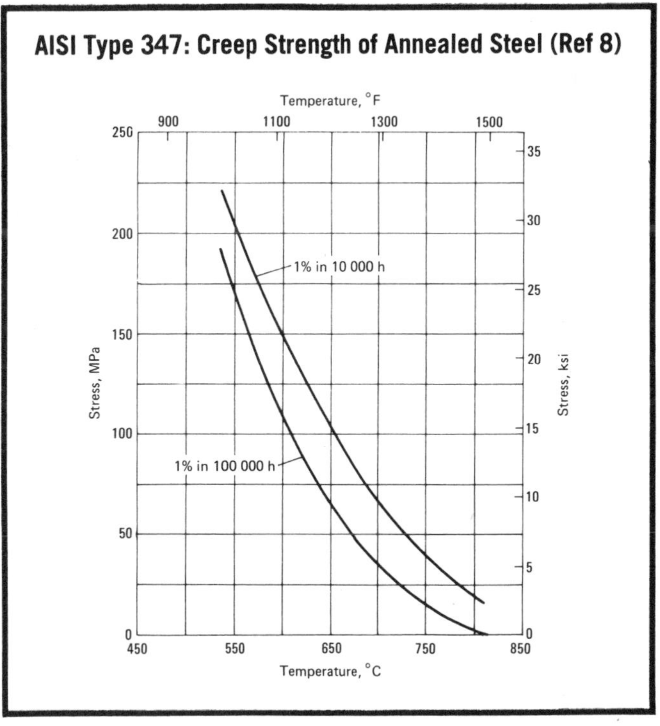

AISI Type 384

AISI Type 384: Chemical Composition

Element	Composition, max %
Carbon	0.080
Manganese	2.0
Phosphorus	0.045
Sulfur	0.030
Silicon	1.0
Chromium	15-17
Nickel	17-19

Characteristics. AISI type 384 is a chromium-nickel, austenitic, stainless steel which has a lower cold work-hardening rate than type 305. It is used when severe cold heading of cold forming operations are required.

Type 384 resists nitric acid well and sulfuric acid moderately, but will not resist hydrochloric and other halogen acids. It is subject to intergranular corrosion if heated or cooled slowly in the 425 to 900 °C (800 to 1650 °F) range. If intergranular corrosion occurs, the steel should have a corrective anneal by water quenching from at least 1040 °C (1900 °F). The safe scaling temperature for type 384 in continuous service is 870 °C (1600 °F).

This alloy can be readily cold worked by blanking, forming, extruding, coining, and swaging. In the annealed condition, type 384 machines with a tough, stringy chip. Machinability of type 384 is about 40% of that for AISI 1212. Cold headed parts can be machined more easily because of the hardness developed by cold working.

Type 384 can be brazed or welded using any of the electric welding processes. If filler metal is required, type 308 is suggested. Annealing must follow welding to prevent intergranular corrosion.

Typical Uses. Applications of AISI type 384 include fasteners; cold headed bolts; screws, upset nuts, and instrument parts; and parts involving severe coining, extrusion, or swaging.

AISI Type 384: Similar Steels (U.S. and/or Foreign). UNS S38400; ASTM A493; SAE J405 (30384)

Physical Properties

AISI Type 384: Thermal Treatment Temperatures

Treatment	Temperature range °C	°F
Forging, start	1150-1230	2100-2250
Forging, finish	925 min	1700 min
Annealing(a)	1040-1150	1900-2100
Hardening(b)	...	...

(a) Water quench. (b) Hardenable by cold work only

AISI Type 384: Selected Physical Properties (Ref 6)

Melting range	1400-1455 °C (2550-2650 °F)
Density	8.0 g/cm³ (0.29 lb/in.³)
Specific heat at 0-100 °C (32-212 °F)	500 J/kg·K (0.12 Btu/lb·°F)
Electrical resistivity at 20 °C (68 °F)	79.1 nΩ·m

AISI Type 384: Average Coefficients of Linear Thermal Expansion (Ref 6)

Temperature range °C	°F	Coefficient µm/m·K	µin./in.·°F
0-100	32-212	17.3	9.6
0-260	32-500	17.8	9.9
0-540	32-1000	18.4	10.2
0-650	32-1200	18.7	10.4

AISI Type 384: Thermal Conductivity (Ref 6)

Temperature °C	°F	Thermal conductivity W/m·K	Btu/ft·h·°F
100	212	16.3	9.4
500	930	21.5	12.4

Mechanical Properties

AISI Type 384: Representative Tensile Properties (Ref 6)

Test specimens were wire

Treatment	Diameter mm	in.	Tensile strength MPa	ksi	Yield strength(a) MPa	ksi	Elongation(b), %	Reduction in area, %	Hardness
Annealed at 1040 °C (1900 °F)	13.0	0.50	515	75	240	35	55	72	70 HRB
Lightly drafted, as for cold heading wire	5.08	0.20	540	78	310	45	...	...	...

(a) 0.2% offset. (b) Gage length is 4 × diam for 3.18-mm (0.125-in.) diam and larger; for smaller diameters, the gage length is 254 mm (10.0 in.)

AISI Type 384: Tensile Properties at Elevated Temperatures (Ref 9)

Test temperature °C	°F	Tensile strength MPa	ksi	Yield strength(a) MPa	ksi	Elongation(b), %	Reduction in area, %	Hardness(c), HB
Room temperature		510	74	205	30	52.5	74.6	126
120	250	470	68	170	25	49.4	75.4	126
260	500	415	60	145	21	47.3	73.1	126
400	750	405	59	150	22	45.8	70.2	126
540	1000	385	56	...	...	43.2	70.2	126

(a) 0.2% offset. (b) In 50 mm (2 in.). (c) Before test

Machining Data

For machining data on AISI type 384, refer to the preceding machining tables for AISI type 201.

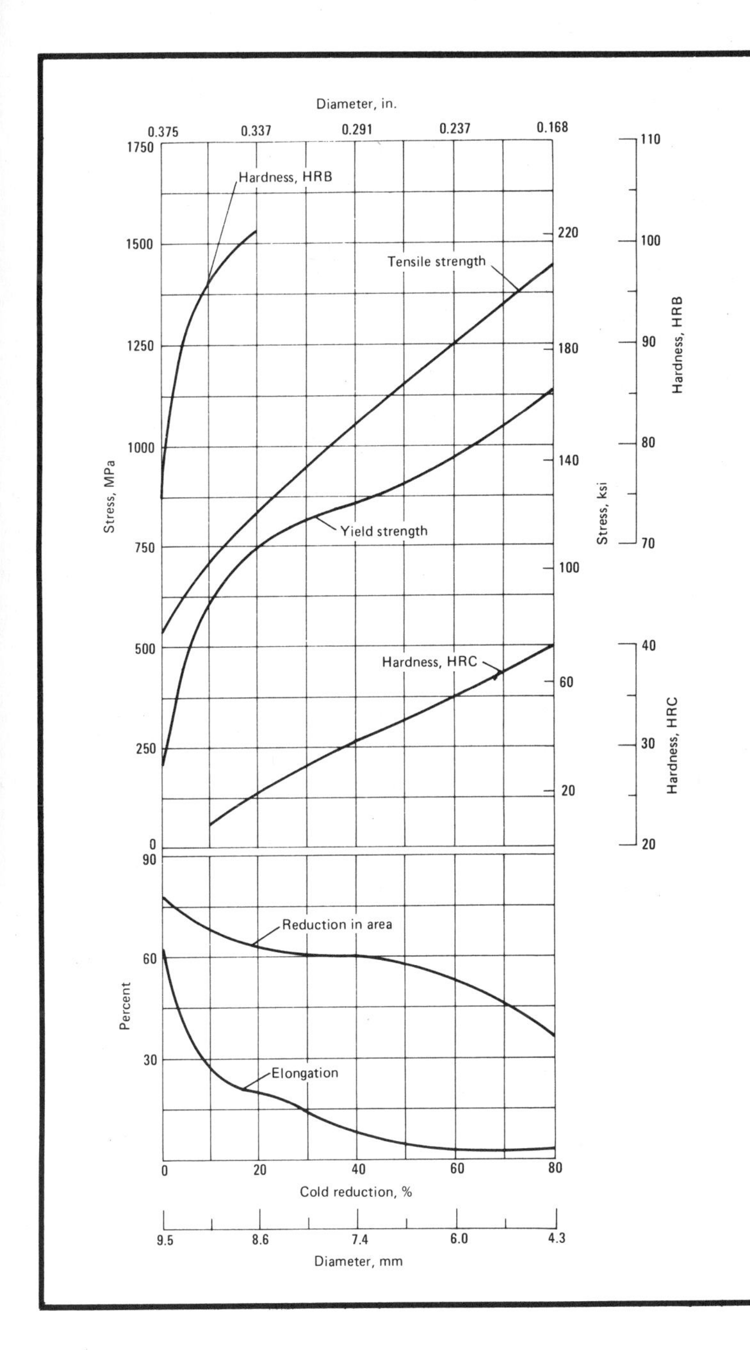

AISI Type 384: Effect of Cold Reduction on Tensile Properties. Annealed at 1065 °C (1950 °F); water quenched. Specimens were 9.53-mm (0.375-in.) round, unstraightened and untempered. Tests were conducted using specimens machined to English units. Elongation was measured in 50 mm (2 in.); yield strength at 0.2% offset. (Ref 10)

AISI Type 403

AISI Type 403: Chemical Composition

Element	Composition, max %
Carbon	0.15
Manganese	1.00
Phosphorus	0.04
Sulfur	0.03
Silicon	0.50
Chromium	11.5-13.0

Characteristics. AISI type 403 is hardenable (martensitic), chromium stainless steel. This selected quality steel is manufactured for the production of turbine blades and very highly stressed parts. It is magnetic in all conditions.

The type 403 alloy may be cold worked by practically all methods of cold forming. Severe forming should be followed by a stress relief to prevent cracking. The chips formed in machining are brittle and stringy. For most applications, material that has been heat treated to hardnesses of 200 to 240 HB, or annealed and cold drawn, is preferred to material in the fully annealed condition. Machinability of type 403 is about 55% of that for free-machining AISI 1212.

This stainless steel can be satisfactorily welded, but is susceptible to air hardening and cracking unless the material is preheated to a minimum temperature of 150 °C (300 °F) prior to welding. When welding type 403, filler metal type 410 should be used.

Type 403 is adaptable to applications where maximum resistance to corrosion is not required. It is resistant to the corrosive action of the atmosphere, fresh water and steam, and the milder acids and alkalis. This grade has good scaling resistance up to 675 °C (1250 °F) in continuous service. Scaling becomes excessive above about 760 °C (1400 °F).

Typical Uses. Applications of AISI type 403 include steam turbine blades and parts; gas turbine blades and jet engine parts; valve parts; cutlery; fasteners; furnace parts; and burners operating below 650 °C (1200 °F).

AISI Type 403: Similar Steels (U.S. and/or Foreign). UNS S40300; AMS 5611, 5612; ASTM A176, A276, A314, A473, A479, A511, A580; FED QQ-S-763; MIL SPEC MIL-S-862; SAE J405 (51403); (W. Ger.) DIN 1.4024; (Fr.) AFNOR Z 12 C 13 M; (Jap.) JIS SUS 416, SUS 403; (U.K.) B.S. 420 S 29, En. 56 B

Physical Properties

AISI Type 403: Thermal Treatment Temperatures

Treatment	Temperature range °C	°F
Forging, start	1095-1205	2000-2200
Forging, finish	760 min	1400 min
Full annealing(a)	815-900	1500-1650
Process annealing(b)	650-760	1200-1400
Hardening(c)	925-1010	1700-1850
Tempering	205-760	400-1400

(a) Cool slowly or furnace cool. (b) Air cool. (c) Air cool or oil quench

AISI Type 403: Thermal Conductivity (Ref 6)

Temperature °C	°F	Thermal conductivity W/m·K	Btu/ft·h·°F
100	212	24.9	14.4
500	930	28.7	16.6

AISI Type 403: Selected Physical Properties (Ref 6)

Melting range	1480-1530 °C (2700-2790 °F)
Density	7.8 g/cm^3 (0.28 lb/in.^{3})
Specific heat at 0-100 °C (32-212 °F)	460 J/kg·K (0.11 Btu/lb·°F)
Electrical resistivity at 20 °C (68 °F)	57 nΩ·m

AISI Type 403: Average Coefficients of Linear Thermal Expansion (Ref 6)

Temperature range °C	°F	Coefficient μm/m·K	μin./in.·°F
0-100	32-212	9.9	5.5
0-315	32-600	11.3	6.3
0-540	32-1000	11.5	6.4
0-650	32-1200	11.7	6.5

Mechanical Properties

AISI Type 403: Representative Tensile Properties (Ref 6)

Product form	Treatment	Tensile strength MPa	ksi	Yield strength(a) MPa	ksi	Elongation(b), %	Reduction in area, %	Hardness, HRB
Sheet	Annealed	485	70	310	45	25	...	80
Strip	Annealed	485	70	310	45	25	...	80
Wire, 6.4-mm (0.25-in.) diam	Soft tempered	655	95	550	80	15	60	92
Bar	Annealed	515	75	275	40	35	70	82
	Tempered	765	111	585	85	23	67	97

(a) 0.2% offset. (b) In 50 mm (2 in.), except for wire. For wire 3.18-mm (0.125-in.) diam and larger, gage length is 4 × diam; for smaller diameters, the gage length is 254 mm (10.0 in.)

AISI Type 403: Endurance Limit

Test specimens were 25-mm (1.0-in.) diam

Condition	Stress MPa	ksi
Annealed	275	40
Tempered	380	55

AISI Type 403: Modulus of Elasticity (Ref 10)

Stressed in:	Modulus GPa	10^6 psi
Tension	200	29
Torsion	76	11

AISI Type 403: Stress-Rupture and Creep Properties (Ref 10)

Test temperature °C	°F	Stress for rupture in: 1000 h MPa	1000 h ksi	10 000 h MPa	10 000 h ksi	Stress for creep rate of 1% in 10 000 h MPa	ksi
425	800	...	...	...	...	296	43.0
480	900	234	34.0	179	26.0	200	29.0
540	1000	131	19.0	90	13.0	63	9.2
595	1100	69	10.0	48	6.9	29	4.2
650	1200	...	...	...	...	14	2.0
705	1300	...	...	...	...	7	1.0

AISI Type 403: Izod Impact Energy

Test specimens were 25-mm (1.0-in.) diam

Condition	Energy J	ft·lb
Annealed	122	90
Tempered	102	75

Machining Data (Ref 2)

AISI Type 403: Planing

Tool material	Depth of cut mm	Depth of cut in.	Speed m/min	Speed ft/min	Feed/stroke mm	Feed/stroke in.
Annealed, with hardness of 135 to 175 HB						
M2 and M3 high speed steel	0.1	0.005	9	30	(a)	(a)
	2.5	0.100	12	40	1.25	0.050
	12	0.500	8	25	1.50	0.060
C-6 carbide	0.1	0.005	46	150	(a)	(a)
	2.5	0.100	49	160	2.05	0.080
Annealed, with hardness of 175 to 225 HB						
M2 and M3 high speed steel	0.1	0.005	8	25	(a)	(a)
	2.5	0.100	11	35	1.25	0.050
	12	0.500	8	25	1.50	0.060
C-6 carbide	0.1	0.005	43	140	(a)	(a)
	2.5	0.100	46	150	2.05	0.080
Quenched and tempered, with hardness of 275 to 325 HB						
M2 and M3 high speed steel	0.1	0.005	5	15	(a)	(a)
	2.5	0.100	8	25	0.75	0.030
	12	0.500	5	15	1.15	0.045
C-6 carbide	0.1	0.005	38	125	(a)	(a)
	2.5	0.100	43	140	1.50	0.060

(a) Feed is 75% the width of the square nose finishing tool

AISI Type 403: Effect of Tempering Temperature on Tensile Properties. Heat treated at 980 °C (1800 °F); oil quenched; tempered for 3 h at temperature given. Test specimens were heat treated in 25-mm (1.0-in.) round; tensile specimens were 13-mm (0.505-in.) diam; Izod notched specimens were 10.0-mm (0.394-in.) square. Tests were conducted using specimens machined to English units. Impact energy test used Izod specimens. Elongation was measured in 50 mm (2 in.); yield strength at 0.2% offset. (Ref 10)

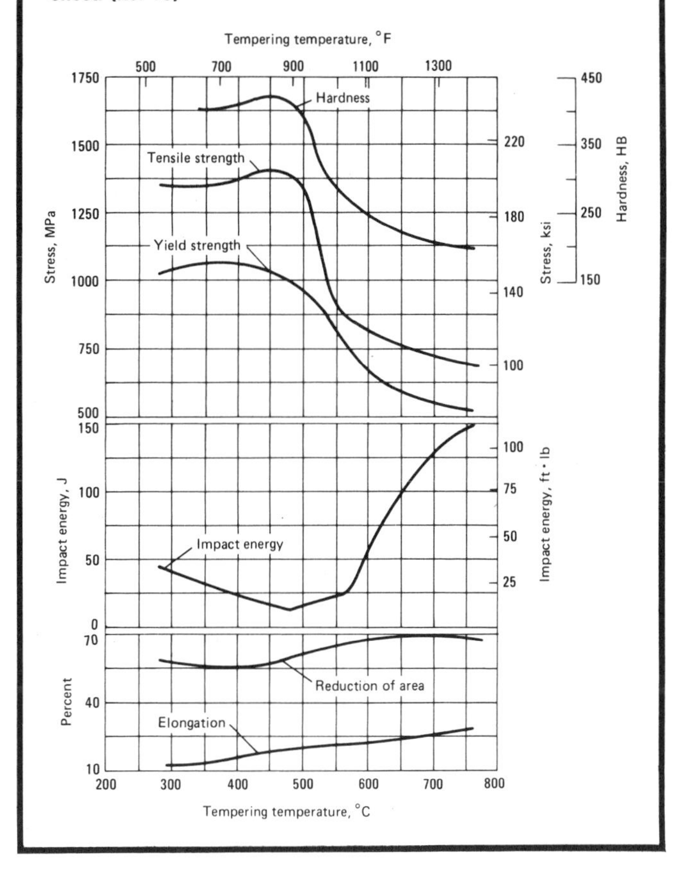

AISI Type 403: Turning (Single Point and Box Tools)

Depth of cut mm	Depth of cut in.	High speed steel Speed m/min	High speed steel Speed ft/min	High speed steel Feed mm/rev	High speed steel Feed in./rev	Uncoated carbide Speed, brazed m/min	Uncoated carbide Speed, brazed ft/min	Uncoated carbide Speed, inserted m/min	Uncoated carbide Speed, inserted ft/min	Uncoated carbide Feed mm/rev	Uncoated carbide Feed in./rev	Coated carbide Speed m/min	Coated carbide Speed ft/min	Coated carbide Feed mm/rev	Coated carbide Feed in./rev
Annealed, with hardness of 135 to 175 HB															
1	0.040	47(a)	155(a)	0.18	0.007	145(b)	475(b)	190(b)	620(b)	0.18	0.007	245(c)	800(c)	0.18	0.007
4	0.150	38(a)	125(a)	0.40	0.015	120(d)	400(d)	145(d)	480(d)	0.40	0.015	190(e)	625(e)	0.40	0.015
8	0.300	30(a)	100(a)	0.50	0.020	100(d)	320(d)	115(d)	380(d)	0.50	0.030	150(e)	500(e)	0.50	0.020
16	0.625	24(a)	80(a)	0.75	0.030	73(d)	240(d)	90(d)	300(d)	0.75	0.040	...	...	...	...
Annealed, with hardness of 175 to 225 HB															
1	0.040	44(a)	145(a)	0.18	0.007	140(b)	460(b)	175(b)	570(b)	0.18	0.007	260(c)	850(c)	0.18	0.007
4	0.150	35(a)	115(a)	0.40	0.015	115(d)	385(d)	135(d)	450(d)	0.40	0.015	170(e)	550(e)	0.40	0.015
8	0.300	27(a)	90(a)	0.50	0.020	90(d)	300(d)	105(d)	350(d)	0.75	0.030	135(e)	450(e)	0.50	0.020
16	0.625	21(a)	70(a)	0.75	0.030	72(d)	235(d)	81(d)	265(d)	1.00	0.040	...	...	...	...
Quenched and tempered, with hardness of 275 to 325 HB															
1	0.040	29(f)	95(f)	0.18	0.007	110(b)	360(b)	140(b)	465(b)	0.18	0.007	215(c)	700(c)	0.18	0.007
4	0.150	23(f)	75(f)	0.40	0.015	85(d)	280(d)	110(d)	360(d)	0.40	0.015	135(e)	450(e)	0.40	0.015
8	0.300	18(f)	60(f)	0.50	0.020	69(d)	225(d)	85(d)	280(d)	0.50	0.020	115(e)	375(e)	0.50	0.020
Quenched and tempered, with hardness of 375 to 425 HB															
1	0.040	20(f)	65(f)	0.18	0.007	88(b)	290(b)	100(b)	320(b)	0.18	0.007	145(c)	475(c)	0.18	0.007
4	0.150	15(f)	50(f)	0.40	0.015	69(d)	225(d)	76(d)	250(d)	0.40	0.015	90(e)	300(e)	0.40	0.015
8	0.300	12(f)	40(f)	0.50	0.020	55(d)	180(d)	60(d)	200(d)	0.50	0.020	76(e)	250(e)	0.50	0.020

(a) High speed steel tool material: M2 or M3. (b) Carbide tool material: C-7. (c) Carbide tool material: CC-7. (d) Carbide tool material: C-6. (e) Carbide tool material: CC-6. (f) Any premium high speed steel tool material (T15, M33, M41 to M47)

Machining Data (Ref 2) (continued)

AISI Type 403: Turning (Cutoff and Form Tools)

Tool material	Speed, m/min (ft/min)	Feed per revolution for cutoff tool width of: 1.5 mm (0.062 in.) mm (in.)	3 mm (0.125 in.) mm (in.)	6 mm (0.25 in.) mm (in.)	Feed per revolution for form tool width of: 12 mm (0.5 in.) mm (in.)	18 mm (0.75 in.) mm (in.)	25 mm (1 in.) mm (in.)	35 mm (1.5 in.) mm (in.)	50 mm (2 in.) mm (in.)
Annealed, with hardness of 135 to 175 HB									
M2 and M3 high speed steel	29 (95)	0.038 (0.0015)	0.050 (0.002)	0.063 (0.0025)	0.050 (0.002)	0.046 (0.0018)	0.038 (0.0015)	0.030 (0.0012)	0.025 (0.001)
C-6 carbide	100 (325)	0.038 (0.0015)	0.050 (0.002)	0.063 (0.0025)	0.050 (0.002)	0.046 (0.0018)	0.038 (0.0015)	0.030 (0.0012)	0.025 (0.001)
Annealed, with hardness of 175 to 225 HB									
M2 and M3 high speed steel	26 (85)	0.038 (0.0015)	0.050 (0.002)	0.063 (0.0025)	0.050 (0.002)	0.046 (0.0018)	0.038 (0.0015)	0.030 (0.0012)	0.025 (0.001)
C-6 carbide	90 (300)	0.038 (0.0015)	0.050 (0.002)	0.063 (0.0025)	0.050 (0.002)	0.046 (0.0018)	0.038 (0.0015)	0.030 (0.0012)	0.025 (0.001)
Quenched and tempered, with hardness of 275 to 325 HB									
Any premium high speed steel (T15, M33, M41-M47)	17 (55)	0.025 (0.001)	0.030 (0.0012)	0.038 (0.0015)	0.038 (0.0015)	0.030 (0.0012)	0.025 (0.001)	0.023 (0.0009)	0.020 (0.0008)
C-6 carbide	56 (185)	0.025 (0.001)	0.030 (0.0012)	0.038 (0.0015)	0.038 (0.0015)	0.030 (0.0012)	0.025 (0.001)	0.023 (0.0009)	0.020 (0.0008)
Quenched and tempered, with hardness of 375 to 425 HB									
Any premium high speed steel (T115, M33, M41-M47)	12 (40)	0.020 (0.0008)	0.025 (0.001)	0.030 (0.0012)	0.030 (0.0012)	0.025 (0.001)	0.023 (0.0009)	0.020 (0.0008)	0.018 (0.0007)
C-6 carbide	63 (175)	0.020 (0.0008)	0.025 (0.001)	0.030 (0.0012)	0.030 (0.0012)	0.025 (0.001)	0.023 (0.0009)	0.020 (0.0008)	0.018 (0.0007)

AISI Type 403: Boring

Depth of cut		High speed steel				Uncoated carbide						Coated carbide			
		Speed		Feed		Speed, brazed		Speed, inserted		Feed		Speed		Feed	
mm	in.	m/min	ft/min	mm/rev	in./rev	m/min	ft/min	m/min	ft/min	mm/rev	in./rev	m/min	ft/min	mm/rev	in./rev
Annealed, with hardness of 135 to 175 HB															
0.25	0.010	47(a)	155(a)	0.075	0.003	145(b)	470(b)	170(b)	550(b)	0.075	0.003	210(c)	685(c)	0.075	0.003
1.25	0.050	38(a)	125(a)	0.13	0.005	115(d)	370(d)	135(d)	435(d)	0.13	0.005	170(e)	550(e)	0.13	0.005
2.50	0.100	30(a)	100(a)	0.30	0.012	87(f)	285(f)	100(f)	335(f)	0.30	0.012	135(g)	440(g)	0.30	0.012
Annealed, with hardness of 175 to 225 HB															
0.25	0.010	44(a)	145(a)	0.075	0.003	130(d)	425(d)	150(d)	500(d)	0.075	0.003	230(e)	750(e)	0.075	0.003
1.25	0.050	35(a)	115(a)	0.13	0.005	105(f)	340(f)	120(f)	400(f)	0.13	0.005	185(g)	600(g)	0.13	0.005
2.50	0.100	27(a)	90(a)	0.30	0.012	81(f)	265(f)	95(f)	315(f)	0.30	0.012	115(g)	375(g)	0.30	0.012
Quenched and tempered, with hardness of 275 to 325 HB															
0.25	0.010	29(h)	95(h)	0.075	0.003	105(d)	345(d)	125(d)	405(d)	0.075	0.003	190(e)	625(e)	0.075	0.003
1.25	0.050	23(h)	75(h)	0.13	0.005	84(f)	275(f)	100(f)	325(f)	0.13	0.005	150(g)	500(g)	0.13	0.005
2.50	0.100	18(h)	60(h)	0.30	0.012	66(f)	215(f)	76(f)	250(f)	0.30	0.012	100(g)	325(g)	0.30	0.012
Quenched and tempered, with hardness of 375 to 425 HB															
0.25	0.010	20(h)	65(h)	0.075	0.003	72(d)	235(d)	84(d)	280(d)	0.075	0.003	120(e)	400(e)	0.075	0.003
1.25	0.050	15(h)	50(h)	0.13	0.005	58(f)	190(f)	69(f)	225(f)	0.13	0.005	100(g)	325(g)	0.13	0.005
2.50	0.100	12(h)	40(h)	0.30	0.012	46(f)	150(f)	53(f)	175(f)	0.30	0.012	60(g)	200(g)	0.30	0.012

(a) High speed steel tool material: M2 or M3. (b) Carbide tool material: C-8. (c) Carbide tool material: CC-8. (d) Carbide tool material: C-7. (e) Carbide tool material: CC-7. (f) Carbide tool material: C-6. (g) Carbide tool material: CC-6. (h) Any premium high speed steel tool material (T15, M33, M41 to M47)

AISI Type 403: End Milling (Profiling)

Tool material	Depth of cut		Speed		Feed per tooth for cutter diameter of: 10 mm (0.375 in.)		12 mm (0.5 in.)		18 mm (0.75 in.)		25-50 mm (1-2 in.)	
	mm	in.	m/min	ft/min	mm	in.	mm	in.	mm	in.	mm	in.
Annealed, with hardness of 135 to 175 HB												
M2, M3, and M7 high speed steel	0.5	0.020	44	145	0.025	0.001	0.050	0.002	0.10	0.004	0.13	0.005
	1.5	0.060	34	110	0.050	0.002	0.075	0.003	0.13	0.005	0.15	0.006
	diam/4	diam/4	29	95	0.038	0.0015	0.050	0.002	0.102	0.004	0.13	0.005
	diam/2	diam/2	26	85	0.025	0.001	0.038	0.0015	0.075	0.003	0.102	0.004
C-5 carbide	0.5	0.020	140	455	0.013	0.0005	0.025	0.001	0.075	0.003	0.13	0.005
	1.5	0.060	105	350	0.025	0.001	0.050	0.002	0.102	0.004	0.15	0.006
	diam/4	diam/4	90	295	0.025	0.001	0.038	0.0015	0.063	0.0025	0.102	0.004
	diam/2	diam/2	84	275	0.013	0.0005	0.025	0.001	0.050	0.002	0.075	0.003
Annealed, with hardness of 175 to 225 HB												
M2, M3, and M7 high speed steel	0.5	0.020	40	130	0.025	0.001	0.050	0.002	0.075	0.003	0.102	0.004
	1.5	0.060	30	100	0.050	0.002	0.075	0.003	0.102	0.004	0.13	0.005
	diam/4	diam/4	27	90	0.025	0.001	0.050	0.002	0.075	0.003	0.102	0.004
	diam/2	diam/2	23	75	0.025	0.001	0.025	0.001	0.050	0.002	0.075	0.003
C-5 carbide	0.5	0.020	120	390	0.013	0.0005	0.025	0.001	0.075	0.003	0.13	0.005
	1.5	0.060	90	300	0.025	0.001	0.050	0.002	0.102	0.004	0.15	0.006
	diam/4	diam/4	78	255	0.025	0.001	0.038	0.0015	0.075	0.003	0.102	0.004
	diam/2	diam/2	72	235	0.013	0.0005	0.025	0.001	0.050	0.002	0.075	0.003
Quenched and tempered, with hardness of 275 to 325 HB												
M2, M3, and M7 high speed steel	0.5	0.020	24	80	0.018	0.0007	0.025	0.001	0.050	0.002	0.075	0.003
	1.5	0.060	18	60	0.025	0.001	0.050	0.002	0.075	0.003	0.102	0.004
	diam/4	diam/4	15	50	0.018	0.0007	0.025	0.001	0.050	0.002	0.075	0.003
	diam/2	diam/2	14	45	0.013	0.0005	0.018	0.0007	0.038	0.0015	0.038	0.0015
C-5 carbide	0.5	0.020	90	300	0.013	0.0005	0.025	0.001	0.075	0.003	0.13	0.005
	1.5	0.060	70	230	0.025	0.001	0.050	0.002	0.102	0.004	0.15	0.006
	diam/4	diam/4	60	200	0.025	0.001	0.038	0.0015	0.075	0.003	0.102	0.004
	diam/2	diam/2	55	180	0.013	0.0005	0.025	0.001	0.050	0.002	0.075	0.003
Quenched and tempered, with hardness of 375 to 425 HB												
Any premium high speed steel tool material (T15, M33, M41-M47)	0.5	0.020	20	65	0.013	0.0005	0.018	0.0007	0.025	0.001	0.050	0.002
	1.5	0.060	17	55	0.013	0.0005	0.025	0.001	0.050	0.002	0.075	0.003
	diam/4	diam/4	15	50	0.013	0.0005	0.013	0.0005	0.038	0.0015	0.063	0.0025
	diam/2	diam/2	12	40	...	...	...	...	0.025	0.001	0.050	0.002
C-5 carbide	0.5	0.020	76	250	0.013	0.0005	0.025	0.001	0.050	0.002	0.075	0.003
	1.5	0.060	58	190	0.025	0.001	0.050	0.002	0.075	0.003	0.102	0.004
	diam/4	diam/4	50	165	0.025	0.001	0.038	0.0015	0.063	0.0025	0.075	0.003
	diam/2	diam/2	46	150	0.013	0.0005	0.025	0.001	0.050	0.002	0.063	0.0025

AISI Type 403: Face Milling

Depth of cut		High speed steel: Speed		High speed steel: Feed/tooth		Uncoated carbide: Speed, brazed		Uncoated carbide: Speed, inserted		Uncoated carbide: Feed/tooth		Coated carbide: Speed		Coated carbide: Feed/tooth	
mm	in.	m/min	ft/min	mm	in.	m/min	ft/min	m/min	ft/min	mm	in.	m/min	ft/min	mm	in.
Annealed, with hardness of 135 to 175 HB															
1	0.040	49(a)	160(a)	0.20	0.008	150(b)	500(b)	165(b)	540(b)	0.20	0.008	245(c)	800(c)	0.20	0.008
4	0.150	38(a)	125(a)	0.30	0.012	120(b)	390(b)	145(b)	480(b)	0.30	0.012	160(c)	525(c)	0.30	0.012
8	0.300	30(a)	100(a)	0.40	0.016	66(d)	215(d)	79(d)	260(d)	0.40	0.016	100(e)	325(e)	0.40	0.016
Annealed, with hardness of 175 to 225 HB															
1	0.040	44(a)	145(a)	0.20	0.008	145(b)	480(b)	155(b)	510(b)	0.20	0.008	235(c)	765(c)	0.20	0.008
4	0.150	35(a)	115(a)	0.30	0.012	110(b)	365(b)	115(b)	380(b)	0.30	0.012	150(c)	500(c)	0.30	0.012
8	0.300	27(a)	90(a)	0.40	0.016	73(d)	240(d)	90(d)	295(d)	0.40	0.016	115(e)	375(e)	0.40	0.016
Quenched and tempered, with hardness of 275 to 325 HB															
1	0.040	30(f)	100(f)	0.15	0.006	115(b)	370(b)	115(b)	410(b)	0.15	0.006	190(c)	625(c)	0.13	0.005
4	0.150	23(f)	75(f)	0.23	0.009	85(b)	280(b)	85(b)	310(b)	0.20	0.008	120(c)	400(c)	0.18	0.007
8	0.300	18(f)	60(f)	0.30	0.012	59(d)	195(d)	73(d)	240(d)	0.25	0.010	95(e)	315(e)	0.23	0.009
Quenched and tempered, with hardness of 275 to 325 HB															
1	0.040	18(f)	60(f)	0.10	0.004	55(b)	180(b)	60(b)	200(b)	0.10	0.004	90(c)	300(c)	0.075	0.003
4	0.150	14(f)	45(f)	0.15	0.006	41(b)	135(b)	46(b)	150(b)	0.15	0.006	60(c)	200(c)	0.13	0.005
8	0.300	11(f)	35(f)	0.20	0.008	29(d)	95(d)	35(d)	115(d)	0.20	0.008	46(e)	150(e)	0.18	0.007

(a) High speed steel tool material: M2 or M7. (b) Carbide tool material: C-6. (c) Carbide tool material: CC-6. (d) Carbide tool material: C-5. (e) Carbide tool material: CC-5. (f) Any premium high speed steel tool material (T15, M33, M41 to M47)

Machining Data (Ref 2) (continued)

AISI Type 403: Drilling

Tool material	Speed, m/min (ft/min)	Feed per revolution for nominal hole diameter of: 1.5 mm (0.062 in.) mm (in.)	3 mm (0.125 in.) mm (in.)	6 mm (0.25 in.) mm (in.)	12 mm (0.5 in.) mm (in.)	18 mm (0.75 in.) mm (in.)	25 mm (1 in.) mm (in.)	35 mm (1.5 in.) mm (in.)	50 mm (2 in.) mm (in.)
Annealed, with hardness of 135 to 175 HB									
M10, M7, and high speed steel	17 (55)	0.025 (0.001)	... (...)	... (...)	... (...)	... (...)	... (...)	... (...)	... (...)
	23 (75)	... (...)	0.075 (0.003)	0.15 (0.006)	0.25 (0.010)	0.33 (0.013)	0.40 (0.016)	0.55 (0.021)	0.65 (0.025)
Annealed, with hardness of 175 to 225 HB									
M10, M7, and M1 high speed steel	20 (65)	0.025 (0.001)	0.075 (0.003)	0.15 (0.006)	0.25 (0.010)	0.33 (0.013)	0.40 (0.016)	0.55 (0.021)	0.65 (0.025)
Quenched and tempered, with hardness of 275 to 325 HB									
M10, M7, and M1 high speed steel	17 (55)	... (...)	0.075 (0.003)	0.102 (0.004)	0.18 (0.007)	0.25 (0.010)	0.30 (0.012)	0.40 (0.015)	0.45 (0.018)
Quenched and tempered, with hardness of 375 to 425 HB									
Any premium high coated steel (T15, M33, M41-M47)	14 (45)	... (...)	0.025 (0.001)	0.050 (0.002)	0.102 (0.004)	0.102 (0.004)	0.102 (0.004)	0.102 (0.004)	0.102 (0.004)

AISI Type 403: Cylindrical and Internal Grinding

Operation	Wheel speed m/s	Wheel speed ft/min	Work speed m/min	Work speed ft/min	Infeed on diameter mm/pass	Infeed on diameter in./pass	Traverse(a)	Wheel identification
Annealed, with hardness of 135 to 275 HB								
Cylindrical grinding								
Roughing	28-33	5500-6500	15-30	50-100	0.050	0.002	1/4	A60JV
Finishing	28-33	5500-6500	15-30	50-100	0.013 max	0.0005 max	1/8	A60JV
Internal grinding								
Roughing	25-33	5000-6500	23-60	75-200	0.013	0.0005	1/3	A60KV
Finishing	25-33	5000-6500	23-60	75-200	0.005 max	0.0002 max	1/6	A60KV
Quenched and tempered, with hardness greater than 275 HB								
Cylindrical grinding								
Roughing	28-33	5500-6500	15-30	50-100	0.050	0.002	1/4	A60IV
Finishing	28-33	5500-6500	15-30	50-100	0.013 max	0.0005 max	1/8	A60IV
Internal grinding								
Roughing	25-33	5000-6500	15-46	50-150	0.013	0.0005	1/3	A60JV
Finishing	25-33	5000-6500	15-46	50-150	0.005 max	0.0002 max	1/6	A60JV

(a) Wheel width per revolution of work

AISI Type 403: Surface Grinding with a Horizontal Spindle and Reciprocating Table

Operation	Wheel speed m/s	Wheel speed ft/min	Table speed m/min	Table speed ft/min	Downfeed mm/pass	Downfeed in./pass	Crossfeed mm/pass	Crossfeed in./pass	Wheel identification
Annealed, with hardness of 135 to 275 HB									
Roughing	28-33	5500-6500	15-30	50-100	0.050	0.002	1.25-12.5	0.050-0.500	A46IV
Finishing	28-33	5500-6500	15-30	50-100	0.013 max	0.0005 max	(a)	(a)	A46IV
Quenched or tempered, with hardness greater than 275 HB									
Roughing	28-33	5500-6500	15-30	50-100	0.025	0.001	0.65-6.5	0.025-0.250	A46HV
Finishing	28-33	5500-6500	15-30	50-100	0.008 max	0.0003 max	(b)	(b)	A46HV

(a) Maximum 1/4 of wheel width. (b) Maximum 1/10 of wheel width

Machining Data (Ref 2) (continued)

AISI Type 403: Reaming

Based on 4 flutes for 3- and 6-mm (0.125- and 0.250-in.) reamers, 6 flutes for 12-mm (0.5-in.) reamers, and 8 flutes for 25-mm (1-in.) and larger reamers

Tool material	Speed m/min	Speed ft/min	Feed per revolution for reamer diameter of: 3 mm (0.125 in.) mm	3 mm (0.125 in.) in.	6 mm (0.25 in.) mm	6 mm (0.25 in.) in.	12 mm (0.5 in.) mm	12 mm (0.5 in.) in.	25 mm (1 in.) mm	25 mm (1 in.) in.	35 mm (1.5 in.) mm	35 mm (1.5 in.) in.	50 mm (2 in.) mm	50 mm (2 in.) in.
Annealed, with hardness of 135 to 175 HB														
Roughing														
M1, M2, and M7 high speed steel	24	80	0.075	0.003	0.102	0.004	0.13	0.005	0.20	0.008	0.25	0.010	0.30	0.012
C-2 carbide	29	95	0.102	0.004	0.20	0.008	0.30	0.012	0.40	0.016	0.50	0.020	0.60	0.024
Finishing														
M1, M2, and M7 high speed steel	12	40	0.102	0.004	0.102	0.004	0.15	0.006	0.20	0.008	0.23	0.009	0.25	0.010
C-2 carbide	18	60	0.102	0.004	0.15	0.006	0.20	0.008	0.25	0.010	0.28	0.011	0.30	0.012
Annealed, with hardness of 175 to 225 HB														
Roughing														
M1, M2, and M7 high speed steel	23	75	0.075	0.003	0.102	0.004	0.13	0.005	0.20	0.008	0.25	0.010	0.30	0.012
C-2 carbide	27	90	0.102	0.004	0.20	0.008	0.30	0.012	0.40	0.016	0.50	0.020	0.60	0.024
Finishing														
M1, M2, and M7 high speed steel	12	40	0.102	0.004	0.102	0.004	0.15	0.006	0.20	0.008	0.23	0.009	0.25	0.010
C-2 carbide	18	60	0.102	0.004	0.13	0.005	0.15	0.006	0.20	0.008	0.23	0.009	0.25	0.010
Quenched and tempered, with hardness of 275 to 325 HB														
Roughing														
M1, M2, and M7 high speed steel	18	60	0.075	0.003	0.102	0.004	0.15	0.005	0.20	0.008	0.25	0.010	0.30	0.012
C-2 carbide	23	75	0.102	0.004	0.15	0.006	0.23	0.009	0.40	0.015	0.45	0.018	0.50	0.020
Finishing														
M1, M2, and M7 high speed steel	11	35	0.075	0.003	0.075	0.003	0.102	0.004	0.15	0.006	0.18	0.007	0.20	0.008
C-2 carbide	15	50	0.075	0.003	0.102	0.004	0.13	0.005	0.15	0.006	0.18	0.007	0.20	0.008
Quenched and tempered, with hardness of 375 to 425 HB														
Roughing														
Any premium high speed steel (T15, M33, M41-M47)	14	45	0.050	0.002	0.102	0.004	0.13	0.005	0.20	0.008	0.25	0.010	0.30	0.012
C-2 carbide	18	60	0.075	0.003	0.13	0.005	0.15	0.006	0.20	0.008	0.25	0.010	0.30	0.012
Finishing														
Any premium high speed steel (T15, M33, M41-M47)	9	30	0.075	0.003	0.075	0.003	0.102	0.004	0.15	0.006	0.18	0.007	0.20	0.008
C-2 carbide	12	40	0.075	0.003	0.102	0.004	0.13	0.005	0.15	0.006	0.18	0.007	0.20	0.008

AISI Type 405

AISI Type 405: Chemical Composition

Element	Composition, %
Carbon	0.08 max
Manganese	1.00 max
Phosphorus	0.04 max
Sulfur	0.03 max
Silicon	1.00 max
Chromium	11.5-14.5
Aluminum	0.10-0.30

Characteristics. AISI type 405 is a nonhardening, 12% chromium stainless steel. A small addition of aluminum restricts the formation of austenite at high temperatures so that hardening does not occur upon quenching. This characteristic makes type 405 especially suited for welding. The welds are ductile, and there is no need for controlled cooling rates. This grade may be welded using chromium-nickel 18-8 types or type 430 filler metals. When maximum corrosion resistance is required, a post-weld annealing is recommended. In this instance, stabilized electrodes should be used to avoid grain-boundary carbide precipitation.

Type 405 provides good corrosion resistance and oxidation resistance up to about 650 °C (1200 °F) in applications not requiring the high strength of hardenable grades. It is resistant to the corrosive action of the atmosphere, fresh water, and various alkalis and mild acids.

In the annealed condition, type 405 may be easily formed, drawn, spun, bent, and roll formed. It is machined at speeds about 55% of those used for free-machining AISI 1212.

Typical Uses. Applications of AISI type 405 include welded structures that cannot be annealed after welding, annealing boxes, quenching racks, vessel linings, and precision rolled profiles for steam turbine parts.

AISI Type 405: Similar Steels (U.S. and/or Foreign). UNS S40500; ASME SA240, SA268, SA479; ASTM A176, A240, A268, A276, A314, A473, A479, A511, A580; FED QQ-S-763; MIL SPEC MIL-S-862; SAE J405 (51405); (W. Ger.) DIN 1.4002; (Fr.) AFNOR Z 6 CA 13; (Ital.) UNI X 6 CrAl 13; (Jap.) JIS SUS 405; (U.K.) B.S. 405 S 17

Physical Properties

AISI Type 405: Thermal Treatment Temperatures

Treatment	Temperature range °C	°F
Forging, start	1095-1205	2000-2200
Forging, finish	760 min	1400 min
Annealing (a)	650-760	1200-1400
Hardening	(b)	(b)

(a) Cool in air. (b) Does not harden appreciably with heat treatment

AISI Type 405: Average Coefficients of Linear Thermal Expansion (Ref 6)

Temperature range °C	°F	Coefficient µm/m·K	µin./in.·°F
0-100	32-212	10.8	6.0
0-315	32-600	11.5	6.4
0-540	32-1000	12.1	6.7
0-815	32-1500	13.5	7.5

AISI Type 405: Selected Physical Properties (Ref 6)

Melting range	1480-1530 °C (2700-2790 °F)
Density	7.8 g/cm³ (0.28 lb/in.³)
Specific heat at 0-100 °C (32-212 °F)	460 J/kg·K (0.11 Btu/lb·°F)
Thermal conductivity at 100 °C (212 °F)	27.0 W/m·K (15.6 Btu/ft·h·°F)
Electrical resistivity at 20 °C (68 °F)	60 nΩ·m

Mechanical Properties

AISI Type 405: Representative Tensile Properties (Ref 6)

Product form	Condition	Tensile strength MPa	ksi	Yield strength(a) MPa	ksi	Elongation(b), %	Reduction in area, %	Hardness
Sheet	Annealed	448	65	276	40	25	...	75 HRB
Wire(c)	Soft tempered	621	90	517	75	15	60	...
Plate	Annealed	448	65	276	40	30	...	150 HB
Bar	Annealed	483	70	276	40	30	60	150 HB
	Annealed and cold drawn(d)	586	85	483	70	20	60	185 HB

(a) 0.2% offset. (b) In 50 mm (2 in.), except for wire. For wire 3.18-mm (0.125-in.) diam and larger, gage length is 4 × diam; for smaller diameters, the gage length is 254 mm (10.0 in.). (c) Diameter of 6.4 mm (0.25 in.). (d) Diameter of 25 mm (1 in.)

Mechanical Properties (continued)

AISI Type 405: Typical Short-Time Elevated-Temperature Tensile Properties (Ref 12)

Test temperature °C	°F	Tensile strength MPa	ksi	Yield strength(a) MPa	ksi	Elongation(b), %
Room temperature		469	68.0	276	40.0	30
95	200	400	58.0	200	29.0	28
205	400	372	54.0	165	24.0	22
315	600	352	51.0	152	22.0	21
425	800	321	46.5	138	20.0	22
540	1000	252	36.5	117	17.0	28
650	1200	159	23.0	93	13.5	34
760	1400	59	8.5	31	4.5	68
870	1600	29	4.2	15	2.2	80

(a) 0.2% offset. (b) In 50 mm (2 in.)

AISI Type 405: Selected Mechanical Properties (Ref 7)

Modulus of elasticity, in tension .... 200 GPa (29 × 10^6 psi)

Izod impact properties at 20 °C (70 °F)........................41 J (30 ft·lb)

Machining Data (Ref 2)

AISI Type 405: Planing

Test specimens were in the annealed condition, with hardness of 135 to 185 HB

Tool material	Depth of cut mm	in.	Speed m/min	ft/min	Feed/stroke mm	in.
M2 and M3 high speed steel	0.1	0.005	9	30	(a)	(a)
	2.5	0.100	12	40	1.25	0.050
	12.0	0.500	8	25	1.50	0.060
C-6 carbide	0.1	0.005	46	150	(a)	(a)
	2.5	0.100	49	160	1.50	0.060

(a) Feed is 75% the width of the square nose finishing tool

AISI Type 405: Face Milling

Test specimens were in the annealed condition, with hardness of 135 to 185 HB

Depth of cut		M2 and M7 high speed steel				Uncoated carbide						Coated carbide			
		Speed		Feed/tooth		Speed, brazed		Speed, inserted		Feed/tooth		Speed		Feed/tooth	
mm	in.	m/min	ft/min	mm	in.	m/min	ft/min	m/min	ft/min	mm	in.	m/min	ft/min	mm	in.
1	0.040	58	190	0.20	0.008	170(a)	565(a)	190(a)	620(a)	0.20	0.008	280(b)	925(b)	0.20	0.008
4	0.150	44	145	0.30	0.012	130(a)	425(a)	145(a)	470(a)	0.30	0.012	185(b)	600(b)	0.30	0.012
8	0.300	35	115	0.40	0.016	90(c)	300(c)	110(c)	365(c)	0.40	0.016	145(d)	475(d)	0.40	0.016

(a) Carbide tool material: C-6. (b) Carbide tool material: CC-6. (c) Carbide tool material: C-5. (d) Carbide tool material: CC-5

AISI Type 405: Turning (Single Point and Box Tools)

Test specimens were in the annealed condition, with hardness of 135 to 185 HB

Depth of cut		M2 and M3 high speed steel				Uncoated carbide						Coated carbide			
		Speed		Feed		Speed, brazed		Speed, inserted		Feed		Speed		Feed	
mm	in.	m/min	ft/min	mm/rev	in./rev	m/min	ft/min	m/min	ft/min	mm/rev	in./rev	m/min	ft/min	mm/rev	in./rev
1	0.040	46	150	0.18	0.007	175(a)	575(a)	200(a)	650(a)	0.18	0.007	260(b)	850(b)	0.18	0.007
4	0.150	37	120	0.40	0.015	135(c)	450(c)	150(c)	500(c)	0.40	0.015	200(d)	650(d)	0.40	0.015
8	0.300	29	95	0.50	0.020	105(c)	350(c)	120(c)	400(c)	0.75	0.030	160(d)	525(d)	0.50	0.020
16	0.625	23	75	0.75	0.030	84(c)	275(c)	95(c)	310(c)	1.00	0.040	...	...	...	...

(a) Carbide tool material: C-7. (b) Carbide tool material: CC-7. (c) Carbide tool material: C-6. (d) Carbide tool material: CC-6

AISI Type 405: Turning (Cutoff and Form Tools)

Test specimens were in the annealed condition, with hardness of 135 to 185 HB

Tool material	Speed, m/min (ft/min)	Feed per revolution for cutoff tool width of: 1.5 mm (0.062 in.) mm (in.)	3 mm (0.125 in.) mm (in.)	6 mm (0.25 in.) mm (in.)	Feed per revolution for form tool width of: 12 mm (0.5 in.) mm (in.)	18 mm (0.75 in.) mm (in.)	25 mm (1 in.) mm (in.)	35 mm (1.5 in.) mm (in.)	50 mm (2 in.) mm (in.)
M2 and M3 high speed steel	27	0.028	0.038	0.050	0.050	0.046	0.038	0.030	0.025
	(90)	(0.0011)	(0.0015)	(0.002)	(0.002)	(0.0018)	(0.0015)	(0.0012)	(0.001)
C-6 carbide	90	0.028	0.038	0.050	0.050	0.046	0.038	0.030	0.025
	(300)	(0.0011)	(0.0015)	(0.002)	(0.002)	(0.0018)	(0.0015)	(0.0012)	(0.001)

Machining Data (Ref 2) (continued)

AISI Type 405: End Milling (Profiling)

Test specimens were in the annealed condition, with hardness of 135 to 185 HB

Tool material	Depth of cut		Speed		Feed per tooth for cutter diameter of: 10 mm (0.375 in.)		12 mm (0.5 in.)		18 mm (0.75 in.)		25-50 mm (1-2 in.)	
	mm	in.	m/min	ft/min	mm	in.	mm	in.	mm	in.	mm	in.
M2, M3, and M7 high speed steel	0.5	0.020	44	145	0.025	0.001	0.050	0.002	0.102	0.004	0.13	0.005
	1.5	0.060	34	110	0.050	0.002	0.075	0.003	0.130	0.005	0.15	0.006
	diam/4	diam/4	29	95	0.025	0.001	0.050	0.002	0.102	0.004	0.13	0.005
	diam/2	diam/2	26	85	0.025	0.001	0.038	0.0015	0.075	0.003	0.102	0.004
C-5 carbide	0.5	0.020	140	455	0.013	0.0005	0.025	0.001	0.075	0.003	0.13	0.005
	1.5	0.060	105	350	0.025	0.001	0.050	0.002	0.102	0.004	0.15	0.006
	diam/4	diam/4	90	295	0.025	0.001	0.038	0.0015	0.075	0.003	0.102	0.004
	diam/2	diam/2	84	275	0.018	0.0007	0.025	0.001	0.050	0.002	0.075	0.003

AISI Type 405: Boring

Test specimens were in the annealed condition, with hardness of 135 to 185 HB

Depth of cut		M2 and M3 high speed steel: Speed		Feed		Uncoated carbide: Speed, brazed		Speed, inserted		Feed		Coated carbide: Speed		Feed	
mm	in.	m/min	ft/min	mm/rev	in./rev	m/min	ft/min	m/min	ft/min	mm/rev	in./rev	m/min	ft/min	mm/rev	in./rev
0.25	0.010	46	150	0.075	0.003	150(a)	490(a)	175(a)	575(a)	0.075	0.003	230(b)	750(b)	0.075	0.003
1.25	0.050	37	120	0.130	0.005	115(c)	385(c)	140(c)	455(c)	0.130	0.005	185(d)	600(d)	0.130	0.005
2.50	0.100	29	95	0.300	0.012	90(e)	300(e)	105(e)	350(e)	0.300	0.012	135(f)	450(f)	0.300	0.012

(a) Carbide tool material: C-8. (b) Carbide tool material: CC-8. (c) Carbide tool material: C-7. (d) Carbide tool material: CC-7. (e) Carbide tool material: C-6. (f) Carbide tool material: CC-6

AISI Type 405: Drilling

Test specimens were in the annealed condition, with hardness of 135 to 185 HB

Tool material	Speed, m/min (ft/min)	Feed per revolution for nominal hole diameter of: 1.5 mm (0.062 in.) mm (in.)	3 mm (0.125 in.) mm (in.)	6 mm (0.25 in.) mm (in.)	12 mm (0.5 in.) mm (in.)	18 mm (0.75 in.) mm (in.)	25 mm (1 in.) mm (in.)	35 mm (1.5 in.) mm (in.)	50 mm (2 in.) mm (in.)
M10, M7, and M1 high speed	20	0.025	0.050	0.102	0.18	0.25	0.30	0.38	0.45
	(65)	(0.001)	(0.002)	(0.004)	(0.007)	(0.010)	(0.012)	(0.015)	(0.018)

AISI Type 405: Reaming

Test specimens were in the annealed condition, with hardness of 135 to 185 HB; based on 4 flutes for 3- and 6-mm (0.125- and 0.25-in.) reamers, 6 flutes for 12-mm (0.5-in.) reamers, and 8 flutes for 25-mm (1-in.) and larger reamers

Tool material	Speed		Feed per revolution for reamer diameter of: 3 mm (0.125 in.)		6 mm (0.25 in.)		12 mm (0.5 in.)		25 mm (1 in.)		35 mm (1.5 in.)		50 mm (2 in.)	
	m/min	ft/min	mm	in.	mm	in.	mm	in.	mm	in.	mm	in.	mm	in.
Roughing														
M1, M2, and M7 high speed steel	23	75	0.075	0.003	0.102	0.004	0.13	0.005	0.20	0.008	0.25	0.010	0.30	0.012
C-2 carbide	27	90	0.102	0.004	0.20	0.008	0.30	0.012	0.40	0.016	0.50	0.020	0.60	0.024
Finishing														
M1, M2, and M7 high speed steel	11	35	0.102	0.004	0.102	0.004	0.15	0.006	0.20	0.008	0.23	0.009	0.25	0.010
C-2 carbide	15	50	0.102	0.004	0.15	0.006	0.20	0.008	0.25	0.010	0.28	0.011	0.30	0.012

AISI Type 405: Cylindrical and Internal Grinding

Test specimens were in the annealed condition, with hardness of 135 to 185 HB

Operation	Wheel speed m/s	Wheel speed ft/min	Work speed m/min	Work speed ft/min	Infeed on diameter mm/pass	Infeed on diameter in./pass	Traverse(a)	Wheel identification
Cylindrical grinding								
Roughing	28-33	5500-6500	15-30	50-100	0.050	0.002	1/2	A60JV
Finishing	28-33	5500-6500	15-30	50-100	0.013 max	0.0005 max	1/6	A60JV
Internal grinding								
Roughing	25-33	5000-6500	23-60	75-200	0.013	0.0005	1/3	A60KV
Finishing	25-33	5000-6500	23-60	75-200	0.005 max	0.0002 max	1/6	A60KV

(a) Wheel width per revolution of work

AISI Type 405: Surface Grinding With a Horizontal Spindle and Reciprocating Table

Test specimens were in the annealed condition, with hardness of 135 to 185 HB

Operation	Wheel speed m/s	Wheel speed ft/min	Table speed m/min	Table speed ft/min	Downfeed mm/pass	Downfeed in./pass	Crossfeed mm/pass	Crossfeed in./pass	Wheel identification
Roughing	28-33	5500-6500	15-30	50-100	0.050	0.002	1.25-12.5	0.050-0.500	A46IV
Finishing	28-33	5500-6500	15-30	50-100	0.013 max	0.0005 max	(a)	(a)	A46IV

(a) Maximum 1/4 of wheel width

AISI Type 409

AISI Type 409: Chemical Composition

Element	Composition, %
Carbon	0.080 max
Manganese	1.000 max
Phosphorus	0.045 max
Sulfur	0.045 max
Silicon	1.000 max
Chromium	10.50-11.75
Titanium	(a)

(a) Composition ranges from six times the carbon content to 0.75%

Characteristics. AISI type 409 is a general-purpose, construction, chromium stainless steel. It is primarily intended for automotive exhaust systems, as well as structural and other applications, where appearance is secondary to mechanical and corrosion-resistant properties.

In general, the corrosion resistance of type 409 is about the same as that of type 410. Type 409 tolerates products such as neutral salts (other than chlorides), dilute alkaline or ammoniacal solutions, and cold dilute acidic solutions. This steel also exhibits superior resistance to the acid-corrosive conditions in automotive exhaust systems and demonstrates negligible general corrosion from exposure to fresh water.

Type 409 stainless steel starts to exhibit destructive scaling in air at approximately 705 °C (1300 °F). This temperature is considered the general maximum-service temperature for continuous service in air.

This alloy has good fabricating characteristics and can be stamped and formed without difficulty. Punch and die radii for type 409 steel should be as generous as conditions permit. These radii should be approximately five to eight times the metal thickness for the die and a minimum of four times the metal thickness for the punch. Die clearance for stamping should be a minimum of 30 to 40% of the metal thickness.

Type 409 has welding characteristics similar to other ferritic stainless steels and is easily welded by the standard welding procedures recommended for stainless steels. This alloy does not air-harden and is not prone to cracking during cooling; consequently, preheating and postheating are not necessary.

Typical Uses. Applications of AISI type 409 include automotive exhaust systems, cold water storage tanks, transformer cases, culverts, shipping containers, and farm equipment.

AISI Type 409: Similar Steels (U.S. and/or Foreign). UNS S40900; ASME SA268; ASTM A176, A268, A651; SAE J405 (51409); (W. Ger.) DIN 1.4512; (Jap.) JIS SUH 409; (U.K.) B.S. 409 S 17

Physical Properties

AISI Type 409: Thermal Conductivity (Ref 8)

Temperature °C	Temperature °F	Thermal conductivity W/m·K	Thermal conductivity Btu/ft·h·°F
95	200	24.9	14.4
540	1000	28.6	16.5

AISI Type 409: Selected Physical Properties (Ref 8)

Melting range	1425-1510 °C (2600-2750 °F)
Density	7.64 g/cm³ (0.276 lb/in.³)
Specific heat at 0-100 °C (32-212 °F)	460 J/kg·K (0.11 Btu/lb·°F)
Electrical resistivity at 20 °C (68 °F)	60 nΩ·m

Physical Properties (continued)

AISI Type 409: Thermal Treatment Temperatures

Treatment	Temperature range °C	°F
Annealing(a)	870-900	1600-1650
Hardening(b)	...	...

(a) Air cool. (b) Not hardenable by heat treatment

AISI Type 409: Average Coefficients of Linear Thermal Expansion (Ref 8)

Temperature range °C	°F	Coefficient µm/m·K	µin./in.·°F
20-100	68-212	11.7	6.5
20-260	68-500	11.9	6.6
20-480	68-900	12.4	6.9
20-650	68-1200	12.9	7.2
20-815	68-1500	13.5	7.5

Mechanical Properties

AISI Type 409: Representative Tensile Properties (Ref 6)

Test specimens were in the annealed condition

Product form	Tensile strength MPa	ksi	Yield strength(a) MPa	ksi	Elongation(b), %	Hardness, HRB
Sheet, strip, plate, and bar	450	65	240	35	25	75

(a) 0.2% offset. (b) In 50 mm (2 in.)

AISI Type 409: Tensile Properties at Elevated Temperatures (Ref 8)

Test temperature °C	°F	Tensile strength MPa	ksi	Yield strength(a) MPa	ksi	Elongation(b), %
27	80	448	65.0	238	34.5	32.5
93	200	406	58.9	207	30.0	29.5
205	400	365	53.0	172	25.0	25.5
315	600	334	48.5	172	25.0	22.5
425	800	326	47.3	165	24.0	21.0
540	1000	254	36.9	136	19.7	24.0
650	1200	148	21.4	92	13.4	44.0
760	1400	61	8.8	39	5.7	67.5
870	1600	30	4.3	19	2.7	75.5

(a) 0.2% offset. (b) In 50 mm (2 in.)

Machining Data

For machining data on AISI type 409, refer to the preceding machining tables for AISI type 405.

AISI Types 410 and 410S

AISI Types 410 and 410S: Chemical Composition

Element	Composition, % Type 410	Type 410S
Carbon	0.15 max	0.08 max
Manganese	1.00 max	1.00 max
Phosphorus	0.04 max	0.04 max
Sulfur	0.03 max	0.03 max
Silicon	1.00 max	1.00 max
Chromium	11.5-13.5	11.5-13.5

Characteristics. AISI type 410 is the basic, hardenable, martensitic, stainless alloy. It can be heat treated to a wide range of mechanical properties. In both the annealed and heat treated conditions, type 410 provides good corrosion resistance to many industrial and domestic environments, as well as potable and mine waters. Type 410S is a low-carbon modification of type 410.

The safe temperature for continuous service of type 410 is 650 °C (1200 °F). This steel will discolor slightly at about 230 °C (450 °F), but does not scale until the temperature exceeds 650 °C (1200 °F).

In the annealed condition, type 410 can be blanked, drawn, formed, and cold headed. When type 410 is machined in the dead soft condition, it is tough and draggy and the chips tend to build up in the tool. Better finishes are obtained in the cold drawn or heat treated condition. The machinability rating of type 410 is about 55% of that for AISI 1212.

Type 410 can be welded satisfactorily. When filler metal is used, it should be of similar analysis. Parts should be preheated to a minimum temperature of 205 °C (400 °F) before welding to prevent cracking. Because type 410 air hardens, it should be annealed immediately after welding to ensure uniform ductility.

Typical Uses. Applications of AISI types 410 and 410S include steam turbine buckets, blades, bucket covers, gas turbine compressor blades, nuclear reactor control rod mechanisms, valves, fasteners, shafting, pump parts, petrochemical equipment, and machine parts.

AISI Type 410: Similar Steels (U.S. and/or Foreign). UNS S41000; AMS 5504, 5505, 5591, 5613, 5776, 5821; ASME SA194 (6), SA240, SA268, SA479; ASTM A176, A193, A194, A240, A276, A314, A473, A479, A493, A511, A580; FED QQ-S-763, QQ-W-423; MIL SPEC MIL-S-862; SAE J405 (51410), J412 (51410); (W. Ger.) DIN 1.4006; (Fr.) AFNOR Z 10 C 13, Z 10 C 14, Z 12 C 13; (Ital.) UNI X 12 Cr 13; (Jap.) JIS SUS 410; (Swed.) SS_{14} 2302; (U.K.) B.S. 410 S 21, En. 56 A, ANC 1 Grade A

AISI Type 410S: Similar Steels (U.S. and/or Foreign). UNS S41008; ASME SA240; ASTM A176, A240, A473; (W. Ger.) DIN 1.4001; (Jap.) JIS SUS 410 S; (U.K.) B.S. 403 S 17

Physical Properties

AISI Types 410, 410S: Thermal Treatment Temperatures

Treatment	Temperature °C	°F
Forging, start	1095-1205	2000-2200
Forging, finish	815 min	1500 min
Full annealing(a)	815-900	1500-1650
Process annealing(b)	650-760	1200-1400
Hardening(c)	925-1010	1700-1850
Tempering(d)	205-760	400-1400

(a) Hold for 1 h per 25 mm (1 in.) of thickness; slow cool to 595 °C (1100 °F); air cool. (b) Hold at temperature to achieve uniform heat distribution; air cool. (c) Hold at temperature 15 to 30 min; quench in oil. (d) Tempering temperature depends on hardness, strength, toughness, or corrosion resistance desired

AISI Types 410, 410S: Thermal Conductivity (Ref 7)

Temperature °C	°F	Thermal conductivity W/m·K	Btu/ft·h·°F
100	212	24.9	14.4
500	930	28.7	16.6

AISI Types 410, 410S: Average Coefficients of Linear Thermal Expansion (Ref 7)

Temperature range °C	°F	Coefficient μm/m·K	μin./in.·°F
0-100	32-212	9.9	5.5
0-315	32-600	11.0	6.1
0-540	32-1000	11.5	6.4
0-650	32-1200	11.7	6.5

AISI Types 410, 410S: Selected Physical Properties (Ref 7)

Melting range	1480-1530 °C (2700-2790 °F)
Density	7.8 g/cm³ (0.28 lb/in.³)
Specific heat at 0-100 °C (32-212 °F)	460 J/kg·K (0.11 Btu/lb·°F)
Electrical resistivity at 20 °C (68 °F)	57 nΩ·m
Electrical resistivity at 650 °C (1200 °F)	108 nΩ·m

Mechanical Properties

AISI Types 410, 410S: Tensile Properties at Elevated Temperatures (Ref 7)

Test specimens were 16-mm (0.625-in.) diam bar; heated to 980 °C (1800 °F) for ½ h; oil quenched; tempered for 2 h at 28 °C (50 °F) above test temperature

Test temperature °C	°F	Tensile strength MPa	ksi	Yield strength(a) MPa	ksi	Elongation(b), %	Reduction in area, %	Hardness(c), HRC
21	70	1525	221	1225	177.5	14.5	63.5	45
205	400	1475	214	1005	146	11	51	43
315	600	1470	213	961	139.5	18	57	43
425	800	1340	194	920	133.5	18.5	59	43.5
480	900	1150	166.5	835	121	14	57	...
540	1000	605	87.5	565	82	21.5	81.5	...
595	1100	440	64	395	57.5	25.5	87	...
650	1200	300	43.5	270	39.5	29.5	96.5	...
705	1300	195	28.5	165	24	34	91.5	...

(a) 0.2% offset. (b) In 50 mm (2 in.). (c) Before testing

AISI Types 410, 410S: Tensile Properties at Elevated Temperatures after Tempering at 605 °C (1125 °F) (Ref 7)

Test specimens were 16-mm (0.625-in.) diam bar; heated to 980 °C (1800 °F) for ½ h; oil quenched; tempered at 650 °C (1200 °F) for 2 h; air cooled; hardness of 24.5 HRC

Test temperature °C	°F	Tensile strength MPa	ksi	Yield strength(a) MPa	ksi	Elongation(b), %	Reduction in area, %
21	70	834	121	721	104.5	21.5	68.5
205	400	741	107.5	650	94.5	18	69
315	600	696	101	615	89.5	16	70
425	800	635	92	570	83	17	68
480	900	585	84.5	525	76.5	18.5	73.5
540	1000	495	72	470	68	22.5	81.5
595	1100	405	58.5	385	55.5	25.5	88
650	1200	305	44	280	40.5	29	90

(a) 0.2% offset. (b) In 50 mm (2 in.)

Machining Data

For machining data on AISI types 410 and 410S, refer to the preceding machining tables for AISI type 403.

Mechanical Properties (continued)

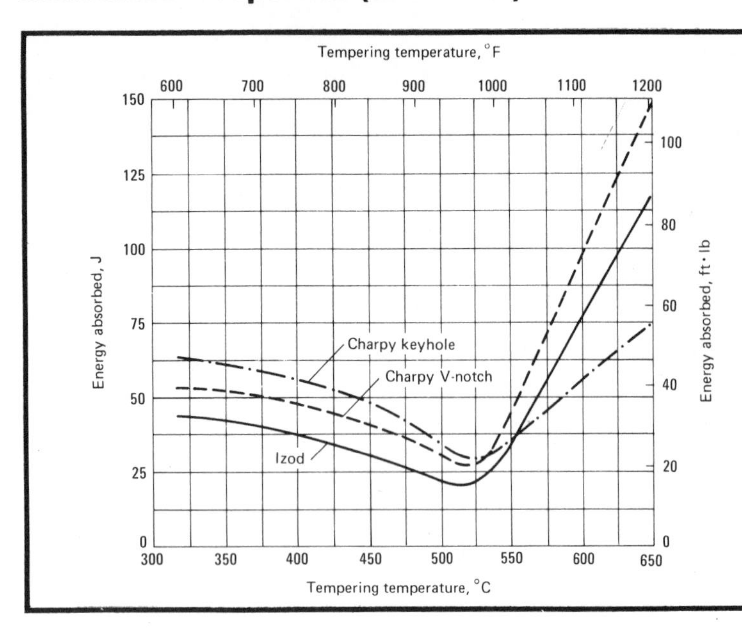

AISI Type 410: Effect of Tempering Temperature on Impact Properties. Impact energy tests used Izod, Charpy V-notch, or Charpy keyhole as indicated. Heat treatment of test specimens: heated to and held at 955 °C (1750 °F) for 1 h; oil quenched, tempered for 4 h at temperature given; air cooled. (Ref 8)

AISI Types 410, 410S: Tensile Properties at Elevated Temperatures after Tempering at 540 °C (1000 °F) (Ref 7)

Test specimens were 16-mm (0.625-in.) diam bar; heated to 980 °C (1800 °F) for ½ h; oil quenched; tempered at 540 °C (1000 °F) for 2 h; air cooled; hardness of 35.5 HRC

Test temperature °C	Test temperature °F	Tensile strength MPa	Tensile strength ksi	Yield strength(a) MPa	Yield strength(a) ksi	Elongation(b), %	Reduction in area, %
21	70	1085	157.5	1005	145.5	13	69.5
205	400	1050	152	927	134.5	11	69.5
315	600	1005	145.5	838	121.5	10.5	65.5
425	800	896	130	758	110	12	70
540	1000	700	101.5	645	93.5	16	77.5
650	1200	275	40	260	38	35	91
760	1400	105	15	90	13	54	96
870	1600	96	14	66	9.5	81	77

(a) 0.2% offset. (b) In 50 mm (2 in.)

AISI Types 410, 410S: Tensile Properties at Elevated Temperatures after Tempering at 605 °C (1125 °F) (Ref 7)

Test specimens were 16-mm (0.625-in.) diam bar; heated to 980 °C (1800 °F) for ½ h; oil quenched; tempered at 605 °C (1125 °F) for 2 h; air cooled; hardness of 28.5 HRC

Test temperature °C	Test temperature °F	Tensile strength MPa	Tensile strength ksi	Yield strength(a) MPa	Yield strength(a) ksi	Elongation(b), %	Reduction in area, %
21	70	924	134	807	117	20	68.5
205	400	817	118.5	727	105.5	17.5	70.5
315	600	772	112	689	100	16.5	67.5
425	800	724	105	650	94.5	17	65.5
480	900	625	91	585	85	18.5	74.5
540	1000	560	81	525	76	22	82
595	1100	450	65	415	60	25	87

(a) 0.2% offset. (b) In 50 mm (2 in.)

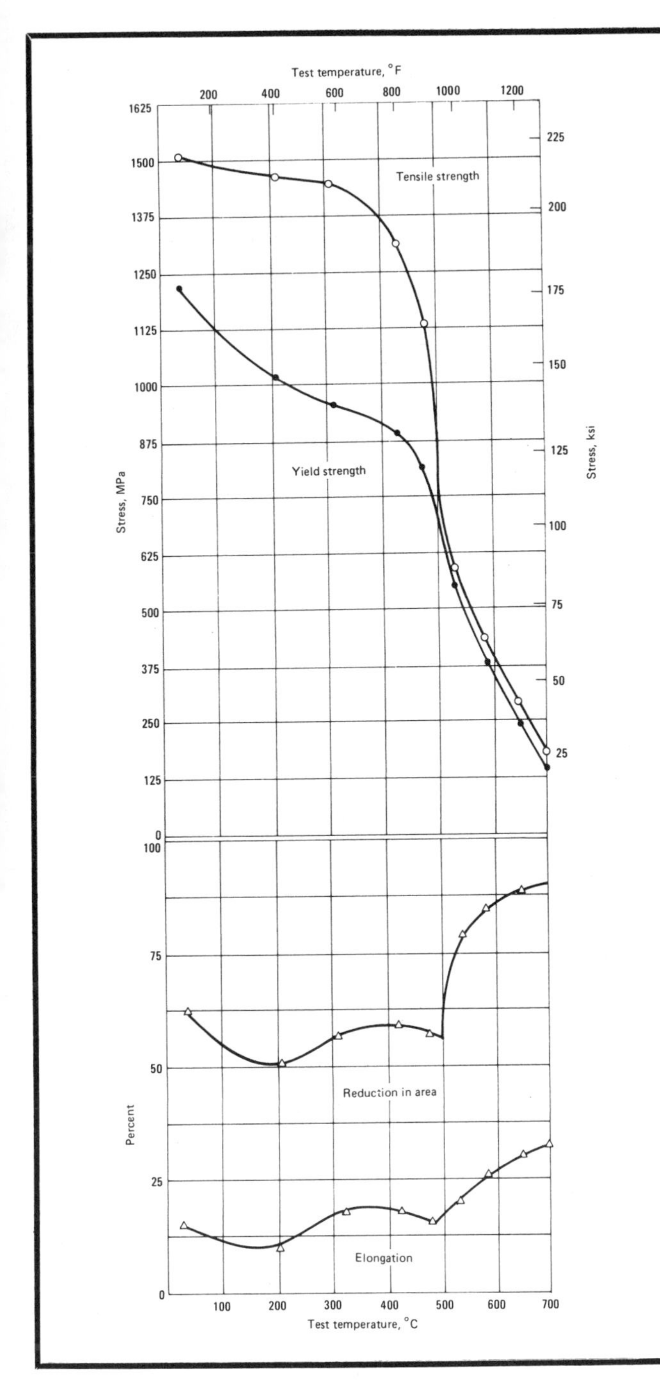

AISI Type 410: Effect of Elevated Temperature on Tensile Properties. Hardened at 980 °C (1800 °F) for ½ h; oil quenched; tempered for 2 h at 28 °C (50 °F) above test temperature. Specimens were 16-mm (0.625-in.) round bar. Tests were conducted using specimens machined to English units. Elongation was measured in 50 mm (2 in.); yield strength at 0.2% offset. (Ref 7)

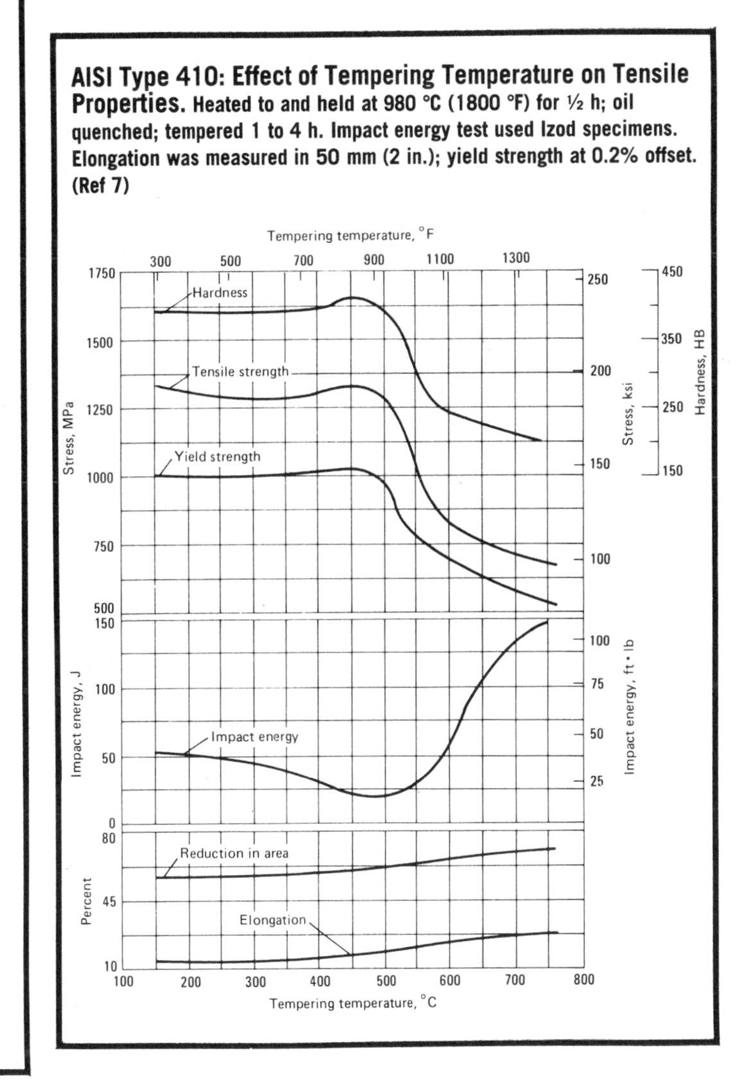

AISI Type 410: Effect of Tempering Temperature on Tensile Properties. Heated to and held at 980 °C (1800 °F) for ½ h; oil quenched; tempered 1 to 4 h. Impact energy test used Izod specimens. Elongation was measured in 50 mm (2 in.); yield strength at 0.2% offset. (Ref 7)

AISI Types 410, 410S: Stress-Rupture and Creep Properties (Ref 7)

Test temperature		Stress for rupture in:						Stress for secondary creep rate of(a):			
		100 h		1000 h		10 000 h(a)		1% in 10 000 h		1% in 100 000 h	
°C	°F	MPa	ksi	MPa	ksi	MPa	ksi	MPa	ksi	MPa	ksi
480	900	...	...	265	38.5	...	...	...	...	110	16.0
540	1000	234	34.0	159	23.0	110	16.0	76	11.0	41	6.0
595	1100	134	19.5	86	12.5	57	8.3	34	5.0	21	3.0
650	1200	74	10.7	46	6.6	28	4.0	21	3.0	10	1.5
705	1300	39	5.7	23	3.3	...	...	...	...	...	...

(a) Extrapolated data

AISI Type 414

AISI Type 414: Chemical Composition

Element	Composition, %
Carbon	0.15 max
Manganese	1.00 max
Phosphorus	0.04 max
Sulfur	0.03 max
Silicon	1.00 max
Chromium	11.5-13.5
Nickel	1.25-2.5

Characteristics. AISI type 414 is a hardenable (martensitic), nickel-bearing, chromium steel. This grade can be heat treated to higher tensile and impact strength than type 410. It will resist corrosion from mild atmospheres, fresh water, steam, ammonia, many petroleum products and organic materials, and several mild acid environments.

Type 414 is rarely considered for high-temperature service. It will discolor at about 230 °C (450 °F), but scale does not form until the temperature exceeds 650 °C (1200 °F).

This stainless steel can be moderately cold formed in the subcritical annealed condition. For cold heading operations, the material can be supplied in the cycle annealed, minimum hardness condition. In the annealed condition, type 414 will tend to tear and drag due to the tendency of the chip to build up on the tool. The machinability rating of this steel is about 45% of that for free-machining AISI 1212.

Type 414 can be welded satisfactorily, but the weld air hardens and requires careful handling to prevent cracking. Recommended treatments include heating to 205 °C (400 °F) minimum prior to welding, and annealing immediately after welding. Electrodes of composition similar to type 410 should be used.

Typical Uses. Applications of AISI type 414 include beater bars, fasteners, gage parts, mild springs, mining equipment, scissors, scraper knives, shafts, spindles, and valve seats.

AISI Type 414: Similar Steels (U.S. and/or Foreign). UNS S41400; AMS 5615; ASTM A276, A314, A473, A511, A580; FED QQ-S-763; SAE J405 (51414)

Physical Properties

AISI Type 414: Selected Physical Properties (Ref 7)

Density	7.8 g/cm^3 (0.28 lb/in.^{3})
Specific heat at 0-100 °C (32-212 °F)	460 J/kg·K (0.11 Btu/lb·°F)
Electrical resistivity at 20 °C (68 °F)	72 nΩ·m

AISI Type 414: Thermal Treatment Temperatures

Treatment	Temperature range °C	°F
Forging, start	1150-1205	2100-2200
Forging, finish	900 min	1650 min
Subcritical annealing(a)	650-705	1200-1300
Cycle annealing	(b)	(b)
Hardening(c)	980-1040	1800-1900
Tempering(d)	205-705	400-1300

(a) Hold at temperature 4 to 8 h; air cool. (b) Heat to 760 °C (1400 °F); hold 20 h; cool to below 315 °C (600 °F); reheat to 675 °C (1250 °F); hold 20 h; air cool. (c) Hold 15 to 30 min; oil or air quench, depending on section thickness. (d) Temperature depends upon hardness, strength, toughness, or corrosion resistance required

AISI Type 414: Average Coefficients of Linear Thermal Expansion (Ref 7)

Temperature range °C	°F	Coefficient µm/m·K	µin./in.·°F
0-100	32-212	10.4	5.8
0-315	32-600	11.0	6.1
0-540	32-1000	12.1	6.7
0-650	32-1200	12.4	6.9

AISI Type 414: Thermal Conductivity (Ref 7)

Temperature °C	°F	Thermal conductivity W/m·K	Btu/ft·h·°F
100	212	24.9	14.4
500	930	28.7	16.6

Mechanical Properties

AISI Type 414: Tensile Properties at Elevated Temperatures (Ref 7)

Test specimens were 16-mm (0.625-in.) diam bar; heated to 980 °C (1800 °F) for ½ h; oil quenched; tempered at 28 °C (50 °F) above test temperature

Test temperature °C	°F	Tensile strength MPa	ksi	Yield strength(a) MPa	ksi	Elongation(b), %	Reduction in area, %	Hardness(c), HRC
21	70	1795	260	1013	147	15	49	48
205	400	1500	217.5	1017	147.5	14.5	45.5	45.5
315	600	1430	207.5	960	139.5	18	55.5	44.5
425	800	1385	201	985	142.5	16.5	55.5	47.5
480	900	1195	173.5	885	128	14.5	55	44.5
540	1000	620	90	560	81.5	21	75.5	36.5
595	1100	445	64.5	395	57.5	24.5	84	33
650	1200	275	40	230	33.5	33.5	91	32
705	1300	185	27	145	21	42.5	93.5	29

(a) 0.2% offset. (b) In 50 mm (2 in.). (c) Before testing

AISI Type 414: Izod Impact Properties (Ref 7)

Condition	Energy J	ft·lb
Annealed	70-110	50-80
Hardened and tempered	30-80	20-60

AISI Type 414: Modulus of Elasticity (Ref 7)

Stressed in:	Modulus GPa	10^6 psi
Tension	200	29

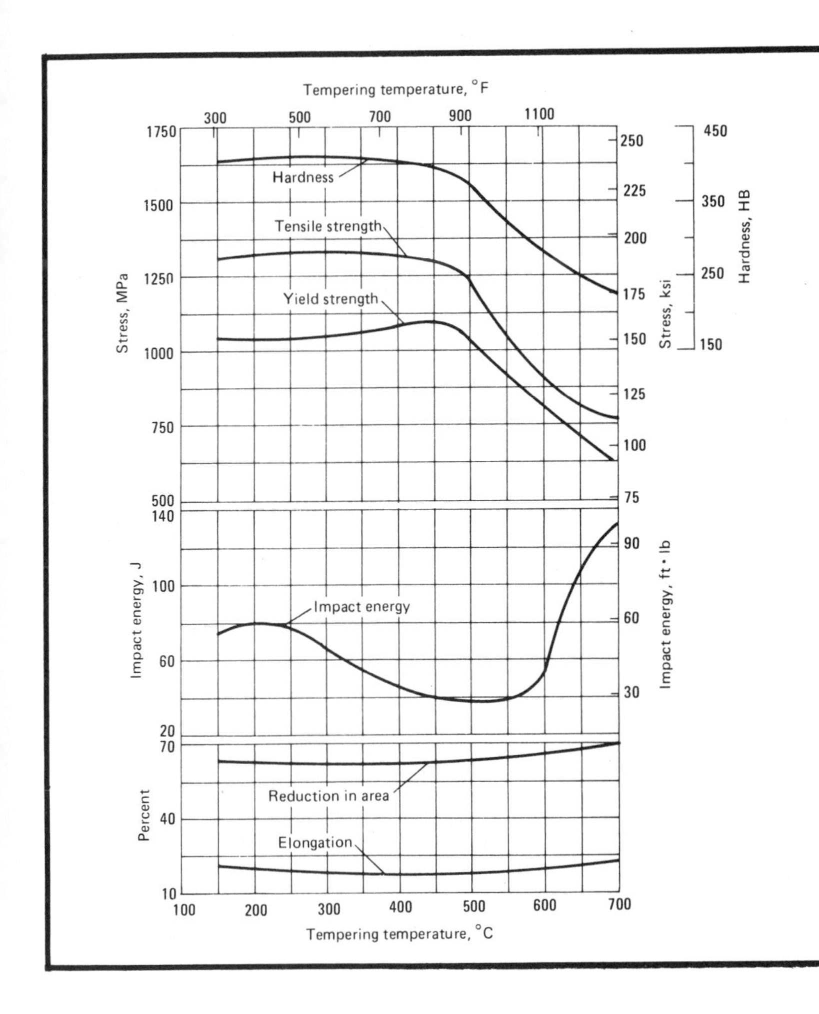

AISI Type 414: Effect of Tempering Temperature on Tensile Properties. Impact energy test used Izod specimens. Elongation was measured in 50 mm (2 in.); yield strength at 0.2% offset. (Ref 8)

AISI Type 414: Representative Tensile Properties (Ref 6)

Product form	Treatment	Tensile strength MPa	Tensile strength ksi	Yield strength(a) MPa	Yield strength(a) ksi	Elongation(b), %	Reduction in area, %	Hardness
Sheet	Annealed	825	120	725	105	15	...	98 HRB
Strip	Annealed	825	120	725	105	15	...	98 HRB
Wire, 6.4-mm (0.25-in.) diam	Soft tempered	930	135	795	115	10	55	29 HRC
Plate	Annealed	795	115	620	90	20	60	235 HB
Bar	Annealed and cold drawn	795	115	620	90	20	60	235 HB
		895	130	795	115	15	58	270 HB

(a) 0.2% offset. (b) In 50 mm (2 in.), except for wire. For wire 3.18-mm (0.125-in.) diam and larger, gage length is 4 × diam; for smaller diameters, the gage length is 254 mm (10.0 in.)

AISI Type 414: Tensile Properties at Elevated Temperatures After Tempering at 650 °C (1200 °F) (Ref 7)

Test specimens were 16-mm (0.625-in.) diam bar; heated to 980 °C (1800 °F) for ½ h; oil quenched; tempered at 650 °C (1200 °F) for 2 h; air cooled; hardness of 30.5 HRC

Test temperature °C	Test temperature °F	Tensile strength MPa	Tensile strength ksi	Yield strength(a) MPa	Yield strength(a) ksi	Elongation(b), %	Reduction in area, %
21	70	1005	145.5	800	116	19	58
205	400	903	131	721	104.5	17	60.5
315	600	879	127.5	680	98.5	15.5	55
425	800	817	118.5	630	91.5	15.5	55.5
480	900	703	102	595	86	17	68
540	1000	550	79.5	495	71.5	21.5	79
595	1100	420	61	385	55.5	25.5	85.5
650	1200	285	41.5	250	36.5	29	90.5

(a) 0.2% offset. (b) In 50 mm (2 in.)

Machining Data (Ref 2)

AISI Type 414: Turning (Single Point and Box Tools)

Depth of cut		High speed steel				Uncoated carbide						Coated carbide			
		Speed		Feed		Speed, brazed		Speed, inserted		Feed		Speed		Feed	
mm	in.	m/min	ft/min	mm/rev	in./rev	m/min	ft/min	m/min	ft/min	mm/rev	in./rev	m/min	ft/min	mm/rev	in./rev
Annealed, with hardness of 225 to 275 HB															
1	0.040	30(a)	100(a)	0.18	0.007	105(b)	350(b)	135(b)	450(b)	0.18	0.007	175(c)	575(c)	0.18	0.007
4	0.150	23(a)	75(a)	0.40	0.015	84(d)	275(d)	105(d)	350(d)	0.40	0.015	135(e)	450(e)	0.40	0.015
8	0.300	18(a)	60(a)	0.50	0.020	67(d)	220(d)	84(d)	275(d)	0.75	0.030	105(e)	350(e)	0.50	0.020
16	0.625	14(a)	45(a)	0.75	0.030	52(d)	170(d)	69(d)	225(d)	1.00	0.040	...	...	...	...
Quenched and tempered, with hardness of 275 to 325 HB															
1	0.040	23(f)	75(f)	0.18	0.007	88(b)	290(b)	115(b)	375(b)	0.18	0.007	120(c)	400(c)	0.18	0.007
4	0.150	18(f)	60(f)	0.40	0.015	69(d)	225(d)	90(c)	295(c)	0.40	0.015	90(e)	300(e)	0.40	0.015
8	0.300	14(f)	45(f)	0.50	0.020	55(d)	180(d)	70(c)	230(c)	0.50	0.020	76(e)	250(e)	0.50	0.020
Quenched and tempered, with hardness of 375 to 425 HB															
1	0.040	18(f)	60(f)	0.18	0.007	84(b)	275(b)	90(b)	300(b)	0.18	0.007	120(c)	400(c)	0.18	0.007
4	0.150	14(f)	45(f)	0.40	0.015	60(d)	200(d)	72(d)	235(d)	0.40	0.015	90(e)	300(e)	0.40	0.015
8	0.300	11(f)	35(f)	0.50	0.020	49(d)	160(d)	56(d)	185(d)	0.50	0.020	76(e)	250(e)	0.50	0.020

(a) High speed steel tool material: M2 or M3. (b) Carbide tool material: C-7. (c) Carbide tool material: CC-7. (d) Carbide tool material: C-6. (e) Carbide tool material: CC-6. (f) Any premium high speed steel tool material (T15, M33, M41 to M47)

AISI Type 414: Turning (Cutoff and Form Tools)

Tool material	Speed, m/min (ft/min)	Feed per revolution for for cutoff tool width of: 1.5 mm (0.062 in.) mm (in.)	3 mm (0.125 in.) mm (in.)	6 mm (0.25 in.) mm (in.)	Feed per revolution for form tool width of: 12 mm (0.5 in.) mm (in.)	18 mm (0.75 in.) mm (in.)	25 mm (1 in.) mm (in.)	35 mm (1.5 in.) mm (in.)	50 mm (2 in.) mm (in.)
Annealed, with hardness of 225 to 275 HB									
M2 and M3 high speed steel	18 (60)	0.033 (0.0013)	0.046 (0.0018)	0.058 (0.0023)	0.046 (0.0018)	0.041 (0.0016)	0.033 (0.0013)	0.025 (0.001)	0.020 (0.0008)
C-6 carbide	69 (225)	0.033 (0.0013)	0.046 (0.0018)	0.058 (0.0023)	0.046 (0.0018)	0.041 (0.0016)	0.033 (0.0013)	0.025 (0.001)	0.020 (0.0008)
Quenched and tempered, with hardness of 275 to 325 HB									
Any premium high speed steel (T15, M33, M41-M47)	15 (50)	0.025 (0.001)	0.030 (0.0012)	0.038 (0.0015)	0.038 (0.0015)	0.030 (0.0012)	0.025 (0.001)	0.023 (0.009)	0.020 (0.0008)
C-6 carbide	52 (170)	0.025 (0.001)	0.030 (0.0012)	0.038 (0.0015)	0.038 (0.0015)	0.030 (0.0012)	0.025 (0.001)	0.023 (0.009)	0.020 (0.0008)
Quenched and tempered, with hardness of 375 to 425 HB									
Any premium high speed steel (T15, M33, M41-M47)	12 (40)	0.020 (0.008)	0.025 (0.001)	0.030 (0.0012)	0.030 (0.0012)	0.025 (0.0010)	0.023 (0.0009)	0.020 (0.008)	0.018 (0.0007)
C-6 carbide	34 (110)	0.020 (0.008)	0.025 (0.001)	0.030 (0.0012)	0.030 (0.0012)	0.025 (0.0010)	0.023 (0.0009)	0.020 (0.008)	0.018 (0.0007)

AISI Type 414: Face Milling

Depth of cut		High speed steel				Uncoated carbide						Coated carbide			
		Speed		Feed/tooth		Speed, brazed		Speed, inserted		Feed/tooth		Speed		Feed/tooth	
mm	in.	m/min	ft/min	mm	in.	m/min	ft/min	m/min	ft/min	mm	in.	m/min	ft/min	mm	in.
Annealed, with hardness of 225 to 275 HB															
1	0.040	37(a)	120(a)	0.15	0.006	140(b)	465(b)	155(b)	510(b)	0.18	0.007	230(c)	750(c)	0.18	0.007
4	0.150	27(a)	90(a)	0.25	0.010	105(b)	350(b)	115(b)	385(b)	0.25	0.010	150(c)	500(c)	0.25	0.010
8	0.300	21(a)	70(a)	0.36	0.014	75(d)	245(d)	90(d)	300(d)	0.36	0.014	120(e)	400(e)	0.36	0.014
Quenched and tempered, with hardness of 275 to 325 HB															
1	0.040	29(f)	95(f)	0.15	0.006	120(b)	400(b)	135(b)	440(b)	0.15	0.006	200(c)	650(c)	0.13	0.005
4	0.150	21(f)	70(f)	0.23	0.009	90(b)	300(b)	100(b)	330(b)	0.20	0.008	130(c)	425(c)	0.18	0.007
8	0.300	17(f)	55(f)	0.30	0.012	64(d)	210(d)	70(d)	225(d)	0.25	0.010	100(e)	325(e)	0.23	0.009
Quenched and tempered, with hardness of 375 to 425 HB															
1	0.040	18(f)	60(f)	0.10	0.004	60(b)	200(b)	67(b)	220(b)	0.10	0.004	100(c)	325(c)	0.075	0.003
4	0.150	14(f)	45(f)	0.15	0.006	46(b)	150(b)	52(b)	170(b)	0.15	0.006	69(c)	225(c)	0.13	0.005
8	0.300	11(f)	35(f)	0.20	0.008	32(b)	105(b)	40(b)	13(b)	0.20	0.008	52(c)	170(c)	0.18	0.007

(a) High speed steel tool material: M2 or M7. (b) Carbide tool material: C-6. (c) Carbide tool material: CC-6. (d) Carbide tool material: C-5. (e) Carbide tool material: CC-5. (f) Any premium high speed steel tool material (T15, M33, M41 to M47)

AISI Type 414: End Milling (Profiling)

Tool material	Depth of cut mm	Depth of cut in.	Speed m/min	Speed ft/min	10 mm (0.375 in.) mm	10 mm (0.375 in.) in.	12 mm (0.5 in.) mm	12 mm (0.5 in.) in.	18 mm (0.75 in.) mm	18 mm (0.75 in.) in.	25-50 mm (1-2 in.) mm	25-50 mm (1-2 in.) in.
					Feed per tooth for cutter diameter of:							
Annealed, with hardness of 225 to 275 HB												
M2, M3, and M7 high speed steel	0.5	0.020	30	100	0.025	0.001	0.050	0.002	0.075	0.003	0.102	0.004
	1.5	0.060	23	75	0.050	0.002	0.075	0.003	0.102	0.004	0.15	0.005
	diam/4	diam/4	20	65	0.025	0.001	0.050	0.002	0.075	0.003	0.102	0.004
	diam/2	diam/2	17	55	0.025	0.001	0.025	0.001	0.050	0.002	0.075	0.003
C-5 carbide	0.5	0.020	100	325	0.013	0.0005	0.025	0.001	0.075	0.003	0.13	0.005
	1.5	0.060	75	245	0.025	0.001	0.050	0.002	0.102	0.004	0.15	0.006
	diam/4	diam/4	64	210	0.025	0.001	0.038	0.0015	0.075	0.003	0.102	0.004
	diam/2	diam/2	60	200	0.013	0.0005	0.025	0.001	0.050	0.002	0.075	0.003
Quenched and tempered, with hardness of 275 to 325 HB												
M2, M3, and M7 high speed steel	0.5	0.020	23	75	0.018	0.0007	0.025	0.001	0.050	0.002	0.075	0.003
	1.5	0.060	18	60	0.025	0.001	0.050	0.002	0.075	0.003	0.102	0.004
	diam/4	diam/4	15	50	0.018	0.0007	0.025	0.001	0.050	0.002	0.075	0.003
	diam/2	diam/2	14	45	0.013	0.0005	0.018	0.0007	0.038	0.0015	0.063	0.0025
C-5 carbide	0.5	0.020	88	290	0.013	0.0005	0.025	0.001	0.075	0.003	0.13	0.005
	1.5	0.060	69	225	0.025	0.001	0.050	0.002	0.102	0.004	0.15	0.006
	diam/4	diam/4	59	195	0.025	0.001	0.038	0.0015	0.075	0.003	0.102	0.004
	diam/2	diam/2	53	175	0.013	0.0005	0.025	0.001	0.050	0.002	0.075	0.003
Quenched and tempered, with hardness of 375 to 425 HB												
Any premium high speed steel (T15, M33, M41-M47)	0.5	0.020	60	60	0.013	0.0005	0.018	0.0007	0.025	0.001	0.050	0.002
	1.5	0.060	50	50	0.013	0.0005	0.025	0.001	0.050	0.002	0.075	0.003
	diam/4	diam/4	40	40	0.013	0.0005	0.013	0.0005	0.038	0.0015	0.063	0.0025
	diam/2	diam/2	35	35	...	...	...	...	0.025	0.001	0.050	0.002
C-5 carbide	0.5	0.020	275	275	0.013	0.0005	0.025	0.001	0.050	0.002	0.075	0.003
	1.5	0.060	175	175	0.025	0.001	0.050	0.002	0.075	0.003	0.102	0.004
	diam/4	diam/4	145	145	0.025	0.001	0.038	0.0015	0.063	0.0025	0.089	0.0035
	diam/2	diam/2	135	135	0.013	0.0005	0.025	0.001	0.050	0.002	0.075	0.003

AISI Type 414: Boring

Depth of cut mm	Depth of cut in.	High speed steel Speed m/min	High speed steel Speed ft/min	High speed steel Feed mm/rev	High speed steel Feed in./rev	Uncoated carbide Speed, brazed m/min	Uncoated carbide Speed, brazed ft/min	Uncoated carbide Speed, inserted m/min	Uncoated carbide Speed, inserted ft/min	Uncoated carbide Feed mm/rev	Uncoated carbide Feed in./rev	Coated carbide Speed m/min	Coated carbide Speed ft/min	Coated carbide Feed mm/rev	Coated carbide Feed in./rev
Annealed, with hardness of 225 to 275 HB															
0.25	0.010	30(a)	100(a)	0.075	0.003	105(b)	345(b)	125(b)	405(b)	0.075	0.003	150(c)	500(c)	0.075	0.003
1.25	0.050	24(a)	80(a)	0.13	0.005	84(d)	275(d)	100(d)	325(d)	0.13	0.005	120(e)	400(e)	0.13	0.005
2.5	0.100	18(a)	60(a)	0.30	0.012	64(d)	210(d)	76(d)	250(d)	0.30	0.012	100(e)	325(e)	0.30	0.012
Quenched and tempered, with hardness of 275 to 325 HB															
0.25	0.010	23(f)	75(f)	0.075	0.003	81(b)	265(b)	95(b)	315(b)	0.075	0.003	135(c)	450(c)	0.075	0.003
1.25	0.050	18(f)	60(f)	0.13	0.005	64(d)	210(d)	76(d)	250(d)	0.13	0.005	105(e)	350(e)	0.13	0.005
2.5	0.100	15(f)	50(f)	0.30	0.012	52(d)	170(d)	60(d)	200(d)	0.30	0.012	76(e)	250(e)	0.30	0.012
Quenched and tempered, with hardness of 375 to 425 HB															
0.25	0.010	20(f)	65(f)	0.075	0.003	64(b)	210(b)	76(b)	250(b)	0.075	0.003	105(c)	350(c)	0.075	0.003
1.25	0.050	15(f)	50(f)	0.13	0.005	52(d)	170(d)	60(d)	200(d)	0.13	0.005	84(e)	275(e)	0.13	0.005
2.5	0.100	11(f)	35(f)	0.30	0.012	41(d)	135(d)	49(d)	160(d)	0.30	0.012	64(e)	210(e)	0.30	0.012

(a) High speed steel tool material: M2 or M3. (b) Carbide tool material: C-7. (c) Carbide tool material: CC-7. (d) Carbide tool material: C-6. (e) Carbide tool material: CC-6. (f) Any premium high speed steel tool material (T15, M33, M41 to M47)

AISI Type 414: Surface Grinding with a Horizontal Spindle and Reciprocating Table

Test specimens were annealed or cold drawn, with hardness of 135 to 275 HB

Operation	Wheel speed m/s	Wheel speed ft/min	Table speed m/min	Table speed ft/min	Downfeed mm/pass	Downfeed in./pass	Crossfeed mm/pass	Crossfeed in./pass	Wheel identification
Annealed, with hardness of 135 to 275 HB									
Roughing	28-33	5500-6500	15-30	50-100	0.050	0.002	1.25-12.5	0.050-0.500	A46IV
Finishing	28-33	5500-6500	15-30	50-100	0.013 max	0.0005 max	(a)	(a)	A46IV
Quenched and tempered, with hardness greater than 275 HB									
Roughing	28-33	5500-6500	15-30	50-100	0.025	0.001	0.65-6.5	0.025-0.250	A46HV
Finishing	28-33	5500-6500	15-30	50-100	0.008 max	0.0003 max	(b)	(b)	A46HV

(a) Maximum ¼ of wheel width. (b) Maximum 1⁄10 of wheel width

Machining Data (Ref 2) (continued)

AISI Type 414: Reaming

Based on 4 flutes for 3- and 6-mm (0.125- and 0.25-in.) reamers, 6 flutes for 12-mm (0.5-in.) reamers, and 8 flutes for 25-mm (1-in.) and larger reamers

Tool material	Speed m/min	Speed ft/min	Feed per revolution for reamer diameter of: 3 mm (0.125 in.) mm	3 mm in.	6 mm (0.25 in.) mm	6 mm in.	12 mm (0.5 in.) mm	12 mm in.	25 mm (1 in.) mm	25 mm in.	35 mm (1.5 in.) mm	35 mm in.	50 mm (2 in.) mm	50 mm in.
Annealed, with hardness of 225 to 275 HB														
Roughing														
M1, M2, and M7 high speed steel	20	65	0.075	0.003	0.102	0.004	0.13	0.005	0.20	0.008	0.25	0.010	0.30	0.012
C-2 carbide	24	80	0.102	0.004	0.20	0.008	0.30	0.012	0.40	0.016	0.50	0.020	0.60	0.024
Finishing														
M1, M2, and M7 high speed steel	11	35	0.102	0.004	0.102	0.004	0.15	0.006	0.20	0.008	0.23	0.009	0.25	0.010
C-2 carbide	15	50	0.102	0.004	0.13	0.005	0.15	0.006	0.20	0.008	0.23	0.009	0.25	0.010
Quenched and tempered, with hardness of 275 to 325 HB														
Roughing														
M1, M2, and M7 high speed steel	15	50	0.075	0.003	0.102	0.004	0.13	0.005	0.20	0.008	0.25	0.010	0.30	0.012
C-2 carbide	20	65	0.102	0.004	0.15	0.006	0.23	0.009	0.40	0.015	0.45	0.018	0.50	0.020
Finishing														
M1, M2, and M7 high speed steel	11	35	0.075	0.003	0.075	0.003	0.102	0.004	0.15	0.006	0.18	0.007	0.20	0.008
C-2 carbide	15	50	0.075	0.003	0.102	0.004	0.13	0.005	0.15	0.006	0.18	0.007	0.20	0.008
Quenched and tempered, with hardness of 375 to 425 HB														
Roughing														
Any premium high speed steel (T15, M33, M41-M47)	9	30	0.050	0.002	0.102	0.004	0.13	0.005	0.20	0.008	0.25	0.010	0.30	0.012
C-2 carbide	14	45	0.075	0.003	0.13	0.005	0.15	0.006	0.20	0.008	0.25	0.010	0.30	0.012
Finishing														
Any premium high speed steel (T15, M33, M41-M47)	9	30	0.075	0.003	0.075	0.003	0.102	0.004	0.15	0.006	0.18	0.007	0.20	0.008
C-2 carbide	12	40	0.075	0.003	0.102	0.004	0.13	0.005	0.15	0.006	0.18	0.007	0.20	0.008

AISI Type 414: Planing

Tool material	Depth of cut mm	Depth of cut in.	Speed m/min	Speed ft/min	Feed/stroke mm	Feed/stroke in.
Annealed, with hardness of 225 to 275 HB						
M2 and M3 high speed steel	0.1	0.005	8	25	(a)	(a)
	2.5	0.100	11	35	0.75	0.030
	12	0.500	8	25	1.15	0.045
C-6 carbide	0.1	0.005	38	125	(a)	(a)
	2.5	0.100	43	140	1.50	0.060
Quenched and tempered, with hardness of 275 to 325 HB						
M2 and M3 high speed steel	0.1	0.005	6	20	(a)	(a)
	2.5	0.100	8	25	0.75	0.030
	12	0.500	5	15	1.15	0.045
C-6 carbide	0.1	0.005	27	90	(a)	(a)
	2.5	0.100	30	100	1.50	0.060

(a) Feed is 75% the width of the square nose finishing tool

AISI Type 414: Drilling

Tool material	Speed, m/min (ft/min)	Feed per revolution for nominal hole diameter of: 1.5 mm (0.062 in.) mm (in.)	3 mm (0.125 in.) mm (in.)	6 mm (0.25 in.) mm (in.)	12 mm (0.5 in.) mm (in.)	18 mm (0.75 in.) mm (in.)	25 mm (1 in.) mm (in.)	35 mm (1.5 in.) mm (in.)	50 mm (2 in.) mm (in.)
Annealed, with hardness of 225 to 275 HB									
M10, M7, and M1 high speed steel	17 (55)	0.025 (0.001)	0.075 (0.003)	0.130 (0.005)	0.18 (0.007)	0.25 (0.010)	0.30 (0.012)	0.40 (0.015)	0.45 (0.018)
Quenched and tempered, with hardness of 275 to 325 HB									
M10, M7, and M1 high speed steel	15 (50)	... (...)	0.075 (0.003)	0.102 (0.004)	0.18 (0.007)	0.23 (0.009)	0.28 (0.011)	0.33 (0.013)	0.40 (0.015)

AISI Type 414: Cylindrical and Internal Grinding

Operation	Wheel speed m/s	Wheel speed ft/min	Work speed m/min	Work speed ft/min	Infeed on diameter mm/pass	Infeed on diameter in./pass	Traverse(a)	Wheel identification
Annealed, with hardness of 135 to 275 HB								
Cylindrical grinding								
Roughing	28-33	5500-6500	15-30	50-100	0.050	0.002	1/4	A60JV
Finishing	28-33	5500-6500	15-30	50-100	0.013 max	0.0005 max	1/8	A60JV
Internal grinding								
Roughing	25-33	5000-6500	23-60	75-200	0.013	0.0005	1/3	A60KV
Finishing	25-33	5000-6500	23-60	75-200	0.005 max	0.0002 max	1/6	A60KV
Quenched and tempered, with hardness greater than 275 HB								
Cylindrical grinding								
Roughing	28-33	5500-6500	15-30	50-100	0.050	0.002	1/4	A60IV
Finishing	28-33	5500-6500	15-30	50-100	0.013 max	0.0005 max	1/8	A60IV
Internal grinding								
Roughing	25-33	5000-6500	15-46	50-150	0.013	0.0005	1/3	A60JV
Finishing	25-33	5000-6500	15-46	50-150	0.005 max	0.0002 max	1/6	A60JV

(a) Wheel width per revolution of work

AISI Types 416, 416Se

AISI Types 416, 416Se: Chemical Composition

Element	Composition, % Type 414	Type 414Se
Carbon	0.15 max	0.15 max
Manganese	1.25 max	1.25 max
Phosphorus	0.06 max	0.06 max
Sulfur	0.15 max	0.06 max
Silicon	1.00 max	1.00 max
Chromium	12-14 max	12-14 max
Molybdenum(a)	0.60 max	...
Selenium	...	0.15 min

(a) Optional

Characteristics. AISI types 416 and 416Se are corrosion-resistant chromium steels with added elements to improve machining and nonseizing characteristics. These grades are the most readily machinable of all the stainless steels and are suitable for use in automatic screw machines. Both types are magnetic in all conditions.

Type 416 is satisfactory for only a limited amount of cold working; therefore machining is used most often for fabricating parts from this grade. Cold upsetting, flaring, or swaging may result in splitting. Although type 416 is forgeable, it is not recommended for parts requiring upsetting operations. Forging grade type 416 is recommended for most forging operations.

Types 416 and 416Se are not recommended in applications that involve welding. When welding is attempted, carefully controlled practices are required. Electrodes of types 410, 308, or 310 may be used, depending upon the type of weld zone desired.

Types 416 and 416Se have a general machinability rating of about 90% based on free-machining AISI 1212. These grades are resistant to the corrosive action of the atmosphere, fresh water, various alkalis, and mild acids. They have good oxidation resistance up to 650 °C (1200 °F) in continuous service. Scaling becomes excessive about 760 °C (1400 °F) in intermittent service.

Typical Uses. Applications of AISI types 416 and 416Se include nongalling, nonseizing, corrosion-resistant parts machined on automatic screw machines, such as screws, bolts, nuts, and studs; rivets; automobile trim; appliance parts; shafts and pump components; valve parts; and hardware.

AISI Type 416: Similar Steels (U.S. and/or Foreign). UNS S41600; ASME SA194; ASTM A194, A314, A473, A581, A582; FED QQ-W-423; MIL SPEC MIL-S-862; SAE J405 (51416); (W. Ger.) DIN 1.4005; (Ital.) UNI X 12 CrS 13; (U.K.) B.S. 416 S 21

AISI Type 416Se: Similar Steels (U.S. and/or Foreign). UNS S41623; AMS 5610; ASME SA194; ASTM A194, A314, A473, A511, A581, A582; MIL SPEC MIL-S-862; SAE J405 (51416Se)

Physical Properties

AISI Types 416, 416Se: Thermal Treatment Temperatures

Treatment	Temperature range °C	°F
Forging, start	1150-1205	2100-2200
Forging, finish	815 min	1500 min
Full annealing(a)	815-900	1500-1650
Process annealing(b)	650-760	1200-1400
Hardening(c)	925-1010	1700-1850
Tempering(d)	230-730	450-1350

(a) Hold for 1 h per 25 mm (1.0 in.) of thickness; air cool. (b) Hold at temperature for uniform heat distribution; cool in air. (c) Hold at temperature for 15 to 30 min; quench in air or oil. (d) Temperature depends on the hardness, strength, toughness, or corrosion resistance required. Holding at 230 to 400 °C (450 to 750 °F) for 1 to 3 h slightly reduces hardness and improves toughness without affecting corrosion resistance. Tempering between 400 to 565 °C (750 to 1050 °F) is usually avoided because of lowered impact toughness and corrosion resistance

AISI Types 416, 416Se: Thermal Conductivity (Ref 6)

Temperature °C	°F	Thermal conductivity W/m·K	Btu/ft·h·°F
100	212	24.9	14.4
500	930	28.7	16.6

AISI Types 416, 416Se: Selected Physical Properties (Ref 6)

Melting range	1480-1530 °C (2700-2790 °F)
Density	7.8 g/cm³ (0.28 lb/in.³)
Specific heat at 0-100 °C (32-212 °F)	460 J/kg·K (0.11 Btu/lb·°F)
Electrical resistivity at 20 °C (68 °F)	57 nΩ·m
Electrical resistivity at 650 °C (1200 °F)	108 nΩ·m

AISI Types 416, 416Se: Average Coefficients of Linear Thermal Expansion (Ref 6)

Temperature range °C	°F	Coefficient μm/m·K	μin./in.·°F
0-100	32-212	9.9	5.5
0-315	32-600	11.0	6.1
0-540	32-1000	11.5	6.4
0-650	32-1200	11.7	6.5

Mechanical Properties

AISI Types 416, 416Se: Effect of Tempering Temperature on Tensile Properties (Ref 7)

Test specimens were 25-mm (1.0-in.) diam bar; oil quenched from 995 °C (1825 °F) and held at tempering temperature for 2 h

Test temperature °C	°F	Tensile strength MPa	ksi	Yield strength(a) MPa	ksi	Elongation(b), %	Reduction in area, %	Hardness, HRC	Izod impact energy J	ft·lb
150	300	1310	190	962	139.5	9	27	39.5	20.3	15.0
260	500	1295	187.5	962	139.5	9.5	23.5	39	20.3	15.0
370	700	1260	182.5	962	139.5	14	46	39	20.3	15.0
480	900	1185	172	983	142.5	16	45	39	20.3	15.0
540	1000	1080	157	895	130	13	45	34	27.3	20.1
595	1100	865	125.5	740	107	17	53	26	54.4	40.1
650	1200	796	115.5	670	97	17.5	54.5	23.5	61.1	45.1
705	1300	741	107.5	600	87	20	57	20	75.0	55.3

(a) 0.2% offset. (b) In 50 mm (2 in.)

AISI Types 416, 416Se: Izod Impact Properties and Endurance Limits (Ref 6)

Test specimens were 25-mm (1.0-in.) diam bar

Condition	Impact energy J	ft·lb	Endurance limit MPa	ksi
Annealed	95	70	275	40
Tempered	34	25	380	55
Tempered and cold drawn	27	20	365	53

AISI Types 416, 416Se: Modulus of Elasticity (Ref 7)

Stressed in:	Modulus GPa	10^6 psi
Tension	200	29
Torsion	83	12

Mechanical Properties (continued)

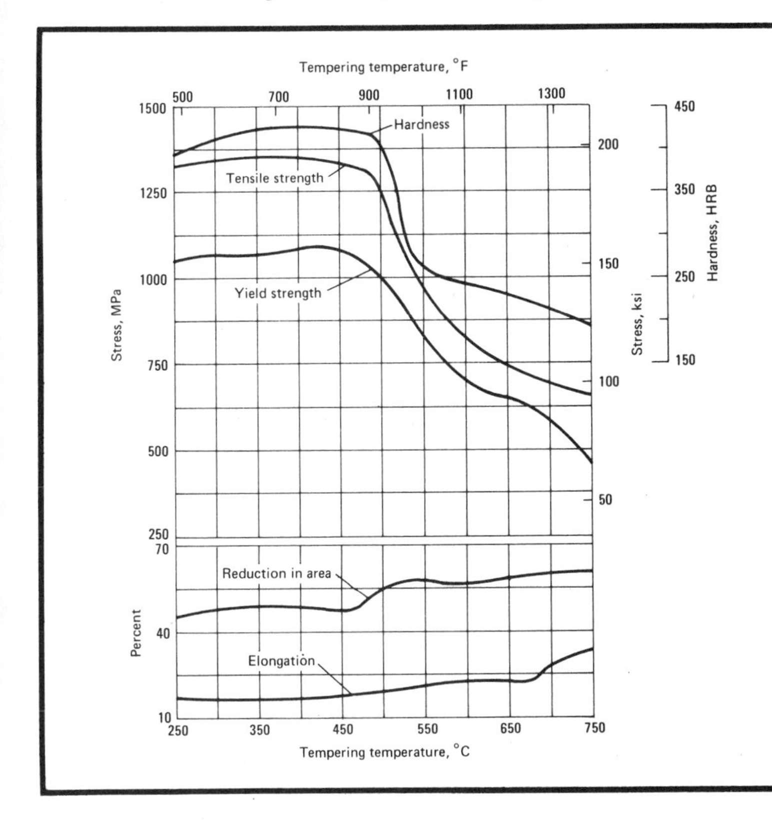

AISI Type 416: Effect of Tempering Temperature on Tensile Properties. Heat treated at 980 °C (1800 °F); oil quenched; tempered for 4 h at temperature given. Elongation was measured in 50 mm (2 in.); yield strength at 0.2% offset. (Ref 10)

AISI Types 416, 416Se: Representative Tensile Properties (Ref 6)

Condition	Tensile strength MPa	Tensile strength ksi	Yield strength(a) MPa	Yield strength(a) ksi	Elongation(b), %	Reduction in area, %	Hardness
Wire, 6.4-mm (0.25-in.) diam							
Annealed	515	75	275	40	20	60	82 HRB
Tempered and cold drawn	620	90	550	80	10	50	92 HRB
Heat treated	930	135	725	105	5	40	29 HRC
Bar							
Annealed	515	75	275	40	30	60	82 HRB
Tempered	760	110	585	85	18	55	97 HRB
Tempered and cold drawn(c)	690	100	585	85	13	50	94 HRB

(a) 0.2% offset. (b) In 50 mm (2 in.), except for wire. For wire 3.18-mm (0.125-in.) diam and larger, gage length is 4 × diam; for smaller diameters, the gage length is 254 mm (10.0 in.). (c) Diameter of 25 mm (1 in.)

AISI Type 416: Creep Rates (Ref 10)

Creep rate of 1% in 10 000 h

Test temperature °C	Test temperature °F	Stress MPa	Stress ksi
540	1000	63	9.2
595	1100	29	4.2
650	1200	14	2.0
705	1300	7	1.0

Machining Data (Ref 2)

AISI Types 416, 416Se: Surface Grinding with a Horizontal Spindle and Reciprocating Table

Operation	Wheel speed m/s	Wheel speed ft/min	Table speed m/min	Table speed ft/min	Downfeed mm/pass	Downfeed in./pass	Crossfeed mm/pass	Crossfeed in./pass	Wheel identification
Annealed or cold drawn, with hardness of 135 to 240 HB									
Roughing	28-33	5500-6500	15-30	50-100	0.050	0.002	1.25-12.5	0.050-0.500	A46IV
Finishing	28-33	5500-6500	15-30	50-100	0.013 max	0.0005 max	(a)	(a)	A46IV
Quenched and tempered, with hardness greater than 275 HB									
Roughing	28-33	5500-6500	15-30	50-100	0.025	0.001	0.65-6.5	0.025-0.250	A46HV
Finishing	28-33	5500-6500	15-30	50-100	0.013 max	0.0015 max	(b)	(b)	A46HV

(a) Maximum 1/4 of wheel width. (b) Maximum 1/10 of wheel width

Machining Data (Ref 2) (continued)

AISI Types 416, 416Se: Cylindrical and Internal Grinding

Operation	Wheel speed m/s	Wheel speed ft/min	Work speed m/min	Work speed ft/min	Infeed on diameter mm/pass	Infeed on diameter in./pass	Traverse(a)	Wheel identification
Annealed or cold drawn, with hardness of 135 to 240 HB								
Cylindrical grinding								
Roughing	28-33	5500-6500	15-30	50-100	0.050	0.002	1/4	A60JV
Finishing	28-33	5500-6500	15-30	50-100	0.013 max	0.0005 max	1/8	A60JV
Internal grinding								
Roughing	25-33	5000-6500	23-60	75-200	0.013	0.0005	1/3	A60KV
Finishing	25-33	5000-6500	23-60	75-200	0.005 max	0.0002 max	1/6	A60KV
Quenched and tempered, with hardness greater than 275 HB								
Cylindrical grinding								
Roughing	28-33	5500-6500	15-30	50-100	0.050	0.002	1/4	A60IV
Finishing	28-33	5500-6500	15-30	50-100	0.013 max	0.0005 max	1/8	A60IV
Internal grinding								
Roughing	25-33	5000-6500	15-46	50-150	0.013	0.0005	1/3	A60JV
Finishing	25-33	5000-6500	15-46	50-150	0.005 max	0.0002 max	1/6	A60JV

(a) Wheel width per revolution of work

AISI Types 416, 416Se: Boring

Depth of cut		High speed steel				Uncoated carbide						Coated carbide			
		Speed		Feed		Speed, brazed		Speed, inserted		Feed		Speed		Feed	
mm	in.	m/min	ft/min	mm/rev	in./rev	m/min	ft/min	m/min	ft/min	mm/rev	in./rev	m/min	ft/min	mm/rev	in./rev
Annealed, with hardness of 135 to 185 HB															
0.25	0.010	60(a)	200(a)	0.075	0.003	155(b)	510(b)	185(b)	600(b)	0.075	0.003	235(c)	775(c)	0.075	0.003
1.25	0.050	49(a)	160(a)	0.13	0.005	125(d)	410(d)	145(d)	475(d)	0.13	0.005	185(e)	615(e)	0.13	0.005
2.5	0.100	41(a)	135(a)	0.30	0.012	105(f)	350(f)	30(f)	420(f)	0.30	0.012	165(g)	545(g)	0.30	0.012
Annealed or cold drawn, with hardness of 185 to 240 HB															
0.25	0.010	53(a)	175(a)	0.075	0.003	135(d)	450(d)	160(d)	525(d)	0.075	0.003	205(e)	675(e)	0.075	0.003
1.25	0.050	43(a)	140(a)	0.13	0.005	115(f)	375(f)	130(f)	425(f)	0.13	0.005	170(g)	550(g)	0.13	0.005
2.5	0.100	37(a)	120(a)	0.30	0.012	100(f)	325(f)	115(f)	375(f)	0.30	0.012	145(g)	475(g)	0.30	0.012
Quenched and tempered, with hardness of 275 to 325 HB															
0.25	0.010	30(h)	100(h)	0.075	0.003	105(d)	340(d)	120(d)	400(d)	0.075	0.003	150(e)	500(e)	0.075	0.003
1.25	0.050	24(h)	80(h)	0.13	0.005	84(f)	275(f)	100(f)	325(f)	0.13	0.005	120(g)	400(g)	0.13	0.005
2.5	0.100	20(h)	65(h)	0.30	0.012	76(f)	250(f)	84(f)	275(f)	0.30	0.012	115(g)	375(g)	0.30	0.012
Quenched and tempered, with hardness of 375 to 425 HB															
0.25	0.010	20(h)	65(h)	0.050	0.002	46(d)	150(d)	56(d)	185(d)	0.050	0.002	84(e)	275(e)	0.050	0.002
1.25	0.050	15(h)	50(h)	0.102	0.004	38(f)	125(f)	46(f)	150(f)	0.102	0.004	66(g)	215(g)	0.102	0.004
2.5	0.100	12(h)	40(h)	0.20	0.008	25(f)	115(f)	41(f)	135(f)	0.20	0.008	53(g)	175(g)	0.20	0.008

(a) High speed steel tool material: M2 or M3. (b) Carbide tool material: C-8. (c) Carbide tool material: CC-8. (d) Carbide tool material: C-7. (e) Carbide tool material: CC-7. (f) Carbide tool material: C-6. (g) Carbide tool material: CC-6. (h) Any premium high speed steel tool material (T15, M33, M41 to M47)

AISI Types 416, 416Se: Face Milling

Depth of cut		High speed steel				Uncoated carbide						Coated carbide			
		Speed		Feed/tooth		Speed, brazed		Speed, inserted		Feed/tooth		Speed		Feed/tooth	
mm	in.	m/min	ft/min	mm	in.	m/min	ft/min	m/min	ft/min	mm	in.	m/min	ft/min	mm	in.
Annealed, with hardness of 135 to 185 HB															
1	0.040	70(a)	230(a)	0.20	0.008	190(b)	620(b)	205(b)	675(b)	0.20	0.008	305(c)	1000(c)	0.20	0.008
4	0.150	53(a)	175(a)	0.30	0.012	135(b)	450(b)	150(b)	500(b)	0.30	0.012	200(c)	650(c)	0.30	0.012
8	0.300	41(a)	135(a)	0.40	0.016	100(d)	325(d)	120(d)	400(d)	0.40	0.016	160(e)	525(e)	0.40	0.016
Annealed or cold drawn, with hardness of 185 to 240 HB															
1	0.040	53(a)	175(a)	0.20	0.008	170(b)	550(b)	185(b)	610(b)	0.18	0.007	280(c)	915(c)	0.18	0.007
4	0.150	41(a)	135(a)	0.30	0.012	120(b)	400(b)	145(b)	475(b)	0.25	0.010	185(c)	600(c)	0.25	0.010
8	0.300	32(a)	105(a)	0.40	0.016	90(d)	300(d)	115(d)	370(d)	0.36	0.014	145(e)	480(e)	0.36	0.014
Quenched and tempered, with hardness of 275 to 375 HB															
1	0.040	32f)	105f)	0.15	0.006	120(b)	400(b)	135(b)	440(b)	0.15	0.006	200(c)	650(c)	0.13	0.005
4	0.150	27f)	90f)	0.23	0.009	90(b)	300(b)	105(b)	350(b)	0.20	0.008	140(c)	455(c)	0.18	0.007
8	0.300	21f)	70f)	0.30	0.012	69(d)	205(d)	84(d)	275(d)	0.25	0.010	110(e)	355(e)	0.23	0.009
Quenched and tempered, with hardness of 375 to 425 HB															
1	0.040	18f)	60f)	0.102	0.004	67(b)	220(b)	76(b)	250(b)	0.102	0.004	115(c)	375(c)	0.075	0.003
4	0.150	14f)	45f)	0.15	0.006	60(b)	200(b)	59(b)	225(b)	0.15	0.006	90(c)	300(c)	0.13	0.005
8	0.300	11f)	35f)	0.20	0.008	43(d)	140(d)	53(d)	175(d)	0.20	0.008	69(e)	325(e)	0.18	0.007

(a) High speed steel tool material: M2 or M7. (b) Carbide tool material: C-6. (c) Carbide tool material: CC-6. (d) Carbide tool material: C-5. (e) Carbide tool material: CC-5. (f) Use any premium high speed steel tool material (T15, M33, M41 to M47)

AISI Types 416, 416Se: Reaming

Based on 4 flutes for 3- and 6-mm (0.125- and 0.250-in.) reamers, 6 flutes for 12-mm (0.5-in.) reamers, and 8 flutes for 25-mm (1-in.) and larger reamers

Tool material	Speed m/min	Speed ft/min	Feed per revolution for reamer diameter of: 3 mm (0.125 in.) mm	3 mm (0.125 in.) in.	6 mm (0.25 in.) mm	6 mm (0.25 in.) in.	12 mm (0.5 in.) mm	12 mm (0.5 in.) in.	25 mm (1 in.) mm	25 mm (1 in.) in.	35 mm (1.5 in.) mm	35 mm (1.5 in.) in.	50 mm (2 in.) mm	50 mm (2 in.) in.
Annealed, with hardness of 135 to 185 HB														
Roughing														
M1, M2, and M7 high speed steel	40	130	0.075	0.003	0.13	0.005	0.20	0.008	0.25	0.010	0.30	0.012	0.40	0.015
C-2 carbide	46	150	0.102	0.004	0.20	0.008	0.30	0.012	0.40	0.016	0.50	0.020	0.60	0.024
Finishing														
M1, M2, and M7 high speed steel	18	60	0.102	0.004	0.102	0.004	0.15	0.006	0.20	0.008	0.25	0.010	0.30	0.012
C-2 carbide	21	70	0.102	0.004	0.15	0.006	0.20	0.008	0.25	0.010	0.28	0.011	0.30	0.012
Annealed or cold drawn, with hardness of 185 to 240 HB														
Roughing														
M1, M2, and M7 high speed steel	34	110	0.075	0.003	0.13	0.005	0.20	0.008	0.25	0.010	0.30	0.012	0.40	0.015
C-2 carbide	38	125	0.102	0.004	0.20	0.008	0.30	0.012	0.40	0.016	0.50	0.020	0.60	0.024
Finishing														
M1, M2, and M7 high speed steel	15	50	0.102	0.004	0.102	0.004	0.15	0.006	0.20	0.008	0.20	0.010	0.30	0.012
C-2 carbide	20	65	0.102	0.004	0.15	0.006	0.20	0.008	0.25	0.010	0.28	0.011	0.30	0.012
Quenched and tempered, with hardness of 275 to 325 HB														
Roughing														
M1, M2, and M7 high speed steel	20	65	0.075	0.003	0.13	0.005	0.20	0.008	0.25	0.010	0.30	0.012	0.40	0.015
C-2 carbide	24	80	0.102	0.004	0.15	0.006	0.23	0.009	0.40	0.015	0.45	0.018	0.50	0.020
Finishing														
M1, M2, and M7 high speed steel	12	40	0.075	0.003	0.102	0.004	0.15	0.006	0.20	0.008	0.23	0.009	0.25	0.010
C-2 carbide	18	80	0.102	0.004	0.13	0.005	0.15	0.006	0.20	0.008	0.23	0.009	0.25	0.010
Quenched and tempered, with hardness of 375 to 425 HB														
Roughing														
Any premium high speed steel (T15, M33, M41-M47)	14	45	0.050	0.002	0.102	0.004	0.13	0.005	0.20	0.008	0.25	0.010	0.30	0.012
C-2 carbide	18	60	0.075	0.003	0.13	0.005	0.15	0.006	0.20	0.008	0.25	0.010	0.30	0.012
Finishing														
Any premium high speed steel (T15, M33, M41-M47)	9	30	0.075	0.003	0.102	0.004	0.13	0.005	0.15	0.006	0.18	0.007	0.20	0.008
C-2 carbide	12	40	0.075	0.003	0.102	0.004	0.13	0.005	0.15	0.006	0.18	0.007	0.20	0.008

AISI Types 416, 416Se: Planing

Tool material	Depth of cut mm	Depth of cut in.	Speed m/min	Speed ft/min	Feed/stroke mm	Feed/stroke in.
Annealed, with hardness of 135 to 185 HB						
M2 and M3 high speed steel	0.1	0.005	15	50	(a)	(a)
	2.5	0.100	21	70	1.25	0.050
	12	0.500	14	45	1.50	0.060
C-6 carbide	0.1	0.005	69	225	(a)	(a)
	2.5	0.100	79	260	1.50	0.060
Annealed or cold drawn, with hardness of 185 to 240 HB						
M2 and M3 high speed steel	0.1	0.005	12	40	(a)	(a)
	2.5	0.100	18	60	0.075	0.030
	12	0.500	11	35	1.15	0.045
C-6 carbide	0.1	0.005	60	200	(a)	(a)
	2.5	0.100	73	240	1.50	0.060
Quenched and tempered, with hardness of 275 to 325 HB						
M2 and M3 high speed steel	0.1	0.005	9	30	(a)	(a)
	2.5	0.100	12	40	0.075	0.030
	12	0.500	8	25	1.15	0.045
C-6 carbide	0.1	0.005	69	125	(a)	(a)
	2.5	0.100	43	140	1.50	0.060

(a) Feed is 75% the width of the square nose finishing tool

AISI Types 416, 416Se: Turning (Single Point and Box Tools)

Depth of cut		High speed steel				Uncoated carbide						Coated carbide			
		Speed		Feed		Speed, brazed		Speed, inserted		Feed		Speed		Feed	
mm	in.	m/min	ft/min	mm/rev	in./rev	m/min	ft/min	m/min	ft/min	mm/rev	in./rev	m/min	ft/min	mm/rev	in./rev
Annealed, with hardness of 135 to 185 HB															
1	0.040	60(a)	200(a)	0.18	0.007	185(b)	600(b)	205(b)	675(b)	0.18	0.007	265(c)	875(c)	0.18	0.007
4	0.150	52(a)	170(a)	0.40	0.015	160(d)	520(d)	185(d)	600(d)	0.40	0.015	235(e)	775(e)	0.40	0.015
8	0.300	40(a)	130(a)	0.50	0.020	120(d)	400(d)	140(d)	460(d)	0.75	0.030	185(e)	600(e)	0.50	0.020
16	0.625	30(a)	100(a)	0.75	0.030	90(d)	300(d)	110(d)	360(d)	1.00	0.040	...	...	...	...
Annealed or cold drawn, with hardness of 185 to 240 HB															
1	0.040	53(a)	175(a)	0.18	0.007	170(b)	550(b)	185(b)	600(b)	0.18	0.007	235(c)	775(c)	0.18	0.007
4	0.150	46(a)	150(a)	0.40	0.015	145(d)	475(d)	160(d)	525(d)	0.40	0.015	205(e)	675(e)	0.40	0.015
8	0.300	37(a)	120(a)	0.50	0.020	115(d)	370(d)	120(d)	400(d)	0.50	0.020	120(e)	400(e)	0.50	0.020
16	0.625	27(a)	90(a)	0.75	0.030	85(d)	280(d)	90(d)	300(d)	0.75	0.030	...	...	...	...
Quenched and tempered, with hardness of 275 to 325 HB															
1	0.040	30(f)	100(f)	0.18	0.007	120(b)	400(b)	135(b)	450(b)	0.18	0.007	175(c)	575(c)	0.18	0.007
4	0.150	24(f)	80(f)	0.40	0.015	105(d)	350(d)	120(d)	400(d)	0.40	0.015	160(e)	525(e)	0.40	0.015
8	0.300	18(f)	60(f)	0.50	0.020	84(d)	275(d)	95(d)	310(d)	0.50	0.020	120(e)	400(e)	0.50	0.020
Quenched and tempered, with hardness of 375 to 425 HB															
1	0.040	18(f)	60(f)	0.13	0.005	60(b)	200(b)	69(b)	225(b)	0.13	0.005	90(c)	300(c)	0.13	0.005
4	0.150	15(f)	50(f)	0.25	0.010	52(d)	170(d)	58(d)	190(d)	0.25	0.010	76(e)	250(e)	0.25	0.010
8	0.300	12(f)	40(f)	0.40	0.015	40(d)	130(d)	46(d)	150(d)	0.40	0.015	60(e)	200(e)	0.40	0.015

(a) High speed steel tool material: M2 or M3. (b) Carbide tool material: C-7. (c) Carbide tool material: CC-7. (d) Carbide tool material: C-6. (e) Carbide tool material: CC-6. (f) Any premium high speed steel tool material (T15, M33, M41 to M47)

AISI Types 416, 416Se: End Milling (Profiling)

Tool material	Depth of cut		Speed		Feed per tooth for cutter diameter of: 10 mm (0.375 in.)		12 mm (0.5 in.)		18 mm (0.75 in.)		25-50 mm (1-2 in.)	
	mm	in.	m/min	ft/min	mm	in.	mm	in.	mm	in.	mm	in.
Annealed, with hardness of 135 to 185 HB												
M2, M3, and M7 high speed steel	0.5	0.020	55	180	0.025	0.001	0.050	0.002	0.102	0.004	0.13	0.005
	1.5	0.060	43	140	0.050	0.002	0.075	0.003	0.13	0.005	0.15	0.006
	diam/4	diam/4	38	125	0.038	0.0015	0.050	0.002	0.102	0.004	0.13	0.005
	diam/2	diam/2	34	110	0.025	0.001	0.038	0.0015	0.075	0.003	0.10	0.004
C-5 carbide	0.5	0.020	160	520	0.013	0.0005	0.050	0.002	0.102	0.004	0.15	0.006
	1.5	0.060	120	400	0.025	0.001	0.075	0.003	0.13	0.005	0.18	0.007
	diam/4	diam/4	105	350	0.085	0.001	0.050	0.002	0.102	0.004	0.15	0.006
	diam/2	diam/2	100	320	0.013	0.0005	0.025	0.001	0.075	0.003	0.13	0.005
Annealed or cold drawn, with hardness of 185 to 240 HB												
M2, M3, and M7 high speed steel	0.5	0.020	49	160	0.025	0.001	0.050	0.002	0.075	0.003	0.102	0.004
	1.5	0.060	37	120	0.050	0.002	0.075	0.003	0.102	0.004	0.13	0.005
	diam/4	diam/4	30	100	0.038	0.0015	0.050	0.002	0.075	0.003	0.102	0.004
	diam/2	diam/2	29	95	0.025	0.001	0.038	0.0015	0.050	0.002	0.075	0.003
C-5 carbide	0.5	0.020	140	455	0.013	0.0005	0.025	0.001	0.102	0.004	0.15	0.006
	1.5	0.060	105	350	0.025	0.001	0.050	0.002	0.13	0.005	0.18	0.007
	diam/4	diam/4	90	295	0.025	0.001	0.038	0.0015	0.102	0.004	0.13	0.005
	diam/2	diam/2	84	275	0.013	0.0005	0.025	0.001	0.075	0.003	0.102	0.004
Quenched and tempered, with hardness of 275 to 325 HB												
M2, M3, and M7 high speed steel	0.5	0.020	30	100	0.018	0.0007	0.038	0.0015	0.075	0.003	0.102	0.004
	1.5	0.060	23	75	0.025	0.001	0.050	0.002	0.102	0.004	0.13	0.005
	diam/4	diam/4	20	65	0.018	0.0007	0.038	0.0015	0.075	0.003	0.102	0.004
	diam/2	diam/2	17	55	0.013	0.0005	0.025	0.001	0.050	0.002	0.075	0.003
C-5 carbide	0.5	0.020	110	360	0.013	0.0005	0.025	0.001	0.050	0.002	0.102	0.004
	1.5	0.060	82	270	0.025	0.001	0.050	0.002	0.075	0.003	0.13	0.005
	diam/4	diam/4	72	235	0.025	0.001	0.038	0.0015	0.050	0.002	0.102	0.004
	diam/2	diam/2	67	220	0.013	0.0005	0.050	0.001	0.075	0.003	0.075	0.003
Quenched and tempered, with hardness of 375 to 425 HB												
Any premium high speed steel (T15, M33, M41-M47)	0.5	0.020	15	50	0.013	0.0005	0.018	0.0007	0.025	0.001	0.050	0.002
	1.5	0.060	11	35	0.013	0.0005	0.025	0.001	0.050	0.002	0.075	0.003
	diam/4	diam/4	9	30	0.013	0.0005	0.013	0.0005	0.038	0.0015	0.063	0.0025
	diam/2	diam/2	8	25	...	...	...	...	0.025	0.001	0.050	0.002
C-5 carbide	0.5	0.020	76	250	0.013	0.0005	0.025	0.001	0.050	0.002	0.075	0.003
	1.5	0.060	58	190	0.025	0.001	0.050	0.002	0.075	0.003	0.102	0.004
	diam/4	diam/4	50	165	0.025	0.001	0.038	0.0015	0.063	0.0025	0.075	0.003
	diam/2	diam/2	46	150	0.013	0.0005	0.025	0.001	0.050	0.002	0.13	0.005

Machining Data (Ref 2) (continued)

AISI Types 416, 416Se: Turning (Cutoff and Form Tools)

Tool material	Speed, m/min (ft/min)	Feed per revolution for cutoff tool width of: 1.5 mm (0.062 in.) mm (in.)	3 mm (0.125 in.) mm (in.)	6 mm (0.25 in.) mm (in.)	Feed per revolution for form tool width of: 12 mm (0.5 in.) mm (in.)	18 mm (0.75 in.) mm (in.)	25 mm (1 in.) mm (in.)	35 mm (1.5 in.) mm (in.)	50 mm (2 in.) mm (in.)
Annealed, with hardness of 135 to 185 HB									
M2 and M3 high speed steel	43 (140)	0.050 (0.002)	0.063 (0.0025)	0.075 (0.003)	0.063 (0.0025)	0.050 (0.002)	0.046 (0.0018)	0.040 (0.0015)	0.025 (0.001)
C-6 carbide	115 (375)	0.050 (0.002)	0.063 (0.0025)	0.075 (0.003)	0.063 (0.0025)	0.050 (0.002)	0.046 (0.0018)	0.040 (0.0015)	0.025 (0.001)
Annealed or cold drawn, with hardness of 185 to 240 HB									
M2 and M3 high speed steel	38 (125)	0.038 (0.0015)	0.050 (0.002)	0.063 (0.0025)	0.063 (0.0025)	0.050 (0.002)	0.046 (0.0018)	0.040 (0.0015)	0.025 (0.001)
C-6 carbide	105 (340)	0.038 (0.0015)	0.050 (0.002)	0.063 (0.0025)	0.063 (0.0025)	0.050 (0.002)	0.046 (0.0018)	0.040 (0.0015)	0.025 (0.001)
Quenched and tempered, with hardness of 275 to 325 HB									
Any premium high speed steel (T15, M33, M41-M47)	24 (80)	0.025 (0.001)	0.030 (0.0012)	0.040 (0.0015)	0.040 (0.0015)	0.030 (0.0012)	0.025 (0.001)	0.020 (0.0008)	0.018 (0.0007)
C-6 carbide	78 (255)	0.025 (0.001)	0.030 (0.0012)	0.040 (0.0015)	0.040 (0.0015)	0.030 (0.0012)	0.025 (0.001)	0.020 (0.0008)	0.018 (0.0007)
Quenched and tempered, with hardness of 375 to 425 HB									
Any premium high speed steel (T15, M33, M41-M47)	12 (40)	0.023 (0.0009)	0.028 (0.0011)	0.036 (0.0014)	0.036 (0.0014)	0.028 (0.0011)	0.023 (0.0009)	0.018 (0.0007)	0.015 (0.0006)
C-6 carbide	34 (110)	0.023 (0.0009)	0.028 (0.0011)	0.036 (0.0014)	0.036 (0.0014)	0.028 (0.0011)	0.023 (0.0009)	0.018 (0.0007)	0.015 (0.0006)

AISI Types 416, 416Se: Drilling

Tool material	Speed, m/min (ft/min)	Feed per revolution for nominal hole diameter of: 1.5 mm (0.062 in.) mm (in.)	3 mm (0.125 in.) mm (in.)	6 mm (0.25 in.) mm (in.)	12 mm (0.5 in.) mm (in.)	18 mm (0.75 in.) mm (in.)	25 mm (1 in.) mm (in.)	35 mm (1.5 in.) mm (in.)	50 mm (2 in.) mm (in.)
Annealed, with hardness of 135 to 185 HB									
M10, M7, and M1 high speed steel	44 (145)	0.025 (0.001)	0.075 (0.003)	0.15 (0.006)	0.25 (0.010)	0.36 (0.014)	0.45 (0.017)	0.55 (0.021)	0.65 (0.025)
Annealed or cold drawn, with hardness of 185 to 240 HB									
M10, M7, and M1 high speed steel	30 (100)	0.025 (0.001)	... (...)	... (...)	... (...)	... (...)	... (...)	... (...)	... (...)
	38 (125)	... (...)	0.075 (0.003)	0.15 (0.006)	0.25 (0.010)	0.36 (0.014)	0.45 (0.017)	0.55 (0.021)	0.65 (0.025)
Quenched and tempered, with hardness of 275 to 325 HB									
M10, M7, and M1 high speed steel	15 (50)	0.025 (0.011)	... (...)	... (...)	... (...)	... (...)	... (...)	... (...)	... (...)
	21 (70)	... (...)	0.075 (0.003)	0.13 (0.005)	0.18 (0.007)	0.23 (0.009)	0.26 (0.011)	0.36 (0.014)	0.45 (0.018)
Quenched and tempered, with hardness of 375 to 425 HB									
M10, M7, and M1 high speed steel	14 (45)	... (...)	0.050 (0.002)	0.075 (0.003)	0.13 (0.005)	0.18 (0.007)	0.23 (0.009)	0.30 (0.012)	0.40 (0.015)

AISI Type 420

AISI Type 420: Chemical Composition

Element	Composition, %
Carbon	0.15 min
Manganese	1.00 max
Phosphorus	0.04 max
Sulfur	0.03 max
Silicon	1.00 max
Chromium	12-14

Characteristics. AISI type 420 is a high-carbon member of the hardenable (martensitic), 12% chromium family of stainless steels. The higher carbon content provides higher heat treated hardness than is characteristic of type 410. In the hardened condition, type 420 exhibits high strength and wear resistance. It develops maximum corrosion resistance when hardened and polished. Because much lower resistance is shown by annealed material and material having its surface contaminated by foreign particles, it is advisable to passivate final parts made from type 420 steel.

This stainless steel can be moderately cold worked when in the fully annealed condition. It is usually machined in the fully annealed condition at speeds about 50% of those used for free-machining AISI 1212.

Type 420 should be preheated to 230 °C (450 °F), and immediately after welding should be annealed for 6 to 8 h at 705 to 760 °C (1300 to 1400 °F) and air cooled. Type 420 electrodes are used when similar properties in the weld and base metal are required; otherwise types 309 or 310 are used.

This steel has good scaling resistance up to about 650 °C (1200 °F) in continuous service. Scaling becomes excessive above 760 °C (1400 °F).

Typical Uses. Applications of AISI type 420 include cutlery, hand tools, dental and surgical instruments, valve trim and parts, shafts, and plastic mold steel.

AISI Type 420: Similar Steels (U.S. and/or Foreign). UNS S42000; AMS 5506, 5621; ASTM A276, A314, A473, A580; FED QQ-S-763, QQ-S-766, QQ-W-423; MIL SPEC MIL-S-862; SAE J405 (51420); (W. Ger.) 1.4021; (Fr.) AFNOR Z 20 C 13; (Ital.) UNI X 20 Cr 13; (Jap.) JIS SUS 420 J1; (Swed.) SS_{14} 2303; (U.K.) B.S. 420 S 37, CDS-18

Physical Properties

AISI Type 420: Selected Physical Properties (Ref 6)

Melting range	1455-1510 °C (2650-2750 °F)
Density	7.8 g/cm³ (0.28 lb/in.³)
Specific heat at 0-100 °C (32-212 °F)	460 J/kg·K (0.11 Btu/lb·°F)
Thermal conductivity at 100 °C (212 °F)	24.9 W/m·K (14.4 Btu/ft·h·°F)
Electrical resistivity at 20 °C (68 °F)	55 nΩ·m

AISI Type 420: Average Coefficients of Linear Thermal Expansion (Ref 6)

Temperature range		Coefficient	
°C	°F	μm/m·K	μin./in.·°F
0-100	32-212	10.3	5.7
0-315	32-600	10.8	6.0
0-540	32-1000	11.7	6.5
0-650	32-1200	12.2	6.8

AISI Type 420: Thermal Treatment Temperatures

Treatment	Temperature range °C	°F
Forging, preheat	760-815	1400-1500
Forging, start	1095-1230	2000-2250
Forging, finish	900 min	1650 min
Full annealing(a)	840-900	1550-1650
Process annealing(b)	730-790	1350-1450
Hardening(c)	1010-1065	1850-1950
Tempering(d)	150-370	300-700

(a) Cool slowly in furnace. (b) Furnace cool or cool in air. (c) Soak at temperature and quench in warm oil. (d) Temper at 150 to 205 °C (300 to 400 °F) for maximum hardness and corrosion resistance

Mechanical Properties

AISI Type 420: Tensile Properties at Elevated Temperatures

Test specimens were 16.0-mm (0.625-in.) diam bar; austenitized at 980 °C or 1800 °F for ½ h; oil quenched; tempered for 2 h at 28 °C (50 °F) above test temperature

Test temperature		Tensile strength		Yield strength(a)		Elongation(b),	Reduction	Hardness	
°C	°F	MPa	ksi	MPa	ksi	%	in area, %	HRC(c)	HRC(d)
21	70	2025	294	1360	197	2.5	...	57	...
205	400	1820	264	1085	157.5	11.5	36	51.5	51
315	600	1705	247.5	1040	150.5	13.5	34.5	50.5	49
425	800	1715	248.5	1155	167.5	12.5	35.5	51.5	52
480	900	1415	205	1095	159	9	28.5	52.5	50
540	1000	660	96	585	84.5	20.5	73.5	36	35.5
595	1100	450	65	380	55	26	83	31	30.5
650	1200	290	42	240	34.5	31.5	90	26.5	26
705	1300	170	25	115	16.5	36	93.5	23.5	21

(a) 0.2% offset. (b) In 50 mm (2 in.). (c) Before testing. (d) After testing

AISI Type 420: Modulus of Elasticity (Ref 7)

Stressed in:	Modulus GPa	10^6 psi
Tension	200	29.0
Torsion	80.7	11.7

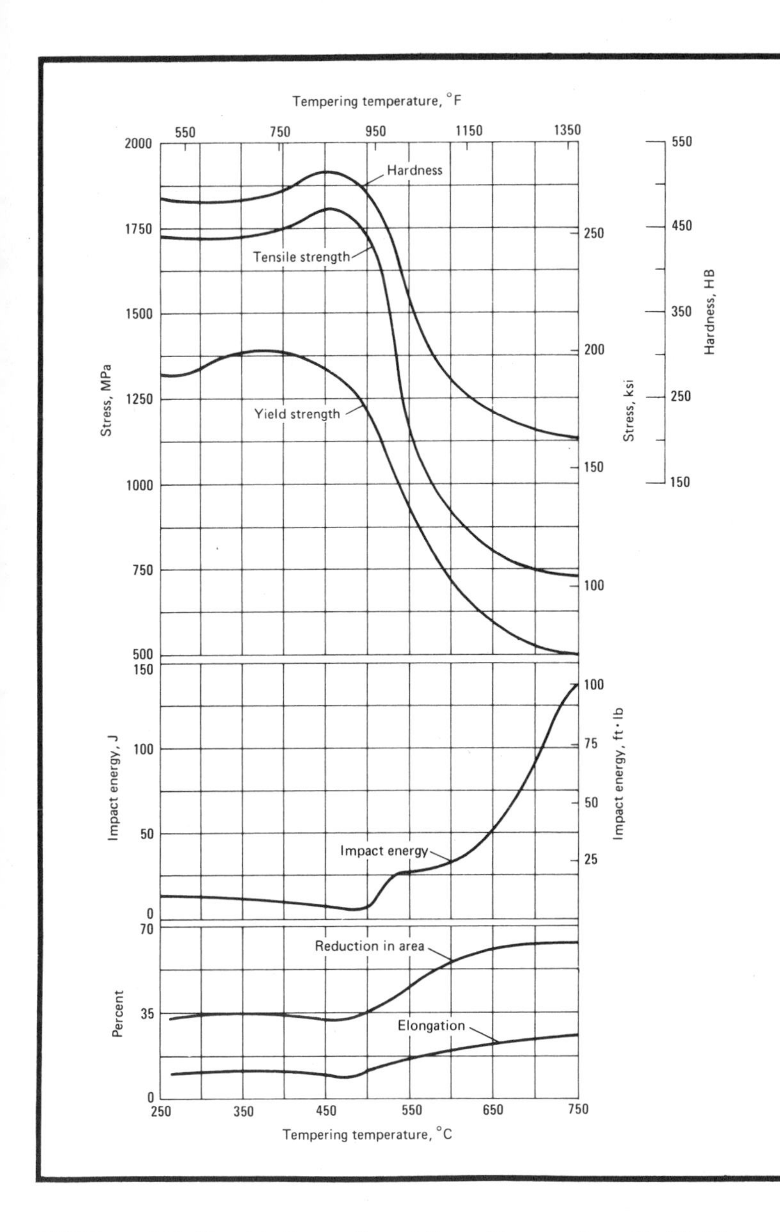

AISI Type 420: Effect of Tempering Temperature on Tensile Properties. Heated to 1010 °C (1850 °F); oil quenched; tempered at temperature indicated for 5 h. Heat treated as 25-mm (1-in.) diam rod; tensile test specimens were 12.8-mm (0.505-in.) diam; Izod impact test specimens were 10.0-mm (0.394-in.) square. Tests were conducted using specimens machined to English units. Elongation was measured in 50 mm (2 in.); yield strength at 0.2% offset. (Ref 10)

AISI Type 420: Tensile Properties at Room Temperature (Ref 6, 7)

Product form	Condition	Tensile strength MPa	Tensile strength ksi	Yield strength(a) MPa	Yield strength(a) ksi	Elongation(b), %	Reduction in area, %	Hardness
Strip	Annealed	655	95	345	50	20	...	92 HRB
Wire	Annealed	655	95	345	50	20	65	92 HRB
	Soft bright tempered	760	110	585	85	15	55	97 HRB
Bar	Annealed	655	95	345	50	25	55	195 HB
	Annealed and cold drawn	725	105	585	85	17	50	215 HB
	Hardened and tempered	760-1720	110-250	515-1380	75-200	22-7	60-25	500-205 HB

(a) 0.2% offset. (b) In 50 mm (2 in.), except for wire. For wire 3.18-mm (0.125-in.) diam and larger, gage length is 4 × diam; for smaller diameters, the gage length is 254 mm (10.0 in.)

AISI Type 420: Impact Properties at Room Temperature (Ref 7)

Condition	Energy J	Energy ft · lb	Hardness, HB
Annealed	100	75	175
Quenched and tempered	9-18	7-13	477

Machining Data

For machining data on AISI type 420, refer to the preceding machining tables for AISI types 416 and 416Se.

AISI Types 420F, 420F(Se)

AISI Types 420F, 420F(Se): Chemical Composition

Element	Composition, % Type 420F	Type 420F(Se)
Carbon	0.15 min	0.15 min
Manganese	1.00 max	1.00 max
Phosphorus	0.04 max	0.04 max
Sulfur	0.15 min	0.03 max
Silicon	1.00 max	1.00 max
Chromium	12-14	12-14
Molybdenum(a)	0.60 max	0.60 max
Selenium	···	0.15 min

(a) Optional

Characteristics. AISI types 420F and 420F(Se) are hardenable (martensitic) stainless steels with elements added to improve machining and nonseizing characteristics. The optimum corrosion-resisting qualities of these grades are obtained in the hardened and tempered condition. Both types are magnetic in all conditions.

Types 420F and 420F(Se) have corrosion resistance similar to type 416; they will resist corrosion from mild atmospheres, fresh water, ammonia, many petroleum products, and several mild acid environments.

Types 420F and 420F(Se) are not often used for service at elevated temperatures, because corrosion resistance of these alloys is reduced when heated above 425 °C (800 °F). Appreciable scale will form if temperatures exceed 650 °C (1200 °F). Therefore, types 420F and 420F(Se) are not recommended for plastic mold cavities where a high mirror-finish is essential.

These steels will withstand moderate cold work in the annealed condition, but are not recommended for cold upsetting. In the fully annealed condition, the machinability rating of types 420F and 420F(Se) is about 55% of that for free-machining AISI 1212. Types 420F and 420F(Se) are not recommended for welding.

Typical Uses. Applications of AISI types 420F and 420F (Se) include valve trim, gears, cams, pivots, and dental and surgical instruments.

AISI Type 420F: Similar Steels (U.S. and/or Foreign). UNS S42020; AMS 5620; SAE J405 (51420F)

AISI Type 420F(Se): Similar Steels (U.S. and/or Foreign). UNS S42023; AMS 5620; ASTM A582, SAE J405 (51420FSe)

Physical Properties

AISI Type 420F: Selected Physical Properties (Ref 6)

Melting range	1455-1510 °C (2650-2750 °F)
Density	7.8 g/cm³ (0.28 lb/in.³)
Specific heat at 0-100 °C (32-212 °F)	460 J/kg·K (0.11 Btu/lb·°F)
Thermal conductivity at 27 °C (80 °F)	25.1 W/m·K (14.5 Btu/ft·h·°F)
Electrical resistivity at 20 °C (68 °F)	55 nΩ·m

AISI Type 420F: Average Coefficients of Linear Thermal Expansion (Ref 6)

Temperature °C	°F	Coefficient μm/m·K	μin./in.·°F
0-100	32-212	10.3	5.7
0-650	32-1200	12.2	6.8

AISI Types 420F, 420F(Se): Thermal Treatment Temperatures

Treatment	Temperature range °C	°F
Forging, preheat	760-815	1400-1500
Forging, start	1095-1205	2000-2200
Forging, finish	900 min	1650 min
Full annealing(a)	845-900	1550-1650
Process annealing(b)	730-790	1350-1450
Hardening(c)	1010-1065	1850-1950
Tempering(d)	150-370	300-700

(a) Cool slowly in furnace; hardness of 179 HB. (b) Furnace cool or cool in air; hardness of 196 HB. (c) Soak at temperature; quench in warm oil. (d) Temper at 150 to 205 °C (300 to 400 °F) for maximum toughness. Tempering above 425 °C (800 °F) after hardening impairs corrosion resistance

Mechanical Properties

AISI Types 420F, 420F(Se): Representative Tensile Properties (Ref 6)

Product form	Condition	Tensile strength MPa	ksi	Yield strength(a) MPa	ksi	Elongation(b), %	Reduction in area, %	Hardness
Bar	Annealed	655	95	380	55	22	50	220 HB
	Annealed and cold drawn	760	110	690	100	14	40	228 HB
Wire	Annealed and cold drawn	690	100	550	80	15	45	99 HRB

(a) 0.2% offset. (b) In 50 mm (2 in.), except for wire. For wire 3.18-mm (0.125-in.) diam and larger, gage length is 4 × diam; for smaller diameters, the gage length is 254 mm (10.0 in.)

Machining Data

For machining data on AISI types 420F and 420F(Se), refer to the preceding machining tables for AISI types 416 and 416Se.

AISI Type 422

AISI Type 422: Chemical Composition

Element	Composition, %
Carbon	0.20-0.25
Manganese	1.00 max
Phosphorus	0.025 max
Sulfur	0.025 max
Silicon	0.075 max
Chromium	11-13
Nickel	0.50-1.00
Molybdenum	0.75-1.25
Vanadium	0.15-0.30
Tungsten	0.75-1.25

Characteristics. AISI type 422 is a hardenable, martensitic, stainless steel designed for use at temperatures up to 650 °C (1200 °F). It has excellent creep-rupture properties in the range of 315 to 650 °C (600 to 1200 °F) and can be heat treated to various high-strength levels, depending on the section, form, and application involved. Type 422 offers high strength/weight ratios from room temperature to 650 °C (1200 °F).

Composition of this steel is balanced to provide a microstructure that is virtually all tempered martensite after heat treatment. This structure ensures uniformity of mechanical properties with a minimum of directionality.

Type 422 is adaptable to applications where maximum resistance to corrosion is not required. It is resistant to the corrosive action of the atmosphere, fresh water, steam, and some of the milder acids and alkalis. This stainless steel has good resistance to scaling or oxidation for continuous service at temperatures up to 675 °C (1250 °F).

Chips formed in machining the type 422 alloy are brittle and stringy with a tendency to gall or build up on the cutting edges and radii of the tool. Material in the subcritical-annealed condition is preferred for most machining applications. The machinability rating of type 422 is about 40% of that for free-machining AISI 1212.

Type 422 may be satisfactorily welded by the gas tungsten arc and gas metal-arc welding processes. Type 422 filler metal should be used to obtain optimum results. Because type 422 is air-hardening, it should be preheated to a minimum temperature of 175 °C (350 °F). If the weldment is not immediately hardened and tempered, it should be stress relieved by postheating at 650 to 675 °C (1200 to 1250 °F) and air cooling.

Typical Uses. Applications of AISI type 422 include compressor and turbine blades, and wheels, rotors, high-temperature bolting, valve parts, and structural components for aircraft and missiles.

AISI Type 422: Similar Steels (U.S. and/or Foreign). UNS S42200; AMS 5655; ASTM A565; SAE J467; (W. Ger.) DIN 1.4935; (Jap.) JIS SUH 616

Physical Properties

AISI Type 422: Selected Physical Properties (Ref 6)

Melting range	1470-1480 °C (2675-2700 °F)
Density	7.8 g/cm³ (0.28 lb/in.³)
Specific heat at 0-100 °C (32-212 °F)	460 J/kg·K (0.11 Btu/lb·°F)

AISI Type 422: Thermal Treatment Temperatures

Treatment	Temperature range °C	°F
Forging, preheat	650-760	1200-1400
Forging, start	1065-1105	1950-2025
Forging, finish	845 min	1550 min
Full annealing(a)	870-900	1600-1650
Subcritical annealing(b)	760-775	1400-1425
Hardening(c)	1040	1900
Tempering(d)	315-650	600-1200

(a) Hold at temperature 1½ h per 25 mm (1.0 in.) of cross section; cool slowly to 705 °C (1300 °F); hold 6 h; slow cool in furnace to 540 °C (1000 °F); air cool. (b) Hold at temperature 6 to 8 h; furnace cool to 650 °C (1200 °F); air cool. (c) Oil quench. (d) Temper for at least 2 h; air cool or oil quench. A temperature of 525 °C (980 °F) is used to obtain highest strength; 650 °C (1200 °F) or higher is used for long-time elevated temperature service

AISI Type 422: Thermal Conductivity (Ref 6)

Temperature °C	°F	Thermal conductivity W/m·K	Btu/ft·h·°F
100	212	23.9	13.8
500	930	27.3	15.8

AISI Type 422: Average Coefficients of Linear Thermal Expansion (Ref 6)

Temperature range °C	°F	Coefficient μm/m·K	μin./in.·°F
0-100	32-212	11.2	6.2
0-315	32-600	11.3	6.3
0-540	32-1000	11.9	6.6
0-650	32-1200	12.1	6.7

Machining Data

For machining data on AISI type 422, refer to the preceding machining tables for AISI type 403.

Mechanical Properties

AISI Type 422: Stress-Rupture Strength and Creep Rates (Ref 7)

Test specimens were austenitized at 1040 °C (1900 °F); oil quenched; tempered at 650 °C (1200 °F) for 2 h; air cooled

Test temperature		Stress to produce rupture in: 10 h		100 h		1000 h		10 000 h		Stress for minimum creep rate of: 1% in 1000 h		1% in 10 000 h		1% in 100 000 h	
°C	°F	MPa	ksi	MPa	ksi	MPa	ksi	MPa	ksi	MPa	ksi	MPa	ksi	MPa	ksi
540	1000	490	71	435	63	390	57	340	49	360	52	270	39	180	26
595	1100	365	53	320	46	255	37	160	23	235	34	140	20	...	...
650	1200	385	56	170	25	120	17	...	...	...	...	...	...	...	...

AISI Type 422: Short-Time Elevated-Temperature Tensile Properties (Ref 10)

Test specimens were austenitized at 1040 °C (1900 °F) for 1 h; oil quenched; tempered for 2 h at temperature indicated

Test temperature °C	°F	Tempering temperature °C	°F	Tensile strength MPa	ksi	Yield strength(a) MPa	ksi	Elongation(b), %	Reduction in area, %
315	600	425	800	1585	230	1055	153	16	13
425	800	425	800	1490	216	951	138	14	10-24
425	800	455	850	1490	216	979	142	19	55
425	800	525	980	1510	219	1080	157	16	52
480	900	525	980	1440	209	1040	151	14	46
540	1000	525	980	1220	177	903	131	12	43
650	1200	650	1200	390	57	390	57	21	85

(a) 0.2% offset. (b) In 36 mm (1.4 in.)

AISI Type 422: Low- and Elevated-Temperature, Charpy V-Notch Impact Properties (Ref 10)

Test specimens were austenitized at 1040 °C (1900 °F) for 1 h; oil quenched; tempered at 525 °C (980 °F) for 2 h; air cooled

Test temperature °C	°F	Impact energy J	ft·lb
−73	−100	4	3
		5	4
−17	0	5	4
		7	5
Room temperature		8	6
		10	7
71	160	14	10
		15	11
100	212	18	13
		20	15
125	260	22	16
		23	17
140	280	28	21
150	300	31	23
		37	27
170	340	38	28
		39	29
205	400	43	32
		52	38

AISI Type 422: Effect of Tempering Temperature on Room- and Elevated-Temperature Tensile Properties (Ref 7)

Test specimens were austenitized at 1040 °C (1900 °F) for 1 h; oil quenched; tempered for 2 h at temperature indicated

Test temperature °C	°F	Tempering temperature °C	°F	Tensile strength MPa	ksi	Yield strength(a) MPa	ksi	Elongation(b), %	Reduction in area, %
21	70	315	600	1580	229	1000	145	16	56
21	70	425	800	1635	237	1160	168	16	53
21	70	480	900	1675	243	1220	177	16	50
21	70	540	1000	1610	233	1005	146	16	54
21	70	650	1200	1030	149	862	125	19	52
315	600	425	800	1585	230	1055	153	16	13
425	800	425	800	1490	216	951	138	14	17
425	800	455	850	1490	216	979	142	19	55
540	1000	540	1000	1165	169	869	126	16	56
650	1200	650	1200	390	57	320	46	21	85

(a) 0.2% offset. (b) In 36 mm (1.4 in.)

AISI Type 422: Effect of Tempering on Room-Temperature Impact Properties (Ref 10)

Test specimens were austenitized at 1040 °C (1900 °F) for 1 h; oil quenched; tempered for 2 h at temperature indicated

Tempering temperature °C	°F	Hardness, HRC	Charpy V-notch impact energy J	ft·lb
370	700	48	12	9
425	800	48	9	7
480	900	50	7	5
525	980	49	8	6
650	1200	34	20	15
650(a)	1200(a)	34	26	19
675(a)	1250(a)	33	33	24
705(a)	1300(a)	29	76	56
730(a)	1350(a)	25	121	89
760	1400	26	52	38

(a) Double tempered

AISI Type 422: Modulus of Elasticity (Ref 10)

Temperature		Stressed in: Tension		Torsion	
°C	°F	GPa	10^6 psi	GPa	10^6 psi
24	75	199.2	28.9	82.7	12.0
260	500	189.6	27.5	79.3	11.5
425	800	166.6	24.2	72.4	10.5
595	1100	141.4	20.5	62.0	9.0

AISI Types 430F, 430F(Se)

AISI Types 430F, 430F(Se): Chemical Composition

Element	Composition, % Type 430F	Type 430F(Se)
Carbon	0.12 max	0.12 max
Manganese	1.25 max	1.25 max
Potassium	0.06 max	0.06 max
Sulfur	0.15 min	0.06 max
Silicon	1.00 max	1.00 max
Chromium	14-18	14-18
Molybdenum(a)	0.60 max	...
Selenium	...	0.15 min

(a) Optional

Characteristics. AISI types 430F and 430F(Se) are free-machining modifications of type 430. They are suitable for the production of parts in automatic screw machines.

These stainless steels will withstand moderate cold work, but are not recommended for cold upsetting. Parts that are machined to shape are the primary applications of types 430F and 430F(Se). The machinability rating of these steels is about 80 to 90% of that for AISI 1212. Types 430F and 430F(Se) are not usually recommended for welding.

These alloys resist corrosion from the atmosphere, fresh water and steam, foodstuffs, dairy products, nitric acid, and many petroleum products and organic materials. Their resistance to chloride stress-corrosion cracking at elevated temperatures is higher than that of austenitic types 304 or 316. For best resistance, finished parts should be passivated.

The safe scaling temperature for types 430F and 430F(Se) in continuous service is 815 °C (1600 °F), and for intermittent service up to 870 °C (1600 °F).

Typical Uses. Applications of AISI types 430F and 430F (Se) include bolts, nuts, studs, aircraft fittings, and valve parts and hardware that require considerable machining. These alloys are not usually recommended for vessels containing gases or liquids under high pressure.

AISI Type 430F: Similar Steels (U.S. and/or Foreign). UNS S43020; ASTM A314, A473, A581, A582; MIL SPEC MIL-S-862; SAE J405 (51430F); (W. Ger.) DIN 1.4104; (Fr.) AFNOR Z 10 CF 17; (Ital.) UNI X 10 CrS 17; (Jap.) JIS SUS 430 F; (Swed.) SS_{14} 2383

AISI Type 430F(Se): Similar Steels (U.S. and/or Foreign). UNS S43023; ASTM A314, A473, A581, A582; MIL SPEC MIL-S-862; SAE J405 (51430F-Se)

Mechanical Properties

AISI Type 430F: Tensile Properties at Room Temperature (Ref 6)

Test specimens were in the annealed condition

Product form	Treatment	Tensile strength MPa	Tensile strength ksi	Yield strength(a) MPa	Yield strength(a) ksi	Elongation(b), %	Reduction in area, %	Hardness
Wire, 6.4-mm (0.25-in.) diam	Soft tempered	655	95	585	85	10	50	92 HRB
Bar	Annealed	550	80	380	55	25	60	170 HB
	Annealed and cold drawn(c)	620	90	550	80	15	55	190 HB

(a) 0.2% offset. (b) In 50 mm (2 in.), except for wire. For wire 3.18-mm (0.125-in.) diam and larger, gage length is 4 × diam; for smaller diameters, the gage length is 254 mm (10.0 in.). (c) Diameter of 25 mm (1 in.)

AISI Type 430F: Tensile Properties at Elevated Temperatures

Annealed bar were held at temperature for 1 h and pulled at the rate of 1.3 mm (0.05 in.) per minute

Test temperature °C	°F	Tensile strength MPa	ksi	Elongation(a), %	Reduction in area, %
21	70	531	77	32	74
93	200	496	72	30	74
205	400	462	67	27	76
315	600	441	64	26	75
425	800	386	56	29	75
535	1000	248	36	35	84
650	1200	131	19	61	97
760	1400	48	7	70	99

(a) In 50 mm (2 in.)

Physical Properties

AISI Types 430F, 430F(Se): Thermal Treatment Temperatures

Treatment	Temperature range °C	°F
Forging, start	1065-1150	1950-2100
Forging, finish	815 min	1500 min
Full annealing(a)	790-815	1450-1500
Process annealing(b)	675-760	1250-1400
Hardening(c)	...	...

(a) Furnace cool to 595 °C (1100 °F); air cool. (b) Air cool. (c) Not appreciably hardenable

For additional data on physical properties of AISI types 430F and 430F(Se), refer to the preceding tables for AISI type 430.

Machining Data (Ref 2)

AISI Types 430F, 430F(Se): Turning (Single Point and Box Tools)

Test specimens were in the annealed condition, with hardness of 135 to 185 HB

Depth of cut		M2 and M3 high speed steels				Uncoated carbide						Coated carbide			
		Speed		Feed		Speed, brazed		Speed, inserted		Feed		Speed		Feed	
mm	in.	m/min	ft/min	mm/rev	in./rev	m/min	ft/min	m/min	ft/min	mm/rev	in./rev	m/min	ft/min	mm/rev	in./rev
1	0.040	64	210	0.18	0.007	185(a)	600(a)	205(a)	675(a)	0.18	0.007	265(b)	875(b)	0.18	0.007
4	0.150	55	180	0.38	0.015	160(c)	525(c)	185(c)	600(c)	0.38	0.015	235(d)	775(d)	0.38	0.015
8	0.300	43	140	0.50	0.020	120(c)	400(c)	140(c)	460(c)	0.75	0.030	185(d)	600(d)	0.50	0.020
16	0.625	34	110	0.75	0.030	90(c)	300(c)	110(c)	360(c)	1.00	0.040	...	...	...	...

(a) Carbide tool material: C-7. (b) Carbide tool material: CC-7. (c) Carbide tool material: C-6. (d) Carbide tool material: CC-6

AISI Types 430F, 430F(Se): Turning (Cutoff and Form Tools)

Test specimens were in the annealed condition, with hardness of 135 to 185 HB

Tool material	Speed, m/min (ft/min)	Feed per revolution for cutoff tool width of: 1.5 mm (0.0062 in.) mm (in.)	3 mm (0.125 in.) mm (in.)	6 mm (0.25 in.) mm (in.)	Feed per revolution for form tool width of: 12 mm (0.5 in.) mm (in.)	18 mm (0.75 in.) mm (in.)	25 mm (1 in.) mm (in.)	35 mm (1.5 in.) mm (in.)	50 mm (2 in.) mm (in.)
M2 and M3 high speed steels	46 (150)	0.038 (0.0015)	0.050 (0.002)	0.061 (0.0024)	0.046 (0.0018)	0.041 (0.0016)	0.036 (0.0014)	0.033 (.0013)	0.028 (.0011)
C-6 carbide	120 (400)	0.038 (0.0015)	0.050 (0.002)	0.061 (0.0024)	0.046 (0.0018)	0.041 (0.0016)	0.036 (0.0014)	0.033 (.0013)	0.028 (.0011)

AISI Types 430F, 430F(Se): Face Milling

Test specimens were in the annealed condition, with hardness of 135 to 185 HB

Depth of cut		M2 and M7 high speed steels				Uncoated carbide						Coated carbide			
		Speed		Feed/tooth		Speed, brazed		Speed, inserted		Feed/tooth		Speed		Feed/tooth	
mm	in.	m/min	ft/min	mm	in.	m/min	ft/min	m/min	ft/min	mm	in.	m/min	ft/min	mm	in.
1	0.040	76	250	0.20	0.008	190(a)	625(a)	210(a)	690(a)	0.20	0.008	310(b)	1025(b)	0.20	0.008
4	0.150	58	190	0.30	0.012	145(a)	470(a)	160(a)	420(a)	0.30	0.012	205(b)	675(b)	0.30	0.012
8	0.300	46	150	0.40	0.016	100(c)	325(c)	120(c)	400(c)	0.40	0.016	160(d)	525(d)	0.40	0.016

(a) Carbide tool material: C-6. (b) Carbide tool material: CC-6. (c) Carbide tool material: C-5. (d) Carbide tool material: CC-5

AISI Types 430F, 430F(Se): End Milling (Profiling)

Test specimens were in the annealed condition, with hardness of 135 to 185 HB

Tool material	Depth of cut		Speed		Feed per tooth for cutter diameter of: 10 mm (0.375 in.)		12 mm (0.5 in.)		18 mm (0.75 in.)		25-50 mm (1-2 in.)	
	mm	in.	m/min	ft/min	mm	in.	mm	in.	mm	in.	mm	in.
M2, M3, and M7 high speed steels	0.5	0.020	55	180	0.025	0.001	0.050	0.002	0.102	0.004	0.13	0.005
	1.5	0.060	43	140	0.050	0.002	0.075	0.003	0.13	0.005	0.15	0.006
	diam/4	diam/4	38	125	0.025	0.001	0.050	0.002	0.102	0.004	0.13	0.005
	diam/2	diam/2	34	110	0.025	0.001	0.038	0.0015	0.075	0.003	0.102	0.004
C-5 carbide	0.5	0.020	160	520	0.025	0.001	0.038	0.0015	0.102	0.004	0.13	0.005
	1.5	0.060	120	400	0.038	0.0015	0.050	0.002	0.13	0.005	0.15	0.006
	diam/4	diam/4	105	350	0.038	0.0015	0.050	0.002	0.102	0.004	0.13	0.005
	diam/2	diam/2	100	320	0.025	0.001	0.038	0.0015	0.075	0.003	0.102	0.004

AISI Types 430F, 430F(Se): Planing

Test specimens were in the annealed condition, with hardness of 135 to 185 HB

Tool material	Depth of cut		Speed		Feed/stroke	
	mm	in.	m/min	ft/min	mm	in.
M2 and M3 high speed steels	0.1	0.005	15	50	(a)	(a)
	2.5	0.100	21	70	1.25	0.050
	12	0.500	14	45	1.50	0.060
C-6 carbide	0.1	0.005	69	225	(a)	(a)
	2.5	0.100	79	260	2.05	0.080

(a) Feed is 75% the width of the square nose finishing tool

Machining Data (Ref 2) (continued)

AISI Types 430F, 430F(Se): Drilling

Test specimens were in the annealed condition, with hardness of 135 to 185 HB

Tool material	Speed, m/min (ft/min)	Feed per revolution for nominal hole diameter of: 1.5 mm (0.062 in.) mm (in.)	3 mm (0.125 in.) mm (in.)	6 mm (0.25 in.) mm (in.)	12 mm (0.5 in.) mm (in.)	18 mm (0.75 in.) mm (in.)	25 mm (1 in.) mm (in.)	35 mm (1.5 in.) mm (in.)	50 mm (2 in.) mm (in.)
M10, M7, and M1 high speed steels	49 (160)	0.025 (0.001)	0.075 (0.003)	0.15 (0.006)	0.25 (0.010)	0.36 (0.014)	0.45 (0.017)	0.55 (0.021)	0.65 (0.025)

AISI Types 430F, 430F(Se): Boring

Test specimens were in the annealed condition, with hardness of 135 to 185 HB

Depth of cut		M2 and M3 high speed steels				Uncoated carbide						Coated carbide			
		Speed		Feed		Speed, brazed		Speed, inserted		Feed		Speed		Feed	
mm	in.	m/min	ft/min	mm/rev	in./rev	m/min	ft/min	m/min	ft/min	mm/rev	in./rev	m/min	ft/min	mm/rev	in./rev
0.25	0.010	60	205	0.075	0.003	155(a)	515(a)	185(a)	600(a)	0.075	0.003	235(b)	775(b)	0.075	0.003
1.25	0.050	50	165	0.130	0.005	120(c)	400(c)	145(c)	475(c)	0.130	0.005	185(d)	615(d)	0.130	0.005
2.50	0.100	44	145	0.300	0.012	105(e)	350(e)	130(e)	425(e)	0.300	0.012	165(f)	545(f)	0.300	0.012

(a) Carbide tool material: C-8. (b) Carbide tool material: CC-8. (c) Carbide tool material: C-7. (d) Carbide tool material: CC-7. (e) Carbide tool material: C-6. (f) Carbide tool material: CC-6

AISI Types 430F, 430F(Se): Reaming

Test specimens were in the annealed condition,with hardness of 135 to 185 HB; based on 4 flutes for 3- and 6-mm (0.125- and 0.25-in.) reamers, 6 flutes for 12-mm (0.5-in.) reamers, and 8 flutes for 25-mm (1-in.) and larger reamers

Tool material	Speed		Feed per revolution for reamer diameter of: 3 mm (0.125 in.)		6 mm (0.25 in.)		12 mm (0.5 in.)		25 mm (1 in.)		35 mm (1.5 in.)		50 mm (2 in.)	
	m/min	ft/min	mm	in.	mm	in.	mm	in.	mm	in.	mm	in.	mm	in.
Roughing														
M1, M2, and M7 high speed steels	40	130	0.075	0.003	0.13	0.005	0.20	0.008	0.25	0.010	0.30	0.012	0.40	0.015
C-2 carbide	46	150	0.102	0.004	0.20	0.008	0.30	0.012	0.40	0.016	0.50	0.20	0.60	0.024
Finishing														
M1, M2, and M7 high speed steels	18	60	0.102	0.004	0.102	0.004	0.15	0.006	0.20	0.008	0.25	0.010	0.30	0.012
C-2 carbide	21	70	0.102	0.004	0.15	0.006	0.20	0.008	0.25	0.010	0.28	0.011	0.30	0.012

AISI Types 430F, 430F(Se): Cylindrical and Internal Grinding

Test specimens were in the annealed condition, with hardness of 135 to 185 HB

Operation	Wheel speed m/s	ft/min	Work speed m/min	ft/min	Infeed on diameter mm/pass	in./pass	Traverse(a)	Wheel identification
Cylindrical grinding								
Roughing	28-33	5500-6500	15-30	50-100	0.050	0.002	½	A60JV
Finishing	28-33	5500-6500	15-30	50-100	0.013 max	0.0005 max	⅙	A60JV
Internal grinding								
Roughing	25-33	5000-6500	23-60	75-200	0.013	0.0005	⅓	A60KV
Finishing	25-33	5000-6500	23-60	75-200	0.005 max	0.0002 max	⅙	A60KV

(a) Wheel width per revolution of work

AISI Types 430F, 430F(Se): Surface Grinding with a Horizontal Spindle and Reciprocating Table

Test specimens were in the annealed condition, with hardness of 135 to 185 HB

Operation	Wheel speed m/s	ft/min	Table speed m/min	ft/min	Downfeed mm/pass	in./pass	Crossfeed mm/pass	in./pass	Wheel identification
Roughing	28-33	5500-6500	15-30	50-100	0.050	0.002	1.25-12.5	0.050-0.500	A46IV
Finishing	28-33	5500-6500	15-30	50-100	0.013 max	0.0005 max	(a)	(a)	A46IV

(a) Maximum ¼ of wheel width

AISI Types 430Ti, S43035

AISI Types 430Ti, S43035: Chemical Composition

Element	Composition, % Type 430Ti	Type S43035(a)
Carbon	0.12 max	0.07 max
Manganese	1.00 max	1.00 max
Phosphorus	0.040 max	0.040 max
Sulfur	0.030 max	0.030 max
Silicon	1.00 max	1.00 max
Chromium	16-18	17-19
Titanium	(b)	(c)

(a) Composition includes 0.15 max aluminum and 0.50 max nickel. (b) Minimum of six times the carbon content. (c) Twelve times the carbon content, up to 1.10 max

Characteristics. AISI types 430Ti and S43035 are ferritic stainless steels specifically designed for ease of welding and as-welded corrosion resistance to mild environments. The addition of titanium, or titanium and aluminum, decreases the amount of martensite which forms at ferritic grain boundaries during cooling from the welding heat.

All conventional brazing and soldering methods will join type 430Ti and S43035 steels. The use of soldering fluxes specifically designed for stainless steels is essential to good solder flow and adhesion. To avoid flux corrosion, the employment of phosphoric acid-based flux is mandatory, especially in plumbing installations where the inside walls of the tubes cannot be cleaned and where flux residues remain in contact for weeks or months prior to flushing with water. The selection of the brazing alloy should be made with the ultimate product service in mind.

Types 430Ti and S43035 must be welded only with inert-gas-shielding techniques, such as the gas tungsten-arc or gas metal-arc processes. If filler metal is added to the weld deposit, weld wire of the same composition as the base metal must be used to maintain corrosion resistance in the weld metal. The weld metal will be free of the martensite normally associated with ferritic stainless steels, and the heat-affected zone will be immune to intergranular attack in mildly corrosive environments. However, postweld annealing is desirable with types 430Ti and S43035 to promote maximum corrosion resistance.

These steels resist oxidation at temperatures up to 815 °C (1500 °F) for continuous service, and up to 870 °C (1600 °F) for intermittent service.

Low yield strengths and work-hardening rates, coupled with good ductility, are characteristics which allow these alloys to be drawn into hemispherical tank heads with little difficulty. Welded tube is easily bent and flared with conventional tubing tools and techniques. The machinability rating of types 430Ti and S43035 is about 60% of that for free-machining AISI 1212.

Typical Uses. Applications of AISI types 430Ti and S43035 include domestic hot water tanks and water systems, hot water holding tanks, and weldments where postweld annealing is difficult.

AISI Type 430Ti: Similar Steels (U.S. and/or Foreign). UNS S43036; ASTM A268, A554, A651; (W. Ger.) DIN 1.4510

AISI Type S43035: Similar Steels (U.S. and/or Foreign). UNS S43035; ASME SA240 (XM-8), SA268 (XM-8), SA479 (XM-8); ASTM A240 (XM-8), A268 (XM-8), A479 (XM-8), A651 (XM-8)

Physical Properties

AISI Types 430Ti, S43035: Selected Physical Properties (Ref 8)

Specific heat at 0-100 °C (32-212 °F)	460 J/kg·K (0.11 Btu/lb·°F)
Thermal conductivity at 100 °C (212 °F)	24.2 W/m·K (14.0 Btu/ft·h·°F)

AISI Types 430TI, S43035: Average Coefficients of Linear Thermal Expansion (Ref 8)

Temperature range °C	°F	Coefficient μm/m·K	μin./in.·°F
20-100	68-212	10.1	5.6
20-500	68-930	11.5	6.4
20-800	68-1470	12.4	6.9

AISI Types 430Ti, S43035: Thermal Treatment Temperatures

Treatment	Temperature range °C	°F
Forging, start	1040-1095	1900-2000
Forging, finish	760 min	1400 min
Annealing(a)	790-870	1450-1600
Hardening(b)	...	...

(a) Air cool. (b) Not appreciably hardenable

Machining Data

For machining data on AISI types 430Ti and S43035, refer to the preceding machining tables for AISI type 405.

Mechanical Properties

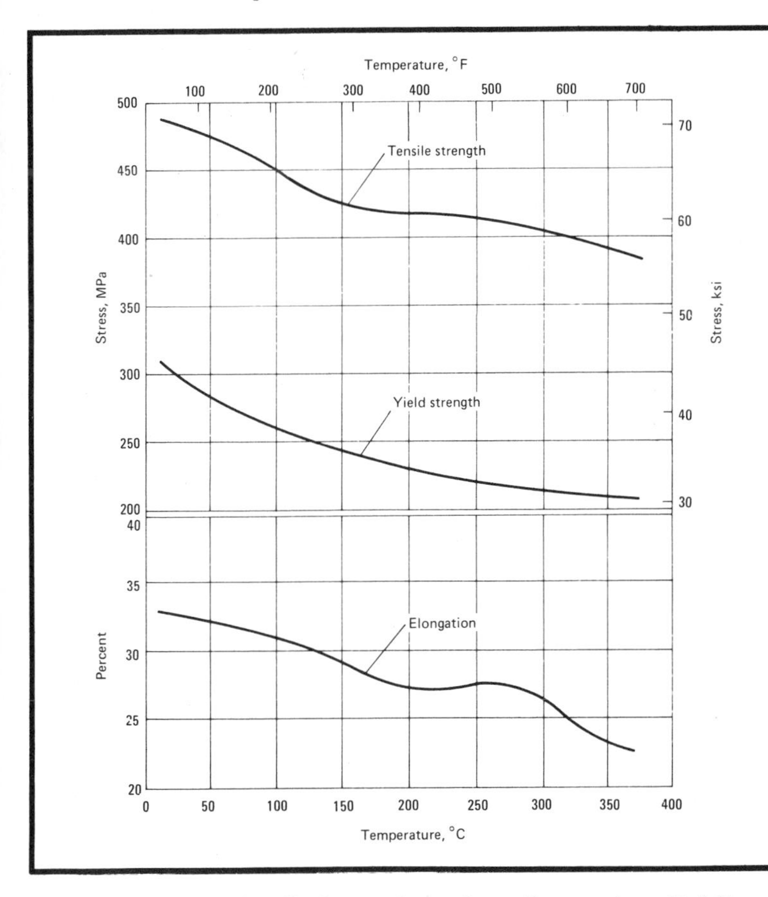

AISI Type S43035: Effect of Temperature on Tensile Properties. Elongation was measured in 50 mm (2 in.). (Ref 8)

AISI Type S43035: Tensile Properties at Room Temperature (Ref 8)

Test specimens were in the annealed condition

Material thickness mm	in.	Tensile strength MPa	ksi	Yield strength(a) MPa	ksi	Elongation(b), %	Hardness, HRB
1.6	0.062	450-515	65-75	275-380	40-55	28-31	78-86
6.35	0.250	515-585	75-85	345-415	50-60	24-29	78-86

(a) 0.2% offset. (b) In 50 mm (2 in.)

AISI Types 430, 434, 435, 436

AISI Types 430, 434, 435, 436: Chemical Composition

Element	Composition, % Type 430	Type 434	Type 435	Type 436
Carbon, max	0.12	0.12	0.12	0.12
Manganese, max	1.00	1.00	1.00	1.00
Phosphorus, max	0.04	0.04	0.04	0.04
Sulfur, max	0.03	0.03	0.03	0.03
Silicon, max	1.00	1.00	1.00	1.00
Chromium	4-18	14-18	14-18	14-18
Molybdenum	···	0.75-1.25	···	0.75-1.25
Niobium plus tantalum	···	···	0.40-0.60	0.40-0.60

Characteristics. AISI type 430 is a nonhardening (ferritic), straight chromium, stainless steel with superior corrosion and heat resistance compared to type 410. This steel is magnetic in all conditions. Type 430 most resembles the 18-8 chromium-nickel grades in fabrication and service performance. Because of excellent corrosion resistance, type 430 is chosen for many appliances, decorative trim, and chemical handling equipment.

Roping is a problem often encountered when type 430 is stretch bent or deep drawn; irregularities develop and appear on the surface as parallel ridges or ropes. Extra finishing steps are required to remove these imperfections.

AISI type 434 contains an addition of molybdenum (1.00% nominal) to produce a grade with increased corrosion resistance and improved resistance to pitting attack from various de-icing chemicals. However, this grade exhibits slightly more tendency than type 430 to rope during forming.

AISI type 435 contains an addition of niobium (0.45% nominal) to produce a grade with improved resistance to roping.

AISI type 436 contains the molybdenum addition of type 434 and the niobium addition of type 435. The result is a stainless steel that exhibits both improved resistance to corrosion and a minimum roping tendency when compared to AISI type 430 stainless steel.

The general comments and physical and mechanical properties that follow apply to the various modifications as well as to type 430. However, some differences do exist and the producer should be consulted when considering use of these steels.

Type 430 is easily cold formed by methods such as cold rolling, bending, drawing, pressing, and heading. Type 435 is well suited for the more severe stretch forming and deep drawing. All four alloys are available in a buffed mirror-bright finish on both sides. If protected during forming, the need for final polishing is reduced, and in some cases eliminated.

Type 430 can be resistance welded, and arc welded using type 430 electrodes. The high heat of welding may cause grain coarsening, embrittlement at room temperature, and a lowered resistance to corrosion. A common practice is to anneal at 760 °C (1400 °F) after welding to restore ductility and corrosion resistance. Heating to temperatures in the range of 150 to 205 °C (300 to 400 °F) prior to heating is recommended.

The general machinability of type 430 is about 55% based on free-machining AISI 1212. Types 430, 434, 435, and 436 have good resistance to scaling up to about 760 to 815 °C (1400 to 1500 °F) in continuous service, and up to about 870 °C (1600 °F) in intermittent service.

Typical Uses. Applications of AISI types 430, 434, 435, and 436 include architectural and automotive trim, appliance parts, bathroom fixtures, oil burner parts, cold headed fasteners, nitrogen fixation equipment, television cones, and nitric acid storage tanks and tank cars.

AISI Type 430: Similar Steels (U.S. and/or Foreign). UNS S43000; AMS 5503, 5627; ASME SA182, SA240, SA268, SA479; ASTM A176, A182, A240, A268, A276, A314, A473, A479, A493, A511, A554, A580, A651; FED QQ-S-763, QQ-S-766, QQ-W-423, STD-66; MIL SPEC MIL-S-862; SAE J405 (51430); (W. Ger.) DIN 1.4016; (Fr.) AFNOR Z 8 C 17; (Ital.) UNI X 8 Cr 17; (Jap.) JIS SUS 430; (Swed.) SS_{14} 2320; (U.K.) B.S. 430 S 15

AISI Type 434: Similar Steels (U.S. and/or Foreign). UNS S43400; ASTM A651; SAE J405 (51434); (W. Ger.) DIN 1.4113; (Fr.) AFNOR Z 8 CD 17.01; (Ital.) UNI X 8 CrMo 17; (Jap.) JIS SUS 434; (Swed.) SS_{14} 2325; (U.K.) B.S. 434 S 19

AISI Type 435: Similar Steels (U.S. and/or Foreign). None listed

AISI Type 436: Similar Steels (U.S. and/or Foreign). UNS S43600; SAE J405 (51436)

Mechanical Properties

AISI Types 430, 434, 435, 436: Stress-Rupture and Creep Rate Properties (Ref 7)

Test temperature		Stress for rupture in:				Stress for secondary creep rate of(a):			
		1000 h		10 000 h(a)		1% in 10 000 h		1% in 100 000 h	
°C	°F	MPa	ksi	MPa	ksi	MPa	ksi	MPa	ksi
480	900	205	30	165	24	...	...	...	...
500	1000	120	17	90	13	59	8.5	45	6.5
595	1100	60	9	40	6	30	4.3	22	3.2
650	1200	35	5	20	3	15	2.2	12	1.7
705	1300	...	...	...	...	9	1.3	...	...

(a) Extrapolated data

AISI Types 430, 434, 435, 436: Tensile Properties at Elevated Temperatures (Ref 7)

Test specimens were 16-mm (0.63-in.) diam bar

Test temperature		Tensile strength		Yield strength(a)		Elongation(b),	Reduction
°C	°F	MPa	ksi	MPa	ksi	%	in area, %
Annealed at 815 °C (1500 °F) and cold worked to 89 HRB							
21	70	538	78.0	441	64.0	32.0	77.0
205	400	459	66.5	390	56.5	28.0	77.5
425	800	379	55.0	317	46.0	28.5	78.0
540	1000	272	39.5	234	34.0	30.0	84.5
650	1200	134	19.5	114	16.5	43.5	95.5
760	1400	66	9.5	52	7.5	70.5	98.0
870	1600	34	5.0	28	4.0	95.5	98.5
Annealed							
21	70	517	75.0	310	45.0	32.0	...
205	400	441	64.0	265	38.5	30.0	...
425	800	365	53.0	231	33.5	29.0	...
540	1000	269	39.0	169	24.5	40.0	...
650	1200	134	19.5	89.6	13.0	56.0	...
760	1400	66	9.5	41	6.0	74.0	...

(a) 0.2% offset. (b) In 50 mm (2 in.)

AISI Types 430, 434, 435, 436: Representative Tensile Properties (Ref 6)

Product form	Condition	Tensile strength MPa	Tensile strength ksi	Yield strength(a) MPa	Yield strength(a) ksi	Elongation(b), %	Reduction in area, %	Hardness
Sheet	Annealed	517	75	345	50	25	...	85 HRB
Strip	Annealed	517	75	345	50	25	...	85 HRB
Wire, 6.4 mm (0.25 in.) diam	Annealed	483	70	276	40	35	70	82 HRB
	Soft tempered	586	85	483	70	15	70	90 HRB
Plate	Annealed	517	75	276	40	30	...	160 HB
Bar	Annealed	517	75	310	45	30	65	155 HB
	Annealed and cold drawn(c)	586	85	483	70	20	65	185 HB

(a) 0.2% offset. (b) In 50 mm (2 in.), except for wire. For wire 3.18-mm (0.125-in.) diam and larger, gage length is 4 × diam; for smaller diameters, the gage length is 254 mm (10.0 in.). (c) Diameter of 25 mm (1 in.)

AISI Types 430, 434, 435, 436: Effect of Cold Reduction on Tensile Properties. Test specimens were in the annealed condition, 9.53-mm (0.375-in.) round, unstraightened. Tests were conducted using specimens machined to English units. Elongation was measured in 50 mm (2 in.); yield strength at 0.2% offset. (Ref 7)

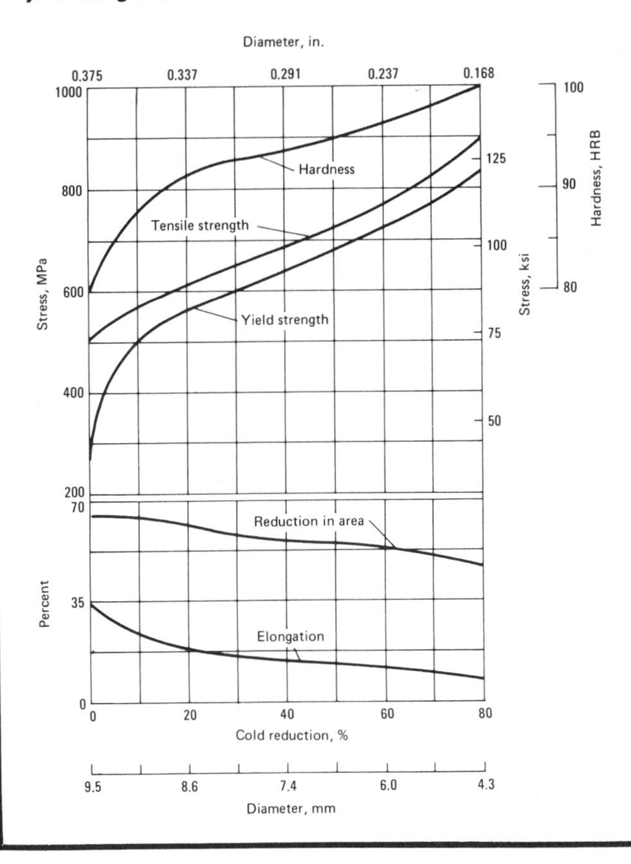

Physical Properties

AISI Types 430, 434, 435, 436: Selected Physical Properties (Ref 6)

Melting range	1425-1510 °C (2600-2750 °F)
Density	7.8 g/cm³ (0.28 lb/in.³)
Specific heat at 0-100 °C (32-212 °F)	460 J/kg·K (0.11 Btu/lb·°F)
Electrical resistivity at 20 °C (68 °F)	60 nΩ·m

AISI Types 430, 434, 435, 436: Thermal Treatment Temperatures

Treatment	Temperature range °C	Temperature range °F
Forging, start	1040-1120	1900-2050
Forging, finish	815 min	1500 min
Annealing(a)	760-815	1400-1500
Hardening(b)	...	...

(a) Air cool. Annealing scale should be removed by pickling in 50% hydrochloric acid heated to 60 to 65 °C (140 to 150 °F). Wash and passivate. (b) Not appreciably hardenable

AISI Types 430, 434, 435, 436: Thermal Conductivity

Temperature °C	Temperature °F	Thermal conductivity W/m·K	Thermal conductivity Btu/ft·h·°F
100	212	26.1	15.1
500	930	26.3	15.2

AISI Types 430, 434, 435, 436: Average Coefficients of Linear Thermal Expansion (Ref 6)

Temperature range °C	Temperature range °F	Coefficient µm/m·K	Coefficient µin./in.·°F
0-100	32-212	10.4	5.8
0-315	32-600	11.0	6.1
0-540	32-1000	11.3	6.3
0-650	31-1200	11.9	6.6
0-815	32-1500	12.4	6.9

Machining Data

For machining data on AISI types 430, 434, 435, and 436, refer to the preceding machining tables for AISI type 405.

AISI Type 431

AISI Type 431: Chemical Composition

Element	Composition, %
Carbon	0.20 max
Manganese	1.00 max
Phosphorus	0.40 max
Sulfur	0.03 max
Silicon	1.00 max
Chromium	15-17
Nickel	1.25-2.50

Characteristics. AISI type 431 is a higher chromium version of the nickel-bearing, hardenable, stainless steel, type 414. It is the most corrosion-resistant grade of the standard, quench-hardenable, stainless steels. This grade is heat treatable to obtain the characteristics of high hardness and strength levels combined with excellent toughness.

Type 431 is resistant to the corrosive action of the atmosphere, various alkalis, and mild acids. It has better resistance to corrosion from marine atmosphere, and is considered to have better resistance to stress corrosion than other martensitic stainless steels. This stainless steel also has good oxidation resistance up to about 815 °C (1500 °F) in continuous service. Scaling of type 431 becomes excessive above temperatures of about 870 °C (1600 °F).

This steel can be moderately cold worked in the subcritical annealed condition. For cold heading operations, type 431 material can be supplied in the cyclic-annealed, minimum hardness condition. In the annealed condition, this steel has a machinability rating of about 45% of that for AISI 1212.

Type 431 can be welded satisfactorily, but the welds air harden and require careful handling to prevent cracking. Treatment recommendations are heating to 205 °C (400 °F) prior to welding, and annealing immediately after welding. Electrodes of similar composition should be used with this grade to ensure similar properties. If less strong but more ductile welds are required, type 310 electrodes should be used.

Typical Uses. Applications of AISI type 431 include aircraft fittings, beater bars, fasteners, conveyor parts, valve parts, pump shafts, and marine hardware.

AISI Type 431: Similar Steels (U.S. and/or Foreign). UNS S43100; AMS 5628; ASTM A276, A314, A473, A493, A579 (63), A580; MIL SPEC MIL-S-862; SAE J405 (51431); (W. Ger.) DIN 1.4057; (Fr.) AFNOR Z 6 CNU 17.04, Z 15 CN 16.02; (Ital.) UNI X 16 CrNi 16; (Jap.) JIS SUS 431; (U.K.) B.S. 431 S 29, 5 S. 80

Physical Properties

AISI Type 431: Selected Physical Properties (Ref 7)

Density	7.8 g/cm³ (0.28 lb/in.³)
Specific heat at 0-100 °C (32-212 °F)	460 J/kg·K (0.11 Btu/lb·°F)
Thermal conductivity at 100 °C (212 °F)	20.2 W/m·K (11.7 Btu/ft·h·°F)
Electrical resistivity at 20 °C (68 °F)	72 nΩ·m

AISI Type 431: Average Coefficients of Linear Thermal Expansion (Ref 7)

Temperature range °C	Temperature range °F	Coefficient μm/m·K	Coefficient μin./in.·°F
0-100	32-212	10.1	5.6
0-315	32-600	12.1	6.7
0-540	32-1000	11.9	6.6
0-650	32-1200	12.2	6.8

AISI Type 431: Thermal Treatment Temperatures

Treatment	Temperature range °C	Temperature range °F
Forging, start	1040-1120	1900-2050
Forging, finish	900 min	1650 min
Subcriticalannealing(a)	620-665	1150-1225
Cycle annealing	(b)	(b)
Hardening(c)	980-1065	1800-1950
Tempering(d)	230-705	450-1300

(a) Soak at temperature 4 to 8 h; air cool. (b) Heat to 760 °C (1400 °F); hold 20 h; cool below 315 °C (600 °F); reheat to 675 °C (1250 °F); hold 20 h; air cool. (c) Hold at temperature 15 to 30 min; oil quench. (d) For maximum corrosion resistance, temper at 230 to 395 °C (450 to 745 °F) for 1 to 3 h. Tempering at temperatures in the 400 to 565 °C (750 to 1050 °F) range should be avoided because of lowered impact toughness and corrosion resistance

Mechanical Properties

AISI Type 431: Representative Tensile Properties (Ref 6)

Product form	Condition	Tensile strength MPa	Tensile strength ksi	Yield strength(a) MPa	Yield strength(a) ksi	Elongation(b), %	Reduction in area, %	Hardness
Wire	Soft tempered	930	135	795	115	10	50	29 HRC
Bar	Annealed	860	125	655	95	20	55	260 HB
	Annealed and cold drawn	895	130	760	110	15	35	270 HB

(a) 0.2% offset. (b) In 50 mm (2 in.), except for wire. For wire 3.18-mm (0.125-in.) diam and larger, gage length is 4 × diam; for smaller diameters, the gage length is 254 mm (10.0 in.)

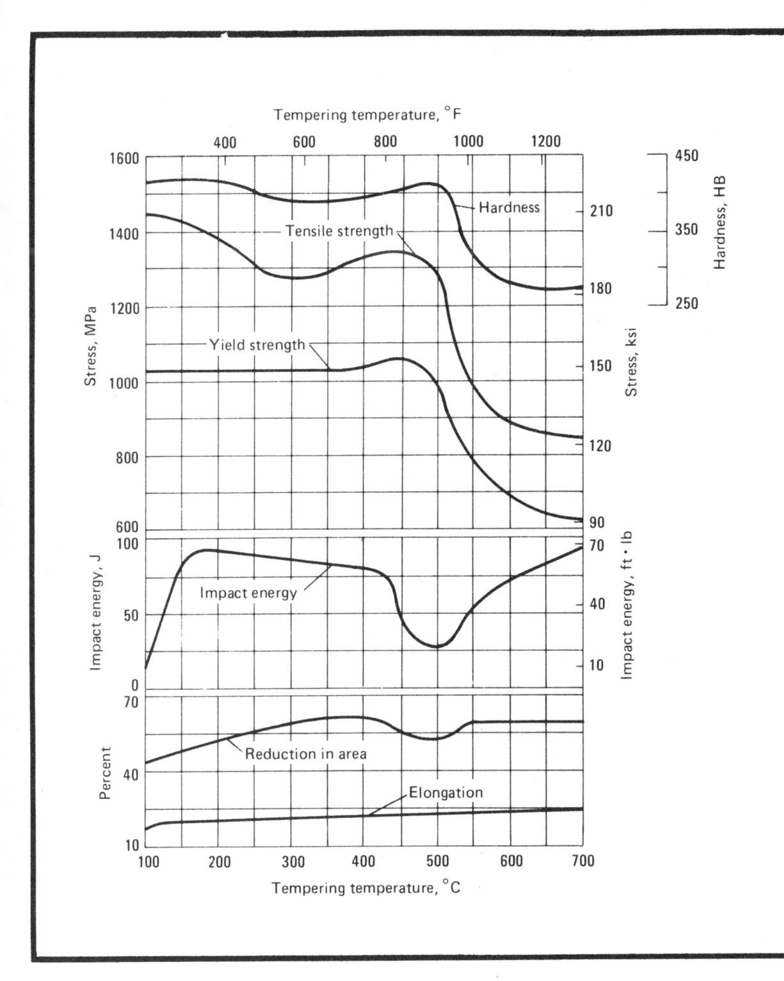

AISI Type 431: Effect of Tempering Temperature on Tensile Properties. Heated to 1040 °C (1900 °F); oil quenched; tempered for 3 h at temperature given. Specimens were heat treated as 25-mm (1-in.) round; tensile test specimens were 12.8-mm (0.505-in.) diam; Izod notched impact test specimens were 10.0-mm (0.394-in.) square. Tests were conducted using specimens machined to English units. Impact energy test used Izod specimens. Elongation was measured in 50 mm (2 in.); yield strength at 0.2% offset. (Ref 10)

AISI Type 431: Tensile Properties at Elevated Temperatures After Tempering at 650 °C (1200 °F)

Test specimens were 16-mm (0.625-in.) diam bar; austenitized at 980 °C (1800 °F) for ½ h; tempered at 650 °C (1200 °F) for 2 h; air cooled; hardness of 25.5 HRC

Test temperature °C	°F	Tensile strength MPa	ksi	Yield strength(a) MPa	ksi	Elongation(b), %	Reduction in area, %
21	70	831	120.5	738	107	20	63.5
205	400	727	105.5	650	94	17.5	66
315	600	685	99	610	88.5	15.5	61.5
425	800	625	91	570	82.5	15.5	61
480	900	560	81.5	510	74	17.5	68.5
540	1000	485	70	470	68	23	77.5
595	1100	385	56	355	51.5	22	83
650	1200	295	42.5	270	39.5	25	86

(a) 0.2% offset. (b) In 50 mm (2 in.)

AISI Type 431: Tensile Properties at Elevated Temperatures After Tempering at Various Temperatures (Ref 7)

Test specimens were 16-mm (0.625-in.) diam bar; austenitized at 980 °C (1800 °F) for ½ h; oil quenched; tempered at 28 °C (50 °F) above test temperature for 2 h

Test temperature °C	°F	Tensile strength MPa	ksi	Yield strength(a) MPa	ksi	Elongation(b), %	Reduction in area, %	Hardness(c), HRC
21	70	1435	208	1140	165.5	16.5	59	45
205	400	1350	196	865	125.5	14	44.5	40.5
315	600	1365	198	995	144	15	44.5	41.5
425	800	1280	185.5	915	133	17	55	42.5
480	900	1035	150	760	110.5	13.5	53.5	42.5
540	1000	565	82	525	76.5	20.5	75	31.5
595	1100	415	60.5	360	52.5	23.5	82	28
650	1200	285	41	260	38	26	86.5	24.5
705	1300	170	25	150	21.5	29	91	23.5

(a) 0.2% offset. (b) In 50 mm (2 in.). (c) Before testing

AISI Type 431: Izod Impact Properties (Ref 7)

Condition	Energy J	ft·lb
Annealed	70-110	50-80
Quenched and tempered	30-80	20-60

AISI Type 431: Modulus of Elasticity (Ref 7)

Stressed in:	Modulus GPa	10^6 psi
Tension	200	29

Machining Data

For machining data on AISI type 431, refer to the preceding machining tables for AISI type 414.

AISI Type 440A

AISI Type 440A: Chemical Composition

Element	Composition, %
Carbon	0.60-0.75
Manganese	1.00 max
Phosphorus	0.04 max
Sulfur	0.03 max
Silicon	1.00 max
Chromium	16-18
Molybdenum	0.75 max

Characteristics. AISI type 440A is a high-carbon, high-chromium, stainless steel designed to provide stainless properties with excellent hardness. This steel attains a hardness of 56 HRC and maximum toughness when heat treated.

Generally type 440A has corrosion resistance similar to type 410. This alloy can be used in the annealed, hardened, or hardened and tempered condition. Optimum corrosion resistance is obtained in the as-hardened condition or with a temper below about 425 °C (800 °F). After heat treatment, parts must be pickled, polished, or ground to remove all scale. After pickling, the parts should be baked at 120 to 150 °C (250 to 300 °F) to remove acid brittleness.

Because corrosion resistance, as well as hardness and strength, are lowered by exposure above about 425 °C (800 °F), type 440A is not usually employed for high temperature service. This steel scales appreciably at temperatures above 760 °C (1400 °F).

When in the fully annealed condition, type 440A can be moderately cold worked and it machines best in the fully annealed condition. The machinability rating of this grade is about 45% of that for AISI 1212.

Type 440A is seldom welded. However, by preheating the parts to 150 to 205 °C (300 to 400 °F), then annealing 6 to 8 h at 730 to 760 °C (1350 to 1400 °F) and air cooling, satisfactory welds can be produced. When required, filler metal should be of a similar composition to ensure matching properties.

Typical Uses. Applications of AISI type 440A include pivot pins, dental and surgical instruments, cutlery, and valve parts.

AISI Type 440A: Similar Steels (U.S. and/or Foreign). UNS S44002; AMS 5631; ASTM A276, A314, A473, A511, A580; FED QQ-S-763; MIL SPEC MIL-S-862; SAE J405 (51440A)

Physical Properties

AISI Type 440A: Selected Physical Properties (Ref 6, 8)

Melting range	1370-1510 °C (2500-2750 °F)
Density	7.8 g/cm^3 (0.28 lb/in.^{3})
Specific heat at 0-100 °C (32-212 °F)	460 J/kg·K (0.11 Btu/lb·°F)
Thermal conductivity at 100 °C (212 °F)	24.2 W/m·K (14.0 Btu/ft·h·°F)
Electrical resistivity at 20 °C (68 °F)	60 nΩ·m

AISI Type 440A: Average Coefficients of Linear Thermal Expansion (Ref 6, 8)

Temperature range °C	°F	Coefficient μm/m·K	μin./in.·°F
20-200	68-390	10.3-11.2	5.7-6.2
20-600	68-1110	11.7-12.6	6.5-7.0

AISI Type 440A: Thermal Treatment Temperatures

Treatment	Temperature range °C	°F
Forging, preheat	760-815	1400-1500
Forging, start	1040-1205	1900-2200
Forging, finish	925 min	1700 min
Full annealing(a)	845-900	1550-1650
Process annealing(a)	730-790	1350-1450
Hardening(b)	1010-1065	1850-1950
Hardening, preheat	540-790	1000-1450
Tempering	150-370	300-700

(a) Hold at temperature; furnace cool. (b) Hold at temperature; quench in warm oil or air cool. Preheating is recommended

Mechanical Properties

AISI Type 440A: Izod Impact Properties (Ref 7)

Condition	Energy J	ft·lb	Hardness, HB
Annealed	20	15	215
Quenched and tempered	7	5	560

AISI Type 440A: Rockwell C Hardness (Ref 7)

Oil quenched at 1010 to 1040 °C (1850 to 1900 °F); tempered for 1 h

Tempering temperature °C	°F	Hardness, HRC
As quenched		57
150	300	55.5
205	400	54
260	500	52.5
315	600	51.5
370	700	50.5
425	800	50
540	1000	46

AISI Type 440A: Modulus of Elasticity (Ref 7)

Stressed in:	Modulus GPa	10^6 psi
Tension	200	29.0

AISI Type 440A: Representative Tensile Properties (Ref 6)

Product form	Condition	Tensile strength MPa	Tensile strength ksi	Yield strength(a) MPa	Yield strength(a) ksi	Elongation(b), %	Reduction in area, %	Hardness
Strip	Annealed	690	100	415	60	20	...	95 HRB
Wire, 6.4-mm (0.25-in.) diam	Annealed	725	105	415	60	18	55	95 HRB
	Soft tempered	795	115	585	85	10	35	99 HRB
Bar	Annealed	725	105	415	60	20	45	95 HRB
	Annealed and cold drawn(c)	795	115	620	90	12	20	99 HRB
	Hardened and tempered at 315 °C (600 °F)	1790	260	1650	240	5	20	510 HB

(a) 0.2% offset. (b) In 50 mm (2 in.), except for wire. For wire 3.18-mm (0.125-in.) diam and larger, gage length is 4 × diam; for smaller diameters, the gage length is 254 mm (10.0 in.). (c) Diameter of 25 mm (1.0 in.)

Machining Data (Ref 2)

AISI Type 440A: Turning (Single Point and Box Tools)

Depth of cut mm	Depth of cut in.	High speed steel(a) Speed m/min	ft/min	Feed mm/rev	Feed in./rev	Uncoated carbide Speed, brazed m/min	ft/min	Speed, inserted m/min	ft/min	Feed mm/rev	Feed in./rev	Coated carbide Speed m/min	ft/min	Feed mm/rev	Feed in./rev
Annealed, with hardness of 225 to 275 HB															
1	0.040	24	80	0.18	0.007	105(b)	350(b)	120(b)	400(b)	0.18	0.007	160(c)	525(c)	0.18	0.007
4	0.150	20	65	0.40	0.015	84(d)	275(d)	95(d)	310(d)	0.40	0.015	120(e)	400(e)	0.40	0.015
8	0.300	15	50	0.50	0.020	69(d)	225(d)	76(d)	250(d)	0.075	0.030	100(e)	325(e)	0.50	0.020
16	0.625	11	35	0.075	0.030	53(d)	175(d)	60(d)	200(d)	0.102	0.040	...	...	...	...
Quenched and tempered, with hardness of 275 to 325 HB															
1	0.040	20	65	0.18	0.007	87(b)	285(b)	95(b)	315(b)	0.18	0.007	120(c)	400(c)	0.18	0.007
4	0.150	15	50	0.40	0.15	69(d)	225(d)	75(d)	245(d)	0.40	0.015	100(e)	325(e)	0.40	0.015
8	0.300	12	40	0.50	0.20	55(d)	180(d)	58(d)	190(d)	0.50	0.020	76(e)	250(e)	0.50	0.020
Quenched and tempered, with hardness of 375 to 475 HB															
1	0.040	7	55	0.13	0.005	58(b)	190(b)	67(b)	220(b)	0.13	0.005	...	...	...	...
4	0.150	12	40	0.25	0.010	43(d)	140(d)	53(d)	175(d)	0.25	0.010	...	...	...	...
8	0.300	9	30	0.40	0.015	34(d)	110(d)	41(d)	135(d)	0.40	0.015	...	...	...	...
Quenched and tempered, with hardness of 48 to 52 HRC															
1	0.040	11	35	0.075	0.003	40(f)	130(f)	46(f)	150(f)	0.13	0.005	...	...	...	...
4	0.150	8	25	0.18	0.007	30(f)	100(f)	37(f)	120(f)	0.25	0.010	...	...	...	...
8	0.300	6	20	0.25	0.010	24(d)	80(d)	27(d)	90(d)	0.40	0.015	...	...	...	...

(a) Any premium high speed steel tool material (T15, M33, M41 to M47). (b) Carbide tool material: C-7. (c) Carbide tool material: CC-7. (d) Carbide tool material: C-6. (e) Carbide tool material: CC-6. (f) Carbide tool material: C-8

AISI Type 440A: Boring

Depth of cut mm	Depth of cut in.	High speed steel(a) Speed m/min	ft/min	Feed mm/rev	Feed in./rev	Uncoated carbide Speed, brazed m/min	ft/min	Speed, inserted m/min	ft/min	Feed mm/rev	Feed in./rev	Coated carbide Speed m/min	ft/min	Feed mm/rev	Feed in./rev
Annealed, with hardness of 225 to 275 HB															
0.25	0.010	24	80	0.075	0.003	90(b)	300(b)	105(b)	350(b)	0.075	0.003	130(c)	450(c)	0.075	0.003
1.25	0.050	20	65	0.13	0.005	72(d)	235(d)	84(d)	275(d)	0.13	0.005	110(e)	360(e)	0.13	0.005
2.5	0.100	15	50	0.30	0.012	56(d)	185(d)	66(d)	215(d)	0.30	0.012	84(e)	275(e)	0.30	0.012
Quenched and tempered, with hardness of 275 to 325 HB															
0.25	0.010	20	65	0.075	0.003	73(b)	240(b)	85(b)	280(b)	0.075	0.003	76(c)	350(c)	0.075	0.003
1.25	0.050	15	50	0.13	0.005	58(d)	190(d)	69(d)	225(d)	0.13	0.005	84(e)	275(e)	0.13	0.005
2.5	0.100	12	40	0.30	0.012	44(d)	145(d)	52(d)	170(d)	0.30	0.012	69(e)	225(e)	0.30	0.012
Quenched and tempered, with hardness of 375 to 425 HB															
0.25	0.010	17	55	0.050	0.002	50(b)	165(b)	59(b)	195(b)	0.050	0.002	...	...	...	...
1.25	0.050	14	45	0.102	0.004	40(d)	130(d)	47(d)	155(d)	0.102	0.004	...	...	...	...
2.5	0.100	9	30	0.20	0.008	32(d)	105(d)	38(d)	125(d)	0.20	0.008	...	...	...	...
Quenched and tempered, with hardness of 48 to 52 HRC															
0.25	0.010	12	40	0.050	0.002	34(b)	110(b)	40(b)	130(b)	0.050	0.002	...	...	...	...
1.25	0.050	9	30	0.102	0.004	27(d)	90(d)	32(d)	105(d)	0.102	0.004	...	...	...	...
2.5	0.100	6	20	0.15	0.006	21(d)	70(d)	26(d)	85(d)	0.20	0.008	...	...	...	...

(a) Any premium high speed steel (T15, M33, M41 to M47). (b) Carbide tool material: C-7. (c) Carbide tool material: CC-7. (d) Carbide tool material: C-6. (e) Carbide tool material: CC-6

Machining Data (Ref 2) (continued)

AISI Type 440A: Turning (Cutoff and Form Tools)

Tool material	Speed, m/min (ft/min)	Feed per revolution for cutoff tool width of: 1.5 mm (0.062 in.) mm (in.)	3 mm (0.125 in.) mm (in.)	6 mm (0.25 in.) mm (in.)	Feed per revolution for form tool width of: 12 mm (0.5 in.) mm (in.)	18 mm (0.75 in.) mm (in.)	25 mm (1 in.) mm (in.)	35 mm (1.5 in.) mm (in.)	50 mm (2 in.) mm (in.)
Annealed, with hardness of 225 to 275 HB									
M2 and M3 high speed steel	17 (55)	0.033 (0.0013)	0.046 (0.0018)	0.033 (0.0023)	0.046 (0.0018)	0.041 (0.0016)	0.033 (0.0013)	0.025 (0.001)	0.020 (0.0008)
C-6 carbide	62 (205)	0.033 (0.0013)	0.046 (0.0018)	0.033 (0.0013)	0.046 (0.0018)	0.041 (0.0016)	0.033 (0.0013)	0.025 (0.001)	0.020 (0.0008)
Quenched and tempered, with hardness of 275 to 325 HB									
Any premium high high speed steel (T15, M33, M41-M47)	15 (50)	0.025 (0.001)	0.050 (0.002)	0.038 (0.0015)	0.038 (0.0015)	0.030 (0.0012)	0.025 (0.001)	0.023 (0.0009)	0.020 (0.0008)
C-6 carbide	52 (170)	0.025 (0.001)	0.050 (0.002)	0.038 (0.0015)	0.038 (0.0015)	0.030 (0.0012)	0.025 (0.001)	0.023 (0.0009)	0.020 (0.0008)
Quenched and tempered, with hardness of 375 to 425 HB									
Any premium high speed steel (T15, M33, M41-M47)	12 (40)	0.020 (0.0008)	0.025 (0.001)	0.033 (0.0013)	0.033 (0.0013)	0.025 (0.001)	0.020 (0.0008)	0.018 (0.0007)	0.015 (0.0006)
C-6 carbide	34 (110)	0.020 (0.0008)	0.025 (0.001)	0.033 (0.0013)	0.033 (0.0013)	0.025 (0.001)	0.020 (0.0008)	0.018 (0.0007)	0.015 (0.0006)
Quenched and tempered, with hardness of 48 to 52 HRC									
Any premium high speed steel (T15, M33, M41-M47)	6 (20)	0.018 (0.0007)	0.025 (0.001)	0.033 (0.0013)	0.028 (0.0011)	0.023 (0.0009)	0.020 (0.0008)	0.015 (0.0006)	0.013 (0.0005)
C-2 carbide	20 (65)	0.018 (0.0007)	0.025 (0.001)	0.033 (0.0013)	0.028 (0.0011)	0.023 (0.0009)	0.020 (0.0008)	0.015 (0.0006)	0.013 (0.0005)

AISI Type 440A: Cylindrical and Internal Grinding

Operation	Wheel speed m/s	Wheel speed ft/min	Work speed m/min	Work speed ft/min	Infeed on diameter mm/pass	Infeed on diameter in./pass	Traverse(a)	Wheel identification
Annealed, with hardness of 135 to 275 HB								
Cylindrical grinding								
Roughing	28-33	5500-6500	15-30	50-100	0.050	0.002	¼	A60JV
Finishing	28-33	5500-6500	15-30	50-100	0.013 max	0.0005 max	⅛	A60JV
Internal grinding								
Roughing	25-33	5000-6500	23-60	75-200	0.013	0.0005	⅓	A60KV
Finishing	25-33	5000-6500	23-60	75-200	0.005 max	0.0002 max	⅙	A60KV
Quenched and tempered, with hardness greater than 275 HB								
Cylindrical grinding								
Roughing	28-33	5500-6500	15-30	50-100	0.050	0.002	¼	A60IV
Finishing	28-33	5500-6500	15-30	50-100	0.013 max	0.0005 max	⅛	A60IV
Internal grinding								
Roughing	25-33	5000-6500	15-46	50-100	0.013	0.0005	⅓	A60JV
Finishing	25-33	5000-6500	15-46	50-100	0.005 max	0.0002 max	⅙	A60JV

(a) Wheel width per revolution of work

AISI Type 440A: Surface Grinding with a Horizontal Spindle and Reciprocating Table

Operation	Wheel speed m/s	Wheel speed ft/min	Table speed m/min	Table speed ft/min	Downfeed mm/pass	Downfeed in./pass	Crossfeed mm/pass	Crossfeed in./pass	Wheel identification
Annealed, with hardness of 135 to 275 HB									
Roughing	28-33	5500-6500	15-30	50-100	0.050	0.002	1.25-12.5	0.050-0.500	A46IV
Finishing	28-33	5500-6500	15-30	50-100	0.013 max	0.0005 max	(a)	(a)	A46IV
Quenched and tempered, with hardness greater than 275 HB									
Roughing	28-33	5500-6500	15-30	50-100	0.025	0.001	0.65-6.5	0.025-0.250	A46HV
Finishing	28-33	5500-6500	15-30	50-100	0.008 max	0.0003 max	(b)	(b)	A46HV

(a) Maximum ¼ of wheel width. (b) Maximum ⅒ of wheel width

AISI Type 440A: Reaming

Based on 4 flutes for 3- and 6-mm (0.125- and 0.25-in.) reamers, 6 flutes for 12-mm (0.5-in.) reamers, and 8 flutes for 25-mm (1-in.) and larger reamers

Tool material	Speed m/min	Speed ft/min	3 mm (0.125 in.) mm	3 mm (0.125 in.) in.	6 mm (0.25 in.) mm	6 mm (0.25 in.) in.	12 mm (0.5 in.) mm	12 mm (0.5 in.) in.	25 mm (1 in.) mm	25 mm (1 in.) in.	35 mm (1.5 in.) mm	35 mm (1.5 in.) in.	50 mm (2 in.) mm	50 mm (2 in.) in.
			Feed per revolution for reamer diameter of:											
Annealed, with hardness of 225 to 275 HB														
Roughing														
M1, M2, and M7 high speed steel	17	55	0.050	0.002	0.075	0.003	0.13	0.005	0.20	0.008	0.25	0.010	0.30	0.012
C-2 carbide	21	70	0.102	0.004	0.20	0.008	0.30	0.012	0.40	0.016	0.50	0.020	0.60	0.024
Finishing														
M1, M2, and M7 high speed steel	11	35	0.075	0.003	0.075	0.003	0.102	0.004	0.15	0.006	0.18	0.007	0.20	0.008
C-2 carbide	15	50	0.102	0.004	0.13	0.005	0.150	0.006	0.20	0.008	0.23	0.009	0.25	0.010
Quenched and tempered, with hardness of 275 to 325 HB														
Roughing														
M1, M2, and M7 high speed steel	12	40	0.050	0.002	0.075	0.003	0.130	0.005	0.20	0.008	0.25	0.010	0.30	0.012
C-2 carbide	17	55	0.102	0.004	0.15	0.006	0.230	0.009	0.40	0.015	0.45	0.018	0.50	0.020
Finishing														
M1, M2, and M7 high speed steel	9	30	0.075	0.003	0.075	0.003	0.102	0.004	0.15	0.006	0.18	0.007	0.20	0.008
C-2 carbide	12	40	0.075	0.003	0.102	0.004	0.130	0.005	0.15	0.006	0.18	0.007	0.20	0.008
Quenched and tempered, with hardness of 375 to 425 HB														
Roughing														
Any premium high speed steel (T15, M33, M41-M47)	9	30	0.050	0.002	0.075	0.003	0.102	0.004	0.18	0.007	0.25	0.010	0.30	0.012
C-2 carbide	14	45	0.075	0.003	0.13	0.005	0.150	0.006	0.20	0.008	0.25	0.010	0.30	0.012
Finishing														
Any premium high speed steel (T15, M33, M41-M47)	6	20	0.075	0.003	0.075	0.003	0.102	0.004	0.15	0.006	0.18	0.007	0.20	0.008
C-2 carbide	11	35	0.075	0.003	0.102	0.004	0.130	0.005	0.15	0.006	0.18	0.007	0.20	0.008
Quenched and tempered, with hardness of 48 to 52 HRC														
Roughing														
Any premium high speed steel (T15, M33, M41-M47)	6	20	0.050	0.002	0.075	0.003	0.102	0.004	0.15	0.006	0.20	0.008	0.25	0.010
C-2 carbide	11	35	0.075	0.003	0.102	0.004	0.130	0.005	0.15	0.006	0.20	0.008	0.25	0.010
Finishing														
Any premium high speed steel (T15, M33, M41-M47)	5	15	0.050	0.002	0.075	0.003	0.102	0.004	0.13	0.005	0.15	0.006	0.15	0.006
C-2 carbide	9	30	0.050	0.002	0.075	0.003	0.102	0.004	0.13	0.005	0.15	0.006	0.15	0.006
Quenched and tempered, with hardness of 52 to 54 HRC														
Roughing														
C-2 carbide	8	25	0.050	0.002	0.075	0.003	0.102	0.004	0.13	0.005	0.15	0.006	0.20	0.008
Finishing														
C-2 carbide	6	20	0.050	0.002	0.050	0.002	0.075	0.003	0.102	0.004	0.102	0.004	0.102	0.004

AISI Type 440A: Planing

Tool material	Depth of cut mm	Depth of cut in.	Speed m/min	Speed ft/min	Feed/stroke mm	Feed/stroke in.
Annealed, with hardness of 225 to 275 HB						
M2 and M3 high speed steel	0.1	0.005	6	20	(a)	(a)
	2.5	0.100	8	25	0.75	0.030
	12	0.500	5	15	1.15	0.045
C-6 carbide	0.1	0.005	27	90	(a)	(a)
	2.5	0.100	30	100	1.50	0.060
Quenched and tempered, with hardness of 275 to 325 HB						
M2 and M3 high speed steel	0.1	0.005	6	20	(a)	(a)
	2.5	0.100	8	25	0.75	0.030
	12	0.500	5	15	1.15	0.045
C-6 carbide	0.1	0.005	27	90	(a)	(a)
	2.5	0.100	30	100	1.50	0.060

(a) Feed is 75% the width of the square nose finishing tool

AISI Type 440A: End Milling (Profiling)

Tool material	Depth of cut		Speed		Feed per tooth for cutter diameter of: 10 mm (0375 in.)		12 mm (0.5 in.)		18 mm (0.75 in.)		25-50 mm (1-2 in.)	
	mm	in.	m/min	ft/min	mm	in.	mm	in.	mm	in.	mm	in.
Annealed, with hardness of 225 to 275 HB												
M2, M3, and M7 high speed steel	0.5	0.020	27	90	0.025	0.001	0.050	0.002	0.075	0.003	0.102	0.004
	1.5	0.060	21	70	0.050	0.002	0.075	0.003	0.102	0.004	0.130	0.005
	diam/4	diam/4	18	60	0.025	0.001	0.050	0.002	0.075	0.003	0.102	0.004
	diam/2	diam/2	15	50	0.025	0.001	0.025	0.001	0.050	0.002	0.075	0.003
C-5 carbide	0.5	0.020	95	310	0.013	0.0005	0.025	0.001	0.075	0.003	0.130	0.005
	1.5	0.060	72	235	0.025	0.001	0.050	0.002	0.102	0.004	0.150	0.006
	diam/4	diam/4	62	205	0.025	0.001	0.038	0.0015	0.075	0.003	0.102	0.004
	diam/2	diam/2	58	190	0.013	0.0005	0.025	0.001	0.050	0.002	0.075	0.003
Quenched and tempered, with hardness of 275 to 325 HB												
M2, M3, and M7 high speed steel	0.5	0.020	21	70	0.018	0.0007	0.025	0.001	0.050	0.002	0.075	0.003
	1.5	0.060	17	55	0.025	0.001	0.050	0.002	0.075	0.003	0.102	0.004
	diam/4	diam/4	14	45	0.018	0.0007	0.025	0.001	0.050	0.002	0.075	0.003
	diam/2	diam/2	12	40	0.013	0.0005	0.018	0.0007	0.038	0.0015	0.063	0.0025
C-5 carbide	0.5	0.020	76	250	0.013	0.0005	0.025	0.001	0.075	0.003	0.130	0.005
	1.5	0.060	58	190	0.025	0.001	0.050	0.002	0.102	0.004	0.150	0.006
	diam/4	diam/4	50	165	0.025	0.001	0.038	0.0015	0.075	0.003	0.102	0.004
	diam/2	diam/2	46	150	0.013	0.0005	0.025	0.001	0.050	0.002	0.075	0.003
Quenched and tempered, with hardness of 375 to 425 HB												
Any premium high speed steel (T15, M33, M41-47)	0.5	0.020	18	60	0.013	0.0005	0.018	0.0007	0.025	0.001	0.050	0.002
	1.5	0.060	14	45	0.013	0.0005	0.025	0.001	0.050	0.002	0.075	0.003
	diam/4	diam/4	12	40	0.013	0.0005	0.013	0.0005	0.038	0.0015	0.063	0.0025
	diam/2	diam/2	11	35	...	...	...	...	0.025	0.001	0.050	0.002
C-5 carbide	0.5	0.020	60	200	0.013	0.0005	0.025	0.001	0.025	0.001	0.075	0.003
	1.5	0.060	46	150	0.025	0.001	0.050	0.002	0.050	0.002	0.102	0.004
	diam/4	diam/4	40	130	0.025	0.001	0.038	0.0015	0.038	0.0015	0.075	0.003
	diam/2	diam/2	35	115	0.013	0.0005	0.025	0.001	0.025	0.001	0.063	0.0025
Quenched and tempered, with hardness of 48 to 52 HRC												
Any premium high speed steel (T15, M33, M41-M47)	0.5	0.020	12	40	0.013	0.0005	0.013	0.0005	0.013	0.0005	0.025	0.001
	1.5	0.060	11	35	...	...	0.013	0.0005	0.025	0.001	0.038	0.0015
	diam/4	diam/4	9	30	...	...	...	...	0.013	0.0005	0.025	0.001
	diam/2	diam/2	8	25	...	...	...	...	...	...	...	...
C-5 carbide	0.5	0.020	24	80	0.013	0.0005	0.013	0.0005	0.025	0.001	0.025	0.001
	1.5	0.060	18	60	...	...	0.013	0.0005	0.050	0.002	0.050	0.002
	diam/4	diam/4	15	50	...	...	...	...	0.038	0.0015	0.038	0.0015
	diam/2	diam/2	14	45	...	...	...	...	0.025	0.001	0.025	0.001

AISI Type 440A: Face Milling

Depth of cut		High speed steel: Speed		Feed/tooth		Uncoated carbide: Speed, brazed		Speed, inserted		Feed/tooth		Coated carbide: Speed		Feed/tooth	
mm	in.	m/min	ft/min	mm	in.	m/min	ft/min	m/min	ft/min	mm	in.	m/min	ft/min	mm	in.
Annealed, with hardness of 225 to 275 HB															
1	0.040	32(a)	105(a)	0.15	0.006	130(b)	430(b)	145(b)	475(b)	0.18	0.007	215(c)	700(c)	0.18	0.007
4	0.150	24(a)	80(a)	0.25	0.010	100(b)	335(b)	110(b)	360(b)	0.25	0.010	140(c)	465(c)	0.25	0.010
8	0.300	18(a)	60(a)	0.36	0.014	70(d)	230(d)	85(d)	280(d)	0.36	0.014	110(e)	365(e)	0.36	0.014
Quenched and tempered, with hardness of 275 to 325 HB															
1	0.040	24(f)	80(f)	0.15	0.006	115(b)	370(b)	125(b)	410(b)	0.15	0.006	190(c)	625(c)	0.13	0.005
4	0.150	18(f)	60(f)	0.23	0.009	84(b)	275(b)	95(b)	305(b)	0.20	0.008	120(c)	400(c)	0.18	0.007
8	0.300	14(f)	45(f)	0.30	0.012	58(d)	190(d)	72(d)	235(d)	0.25	0.010	90(e)	300(e)	0.23	0.009
Quenched and tempered, with hardness of 375 to 425 HB															
1	0.040	18(f)	60(f)	0.100	0.004	55(b)	180(b)	60(b)	200(b)	0.102	0.004	...	...	...	...
4	0.150	14(f)	45(f)	0.15	0.006	41(b)	135(b)	46(b)	150(b)	0.15	0.006	...	...	...	...
8	0.300	11(f)	35(f)	0.20	0.008	29(b)	95(b)	35(b)	115(b)	0.20	0.008	...	...	...	...
Quenched and tempered, with hardness of 48 to 52 HRC															
1	0.040	11(f)	35(f)	0.050	0.002	47(b)	155(b)	58(b)	190(b)	0.050	0.002	...	...	...	...
4	0.150	8(f)	25(f)	0.075	0.003	38(b)	125(b)	47(b)	155(b)	0.075	0.003	...	...	...	...
8	0.300	6(f)	20(f)	0.102	0.004	30(b)	100(b)	37(b)	120(b)	0.102	0.004	...	...	...	...

(a) High speed steel tool material: M2 or M7. (b) Carbide tool material: C-6. (c) Carbide tool material: CC-6. (d) Carbide tool material: C-5. (e) Carbide tool material: CC-5. (f) Any premium high speed tool material (T15, M33, M41 to M47)

Machining Data (Ref 2) (continued)

AISI Type 440A: Drilling

Tool material	Speed, m/min (ft/min)	Feed per revolution for nominal hole diameter of: 3 mm (0.125 in.) mm (in.)	6 mm (0.25 in.) mm (in.)	12 mm (0.5 in.) mm (in.)	18 mm (0.75 in.) mm (in.)	25 mm (1 in.) mm (in.)	35 mm (1.5 in.) mm (in.)	50 mm (2 in.) mm (in.)
Annealed, with hardness of 225 to 275 HB								
M10, M7, and M1 high speed steel	15 (50)	0.050 (0.002)	0.102 (0.004)	0.180 (0.007)	0.230 (0.009)	0.280 (0.011)	0.33 (0.013)	0.40 (0.015)
Quenched and tempered, with hardness of 275 to 325 HB								
M10, M7, and M1 high speed steel	12 (40)	0.025 (0.001)	0.075 (0.003)	0.130 (0.005)	0.150 (0.006)	0.180 (0.007)	0.23 (0.009)	0.25 (0.010)
Quenched and tempered, with hardness of 375 to 425 HB								
Any premium high speed steel	8 (25)	0.025 (0.001)	0.050 (0.002)	0.102 (0.004)	0.102 (0.004)	0.102 (0.004)	0.102 (0.004)	0.102 (0.004)
Quenched and tempered, with hardness of 48 to 52 HRC								
Any premium high speed steel	3 (10)	0.025 (0.001)	0.050 (0.002)	0.075 (0.003)	0.075 (0.003)	0.102 (0.004)	0.102 (0.004)	0.102 (0.004)

AISI Type 440B

AISI Type 440B: Chemical Composition

Element	Composition, %
Carbon	0.75-0.95
Manganese	1.00 max
Phosphorus	0.04 max
Sulfur	0.03 max
Silicon	1.00 max
Chromium	16-18
Molybdenum	0.075 max

Characteristics. AISI type 440B is a high-carbon, high-chromium steel designed to provide stainless properties with excellent hardness. This alloy attains a hardness of 58 HRC with excellent toughness when heat treated. Corrosion resistance of type 440B is generally similar to type 410.

Type 440B is always used in the hardened and tempered condition. Optimum corrosion resistance is obtained with a temper below about 425 °C (800 °F). After heat treatment, parts must be pickled, polished, or ground to remove all scale. After pickling, parts should be baked at 120 to 150 °C (250 to 300 °F) to remove acid brittleness.

Because corrosion resistance is impaired when type 440B is used in the annealed condition, or hardened and tempered above 425 °C (800 °F), it is not usually recommended for elevated temperature applications.

Type 440B can be cold formed, headed, and upset in the fully annealed condition. For most machining operations, this steel cuts best in the fully annealed condition. Because the chips are tough and stringy, chip curlers and breakers are important. The machinability rating of this steel is about 40% of that for AISI 1212.

Type 440B is seldom welded. However, by preheating the parts to 150 to 205 °C (300 to 400 °F) before welding, then annealing 6 to 9 h at 730 to 760 °C (1350 to 1400 °F) and air cooling, satisfactory welds can be produced. When filler metal is required, a composition similar to the base metal should be used.

Typical Uses. Applications of AISI type 440B include bearings, cutlery, spatula blades, food processing knives, and hardened balls. The steel makes fair magnets.

AISI Type 440B: Similar Steels (U.S. and/or Foreign). UNS S44003; ASTM A276, A314, A473, A580; FED QQ-S-763; MIL SPEC MIL-S-862; SAE J405 (51440B); (W. Ger.) DIN 1.4112; (Jap.) JIS SUS 440B

Machining Data

For machining data on AISI type 440B, refer to the preceding machining tables for AISI type 440A.

Physical Properties

AISI Type 440B: Thermal Treatment Temperatures

Treatment	Temperature range °C	°F
Forging, preheat	760-815	1400-1500
Forging, start	1040-1175	1900-2150
Forging, finish	925 min	1700 min
Full annealing(a)	845-900	1550-1650
Process annealing(a)	730-790	1350-1450
Hardening(b)	1010-1065	1850-1950
Tempering(c)	150-370	300-700

(a) Hold at temperature for uniform heat distribution; furnace cool. (b) Preheat to 540 to 790 °C (1000 to 1450 °F); hold at temperature; quench in warm oil or air cool. (c) Oil quench

For additional data on physical properties of AISI type 440B, refer to the preceding tables for AISI type 440A.

Mechanical Properties

AISI Type 440B: Representative Tensile Properties (Ref 6)

Condition	Tensile strength MPa	ksi	Yield strength(a) MPa	ksi	Elongation(b), %	Reduction in area, %	Hardness
Wire, 6.4-mm (0.25-in.) diam							
Annealed	740	107	425	62	16	40	96 HRB
Soft bright tempered	795	115	620	90	8	25	99 HRB
Bar							
Annealed	740	107	425	62	18	35	96 HRB
Annealed and cold drawn(c)	825	120	655	95	9	20	23 HRC
Hardened and tempered at 315 °C (600°F)	1930	280	1860	270	3	15	555 HB

(a) 0.2% offset. (b) In 50 mm (2 in.), except for wire. For wire 3.18-mm (0.125-in.) diam and larger, gage length is 4 × diam; for smaller diameters, the gage length is 254 mm (10.0 in.). (c) Diameter of 25 mm (1 in.)

AISI Type 440B: Rockwell C Hardness (Ref 7)

Test specimens were 25-mm (1.0-in.) diam bar; oil quenched from 1040 °C (1900°F); tempered for 1 h

Tempering temperature °C	°F	Hardness, HRC
As quenched		59
150	300	58
205	400	56
260	500	54
315	600	54
370	700	54
425	800	54
480	900	53.5
540	1000	51

AISI Type 440B: Impact Properties at Room Temperature (Ref 7)

Condition	Energy J	ft·lb	Hardness, HB
Annealed	20	15	220
Quenched and tempered	4	3	580

AISI Type 440B: Modulus of Elasticity (Ref 7)

Stressed in:	Modulus GPa	10^6 psi
Tension	200	29

AISI Type 440C

AISI Type 440C: Chemical Composition

Element	Composition, %
Carbon	0.95-1.20
Manganese	1.00 max
Phosphorus	0.040 max
Sulfur	0.030 max
Silicon	1.00 max
Chromium	16-18
Molybdenum	0.75 max

Characteristics. AISI type 440C is a high-carbon, high-chromium steel which, with heat treatment, acquires the highest hardness of any type of corrosion- or heat-resisting steel. It has optimum corrosion resistance in the hardened and tempered condition.

Type 440C resists corrosion in normal domestic environments and very mild industrial environments, including many petroleum products and organic materials. This grade is always used in the hardened and tempered condition. Optimum corrosion resistance is obtained by hardening from 1095 °C (2000 °F) to ensure better carbide solution.

To avoid excessive grain coarsening in this stainless steel, care should be taken to minimize time at 1095 °C (2000 °F). For best resistance, the tempering temperature should be below 425 °C (800 °F). Parts of type 440C must be pickled, ground, or polished after heat treatment to remove all scale.

After pickling, parts should be baked at 120 to 150 °C (250 to 300 °F) to remove acid brittleness.

This steel machines best and can be moderately cold formed or headed when annealed for maximum softness. The machinability rating of type 440C is about 40% of that for AISI 1212.

Satisfactory welds can be produced by preheating to 260 °C (500 °F), maintaining 260 °C (500 °F) during welding, and annealing immediately after welding. For filler metal, a composition similar to the base metal should be used.

Typical Uses. Applications of AISI type 440C include bearing balls and races, cutlery, needle valves, ball check valves, valve seats, pump parts, bushings, and wear-resistant textile machine components.

AISI Type 440C: Similar Steels (U.S. and/or Foreign). UNS S44004; AMS 5618, 5630; ASTM A276, A314, A473, A493, A580; FED QQ-S-763; MIL SPEC MIL-S-862; SAE J405 (51440C); (W. Ger.) DIN 1.4125; (Jap.) JIS SUS 440C

Physical Properties

AISI Type 440C: Thermal Treatment Temperatures

Treatment	Temperature range °C	°F
Forging, preheat	760-815	1400-1500
Forging, start	1040-1150	1900-2100
Forging, finish	925 min	1700 min
Full annealing(a)	845-900	1550-1650
Process annealing(a)	730-790	1350-1450
Hardening(b)	1010-1065	1850-1950
Tempering(c)	150-370	300-700

(a) Hold at temperature for uniform heat distribution; furnace cool. (b) Preheat to 540 to 790 °C (1000 to 1450 °F). Hold at temperature for uniform heat distribution; quench in warm oil or air cool. (c) Temper as soon after quenching as possible

For additional data on physical properties of AISI type 440C, refer to the preceding tables for AISI type 440A.

Mechanical Properties

AISI Type 440C: Representative Tensile Properties (Ref 6)

Product form	Condition	Tensile strength MPa	Tensile strength ksi	Yield strength(a) MPa	Yield strength(a) ksi	Elongation(b), %	Reduction in area, %	Hardness, HRB
Wire, 6.4-mm (0.25-in.) diam	Annealed	760	110	450	65	13	30	97
	Soft tempered	860	125	690	100	6	20	24
Bar	Annealed	760	110	450	65	14	25	97
	Annealed and cold drawn(c)	860	125	690	100	7	20	260
	Hardened and tempered at 315 °C (600 °F)	1970	285	1900	275	2	10	580

(a) 0.2% offset. (b) In 50 mm (2 in.), except for wire. For wire 3.18-mm (0.125-in.) diam and larger, gage length is 4 × diam; for smaller diameters, the gage length is 254 mm (10.0 in.). (c) Diameter of 25 mm (1 in.)

AISI Type 440C: Rockwell C Hardness (Ref 7)

Test specimens were 25-mm (1.0-in.) diam bar; oil quenched from 1040 °C (1900°F); tempered for 1 h

Tempering temperature °C	°F	Hardness, HRC
As quenched		60.0
150	300	59.0
205	400	57.5
260	500	56.0
315	600	55.0
370	700	55.0
425	800	56.0
480	900	57.0
540	1000	52.5
595	1100	43.0

Machining Data

For machining data on AISI type 440C, refer to the preceding machining tables for AISI type 440A.

AISI Types 440F and 440F(Se)

AISI Types 440F and 440F(Se): Chemical Composition

Element	Composition, % Type 440F	Type 440F(Se)
Carbon	0.95-1.20	0.95-1.20
Manganese	1.00 max	1.00 max
Phosphorus	0.04 max	0.04 max
Sulfur	0.05 min	0.03 max
Silicon	1.00 max	1.00 max
Chromium	16-18	16-18
Molybdenum	0.75 max	0.75 max
Selenium	...	0.10 min

Characteristics. AISI types 440F and 440F(Se) are free-machining versions of type 440C. In addition to being free-machining, these steels are easy to grind and polish, and have certain nongalling and nonseizing properties in metal-to-metal contact. After heat treatment, they exhibit a hardness of about 60 HRC.

Types 440F and 440F(Se) are always used in the hardened and tempered condition. Optimum corrosion resistance is obtained with a temper below 425 °C (800 °F).

These stainless steels can be moderately cold worked in the fully annealed condition. However, major uses of these steels are in parts that are machined to shape. Machinability of types 440F and 440F(Se) is about 46% of that for AISI 1212. They machine best in the fully annealed condition.

These steels are not recommended for welding, or for vessels which contain gases or liquids under high pressure.

Typical Uses. Applications of AISI types 440F and 440F(Se) include bearings, nozzles, valve parts, and balls and seats for oil well pumps.

AISI Type 440F: Similar Steels (U.S. and/or Foreign). UNS S44020; AMS 5632 (1); MIL SPEC MIL-S-862; SAE J405 (51440F)

AISI Type 440F(Se): Similar Steels (U.S. and/or Foreign). UNS S44023; AMS 5632 (2); MIL SPEC MIL-S-862; SAE J405 (51440F-Se)

Physical Properties

AISI Types 440F and 440F(Se): Thermal Treatment Temperatures

Treatment	Temperature range °C	°F
Forging, preheat	760-815	1400-1500
Forging, start	1040-1150	1900-2100
Forging, finish	925 min	1700 min
Full annealing(a)	845-870	1550-1600
Process annealing(b)	730-760	1350-1400
Hardening(c)	1010-1065	1850-1950
Tempering(d)	150-425	300-800

(a) Heat uniformly and furnace cool very slowly; hardness of approximately 223 HB. (b) Heat uniformly; air cool; hardness of approximately 225 HB. (c) Heat uniformly; quench in warm oil or cool in air. (d) To remove peak stress but retain maximum hardness, temper for at least 1 h at 150 to 175 °C (300 to 350 °F)

For additional data on physical properties of AISI types 440F and 440F(Se), refer to the preceding tables for AISI type 440A.

Machining Data

For machining data on AISI types 440F and 440F(Se), refer to the preceding machining tables for AISI type 440A.

Mechanical Properties

AISI Types 440F, 440F(Se): Rockwell C Hardness (Ref 9)

Test specimens were 25-mm (1.0-in.) diam bar; oil quenched from 1040 °C (1900 °F); tempered 1 h

Tempering temperature °C	°F	Hardness, HRC
150	300	60
205	400	59
260	500	57
315	600	56
370	700	56
425	800	56

AISI Types 442, 443

AISI Types 442, 443: Chemical Composition

Element	Composition, % Type 442	Type 443
Carbon	0.20 max	0.20 max
Manganese	1.00 max	1.00 max
Phosphorus	0.04 max	0.40 max
Sulfur	0.03 max	0.03 max
Silicon	1.00 max	1.00 max
Chromium	18-23	18-23
Copper	...	0.90-1.25

Characteristics. AISI types 442 and 443 are nonhardenable ferritic stainless steels. High-carbon content in these stainless steels promotes corrosion and heat resistance. They are used principally for the manufacture of parts that must resist scaling at high temperatures, but where strength and toughness are not prime factors. Both types are magnetic in all conditions.

Type 443 is a modification of type 442 which contains 0.90 to 1.25% copper. This copper addition significantly increases resistance to sulfuric acid.

Types 442 and 443 may be cold drawn, stamped, or otherwise formed into most desired shapes. Their cold workability is between that of type 430 and type 446, the latter being more difficult to fabricate. Preheating to 120 to 175 °C (250 to 350 °F) improves formability. Types 442 and 443 are machined at speeds about 55% of those used for free-machining AISI 1212.

To minimize grain coarsening in these alloys, overheating during welding should be avoided. Parts should be preheated to 150 to 205 °C (300 to 400 °F), and after welding should be annealed at 705 to 785 °C (1300 to 1450 °F), then air cooled or water quenched.

Type 442 has good resistance to scaling up to about 955 °C (1750 °F) in continuous service, and to about 1040 °C (1900 °F) for intermittent service. For type 443, the safe scaling temperature for continuous service is 870 °C (1600 °F), and for intermittent exposure up to 980 °C (1800 °F).

These alloys will resist all ordinary corrosion conditions including salt spray, all foodstuffs, most organic chemicals, many fused salts, and molten nonferrous metals. They will not withstand the halogen acids. Type 443 will resist weak sulfuric acid solutions better than the 18-8 stainless steels.

Typical Uses. Applications of AISI types 442 and 443 include high-temperature parts for resistance to oxidation, annealing boxes, oil burner parts, nozzles, and heating elements.

AISI Type 442: Similar Steels (U.S. and/or Foreign). UNS S44200; ASTM A176; SAE J405 (51442)

AISI Type 443: Similar Steels (U.S. and/or Foreign). UNS S44300; ASTM A268, A511

Physical Properties

AISI Type 442: Thermal Treatment Temperatures (Ref 7)

Treatment	Temperature range °C	°F
Forging, start	1040-1065	1900-1950
Forging, finish	870 min	1600 min
Annealing (a)	705-815	1300-1500
Hardening (b)	...	...

(a) Water quench or air cool. Sheet or strip should be held at temperature 3 min for each 2.5 mm or 0.10 in. of thickness. (b) Hardens moderately by cold work only

AISI Type 443: Thermal Treatment Temperatures (Ref 9)

Treatment	Temperature range °C	°F
Forging, preheat	760-815	1400-1500
Forging, start	1040-1120	1900-2050
Forging, finish	900 min	1650 min
Annealing (a)	760-815	1400-1500
Hardening (b)	...	...

(a) Water quench. (b) Hardens moderately by cold work only

AISI Types 442, 443: Thermal Conductivity (Ref 7)

Temperature °C	°F	Thermal conductivity W/m·K	Btu/ft·h·°F
100	212	21.6	12.5
500	930	24.6	14.2

AISI Types 442, 443: Average Coefficients of Linear Thermal Expansion (Ref 7)

Temperature range °C	°F	Coefficient μm/m·K	μin./in.·°F
20-100	68-212	10.1	5.6
20-650	68-1200	12.1	6.7

AISI Types 442, 443: Electrical Resistivity (Ref 7)

AISI type	Temperature °C	°F	Electrical resistivity, nΩ·m
442	20	68	64
443	20	68	68

AISI Types 442, 443: Selected Physical Properties (Ref 7)

Melting range	1425-1510 °C (2600-2750 °F)
Density	7.8 g/cm³ (0.28 lb/in.³)
Specific heat at 0-100 °C (32-212 °F)	460 J/kg·K (0.11 Btu/lb·°F)

Mechanical Properties

AISI Types 442, 443: Tensile Properties at Room Temperature (Ref 7)

Test specimens were in the annealed condition

Product form	Tensile strength MPa	Tensile strength ksi	Yield strength(a) MPa	Yield strength(a) ksi	Elongation(b), %	Reduction in area, %	Hardness
Strip	515	75	310	45	25	...	80 HRB
Bar	515	75	310	45	30	50	160 HB

(a) 0.2% offset. (b) In 50 mm (2 in.)

AISI Types 442, 443: Tensile Properties at Elevated Temperatures (Ref 7)

Test specimens were 16.0-mm (0.63-in.) bar; annealed at 815 °C (1500 °F); cold worked to hardness of 92 HRC

Test temperature °C	Test temperature °F	Tensile strength MPa	Tensile strength ksi	Yield strength(a) MPa	Yield strength(a) ksi	Elongation(b), %	Reduction in area, %
21	70	545	79.0	427	62.0	35.5	79.0
205	400	462	67.0	365	53.0	31.5	78.5
425	800	379	55.0	307	44.5	31.5	80.0
540	1000	262	38.0	221	32.0	34.0	84.5
650	1200	114	16.5	103	15.0	53.5	97.0
760	1400	52	7.5	48	7.0	91.5	98.5
870	1600	31	4.5	24	3.5	87.0	98.5
980	1800	17	2.5	14	2.0	134.0	99.0

(a) 0.2% offset. (b) In 50 mm (2 in.)

AISI Type 443: Short-Time Tensile Tests at Elevated Temperatures (Ref 9)

Test temperature °C	Test temperature °F	Tensile strength MPa	Tensile strength ksi	Yield strength(a) MPa	Yield strength(a) ksi	Elongation(b), %	Reduction in area, %
21	70	621	90	345	50	22	55
425	800	414	60	207	30	19	40
480	900	386	56	186	27	17	43
540	1000	345	50	165	24	18	47
595	1100	303	44	145	21	20	55
650	1200	255	37	124	18	23	70
705	1300	179	26	76	11	41	82
760	1400	110	16	41	6	50	90
815	1500	69	10	21	3	61	93
870	1600	41	6	14	2	65	96

(a) 0.2% offset. (b) In 50 mm (2 in.)

AISI Types 442, 443: Modulus of Elasticity (Ref 7)

Stressed in:	Modulus GPa	Modulus 10^6 psi
Tension	200	29

AISI Types 442, 443: Izod Impact Properties (Ref 7)

Temperature °C	Temperature °F	Energy J	Energy ft·lb
21	70	14	10

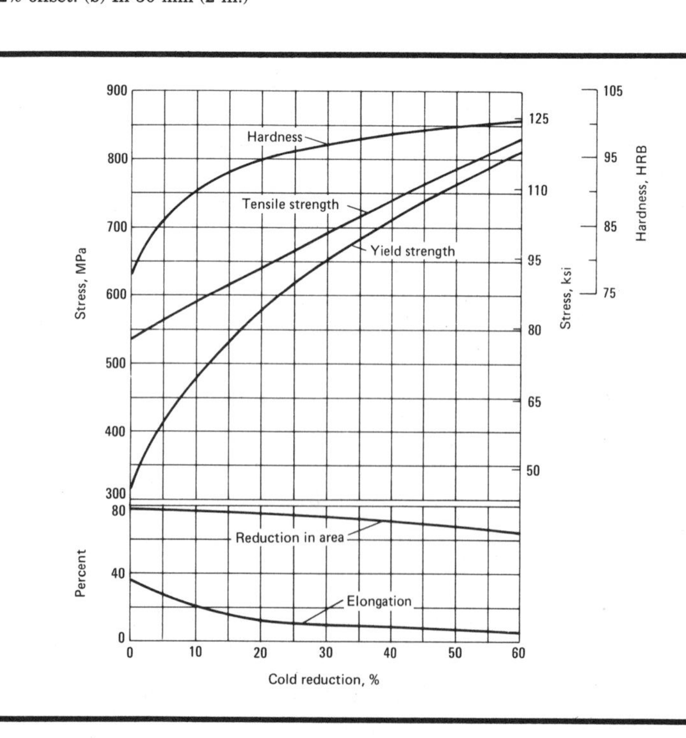

AISI Type 442: Effect of Cold Reduction on Tensile Properties. Elongation was measured in 50 mm (2 in.); yield strength at 0.2% offset. (Ref 7)

Machining Data

For machining data on AISI types 442 and 443, refer to the preceding machining tables for AISI type 405.

AISI Type 442: Effect of Test Temperature on Tensile Properties. Test specimens were 15.9-mm (0.625-in.) round bar in the annealed and cold worked condition, with hardness of 92 HRB. Elongation was measured in 50 mm (2 in.); yield strength at 0.2% offset. (Ref 7)

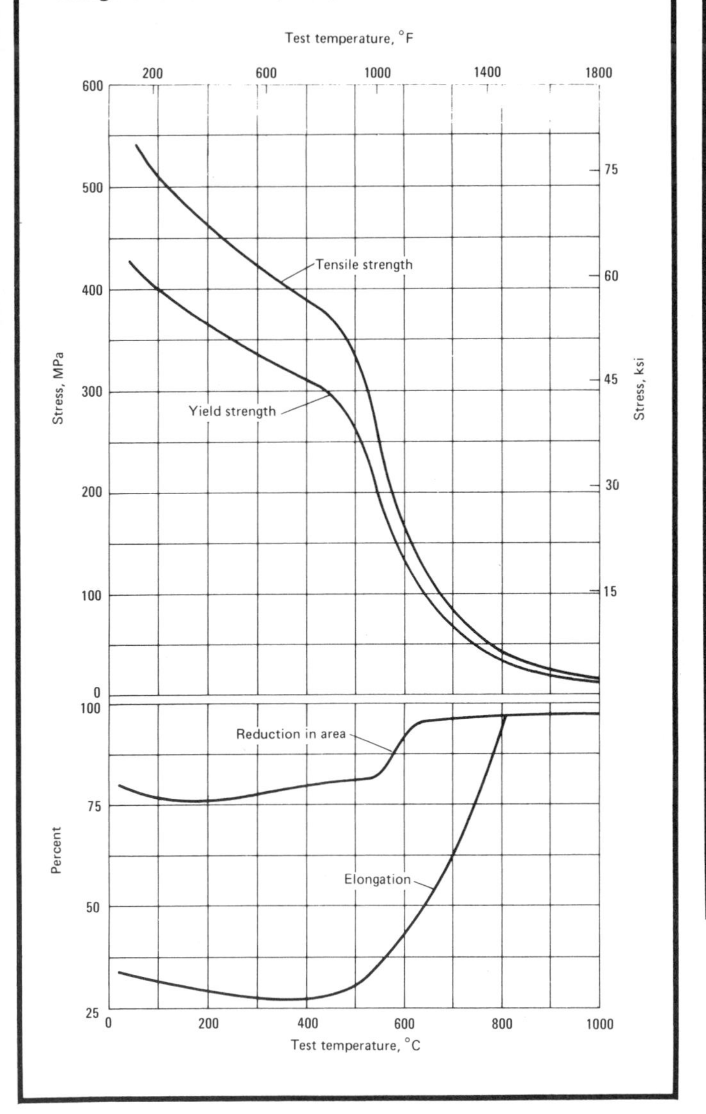

AISI Type 443: Effect of Cold Reduction on Tensile Properties. Test specimens were in the annealed condition, 9.53-mm (0.375-in.) round, unstraightened. Tests were conducted using specimens machined to English units. Elongation was measured in 50 mm (2 in.); yield strength at 0.2% offset. (Ref 10)

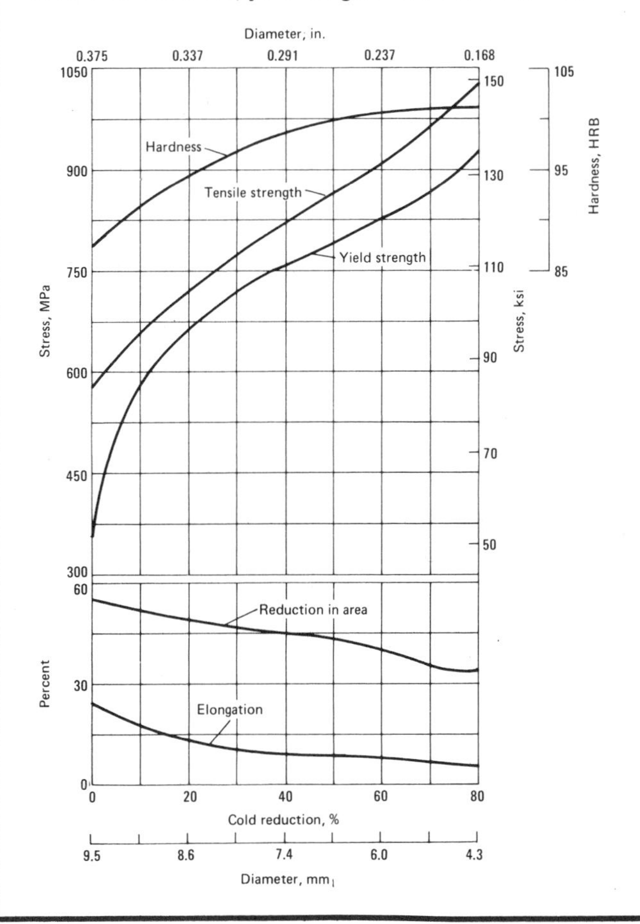

AISI Type 446

AISI Type 446: Chemical Composition

Element	Composition, %
Carbon	0.20 max
Manganese	1.50 max
Phosphorus	0.040 max
Sulfur	0.030 max
Silicon	1.00 max
Chromium	23-27
Nickel	0.25 max

Characteristics. AISI type 446 is a high-chromium, nonhardenable ferritic stainless steel. The chromium content of this steel, nominally 25%, promotes the best resistance to corrosion and scaling of all the standard straight-chromium grades.

Type 446 is well-suited for service in temperatures up to 1095 °C (2000 °F) for parts requiring excellent resistance to oxidation, but where strength and toughness are not prime factors. This grade has good resistance to carburization and combustion gases. Type 446 also resists attack from sulfur-containing gases.

In the annealed condition, type 446 may be drawn, stamped, or otherwise formed to most shapes. Because of room temperature brittleness, preheating to 120 to 175 °C (250 to 350 °F) improves ductility during forming of intricate shapes. Type 446 is machined at speeds about 55% of those for free-machining AISI 1212.

To minimize coarse grains, overheating should be avoided during welding. When using type 446 electrodes, parts should be preheated to 150 to 205 °C (300 to 400 °F), and after welding, should be annealed at 705 to 785 °C (1300 to 1450 °F), then air cooled. Peening of the weld zone prior to annealing will improve ductility. Type 310 electrodes may be used for better ductility.

Typical Uses. Applications of AISI type 446 include high-temperature parts for resistance to oxidation, annealing boxes, heaters, salt bath electrodes, oil burner parts, metal-to-glass seals, and nozzles.

AISI Type 446: Similar Steels (U.S. and/or Foreign). UNS 44600; ASME SA268; ASTM A176, A268, A276, A314, A473, A511, A580; FED QQ-S-763, QQ-S-766; MIL SPEC MIL-S-862; SAE J405 (51446)

Physical Properties

AISI Type 446: Thermal Treatment Temperatures

Treatment	Temperature range °C	°F
Forging, start	1040-1120	1900-2050
Forging, finish	870 min	1600 min
Annealing(a)	730-815	1350-1500
Hardening(b)	...	...

(a) Air cool or water quench. (b) Moderately hardenable by cold work

AISI Type 446: Thermal Conductivity (Ref 6)

Temperature °C	°F	Thermal conductivity W/m·K	Btu/ft·h·°F
100	212	21.6	12.5
500	930	24.6	14.2

AISI Type 446: Average Coefficients of Linear Thermal Expansion (Ref 6)

Temperature range °C	°F	Coefficient μm/m·K	μin./in.·°F
0-100	32-212	10.4	5.8
0-315	32-600	10.8	6.0
0-540	32-1000	11.2	6.2
0-650	32-1200	11.5	6.4
0-980	32-1800	12.1	6.7

AISI Type 446: Selected Physical Properties (Ref 6)

Melting range	1425-1510 °C (2600-2750 °F)
Density	7.8 g/cm^3 (0.28 lb/in.^{3})
Specific heat at 0-100 °C (32-212 °F)	460 J/kg·K (0.11 Btu/lb·°F)
Electrical resistivity at 20 °C (68 °F)	64 nΩ·m

Mechanical Properties

AISI Type 446: Tensile Properties

Product form	Treatment	Tensile strength MPa	ksi	Yield strength(a) MPa	ksi	Elongation(b), %	Reduction in area, %	Hardness, HRB
Sheet	Annealed	550	80	345	50	20	...	83
Strip	Annealed	550	80	345	50	20	...	83
Wire, 6.4-mm (0.25-in.) diam(c)	Soft tempered	655	95	550	80	15	50	92
Plate	Annealed	585	85	380	55	25	...	84
Bar	Annealed	550	80	345	50	25	45	86
	Annealed and cold drawn	585	85	485	70	20	45	90

(a) 0.2% offset. (b) In 50 mm (2 in.), except for wire. For wire 3.18-mm (0.125-in.) diam and larger, gage length is 4 × diam; for smaller diameters, the gage length is 254 mm (10.0 in.). (c) Diameter of 25 mm (1.0 in.)

AISI Type 446: Tensile Properties at Elevated Temperatures

Test specimens were 16-mm (0.63-in.) diam; annealed at 815 °C (1500 °F); cold worked to hardness of 96 HRB

Test temperature °C	°F	Tensile strength MPa	ksi	Yield strength(a) MPa	ksi	Elongation(b), %	Reduction in area, %
21	70	607	88.0	462	67.0	26.0	64.0
205	400	527	76.5	403	58.5	24.0	61.5
425	800	459	66.5	359	52.0	24.0	63.5
540	1000	338	49.0	269	39.0	32.5	71.5
650	1200	128	18.5	107	15.5	48.5	94.0
760	1400	66	9.5	52	7.5	66.0	98.0
870	1600	34	5.0	28	4.0	101.0	98.5
980	1800	17	2.5	14	2.0	135.0	99.0

(a) 0.2% offset. (b) In 50 mm (2 in.)

Mechanical Properties (continued)

AISI Type 446: Effect of Test Temperature on Tensile Properties. Test specimens were 15.9-mm (0.625-in.) round bar in the annealed and cold worked condition, with hardness of 96 HRB. Tests were conducted using specimens machined to English units. Elongation was measured in 50 mm (2 in.); yield strength at 0.2% offset. (Ref 7)

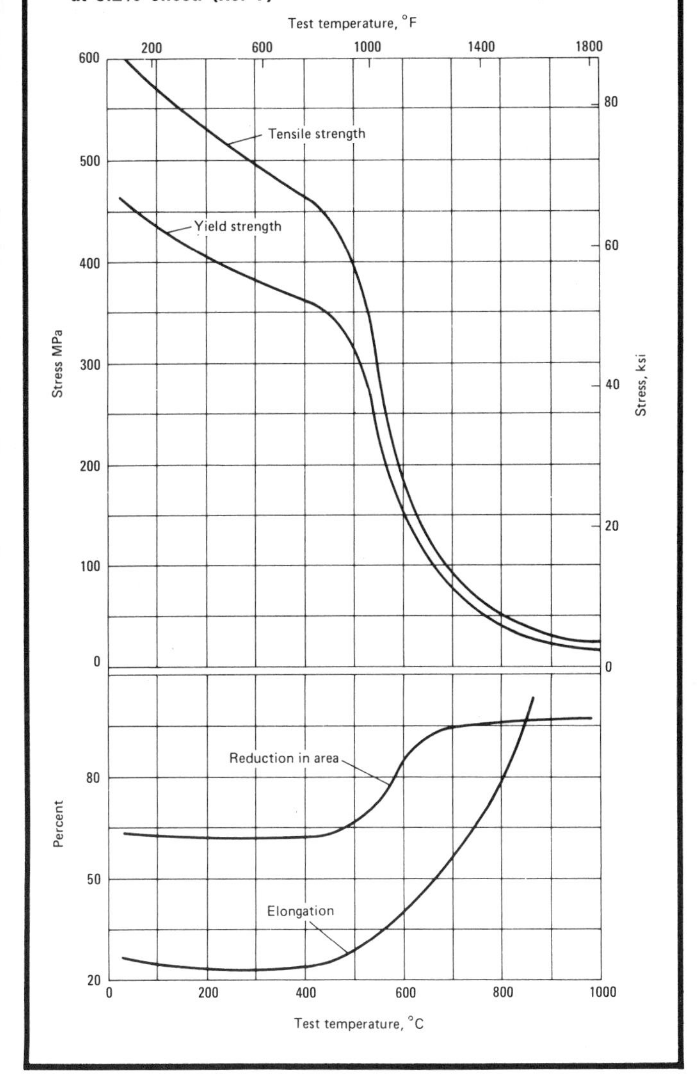

AISI Type 446: Stress-Rupture and Creep Properties (Ref 10)

Test temperature °C	Test temperature °F	Stress for rupture in: 1000 h MPa	1000 h ksi	10 000 h MPa	10 000 h ksi	Stress for creep rate of 1% in 10 000 h MPa	ksi
540	1000	...	...	...	...	41	6.0
595	1100	39	5.6	21	3.0	21	3.0
650	1200	28	4.0	15	2.2	10	1.5
705	1300	19	2.7	11	1.6	5	0.7
760	1400	12	1.8	8	1.1	...	...
815	1500	8	1.2	6	0.8	2	0.3
870	1600	6	0.8	3	0.5	...	...

Machining Data

For machining data on AISI type 446, refer to the preceding machining tables for AISI type 405.

AISI Types 501, 502

AISI Types 501, 502: Chemical Composition

Element	Composition, % Type 501	Type 502
Carbon	0.10 min	0.10 max
Manganese	1.00 max	1.00 max
Phosphorus	0.040 max	0.040 max
Sulfur	0.030 max	0.030 max
Silicon	1.00 max	1.00 max
Chromium	4-6	4-6
Molybdenum	0.40-0.65	0.45-0.65

Characteristics. AISI types 501 and 502 are heat-resisting, chromium steels used for their heat resistance and good mechanical properties at moderately elevated temperatures. When annealed, type 502 has greater ductility and lower tensile strength than type 501. Both steels are magnetic in all conditions.

Typical Uses. Applications of AISI types 501 and 502 include high-temperature bolting materials, forgings for high-temperature and aircraft applications, and plates for pressure vessels.

AISI Type 501: Similar Steels (U.S. and/or Foreign). UNS S50100; AMS 5502, 5602; ASME SA194 (3), SA387 (5); ASTM A193, A194, A314, A387 (5), A473; SAE J405 (51501); (W. Ger.) DIN 1.7362

AISI Type 502: Similar Steels (U.S. and/or Foreign). UNS S50200; AMS 6466, 6467; ASME SA387 (5); ASTM A314, A387 (5), A473; SAE J405 (51502)

Physical Properties

AISI Types 501, 502: Thermal Treatment Temperatures

Treatment	Temperature range °C	°F
Forging, start	1150-1205	2100-2200
Full annealing(a)	830-870	1525-1600
Process annealing(a)	720-745	1325-1375
Hardening(b)	870-925	1600-1700
Tempering(b)	205-760	400-1400

(a) Cool slowly in furnace or air cool. (b) Data for type 501; type 502 is generally used in the annealed condition

AISI Types 501, 502: Thermal Conductivity (Ref 6)

Temperature °C	°F	Thermal conductivity W/m·K	Btu/ft·h·°F
100	212	36.7	21.2
500	930	33.7	19.5

AISI Types 501, 502: Selected Physical Properties (Ref 6)

Melting range	1480-1540 °C (2700-2800 °F)
Density	7.8 g/cm^3 (0.28 lb/in.^{3})
Specific heat at 0-100 °C (32-212 °F)	460 J/kg·K (0.11 Btu/lb·°F)
Electrical resistivity at 20 °C (68 °F)	40 nΩ·m

AISI Types 501, 502: Average Coefficients of Linear Thermal Expansion (Ref 6)

Temperature range °C	°F	Coefficient μm/m·K	μin./in.·°F
0-100	32-212	11.2	6.2
0-315	32-600	12.2	6.8
0-540	32-1000	13.0	7.2
0-650	32-1200	13.1	7.3

Mechanical Properties

AISI Type 501: Representative Tensile Properties (Ref 6)

Product form or treatment	Tensile strength MPa	Tensile strength ksi	Yield strength(a) MPa	Yield strength(a) ksi	Elongation(b), %	Reduction in area, %	Hardness, HB
Test specimens in the annealed condition							
Plate	485	70	205	30	28	65	160
Bar	485	70	205	30	28	65	160
Bar oil quenched from 900 °C (1650 °F)							
Tempered at 540 °C (1000 °F)	1205	175	930	135	15	50	370
Tempered at 595 °C (1100 °F)	965	140	760	110	18	55	290
Tempered at 650 °C (1200 °F)	795	115	620	90	20	60	240

(a) 0.2% offset. (b) In 50 mm (2 in.)

AISI Type 502: Representative Tensile Properties (Ref 6)

Test specimens were in the annealed condition

Product form	Tensile strength MPa	Tensile strength ksi	Yield strength(a) MPa	Yield strength(a) ksi	Elongation(b), %	Reduction in area, %	Hardness
Sheet	480	70	...	...	30	...	75 HRB
Strip	480	70	...	...	30	...	75 HRB
Wire	515	75	205	30	30	...	72 HRB
Plate(c)	450	65	170	25	30	75	150 HB
Bar(c)	450	65	170	25	30	75	150 HB

(a) 0.2% offset. (b) In 50 mm (2 in.), except for wire. For wire 3.18-mm (0.125-in.) diam and larger, gage length is 4 × diam; for smaller diameters, the gage length is 254 mm (10.0 in.). (c) Charpy impact energy of 61 J (45 ft·lb); Izod impact energy of 115 J (85 ft·lb)

Machining Data

For machining data on AISI types 501 and 502, refer to the preceding machining tables for AISI type 403.

AISI Type 615

AISI Type 615: Chemical Composition

Element	Composition, %
Carbon	0.15-0.20
Manganese	0.50 max
Phosphorus	0.040 max
Sulfur	0.030 max
Silicon	0.50 max
Chromium	12-14
Nickel	1.8-2.2
Molybdenum	0.50 max
Tungsten	2.5-3.5

Characteristics. AISI type 615 (S41800, Greek Ascoloy) is a high-strength modification of the hardenable, martensitic, 12% chromium, type 410 stainless steel. Additions of nickel and tungsten account for the improved properties and make this alloy more resistant to tempering temperatures, while retaining both corrosion and oxidation resistance through the useful temperature range of this steel, up to 565 °C (1050 °F).

Furthermore, this alloy has good resistance to stress corrosion cracking, and adequate creep resistance and stress-rupture strength at temperatures through 565 °C (1050 °F).

Type 615 requires higher forming pressures for cold forming than type 410. Although it does not work harden as severely as the 18-8 stainless steels, process annealing may be required. Severely formed parts should be annealed or stress relieved. Annealed sheet and strip are capable of passing the bend test requirement of 105° around a diameter equal to four times the nominal thickness of the material, with the axis of the bend parallel to the rolling direction.

The machinability rating of type 615 is about 55% of that for AISI 1212. Generally, the best machinability is obtained in the fully annealed condition. Because of its lower hardening rate, type 615 machines better than the 18-8 stainless steels.

Because of susceptibility to strain cracking, welding of type 615 is generally not recommended. When welding type 615, preheat temperatures of 205 to 315 °C (400 to 600 °F) and postheat treatments above 540 °C (1000 °F) are essential. Type 310 filler metal is used to obtain sufficient weld ductility. Weldments are subsequently hardened and tempered.

Type 615 is readily brazed. Hardening and tempering is required after copper brazing. When silver brazing, a subsequent stress relief is required; brazing below 760 °C (1400 °F) does not require additional heat treating.

Typical Uses. Applications of AISI type 615 include compressor blades for gas turbines, turbine discs in the lower temperature stages, rings for jet engines, fasteners, and valve parts.

AISI Type 615: Similar Steels (U.S. and/or Foreign). UNS S41800; AMS 5354, 5508, 5615, 5817, 7470; ASTM A565; SAE J467 (Greek Ascoloy)

Physical Properties

AISI Type 615: Thermal Treatment Temperatures

Treatment	Temperature range °C	°F
Forging, start	1175	2150
Forging, finish	925 min	1700 min
Full annealing	790	1450
Process annealing	705	1300
Hardening	955-980	1750-1800
Tempering	540-675	1000-1250

AISI Type 615: Selected Physical Properties (Ref 7)

Melting range	1460-1465 °C (2660-2670 °F)
Density	7.86 g/cm³ (0.284 lb/in.³)
Specific heat at 0-100 °C (32-212 °F)	460 J/kg·K (0.11 Btu/lb·°F)
Electrical resistivity at 20 °C (68 °F)	79 nΩ·m

AISI Type 615: Average Coefficients of Linear Thermal Expansion (Ref 7)

Temperature range °C	°F	Coefficient μm/m·K	μin./in.·°F
25-95	80-200	10.4	5.8
25-205	80-400	11.0	6.1
25-315	80-600	11.2	6.2
25-425	80-800	11.5	6.4
25-540	80-1000	11.7	6.5
25-595	80-1100	11.9	6.6

Mechanical Properties

AISI Type 615: Creep Properties (Ref 7)

Test specimens were hardened at 955 °C (1750 °F); oil quenched; tempered at 565 °C (1050 °F) to hardness of 33.5 to 35.5 HRC

Test temperature		Stress for 0.1% creep in:				Stress for 1% creep in:			
		100 h		150 h		100 h		150 h	
°C	°F	MPa	ksi	MPa	ksi	MPa	ksi	MPa	ksi
370	700	503	73.0	483	70.0	...	...	689	100.0
425	800	324	47.0	303	44.0	...	...	...	...
480	900	228	33.0	214	31.0	552	80.0	538	78.0
540	1000	131	19.0	117	17.0	331	48.0	324	47.0
590	1100	45	6.5	36	5.2	200	29.0	193	28.0

Mechanical Properties (continued)

AISI Type 615: Charpy Keyhole and Izod V-notch Impact Energy at Various Tempering Temperatures (Ref 7)

Test specimens were hardened at 955 °C (1750 °F); oil quenched; tempered as indicated

Tempering temperature °C	°F	Hardness, HRC	Impact energy(a) J	ft · lb
Specimens held 4 h at tempering temperature; air cooled				
260	500	46	26	19
315	600	46	22	16
370	700	46	14	10
425	800	47	12	9
480	900	44	14	10
540	1000	39	23	17
590	1100	33	54	40
650	1200	28	104	77
705	1300	27	88	65
Specimens held 2 h at tempering temperature; air cooled				
510	950	...	22	16
540	1000	...	24	18
565	1050	...	27	20
590	1100	...	28	21
650	1200	...	41	30

(a) For test specimens held 4 h at tempering temperature, impact energy was measured using Izod V-notch; for specimens held 2 h, Charpy keyhole was used

AISI Type 615: Tensile Properties at Elevated Temperature (Ref 7)

Test specimens were hardened at 980 °C (1800 °F); held for ½ h; oil quenched; tempered at temperature indicated; held for 2 h; air cooled

Tempering temperature °C	°F	Test temperature °C	°F	Tensile strength MPa	ksi	Yield strength(a) MPa	ksi	Elongation(b), %	Reduction in area, %
565	1050	425	800	929	134.8	839	121.7	13.3	67.3
		540	1000	711	103.1	673	97.6	17.1	71.4
		650	1200	319	46.2	303	44.0	29.4	89.6
		760	1400	177	25.6	147	21.3	84.6	89.9
650	1200	425	800	817	118.5	776	112.5	16.0	55.0
		540	1000	621	90.0	614	89.0	24.0	72.5
		650	1200	355	51.5	348	50.5	30.0	86.0

(a) 0.2% offset. (b) In 50 mm (2 in.)

AISI Type 615: Stress-Rupture Strength (Ref 7)

Test temperature °C	°F	Stress for rupture in: 10 h MPa	10 h ksi	100 h MPa	100 h ksi	1000 h MPa	1000 h ksi	10 000 h(a) MPa	10 000 h(a) ksi
Hardened at 980 °C (1800 °F); held ½ h; oil quenched; tempered at 680 °C (1260 °F); held 1½ h; air cooled									
480	900	448	65	379	55.0	331	48.0	290	42.0
540	1000	331	48	290	42.0	248	36.0	210	30.5
590	1100	262	38	216	31.3	172	25.0	128	18.5
650	1200	186	27	141	20.5	79	11.5	41	6.0
Hardened at 955 °C (1750 °F); oil quenched; tempered at 650 °C (1200 °F) to hardness of 27-29 HRC									
480	900	...	...	386	56.0	345	50.0	...	...
540	1000	...	...	338	49.0	241	35.0	...	...
590	1100	...	...	207	30.0	138	20.0	...	...
650	1200	...	...	131	19.0	69	10.0	...	...

(a) Extrapolated data

AISI Type 615: Typical Tensile Properties at Room Temperature (Ref 7)

Tempering temperature °C	°F	Tensile strength MPa	ksi	Yield strength 0.02% offset MPa	ksi	0.2% offset MPa	ksi	Elongation(a), %	Reduction in area, %
Harden at 955 °C (1750 °F); oil quench; temper 4 h									
260	500	1440	209	1035	150	1205	175	18	53
315	600	1460	212	1105	160	1240	180	17	55
370	700	1495	217	1140	165	1260	183	16	56
425	800	1495	217	1160	168	1275	185	16	57
480	900	1360	197	1090	158	1140	165	17	58
540	1000	1140	165	945	137	970	141	18	59
590	1100	1000	145	825	120	850	123	18	60
650	1200	930	135	740	107	745	108	20	61
705	1300	860	125	670	97	675	98	22	62
Harden at 980 °C (1800 °F); oil quench; temper 4 h									
565	1050	1170	170	...	...	1035	150	13.3	64
650	1200	980	142	...	...	780	113	19.5	60

(a) In 50 mm (2 in.)

Machining Data

For machining data on AISI type 615, refer to the preceding machining tables for AISI type 414.

AISI Type 630

AISI Type 630: Chemical Composition

Element	Composition, %
Carbon	0.07 max
Manganese	1.00 max
Phosphorus	0.040 max
Sulfur	0.030 max
Silicon	1.00 max
Chromium	15.5-17.5
Nickel	3-5
Copper	3-5
Niobium and tantalum	0.15-0.45

Characteristics. AISI type 630 (S17400) is a precipitation-hardening stainless steel which is capable of high strength and hardness levels after a relatively simple heat treatment procedure. This grade is martensitic and magnetic in both the solution-treated and precipitation-hardened conditions.

Type 630 has high resistance to crack propagation. Transverse properties and corrosion resistance are normally better than the regular martensitic chromium-type stainless steel. Because of the single low-temperature, precipitation-hardening treatment of this grade, 480 to 620 °C (900 to 1150 °F), scaling and distortion are virtually eliminated. Therefore, the material can be finish machined to close tolerances prior to heat treatment.

Although type 630 can be cold worked, the amount is limited by the high initial yield strength. This grade is readily machined in both the solution-treated and various age-hardened conditions. In the solution-treated condition, type 630 machines similarly to types 302 and 304. Machinability will increase as the hardening temperature is increased.

This stainless steel can be welded by all methods with no preheating required. Properties in the weld comparable to those of the base metal can be obtained by suitable postweld heat treatment.

The amount of contraction in hardening solution-treated material, condition A, to condition H900 is about 0.010 to 0.015 mm per 25 mm (0.0004 to 0.0006 in. per 1 in.). Condition A material, when hardened to condition H1150, will contract approximately 0.023 to 0.030 mm per 25 mm (0.0009 to 0.0012 in. per 1 in.).

Typical Uses. Applications of AISI type 630 include oil field valve parts, chemical process equipment, aircraft fittings, fasteners, pump shafts, nuclear reactor components, gears, paper mill equipment, missile fittings, and jet engine parts.

AISI Type 630: Similar Steels (U.S. and/or Foreign). UNS S17400; AMS 5342, 5343, 5344, 5355, 5604, 5622, 5643, 5825; ASME SA564, SA705; ASTM A564, A693, A705; MIL SPEC MIL-C-24111, MIL-S-81506, MIL-S-81591; SAE J467 (17-4PH)

Physical Properties

AISI Type 630: Selected Physical Properties in the H900 Condition (Ref 6, 12)

Melting range	1405-1440 °C (2560-2625 °F)
Density	7.81 g/cm³ (0.282 lb/in.³)
Specific heat at 0-100 °C (32-212 °F)	460 J/kg·K (0.11 Btu/lb·°F)
Electrical resistivity at 20 °C (70 °F)	77 nΩ·m

AISI Type 630: Average Coefficients of Linear Thermal Expansion (Ref 6, 12)

Temperature range °C	Temperature range °F	Coefficient µm/m·K	Coefficient µin./in.·°F
−73-20	−100-70	10.4	5.8
20-95	70-200	10.8	6.0
20-205	70-400	10.8	6.0
20-315	70-600	11.3	6.3
20-425	70-800	11.7	6.5

AISI Type 630: Thermal Treatment Temperatures

Treatment	Temperature range °C	Temperature range °F
Forging, start	1175-1230	2150-2250
Forging, finish	1010 min	1850 min
Solution treating, condition A	1025-1050	1875-1925
Hardening	480-620	900-1150

AISI Type 630: Thermal Conductivity in the H900 Condition (Ref 12)

Temperature °C	Temperature °F	Thermal conductivity W/m·K	Thermal conductivity Btu/ft·h·°F
150	300	17.8	10.3
260	500	19.5	11.3
460	860	22.5	13.0
480	900	22.7	13.1

Mechanical Properties

AISI Type 630: Typical Mechanical Properties in Various Conditions (Ref 9)

Condition	Tensile strength MPa	Tensile strength ksi	Yield strength(a) MPa	Yield strength(a) ksi	Elongation(b), %	Reduction in area, %	Hardness, HB	Impact energy(c) J	Impact energy(c) ft·lb
H900	1365	198	1260	183	15	52	420	21	16
H1025	1160	168	1115	162	16	58	352	54	40
H1075	1130	164	1020	148	17	59	341	61	45
H1150	992	144	869	126	20	60	311	75	55
H1150M	848	123	600	87	22	66	377	136	100

(a) 0.2% offset. (b) In 50 mm (2 in.). (c) Charpy V-notch

Mechanical Properties (continued)

AISI Type 630: Stress-Rupture and Creep Properties (Ref 9)

Test temperature °C	°F	Condition	Stress to produce rupture in: 100 h MPa	100 h ksi	1000 h MPa	1000 h ksi	Stress for creep rate of: 0.1% in 1000 h MPa	0.1% in 1000 h ksi	0.01% in 1000 h MPa	0.01% in 1000 h ksi
315	600	H900	...	...	...	...	931	135	862	125
330	625	H900	1117	162	1082	157	...	...	...	...
		H1075	945	137	924	134	...	...	...	...
370	700	H900	1076	156	1034	150	724	105	689	100
		H1075	869	126	848	123	...	...	...	...
425	800	H900	965	140	883	128	414	60	296	43
		H1075	745	108	710	103	...	...	...	...
480	900	H900	...	...	...	...	159	23	...	...

AISI Type 630: Typical Mechanical Properties in the Longitudinal Direction (Ref 12)

Condition	Tensile strength MPa	Tensile strength ksi	Yield strength(a) MPa	Yield strength(a) ksi	Elongation(b), %	Reduction in area, %	Hardness, HB	Impact energy(c) J	Impact energy(c) ft·lb
H900	1380	200	1275	185	14	50	420	20	15
H925	1310	190	1205	175	14	54	409	34	25
H1025	1170	170	1135	165	15	56	352	47	35
H1075	1135	165	1035	150	16	58	341	54	40
H1100	1035	150	931	135	17	58	332	61	45
H1150	1000	145	862	125	19	60	311	68	50
H1150M	862	125	585	85	22	68	277	136	100

(a) 0.2% offset. (b) In 50 mm (2 in.). (c) Charpy V-notch

AISI Type 630: Impact Strength at Subzero Temperatures (Ref 12)

Test temperature °C	°F	Charpy V-notch impact energy for heat treat condition of: H925 J	H925 ft·lb	H1025 J	H1025 ft·lb	H1150 J	H1150 ft·lb	H1150M(a) J	H1150M(a) ft·lb	H1150M(b) J	H1150M(b) ft·lb
24	75	41	30	100	75	130	95	140	105	130	95
−12	10	22	16	79	58	125	93	...	...	115	85
−40	−40	12	9	54	40	105	76	...	...	100	75
−79	−110	7.5	5.5	20	15	65	48	...	...	90	65
−195	−320	4.8	3.5	6.1	4.5	8.8	6.5	38	28	7	5

(a) Test samples from 25-mm (1-in.) diam bar, longitudinal direction. (b) Test samples from 100-mm (4-in.) diam bar, longitudinal direction

AISI Type 630: Short-Time Tensile Properties at Elevated Temperatures (Ref 12)

Test temperature °C	°F	Condition	Tensile strength MPa	Tensile strength ksi	Yield strength(a) MPa	Yield strength(a) ksi	Elongation(b), %	Reduction in area, %
24	75	H900	1400	203	1280	186	11	50
		H925	1315	191	1255	182	18	54
		H1025	1200	174	1160	168	15	55
		H1075	1130	164	1095	159	16	54
		H1150	965	140	900	129	17	61
315	600	H900	1190	173	1035	150	10	31
		H925	1135	165	1000	145	12	32
		H1025	1005	146	931	135	12	42
		H1075	950	138	910	132	9	38
		H1150	855	124	827	120	12	54
370	700	H900	1165	169	1005	146	8	25
		H925	1110	161	980	142	12	33
		H1025	979	142	903	131	10	38
		H1075	924	134	876	127	9	33
		H1150	827	120	786	114	12	52
425	800	H900	1115	162	972	141	10	21
		H925	1070	155	958	139	10	34
		H1025	945	137	882	128	11	39
		H1075	883	128	834	121	10	30
		H1150	800	116	772	112	13	43
480	900	H900	1025	149	910	132	10	30
		H925	1000	145	883	128	10	35
		H1025	869	126	814	118	12	39
		H1075	786	114	758	110	11	38
		H1150	752	109	717	104	13	51
540	1000	H900	820	119	731	106	15	46
		H925	800	116	710	103	16	45
		H1025	731	106	696	101	15	43
		H1075	680	99	650	94	16	55
		H1150	660	96	640	93	15	55

(a) 0.2% offset. (b) In 50 mm (2 in.)

Mechanical Properties (continued)

AISI Type 630: Stress-Rupture Strength (Ref 12)

Test temperature °C	°F	Time to rupture, h	Condition	Strength MPa	Strength ksi	Elongation at rupture(a), %	Reduction in area, %
330	625	100	H925	1125	163	3.0	13.0
			H1075	945	137	3.5	4.5
			H1150	850	123	5.5	17.5
		1000	H925	1105	160	2.5	12.0
			H1075	925	134	3.0	14.0
			H1150	840	122	4.5	16.5
370	700	100	H900	1075	156	3.0	7.0
			H925	1060	154	3.0	13.5
			H1075	869	126	4.0	15.5
			H1150	786	114	6.5	19.0
		1000	H900	1035	150	2.0	6.0
			H925	1040	151	2.5	12.5
			H1075	848	123	3.5	15.0
			H1150	765	111	5.5	18.0
425	800	100	H900	965	140	4.0	8.0
			H925	883	128	3.5	13.5
			H1075	745	108	6.0	16.0
			H1150	689	100	6.5	25.5
		1000	H900	883	128	4.0	6.0
			H925	834	121	4.3	13.0
			H1075	710	103	5.5	15.0
			H1150	650	94	6.0	20.0
480	900	100	H900	655	95	5.0	9.0
			H1150	550	80	9.0	40.0
		1000	H900	415	60	12.0	25.0
			H1150	490	71	9.0	36.0

(a) In 50 mm (2 in.)

AISI Type 630: Properties and Formability of Sheet and Strip at Room Temperature (Ref 12)

Condition	Tensile strength MPa	Tensile strength ksi	Yield strength(a) MPa	Yield strength(a) ksi	Elongation(b), %	Hardness, HRC	Olsen cup draw mm	Olsen cup draw in.
A	1079	156.5	758.4	110.0	6.7	33.0	8.38	0.330
H900 for 1 h	1382	200.5	1253	181.7	10.5	44.5	6.35	0.250
H1150 for 4 h	996.3	144.5	915.6	132.8	13.2	32.5	8.38	0.330
Aged at 705 °C (1300 °F) for 2 h	983.9	142.7	713.6	103.5	9.5	30.0	6.30	0.248
Aged at 760 °C (1400 °F) for 2 h	1024	148.5	806.7	117.0	8.7	32.0	7.85	0.309
H1150M	903.2	131.0	748.1	108.5	16.2	27.5	8.31	0.327

(a) 0.2% offset. (b) In 50 mm (2 in.)

AISI Type 630: Bend Test Data (Ref 12)
t is stock thickness

Condition	Minimum bend radius for bend of: Longitudinal 90°	Longitudinal 135°	Longitudinal 180°	Transverse 90°	Transverse 135°	Transverse 180°
A	3*t*	3*t*	6*t*	4*t*	5*t*	9*t*
H900	3*t*	3*t*	5*t*	4*t*	6*t*	9*t*
H925	2*t*	3*t*	5*t*	4*t*	6*t*	9*t*
H1025	2*t*	3*t*	4*t*	4*t*	6*t*	7*t*
H1075	2*t*	3*t*	4*t*	4*t*	4*t*	7*t*
H1150	2*t*	2*t*	4*t*	2*t*	3*t*	6*t*

Machining Data (Ref 2)

AISI Type 630: Surface Grinding With a Horizontal Spindle and Reciprocating Table

Operation	Wheel speed m/s	Wheel speed ft/min	Table speed m/min	Table speed ft/min	Downfeed mm/pass	Downfeed in./pass	Crossfeed mm/pass	Crossfeed in./pass	Wheel identification
Solution treated, with hardness of 150 to 200 HB									
Roughing	28-33	5500-6500	15-30	50-100	0.050	0.002	1.25-12.5	0.050-0.500	A46IV
Finishing	28-33	5500-6500	15-30	50-100	0.013 max	0.0005 max	(a)	(a)	A46IV
Solution treated or hardened, with hardness of 275 to 440 HB									
Roughing	28-33	5500-6500	15-30	50-100	0.025	0.001	0.65-6.5	0.025-0.250	A46HV
Finishing	28-33	5500-6500	15-30	50-100	0.013 max	0.0005 max	(b)	(b)	A46HV

(a) Maximum 1/4 of wheel width. (b) Maximum 1/10 of wheel width

AISI Type 630: Reaming

Based on 4 flutes for 3- and 6-mm (0.125- and 0.25-in.) reamers, 6 flutes for 12-mm (0.5-in.) reamers, and 8 flutes for 25-mm (1-in.) and larger reamers

Tool material	Speed		Feed per revolution for reamer diameter of: 3 mm (0.125 in.)		6 mm (0.25 in.)		12 mm (0.5 in.)		25 mm (1 in.)		35 mm (1.5 in.)		50 mm (2 in.)	
	m/min	ft/min	mm	in.	mm	in.	mm	in.	mm	in.	mm	in.	mm	in.
Solution treated, with hardness of 150 to 200 HB														
Roughing														
M1, M2, and M7 high speed steel	17	55	0.075	0.003	0.13	0.005	0.20	0.008	0.25	0.010	0.30	0.012	0.40	0.015
C-2 carbide	21	70	0.102	0.004	0.20	0.008	0.30	0.012	0.40	0.016	0.50	0.020	0.60	0.024
Finishing														
M1, M2, and M7 high speed steel	11	35	0.102	0.004	0.102	0.004	0.15	0.006	0.20	0.008	0.23	0.009	0.25	0.010
C-2 carbide	15	50	0.102	0.004	0.15	0.006	0.20	0.008	0.25	0.010	0.28	0.011	0.30	0.012
Solution treated or hardened, with hardness of 275 to 325 HB														
Roughing														
M1, M2, and M7 high speed steel	12	40	0.075	0.003	0.13	0.005	0.20	0.008	0.25	0.010	0.30	0.012	0.40	0.015
C-2 carbide	17	55	0.102	0.004	0.15	0.006	0.23	0.009	0.40	0.015	0.45	0.018	0.50	0.020
Finishing														
M1, M2, and M7 high speed steel	9	30	0.075	0.003	0.075	0.003	0.102	0.004	0.15	0.006	0.18	0.007	0.20	0.008
C-2 carbide	12	40	0.075	0.003	0.102	0.004	0.13	0.005	0.15	0.006	0.18	0.007	0.20	0.008
Solution treated or hardened, with hardness of 325 to 375 HB														
Roughing														
M1, M2, and M7 high speed steel	11	35	0.050	0.002	0.102	0.004	0.15	0.006	0.20	0.008	0.28	0.011	0.30	0.012
C-2 carbide	15	50	0.102	0.004	0.15	0.006	0.20	0.008	0.30	0.012	0.36	0.014	0.40	0.016
Finishing														
M1, M2, and M7 high speed steel	6	20	0.075	0.003	0.075	0.003	0.102	0.004	0.15	0.006	0.18	0.007	0.20	0.008
C-2 carbide	11	35	0.075	0.003	0.102	0.004	0.13	0.005	0.15	0.006	0.18	0.007	0.20	0.008
Hardened, with hardness of 375 to 440 HB														
Roughing														
Any premium high speed steel (T15, M33, M41-M47)	9	30	0.050	0.002	0.075	0.003	0.102	0.004	0.18	0.007	0.25	0.010	0.30	0.012
C-2 carbide	14	45	0.075	0.003	0.13	0.005	0.15	0.006	0.20	0.008	0.25	0.010	0.30	0.012
Finishing														
Any premium high speed steel (T15, M33, M41-M47)	5	15	0.050	0.002	0.075	0.003	0.102	0.004	0.13	0.005	0.15	0.006	0.18	0.007
C-2 carbide	11	35	0.075	0.003	0.102	0.004	0.13	0.005	0.15	0.006	0.18	0.007	0.20	0.008

AISI Type 630: Cylindrical and Internal Grinding

Operation	Wheel speed		Work speed		Infeed on diameter		Traverse(a)	Wheel identification
	m/s	ft/min	m/min	ft/min	mm/pass	in./pass		
Solution treated, with hardness of 150 to 200 HB								
Cylindrical grinding								
Roughing	28-33	5500-6500	21-30	70-100	0.050	0.002	1/4	A60JV
Finishing	28-33	5500-6500	21-30	70-100	0.013 max	0.0005 max	1/8	A60JV
Internal grinding								
Roughing	25-33	5000-6500	23-60	75-200	0.013	0.0005	1/3	A60KV
Finishing	25-33	5000-6500	23-60	75-200	0.005 max	0.0002 max	1/6	A60KV
Solution treated or hardened, with hardness of 275 to 440 HB								
Cylindrical grinding								
Roughing	28-33	5500-6500	21-30	70-100	0.050	0.002	1/4	A60IV
Finishing	28-33	5500-6500	21-30	70-100	0.013 max	0.0005 max	1/8	A60IV
Internal grinding								
Roughing	25-33	5000-6500	23-60	75-100	0.013	0.0005	1/3	A60JV
Finishing	25-33	5000-6500	23-60	75-100	0.005 max	0.0002 max	1/6	A60JV

(a) Wheel width per revolution of work

Machining Data (Ref 8) (continued)

AISI Type 630: Turning (Single Point and Box Tools)

Depth of cut		High speed steel(a)				Uncoated carbide						Coated carbide			
		Speed		Feed		Speed, brazed		Speed, inserted		Feed		Speed		Feed	
mm	in.	m/min	ft/min	mm/rev	in./rev	m/min	ft/min	m/min	ft/min	mm/rev	in./rev	m/min	ft/min	mm/rev	in./rev
Solution treated, with hardness of 150 to 200 HB															
1	0.040	29(a)	95(a)	0.18	0.007	105(b)	350(b)	120(b)	400(b)	0.18	0.007	160(c)	525(c)	0.18	0.007
4	0.150	24(a)	80(a)	0.40	0.015	90(d)	300(d)	105(d)	350(d)	0.40	0.015	135(e)	450(e)	0.40	0.015
8	0.300	18(a)	60(a)	0.50	0.020	73(d)	240(d)	84(d)	275(d)	0.75	0.030	105(e)	350(e)	0.50	0.020
16	0.625	14(a)	45(a)	0.75	0.030	55(d)	180(d)	64(d)	210(d)	1.00	0.040	...	...	...	...
Solution treated or hardened, with hardness of 275 to 325 HB															
1	0.040	23(f)	75(f)	0.18	0.007	90(b)	300(b)	105(b)	350(b)	0.18	0.007	135(c)	450(c)	0.18	0.007
4	0.150	18(f)	60(f)	0.40	0.015	76(d)	250(d)	90(d)	300(d)	0.40	0.015	120(e)	400(e)	0.40	0.015
8	0.300	14(f)	45(f)	0.50	0.020	60(d)	200(d)	73(d)	240(d)	0.50	0.020	90(e)	300(e)	0.50	0.020
Solution treated or hardened, with hardness of 325 to 375 HB															
1	0.040	21(f)	70(f)	0.18	0.007	88(b)	290(b)	100(b)	325(b)	0.13	0.005	...	...	...	...
4	0.150	17(f)	55(f)	0.40	0.015	75(d)	245(d)	84(d)	275(d)	0.25	0.010	...	...	...	...
8	0.300	12(f)	40(f)	0.50	0.020	60(d)	200(d)	69(d)	225(d)	0.40	0.015	...	...	...	...
Hardened, with hardness of 325 to 440															
2	0.040	14(f)	45(f)	0.13	0.005	58(b)	190(b)	69(b)	225(b)	0.13	0.005	...	...	...	...
4	0.150	9(f)	30(f)	0.25	0.010	49(d)	160(d)	58(d)	190(d)	0.25	0.010	...	...	...	...
8	0.300	8(f)	25(f)	0.40	0.015	40(d)	130(d)	47(d)	150(d)	0.40	0.015	...	...	...	...

(a) High speed tool material: M2 or M3. (b) Carbide tool material: C-7. (c) Carbide tool material: CC-7. (d) Carbide tool material: C-6. (e) Carbide tool material: CC-6. (f) Any premium high speed steel tool material (T15, M33, M41 to M47)

AISI Type 630: Turning (Cutoff and Form Tools)

		Feed per revolution for cutoff tool width of:			Feed per revolution for form tool width of:				
Tool material	Speed, m/min (ft/min)	1.5 mm (0.062 in.) mm (in.)	3 mm (0.125 in.) mm (in.)	6 mm (0.25 in.) mm (in.)	12 mm (0.5 in.) mm (in.)	18 mm (0.75 in.) mm (in.)	25 mm (1 in.) mm (in.)	35 mm (1.5 in.) mm (in.)	50 mm (2 in.) mm (in.)
Solution treated, with hardness of 150 to 200 HB									
M2 and M3 high speed steel	18	0.030	0.038	0.050	0.050	0.046	0.041	0.033	0.028
	(60)	(0.0012)	(0.0015)	(0.002)	(0.002)	(0.0018)	(0.0016)	(0.0013)	(0.0011)
C-6 carbide	62	0.030	0.038	0.050	0.050	0.046	0.041	0.033	0.028
	(205)	(0.0012)	(0.0015)	(0.002)	(0.002)	(0.0018)	(0.0016)	(0.0013)	(0.0011)
Solution treated or hardened, with hardness of 275 to 325 HB									
Any premium high speed steel (T15, M33, M41-M47)	17	0.025	0.033	0.046	0.046	0.041	0.036	0.028	0.023
	(55)	(0.001)	(0.0013)	(0.0018)	(0.0018)	(0.0016)	(0.0014)	(0.0011)	(0.0009)
C-6 carbide	58	0.025	0.033	0.046	0.046	0.041	0.036	0.028	0.023
	(190)	(0.001)	(0.0013)	(0.0018)	(0.0018)	(0.0016)	(0.0014)	(0.0011)	(0.0009)
Solution treated or hardened, with hardness of 325 to 375 HB									
Any premium high speed steel (T15, M33, M41-M47)	12	0.020	0.028	0.041	0.041	0.036	0.030	0.023	0.018
	(40)	(0.0008)	(0.0011)	(0.0016)	(0.0016)	(0.0014)	(0.0012)	(0.0009)	(0.0007)
C-6 carbide	52	0.020	0.028	0.041	0.041	0.036	0.030	0.023	0.018
	(170)	(0.0008)	(0.0011)	(0.0016)	(0.0016)	(0.0014)	(0.0012)	(0.0009)	(0.0007)
Hardened, with hardness of 375 to 440 HB									
Any premium high speed steel (T15, M33, M41-M47)	9	0.015	0.023	0.036	0.036	0.030	0.025	0.018	0.015
	(30)	(0.0006)	(0.0009)	(0.0014)	(0.0014)	(0.0012)	(0.001)	(0.0007)	(0.0006)
C-6 carbide	34	0.015	0.023	0.036	0.036	0.030	0.025	0.018	0.015
	(110)	(0.0006)	(0.0009)	(0.0014)	(0.0014)	(0.0012)	(0.001)	(0.0007)	(0.0006)

Machining Data (Ref 8) (continued)

AISI Type 630: Face Milling

Depth of cut		High speed steel				Uncoated carbide						Coated carbide			
		Speed		Feed/tooth		Speed, brazed		Speed, inserted		Feed/tooth		Speed		Feed/tooth	
mm	in.	m/min	ft/min	mm	in.	m/min	ft/min	m/min	ft/min	mm	in.	m/min	ft/min	mm	in.
Solution treated with hardness of 150 to 200 HB															
1	0.040	34(a)	110(a)	0.20	0.008	115(b)	375(b)	120(b)	390(b)	0.20	0.008	175(c)	575(c)	0.20	0.008
4	0.150	26(a)	85(a)	0.30	0.012	105(b)	350(b)	115(b)	375(b)	0.30	0.012	145(c)	475(c)	0.30	0.012
8	0.300	20(a)	65(a)	0.40	0.016	72(d)	235(d)	88(d)	290(d)	0.40	0.016	115(e)	375(e)	0.40	0.016
Solution treated or hardened with hardness of 275 to 325 HB															
1	0.040	24(f)	80(f)	0.15	0.006	84(b)	275(b)	88(b)	290(b)	0.15	0.006	130(c)	425(c)	0.13	0.005
4	0.150	18(f)	60(f)	0.23	0.009	76(b)	250(b)	84(b)	275(b)	0.20	0.008	105(c)	350(c)	0.18	0.007
8	0.300	14(f)	45(f)	0.30	0.012	53(d)	175(d)	66(d)	215(d)	0.25	0.010	84(e)	275(e)	0.23	0.009
Solution treated or hardened with hardness of 325 to 375 HB															
1	0.040	21(f)	75(f)	0.13	0.005	67(b)	220(b)	79(b)	260(b)	0.13	0.005	...	...	...	...
4	0.150	17(f)	55(f)	0.20	0.008	60(b)	200(b)	67(b)	220(b)	0.18	0.007	...	...	...	...
8	0.300	12(f)	40(f)	0.25	0.010	43(d)	140(d)	52(d)	170(d)	0.23	0.009	...	...	...	...
Hardened with hardness of 375 to 440 HB															
1	0.040	18(f)	60(f)	0.102	0.004	43(b)	140(b)	58(b)	190(b)	0.102	0.004	...	...	...	...
4	0.150	14(f)	45(f)	0.15	0.006	40(b)	130(b)	43(b)	140(b)	0.15	0.006	...	...	...	...
8	0.300	11(f)	35(f)	0.20	0.008	27(b)	90(b)	34(b)	110(b)	0.20	0.008	...	...	...	...

(a) High speed steel tool material: M2 or M7. (b) Carbide tool material: C-6. (c) Carbide tool material: CC-6. (d) Carbide tool material: C-5. (e) Carbide tool material: CC-5. (f) Any premium high speed tool material (T15, M33, M41 to M47)

AISI Type 630: End Milling (Profiling)

Tool material	Depth of cut		Speed		Feed per tooth for cutter diameter of: 10 mm (0.375 in.)		12 mm (0.5 in.)		18 mm (0.75 in.)		25-50 mm (1-2 in.)	
	mm	in.	m/min	ft/min	mm	in.	mm	in.	mm	in.	mm	in.
Solution treated, with hardness of 150 to 200 HB												
M2, M3, and M7 high speed steel	0.5	0.020	30	100	0.013	0.0005	0.025	0.001	0.050	0.002	0.075	0.003
	1.5	0.060	23	75	0.025	0.001	0.050	0.002	0.075	0.003	0.102	0.004
	diam/4	diam/4	20	65	0.018	0.0007	0.025	0.001	0.050	0.002	0.075	0.003
	diam/2	diam/2	17	55	0.013	0.0005	0.018	0.0007	0.025	0.001	0.050	0.002
C-2 carbide	0.5	0.020	110	360	0.013	0.0005	0.025	0.001	0.075	0.003	0.13	0.005
	1.5	0.060	82	270	0.025	0.001	0.050	0.002	0.102	0.004	0.15	0.006
	diam/4	diam/4	72	235	0.025	0.001	0.038	0.0015	0.075	0.003	0.102	0.004
	diam/2	diam/2	67	220	0.013	0.0005	0.025	0.001	0.050	0.002	0.075	0.003
Solution treated or hardened, with hardness of 275 to 325 HB												
M2, M3, and M7 high speed steel	0.5	0.020	26	85	0.013	0.0005	0.025	0.001	0.050	0.002	0.075	0.003
	1.5	0.060	20	65	0.025	0.001	0.050	0.002	0.075	0.003	0.102	0.004
	diam/4	diam/4	17	55	0.018	0.0007	0.025	0.001	0.050	0.002	0.075	0.003
	diam/2	diam/2	15	50	0.013	0.0005	0.018	0.0007	0.025	0.001	0.050	0.002
C-2 carbide	0.5	0.020	82	270	0.013	0.0005	0.025	0.001	0.075	0.003	0.13	0.005
	1.5	0.060	64	210	0.025	0.001	0.050	0.002	0.102	0.004	0.15	0.006
	diam/4	diam/4	55	180	0.025	0.001	0.038	0.0015	0.075	0.003	0.102	0.004
	diam/2	diam/2	50	165	0.013	0.0005	0.025	0.001	0.050	0.002	0.075	0.003
Solution treated or hardened, with hardness of 325 to 375 HB												
M2, M3, and M7 high speed steel	0.5	0.020	24	80	0.013	0.0005	0.018	0.0007	0.025	0.001	0.050	0.002
	1.5	0.060	18	60	0.013	0.0005	0.025	0.001	0.050	0.002	0.075	0.003
	diam/4	diam/4	15	50	0.013	0.0005	0.018	0.0007	0.038	0.0015	0.050	0.002
	diam/2	diam/2	14	45	0.013	0.0005	0.013	0.0005	0.025	0.001	0.038	0.0015
C-2 carbide	0.5	0.020	79	260	0.013	0.0005	0.025	0.001	0.050	0.002	0.075	0.003
	1.5	0.060	60	200	0.025	0.001	0.050	0.002	0.075	0.003	0.102	0.004
	diam/4	diam/4	52	170	0.025	0.001	0.038	0.0015	0.063	0.0025	0.089	0.0035
	diam/2	diam/2	50	165	0.013	0.0005	0.025	0.001	0.050	0.002	0.075	0.003
Hardened, with hardness of 375 to 440 HB												
Any premium high speed steel (T15, M33, M41-M47)	0.5	0.020	21	70	0.013	0.0005	0.018	0.0007	0.025	0.0010	0.050	0.002
	1.5	0.060	17	55	0.013	0.0005	0.025	0.001	0.050	0.002	0.075	0.003
	diam/4	diam/4	14	45	0.013	0.0005	0.018	0.0007	0.038	0.0015	0.050	0.002
	diam/2	diam/2	12	40	...	...	0.013	0.0005	0.018	0.0007	0.025	0.001
C-2 carbide	0.5	0.020	34	110	0.013	0.0005	0.025	0.001	0.050	0.002	0.075	0.003
	1.5	0.060	26	85	0.025	0.001	0.050	0.002	0.075	0.003	0.102	0.004
	diam/4	diam/4	21	70	0.025	0.001	0.038	0.0015	0.063	0.0025	0.089	0.0035
	diam/2	diam/2	20	65	0.013	0.0005	0.025	0.001	0.050	0.002	0.075	0.003

Machining Data (Ref 8) (continued)

AISI Type 630: Boring

Depth of cut mm	in.	High speed steel: Speed m/min	ft/min	Feed mm/rev	in./rev	Uncoated carbide: Speed, brazed m/min	ft/min	Speed, inserted m/min	ft/min	Feed mm/rev	in./rev	Coated carbide: Speed m/min	ft/min	Feed mm/rev	in./rev
Solution treated, with hardness of 150 to 200 HB															
0.25	0.010	29(a)	95(a)	0.075	0.003	90(b)	295(b)	105(b)	345(b)	0.075	0.003	135(c)	440(c)	0.075	0.003
1.25	0.050	23(a)	75(a)	0.13	0.005	72(d)	235(d)	84(d)	275(d)	0.13	0.005	115(e)	375(e)	0.13	0.005
2.5	0.100	20(a)	65(a)	0.30	0.012	64(f)	210(f)	75(f)	245(f)	0.30	0.012	95(g)	315(g)	0.30	0.012
Solution treated or hardened, with hardness of 275 to 325 HB															
0.25	0.010	23(h)	75(h)	0.075	0.003	81(d)	265(d)	95(d)	310(d)	0.075	0.003	120(e)	400(e)	0.075	0.003
1.25	0.050	18(h)	60(h)	0.13	0.005	64(f)	210(f)	75(f)	245(f)	0.13	0.005	95(g)	315(g)	0.13	0.005
2.5	0.100	15(h)	50(h)	0.30	0.012	55(f)	180(f)	64(f)	210(f)	0.30	0.012	84(g)	275(g)	0.30	0.012
Solution treated or hardened, with hardness of 325 to 375 HB															
0.25	0.010	21(h)	70(h)	0.075	0.003	72(d)	235(d)	85(d)	280(d)	0.050	0.002	...	...	...	...
1.25	0.050	17(h)	55(h)	0.13	0.005	58(f)	190(f)	69(f)	225(f)	0.102	0.004	...	...	...	...
2.5	0.100	14(h)	45(h)	0.30	0.012	49(f)	160(f)	58(f)	190(f)	0.200	0.008	...	...	...	...
Hardened, with hardness of 375 to 440 HB															
0.25	0.010	14(h)	45(h)	0.050	0.002	50(d)	165(d)	59(d)	195(d)	0.050	0.002	...	...	...	...
1.25	0.050	11(h)	35(h)	0.102	0.004	40(f)	130(f)	47(f)	155(f)	0.102	0.004	...	...	...	...
2.5	0.100	8(h)	25(h)	0.20	0.008	35(f)	115(f)	41(f)	135(f)	0.200	0.008	...	...	...	...

(a) High speed steel tool material: M2 or M3. (b) Carbide tool material: C-8. (c) Carbide tool material: CC-8. (d) Carbide tool material: C-7. (e) Carbide tool material: CC-7. (f) Carbide tool material: C-6. (g) Carbide tool material: CC-6. (h) Any premium high speed steel tool material (T15, M33, M41 to M47)

AISI Type 630 Drilling

Tool material	Speed, m/min (ft/min)	Feed per revolution for nominal hole diameter of: 1.5 mm (0.062 in.) mm (in.)	3 mm (0.125 in.) mm (in.)	6 mm (0.25 in.) mm (in.)	12 mm (0.5 in.) mm (in.)	18 mm (0.75 in.) mm (in.)	25 mm (1 in.) mm (in.)	35 mm (1.5 in.) mm (in.)	50 mm (2 in.) mm (in.)
Solution treated, with hardness of 150 to 200 HB									
M10, M7, and M1 high speed steel	15 (50)	0.025 (0.001)	0.050 (0.002)	0.102 (0.004)	0.18 (0.007)	0.20 (0.008)	0.25 (0.010)	0.30 (0.012)	0.40 (0.015)
Solution treated or hardened, with hardness of 275 to 325 HB									
M10, M7, and M1 high speed steel	14 (45)	... (...)	0.050 (0.002)	0.102 (0.004)	0.18 (0.007)	0.20 (0.008)	0.25 (0.010)	0.30 (0.012)	0.40 (0.015)
Solution treated or hardened, with hardness of 325 to 375 HB									
Any premium high speed steel (T15, M33, M41-M47)	11 (35)	... (...)	0.050 (0.002)	0.102 (0.004)	0.15 (0.006)	0.20 (0.008)	0.23 (0.009)	0.28 (0.011)	0.30 (0.012)
Hardened, with hardness of 375 to 440 HB									
Any premium high speed steel (T15, M33, M41-M47)	8 (25)	... (...)	0.025 (0.001)	0.050 (0.002)	0.075 (0.003)	0.102 (0.004)	0.102 (0.004)	0.102 (0.004)	0.102 (0.004)

AISI Type 630: Planing

Tool material	Depth of cut mm	in.	Speed m/min	ft/min	Feed/stroke mm	in.
Solution treated, with hardness of 150 to 200 HB						
M2 or M3 high speed steel	0.1	0.005	5	15	(a)	(a)
	2.5	0.100	8	25	0.75	0.030
	12	0.500	5	15	1.15	0.045
C-6 carbide	0.1	0.005	38	125	(a)	(a)
	2.5	0.100	43	140	1.50	0.060

Tool material	Depth of cut mm	in.	Speed m/min	ft/min	Feed/stroke mm	in.
Solution treated or hardened, with hardness of 275 to 325 HB						
M2 or M3 high speed steel	0.1	0.005	5	15	(a)	(a)
	2.5	0.100	6	20	0.50	0.020
	12	0.500	5	15	0.75	0.030
C-6 carbide	0.1	0.005	34	110	(a)	(a)
	2.5	0.100	38	125	1.50	0.060

(a) Feed is 75% the width of the square nose finishing tool

AISI Type 631

AISI Type 631: Chemical Composition

Element	Composition, %
Carbon	0.09
Manganese	1.00
Phosphorus	0.040
Sulfur	0.040
Silicon	0.040
Chromium	16-18
Nickel	6.50-7.75
Aluminum	0.75-1.50

Characteristics. AISI type 631 (S17700) is a precipitation-hardening stainless steel that provides high strength and hardness, excellent fatigue properties, good corrosion resistance, and minimum distortion on heat treatment. It is easily formed in the solution-treated (annealed) condition, then hardened to high-strength levels by simple heat treatments.

In the heat treated conditions, type 631 has good mechanical properties at temperatures up to 480 °C (900 °F). Corrosion resistance of this grade in conditions TH1050 and RH950 is superior to that of the hardenable chromium types, and in some environments, is similar to types 302 and 304. Spring wire is produced by severe cold drawing which can be hardened by a single-step, low-temperature heat treatment.

Type 631 in condition A, mill annealed, can be formed similarly to type 301 stainless steel. It work hardens rapidly, and may require intermediate annealing when deep drawing or forming intricate parts.

Type 631 is readily welded by many of the arc and resistance processes applicable to stainless steel. Preheating and postannealing practices are not necessary. Inert gas shielding should be used to avoid oxidation of aluminum during fusion welding.

Machining characteristics of type 631 are similar to those of type 302 stainless steel. Normal practice for machining materials in the hardness range of 275 to 400 HB should be followed.

Typical Uses. Applications of AISI type 631 include aircraft structural parts; flat and round wire springs; and drawn, bent or formed parts.

AISI Type 631: Similar Steels (U.S. and/or Foreign). UNS S17700; AMS 5528, 5529, 5568, 5644, 5673, 5678, 5824; ASME SA705; ASTM A313, A564, A579 (62), A693, A705; MIL SPEC MIL-S-25043, MIL-W-46078; SAE J217, J467 (17-7PH); (W. Ger.) DIN 1.4568

Physical Properties

AISI Type 631: Thermal Treatment Temperatures

Treatment	Temperature range °C	Temperature range °F
Forging	1175-1230	2150-2250
Solution treatment, condition A	1040-1045	1890-1910
Transformation to condition C	(a)	(a)
Transformation to condition T(b)	760	1400
Transformation to condition R(c)	950	1750

(a) Cold rolled or cold drawn. (b) Air cool to 16 °C (60 °F). (c) Plus −73 °C (−100 °F)

AISI Type 631: Average Coefficients of Linear Thermal Expansion (Ref 12)

Condition	Temperature range °C	Temperature range °F	Coefficient μm/m·K	Coefficient μin./in.·°F
A	21-93	70-200	15.3	8.5
	21-205	70-400	16.2	9.0
	21-315	70-600	17.1	9.5
	21-425	70-800	17.3	9.6
TH1050	21-93	70-200	10.1	5.6
	21-205	70-400	11.0	6.1
	21-315	70-600	11.3	6.3
	21-425	70-800	11.9	6.6
RH950	21-93	70-200	10.3	5.7
	21-205	70-400	11.9	6.6
	21-315	70-600	12.2	6.8
	21-425	70-800	12.4	6.9
CH900	21-93	70-200	11.0	6.1
	21-205	70-400	11.2	6.2
	21-315	70-600	11.5	6.4
	21-425	70-800	11.9	6.6

AISI Type 631: Thermal Conductivity (Ref 12)

Condition	Temperature °C	Temperature °F	Thermal conductivity W/m·K	Thermal conductivity Btu/ft·h·°F
TH1050	150	300	17.0	9.8
	260	500	18.5	10.7
	450	840	21.1	12.2
	480	900	21.1	12.2
RH960(a)	150	300	17.0	9.8
	260	500	18.5	10.7
	450	840	21.1	12.2
	480	900	21.1	12.2
CH900	150	300	16.4	9.5
	260	500	18.3	10.6
	450	840	21.6	12.5
	480	900	21.8	12.6

(a) Estimated values

AISI Type 631: Selected Physical Properties at Various Heat Treated Conditions (Ref 12)

Melting range	1405-1440 °C (2560-2625 °F)
Specific heat at 0-100 °C (32-212 °F)	460 J/kg·K (0.11 Btu/lb·°F)

AISI Type 631: Density (Ref 12)

Condition	Density g/cm³	Density lb/in.³
A	7.81	0.282
TH1050	7.64	0.276
RH950	7.64	0.276
CH900	7.67	0.277

AISI Type 631: Electrical Resistivity at Room Temperature (Ref 12)

Condition	Electrical resistivity, nΩ·m
A	80
TH1050	82
RH950	83
CH900	83.8

Mechanical Properties

AISI Type 631: Representative Tensile Properties (Ref 6)

Condition	Tensile strength MPa	Tensile strength ksi	Yield strength(a) MPa	Yield strength(a) ksi	Elongation(b), %	Reduction in area, %	Hardness
Sheet							
Solution treated	895	130	275	40	35	...	85 HRB
CH900	1825	265	1795	260	2	...	43 HRC
TH1050	1380	200	...	...	...	...	...
Plate and bar							
Solution treated	895	130	275	40	10	30	90 HRB
RH950	1380	200	1205	175	10	30	44 HRC
TH1050	1205	175	1070	155	12	34	42 HRC

(a) 0.2% offset. (b) In 50 mm (2 in.)

AISI Type 631: Properties Acceptable for Material Specification of Cold Drawn Wire (Ref 12)

Wire diameter		Tensile strength			
		Condition C		Condition CH900	
mm	in.	MPa	ksi	MPa	ksi
0.762-1.041	0.030-0.041	1790-2000	260-290	2205-2415	320-350
1.067-1.295	0.042-0.051	1760-1965	255-285	2140-2345	310-340
1.321-1.549	0.052-0.061	1725-1930	250-280	2105-2310	305-335
1.575-1.803	0.062-0.071	1670-1875	242-272	2050-2255	297-327
1.829-2.184	0.072-0.086	1655-1860	240-270	2015-2220	292-322
2.210-2.286	0.087-0.090	1585-1795	230-260	1945-2050	282-312
2.311-2.540	0.091-0.100	1565-1770	227-257	1925-2130	279-309
2.565-2.692	0.101-0.106	1540-1745	223-253	1900-2100	274-304
2.718-3.302	0.107-0.130	1525-1730	221-251	1875-2080	272-302
3.327-3.505	0.131-0.138	1480-1690	215-245	1795-2000	260-290
3.531-3.708	0.139-0.146	1470-1675	213-243	1780-1985	258-288
3.743-4.115	0.147-0.162	1455-1660	211-241	1765-1970	256-286
4.140-4.572	0.163-0.182	1440-1650	209-239	1750-1960	254-284
4.597-5.258	0.181-0.207	1430-1635	207-237	1740-1945	252-282
5.283-5.715	0.208-0.225	1400-1605	203-233	1710-1915	248-278
5.740-7.772	0.226-0.306	1365-1570	198-228	1660-1875	242-272
7.798-8.407	0.307-0.331	1325-1530	192-222	1620-1825	235-265

AISI Type 631: Typical Elevated-Temperature Properties of Sheet and Strip (Ref 12)

Test temperature °C	Test temperature °F	Condition	Tensile strength MPa	Tensile strength ksi	Yield strength(a) MPa	Yield strength(a) ksi	Elongation(b), %
24	75	TH1050	1330	193	1255	182	10.0
		RH950	1585	230	1495	217	6.0
		CH900	1805	262	1695	246	5.0
150	300	TH1050	1235	179	1170	170	8.0
		RH950	1435	208	1325	192	4.5
		CH900	1710	248	1614	234	4.0
260	500	TH1050	1150	167	1105	160	4.5
		RH950	1345	195	1215	176	4.5
		CH900	1570	228	1475	214	3.0
315	600	TH1050	1115	162	1070	155	4.0
		RH950	1305	189	1165	169	5.0
		CH900	1530	222	1405	204	3.0
370	700	TH1050	1075	156	1005	146	4.5
		RH950	1250	181	1115	162	7.0
425	800	TH1050	986	143	896	130	6.2
		RH950	1105	160	945	137	12.0
		CH900	1425	207	1215	176	5.0
480	900	TH1050	855	124	689	100	10.0
		RH950	915	133	786	114	15.0
		CH900	1260	183	993	144	6.0

(a) 0.2% offset. (b) In 50 mm (2 in.)

Mechanical Properties (continued)

AISI Type 631: Stress-Rupture and Creep-Strength Properties of Sheet and Strip (Ref 12)

Test temperature			Stress ro rupture in:				Stress to produce permanent deformation of:			
			100 h		1000 h		0.1% in 1000 h		0.2% in 1000 h	
°C	°F	Condition	MPa	ksi	MPa	ksi	MPa	ksi	MPa	ksi
315	600	TH1050	1170	170	1090	158	600	87	724	105
		RH950	1295	188	1240	180	725	105	869	126
		CH900	1515	220	1490	216	...	...	...	...
370	700	TH1050	895	130	840	122	395	57	48	70
		RH950	1165	169	100	146	415	60	600	87
		CH900	1340	194	1240	180	...	...	...	...
425	800	TH1050	758	110	620	90	275	40	310	45
		RH950	779	113	635	92	215	31	250	36
		CH900	931	135	505	73	...	...	...	...
480	900	TH1050	540	78	360	52	105	15	125	18
		RH950	420	61	305	44	83	12	97	14
		CH900	365	53	250	36	...	...	...	...

AISI Type 631, Condition CH900: Fatigue Properties of Helical Compression Springs (Ref 12)

Wire diameter		Surface	Fatigue strength	
mm	in.	condition	MPa	ksi
1.70	0.067	Cold drawn	585	85
		Electropolished	725	105
2.16	0.085	Cold drawn	585	85
2.67	0.105	Cold drawn	585	85
		Shot peened	825	120
		Electropolished, shot peened	895	130
3.18	0.125	Cold drawn	505	73
		Electropolished	585	85

Machining Data

For machining data on AISI type 631, refer to the preceding tables for AISI type 630.

AISI Type 632

AISI Type 632: Chemical Composition

Element	Composition, %
Carbon	0.09 max
Manganese	1.00 max
Phosphorus	0.04 max
Sulfur	0.03 max
Silicon	1.00 max
Chromium	14-16
Nickel	6.50-7.75
Molybdenum	2-3
Aluminum	0.75-1.50

Characteristics. AISI type 632 (S15700) is a semiaustenitic, precipitation-hardening, stainless steel that provides high strength and hardness, good corrosion resistance, and minimum distortion on heat treatment. It is easily formed in the annealed condition and develops an effective balance of properties through simple heat treatments. For applications requiring exceptionally high strength, cold reduced type 632, condition CH900, is ideal and can be used where limited ductility and workability are permissible.

In its heat-treated condition, type 632 stainless steel provides excellent mechanical properties at temperatures up to 480 °C (900 °F). Corrosion resistance of this grade is superior to that of the hardenable chromium types, in some environments approximating that of type 304. Type 632 has the greatest resistance to stress corrosion cracking in condition CH900. Condition TH1050 is somewhat less resistant than condition CH900, but appears to be more resistant to stress cracking than condition RH950.

Type 632 in condition A can be formed comparably to type 301 stainless steel. It work hardens rapidly, and may require intermediate annealing when deep drawing or forming intricate parts.

Type 632 is readily welded by many of the arc and resistance welding processes applicable to stainless steels. Preheating and postannealing practices are not necessary. The major precaution advised is the use of gas shielding to avoid oxidation of aluminum during fusion welding.

Typical Uses. Applications of AISI type 632 include retaining rings, springs, diaphragms, aircraft bulkheads, welded and brazed honeycomb paneling, and other aircraft components that require high strength at elevated temperatures.

AISI Type 632: Similar Steels (U.S. and/or Foreign). UNS S15700; AMS 5520, 5657, 5812, 5813; ASTM A564, A579, A693, A705; SAE J467 (PH15-7-Mo); (W. Ger.) DIN 1.4532

Physical Properties

AISI Type 632: Thermal Treatment Temperatures

Treatment	Temperature range °C	°F
Annealing, condition A	1050-1080	1925-1975
Austenite conditioning	745-775	1375-1425
Hardening	510-565	950-1050

AISI Type 632: Thermal Conductivity

Temperature °C	°F	Thermal conductivity W/m·K	Btu/ft·h·°F
21	70	15.1	8.7
95	200	16.1	9.3
205	400	17.6	10.2
315	600	19.2	11.1
425	800	20.8	12.0
480	900	21.6	12.5

AISI Type 632: Average Coefficients of Linear Thermal Expansion (Ref 12)

Temperature range °C	°F	Coefficient μm/m·K	μin./in.·°F
21-93	70-200	9.0	5.0
21-205	70-400	9.7	5.4
21-316	70-600	10.1	5.6
21-425	70-800	10.6	5.9
21-480	70-900	10.8	6.0
21-540	70-1000	11.1	6.1

AISI Type 632: Selected Physical Properties in the RH950 Condition (Ref 12)

Density	7.67 g/cm³ (0.277 lb/in.³)
Electrical resistivity at 20 °C (68 °F)	83 nΩ·m

Mechanical Properties

AISI Type 632: Stress-Rupture and Creep Properties at Elevated Temperatures (Ref 12)

Test temperature °C	°F	Condition	Stress to produce rupture in: 100 h MPa	100 h ksi	1000 h MPa	1000 h ksi	Stress to produce permanent deformation of: 1% in 1000 h MPa	1% in 1000 h ksi	2% in 1000 h MPa	2% in 1000 h ksi
315	600	RH950	1395	202	1380	200	906.6	131.5	1035	150.1
		TH1050	1235	179	1225	178	...	...	...	...
370	700	RH950	1330	193	1315	191	830.8	120.5	979.0	142.0
		TH1050	1110	161	1040	151	...	...	...	...
425	800	RH950	1200	174	1180	171	655	95.0	752.9	109.2
		TH1050	960	139	945	137	...	...	...	...
480	900	RH950	860	125	745	108	248	36.0	279	40.5
		TH1050	745	108	675	98	...	...	...	...

AISI Type 632: Short-Time Properties at Elevated Temperatures (Ref 12)

Test temperature °C	°F	Condition	Tensile strength MPa	Tensile strength ksi	Yield strength(a) MPa	Yield strength(a) ksi	Elongation(b), %	Shear strength MPa	Shear strength ksi
24	1075	TH1050	1462	212	1413	205	7.0	986	143
		RH950	1634	237	1517	220	5.0	1117	162
		CH900, Longitudinal	1744	253	1675	243	3.0	...	...
		CH900, Transverse	1800	261	1758	255	3.0	...	...
150	300	TH1050	1379	200	1344	195	4.5	896	130
		RH950	1517	220	1379	200	4.0	1000	145
		CH900, Longitudinal	1655	240	1551	225	2.0	...	...
		CH900, Transverse	1779	258	1606	233	2.0	...	...
315	600	TH1050	1255	182	1886	172	4.5	800	116
		RH950	1379	200	1200	174	5.0	883	128
		CH900, Longitudinal	1517	220	1407	204	1.5	...	...
		CH900, Transverse	1641	238	1455	211	1.5	...	...
370	700	TH1050	1207	175	1131	164	6.0	758	110
		RH950	1344	195	1138	165	6.0	855	124
		CH900, Longitudinal	1434	208	1331	193	1.5	...	...
		CH900, Transverse	1572	228	1379	200	1.5	...	...
425	800	TH1050	1124	163	1034	150	9.0	717	104
		RH950	1255	182	1034	150	8.0	800	116
		CH900, Longitudinal	1372	199	1248	181	1.5	...	...
		CH900, Transverse	1510	219	1310	190	1.5	...	...
480	900	TH1050	979	142	876	127	14.0	662	96
		RH950	1103	160	896	130	10.0	710	103
		CH900, Longitudinal	1269	184	1138	165	2.5	...	...
		CH900, Transverse	1393	202	1207	175	2.5	...	...
540	1000	TH1050	793	115	724	105	19.0	552	80
		RH950	896	130	724	105	14.0	607	88
		CH900, Longitudinal	1089	158	903	131	4.0	...	...
		CH900, Transverse	1193	173	986	143	4.0	...	...

(a) 0.2% offset. (b) In 50 mm (2 in.)

Mechanical Properties (continued)

AISI Type 632: Typical Tensile Properties at Room Temperature (Ref 12)

Condition	Tensile strength MPa	Tensile strength ksi	Yield strength(a) MPa	Yield strength(a) ksi	Elongation(b), %	Hardness
A	896	130	379	55	35	88 HRB
T	1000	145	621	90	7	28 HRC
TH1050	1448	210	1379	200	7	24 HRC
A1750	1034	150	379	55	12	85 HRB
R100	1241	180	862	125	7	40 HRC
RH950	1655	240	1551	225	6	48 HRC
C	1517	220	1310	190	5	45 HRC
CH900	1827	265	1793	260	2	50 HRC

(a) 0.2% offset. (b) In 50 mm (2 in.)

Machining Data

For machining data on AISI type 632, refer to the preceding machining tables for AISI type 630.

AISI Type 633

AISI Type 633: Chemical Composition

Element	Composition, %
Carbon	0.07-0.11
Manganese	0.50-1.25
Phosphorus	0.04 max
Sulfur	0.03 max
Silicon	0.50 max
Chromium	16-17
Nickel	4-5
Molybdenum	2.50-3.25
Nitrogen	0.07-0.13

Characteristics. AISI type 633 (S35000) is a chromium-nickel-molybdenum stainless steel which can be hardened by martensitic transformation and/or precipitation hardening. Depending on heat treatment, type 633 may have an austenitic structure for optimal formability, or a martensitic structure with strengths comparable to martensitic steels. This alloy contains some delta ferrite, normally about 5 to 10% of the phase.

Type 633 has corrosion resistance superior to other quench-hardenable, martensitic stainless steels. Material in the double-aged, or equalized and overtempered condition, is susceptible to intergranular corrosion because of the precipitation of chromium carbides. When this alloy is hardened by treatments employing subzero cooling, it is not subject to intergranular attack.

In the annealed condition, type 633 is austenitic and has cold forming characteristics similar to the austenitic stainless steels. It has a high rate of work hardening, and cold forming will cause martensite formation in proportion to the amount of deformation. In the hardened condition, this alloy has sufficient ductility for limited forming or straightening operations.

Optimal machinability of type 633 is obtained in the equalized and overtempered condition. Machining in the annealed condition is not recommended.

Inert-gas-shielded or shielded-metal-arc welding of type 633 can be used to produce sound welds with good ductility. For optimum properties, postweld reannealing at 930 °C (1710 °F) is required. For fusion welding of type 633, filler metal of the same composition as the base metal is recommended. Rod of types 308, 309, or 310 can be used for increased ductility. Type 633 can be brazed with the common silver-base or nickel-base brazing alloys.

Typical Uses. Applications of AISI type 633 include gas turbine compressor blades, discs, rotors, shafts, and similar parts where high strength at room and intermediate temperatures is required.

AISI Type 633: Similar Steels (U.S. and/or Foreign). UNS S35000; AMS 5546, 5548, 5554, 5745, 5774, 5775; ASTM A579 (61), A693; SAE J467 (AM-350)

Physical Properties

AISI 633: Thermal Treatment Temperatures

Treatment	Temperature range °C	Temperature range °F
Forging and annealing		
Forging, start	1175 max	2150 max
Forging, finish	925-980	1700-1800
Annealing	1010-1065	1850-1950
Hardening		
Subzero cooling, conditioning	920-945	1685-1735
Subzero cooling, tempering	455 or 540	850 or 1000
Double aging	730-760	1350-1400
Double aging, reheat	440-470	825-875
Equalize and overtemper	745-800	1375-1475
	540-595	1000-1100

AISI Type 633: Average Coefficients of Linear Thermal Expansion (Ref 9)

Test specimens were subzero cooled; tempered at 455 °C (850 °F)

Temperature range °C	Temperature range °F	Coefficient μm/m·K	Coefficient μin./in.·°F
20-100	68-212	11.3	6.3
20-300	68-570	12.2	6.8
20-400	68-750	12.6	7.0
20-500	68-930	13.0	7.2
20-620	68-1150	13.0	7.2
20-730	68-1350	12.1	6.7
20-815	68-1500	12.6	7.0
20-925	68-1700	13.5	7.5

Physical Properties (continued)

AISI Type 633: Selected Physical Properties (Ref 9)

Melting range .... 1370-1400 °C (2500-2550 °F)
Density in the annealed condition................. 8.0 g/cm³ (0.29 lb/in.³)

AISI Type 633: Electrical Resistivity (Ref 9)

Test specimens were subzero cooled; tempered at 455 °C (850 °F)

Temperature °C	°F	Electrical resistivity, nΩ·m
27	80	79
280	540	94
730	1350	115

AISI Type 633: Thermal Conductivity (Ref 9)

Test specimens were subzero cooled; tempered at 455 °C (850 °F)

Temperature °C	°F	Thermal conductivity W/m·K	Btu/ft·h·°F
38	100	14.5	8.4
93	200	15.4	8.9
150	300	16.2	9.4
205	400	17.0	9.8
260	500	17.8	10.3
315	600	18.7	10.8
370	700	19.6	11.3
425	800	20.2	11.7
480	900	21.1	12.2

Mechanical Properties

AISI Type 633: Elevated-Temperature Properties (Ref 9)

Test specimens were subzero cooled; tempered at 455 °C (850 °F)

Test temperature °C	°F	Tensile strength MPa	ksi	Yield strength(a) MPa	ksi	Elongation(b), %
27	80	1400	203	1170	170	13.0
205	400	1295	188	970	141	8.5
315	600	1305	189	940	136	7.0
370	700	1310	190	880	128	8.0
425	800	1280	186	860	125	9.5
580	900	1145	166	765	111	9.0
640	1000	730	106	585	85	16.0

(a) 0.2% offset. (b) In 50 mm (2 in.)

AISI Type 633: Charpy V-Notch Impact Strength (Ref 9)

Test specimens were subzero cooled and tempered

Test temperature °C	°F	Tempering temperature °C	°F	Impact energy J	ft·lb
−195	−320	455	850	5	4
		540	1000	8	6
−73	−100	455	850	11	8
		540	1000	...	...
21	70	455	850	19	14
		540	1000	34	25
100	212	455	850	33	24

AISI Type 633: Effect of Long-Time Exposure Under Stress at Elevated Temperatures on Room-Temperature Properties (Ref 9)

Test specimens were subzero cooled; tempered at 455 °C (850 °F)

Exposure stress MPa	ksi	Tensile strength MPa	ksi	Yield strength(a) MPa	ksi	Elongation(b), %
Exposed at room temperature						
...	...	1385	201	1090	158	12.0
Exposed at 315 °C (600 °F) for 1000 h						
410	60	1365	198	1115	162	14.0
620	90	1395	202	1220	177	13.0
970	140	1405	204	1385	201	12.0
Exposed at 370 °C (700 °F) for 1000 h						
410	60	1405	204	1165	169	11.0
620	90	1420	206	1240	180	11.0
1030	150	1570	228	1565	227	5.0
Exposed at 425 °C (800 °F) for 1000 h						
410	60	1515	220	1310	190	7.0
620	90	1475	214	1325	192	7.5
900	130	1515	220	1460	212	4.5(c)

(a) 0.2% offset. (b) In 50 mm (2 in.). (c) Broke outside gage marks

AISI Type 633: Stress-Rupture Properties (Ref 9)

Test specimens were subzero cooled and tempered

Test temperature °C	°F	Tempering temperature °C	°F	Stress for rupture in: 10 h MPa	10 h ksi	100 h MPa	100 h ksi	1000 h MPa	1000 h ksi
425	800	455	850	1295	188	1280	186	1260	183
		540	1000	910	132	895	130	875	127
480	900	455	850	965	140	815	118	655	95
		540	1000	760	110	710	103	675	98

Mechanical Properties (continued)

AISI Type 633: Typical Tensile Properties at Room Temperature (Ref 9)

Treatment	Tensile strength MPa	Tensile strength ksi	Yield strength(a) MPa	Yield strength(a) ksi	Elongation(b), %	Reduction in area, %	Hardness
Subzero cooled and tempered at 455 °C (850 °F)	1365	198	1115	162	15	49	48 HRC
Subzero cooled and tempered at 540 °C (1000 °F)	1125	163	1035	150	22	53	38 HRC
Double aged	1180	171	980	142	12	...	40 HRC
Annealed	1105	160	415	60	30	...	95 HRB

(a) 0.2% offset. (b) In 50 mm (2 in.)

Machining Data

For machining data on AISI type 633, refer to the preceding machining table for AISI type 630.

AISI Type 634

AISI Type 634: Chemical Composition

Element	Composition, %
Carbon	0.10-0.15
Manganese	0.50-1.25
Phosphorus	0.040 max
Sulfur	0.030 max
Silicon	0.50 max
Chromium	15-16
Nickel	4-5
Molybdenum	2.50-3.25
Nitrogen	0.07-0.13

Characteristics. AISI type 634 (S35500) is a chromium-nickel-molybdenum stainless steel which can be hardened by martensitic transformation and/or precipitation hardening. In the annealed condition, the alloy is austenitic and exhibits properties similar to the austenitic stainless steels. This condition offers good formability and weldability relative to ferritic or martensitic materials. After fabrication, appropriate heat treatments are employed to harden the alloy to high strengths similar to the hardenable martensitic stainless steels.

Type 634 offers good resistance to atmospheric corrosion and to a number of other mild chemical environments. Material in the double-aged or equalized and overtempered condition is susceptible to intergranular corrosion because of grain-boundary precipitation of carbides. When hardened by subzero cooling, this alloy is not subject to intergranular attack.

The machinability rating of type 634 is 30% in the annealed condition and about 25% in the hardened condition when compared to 100% machinability rating of free-machining AISI 1212.

Type 634 may be welded by the inert-gas-shielded, covered-electrode, or submerged-arc welding processes. Filler metal should be of the same composition as the base metal.

In the fully heat treated conditions type 634 can be limited cold formed, by cold forming operations such as bending, dimpling, and straightening. Forming characteristics in the annealed condition are similar to those of the austenitic stainless steels. Forming at a temperature of 150 °C (300 °F) may be used to take advantage of lower strength and lower work hardening rates at this temperature.

Typical Uses. Applications of AISI type 634 include structural components for high-speed aircraft, jet engine components, and high-strength, cold drawn wire and springs.

AISI Type 634: Similar Steels (U.S. and/or Foreign). UNS S35500; AMS 5547, 5549, 5594, 5742, 5743, 5744, 5780, 5781; ASTM A564, A579, A693, A705; SAE J467 (AM-355)

Physical Properties

AISI Type 634: Average Coefficients of Linear Thermal Expansion (Ref 7)

Temperature range °C	Temperature range °F	Coefficient: Hardened µm/m·K	Coefficient: Hardened µin./in.·°F	Coefficient: Subzero cooled and tempered µm/m·K	Coefficient: Subzero cooled and tempered µin./in.·°F
25-100	80-212	17.1	9.5	11.5	6.4
25-200	80-390	18.1	10.1	11.9	6.6
25-310	80-590	18.4	10.2	12.2	6.8
25-400	80-750	18.9	10.5	12.6	7.0
25-500	80-930	19.3	10.7	13.0	7.2
25-600	80-1110	19.8	11.0	13.0	7.2
25-1000	80-1830	20.3	11.3	13.3	7.4

AISI Type 634: Electrical Resistivity

Test specimens were subzero cooled and tempered at 455 °C (850 °F)

Temperature °C	Temperature °F	Electrical resistivity, nΩ·m
25	80	75.7
100	212	79.7
245	420	86.8
390	735	94.8
565	1050	103.6

Machining Data

For machining data on AISI type 634, refer to the preceding machining tables for AISI type 630.

Physical Properties (continued)

AISI Type 634: Thermal Treatment Temperatures

Treatment	Temperature range °C	°F
Forging		
Forging, preheat	870	1600
Forging, start	1175 max	2150 max
Forging, finish	925-980	1700-1800
Annealing		
Annealing	1010-1040	1850-1900
Hardening		
Conditioning	930-955	1710-1750
Subzero cooling	−73	−100
Subzero cooling, tempering	455-540	850-1000
Double aging	730-760	1350-1400
Equalize and overtemper	745-800	1375-1475
	540-595	1000-1100

AISI Type 634: Thermal Conductivity (Ref 7)

Temperature °C	°F	Thermal conductivity W/m·K	Btu/ft·h·°F
38	100	15.1	8.7
150	300	16.4	9.5
260	500	17.8	10.3
370	700	19.4	11.2
480	900	20.7	12.0

AISI Type 634: Selected Physical Properties (Ref 7)

Melting range	1370-1400 °C (2500-2550 °F)
Density	7.8 g/cm^3 (0.28 lb/in.^{3})
Specific heat at 0-100 °C (32-212 °F)	502 J/kg·K (0.12 Btu/lb·°F)

Mechanical Properties

AISI Type 634: Room-Temperature Properties of Bar After Exposure to Elevated Temperature (Ref 7)

Test specimens were 9.53-mm (0.375-in.) diam

Specimen mm	in.	Exposure temperature °C	°F	Exposure time, h	Tensile strength MPa	ksi	Yield strength(a) MPa	ksi	Elongation(b), %	Reduction in area, %
Subzero cooled and tempered at 540 °C (1000 °F)										
133(c)	5.25(c)	...	...	...	1158	168.0	1103	160	18	52
		260	500	1000	1169	169.5	1110	161	18	54
		370	700	1000	1344	195.0	1289	187	15	47
102(c)	4.00(c)	...	...	...	1134	164.5	1076	156.0	18	57
		260	500	100	1124	163.0	1072	155.5	19	58
		260	500	1000	1127	163.5	1096	159.0	18	58
		370	700	100	1186	172.0	1141	165.5	18	54
		370	700	1000	1227	178.0	1207	175.0	17	54
25(d)	1.00(d)	...	...	...	1148	166.5	1100	159.5	19	57
		455	850	24	1217	176.5	1172	170.0	19	59
		455	850	100	1269	184.0	1241	180.0	18	55
		455	850	1000	1376	199.5	1300	188.5	17	49
Subzero cooled and tempered at 455 °C (850 °F)										
25(d)	1.00(d)	...	...	...	1407	204.0	1207	175.0	16	50
		455	850	24	1396	202.5	1245	180.5	16	52
		455	850	100	1413	205.0	1310	190.0	15	50
		455	850	1000	1441	209.0	1386	201.0	16	50

(a) 0.2% offset. (b) Gage length is 4 × diam of the test specimen. (c) Square. (d) Round

AISI Type 634: Charpy V-Notch Impact Properties (Ref 7)

Test specimens were subzero cooled and tempered

Test temperature °C	°F	Tempering temperature °C	°F	Impact energy J	ft·lb
−195	−320	455	850	4	3
		540	1000	14	10
−73	−100	455	850	14	10
		540	1000	34	25
−12	10	455	850	20	15
		540	1000	52	38
21	70	455	850	23	17
		540	1000	62	46
100	212	455	850	26	19
		540	1000	68	50

AISI Type 634: Creep Strength of Sheet (Ref 7)

Test specimens were subzero cooled and tempered at 455 °C (850 °F)

Test temperature °C	°F	Stress for permanent deformation of: 0.1% in 1000 h MPa	ksi	0.2% in 1000 h MPa	ksi
315	600	848	123	1000	145
370	700	724	105	862	125
425	800	421	61	627	91

Mechanical Properties (continued)

AISI Type 634: Room-Temperature Tensile Properties of Sheet and Bar After Exposure to Elevated Temperature Under Stress (Ref 7)

Specimen mm	Specimen in.	Exposure temperature °C	Exposure temperature °F	Exposure stress MPa	Exposure stress ksi	Exposure time, h	Tensile strength MPa	Tensile strength ksi	Yield strength(a) MPa	Yield strength(a) ksi	Elongation(b), %	Reduction in area, %
Subzero cooled and tempered at 455 °C (850 °F)												
Sheet												
0.69	0.027	...	...	...	...	...	1520	220.5	1286	186.5	8	...
		315	600	689	100	502	1617	234.5	1424	206.5	11	...
		315	600	862	125	517	1620	235.0	1469	213.0	7	...
1.27	0.050	...	...	...	...	...	1458	211.5	1169	169.5	12	...
		315	600	455	66.0	1001	1472	213.5	1186	172.0	12	...
		315	600	683	99.0	1001	1475	214.0	1238	179.5	11	...
		315	600	1024	148.5	1001	1469	213.0	1424	206.5	8	...
		370	700	445	64.5	1003	1496	217.0	1227	178.0	11	...
		370	700	669	97.0	1072	1507	218.5	1300	188.5	12	...
		370	700	1003	145.5	1072	1503	218.0	1496	217.0	6	...
		425	800	427	62.0	1072	1572	228.0	1379	200.0	13	...
		425	800	641	93.0	1016	1548	224.5	1410	204.5	13	...
		425	800	965	140.0	1007	1565	227.0	1524	221.0	9	...
Round bar												
25	1.0	...	...	...	...	...	1462	212.0	1155	167.5	21	54
		315	600	710.1	103.0	1000	1465	212.5	1265	183.5	9	54
		370	700	717.0	104.0	1000	1500	217.5	1324	192.0	9	43
Subzero cooled and tempered at 540 °C (1000 °F)												
Round bar												
25	1.0	...	...	...	...	...	1276	185.0	1134	164.5	16	57
		370	700	676	98.0	5400	1489	216.0	1382	200.5	15	61
		...	...	...	...	...	1213	176.0	1065	154.5	18	60
		370	700	641	93.0	5400	1403	203.5	1096	159.0	16	51
		...	...	...	...	...	1317	191.0	1213	176.0	9	54
		(c)	(c)	413	60.0	(c)	1427	207.0	1365	198.0	11	53
		...	...	...	...	...	1372	199.0	1255	182.0	7	53
		(c)	(c)	413	60.0	(c)	1544	224.0	1482	215.0	5	50
		(c)	(c)	827	120.0	(c)	1551	225.0	...	...	5	49

(a) 0.2% offset. (b) In 50 mm (2 in.). (c) Cycled daily for 15 days between −73 °C and 425 °C (−100 °F and 800 °F); held at 425 °C (800 °F) about 15 h during each cycle

AISI Type 634: Stress-Rupture Properties for Flat Rolled Bar Stock (Ref 7)

Condition	Test temperature °C	Test temperature °F	Stress for rupture in: 10 h MPa	10 h ksi	100 h MPa	100 h ksi	1000 h MPa	1000 h ksi
Subzero cooled and tempered at 455 °C (850 °F)	425	800	1296	188	1276	185	1255	182
	480	900	972	141	827	120	676	98
	540	1000	607	88	496	72	400	58
Subzero cooled and tempered at 540 °C (1000 °F)	425	800	965	140	951	138	931	135
	480	900	758	110	724	105	683	99
	540	1000	579	84	490	71	414	60

AISI Type 651

AISI Type 651: Chemical Composition

Element	Composition, %
Carbon	0.28-0.35
Manganese	0.75-1.50
Phosphorus	0.040 max
Sulfur	0.030 max
Silicon	0.30-0.80
Chromium	18-21
Nickel	8-11
Molybdenum	1.00-1.75
Copper	0.50 max
Niobium	0.25-0.60
Titanium	0.10-0.35
Tungsten	1.00-1.75

Characteristics. AISI type 651 (K63198) is a 19% chromium, 9% nickel, iron-base, austenitic alloy. This steel contains relatively small amounts of alloying elements to provide an economical, high-strength, heat-resisting alloy. This grade is not hardenable by heat treatment. Bar and forgings are often warm formed to higher strength levels than are obtainable in the annealed (solution-treated) condition. Warm working is suitable for increasing strength up to the 650 to 705 °C (1200 to 1300 °F) service temperature area.

Corrosion resistance is similar to that of the higher carbon, nonstabilized, austenitic stainless steels. When annealed and rapidly quenched or otherwise properly heat treated to avoid sensitization, type 651 exhibits ample corrosion resistance. Oxidation resistance is good up to 955 °C (1750 °F) in continuous service, and up to 780 °C (1450 °F) for intermittent service.

Type 651 is readily cold formed in the annealed condition at a hardness of about 88 to 95 HRB. Cold rolled and stress relieved material is not suitable for severe cold forming. This alloy work hardens rapidly, and will probably require annealing at some stage of forming.

The machinability rating of this alloy is 45% of that for AISI 1212. Type 651 is most readily machined when cold drawn 15 to 20% and stress relieved to a hardness of 220 to 240 HB. Warm working followed by stress relieving to a similar hardness also provides good machinability.

Type 651 is readily welded using the common welding processes. Filler metal should be of the same composition as the base metal. It is recommended that fusion-welded components be stress relieved at 900 to 980 °C (1650 to 1800 °F). If the part is to be employed at temperatures within the carbide precipitation range, the 900 °C (1650 °F) temperature should be used.

Typical Uses. Applications of AISI type 651 include rotor forgings, buckets, bolts, aircraft exhaust manifolds, collector rings, jet engine and gas turbine tail cones, and similar heat-resisting assemblies.

AISI Type 651: Similar Steels (U.S. and/or Foreign). UNS K63198; AMS 5369, 5526, 5527, 5579, 5720, 5721, 5722; ASTM A453, A457, A458, A477; SAE J467 (19-9 DL); MIL SPEC MIL-S-46042

Physical Properties

AISI Type 651: Thermal Treatment Temperatures

Treatment	Temperature °C	Temperature °F
Forging, start	1175	2150
Forging, finish	650 min	1200 min
Warm working	650-815	1200-1500

AISI Type 651: Thermal Conductivity (Ref 7)

Temperature °C	Temperature °F	Thermal conductivity W/m·K	Thermal conductivity Btu/ft·h·°F
21	70	13.5	7.8
425	800	18.5	10.7
650	1200	21.3	12.3

AISI Type 651: Average Coefficients of Linear Thermal Expansion (Ref 7)

Temperature range °C	Temperature range °F	Coefficient μm/m·K	Coefficient μin./in.·°F
21-93	70-200	15.3	8.5
21-205	70-400	16.3	9.1
21-315	70-600	16.7	9.3
21-425	70-800	17.3	9.6
21-540	70-1000	17.6	9.8
21-815	70-1500	18.0	10.0

AISI Type 651: Selected Physical Properties (Ref 7)

Melting range	1420-1435 °C (2590-2615 °F)
Density	7.94 g/cm³ (0.287 lb/in.³)
Specific heat at 0-100 °C (32-212 °F)	420 J/kg·K (0.10 Btu/lb·°F)

Mechanical Properties

AISI Type 651: Stress-Rupture Strength and Creep Rate (Ref 7)

Test temperature °C	Test temperature °F	Stress to product rupture in: 10 h MPa	10 h ksi	100 h MPa	100 h ksi	1000 h MPa	1000 h ksi	10 000 h MPa	10 000 h ksi	Stress for minimum creep rate of: MPa	ksi	1% in 100 000 h MPa	1% in 100 000 h ksi
19- to 76-mm (0.75 to 3-in.) diam bar, warm worked and stress relieved													
425	800	...	...	...	...	...	...	...	...	524	76.0	228	33.0
540	1000	503	73.0	441	64.0	386	56.0	331	48.0	276	40.0	131	19.0
650	1200	345	50.0	303	44.0	255	37.0	214	31.0	134	19.5	68	9.8
19- to 76-mm (0.75 to 3-in.) diam bar, solution treated and aged													
730	1350	210	30.5	155	22.5	117	17.0	86	12.5	68.9	10.0	36	5.2
815	1500	138	20.0	89.6	13.0	59	8.6	39	5.6	38	5.5	...	...

Mechanical Properties (continued)

AISI Type 651: Typical Tensile Properties at Room and Elevated Temperatures (Ref 7)

Test specimens were 60-mm (2.4-in.) diam bar; warm rolled and stress relieved at 730 °C (1350 °F) for 8 h, with hardness of 229 to 241 HB

Test temperature °C	°F	Tensile strength MPa	ksi	Yield strength(a) MPa	ksi	Elongation(b), %	Reduction in area, %
24	75	838	121.5	579	84.0	43.0	55.5
150	300	703	102.0	517	75.0	30.0	52.5
260	500	707	102.5	496	72.0	27.5	49.5
370	700	672	97.5	455	66.0	31.5	49.5
480	900	634	92.0	427	62.0	29.5	48.5
540	1000	614	89.0	396	57.5	29.5	46.0
650	1200	534	77.5	362	52.5	30.5	47.5
705	1300	459	66.5	352	51.0	29.0	49.0
760	1400	365	53.0	317	46.0	30.5	56.5
815	1500	331	48.0	279	40.5	29.0	62.0

(a) 0.2% offset. (b) In 50 mm (2 in.)

AISI Type 651: Stress-Rupture Strength of Cold Rolled Sheet (Ref 7)

Test specimens were 1-mm (0.040-in.) thick

Treatment	Test temperature °C	°F	Stress to rupture in: 10 h MPa	10 h ksi	100 h MPa	100 h ksi
Cold rolled; heated to 980 °C (1800 °F); annealed at 650 °C (1200 °F) for 5 min	650	1200	276	40.0	217	31.5
Cold rolled; heated to 900 °C (1650 °F); annealed at 650 °C (1200 °F) for 30 min	650	1200	259	37.5	193	28.0

AISI Type 651: Charpy V-Notch Impact Properties (Ref 7)

Treatment	Hardness, HB	Impact energy at test temperature of: −195 °C (−323 °F) J	ft·lb	−76 °C (−104 °F) J	ft·lb	24 °C (76 °F) J	ft·lb
Hot rolled and stress relieved at 650 °C (1200 °F)(a)	228	...	...	...	...	62	46
Annealed at 980 °C (1800 °F); air cooled	196	34	25	52	38	72	53
Warm worked and stress relieved	311	15	11	26	19	37	27
Cold drawn and stress relieved	286	18	13	26	19	30	22

(a) Impact energy at 260 °C (500 °F), 72 J (53 ft·lb); at 540 °C (1000 °F), 71 J (52 ft·lb); at 650 °C (1200 °F), 76 J (56 ft·lb); at 815 °C (1500 °F), 81 J (60 ft·lb)

AISI Type 651: Typical Properties of Cold Rolled Sheet at Room and Elevated Temperatures (Ref 7)

Test specimens were 0.89- to 1.3-mm (0.035- to 0.050-in.) thick

Test temperature °C	°F	Tensile strength MPa	ksi	Yield strength(a) MPa	ksi	Elongation(b), %
Annealed at 980 °C (1800 °F); air cooled; hardness of 88 HRB						
24	75	752	109.0	365	53.0	39.0
150	300	600	87.0	317	46.0	30.0
315	600	583	84.5	296	43.0	25.0
425	800	572	83.0	272	39.5	25.0
480	900	555	80.5	269	39.0	24.0
650	1200	434	63.0	241	35.0	25.0
815	1500	234	34.0	172	25.0	47.0
Cold rolled and stress relieved, with hardness of 35 HRC						
24	75	965	140.0	772	112.0	19.5
425	800	827	120.0	724	105.0	6.5

(a) 0.2% offset. (b) In 50 mm (2 in.)

Machining Data (Ref 2)

AISI Type 651: Turning (Cutoff and Form Tools)

Tool material	Speed, m/min (ft/min)	Feed per revolution for cutoff tool width of: 1.5 mm (0.062 in.) mm (in.)	3 mm (0.125 in.) mm (in.)	6 mm (0.25 in.) mm (in.)	Feed per revolution for form tool width of: 12 mm (0.5 in.) mm (in.)	18 mm (0.75 in.) mm (in.)	25 mm (1 in.) mm (in.)	35 mm (1.5 in.) mm (in.)	50 mm (2 in.) mm (in.)
Solution treated, with hardness of 180 to 230 HB									
Any premium high speed steel (T15, M33, M41-M47)	8 (25)	0.050 (0.002)	0.102 (0.004)	0.13 (0.005)	0.075 (0.003)	0.063 (0.0025)	0.050 (0.002)	0.038 (0.0015)	0.025 (0.001)
C-2 carbide	29 (95)	0.050 (0.002)	0.102 (0.004)	0.13 (0.005)	0.075 (0.003)	0.063 (0.0025)	0.050 (0.002)	0.038 (0.0015)	0.025 (0.001)
Solution treated and aged, with hardness of 250 to 320 HB									
Any premium high speed steel (T15, M33, M41-M47)	6 (20)	0.050 (0.002)	0.102 (0.004)	0.13 (0.005)	0.075 (0.003)	0.063 (0.0025)	0.050 (0.002)	0.038 (0.0015)	0.025 (0.001)
C-2 carbide	24 (80)	0.050 (0.002)	0.102 (0.004)	0.13 (0.005)	0.075 (0.003)	0.063 (0.0025)	0.050 (0.002)	0.038 (0.0015)	0.025 (0.001)

AISI Type 651: End Milling (Profiling)

Tool material	Depth of cut mm	Depth of cut in.	Speed m/min	Speed ft/min	Feed per tooth for cutter diameter of: 10 mm (0.375 in.) mm	10 mm (0.375 in.) in.	12 mm (0.5 in.) mm	12 mm (0.5 in.) in.	18 mm (0.75 in.) mm	18 mm (0.75 in.) in.	25-50 mm (1-2 in.) mm	25-50 mm (1-2 in.) in.
Solution treated, with hardness of 180 to 230 HB												
Any premium high speed steel (T15, M33, M41-M47)	0.5	0.020	12	40	0.025	0.001	0.025	0.001	0.050	0.002	0.050	0.002
	1.5	0.060	9	30	0.050	0.002	0.050	0.002	0.075	0.003	0.102	0.004
	diam/4	diam/4	8	25	0.038	0.0015	0.038	0.0015	0.050	0.002	0.075	0.003
	diam/2	diam/2	6	20	0.025	0.001	0.025	0.001	0.050	0.002	0.063	0.0025
C-2 carbide	0.5	0.020	46	150	0.025	0.001	0.025	0.001	0.050	0.002	0.050	0.002
	1.5	0.060	34	110	0.025	0.001	0.050	0.002	0.075	0.003	0.102	0.004
	diam/4	diam/4	29	95	0.025	0.001	0.038	0.0015	0.050	0.002	0.063	0.0025
	diam/2	diam/2	27	90	...	...	...	...	0.038	0.0015	0.005	0.002
Solution treated and aged, with hardness of 250 to 320 HB												
Any premium high speed steel (T15, M33, M41-M47)	0.5	0.020	9	30	0.025	0.001	0.025	0.001	0.050	0.002	0.050	0.002
	1.5	0.060	8	25	0.050	0.002	0.050	0.002	0.075	0.003	0.102	0.004
	diam/4	diam/4	6	20	0.038	0.0015	0.038	0.0015	0.063	0.0025	0.075	0.003
	diam/2	diam/2	5	15	0.025	0.001	0.025	0.001	0.050	0.002	0.063	0.0025
C-2 carbide	0.5	0.020	30	100	0.025	0.001	0.025	0.001	0.050	0.002	0.050	0.002
	1.5	0.060	23	75	0.025	0.001	0.050	0.002	0.075	0.003	0.075	0.003
	diam/4	diam/4	20	65	0.025	0.001	0.038	0.0015	0.050	0.002	0.063	0.0025
	diam/2	diam/2	18	60	...	...	...	...	0.038	0.0015	0.050	0.002

AISI Type 651: Turning (Single Point and Box Tools)

Depth of cut mm	Depth of cut in.	High speed steel(a) Speed m/min	Speed ft/min	Feed mm/rev	Feed in./rev	Uncoated carbide Speed, brazed m/min	Speed, brazed ft/min	Speed, inserted m/min	Speed, inserted ft/min	Feed mm/rev	Feed in./rev
Solution treated, with hardness of 180 to 230 HB											
0.8	0.030	12	45	0.13	0.005	49(b)	160(b)	58(b)	190(b)	0.13	0.005
2.5	0.100	9	35	0.18	0.007	41(c)	135(c)	49(c)	160(c)	0.18	0.007
5.0	0.200	...	...	...	...	30(c)	100(c)	37(c)	120(c)	0.25	0.010
Solution treated and aged, with hardness of 250 to 320 HB											
0.8	0.030	12	40	0.13	0.005	44(b)	145(b)	52(b)	170(b)	0.13	0.005
2.5	0.100	9	30	0.18	0.007	37(c)	120(c)	44(c)	145(c)	0.18	0.007
5.0	0.200	...	...	...	...	26(c)	85(c)	30(c)	100(c)	0.25	0.010

(a) Any premium high speed steel (T15, M33, M41 to M47). (b) Carbide tool material: C-3. (c) Carbide tool material: C-2

Machining Data (Ref 2) (continued)

AISI Type 651: Reaming

Based on 4 flutes for 3- and 6-mm (0.125- and 0.25-in.) reamers, 6 flutes for 12-mm (0.5-in.) reamers, and 8 flutes for 25-mm (1-in.) and larger reamers

Tool material	Speed m/min	Speed ft/min	3 mm (0.125 in.) mm	3 mm (0.125 in.) in.	6 mm (0.25 in.) mm	6 mm (0.25 in.) in.	12 mm (0.5 in.) mm	12 mm (0.5 in.) in.	25 mm (1 in.) mm	25 mm (1 in.) in.	35 mm (1.5 in.) mm	35 mm (1.5 in.) in.	50 mm (2 in.) mm	50 mm (2 in.) in.
			Feed per revolution for reamer diameter of:											
Solution treated, with hardness of 180 to 230 HB														
Roughing														
Any premium high speed steel (T15, M33, M41-M47)	9	30	0.075	0.003	0.15	0.006	0.25	0.010	0.30	0.012	0.36	0.014	0.40	0.016
C-2 carbide	21	70	0.050	0.002	0.15	0.006	0.25	0.010	0.30	0.012	0.36	0.014	0.40	0.016
Finishing														
Any premium high speed steel (T15, M33, M41-M47)	8	25	0.075	0.003	0.102	0.004	0.15	0.006	0.20	0.008	0.25	0.010	0.30	0.012
C-2 carbide	12	40	0.075	0.003	0.102	0.004	0.15	0.006	0.20	0.008	0.25	0.010	0.30	0.012
Solution treated and aged, with hardness of 250 to 320 HB														
Roughing														
Any premium high speed steel (T15, M33, M41-M47)	8	25	0.075	0.003	0.15	0.006	0.25	0.010	0.30	0.012	0.36	0.014	0.40	0.016
C-2 carbide	20	65	0.050	0.002	0.15	0.006	0.25	0.010	0.30	0.012	0.36	0.014	0.40	0.016
Finishing														
Any premium high speed steel (T15, M33, M41-M47)	6	20	0.075	0.003	0.102	0.004	0.15	0.006	0.20	0.008	0.25	0.010	0.30	0.012
C-2 carbide	9	30	0.075	0.003	0.102	0.004	0.15	0.006	0.20	0.008	0.25	0.010	0.30	0.012

AISI Type 651: Drilling

Tool material	Speed m/min	Speed ft/min	3 mm (0.125 in.) mm	3 mm (0.125 in.) in.	6 mm (0.25 in.) mm	6 mm (0.25 in.) in.	12 mm (0.5 in.) mm	12 mm (0.5 in.) in.	18 mm (0.75 in.) mm	18 mm (0.75 in.) in.	25 mm (1 in.) mm	25 mm (1 in.) in.
			Feed per revolution for nominal hole diameter of:									
Solution treated, with hardness of 180 to 230 HB												
Any premium high speed steel (T15, M33, M41-M47)	8	25	0.050	0.002	0.102	0.004	0.15	0.006	0.20	0.008	0.25	0.010
Solution treated and aged, with hardness of 250 to 320 HB												
Any premium high speed steel (T15, M33 M41-M47)	6	20	0.050	0.002	0.102	0.004	0.15	0.006	0.20	0.008	0.20	0.008

AISI Type 651: Boring

Depth of cut mm	Depth of cut in.	High speed steel(a) Speed m/min	High speed steel(a) Speed ft/min	High speed steel(a) Feed mm/rev	High speed steel(a) Feed in./rev	Uncoated carbide Speed, brazed m/min	Uncoated carbide Speed, brazed ft/min	Uncoated carbide Speed, inserted m/min	Uncoated carbide Speed, inserted ft/min	Uncoated carbide Feed mm/rev	Uncoated carbide Feed in./rev
Solution treated, with hardness of 180 to 230 HB											
0.25	0.010	14	45	0.050	0.002	40(b)	130(b)	47(b)	155(b)	0.050	0.002
1.25	0.050	11	35	0.102	0.004	32(c)	105(c)	38(c)	125(c)	0.102	0.004
2.50	0.100	5	15	0.15	0.006	21(c)	70(c)	26(c)	85(c)	0.150	0.006
Solution treated and aged, with hardness of 250 to 320 HB											
0.25	0.010	12	40	0.050	0.002	38(b)	125(b)	44(b)	145(b)	0.050	0.002
1.25	0.050	9	30	0.102	0.004	30(c)	100(c)	35(c)	115(c)	0.102	0.004
2.50	0.100	3	10	0.15	0.006	18(c)	60(c)	21(c)	70(c)	0.150	0.006

(a) Any premium high speed steel (T15, M33, M41 to M47). (b) Carbide tool material: C-3. (c) Carbide tool material: C-2

Machining Data (Ref 2) (continued)

AISI Type 651: Surface Grinding with a Horizontal Spindle and Reciprocating Table

Operation	Wheel speed m/s	Wheel speed ft/min	Table speed m/min	Table speed ft/min	Downfeed mm/pass	Downfeed in./pass	Crossfeed mm/pass	Crossfeed in./pass	Wheel identification
Solution treated, with hardness of 180 to 230 HB									
Roughing	15-20	3000-4000	15-30	50-100	0.025	0.001	0.65-6.5	0.025-0.250	A46HV
Finishing	15-20	3000-4000	15-30	50-100	0.013 max	0.0005 max	(a)	(a)	A46HV
Solution treated, with hardness of 250 to 320 HB									
Roughing	15-20	3000-4000	15-30	50-100	0.025	0.001	0.65-6.5	0.025-0.250	A46HV
Finishing	15-20	3000-4000	15-30	50-100	0.013 max	0.0005 max	(a)	(a)	A46HV

(a) Maximum 1/10 of wheel width

AISI Type 651: Cylindrical and Internal Grinding

Operation	Wheel speed m/s	Wheel speed ft/min	Work speed m/min	Work speed ft/min	Infeed on diameter mm/pass	Infeed on diameter in./pass	Traverse(a)	Wheel identification
Solution treated, with hardness of 180 to 230 HB								
Cylindrical grinding								
Roughing	15-20	3000-4000	15-46	50-150	0.013	0.0005	1/3	A60JV
Finishing	15-20	3000-4000	15-46	50-150	0.005 max	0.0002 max	1/6	A60JV
Internal grinding								
Roughing	15-20	3000-4000	15-46	50-150	0.013	0.0005	1/3	A60JV
Finishing	15-20	3000-4000	15-46	50-150	0.005 max	0.0002 max	1/6	A60JV
Solution treated and aged, with hardness of 250 to 320 HB								
Cylindrical grinding								
Roughing	15-20	3000-4000	15-46	50-150	0.013	0.0005	1/3	A60JV
Finishing	15-20	3000-4000	15-46	50-150	0.005 max	0.0002 max	1/6	A60JV
Internal grinding								
Roughing	15-20	3000-4000	15-46	50-150	0.013	0.0005	1/3	A60JV
Finishing	15-20	3000-4000	15-46	50-150	0.005 max	0.0002 max	1/6	A60JV

(a) Wheel width per revolution of work

AISI Type 651: Face Milling

Depth of cut mm	Depth of cut in.	High speed steel(a) Speed m/min	High speed steel(a) Speed ft/min	High speed steel(a) Feed/tooth mm	High speed steel(a) Feed/tooth in.	Uncoated carbide Speed, brazed m/min	Uncoated carbide Speed, brazed ft/min	Uncoated carbide Speed, inserted m/min	Uncoated carbide Speed, inserted ft/min	Uncoated carbide Feed/tooth mm	Uncoated carbide Feed/tooth in.
Solution treated, with hardness of 180 to 230 HB											
1	0.040	18	60	0.13	0.005	24	80	30	100	0.15	0.006
4	0.150	9	30	0.20	0.008	20	65	24	80	0.20	0.008
8	0.300	5	15	0.25	0.010	14	45	18	60	0.25	0.010
Solution treated and aged, with hardness of 250 to 320 HB											
1	0.040	12	40	0.102	0.004	18	60	21	70	0.15	0.006
4	0.150	6	20	0.15	0.006	17	55	20	65	0.20	0.008
8	0.300	5	15	0.20	0.008	11	35	12	40	0.25	0.010

(a) Any premium high speed steel (T15, M33, M41 to M47)

AISI Type 660

AISI Type 660: Chemical Composition

Element	Composition, %
Carbon	0.08 max
Manganese	2.00 max
Phosphorus	0.040 max
Sulfur	0.030 max
Silicon	1.00 max
Chromium	13.5-16
Copper	0.25 max
Nickel	24-27
Titanium	1.90-2.35
Aluminum	0.035 max
Vanadium	0.10-0.50
Boron	0.001-0.010

Characteristics. AISI type 660 (K66268) is a corrosion- and heat-resistance, austenitic, iron-nickel-chromium steel that is age hardenable because of titanium and nickel in the content. This steel is recommended for applications requiring high strength and corrosion resistance at temperatures up to 705 °C (1300 °F), and for lower stress applications at higher temperatures. The type 660 alloy is also used in high-strength nonmagnetic applications because it remains essentially nonmagnetic even after severe cold working.

Type 660 has good oxidation resistance for continuous service up to 815 °C (1500 °F), and intermittent service up to 980 °C (1800 °F). In the solution-treated condition, type 660 work hardens at about the same rate as type 310 stainless steel. Due to the work-hardening effects during cold drawing and other cold forming operations, intermediate anneals at 980 °C (1800 °F) are required.

Because it is gummy in the soft, solution-treated condition, type 660 is generally machined after it has been partially or fully aged. Material cold worked after solution treating also exhibits good machining characteristics.

Welding of type 660 is performed preferably in the solution-treated condition. This alloy is susceptible to hot cracking, particularly in the aged condition. The inert gas and shielded-metal-arc methods should be used to prevent loss of titanium and hardenability. Resistance seam and spot welds can be made using high current and high electrode pressures. Flash welding can be performed on nearly all section sizes.

Type 660 can be successfully brazed in a pure, dry, hydrogen atmosphere, or in a vacuum. Nickel plating before brazing is helpful in promoting wetting. Improvements in ductility are accomplished by solution heat treating at 900 or 1010 °C (1650 or 1800 °F) after the brazing cycle.

Typical Uses. Applications of AISI type 660 include turbine shafts, housings and castings, afterburner and tail cone parts, welded assemblies, bolts, nuts, springs, and various other hardware in jet engines, gas turbines, and turbo superchargers.

AISI Type 660: Similar Steels (U.S. and/or Foreign). UNS K66286; AMS 5525, 5731, 5732, 5734, 5735, 5736, 5737, 5804, 5805; ASME SA638; ASTM A453, A638; SAE J467 (A286); (W. Ger.) DIN 1.4980

Physical Properties

AISI Type 660: Thermal Conductivity (Ref 7)

Temperature °C	Temperature °F	Thermal conductivity W/m·K	Thermal conductivity Btu/ft·h·°F
21	70	12.6	7.3
150	300	15.1	8.7
425	800	20.2	11.7
600	1110	23.79	13.75
650	1200	24.7	14.3

AISI Type 660: Selected Physical Properties (Ref 7)

Melting range	1370-1425 °C (2500-2600 °F)
Density	7.92 g/cm³ (0.286 lb/in.³)
Specific heat at 0-100 °C (32-212 °F)	460 J/kg·K (0.11 Btu/ft·h·°F)

AISI Type 660: Electrical Resistivity (Ref 7)

Temperature °C	Temperature °F	Electrical resistivity, nΩ·m
30	87	91.0
540	1000	115.6
650	1200	118.8
730	1350	120.1
815	1500	122.4

AISI Type 660: Average Coefficients of Linear Thermal Expansion (Ref 7)

Temperature range		Coefficient			
		Solution treated		Solution treated and aged	
°C	°F	μm/m·K	μin./in.·°F	μm/m·K	μin./in.·°F
21-100	70-212	16.9	9.4	16.7	9.3
21-500	70-930	17.4	9.67	17.6	9.8
21-650	70-1200	17.6	9.79	...	...
21-750	70-1380	17.7	9.81	18.5	10.3
21-900	70-1650	19.4	10.8	...	...

AISI Type 660: Thermal Treatment Temperatures

Treatment	Temperature range °C	Temperature range °F
Forging, start	1095-1175	2000-2150
Forging, finish	925 min	1700 min
Solution treatment	900 or 1010	1650 or 1800
Hardening(a)	720	1325

(a) This alloy can also be cold worked and aged to produce very high strengths

Mechanical Properties

AISI Type 660: Short-Time Tensile Properties (Ref 8)

Test temperature °C	°F	Tensile strength MPa	ksi	Yield strength(a) 0.02% offset MPa	ksi	0.2% offset MPa	ksi	Elongation(a), %	Reduction in area, %
−195	−320	1441	209.0	...	...	889.4	129.0	35.0	...
−75	−100	1224	177.5	...	...	810.1	117.5	30.0	...
20	68	1007	146.0	569	82.5	703.2	102.0	25.0	36.8
205	400	1000	145.0	524	76.0	645	93.5	21.5	52.8
370	700	948.0	137.5	490	71.0	645	93.5	22.0	45.0
425	800	951.4	138.0	496	72.0	645	93.5	18.5	35.0
540	1000	903.2	131.0	427	62.0	603	87.5	18.5	31.2
595	1100	841.2	122.0	445	64.5	621	90.0	21.0	23.0
650	1200	713.6	103.5	431	62.5	607	88.0	13.0	14.5
705	1300	596	86.5	472	68.5	...	...	11.0	9.6
760	1400	441	64.0	307	44.5	...	...	18.5	23.4
815	1500	252	36.5	213	31.0	...	...	68.5	37.5

(a) In 50 mm (2 in.)

AISI Type 660: Tensile Properties of Cold Drawn and Aged Wire at Room Temperature (Ref 8)

Test specimens were cold drawn and aged for 16 h at temperature indicated

Aging temperature °C	°F	Tensile strength MPa	ksi	Yield strength(a) MPa	ksi	Elongation(b), %	Reduction in area, %
Wire diameter of 5.61 mm (0.221 in.), reduced in area by 22% from original diameter							
None	None	869	126	...	...	...	...
425	800	951	138	883	128	19.5	51.0
480	900	951	138	869	126	18.0	58.0
540	1000	992	144	855	124	21.5	58.0
595	1100	1158	168	1062	154	21.5	52.5
650	1200	1248	181	1124	163	16.0	45.0
705	1300	1296	188	1138	165	14.0	35.0
760	1400	1200	174	938	136	17.5	39.0
Wire diameter of 4.45 mm (0.175 in.), reduced in area by 51% from original diameter							
None	None	1124	163	...	...	...	...
425	800	1220	177	1179	171	11.0	58.0
480	900	1262	183	1145	166	9.0	60.0
540	1000	1317	191	1179	171	11.0	53.5
595	1100	1393	202	1234	179	9.0	55.0
650	1200	1531	222	1372	199	9.0	48.0
705	1300	1413	205	1186	172	15.0	30.0
760	1400	1131	164	834	121	20.0	38.5
Wire diameter of 2.84 mm (0.112 in.), reduced in area by 80% from original diameter							
None	None	1310	190	...	...	...	...
425	800	1482	215	1351	196	9.5	40.0
480	900	1565	227	1420	206	9.5	31.0
540	1000	...	...	...	...	...	...
595	1100	1641	238	1475	214	6.0	5.5
650	1200	1717	249	1589	230	3.0	12.0
705	1300	1269	184	1034	150	13.0	26.5

(a) 0.2% offset. (b) Gage length is 4 × diam of the test specimen

AISI Type 660: Stress-Rupture Properties (Ref 7)

Test temperature °C	°F	Stress for rupture in: 100 h MPa	ksi	1000 h MPa	ksi	10 000 h(a) MPa	ksi	Elongation(b), % 100 h	1000 h
540	1000	689	100	600	87	525	76	3	3
595	1100	565	82	490	71	370	54	3	3
650	1200	425	62	315	46	235	34	5	9
705	1300	305	44	205	30	105	15	12	24
730	1350	240	35	150	22	...	...	28	35
815	1500	90	13	...	...	...	...	55	...

(a) Extrapolated data. (b) Gage length is 4 × diam of the test specimen

AISI Type 660: Creep-Strength Data (Ref 7)

Test temperature °C	°F	Stress for creep rate of: 0.5% in 100 h MPa	ksi	1% in 100 h MPa	ksi	0.5% in 1000 h MPa	ksi	1% in 1000 h MPa	ksi
540	1000	560	81	634	92.0	540	78	586	85.0
595	1100	525	76	552	80.0	470	68	483	70.0
650	1200	365	53	414	60.0	240	35	283	41.0
705	1300	205	30	245	35.5	...	...	155	22.5

AISI Type 660: Charpy V-Notch Impact Energy (Ref 8)

Test specimens were 22-mm (0.875-in.) diam bar; solution treated at 980 °C (1800 °F); oil quenched; heated to 720 °C (1325 °F); held for 16 h; air cooled

Test temperature °C	°F	Impact energy J	ft·lb
−195	−320	77.3	57.0
−75	−100	92.2	68.0
27	80	86.8	64.0
210	410	80.7	59.5
430	810	69.8	51.5
545	1010	61.7	45.5
600	1115	59.7	44.0
655	1215	47.5	35.0
675	1250	55.6	41.0
690	1275	50.8	37.5
712	1315	59.7	44.0
740	1365	81.3	60.0
790	1450	70.5	52.0

Machining Data

For machining data on AISI type 660, refer to the preceding machining tables for AISI type 651.

AISI Type 661

AISI Type 661: Chemical Composition

Element	Composition, %
Carbon	0.08-0.16
Manganese	1.00-2.00
Phosphorus	0.040 max
Sulfur	0.030 max
Silicon	1.00 max
Chromium	20.00-22.50
Nickel	19.00-21.00
Molybdenum	2.50-3.50
Niobium and tantalum	0.75-1.25
Tungsten	2.00-3.00
Nitrogen	0.20 max
Cobalt	18.50-21.00

Characteristics. AISI type 661 (N-155, R30155) is an iron-base, austenitic alloy containing approximately 20% each of chromium, nickel, and cobalt, and lesser amounts of molybdenum, tungsten, and niobium. This stainless steel is primarily a solid-solution strengthened alloy, but, to a limited extent, does respond to precipitation-aging treatments. Type 661 is used over a wide range of temperatures from subzero to about the 980 to 1040 °C (1800 to 1900 °F) range.

Maximum corrosion resistance is obtained in the solution-treated condition. Stress relieving is detrimental to corrosion resistance because of the precipitation of carbides. In the solution-treated condition, the resistance of type 661 to weak hydrochloric acid and sulfuric acid is better than that for the austenitic stainless steels, whereas resistance to nitric acid is about equal.

Resistance to salt spray and to all atmospheres encountered in jet engine and turbo supercharger operations is good. Oxidation resistance is good for continuous service up to 1040 °C (1900 °F) and up to 870 °C (1600 °F) in intermittent service.

The machinability rating for type 661 is about 18% compared to AISI 1212. Type 661 can be readily cold formed by operations such as drawing, hydroforming, spinning, bending, and roll forming. In the annealed condition, sheet and strip form without developing an orange-peel surface. Reductions between about 2 to 6% should be avoided, if possible, to prevent large grains from developing during subsequent annealing. This alloy work hardens rapidly and usually requires annealing at 1175 °C (2150 °F).

Type 661 is welded by the shielded-metal-arc, shielded-inert-gas, and resistance welding methods. Filler metal should be the same composition as that of the base metal, or may be Hastelloy W, the latter being preferred for joining type 661 to other nickel- or iron-base super alloys.

Typical Uses. Applications of AISI type 661 include turbine rotors, shafts and blades, exhaust valves, bolts, rivets, tail pipes and cones, afterburner parts, exhaust manifolds, nozzles, and combustion chambers.

AISI Type 661: Similar Steels (U.S. and/or Foreign). UNS R30155; AMS 5376, 5531, 5532, 5585, 5768, 5769, 5794, 5795; ANSI G81.40; ASTM A639, A567; (W. Ger.) DIN 1.4971

Physical Properties

AISI Type 661: Thermal Treatment Temperatures

Treatment	Temperature range °C	°F
Forging, start	1175 max	2150 max
Forging, finish	955 max	1750 max
Solution treatment	1175	2150
Precipitation treatment	650-900	1200-1650

AISI Type 661: Thermal Conductivity (Ref 7)

Temperature °C	°F	Thermal conductivity W/m·K	Btu/ft·h·°F
21	70	12.3	7.1
95	200	13.1	7.6
205	400	14.5	8.4
315	600	16.1	9.3
425	800	17.6	10.2
540	1000	19.2	11.1
650	1200	20.8	12.0

AISI Type 661: Selected Physical Properties (Ref 7)

Melting range	1300-1355 °C (2375-2475 °F)
Density	8.25 g/cm^3 (0.298 lb/in.^{3})
Specific heat at 21-100 °C (70-212 °F)	435 J/kg·K (0.104 Btu/lb·°F)

AISI Type 661: Average Coefficients of Linear Thermal Expansion (Ref 7)

Temperature range °C	°F	Coefficient μm/m·K	μin./in.·°F
21-95	70-200	14.0	7.8
21-205	70-400	15.3	8.5
21-315	70-600	15.3	8.5
21-540	70-1000	16.4	9.1
21-650	70-1200	17.1	9.5
21-760	70-1400	17.5	9.7
21-870	70-1600	17.8	9.9
21-980	70-1800	18.2	10.1

AISI Type 661: Electrical Resistivity (Ref 7)

Temperature °C	°F	Electrical resistivity, nΩ·m
25	77	94.2
200	390	101.7
400	750	109.2
600	1110	115.0
800	1470	118.0
1000	1830	122.7

Mechanical Properties

AISI Type 661: Tensile Properties of Bar at Room and Subzero Temperatures (Ref 7)

Test specimens were 16-mm (0.63-in.) diam bar; heated to 1190 °C (2175 °F) for 1 h; air cooled; heated to 730 °C (1350 °F) for 16 h; air cooled

Test temperature °C	°F	Tensile strength MPa	ksi	Yield strength(a) MPa	ksi	Elongation(b), %	Reduction in area, %
24	75	823.9	119.5	362	52.5	43	40
−18	0	882.5	128.0	393	57.0	42	45
−73	−100	968.7	140.5	483	70.1	36	29
−130	−200	1069	155.0	565	82.0	29	25
−185	−300	1131	164.0	710	103.0	22	20

(a) 0.2% offset. (b) In 25 mm (1 in.)

AISI Type 661: Tensile Properties of Bar at Elevated Temperatures (Ref 7)

Test specimens were 33-mm (1.3-in.) square bar; heated to 1175 °C (2150 °F) for 1 h; water quenched; heated to 815 °C (1500 °F) for 4 h; air cooled

Test temperature		Tensile strength		Yield strength				Elongation(a),	Reduction
				0.02% offset		0.2% offset			
°C	°F	MPa	ksi	MPa	ksi	MPa	ksi	%	in area, %
540	1000	621	90.0	217	31.5	265	38.5	47	48.5
650	1200	562	81.5	217	31.5	248	36.0	44	46.0
760	1400	403	58.5	186	27.0	238	34.5	36	36.0
870	1600	248	36.0	193	28.0	224	32.5	44	42.0

(a) In 50 mm (2 in.)

AISI Type 661: Creep Rates of Bar and Sheet (Ref 7)

Test specimens were solution treated at 1175 °C (2150 °F); water quenched

Test temperature		Stress for secondary creep rate of:							
		1% in 100 h		1% in 1000 h		1% in 10 000 h		1% in 100 000 h(a)	
°C	°F	MPa	ksi	MPa	ksi	MPa	ksi	MPa	ksi
650	1200	276	40.0	234	34.0	200	29.0	152	22.0
705	1300	...	...	...	...	121	17.5	97	14.0
730	1350	155	22.5	128	18.5	96.5	14.0	79.2	11.5
815	1500	100	14.5	79.3	11.5	62	9.0	45	6.5
900	1650	48	7.0	34	5.0	24	3.5	17	2.5

(a) Extrapolated data

Mechanical Properties (continued)

AISI Type 661: Stress-Rupture Properties of Bar (Ref 7)

Test specimens were 33-mm (1.3-in.) square, hot rolled bar; heat treated at 1175 °C (2150 °F) for 1 h; water quenched; plus 815 °C (1500 °F) for 4 h; air cooled

Test temperature		Stress for rupture in:							
		10 h		100 h		1000 h		10 000 h(a)	
°C	°F	MPa	ksi	MPa	ksi	MPa	ksi	MPa	ksi
650	1200	427	62.0	345	50.0	276	40.0	217	31.5
705	1300	324	47.0	259	37.5	200	29.0	148	21.5
760	1400	245	35.5	186	27.0	138	20.0	100	14.5
815	1500	176	25.5	128	18.5	90	13.0	64	9.3
870	1600	128	18.5	88.3	12.8	60	8.7	40	5.8
925	1700	86.2	12.5	60	8.7	38	5.5	24	3.5

(a) Extrapolated data

AISI Type 661: Stress-Rupture Properties of Sheet (Ref 7)

Test specimens were 1.9-mm (0.075-in.) thick; heat treated to 1175 °C (2150 °F) for 10 min; water quenched

Test temperature		Stress for rupture in:							
		10 h		100 h		1000 h		10 000 h(a)	
°C	°F	MPa	ksi	MPa	ksi	MPa	ksi	MPa	ksi
650	1200	365	53.0	317	46.0	265	38.5	207	30.0
705	1300	307	44.5	245	35.5	186	27.0	138	20.0
760	1400	231	33.5	172	25.0	128	18.5	93.1	13.5
815	1500	169	24.5	121	17.5	86.2	12.5	63	9.1
870	1600	117	17.0	84.8	12.3	59	8.5	40	5.8
925	1700	82.7	12.0	59	8.5	38	5.5	...	...

(a) Extrapolated data

AISI Type 661: Tensile Properties of Sheet at Room and Elevated Temperatures (Ref 7)

Test specimens were 0.075- to 3.4-mm (0.030- to 0.135-in.) thick sheet; heated to 1175 °C (2150 °F) for 10 min; water quenched

Test temperature		Tensile strength		Yield strength(a)		Elongation(b),
°C	°F	MPa	ksi	MPa	ksi	%
Room temperature		820.5	119.0	407	59.0	42.5
425	800	689.5	100.0	272	39.5	42.5
540	1000	676	98.0	248	36.0	42.5
595	1100	655	95.0	269	39.0	40.0
650	1200	600	87.0	255	37.0	37.5
705	1300	514	74.5	259	37.5	31.5
760	1400	452	65.5	234	34.0	25.5
815	1500	348	50.5	210	30.5	38.0
870	1600	293	42.5	179	26.0	29.5
925	1700	217	31.5	131	19.0	50.0
980	1800	162	23.5	93.1	13.5	39.5

(a) 0.2% offset. (b) In 50 mm (2 in.)

AISI Type 661: Charpy V-Notch Impact Properties (Ref 7)

Test specimens were 16-mm (0.63-in.) diam bar; heat treated at 1190 °C (2175 °F) for 1 h; air cooled; plus 730 °C (1350 °F) for 16 h; air cooled

Test temperature		Impact energy	
°C	°F	J	ft·lb
−185	−300	22	16
−130	−200	43	32
−73	−100	54	40
−18	0	68	50
24	75	81	60
205	400	104	77
425	800	108	80
540	1000	108	80
650	1200	89	66
760	1400	87	64
870	1600	87	64

Machining Data

For machining data on AISI type 661, refer to the preceding machining tables for AISI type 651.

Type N08366

Type N08366 Chemical Composition

Element	Composition, %
Carbon	0.03 max
Manganese	1.50 max
Phosphorus	0.030 max
Sulfur	0.010 max
Silicon	0.50 max
Chromium	20-22
Nickel	23.5-25.5
Molybdenum	6-7

Characteristics. Alloy type N08366 is an austenitic, chromium-nickel-molybdenum stainless steel developed for service in chloride and other pitting and crevice-corrosion environments, such as sea water. This steel is austenitic in the fully annealed condition.

Alloy N08366 is resistant to atmospheric corrosion and similar mild types of corrosion. In certain strongly oxidizing acids, such as nitric acid, this steel is less resistant than any of the standard chromium-nickel types. Type N08366 steel provides greater resistance to chloride pitting and crevice corrosion than types 316 or 304.

The low carbon content (0.03% max) of alloy N08366 provides resistance to intergranular attack because of precipitation of carbides from short exposures to the carbide-precipitation temperatures, such as welding. When held for longer periods in the 425 to 815 °C (800 to 1500 °F) range, a molybdenum-containing phase, such as sigma, precipitates. This phase is the principal precipitate formed on slow cooling from annealing.

Typical Uses. Applications of type N08366 include heat exchanger and condenser tubes exposed to sea water.

Type N08366: Similar Steels (U.S. and/or Foreign). UNS N08366; AL-6X (Allegheny Ludlum Steel Corp); 6X (Al Tech Specialty Steel Corp.)

Physical Properties

Type N08366: Selected Physical Properties (Ref 8)

Melting range	1370 °C (2500 °F)
Density	8.11 g/cm³ (0.293 lb/in.³)
Electrical resistivity at 20 °C (68 °F)	95 nΩ·m
Thermal conductivity at 20-100 °C (68-212 °F)	13.81 W/m·K (7.98 Btu/ft·h·°F)

Type N08366: Thermal Treatment Temperatures

Treatment	Temperature range °C	°F
Forging, start	1205-1260	2200-2300
Forging, finish	925-980	1700-1800
Annealing	1095-1120	2000-2050
Hardening(a)	...	...

(a) Not hardenable by heat treatment

Type N08366: Average Coefficients of Linear Thermal Expansion (Ref 8)

Temperature range °C	°F	Coefficient µm/m·K	µin./in.·°F
20-100	68-212	15.3	8.5
20-500	68-930	16.0	8.9
20-1000	68-1830	18.0	10.0

Mechanical Properties

Type N08366: Effect of Cold Reduction on Tensile Properties of Sheet Material (Ref 8)

Cold reduction, %	Tensile strength MPa	ksi	Yield strength(a) MPa	ksi	Elongation(b), %	Hardness
Sheet thickness of 1.65 mm (0.065 in.)						
Annealed	641	93.0	291	42.2	46.0	80 HRB
20	894	129.7	795	115.3	14.0	25 HRC
50	1117	162.0	1044	151.4	4.5	...
60	1200	174.1	1095	158.8	4.0	33 HRC
Sheet thickness of 1.02 mm (0.040 in.)						
Annealed	583	84.6	262	38.0	61.0	80 HRB
10	652	94.5	492	71.3	40.0	...
20	769	111.6	680	98.6	21.0	25 HRC
31	917	133.0	824	119.5	11.0	...
49	1020	148.0	933	135.3	6.0	...
60	1169	169.6	1036	150.3	3.5	80 HRC

(a) 0.2% offset. (b) In 50 mm (2 in.)

Type N08366: Typical Tensile Properties of Strip (Ref 8)

Condition	Tensile strength MPa	ksi	Yield strength(a) MPa	ksi	Elongation(b), %	Reduction in area, %	Modulus of elasticity GPa	10⁶ psi
Annealed	620	90	276	40	45	60	200	29

(a) 0.2% offset. (b) In 50 mm (2 in.)

AISI Type S13800

AISI Type S13800: Chemical Composition

Element	Composition, %
Carbon	0.05 max
Manganese	0.10 max
Phosphorus	0.010 max
Sulfur	0.008 max
Silicon	0.10 max
Chromium	12.25-13.25
Nickel	7.5-8.5
Aluminum	0.90-1.35
Molybdenum	2.0-2.5
Nitrogen	0.010 max

Characteristics. AISI type S13800 is a martensitic, precipitation hardening (maraging), stainless steel that combines high strength and hardness with good corrosion resistance. This alloy exhibits good ductility and toughness in large sections, in both the transverse and longitudinal directions. Compared to other ferrous-base materials, this alloy offers a high level of useful mechanical properties under severe environmental conditions.

Type S13800 has good fabricating characteristics and can be age-hardened by a single low-temperature treatment. Cold work prior to aging increases aging, especially in the lower aging temperatures.

Rusting resistance of type S13800 steel is similar to that of type 304. In strongly oxidizing and reducing acids, and in atmospheric exposures, the general corrosion resistance of type S13800 approaches that of type 304. The general level of corrosion resistance of type S13800 steel is greatest in the fully hardened condition and decreases slightly as the aging temperature is increased.

Type S13800 can be machined in both the solution-treated and various age-hardened conditions. The machinability rating of this grade is about 35 to 40% of that for free-machining AISI 1212. It can be satisfactorily welded by all methods. No preheating is required. For achieving properties in the weld comparable to those of the base metal, a postweld solution heat treatment should be used prior to aging.

Typical Uses. Applications of AISI types S13800 include valve parts, fittings, cold headed and machined fasteners, shafts, landing gear parts, pins, lockwashers, aircraft components, and nuclear reactor components.

AISI Type S13800: Similar Steels (U.S. and/or Foreign). UNS S13800; AMS 5629, 5840; ASME SA705 (XM-13); ASTM A564 (XM-13), A693 (XM-13), A705 (XM-13)

Physical Properties

AISI Type S13800: Thermal Treatment Temperatures

Treatment	Temperature range °C	°F
Forging, start	1175-1205	2150-2200
Forging, finish	955 min	1750 min
Solution treatment(a)	915-935	1685-1715
Hardening(b)	510-620	950-1150

(a) Referred to as condition A, or as supplied from the mill. Time of solution treatment depends on section thickness. Air cool or oil quench below 15 °C (60 °F). (b) Temperature depends on properties desired

AISI Type S13800: Selected Physical Properties in the H950 Condition (Ref 6)

Melting range	1405 to 1440 °C (2560 to 2625 °F)
Density	7.8 g/cm³ (0.28 lb/in.³)
Specific heat at 0-100 °C (32-212 °F)	460 J/kg·K (0.11 Btu/lb·°F)

AISI Type S13800: Electrical Resistivity for Condition A (Ref 6)

Temperature °C	°F	Electrical resistivity, nΩ·m
25	77	100.1
100	212	101.9
200	390	104.6
315	600	106.1
425	800	108.1
540	1000	109.1
595	1110	109.5

AISI Type S13800: Average Coefficients of Linear Thermal Expansion (Ref 6)

Temperature range °C	°F	Coefficient µm/m·K	µin./in.·°F
21-95	70-200	10.6	5.9
21-205	70-400	10.8	6.0
21-315	70-600	11.2	6.2
21-425	70-800	11.3	6.3
21-540	70-1000	11.9	6.6

AISI Type S13800: Thermal Conductivity for Condition A (Ref 6)

Temperature °C	°F	Thermal conductivity W/m·K	Btu/ft·h·°F
100	212	14.0	8.1
200	390	15.7	9.1
315	600	17.8	10.3
425	800	20.4	11.8
540	1000	22.3	12.9
595	1110	22.5	13.0

Machining Data

For machining data on AISI Type S13800, refer to the preceding machining table for AISI type 630.

Mechanical Properties

AISI Type S13800: Typical Longitudinal Mechanical Properties (Ref 9)

Center or intermediate test location

Condition	Tensile strength MPa	Tensile strength ksi	Yield strength(a) MPa	Yield strength(a) ksi	Elongation(b), %	Reduction in area, %	Hardness, HRC	Impact energy(c) J	Impact energy(c) ft·lb	Modulus of elasticity GPa	Modulus of elasticity 10^6 psi
RH950	1620	235	1480	215	12	45	48	27	20	...	...
H950	1550	225	1450	210	12	50	47	41	30	197	28.6
H1000	1480	215	1415	205	13	55	45	54	40	221	32.0
H1050	1310	190	1240	180	15	55	43	81	60	212	30.8
H1100	1105	160	1035	150	18	60	36	49	36	197	28.6
H1150	1000	145	725	105	20	63	33	...	...	230	33.3
H1150M	895	130	585	85	22	70	28	163	120	172	25.0

(a) 0.2% offset. (b) In 50 mm (2 in.). (c) Charpy V-notch

AISI Type S13800: Typical Transverse Mechanical Properties (Ref 9)

Center or intermediate test location

Condition	Tensile strength MPa	Tensile strength ksi	Yield strength(a) MPa	Yield strength(a) ksi	Elongation(b), %	Reduction in area, %	Hardness, HRC
H950	1550	225	1450	210	12	40	47
H1000	1480	215	1415	205	13	50	45
H1050	1310	190	1240	180	15	55	43
H1100	1105	160	1035	150	18	60	36
H1150	1000	145	725	105	20	63	33
H1150M	895	130	585	85	22	70	28

(a) 0.2% offset. (b) In 50 mm (2 in.)

AISI Type S13800: Typical Cryogenic Charpy V-Notch Energy in Longitudinal Values (Ref 9)

Test temperature °C	Test temperature °F	H950 J	H950 ft·lb	H1000 J	H1000 ft·lb	H1150M J	H1150M ft·lb
Room temperature		45	33	52	38	163+	120+
0	32	31	23	41	30	...	...
−54	−65	18	13	27	20	...	...
−73	−100	11	8	22	16	119	88
−115	−175	7	5	16	12	96	71
−140	−220	5	4	9	7	...	...
−195	−320	5	4	5	4	41	30

AISI Type S14800

AISI Type S14800: Chemical Composition

Element	Composition, %
Carbon	0.05
Manganese	1.00
Phosphorus	0.015
Sulfur	0.010
Silicon	1.00
Chromium	13.75-15
Nickel	7.75-8.75
Molybdenum	2-3
Aluminum	0.75-1.5

Characteristics. AISI type S14800 is a precipitation-hardening, chromium-nickel, stainless steel with high-strength properties at both room and elevated temperatures, coupled with high-fracture toughness. In the precipitation-hardened conditions, type S14800 is superior in corrosion resistance to that of the chromium stainless steels. In some environments, it approximates the austenitic chromium-nickel types of stainless steels. This alloy is essentially austenitic in the solution-treated (annealed) condition, but martensitic in the age-hardened condition.

Type S14800 condition A, mill annealed at 980 to 1010 °C (1800 to 1850 °F), can be formed similarly to type 301 stainless steel. It work hardens rapidly, and may require intermediate annealing in drawing deep shapes or forming intricate parts. In condition C, cold rolled at the mill, the material is extremely hard and strong, and fabrication techniques used for such materials must be followed.

Welding of type S14800 is most successful when using the gas tungsten-arc welding process. Shielded-metal-arc welding can be done using a standard chromium-nickel grade of stainless steel. Type S14800 is readily brazed.

Sheets and brazed honeycomb panels of this grade have been machined successfully using chemical milling.

Typical Uses. Applications of AISI type S14800 include pressure tanks, honecomb paneling, and heat shields for aircraft and aerospace applications.

AISI Type S14800: Similar Steels (U. S. and/or Foreign). UNS S14800; AMS 5601, 5603

Physical Properties

AISI Type S14800: Thermal Treatment Temperatures

Treatment	Temperature range °C	°F
Solution treatment, condition A	980-1010	1800-1850
Precipitation hardening	510 or 565	950 or 1050
Condition C(a)	...	...
Hardening, condition CH900	480	900

(a) Cold rolled at mill

AISI Type S14800: Density (Ref 12)

Condition	Density g/cm³	lb/in.³
A	7.83	0.283
SRH950	7.70	0.278

AISI Type S14800: Average Coefficients of Linear Thermal Expansion (Ref 12)

Temperature range °C	°F	Coefficient μm/m·K	μin./in.·°F
21-93	70-200	9.5	5.3
21-205	70-400	10.6	5.9
21-315	70-600	11.2	6.2
21-425	70-800	11.3	6.3
21-540	70-1000	11.5	6.4

Machining Data

For machining data on AISI type S14800, refer to the preceding machining tables for AISI type 630.

Mechanical Properties

AISI Type S14800: Short-Time Elevated Temperature Properties for Condition SRH950 (Ref 12)

Test temperature °C	°F	Longitudinal direction: Tensile strength MPa	ksi	Yield strength(a) MPa	ksi	Elongation(b), %	Tranverse direction: Tensile strength MPa	ksi	Yield strength(a) MPa	ksi	Elongation(b), %
−73	−100	1840	267	1725	250	12	1850	268	1780	258	8
27	80	1615	234	1505	218	5	1670	242	1570	228	5
260	500	1435	208	1255	182	4	1460	212	1315	191	3
290	550	1415	205	1240	180	4	1455	211	1330	193	3
315	600	1370	199	1215	176	4	1425	207	1280	186	3
345	650	1370	199	1205	175	5	1405	204	1255	182	3
425	800	1240	180	1055	153	8	1290	187	1095	159	6
540	1000	910	132	730	106	18	930	135	760	110	17

(a) 0.2% offset. (b) In 50 mm (2 in.)

AISI Type S14800: Elevated-Temperature Stability for Condition SRH1050 (Ref 12, 13)

Test specimens were vacuum induction melted; tested in transverse direction; 25-mm (1.0-in.) wide, 6.4-mm (0.025-in.) thick, with an edge notch-root radius of 0.018 mm (0.0007 in.)

Exposure prior to testing	Test temperature °C	°F	Tensile strength MPa	ksi	Yield strength(a) MPa	ksi	Elongation(b), %	Notch strength MPa	ksi	Ratio NS/YS(c)	Ratio NS/TS(d)
Not exposed	−79	−110	1620	234.9	1571	227.9	12.0	1706	247.5	1.09	1.05
	24	75	1502	217.9	1447	209.8	6.0	1592	230.9	1.10	1.06
	345	650	1252	181.6	1046	151.7	5.0	1065	154.4	1.02	0.85
345 °C (650 °F) for 1000 h; 275-MPa (40-ksi) load	−79	−110	1744	252.9	1680	243.7	11.0	1679	243.5	1.00	0.96
	24	75	1617	234.5	1553	225.3	5.0	1502	217.9	0.97	0.93
	345	650	1308	189.7	1229	178.2	4.0	1149	166.7	0.94	0.88

(a) 0.2% offset. (b) In 50 mm (2 in.). (c) Notch strength to yield strength. (d) Notch strength to tensile strength

Mechanical Properties (continued)

AISI Type S14800: Effect of Cold Reduction on Properties of Material in Conditions C and CH900 (Ref 12)

	Condition C						Condition CH900					
	Tensile strength		Yield strength(a)		Elonga-tion(b),		Tensile strength		Yield strength(a)		Elonga-tion(b),	
Cold reduction, %	MPa	ksi	MPa	ksi	%	Hardness	MPa	ksi	MPa	ksi	%	Hardness
6.3	903	131	421	61	23.0	90.0 HRB	889	129	490	71	28.5	88.5 HRB
11.3	97	141	448	65	16.5	21.5 HRC	896	130	648	94	23.5	24.5 HRC
22.2	1124	163	538	78	11.0	33.0 HRC	1089	158	931	135	15.5	39.5 HRC
29.2	1158	168	752	109	10.0	38.0 HRC	1434	208	1420	206	5.0	47.5 HRC
39.2	1227	178	1034	150	8.0	41.0 HRC	1737	252	1724	250	2.0	50.5 HRC
49.9	1331	193	1220	177	2.0	43.0 HRC	1910	277	1903	276	1.0	53.0 HRC
59.7	1441	209	1427	207	1.0	43.5 HRC	1999	290	1999	290	1.0	54.0 HRC
72.5	1551	225	1531	222	1.0	45.0 HRC	2137	310	2103	305	0	55.5 HRC

(a) 0.2% offset. (b) In 50 mm (2 in.)

AISI Type S14800: Typical Properties in Conditions A, SRH950, and SRH1050 (Ref 12)

	Tensile strength		Yield strength(a)		Elongation(b),	
Condition	MPa	ksi	MPa	ksi	%	Hardness
A	1035	150	450	65	20	100 HRB
SRH950	1515	220	1310	190	3	45-51 HRC
SRH1050	1380	200	1240	180	3	38-45 HRC

(a) 0.2% offset. (b) In 50 mm (2 in.)

AISI Type S15500

AISI Type S15500: Chemical Composition

Element	Composition, %
Carbon	0.07 max
Manganese	1.00 max
Phosphorus	0.040 max
Sulfur	0.030 max
Silicon	1.00 max
Chromium	14-15.5
Nickel	3.5-5.5
Copper	2.5-4.5
Niobium and tantalum	0.15-0.45

Characteristics. AISI type S15500 is a martensitic, precipitation hardening (maraging), stainless steel that combines high strength and hardness with excellent corrosion resistance. It can be hardened by a single-step, low temperature heat treatment that virtually eliminates scaling and distortion, plus gives type S15500 good forgeability and good transverse mechanical properties.

The general corrosion resistance of type S15500 is comparable to that of type 304, in most media. Good resistance to stress corrosion cracking is obtained by hardening at temperatures of 550 °C (1025 °F) and higher.

This alloy is generally used in the form of bar and forgings which do not require much forming. It is readily machined in both the solution-treated and various age-hardened conditions. Type S15500 machines similarly to types 302 and 304. Machinability will improve as the hardening temperature is increased.

Type S15500 can be welded satisfactorily by all methods. No preheating is required. Properties in the weld comparable to those of the base metal can be obtained by suitable postweld heat treatment.

Typical Uses. Applications of AISI type S15500 include valve parts, fittings and fasteners, shafts, gears, chemical process equipment, paper mill equipment, aircraft components, and nuclear reactor components.

AISI Type S15500: Similar Steels (U.S. and/or Foreign). UNS S15500; AMS 5658, 5659, 5862; ASME SA705 (XM-12); ASTM A564 (XM-12), A693 (XM-12), A 705 (XM-12)

Physical Properties

AISI Type S15500: Selected Physical Properties in the H900 Condition (Ref 6)

Solution treated plus 480 °C (900 °F) for 1 h; air cool

Melting range	1405-1440 °C (2560-2625 °F)
Density	7.8 g/cm³ (0.28 lb/in.³)
Specific heat	420 J/kg·K (0.10 Btu/lb·°F)
Electrical resistivity at 20 °C (68 °F)	77 nΩ·m

AISI Type S15500: Thermal Conductivity

Temperature		Thermal conductivity	
°C	°F	W/m·K	Btu/ft·h·°F
150	300	17.8	10.3
260	500	19.5	11.3
460	860	22.5	13.0
480	900	22.7	13.1

Physical Properties (continued)

AISI Type S15500: Thermal Treatment Temperatures

Treatment	Temperature range °C	°F
Forging, start	1175-1205	2150-2200
Forging, finish	1010 min	1850 min
Solution treatment(a)	1025-1050	1875-1925
Hardening(b)	480-620	900-1150

(a) Cool rapidly to room temperature. (b) Temperature depends upon desired strength and toughness

AISI Type S15500: Average Coefficients of Linear Thermal Expansion (Ref 6)

Temperature range °C	°F	Coefficient µm/m·K	µin./in.·°F
−73-21	−100-70	10.4	5.8
21-93	70-200	10.8	6.0
21-205	70-400	10.8	6.0
21-315	70-600	11.3	6.3
21-425	70-800	11.7	6.5

Mechanical Properties

AISI Type S15500: Typical Mechanical Properties in the Transverse Direction, at an Intermediate Location (Ref 10)

Condition	Tensile strength MPa	Tensile strength ksi	Yield strength(a) MPa	Yield strength(a) ksi	Elongation(b), %	Reduction in area, %	Hardness, HB	Impact energy(c) J	Impact energy(c) ft·lb
H900	1380	200	1275	185	10	30	420	10	7
H925	1310	190	1205	175	11	35	409	23	17
H1025	1170	170	1140	165	12	42	352	37	27
H1075	1140	165	1035	150	13	43	341	41	30
H1100	1035	150	930	135	14	44	332	41	30
H1150	1000	145	860	125	15	45	311	68	50
H1150M	860	125	585	85	18	50	277	136	100

(a) 0.2% offset. (b) In 50 mm (2 in.). (c) Charpy V-notch

AISI Type S15500: Short-Time Tensile Properties at Various Test Temperatures and Conditions (Ref 12)

Test temperature °C	°F	Condition	Tensile strength MPa	Tensile strength ksi	Yield strength(a) MPa	Yield strength(a) ksi	Elongation(b), %	Reduction in area, %
24	75	H925	1317	191	1213	176	16	59
		H1025	1145	166	1110	161	17	64
		H1100	1069	155	1034	150	19	67
		H1150M	896	130	717	104	23	75
205	400	H925	1158	168	1048	152	15	54
		H1025	1014	147	958	139	15	58
		H1100	952	138	924	134	16	62
		H1150M	765	111	689	100	20	64
315	600	H925	1096	159	965	140	14	59
		H1025	958	139	903	131	14	57
		H1100	910	132	869	126	14	57
		H1150M	717	104	662	96	19	70
425	800	H925	1027	149	1414	126	15	60
		H1025	917	133	820	119	15	60
		H1100	848	123	786	114	14	60
		H1150M	676	98	607	88	17	69
540	1000	H925	758	110	634	92	17	70
		H1025	724	105	627	91	18	70
		H1100	662	96	607	88	18	71
		H1150M	552	80	462	67	20	74
650	1200	H925	400	58	317	46	26	83
		H1025	372	54	283	41	28	83
−195	−320	H1025	1558	226	1524	221	15	55
		H1100	1448	210	1413	205	18	60
		H1150M	1386	201	1007	146	27	65
−73	−100	H925	1462	212	1372	199	17	61
		H1025	1269	184	1234	179	18	67
		H1100	1186	172	1145	166	19	66
		H1150M	1041	151	738	107	25	74

(a) 0.2% offset. (b) Gage length is 4 × diam of the test specimen

AISI Type S15500: Typical Mechanical Properties in the Longitudinal Direction, at an Intermediate Location (Ref 10)

Condition	Tensile strength MPa	Tensile strength ksi	Yield strength(a) MPa	Yield strength(a) ksi	Elongation(b), %	Reduction in area, %	Hardness, HB	Impact energy(c) J	Impact energy(c) ft·lb
H900	1380	200	1275	185	14	50	420	20	15
H925	1310	190	1205	175	14	54	409	34	25
H1025	1170	170	1140	165	15	56	352	48	35
H1075	1140	165	1035	150	16	58	341	54	40
H1100	1035	150	930	135	17	58	332	61	45
H1150	1000	145	860	125	19	60	311	68	50
H1150M	860	125	585	85	22	68	277	136	100

(a) 0.2% offset. (b) In 50 mm (2 in.). (c) Charpy V-notch

AISI Type S15500: Charpy V-Notch Impact Properties at Subzero Temperatures

Test temperature °C	Test temperature °F	Impact energy for heat treat condition of: H925 J	H925 ft·lb	H1025 J	H1025 ft·lb	H1100 J	H1100 ft·lb	H1150M J	H1150M ft·lb
24	75	79	58	114	84	130	96	236	174
−12	10	38	28	62	46	108	80	233	172
−40	−40	22	16	31	23	73	54	226	167
−73	−100	9	7	12	9	37	27	206	152
−195	−320	...	...	3	2	4.7	3.5	45	33

Machining Data

For machining data on AISI type S15500, refer to the preceding machining tables for AISI type 630.

Type S20910

Type S20910: Chemical Composition

Element	Composition, %
Carbon	0.06 max
Manganese	4-6
Phosphorus	0.04 max
Sulfur	0.03 max
Silicon	1.00 max
Chromium	20.50-23.50
Nickel	11.50-13.50
Molybdenum	1.50-3.00
Nitrogen	0.20-0.40
Niobium	0.10-0.30
Vanadium	0.10-0.30

Characteristics. Type S20910 (Nitronic 50) is an austenitic stainless steel that has greater corrosion resistance than that provided by types 316 and 316L, plus approximately twice the yield strength at room temperature. The steel has very good mechanical properties at both elevated and subzero temperatures.

Type S20910 stainless steel provides adequate corrosion resistance for many applications in the 1065 °C (1950 °F) annealed condition. In very corrosive media, or where material is to be used in the as-welded condition, the 1120 °C (2050 °F) annealed condition should be specified to minimize intergranular attack. This alloy has excellent resistance to sulfide stress cracking in all conditions.

The inert-gas-shielded arc welding processes can be used to weld type S20910. Good weld-joint properties can be obtained without the necessity of preheat or postweld annealing. Filler metal, where used, should match the composition of the base metal for comparable strength and corrosion resistance. Types 308L and 309 produce sound weld joints.

The machinability of S20910 is approximately 21% of that for free-machining AISI 1212. Rigid setups should be used in holding the work and tools. This alloy produces a good surface finish, but because of the work hardening characteristics, tools should not be permitted to slide over the surface.

Typical Uses. Materials of type S20910 have applications in the petroleum, chemical, pulp and paper, textile, food processing, and marine industries. Representative components using the combination of good corrosion resistance and high strength are pumps, valves and fittings, fasteners, cables, chains, screens and wire cloth, marine hardware, boat shafting, heat exchanger parts, springs, and photographic equipment.

Type S20910: Similar Steels (U.S. and/or Foreign). UNS S20910; AMS 5764, 5861; ASME SA182 (XM-19), SA240 (XM-19), SA249 (XM-19), SA312 (XM-19), SA403 (XM-19), SA412 (XM-19), SA479 (XM-19); ASTM A182 (XM-19), A240 (XM-19), A249 (XM-19), A269 (XM-19), A276 (XM-19), A312 (XM-19), A403 (XM-19), A412 (XM-19), A429 (XM-19), A479 (XM-19), A580 (XM-19)

Physical Properties

Type S20910: Average Coefficients of Linear Thermal Expansion (Ref 12)

Test specimens were in the annealed condition

Temperature range °C	°F	Coefficient µm/m·K	µin./in.·°F
20-95	70-200	16.2	9.0
20-205	70-400	16.6	9.2
20-315	70-600	17.3	9.6
20-425	70-800	17.8	9.9
20-540	70-1000	18.4	10.2
20-650	70-1200	18.9	10.5
20-760	70-1400	19.4	10.8
20-870	70-1600	20.0	11.1

Type S20910: Selected Physical Properties (Ref 12)

Density	7.89 g/cm³ (0.285 lb/in.³)
Electrical resistivity at room temperature	82 nΩ·m

Type S20910: Thermal Treatment Temperatures

Treatment	Temperature range °C	°F
Forging	1175-1230	2150-2250
Annealing(a)	1065 or 1120	1950 or 2050
Hardening(b)	...	...

(a) Use 1065 °C (1950 °F) annealed material for greatest strength, and 1120 °C (2050 °F) annealed material for best corrosion resistance. (b) Not hardenable by heat treatment

Type S20910: Thermal Conductivity (Ref 12)

Temperature °C	°F	Thermal conductivity W/m·K	Btu/ft·h·°F
150	300	15.6	9.0
315	600	17.8	10.3
480	900	20.4	11.8
650	1200	23.0	13.3
815	1500	25.3	14.6

Mechanical Properties

Type S20910: Typical Tensile Properties of Annealed and High-Strength Bar at Room Temperature (Ref 12)

Test specimens were 25-mm (1-in.) diam annealed bar and 25- to 50-mm (1- to 2-in.) diam high-strength bar

Condition	Tensile strength MPa	ksi	Yield strength(a) MPa	ksi	Elongation(b), %	Reduction in area, %	Hardness	Impact energy(c) J	ft·lb
Annealed at 1120 °C (2050 °F); water quenched	825	120	415	60	50	70	98 HRB	230	170
Annealed at 1065 °C (1950 °F); water quenched	860	125	450	65	45	65	23 HRC	176	130
High strength, typical	1105	160	1000	145	25	55	32 HRC	122	90
High strength, minimum	930	135	725	105	20	50	26 HRC	...	...

(a) 0.2% offset. (b) In 50 mm (2 in.) or 4 × diam of the test specimen. (c) Charpy V-notch

Type S20910: Typical Impact Properties of Annealed Bar (Ref 12)

Test specimens were 25-mm (1-in.) diam bar; annealed at 1120 °C (2050 °F)

Test temperature °C	°F	Charpy V-notch impact energy, Annealed J	Annealed ft·lb	Simulated HAZ(a) J	Simulated HAZ(a) ft·lb
24	75	230	170	230	170
−73	−100	155	115	155	115
−195	−320	70	50	70	50

(a) Heat treated at 675 °C (1250 °F) for 1 h to simulate the heat-affected zone of heavy weldments

Type S20910: Typical Subzero Impact Properties of High-Strength Bar (Ref 12)

Test temperature °C	°F	Charpy V-notch impact energy for bar diameter of: 25 mm (1 in.) J	25 mm (1 in.) ft·lb	38 mm (1.5 in.) J	38 mm (1.5 in.) ft·lb	50 mm (2 in.) J	50 mm (2 in.) ft·lb
27	80	171	126	203	150	201	148
−60	−75	155	114	178	131	178	131
−130	−200	65	48	84	62	83	61
−195	−320	42	31	56	41	49	36

Type S20910: Typical Tensile Properties of Cold Drawn Wire (Ref 12)

Starting size of test specimens was 6.4-mm (0.25-in.) diam rod; annealed at 1120 °C (2050 °F)

Cold reduction, %	Tensile strength MPa	ksi	Yield strength(a) MPa	ksi	Elongation(b), %	Reduction in area, %
15	1140	165	985	143	23	56
30	1340	194	1200	174	15	49
45	1490	216	1350	196	11	45
60	1615	234	1490	216	9	42
75	1695	246	1615	234	8	39

(a) 0.2% offset. (b) In 50 mm (2 in.) or 4 × diam of the test specimen

Mechanical Properties (continued)

Type S20910: Typical Short-Time Elevated Temperature Tensile Properties (Ref 12)

Condition	Bar diameter mm	Bar diameter in.	Test temperature °C	Test temperature °F	Tensile strength MPa	Tensile strength ksi	Yield strength(a) MPa	Yield strength(a) ksi	Elongation(b), %	Reduction in area, %
Annealed at 1065 °C (1950 °F)	19-32	0.75-1.25	24	75	855	124	538	78	40.5	67.5
			95	200	772	112	455	66	40.5	67.5
			205	400	703	102	400	58	37.5	67.0
			315	600	676	98	372	54	37.5	64.0
			425	800	648	94	345	50	39.5	63.0
			540	1000	614	89	331	48	36.5	62.5
			650	1200	552	80	303	44	36.5	63.0
			730	1350	469	68	290	42	42.5	71.5
			815	1500	345	50	221	32	59.5	85.0
Annealed at 1120 °C (2050 °F)	25-38	1-1.5	24	75	807	117	414	60	45.0	71.0
			95	200	738	107	338	50	43.5	70.5
			205	400	662	96	262	38	43.5	69.5
			316	600	634	92	241	35	42.5	67.5
			425	800	614	89	234	34	43.5	66.0
			540	1000	579	84	221	32	41.0	66.5
			650	1200	510	74	214	31	38.0	64.0
			730	1350	455	66	214	31	37.0	61.5
			815	1500	359	52	207	30	41.0	61.0
High-strength	25-50	1-2	24	75	1034	150	869	126	29.0	64.0
			95	200	931	135	772	112	28.0	65.0
			205	400	855	124	696	101	27.0	63.5
			315	600	807	117	641	93	27.5	61.0
			425	800	765	111	593	86	28.5	61.0
			540	1000	710	103	552	80	27.0	60.5
			595	1100	683	99	531	77	26.0	59.5

(a) 0.2% offset. (b) Gage length is 4 × diam of the test specimen

Type S20910: Typical Stress-Rupture Properties of Annealed and High-Strength Bars (Ref 12)

Condition	Bar diameter mm	Bar diameter in.	Test temperature °C	Test temperature °F	Stress to produce rupture in: 100 h MPa	100 h ksi	1000 h MPa	1000 h ksi	10 000 h(a) MPa	10 000 h(a) ksi
Annealed at 1065 °C (1950 °F)	19-32	0.75-1.25	540	1000	625	91	605	88	496	72
			595	1100	495	72	425	62	324	47
			650	1200	380	55	260	38	152	22
			730	1350	145	21	85	12	41.4	6
			815	1500	70	10	26	3.7	9.0	1.3
Annealed at 1120 °C (2050 °F)	25-50	1-2	540	1000	...	...	...	...	...	...
			595	1100	450	65	370	54	296	43.0
			650	1200	345	50	285	41	224	32.5
			730	1350	200	29	105	15	58.6	8.5
			815	1500	90	13	45	6.5	24.1	3.5
High-strength	25	1	540	1000	675	98	655	95	538	78
			595	1100	585	85	455	66	352	51

(a) Estimated

Machining Data (Ref 2)

Type S20910: Surface Grinding with a Horizontal Spindle and Reciprocating Table

Operation	Wheel speed m/s	Wheel speed ft/min	Table speed m/min	Table speed ft/min	Downfeed mm/pass	Downfeed in./pass	Crossfeed mm/pass	Crossfeed in./pass	Wheel identification
Annealed, with hardness of 210 to 250 HB									
Roughing	28-33	5500-6500	15-30	50-100	0.050	0.002	1.25-12.5	0.050-0.500	C46JV
Finishing	28-33	5500-6500	15-30	50-100	0.013 max	0.0005 max	(a)	(a)	C46JV
Cold drawn, with hardness of 325 to 375 HB									
Roughing	28-33	5500-6500	15-30	50-100	0.025	0.001	0.65-6.5	0.025-0.250	C46HV
Finishing	28-33	5500-6500	15-30	50-100	0.013 max	0.0005 max	(b)	(b)	C46HV

(a) Maximum 1/4 of wheel width. (b) Maximum 1/10 of wheel width

Machining Data (Ref 2) (continued)

Type S20910: Turning (Single Point and Box Tools)

Depth of cut mm	in.	High speed steel Speed m/min	ft/min	Feed mm/rev	in./rev	Uncoated carbide Speed, brazed m/min	ft/min	Speed, inserted m/min	ft/min	Feed mm/rev	in./rev	Coated carbide Speed m/min	ft/min	Feed mm/rev	in./rev
Annealed, with hardness of 210 to 250 HB															
1	0.040	18(a)	60(a)	0.18	0.007	58(b)	190(b)	67(b)	220(b)	0.18	0.007	84(c)	275(c)	0.18	0.007
4	0.150	15(a)	50(a)	0.40	0.015	49(b)	160(b)	58(b)	190(b)	0.40	0.015	76(c)	250(c)	0.40	0.015
8	0.300	12(a)	40(a)	0.50	0.020	38(d)	125(d)	46(d)	150(d)	0.50	0.020	60(e)	200(e)	0.50	0.020
16	0.625	8(a)	25(a)	0.75	0.030	29(d)	95(d)	35(d)	115(d)	0.75	0.030	...	...	...	...
Cold drawn, with hardness of 325 to 375 HB															
1	0.040	15(f)	50(f)	0.18	0.007	46(b)	150(b)	53(b)	175(b)	0.18	0.007	38(c)	125(c)	0.18	0.007
4	0.150	12(f)	40(f)	0.40	0.015	40(b)	130(b)	46(b)	150(b)	0.40	0.015	30(c)	100(c)	0.40	0.015
8	0.300	9(f)	30(f)	0.50	0.020	30(d)	100(d)	37(d)	120(d)	0.50	0.020	...	...	...	...

(a) High speed steel tool material: M2 or M3. (b) Carbide tool material: C-3. (c) Carbide tool material: CC-3. (d) Carbide tool material: C-2. (e) Carbide tool material: CC-2. (f) Any premium high speed steel tool material (T15, M33, M41 to M47)

Type S20910: Turning (Cutoff and Form Tools)

Tool material	Speed, m/min (ft/min)	Feed per revolution for cutoff tool width of: 1.5 mm (0.062 in.) mm (in.)	3 mm (0.125 in.) mm (in.)	6 mm (0.25 in.) mm (in.)	Feed per revolution for form tool width of: 12 mm (0.5 in.) mm (in.)	18 mm (0.75 in.) mm (in.)	25 mm (1 in.) mm (in.)	35 mm (1.5 in.) mm (in.)	50 mm (2 in.) mm (in.)
Annealed, with hardness of 210 to 250 HB									
M2 and M3 high speed steel	14 (45)	0.025 (0.001)	0.030 (0.0012)	0.038 (0.0015)	0.038 (0.0015)	0.033 (0.0013)	0.030 (0.0012)	0.025 (0.001)	0.023 (0.0009)
C-2 carbide	46 (150)	0.025 (0.001)	0.030 (0.0012)	0.038 (0.0015)	0.038 (0.0015)	0.033 (0.0013)	0.030 (0.0012)	0.025 (0.001)	0.023 (0.0009)
Cold drawn, with hardness of 325 to 375 HB									
Any premium high speed steel (T15, M33, M41-M47)	9 (30)	0.020 (0.0008)	0.025 (0.001)	0.033 (0.0013)	0.033 (0.0013)	0.028 (0.0011)	0.025 (0.001)	0.020 (0.0008)	0.018 (0.0007)
C-2 carbide	29 (95)	0.020 (0.0008)	0.025 (0.001)	0.033 (0.0013)	0.033 (0.0013)	0.028 (0.0011)	0.025 (0.001)	0.020 (0.0008)	0.018 (0.0007)

Type S20910: Face Milling

Depth of cut mm	in.	High speed steel(a) Speed m/min	ft/min	Feed/tooth mm	in.	Uncoated C-2 carbide Speed, brazed m/min	ft/min	Speed, inserted m/min	ft/min	Feed/tooth mm	in.	Coated CC-2 carbide Speed m/min	ft/min	Feed/tooth mm	in.
Annealed, with hardness of 210 to 250 HB															
1	0.040	24	80	0.15	0.006	59	195	73	240	0.18	0.007	105	350	0.18	0.007
4	0.150	18	60	0.25	0.010	52	170	64	210	0.25	0.010	84	275	0.25	0.010
8	0.300	14	45	0.36	0.014	41	135	50	165	0.36	0.014	69	225	0.36	0.014
Cold drawn, with hardness of 325 to 375 HB															
1	0.040	20	65	0.13	0.005	47	155	58	190	0.13	0.005	84	275	0.102	0.004
4	0.150	15	50	0.20	0.008	41	135	50	165	0.18	0.007	60	200	0.15	0.006
8	0.300	11	35	0.25	0.010	32	105	40	130	0.23	0.009	46	150	0.20	0.008

(a) High speed steel tool materials: use M2 or M7 for annealed stock and any premium high speed steel (T15, M33, M41 to M47) for cold drawn stock

Machining Data (Ref 2) (continued)

Type S20910: End Milling (Profiling)

Tool material	Depth of cut mm	Depth of cut in.	Speed m/min	Speed ft/min	Feed per tooth for cutter diameter of: 10 mm (0.375 in.) mm	10 mm (0.375 in.) in.	12 mm (0.5 in.) mm	12 mm (0.5 in.) in.	18 mm (0.75 in.) mm	18 mm (0.75 in.) in.	25-50 mm (1-2 in.) mm	25-50 mm (1-2 in.) in.
Annealed, with hardness of 210 to 250 HB												
M2, M3, M7 high speed steel	0.5	0.020	18	60	0.025	0.001	0.050	0.002	0.075	0.003	0.102	0.004
	1.5	0.060	14	45	0.038	0.0015	0.075	0.003	0.102	0.004	0.13	0.005
	diam/4	diam/4	12	40	0.025	0.001	0.038	0.0015	0.075	0.003	0.102	0.004
	diam/2	diam/2	11	35	0.018	0.0007	0.025	0.001	0.050	0.002	0.075	0.003
C-2 carbide	0.5	0.020	67	220	0.013	0.0005	0.025	0.001	0.050	0.002	0.075	0.003
	1.5	0.060	52	170	0.025	0.001	0.050	0.002	0.075	0.003	0.102	0.004
	diam/4	diam/4	43	140	0.025	0.001	0.038	0.0015	0.038	0.0025	0.075	0.003
	diam/2	diam/2	40	130	0.013	0.0005	0.025	0.001	0.050	0.002	0.063	0.0025
Cold drawn, with hardness of 325 to 375 HB												
Any premium high speed steel (T15, M33, M41 to M47)	0.5	0.020	15	50	0.025	0.001	0.050	0.002	0.075	0.003	0.102	0.004
	1.5	0.060	11	35	0.038	0.0015	0.075	0.003	0.102	0.004	0.13	0.005
	diam/4	diam/4	9	30	0.025	0.001	0.038	0.0015	0.075	0.003	0.102	0.004
	diam/2	diam/2	8	25	0.018	0.0007	0.025	0.001	0.050	0.002	0.075	0.003
C-2 carbide	0.5	0.020	53	175	0.013	0.0005	0.025	0.001	0.050	0.002	0.075	0.003
	1.5	0.060	41	135	0.025	0.001	0.050	0.002	0.075	0.003	0.102	0.004
	diam/4	diam/4	35	115	0.025	0.001	0.038	0.0015	0.063	0.0025	0.075	0.003
	diam/2	diam/2	30	100	0.013	0.0005	0.025	0.001	0.050	0.002	0.063	0.0025

Type S20910: Reaming

Based on 4 flutes for 3- and 6-mm (0.125- and 0.25-in.) reamers, 6 flutes for 12-mm (0.5-in.) reamers, and 8 flutes for 25-mm (1-in.) and larger reamers

Tool material	Speed m/min	Speed ft/min	Feed per revolution for reamer diameter of: 3 mm (0.125 in.) mm	3 mm (0.125 in.) in.	6 mm (0.25 in.) mm	6 mm (0.25 in.) in.	12 mm (0.5 in.) mm	12 mm (0.5 in.) in.	25 mm (1 in.) mm	25 mm (1 in.) in.	35 mm (1.5 in.) mm	35 mm (1.5 in.) in.	50 mm (2 in.) mm	50 mm (2 in.) in.
Annealed, with hardness of 210 to 250 HB														
Roughing														
M1, M2 and M7 high speed steel	18	60	0.075	0.003	0.13	0.005	0.20	0.008	0.25	0.010	0.30	0.012	0.40	0.015
C-2 carbide	23	75	0.102	0.004	0.20	0.008	0.30	0.012	0.40	0.016	0.50	0.020	0.60	0.024
Finishing														
M1, M2, and M7 high speed steel	9	30	0.075	0.003	0.075	0.003	0.102	0.004	0.15	0.006	0.18	0.007	0.20	0.008
C-2 carbide	12	40	0.075	0.003	0.102	0.004	0.15	0.006	0.20	0.008	0.23	0.009	0.25	0.010
Cold drawn, with hardness of 325 to 375 HB														
Roughing														
M1, M2, and M7 high speed steel	12	40	0.075	0.003	0.102	0.004	0.13	0.005	0.20	0.008	0.25	0.010	0.30	0.012
C-2 carbide	17	55	0.102	0.004	0.15	0.006	0.23	0.009	0.40	0.015	0.45	0.018	0.50	0.020
Finishing														
M1, M2 and M7 high speed steel	6	20	0.075	0.003	0.075	0.003	0.102	0.004	0.13	0.005	0.15	0.006	0.18	0.007
C-2 carbide	11	35	0.075	0.003	0.102	0.004	0.13	0.005	0.15	0.006	0.18	0.007	0.20	0.008

Machining Data (Ref 2) (continued)

Type S20910: Drilling

Tool material	Speed, m/min (ft/min)	Feed per revolution for nominal hole diameter of: 1.5 mm (0.062 in.) mm (in.)	3 mm (0.125 in.) mm (in.)	6 mm (0.25 in.) mm (in.)	12 mm (0.5 in.) mm (in.)	18 mm (0.75 in.) mm (in.)	25 mm (1 in.) mm (in.)	35 mm (1.5 in.) mm (in.)	50 mm (2 in.) mm (in.)
Annealed, with hardness of 210 to 250 HB									
M10, M7, M1 high speed steel	14 (45)	0.025 (0.001)	0.050 (0.002)	0.102 (0.004)	0.18 (0.007)	0.25 (0.010)	0.30 (0.012)	0.40 (0.015)	0.45 (0.018)
Cold drawn, with hardness of 325 to 375 HB									
Any premium high speed steel (T15, M33, M41-M47)	9 (30)	0.025 (0.001)	0.050 (0.002)	0.102 (0.004)	0.18 (0.007)	0.25 (0.010)	0.30 (0.012)	0.40 (0.015)	0.45 (0.018)

Type S20910: Boring

Depth of cut mm	Depth of cut in.	High speed steel Speed m/min	High speed steel Speed ft/min	High speed steel Feed mm/rev	High speed steel Feed in./rev	Uncoated carbide Speed, brazed m/min	Speed, brazed ft/min	Speed, inserted m/min	Speed, inserted ft/min	Feed mm/rev	Feed in./rev	Coated carbide Speed m/min	Speed ft/min	Feed mm/rev	Feed in./rev
Annealed, with hardness of 210 to 250 HB															
0.25	0.010	20	65(a)	0.075	0.003	50	165(b)	59	195(b)	0.075	0.003	73	240(c)	0.075	0.003
1.25	0.050	15	50(a)	0.13	0.005	40	130(d)	47	155(d)	0.13	0.005	58	190(e)	0.13	0.005
2.5	0.100	12	40(a)	0.30	0.012	35	115(d)	41	135(d)	0.30	0.012	53	175(e)	0.30	0.012
Cold drawn, with hardness of 325 to 375 HB															
0.25	0.010	15	50(f)	0.075	0.003	40	130(b)	47	155(b)	0.075	0.003	64	210(c)	0.075	0.003
1.25	0.050	12	40(f)	0.13	0.005	32	105(d)	38	125(d)	0.13	0.005	53	175(e)	0.13	0.005
2.5	0.100	9	30(f)	0.30	0.012	29	95(d)	34	110(d)	0.30	0.012	46	150(e)	0.30	0.012

(a) High speed steel tool material: M2 or M3. (b) Carbide tool material: C-3. (c) Carbide tool material: CC-3. (d) Carbide tool material: C-2. (e) Carbide tool material: CC-2. (f) Any premium high speed steel tool material (T15, M33, M41 to M47)

Type S20910: Cylindrical and Internal Grinding

Operation	Wheel speed m/s	Wheel speed ft/min	Work speed m/min	Work speed ft/min	Infeed on diameter mm/pass	Infeed on diameter in./pass	Traverse(a)	Wheel identification
Annealed, with hardness of 210 to 250 HB								
Cylindrical grinding								
Roughing	28-33	5500-6500	15-30	50-100	0.050	0.002	1/2	C54JV
Finishing	28-33	5500-6500	15-30	50-100	0.013 max	0.0005 max	1/6	C54JV
Internal grinding								
Roughing	25-33	5000-6500	23-60	75-200	0.013	0.0005	1/3	C60KV
Finishing	25-33	5000-6500	23-60	75-200	0.005 max	0.0002 max	1/6	C60KV
Cold drawn, with hardness of 325 to 375 HB								
Cylindrical grinding								
Roughing	28-33	5500-6500	15-30	50-100	0.050	0.002	1/4	C54IV
Finishing	28-33	5500-6500	15-30	50-100	0.013 max	0.0005 max	1/8	C54IV
Internal grinding								
Roughing	25-33	5000-6500	15-46	50-150	0.013	0.0005	1/3	C60JV
Finishing	25-33	5000-6500	15-46	50-150	0.005 max	0.0002 max	1/6	C60JV

(a) Wheel width per revolution of work

Type S21800

Type S21800: Chemical Composition

Element	Composition, %
Carbon	0.10 max
Manganese	7-9
Phosphorus	0.040 max
Sulfur	0.030 max
Silicon	3.50-4.50
Chromium	16-18
Nickel	8-9
Nitrogen	0.08-0.18

Characteristics. Type S21800 (Nitronic 60) is a galling- and wear-resistant, austenitic stainless steel. Corrosion resistance of this steel is better than that of type 304 in most media and pitting resistance is superior to that of type 316. Room-temperature yield strength is about twice that of type 304.

Type S21800 provides excellent high-temperature oxidation resistance and low-temperature impact resistance. In quiet seawater at ambient temperature, S21800 exhibits much better crevice-corrosion resistance than type 304 and slightly better resistance than type 316.

The machinability rating of type S21800 is 23% of that for AISI 1212. (The machinability rating of type 304 is about 45%.)

Type S21800 is readily welded using the inert-gas-shielded processes. Fusion welds made without filler metal with the gas tungsten-arc process will be sound, with wear characteristics approximating those of the unwelded base metal. Heavy weld deposits made with the gas-metal-arc process will also sound, with tensile strengths slightly above those of the unwelded base metal. Wear properties of these heavy weld deposits are slightly less than those of the base metal. When type S21800 stainless steel is used for weld overlay on most other stainless steels, sound deposits develop with properties about equal to that of an all-weld deposit.

Typical Uses. Applications of type S21800 include valve stems, seats and trim; nuts and bolts; screening; chain drive systems; and pins, bushings, and roller bearings.

Type S21800: Similar Steels (U.S. and/or Foreign). UNS S21800; ASME Code Case 1817; ASTM A193 (grade B8S), A194 (grade 8S), A276, A479, A580

Physical Properties

Type S21800: Selected Physical Properties (Ref 12)

Density	7.61 g/cm³ (0.275 lb/in.³)
Electrical resistivity at room temperature	98.2 nΩ·m
Magnetic permeability in the annealed condition	1.02 max
Magnetic permeability for cold reduction of 10-40%	1.02 max
Magnetic permeability for cold reduction of 50-70%	1.05 max

Type S21800: Thermal Treatment Temperatures

Treatment	Temperature range °C	°F
Forging, preheat	815	1500
Forging, equalize	1095	2000
Forging, forge	1175	2150
Annealing(a)	1065	1950
Hardening(b)	...	...

(a) Follow with rapid cooling. (b) Not hardenable by heat treatment

Type S21800: Average Coefficients of Linear Thermal Expansion (Ref 12)

Temperature range °C	°F	Coefficient μm/m·K	μin./in.·°F
25-95	75-200	15.8	8.8
25-205	75-400	16.6	9.2
25-315	75-600	17.3	9.6
25-425	75-800	17.6	9.8
25-540	75-1000	18.0	10.0
25-650	75-1200	18.5	10.3
25-760	75-1400	18.9	10.5
25-870	75-1600	19.3	10.7
25-980	75-1800	19.8	11.0

Type S21800: Dynamic Coefficients of Friction (Ref 12)

Tested in water at 20 °C (68 °F); self-mated

Alloy	Dynamic coefficient of friction for test stress level, N/mm³, of: 0.8	5.6	14.0	28.0	56.0	112.0
S21800	0.50	0.35	0.38	0.44	0.44	0.44
Stellite 6B	0.30	0.60	0.63	...	...	...
S24100	...	...	0.45	0.53	0.65	0.58

Mechanical Properties

Type S21800: Low-Temperature Mechanical Properties (Ref 12)

Test temperature °C	°F	Tensile strength MPa	ksi	Yield strength(a) MPa	ksi	Elongation(b), %	Reduction in area, %	Fracture strength MPa	ksi	Impact energy(c) J	ft·lb
−24	75	754	109.3	401	58.1	66.4	79.0	2317	336.1	313	231
−18	0	883	128.1	464	67.3	71.3	79.7	2988	433.4	293	216
−73	−100	1023	148.4	537	77.9	70.5	80.9	3083	447.1	267	197
−130	−200	1156	167.6	603	87.4	62.4	78.4	3151	457.0	230	170
−195	−320	1502	217.9	699	101.4	59.5	65.8	409	594.0	187	138
−255	−425	1405	203.8	864	125.3	23.5	26.6	1914	277.6	...	...

(a) 0.2% offset. (b) In 25 mm (1 in.). (c) Charpy V-notch

Type S21800: Typical Elevated-Temperature Tensile Properties of Annealed Bar (Ref 12)

Test specimens were 19- and 25-mm (0.75- and 1-in.) diam bar

Test temperature °C	°F	Tensile strength MPa	ksi	Yield strength(a) MPa	ksi	Elongation(b), %	Reduction in area, %	Hardness, HB
Room temperature		734	106.5	390	56.5	61.7	71.9	200
95	200	640	92.8	306	44.4	63.3	72.4	187
150	300	620	89.9	261	37.8	64.4	73.7	...
205	400	582	84.4	226	32.8	64.0	73.7	168
260	500	566	82.1	221	32.1	61.5	73.0	...
315	600	555	80.5	205	29.7	59.6	73.1	155
370	700	548	79.5	201	29.2	59.1	72.6	...
425	800	540	78.3	200	29.0	56.5	72.1	148
480	900	532	77.1	195	28.3	53.9	71.6	...
540	1000	520	75.4	193	28.0	52.2	70.4	145
595	1100	494	71.6	198	28.7	48.7	70.0	...
650	1200	459	66.6	194	28.1	48.2	69.6	144
705	1300	407	59.0	189	27.5	41.4	50.0	...
760	1400	343	49.8	174	25.3	47.1	53.9	143
815	1500	255	37.0	164	23.8	72.8	75.0	...
870	1600	208	30.2	113	16.4	72.8	...	110

(a) 0.2% offset. (b) Gage length is 4 × diam of the test specimen

Type S21800: Elevated-Temperature Stress-Rupture Strength of Annealed Bar (Ref 12)

Test specimens were 16- to 25-mm (0.63- to 1-in.) diam bar

Test temperature °C	°F	Stress to produce rupture in: 100 h MPa	ksi	1000 h MPa	ksi	10 000 h MPa	ksi
540	1000	495	72	360	52	24	35
595	1100	340	49	215	31	10	20
650	1200	200	29	115	17	69	10
730	1350	97	14	55	8	...	...
815	1500	46	6.7	30	4	...	...

Type S21800: Typical Tensile Properties of Cold Drawn Bar(a) at Room Temperature (Ref 12)

Reduction, %	Hardness, HRC	Tensile strength MPa	ksi	Yield strength(b) MPa	ksi	Elongation(c), %	Reduction in area, %
10	24	827	120	627	91	51	68
20	31	965	140	772	112	35	65
30	34	1110	161	910	132	26	62
40	37.5	1345	195	1055	153	20	57
50	41	1495	217	1200	174	15	53
60	43	1655	240	1345	195	12	48
70	46	1815	263	1495	217	10	40

(a) Test specimens were 11.2-mm (0.442-in.) diam at start. (b) 0.2% offset. (c) Gage length is 4 × diam of the test specimen

Type S21800: Typical Tensile Properties of Annealed Bar at Room Temperature (Ref 12)

Diameter mm	in.	Hardness	Tensile strength MPa	ksi	Yield strength(a) MPa	ksi	Elongation(b), %	Reduction in area, %
25	1.0	95 HRB	710	103	414	60	64	74
44	1.75	100 HRB	696	101	386	56	62	33
57	2.25	100 HRB	696	101	414	60	60	76
76	3.0	97 HRB	779	113	448	65	55	67
105	4.125	95 HRC	731	106	386	56	57	67

(a) 0.2% offset. (b) Gage length is 4 × diam of the test specimen

Type S21800: Modules of Elasticity (Ref 12)

180 GPa............ 26.2×10^6 psi

Machining Data

For machining data on type S21800, refer to the preceding machining tables for type S20910.

Types S21900, S21904

Types S21900, S21904: Chemical Composition

Element	Composition, % Type S21900	Type S21904
Carbon	0.08 max	0.04 max
Manganese	8-10	8-10
Phosphorus	0.060 max	0.060 max
Sulfur	0.030 max	0.030 max
Silicon	1.00 max	1.00 max
Chromium	18-21	18-21
Nickel	5-7	5-7
Nitrogen	0.15-0.40	0.15-0.40

Characteristics. Types S21900 and S21904 (Nitronic 40) are nitrogen-strengthened, austenitic stainless steels. Type S21904 is a low-carbon modification (0.040%) of type S21900. These steels have high strength levels, excellent corrosion resistance, and low magnetic permeability, even after severe cold working. They are essentially nonmagnetic in both the annealed and cold worked conditions. They have good elevated temperature properties and retain high strength and toughness at subzero temperatures.

In chemical media and industrial and marine atmospheres, the resistance of types S21900 and S21904 to attack is between that of type 304 and type 316. The low-carbon grade, S21904, is resistant to intergranular attack, and can usually be used in the as-welded condition. Stainless steels S21900 and S21904 exhibit good resistance to stress corrosion cracking in hot chloride solutions and in the sensitized and cold rolled conditions.

These alloys may be successfully welded by the shielded-metal-arc and inert-gas-shielded techniques. Filler metal should be of the same composition as the base metal. Standard stainless steel electrode types may be substituted, but their selection should be based on the properties required.

The machinability rating of types S21900 and S21904 is about 30% of that for free-machining AISI 1212. These steels may be cold formed, but due to their high yield strength, forming forces are somewhat higher than those needed for other austenitic stainless steels, such as type 304.

Typical Uses. Types S21900 and S21904 are primarily used in aircraft applications, such as ducting and bellows systems, tail pipes and exhaust systems, clamps, fasteners, flanges, and hydraulic tubing.

Type S21900: Similar Steels (U.S. and/or Foreign). UNS S21900; AMS 5561; ASTM A269 (XM-10), A276 (XM-10), A314 (XM-10), A412 (XM-10), A473 (XM-10), A580 (XM-10)

Type S21904: Similar Steels (U.S. and/or Foreign). UNS S21904; AMS 5595, 5656; ASME SA412; ASTM A269 (XM-11), A276 (XM-11), A314 (XM-11), A412 (XM-11), A473 (XM-11), A580 (XM-11)

Physical Properties

Types S21900, S21904: Thermal Treatment Temperatures

Treatment	Temperature range °C	°F
Forging, preheat	870	1600
Forging, start	1205	2200
Forging, finish	925 min	1700 min
Annealing and stress relieving(a)	1065-1120	1950-2050
Hardening(b)	...	...

(a) Follow with rapid air cooling for sheet and strip; use water or oil quench for heavier sections. (b) Hardenable by cold work only

Types S21900, S21904: Thermal Conductivity (Ref 12)

Temperature °C	°F	Thermal conductivity W/m·K	Btu/ft·h·°F
−180	−290	7.8	4.5
−73	−100	10.9	6.3
95	200	13.8	8.0
205	400	16.1	9.3
315	600	18.2	10.5
425	800	20.2	11.7
540	1000	22.5	13.0
650	1200	24.7	14.3
760	1400	26.8	15.5
870	1600	28.9	16.7

Types S21900, S21904: Average Coefficients of Linear Thermal Expansion (Ref 12)

Test specimens were in the annealed condition

Temperature range °C	°F	Coefficient μm/m·K	μin./in.·°F
25-95	80-200	16.7	9.3
25-205	80-400	17.3	9.6
25-315	80-600	18.2	10.1
25-540	80-1000	19.1	10.6
25-760	80-1400	20.0	11.1
25-870	80-1600	20.2	11.2
25-980	80-1800	20.5	11.4

Types S21900, S21904: Selected Physical Properties (Ref 12)

Density	7.83 g/cm³ (0.283 lb/in.³)
Electrical resistivity at room temperature	73 nΩ·m

Mechanical Properties

Types S21900, S21904: Typical Tensile Properties of Cold Drawn Wire (Ref 12)

Stress relieving heat treatment	Tensile strength MPa	Tensile strength ksi	Yield strength(a) MPa	Yield strength(a) ksi	Elongation(b), %	Reduction in area, %	Hardness, HRC	Relaxation, loss in load(c), %
15% final cold reduction								
None	883	128	745	108	56	70	...	...
540 °C (1000 °F) for 2 h; air cool	883	128	703	102	56	68	21	10.2
620 °C (1150 °F) for 2 h; air cool	862	125	635	92	45	67	20	10.0
705 °C (1300 °F) for 2 h; air cool	841	122	640	93	60	65	20	6.8
30% final cold reduction								
None	1235	179	1115	162	28	57	40	...
540 °C (1000 °F) for 2 h; air cool	1235	179	1115	162	28	57	39	9.1
620 °C (1150 °F) for 2 h; air cool	1145	166	1025	149	28	54	33	6.8
705 °C (1300 °F) for 2 h; air cool	1095	159	965	140	40	57	30	7.3

(a) 0.2% offset. (b) Gage length is 4 × diam of the test specimen. (c) Starting condition: compression spring stress at 275 MPa (40 ksi) at room temperature; exposed at 455 °C (850 °F) for 5 days

Types S21900, S21904: Typical Properties at Cryogenic Temperatures (Ref 12)

Test specimens were 120-mm (4.75-in.) thick slab in the annealed condition; tested in the transverse direction

Test temperature °C	Test temperature °F	Tensile strength MPa	Tensile strength ksi	Yield strength(a) MPa	Yield strength(a) ksi	Elongation(b), %	Reduction in area, %	Charpy V-notch J	Charpy V-notch ft·lb	Charpy Keyhole J	Charpy Keyhole ft·lb
24	75	710	103	400	58	50	70.0	278	205	...	...
−79	−110	924	134	600	87	59	71.0	198	146	81	60
−195	−320	1400	203	1035	150	...	24.0	88	65	50	37
−255	−425	1690	245	1350	196	15	20.5	72	53	46	34

(a) 0.2% offset. (b) In 25 mm (1 in.)

Types S21900, S21904: Short-Time Elevated-Temperature Tensile Properties of Annealed Sheet and Strip (Ref 12)

Test temperature °C	Test temperature °F	Tensile strength MPa	Tensile strength ksi	Yield strength(a) MPa	Yield strength(a) ksi	Elongation(b), %
24	75	765	111.0	444	64.4	42.5
95	200	693	100.5	353	51.2	41.5
205	400	616	89.4	289	41.9	40.0
315	600	594	86.1	259	37.5	33.0(c)
425	800	549	79.6	221	32.0	33.0(c)
480	900	518	75.1	202	29.3	41.0
540	1000	490	71.1	203	29.4	35.0
595	1100	467	67.8	193	28.0	30.0
650	1200	416	60.3	184	26.7	26.0
705	1300	354	51.3	185	26.8	24.0

(a) 0.2% offset. (b) In 50 mm (2 in.). (c) Broke on gage mark

Types S21900, S21904: Stress-Rupture Properties of Annealed Bar and Sheet (Ref 12)

Test temperature °C	Test temperature °F	Stress to produce rupture in 100 h, MPa	100 h, ksi	1000 h, MPa	1000 h, ksi
Bar					
650	1200	231	33.5	190	27.5
730	1350	145	21.0	93	13.5
815	1500	70	10.2	42	6.1
Sheet					
650	1200	221	32.0	148	21.5
815	1500	44	6.4	19	2.8

Types S21900, S21904: Tensile Properties of Annealed Bar at Room Temperature (Ref 12)

Section size mm	Section size in.	Test direction	Tensile strength MPa	Tensile strength ksi	Yield strength(a) MPa	Yield strength(a) ksi	Elongation(b), %	Reduction in area, %
152 × 152	6 × 6	Longitudinal	685	99	450	65	48	70
		Transverse	670	97	370	54	32	47
102 × 102	4 × 4	Longitudinal	675	98	400	58	55	75
64 × 112	2.5 × 4.5	Longitudinal	690	100	470	68	48	70
25 round	1 round	Longitudinal	690	100	395	57	53	75

(a) 0.2% offset. (b) In 50 mm (2 in.)

Types S21900, S21904: Properties of Welded Joints In Sheet and Strip at Elevated Temperatures (Ref 12)

Test temperature °C	°F	Tensile strength MPa	ksi	Yield strength(a) MPa	ksi	Elongation, % In 50 mm (2 in.)	In 13 mm (0.5 in.)
24	75	794	115.2	459	66.6	37.0	36.0
425	800	553	80.2	231	33.5	37.5	...
480	900	536	77.8	227	32.9	33.0	...
540	1000	507	73.5	216	31.4	28.5	...
595	1100	461	66.8	199	28.9	24.5	25.0
650	1200	365	52.9	192	27.9	20.0	20.0
705	1300	332	48.2	183	26.5	11.0	12.5

(a) 0.2% offset

Types S21900, S21904: Effect of Cold Reduction and Stress Relieving on Sheet and Strip (Ref 12)

Stress relieving heat treatment	Cold reduction, %	Tensile strength MPa	ksi	Yield strength(a) MPa	ksi	Elongation(b), %	Hardness
Tested at room temperature							
None	0	794	115.2	482	69.9	44.5	94 HRB
	15	978	141.9	905	131.2	22.5	34 HRC
	30	1222	177.2	1099	159.4	12.0	40 HRC
480 °C (900 °F) for 1 h	0	803	116.5	494	71.6	44.0	94 HRB
	15	1007	146.0	912	132.3	23.0	34 HRC
	30	1278	185.4	122	178.3	11.0	42 HRC
675 °C (1250 °F) for 1 h	0	798	115.8	492	71.4	44.0	94 HRB
	15	977	141.7	814	118.0	26.0	32 HRC
	30	1151	167.0	106	154.2	18.5	39 HRC
Tested at 480 °C (900 °F)							
None	0	537	77.9	234	33.9	41.5	...
	15	654	94.9	529	76.7	18.0	...
	30	836	121.2	717	104.0	7.0	...
675 °C (1250 °F) for 1 h	0	538	78.1	239	34.7	42.0	...
	15	631	91.5	490	71.0	21.5	...
	30	774	112.2	669	97.0	13.0	...

(a) 0.2% offset. (b) In 50 mm (2 in.)

Types S21900, S21904: Tensile Properties of Cold Rolled Sheet at Short-Time Elevated Temperatures (Ref 12)

Test temperature °C	°F	Tensile strength MPa	ksi	Yield strength(a) MPa	ksi	Elongation(b), %	Hardness, HRC
10% cold reduction							
24	75	870	126.2	630	91.4	27.0	26
95	200	779	113.0	611	88.6	28.0	...
205	400	696	100.9	500	72.5	25.0	...
315	600	681	98.7	470	68.1	25.5	...
425	800	624	90.5	430	62.4	26.5	...
540	1000	552	80.0	377	54.7	24.0	...
650	1200	469	68.0	351	50.9	17.0	...
25% cold reduction							
24	75	1097	159.1	876	127.0	14.5	35
95	200	998	144.8	881	127.8	12.0	...
205	400	931	135.1	836	121.3	5.5	...
315	600	877	127.2	759	110.1	6.0	...
425	800	798	115.8	667	96.8	10.5	...
540	1000	715	103.7	601	87.1	9.0	...
650	1200	592	85.8	500	72.5	8.0	...
60% cold reduction							
24	75	1407	204.0	1162	168.6	6.0	43
95	200	1321	191.6	1189	172.5	4.5	...
205	400	1256	182.2	1122	162.8	3.5	...
315	600	1214	176.1	1045	151.6	4.0	...

(a) 0.2% offset. (b) In 50 mm (2 in.)

Machining Data

For machining data on types S21900 and S21904 refer to the preceding machining tables for type S20910.

Type S24000

Type S24000: Chemical Composition

Element	Composition, %
Carbon	0.08 max
Manganese	11.5-14.5
Phosphorus	0.060 max
Sulfur	0.030 max
Silicon	1.00 max
Chromium	17-19
Nickel	2.50-3.75
Nitrogen	0.20-0.40

Characteristics. Type S24000 (Nitronic 33) is a low-nickel, austenitic, stainless steel with a yield strength in the annealed condition approximately twice that of type 304 stainless steel. Resistance of type S24000 to stress-corrosion cracking is better than that of type 304. The alloy has low magnetic permeability. After severe cold working, type S24000 has excellent strength and ductility at cryogenic temperatures and resistance to wear and galling that is superior to the standard austenitic stainless steels.

Type S24000 stainless steel may exhibit stress-corrosion cracking in hot chloride environments under certain conditions. It is superior to type 304 at low stress levels and equal to type 304 at high stress levels. This alloy exhibits excellent resistance to cracking polythionic acids ($H_sS_xO_6$, where x is usually 3, 4, or 5). These acids can form in petroleum refinery units, particularly desulfurizers, during shutdown. Type S24000 steel is usually not subject to intergranular attack.

The machinability rating of type S24000 is about 30% of that for AISI 1212. Tools should not be permitted to dwell or ride on the work surface, which will result in burnishing or work hardening.

Type S24000 stainless steel can be welded satisfactorily with most of the fusion welding processes. Filler metal should be Nitronic 35W, type 308L or 312.

Typical Uses. Applications of type S24000 include heat exchangers; process vessels, pipe, and tubing; riser pipe for underground power transmission; and cryogenic tanks, valves, and piping.

Type S24000: Similar Steels (U.S. and/or Foreign). UNS S24000; ASME SA240 (XM-29), SA249 (XM-29), SA312 (XM-29), SA688 (XM-29); ASTM A240 (XM-29), A249 (XM-29), A269 (XM-29), A276 (XM-29), A312 (XM-29), A313 (XM-29), A358 (XM-29), A412 (XM-29), A479 (XM-29), A580 (XM-29), A688 (XM-29)

Physical Properties

Type S24000: Average Coefficients of Linear Thermal Expansion (Ref 12)

Temperature range °C	°F	Coefficient µm/m·K	µin./in.·°F
25-93	78-200	16.0	8.9
25-205	78-400	16.6	9.2
25-315	78-600	17.5	9.7
25-425	78-800	18.2	10.1
25-540	78-1000	18.7	10.4
25-650	78-1200	19.4	10.8
25-760	78-1400	20.2	11.2
25-870	78-1600	20.5	11.4
25-980	78-1800	21.1	11.7

Type S24000: Thermal Treatment Temperatures

Treatment	Temperature range °C	°F
Annealing	1040-1095	1900-2000
Hardening(a)	...	...

(a) Hardenable by cold work only

Type S24000: Thermal Conductivity at Elevated Temperatures (Ref 12)

Temperature °C	°F	Thermal conductivity W/m·K	Btu/ft·h·°F
100	212	15.9	9.2
200	390	17.3	10.0
300	570	18.9	10.9
400	750	20.4	11.8
500	930	22.0	12.7
600	1110	23.4	13.5
700	1290	24.7	14.3
800	1470	26.1	15.1

Type S24000: Electrical Resistivity (Ref 12)

Temperature °C	°F	Electrical resistivity, nΩ·m
25	77	74
100	212	80
200	390	87
400	750	100
600	1110	111
800	1470	119

Type S24000: Density (Ref 12)

7.75 g/cm^3 .... 0.280 lb/in.^{3}

Machining Data

For machining data on type S24000, refer to preceding machining tables for type S20910.

Mechanical Properties

Type S24000: Typical Mechanical Properties of Annealed Sheet at Cryogenic Temperatures (Ref 12)

Test temperature °C	°F	Tensile strength MPa	ksi	Yield strength(a) MPa	ksi	Elongation(b), %
−18	0	1007	146.1	588	85.3	64.5
−46	−50	1048	152.0	647	93.9	63.0
−73	−100	1145	166.0	718	104.1	60.5
−100	−150	1237	179.4	796	115.5	55.0
−130	−200	1345	195.1	907	131.5	49.5
−150	−242	1437	208.4	1009	146.3	42.5
−195	−320	1607	233.1	1209	175.3	20.0

(a) 0.2% offset. (b) In 50 mm (2 in.)

Type S24000: Short-Time Elevated Temperature Properties (Ref 12)

Test specimens were 1.3-mm (0.050-in.) thick sheet in the annealed condition

Test temperature °C	°F	Tensile strength MPa	ksi	Yield strength(a) MPa	ksi	Elongation(b), %
24	75	807	117	495	72	49
93	200	731	106	405	59	50
205	400	640	93	310	45	44
315	600	625	91	270	39	44
425	800	565	82	250	36	49
540	1000	510	74	220	32	39

(a) 0.2% offset. (b) In 50 mm (2 in.)

Type S24000: Typical Properties of Cold Reduced Sheet (Ref 12)

Cold reduction, %	Tensile strength MPa	ksi	Yield strength(a) MPa	ksi	Elongation(b), %	Hardness
0	793	115	470	68	51.0	95 HRB
10	917	133	724	105	32.0	30 HRC
20	1105	160	965	140	18.0	37 HRC
30	1270	184	1150	167	10.0	41 HRC
40	1380	200	1250	181	7.5	42 HRC
50	1450	210	1315	191	6.5	44 HRC
60	1530	222	1370	199	5.0	45 HRC

(a) 0.2% offset. (b) In 50 mm (2 in.)

Types S24000: Minimum Properties Acceptable for Material Specification (Ref 12)

Product	Tensile strength MPa	ksi	Yield strength(a) MPa	ksi	Elongation(b), %	Hardness HRB
Sheet, strip	725	105	380	55	40	100 max
Plate	690	100	380	55	40	100 max
Annealed bar up to 200-mm (8-in.) diam	690	100	380	55	30	92

(a) 0.2% offset. (b) In 50 mm (2 in.) or 4 × diam of the test specimen

Type S24000: Stress-Rupture Strength (Ref 12)

Test specimens were 1.57-mm (0.062-in.) thick, mill-annealed sheet

Test temperature °C	°F	Stress to rupture in: 100 h MPa	ksi	1000 h MPa	ksi	10 000 h(a) MPa	ksi
480	900	538(a)	78.0(a)	434	63.0	359	52.0
540	1000	400	58.0	324	47.0	259	37.5
595	1100	338	49.0	265	38.5	214	31.0
650	1200	248	36.0	162	23.5	100	14.5
730	1350	103	15.0	52	7.6	28	4.0
815	1500	47	6.8	23(a)	3.3(a)	11	1.6

(a) Extrapolated data

Type S24100

Type S24100: Chemical Composition

Element	Composition, %
Carbon	0.15 max
Manganese	11-14
Phosphorus	0.060 max
Sulfur	0.030 max
Silicon	1.00 max
Chromium	16.5-19.5
Nickel	0.50-2.50
Nitrogen	0.20-0.45

Characteristics. Type S24100 (Nitronic 32) is a low-nickel, austenitic stainless steel that provides approximately twice the yield strength of type 304, as well as comparable corrosion resistance. The high work hardenability of this steel permits cold drawing to high-strength levels while maintaining good ductility. Type S24100 exhibits excellent wear- and galling-resistance in standard tests for these characteristics.

This stainless steel resists staining and rusting in 5% sodium chloride (NaCl) fog and high-humidity atmospheres. Weldments in their as-welded condition also resist corrosion in 5% NaCl fog. The corrosion resistance of type S24100 in weak acid solutions and pitting media approaches that of type 304 stainless steel. In more aggressive media, type S24100 stainless steel is somewhat less corrosion resistant than type 304.

Type S24100 is comparable to type 302, with respect to intergranular attack. But because of higher carbon content, it is not as resistant to intergranular attack as type 304. Weldments in their as-welded condition exhibit only a moderate corrosive attack in the sensitized area in boiling 65% nitric acid (HNO_3). Type S24100 steel may exhibit stress-corrosion cracking in hot chloride environments under certain conditions.

This steel is readily welded in all forms. Weld joints in the as-welded condition have strength equivalent to the unwelded base metal. Type S24100 has a machinability rating about 30% of the rating for AISI 1212. Tools should not be allowed to dwell or ride on the work surface, which results in burnishing or work-hardening.

Typical Uses. Applications of type S24100 include high-strength shafting and bolting, clamps for pole line hardware, concrete reinforcing accessories, abrasion and corrosion-resistant screens, high-strength nonmagnetic springs, wire forms, racks, and cages.

Type S24100: Similar Steels (U.S. and/or Foreign). UNS S24100; ASTM A276 (XM-28), A313 (XM-28), A580 (XM-28)

Physical Properties

Type S24100: Thermal Treatment Temperatures

Treatment	Temperature range °C	°F
Forging	1150-1205	2100-2200
Annealing(a)	1040	1900
Hardening(b)	...	...

(a) Follow with rapid cooling. (b) Hardenable by cold work only

Type S24100: Selected Physical Properties (Ref 12)

Density	7.78 g/cm³ (0.281 lb/in.³)
Magnetic permeability, when H is 200 Oe, in the annealed condition	1.008
Magnetic permeability, when H is 200 Oe, in the 70% cold drawn condition	1.011

Type S24100: Average Coefficients of Linear Thermal Expansion (Ref 12)

Temperature range °C	°F	Coefficient μm/m·K	μin./in.·°F
24-95	75-200	16.2	9.0
24-205	75-400	16.9	9.4
24-315	75-600	17.6	9.8
24-425	75-800	18.0	10.0
24-540	75-1000	18.5	10.3
24-650	75-1200	19.3	10.7
24-760	75-1400	19.8	11.0
24-870	75-1600	20.2	11.2
24-980	75-1800	20.9	11.6

Machining Data

For machining data on type S24100, refer to the preceding machining tables for type S20910.

Mechanical Properties

Type S24100: Typical Tensile Properties at Room Temperature (Ref 12)

Test specimens were in the annealed condition

Bar diameter		Tensile strength		Yield strength(a)		Elongation(b),	Reduction	Hardness,	Impact energy(c)	
mm	in.	MPa	ksi	MPa	ksi	%	in area, %	HRB	J	ft·lb
25	1	793	115	415	60	55	70	96	310	230

(a) 0.2% offset. (b) Gage length is 4 × diam of the test specimen. (c) Charpy V-notch

Type S24100: Typical Short-Time Elevated Temperature Tensile Properties (Ref 12)

Test specimens were 25-mm (1-in.) diam, mill-annealed bar

Test temperature		Tensile strength		Yield strength(a)		Elongation(b),	Reduction
°C	°F	MPa	ksi	MPa	ksi	%	in area, %
27	80	814	118	485	70	53	70
95	200	717	104	395	57	52	75
205	400	650	94	305	44	50	75
315	600	615	89	270	39	50	75
425	800	570	83	250	36	50	76
540	1000	525	76	235	34	47	76
650	1200	455	66	215	31	39	67
760	1400	345	50	200	29	31	38

(a) 0.2% offset. (b) Gage length is 4× diam of the test specimen

Type S24100: Typical Mechanical Properties of Cold Drawn Wire (Ref 12)

Test specimens were 6.4-mm (0.25-in.) diam rod in the annealed condition

Cold	Tensile strength		Yield strength(a)		Elongation(b),	Reduction	Hardness,
reduction, %	MPa	ksi	MPa	ksi	%	in area, %	HRC
0	827	120	450	65	55.0	70	...
10	1000	145	758	110	43.0	65	23.0
20	1185	172	931	135	28.5	60	27.5
30	1310	190	1090	158	18.5	55	32.0
40	1450	210	1240	180	12.5	50	37.0
50	1640	238	1405	204	10.0	45	40.0
60	1825	265	1545	224	9.5	40	44.0

(a) 0.2% offset. (b) In 50 mm (2 in.) or 4 × diam of the test specimen

Type S30430

Type S30430: Chemical Composition

Element	Composition, max %
Carbon	0.080
Manganese	2.0
Phosphorus	0.045
Sulfur	0.030
Silicon	1.0
Chromium	17-19
Nickel	8-10
Copper	3-4

Characteristics. Type S30430 is an austenitic stainless steel exhibiting a lower rate of work hardening than type 305. It is nonmagnetic in the annealed condition and becomes only very faintly magnetic after severe cold working.

Type S30430 exhibits the same corrosion resistance as type 304. Because this grade is subject to intergranular carbide precipitation when exposed in the temperature range of 480 to 870 °C (900 to 1600 °F), it is not recommended for use in the as-welded or as-forged condition. This steel has a safe scaling temperature in continuous exposure of 870 °C (1600 °F).

Typical Uses. Type S30430 is used for severe cold heading applications such as cold formed nuts and recessed head fasteners.

Type S30430: Similar Steels (U.S. and/or Foreign). UNS S30430; ASTM A493 (XM-7)

Physical Properties

Type S30430: Thermal Treatment Temperatures

Treatment	Temperature range °C	°F
Forging, start	1150-1260	2100-2300
Forging, finish	925 min	1700 min
Annealing(a)	1010-1120	1850-2050
Hardening(b)	...	...

(a) Although small forgings can be air cooled, better corrosion resistance can be obtained by water quenching; large forgings should be annealed after forging. (b) Hardenable by cold work only

Type S30430: Selected Physical Properties (Ref 6)

Melting range	1400-1455 °C (2550-2650 °F)
Density	8.00 g/cm³ (0.289 lb/in.³)
Specific heat at 0-100 °C (32-212 °F)	500 J/kg·K (0.12 Btu/lb·°F)
Electrical resistivity at 20 °C (68 °F)	72 nΩ·m

Type S30430: Average Coefficients of Linear Thermal Expansion (Ref 6)

Temperature range °C	°F	Coefficient μm/m·K	μin./in.·°F
0-100	32-212	17.3	9.6
0-315	32-600	17.8	9.9
0-650	32-1200	18.7	10.4

Type S30430: Thermal Conductivity (Ref 6)

Temperature °C	°F	Thermal conductivity W/m·K	Btu/ft·h·°F
100	212	16.3	9.4
500	930	21.5	12.4

Mechanical Properties

Type S30430: Tensile Properties (Ref 6)

Test specimens were 2.54- to 25.4-mm (0.1- to 1.0-in.) diam wire

Condition	Tensile strength MPa	ksi	Yield strength(a) MPa	ksi	Elongation(b), %	Reduction in area, %	Hardness	Charpy impact energy J	ft·lb
Annealed	505	73	215	31	70	80	70 HRB	325	240
Soft temper	560	81	380	55	...	...	...	...	...

(a) 0.2% offset. (b) In 50 mm (2 in.)

Type S30430: Modulus of Elasticity (Ref 6)

Stressed in:	Modulus GPa	10^6 psi
Tension	193	28.0
Torsion	86	12.5

Machining Data

For machining data on type S30430, refer to the preceding machining tables for AISI type 201.

Type S44625

Type S44625: Chemical Composition

Element	Composition, %
Carbon	0.010 max
Manganese	0.40 max
Phosphorus	0.020 max
Sulfur	0.020 max
Silicon	0.40 max
Chromium	25-27.5
Nickel	0.50 max
Molybdenum	0.75-1.50
Nitrogen	0.015 max
Niobium	0.05-0.20
Copper	0.20 max
Nickel and copper	0.50 max

Characteristics. Type S44625 (XM-27) is a ferritic, chromium-molybdenum stainless steel with good fabrication characteristics. This alloy provides good resistance to stress-corrosion cracking, pitting, and crevice corrosion. It has excellent resistance to sodium hydroxide, nitric acid, amines, ammonium carbamate, and organic acids, such as formic, oxalic, and acetic. This steel is highly resistant to intergranular corrosion and is not sensitized by thermal cycles experienced in welding operations.

Type S44625 is subject to 475 °C (885 °F) embrittlement on extended exposure to 370 to 570 °C (700 to 1060 °F), and the strength decreases rapidly above 540 °C (1000 °F).

In machining operations, the low work-hardening rate and excellent ductility tend to produce long, stringy chips. Procedures for machining type 446 ferritic stainless steel should be followed. The formability and deep-drawing characteristics are similar to carbon steel and slightly better than those for type 304 stainless steel. The minimum bend radius for tubing is two times the tube diameter. Roller expansion into tube sheets presents no difficulties.

Type S44625 may be joined by gas tungsten-arc and shielded-metal-arc welding. Filler metals which may be used include type 310 and 312 stainless steels and the nickel-chromium and nickel-chromium-molybdenum alloys.

Typical Uses. Specific uses of type S44625 include heat exchangers, preheaters, tanks, pressure vessels, piping, heat recuperators, ducting, tube sheets, reboilers, and thermocouple shields. Other applications include chemical process plants, oil refineries, food processing plants, and high-temperature oxidation and sulfidation environments.

Type S44625: Similar Steels (U.S. and/or Foreign). UNS S44625; ASME SA240 (XM-27), SA268 (XM-27), SA479 (XM-27), SA737 (XM-27); ASTM A176 (XM-27), A240 (XM-27), A268 (XM-27), A276 (XM-27), A314 (XM-27), A479 (XM-27), A731 (XM-27)

Physical Properties

Type S44625: Selected Physical Properties (Ref 8)

Density	7.8 g/cm³ (0.28 lb/in.³)
Specific heat at 0-100 °C (32-212 °F)	427 J/kg·K (0.102 Btu/lb·°F)
Electrical resistivity at room temperature	52-57 nΩ·m
Average coefficient of linear thermal expansion at 21-100 °C (70-212 °F)	9.9 μm/m·K (5.5 μin./in.·°F)

Type S44625: Thermal Conductivity (Ref 8)

Temperature °C	Temperature °F	Thermal conductivity W/m·K	Thermal conductivity Btu/ft·h·°F
Room temperature		20.33	11.75
260	500	16.8	9.7

Mechanical Properties

Type S44625, AISI Types 304, 316: Corrosion in Boiling Organic Acids (Ref 8)

Alloy grade	45% formic mm/yr	45% formic mil/yr	88% formic mm/yr	88% formic mil/yr	10% oxalic mm/yr	10% oxalic mil/yr	20% acetic mm/yr	20% acetic mil/yr	99% acetic mm/yr	99% acetic mil/yr
Type S44625	0.08	3	<0.0025	<0.1	0.08	3	0.025	1	0.01	0.5
AISI type 304	1.2	48	2.44	96.0	1.2	48	0.05	2	0.46	18.0
AISI type 316	0.30	12	0.23	9.0	1.0	40	0.025	1	0.05	2.0

(Column group: Corrosion rate for boiling organic acids)

Type S44625, Carbon Steel, AISI Type 304: Comparison of Sheet Formability (Ref 8)

Alloy grade	Olsen cup height mm	Olsen cup height in.	Strain-hardening exponent, n	Average strain ratio, r	Limiting draw ratio, *LDR*
Type S44625	10.7-11.4	0.42-0.45	0.17-0.18	1.31-1.45	2.34
Carbon steels	10.2-11.4	0.40-0.45	0.22	1.00-1.80	2.15-2.50
AISI type 304	12.7-14.2	0.50-0.56	0.45-0.50	0.90-1.00	2.16

Type S44625: Typical Tensile Properties of Annealed Sheet, Plate, and Tube (Ref 8)

Product form	Tensile strength MPa	Tensile strength ksi	Yield strength(a) MPa	Yield strength(a) ksi	Elongation(b), %	Hardness, HB
Sheet and plate						
Typical	485	70	345	50	30	83
ASTM A240, min	450	65	275	40	22	90 max
Tubing						
Typical	515	75	415	60	25	88
ASTM A268, min	450	65	275	40	20	90 max

(a) 0.2% offset. (b) In 50 mm (2 in.)

Mechanical Properties (continued)

Type S44625: Effect of Extended Exposure on Tensile Properties (Ref 7)

Extended exposure was 10 000 h

Thermal exposure °C	°F	Tensile strength MPa	ksi	Yield strength(a) MPa	ksi	Elongation(b), %	Hardness, HV
345	650	454	65.9	359	52.0	35	160
370	700	531	77.0	395	57.0	32	176
400	750	617	89.5	598	86.8	3	251

(a) 0.2% offset. (b) In 50 mm (2 in.)

Type S44625: Effect of Cold Work on Tensile Properties of Annealed Strip (Ref 8)

Reduction, %	Tensile strength MPa	ksi	Yield strength(a) MPa	ksi	Elongation(b), %	Hardness, HB
0	485	70	345	50	35.0	83
20	655	95	605	88	8.0	97
40	760	110	745	108	6.0	101
60	855	124	800	116	4.0	104
80	905	131	862	125	2.5	104

(a) 0.2% offset. (b) In 50 mm (2 in.)

Type S44625, AISI Type 316: Corrosion in Boiling, Dilute Sulfuric Acid Solutions (Ref 8)

Alloy grade	Nonactivated solution mm/yr	mil/yr	Activated solution mm/yr	mil/yr
Corrosion rate of 1% H_2SO_4				
Type S44625	0.02	0.7	13.7	541
AISI type 316	0.551	21.7	0.66	26
Corrosion rate of 5% H_2SO_4				
Type S44625	0.36	14	76.7	3020
AISI type 316	2.5	98	2.72	107

Machining Data

For machining data on type S44625, refer to the preceding machining tables for AISI type 405.

Type S45000

Type S45000: Chemical Composition

Element	Composition, %
Carbon	0.05 max
Manganese	1.00 max
Phosphorus	0.030 max
Sulfur	0.030 max
Silicon	1.00 max
Chromium	14-16
Nickel	5-7
Molybdenum	0.50-1.00
Copper	1.25-1.75
Niobium(a)	...

(a) Minimum content is eight times the carbon content

Characteristics. Type S45000 (Custom 450) is a martensitic, age-hardenable stainless steel which exhibits very good corrosion resistance with moderate strength. This alloy has a yield strength somewhat greater than 690 MPa (100 ksi) in the annealed condition, but is easily fabricated. A single-step aging treatment develops higher strength with good ductility and toughness.

Type S45000 will resist atmospheric corrosion, including salt water atmospheres. It shows excellent resistance to rusting and pitting in 5 and 20% salt spray at 35 °C (95 °F), and to oxidation up to 650 °C (1200 °F). The maximum service temperature for continuous service is 425 °C (800 °F).

The work-hardening rate of S45000 is relatively low, permitting a good deal of cold reduction without intermediate annealing. Deep drawing or stretching operations with sharp bends which produce localized elongation are to be avoided. This alloy can be machined using the same practices employed with other martensitic stainless steels at comparable hardness levels.

Type S45000 can be welded using most fusion and resistance welding methods. No preheating is required to prevent cracking during welding. Where optimal corrosion resistance is required, parts should be annealed after welding. Similar filler metal should be used to maintain optimum weld metal strength. Brazing temperature should coincide with the annealing temperature range so that reannealing is not necessary.

Type S45000: Similar Steels (U.S. and/or Foreign). UNS S45000; AMS 5763; ASME SA564 (XM-25), SA705 (XM-25); ASTM A564 (XM-25), A693 (XM-25), A705 (XM-25)

Physical Properties

Type S45000: Average Coefficients of Linear Thermal Expansion (Ref 9)

Temperature range °C	Temperature range °F	Coefficient: Condition A(a) μm/m·K	Condition A(a) μin./in.·°F	Condition B(b) μm/m·K	Condition B(b) μin./in.·°F
24-95	75-200	10.6	5.9	10.8	6.0
24-150	75-300	10.1	5.6	10.4	5.8
24-205	75-400	10.3	5.7	10.6	5.9
24-260	75-500	10.4	5.8	11.0	6.1
24-315	75-600	10.6	5.9	11.2	6.2
24-370	75-700	10.8	6.0	11.3	6.3
24-425	75-800	11.0	6.1	11.5	6.4
24-480	75-900	11.1	6.2	11.7	6.5
24-540	75-1000	11.0	6.1	11.75	6.53
24-595	75-1100	11.2	6.2	11.75	6.53

(a) Annealed at 1040 °C (1900 °F); water quenched. (b) Annealed and aged at 480 °C (900 °F) for 4 h; air cooled

Type S45000: Thermal Treatment Temperatures

Treatment	Temperature range °C	Temperature range °F
Forging, start	1150-1175	2100-2150
Forging, finish	1040 min	1900
Solution treatment	1025-1050	1875-1925
Hardening (aging)	480-565	900-1050
Overaging	620 max	1150 max

Type S45000: Density (Ref 9)

7.8 g/cm^3 .... 0.28 lb/in.^{3}

Machining Data

For machining data on type S45000, refer to the preceding machining tables for AISI type 630.

Mechanical Properties

Type S45000: Typical Tensile Properties of 25-mm (1.0 in.) Diam Bar at Room Temperature (Ref 9)

Treatment	Tensile strength MPa	Tensile strength ksi	Yield strength(a) MPa	Yield strength(a) ksi	Notch tensile strength(b) MPa	Notch tensile strength(b) ksi	Elongation(c), %	Reduction in area, %	Hardness, HRC	Impact energy(d) J	Impact energy(d) ft·lb
Solution annealed	986	143	814	118	1525	221	13.3	48.0	28.0	125, 133, 141	92, 98, 104
	972	141	814	118	1515	220	13.5	52.4	...	...	...
Aged at 455 °C (850 °F)	1350	196	1280	186	2050	297	14.5	54.5	44.0	24, 27, 31	18, 20, 23
	1345	195	1260	183	2050	297	14.7	53.5	...	...	...
Aged at 480 °C (900 °F)	1345	195	1280	186	2050	297	13.9	56.7	42.5	52, 56, 57	38, 41, 42
	1350	196	1305	189	2060	299	14.8	54.5	...	...	...
Aged at 510 °C (950 °F)	1290	187	1270	184	1985	288	15.6	58.8	41.5	53, 66, 71	39, 49, 52
	1290	187	1270	184	1980	287	15.7	57.7	...	...	...
Aged at 540 °C (1000 °F)	1195	173	1165	169	1890	274	16.8	61.7	39.0	68, 71, 71	50, 52, 52
	1185	172	1165	169	1870	271	17.7	64.6	...	...	...
Aged at 565 °C (1050 °F)	1095	159	1050	152	1760	255	19.8	64.6	37.0	91, 95	67, 70
	1105	160	1050	152	1760	255	19.5	66.5	...	...	...
Aged at 620 °C (1150 °F)	985	143	640	93	1425	207	22.6	69.0	28.0	122, 130, 141	90, 96, 104
	970	141	620	90	1450	210	24.1	69.9	...	...	...

(a) 0.2% offset. (b) The stress concentration factor, K_t, is 10. (c) Gage length is 4 × diam of the test specimen. (d) Charpy V-notch

Type S45000: Effect of Test and Aging Temperatures on Tensile Properties of 25-mm (1-in.) Diam Bar (Ref 9)

Test temperature °C	Test temperature °F	Aging temperature °C	Aging temperature °F	Tensile strength MPa	Tensile strength ksi	Yield strength(a) MPa	Yield strength(a) ksi	Notch tensile strength(b) MPa	Notch tensile strength(b) ksi	Elongation(c), %	Reduction in area, %	Impact energy(d) J	Impact energy(d) ft·lb
−195	−320	480	900	1795	260	1715	249	585	85	4.5	7.8	1.4	1
		565	1050	1540	223	1415	205	1560	226	22.0	57.7	7	5
		620	1150	1510	219	938	136	1715	249	29.8	55.3	49	36
−73	−100	480	900	1490	216	1425	207	1770	257	15.7	55.5	5.4	4
		565	1050	1240	180	1150	167	1950	283	21.1	64.6	56	41
		620	1150	1145	166	660	96	1655	240	25.2	66.6	89	66
−18	0	480	900	1415	205	1340	194	2110	306	15.2	56.7	22	16
		565	1050	1170	170	1105	160	1840	267	20.6	65.5	87	64
		620	1150	1060	154	640	93	1515	220	24.3	69.0	115	85
315	600	480	900	1105	160	951	138	...	...	12.0	47.7	54	40
		565	1050	1050	152	1005	146	...	...	12.4	49.3	68	50
		620	1150	917	133	862	125	...	...	14.1	53.7	111	82
425	800	480	900	1035	150	903	131	...	...	12.4	44.5	57	42
		565	1050	986	143	896	130	...	...	12.2	44.5	73	54
		620	1150	834	121	793	115	...	...	13.4	49.1	111	82
565	1050	480	900	580	84	525	76	...	...	24.0	74.6	89	66
		565	1050	585	85	540	78	...	...	26.5	73.7	91	67
		620	1150	540	78	485	70	...	...	30.0	76.6	113	83

(a) 0.2 offset. (b) The stress concentration factor, K_t, is 10.(c) Gage length is 4 × diam of the test specimen. (d) Charpy V-notch

Type S45500

Type S45500: Chemical Composition

Element	Composition, %
Carbon	0.05 max
Manganese	0.50 max
Phosphorus	0.040 max
Sulfur	0.030 max
Silicon	0.50 max
Chromium	11.0-12.5
Nickel	7.5-9.5
Molybdenum	0.50 max
Niobium	0.10-0.50
Copper	1.5-2.5
Titanium	0.80-1.4

Characteristics. Type S45500 (Custom 455) is a martensitic, age-hardenable stainless steel that is relatively soft and easily formable in the annealed condition. A single-step, age-hardening treatment develops high yield strengths with good ductility and toughness.

This alloy resists staining in normal air atmospheres, and shows no corrosion in fresh water. It has good resistance to pitting in 5% salt spray at 35 °C (95 °F) and in 5% ferric chloride at room temperature. In mild chemical environments, the level of general corrosion resistance is higher than that of 12% chromium steels (type 410) and approaches that of 17% chromium steels (type 430). Type S45500 has inherently good resistance to stress-corrosion cracking. This resistance improves as aging temperature is increased.

Type S45500 has good scaling resistance up to about 650 °C (1200 °F). It is suggested for use in service temperatures up to 425 °C (800 °F).

In the annealed condition, type S45500 may be cold drawn and cold rolled without intermediate annealing. Deep drawing or stretching operations will require intermediate anneals because elongation tends to be localized. Cold heading and warm forming operations are also easily performed. Type S45500 can be machined in the annealed condition, and welded in the same manner as other stainless steels.

Type S45500: Similar Steels (U.S. and/or Foreign). UNS S45500; AMS 5578, 5617, 5672, 5860; ASTM A313 (XM-16), A564 (XM-16), A693 (XM-16), A705 (XM-16)

Physical Properties

Type S45500: Average Coefficients of Linear Thermal Expansion (Ref 9)

Temperature range °C	Temperature range °F	Coefficient μm/m·K	Coefficient μin./in.·°F
22-95	72-200	10.6	5.9
22-150	72-300	10.8	6.0
22-205	72-400	11.2	6.2
22-260	72-500	11.7	6.5
22-315	72-600	12.1	6.7

Type S45500: Thermal Treatment Temperatures

Treatment	Temperature range °C	Temperature range °F
Forging, start	1040-1150	1900-2100
Forging, finish	815-925	1500-1700
Solution annealing	815-845	1500-1550
Hardening	480-565	900-1050

Type S45500: Thermal Conductivity (Ref 9)

Temperature °C	Temperature °F	Thermal conductivity W/m·K	Thermal conductivity Btu/ft·h·°F
100	212	18.0	10.4
200	390	19.7	11.4
300	570	21.3	12.3
400	750	23.4	13.5
500	930	24.7	14.3

Type S45500: Electrical Resistivity (Ref 9)

Treatment	Electrical resistivity, nΩ·m
Annealed	90
Annealed and aged at 510 °C (950 °F) for 4 h; air cooled	75

Type S45500: Density (Ref 9)

7.8 g/cm^3 .... 0.28 lb/in.^{3}

Machining Data

For machining data on type S45500, refer to the preceding machining tables for AISI type 630.

Mechanical Properties

Type S45500: Typical Mechanical Properties of Bar at Room Temperature (Ref 9)

Treatment	Size round mm	Size round in.	Tensile strength MPa	Tensile strength ksi	Yield strength(a) MPa	Yield strength(a) ksi	Notch tensile strength(b) MPa	Notch tensile strength(b) ksi	Elongation(c), %	Reduction in area, %	Hardness, HRC	Impact energy(d) J	Impact energy(d) ft·lb
Annealed at 815 °C (1500 °F) for ½ h; water quenched	100	4	965	140	795	115	...	...	12	50	31	...	...
	25	1	1000	145	795	115	1585	230	14	60	31	95	70
Hardened at 480 °C (900 °F) for 4 h; air cooled	100	4	...	...	...	...	...	...	...	...	...	...	...
	25	1	1725	250	1670	245	1795	260	10	45	49	12	9
Hardened at 510 °C (950 °F) for 4 h; air cooled	100	4	1585	230	1515	220	1725	250	10	45	48	11	8
	25	1	1620	235	1550	225	2070	300	12	50	48	19	14
Hardened at 540 °C (1000 °F) for 4 h; air cooled	100	4	1415	205	1345	195	1725	250	12	45	45	16	12
	25	1	1450	210	1340	200	2000	290	14	55	45	27	20
Hardened at 565 °C (1050 °F) for 4 h; air cooled	100	4	1310	190	1205	175	1725	250	14	50	40	34	25
	25	1	1310	190	1205	175	1795	260	15	55	40	47	35

(a) 0.2% offset. (b) The stress concentration factor, K_t, is 10. (c) Gage length is 4 × diam of the test specimen. (d) Charpy V-notch

Type S45500: Typical Tensile Properties of Strip at Room Temperature (Ref 9)

Treatment	Thickness mm	Thickness in.	Tensile strength MPa	Tensile strength ksi	Yield strength(a) MPa	Yield strength(a) ksi	Elongation(b), %	Elongation(c), %	Reduction in area, %	Hardness HRC
Annealed at 815 °C (1500 °F) for ½ h; water quenched	4.1	0.16	1105	160	930	135	18	8	54	33
	1.3	0.05	1105	160	1035	150	10	6	...	34
Hardened at 480 °C (900 °F) for 4 h; air cooled	4.1	0.16	1795	260	1725	250	8	3	25	31
	1.3	0.05	1795	260	1725	250	6	3	...	51
Hardened at 510 °C (950 °F) for 4 h; air cooled	4.1	0.16	1725	250	1655	240	10	4	40	48
	1.3	0.05	1725	250	1655	240	8	4	...	47
Hardened at 540 °C (1000 °F) for 4 h; air cooled	4.1	0.16	1515	220	1450	210	14	6	45	46
	1.3	0.05	1515	220	1450	210	12	5	...	44

(a) 0.2% offset. (b) In 25 mm (1 in.). (c) In 50 mm (2 in.)

Type S45500: Typical Tensile Properties at Elevated Temperature (Ref 9)

Test specimens were hardened at 815 °C (1500 °F) for ½ h; water quenched; aged at 480 °C (900 °F) for 4 h; air cooled

Test temperature °C	Test temperature °F	Tensile strength MPa	Tensile strength ksi	Yield strength(a) MPa	Yield strength(a) ksi	Elongation(b), %	Reduction in area, %	Hardness(c), HRC
Room temperature		1690	245	1635	237	11	48	49
260	500	1475	214	1370	199	10	49	49
315	600	1405	204	1295	188	11	50	49
370	700	1345	195	1240	180	12	52	49
425	800	1240	180	1145	166	14	50	49

(a) 0.2% offset. (b) Gage length is 4 × diam of the test specimen. (c) After testing

Ferralium® Alloy 255

Ferralium® Alloy 255: Chemical Composition (Ref 14)

Compositions given are nominal

Element	Composition, %
Carbon	0.04 max
Manganese	0.80
Silicon	0.45
Chromium	26
Nickel	5.5
Molybdenum	3.3
Nitrogen	0.17
Copper	1.7

Characteristics. Ferralium® alloy 255 is a patented, ferritic-austenitic stainless steel which combines high-mechanical strength, ductility, and hardness with resistance to corrosion and erosion.

Under most service conditions, alloy 255 has excellent resistance to sulfuric, phosphoric, nitric, and other acids and salts. It is also highly resistant to acetic, formic, and other organic acids and compounds. Alloy 255 is generally not suitable for handling hydrochloric acid and other severely reducing acids and compounds.

This steel has good resistance to stress-corrosion cracking, crevice corrosion, and pitting when compared to types 304 and 316 austenitic stainless steels. It is highly resistant to stress-corrosion cracking in sodium chloride and sea water.

Ferralium® alloy 255 is a duplex alloy with approximately equal portions of austenite- and ferrite-matrix phases. It contains primary $M_{23}C_6$ carbide in the annealed condition and forms secondary $M_{23}C_6$ carbides when exposed in the 540 to 1095 °C (1000 to 2000 °F) temperature range. Aging or use of this alloy in the 760 to 870 °C (1400 to 1600 °F) range should be avoided, because a large amount of sigma phase is produced. The presence of sigma phase greatly reduces the ductility of this alloy.

Ferralium® alloy 255 can be readily welded using conventional techniques. It can be machined in either the solution-treated or hardened condition using the same techniques as for austenitic stainless steels.

Typical Uses. Applications of Ferralium® alloy 255 include pumps, agitators, valves, and piping for handling wet phosphoric acid slurries; marine scrubbers, pumps, and mechanical seals; and equipment for handling hot organic acids, fatty acids, and solutions of nitric and hydrofluoric acids.

Physical Properties (Ref 14)

Ferralium® Alloy 255: Heat Treatment. Ferralium Alloy 255 is solution-treated at 1040 to 1120 °C (1900 to 2050 °F) and rapidly cooled. The alloy can be hardened by a subsequent heat treatment at 510 °C (950 °F) for 4 h, then air cooled.

Ferralium® Alloy 255: Selected Physical Properties

Density at 20 °C (68 °F)	7.81 g/cm³ (0.282 lb/in.³)
Electrical resistivity at 25 °C (77 °F)	0.84 μΩ · m
Thermal conductivity at 100 °C (212 °F)	15.1 W/m·K (8.75 Btu/ft·h·°F)

Ferralium® Alloy 255: Mean Coefficients of Linear Thermal Expansion

Temperature range °C	°F	Coefficient μm/m·K	μin./in. · °F
20-93	68-200	11.0	6.1
20-205	68-400	11.7	6.5
20-315	68-600	12.1	6.7
20-425	68-800	12.4	6.9
20-540	68-1000	13.0	7.2

Ferralium® Alloy 255: Average Tensile Properties and Modulus of Elasticity at Elevated Temperatures

Test specimens were 10.2- to 19-mm (0.400- to 0.75-in.) thick plate; heat treated at 1120 °C (2050 °F); cooled rapidly

Test temperature °C	°F	Tensile strength MPa	ksi	Yield strength(a) MPa	ksi	Elongation(b), %	Modulus of elasticity GPa	psi × 10⁶
Room temperature		867.4	125.8	674	97.8	27	210	30.5
95	200	786.0	114.0	594	86.1	27	206	29.9
205	400	704.0	102.1	515	74.7	27	198	28.7
315	600	706.7	102.5	494	71.7	29	189	27.4
425	800	694.3	100.7	487	70.7	26	179	25.9
540	1000	580	84.1	425	61.7	28	165	24.0

(a) 0.2% offset. (b) In 50 mm (2 in.)

Mechanical Properties (Ref 14) (continued)

Ferralium® Alloy 255: Typical Charpy V-Notch Properties at Room Temperature

Test specimens were 12.7 mm (0.500 in.) thick plate

Condition	Aging time, h	Impact energy for specimen: A(a) J	A(a) ft·lb	B(b) J	B(b) ft·lb	C(c) J	C(c) ft·lb	D(d) J	D(d) ft·lb
Solution heat treated at 1120 °C (2050 °F)	0	190	140	268	198	126	93	207	153
Aged at 205 °C (400 °F)	1	197	145	289	213	118	87	213	157
	10	201	148	256	189	137	101	230	170
	100	226	167	236	174	141	104	209	154
Aged at 315 °C (600 °F)	1	218	161	317	234	178	131	199	147
	10	219	162	320	236	155	114	190	140
	100	191	141	305	225	172	127	178	131
Aged at 425 °C (800 °F)	1	148	109	240	177	169	125	149	110
	10	60	44	258	190	41	30	92	68
	100	54	40	256	189	38	28	75	55
Aged at 540 °C (1000 °F)	1	114	84	183	135	164	121	123	91
	10	75	55	87	64	175	129	156	115
	100	56	41	142	105	41	30	85	63

(a) Longitudinal specimen with notch perpendicular to plate surface. (b) Longitudinal specimen with notch parallel to plate surface. (c) Transverse specimen with notch perpendicular to plate surface. (d) Transverse specimen with notch parallel to plate surface

Mechanical Properties (Ref 14)

Ferralium® Alloy 255: Average Tensile Properties and Hardness at Room Temperature

Product form	Thickness mm	Thickness in.	Tensile strength MPa	Tensile strength ksi	Yield strength(a) MPa	Yield strength(a) ksi	Elongation(b), %	Hardness
Solution heat treated at 1120 °C (2050 °F); cooled rapidly								
Sheet	0.84-2.1	0.033-0.083	896	130	683	99	25	24 HRC
	3.18	0.125	855	124	648	94	25	21 HRC
	4.75	0.187	869	126	662	96	25	23 HRC
Plate	6.4	0.250	841	122	641	93	27	24 HRC
	9.53	0.375	841	122	627	91	31	23 HRC
	12.7	0.500	834	121	600	87	34	97 HRB
	19-25	0.75-1.00	800	116	600	87	31	95 HRB
Solution heat treated; hardened for 4 h at 510 °C (950 °F)								
Sheet	0.84-2.1	0.033-0.083	979	142	772	112	22	...
	3.18	0.125	903	131	703	102	24	...
	4.75	0.187	910	132	689	100	24	...
Plate	6.4	0.250	855	129	683	99	26	35 HRC
	9.53	0.375	883	128	683	99	30	...
	12.7	0.500	876	127	641	93	31	32 HRC
	19-25	0.75-1.00	938	136	717	104	28	...

(a) 0.2% offset. (b) In 50 mm (2 in.)

Machining Data

For machining data on Ferralium® Alloy 255, refer to the preceding machining tables for AISI type 301.

Tool Steels

The classification of tool steels indicates an extra high-quality product, not exclusively used as tool material. Tool steels are closely controlled and inspected throughout manufacture, and are made and sold in relatively small lots. Thus, they differ from machinery and structural steels, not so much by final application, but by the degree of quality control exercised and quantities involved.

Tool steels may also be considered as a group of special grades which, after proper heat treatment, will provide the combination of properties required for cutting tool and die applications. Sometimes, only one or two tool steel compositions will be satisfactory for a given application, whereas in the manufacture of machinery parts, a variety of machinery steels might be used interchangeably without materially affecting the service life of the part. Thus, because most grades have been developed to meet specific requirements, tool steels are much more specialized than machinery or structural steels.

An important characteristic of tool steels as a group is that they generally tend to have excess carbides in their structure, making them harder and more wear resistant.

The American Iron and Steel Institute (AISI) published an identification and type classification of tool steels in which the steels are divided into the following major categories, each with an identifying letter symbol or symbols:

- Water-hardening tool steels, W
- Shock-resisting tool steels, S
- Cold work tool steels:
  - Oil-hardening types, O
  - Medium-alloy, air-hardening types, A
  - High-carbon, high-chromium types, D
- Special-purpose tool steels:
  - Low-alloy types, L
  - Carbon-tungsten types, F
  - Mold steels, P; P1 to P19 inclusive for low carbon types; P20 to P39 inclusive for other types
- Hot work tool steels:
  - Chromium-base types, H1-H29 inclusive
  - Tungsten-base types, H-30-H39 inclusive
  - Molybdenum-base types, H40-H59 inclusive
- High speed tool steels:
  - Tungsten-base types, T
  - Molybdenum-base types, M

End use or common properties is the criterion of comparison for each category, excluding water-hardening steels. The water-hardening tool steels are essentially carbon steels and are classified together because of a common manner of heat treatment.

Within the various groups there may be subdivisions. Thus, within the cold work tool steels category there are three subdivisions identified by the letters O, A, and D, which are based on the method of heat treatment, or the similarity of chemical composition and properties. These subdivisions are oil-hardening and medium-alloy, air-hardening types, which are based on quenching method, and high-carbon, high-chromium types, which are based on similar properties.

REFERENCES

1. *1981 SAE Handbook,* Part 1, Society of Automotive Engineers, Inc., Warrendale, PA, 1981
2. *Machining Data Handbook,* 3rd ed., Vol 1 and 2, Machinability Data Center, Metcut Research Associates, Inc., Cincinnati, OH, 1980
3. G.A. Roberts and R.A. Cary, *Tool Steels,* 4th ed., American Society for Metals, 1980
4. Bethlehem Steel Corp., Bethlehem, PA 18016
5. Carpenter Technology Corp., 101 W. Bern St., Reading, PA 19603
6. Crucible Specialty Metals Div., P.O. Box 977, Syracuse, NY 13201
7. Jessop Steel Co., wholly owned subsidiary of Athlone Industries, Inc., Washington, PA 15301
8. Teledyne Vasco, P.O. Box 151, Latrobe, PA 15650
9. The International Nickel Co., One New York Plaza, New York, NY 10004

Water-Hardening Tool Steels (W)

Water-Hardening Tool Steels (W): Chemical Composition

Compositions also include the following elements in maximum percentages of: 0.15 tungsten, 0.10 molybdenum, 0.20 copper, and 0.20 nickel

Steel type	Chemical composition, % C	Mn	P max	S max	Si	Cr	V
W1	0.60-1.40(a)	0.10-0.40	0.025	0.025	0.10-0.40	0.15 max	0.10 max
W2	0.60-1.40(a)	0.10-0.40	0.030	0.030	0.10-0.40	0.15 max	0.15-0.35
W5	1.05-1.15	0.10-0.40	0.030	0.030	0.10-0.40	0.40-0.60	0.10 max

(a) Various carbon contents are available; most frequently offered in the 0.90 to 1.10% carbon range

Characteristics. The water-hardening tool steels, listed under the letter symbol W, are essentially carbon steels, and are among the least expensive of tool steels. They must be water quenched to attain the necessary hardness, and except in very small sizes, will harden with a hard case and a soft core. They may be used for a wide variety of tools, but their limitations must be recognized, and discussion of the application with the producer is strongly recommended.

SAE Standard J438 (Ref 1) lists four grades in which water-hardening tool steels are usually available:

- *Special, grade 1:* the highest-quality, water-hardening carbon tool steel, controlled for hardenability and subject to the most rigid tests to ensure maximum uniformity in performance, with a composition held to closest limits
- *Extra, grade 2:* a high-quality, water-hardening carbon tool steel, controlled for hardenability and subject to tests to ensure good service for general application
- *Standard, grade 3:* a good-quality, water-hardening carbon tool steel recommended for applications where some latitude with respect to uniformity is permissible
- *Commercial, grade 4:* a commercial-quality, water hardening carbon tool steel, not controlled for hardenability, not subject to special tests

Microscopic, macroscopic, and hardenability requirements are not specified for standard- and commercial-quality carbon tool steels but do apply to special- and extra-quality material.

The following limits on composition are generally required for standard and commercial grades:

Grade	Chemical composition, max % Mn	Si	Cr
Standard	0.35	0.35	0.15
Commercial	0.35	0.35	0.20

The total of manganese, silicon, and chromium should not exceed 0.75%

In comparison with other tool steels, the water-hardening tool steels have poor nondeforming properties, fair safety in hardening, shallow depth of hardening, good toughness, poor resistance to the softening effect of heat, fair to good wear resistance, and the best machinability. The water-hardening tool steels have a machinability rating of 100%, as a basis for comparison with other tool steels. When compared with free-machining AISI 1212 steel, the machinability rating of W tool steels is 40%.

Typical Uses. Water-hardening tool steels are available as hot and cold finished bar, plate, sheet, rod, wire, and forgings. These products are normally fabricated into tools, dies, and fixtures.

AISI Type W1: Similar Steels (U.S. and/or Foreign). UNS T72301; ASTM A686; SAE J437 (W108), (W109), (W110), (W112), J438 (W108), (W109), (W110), (W112)

AISI Type W2: Similar Steels (U.S. and/or Foreign). UNS T72302; SAE J437 (W209), (W210), J438 (W209), (W210)

AISI Type W5: Similar Steels (U.S. and/or Foreign). UNS T72305; ASTM A686

Water-Hardening Tool Steels: Approximate Critical Points (Ref 4)

Steel type	Transformation point	Temperature range(a) °C	°F
W1, 0.80% C	Ac range	745-760	1370-1400
	Ar range	699-696	1290-1285
W1, 1.05% C	Ac range	745-765	1370-1410
	Ar range	701-699	1295-1290
W2, 1.05% C	Ac range	750-765	1380-1410
	Ar range	699-693	1290-1280

(a) When heating or cooling at a rate of 205 °C (400 °F) per hour

Physical Properties

Water-Hardening Tool Steels (W): Average Coefficients of Linear Thermal Expansion (Ref 9)

Temperature range °C	°F	Coefficient μm/m · K	μin./in. · °F
20-100	68-212	10.4	5.76
20-200	68-390	11.0	6.13
20-300	68-570	12.2	6.80
20-400	68-750	13.1	7.28
20-500	68-930	13.8	7.64
20-600	68-1110	14.2	7.90
20-700	68-1290	14.6	8.11

Water-Hardening Tool Steels (W): Selected Physical Properties (Ref 5)

Density	7.83 g/cm³ (0.283 lb/in.³)
Specific heat at 0-100 °C (32-212 °F)	461 J/kg · K (0.11 Btu/lb · °F)
Electrical resistivity at 20 °C (68 °F)	30 nΩ · m

Physical Properties

Water-Hardening Tool Steels (W): Thermal Treatment Temperatures

Treatment	Temperature range °C	°F	Treatment	Temperature range °C	°F
Forging, preheat	790	1450	Annealing(a)	760-790	1400-1450
Forging, start	980-1065	1800-1950	Hardening, preheat(b)	565-650	1050-1200
Forging, finish	815	1500	Hardening, quench from W1 (c)	770-790	1420-1450
Normalizing, preheat	790	1450	Hardening, quench from W2, W5 (c)	770-815	1420-1500
Normalizing, hold at	815	1500	Tempering(d)	175-275	350-525

(a) Maximum cooling rate of 42 °C (75 °F) per hour; hardness of 159 to 202 HB or 84 to 94 HRB. (b) For large tools and tools with intricate sections. (c) Hardness after quenching of 65 to 67 HRC. (d) Hardness after tempering of 65 to 56 HRC

Mechanical Properties

Water-Hardening Tool Steels (W): Modulus of Elasticity (Ref 9)

205 GPa............... 30×10^6 psi

Water-Hardening Tool Steels (W): Effect of Tempering Temperature on Ultimate Compressive Strength (Ref 4)

Tempering temperature °C	°F	Ultimate compressive strength MPa	ksi
As quenched		3590	521
150	300	3765	546
205	400	2910	422
260	500	2605	378
315	600	2350	341
370	700	1875	272
425	800	1595	231
480	900	1150	167
540	1000	1060	154
595	1100	1035	150
650	1200	760	110

Type W1 Water-Hardening Tool Steels: Effect of Tempering Temperature on Hardness

Test specimens were brine quenched from 790 °C (1450 °F); tempered 1 h

Tempering temperature °C	°F	Hardness, HRC	Equivalent scleroscope hardness
As hardened		66-67	96
95	200	66-67	96
150	300	64-65	92
175	350	63-64	90
190	375	62-63	88
205	400	61-62	86
260	500	58-59	82
315	600	54-55	74
370	700	50-51	68
425	800	46-47	62

Water-Hardening Tool Steels: Effect of Tempering Temperature on Impact Strength.

Tests were conducted using unnotched Charpy specimens for carbon tool steel:
—— 0.78% carbon, water quenched at 790 °C (1450 °F);
--- 1.08% carbon, water quenched at 775 °C (1425 °F);
— — 1.0% carbon, water quenched at 790 °C (1450 °F). (Ref 4)

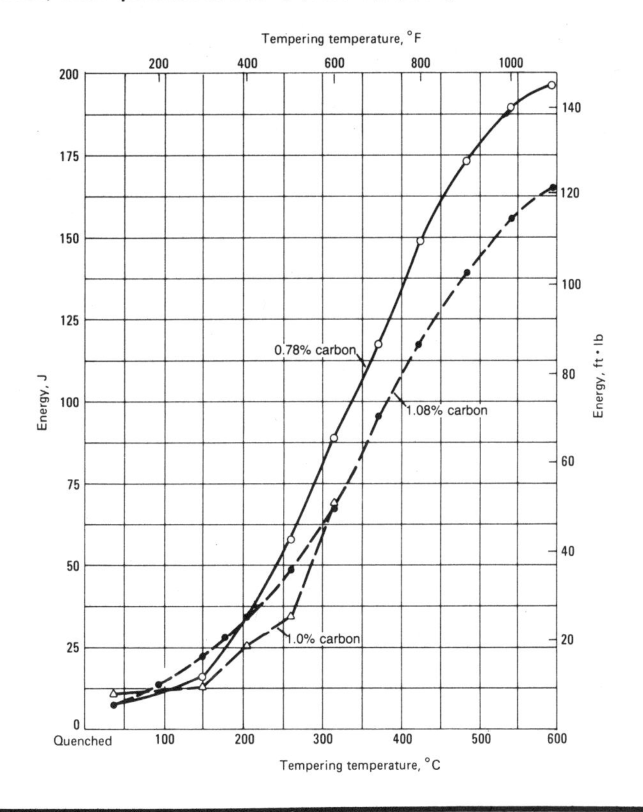

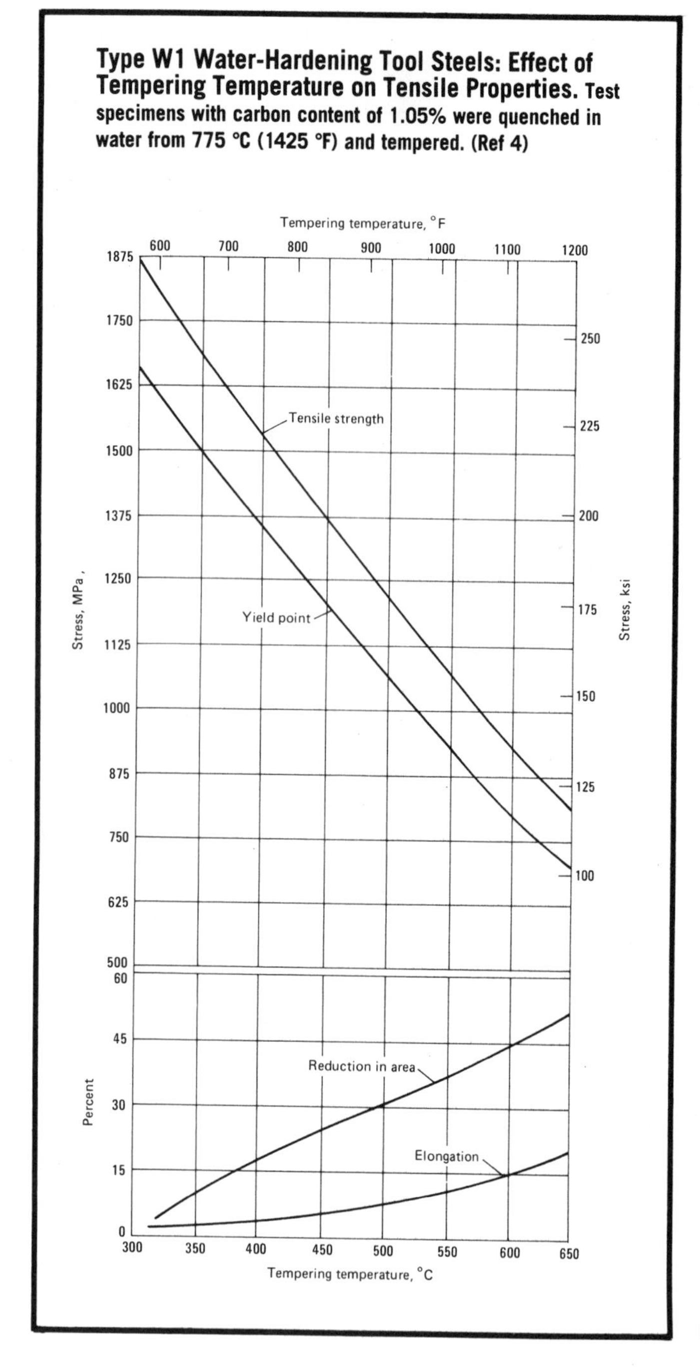

Type W1 Water-Hardening Tool Steels: Effect of Tempering Temperature on Tensile Properties. Test specimens with carbon content of 1.05% were quenched in water from 775 °C (1425 °F) and tempered. (Ref 4)

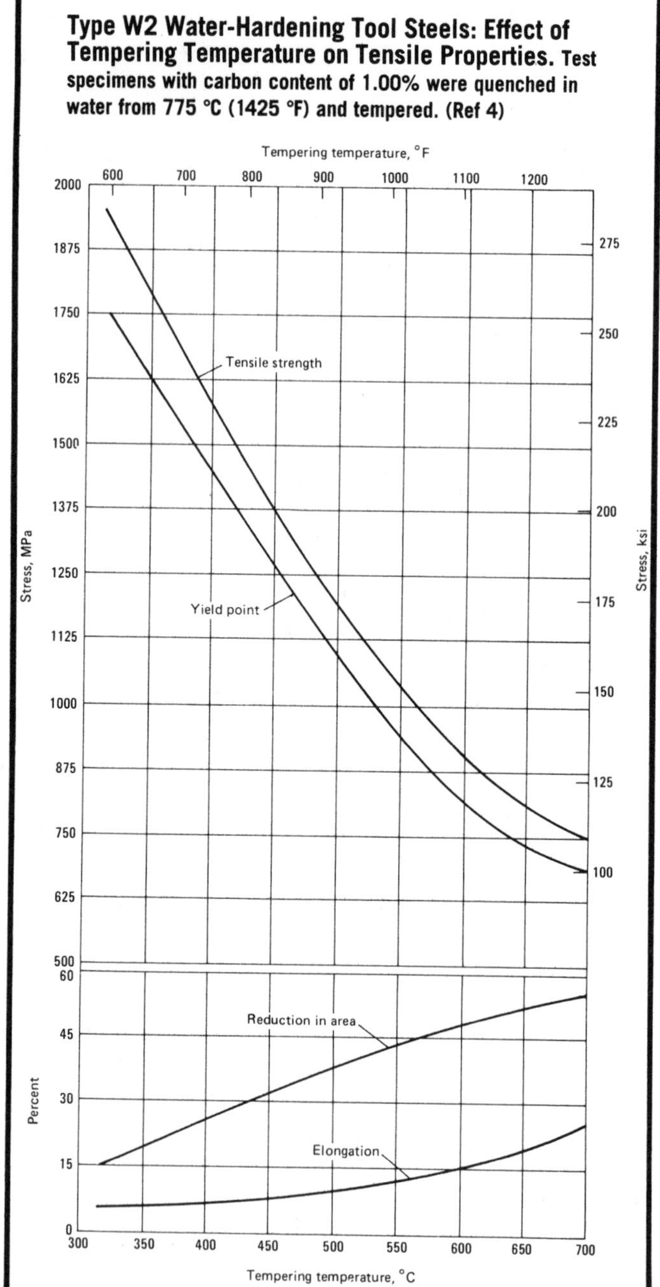

Type W2 Water-Hardening Tool Steels: Effect of Tempering Temperature on Tensile Properties. Test specimens with carbon content of 1.00% were quenched in water from 775 °C (1425 °F) and tempered. (Ref 4)

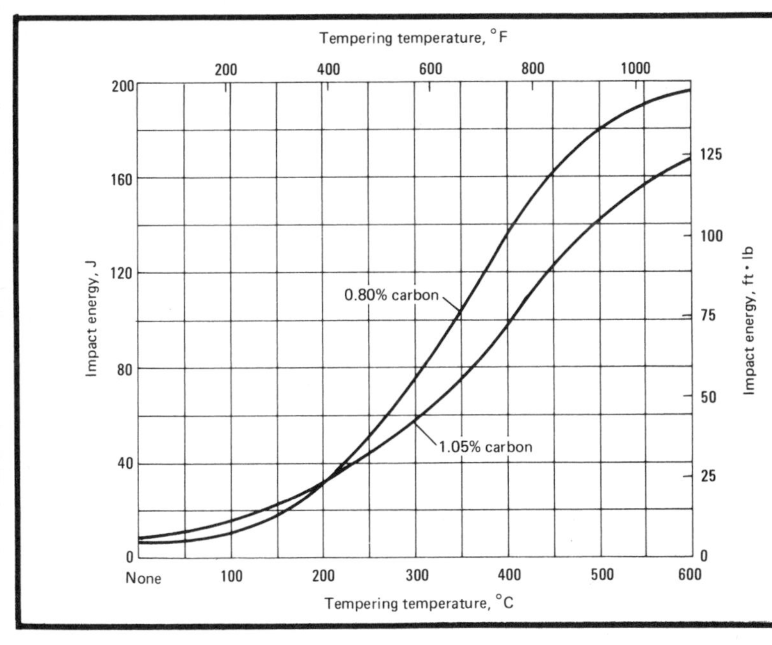

Type W1 Water-Hardening Tool Steels: Effect of Tempering Temperature on Impact Properties. Test specimens with carbon compositions of 0.80% and 1.05% were quenched in water from 775 °C (1425 °F) and tempered. Impact energy tests used unnotched Charpy specimens. (Ref 4)

Type W2 Water-Hardening Tool Steels: Effect of Tempering Temperature on Impact Values. Test specimens with carbon content of 1.00% were quenched in water from 775 °C (1425 °F) and tempered. Impact energy tests used unnotched Charpy specimens. (Ref 4)

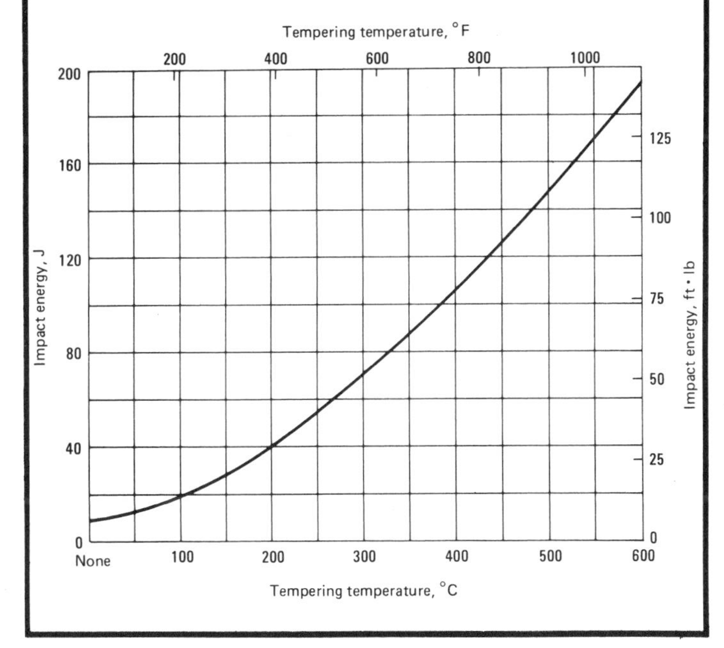

Type W1 Water-Hardening Tool Steels: Relationship of Hardness and Tempering Temperature. Test specimens with carbon content of 1.00% were austenitized at 785 °C (1450 °F) and quenched in water. (Ref 8)

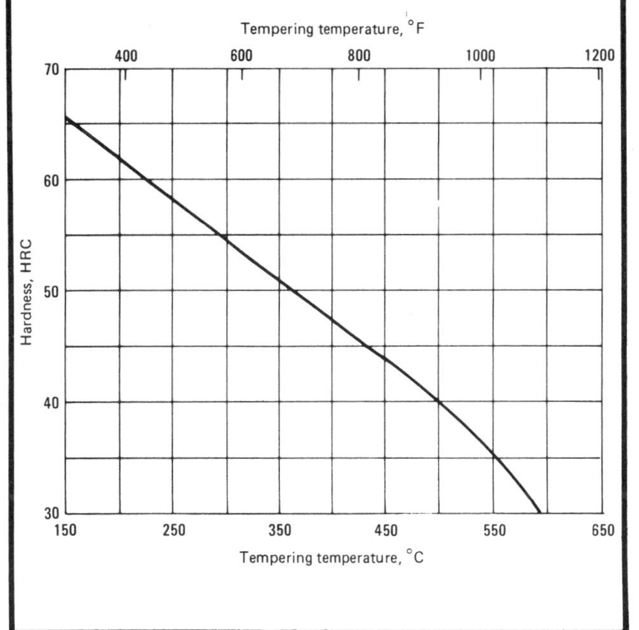

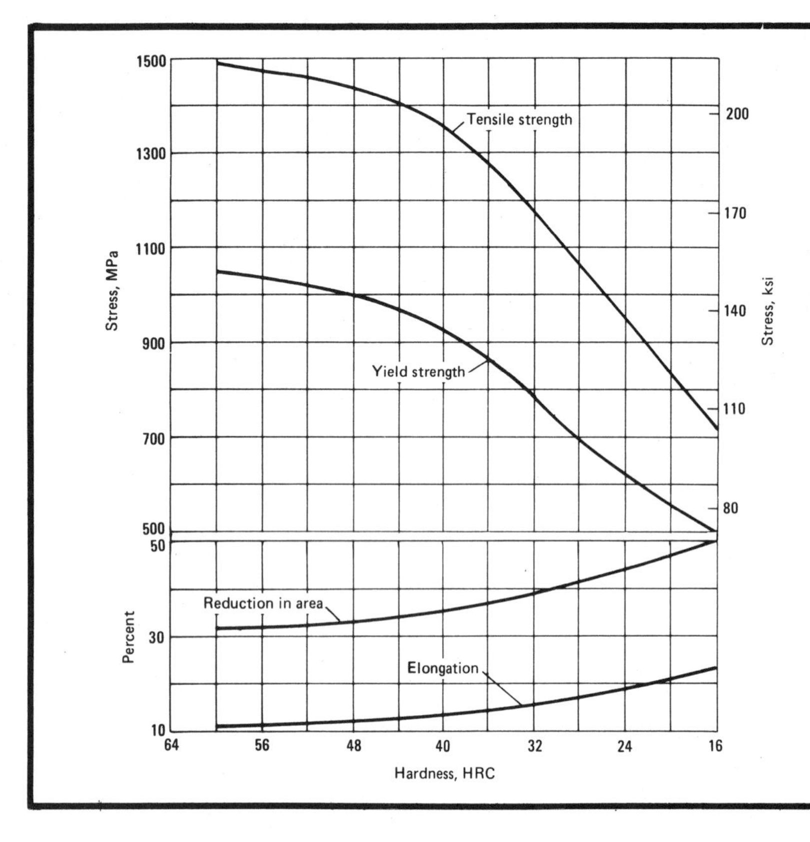

Type W1 Water-Hardening Tool Steels: Relationship of Mechanical Properties and Hardness. Test specimens with a carbon content of 1.20% were austenitized at 785 to 815 °C (1450 to 1500 °F) and quenched in a brine solution. Elongation was measured in 50 mm (2 in.). (Ref 5)

Shock-Resisting Tool Steels (S)

Shock-Resisting Tool Steels (S): Chemical Composition

Steel type	Chemical composition, %								
	C	Mn	P max	S max	Si	Cr	Mo	V	W
S1	0.40-0.55	0.10-0.40	0.030	0.03	0.15-1.20	1.00-1.80	0.50 max	0.15-0.30	1.50-3.00
S2	0.40-0.55	0.30-0.50	0.030	0.03	0.90-1.20	...	0.30-0.50	0.50 max	...
S4	0.50-0.65	0.60-0.95	0.030	0.03	1.75-2.25	0.35 max	...	0.35 max	...
S5	0.50-0.65	0.60-1.00	0.030	0.03	1.75-2.25	0.35 max	0.20-1.35	0.35 max	...
S6	0.40-0.50	1.20-1.50	0.030	0.03	2.00-2.50	1.20-1.50	0.30-0.50	0.20-0.40	...
S7	0.45-0.55	0.20-0.80	0.030	0.03	0.20-1.00	3.00-3.50	1.30-1.80	0.20-0.30(a)	...

(a) Optional

Characteristics. The shock-resisting tool steels are designated by the letter symbol S. There is considerable variation in their alloy content and hardenability, but all are intended for applications requiring high toughness and resistance to shock loading.

Type S1 is a versatile chromium-tungsten tool steel that provides excellent service in both hot and cold work shock applications. This steel exhibits good wear resistance and hot hardness value.

Type S2 is an extremely tough, strong steel that is recommended for applications where toughness is required above all other qualities. It has a hard case and tough core after heat treatment.

Type S4 is a tough alloy tool steel with good resistance to battering and shock at a relatively high hardness.

Type S5 is an extremely tough silicon-manganese tool steel with outstanding shock and abrasion resistance. This steel has greater hardenability than type S4. The depth of hardening of S5 when quenched in oil is approximately equal to that of type S4 which has been water or brine quenched from the same temperature.

Type S6 has good deep-hardening characteristics, plus excellent safety in hardening. Furthermore, it affords superior toughness and wear resistance in the hardness range of 54 to 56 HRC.

Type S7 is an air-hardening steel with high impact and shock resistance. This grade also has excellent toughness combined with good resistance to softening at temperatures up to 540 °C (1000 °F). Air hardening makes type S7 more dimensionally stable and safer to heat treat than other type S steels.

The shock-resisting steels can be welded using the inert-gas-shielded and shielded-metal-arc processes. Filler metal similar to the composition of the alloy should be used. Preheating and postheating are recommended when welding these steels.

The shock resisting steels have a machinability rating of about 75% of that for a 1% carbon tool steel.

Typical Uses. The shock-resisting tool steels have the following uses: heavy duty blanking and forming dies, punches, chisels, shear blades, slitter knives, cold striking dies, stamps, and hot work applications, such as headers, piercers, forming tools, shear blades, and drop forge die inserts.

AISI Type S1: Similar Steels (U.S. and/or Foreign). UNS T41901; ASTM A681; FED QQ-T-570; SAE J438; (W. Ger.) DIN 1.2550; (Fr.) AFNOR 55 WC 20; (Ital.) UNI 58 WCr 9 KU

AISI Type S2: Similar Steels (U.S. and/or Foreign). UNS T41902; ASTM A681; FED QQ-T-570; SAE J437, J438

AISI Type S4: Similar Steels (U.S. and/or Foreign). UNS T41904; ASTM A681; FED QQ-T-570

AISI Type S5: Similar Steels (U.S. and/or Foreign). UNS T41905; ASTM A681; FED QQ-T-570; SAE J437, J438

AISI Type S6: Similar Steels (U.S. and/or Foreign). UNS T41906; ASTM A681; FED QQ-T-570

AISI Type S7: Similar Steels (U.S. and/or Foreign). UNS T41907; ASTM A681

Physical Properties

Shock-Resisting Tool Steels (S): Thermal Treatment Temperatures

All grades should be cooled slowly from the annealing temperatures

Treatment	Temperature range °C	°F
Type S1		
Forging, start	1095-1150	2000-2100
Forging, finish	870 min	1600 min
Annealing	780-815	1440-1500
Hardening, preheat	760-790	1400-1450
Hardening, oil quench from	925-980	1700-1800
Tempering, for cold work	150-260	300-500
Tempering, for hot work	540-650	1000-1200
Type S2		
Forging, start	980-1065	1800-1950
Forging, finish	870 min	1600 min
Annealing	745-800	1375-1475
Hardening	845	1550
Tempering	150-480	300-900
Type S4		
Forging, start	1010-1050	1850-1925
Forging, finish	845 min	1550 min
Annealing	760-775	1400-1425
Hardening, water quench from	845-870	1550-1600
Hardening, oil quench from	870-900	1600-1650
Tempering	175-480	350-900
Type S5		
Forging, start	1010-1095	1850-2000
Forging, finish	900 min	1650 min
Annealing	730-775	1350-1425
Hardening	845-870	1550-1600
Tempering	175-480	350-900
Type S6		
Forging, start	1010-1095	1850-2000
Forging, finish	900 min	1650 min
Annealing	800-830	1475-1525
Hardening	915-955	1675-1750
Tempering	205-315	400-600
Type S7		
Forging, start	1095-1120	2000-2050
Forging, finish	925 min	1700 min
Annealing	815-870	1500-1550
Hardening, preheat	650-705	1200-1300
Hardening, oil quench from	940-955	1725-1750

Type S7 Shock-Resisting Tool Steels: Density

7.83 g/cm³ ....0.283 lb/in.³

Type S2 Shock-Resisting Tool Steels: Average Coefficients of Linear Thermal Expansion

Temperature range °C	°F	Coefficient μm/m·K	μin./in.·°F
20-100	68-212	10.9	6.04
20-200	68-390	11.9	6.63
20-300	68-570	13.0	7.24
20-400	68-750	13.6	7.54
20-500	68-930	14.1	7.83
20-600	68-1110	14.4	8.01
20-700	68-1290	14.6	8.12
20-750	68-1380	14.8	8.22

Type S7 Shock-Resisting Tool Steels: Average Coefficients of Linear Thermal Expansion

Temperature range °C	°F	Coefficient μm/m·K	μin./in.·°F
25-200	77-390	12.59	6.99
25-300	77-570	12.99	7.22
25-400	77-750	13.33	7.41
25-500	77-930	13.68	7.60
25-600	77-1110	14.01	7.78
25-700	77-1290	14.27	7.93

Mechanical Properties

Shock-Resisting Tool Steels (S): Effect of Tempering Temperature on Hardness (Ref 4, 5, 6)

Tempering temperature °C	°F	Hardness, HRC S1 (a)	S2 (b)	S4 (c)	S4 (d)	S5 (e)	S7 (f)
As quenched		61	61-63	64-65	61-63	61-62	59-61
150	300	58	59-60	60-61	60-61	60	58-60
205	400	56	57-58	59-60	58-60	59	56-58
260	500	54	55-56	59-60	58-59	59	54-56
315	600	53	54-55	57-58	57-58	58	53-55
370	700	53	52-53	56-57	56-57	57	52-54
425	800	50	49	51-52	51-52	54	52-54
480	900	48	45	48-49	47-48	51	51-53
540	1000	47	42-43	44-45	44-45	49	50-52
595	1100	47	38	...	...	44	46-48
650	1200	42	31	...	...	37	40-42
705	1300	...	23-24	...	...	31	33-35

(a) Oil quenched from 955 °C (1750 °F); tempered 1 h. (b) Brine or oil quenched from 845 to 870 °C (1550 to 1600 °F); tempered 1 h. (c) Water quenched from 870 °C (1600 °F); tempered 2 h. (d) Oil quenched from 900 °C (1650 °F); tempered 2 h. (e) Oil quenched from 870 °C (1600 °F). (f) Air quenched from 940 °C (1725 °F)

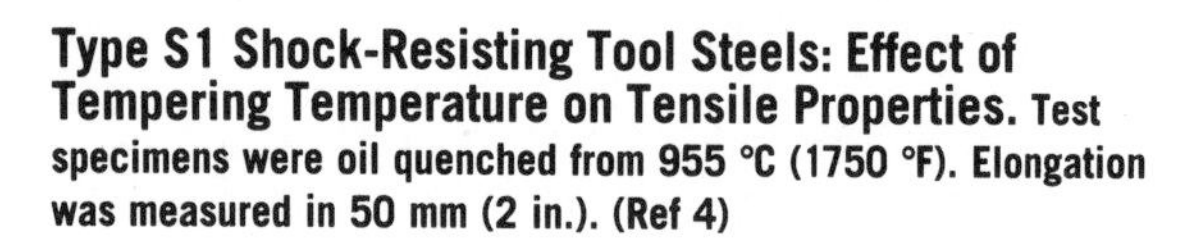

Type S1 Shock-Resisting Tool Steels: Effect of Tempering Temperature on Tensile Properties. Test specimens were oil quenched from 955 °C (1750 °F). Elongation was measured in 50 mm (2 in.). (Ref 4)

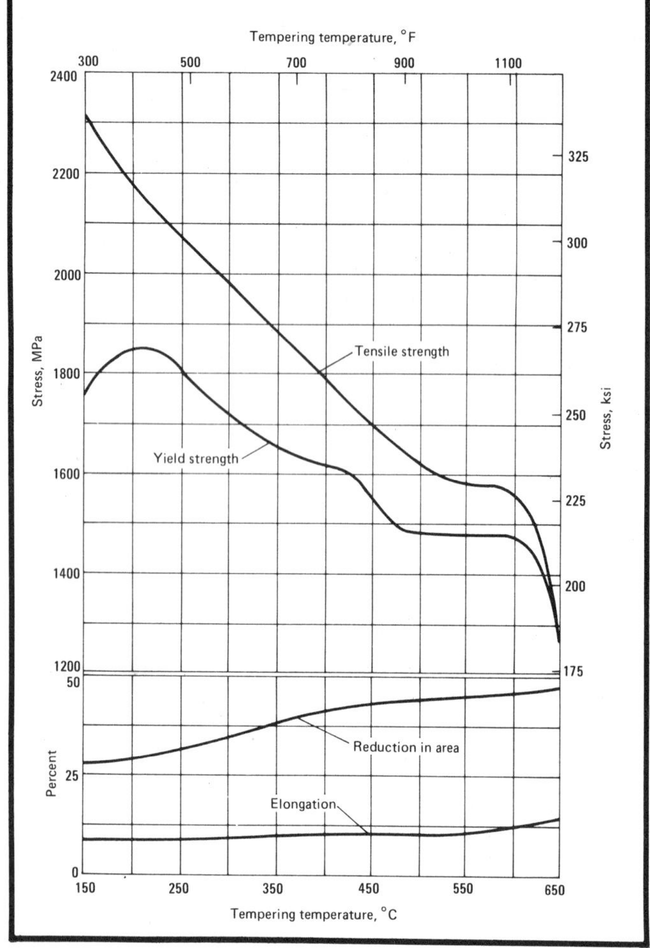

Type S1 Shock-Resisting Tool Steels: Effect of Tempering Temperature on Toughness. (1) unnotched Charpy test: specimens with compositions of 0.50% carbon, 0.75% silicon, 2.50% tungsten, 1.15% chromium, and 0.20% vanadium were oil quenched from 955 °C (1750 °F) (2) unnotched Charpy test: specimens with compositions of 0.45% carbon, 2.40% tungsten, 1.50% chromium, and 0.25% vanadium were oil quenched at 925 °C (1700 °F). (3) torsion impact test: specimens with compositions of 0.45% carbon, 2.40% tungsten, 1.50% chromium, and 0.25% vanadium were oil quenched at 925 °C (1700 °F). (Ref 4)

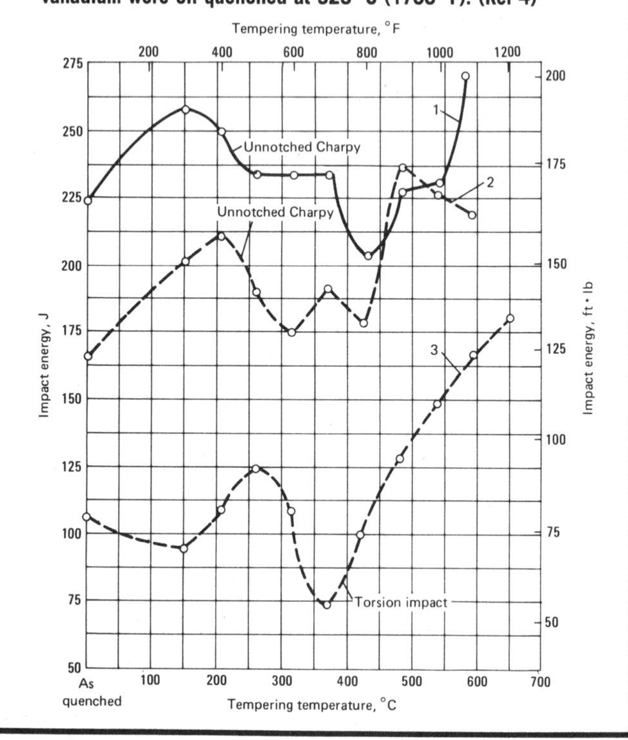

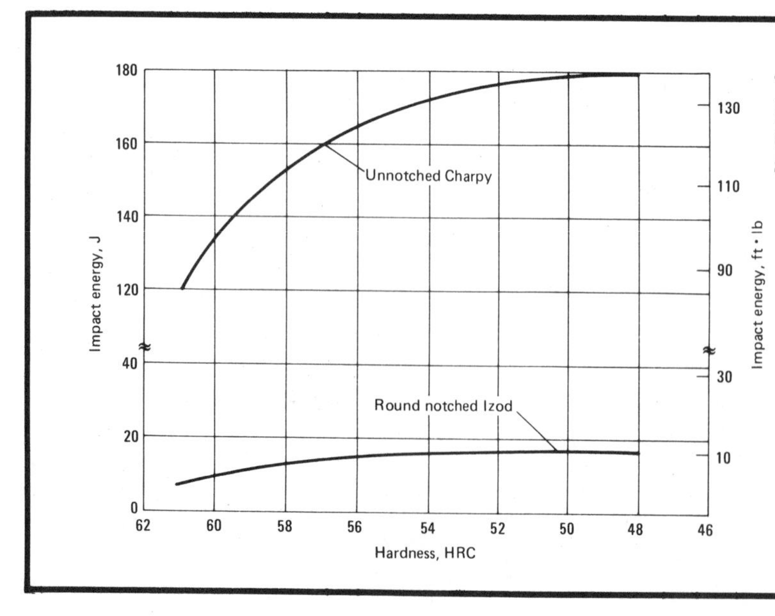

Type S5 Shock-Resisting Tool Steels: Relationship of Impact Properties and Hardness. Round notched Izod specimens were austenitized at 855 to 870 °C (1575 to 1600 °F); then quenched in oil. (Ref 5)

Type S2 Shock-Resisting Tool Steels: Relationship of Tensile Properties and Hardness. Specimens were austenitized at 845 °C (1550 °F); then quenched in brine. Elongation was measured in 50 mm (2 in.); yield strength at 0.2% offset. (Ref 5)

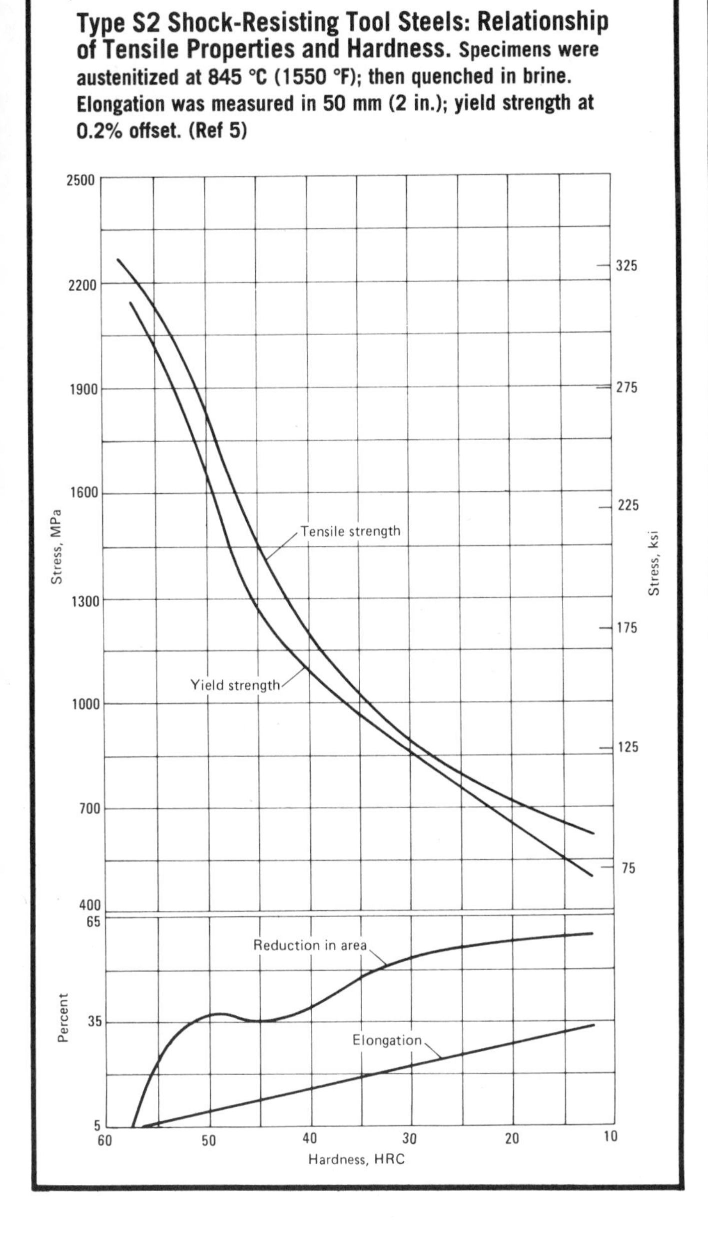

Type S5 Shock-Resisting Tool Steels: Relationship of Tensile Properties and Hardness. Specimens were austenitized at 855 to 870 °C (1575 to 1600 °F); then quenched in oil. Elongation was measured in 50 mm (2 in.). (Ref 5)

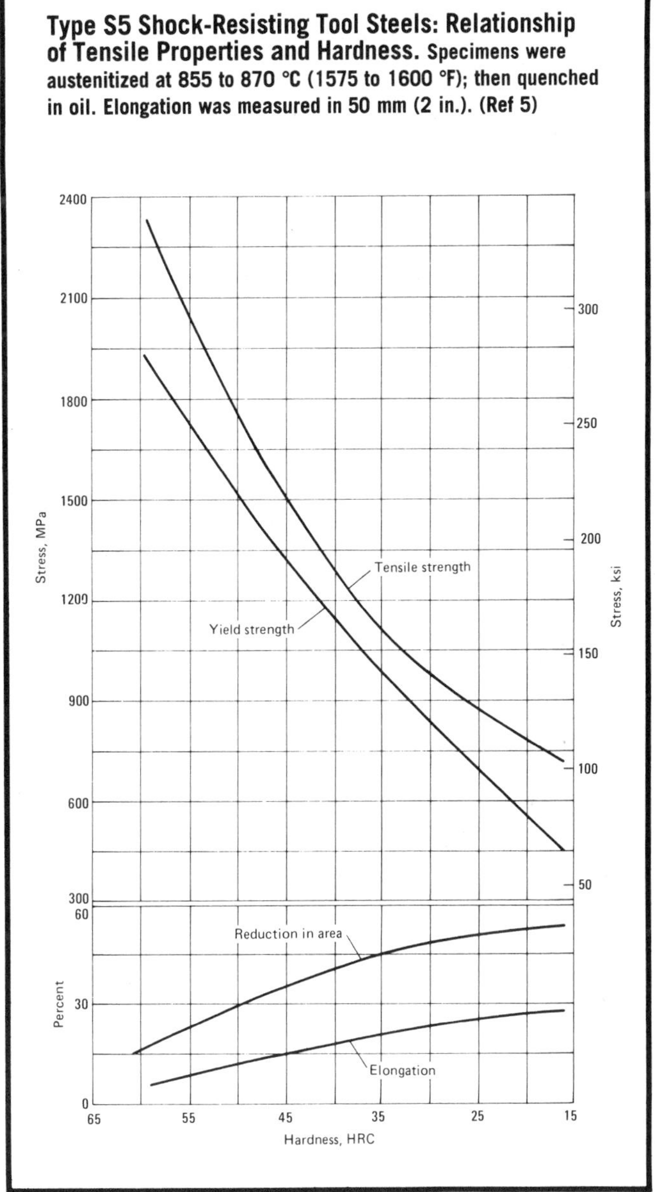

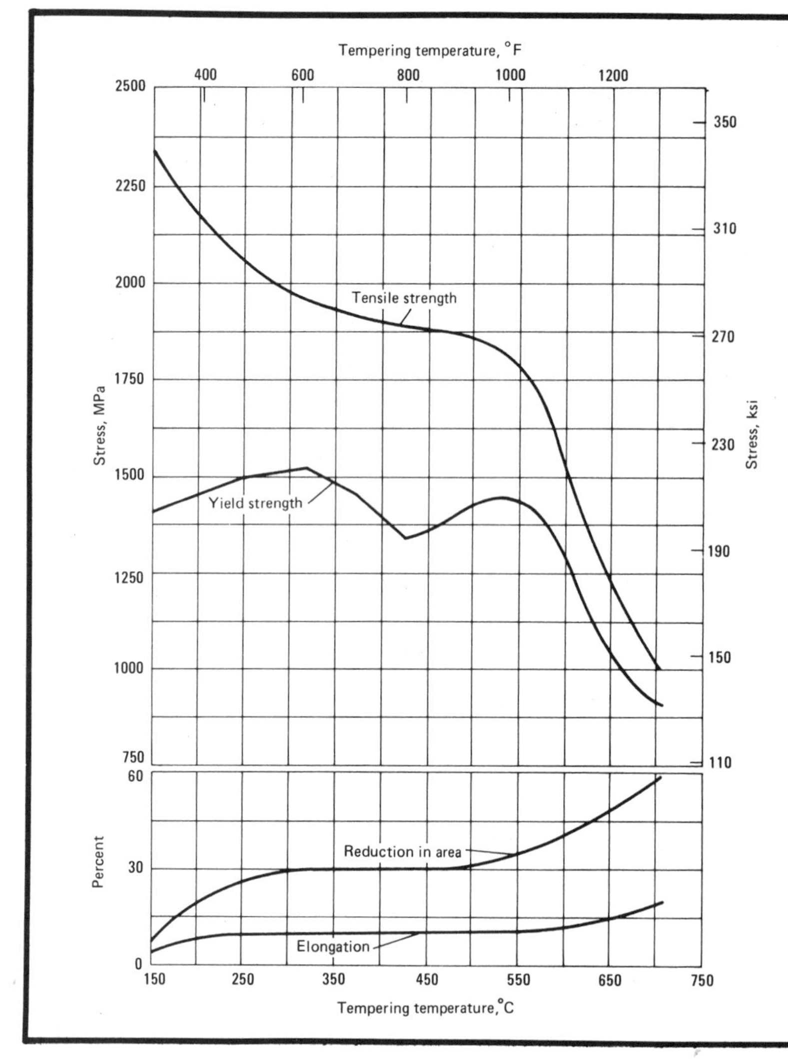

Type S7 Shock-Resisting Tool Steels: Effect of Tempering Temperature on Tensile Properties. Test specimens were quenched in still air from 940 °C (1725 °F) and tempered. Elongation was measured in 50 mm (2 in.); yield strength at 0.2% offset. (Ref 4)

Oil-Hardening Tool Steels (O)

Oil-Hardening Tool Steels (O): Chemical Composition

Steel type	Chemical composition, %								
	C	Mn	P max	S max	Si max	Cr	Mo	V max	W
O1	0.85-1.00	1.00-1.40	0.030	0.030	0.50	0.40-0.60	...	0.30	0.40-0.60
O2	0.85-0.95	1.40-1.80	0.030	0.030	0.50	0.35 max	0.30 max	0.30	...
O6 (a)	1.25-1.55	0.30-1.10	0.030	0.030	0.55-1.50	0.30 max	0.20-0.30	...	...
O7	1.10-1.30	1.00 max	0.030	0.030	0.60	0.35-0.85	0.30 max	0.40	1.00-2.00

(a) Contains free graphite in the microstructure to improve machinability

Characteristics. The oil-hardening tool steels are a subcategory of the cold work tool steels, designated by the letter symbol O. These low-alloy types must be oil quenched in heat treatment. Sizes larger than approximately 50 to 64 mm (2.0 to 2.5 in.) may exhibit lower hardness in the interior.

Type O1 is a general-purpose, oil-hardening, nondeforming alloy tool steel suitable for applications where extreme dimensional accuracy is required. Normal care in heat treatment gives good results in hardening and produces small dimensional changes. This grade has good abrasion-resistance, and sufficient toughness for normal tool and die applications. The machinability rating of type O1 is between 65 and 75% of a 1% carbon, water-hardening tool steel.

Type O2 is safe to harden, even in the most intricate sections. It shows very little size change or warpage during the hardening operation, less than that for type O1. The machinability rating of this grade is between 90 and 100% of a 1% carbon water-hardening steel.

Type O6 is a medium-alloy, oil-hardening tool steel. In the annealed condition about one third of the carbon is present as graphitic carbon, and the remainder as combined carbon in the form of carbides. Type O6 has a machinability rating of 125%, compared with a 1% carbon tool steel.

Type O7 has a tungsten addition which promotes harder carbides and a higher carbon content. This grade maintains a keen cutting edge. The machinability rating of annealed type O7 is 85% of a 1% carbon water-hardening tool steel.

Typical Uses. Oil-hardening tool steel types O1 and O2 are used for blanking, forming, drawing, molding, and trimming dies; spindles; thread gages; collets; and stamps. Type O6 is used for ring and plug gages, drawing dies, perforating dies, and punches where excellent wear resistance is desired. Type O7 is used for punches, dies, and knives where a keen cutting edge is required.

AISI Type O1: Similar Steels (U.S. and/or Foreign). UNS T31501; ASTM A681; FED QQ-T-570; SAE J437, J438; (W. Ger.) DIN 1.2510; (U.K.) B.S. BO 1

AISI Type O2: Similar Steels (U.S. and/or Foreign). UNS T31502; ASTM A681; FED QQ-T-570; SAE J437, J438; (W. Ger.) DIN 1.2842; (Fr.) AFNOR 90 MV 8; (Ital.) UNI 88 MnV 8 KU; (U.K.) B.S. BO 2

AISI Type O6: Similar Steels (U.S. and/or Foreign). UNS T31506; ASTM A681; FED QQ-T-570; SAE J437, J438

AISI Type O7: Similar Steels (U.S. and/or Foreign). UNS T31507; ASTM A681; FED QQ-T-570

Oil-Hardening Tool Steels (O): Approximate Critical Points (Ref 5)

Steel type	Transformation point	Temperature range °C	°F
O1	Ac range	755-790	1390-1450
	Ar range	695-680	1280-1260
O2	Ac range	725-750	1340-1380
O6	Ac range	760-770	1400-1420
	Ar range	725-695	1340-1280

Physical Properties

Oil-Hardening Tool Steels (O): Thermal Treatment Temperatures

Treatment	Temperature range °C	°F
Types O1, O6		
Forging, start	980-1065	1800-1950
Forging, finish	870 min	1600 min
Annealing	760-790	1400-1450
Hardening	790-815	1450-1500
Tempering	175-260	350-500
Type O2		
Forging, start	980-1050	1800-1925
Forging, finish	815 min	1500 min
Annealing	745-770	1375-1425
Hardening	760-800	1400-1475
Tempering	175-260	350-500
Type O7		
Forging, start	980-1090	1800-2000
Forging, finish	925 min	1700 min
Annealing	790-815	1450-1500
Hardening, water quench from	790-830	1450-1525
Hardening, oil quench from	815-885	1500-1625
Tempering	150-260	300-500

Type O1 Oil-Hardening Tool Steels: Average Coefficients of Linear Thermal Expansion (Ref 5)

Temperature range °C	°F	Coefficient μm/m·K	μin./in.·°F
38-260	100-500	10.8	6.0
38-425	100-800	12.8	7.1
38-540	100-1000	14.0	7.8
38-650	100-1200	14.4	8.0

Type O6 Oil-Hardening Tool Steels: Average Coefficients of Linear Thermal Expansion (Ref 5)

Temperature range °C	°F	Coefficient μm/m·K	μin./in.·°F
25-250	77-480	12.48	6.9
25-300	77-570	12.86	7.1
25-350	77-660	13.17	7.3
25-400	77-750	13.43	7.5
25-450	77-840	13.65	7.6
25-500	77-930	13.86	7.7

Type O2 Oil-Hardening Tool Steels: Average Coefficients of Linear Thermal Expansion (Ref 5)

Temperature range °C	°F	Coefficient μm/m·K	μin./in.·°F
25-40	77-100	10.3	5.7
25-95	77-200	11.7	6.5
25-150	77-300	12.6	7.0
25-205	77-400	12.6	7.0
25-260	77-500	13.0	7.2
25-315	77-600	13.3	7.4
25-370	77-700	13.9	7.7
25-425	77-800	13.9	7.7
25-480	77-900	14.4	8.0
25-540	77-1000	14.6	8.1
25-595	77-1100	14.9	8.3
25-650	77-1200	15.1	8.4

Oil-Hardening Tool Steels (O): Density (Ref 5)

Steel type	Density g/cm³	lb/in.³
O1	7.83	0.283
O2	7.61	0.275
O6	7.67	0.277

Mechanical Properties

Oil-Hardening Tool Steels (O): Modulus of Elasticity (Ref 5)

Steel type	Modulus GPa	10^6 psi
O1	214	31.0
O2	209	30.3

Mechanical Properties (continued)

Type O1 Oil-Hardening Tool Steels: Effect of Tempering Temperature on Hardness (Ref 5)

Test specimens were oil quenched from 800 °C (1475 °F); tempered 1 h at heat

Tempering temperature °C	°F	Hardness, HRC
As hardened		63-65
95	200	63-65
150	300	63-65
175	350	62-64
205	400	60-63
230	450	60-62
260	500	58-60
315	600	55-57
370	700	52-54
425	800	48-50
480	900	44-47
540	1000	40-44

Type O6 Oil-Hardening Tool Steels: Effect of Tempering Temperature on Hardness (Ref 5)

Test specimens were oil quenched from 800 °C (1475 °F); tempered 1 h at temperature indicated

Tempering temperature °C	°F	Hardness, HRC
As hardened		65.0
120	250	64.0
150	300	63.0
175	350	62.0
190	375	61.0
205	400	61.0
230	450	60.0
260	500	59.5
315	600	58.5
370	700	56.0
425	800	51.0
480	900	47.0
540	1000	42.5
595	1100	37.5
605	1200	31.0

Type O1 Oil-Hardening Tool Steels: End-Quench Hardenability. Test specimens were austenitized at 800 °C (1475 °F), oil quenched; then tempered for a minimum of 1 h to desired hardness. (Ref 5)

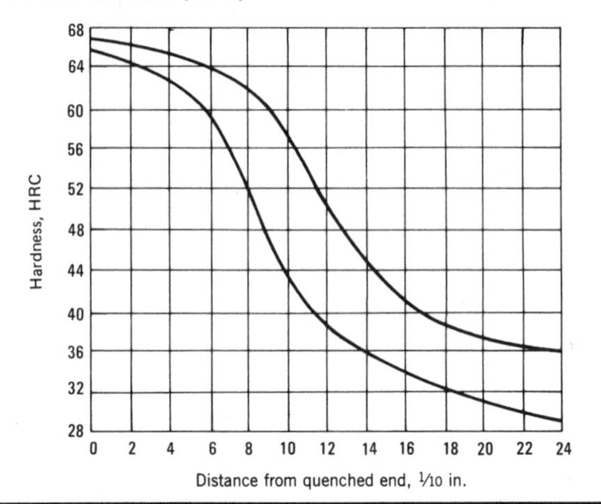

Type O2 Oil-Hardening Tool Steels: Effect of Tempering Temperature on Hardness (Ref 5)

Test specimens were oil quenched from 800 °C (1475 °F); tempered for 1 h at temperature indicated

Tempering temperature °C	°F	Hardness, HRC
As quenched		63-65
95	200	63-65
150	300	63-64
175	350	62-63
190	375	61-62
205	400	60-61
260	500	58-59
315	600	54-55
370	700	50-51
425	800	47-48
480	900	43-44
540	1000	33-37

Type O6 Oil-Hardening Tool Steels: End-Quench Hardenability. Test specimens were preheated to 675 °C (1250 °F); austenitized at 785 to 815 °C (1450 to 1500 °F) for 1 h; then quenched in oil. (Ref 4)

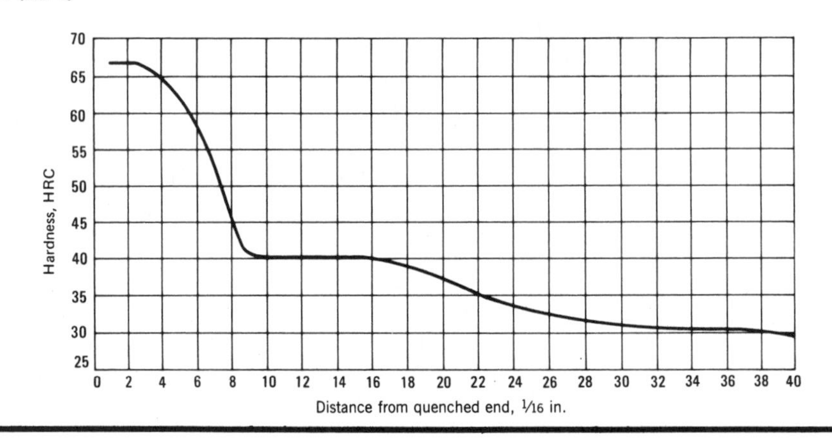

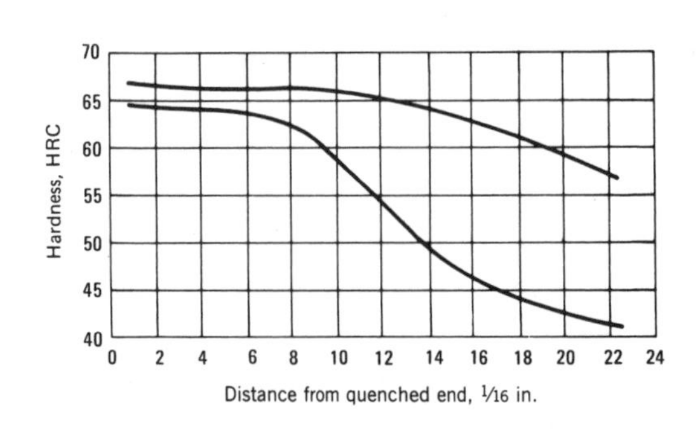

Type O2 Oil-Hardening Tool Steels: End-Quench Hardenability. Test specimens were austenitized at 775 to 830 °C (1425 to 1525 °F); oil quenched; then tempered for 1 h to desired hardness. (Ref 5)

Type O1 Oil-Hardening Tool Steels: Effect of Tempering Temperature on Size Change. Specimens 25-mm (1-in.) diam by 102-mm (4-in.) long were austenitized at 800 °C (1475 °F) for 20 min; oil quenched; and tempered for 1 h at temperature indicated. (Ref 5)

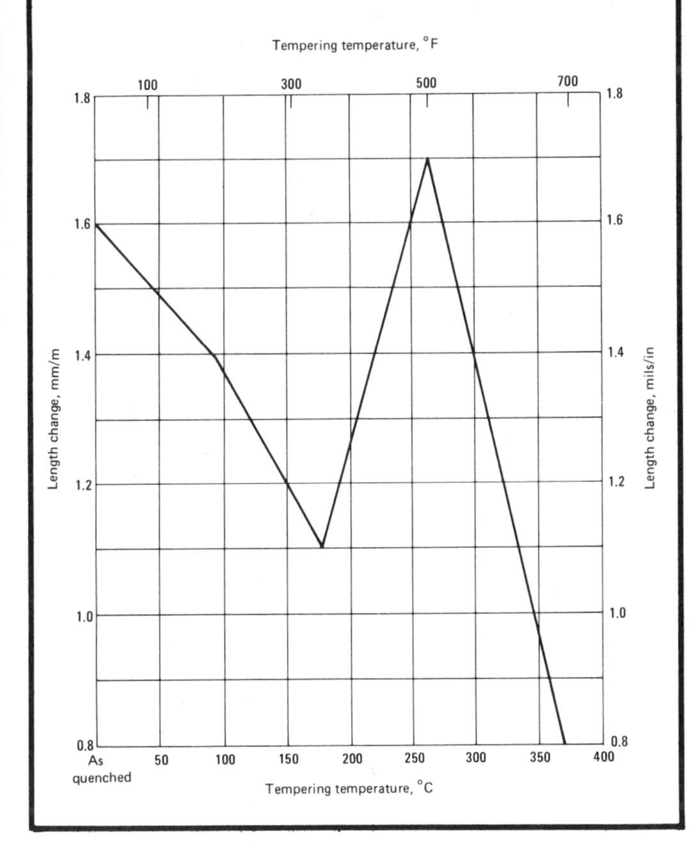

Type O1 Oil-Hardening Tool Steels: Relationship of Tensile Properties and Hardness. Specimens were austenitized at 800 to 815 °C (1475 to 1500 °F); oil quenched; and tempered a minimum of 1 h to desired hardness. Impact energy tests used unnotched Izod specimens. Elongation was measured in 50 mm (2 in.). (Ref 5)

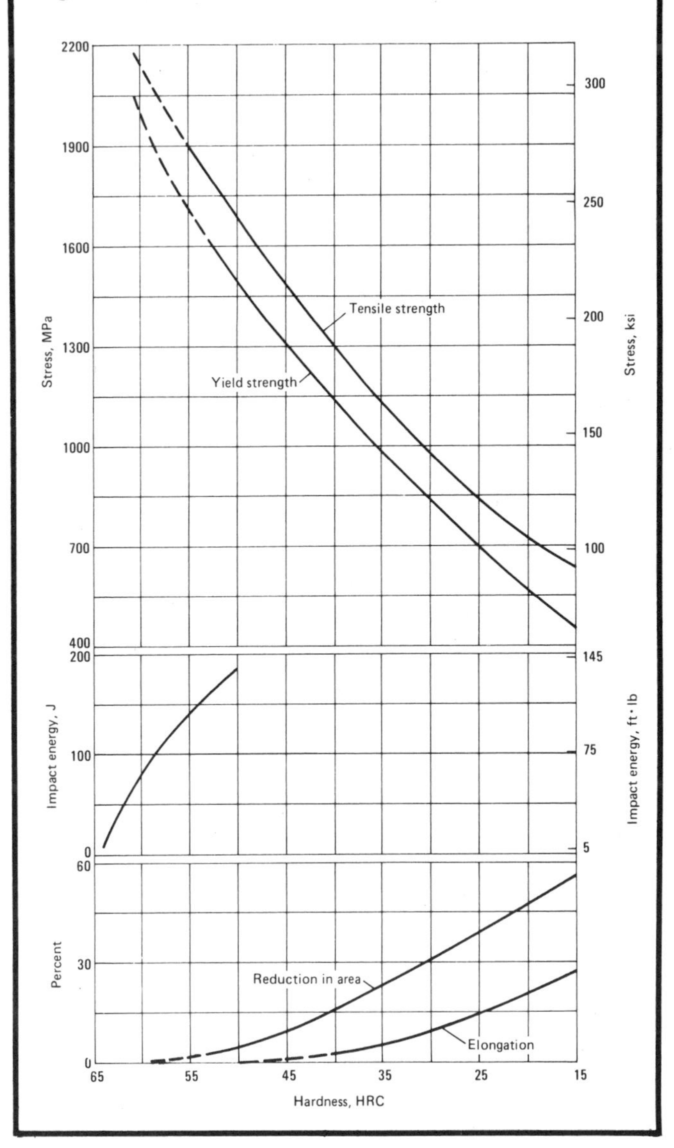

Type O2 Oil-Hardening Tool Steels: Relationship of Tensile Properties and Hardness. Test specimens were austenitized at 775 to 830 °C (1425 to 1525 °F); oil quenched; then tempered for 1 h to desired hardness. Impact energy tests used unnotched Izod specimens. Elongation was measured in 50 mm (2 in.). (Ref 5)

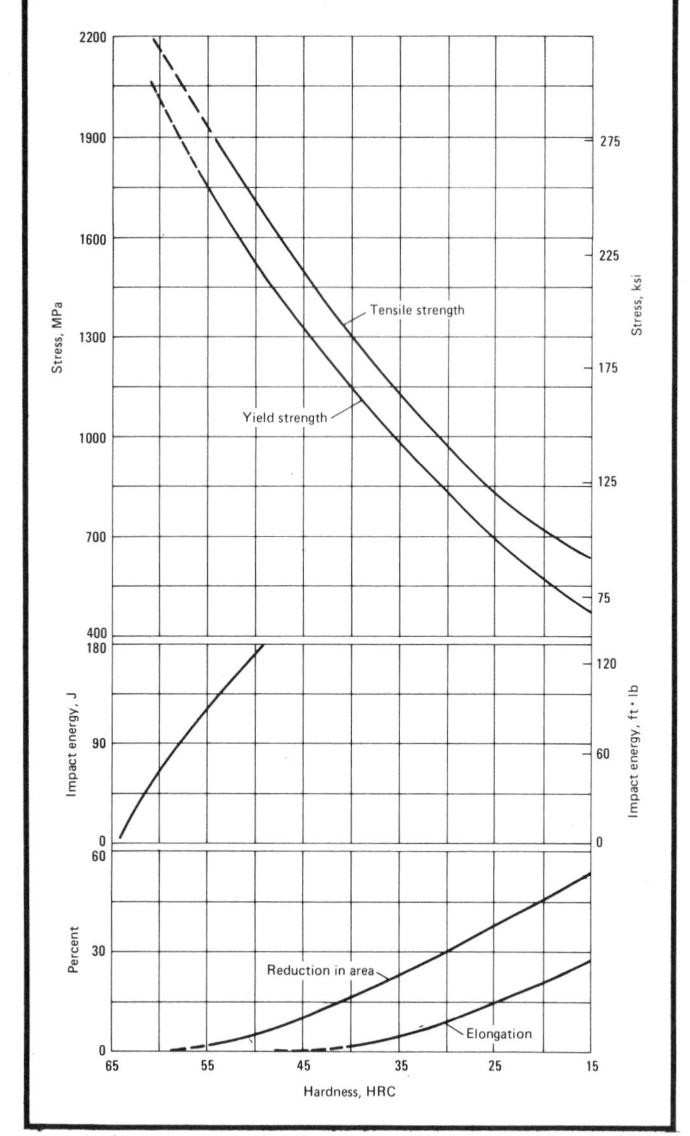

Type O2 Oil-Hardening Tool Steels: Effect of Tempering Temperature on Size Change. Test specimens were 25-mm (1.0-in.) round; oil quenched from 800 °C (1475 °F); tempered 1 h at heat. (Ref 5)

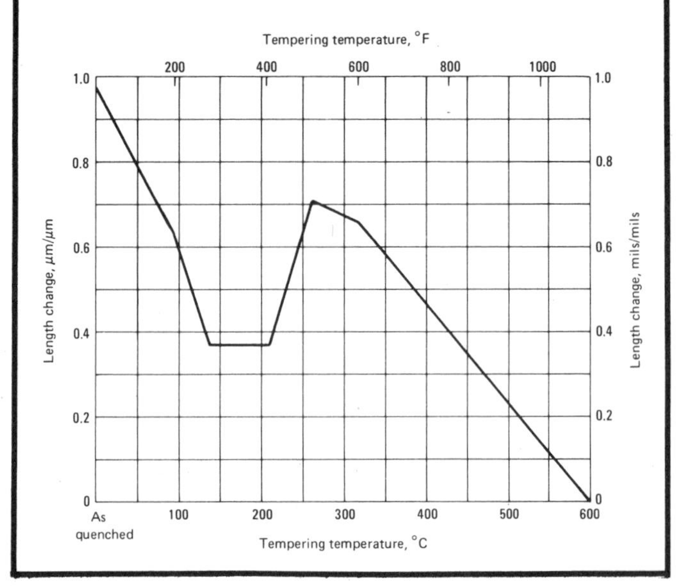

Type O6 Oil-Hardening Tool Steels: Effect of Tempering Temperature on Size Change. Test specimens were austenitized in salt at 800 °C (1475 °F) for 25 min; oil quenched; and tempered for 1 h at temperature. Size change samples were 19-mm (0.75-in.) round by 50-mm (2-in.) long. In larger sample sizes, expansion characteristics may vary from the data shown. (Ref 5)

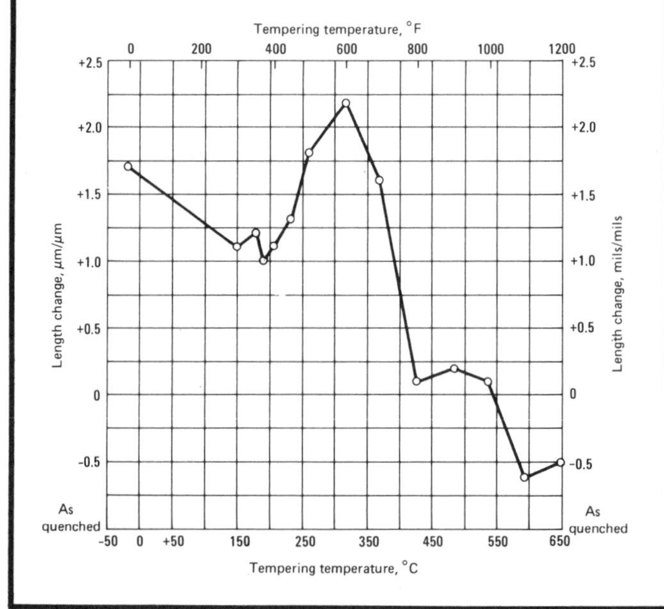

Medium-Alloy, Air-Hardening Tool Steels (A)

Medium-Alloy, Air-Hardening Tool Steels (A): Chemical Composition

Some types can be produced with a sulfur addition to improve machinability

Steel type	Chemical composition, % C	Mn	P max	S max	Si	Cr	Ni	Mo	W	V
A2	0.95-1.05	1.00 max	0.030	0.030	0.50 max	4.75-5.50	...	0.90-1.40	...	0.15-0.50
A3	1.20-1.30	0.40-0.60	0.030	0.030	0.50 max	4.75-5.50	...	0.90-1.40	...	0.80-1.40
A4	0.95-1.05	1.80-2.20	0.030	0.030	0.50 max	0.90-2.20	...	0.90-1.40	...	...
A5	0.95-1.05	2.80-3.20	0.030	0.030	0.50 max	0.90-1.20	...	0.90-1.40	...	...
A6	0.65-0.75	1.80-2.50	0.030	0.030	0.50 max	0.90-1.20	...	0.90-1.20	...	...
A7	2.00-2.85	0.80 max	0.030	0.030	0.50 max	5.00-5.75	...	0.90-1.40	0.50-1.50(a)	3.90-5.15
A8	0.50-0.60	0.50 max	0.030	0.030	0.75-1.10	4.75-5.50	...	1.15-1.65	1.00-1.50	...
A9	0.45-0.55	0.50 max	0.030	0.030	0.95-1.15	4.75-5.50	1.25-1.75	1.30-1.80	...	0.80-1.40
A10(b)	1.25-1.50	1.60-2.10	0.030	0.030	1.00-1.50	...	1.55-2.05	1.25-1.75	...	...

(a) Optional. (b) Contains free graphite in the microstructure to improve machinability

Characteristics. The medium-alloy, air-hardening tool steels are subcategories of the cold work tool steels which are designated by the letter symbol A. They include a wide range of carbon and alloy contents, but all have high hardenability and exhibit a high degree of dimensional stability in heat treatment. This group of steels also exhibits good wear resistance, fatigue life, toughness, and deep-hardening qualities.

The low carbon types A8 and A9 offer greater shock resistance than the other steels in this group, but are lower in their resistance to wear. Type A7, which has high carbon and vanadium, exhibits maximum abrasion resistance, but should be restricted to applications where toughness is not a prime consideration.

The machinability rating of the medium-alloy, air hardening tool steels is about 65% of that for the carbon tool steels. The rating for grade A4 is about 20% higher.

Atomic hydrogen, oxy-acetylene, gas tungsten-arc, and shielded-metal-arc welding processes are usually acceptable for use on the air-hardening tool steels. These steels should be preheated to 150 to 260 °C (300 to 500 °F) prior to welding. After welding, annealed material should be reannealed. Hardened material should be heated to 205 to 315 or 345 °C (400 to 600 or 650 °F). These steels may be nitrided to produce an exceedingly hard surface.

Typical Uses. Applications of the medium-alloy, air hardening tool steels include cold forming, blanking and bending dies, forming rolls, drill bushings, knurling tools, master dies and gages, and similar uses where low distortion in heat treatment, or good wear resistance, is required.

AISI Type A2: Similar Steels (U.S. and/or Foreign). UNS T30102; ASTM A681; FED QQ-T-570; SAE J437, J438; (W. Ger.) DIN 1.2363; (Fr.) AFNOR Z 100 CDV 5; (Jap.) JIS SKD 12; (Swed.) SS_{14} 2260; (U.K) B.S. BA 2

AISI Type A3: Similar Steels (U.S. and/or Foreign). UNS T30103; ASTM A681; FED QQ-T-570

AISI Type A4: Similar Steels (U.S. and/or Foreign). UNS T30104; ASTM A681; FED QQ-T-570

AISI Type A5: Similar Steels (U.S. and/or Foreign). UNS T30105; ASTM A681; FED QQ-T-570

AISI Type A6: Similar Steels (U.S. and/or Foreign). UNS T30106; ASTM A681; FED QQ-T-570

AISI Type A7: Similar Steels (U.S. and/or Foreign). UNS T30107; ASTM A681; FED QQ-T-570

AISI Type A8: Similar Steels (U.S. and/or Foreign). UNS T30108; ASTM A681; FED QQ-T-570

AISI Type A9: Similar Steels (U.S. and/or Foreign). UNS T30109; ASTM A681; FED QQ-T-570

AISI Type A10: Similar Steels (U.S. and/or Foreign). UNS T30110; ASTM A681; FED QQ-T-570

Medium-Alloy, Air-Hardening Tool Steels (A): Approximate Critical Points (Ref 4)

Steel type	Transformation point	Temperature range °C	°F
A2	Ac range	800-840	1475-1540
	Ar range	710-355	1310-670
A4	Ac range	755-795	1390-1460
	Ar range	705-665	1300-1230
A6	Ac range	740-770	1360-1420
	Ar range	670-645	1240-1190

Physical Properties

Type A2 Medium-Alloy, Air-Hardening Tool Steels: Average Coefficients of Linear Thermal Expansion (Ref 5)

Temperature range °C	°F	Coefficient μm/m·K	μin./in.·°F
20-100	68-212	10.7	5.96
20-200	68-390	12.0	6.64
20-300	68-570	12.7	7.05
20-400	68-750	13.2	7.36
20-500	68-930	13.7	7.60
20-600	68-1110	14.0	7.75
20-700	68-1290	14.3	7.92
20-750	68-1380	14.4	7.98

Physical Properties (continued)

Medium-Alloy, Air-Hardening Tool Steels (A): Thermal Treatment Temperatures

Always anneal after forging and before machining or hardening; cool slowly from annealing temperature; when hardening, most grades should be cooled below 65 °C (150 °F), then tempered immediately; the usual tempering temperature is 150 to 260 °C (300 to 500 °F)

Treatment	Temperature range °C	°F
Type A2		
Forging, preheat	675	1250
Forging, start	1065-1150	1950-2100
Forging, finish	870	1600
Annealing	845-900	1550-1650
Hardening, preheat	790-845	1450-1550
Hardening, quench from	955-995	1750-1825
Type A4		
Forging, preheat	650	1200
Forging, start	1065-1095	1950-2000
Forging, finish	925	1700
Annealing	760-775	1400-1425
Hardening, quench from	830-855	1525-1575
Type A6		
Forging, start(a)	1040-1150	1900-2100
Forging, finish	870-980	1600-1800
Annealing	845-900	1550-1650
Hardening, quench from	830-870	1525-1600
Type A7		
Annealing	845-900	1550-1650
Hardening, preheat	790-815	1450-1500
Hardening, quench from	955-980	1750-1800

(a) Heat slowly to a uniform temperature

Type A6 Medium-Alloy, Air-Hardening Tool Steels: Average Coefficients of Linear Thermal Expansion (Ref 5)

Temperature range °C	°F	Coefficient μm/m·K	μin./in.·°F
20-100	68-212	11.8	6.57
20-200	68-390	12.4	6.91
20-300	68-570	13.2	7.32
20-400	68-750	13.7	7.59
20-500	68-930	14.1	7.86
20-600	68-1110	14.4	8.01
20-700	68-1290	14.7	8.19

Medium-Alloy, Air-Hardening Tool Steels (A): Density (Ref 5)

Steel type	Density g/cm³	lb/in.³
A2	7.86	0.284
A6	8.03	0.290

Mechanical Properties

Medium-Alloy, Air-Hardening Tool Steels (A): Modulus of Elasticity (Ref 5)

Steel type	Modulus GPa	10^6 psi
A2	203	29.5
A6	200	29.0

Type A2 Medium-Alloy, Air-Hardening Tool Steels: Effect of Tempering Temperature on Hardness (Ref 5)

Air quenched from 970 °C (1775 °F); tempered 1 h at temperature indicated

Tempering temperature °C	°F	Hardness, HRC
As hardened		63-64
150	300	63-64
175	350	61-63
205	400	60-62
230	450	59-61
260	500	58-60
315	600	57-59
370	700	57-59
425	800	57-59
480	900	57-59
540	1000	56-58
595	1100	50-51
650	1200	44-45

Type A2 Medium-Alloy, Air-Hardening Tool Steels: Typical Hardness and Charpy V-Notch Impact Properties (Ref 6)

Temperature °C	°F	Hardness, HRC	Impact energy J	ft·lb
As air hardened		63-65	...	...
205	400	60-62	42	31
260	500	59-61	56	41
315	600	58-60	50	37
370	700	57-59	45	33
425	800	57-59	42	31
480	900	57-59	39	29
540	1000	56-58	56	41
595	1100	50-52	...	...
650	1200	42-44	...	...

Medium-Alloy, Air-Hardening Tool Steels (A): Effect of Tempering Temperature on Hardness (Ref 4, 7)

Temperature °C	°F	Hardness, HRC A2 (a)	A4 (b)	A6 (c)	A7 (d)
As quenched		62-64	63	61-62	65-66.5
95	200	...	...	61-62	...
150	300	...	61	60-61	64-65
175	350	...	...	59-60	...
205	400	60-62	60	58-59	63-65
260	500	58-60	59	56-57	62-64
315	600	58-60	58	55-56	61-63
370	700	57-59	57	54-55	60-62
425	800	56-58	56	52-53	59-61
480	900	56-58	54	50-51	58-60
540	1000	55-57	53	48-49	56-58
595	1100	...	50	...	53-55
650	1200	...	40	...	44-46

(a) Air quenched from 980 °C (1800 °F); tempered 2.5 h. (b) Air quenched from 845 °C (1550 °F). (c) Air quenched from 845 °C (1550 °F); tempered 2 h. (d) Air quenched from 955 °C (1750 °F); tempered 3 h

Type A2: Medium-Alloy, Air-Hardening Tool Steels: Dimensional Change During Heat Treatment (Ref 6)

Test specimens were 19-mm (0.75-in.) diam bar austenitized at 970 °C (1775 °F) for ½ h; air cooled; tempered at temperature indicated

Tempering temperature °C	°F	Hardness, HRC	Dimensional change, %
As air quenched		64.5	+0.10
95	200	64.0	+0.10
150	300	62.5	+0.08
205	400	61.0	+0.07
260	500	59.0	+0.10
315	600	58.0	+0.09

Type A2 Medium-Alloy, Air-Hardening Tool Steels: Relationship of Tensile Properties and Hardness. Test specimens were austenitized at 940 to 970 °C (1725 to 1775 °F); air cooled; and tempered to desired hardness. Impact energy tests used unnotched Izod specimens. Elongation was measured in 50 mm (2 in.); yield strength at 0.02% offset. (Ref 5)

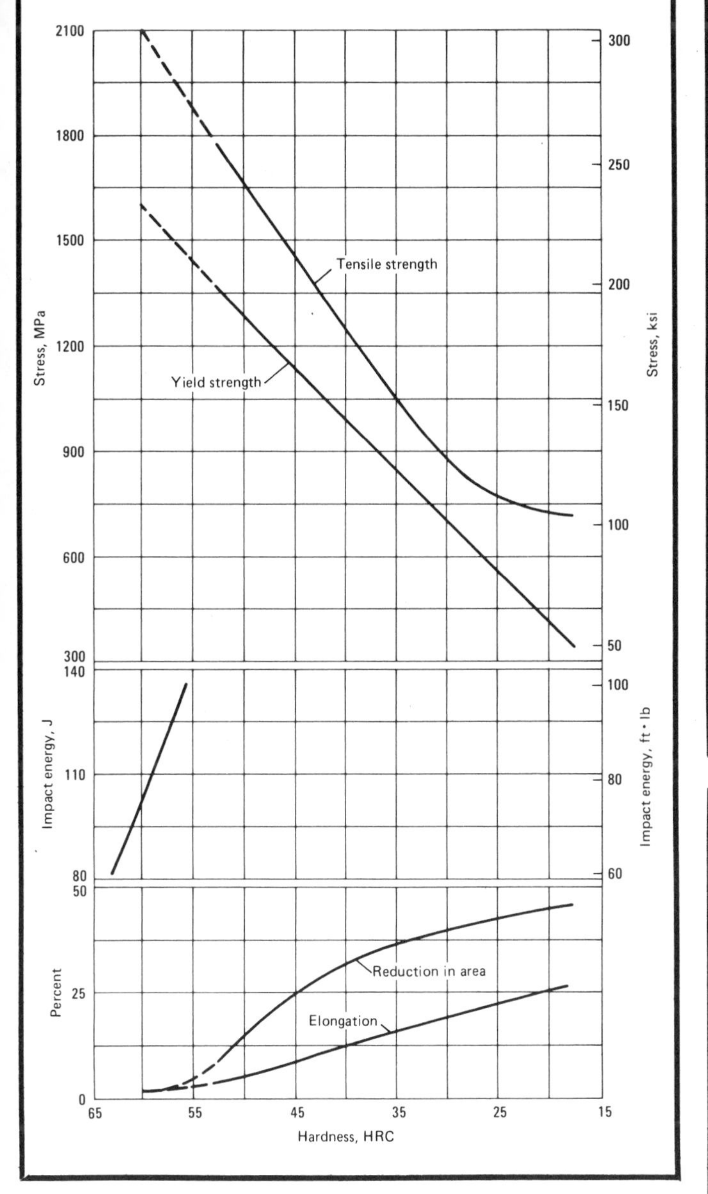

Type A6 Medium-Alloy, Air-Hardening Tool Steels: Relationship of Impact Properties and Hardness (Ref 5)

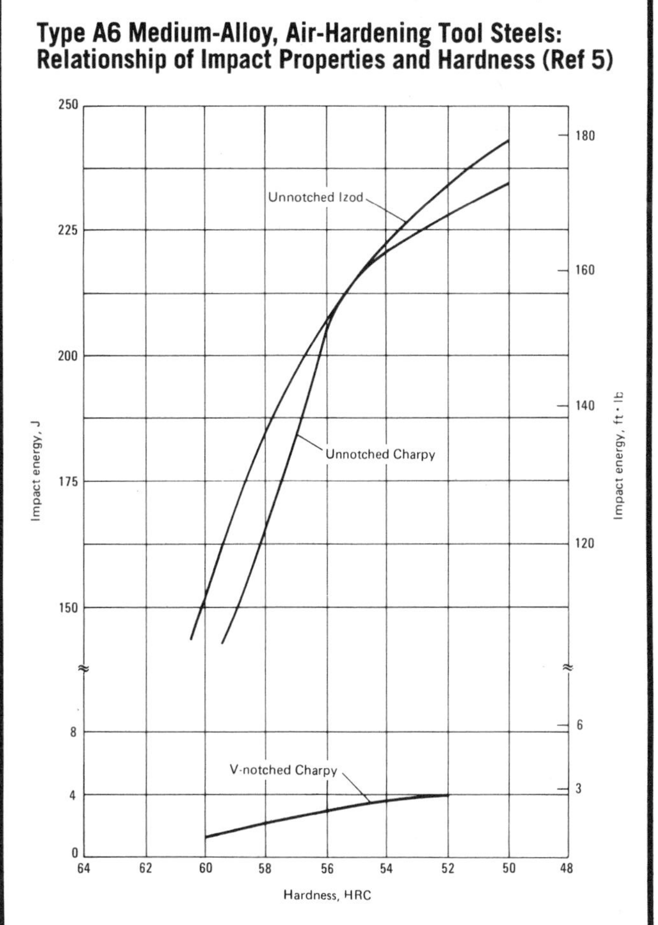

Type A6 Medium-Alloy, Air-Hardening Tool Steels: Relationship of Tensile Properties and Hardness. Test specimens were austenitized at 830 to 870 °C (1525 to 1600 °F); air cooled; then tempered to desired hardness. Elongation was measured in 50 mm (2 in.). (Ref 5)

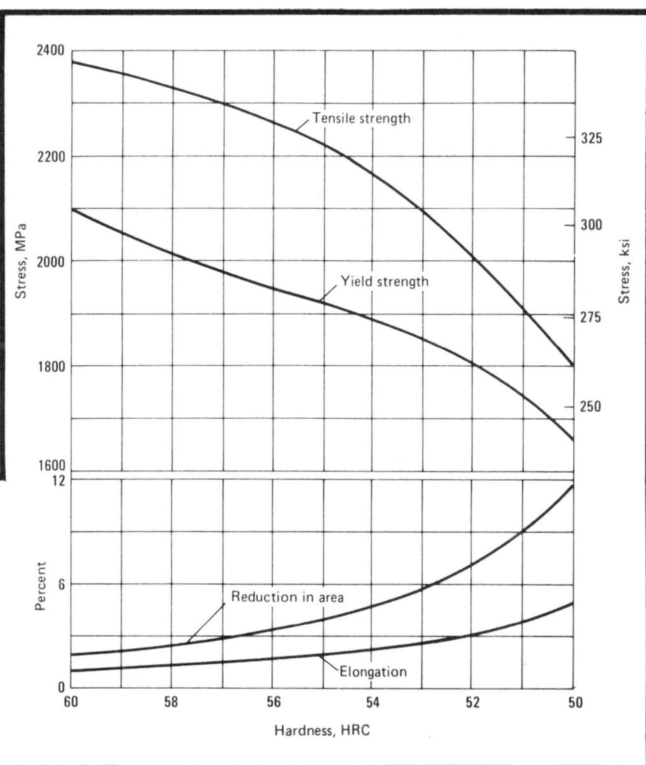

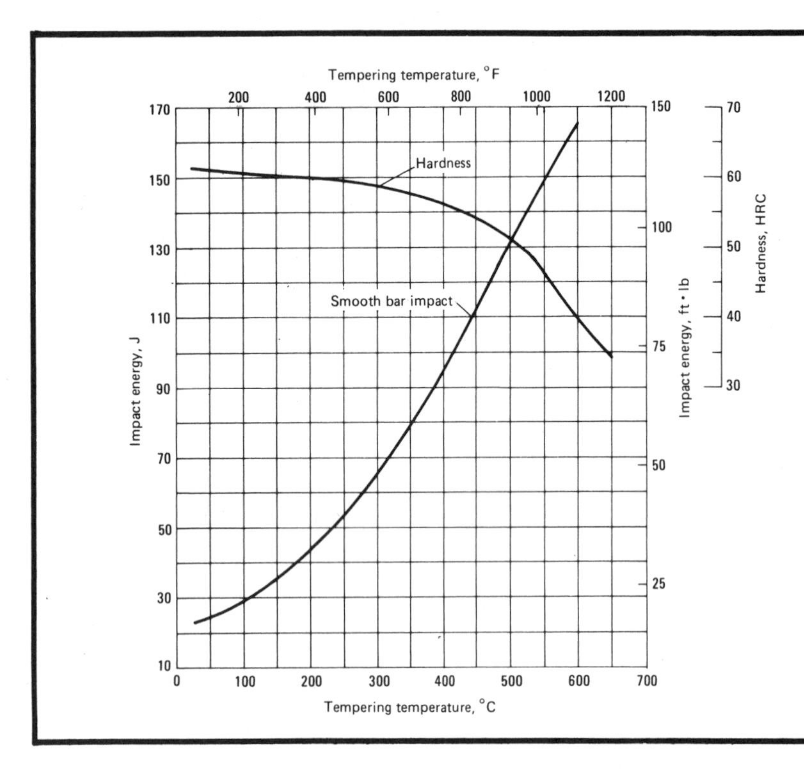

Type A10 Medium-Alloy, Air-Hardening Tool Steels: Effect of Tempering on Hardness and Smooth Bar Impact Toughness. Test specimen composition: 1.35% carbon, 1.80% manganese, 1.20% silicon, 1.85% nickel, 0.20% chromium, and 1.50% molybdenum; hardened at 795 °C (1460 °F); air cooled. Impact tests used smooth bar Izod specimens. (Ref 9)

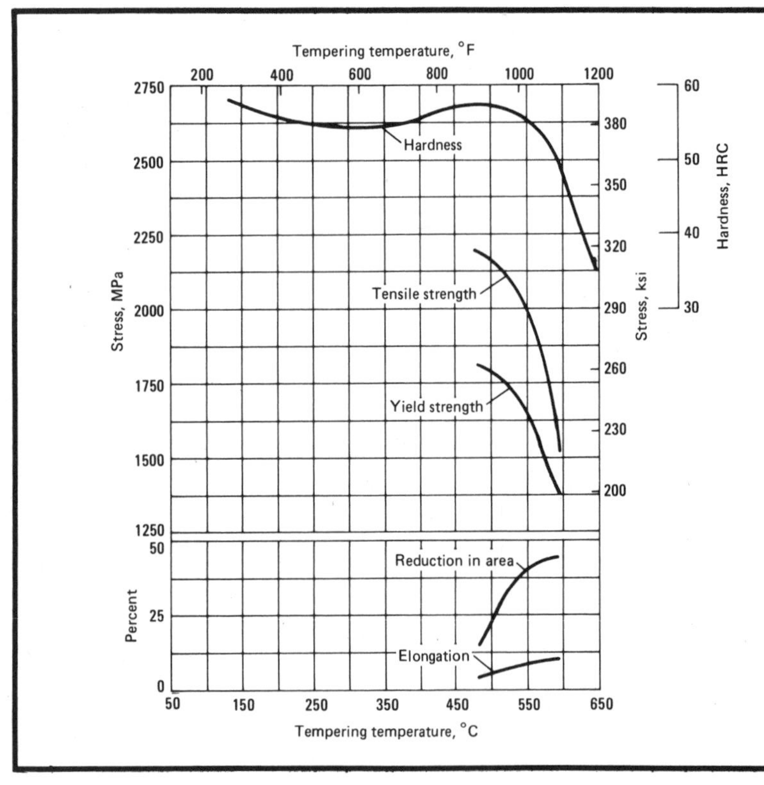

Type A9 Medium-Alloy, Air-Hardening Tool Steels: Effect of Tempering Temperature on Tensile Properties. Test specimen composition: 0.50% carbon, 0.40% manganese, 1.00% silicon, 1.50% nickel, 5.00% chromium, 1.40% molybdenum, and 1.00% vanadium; hardened at 1010 °C (1850 °F); air cooled; double tempered. Elongation was measured in 50 mm (2 in.). (Ref 9)

High-Carbon, High-Chromium Tool Steels (D)

High-Carbon, High-Chromium Tool Steels (D): Chemical Composition

Steel type	Chemical composition, %									
	C	Mn max	P max	S max	Si max	Cr	Mo	V	Co	W
D2	1.40-1.60	0.60	0.030	0.030	0.60	11-13	0.70-1.20	1.10 max	1.00 max	...
D3	2.00-2.35	0.60	0.030	0.030	0.60	11-13	...	1.00 max	...	1.00 max
D4	2.05-2.40	0.60	0.030	0.030	0.60	11-13	0.70-1.20	1.00 max	...	...
D5	1.40-1.60	0.60	0.030	0.030	0.60	11-13	0.70-1.20	1.00 max	2.50-3.50	...
D7	2.15-2.60	0.60	0.030	0.030	0.60	11.5-13.5	0.70-1.20	3.80-4.40	...	...

Characteristics. High-carbon, high-chromium tool steels are subcategories of cold work tool steels which are designated by the letter symbol D. They are all characterized by high carbon content, from 1.40 to 2.60%, and nominally 12% chromium. The types containing molybdenum are air hardening, and therefore offer a high degree of dimensional stability in heat treatment. Although these steels can be quenched in oil, the movement is slightly greater when oil quenched. The D series exhibits high wear resistance, which increases with greater carbon and vanadium content.

The high-carbon, high-chromium tool steels are essentially nondeforming during heat treatment. For example, a piece of AISI D3 25-mm (1-in.) long would expand about 0.02-mm (0.0008 in.) in hardening, but when tempered at 205 °C (400 °F), it returns to within 0.005 mm (0.0002 in.) of its original length.

AISI type D steels may be satisfactorily welded using the atomic hydrogen, oxy-acetylene, gas tungsten-arc, and shielded-metal-arc processes. Preheating and postheating are usually required. When welding hardened material, the preheating or postheating temperature must not exceed the tempering temperature. The machinability rating of these steels is 40 to 60% of that for the 1% water-hardening tool steels.

Typical Uses. Applications of high-carbon, high-chromium steels include spindles, hobs, cold rolls, slitting cutters, blanking dies, forming dies, coining dies, bushings, taps, broaches, sand blast nozzles, brick molds, and plug and ring gages.

AISI Type D2: Similar Steels (U.S. and/or Foreign). UNS T30402; ASTM A681; FED QQ-T-570; SAE J437, J438; (W. Ger.) DIN 1.2379; (Ital.) UNI X 150 CrMo 12 KU; (U.K.) B.S. BD 2

AISI Type D3: Similar Steels (U.S. and/or Foreign). UNS T30403; ASTM A681; FED QQ-T-570; SAE J437 (D2), J438 (D2); (W. Ger.) DIN 1.2080; (Fr.) AFNOR Z 200 C 12; (Ital.) UNI X 210 Cr 13 KU; (Jap.) JIS SKD 1; (U.K.) B.S. BD 3

AISI Type D4: Similar Steels (U.S. and/or Foreign). UNS T30404; ASTM A681; FED QQ-T-570

AISI Type D5: Similar Steels (U.S. and/or Foreign). UNS T30405; ASTM A681; FED QQ-T-570; SAE J437, J438

AISI Type D7: Similar Steels (U.S. and/or Foreign). UNS T30407; ASTM A681; FED QQ-T-570; SAE J437, J438

Physical Properties

Type D3 High-Carbon, High-Chromium Tool Steels: Average Coefficients of Linear Thermal Expansion (Ref 5)

Temperature range °C	°F	Coefficient μm/m·K	μin./in.·°F
20-100	68-212	10.7	5.93
20-200	68-390	11.6	6.45
20-300	68-570	12.1	6.73
20-400	68-750	12.4	6.89
20-500	68-930	12.8	7.09
20-600	68-1110	12.9	7.14
20-700	68-1290	13.0	7.21
20-750	68-1380	13.1	7.26

Type D2 High-Carbon, High-Chromium Tool Steels: Average Coefficients of Linear Thermal Expansion (Ref 5)

Temperature range °C	°F	Coefficient μm/m·K	μin./in.·°F
20-100	68-212	10.5	5.81
20-200	68-390	11.3	6.29
20-300	68-570	11.8	6.56
20-400	68-750	12.2	6.76
20-500	68-930	12.5	6.93
20-600	68-1110	12.6	7.00
20-700	68-1290	12.8	7.09
20-750	68-1380	...	...
20-800	68-1470	13.0	7.24

High-Carbon, High-Chromium Tool Steels (D): Thermal Treatment Temperatures

Treatment	Temperature range °C	°F
Forging, start	1040-1095	1900-2000
Forging, finish	925 min	1700 min
Annealing (a)	870-900	1600-1650
Hardening, preheat	790-815	1450-1500
Hardening, oil quench from	955-980	1750-1800
Hardening, air quench from	980-1010	1800-1850
Tempering (b)	95-650	200-1200

(a) Cool slowly in furnace. (b) Temperature depends on hardness and toughness required

Type D3 High-Carbon, High-Chromium Tool Steels: Density (Ref 5)

7.86 g/cm³ .... 0.284 lb/in.³

Mechanical Properties

High-Carbon, High-Chromium Tool Steels (D): Effect of Tempering Temperature on Hardness (Ref 5, 6)

Tempering temperature °C	°F	Hardness, HRC D2 (a)	D3 (b)	D4 (c)
As quenched		62-63	65-66	64-66
95	200	61-62	64-66	...
150	300	...	63-64	...
205	400	59-60	62-63	61-63
230	450	59-60	...	...
260	500	...	...	59-61
290	550	56-57	...	...
315	600	...	59-60	58-60
370	700	56-57	...	58-60
425	800	56-57	58-59	58-60
480	900	58-59	...	58-60
540	1000	59-60	51-53	55-57
595	1100	50-55	...	50-52
650	1200	44-45	37-39	42-44
705	1300	...	32-34	...

(a) Oil quenched from 1010 °C (1850 °F); tempered 1 h. (b) Oil quenched from 970 °C (1775 °F). (c) Air quenched from 995 °C (1825 °F); tempered 3 h

Types D2, D3 High-Carbon, High-Chromium Tool Steels: Effect of Tempering Temperature on Impact Energy.

Type D2 test specimens were air cooled at 1010 to 1025 °C (1850 to 1875 °F). Type D3 test specimens were oil quenched at 980 °C (1800 °F). Impact energy tests used unnotched Izod specimens. (Ref 3)

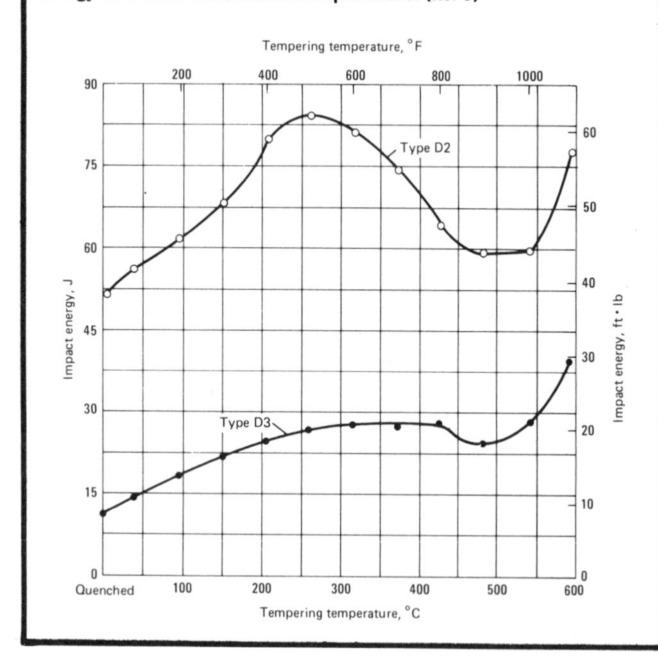

Type D3 High-Carbon, High-Chromium Tool Steels: Modulus of Elasticity (Ref 5)

20 GPa .... 30×10^6 psi

Types D2, D3 High-Carbon, High-Chromium Tool Steels: Effect of Tempering Temperature on Torsion Impact Energy.

Type D2 test specimen composition: 1.60% carbon, 13.00% chromium, 0.75% molybdenum, and 0.27% vanadium; air quenched at 980 °C (1800 °F). Type D3 test specimen composition: 2.10% carbon, 12.50% chromium, and 0.50% nickel; oil quenched at 970 °C (1775 °F). The absolute magnitude of impact energy should not be compared because the steels were tested under different conditions. (Ref 3)

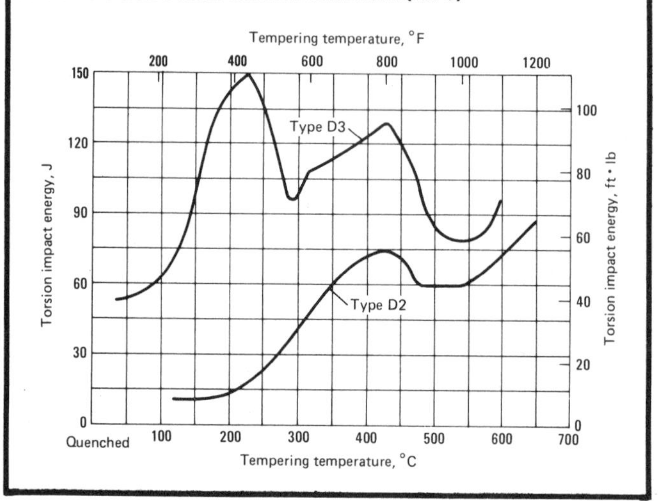

Types D3, D2 High-Carbon, High-Chromium Tool Steels: Comparison of the Ductility in a Static Torsion Test of Materials Quenched to Maximum Hardness. Type D3 test specimens were oil quenched from 970 °C (1775 °F). Type D2 test specimens were air cooled from 1010 °C (1850 °F). Test specimens of both types were tempered at the following temperatures: —— 175 °C (350 °F); — — 290 °C (550 °F); - - - 400 °C (750 °F). The Rockwell C hardness after tempering is given for each tempering temperature. (Ref 8)

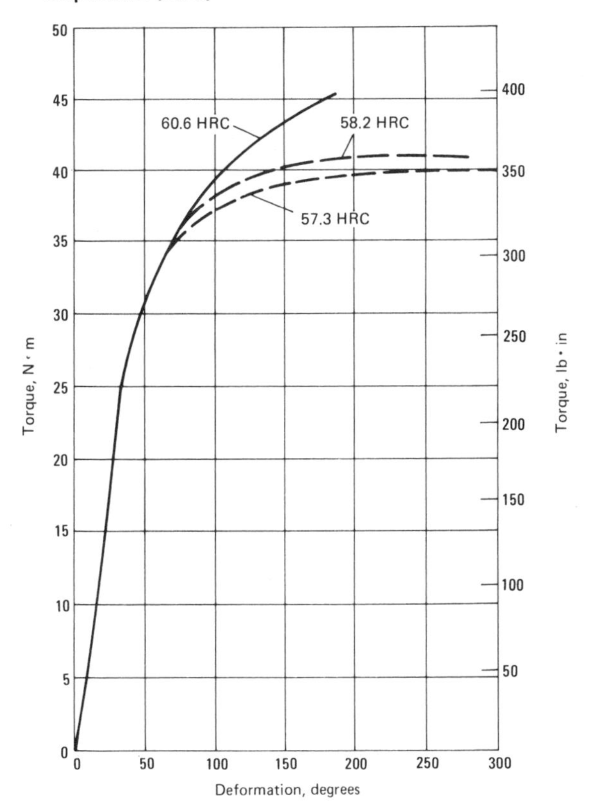

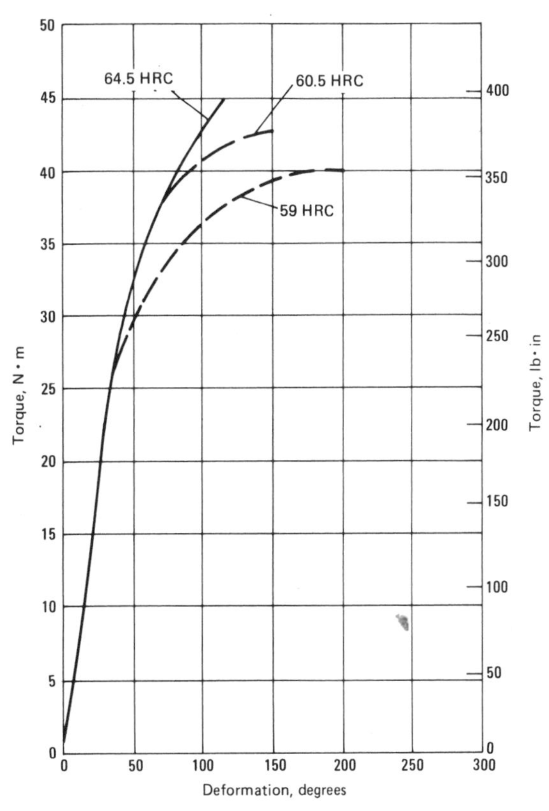

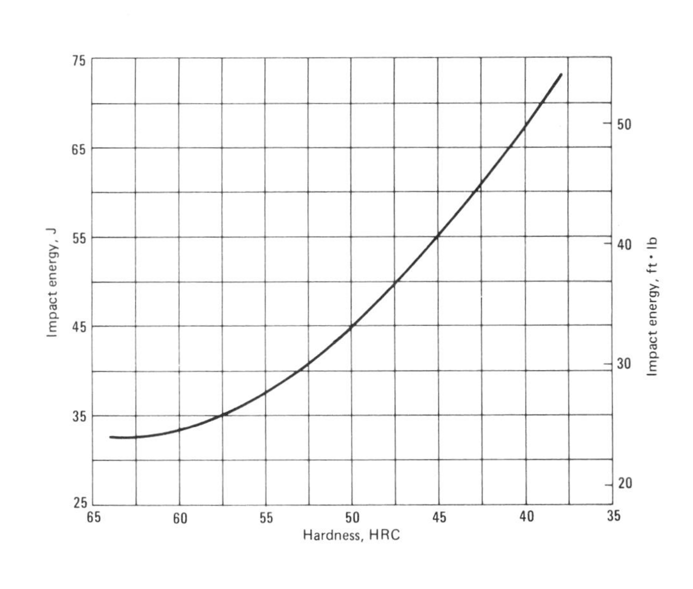

Type D3 High-Carbon, High-Chromium Tool Steels: Effect of Hardness on Impact Properties. Tests used unnotched Izod specimens. (Ref 5)

Types D2, D3 High-Carbon, High-Chromium Tool Steels: Effect of Carbon Content on the Ultimate Compressive Strength and Impact Energy.
—— type D2 test specimens with composition of 1.55% carbon, 12.0% chromium, 0.80% molybdenum, and 0.25% vanadium; tempered at 1010 °C (1850 °F) and quenched in air.
– – – type D3 test specimens with composition of 2.10% carbon, 12.0% chromium, and 1.0% vanadium; tempered at 980 °C (1800 °F) and quenched in oil.
—– type D3 test specimens with composition of 2.40% carbon and 11.5% chromium; tempered at 955 °C (1750 °F) and quenched in oil. Impact energy tests used unnotched Izod test specimens. Hardness after treatments was measured in Rockwell C; HRC values are given at data points. (Ref 3)

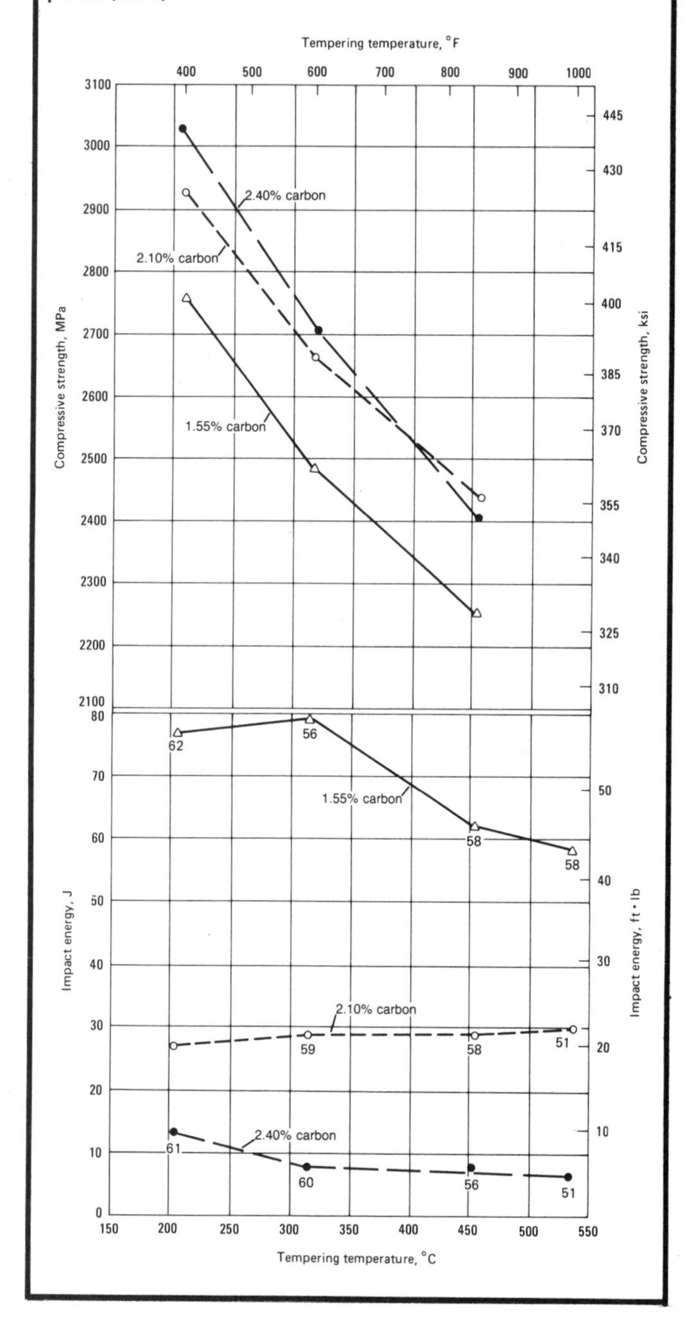

Type D2 High-Carbon, High-Chromium Tool Steels: Effect of Tempering on Rockwell C Hardness and Compressive Stress. Oversize test specimens 12.8-mm (0.505-in.) diam by 19.1-mm (0.750-in.) long were austenitized at 1010 °C (1850 °F) and tempered at 56 °C (100 °F) intervals. (Ref 4)

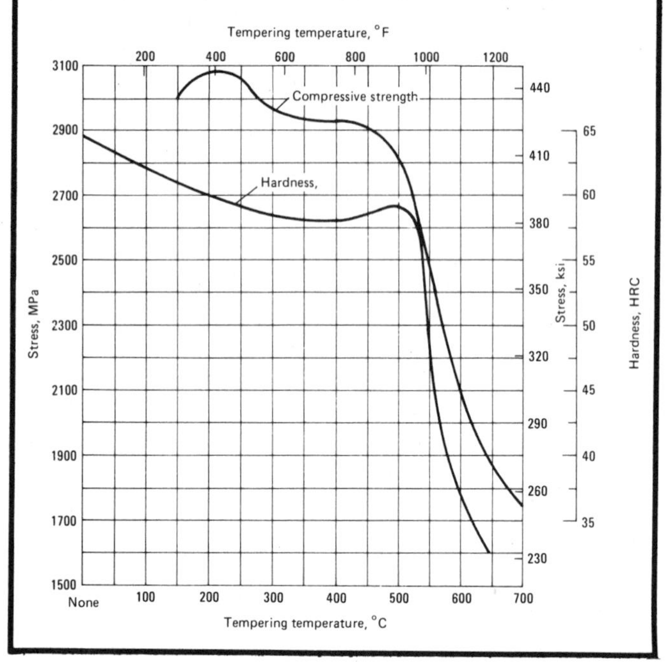

Mold Steels (P)

Mold Steels (P): Chemical Composition

Steel type	Chemical composition, % C	Mn	P max	S max	Si	Cr	Mo	Ni
P2	0.10 max	0.10-0.40	0.030	0.030	0.10-0.40	0.75-1.25	0.15-0.40	0.10-0.50
P3	0.10 max	0.20-0.60	0.030	0.030	0.40 max	0.40-0.75	···	1.00-1.50
P4	0.12 max	0.20-0.60	0.030	0.030	0.10-0.40	4.00-5.25	0.40-1.00	···
P5	0.10 max	0.20-0.60	0.030	0.030	0.40 max	2.00-2.50	···	0.35 max
P6	0.05-0.15	0.20-0.60	0.030	0.030	0.40 max	1.25-1.75	···	3.25-3.75
P20	0.28-0.40	0.60-1.00	0.030	0.030	0.20-0.80	1.40-2.00	0.30-0.65	···
P21 (a)	0.18-0.22	0.20-0.40	0.030	0.030	0.20-0.40	0.20-0.30	0.20-0.40	4.00-4.25

(a) Also contains 1.05 to 1.25% aluminum and 0.15 to 0.25% vanadium

Characteristics. The mold steels, designated by the letter symbol P, are generally intended for mold applications. Types P2 to P6 are very low in carbon content, and are usually supplied at very low hardness to facilitate cold hubbing of the impression. They are then carburized to develop the required surface properties for injection and compression molds for plastics. Types P20 and P21 are usually supplied in the prehardened condition, so that the cavity can be machined and the mold placed directly in service.

Also used as mold materials are stainless steels type 420, low-carbon (0.03%) type 414, and AISI A6 air-hardening tool steel. The latter is a graphitic steel which provides good machining characteristics when producing intricate cavities and cores.

Machinability of alloys P2, P3, and P4 is 80 to 90% of a 1% water-hardening carbon tool steel. Other approximate machinability ratings are: 60% for type P5, 40% for type P6, and 65% for types P20 and P21.

Electrical discharge machining (EDM) is widely used on prehardened P20 and P21, types 414 and 420 stainless steels, and AISI A6. Because this method of machining leaves a rehardened surface layer on the steel, certain precautions should be taken. It is advisable to remove this layer, or at least temper at 540 °C (1000 °F) subsequent to EDM machining, to transform this white surface layer.

Typical Uses. Applications of mold steels include cavities and cores for plastic molds, zinc die casting dies, and holder blocks.

AISI Type P2: Similar Steels (U.S. and/or Foreign). UNS T51602; ASTM A681

AISI Type P3: Similar Steels (U.S. and/or Foreign). UNS T51603; ASTM A681

AISI Type P4: Similar Steels (U.S. and/or Foreign). UNS T51604; ASTM A681; (W. Ger.) DIN 1.2341

AISI Type P5: Similar Steels (U.S. and/or Foreign). UNS T51605; ASTM A681

AISI Type P6: Similar Steels (U.S. and/or Foreign). UNS T51606; ASTM A681; (W. Ger.) DIN 1.2735

AISI Type P20: Similar Steels (U.S. and/or Foreign). UNS T51620; ASTM A681; (W. Ger.) DIN 1.2330

AISI Type P21: Similar Steels (U.S. and/or Foreign). UNS T51621; ASTM A681

Physical Properties

Mold Steels (P): Thermal Treatment Temperatures

Dies and molds should be slow cooled after annealing; tempering temperature is usually 28 °C (50 °F) above operating temperature

Treatment	Temperature range °C	°F
Type P4		
Forging, start	1095-1120	2000-2050
Forging, finish	870 min	1600 min
Normalizing(a)	···	···
Full annealing	870	1600
Process annealing	760	1400
Carburizing	915-925	1675-1700
Hardening	940-955	1725-1750
Tempering, plastic molds	425 max	800 max
Tempering, die-cast dies	425 min	800 min
Type P5		
Forging	1065-1205	1950-2200
Normalizing	925-955	1700-1750
Full annealing	815-870	1500-1600
Process annealing	705-730	1300-1350
Carburizing	870-900	1600-1650
Hardening	845-870	1550-1600
Tempering	150-260	300-500
Type P6		
Forging	1150 max	2100 max
Normalizing	900-955	1650-1750
Annealing	675-695	1250-1280
Stress relieving	595	1100
Carburizing, cyanide	790	1450
Carburizing, pack and gas	815-870	1500-1600
Hardening	775-815	1425-1500
Cold treating	−73	−100
Tempering	150-260	300-500
Type P20		
Forging, start	1040-1095	1900-2000
Forging, finish	925 min	1700 min
Normalizing(a)	···	···
Annealing	790-815	1450-1500
Stress relieving	540	1000
Carburizing	870-925	1600-1700
Hardening	815-845	1500-1550
Tempering	150-315	300-600
Nitriding	540-550	1000-1025
Type P21		
Forging, start	1095-1150	2000-2100
Forging, finish	925 min	1700 min
Hardening	705-730	1300-1350
Tempering	510-550	950-1025

(a) Not recommended

Mechanical Properties

Type P4 Mold Steel: Typical Tensile Properties (Ref 5)

Test specimens were air cooled from 955 °C (1750 °F); tempered at 425 °C (800 °F)

Test temperature °C	°F	Tensile strength MPa	ksi	Yield strength(a) MPa	ksi	Elongation(b), %	Reduction in area, %	Hardness, HRC	Impact energy(c) J	ft·lb
Room temperature		1290	187	1035	150	15.8	53.4	42	43	32
425	800	1260	183	1005	146	19.4	55.8	...	54	40

(a) 0.2% offset. (b) In 50 mm (2 in.). (c) Charpy V-notch

Type P4 Mold Steels: Typical Short-Time Elevated Temperature Properties

Test specimens were air cooled from 955 °C (1750 °F); tempered at 540 °C (1000 °F)

Test temperature °C	°F	Tensile strength MPa	ksi	Yield strength MPa	ksi	Elongation(a), %	Reduction in area, %
Room temperature		1213	176	951	138	18.4	64.0
100	212	1200	174	1014	147	18.4	64.0
205	400	1158	168	1014	147	17.4	63.5
315	600	1179	171	986	143	15.6	58.4
427	800	1151	167	1000	145	15.8	62.9
540	1000	1020	148	910	132	15.9	59.9

(a) In 50 mm (2 in.)

Type P4 Mold Steels: Effect of Tempering Temperature on Hardness of Die-Casting Dies (Ref 5)

Test specimens were 38.1-mm (1.50-in.) round packed in new gray cast iron chips; cooled in air from 955 °C (1750 °F)

Tempering temperature °C	°F	Surface hardness, HRC	Core hardness, HRC
As treated		52-54	36-38
150	300	50-53	36-38
205	400	50-53	38-39
260	500	50-51	37-38
315	600	49-50	37-38
370	700	49-50	37-38
425	800	49-50	38-39
480	900	50-51	38-39
540	1000	48-49	39-40
595	1100	42-44	37-38
650	1200	32-34	21-22

Type P20 Mold Steels: Effect of Tempering Temperature on Tensile Properties (Ref 4)

Test specimens were 25-mm (1-in.) diam; oil quenched from 845 °C (1555 °F)

Tempering temperature °C	°F	Tensile strength MPa	ksi	Yield strength(a) MPa	ksi	Elongation(b), %	Reduction in area, %	Hardness, HRC
150	300	1960	284	1360	197	11.0	35.0	54
205	400	1905	276	1480	215	11.5	36.6	53
260	500	1805	262	1490	216	11.5	40.4	52
315	600	1750	254	1470	213	12.0	43.7	50
370	700	1670	242	1425	207	12.0	46.3	48
425	800	1545	224	1360	197	12.0	47.4	46
480	900	1425	207	1290	187	13.5	50.8	44
540	1000	1325	192	1190	173	14.5	50.6	41
565	1050	1240	180	1095	159	16.5	54.1	38
595	1100	1160	168	1040	151	16.0	58.3	35
620	1150	995	144	895	130	17.5	59.3	30
650	1200	895	130	785	114	18.5	62.3	26

(a) 0.2% offset. (b) In 50 mm (2 in.)

Type P20 Mold Steels: Effect of Tempering Temperature on Hardness (Ref 6)

Test specimens were 100 × 100-mm (4 × 4-in.) square sections; oil quenched from 815 °C (1500 °F); tempered 4 h at temperature indicated

Tempering temperature °C	°F	Hardness, HRC
As hardened		50-51
205	400	48-49
260	500	47-48
315	600	46-47
370	700	44-45
425	800	43-44
480	900	39-40
540	1000	39-40
565	1050	36-37
595	1100	33-34
620	1150	30-31
650	1200	26-27
675	1250	20-22

Type P4 Mold Steels: Effect of Tempering Temperature on Hardness of Plastic Molds (Ref 5)

Tempering temperature °C	°F	Case hardness, HRC	Core hardness, HRC
As treated		61-62	36-38
95	200	62-63	36-38
150	300	61-62	36-38
205	400	59-60	38-39
315	600	57-58	37-38
425	800	54-56	38-39
540	1000	54-55	39-40

Type P5 Mold Steels: Effect of Tempering Temperature on Case Hardness of Carburized and Hardened Material (Ref 5)

Tempering temperature °C	°F	Hardness, HRC Oil quench(a)	Water quench(b)
As quenched		65-67	66-67
95	200	64-65	66
150	300	63-64	65
204	400	60-61	63
260	500	58-59	60

(a) At 855 °C (1575 °F). (b) At 790 °C (1450 °F)

Mechanical Properties (continued)

Type P6 Mold Steels: Average Mechanical Properties of Case and Core of Case-Hardened Material (Ref 5)

Treatment	Case hardness, HRC	Case depth		Core Properties: Tensile strength		Yield strength		Elongation(a), %	Reduction in area, %	Hardness, HRC	Impact energy(b)	
		mm	in.	MPa	ksi	MPa	ksi				J	ft·lb
Carburized at 870-900 °C (1600-1650 °F); pack or air cooled; oil quenched from 815 °C (1500 °F); reheated to 775 °C (1425 °F); oil quenched; tempered at 95-150 °C (200-300 °F)	62	2.0	0.078	1140	165	930	135	17.0	55.6	37	35	26
Carburized at 845-870 °C (1550-1600 °F); pack or air cooled; oil quenched from 845 °C (1550 °F); tempered at 95-150 °C (200-300 °F)	61	2.0	0.078	1140	165	930	135	16.0	53.0	37	33	24
Carburized at 845-870 °C (1550-1600 °F); quenched directly in oil; tempered at 95-150 °C (200-300 °F)	61	1.6	0.063	1170	170	965	140	14.0	48.0	38	30	22

(a) In 50 mm (2 in.). (b) Izod impact energy for case and core combined

Type P6 Mold Steels: Effect of Tempering Temperature on Case Hardness (Ref 5)

Test specimens were carburized at 870 °C (1600 °F) for 8 h; air cooled; hardened at 790 °C (1450 °F); oil quenched; cold treated at −84 °C (−120 °F)

Tempering temperature °C	°F	Hardness, HRC
As quenched		64-65
150	300	63
175	350	61-62
205	400	61
230	450	60
260	500	59-60
290	550	58-59

Type P6 Mold Steels: Effect of Tempering on the Core or Uncarburized Samples (Ref 5)

Test specimens were oil quenched from 790 to 870 °C (1450 to 1600 °F)

Tempering temperature °C	°F	Hardness, HRC
As quenched		39-41
205	400	39-41
260	500	39
315	600	37
425	800	35
480	900	32
540	1000	27
595	1100	21
650	1200	18

Type P20 Mold Steels: Effect of Tempering Temperature on Hardness (Ref 6)

Test specimens were 100-mm (4-in.) diam bar; gas carburized at 870 °C (1600 °F); furnace cooled to 800 °C (1475 °F); oil quenched and double tempered 4 h each time at temperature indicated; larger sections than those tested will show lower hardness values; smaller sections will show higher hardness values

Tempering temperature °C	°F	Case hardness, HRC	Core hardness, HRC
315	600	57-58	47-48
345	650	57-58	46-47
370	700	55-56	45-46
400	750	54-55	44-45
425	800	53-55	43-44
480	900	52-53	39-40

Type P5 Mold Steels: End-Quench Hardenability. Test specimens were carburized at 925 °C (1700 °F). (Ref 5)

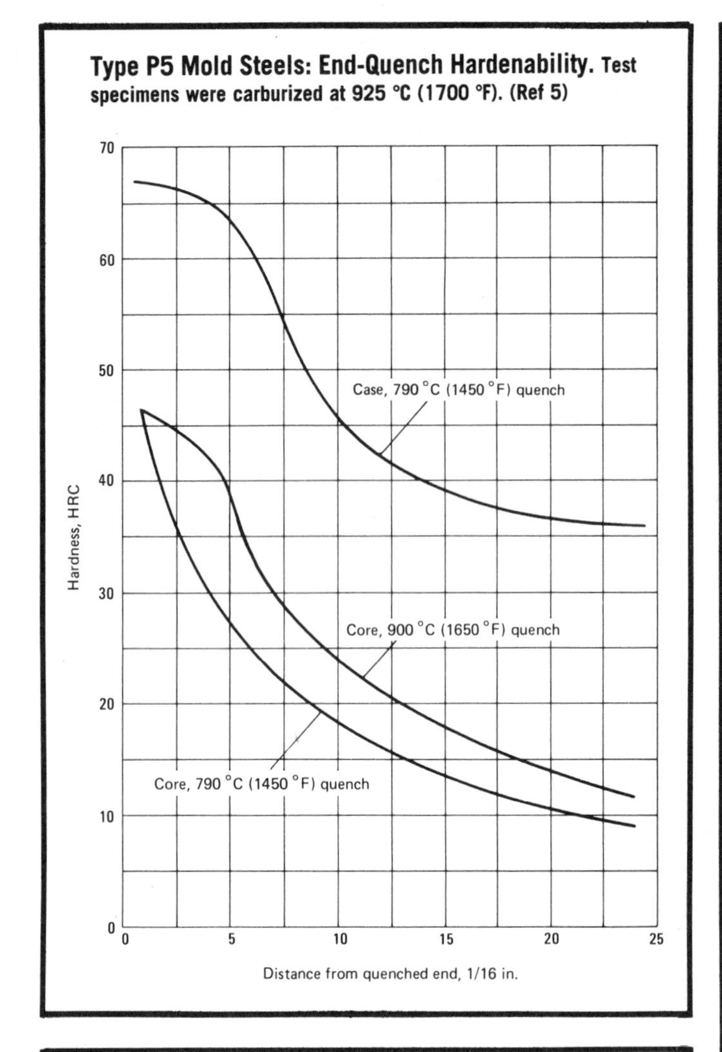

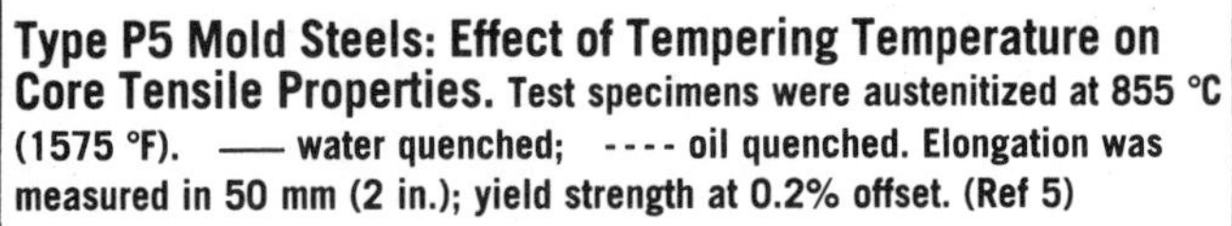
Type P5 Mold Steels: Effect of Tempering Temperature on Core Tensile Properties. Test specimens were austenitized at 855 °C (1575 °F). —— water quenched; ---- oil quenched. Elongation was measured in 50 mm (2 in.); yield strength at 0.2% offset. (Ref 5)

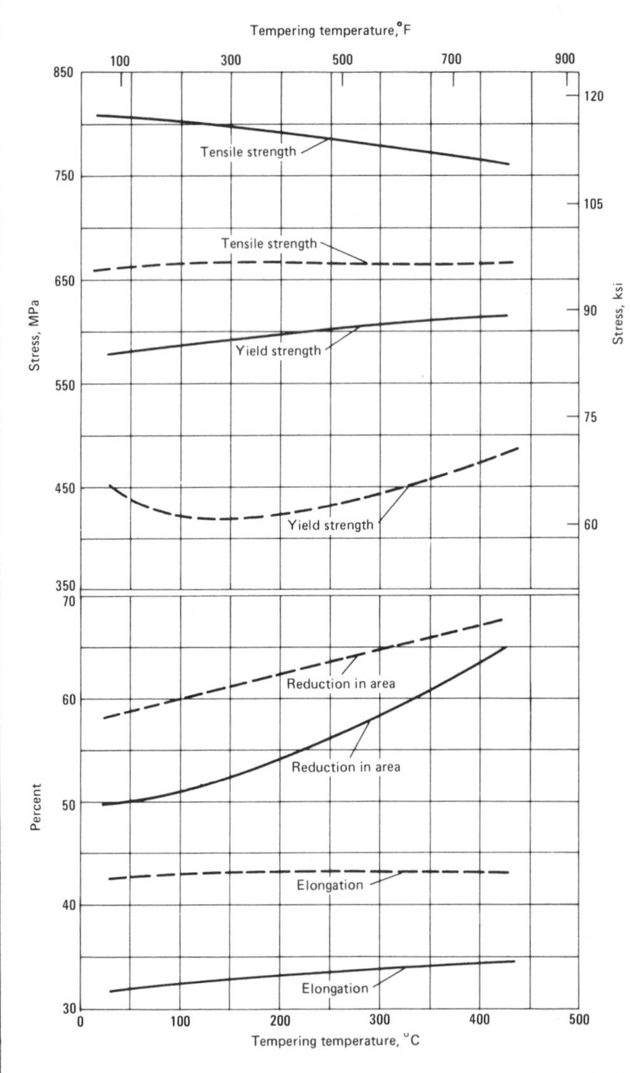

Type P6 Mold Steels: Relationship of Core Tensile Properties and Hardness. Elongation was measured in 50 mm (2 in.). (Ref 5)

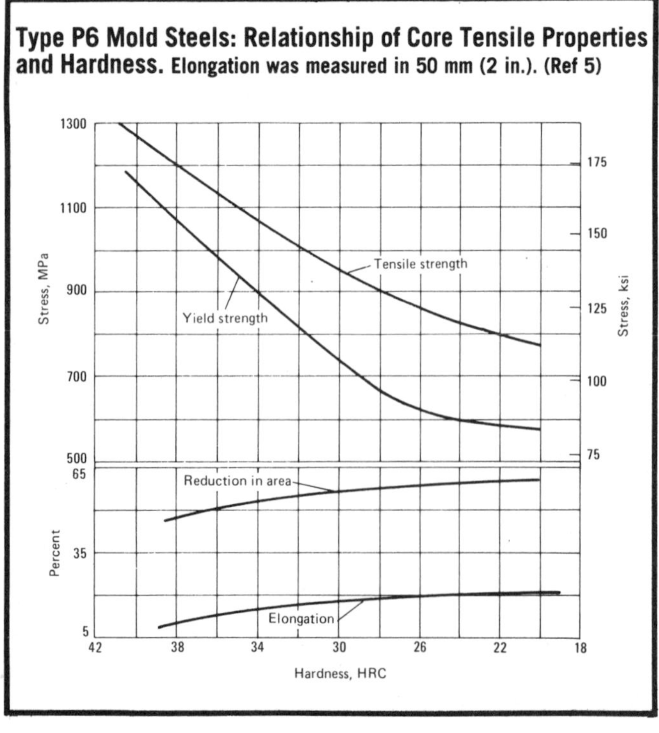

Hot Work Tool Steels (H)

Hot Work Tool Steels (H): Chemical Composition

Steel type	C	Cr	Mo	W	V
Chromium types					
H10	0.35-0.45	3.00-3.75	2.00-3.00	...	0.25-0.75
H11	0.33-0.43	4.75-5.50	1.10-1.60	...	0.30-0.60
H12	0.30-0.40	4.75-5.50	1.25-1.75	1.00-1.70	0.50 max
H13	0.32-0.40	4.75-5.50	1.10-1.75	...	0.80-1.20
H14	0.35-0.45	4.75-5.50	0.20-0.50	4.00-5.25	...
H19(a)	0.32-0.45	4.00-4.50	0.20-0.50	3.75-4.50	1.75-2.20
Tungsten types					
H21	0.32-0.45	4.00-4.50	0.20-0.50	3.75-4.50	1.75-2.20
H22	0.30-0.40	1.75-3.75	0.15-0.40	10.00-11.75	0.25-0.50(b)
H23	0.25-0.35	11.00-12.75	...	11.00-12.75	0.75-1.25(b)
H24	0.42-0.53	2.50-3.50	...	14.00-16.00	0.40-0.60(b)
H25	0.22-0.32	3.75-4.50	...	14.00-16.00	0.40-0.60(b)
H26	0.45-0.55	3.75-4.50	...	17.25-19.00	0.75-1.25
Molybdenum types					
H41	0.60-0.75	3.50-4.00	8.20-9.20	1.40-2.10	1.00-1.30
H42	0.55-0.70	3.75-4.50	4.50-5.50	5.50-6.75	1.75-2.20
H43	0.50-0.65	3.75-4.50	7.75-8.50	...	1.80-2.20

(a) Contains 4.00 to 4.50% cobalt. (b) Optional

Characteristics. The hot work tool steels are designated by the letter symbol H. Types H1 through H19 hot work steels are referred to as chromium types, although there may be other significant alloying elements present. The types containing molybdenum, H10, H11, H12, and H13, are the most widely used of all the hot work steels, and are characterized by high hardenability and excellent toughness. The chromium-tungsten type H14 and the chromium-tungsten-cobalt type H19 offer greater resistance to softening, but are less ductile in service.

In addition to hot work tooling, some of these hot work steels are used for cold work applications requiring exceptional toughness at relatively high hardness levels. The tungsten types are intended for those hot work applications where resistance to the softening effect of elevated temperature is of greatest importance and a lesser degree of toughness can be tolerated.

The molybdenum types are low-carbon modifications of molybdenum high speed steels. They offer excellent resistance to the softening effect of elevated temperature, but like the tungsten types, should be restricted to those applications where less ductility is acceptable. They generally are not readily available except on a mill delivery basis.

The machinability of hot work steels is 55 to 65% for types H11 and H12, 45 to 55% for type H13, 60 to 70% for type H19, and 40 to 50% for type H21. These ratings are based on a 1% carbon steel with a machinability rating of 100%.

Hot work tool steels can be welded using the oxy-acetylene, inert-shielded-gas, and shielded-metal-arc processes. Preheating and postheating are recommended. Filler metals should be similar in composition to the base metal.

Typical Uses. Applications of hot work tool steels include tools which require resistance to softening at elevated temperatures, such as forging dies and punches; aluminum, magnesium, and zinc die-casting dies; aluminum extrusion dies; hot heading dies; piercing and forming punches; forging dies; cold work dies where extreme toughness is required, such as bulldozer dies; dies with deep recesses and sharp corners; mandrels for tube extrusion on brass, cupronickel, aluminum, and magnesium; and nut piercers and punches.

AISI Type H10: Similar Steels (U.S. and/or Foreign). UNS T20810; ASTM A681; FED QQ-T-570; (W. Ger.) DIN 1.2365; (Fr.) AFNOR 32 DCV 28

AISI Type H11: Similar Steels (U.S. and/or Foreign). UNS T20811; AMS 6437, 6485, 6487, 6488; ASTM A681; FED QQ-T-570; SAE J437, J438, J467; (W. Ger.) DIN 1.2343; (Fr.) AFNOR Z 38 CDV 5; (Ital.) UNI X 35 CrMo 05 KU; (Jap.) JIS SKD 6; (U.K.) B.S. BH 11

AISI Type H12: Similar Steels (U.S. and/or Foreign). UNS T20812; ASTM A681; FED QQ-T-570; SAE J437, J438, J467; (W. Ger.) DIN 1.2606; (Ital.) UNI X 35 CrMoW 05 KU; (Jap.) JIS SKD 62; (U.K.) B.S. BH 12

AISI Type H13: Similar Steels (U.S. and/or Foreign). UNS T20813; ASTM A681; FED QQ-T-570; SAE J437, J438, J467; (W. Ger.) DIN 1.2344; (Fr.) AFNOR Z 40 COV 5; (Ital.) UNI X 35 CrMoV 05 KU; (Jap.) JIS SKD 61; (Swed.) SS_{14} 2242; (U.K.) B.S. BH 13

AISI Type H14: Similar Steels (U.S. and/or Foreign). UNS T20814; ASTM A681; FED QQ-T-570

AISI Type H19: Similar Steels (U.S. and/or Foreign). UNS T20819; ASTM A681; FED QQ-T-570

AISI Type H21: Similar Steels (U.S. and/or Foreign). UNS T20821; ASTM A681; FED QQ-T-570; SAE J437, J438; (W. Ger.) DIN 1.2581; (Fr.) AFNOR Z 30 WCV 9; (Ital.) UNI X 28 W 09 KU; (Jap.) JIS SKD 5; (U.K.) B.S. BH 21

AISI Type H22: Similar Steels (U.S. and/or Foreign). UNS T20822; ASTM A681; FED QQ-T-570

AISI Type H23: Similar Steels (U.S. and/or Foreign). UNS T20823; ASTM A681; FED QQ-T-570; (W. Ger.) DIN 1.2625

AISI Type H24: Similar Steels (U.S. and/or Foreign). UNS T20824; ASTM A681; FED QQ-T-570

AISI Type H25: Similar Steels (U.S. and/or Foreign). UNS T20825; ASTM A681; FED QQ-T-570

AISI Type H26: Similar Steels (U.S. and/or Foreign). UNS T20826; ASTM A681; FED QQ-T-570

AISI Type H41: Similar Steels (U.S. and/or Foreign). UNS T20841; ASTM A681; FED QQ-T-570; (W. Ger.) 1.3346; (Fr.) AFNOR Z 85 DCWV 08-04-02-01; (Ital.) UNI X 82 MoW 09 KU; (U.K.) B.S. BM 1

AISI Type H42: Similar Steels (U.S. and/or Foreign). UNS T20842; ASTM A681; FED QQ-T-570

AISI Type H43: Similar Steels (U.S. and/or Foreign). UNS T20843; ASTM A681; FED QQ-T-570

Physical Properties

Hot Work Tool Steels (H): Thermal Treatment Temperatures

Treatment	Temperature range °C	°F
Types H11, H12, H13, H19, H21		
Forging, start	1095-1135	2000-2075
Forging, finish	925 min	1700 min
Annealing	845-870	1550-1600
Stress relieving	650-675	1200-1250
Hardening	1010-1025	1850-1875
Tempering	(a)	(a)
Type H42		
Forging, start	1040-1120	1900-2050
Forging, finish	980 min	1800 min
Annealing	845-870	1550-1600
Hardening, preheat	790-815	1450-1500
Hardening, quench from	1120-1230	2050-2250
Tempering(a)	540-675	1000-1250

(a) Temperature depends on hardness and other requirements

Hot Work Tool Steels (H): Average Coefficients of Linear Thermal Expansion (Ref 5, 6)

Temperature range °C	°F	Coefficient μm/m·K	μin./in.·°F
Types H11, H13			
25-95	80-200	11.0	6.1
25-205	80-400	11.5	6.4
25-425	80-800	12.2	6.8
25-540	80-1000	12.4	6.9
25-650	80-1200	13.1	7.3
25-790	80-1450	13.5	7.5
260-650	500-1200	14.0	7.8
260-790	500-1450	14.4	8.0
425-650	800-1200	14.6	8.1
425-790	800-1450	14.8	8.2
Type H12			
25-95	80-200	11.0	6.1
25-205	80-400	11.7	6.5
25-425	80-800	12.6	7.0
25-675	80-1250	13.5	7.5
25-790	80-1450	15.1	8.4
Type H21			
25-95	80-200	10.3	5.7
25-205	80-400	11.0	6.1
25-425	80-800	12.2	6.8
25-650	80-1200	13.1	7.3
25-790	80-1450	13.7	7.6

Hot Work Tool Steels (H): Mean Apparent Specific Heat (Ref 5, 6)

Steel type	Temperature range °C	°F	Specific heat J/kg·K	Btu/lb·°F
H11, H13	0-100	32-212	460	0.11
H21	480	800	628	0.15

Hot Work Tool Steels (H): Density (Ref 5, 6)

Steel type	Density g/cm³	lb/in.³
H11, H13	7.8	0.28
H12	7.81	0.282
H21	8.19	0.296

Hot Work Tool Steels (H): Thermal Conductivity (Ref 5, 6)

Temperature °C	°F	Thermal conductivity W/m·K	Btu/ft·h·°F
Types H11, H13			
215	420	24.6	14.2
350	660	24.4	14.1
475	890	24.3	14.0
605	1120	24.7	14.3
Type H12			
25	80	17	10.1
205	400	23	13.3
425	800	25	14.5
650	1200	26	15.3
Type H21			
25	80	25.1	14.5
205	400	29.7	17.2
425	800	29.7	17.2
650	1200	28.9	16.7
760	1400	28.0	16.2

Mechanical Properties

Hot Work Tool Steels (H): Effect of Tempering Temperature on Hardness (Ref 4, 5, 6)

Tempering temperature °C	°F	Hardness, HRC H11 (a)	H12 (b)	H13 (c)	H19 (d)	H21 (e)	H21 (f)
As quenched		56-57	54-56	51-53	···	53	52
150	300	55-56	···	···	···	52	51
205	400	55-56	···	···	···	···	···
260	500	55-56	53	···	···	···	···
315	600	54-55	···	51-53	···	···	···
370	700	54-55	···	···	···	48	47
425	800	54-55	53	51-53	···	···	···
480	900	55-56	52-54	51-53	···	49	48
510	950	···	···	52-54	58-59	···	···
540	1000	56-57	55-56	52-54	57-58	51	59
565	1050	···	52	51-53	55-56	52	50
595	1100	45-50	48	49-51	52-53	51	50.5
620	1150	···	44	45-47	50-51	50	51
650	1200	33-38	40	39-41	45-46	48	47
675	1250	···	35	31-33	39-40	39	38
705	1300	25-30	29-30	28-30	34-35	34	34

(a) Air or oil quenched from 995 to 1025 °C (1825 to 1875 °F). (b) Air quenched from 1010 °C (1850 °F); tempered 2 h. (c) Air quenched from 1025 °C (1875 °F) or oil quenched from 1010 °C (1850 °F); tempered 1 h. (d) Oil quenched from 1175 °C (2150 °F); double tempered for 2 h each time. (e) Oil quenched from 1175 °C (2150 °F); tempered 2 h. (f) Air quenched from 1175 °C (2150 °F); tempered 2 h

Hot Work Tool Steels (H): Effect of Hardness and Temperature on Charpy V-Notch Impact Values (Ref 6)

Oversized Charpy specimens were austenitized at 1010 °C (1850 °F) for 1 h; air cooled and double tempered to the indicated hardness, prior to grinding to finished size

Original hardness, HRC	Impact energy at temperature of: 25 °C (80 °F) J	ft·lb	260 °C (500 °F) J	ft·lb	540 °C (1000 °F) J	ft·lb	595 °C (1100 °F) J	ft·lb
Type H11								
54-55	18	13	24	18	27	20	···	···
48-49	24	18	38	28	42	31	41	30
42-44	31	23	58	43	64	47	···	···
Type H12								
50	7	5	16	12	16	12	24	18
44	20	15	34	25	41	30	46	34
39	19	14	30	22	64	47	84	62
36	20	15	30	22	68	50	81	60

Mechanical Properties (continued)

Hot Work Tool Steels (H): Effect of Hardness and Temperature on Charpy V-Notch Impact Values

Test specimens were 16-mm (0.63-in.) square bar; austenitized at 1175 °C (2150 °F); oil quenched and tempered to the hardness indicated

Hardness, HRC	Impact energy at test temperature of: 27 °C (80 °F) J	ft·lb	150 °C (300 °F) J	ft·lb	260 °C (500 °F) J	ft·lb	425 °C (800 °F) J	ft·lb	540 °C (1000 °F) J	ft·lb
Type H19										
48-50	8	6	16	12	18	13	19	14	20	15
44-46	15	11	18	13	23	17	26	19	26	19
38-40	15	11	24	18	31	23	37	27	41	30
Type H21										
50	7	5	9	7	16	12	22	16	...	...
44	8	6	14	10	2	16	26	19	...	...
40	8	6	16	12	39	29	47	35	...	...
38	11	8	16	12	43	32	47	35	...	...

Type H13 Hot Work Tool Steels: Charpy V-Notch Impact Properties (Ref 6)

Test specimens were air cooled from 1010 °C (1850 °F) and double tempered 2 + 2 h at the temperature indicated

Tempering temperature °C	°F	Hardness, HRC	Impact energy at temperature of: −75 °C (−100 °F) J	ft·lb	Room J	ft·lb	260 °C (500 °F) J	ft·lb	540 °C (1000 °F) J	ft·lb	565 °C (1050 °F) J	ft·lb	595 °C (1100 °F) J	ft·lb
525	975	54	7	5	14	10	27	20	31	23	...	...	...	...
565	1050	52	7	5	14	10	30	22	34	25	34	25	...	...
605	1125	47	8	6	24	18	41	30	45	33	...	...	43	32
615	1140	43	9	7	24	18	52	38	60	44	...	...	57	42

Hot Work Tool Steels (H): Effect of Tempering Temperature on Hardness of Alloys Quenched in Oil from Various Temperatures (Ref 7)

Tempering temperature °C	°F	Hardness after quenching in oil from: 1095 °C (2000 °F) HB	HRC	1150 °C (2100 °F) HB	HRC	1205 °C (2200 °F) HB	HRC
Type H21							
425	800	460	48	496	51	552	53
480	900	484	50	496	51	552	53
540	1000	509	52	509	52	552	53
595	1100	426	45	437	46	448	47
650	1200	352	38	352	38	352	38
Type H22							
425	800	484	50	496	51	496	51
480	900	522	53	522	53	534	54
540	1000	534	54	547	55	560	56
595	1100	509	52	534	54	560	56
650	1200	484	50	509	52	522	53
Type H24							
425	800	547	55	587	58	587	58
480	900	534	54	573	57	587	58
540	1000	534	54	547	55	587	58
595	1100	484	50	509	52	547	55
650	1200	393	42	415	44	448	47

Hot Work Tool Steels (H): Modulus of Elasticity

Temperature °C	°F	Modulus GPa	10^6 psi
Types H11, H13			
20	70	210	30.5
150	300	192	27.8
260	500	180	26.1
345	650	191	27.7
425	800	188	27.3
480	900	186	27.0
540	1000	157	22.7
650	1200	114	16.5
Type H12			
...	...	207	30

Hot Work Tool Steels (H): Effect of Tempering Temperature on Hardness of Alloys Quenched in Air (Ref 7)

Tempering temperature °C	°F	Hardness after quenching in air from: 1095 °C (2000 °F) HB	HRC	1205 °C (2200 °F) HB	HRC
Type H21					
425	800	415	44	437	46
480	900	437	46	472	49
540	1000	509	52	534	54
595	1100	460	48	522	53
650	1200	342	37	362	39
Type H22					
425	800	426	45	460	48
480	900	460	48	472	49
540	1000	509	52	522	53
595	1100	484	50	534	54
650	1200	460	48	496	51

Mechanical Properties (continued)

Type H42 Hot Work Tool Steel: Effect of Tempering Temperature on Hardness (Ref 7)

Test specimens were double tempered 2 + 2 h at temperature indicated; quenched in air

Hardening temperature °C	°F	As quenched hardness, HRC	540 °C (1000 °F)	550 °C (1025 °F)	565 °C (1050 °F)	595 °C (1100 °F)	620 °C (1150 °F)	650 °C (1200 °F)
			Hardness, HRC, after double tempering at temperature of:					
1175	2150	58	60	59	59	56	54	50
1205	2200	59	61	60	59	57	54	49
1220	2225	58-59	60	60	59	57	54	49
1230	2250	58-59	60	60	59	56	53	48

Types H11, H13 Hot Work Tool Steels: Effect of Elevated Temperature on Tensile Strength. Hardness was measured using the Rockwell C hardness test. (Ref 5)

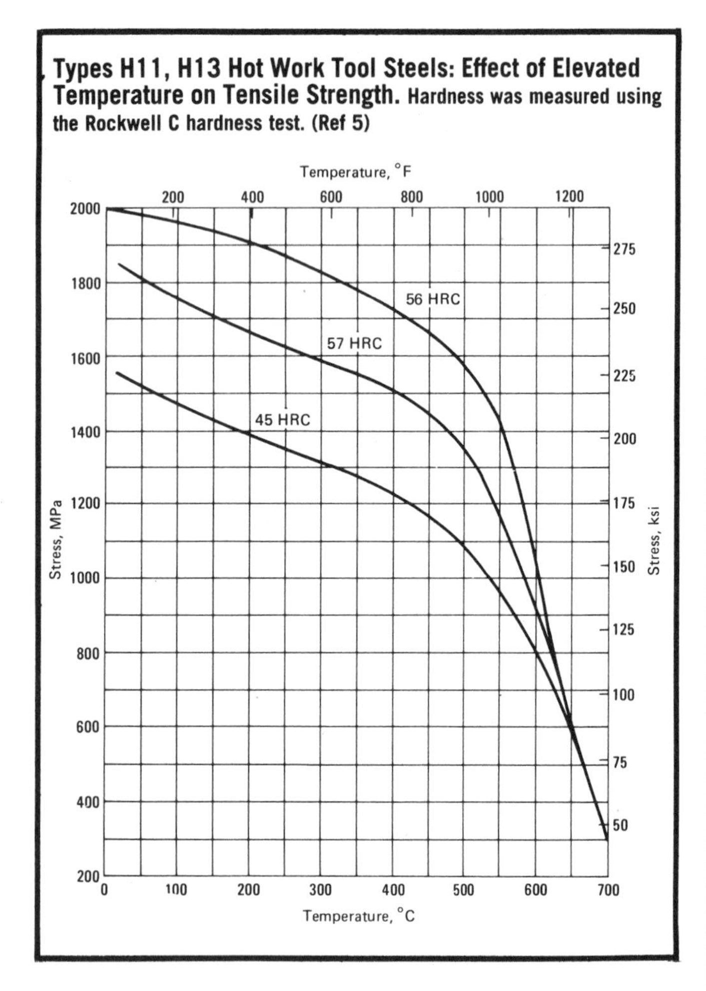

Types H11, H13 Hot Work Tool Steels: Effect of Temperature on Hot Brinell Hardness (Ref 5)

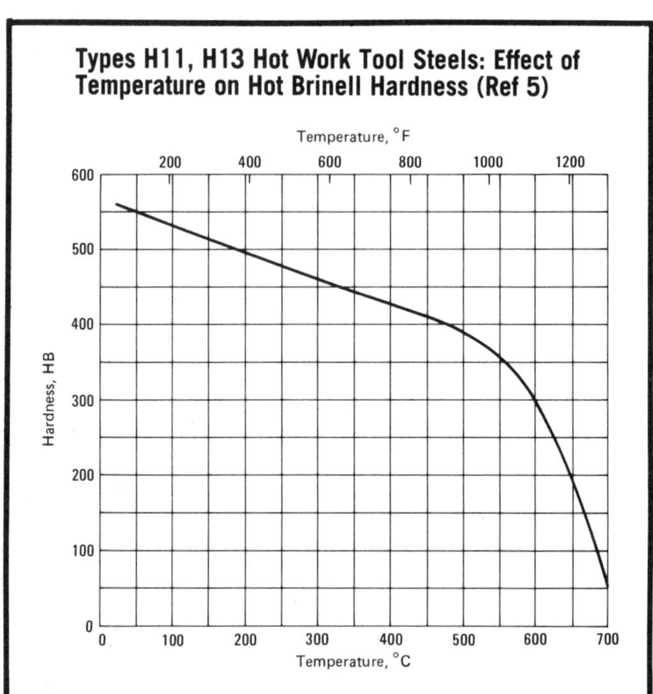

Type H12 Hot Work Tool Steels: Effect of Elevated Temperature on Tensile Strength. Hardness was measured using Rockwell C hardness test. (Ref 5)

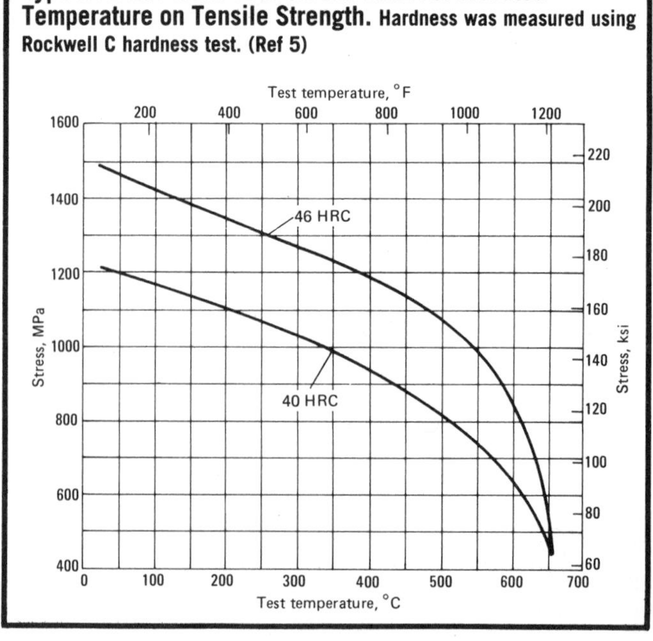

Types H11, H13 Hot Work Tool Steels: Relationship of Tensile Properties and Hardness.

Impact energy tests used Charpy V-notch specimens. Elongation was measured in 50 mm (2 in.); yield strength at 0.2% offset. (Ref 5)

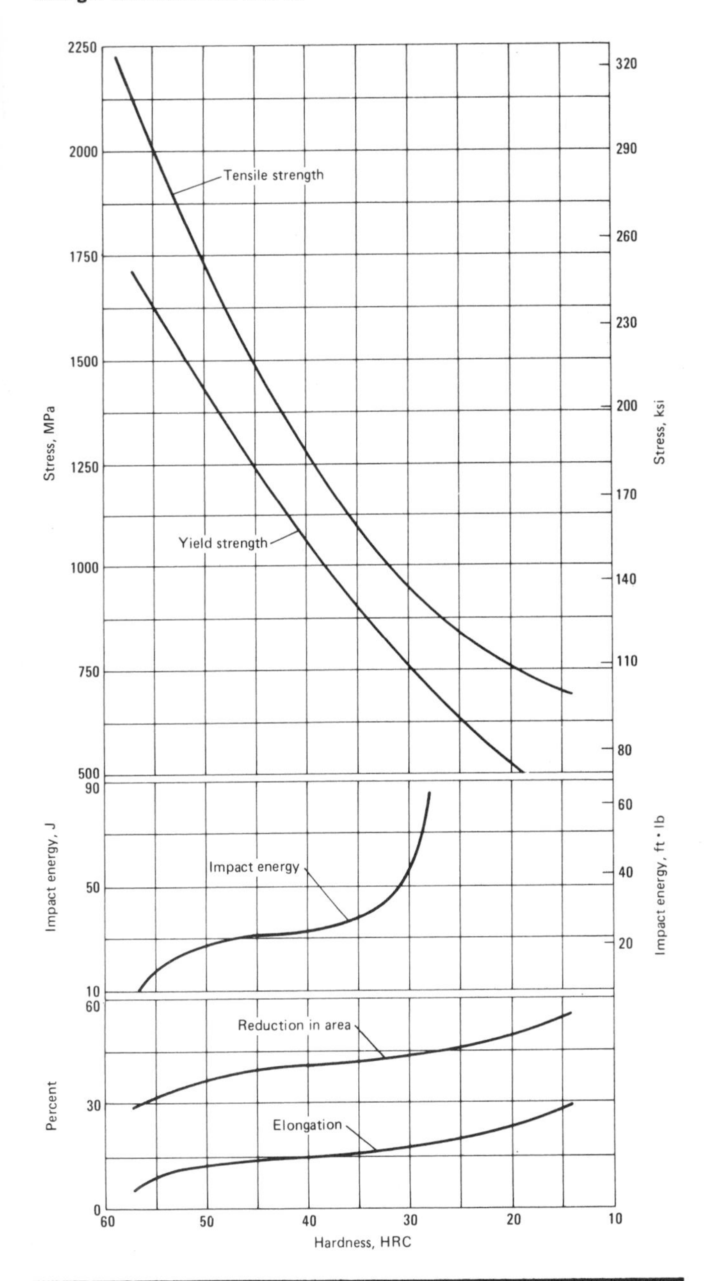

Type H12 Hot Work Tool Steels: Relationship of Tensile Properties and Hardness.

Elongation was measured in 50 mm (2 in.). (Ref 5)

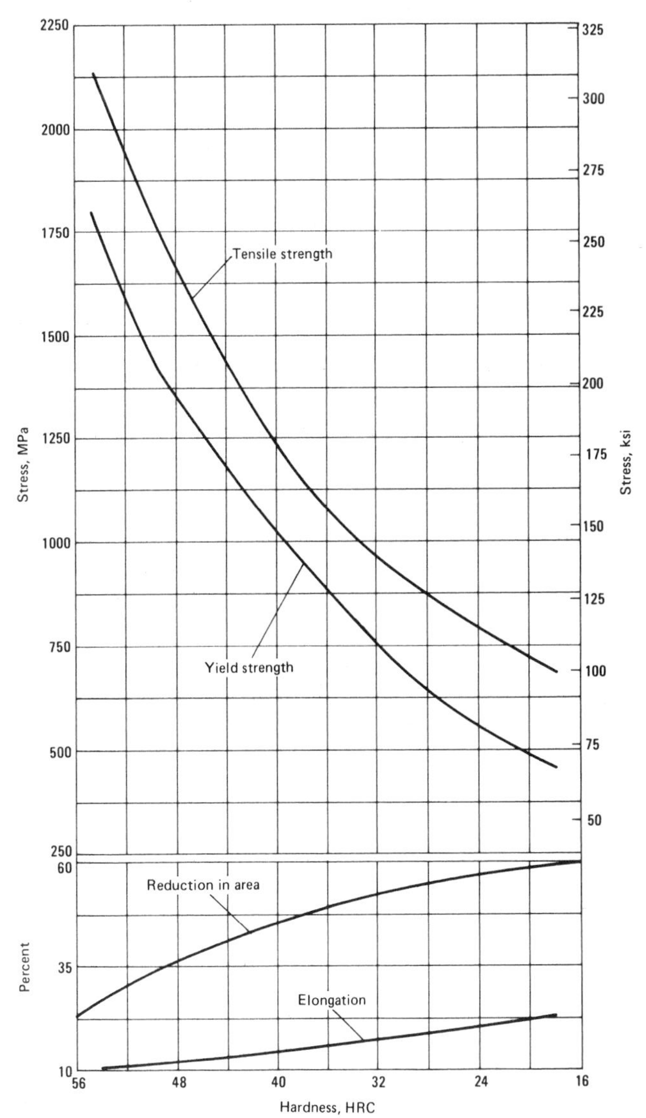

Type H12 Hot Work Tool Steels: Effect of Temperature on Hot Brinell Hardness. Test specimens were air quenched from 1010 °C (1850 °F). Rockwell C hardness in the as-quenched condition was 52 to 54 HRC.

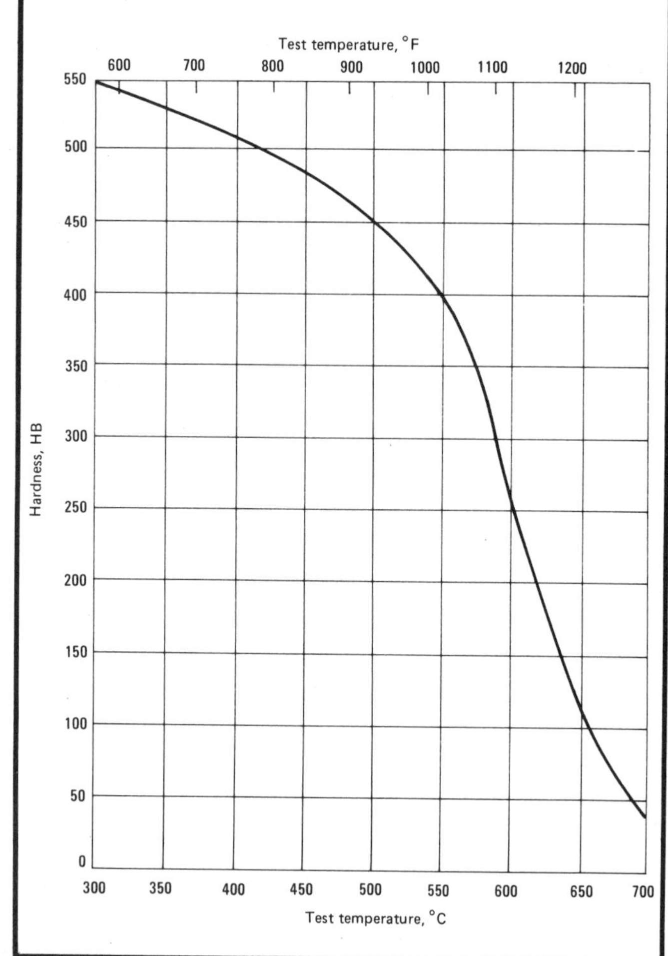

Type H21 Hot Work Tool Steels: Relationship of Hardness to Tensile and Yield Strength. Yield strength was measured at 0.2% offset. (Ref 5)

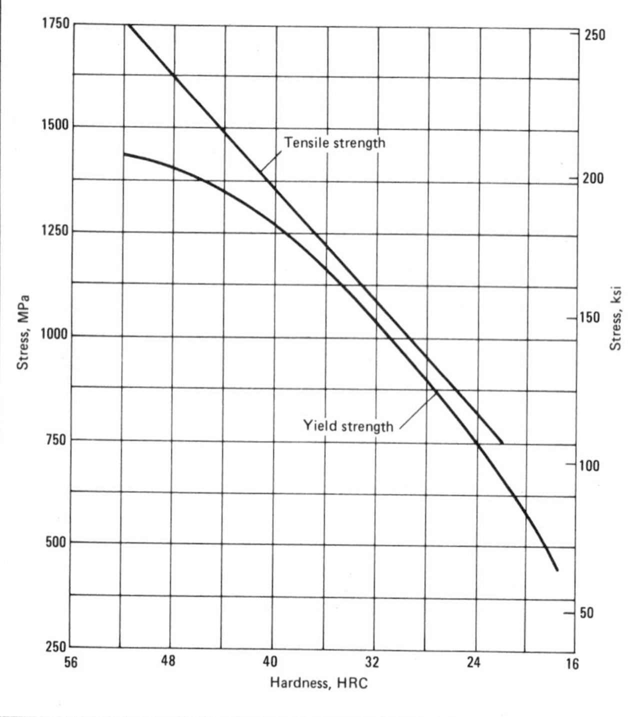

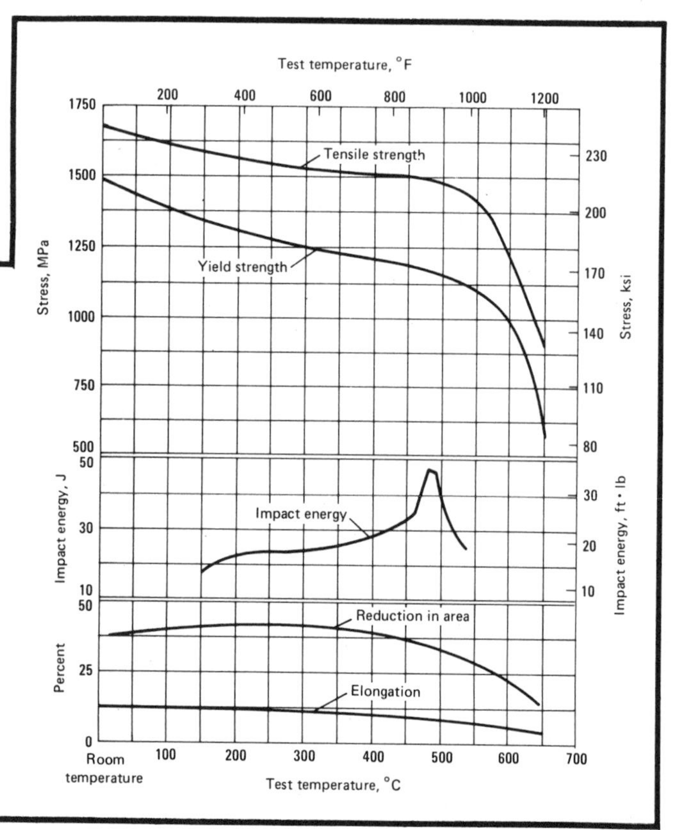

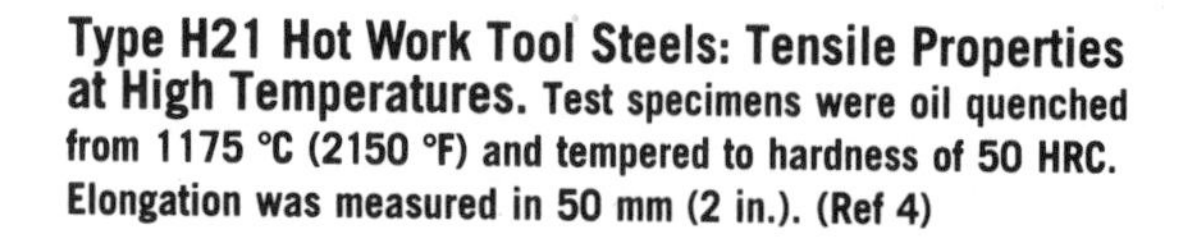

Type H21 Hot Work Tool Steels: Tensile Properties at High Temperatures. Test specimens were oil quenched from 1175 °C (2150 °F) and tempered to hardness of 50 HRC. Elongation was measured in 50 mm (2 in.). (Ref 4)

High Speed Steels (T)

High Speed Steels (T): Chemical Composition

Steel type	C	Cr	Mo	Co	V	W
	Chemical composition, %					
T1	0.65-0.80	3.75-4.50	...	...	0.90-1.30	17.25-18.75
T2	0.80-0.90	3.75-4.50	1.00 max	...	1.80-2.40	17.50-19.00
T4	0.70-0.80	3.75-4.50	0.40-1.00	4.25-5.75	0.80-1.20	17.50-19.00
T5	0.75-0.85	3.75-5.00	0.50-1.25	7.00-9.50	1.80-2.40	17.50-19.00
T6	0.75-0.85	4.00-4.75	0.40-1.00	11.00-13.00	1.50-2.10	18.50-21.00
T8	0.75-0.85	3.75-4.50	0.40-1.00	4.25-5.75	1.80-2.40	13.25-14.75
T15	1.50-1.60	3.75-5.00	1.00 max	4.75-5.25	4.50-5.25	12.00-13.00

Characteristics. The high speed steels designated by the letter symbol T have tungsten as the principal alloying element. They are more resistant to decarburization in heat treatment than the molybdenum (M) types, and usually are hardened from higher temperatures. The type T high speed steels have high abrasion resistance and are recommended for light, fine cuts on hard materials at high speeds, where good finishes are required.

The type T high speed steels containing cobalt have greater red hardness and good wear resistance, but slightly less toughness than those steels without cobalt.

Typical Uses. Applications of the type T high speed steels include single point tools, milling cutters, drills, taps and reamers, gear cutters, broaches, saw blades, woodworking tools, hot forming punches and dies, blanking dies, slitters, trim dies, powder compacting dies, cold extrusion punches, thread rolling dies, and ball and roller bearings.

AISI Type T1: Similar Steels (U.S. and/or Foreign). UNS T12001; AMS 5626; ASTM A600; FED QQ-T-590; SAE J437, J438; (W. Ger.) DIN 1.3355; (Fr.) AFNOR Z 80 WCV 18-04-01; (Ital.) UNI X 75 W 18 KU; (Jap.) JIS SKH 2; (U.K.) B.S. BT 1

AISI Type T2: Similar Steels (U.S. and/or Foreign). UNS T12002; ASTM A600; FED QQ-T-590; SAE J437, J438

AISI Type T4: Similar Steels (U.S. and/or Foreign). UNS T12004; ASTM A600; FED QQ-T-590; SAE J437, J438; (W. Ger.) DIN 1.3255; (Fr.) AFNOR Z 80 WKCV 18-05-04-01; (Ital.) UNI X 78 WCo 1805 KU; (Jap.) JIS SKH 3; (U.K.) B.S. BT 4

AISI Type T5: Similar Steels (U.S. and/or Foreign). UNS T12005; ASTM A600; FED QQ-T-590; SAE J437, J438; (W. Ger.) DIN 1.3265; (Ital.) UNI X 80 WCo 1810 KU; (Jap.) JIS SKH 4A; (U.K.) B.S. BT 5

AISI Type T6: Similar Steels (U.S. and/or Foreign). UNS T12006; ASTM A600; FED QQ-T-590

AISI Type T8: Similar Steels (U.S. and/or Foreign). UNS T12008; ASTM A600; FED QQ-T-590; SAE J437, J438

AISI Type T15: Similar Steels (U.S. and/or Foreign). UNS T12015; ASTM A600; FED QQ-T-590; (W. Ger.) DIN 1.3202; (Ital.) UNI X 150 WCoV 130505 KU; (U.K.) B.S. BT 15

Physical Properties

High Speed Steels (T): Thermal Treatment Temperatures

Steel type	Starting forging temperature(a)		Annealing temperature		Annealed hardness, HB	Hardening temperature		Tempering temperature		Tempered hardness, HRC
	°C	°F	°C	°F		°C	°F	°C	°F	
T1	1065-1120	1950-2050	845-870	1550-1600	217-255	1260-1315	2300-2400	550-595	1025-1100	65-60
T2	1065-1120	1950-2050	870-900	1600-1650	223-255	1260-1300	2300-2375	540-595	1000-1100	66-61
T4	1105-1165	2025-2125	870-900	1600-1650	229-269	1260-1300	2300-2375	540-595	1000-1100	66-62
T5	1040-1095	1900-2000	870-900	1600-1650	235-285	1260-1280	2300-2340	540-595	1000-1100	65-60
T6	...	...	870-900	1600-1650	248-302	1260-1300	2325-2375	540-595	1000-1100	65-60
T8	1105-1165	2025-2125	870-900	1600-1650	229-255	1260-1300	2300-2375	540-595	1000-1100	65-60
T15	1095-1150	2000-2100	870-900	1600-1650	241-277	1205-1260	2200-2300	540-650	1000-1200	68-63

(a) Stop forging at a minimum temperature of 925 °C (1700 °F)

Mechanical Properties

Types T2, T4, T8 High Speed Steels: Effect of Tempering Temperature on Hardness (Ref 6, 7)

Tempering temperature °C	°F	Hardness, HRC T2 (a)	T4 (b)	T5 (b)	T8 (c)
As hardened		63.5-65	64-66	64-66	64-66
565	1050	64-66	64-66	64-66	64-66
580	1075	64-66	64-65.5	64-65.5	64-66
595	1100	62.5-64.5	62.5-64.5	62.5-64.5	63-65

(a) Oil quenched from 1290 °C (2350 °F); double tempered for 2 h as indicated. (b) Oil quenched from 1300 °C (2375 °F); double tempered for 2 h as indicated. (c) Oil quenched from 1290 °C (2350 °F); double tempered for 2 h as indicated

High Speed Steels (T): Hot Hardness (Mutual Indentation Method) After Treatment for Maximum Hardness Before Testing (Ref 3)

Steel type	Hardness at test temperature of: 595 °C (1100 °F), HB	650 °C (1200 °F), HB	Room temperature, HRC
T1	480	337	65.5
T2	490	355	66.0
T5	520	400	65.0
T7	440	340	65.5

Mechanical Properties (continued)

Type T1 High Speed Steel: Effect of Tempering Temperature on Hardness After Oil and Still-Air Quenching (Ref 4)

Test specimens were 25-mm (1-in.) diam, 64-mm (2.5-in.) long; tempered 2 h at temperature indicated

Tempering temperature °C	°F	Hardness, HRC, after quenching at temperature of: 1150 °C (2100 °F)	1205 °C (2200 °F)	1260 °C (2300 °F)	1290 °C (2350 °F)	1315 °C (2400 °F)
Oil quenched						
As quenched		63	65	65	66	66
150	300	63	65	65	65	65
205	400	62	63	64	64	64
260	500	60	61	62	62	63
315	600	60	61	61	61	62
370	700	59	60	61	61	61
425	800	59	60	61	61	61
480	900	60	61	62	63	63
540	1000	61	62	64	65	65
565	1050	61	62	64	65	65
595	1100	61	62	64	64	64
650	1200	53	55	57	57	58
705	1300	44	45	47	47	47
760	1400	33	36	36	37	37
Still-air quenched						
As quenched		63	64	64	64	64
150	300	63	64	64	64	64
205	400	62	63	64	64	64
260	500	59	60	61	63	63
315	600	59	61	61	62	62
370	700	59	61	62	62	62
425	800	59	61	61	62	62
480	900	59	61	62	62	62
540	1000	59	62	63	63	63
565	1050	59	60	62	63	63
595	1100	57	60	61	62	62
650	1200	52	55	55	55	56
705	1300	43	45	43	43	43
760	1400	26	28	31	31	31

High Speed Steels (T): Hot Hardness (Ref 3)

Steel type	Hardness at 21 °C (70 °F) before tests, HRC	Hardness, HRC, at test temperature of: 540 °C (1000 °F)	565 °C (1050 °F)	595 °C (1100 °F)
T1	65.2	58.3	57.2	55.8
T2	66.3	60.2	57.8	56.2
T4	66.2	60.2	59.5	58.7
T6	67.2	62.0	61.4	59.6
T7	64.0	58.0	56.5	54.5

Types T1, T4, T5, T6, T15 High Speed Steels: Relationship of Impact Properties and Hardness.

Test specimens were austenitized at the following temperatures: T1 at 1315 °C (2400 °F); T4, T5, and T6 at 1290 °C (2350 °F); T15 at 1250 °C (2285 °F). Impact energy tests used unnotched Izod specimens. (Ref 3)

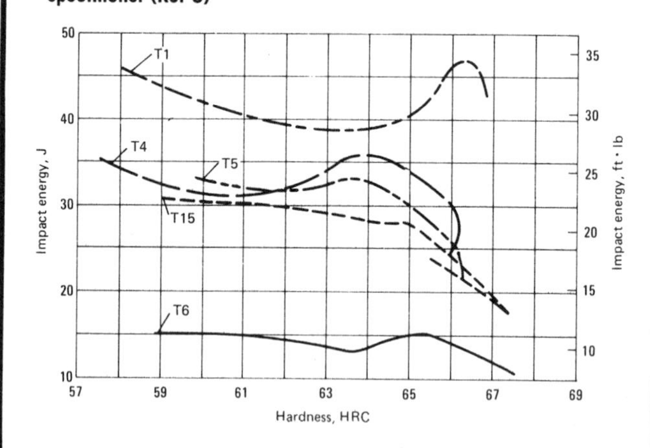

Type T1 High Speed Steels: Relationship of Impact Properties to Hardness.

Impact energy tests used unnotched Izod specimens. Specimens were hardened at various temperatures as indicated. (Ref 3)

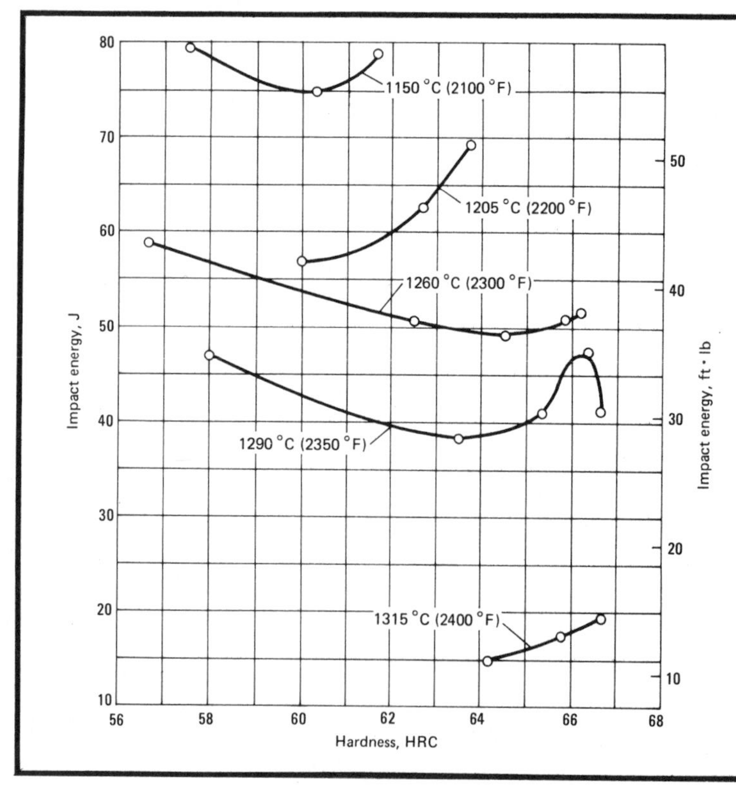

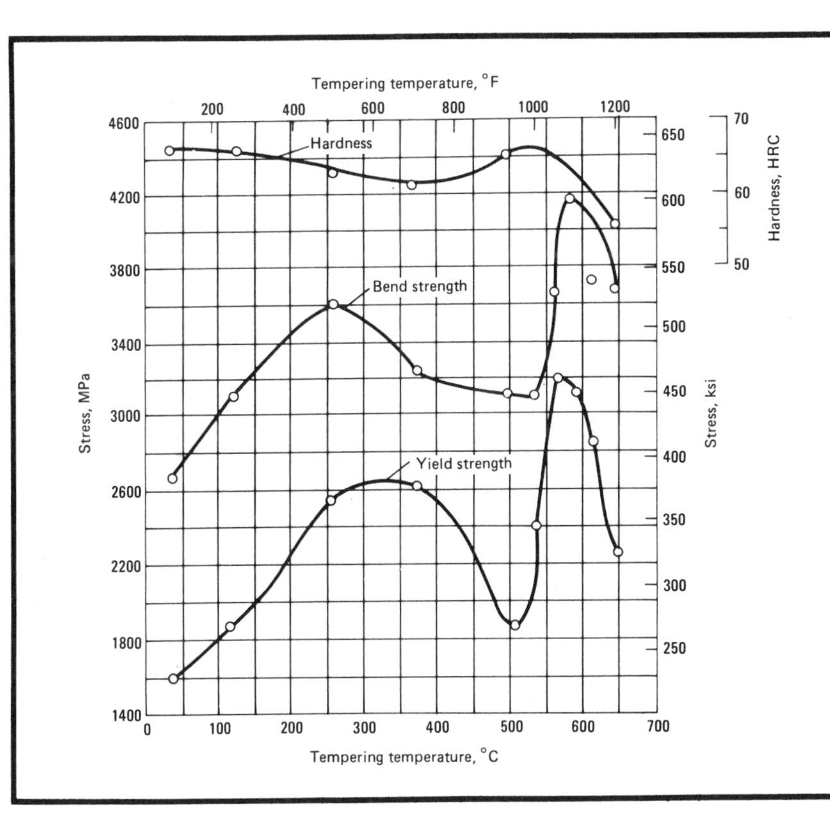

Type T1 High Speed Steels: Effect of Tempering Temperature on Yield Strength, Bend Strength, and Hardness. Test specimens were austenitized at 1290 °C (2350 °F). (Ref 3)

High Speed Steels (M)

High Speed Steels (M): Chemical Composition

Steel type	Chemical composition, % C	Mn	Si	Cr	Mo	V	W	Co
M1	0.78-0.84	0.15-0.40	0.20-0.45	3.50-4.00	8.20-9.20	1.00-1.30	1.40-2.10	...
M2	0.78-1.05	0.15-0.40	0.20-0.45	3.75-4.50	4.50-5.50	1.75-2.20	5.50-6.75	...
M3 (class 1)	1.00-1.10	0.15-0.40	0.20-0.45	3.75-4.50	4.75-6.50	2.25-2.75	5.00-6.75	...
M3 (class 2)	1.15-1.25	0.15-0.40	0.20-0.45	3.75-4.50	4.75-6.50	2.75-3.25	5.00-6.75	...
M4	1.25-1.40	0.15-0.40	0.20-0.45	3.75-4.75	4.25-5.50	3.75-4.50	5.25-6.50	...
M6	0.75-0.85	0.15-0.40	0.20-0.45	3.75-4.75	4.50-5.50	1.30-1.70	3.75-4.75	11.0-13.0
M7	0.98-1.05	0.15-0.40	0.20-0.50	3.50-4.00	8.40-9.10	1.75-2.25	1.40-2.10	...
M10	0.84-1.05	0.15-0.40	0.20-0.45	3.75-4.50	7.75-8.50	1.80-2.20	...	...
M30	0.75-0.85	0.15-0.40	0.20-0.45	3.50-4.25	7.75-9.00	1.00-1.40	1.30-2.30	4.50-5.50
M33	0.85-0.92	0.15-0.40	0.20-0.45	3.50-4.00	9.00-10.00	1.00-1.35	1.40-2.10	7.75-8.75
M34	0.85-0.92	0.15-0.40	0.20-0.45	3.50-4.00	7.75-9.20	1.90-2.30	1.40-2.10	7.75-8.75
M36	0.80-0.90	0.15-0.40	0.20-0.45	3.75-4.50	4.50-5.50	1.75-2.25	5.50-6.50	7.75-8.75
M41	1.05-1.15	0.20-0.60	0.15-0.50	3.75-4.50	3.25-4.25	1.75-2.25	6.50-7.25	4.75-5.75
M42	1.05-1.15	0.15-0.40	0.15-0.50	3.50-4.25	9.00-10.00	0.95-1.35	1.15-1.85	7.75-8.75
M43	1.15-1.25	0.20-0.40	0.15-0.50	3.50-4.25	7.50-8.50	1.50-1.75	2.25-3.00	7.75-8.75
M44	1.10-1.20	0.20-0.40	0.30-0.55	4.00-4.75	6.00-7.00	1.85-2.20	5.00-5.75	11.00-12.25
M46	1.22-1.30	0.20-0.40	0.40-0.60	3.70-4.20	8.00-8.50	3.00-3.30	1.90-2.20	7.80-8.80
M47	1.05-1.15	0.15-0.40	0.20-0.45	3.50-4.00	9.25-10.00	1.15-1.35	1.30-1.80	4.75-5.25

Characteristics and Typical Uses. The high speed steels designated by the letter symbol M are generally considered to have molybdenum as the principal alloying element, although several contain an equal or slightly greater amount of elements such as tungsten or cobalt. Those types with higher carbon and vanadium contents generally offer improved abrasion resistance, but machinability and grindability may be adversely affected. The series beginning with M41 is characterized by the capability of attaining exceptionally high hardness in heat treatment.

In addition to being used as cutting tools, some of these high speed steels are successfully used for such cold work applications as cold header die inserts, thread rolling dies, and blanking dies. For such applications, in order to increase toughness, the high speed steels are frequently hardened from a lower temperature than that used for cutting tools. It should be recognized that because of differences in properties, availability, and economic considerations, all of these steels should not be considered for each application.

The machinability rating for the high speed steels is about 35 to 45% of that for a 1% carbon tool steel, except M2 and M7 which have a rating of about 60%.

AISI Type M1: Similar Steels (U.S. and/or Foreign). UNS T11301; ASTM A600; FED QQ-T-590; SAE J437, J438; (W. Ger.) DIN 1.3346; (Fr.) AFNOR Z 85 DCWV 08-04-02-01; (Ital.) UNI X 82 MoW 09 KU; (U.K.) B.S. BM 1

AISI Type M2: Similar Steels (U.S. and/or Foreign). UNS T11302; ASTM A600; FED QQ-T-590; SAE J437, J438; (W. Ger.) DIN 1.3343; (Fr.) AFNOR Z 85 WDCV 06-05-04-02; (Ital.) UNI X 82 WMo 0605 KU; (Jap.) JIS SKH 9; (Swed.) SS_{14} 2722; (U.K.) B.S. BM 2

AISI Type M4: Similar Steels (U.S. and/or Foreign). UNS T11304; ASTM A600; FED QQ-T-590; SAE J437, J438

AISI Type M6: Similar Steels (U.S. and/or Foreign). UNS T11306; ASTM A600; FED QQ-T-590

AISI Type M7: Similar Steels (U.S. and/or Foreign). UNS T11307; ASTM A600; FED QQ-T-590; (W. Ger.) DIN 1.3348

AISI Type M10: Similar Steels (U.S. and/or Foreign). UNS T11310; ASTM A600; FED QQ-T-590

AISI Type M3 (Class 1): Similar Steels (U.S. and/or Foreign). UNS T11313; ASTM A600; FED QQ-T-590; SAE J437, J438; (W. Ger.) DIN 1.3342; (Fr.) AFNOR Z 90 WDCV 06-05-04-02

AISI Type M3 (Class 2): Similar Steels (U.S. and/or Foreign). UNS T11323; ASTM A600; FED QQ-T-590; SAE J437, J438; (W. Ger.) DIN 1.3344; (Fr.) AFNOR Z 120 WDCV 06-05-04-03, Z 130 WDCV 06-05-04-04; (Jap.) JIS SKH 52, SKH 53

AISI Type M30: Similar Steels (U.S. and/or Foreign). UNS T11330; ASTM A600; FED QQ-T-590

AISI Type M33: Similar Steels (U.S. and/or Foreign). UNS T11333; ASTM A600; FED QQ-T-590; (W. Ger.) DIN 1.3249; (U.K.) B.S. BM 34

AISI Type M34: Similar Steels (U.S. and/or Foreign). UNS T11334; ASTM A600; FED QQ-T-590; (W. Ger.) DIN 1.3249; (U.K.) B.S. BM 34

AISI Type M36: Similar Steels (U.S. and/or Foreign). UNS T11336; ASTM A600; FED QQ-T-590

AISI Type M41: Similar Steels (U.S. and/or Foreign). UNS T11341; ASTM A600; FED QQ-T-590; (W. Ger.) DIN 1.3246; (Fr.) AFNOR Z 110 WKCDV 07-05-04-04-02

AISI Type M42: Similar Steels (U.S. and/or Foreign). UNS T11342; ASTM A600; FED QQ-T-590; (W. Ger.) DIN 1.3246; (Fr.) AFNOR Z 110 WKCDV 07-05-04-04-02

AISI Type M43: Similar Steels (U.S. and/or Foreign). UNS T11343; ASTM A600; FED QQ-T-590

AISI Type M44: Similar Steels (U.S. and/or Foreign). UNS T11344; ASTM A600; FED QQ-T-590

AISI Type M46: Similar Steels (U.S. and/or Foreign). UNS T11346; ASTM A600; FED QQ-T-590

AISI Type M47: Similar Steels (U.S. and/or Foreign). UNS T11347; ASTM A600

Physical Properties

High Speed Steels (M): Thermal Treatment Temperatures, Forging and Annealing (Ref 5, 6)

Steel type	Forging temperature: Preheat °C	Preheat °F	Start °C	Start °F	Finish °C	Finish °F	Annealing temperature °C	°F
M1	815-870	1500-1600	1065-1120	1950-2050	925 min	1700 min	845-870	1550-1600
M2	815-870	1500-1600	1065-1120	1950-2050	925 min	1700 min	845-870	1550-1600
M3 (class 1)	...	...	1080-1135	1975-2075	925 min	1700 min	845-870	1550-1600
M3 (class 2)	...	...	1025-1080	1875-1975	925 min	1700 min	845-870	1550-1600
M4	760-815	1400-1500	1120	2050	925 min	1700 min	845-870	1550-1600
M7	815-870	1500-1600	1065-1120	1950-2050	925 min	1700 min	830-855	1525-1575
M10	815-870	1500-1600	1065-1120	1950-2050	925 min	1700 min	845-870	1550-1600
M33	...	...	1080-1135	1975-2075	925 min	1700 min	870-900	1600-1650
M35	...	...	1040-1095	1900-2000	925 min	1700 min	870	1600
M41	...	...	1080-1135	1975-2075	925 min	1700 min	845-870	1550-1600
M42	815-870	1500-1600	1065-1120	1950-2050	980 min	1800 min	845-870	1550-1600

High Speed Steels (M): Thermal Treatment Temperatures, Hardening and Tempering (Ref 5, 6)

Steel type	Hardening temperature: Preheat °C	Preheat °F	Quench from(a) °C	Quench from(a) °F	Salt bath(a) °C	Salt bath(a) °F	Tempering Temperature(b) °C	°F
M1	760-815	1400-1500	1175-1205	2150-2200	540-595	1000-1100	540-650	1000-1200
M2	760-815	1400-1500	1175-1230	2150-2250	540-595	1000-1100	550-565	1025-1050
M3 (class 1)	815-845	1500-1550	1205-1230	2200-2250	540-620	1000-1150	540-595	1000-1100
M3 (class 2)	815-845	1500-1550	1205-1230	2200-2250	540-595	1000-1100	540-595	1000-1100
M4	760-815	1400-1500	1205-1220	2200-2225	540-595	1000-1100	540-650	1000-1200
M7	760-815	1400-1500	1175-1220	2150-2225	540-595	1000-1100	540-650	1000-1200
M10	760-815	1400-1500	1175-1220	2150-2225	540-595	1000-1100	540-650	1000-1200
M33	815-845	1500-1550	1205-1220	2200-2225	540-595	1000-1100	540-580	1000-1075
M35	815-845	1500-1550	1220-1230	2225-2250	540-620	1000-1150	550-565	1025-1050
M41	815-845	1500-1550	1190-1205	2175-2200	540-620	1000-1150	525-595	975-1100
M42	815-870	1500-1600	1175-1205	2150-2200	540-620	1000-1150	510-595	950-1100

(a) High speed steels may be oil quenched from temperature given or quenched in a molten salt bath. When quenched in molten salt, the quenching temperature is about 14 °C (25 °F) higher. (b) Temperatures given are those normally used for best results. Higher or lower temperatures may be used for specific properties

Mechanical Properties

Type M1 High Speed Steels: Effect of Tempering Temperature on Hardness After Quenching in Oil (Ref 4)

Tempering temperature °C	°F	Hardness, HRC, after oil quenching at temperature of: 1150 °C (2100 °F)	1205 °C (2200 °F)	1230 °C (2250 °F)	1245 °C (2275 °F)	1260 °C (2300 °F)
As quenched		64	64	64	63	63
150	300	63	63	63	62	62
205	400	62	62	61	61	61
260	500	60	61	60	59	58
315	600	59	60	60	58	58
370	700	60	60	60	59	59
425	800	60	61	61	60	60
480	900	61	63	63	62	62
540	1000	65	66	65	65	64
550	1025	65	66	66	66	65
565	1050	64	66	67	65	65
580	1075	63	65	67	64	64
595	1100	63	65	64	64	64
620	1150	60	62	62	61	62
650	1200	57	58	58	58	58
705	1300	39	41	43	41	48
760	1400	34	37	38	35	37

Type M3, Class 1 and 2: Effect of Tempering Temperature on Hardness (Ref 6)

Test specimens were oil quenched from 1230 °C (2250 °F) and double tempered for 2 h each time

Tempering temperature °C	°F	Hardness, HRC Class 1	Class 2
As quenched		64-66	65-66
540	1000	65-67	66-67
565	1050	64-66	65-66
595	1100	62-64	63-64

Type M2 High Speed Steels: Effect of Tempering Temperature on Hardness After Quenching in Oil and Air (Ref 4)

Test specimens were oil or air quenched from 1230 °C (2250 °F)

Tempering temperature °C	°F	Oil-quenched hardness, HRC	Air-quenched hardness, HRC
150	300	65	65
205	400	64	63
260	500	63	62.5
315	600	62.5	62.5
370	700	63	62.5
425	800	63.5	63.5
455	850	63.5	63.5
480	900	65	64
510	950	66	65
540	1000	66	65.5
565	1050	66	63.5
595	1100	64.5	61.5
620	1150	62	60
650	1200	53.5	53
705	1300	43	39.5
760	1400	33.5	34

Types M33, M35, M42 High Speed Steels: Effect of Tempering Temperature on Hardness (Ref 6)

Tempering temperature °C	°F	Hardness, HRC M33 (a)	M35 (b)	M35 (c)	M42 (d)
As quenched		64-65	...	...	64-66
525	975	...	...	...	67-69
540	1000	67-68	64.5-66.5	65-67	67-69
550	1025	67-68	64-66	64.5-66.5	66.5-68.5
565	1050	66-68	63.5-65.5	64-66	66-68
580	1075	65-67	...	...	65.5-67.5
595	1100	...	62-64	62.5-64.5	64-66

(a) Oil quenched from 1205 °C (2200 °F) in a salt bath; double tempered for 2 h each time at temperature indicated. (b) Austenitized in an atmosphere furnace at 1220 °C (2225 °F); double tempered for 2 h each time at temperature indicated. (c) Austenitized in an atmosphere furnace at 1230 °C (2250 °F); double tempered 2 h each time at temperature indicated. (d) Austenitized at 1190 °C (2175 °F) in a salt bath; oil quenched; triple tempered for 2 h each time at temperature indicated

High Speed Steels (M): Hot Hardness (Mutual Indentation Method) After Treating for Maximum Hardness Before Testing (Ref 3)

Steel type	Hardness at test temperature of: 595 °C (1100 °F), HB	650 °C (1200 °F), HB	Room temperature, HRC
M1	475	308	66.8
M2	465	342	65.5
M10	458	313	65.0
M36	500	390	66.0

High Speed Steels (M): Hot Hardness (Ref 3)

Steel type	Hardness at 21 °C (70 °F) before tests, HRC	Hardness, HRC, at test temperature of: 540 °C (1000 °F)	565 °C (1050 °F)	595 °C (1100 °F)
M2	65.3	58.0	57.0	55.9
M4	65.4	59.0	57.5	56.7
M10	65.3	58.0	57.0	55.5

Mechanical Properties (continued)

Types M4, M7, M10 High Speed Steels: Effect of Tempering Temperature on Hardness (Ref 5, 6)

Tempering temperature °C	°F	Hardness, HRC M4 (a)	M4 (b)	M7 (c)	M7 (d)	M10 (e)
As quenched		64-66	64-66	...	...	65
425	800	...	...	...	...	61-63
480	900	...	...	...	...	62-63
540	1000	63.5-65.5	64-66	64-66	64.5-66.5	65
550	1025	63-65	64-66	64-66	64.5-66.5	...
565	1050	62-64	63-65	63-65	64-66	64
580	1075	61-63	62-64	62-64	63-65	...
595	1100	59-61	61-63	59-61	60-62	62
620	1150	56-58	58-60	55-57	57-59	...
650	1200	...	...	50-52	51-53	55-56

(a) Austenitized in a salt bath at 1165 °C (2125 °F); oil quenched; triple tempered at temperature indicated for 2 h each time. (b) Austenitized in a salt bath at 1205 °C (2200 °F); oil quenched; triple tempered at temperature indicated for 2 h each time. (c) Austenitized in a salt bath at 1190 °C (2175 °F); oil quenched below 540 °C (1000 °F); air cooled; double tempered for 2 h each time at temperature indicated. (d) Austenitized in a salt bath at 1205 °C (2200 °F); oil quenched to below 540 °C (1000 °F); double tempered for 2 h each time at temperature indicated. (e) Austenitized in a salt bath at 1205 °C (2200 °F); oil quenched; tempered at temperature indicated

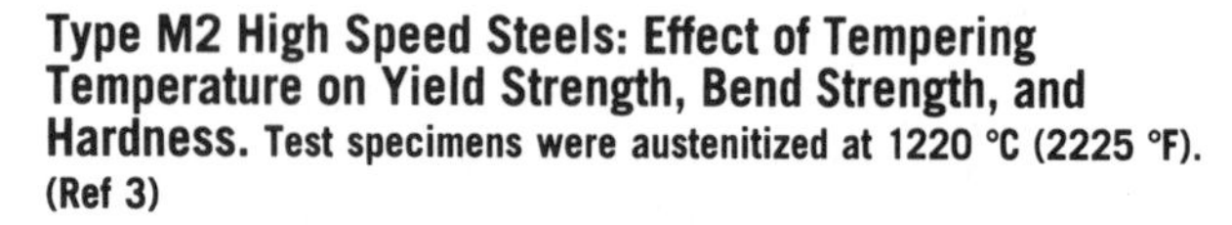

Type M2 High Speed Steels: Effect of Tempering Temperature on Yield Strength, Bend Strength, and Hardness. Test specimens were austenitized at 1220 °C (2225 °F). (Ref 3)

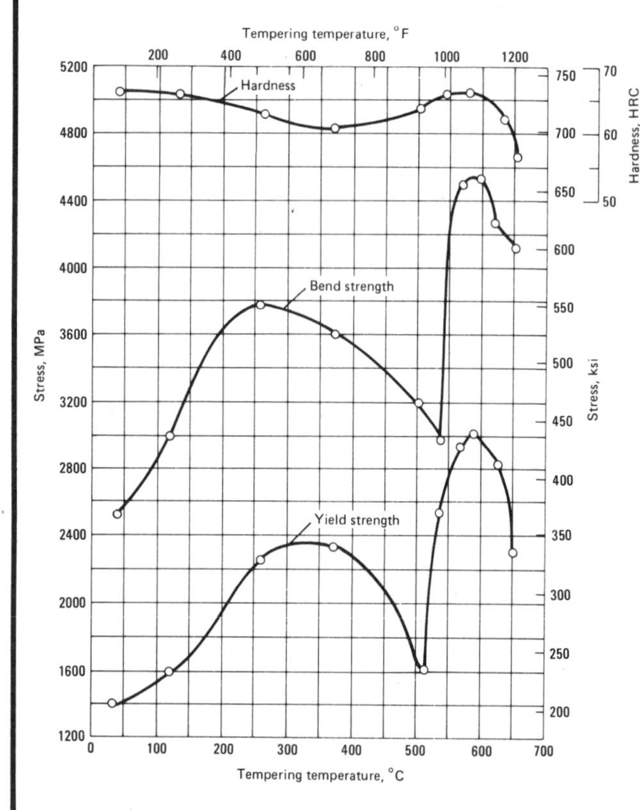

Types M1, M2, M4, M10 High Speed Steels: Relationship of Impact Properties and Hardness. Test specimens were austenitized at 1220 °C (2225 °F). Impact energy tests used unnotched Izod specimens. (Ref 3)

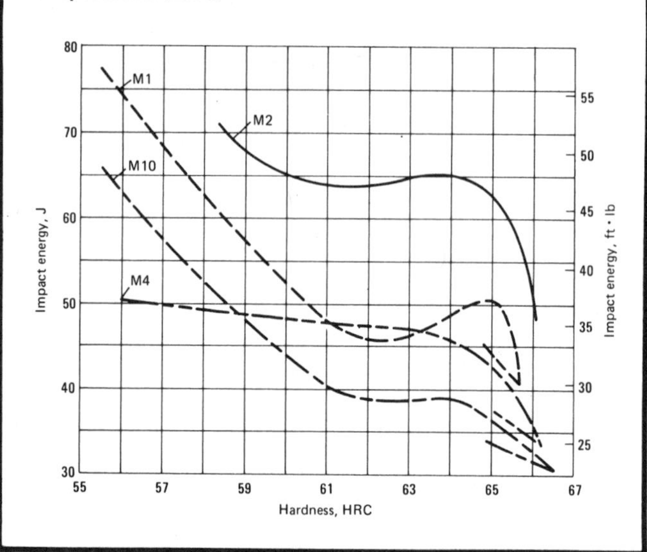

Maraging Steels

Nickel, rather than carbon, is the principal alloying element in the maraging steels. With a maximum carbon content of 0.030%, these steels are essentially carbon-free. They utilize the soft, ductile, iron-nickel martensite, which can be age hardened by additions of other alloying elements. Other major alloying elements in maraging steels are cobalt, molybdenum, aluminum, and titanium.

The most common maraging steels are the four 18Ni grades–200, 250, 300, and 350. The numerical designations generally, but not always, represent the ultimate tensile strength of the grade.

Mechanical properties are usually high for all 18Ni grades of maraging steels. Particularly noteworthy are yield strengths near the ultimate tensile strength. The differences in other mechanical property levels among the four grades permit flexibility in selecting the combination of properties best suited for a given application.

Another characteristic of nickel maraging steels is the hardening reaction of age hardening. In the annealed condition, as furnished to the user, these steels are very tough and relatively soft with a hardness of 30 to 35 HRC. Therefore, they are readily machined and formed.

After machining and forming, aging—a precipitation hardening process which requires no protective atmosphere and relatively low furnace temperatures—produces hardness levels sufficient for many structural and tooling applications. For uses requiring maximum abrasion resistance, maraging steels can be nitrided.

Applications of maraging steels in the aerospace industry include solid-propellant-rocket motor cases, load cells, flexures for guidance mechanisms of missiles, helicopter drive shafts, and aircraft wing components. Tooling applications include aluminum die casting dies, cores, and pins; plastic molding dies; trim knives; cold forming dies; and springs.

Machinability of the maraging steels in the annealed condition is comparable to that of AISI grade 4340 and similar steels at the same hardness level, 30 to 35 HRC. However, when the material is aged (heat treated), the choices of cutting tools and machining conditions become very important. Rigid equipment, firm tool supports, very sharp tools,and an abundance of coolant are essential. Suggested conditions for some machining operations are offered in this section.

Because of the relative softness of the nickel martensite, maraging steels can be readily formed, spun, drawn, or hydroformed while cold, with minimal work hardening.

REFERENCES

1. *Machining Data Handbook,* 3rd ed., Vol 1 and 2, Metcut Research Associates Inc., Cincinnati, OH, 1980
2. VascoMax, Teledyne Vasço, Latrobe, PA, 1977

Machining Data (Ref 1)

Maraging Steels: Turning (Cutoff and Form Tools)

Test specimens were in the annealed condition, with hardness of 275 to 325 HB

Tool material	Speed, m/min (ft/min)	Feed per revolution for cutoff tool width of: 1.5 mm (0.062 in.) mm	in.	3 mm (0.125 in.) mm	in.	6 mm (0.25 in.) mm	in.	Feed per revolution for form tool width of: 12 mm (0.5 in.) mm	in.	18 mm (0.75 in.) mm	in.	25 mm (1 in.) mm	in.	35 mm (1.5 in.) mm	in.	50 mm (2 in.) mm	in.
M2 and M3 high speed steel	17 (55)	0.033	0.0013	0.041	0.0016	0.050	0.002	0.041	0.0016	0.036	0.0014	0.033	0.0013	0.028	0.0011	0.023	0.0009
C-6 carbide	53 (175)	0.033	0.0013	0.041	0.0016	0.050	0.002	0.041	0.0016	0.036	0.0014	0.033	0.0013	0.028	0.0011	0.023	0.0009

Machining Data (Ref 1) (continued)

Maraging Steels: Turning (Single Point and Box Tools)

Depth of cut		High speed steel(a)				Uncoated carbide(b)						Coated carbide(c)			
		Speed		Feed		Speed, brazed		Speed, inserted		Feed		Speed		Feed	
mm	in.	m/min	ft/min	mm/rev	in./rev	m/min	ft/min	m/min	ft/min	mm/rev	in./rev	m/min	ft/min	mm/rev	in./rev
Annealed with a hardness of 275 to 325 HB															
1	0.040	27	90	0.13	0.005	115	380	150	500	0.18	0.007	200	650	0.18	0.007
4	0.150	21	70	0.25	0.010	90	300	120	400	0.40	0.015	160	525	0.40	0.015
8	0.300	15	50	0.40	0.015	70	230	90	300	0.50	0.020	120	400	0.50	0.020
Maraged, with hardness of 50 50 to 52 HRC															
1	0.040	18	60	0.13	0.005	41	135	53	175	0.13	0.005	...	...	...	...
4	0.150	14	45	0.25	0.010	34	110	43	140	0.25	0.010	...	...	...	...

(a) Any premium high speed steel tool material (T15, M33, M41 to M47). (b) Carbide tool material: C-3. (c) Carbide tool material: CC-3

Maraging Steels: Surface Grinding with a Horizontal Spindle and Reciprocating Table

Operation	Wheel speed		Table speed		Downfeed		Crossfeed		Wheel identification
	m/s	ft/min	m/min	ft/min	mm/pass	in./pass	mm/pass	in./pass	
Annealed or maraged, with maximum hardness of 50 HRC									
Roughing	20-30	4000-6000	15-30	50-100	0.075	0.003	1.25-12.5	0.050-0.500	A46JV
Finishing	20-30	4000-6000	15-30	50-100	0.013 max	0.0005 max	(a)	(a)	A46JV
Maraged, with hardness greater than 50 HRC									
Roughing	15-20	3000-4000	15-30	50-100	0.050	0.002	0.65-6.5	0.025-0.250	A46JV
Finishing	15-20	3000-4000	15-30	50-100	0.013 max	0.0005 max	(b)	(b)	A46JV

(a) Maximum ¼ of wheel width. (b) Maximum 1/10 of wheel width

Maraging Steels: Cylindrical and Internal Grinding

Operation	Wheel speed		Work speed		Infeed on diameter		Traverse(a)	Wheel identification
	m/s	ft/min	m/min	ft/min	mm/pass	in./pass		
Cylindrical grinding								
Annealed or maraged, with maximum hardness of 50 HRC								
Roughing	28-33	5500-6500	21-30	70-100	0.050	0.002	½	A60KV
Finishing	28-33	5500-6500	21-30	70-100	0.013 max	0.0005 max	1/6	A60KV
Maraged, with hardness more than 50 HRC								
Roughing	20-28	4000-6500	21-30	70-100	0.050	0.002	¼	A80IV
Finishing	20-28	4000-6500	21-30	70-100	0.013 max	0.0005 max	⅛	A80IV
Internal grinding								
Annealed or maraged, with maximum hardness of 50 HRC								
Roughing	25-33	5000-6500	23-60	75-200	0.013	0.0005	⅓	A60KV
Finishing	25-33	5000-6500	23-60	75-200	0.005 max	0.0002 max	1/6	A60KV
Maraged, with hardness more than 50 HRC								
Roughing	20-28	4000-5500	23-60	75-100	0.013	0.0005	⅓	A80IV
Finishing	20-28	4000-5500	23-60	75-100	0.005 max	0.0002 max	1/6	A80IV

(a) Wheel width per revolution of work

Maraging Steels: Planing

Test specimens were in the annealed condition, with hardness of 275 to 325 HB

Tool material	Depth of cut		Speed		Feed/stroke	
	mm	in.	m/min	ft/min	mm	in.
M2 and M3 high speed steel	0.1	0.005	6	20	(a)	(a)
	2.5	0.100	11	35	0.75	0.030
	12	0.500	5	15	1.15	0.045
C-2 carbide	0.1	0.005	43	140	(a)	(a)
	2.5	0.100	52	170	1.50	0.060
	12	0.500	47	120	1.25	0.050

(a) Feed is 75% the width of the square nose finishing tool

Maraging Steels: Boring

Depth of cut		High speed steel(a)				Uncoated carbide						Coated carbide			
		Speed		Feed		Speed, brazed		Speed, inserted		Feed		Speed		Feed	
mm	in.	m/min	ft/min	mm/rev	in./rev	m/min	ft/min	m/min	ft/min	mm/rev	in./rev	m/min	ft/min	mm/rev	in./rev
Annealed, with hardness of 275 to 325 HB															
0.25	0.010	26	85	0.050	0.002	115(b)	370(b)	135(b)	435(b)	0.075	0.003	175(c)	570(c)	0.075	0.003
1.25	0.050	21	70	0.102	0.004	90(d)	300(d)	105(d)	350(d)	0.13	0.005	140(e)	455(e)	0.130	0.005
2.5	0.100	17	55	0.20	0.008	73(d)	240(d)	85(d)	280(d)	0.30	0.012	110(e)	365(e)	0.300	0.012
Maraged, with hardness of 50 to 52 HRC															
0.25	0.010	18	60	0.050	0.002	40(b)	130(b)	46(b)	150(b)	0.050	0.002	...	...	...	...
1.25	0.050	15	50	0.102	0.004	30(d)	100(d)	37(d)	120(d)	0.102	0.004	...	...	...	...
2.5	0.100	11	35	0.20	0.008	26(d)	85(d)	30(d)	100(d)	0.20	0.008	...	...	...	...

(a) Any premium high speed steel tool material (T15, M33, M41 to M47). (b) Carbide tool material: C-3. (c) Carbide tool material: CC-3. (d) Carbide tool material: C-2. (e) Carbide tool material: CC-2

Maraging Steels: End Milling (Profiling)

Tool material	Depth of cut		Speed		Feed per tooth for cutter diameter of: 10 mm (0.375 in.)		12 mm (0.5 in.)		18 mm (0.75 in.)		25-50 mm (1-2 in.)	
	mm	in.	m/min	ft/min	mm	in.	mm	in.	mm	in.	mm	in.
Annealed, with hardness of 275 to 325 HB												
M2, M3, and M7 high speed steel	0.5	0.020	29	95	0.018	0.0007	0.025	0.001	0.050	0.002	0.075	0.003
	1.5	0.060	21	70	0.025	0.001	0.050	0.002	0.075	0.003	0.102	0.004
	diam/4	diam/4	18	60	0.018	0.0007	0.025	0.001	0.050	0.002	0.075	0.003
	diam/2	diam/2	15	50	0.013	0.0005	0.013	0.0005	0.025	0.001	0.050	0.002
C-5 carbide	0.5	0.020	105	350	0.025	0.001	0.038	0.0015	0.075	0.003	0.102	0.004
	1.5	0.060	81	265	0.025	0.001	0.050	0.002	0.102	0.004	0.13	0.005
	diam/4	diam/4	70	230	0.025	0.001	0.038	0.0015	0.075	0.003	0.102	0.004
	diam/2	diam/2	66	215	0.018	0.0007	0.025	0.001	0.050	0.002	0.075	0.003
Maraged, with hardness of 50 to 52 HRC												
Any premium high speed steel (T15, M33, M41-M47)	0.5	0.020	9	30	0.013	0.0005	0.025	0.001	0.025	0.001	0.025	0.001
	1.5	0.060	8	25	...	...	0.025	0.001	0.025	0.001	0.025	0.001
	diam/4	diam/4	6	20	...	...	0.013	0.0005	0.013	0.0005	0.025	0.001
C-5 carbide	0.5	0.020	30	100	0.013	0.0005	0.013	0.0005	0.038	0.0015	0.050	0.002
	1.5	0.060	23	75	...	...	0.025	0.001	0.050	0.002	0.075	0.003
	diam/4	diam/4	20	65	...	...	...	...	0.050	0.002	0.075	0.003
	diam/2	diam/2	18	60	...	...	...	...	0.038	0.0015	0.050	0.002

Maraging Steels: Reaming

Based on 4 flutes for 3- and 6-mm (0.125- and 0.25-in.) reamers, 6 flutes for 12-mm (0.5-in.) reamers, and 8 flutes for 25-mm (1-in.) and larger reamers

Tool material	Speed		Feed per revolution for reamer diameter of: 3 mm (0.125 in.)		6 mm (0.25 in.)		12 mm (0.5 in.)		25 mm (1 in.)		35 mm (1.5 in.)		50 mm (2 in.)	
	m/min	ft/min	mm	in.	mm	in.	mm	in.	mm	in.	mm	in.	mm	in.
Annealed, with hardness of 275 to 325 HB														
Roughing														
M1, M2, and M7 high speed steel	17	55	0.075	0.003	0.13	0.005	0.20	0.008	0.30	0.012	0.40	0.015	0.50	0.020
C-2 carbide	21	70	0.102	0.004	0.15	0.006	0.20	0.008	0.30	0.012	0.40	0.015	0.50	0.020
Finishing														
M1, M2, and M7 high speed steel	8	25	0.075	0.003	0.15	0.006	0.20	0.008	0.30	0.012	0.40	0.015	0.50	0.020
C-2 carbide	11	35	0.102	0.004	0.15	0.006	0.20	0.008	0.30	0.012	0.40	0.015	0.50	0.020
Maraged, with hardness of 50 to 52 HRC														
Roughing														
Any premium high speed steel (T15, M33, M41-M47)	5	15	0.025	0.001	0.025	0.001	0.038	0.0015	0.050	0.002	0.050	0.002	0.050	0.002
C-2 carbide	8	25	0.050	0.002	0.075	0.003	0.102	0.004	0.13	0.005	0.15	0.006	0.15	0.006
Finishing														
Any premium high speed steel (T15, M33, M41-M47)	3	10	0.038	0.0015	0.050	0.002	0.075	0.003	0.102	0.004	0.102	0.004	0.102	0.004
C-2 carbide	6	20	0.050	0.002	0.075	0.003	0.102	0.004	0.13	0.005	0.15	0.006	0.15	0.006

Machining Data (Ref 1) (continued)

Maraging Steels: Drilling

Tool material	Speed m/min	Speed ft/min	3 mm (0.125 in.) mm	3 mm (0.125 in.) in.	6 mm (0.25 in.) mm	6 mm (0.25 in.) in.	12 mm (0.5 in.) mm	12 mm (0.5 in.) in.	18 mm (0.75 in.) mm	18 mm (0.75 in.) in.	25 mm (1 in.) mm	25 mm (1 in.) in.	35 mm (1.5 in.) mm	35 mm (1.5 in.) in.	50 mm (2 in.) mm	50 mm (2 in.) in.
			Feed per revolution for nominal hole diameter of:													
Annealed, with hardness of 275 to 325 HB																
M10, M7, and M1 high speed steel	17	55	0.075	0.003	0.13	0.005	0.18	0.007	0.23	0.009	0.25	0.010	0.33	0.013	0.40	0.015
Maraged, with hardness of 50 to 52 HRC																
Any premium high speed steel (T15, M33, M41-M47)	6	20	0.050	0.002	0.075	0.003	0.102	0.004	0.102	0.004	0.102	0.004	0.102	0.004	0.102	0.004

Maraging Steels: Face Milling

Depth of cut mm	Depth of cut in.	High speed steel(a) Speed m/min	High speed steel(a) Speed ft/min	High speed steel(a) Feed/tooth mm	High speed steel(a) Feed/tooth in.	Uncoated carbide(b) Speed, brazed m/min	Uncoated carbide(b) Speed, brazed ft/min	Uncoated carbide(b) Speed, inserted m/min	Uncoated carbide(b) Speed, inserted ft/min	Uncoated carbide(b) Feed/tooth mm	Uncoated carbide(b) Feed/tooth in.	Coated carbide(c) Speed m/min	Coated carbide(c) Speed ft/min	Coated carbide(c) Feed/tooth mm	Coated carbide(c) Feed/tooth in.
Annealed, with hardness of 275 to 325 HB															
1	0.040	34	110	0.15	0.006	85	280	95	310	0.15	0.006	140	465	0.13	0.005
4	0.150	26	85	0.23	0.009	79	260	88	290	0.20	0.008	115	375	0.18	0.007
8	0.300	20	65	0.30	0.012	56	185	69	225	0.25	0.010	88	290	0.23	0.009
Maraged, with hardness of 50 to 52 HRC															
1	0.040	12	40	0.050	0.002	47	155	58	190	0.050	0.002	...	...	...	...
4	0.150	9	30	0.075	0.003	38	125	47	155	0.075	0.003	...	...	...	...
8	0.300	6	20	0.102	0.004	30	100	37	120	0.102	0.004	...	...	...	...

(a) Any premium high speed steel tool material (T15, M33, M41 to M47). (b) Carbide tool material: C-2. (c) Carbide tool material: CC-2

Grade 18Ni(200)

18Ni(200): Chemical Composition (Ref 2)

Element	Composition, %	Element	Composition, %
Nickel	18.50	Manganese	0.10 max
Cobalt	8.50	Carbon	0.03 max
Molybdenum	3.25	Sulfur	0.01 max
Titanium	0.20	Phosphorus	0.01 max
Aluminum	0.10	Zirconium	0.01
Silicon	0.10 max	Boron	0.003

18Ni(200): Similar Steels (U.S. and/or Foreign). UNS K92810; ASTM A538 (A), A579 grade 71

Physical Properties (Ref 2)

18Ni(200): Average Coefficients of Linear Thermal Expansion

Temperature range °C	Temperature range °F	Coefficient μm/m·K	Coefficient μin./in.·°F
21-480	70-900	10.1	5.6

18Ni(200): Density

8.00 g/cm³ ....0.289 lb/in.³

18Ni(200): Thermal Conductivity

Temperature °C	Temperature °F	Thermal conductivity W/m·K	Thermal conductivity Btu/ft·h·°F
20	68	19.6	11.3
50	120	20.1	11.6
100	212	21.2	12.1

Mechanical Properties (Ref 2)

18Ni(200): Effect of Aging Time on Tensile Properties. Bar of small cross section solution annealed at 815 °C (1500 °F) for 1 h, air cooled, and aged at 480 °C (900 °F) for the times indicated. Elongation measured in gage length of $4.5\sqrt{\text{area}}$; yield strength at 0.2% offset

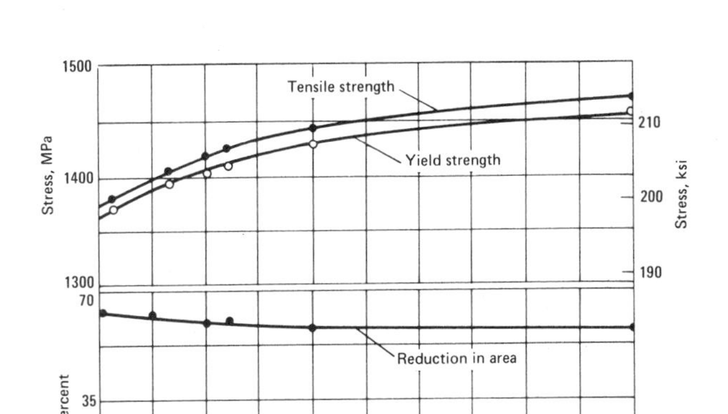

18Ni(200): Effect of Test Temperature on Charpy V-Notch Impact Energy. Bar of small cross section solution annealed at 815 °C (1500 °F) for 30 min, air cooled, and aged at 480 °C (900 °F) for 3 h

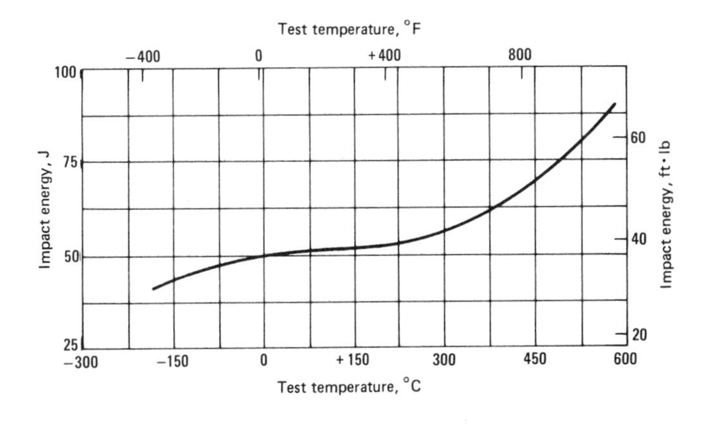

18Ni(250): Tensile Properties

Stock shape	Stock size mm	Stock size in.	Tensile strength MPa	Tensile strength ksi	Yield strength(a) MPa	Yield strength(a) ksi	Elongation(b), %	Reduction in area, %	Hardness, HRC	Direction of testing
Nominal annealed properties										
(c)	(c)	(c)	965	140	655	95	17	75	30	...
Nominal room temperature properties after aging										
Round bar	16	0.63	1824	264.5	1764	255.8	11.5	57.9	51.3	Longitudinal
Round bar	32	1.25	1844	268.5	1784	258.8	11	56.5	51.8	Longitudinal
Round bar	75	3.	1750	253.5	1712	248.3	11	53.4	50.4	Longitudinal
Square bar	150	6.	1731	251.0	1693	245.8	10	46.7	50.8	Longitudinal
Square bar	150	6.	1723	249.9	1690	245.2	8.1	30.3	50.3	Transverse
Sheet	6	0.25	1874	271.9	1832	265.7	8	40.8	50.6	Transverse

(a) 0.2% offset. (b) In gage length of $4.5\sqrt{\text{area}}$. (c) Not applicable

18Ni(200): Effect of Stress Concentration Factor on Tensile Properties

Stress concentration factor is K_t; solution annealed at 815 °C (1500 °F) for 1 h, air cooled, and aged at 480 °C (900 °F) for 3 h

K_t	Notch tensile strength: Average MPa	Average ksi	Range MPa	Range ksi	Notch-to-smooth tensile strength ratio(a)
2.0	2226	322.9	2179-2298	316.0-333.3	1.52
3.0	2256	327.2	2231-2306	323.6-334.5	1.54
5.0	2246	325.8	2208-2265	320.3-328.5	1.54
6.25	2269	329.1	2237-2349	324.4-340.7	1.55
7.0	2273	329.7	2204-2338	319.7-339.1	1.55
9.0	2266	328.6	2244-2300	325.5-333.6	1.55

(a) Based on smooth bar tensile strength of 1462 MPa (212.0 ksi)

Mechanical Properties (Ref 2) (continued)

18Ni(200): R.R. Moore Rotating Beam Fatigue Tests on Production Bar Stock.

Bar of small cross section solution annealed at 815 °C (1500 °F) for 30 min, air cooled, and aged at 480 °C (900 °F) for 3 h

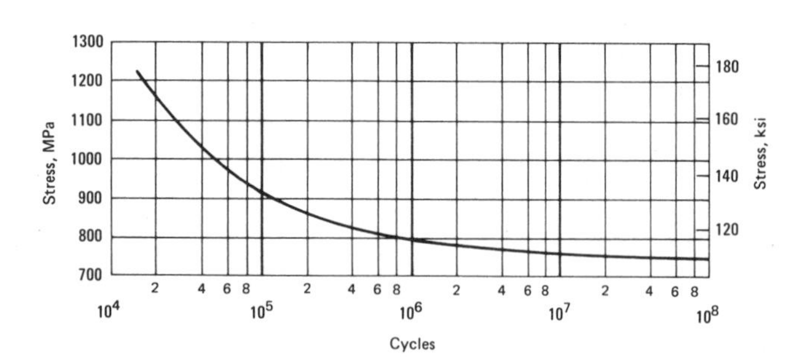

18Ni(200): Compressive Strength

Solution annealed at 815 °C (1500 °F) for 30 min, air cooled, and aged at 480 °C (900 °F) for 3 h; average of four tests per condition

Condition	Compressive strength: Proportional limit, MPa	Proportional limit, ksi	Yield strength(a), MPa	Yield strength(a), ksi	Hardness, HRC
Solution annealed	724	105.0	1000	145.0	28
Aged	1265	183.4	1469	213.0	43

(a) 0.2% offset

18Ni(200): Modulus of Elasticity

183 GPa .... 26.6 10^6 psi

18Ni(200): Effect of Testing Temperature on Mechanical Properties

Solution annealed at 815 °C (1500 °F) for 1 h, air cooled, and aged at 480 °C (900 °F) for 3 h

Test temperature, °C	°F	Tensile strength, MPa	Tensile strength, ksi	Yield strength(a), MPa	Yield strength(a), ksi	Elongation(b), %	Reduction in area, %
315	600	1217	176.5	1141	165.5	12.5	60
425	800	1154	167.4	1059	153.6	14	61
480	900	1044	151.4	977	141.7	18	66.3
510	950	953	138.2	876	127.1	18.5	69.6
540	1000	840	121.9	743	107.7	24	73.2

(a) 0.2% offset. (b) In gage length of $4.5\sqrt{area}$

Grade 18Ni(250)

18Ni(250): Chemical Composition (Ref 2)

Element	Composition, %	Element	Composition, %
Nickel	18.50	Manganese	0.10 max
Cobalt	7.50	Carbon	0.03 max
Molybdenum	4.80	Sulfur	0.01 max
Titanium	0.40	Phosphorus	0.01 max
Aluminum	0.10	Zirconium	0.01
Silicon	0.10 max	Boron	0.003

18Ni(250): Similar Steels (U.S. and/or Foreign). UNS K92890; ASTM A538 (B), A579 grade 72

Physical Properties (Ref 2)

18Ni(250): Average Coefficients of Linear Thermal Expansion

Temperature range, °C	°F	Coefficient, µm/m·K	µin./in.·°F
21-480	70-900	10.1	5.6

18Ni(250): Density

8.00 g/cm³ .... 0.289 lb/in.³

18Ni(250): Thermal Conductivity

Temperature, °C	°F	Thermal conductivity, W/m·K	Btu/ft·h·°F
20	68	25.3	14.6
50	120	25.8	14.9
100	212	27.0	15.6

Mechanical Properties (Ref 2)

18Ni(250): Effect of Test Temperature on Charpy V-Notch Impact Energy. Bar of small cross section solution annealed at 815 °C (1500 °F) for 30 min, air cooled, and aged at 480 °C (900 °F) for 3 h

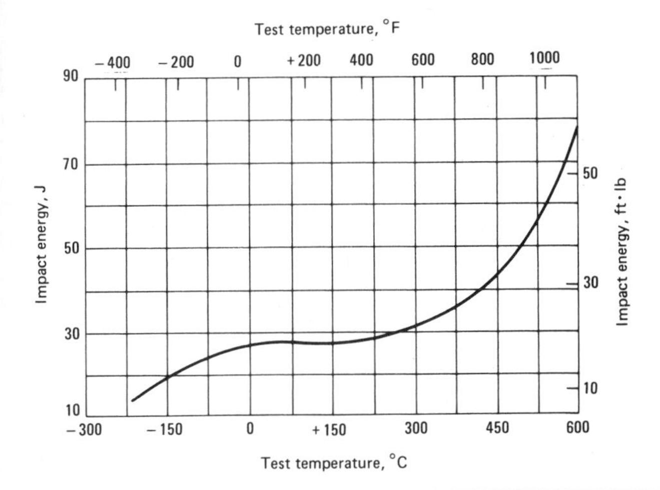

18Ni(250): R.R. Moore Rotating Beam Fatigue Tests on Production Bar Stock. Bar of small cross section solution annealed at 815 °C (1500 °F) for 30 min, air cooled, and aged at 480 °C (900 °F) for 3 h

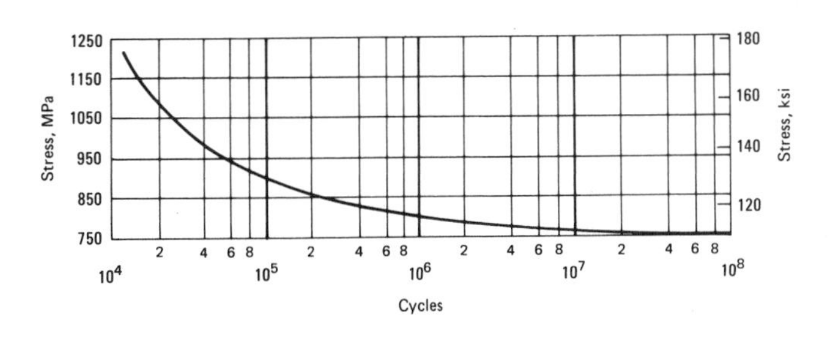

18Ni(250): Tensile Properties

Stock shape	Stock size mm	Stock size in.	Tensile strength MPa	Tensile strength ksi	Yield strength(a) MPa	Yield strength(a) ksi	Elongation(b), %	Reduction in area, %	Hardness, HRC	Direction of testing
Nominal annealed properties										
(c)	(c)	(c)	965	140	655	95	17	75	30	...
Nominal room temperature properties after aging										
Round bar	16	0.63	1824	264.5	1764	255.8	11.5	57.9	51.3	Longitudinal
Round bar	32	1.25	1844	268.5	1784	258.8	11	56.5	51.8	Longitudinal
Round bar	75	3	1750	253.5	1712	248.3	11	53.4	50.4	Longitudinal
Square bar	150	6	1731	251.0	1693	245.8	10	46.7	50.8	Longitudinal
Square bar	150	6	1723	249.9	1690	245.2	8.1	30.3	50.3	Transverse
Sheet	6	0.25	1874	271.9	1832	265.7	8	40.8	50.6	Transverse

(a) 0.2% offset. (b) In gage length of $4.5\sqrt{area}$. (c) Not applicable

18Ni(250): Effects of Various Aging Treatments on Tensile Properties of Sheet

Test specimens were 3.18-mm (0.125-in.) thick standard ASTM sheet; solution annealed at 815 °C (1500 °F) for 30 min, air cooled, and aged as indicated

Aging temperature °C	Aging temperature °F	Aging time, h	Yield strength(a) MPa	Yield strength(a) ksi	Tensile strength MPa	Tensile strength ksi	Elongation, % 25 mm (1 in.)	Elongation, % 50 mm (2 in.)	Reduction in area, %
455	850	3	1770	256.7	1810	262.5	9.7	4.7	47.6
480	900	1	1769	256.6	1824	264.5	9.4	4.8	48.7
480	900	3	1932	280.2	1979	287.0	8.8	4.5	44.9
480	900	6	1844	267.5	1931	280.1	8.5	4.2	44.7
510	950	3	1809	262.3	1865	270.5	9.7	4.5	47.2

(a) 0.2% offset

18Ni(250): Effect of Test Temperature on Tensile Properties

Solution annealed at 815 °C (1500 °F) for 1 h, air cooled, and aged at 480 °C (900 °F) for 3 h

Reheat temperature °C	Reheat temperature °F	Tensile strength MPa	Tensile strength ksi	Yield strength(a) MPa	Yield strength(a) ksi	Elongation(b), %	Reduction in area, %
315	600	1609	233.4	1548	224.5	11.5	56
425	800	1524	221	1453	210.8	12	56.1
480	900	1379	200	1276	185.1	16.5	64.6
540	1000	1029	149.2	890	129.1	23	79.2

(a) 0.2% offset. (b) In gage length of $4.5\sqrt{area}$

Mechanical Properties (Ref 2) (continued)

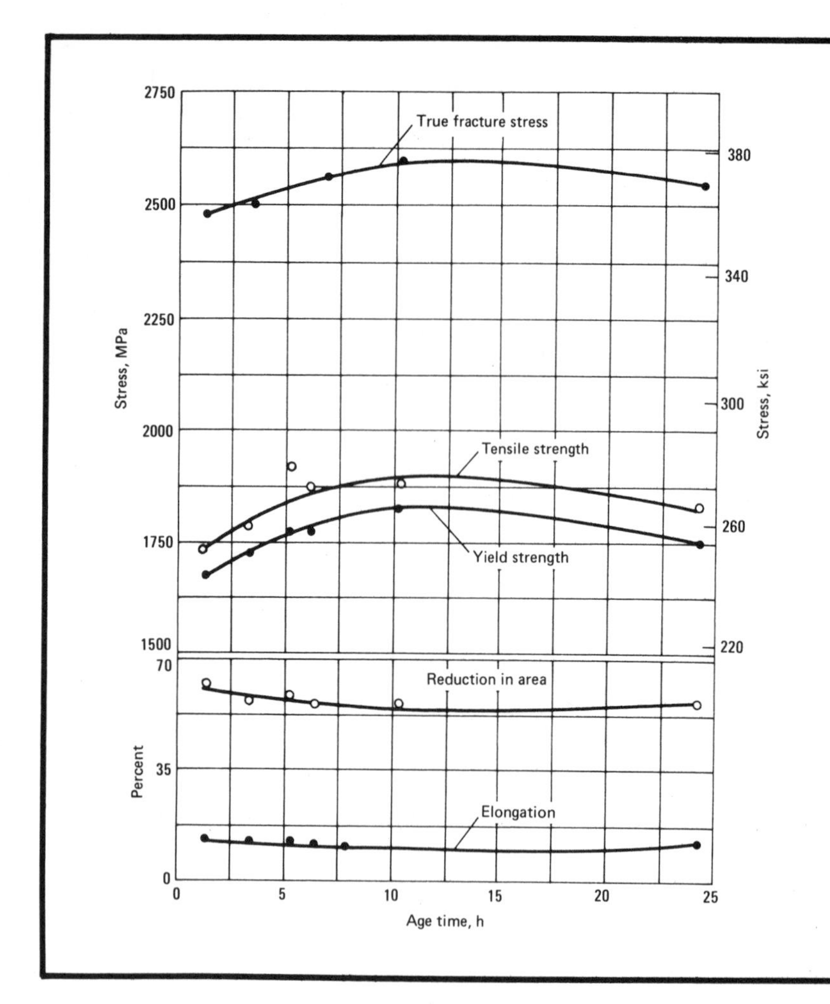

18Ni(250): Effect of Aging Time on Tensile Properties. Bar os small cross section solution annealed at 815 °C (1500 °F) for 1 h, air cooled, and aged at 480 °C (900 °F) for the times indicated. Elongation measured in gage length of $4.5\sqrt{area}$; yield strength at 0.2% offset

18Ni(250): Compressive Strength

Solution annealed at 815 °C (1500 °F) for 30 min, air cooled, and aged at 480 °C (900 °F) for 3 h; average of three tests per condition

Condition	Compressive strength: Proportional limit, MPa	Proportional limit, ksi	Yield strength(a), MPa	Yield strength(a), ksi	Hardness, HRC
Solution annealed	724	105.0	1027	149.0	29.0
Aged	1664	241.3	1931	280.0	51.0

(a) 0.2% offset

18Ni(250): Modulus of Elasticity

190 GPa	27 10^6 psi

18Ni(250): Effect of Stress Concentration Factor on Tensile Properties

Stress concentration factor is K_t; solution annealed at 815 °C (1500 °F) for 1 h, air cooled, and aged at 480 °C (900 °F) for 3 h

K_t	Notch tensile strength: Average, MPa	Average, ksi	Range, MPa	Range, ksi	Notch-to-smooth tensile strength ratio(a)
2.0	2784	403.8	2769-2802	401.6-406.4	1.49
3.0	2751	399.0	2714-2775	393.6-402.4	1.48
5.0	2629	381.3	2598-2664	376.7-386.3	1.41
6.25	2659	385.7	2647-2703	383.9-392.0	1.43
7.0	2603	377.5	2592-2636	375.9-382.3	1.40
9.0	2625	380.7	2603-2647	377.5-383.9	1.41

(a) Based on smooth bar tensile strength of 1862 MPa (270.1 ksi)

Grade 18Ni(300)

18Ni(300): Chemical Composition (Ref 2)

Element	Composition, %	Element	Composition, %
Nickel	18.50	Manganese	0.10 max
Cobalt	9.00	Carbon	0.03 max
Molybdenum	4.80	Sulfur	0.01 max
Titanium	0.60	Phosphorus	0.01 max
Aluminum	0.10	Zirconium	0.01
Silicon	0.10 max	Boron	0.003

18Ni(300): Similar Steels (U.S. and/or Foreign). UNS K93120; ASTM A538 (C), A579 grade 73

Physical Properties (Ref 2)

18Ni(300): Average Coefficients of Linear Thermal Expansion

Temperature range °C	°F	Coefficient μm/m·K	μin./in.·°F
21-480	70-900	10.1	5.6

18Ni(300): Density

8.00 g/cm³ .... 0.289 lb/in.³

18Ni(300): Thermal Conductivity

Temperature °C	°F	Thermal conductivity W/m·K	Btu/ft·h·°F
20	68	25.3	14.6
50	120	25.8	14.9
100	212	27.0	15.6

Mechanical Properties (Ref 2) (continued)

18Ni(300): Effect of Test Temperature on Charpy V-Notch Impact Energy.

Bar of small cross section solution annealed at 815 °C (1500 °F) for 30 min, air cooled, and aged at 480 °C (900 °F) for 3 h

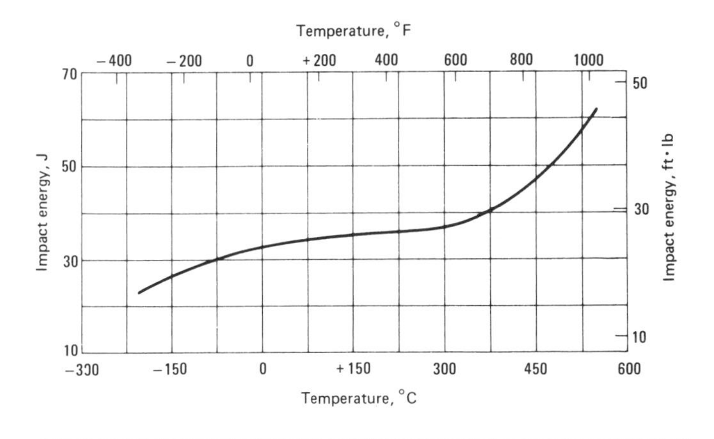

18Ni(300): R.R. Moore Rotating Beam Fatigue Tests on Production Bar Stock.

Bar of small cross section solution annealed at 815 °C (1500 °F) for 30 min, air cooled, and aged at 480 °C (900 °F) for 3 h

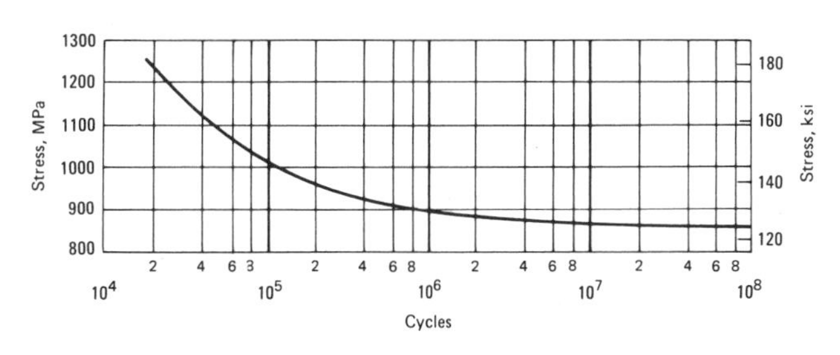

18Ni(300): Effect of Test Temperature on Tensile Properties

Solution annealed at 815 °C (1500 °F) for 1 h, air cooled, and aged at 480 °C (900 °F) for 3 h

Test temperature °C	°F	Tensile strength MPa	ksi	Yield strength(a) MPa	ksi	Elongation(b), %	Reduction in area, %
315	600	1771	256.8	1693	245.6	12	61.8
425	800	1655	240.1	1570	227.7	14	61.3
480	900	1454	210.9	1343	194.8	17.3	68.4
510	950	1304	189.1	1192	172.9	22	76.5
540	1000	1158	168	1056	153.2	24	77.2

(a) 0.2% offset. (b) In gage length of $4.5\sqrt{area}$

Mechanical Properties (Ref 2) (continued)

18Ni(300): Tensile Properties

Stock shape	Stock size mm	Stock size in.	Tensile strength MPa	Tensile strength ksi	Yield strength(a) MPa	Yield strength(a) ksi	Elongation(b), %	Reduction in area, %	Hardness, HRC	Direction of testing
Nominal annealed properties										
(c)	(c)	(c)	1034	150	758	110	18	72	32	...
Nominal room temperature properties after aging										
Round bar	16	0.63	2027	294	2000	290	11.8	56.6	54.3	Longitudinal
Round bar	32	1.25	2041	296	2020	293	11.6	55.8	54.7	Longitudinal
Round bar	75	3	2025	293.7	1977	286.8	10.3	46.6	54	Longitudinal
Square bar	150	6	1962	284.6	1915	277.8	9.8	43.9	53.9	Longitudinal
Square bar	150	6	1953	283.2	1911	277.1	6.6	28.4	54.3	Transverse
Sheet	6	0.25	2169	314.6	2135	309.7	7.7	35	55.1	Transverse

(a) 0.2% offset. (b) In gage length of $4.5\sqrt{area}$. (c) Not applicable

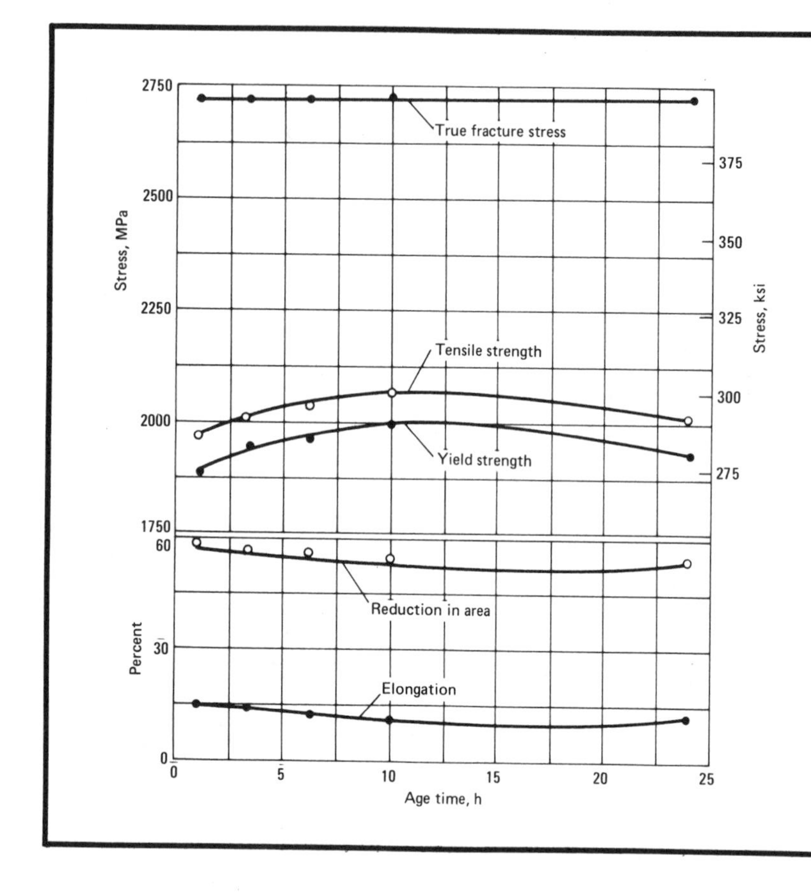

18Ni(300): Effect of Aging Time on Tensile Properties. Bar of small cross section solution annealed at 815 °C (1500 °F) for 30 min, air cooled, and aged at 480 °C (900 °F) for the times indicated. Elongation measured in gage length of $4.5\sqrt{area}$; yield strength at 0.2% offset

18Ni(300): Effect of Stress Concentration Factor on Tensile Properties

Stress concentration factor is K_t; solution annealed at 815 °C (1500 °F) for 1 h, air cooled, and aged at 480 °C (900 °F) for 3 h

K_t	Notch tensile strength, Average MPa	Average ksi	Range MPa	Range ksi	Notch-to-smooth tensile strength ratio(a)
2.0	2937	426.0	2914-2981	422.6-432.3	1.45
3.0	2899	420.5	2892-2908	419.4-421.8	1.43
5.0	2881	417.9	2836-2947	411.3-427.4	1.42
6.25	2885	418.4	2847-2919	412.9-423.4	1.42
7.0	2854	414.0	2785-2936	403.9-425.8	1.41
9.0	2898	420.3	2836-2919	411.3-423.4	1.43

(a) Based on smooth bar tensile strength of 2024 MPa (293.5 ksi)

Mechanical Properties (Ref 2) (continued)

18Ni(300): Effect of Sheet Thickness on Tensile Properties

Test specimens were standard ASTM sheet; solution annealed at 815 °C (1500 °F) for 15 min, air cooled, and aged at 480 °C (900 °F) for 3 h; tests were conducted using test specimens machined to English units

Sheet thickness		Yield strength(a)		Tensile strength		Elongation(b), %	
mm	in.	MPa	ksi	MPa	ksi	25 mm (1 in.)	50 mm (2 in.)
6.35	0.250	2173	315.1	2212	320.8	9.0	5.0
3.18	0.125	2164	313.9	2184	316.8	6.8	3.4
2.29	0.090	2125	308.2	2156	312.7	6.0	3.2
1.65	0.065	2078	301.4	2118	307.2	5.0	3.0
1.14	0.045	2013	291.9	2034	295.0	4.0	2.0
0.64	0.025	2027	294.0	2041	296.0	2.0	1.0

(a) 0.2% offset. (b) Because the change in elongation with thickness is caused by the changing geometry of the test specimen, not a change in material ductility, a gage length of $4.5\sqrt{area}$ should be used for correct elongation measurements, rather than a fixed 25- or 50-mm (1- or 2-in.) gage length

18Ni(300): Compressive Strength

Solution annealed at 815 °C (1500 °F) for 30 min, air cooled, and aged at 480 °C (900 °F) for 3 h; average of three tests per condition

Condition	Compressive strength: Proportional limit, MPa	ksi	Yield strength(a), MPa	ksi	Hardness, HRC
Solution annealed	724	105.0	1034	150.0	31.0
Aged	1875	272.0	2189	317.5	53.5

(a) 0.2% offset

18Ni(300): Modulus of Elasticity

190 GPa .... 27.5 10^6 psi

18Ni(300): Effects of Various Aging Treatments on Tensile Properties of Sheet

Test specimens were 3.18-mm (0.125-in.) thick standard ASTM sheet; solution annealed at 815 °C (1500 °F) for 30 min, air cooled, and aged as indicated

Aging temperature		Aging time, h	Yield strength(a)		Tensile strength		Elongation, %		Reduction in area, %
°C	°F		MPa	ksi	MPa	ksi	25 mm (1 in.)	50 mm (2 in.)	
455	850	3	2033	294.8	2134	309.5	7.0	3.5	34.2
480	900	1	2047	296.9	2115	306.7	8.2	4.2	38.6
480	900	3	2164	313.9	2184	316.8	6.8	3.4	32.5
480	900	6	2166	314.2	2215	321.2	7.5	3.7	33.2
510	950	3	2107	305.6	2124	308.1	8.0	4.0	33.6

(a) 0.2% offset

Grade 18Ni(350)

18Ni(350): Chemical Composition (Ref 2)

Element	Composition, %	Element	Composition, %
Nickel	18.50	Managanese	0.10 max
Cobalt	12.00	Carbon	0.03 max
Molybdenum	4.80	Sulfur	0.01 max
Titanium	1.40	Phosphorus	0.01 max
Aluminum	0.10	Zirconium	0.01
Silicon	0.10 max	Boron	0.003

18Ni(350): Similar Steels (U.S. and/or Foreign). No public specifications

Physical Properties (Ref 2)

18Ni(350): Average Coefficients of Linear Thermal Expansion

Temperature range °C	°F	Coefficient µm/m·K	µin./in.·°F
21-480	70-900	11.3	6.3

18Ni(350): Density

8.08 g/cm³ .... 0.292 lb/in.³

18Ni(350): Thermal Conductivity

Temperature °C	°F	Thermal conductivity W/m·K	Btu/ft·h·°F
20	68	25.5	14.6
50	120	25.8	14.9
100	210	27.0	15.6

Mechanical Properties (Ref 2)

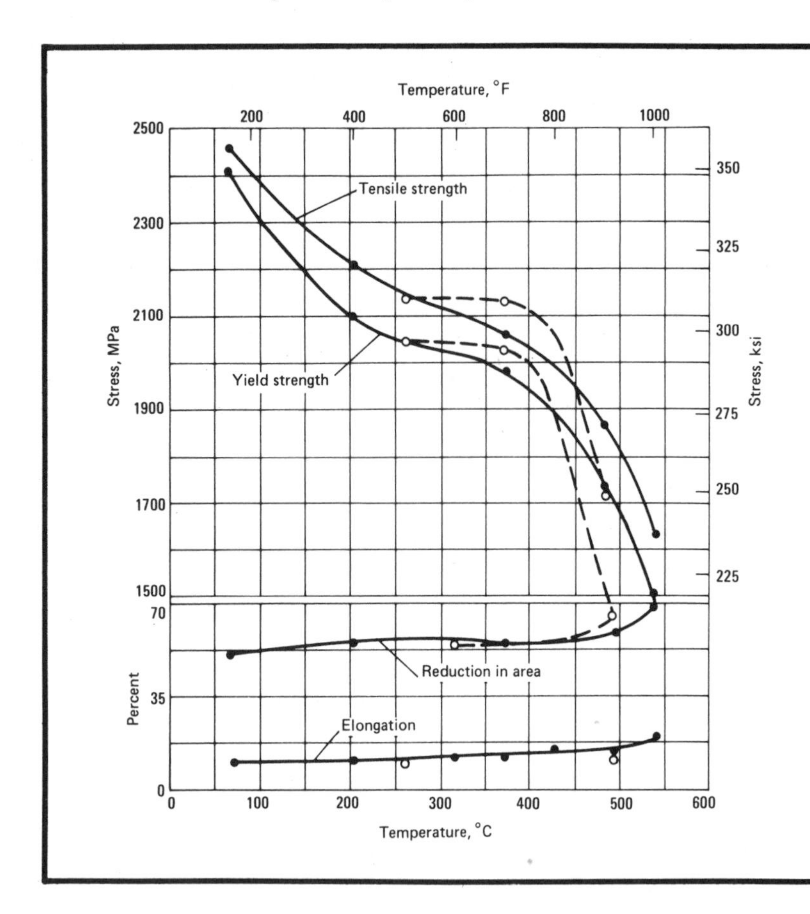

18Ni(350): Tensile Properties as a Function of Test Exposure Temperature. —— exposed for a short time; – – – exposed for 200 h. Bar of small cross section solution annealed at 800 °C (1475 °F) for 1 h, air cooled, and aged at 510 °C (950 °F) for 3 h. Elongation in gage length of $4.5\sqrt{area}$

18Ni(350): Effect of Test Temperature on Charpy V-Notch Impact Energy. Bar of small cross section solution annealed at 815 °C (1500 °F) for 30 min, air cooled, and aged at 480 °C (900 °F) for 3 h

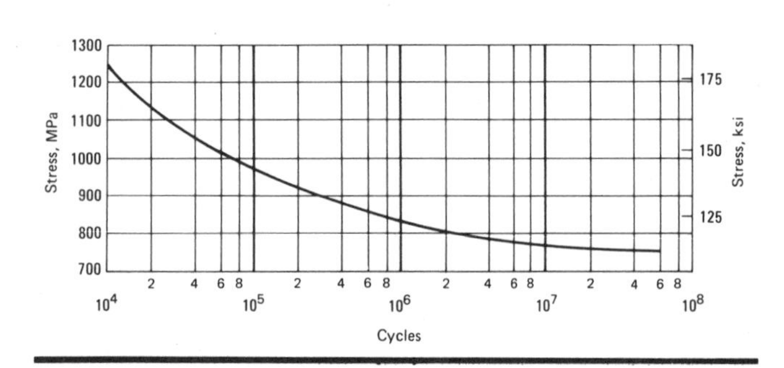

18Ni(350): R.R. Moore Rotating Beam Fatigue Tests on Production Bar Stock. Bar of small cross section solution annealed at 815 °C (1500 °F) for 30 min, air cooled, and aged at 480 °C (900 °F) for 3 h

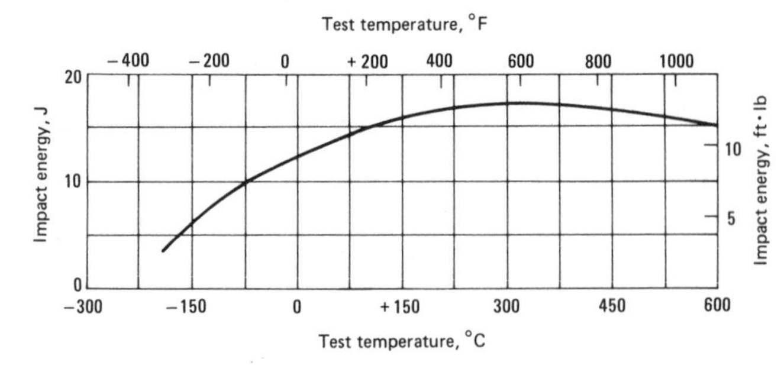

18Ni(350): Compressive Strength

Solution annealed at 815 °C (1500 °F) for 30 min, air cooled, and aged at 480 °C (900 °F) for 3 h; average of three tests per condition

Condition	Compressive strength: Proportional limit, MPa	Proportional limit, ksi	Yield strength(a), MPa	Yield strength(a), ksi	Hardness, HRC
Solution annealed	745	108.0	1107	160.5	34.3
Aged	2408	349.3	2676	388.1	59.6

(a) 0.2% offset

18Ni(350): Modulus of Elasticity

200 GPa . . . . 29 10^6 psi

Mechanical Properties (Ref 2)

18Ni(350): Tensile Properties

Stock shape	Stock size mm	Stock size in.	Tensile strength MPa	Tensile strength ksi	Yield strength(a) MPa	Yield strength(a) ksi	Elongation(b), %	Reduction in area, %	Hardness, HRC	Direction of testing
Nominal annealed properties										
(c)	(c)	(c)	1140	165	827	120	18	70	35	...
Nominal room temperature properties after aging										
Round bar	16	0.63	2415	350.2	2363	342.7	7.5	35.4	57.8	Longitudinal
Round bar	32	1.25	2391	346.8	2348	340.6	7.6	33.8	58.4	Longitudinal
Round bar	75	3	2359	342.2	2320	336.5	6.2	28.6	58.2	Longitudinal
Sheet	6	0.25	2451	355.5	2395	347.3	3	15.4	57.7	Transverse

(a) 0.2% offset. (b) In gage length of $4.5\sqrt{area}$. (c) Not applicable

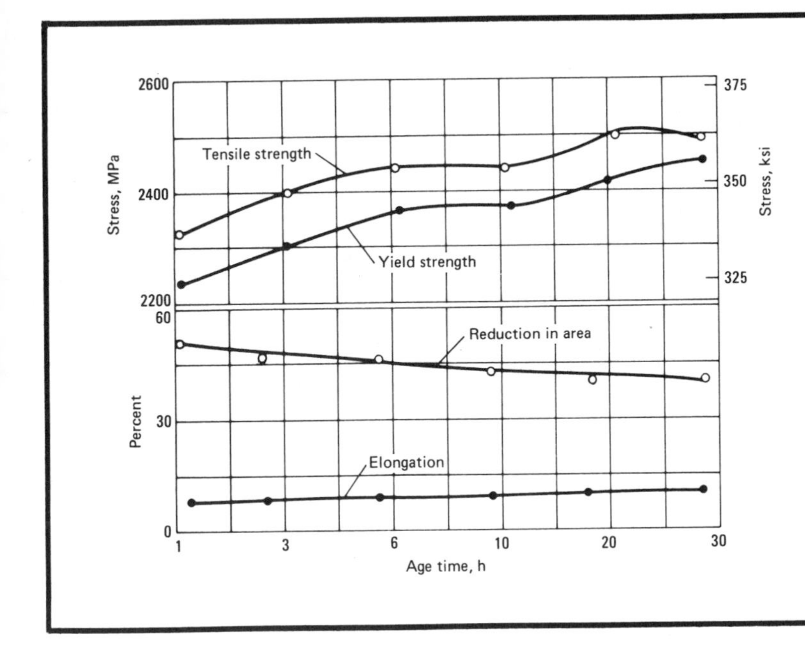

18Ni(250): Effect of Aging Time on Tensile Properties. Bar os small cross section solution annealed at 815 °C (1500 °F) for 1 h, air cooled, and aged at 480 °C (900 °F) for the times indicated. Elongation measured in gage length of $4.5\sqrt{area}$; yield strength at 0.2% offset

18Ni(350): Effect of Test Temperature on Tensile Properties

Solution annealed at 815 °C (1500 °F) for 1 h, air cooled, and aged at 480 °C (900 °F) for 6 h

Test temperature °C	Test temperature °F	Tensile strength MPa	Tensile strength ksi	Yield strength(a) MPa	Yield strength(a) ksi	Elongation(b), %	Reduction in area, %
315	600	2139	310.2	2037	295.4	12.3	54.9
425	800	1988	288.4	1912	277.3	15.6	57.6
480	900	1864	270.4	1737	251.9	17.4	60.3
540	1000	1611	233.6	1460	211.8	20	70.9

(a) 0.2% offset. (b) In gage length of $4.5\sqrt{area}$

18Ni(350): Effect of Stress Concentration Factor

Stress concentration factor is K_t; solution annealed at 815 °C (1500 °F) for 1 h, air cooled, and aged at 480 °C (900 °F) for 3 h

K_t	Notch tensile strength: Average MPa	Average ksi	Range MPa	Range ksi	Notch-to-smooth tensile strength ratio(a)
2.00	2990	433.7	2947-3015	427.4-437.3	1.20
6.25	2305	334.3	2287-2328	331.7-337.6	0.93
9.00	2296	333.0	2253-2335	326.7-338.6	0.92

(a) Based on smooth bar tensile strength of 2501 MPa (362.8 ksi)

ASTM Steels

The steels listed by the Standard Specifications of the American Society for Testing and Materials (ASTM) are those carbon and alloy steels that are commonly used, but have not been assigned AISI/SAE numbers. ASTM steels are available as structural shapes, plate, and bar intended for riveted, bolted, or welded construction of bridges and buildings, and for ship construction. ASTM grades of hot rolled commercial quality, and hot and cold rolled structural and drawing quality carbon steel sheet and strip are used for fabricating industrial parts by bending, forming, drawing, and welding.

ASTM Standard Specifications also include low-carbon sheet and coiled steel used in porcelain enameling and for coated stock. The coated steels are aluminum coated, lead-tin (terne) coated, electrodeposited tin or zinc coated, and galvanized (hot dipped, zinc coated) sheet and coil.

Carbon and alloy steel spring wire and carbon steel used in the manufacture of electric-resistance-welded (ERW), arc-welded, and seamless pipe and tubing follow ASTM specifications. ASTM specifications for carbon and alloy steels that apply to the very specialized category of pressure vessels are not included in this handbook. Pressure vessel steels are often specified by either ASTM or the American Society of Mechanical Engineers (ASME).

REFERENCES

1. Steel Piping, Tubing, and Fittings, *Annual Book of ASTM Standards,* Part 1, American Society for Testing and Materials, Philadelphia, PA, 1981, 1982
2. Steel Plate, Sheet, Strip, and Wire; Metallic Coated Products; Fences, Part 3, *Annual Book of ASTM Standards,* American Society for Testing and Materials, Philadelphia, PA, 1981, 1982
3. Structural Steel; Concrete Reinforcing Steel; Pressure Vessel Plate and Forgings; Steel Rails, Wheels, and Tires; Fasteners, Part 4, *Annual Book of ASTM Standards,* American Society for Testing and Materials, Philadelphia, PA, 1981, 1982
4. Steel Bars, Chain, and Springs; Bearing Steel; Steel Forgings, Part 5, *Annual Book of ASTM Standards,* American Society for Testing and Materials, Philadelphia, PA, 1981, 1982

ASTM A36

ASTM A36: Chemical Composition

Product form	C	Mn	P max	S max	Si	Cu(a)
	Chemical composition, %					
Shape	0.26 max	0.85-1.35(b)	0.04	0.05	0.15-0.40(b)	0.20
Plate	0.25-0.29	0.85-1.20	0.04	0.05	0.15-0.40	0.20
Bar	0.26-0.29	0.60-0.90	0.04	0.05	...	0.20

(a) Minimum copper content when copper steel is specified. (b) Required for shapes over 634 kg/m (426 lb/ft)

Typical uses of ASTM A36 steel include bridges, buildings, and general structural purposes.

ASTM A36: Similar Steels (U.S. and/or Foreign). UNS K02600; ASTM A36 (shapes)

Characteristics. ASTM A36 describes carbon steel shapes, plate, and bar of structural quality for use in riveted, bolted, or welded construction of bridges and buildings, and for general structural purposes. When this steel is used in welded construction, welding procedure should be suitable for the steel and the intended service.

Unless otherwise provided in the order, the specifications listed in the table on material specifications will govern the delivery of otherwise unspecified accessory materials when included with material purchased under this specification.

ASTM A36: Material Specifications

Material	ASTM designation(a)
Plate to be bent or formed cold	A283, grade C
Steel rivets	A502, grade I
Bolts and nuts	A307, A325
Cast steel	A27, grade 65-35
Forgings (carbon steel)	A668, class D
Hot rolled sheets	A570, grade D
Hot rolled strip	A570, grade D
Cold formed tubing	A500, grade B
Hot formed tubing	A501

(a) Specifications of the American Society for Testing and Materials

Mechanical Properties

ASTM A36: Tensile Properties for Plate, Shapes, and Bar

Product form	Tensile strength MPa	Tensile strength ksi	Yield point MPa	Yield point ksi	Elongation, min % In 200 mm (8 in.)(a)	Elongation, min % In 50 mm (2 in.)
Plate(b)	400-550	58-80	250	36	20	23
Bar	400-550	58-80	250	36	20	23
Shapes(c)	400-550	58-80	250	36	20	21

(a) For material less than 8 mm (0.3 in.) in thickness or diameter, a deduction from the percentage of elongation in 200 mm (8 in.) shall be 1.25% for each 0.8 mm (0.03 in.) of the specified thickness or diameter below 8 mm (0.3 in.). (b) For plates wider than 590 mm (24 in.), the test is taken in the transverse direction. Yield point for plates larger than 200 mm (8 in.) in thickness is 220 MPa (32 ksi). (c) For wide flange shapes larger than 634 kg/m (426 lb/ft) the minimum tensile strength is 400 MPa (58 ksi); elongation in 50 mm (2 in.) is 19% minimum

ASTM A131

ASTM A131: Chemical Composition

Nickel, chromium, molybdenum, copper, niobium, vanadium, or aluminum contents need not be reported on the mill sheet unless intentionally added

Grade	C max	Mn	P max	S max	Si
A (a)	0.23(b)	(b)	0.05	0.05	...
B (a)	0.21	0.80-1.10	0.04	0.04	0.35 max
D (a)(c)	0.21	0.70-1.35	0.04	0.04	0.10-0.35
E (a)	0.18	0.70-1.35	0.04	0.04	0.10-0.35
CS (a)	0.16	1.00-1.35	0.04	0.04	0.10-0.35
DS (a)	0.16	1.00-1.35	0.04	0.04	0.10-0.35
AH32 (d), AH36 (d), DH32, DH36, EH32, EH36	0.18	0.90-1.60	0.04	0.05	0.10-0.50(d)

(a) For grades A through DS, the carbon content plus 1/6 of the manganese content shall not exceed 0.40%; the upper limit of the manganese may be exceeded up to 1.65% if the carbon content does not exceed 0.40%. (b) A maximum carbon content of 0.26% is acceptable for plate equal to or less than 13 mm (0.50 in.) and all thicknesses of shapes. Plate larger than 13 mm (0.50 in.) in thickness shall have a minimum manganese content not less than 2.5 times the carbon content. (c) For material equal to or less than 25 mm (1.0 in.) in thickness, 0.60% manganese content is acceptable. The lower limit of silicon does not apply to semikilled steel. (d) Materials as large as 13 mm (0.50 in.) in thickness may be semikilled, in which case the 0.10% minimum silicon does not apply

ASTM A131: Similar Steels (U.S. and/or Foreign):

- UNS K01601; ASTM A131 (CS, DS)
- UNS K01801; ASTM A131 (E)
- UNS K01805; ASTM A131 (AH32, DH32, EH32)
- UNS K01806; ASTM A131 (AH36, DH36, EH36)
- UNS K02101; ASTM A131 (D)
- UNS K02102; ASTM A131 (B); MIL SPEC MIL-S-22698 (B), MIL-S-23495
- UNS K02300; ASTM A131 (A)

Characteristics and Typical Uses. ASTM A131 describes structural steel shapes, plate, bar, and rivets intended for use in ship construction. The material is available in two categories: ordinary strength, grades A, B, D, DS, CS, and E, with a specified minimum yield point of 235 MPa (34 ksi); and higher strength, grades AH, DH, and EH, with specified minimum yield points of either 315 MPa (45.5 ksi) or 350 MPa (51 ksi). The maximum thickness of plate for all grades is 50 mm (2 in.). The values stated in English units are to be regarded as the standard.

Except for grade A steel up to and including 13 mm (0.50 in.) in thickness, rimming-type steels shall not be applied. Grades AH32 and AH36 as large as 13 mm (0.50 in.) in thickness may be semikilled.

Fine-grain practice is used to make grades D, DS, CS, E, DH32, DH36, EH32, and EH36. For ordinary-strength grades, aluminum is used to obtain grain refinement. For the higher strength grades, aluminum, vanadium, or niobium may be used for grain refinement. Grade D material 35.0 mm (1.38 in.) and under in thickness, at the option of the manufacturer, may be semikilled and exempt from the fine austenitic grain-size requirement.

Physical Properties

ASTM A131: Normalizing of Grades AH32, AH36, DH32, DH36

When niobium or vanadium is used in combination with aluminum, heat treatment requirements for niobium or vanadium apply

ASTM grade	Normalizing requirements
Aluminum-treated steels	
AH32, AH36	Not required
DH32, DH36	When larger than 25 mm (1 in.)
Niobium-treated steels	
AH32, AH36	When larger than 13 mm (0.50 in.)
DH32, DH36	When larger than 13 mm (0.50 in.)
Vanadium-treated steels	
AH32, AH36	Not required
DH32, DH36	When larger than 19 mm (0.75 in.)

ASTM A131: Thermal Treatment. Plate in all thicknesses ordered to grades CS, E, EH32, and EH36, and grade D plate larger than 35 mm (1.38 in.) are normalized. When grade D steel is furnished semikilled, it is normalized when larger than 25 mm (1 in.) in thickness. Control rolling of grade D may be substituted for normalizing.

Mechanical Properties

ASTM A131: Charpy V-Notch Impact Properties

Grade	Test temperature °C	Test temperature °F	Charpy V-notch energy, avg min, Longitudinal J	Longitudinal ft·lb	Transverse J	Transverse ft·lb
B (a)	0	32	25	20	19	14
D (b)	−10	14	25	20	19	14
E	−40	−40	25	20	19	14
DH32 (c)	−20	−4	35	25	23	17
EH32	−40	−40	35	25	23	17
DH36 (c)	−20	−4	35	25	23	17
EH36	−40	−40	35	25	23	17

(a) Larger than 25 mm (1.0 in.) in thickness. (b) Toughness tests not required on normalized material when produced fully killed and with a fine, austenitic grain size. (c) Toughness tests not required on normalized material

ASTM A131: Tensile Properties

Grade	Tensile strength MPa	Tensile strength ksi	Yield point MPa	Yield point ksi	Elongation, % In 200 mm (8 in.)(a)	Elongation, % In 50 mm (2 in.)
A (b), B, D, DS, and E structural steel	400-490	58-71	235	34.0	21	24
Rivet steel and steel for cold flanging	380-450	55-65	205	30.0	23	26
AH32, DH32, EH32	470-585	68-85	315	45.5	19	22
AH36, DH36, EH36	490-620	71-90	350	51.0	19	22

(a) For material less than 8 mm (0.3 in.) in thickness or diameter, a deduction of 1.25 percentage points may be made for each decrease of 0.8 mm (0.03 in.) of the specified thickness or diameter. (b) A tensile strength range from 400 to 550 MPa (58 to 80 ksi) may be applied to grade A shapes. When grade A structural steel is larger than 25.4 mm (1.0 in.) in thickness, the minimum yield point may be reduced to 220 MPa (32 ksi)

ASTM A135

ASTM A135: Chemical Composition

Grade	Chemical composition, max % C	Mn	P	S
A	0.25	0.95	0.050	0.060
B	0.30	1.20	0.050	0.060

Characteristics and Typical Uses. ASTM A135 describes two grades of electric-resistance welded steel pipe in nominal sizes 50 to 760 mm (2 to 30 in.) inclusive, with nominal (average) wall thickness up to 12.7 mm (0.500 in.) inclusive, and in nominal sizes 19 to 127 mm (0.75 to 5 in.) inclusive with nominal (average) wall thickness of 2.11 to 3.40 mm (0.083 to 0.134 in.), depending on size. The values stated in English units are to be regarded as the standard.

The pipe may be furnished as either nonexpanded or cold expanded, at the option of the manufacturer. When pipe is cold expanded, the amount of expansion is not to exceed 1.5% of the outside diameter pipe size. ASTM A135 pipe is intended for conveying liquid, gas, or vapor. Only grade A is adapted for flanging and bending.

Flattening Test. Flattening tests are conducted to verify the integrity of the welds for ASTM A135 steel pipe. In a flattening test, a specimen at least 100 mm (4 in.) in length is flattened between parallel plates in three steps. During the first step, which is a test for ductility of the weld, no cracks or breaks on the inside or outside surfaces shall occur until the distance between the plates is less than two thirds of the original outside diameter of the pipe.

During the second step, which is a continuation of the first step and a test for ductility exclusive of the weld, no cracks or breaks on the inside or outside surfaces shall occur until the distance between the plates is less than one third of the original outside diameter of the pipe, but not less than five times the wall thickness of the pipe.

During the third step of a flattening test, which is a test for soundness, the flattening shall be continued until the specimen breaks, or the opposite walls of the pipe meet. Evidence of laminated or unsound material, or of incomplete weld, that is revealed anytime during the entire flattening test is cause for rejection.

For pipe produced in single lengths, the flattening test is made on both crop ends cut from each length of pipe. Tests from each end are made alternately with the weld at 0 and 90° from the direction of the line of force. For pipe produced in multiple lengths, the flattening test is made on crop ends which represent the front and back of each coil with the weld at 90° from the direction of the line of force, and on two intermediate rings representing each coil with the weld 0° from the direction of the line of force.

Hydrostatic Test. Each length of pipe is tested at the mill to a hydrostatic pressure calculated from the following equation, provided that the specified test pressure does not exceed 17.2 MPa (2500 psi):

$$P = \frac{2\,St}{D}$$

where P is minimum hydrostatic test pressure measured in pounds per square inch; S is allowable fiber stress, 16 to 18 ksi for grade A and 20 to 22 ksi for grade B (in no case is the stress produced to exceed 80% of the specified yield point); t is specified wall thickness, measured in inches; D is specified outside diameter, measured in inches.

The hydrostatic pressure is maintained for not less than 5 s. For pipe with wall thicknesses greater than 3.91 mm (0.154 in.), the pipe is to be jarred near both ends with a 0.9-kg (2-lb) steel hammer or its equivalent.

Electromagnetic (eddy current) or ultrasonic tests may be used as an alternate to the hydrostatic test for Schedule 10 pipe in nominal pipe sizes 19 to 127 mm (0.75 to 5 in.), and when accepted by the purchaser. These tests are made in accordance with ASTM Recommended Practice for E213, E273, or E309.

ASTM A135: Similar Steels (U.S. and/or Foreign). None listed

Mechanical Properties

ASTM A135: Tensile Properties of Electric-Resistance-Welded Steel Pipe

Grade	Tensile strength, MPa	Tensile strength, ksi	Yield strength, MPa	Yield strength, ksi	Elongation(a), %
A	330	48	205	30	35
B	415	60	240	35	30

(a) In 50 mm (2 in.), the basic minimum elongation for walls 7.9 mm (0.31 in.) and larger in thickness, longitudinal strip tests, and for all small sizes tested in full section. For longitudinal strip tests, the width of the gage section is 38 mm (1.5 in.). For each decrease of 0.8 mm (0.03 in.) in wall thickness smaller than 7.9 mm (0.31 in.), a deduction of 1.75% for grade A and 1.50% for grade B is made from the basic minimum elongation. Minimum elongation may be calculated using the following equations: for grade A, $E = 56t + 17.50$, and for grade B, $E = 48t + 15.00$, where E is elongation in 50 mm (2 in.); t is actual thickness of specimen measured in inches

ASTM A139

ASTM A139: Chemical Composition

Grade	C	Mn	P	S
	Chemical composition, max %			
A	...	1.00	0.04	0.050
B	0.30	1.00	0.04	0.050
C	0.30	1.20	0.04	0.050
D	0.30	1.30	0.04	0.050
E	0.30	1.40	0.04	0.050

Characteristics and Typical Uses. ASTM A139 describes five grades of electric-fusion, (arc-) welded, straight-seam or spiral-seam, steel pipe 10 to 2340 mm (4 to 92 in.) in diameter inclusive with nominal (average) wall thicknesses up to 25 mm (1 in.) inclusive. The values stated in English units are to be regarded as the standard.

The five grades of ASTM A139 steel are pipe mill grades with mechanical properties that differ from standard plate grades. The pipe is intended for conveying liquid, gas, or vapor. ASTM A139 steel is made by the open-hearth, basic-oxygen, or electric-furnace process.

The longitudinal edges of the steel are shaped to give satisfactory results with the particular welding process employed. Welds are made by automatic means, except tack welds, if used, and are of uniform width and height for the entire length of the pipe. All longitudinal seams, spiral seams, and shop girth seams of ASTM A139 steel pipe are butt welded.

Hydrostatic Test. Each length of pipe is tested by the manufacturer to a hydrostatic pressure that will produce a stress in the pipe of not less than 60% nor more than 85% of the minimum specified yield point at room temperature.

The pressure is determined by the following equation:

$$P = \frac{2\,St}{D}$$

where P is hydrostatic pressure, measured in pounds per square inch, not to exceed 19 MPa (2800 psi); S is 0.60 to 0.85 times the minimum specified yield point of the grade of steel used, measured in pounds per square inch; t is specified wall thickness, measured in inches; D is specified outside diameter, measured in inches.

The test pressure is held for not less than 5 s. Pipe with wall thicknesses larger than 3.91 mm (0.154 in.) shall be struck, while under pressure, with a 0.9-kg (2-lb) hammer, or its equivalent, near the weld at both ends of the pipe.

ASTM A139: Similar Steels (U.S. and/or Foreign):

- UNS K03003; ASTM A139 (B)
- UNS K03004; ASTM A139 (C)
- UNS K03010; ASTM A139 (D)
- UNS K03012; ASTM A139 (E)

Mechanical Properties

ASTM A139: Tensile Properties of Electric-Fusion, Arc-Welded Steel Pipe

Grade	Tensile strength MPa	Tensile strength ksi	Yield point MPa	Yield point ksi	Elongation(a), min %	Reduction in elongation(b), %
A	330	48	205	30	35	1.75
B	415	60	240	35	30	1.50
C	415	60	290	42	25	1.25
D	415	60	315	46	23	1.50
E	455	66	360	52	22	2.00

(a) In 50 mm (2 in.), the basic elongation for wall 7.9 mm (0.31 in.) and larger in thickness for longitudinal strip tests. (b) For longitudinal strip tests, a deduction for each 0.8-mm (0.03-in.) decrease in wall thickness below 7.9 mm (0.31 in.) is taken from the basic minimum elongation of the listed percentage points

ASTM A227, A227M

ASTM A227, A227M: Chemical Composition

In any one lot, carbon may vary not more than 0.13% and manganese not more than 0.30%

C	Mn	P	S	Si
0.45-0.85	0.30-1.30	0.040 max	0.050 max	0.15-0.35

Characteristics and Typical Uses. ASTM A227 describes two classes of round, hard drawn, spring steel wire with properties and quality suited for the manufacture of mechanical springs and wire forms. ASTM A227M is a metric companion to ASTM A227. These steels are made by the open-hearth, basic-oxygen, or electric-furnace process. A sufficient discard is made to assure freedom from injurious piping and undue segregation. The wire is cold drawn to produce the desired mechanical properties.

Test specimens taken from either end of the coils are required to meet the tensile and wrap test requirements given in the accompanying tables. The surface of the wire as received is to be smooth and free of rust. No serious die marks, scratches, or seams may be present.

ASTM A227, A227M: Similar Steels (U.S. and/or Foreign). UNS K06501; ASTM A227

Mechanical Properties

ASTM A227, A227M: Wrap Test Requirements

Wire diameter, mm	Wire diameter, in.	Ratio of mandrel to wire diameter(a) Grade I	Grade II
0.50 to 4.0 incl	0.020 to 0.162 incl	1	2
Over 4.0 to 8.0 incl(a)	Over 0.162 to 0.312 incl(b)	2	4

(a) Wrap tests on wire over 8.5 mm in diameter are not applicable because conventional methods will not accommodate wire over 8.0 mm diameter. (b) Wrap tests on wire over 0.312 in. in diameter are not applicable because conventional methods will not accommodate wire over 0.312 in. in diameter

Mechanical Properties (continued)

ASTM A227M: Tensile Requirements, Metric Units

Diameters were based on a preferred number series for metric sizes

Diameter, mm	Tensile strength, MPa Class I min	Class I max	Class II min	Class II max
0.50	1960	2240	2240	2520
0.60	1920	2200	2200	2480
0.70	1870	2140	2140	2410
0.80	1830	2100	2100	2730
1.0	1770	2040	2040	2310
1.6	1640	1880	1880	2120
2.0	1580	1810	1810	2040
3.0	1460	1680	1680	1900
4.0	1380	1590	1590	1800
5.0	1320	1510	1510	1700
6.0	1280	1470	1470	1650
7.0	1220	1410	1410	1600
8.0	1190	1370	1370	1550
10.0	1130	1310	...	...
13.0	1070	1240	...	...
16.0	1010	1170	...	...

ASTM A227: Tensile Requirements, English Units

Tensile strength values for intermediate diameters may be interpolated

Diameter, in.	Tensile strength, ksi Class I min	Class I max	Class II min	Class II max
0.020	283	323	324	364
0.026	275	315	316	356
0.032	266	306	307	347
0.041	255	293	294	332
0.054	243	279	280	316
0.072	232	266	267	301
0.092	220	253	254	287
0.120	210	241	242	273
0.148	203	234	231	261
0.177	195	225	226	256
0.207	190	218	219	247
0.250	182	210	211	239
0.375	167	193	194	220
0.500	156	180	181	205
0.625	147	170	171	194

ASTM A228

ASTM A228: Chemical Composition

C	Mn	P	S	Si
0.70-1.00	0.20-0.60	0.025 max	0.030 max	0.10-0.30

Characteristics and Typical Uses. ASTM A228 describes a high-quality, round, hard drawn, music steel, spring wire, that is uniform in mechanical properties. This wire is intended for use in manufacturing springs subject to high stresses or requiring good fatigue properties.

ASTM A228 steel is made by the open-hearth, basic-oxygen, or electric-furnace process. A sufficient discard is made to ensure freedom from injurious piping and undue segregation. The wire is cold drawn to produce the desired mechanical properties.

Music wire is supplied with many different types of finishes, including bright, phosphate, tin, cadmium, and others. The finish desired is specified on the purchase order.

The surface of the wire must be smooth and free from defects, such as seams, pits, die marks, and other defects tending to impair the use of the wire for springs. Surface requirements are negotiated when the wire is ordered.

ASTM A228: Similar Steels (U.S. and/or Foreign):

- UNS K08500; ASTM A228
- UNS G10860; AISI 1086; AMS 5112E; ASTM A29, A228, A682; FED QQ-S-700 (C1086); SAE J403, J412, J414; (W. Ger.) DIN 1.1269; (Fr.) AFNOR XC 90

Mechanical Properties

ASTM A228: Tensile Requirements For Steel Music Wire, Spring Quality

The values given in English units are to be regarded as the standard for ASTM A228; tensile strength values for intermediate diameter may be interpolated

Wire diameter mm	Wire diameter in.	Tensile strength Minimum MPa	Minimum ksi	Maximum MPa	Maximum ksi
0.1	0.004	3030	439	3340	485
0.2	0.008	2750	399	3040	441
0.3	0.012	2600	377	2880	417
0.4	0.016	2500	362	2760	400
0.61	0.024	2350	341	2600	377
0.81	0.032	2250	327	2490	361
1.0	0.040	2170	315	2410	349
1.3	0.051	2090	303	2310	335
1.6	0.063	2020	293	2230	324
2.0	0.080	1940	282	2150	312
2.54	0.100	1870	271	2070	300
3.18	0.125	1800	261	1990	288
3.56	0.140	1770	256	1950	283
3.81	0.150	1740	253	1920	279
4.50	0.177	1690	245	1860	270
5.26	0.207	1640	238	1820	264
6.35	0.250	1590	230	1760	255

ASTM A229

ASTM A229: Chemical Composition

Chemical composition, %				
C	Mn	P	S	Si
0.55-0.85	0.30-1.20(a)	0.040	0.030	0.10-0.35

(a) Generally, 0.80 to 1.20 manganese for 4.88-mm (0.192-in.) diam and larger; 0.30 to 0.90 manganese for diameters less than 4.88 mm (0.192 in.)

Characteristics and Typical Uses. ASTM A229 describes two classes of oil-tempered spring wire. The material is similar to AISI/SAE 1065 and is used in the pretempered condition for manufacturing mechanical springs and wire forms. ASTM A229 steel is made by the open-hearth, basic-oxygen, or electric-furnace process. The wire is drawn to diameter, then hardened and tempered to produce the desired mechanical properties.

In addition to tensile requirements, the wire is subjected to a wrap test. Wire 4.11 mm (0.162 in.) and smaller in diameter is to be wound on itself as an arbor without breakage. Wire up to and including 7.92 mm (0.312 in.) diam is to be wound on a mandrel twice the diameter without breakage. Wrap test on wire larger than 7.92 mm (0.312 in.) is not applicable. The values stated in English units are to be regarded as the standard.

ASTM A229: Similar Steels (U.S. and/or Foreign):

- UNS K07001; ASTM A229
- UNS G10650; AISI 1065; ASTM A229, A682; FED QQ-S-700 (C1065); MIL SPEC MIL-S-46049, MIL-S-46409; SAE J403, J412, J414; (W. Ger.) DIN 1.1230

Mechanical Properties

ASTM A229: Tensile Properties of Oil-Tempered Steel Spring Wire

Wire diameter		Class I tensile strength				Class II tensile strength			
		Minimum		Maximum		Minimum		Maximum	
mm	in.	MPa	ksi	MPa	ksi	MPa	ksi	MPa	ksi
0.51	0.020	2020	293	2230	323	2230	324	2440	354
0.66	0.026	1970	286	2180	316	2190	317	2390	347
0.81	0.032	1930	280	2140	310	2140	311	2350	341
1.0	0.041	1830	266	2040	296	2050	297	2250	327
1.4	0.054	1740	253	1950	283	1960	284	2160	314
1.6	0.062	1700	247	1910	277	1920	278	2120	308
2.0	0.080	1620	235	1830	265	1830	266	2040	296
3.05	0.120	1520	220	1720	250	1730	251	1940	281
4.11	0.162	1410	205	1590	230	1590	231	1770	256
5.26	0.207	1310	190	1480	215	1490	216	1660	241
6.35	0.250	1280	185	1450	210	1450	211	1630	236
7.92	0.312	1260	183	1430	208	1440	209	1610	234
9.53	0.375	1240	180	1410	204	1420	206	1590	231
12.7	0.500	1170	170	1340	195	1350	196	1520	221
15.9	0.625	1140	165	1310	190	1320	191	1490	216

ASTM A230

ASTM A230: Chemical Composition

Carbon and manganese may be varied by the manufacturer provided that the mechanical properties specified are maintained

Chemical composition, %				
C	Mn	P	S	Si
0.60-0.75	0.60-0.90	0.025 max	0.030 max	0.15-0.35

Characteristics and Typical Uses. ASTM A230 describes the best quality, hard drawn, round, carbon steel wire. This product is used for manufacturing engine valve springs, and other springs which require high fatigue properties.

ASTM A230 steel is made by the electric-furnace, open-hearth, or basic-oxygen process. Sufficient discard is made to ensure freedom from injurious piping and undue segregation. The material is properly patented, or control cooled and cold drawn, to produce the specified mechanical properties. The wire must be of uniform quality and the surface free from imperfections, such as seams, pits, die marks, scratches, and other imperfections tending to impair the fatigue value of the springs.

In addition to tensile requirements, the material must pass a wrap test. Wire 4.11 mm (0.162 in.) and smaller in diameter must wind on itself as an arbor without breakage. Wire up to and including 6.35 mm (0.250 in.) in diameter must wind without breakage on a mandrel twice the diameter of the wire. Wrap test on wires larger than 7.92-mm (0.312-in.) diam is not applicable. The values stated in English units are to be regarded as the standard.

ASTM A230: Similar Steels (U.S. and/or Foreign):

- UNS K06701; ASTM A230; SAE J172
- UNS G10640; AISI 1064; ASTM A26, A29, A57, A230, A682; SAE J403, J412, J414; (W. Ger.) DIN 1.1221; (Fr.) AFNOR XC 60, XC 65; (Swed.) SS_{14} 1665

Mechanical Properties

ASTM A230: Tensile Properties of Oil-Tempered, Valve, Spring-Quality, Carbon Steel Wire

Wire diameter, mm	Wire diameter, in.	Tensile strength Minimum MPa	Minimum ksi	Maximum MPa	Maximum ksi	Reduction in area, min %
1.6-2.3	0.062-0.092	1650	240	1790	260	(a)
2.4-3.25	0.093-0.128	1620	235	1760	255	40
3.28-4.11	0.129-0.162	1590	230	1720	250	40
4.14-4.88	0.163-0.192	1550	225	1690	245	40
4.90-5.72	0.193-0.225	1520	220	1650	240	40
5.74-6.35	0.226-0.250	1480	215	1620	235	40

(a) Reduction in area does not apply to wire 2.34-mm (0.092-in.) diam and smaller

ASTM A231, A232

ASTM A231, A232: Chemical Composition

Grade	C	Mn	P max	S max	Si	Cr	V min
A231	0.48-0.53	0.70-0.90	0.040	0.040	0.20-0.35	0.20-0.35	0.15
A232	0.48-0.53	0.70-0.90	0.020	0.035	0.20-0.35	0.20-0.35	0.15

Characteristics and Typical Uses. ASTM A231 describes round, chromium-vanadium, alloy steel, spring wire with qualities and properties intended for manufacturing springs used at moderately elevated temperatures.

ASTM A232 designates the best quality of round, chromium-vanadium, alloy steel, valve-spring wire. This material is uniform in quality and temper and is intended for manufacturing valve springs and other springs requiring high fatigue properties, when used at moderately elevated temperatures.

ASTM A231 or A232 wire is supplied in either the annealed and cold drawn, or oil-tempered condition, as specified. The surface of the wire is furnished free of imperfections, such as seams, pits, die marks, scratches, and other imperfections which tend to impair the fatigue value of the springs.

Transverse sections of the valve-spring-quality wire when properly mounted, polished, and etched shall show no completely decarburized (carbon-free) areas when examined at a magnification of 100 diameters. Partial decarburization shall not exceed a depth of 0.025 mm (0.001 in.) on wire 4.88 mm (0.192 in.) and smaller in diameter, or 0.0381 mm (0.0015 in.) on wire larger than 4.88 mm (0.192 in.) in diameter. The values stated in English units are to be regarded as the standard.

ASTM A231: Similar Steels (U.S. and/or Foreign). UNS K15048

ASTM A232: Similar Steels (U.S. and/or Foreign). UNS K15047; FED QQ-W-412

Mechanical Properties

ASTM A231, A232: Tensile Properties of Chromium-Vanadium Steel Spring and Valve Spring Wire in the Oil-Tempered Condition

Values for intermediate sizes may be interpolated

Wire bracket, mm	Wire bracket, in.	Tensile strength Minimum MPa	Minimum ksi	Maximum MPa	Maximum ksi	Reduction in area, min %
0.51	0.020	2070	300	2240	325	(a)
0.81	0.032	2000	290	2170	315	(a)
1.0	0.041	1930	280	2100	305	(a)
1.4	0.054	1860	270	2030	295	(a)
1.6	0.062	1830	265	2000	290	(a)
2.0	0.080	1760	255	1900	275	(a)
2.67	0.105	1690	245	1830	265	45
3.43	0.135	1620	235	1760	255	45
4.11	0.162	1550	225	1690	245	40
4.88	0.192	1520	220	1650	240	40
6.20	0.244	1450	210	1590	230	40
7.19	0.283	1410	205	1550	225	40
7.92	0.312	1400	203	1540	223	40
9.53	0.375	1380	200	1520	220	40
11.1	0.438	1340	195	1480	215	40
12.7	0.500	1310	190	1450	210	40

(a) Reduction in area test not applicable

ASTM A231, A232: Wrap Test Requirements

Wire to wind around mandrel without breakage

Wire diameter, mm	Wire diameter, in.	Ratio of mandrel to wire diameter
4.11 and smaller	0.162 and smaller	1
Larger than 4.11 to 7.92 incl	Larger than 0.162 to 0.312 incl	2
Larger than 7.92	Larger than 0.312	(a)

(a) Wrap test not applicable

ASTM A283

ASTM A283: Chemical Composition

Chemical composition(a), %		
P	S	Cu(b)
0.04 max	0.05 max	0.20 min

(a) Heat analysis. (b) When copper steel is specified

Characteristics and Typical Uses. ASTM A283 designates low and intermediate tensile strength, carbon steel plate, shapes, and bar. Requirements for four grades, A, B, C, and D, are given for carbon steel plate of structural quality for general application. Requirements are also given for two grades, C and D, for carbon steel shapes and bar. The values stated in English units are to be regarded as the standard.

Steel for the ASTM A283 specification is made by the open-hearth, basic-oxygen, or electric-furnace processes.

ASTM A283: Similar Steels (U.S. and/or Foreign). None listed

Mechanical Properties

ASTM A283: Tensile Properties

For plates wider than 610 mm (24 in.), the test specimen is taken in the transverse direction

Grade	Tensile strength		Yield point		Elongation(a), %	
	MPa	ksi	MPa	ksi	In 200 mm (8 in.)(b)	In 50 mm (2 in.)
A	310-380	45-55	165	24	27	30
B	345-415	50-60	185	27	25	28
C	380-450	55-65	205	30	22	25
D(c)	415-495	60-72	230	33	20	23

(a) For plates wider than 610 mm (24 in.), the elongation requirement is reduced by two percentage points. (b) For material less than 8 mm (0.3 in.) in thickness, a deduction of 1.25 percentage points is made for each decrease of 0.8 mm (0.03 in.) of the specified thickness below 8 mm (0.3 in.). (c) For material over 38 mm (1.5 in.) in thickness, the upper limit is 515 MPa (75 ksi)

ASTM A284

ASTM A284: Chemical Composition

Grade	Chemical composition(a), %				
	C(b)	Mn max	P max	S max	Si
C	0.24-0.36	0.90	0.04	0.05	0.15-0.40
D	0.27-0.35	0.90	0.04	0.05	0.15-0.40

(a) Heat analysis. (b) Carbon content increases with plate thickness

Characteristics and Typical Uses. ASTM A284 describes carbon steel plate subject to mechanical test requirements that are intended for machine parts and general construction by gas cutting, welding, or other means.

A definite silicon content is specified in order to limit the carbon content to the lowest practicable amount consistent with the specified tensile strength and the thickness of the material. Welding technique is important and should be in accordance with approved methods.

The maximum thickness for plate delivered under this specification is 305 mm (12 in.) for plate of grade C, and 200 mm (8 in.) for plate of grade D. The values stated in English units are to be regarded as the standard.

ASTM A284 steel may be made by the open-hearth, basic-oxygen, or electric-furnace process.

ASTM A284: Similar Steels (U.S. and/or Foreign):

- UNS K01804; ASTM A284 (A)
- UNS K02001; ASTM A284 (B), A515 (55)
- UNS K02401; ASTM A284 (C), A515 (60); MIL SPEC MIL-S-23495 (C)

Mechanical Properties

ASTM A284: Tensile Properties

For plate wider than 610 mm (24 in.), the test specimen is taken in the transverse direction, and the elongation requirement is reduced two percentage points

Grade	Tensile strength, min		Yield strength, min		Elongation, %	
	MPa	ksi	MPa	ksi	In 200 mm (8 in.)(a)	In 50 mm (2 in.)
C	415	60	205	30	21	25
D	415	60	230	33	21	24

(a) For material less than 8 mm (0.3 in.) in thickness, a deduction of 1.25 percentage points is made for each decrease of 0.8 mm (0.03 in.) of the specified thickness below 8 mm (0.3 in.)

Physical Properties

ASTM A284: Thermal Treatment. Plate larger than 100 mm (4 in.) in thickness shall be treated to produce grain refinement either by normalizing or heating uniformly for hot forming. If the required treatment is to be obtained in conjunction with the hot forming operation, the temperature to which plate is heated shall be equivalent to, but not significantly exceeding the normalizing temperature.

ASTM A308

Characteristics and Typical Uses. ASTM A308 describes cold rolled sheet steel in coils and cut lengths, coated with a lead-tin alloy (terne alloy) by the hot dip process. The minimum tin content is 3%; however, the use of higher percentages of tin will vary depending on the type of coating equipment used. Base steel compositions which are applicable to ASTM A308 are those for ASTM A366, A568, A611, A619, and A620.

The ASTM A308 product is commonly known as long terne and is used where ease of solderability and a degree of corrosion resistance are desirable. It may also be used for stampings where difficulties in drawing may be lessened by the coating which acts as a lubricant in the die.

ASTM A308, long terne, is available in the following quality designations: commercial quality; drawing quality; drawing quality, special killed; and structural quality. The specification does not cover the product known as short terne plate which is described in "Federal Specifications QQ-T-191b and QQ-T-201b".

For qualities other than structural quality, the bend test specimen of coated metal shall be capable of being bent 180° flat on itself in any direction without loss of adherence of the coating on the outside of the bend only.

ASTM A308: Similar Steels (U.S. and/or Foreign). None listed

Mechanical Properties

ASTM A308: Coating Designations and Minimum Coating Test Limits

The coating designation number is the term by which this product is specified; the weight of the coating in g/m² (oz/ft²) of sheet refers to the total coating on both surfaces; the values stated in English units are to be regarded as standard

Coating designation	Minimum coating check limits: Triple-spot test, g/m²	oz/ft²	Single-spot test, g/m²	oz/ft²
LT01	No check		No check	
LT25	75	0.25	60	0.20
LT35	105	0.35	75	0.25
LT40	120	0.40	90	0.30
LT55	170	0.55	120	0.40
LT85	260	0.85	215	0.70
LT110	335	1.10	275	0.90

ASTM A366, A569

ASTM A366, A569: Chemical Composition

Chemical composition, %: C	Mn	P	S	Cu(a)
0.15 max	0.60 max	0.035 max	0.040 max	0.20

(a) Minimum when copper is specified

Characteristics and Typical Uses. ASTM A366 describes cold rolled carbon steel sheet of commercial quality. ASTM A569 designates hot rolled carbon steel sheet and strip of commercial quality, in coils and cut lengths, with a maximum carbon content of 0.15%.

ASTM A366 and A569 steels are intended for parts where bending, moderate forming or drawing, and welding may be involved. Unless otherwise specified, hot rolled material is furnished in the as-rolled condition, not annealed or pickled. The steels are capable of being bent, at room temperature, in any direction through 180° flat on itself, without cracking on the outside of the bent portion.

Hot rolled sheet stock is supplied with mill edge or cut edge. Strip stock is supplied with mill edge or slit (cut) edge. Narrow widths cut from wide cold rolled sheet are not strip, unless they qualify as strip because of thickness, special finish, special edge, or special temper.

ASTM A366, A569: Similar Steels (U.S. and/or Foreign). None listed

ASTM A401

ASTM A401: Chemical Composition

Chemical composition, %: C	Mn	P max	S max	Si	Cr
0.51-0.59	0.60-0.80	0.035	0.040	1.20-1.60	0.60-0.80

Characteristics and Typical Uses. ASTM A401 describes round, chromium-silicon, alloy steel spring wire with properties intended for manufacturing springs that are resistant to set when used at moderately elevated temperatures. This wire is furnished in either the annealed and cold drawn, or oil-tempered condition.

The surface of ASTM A401 wire as received shall be free of rust and excessive scale. No serious die marks, scratches, or seams may be present. Based upon examination of an etched transverse specimen, seams are not to exceed 3.5% of the wire diameter, or 0.254 mm (0.010 in.), whichever is smaller, as measured on a transverse section. The values stated in English units are to be regarded as the standard.

A401 steel may be made by the open-hearth, basic-oxygen or electric-furnace process.

ASTM A401: Similar Steels (U.S. and/or Foreign). UNS K15590; ASTM A401; FED QQ-W-412 (II)

Mechanical Properties

ASTM A401: Tensile Properties of Chromium-Silicon Steel Spring Wire

Values for intermediate sizes may be interpolated

Wire diameter		Tensile strength				Reduction in area, min %
		Minimum		Maximum		
mm	in.	MPa	ksi	MPa	ksi	
0.81	0.032	2070	300	2240	325	(a)
1.0	0.041	2050	298	2230	323	(a)
1.4	0.054	2010	292	2190	317	(a)
1.6	0.062	2000	290	2170	315	(a)
2.3	0.092	1970	285	2140	310	45
3.05	0.120	1900	275	2070	300	45
3.43	0.135	1860	270	2030	295	40
4.11	0.162	1830	265	2000	290	40
4.50	0.177	1790	260	1970	285	40
4.88	0.192	1790	260	1950	283	40
5.56	0.219	1760	255	1920	278	40
6.35	0.250	1720	250	1900	275	40
7.92	0.312	1690	245	1860	270	40
9.53	0.375	1660	240	1830	265	40
11.1	0.438	1620	235	1790	260	40

(a) Reduction in area test not applicable

ASTM A401: Wrap Test Requirements

Wire to wind around mandrel without breakage

Wire diameter		Ratio of mandrel to wire diameter
mm	in.	
4.11 and smaller	0.162 and smaller	1
Larger than 4.11 to 7.92 incl	Larger than 0.162 to 0.312 incl.	2
Larger than 7.92	Larger than 0.312	(a)

(a) Wrap test not applicable

ASTM A424

ASTM A424: Chemical Composition

Steel type	Chemical composition, max %			
	C	Mn	P	S(a)
I	0.008	0.60	...	0.040
II, A	0.04	0.12	0.015	0.040
II, B	0.08	0.20	0.015	0.040

(a) 0.035 max for drawing quality steel

Characteristics and Typical Uses. ASTM A424 describes sheet steel in cut lengths and coils for porcelain enameling. This material is chemically constituted and processed to make it suitable for the fabricating and enameling requirements of articles for porcelain enameling under proper conditions. It is furnished as type I; type II, composition A; type II, composition B; and in three qualities: commercial, drawing, and drawing, special killed.

Type I has an extremely low carbon level commonly produced by decarburizing in an open-coil process, in which the coil laps are separated for easy flow of annealing gases. This material is suitable for direct-cover-coat enameling practice, but this requirement must be specified by the purchaser. Type I material is also suitable for ground- and cover-coat enameling practice. It has good sag resistance and good formability.

Type II has moderately low carbon and manganese levels as produced in the melting operation. This material is suitable for ground- and cover-coat enameling practice. Type II, composition A, is intended for use where resistance to sag is of prime importance. Type II, composition B, is intended for use where formability is of prime importance.

Commercial quality steel is intended for parts where moderate forming, moderate drawing, or bending may be involved. Drawing quality is intended for fabricating identified parts where drawing or severe forming may be involved. Drawing quality, special killed steel is intended for fabricating identified parts where the draw is particularly severe or where the material must be essentially free of changes in mechanical properties over a period of time.

ASTM A424: Similar Steels (U.S. and/or Foreign). None listed

Mechanical Properties

ASTM A424: Typical Mechanical Properties

Steel type	Tensile strength		Yield strength		Total elongation, %	Hardness, HRB
	MPa	ksi	MPa	ksi		
Type I						
Direct cover coat	310	45	170	25	42	40
Commercial quality	295	43	230	33	37	40
Drawing quality	275	40	170	25	45	35
Type II, A or B						
Commercial quality	345	50	240	35	28	50
Drawing quality	295	43	200	29	38	37

ASTM A446

ASTM A446: Chemical Composition

Grade	C	Chemical composition, max % P	S	Cu(a)
A	0.20	0.04	0.04	0.20
B	0.20	0.10	0.04	0.20
C	0.25	0.10	0.04	0.20
D	0.40	0.20	0.04	0.20
E	0.20	0.04	0.04	0.20
F	0.50	0.04	0.04	0.20

(a) Minimum when copper steel is specified

Characteristics and Typical Uses. ASTM A446 describes steel sheet of structural quality in coils and cut lengths that are zinc-coated (galvanized). Sheet of this quality is intended for use when properties or values indicated by tension, hardness, or other commonly accepted mechanical tests or structural properties, are specified or required. ASTM A446 structural quality sheet can be produced in six grades, A through F, according to the base metal mechanical requirements. Structural quality galvanized steel sheet is produced with any of the types of coatings and coating designations listed in the latest revision of "ASTM A525".

The base metal bend test specimen shall be capable of being bent at room temperature through 180° in the longitudinal or transverse direction, without cracking of the base metal on the outside of the bent portion. The inside diameter of the bend shall have a relation to the thickness as follows:

Grade	Ratio of bend diameter to specimen thickness
A	1.5
B	2
C	2.5

Bend metal tests are not required for grades D, E, and F.

The values stated in English units are to be regarded as the standard.

ASTM A446: Similar Steels (U.S. and/or Foreign). None listed

Mechanical Properties

ASTM A446: Mechanical Requirements for Base Metal

Grade	Tensile strength MPa	Tensile strength ksi	Yield point MPa	Yield point ksi	Elongation(a), min %
A	310	45	230	33	20
B	360	52	255	37	18
C	380	55	275	40	16
D	450	65	345	50	12
E	565	82	550	80(b)	...
F	485	70	345	50	12

(a) In 50 mm (2 in.). (b) The yield point approaches the tensile strength; because there is no halt in the gage or drop in the beam, the yield point should be taken as the stress at 0.5% elongation under load. If the hardness is 85 HRB or higher, no tension test is required

ASTM A446: Coating Bend Test

Coating designation	Ratio of bend diameter to specimen thickness Grade A	Grade B	Grade C
G 235	3	3	3
G 210	2	2	2.5
G 185	2	2	2.5
G 165	2	2	2.5
G 140	2	2	2.5
G 115	1.5	2	2.5
G 90	1.5	2	2.5
G 60	1.5	2	2.5
G 01	1.5	2	2.5

ASTM A463

ASTM A463: Chemical Composition

Quality	C	Chemical composition, max % Mn	P	S
Commercial quality	0.15	0.60	0.035	0.040
Drawing quality or drawing quality, special killed	0.10	0.50	0.025	0.035

Characteristics. ASTM A463 describes aluminum-coated steel sheet, type 1, in coils and cut sheet coated with an aluminum-silicon alloy by the hot dip process. The aluminum-silicon alloy for coating has 5 to 11% silicon added to promote better adherence.

Aluminum-coated steel sheet is available in commercial quality; drawing quality; drawing quality, special killed; and structural quality. Type 2 steel sheet, not covered by this specification, is hot dipped in a molten bath of commercially pure aluminum. The values stated in English units are to be regarded as the standard for ASTM A463.

Aluminum-coated steel can be cold bent 180° over a radius equal to twice the thickness in any direction without flaking, on the outside of the bend only. ASTM A463 material can be joined by any of the welding processes suitable for steel, but the procedures must be adapted to the aluminum coating. Inert gas shielding protects the coating and prevents formation of aluminum oxide. However, excessive alloying of weld metal must be minimized.

Shielded-metal-arc welding using low-hydrogen electrodes is satisfactory. An electrode specifically designed to accommodate the alumina slag produces high-quality deposits from the standpoint of pinholes and aluminum oxide coatings. Oxyacetylene welding processes are not ordinarily recommended for aluminum-coated steel, because they tend to damage the aluminum coating more than other types of welding.

Soldering and brazing aluminum-coated steel present the same difficulties encountered with these processes on solid aluminum. The protective oxide film on the surface of the aluminum coating must be removed to establish metal-to-metal contact between the molten solder or braze metal and the coated base metal.

Typical Uses. Aluminum-coated steel, type 1 of ASTM A463, is used for combustion chambers, heat exchangers, automotive exhaust systems, outdoor grills, heat reflectors, ovens, and appliances. Type 2 steel has superior resistance to atmospheric corrosion and is used for metal roofing and other exposed products.

ASTM A463: Similar Steels (U.S. and/or Foreign). None listed

Physical Properties

ASTM A463: Thickness and Weight Equivalents of Aluminum-Coated Steel Sheet, Type 1

Thickness		Weight	
mm	in.	kg/m²	lb/ft²
0.38	0.015	2.79	0.571
0.51	0.020	3.78	0.775
0.64	0.025	4.78	0.979
0.76	0.030	5.78	1.183
0.89	0.035	6.77	1.387
1.02	0.040	7.77	1.591
1.14	0.045	8.76	1.795
1.27	0.050	9.76	1.999
1.40	0.055	10.76	2.203
1.52	0.060	11.75	2.407
1.65	0.065	12.75	2.611
1.78	0.070	13.74	2.815
1.9	0.075	14.74	3.019
2.03	0.080	15.73	3.223
2.16	0.085	16.73	3.427
2.29	0.090	17.73	3.631
2.41	0.095	18.72	3.835
2.54	0.100	19.72	4.039
2.68	0.105	20.71	4.243
2.79	0.110	21.71	4.447
2.92	0.115	22.71	4.651
3.02	0.119	23.50	4.814

ASTM A463: Coating Designations and Minimum Coating Test Limits for Aluminum-Coated Steel Sheet, Type 1

The weight of coating and coating thickness refers to the total coating on both surfaces; the coating designation is the term by which this product is specified

Coating designation	Triple-spot test(a) g/m²	Triple-spot test(a) oz/ft²	Single-spot test(a) g/m²	Single-spot test(a) oz/ft²	Coating thickness(b) mm	Coating thickness(b) in.
T1 40, regular	120	0.40	90	0.30	0.03	0.0013
T1 25, light	75	0.25	60	0.20	0.02	0.0008

(a) Total both sides. (b) Based on minimum single-spot test using calculation based on aluminum density of 0.27 kg/m³ (0.0975 lb/in.³)

ASTM A485

ASTM A485: Chemical Composition

The following maximum compositions are present in each grade: 0.025% phosphorous; 0.025% sulfur, 0.25% nickel, 0.35% copper

Grade	C	Mn	Si	Cr	Mo
1	0.90-1.05	0.95-1.25	0.45-0.75	0.90-1.20	0.10 max
2	0.85-1.00	1.40-1.70	0.50-0.80	1.40-1.80	0.10 max
3	0.95-1.10	0.65-0.90	0.15-0.35	1.10-1.50	0.20-0.30
4	0.95-1.10	1.05-1.35	0.15-0.35	1.10-1.50	0.45-0.60

(Column headers C through Mo fall under: Chemical composition, %)

ASTM A485: Similar Steels (U. S. and/or Foreign):

- UNS K19195; ASTM A485 (2)
- UNS K19667; ASTM A485 (1)
- UNS K19965; ASTM A485 (3)
- UNS K19990; ASTM A485 (4)

Characteristics and Typical Uses. ASTM A485 describes high-hardenability modifications of high-carbon, chromium-bearing quality, steel billets for rolling or forging, tube rounds, bar, coils, and tube to be used in manufacturing anti-friction bearings. Four typical steel compositions, grades 1, 2, 3, and 4, are covered by this specification.

ASTM A485 steel is made by the electric-furnace process and must be vacuum carbon deoxidized unless otherwise agreed upon between the purchaser and supplier. Where not specified, nickel, copper, and molybdenum are residual elements, and are to be reported in the analysis of the heat if required by the purchaser.

Each heat is to be tested for hardenability. Normalizing, followed by spheroidizing annealing, precedes heating for end quenching. In heating for end quenching, the specimens are held for a minimum of 30 min at 815 °C (1500 °F), plus or minus 4.5 °C (8 °F). End-quench procedure should be in accordance with "ASTM A255".

The applicable end-quench values for hardenability are as follows:

Grade	Minimum hardness at a distance of 1/16 in., HRC 10	20	28
1	46	...	...
2	...	52	32
3	46	...	...
4	...	52	35

ASTM A500

ASTM A500: Chemical Composition

Grade	C	Chemical composition, max % Mn	P	S	Cu(a)
A, B	0.30	...	0.05	0.063	0.18
C	0.27	1.40	0.05	0.063	0.18

(a) Minimum when copper steel is specified

Characteristics and Typical Uses. ASTM A500 describes cold formed welded and seamless carbon steel in round, square, rectangular, or special-shape structural tubing. This tubing is used for welded, riveted, or bolted construction of bridges and buildings, and general structural use. The ASTM A500 welded steel tubing is produced in sizes with a maximum periphery of 1630 mm (64 in.) and a maximum wall thickness of 13 mm (0.50 in.). The seamless tubing is produced with a maximum periphery of 810 mm (32 in.) and a maximum wall thickness of 13 mm (0.50 in.). The values stated in English units are to be regarded as the standard.

ASTM A500 steel is made by the open-hearth, basic-oxygen, or electric-furnace process. The welded tubing is made from flat rolled steel using an automatic welding process which produces a longitudinal weld without addition of filler metal. The longitudinal butt joint of welded tubing is welded across the thickness of the joint in a manner that ensures the structural design strength of the tubing. Structural tubing welded by electric resistance methods is normally furnished without removal of inside flash.

ASTM A500: Similar Steels (U.S. and/or Foreign):

- UNS K03000; ASTM A500 (A, B), A501
- UNS K02705; ASTM A500 (C)

Mechanical Properties

ASTM A500: Tensile Properties of Cold Formed Welded and Seamless Carbon Steel Tubing

Grade	Tensile strength MPa	ksi	Yield strength MPa	ksi	Elongation(a), min %
Round structural tubing					
A	310	45	230	33	25(b)
B	400	58	290	42	23(c)
C	425	62	315	46	21(d)
Shaped structural tubing					
A	310	45	270	39	25(b)
B	400	58	315	46	23(c)
C	425	62	345	50	21(d)

(a) In 50 mm (2 in.). (b) Applies to specified wall thickness of 3.05 mm (0.120 in.) and larger; for thicknesses smaller than 3.05 mm (0.120 in.), the minimum elongation is calculated by the formula $E = 56t + 17.5$, where E is elongation and t is material thickness measured in inches. (c) Applies to specified wall thicknesses 4.57 mm (0.180 in.) and larger; for wall thicknesses smaller than 4.57 mm (0.180 in.), the minimum elongation is calculated by the formula $E = 61t + 12$, where E is elongation and t is material thickness measured in inches. (d) Applies to specified wall thicknesses 3.05 mm (0.120 in.) and larger; for lighter wall thicknesses, elongation is determined by agreement with the manufacturer

ASTM A501

ASTM A501: Chemical Composition

Chemical composition, max %			
C	P	S	Cu(a)
0.30	0.05	0.063	0.18

(a) Minimum when copper steel is specified

Characteristics and Typical Uses. ASTM A501 describes hot formed welded and seamless carbon steel of square, round, rectangular, or special-shape structural tubing. This tubing is intended for welded, riveted, or bolted construction of bridges and buildings, and for general structural purposes.

Square and rectangular tubing is furnished in sizes 25 to 250 mm (1 to 10 in.) across flat sides with wall thicknesses 2.41 to 25.4 mm (0.095 to 1.00 in.), depending on size. Round tubing is furnished in nominal diameters of 13 to 610 mm (0.5 to 24 in.) inclusive, with nominal (average) wall thickness 2.77 to 25.4 mm (0.109 to 1.00 in.), depending on size. The values stated in English units are to be regarded as the standard.

ASTM A501 tubing may be furnished with hot-dipped galvanized coating. This steel is made by the open-hearth, basic-oxygen, or electric-furnace process. Tubing is produced by the seamless or furnace butt welded process (continuous welded). It may also be made by the electric-resistance-welding process and subsequently reheated through the cross section of the tubing, and hot formed by a reducing or shaping process, or both.

Bend Test. Square and rectangular tubing is subjected to a bend test. The bend test specimen is taken longitudinally from the tubing and represents the full wall thickness of material. The sides of the bend test specimen may have the corners rounded to a maximum radius of 1.6 mm (0.63 in.).

The specimen is expected to withstand cold bending through 180°, without cracking on the outside of the bent portion, to an inside diameter which has a relation to the thickness of the specimen as follows:

Material thickness		Ratio of bend diameter to specimen thickness
mm	in.	
19 and smaller	0.75 and smaller	0.5
Larger than 19 to 25.4 incl	Larger than 0.75 to 1.0 incl	1

ASTM A501: Similar Steels (U.S. and/or Foreign). UNS K03000; ASTM A500 (A, B), A501

Mechanical Properties

ASTM A501: Tensile Properties for Hot Formed Welded and Seamless Carbon Steel Tubing

Tensile strength		Yield strength		Elongation(a), %	
MPa	ksi	MPa	ksi	In 200 mm (8 in.)	In 50 mm (2 in.)
400	58	250	36	20(b)	23

(a) Elongation may be determined in a gage length of either 50 mm (2 in.) or 200 mm (8 in.). (b) For material smaller than 7.9 mm (0.31 in.) in thickness, a deduction from the percentage elongation of 1.25% in 200 mm (8 in.) may be made for each decrease of 0.8 mm (0.03 in.) of the specified thickness under 7.9 mm (0.31 in.)

ASTM A514

ASTM A514: Chemical Composition

Grade	Thickness, max mm	Thickness, max in.	C	Mn	Si	Ni	Cr	Mo	V	Ti	Zr(b)	Cu	B
			Chemical composition(a), %										
A	32	1.25	0.15-0.21	0.80-1.10	0.40-0.80	...	0.50-0.80	0.18-0.28	...	...	0.05-0.15	...	0.0025 max
B	32	1.25	0.12-0.21	0.70-1.00	0.20-0.35	...	0.40-0.65	0.15-0.25	0.03-0.08	0.01-0.03	...	...	0.0005-0.005
C	32	1.25	0.10-0.20	1.10-1.50	0.15-0.30	...	...	0.20-0.30	...	...	...	...	0.001-0.005
D	32	1.25	0.13-0.20	0.40-0.70	0.20-0.35	...	0.85-1.20	0.15-0.25	(c)	0.04-0.10	...	0.20-0.40	0.0015-0.005
E	150	6	0.12-0.20	0.40-0.70	0.20-0.35	...	1.40-2.00	0.40-0.60	(c)	0.04-0.10	...	0.20-0.40	0.0015-0.005
F	100	4	0.10-0.20	0.60-1.00	0.15-0.35	0.70-1.00	0.40-0.65	0.40-0.60	0.03-0.08	...	...	0.15-0.50	0.0005-0.006
G	50	2	0.15-0.21	0.80-1.10	0.50-0.90	...	0.50-0.90	0.40-0.60	...	...	0.05-0.15	...	0.0025 max
H	50	2	0.12-0.21	0.95-1.30	0.20-0.35	0.30-0.70	0.40-0.65	0.20-0.30	0.03-0.08	...	...	...	0.0005-0.005
J	32	1.25	0.12-0.20	0.45-0.70	0.20-0.35	...	...	0.50-0.65	...	...	...	...	0.001-0.005
K	50	2	0.10-0.20	1.10-1.50	0.15-0.30	...	...	0.45-0.55	...	...	...	...	0.001-0.005
L	50	2	0.13-0.20	0.40-0.70	0.20-0.35	...	1.15-1.65	0.25-0.40	(c)	0.04-0.10	...	0.20-0.40	0.0015-0.005
M	50	2	0.12-0.21	0.45-0.70	0.20-0.35	1.20-1.50	...	0.45-0.60	...	...	...	...	0.001-0.005
N	19	0.75	0.15-0.21	0.80-1.10	0.40-0.90	...	0.50-0.80	0.25 max	...	...	0.05-0.15	...	0.0005-0.0025
P	150	6	0.12-0.21	0.45-0.70	0.20-0.35	1.20-1.50	0.85-1.20	0.45-0.60	...	...	...	...	0.001-0.005
Q	150	6	0.14-0.21	0.95-1.30	0.15-0.35	1.20-1.50	1.00-1.50	0.40-0.60	0.03-0.08	...	...	...	...

(a) Heat analysis. (b) Zirconium may be replaced by cerium. When cerium is added, the cerium/sulfur ratio should be approximately 1.5 to 1, based on heat analysis. (c) May be substituted for all or part of titanium content on a one for one basis

Characteristics and Typical Uses. ASTM A514 describes high yield strength, quenched, and tempered alloy steel plate of structural quality that is intended primarily for use in welded bridges and other structures. Plate thicknesses range from 19 to 152 mm (0.75 to 6 in.). The values stated in English units are to be regarded as the standard.

Welding technique is of fundamental importance and must not adversely affect the properties of the plate, especially in the heat-affected zone. Welding procedures are to be suitable for the materials being welded.

ASTM A514 steel is made by open-hearth, basic-oxygen, or electric-furnace processes. Additional refining by vacuum-arc-remelt (VAR) or electroslag-remelt (ESR) is permitted.

ASTM A514: Similar Steels (U.S. and/or Foreign):

- UNS K11511; AMS 6386 (3); ASTM A514 (C), A517 (C)
- UNS K11523; ASTM A514 (C), A517 (C)
- UNS K11576; ASTM A514 (F), A517 (F), A592 (F)
- UNS K11625; AMS 6386 (5); ASTM A514 (J), A517 (J)
- UNS K11646; ASTM A514 (H), A517 (H)
- UNS K11662; AMS 6386 (4); ASTM A514 (D), A517 (D)
- UNS K11682; ASTM A514 (L), A517 (L)
- UNS K11683; ASTM A514 (M), A517 (M)
- UNS K11847; ASTM A514 (N), A517 (N)
- UNS K11856; AMS 6386 (1); ASTM A514 (A), A517 (A), A592 (A)
- UNS K11872; ASTM A514 (G), A517 (G)
- UNS K21604; ASTM A514 (E), A517 (E)
- UNS K21650; ASTM A514 (P), A517 (P)

Physical Properties

ASTM A514: Thermal Treatment. The material is to be heat treated by the manufacturer to conform to the tensile and hardness requirements, by heating to not less than 900 °C (1650 °F), followed by quenching in water or oil, and tempering at not less than 620 °C (1150 °F).

Mechanical Properties

ASTM A514: Tensile and Hardness Requirements

On plate wider than 610 mm (24 in.), the test specimens were taken in the transverse direction; either the ASTM standard, full-thickness, rectangular specimen or the 12.5-mm (0.5-in.) diam specimen may be used for plate larger than 19 to 40 mm (0.75 to 1.5 in.) in thickness

Plate thickness	Tensile strength		Yield strength, min(a)		Elongation(b), %	Reduction in area(c), min %	Hardness(d), HB
	MPa	ksi	MPa	ksi			
To 19 mm (0.75 in.) incl.	760-895	110-130	690	100	18	40(e)	235-293
Larger than 19 mm (0.75 in.) to 64 mm (2.5 in.) incl.	760-895	110-130	690	100	18	40(e), 50(f)	...
Larger than 64 mm (2.5 in.) to 150 mm (6 in.) incl.	690-895	100-130	620	90	16	50(f)	...

(a) Measured at 0.2% offset or 0.5% extension under load. (b) In 50 mm (2 in.) for plates tested in the tranverse direction, the elongation minimum percent is reduced by 2%. When measured on the full thickness by 40-mm (1.5-in.) specimen, the elongation is determined in a 50-mm (2-in.) gage length which includes the fracture and shows the greatest elongation. (c) For plates tested in the transverse direction, the reduction in area minimum requirement is reduced by 5%. (d) For plates 9.5 mm (0.375 in.) and less in thickness, a Brinell hardness test may be used for tension testing each plate, in which case a tension test shall be made from a corner of each of two plates per lot. (e) When measured on the full thickness by 40-mm (1.5-in.) specimen. (f) When measured on the 12.5-mm (0.50-in.) diam specimen

ASTM A525, A525M, A526, A527, A528, A642

ASTM A525, A525M, A526, A527, A528, A642: Chemical Composition

Designation	Quality	C	Mn	P	S	Cu(a)
		Chemical composition, max %				
A526	Commercial	0.15	0.60	0.035	0.040	0.20
A527	Lock-forming	0.15	0.60	0.035	0.040	0.20
A528	Drawing	0.10	0.50	0.025	0.035	...
A642	Drawing, special killed	0.10	0.50	0.025	0.035	...

(a) Minimum when copper steel is specified

ASTM A525, A525M, A526, A527, A528, A642: Similar Steels (U.S. and/or Foreign). None listed

Characteristics and Typical Uses. ASTM A525, A525M, A526, A527, A528, and A642 describe the requirements for steel sheet in cut lengths and coils, which has been zinc-coated (galvanized) on continuous lines by the hot dip process. Galvanized steel sheet is customarily available in commercial quality, A526; lock-forming quality, A527; drawing quality, A528; drawing quality, special killed, A642; and structural quality, A446. The ASTM A525 specification gives the general requirements for galvanized steel sheet; ASTM A525M is a complete metric companion to ASTM A525. The values expressed in English units are to be regarded as the standard for materials other than A525M.

Galvanized steel sheet is produced to various zinc-coating designations, which are designed to provide coating compatible with the service life required. Except for differential-coated sheet, the coating is always expressed as the total coating of both surfaces. Galvanized steel sheet can be produced with the following types of coatings: regular spangle, minimized spangle, iron-zinc alloy, wiped, and differential.

Regular spangle is the result of the unrestricted growth of zinc crystals during normal solidification. The coating designation has a prefix G. Minimized spangle is obtained by treating the regular galvanized sheet during the solidification of the zinc to restrict the normal spangle formation. Minimum spangle is normally produced in coating designations G 90 and lighter. Iron-zinc alloy galvanized sheet is produced by processing the steel through the galvanizing line in a manner which completely alloys the coating. Iron-zinc alloy coating is designated by the prefix A. Coating designation A 60 is commonly known as galvannealed sheet.

Wiped galvanized sheet is produced by wiping down the molten zinc as it leaves the pot. This product has a light iron-zinc alloy coating, is not spangled, and may have striations of free zinc at the surface of the coating. Differentially coated galvanized steel sheet has a specified coating designation on one surface and a significantly lighter specified coating designation on the other surface.

Physical Properties

ASTM A525, A526, A527, A528, A642: Weight of Coating, Total for Both Sides

English units are to be regarded as the standard

Coating designation	Minimum check limit: Triple-spot test, g/m²	Triple-spot test, oz/ft²	Single-spot test, g/m²	Single-spot test, oz/ft²
Regular type of coating (G)				
G 235	717	2.35	610	2.00
G 210	640	2.10	549	1.80
G 185	564	1.85	488	1.60
G 165	503	1.65	427	1.40
G 140	427	1.40	366	1.20
G 115	351	1.15	305	1.00
G 90	275	0.90	244	0.80
G 60	183	0.60	152	0.50
G 30	91	0.30	61	0.20
G 01	No minimum		No minimum	
Alloyed type of coating (A)				
A 60	183	0.60	152	0.50
A 40	122	0.40	91	0.30
A 25	76	0.25	61	0.20
A 01	No minimum		No minimum	

ASTM A525M: Mass of Coating, Total Both Sides in Metric Units

Values given are based on a preferred number series for metric sizes

Coating designation	Minimum check limit: Triple-spot test, g/m²	Single-spot test, g/m²
Zinc type of coating (Z)		
700	700	595
600	600	510
450	450	385
350	350	300
275	275	235
180	180	150
90	90	75
001	(a)	(a)
Iron-zinc alloyed type of coating (ZF)		
180	180	150
100	100	85
75	75	60
001	(a)	(a)

(a) No minimum

Mechanical Properties

ASTM A525M: Coating Bend Tests, Metric Units

Based on number of pieces of same thickness used in coating bend test; bends may be made over mandrels of equivalent thickness with radius proportional to the thickness; values given are based on a preferred number series for metric units

Coating designation	Galvanized sheet thickness: To 1.0 mm incl	Larger than 1.0-2.0 mm incl	Larger than 2.0 mm
Z700	2	3	3
Z600	2	2	2
Z450	1	1	2
Z350	0	0	1
Z275	0	0	1
Z180	0	0	0
Z90	0	0	0
Z001	0	0	0

ASTM A525, A526, A527, A528, A642: Coating Bend Tests, English Units

Based on number of pieces of same thickness used in coating bend test; bends may be made over a mandrel of equivalent thickness with radius proportional to the thickness

Coating designation	Galvanized sheet thickness: 0.1756-0.0748 in.	0.0747-0.0382 in.	0.0381-0.0131 in.
G 235	3	3	2
G 210	2	2	2
G 185	2	2	2
G 165	2	2	2
G 140	2	1	1
G 115	1	0	0
G 90	1	0	0
G 60	0	0	0
G 01	0	0	0

ASTM A529

ASTM A529: Chemical Composition

Chemical composition, %				
C	Mn	P	S	Cu(a)
0.27 max	1.20 max	0.040 max	0.05 max	0.20 min

(a) When copper is specified

Characteristics and Typical Uses. ASTM A529 describes carbon steel plate and bar 12.7 mm (0.50 in.) and smaller in thickness or diameter and group 1 shapes shown in Table A of "ASTM Specification A6; General Requirements for Rolled Steel Plates, Shapes, Sheet Piling, and Bars for Structural Use". The values stated in English units are to be regarded as the standard.

ASTM A529 steel is of structural quality for use in buildings and similar riveted, bolted, or welded construction. When used in welded construction, welding procedures should be suitable for the steel and intended use.

This steel may be made by the open-hearth, basic-oxygen, or electric-furnace process.

ASTM A529: Similar Steels (U.S. and/or Foreign). UNS K02703

Mechanical Properties

ASTM A529: Tensile Properties

For plates wider than 610 mm (24 in.), the test specimen is taken in the transverse direction and the elongation requirement is reduced two percentage points

Tensile strength		Yield point		Elongation(a),
MPa	ksi	MPa	ksi	min %
415-585	60-85	290	42	19

(a) In 200 mm (8 in.) for material less than 8 mm (0.3 in.) in thickness or diameter, a deduction of 1.25 percentage points is made for each decrease of 0.8 mm (0.03 in.) of the specified thickness or diameter

ASTM A570

ASTM A570: Chemical Composition

	Chemical composition, %				
Grade	C max	Mn max	P max	S max	Cu(a)
30, 33, 36, 40	0.25	0.90	0.040	0.05	0.20
45, 50	0.25	1.35	0.040	0.05	0.20

(a) Minimum when copper is specified

Characteristics and Typical Uses. ASTM A570 describes hot rolled carbon steel sheet and strip of structural quality in cut lengths and coils. This material is intended for structural purposes where mechanical test values are required. It is available in a maximum thickness of 5.8 mm (0.2299 in.) in six grades: 30, 33, 36, 40, 45, and 50. Each grade number signifies the minimum yield point of that grade. The values stated in English units are to be regarded as the standard.

ASTM A570 material can be bent at room temperature in any direction through 180° without cracking on the outside of the bent portion, to an inside diameter that has a relation to the thickness of the specimen as follows:

Grade	Ratio of bend diameter to specimen thickness
30	1
33	1.5
36	1.5
40	2
45	2.5
50	3

Tensile test specimens were taken in the longitudinal direction.

ASTM A570: Similar Steels (U.S. and/or Foreign):

- UNS K02500; ASTM A570 (A, B, C)
- UNS K02502; ASTM A570 (D, E)

Mechanical Properties

ASTM A570: Tensile Properties of Specimens Taken in the Longitudinal Direction

Elongation varies with stock thickness; the larger values relate to the thicker material

	Tensile strength, min		Yield point, min		Elongation, %	
Grade	MPa	ksi	MPa	ksi	In 200 mm (8 in.)	In 50 mm (2 in.)
30	340	49	205	30	21-25	17-19
33	360	52	230	33	18-23	16-18
36	365	53	250	36	17-22	15-17
40	380	55	275	40	15-21	14-16
45	415	60	310	45	13-19	12-14
50	450	65	345	50	11-17	10-12

ASTM A573

ASTM A573: Chemical Composition

Grade	C	Chemical composition(a), % Mn	P max	S max	Si
58	0.23 max	0.60-0.90(b)	0.04	0.05	0.10-0.35
65	0.24-0.26	0.85-1.20	0.04	0.05	0.15-0.40
70	0.27-0.28	0.85-1.20	0.04	0.05	0.15-0.40

(a) Heat analysis. (b) The upper limit of manganese may be exceeded if the carbon content plus 1/6 manganese content does not exceed 0.40% based on heat analysis

Characteristics and Typical Uses. ASTM A573 describes structural quality carbon-manganese-silicon steel plate of three tensile strength ranges. These steel materials are intended primarily for service at atmospheric temperatures where improved notch toughness is important. Plate covered by this specification has a maximum thickness of 38 mm (1.5 in.).

ASTM A573 steel is intended for fusion welding. The welding procedures used should be in accordance with approved methods. This steel may be made by the open-hearth, basic-oxygen, or electric-furnace process.

ASTM A573: Similar Steels (U.S. and/or Foreign):

- UNS K02301; ASTM A573 (58)
- UNS K02404; ASTM A573 (65)
- UNS K02701; ASTM A573 (70)

Mechanical Properties

ASTM A573: Tensile Properties

For plates wider than 610 mm (24 in.), the test specimen is taken in the transverse direction and the elongation requirement is reduced two percentage points

Grade	Tensile strength MPa	ksi	Yield strength, min MPa	ksi	Elongation(a), min %
58	400-490	58-71	220	32	21
65	450-530	65-77	240	35	20
70	485-620	70-90	290	42	18

(a) In 200 mm (8 in.)

ASTM A591

ASTM A591: Chemical Composition

Quality	C	Chemical composition, max % Mn	P	S
Commercial	0.15	0.60	0.035	0.040
Drawing	0.10	0.50	0.025	0.035

Characteristics and Typical Uses. ASTM A591 describes cold rolled steel sheet in cut lengths and coils, zinc-coated by electrodeposition. This product is intended for the manufacture of formed or miscellaneous parts and can be furnished in the chemically treated condition to make it more suitable for painting. ASTM A591 material is produced in light coating weights and is not intended to withstand outside exposure without chemical treating and painting.

Electrolytic zinc-coated sheet is customarily available in widths of 13 to 1520 mm (0.5 to 60 in.) and in thicknesses of 0.229 to 2.75 mm (0.009 to 0.1083 in.). It is produced as commercial quality; drawing quality; drawing quality, special killed; or structural quality steel. The values stated in English units are to be regarded as the standard.

Except for thickness tolerances, material furnished under this specification conforms to the applicable requirements: of "ASTM A568, Steel, Carbon and High-Strength Low-Alloy Hot-Rolled Sheet, Hot-Rolled Strip, and Cold-Rolled Sheet, General Requirements".

The coatings should not show flaking on any of the qualities ordered. The coating bend test is made in accordance with the restrictions noted in the base metal bend test of "ASTM A525, Steel Sheet, Zinc-Coated (Galvanized) by the Hot-Dip Process, General Requirements".

Drawability and formability of electrolytic zinc-coated steel, ASTM A591, is very good with very little die wear and die pickup encountered. Resistance welding requires an increase in time and force of 20 to 40% compared to bare steel. Fusion and arc weld quality is satisfactory, with some damage to coating continuity. Solderability and paintability are good. The surfaces should be alkaline cleaned, rinsed, zinc-phosphate coated, and chromate rinsed before painting.

ASTM A591: Similar Steels (U.S. and/or Foreign). None listed

Physical Properties

ASTM A591: Ordered Coating and Minimum Coating Test Limits

Coating class	Minimum check limit: Decimal equivalents(a) mm	in.	Coating weight(b) g/m²	oz/ft²
Triple-spot test				
A	None	None	None	None
B	0.00165	0.000065	22.9	0.075
C	0.00356	0.000140	50.4	0.165
Single-spot test				
A	None	None	None	None
B	0.00152	0.00006	21.3	0.070
C	0.00318	0.000125	45.8	0.150

(a) One side. (b) Total for both sides

ASTM A599

Characteristics and Typical Uses. ASTM A599 describes cold rolled steel sheet in cut lengths or coils, which has been tin coated by electrodeposition. This product is commonly known as electrolytic tin-coated sheet, and is used where solderability is desirable, appearance is important, or a degree of corrosion resistance, under specific conditions, is advantageous. The base steel compositions which are applicable for ASTM A599 are those for ASTM A366, A568, A611, A619, and A620.

ASTM A599 sheet is customarily available in commercial quality; drawing quality; drawing quality, special killed; and structural quality. The tin coating may be ordered as unmelted or melted.

Unmelted tin coating is tin coated by electrodeposition on a base steel which normally has a dull, blasted-roll surface texture. The deposited tin also has a dull gray appearance. Melted tin is tin coated by electrodeposition on a base steel which normally has a ground-roll finish. The tin coating is then melted to reflow the tin. The resulting coating has a brighter appearance than unmelted tin. An iron-tin alloy layer is developed during the melting operation, thus reducing the amount of free tin available.

For all qualities, except structural quality, the material shall be capable of being bent at room temperature in any direction through 180° flat on itself without fracturing of the base metal or flaking the coating.

ASTM A599: Similar Steels (U.S. and/or Foreign). None listed

Mechanical Properties

ASTM A599: Coating Designations and Minimum Coating Test Limits, Total Both Sides

The values stated in English units are to be regarded as the standard

Coating designation	Nominal coating, g/m²	Nominal coating, oz/ft²	Triple-spot g/m²	Triple-spot oz/ft²	Single-spot g/m²	Single-spot oz/ft²
			Minimum coating test limits			
25	5.6	0.018	3.7	0.012	2.8	0.009
50	11.2	0.037	7.3	0.024	5.6	0.018
75	16.8	0.055	11.0	0.036	8.2	0.027
100	22.4	0.073	14.6	0.048	11.0	0.036
125	28.0	0.092	18.3	0.060	13.8	0.045

ASTM A611

ASTM A611: Chemical Composition

Grade	C	Mn	P	S	Cu(a)
	Chemical composition, max %				
A, B, C, E	0.20	0.60	0.04	0.04	0.20
D	0.20	0.90	0.04	0.04	0.20

(a) Minimum where copper steel is specified

Characteristics and Typical Uses. ASTM A611 describes cold rolled carbon steel sheet of structural quality in cut lengths or coils. This material is available in five strength levels, categorized in grades A through E. Grades A, B, C, and D have moderate ductility; grade E is a full-hard product with no specified minimum elongation. ASTM A611 material is furnished in a matte (dull) finish unless otherwise ordered. The sheet may be oiled or dry, as specified.

ASTM A611 steel can be bent, at room temperature, in any direction through 180° without cracking on the outside of the bent portion, to an inside diameter which has a relation to the thickness of the specimen as follows:

Grade	Ratio of bend diameter to specimen thickness
A	0
B	1
C	1.5
D	2
E	Not applicable

ASTM A611: Similar Steels (U.S. and/or Foreign). None listed

Mechanical Properties

ASTM A611: Tensile Properties Taken on Longitudinal Test Specimens

The values stated in English units are to be regarded as the standard

Grade	Tensile strength, min MPa	Tensile strength, min ksi	Yield point, min MPa	Yield point, min ksi	Elongation(a), min %
A	290	42	170	25	26
B	310	45	205	30	24
C	330	48	230	33	22
D	275	40	360	52	20
E	565	82	550	80(b)	...

(a) In 50 mm (2 in.). (b) On this full-hard product, the yield point approaches the tensile strength; because there is no halt in the gage or drop in the beam, the yield point is taken as the stress at 0.5% elongation, under load

ASTM A619, A620

ASTM A619, A620: Chemical Composition

Chemical composition, %				
C	Mn	P	S	Al
0.10 max	0.50 max	0.025 max	0.035 max	(a)

(a) If aluminum is used as the oxidizing agent, in ASTM A620, the total aluminum content by product analysis is usually in excess of 0.010%

Characteristics and Typical Uses. ASTM A619 describes cold rolled carbon steel sheet of drawing quality in coils and cut lengths. The material is intended for fabricating identified parts where drawing or severe forming may be involved.

ASTM A620 designates cold rolled carbon steel that is drawing quality, special killed material. This material is intended for fabricating identified parts where particularly severe drawing or forming, or essential freedom from aging is required.

These steels are furnished in two class standards. Class 1 material is intended for applications where surface appearance is of primary importance, such as exposed parts. This class will meet requirements for controlled surface texture, surface quality, and flatness. It is normally processed to be free of stretcher strains and fluting.

Class 2 material is intended for applications where surface appearance is not of primary importance. Limitations on degree and frequency of surface imperfections, as well as restrictions on texture and flatness, are not applicable. This material may have coil breaks and a tendency toward fluting and stretcher straining.

ASTM A619, A620: Similar Steels (U.S. and/or Foreign). None listed

ASTM A621, A622

ASTM A621, A622: Chemical Composition

Chemical composition, %				
C	Mn	P	S	Al
0.10 max	0.50 max	0.025 max	0.025 max	(a)

(a) If aluminum is used as the deoxidizing agent, the total aluminum content by product analysis is usually in excess of 0.010%

Characteristics and Typical Uses. ASTM A621 describes hot rolled carbon steel sheet and strip of drawing quality grade in cut lengths or coils. This material is intended for use in fabricating identified parts where drawing or severe forming may be involved, and surface appearance is not of primary importance.

ASTM A622 designates hot rolled carbon steel sheet and strip of drawing quality, special killed grade intended for fabricating identified parts where particularly severe drawing or forming may be involved, and surface appearance is usually not of primary importance.

Drawing quality sheet is manufactured from specially produced or selected steels which have been processed to have good uniform drawing properties for use in fabricating an identified part with severe deformations. Similarly, drawing quality, special killed sheet is manufactured from specially produced or selected killed steels, normally aluminum, that have been processed to have good uniform drawing properties for use in fabricating an identified part with extremely severe deformation.

Unless otherwise specified, the ASTM 621 or 622 material is furnished as-rolled, that is, without removing the hot rolled oxide or scale, and not oiled. When required, pickling or blast cleaning (descaling) of the material may be specified. Pickled or blast-cleaned material is furnished oiled, or when specified, dry.

The manufacturer assumes the responsibility for selection of steel, control of processing, and ability of the material to form specific parts within properly established breakage limits. Deformations of identified parts are assessed by the use of the scribed-square test as described in Appendix A5 of "ASTM Standard Specification A568".

Experience has shown that drawing quality steel is required if the percent increase in area of the most deformed 25-mm (1-in.) square, in a satisfactory untrimmed part, is more than 25% for material 2.09 to 4.75 mm (0.0822 to 0.187 in.) thick, or more than 30% for material larger than 4.75 mm (0.187 in.). Where the draw is particularly severe, or essential freedom from aging is required, drawing quality, special killed steel, ASTM A622, is recommended. The values stated in English units are to be regarded as the standard.

ASTM A621, A622: Similar Steels (U.S. and/or Foreign). None listed

ASTM A678

ASTM A678: Chemical Composition

Boron may be added only by agreement between the producer and purchaser

Grade	C max	Mn(a)	Mn(b)	P max	S max	Si	Cu min(c)
			Chemical composition, %				
A	0.16	0.90-1.50	...	0.040	0.050	0.15-0.50	0.20
B	0.20	0.70-1.35	1.00-1.60	0.040	0.050	0.15-0.50	0.20
C	0.22	1.00-1.60	1.00-1.60	0.040	0.050	0.20-0.50	0.20

(a) For thickness of 38 mm (1.5 in.) and smaller. (b) For thickness larger than 38 mm (1.5 in.) up to 64 mm (2.5 in.) inclusive. (c) When specified

ASTM A678: Similar Steels (U.S. and/or Foreign):

- UNS K01600; ASTM A678 (A)
- UNS K02002; ASTM A678 (B)
- UNS K02204; ASTM A678 (C)

Characteristics and Typical Uses. ASTM A678 describes quenched and tempered carbon steel plate of structural quality for welded, riveted, or bolted construction. The material under this specification is available in three grades with the following maximum thicknesses: grade A, 38 mm (1.5 in.); grade B, 63.4 mm (2.5 in.); grade C, 50.8 mm (2 in.). The values stated in English units are to be regarded as standard.

Special precautions should be followed when welding ASTM A678 material to produce sound welds and to minimize adverse effects on the properties of the plate, especially in the heat-affected zone.

Physical Properties

ASTM A678: Thermal Treatments. The material is treated by the manufacturer by heating to a temperature that produces an austenitic structure, but not to exceed 925 °C (1700 °F); holding a sufficient time to attain uniform heat throughout the material; quenching in a suitable medium; and tempering at not less than 595 °C (1100 °F).

Mechanical Properties

ASTM A678: Tensile Properties

For plates wider than 610 mm (24 in.), the test specimen is taken in the transverse direction, and the elongation requirement is reduced two percentage points

Grade	Tensile strength MPa	Tensile strength ksi	Yield strength(a), min MPa	Yield strength(a), min ksi	Elongation(b), min %
A	485-620	70-90	345	50	22
B	550-690	80-100	415	60	22
C to 19 mm (0.75 in.) incl	655-795	95-115	515	75	19
C larger than 19-38 mm (0.75-1.5 in.) incl	620-760	90-110	485	70	19
C larger than 38-51 mm (1.5 to 2.0 in.) incl	585-725	85-105	450	65	19

(a) Measured at 0.2% offset or 0.5% extension under load. (b) In 50 mm (2 in.). For thicknesses 19 mm (0.75 in.) and under, measured on 38-mm (1.5-in.) wide full-thickness rectangular specimen. The elongation is measured in a 50-mm (2-in.) gage length which includes the fracture and shows the greatest elongation

ASTM A699

ASTM A699: Chemical Composition

C max	Mn(b)	Mn(c)	P max	S max	Si max	Nb	Mo
			Chemical composition(a), %				
0.06	1.20-1.90	1.50-2.20	0.04	0.025	0.40	0.03-0.09	0.25-0.35

(a) Heat analysis. (b) For thickness of 16 mm (0.63 in.) and smaller. (c) For thickness larger than 16 mm (0.63 in.)

ASTM A699: Similar Steels (U.S. and/or Foreign). UNS K10614

Characteristics. ASTM A699 describes a low-carbon, manganese-molybdenum-niobium alloy steel. Four different grades of plate, shapes, and bar are included, providing various combinations of tensile strengths and notch toughness.

Classes 1 and 2 provide minimum yield strength levels of 485 MPa (70 ksi) and 515 MPa (75 ksi), respectively, with no specified notch toughness requirements. Classes 3 and 4 provide the same minimum yield strength levels together with longitudinal Charpy V-notch toughness requirements of 27 J (20 ft·lb) minimum at −46 °C (−50 °F). The values stated in English units are to be regarded as the standard.

Classes 2 and 4 are precipitation hardened to enhance mechanical properties. ASTM A699 steel may be welded, but the welding technique must not adversely affect the properties of the material, especially in the heat-affected zone.

This steel may be made by the open-hearth, basic-oxygen, or electric-furnace process. Further refinement by electroslag remelt (ESR) or vacuum-arc remelt (VAR) is permitted. ASTM A699 steel is made to either semikilled or killed steel practice.

Mechanical Properties

ASTM A699: Tensile Properties

Class	Tensile strength MPa	Tensile strength ksi	Yield strength(a), min MPa	Yield strength(a), min ksi	Elongation, % In 200 mm (8 in.)	Elongation, % In 50 mm (2 in.)
1	620-760	90-110	485	70	12	18
2	620-760	90-110	515	75	12	18
3	585-725	85-105	485	70	12	18
4	585-725	85-105	515	75	12	18

(a) 0.2% offset or 0.5% extension under load

Physical Properties

ASTM A699: Thermal Treatment. Classes 2 and 4 are precipitation heat treated in the temperature range from 540 to 650 °C (1000 to 1200 °F) for a time to be determined by the manufacturer.

ASTM A709

ASTM A709: Chemical Composition

Grade and classification	C	Mn	P max	S max
36; carbon steel shapes (a)(b)	0.26 max	...	0.04	0.05
36; carbon steel plate (b)(c)	0.25-0.29	0.80-1.20	0.04	0.05
36; carbon steel bar(b)	0.26-0.29	0.60-0.90	0.04	0.05
50; high-strength, low-alloy steel(c)(d)	0.23 max	1.35 max(e)	0.04	0.05
50W; high-strength, low-alloy steel(f)	0.20 max	1.35 max	0.04	0.05
100 and 100W; alloy steel	0.10-0.21	0.40-1.50	0.035	0.04

(a) Manganese content of 0.85 to 1.35% and silicon content of 0.15 to 0.40% is required for shapes larger than 634 kg/m (426 lb/ft). (b) Manganese content may be 1.35% max when notch toughness is specified. (c) Silicon content of 0.15 to 0.40% is required for plate larger than 38 mm (1.5 in.) in thickness. (d) For bar larger than 38 mm (1.5 in.) in diameter, thickness or distance between parallel faces is to be made by a killed steel practice. (e) Minimum requirement by heat analysis: 0.80% (0.75% product analysis) for plate thickness larger than 19 mm (0.75 in.); 0.50% (0.45% product analysis) for plate thickness of 9.5 mm (0.38 in.) and smaller, and for all other products. The manganese-to-carbon ratio should not be less than 2 to 1. (f) When normalized, the maximum ladle carbon may be 0.22% and the maximum manganese may be 1.40%

Characteristics and Typical Uses. ASTM A709 describes carbon and high-strength, low-alloy steel for structural shapes, plate, and bar as well as quenched and tempered alloy steel for structural plate, which are intended for use in bridges. Five grades are available in three strength levels, 36, 50, and 100. Grades 50W and 100W have enhanced atmospheric corrosion resistance.

The values stated in English units are to be regarded as the standard. ASTM A709 steels are weldable, but the welding procedures should be suitable for the steel being welded and the use intended.

ASTM A709 steels may be made by the open-hearth, basic-oxygen, or electric-furnace processes. Additional refining by electroslag remelting (ESR) or vacuum-arc remelting (VAR) is permitted. Grades 36 and 50 are not to be made from rimmed or capped steel. Grade 50W is to be made to a killed fine-grain practice. Grades 100 and 100W steel are killed fine grain (ASTM No. 5 or finer) as determined in accordance with "ASTM E112", specifically Plate IV.

ASTM A709: Similar Steels (U.S. and/or Foreign). None listed

Physical Properties

ASTM A709: Thermal Treatment. Grades 100 and 100W steels are heat treated by the manufacturer to conform to the tensile and hardness requirements by heating to not less than 900 °C (1650 °F); quenching in water or oil; and tempering at not less than 620 °C (1150 °F).

Mechanical Properties

ASTM A709: Tensile and Hardness Properties

For plates wider than 610 mm (24 in.), the test specimen is taken in the transverse direction, the elongation requirement is reduced two percentage points, and the reduction in area requirement, where applicable, is reduced five percentage points

Plate thickness	Structural shapes	Tensile strength MPa	Tensile strength ksi	Yield point/ strength(a), min MPa	Yield point/ strength(a), min ksi	Elongation, min %: Plate and bar In 200 mm (8 in.)	Plate and bar In 50 mm (2 in.)	Shapes In 200 mm (8 in.)	Shapes In 50 mm (2 in.)	Reduction in area, %	Hardness, HB
Grade 36											
To 200 mm (8 in.) incl	Up to 634 kg/m (426 lb/ft)	400-550	58-80	250	36	20	23	20	21	...	...
	Larger than 634 kg/m (426 lb/ft)	400 min	58 min	250	36	...	...	20	19	...	...
Larger than 200 mm (8 in.)	...	400 min	58 min	220	32	20	23	...	...	...	...
Grade 50											
To 50 mm (2 in.) incl	Groups 1, 2, 3, and 4	450 min	65 min	345	50	18	21	18	21(b)	...	...
Grade 50W											
To 100 mm (4 in.) incl	Groups 1, 2, 3, 4, and 5	485 min	70 min	345	50	18	21	18	21(c)	...	...
Grades 100 and 100W											
To 64 mm (2.5 in.) incl	...	760-895	110-130	690	100	...	18	...	...	40(d), 50(e)	228-269
Larger than 64 mm (2.5 in.) up to 100 mm (4 in.)	...	690-895	100-130	620	90	...	17	...	...	50(e)	212-269

(a) Measured at 0.2% offset or 0.5% extension under load. (b) 19% for shapes larger than 634 kg/m (426 lb/ft). (c) 18% for wide flange shapes larger than 634 kg/m (426 lb/ft). (d) Measured on 38-mm (1.5-in.) wide, full thickness rectangular, standard ASTM specimen. This is the only specimen to be used for thickness of 19 mm (0.75 in.) and smaller. The reduction in area is measured in a 50 mm (2 in.) gage length that includes the fracture and shows the greatest elongation. (e) Measured on a 13 mm (0.5 in.) diameter standard ASTM specimen. This is the only specimen to be used for thicknesses larger than 38 mm (1.5 in.)

ASTM A710

ASTM A710: Chemical Composition

Grade	C	Mn	P max	S max	Si	Ni	Cr	Mo	Cu	Nb min
	Chemical composition, %									
A	0.07 max	0.40-0.70	0.025	0.025	0.40 max	0.70-1.00	0.60-0.90	0.15-0.25	1.00-1.30	0.02
B	0.06 max	0.40-0.65	0.025	0.025	0.15-0.40	1.20-1.50	...	...	1.00-1.30	0.02

Characteristics and Typical Uses. ASTM A710 describes low-carbon, age-hardening, nickel-copper-molybdenum-niobium, and nickel-copper-molybdenum alloy steel plate, shapes, and bar for general applications. Two different grades and three different conditions are provided.

Grade A provides minimum yield strength levels ranging from 380 to 585 MPa (55 to 85 ksi), depending on thickness and condition. Grade B provides minimum yield strength levels ranging from 515 to 585 MPa (75 to 85 ksi), depending on thickness.

Grade A, class 1 plates have a maximum thickness of 19 mm (0.75 in.). Grade A, classes 2 and 3 are normally limited to a maximum of 100 mm (4 in.). This thickness depends on the capacity of the composition to meet the specified mechanical properties. Bar and shapes are available only as grade A. The values stated in English units are to be regarded as the standard.

ASTM A710 materials are weldable, but the welding technique must not adversely affect the properties of the material, especially in the heat-affected zone.

ASTM A710: Similar Steels (U.S. and/or Foreign):

- UNS K20622; ASTM A710 (B)
- UNS K20747; ASTM A710 (A)

Physical Properties

ASTM A710: Thermal Treatment. Grade A, class 1, and grade B materials are precipitation heat treated at 540 to 650 °C (1000 to 1200 °F) for a time determined by the manufacturer. Grade A, class 2 material is normalized at 870 to 930 °C (1600 to 1700 °F), then precipitation heat treated at 540 to 650 °C (1000 to 1200 °F) for a time determined by the manufacturer. Grade A, class 3 material is water or oil quenched from 870 to 925 °C (1600 to 1700 °F), then precipitation heat treated from 540 to 650 °C (1000 to 1200 °F) for a time determined by the manufacturer.

Mechanical Properties

ASTM A710: Tensile Properties

Thickness	Tensile strength, min		Yield strength(a), min		Elongation(b),
	MPa	ksi	MPa	ksi	%
Grade A, class 1					
Smaller than 6.4-8.0 mm (0.25-0.3 in.) incl.	620	90	585	85	20
Larger than 8-19 mm (0.3-0.75 in.) incl.	620	90	550	80	20
Grade A, class 2					
Smaller than 6.4-25 mm (0.25-1.0 in.) incl	495	72	450	65	20
Larger than 25-50 mm (1-2 in.) incl	495	72	415	60	20
Larger than 50 mm (2 in.)	450	65	380	55	20
Grade A, class 3					
Smaller than 6.4-50 mm (0.25-2 in.) incl.	585	85	515	75	20
Larger than 50 mm (2.0 in.)	515	75	450	65	20
Grade B					
6.4 mm (0.25 in.) and smaller	620	90	585	85	...
Larger than 6.4-9.5 mm (0.25-0.38 in.) incl	620	90	565	82	...
Larger than 9.5-13 mm (0.38-0.50 in.) incl	620	90	550	80	...
Larger than 13-19 mm (0.5-0.75 in.) incl.	605	88	515	75	...

(a) 0.2% offset or 0.5% extension under load. (b) In 50 mm (2 in.)

ASTM A710: Charpy V-notch Impact Properties

Grade	Test temperature		Impact energy, min	
	°C	°F	J	ft·lb
A, class 2	−45	−50	68	50
A, class 3	−62	−80	68	50
B, longitudinal	−35	−32	27	20
B, transverse	−32	−25	20	15

Cross Reference to Steels

FRANCE

Designation	AISI	Page
AFNOR		
20 MC 5	5120	167
20 MC 5	5120H	167
20 NCD 2	8617	188
20 NCD 2	8617H	188
20 NCD 2	8620	189
20 NCD 2	8620H	189
22 NCD 2	8617	188
22 NCD 2	8617H	188
22 NCD 2	8620	189
22 NCD 2	8620H	189
25 CD 4 (S)	4130	108
25 CD 4 (S)	4130H	108
32 C 4	5130H	168
32 C 4	5132	170
32 DCV 28	H10	457
35 CD 4	4135	110
35 CD 4	4135H	110
35 CD 4 TS	4135	110
35 CD 4 TS	4135H	110
35 M 5	1039	26
38 C 4	5132H	170
38 C 4	5135	171
40 CD 4	4137	112
40 CD 4	4137H	112
40 CD 4	4140	113
40 CD 4	4140H	113
40 M 5	1335	81
40 M 5	1335H	81
42 C 2	5140H	172
42 C 2	5150	175
42 C 4	5135H	171
42 C 4	5140	172
42 CD 4	4137	112
42 CD 4	4137H	112
42 CD 4	4140	113
42 CD 4	4140H	113
45 C 2	5140H	172
45 C 2	5150	175
50 CV 4	6150	182
50 CV 4	6150H	182
55 C 3	5155	177
55 C 3	5155H	177
55 S 7	9255	220
55 WC 20	S1	436
60 S 7	9260	222
60 S 7	9260H	222
61 SC 7	9260	222
61 SC 7	9260H	222
90 MV 8	O2	440
100 C 6	E52100	154
CC 20	1020	16
CC 35	1035	21
CC 55	1060	34
XC 10	1010	10
XC 15	1015	11
XC 15	1017	13
XC 18	1015	11
XC 18	1017	13
XC 18 S	1023	20
XC 25	1023	20
XC 32	1034	21
XC 35	1034	21
XC 38	1034	21

FRANCE (continued)

Designation	AISI	Page
AFNOR		
XC 38 TS	1038	26
XC 38 TS	1038H	26
XC 42	1045	30
XC 42	1045H	30
XC 42 TS	1045	30
XC 42 TS	1045H	30
XC 45	1045	30
XC 45	1045H	30
XC 48	1045	30
XC 48	1045H	30
XC 60	1064	40
XC 65	1064	40
XC 68	1070	40
XC 90	1086	488
Z 2 CND 17.12	316L	296
Z 2 CND 19.15	317L	303
Z 6 CA 13	405	324
Z 6 CN 18.09	304	273
Z 6 CND 17.11	316	292
Z 6 CNN6 18.10	347	311
Z 6 CNT 18.10	321	304
Z 6 CNU 17.04	431	356
Z 8 C 17	430	353
Z 8 CD 17.01	434	353
Z 10 C 13	410	328
Z 10 C 14	410	328
Z 10 CF 17	430F	349
Z 10 CNF 18.09	303	265
Z 12 C 13	410	328
Z 12 C 13 M	403	317
Z 12 CN 17.08	301	253
Z 12 CNS 25.20	310	288
Z 12 CNS 25.20	314	291
Z 15 CN 16.02	431	356
Z 15 CN 24.13	309S	285
Z 18 N 5	A2515	94
Z 20 C 13	420	344
Z 30 WCV 9	H21	457
Z 38 CDV 5	H11	457
Z 40 COV 5	H13	457
Z 80 WCV 18-04-01	T1	463
Z 80 WKCV 18-05-04-01	T4	463
Z 85 DCWV 08-04-02-01	H41	457
Z 85 DCWV 08-04-02-01	M1	465
Z 85 WDCV 06-05-04-02	M2	465
Z 90 WDCV 06-05-04-02	M3 (Class 1)	465
Z 100 CDV 5	A2	445
Z 110 WKCDV 07-05-04-04-02	M41	465
Z 110 WKCDV 07-05-04-04-02	M42	465
Z 120 WDCV 06-05-04-03	M3 (Class 2)	465
Z 130 WDCV 06-05-04-04	M3 (Class 2)	465
Z 200 C 12	D3	449

GERMANY (Federal Republic of)

Designation	AISI	Page
DIN		
1.0204	1008	6
1.0402	1020	16
1.0419	1016	13
1.0501	1035	21
1.0601	1060	34
1.0700	1108	45
1.0702	1109	45
1.0711	1212	62
1.0715	1213	62
1.0718	12L13	65
1.0718	12L14	65
1.0904	9255	220
1.0909	9260	222
1.0909	9260H	222
1.0912	1345	87
1.0912	1345H	87
1.1121	1010	10
1.1133	1022	18
1.1141	1015	11
1.1141	1017	13
1.1151	1023	20
1.1157	1039	26
1.1158	1025	20
1.1165	1330	76
1.1165	1330H	76
1.1167	1335	81
1.1167	1335H	81
1.1172	1030	21
1.1176	1038	26
1.1176	1038H	26
1.1181	1034	21
1.1186	1040	26
1.1191	1045	30
1.1191	1045H	30
1.1209	1055	34
1.1210	1050	31
1.1221	1064	40
1.1226	1548	69
1.1230	1065	40
1.1231	1070	40
1.1269	1086	488
1.1273	1090	43
1.1274	1095	43
1.2080	D3	449
1.2330	P20	453
1.2341	P4	453
1.2343	H11	457
1.2344	H13	457
1.2363	A2	445
1.2365	H10	457
1.2379	D2	449
1.2510	O1	440
1.2550	S1	436
1.2581	H21	457
1.2606	H12	457
1.2625	H23	457
1.2735	P6	453
1.2842	O2	440
1.3202	T15	463
1.3246	M41	465
1.3246	M42	465
1.3249	M33	465
1.3249	M34	465

GERMANY (continued)

DIN

Designation	AISI	Page
1.3255	T4	463
1.3265	T5	463
1.3342	M3 (Class 1)	465
1.3343	M2	465
1.3344	M3 (Class 2)	465
1.3346	H41	457
1.3346	M1	465
1.3348	M7	465
1.3355	T1	463
1.3501	E50100	154
1.3503	E51100	154
1.3505	E52100	154
1.4001	410S	329
1.4002	405	324
1.4005	416	337
1.4006	410	328
1.4016	430	353
1.4021	420	344
1.4024	403	317
1.4057	431	356
1.4104	430F	349
1.4112	440B	363
1.4113	434	353
1.4125	440C	364
1.4301	304	273
1.4303	305	281
1.4303	308	283
1.4305	303	265
1.4306	304L	277
1.4310	301	253
1.4401	316	292
1.4404	316L	296
1.4438	317L	303
1.4449	317	301
1.4510	430Ti	352
1.4512	409	327
1.4532	632	384
1.4541	321	304
1.4546	348	311
1.4550	347	311
1.4568	631	382
1.4828	309	285
1.4833	309S	285
1.4841	310	288
1.4841	314	291
1.4935	422	347
1.4971	661	398
1.4980	660	396
1.5069	1340H	83
1.5419	4419	137
1.5419	4419H	137
1.5419	4422	138
1.5680	A2515	94
1.5711	3140	95
1.6511	9840	231
1.6523	8617	188
1.6523	8617H	188
1.6523	8620	189
1.6523	8620H	189
1.6543	8622	193
1.6543	8622H	193
1.6543	8720	215
1.6543	8720H	215
1.6543	8822	218
1.6543	8822H	218
1.6545	8630	197
1.6545	8630H	197
1.6546	8640	202
1.6546	8640H	202
1.6546	8740	216
1.6546	8740H	216
1.6562	E4340	130
1.6562	E4340H	130
1.6565	4340	130
1.6565	4340H	130

GERMANY (continued)

DIN

Designation	AISI	Page
1.6755	4718	146
1.6755	4718H	146
1.7006	5140H	172
1.7006	5150	175
1.7007	50B40	160
1.7007	50B40H	160
1.7030	5130	168
1.7033	5130H	168
1.7033	5132	170
1.7034	5132H	170
1.7034	5135	171
1.7035	5135H	171
1.7035	5140	172
1.7138	50B50	164
1.7138	50B50H	164
1.7147	5120	167
1.7147	5120H	167
1.7176	5155	177
1.7176	5155H	177
1.7218	4130	108
1.7218	4130H	108
1.7220	4135	110
1.7220	4135H	110
1.7223	4142H	117
1.7225	4137	112
1.7225	4137H	112
1.7225	4140	113
1.7225	4140H	113
1.7228	4147	119
1.7228	4147H	119
1.7228	4150	120
1.7228	4150H	120
1.7362	501	371
1.7511	6118	181
1.7511	6118H	181
1.8159	6150	182
1.8159	6150H	182

ITALY

UNI

Designation	AISI	Page
9 SMn 23	1213	62
9 SMnPb 23	12L13	65
9 SMnPb 23	12L14	65
10 S 20	1212	62
20 NiCrMo	8617H	188
20 NiCrMo 2	8617	188
20 NiCrMo 2	8620	189
20 NiCrMo 2	8620H	189
25 CrMo 4	4130	108
25 CrMo 4	4130H	108
25 CrMo 4 KB	4130	108
25 CrMo 4 KB	4130H	108
30 NiCrMo 2 KB	8630	197
30 NiCrMo 2 KB	8630H	197
34 Cr 4 KB	5130H	168
34 Cr 4 KB	5132	170
34 CrMo 4 KB	4135	110
34 CrMo 4 KB	4135H	110
35 CrMo 4	4135	110
35 CrMo 4	4135H	110
35 CrMo 4 F	4135	110
35 CrMo 4 F	4135H	110
38 Cr 4 KB	5132H	170
38 Cr 4 KB	5135	171
38 CrB 1 KB	50B40	160
38 CrB 1 KB	50B40H	160

ITALY (continued)

UNI

Designation	AISI	Page
38 CrMo 4	4142H	117
38 CrMo 4 KB	4137	112
38 CrMo 4 KB	4137H	112
38 CrMo 4 KB	4140	113
38 CrMo 4 KB	4140H	113
38 NiCrMo 4	9840	231
40 Cr 4	5135H	171
40 Cr 4	5140	172
40 CrMo 4	4137	112
40 CrMo 4	4137H	112
40 CrMo 4	4140	113
40 CrMo 4	4140H	113
40 NiCrMo 2 KB	8640	202
40 NiCrMo 2 KB	8640H	202
40 NiCrMo 2 KB	8740	216
40 NiCrMo 2 KB	8740H	216
40 NiCrMo 7	E4340	130
40 NiCrMo 7	E4340H	130
40 NiCrMo 7 KB	E4340	130
40 NiCrMo 7 KB	E4340H	130
41 Cr 4 KB	5135H	171
41 Cr 4 KB	5140	172
50 CrV 4	6150	182
50 CrV 4	6150H	182
55 Si 8	9255	220
58 WCr 9 KU	S1	436
88 MnV 8 KU	O2	440
100 Cr 6	E52100	154
C 20	1020	16
C 35	1035	21
C 60	1060	34
CB 10 FU	1008	6
CB 35	1030	21
G 22 Mn 3	1022	18
G 22 Mo 5	4419	137
G 22 Mo 5	4419H	137
G 22 Mo 5	4422	138
G 40 CrMo 4	4137	112
G 40 CrMo 4	4137H	112
G 40 CrMo 4	4140	113
G 40 CrMo 4	4140H	113
ICL 472 T	321	304
X 2 CrNi 18 11	304L	277
X 2 CrNi 18 11 KG	304L	277
X 2 CrNi 18 11 KW	304L	277
X 2 CrNiMo 17 12	316L	296
X 3 CrNi 18 11	304L	277
X 5 CrNi 18 10	304	273
X 5 CrNiMo 17 12	316	292
X 5 CrNiMo 18 15	317	301
X 6 CrAl 13	405	324
X 6 CrNi 23 14	309S	285
X 6 CrNiTi 18 11	321	304
X 6 CrNiTi 18 11 KG	321	304
X 6 CrNiTi 18 11 KT	321	304
X 6 CrNiTi 18 11 KW	321	304
X 8 Cr 17	430	353
X 8 CrMo 17	434	353
X 8 CrNi 19 10	305	281
8 CrNi 19 10	308	283
X 8 CrNiNb 18 11	347	311
X 10 CrNiS 18 09	303	265
X 10 CrS 17	430F	349
X 12 Cr 13	410	328
X 12 CrNi 17 07	301	253
X 12 CrS 13	416	337
X 16 CrNi 16	431	356
X 16 CrNi 23 14	309	285
X 16 CrNiSi 25 20	310	288
X 16 CrNiSi 25 20	314	291

ITALY (continued)

Designation	AISI	Page
UNI		
X 20 Cr 13	420	344
X 22 CrNi 25 20	310	288
X 22 CrNi 25 20	314	291
X 28 W 09 KU	H21	457
X 35 CrMo 05 KU	H11	457
X 35 CrMoV 05 KU	H13	457
X 35 CrMoW 05 KU	H12	457
X 75 W 18 KU	T1	463
X 78 WCo 1805 KU	T4	463
X 80 WCo 1810 KU	T5	463
X 82 MoW 09 KU	H41	457
X 82 MoW 09 KU	M1	465
X 82 WMo 0605 KU	M2	465
X 150 CrMo 12 KU	D2	449
X 150 WCoV 130505 KU	T15	463
X 210 Cr 13 KU	D3	449

JAPAN

Designation	AISI	Page
JIS		
S 9 CK	1010	10
S 10 C	1010	10
S 12 C	1010	10
S 15 C	1015	11
S 15 C	1017	13
S 15 CK	1015	11
S 15 CK	1017	13
S 17 C	1015	11
S 17 C	1017	13
S 20 C	1023	20
S 20 CK	1023	20
S 22 C	1023	20
S 25 C	1025	20
S 28 C	1025	20
S 38 C	1034	21
S 40 C	1040	26
S 45 C	1045	30
S 45 C	1045H	30
S 48 C	1045	30
S 48 C	1045H	30
S 53 C	1050	31
S 55 C	1050	31
SCCrM 1	4130	108
SCCrM 1	4130H	108
SCCrM 3	4135	110
SCCrM 3	4135H	110
SCM 1	4135	110
SCM 1	4135H	110
SCM 2	4130	108
SCM 2	4130H	108
SCM 4	4137	112
SCM 4	4137H	112
SCM 4	4140	113
SCM 4	4140H	113
SCM 4 H	4137	112
SCM 4 H	4137H	112
SCM 4 H	4140	113
SCM 4 H	4140H	113
SCM 5	4147	119
SCM 5	4147H	119

JAPAN (continued)

Designation	AISI	Page
JIS		
SCM 5	4150	120
SCM 5	4150H	120
SCM 5 H	4147	119
SCM 5 H	4147H	119
SCM 5 H	4150	120
SCM 5 H	4150H	120
SCMn 2	1330	76
SCMn 2	1330H	76
SCMn 3	1335	81
SCMn 3	1335H	81
SCPH 11	4419	137
SCPH 11	4419H	137
SCPH 11	4422	138
SCr 2	5130H	168
SCr 2	5132	170
SCr 2 H	5130H	168
SCr 2 H	5132	170
SCr 3 H	5132H	170
SCr 3 H	5135	171
SCr 4 H	5135H	171
SCr 4 H	5140	172
SCS 19	304L	277
SKD 1	D3	449
SKD 5	H21	457
SKD 6	H11	457
SKD 12	A2	445
SKD 61	H13	457
SKD 62	H12	457
SKH 2	T1	463
SKH 3	T4	463
SKH 4A	T5	463
SKH 9	M2	465
SKH 52	M3 (Class 2)	465
SKH 53	M3 (Class 2)	465
SMn 1 H	1330	76
SMn 1 H	1330H	76
SMn 2	1335	81
SMn 2	1335H	81
SMn 2 H	1335	81
SMn 2 H	1335H	81
SMnC 21	1022	18
SNCM 8	4340	130
SNCM 8	4340H	130
SNCM 21	8617	188
SNCM 21	8617H	188
SNCM 21	8620	189
SNCM 21	8620H	189
SNCM 21 H	8617	188
SNCM 21 H	8617H	188
SNCM 21 H	8620	189
SNCM 21 H	8620H	189
SUH 309	316	292
SUH 310	316L	296
SUH 409	409	327
SUH 616	422	347
SUM 11	1109	45
SUM 12	1109	45
SUM 21	1212	62
SUM 22	1213	62
SUM 22 L	12L13	65
SUM 22 L	12L14	65
SUM 23 L	12L13	65
SUM 24 L	12L13	65
SUM 24 L	12L14	65
SUP 4	1095	43
SUP 10	6150	182
SUP 10	6150H	182
SUP 11	50B50	164
SUP 11	50B50H	164
SUS 301	301	253
SUS 303	303	265
SUS 304	304	273
SUS 304 L	304L	277
SUS 305	305	281
SUS 305	308	283

JAPAN (continued)

Designation	AISI	Page
JIS		
SUS 305 J1	305	281
SUS 305 J1	308	283
SUS 316	316	292
SUS 316 L	316L	296
SUS 317	317	301
SUS 321	321	304
SUS 347	347	311
SUS 403	403	317
SUS 405	405	324
SUS 410	410	328
SUS 410 S	410S	329
SUS 416	403	317
SUS 420 J1	420	344
SUS 430	430	353
SUS 430	430F	349
SUS 431	431	356
SUS 434	434	353
SUS 440 B	440B	363
SUS 440 C	440C	364
SUS Y 310	310	288
SUS Y 310	314	291
SUS Y 316	316	292

SWEDEN

Designation	AISI	Page
SS_{14}		
1370	1015	11
1370	1017	13
1450	1020	16
1550	1035	21
1665	1064	40
1672	1045	30
1672	1045H	30
1678	1064	40
1770	1070	40
1778	1070	40
1870	1095	43
1914	12L13	65
1914	12L14	65
2090	9255	220
2120	1335	81
2120	1335H	81
2225	4130	108
2225	4130H	108
2230	6150	182
2230	6150H	182
2234	4135	110
2234	4135H	110
2242	H13	457
2244	4137	112
2244	4137H	112
2244	4140	113
2244	4140H	113
2260	A2	445
2302	410	328
2303	420	344
2320	430	353
2325	434	353
2332	304	273
2337	321	304
2338	347	311
2346	303	265
2347	316	292
2348	316L	296
2352	304L	277
2367	317L	303
2383	430F	349
2722	M2	465

UNITED KINGDOM

Designation	AISI	Page
B.S.		
040 A 20	1020	16
060 A 35	1035	21
060 A 62	1060	34
060 A 96	1095	43
070 M 20	1020	16
080 A 32	1035	21
080 A 35	1035	21
080 A 37	1035	21
080 A 40	1040	26
080 M 36	1035	21
2 S 93	1040	26
2 S 117	5135H	171
2 S 117	5140	172
2 S 119	4340	130
2 S 119	4340H	130
2 S 130	348	311
2 S 516	1345	87
2 S 516	1345H	87
2 S 517	1345	87
2 S 517	1345H	87
3 S 95	4340	130
3 S 95	4340H	130
5 S 80	431	356
120 M 36	1039	26
150 M 36	1039	26
220 MO7	1213	62
250 A 53	9255	220
250 A 58	9260	222
250 A 58	9260H	222
302 S 17	304	273
303 S 21	303	265
304 S 12	304L	277
304 S 14	304L	277
304 S 15	304	273
304 S 16	304	273
304 S 18	304	273
304 S 22	304L	277
304 S 25	304	273
304 S 40	304	273
310 S 24	310	288
310 S 24	314	291
316 S 12	316L	296
316 S 14	316L	296
316 S 16	316	292
316 S 18	316	292
316 S 22	316L	296
316 S 24	316L	296
316 S 25	316	292
316 S 26	316	292
316 S 29	316L	296
316 S 30	316	292
316 S 30	316L	296
316 S 31	316L	296
316 S 37	316L	296
316 S 40	316	292
316 S 41	316	292
316 S 82	316L	296
317 S 12	317L	303
321 S 12	321	304
321 S 18	321	304
321 S 22	321	304
321 S 27	321	304
321 S 40	321	304
321 S 49	321	304
321 S 50	321	304
321 S 59	321	304
321 S 87	321	304
347 S 17	347	311
347 S 17	348	311
347 S 18	348	311
347 S 40	348	311
403 S 17	410S	329
405 S 17	405	324
409 S 17	409	327
410 S 21	410	328

UNITED KINGDOM (continued)

Designation	AISI	Page
B.S.		
416 S 21	416	337
420 S 29	403	317
420 S 37	420	344
430 S 15	430	353
431 S 29	431	356
434 S 19	434	353
530 A 30	5130	168
530 A 32	5130H	168
530 A 32	5132	170
530 A 36	5132H	170
530 A 36	5135	171
530 A 40	5135H	171
530 A 40	5140	172
530 H 30	5130	168
530 H 32	5130H	168
530 H 32	5132	170
530 H 36	5132H	170
530 H 36	5135	171
530 H 40	5135H	171
530 H 40	5140	172
530 M 40	5135H	171
530 M 40	5140	172
534 A 99	E52100	154
535 A 99	E52100	154
640 M 40	3140	95
708 A 37	4135	110
708 A 37	4135H	110
708 A 42	4137	112
708 A 42	4137H	112
708 A 42	4140	113
708 A 42	4140H	113
708 M 40	4137	112
708 M 40	4137H	112
708 M 40	4140	113
708 M 40	4140H	113
709 A 40	4137H	112
709 M 40	4137	112
709 M 40	4140	113
709 M 40	4140H	113
735 A 50	6150	182
735 A 50	6150H	182
805 A 20	8622	193
805 A 20	8622H	193
805 A 20	8720	215
805 A 20	8720H	215
805 A 20	8822	218
805 A 20	8822H	218
805 H 20	8617	188
805 H 20	8617H	188
805 H 20	8620	189
805 H 20	8620H	189
805 M 20	8617	188
805 M 20	8617H	188
805 M 20	8620	189
805 M 20	8620H	189
816 M 40	9840	231
817 M 40	4340	130
817 M 40	4340H	130
3111 Type 6	4340	130
3111 Type 6	4340H	130
ANC 1 Grade A	410	328
ANC 3 Grade B	347	311
BA 2	A2	445
BD 2	D2	449
BD 3	D3	449
BH 11	H11	457
BH 12	H12	457
BH 13	H13	457
BH 21	H21	457
BM 1	H41	457
BM 1	M1	465
BM 2	M2	465
BM 34	M33	465
BM 34	M34	465
BO 1	O1	440

UNITED KINGDOM (continued)

Designation	AISI	Page
B.S.		
BO 2	O2	440
BT 1	T1	463
BT 4	T4	463
BT 5	T5	463
BT 15	T15	463
CDS-18	420	344
CDS-20	321	304
CDS 105/106	1039	26
CDS 110	4130	108
CDS 110	4130H	108
En. 44 B	1095	43
En. 47	6150	182
En. 47	6150H	182
En. 56 A	410	328
En. 56 B	403	317
En. 58 B	321	304
En. 58 C	321	304
En. 58 E	304	273
En. 58 F	347	311
En. 58 G	347	311
En. 58 H	316	292
S. 139	E4340	130
S. 139	E4340H	130
S. 525	348	311
S. 527	348	311
S. 536	304L	277
S. 537	316L	296
Type 3	5132H	170
Type 3	5135	171
Type 7	8640	202
Type 7	8640H	202
Type 7	8740	216
Type 7	8740H	216
Type 8	E4340	130
Type 8	E4340H	130

UNITED STATES

Designation	Page
AISI	
201	246
202	251
301	253
302	258
302B	262
303	265
303MA	271
303Se	265
304	273
304HN	280
304L	277
304N	280
305	281
308	283
309	285
309S	285
309S(Cb)	285
310	288
310S	288
314	291
316	292
316F	299
316L	296
316N	300
317	301
317L	303
321	304
329	308
330	309

UNITED STATES (continued)

UNITED STATES (continued)

UNITED STATES (continued)

UNITED STATES (continued)

Designation	Page
AISI	
86B45H	206
8650	209
8650H	209
86B50	209
8655	212
8655H	212
8660	213
8660H	213
8720	215
8720H	215
8740	216
8740H	216
8822	218
8822H	218
9255	220
9260	222
9260H	222
94B15	226
94B15H	226
94B17	227
94B17H	227
94B30	229
94B30H	229
94B40	230
9840	231
A2	445
A3	445
A4	445
A5	445
A6	445
A7	445
A8	445
A9	445
A10	445
A2317	88
A2515	94
D2	449
D3	449
D4	449
D5	449
D7	449
E4340	130
E4340H	130
E9310	223
E9310H	223
E50100	154
E51100	154
E52100	154
H10	457
H11	457
H12	457
H13	457
H14	457
H19	457
H21	457
H22	457
H23	457
H24	457
H25	457
H26	457
H41	457
H42	457
H43	457
M1	465
M2	465
M3 (Class 1)	465
M3 (Class 2)	465
M4	465
M6	465
M7	465
M10	465
M30	465
M33	465

UNITED STATES (continued)

Designation	AISI	Page
AISI		
M34		465
M36		465
M41		465
M42		465
M43		465
M44		465
M46		465
M47		465
O1		440
O2		440
O6		440
O7		440
P2		453
P3		453
P4		453
P5		453
P6		453
P20		453
P21		453
S1		436
S2		436
S4		436
S5		436
S6		436
S7		436
S43035		352
T1		463
T2		463
T4		463
T5		463
T6		463
T8		463
T15		463
W1		432
W2		432
W5		432
AMS		
5010 D	1212	62
5024 C	1137	56
5032	1020	16
5040	1010	10
5042	1010	10
5044	1010	10
5045	1020	16
5047	1010	10
5053	1010	10
5060	1015	11
5069	1018	13
5070	1022	18
5075	1025	20
5077	1025	20
5080	1035	21
5082	1035	21
5085	1050	31
5110	1080	41
5112	1090	43
5112 E	1086	488
5115	1070	40
5120 D	1074	40
5121	1095	43
5122	1095	43
5132	1095	43
5331	4340	130
5333	8615	187
5342	630	375
5343	630	375
5344	630	375
5354	615	373
5355	630	375
5369	651	391
5376	661	398
5501	304	273

UNITED STATES (continued)

Designation	AISI	Page
AMS		
5502	501	371
5503	430	353
5504	410	328
5505	410	328
5506	420	344
5507	316L	296
5508	615	373
5510	321	304
5511	304L	277
5512	347	311
5513	304	273
5514	305	281
5515	302	258
5516	302	258
5517	301	253
5518	301	253
5519	301	253
5520	632	384
5521	310S	288
5522	314	291
5523	309S	285
5524	316	292
5525	660	396
5526	651	391
5527	651	391
5528	631	382
5529	631	382
5531	661	398
5532	661	398
5546	633	386
5547	634	388
5548	633	386
5549	634	388
5554	633	386
5556	347	311
5557	321	304
5558	347	311
5559	321	304
5560	304	273
5561		415
5565	304	273
5566	304	273
5567	304	273
5568	631	382
5570	321	304
5571	347	311
5572	310S	288
5573	316	292
5574	309S	285
5575	347	311
5576	321	304
5577	310S	288
5578		426
5579	651	391
5585	661	398
5591	410	328
5592	330	309
5594	634	388
5595		415
5601		403
5602	501	371
5603		403
5604	630	375
5610	416Se	337
5611	403	317
5612	403	317
5613	410	328
5615	414	332
5615	615	373
5617		426
5618	440C	364
5620	420F	346
5620	420F(Se)	346
5621	420	344
5622	630	375

UNITED STATES (continued)

AMS

Designation	AISI	Page
5626	T1	463
5627	430	353
5628	431	356
5629		402
5630	440C	364
5631	440A	358
5632 (Type 1)	440F	366
5632 (Type 2)	440F(Se)	366
5636	302	258
5637	302	258
5639	304	273
5640 (Type 1)	303	265
5640 (Type 2)	303Se	265
5641	303Se	265
5643	630	375
5644	631	382
5645	321	304
5646	347	311
5647	304L	277
5648	316	292
5649	316F	299
5650	309S	285
5651	310S	288
5652	314	291
5653	316L	296
5654	347	311
5655	422	347
5656		415
5657	632	384
5658		405
5659		405
5672		426
5673	631	382
5674	347	311
5678	631	382
5680	347	311
5681	347	311
5685	305	281
5686	305	281
5688	302	258
5689	321	304
5690	316	292
5691	316	292
5694	310	288
5695	310	288
5697	304	273
5716	330	309
5720	651	391
5721	651	391
5722	651	391
5731	660	396
5732	660	396
5734	660	396
5735	660	396
5736	660	396
5737	660	396
5738	303Se	265
5742	634	388
5743	634	388
5744	634	388
5745	633	386
5763		424
5764		407
5768	661	398
5769	661	398
5774	633	386
5775	633	386
5776	410	328
5780	634	388
5781	634	388
5794	661	398
5795	661	398
5804	660	396
5805	660	396

UNITED STATES (continued)

AMS

Designation	AISI	Page
5812	632	384
5813	632	384
5817	615	373
5821	410	328
5824	631	382
5825	630	375
5840		402
5860		426
5861		407
5862		405
6260 F	E9310	223
6265 B	E9310	223
6272	8617	188
6274	8620	189
6275	94B17	227
6275 A	94B15	226
6276	8620	189
6277	8620	189
6280	8630	197
6281	8630	197
6290	4615	140
6294	4620	141
6322	8740	216
6323	8740	216
6325	8740	216
6327	8740	216
6342 C	9840	231
6350	4130	108
6355	8630	197
6356	4130	108
6358	8740	216
6359	4340	130
6360	4130	108
6361	4130	108
6362	4130	108
6365 C	4135	110
6370	4130	108
6371	4130	108
6372 C	4135	110
6373	4130	108
6381	4140	113
6382	4140	113
6386 (Type 1)		497
6386 (Type 3)		497
6386 (Type 4)		497
6386 (Type 5)		497
6390	4140	113
6395	4140	113
6414	4340	130
6415	4340	130
6437	H11	457
6440	E52100	154
6441	E52100	154
6442 B	E50100	154
6443	E51100	154
6444	E52100	154
6446	E51100	154
6447	E52100	154
6448	6150	182
6449	E51100	154
6450	6150	182
6455	6150	182
6466	502	371
6467	502	371
6485	H11	457
6487	H11	457
6488	H11	457
6530	8630	197
6550	8630	197
7240	1060	34
7301	6150	182
7304	1095	43
7470	615	373

UNITED STATES (continued)

ANSI

Designation	AISI	Page
G81.40	661	398

ASME

Designation	AISI	Page
5041	1006	6
Code Case 1817		413
SA182	304	273
SA182	304L	277
SA182	304N	280
SA182	310	288
SA182	316	292
SA182	316L	296
SA182	316N	300
SA182	321	304
SA182	347	311
SA182	348	311
SA182	430	353
SA182 (XM-19)		407
SA193	305	281
SA193	316	292
SA193	321	304
SA193	347	311
SA194	303	265
SA194	303Se	265
SA194	305	281
SA194	316	292
SA194	321	304
SA194	347	311
SA194	416	337
SA194	416Se	337
SA194 (Type 3)	501	371
SA194 (Type 6)	410	328
SA194 (Type 8)	304	273
SA213	304	273
SA213	304L	277
SA213	304N	280
SA213	310	288
SA213	316	292
SA213	316L	296
SA213	316N	300
SA213	321	304
SA213	347	311
SA213	348	311
SA240	302	258
SA240	304	273
SA240	304L	277
SA240	304N	280
SA240	305	281
SA240	309S	285
SA240	310S	288
SA240	316	292
SA240	316L	296
SA240	316N	300
SA240	317	301
SA240	317L	303
SA240	321	304
SA240	347	311
SA240	348	311
SA240	405	324
SA240	410	328
SA240	410S	329
SA240	430	353
SA240 (XM-8)	S43035	352
SA240 (XM-19)		407
SA240 (XM-21)	304HN	280
SA240 (XM-27)		422
SA240 (XM-29)		418
SA249	304	273
SA249	304L	277
SA249	304N	280
SA249	309	285
SA249	310	288
SA249	316	292
SA249	316L	296

UNITED STATES (continued)

Designation	AISI	Page
ASME		
SA249	316N	300
SA249	317	301
SA249	321	304
SA249	347	311
SA249	348	311
SA249 (XM-19)		407
SA249 (XM-29)		418
SA268	329	308
SA268	405	324
SA268	409	327
SA268	410	328
SA268	430	353
SA268	446	369
SA268 (XM-8)	S43035	352
SA268 (XM-27)		422
SA312	304	273
SA312	304L	277
SA312	304N	280
SA312	309	285
SA312	310	288
SA312	316	292
SA312	316L	296
SA312	316N	300
SA312	317	301
SA312	321	304
SA312	347	311
SA312	348	311
SA312 (XM-19)		407
SA312 (XM-29)		418
SA320	303	265
SA320	303Se	265
SA320	316	292
SA320	321	304
SA320 (B8)	304	273
SA320 (B8C)	347	311
SA358	304	273
SA358	304N	280
SA358	309	285
SA358	310	288
SA358	316	292
SA358	316N	300
SA358	321	304
SA358	347	311
SA358	348	311
SA376	304	273
SA376	304N	280
SA376	316	292
SA376	316N	300
SA376	321	304
SA376	347	311
SA376	348	311
SA387 (Type 5)	501	371
SA387 (Type 5)	502	371
SA403	304	273
SA403	304L	277
SA403	309	285
SA403	310	288
SA403	316	292
SA403	316L	296
SA403	316N	300
SA403	317	301
SA403	321	304
SA403	347	311
SA403	348	311
SA403 (XM-19)		407
SA409	304	273
SA409	309	285
SA409	310	288
SA409	316	292
SA409	317	301
SA409	321	304
SA409	347	311
SA409	348	311
SA412		415

UNITED STATES (continued)

Designation	AISI	Page
ASME		
SA412	201	246
SA412 (XM-19)		407
SA430	304	273
SA430	304N	280
SA430	316	292
SA430	316N	300
SA430	321	304
SA430	347	311
SA479	302	258
SA479	304	273
SA479	304L	277
SA479	304N	280
SA479	310S	288
SA479	316	292
SA479	316L	296
SA479	316N	300
SA479	321	304
SA479	347	311
SA479	348	311
SA479	405	324
SA479	410	328
SA479	430	353
SA479 (XM-8)	S43035	352
SA479 (XM-19)		407
SA479 (XM-27)		422
SA564	630	375
SA564 (XM-25)		424
SA638	660	396
SA688	304	273
SA688	304L	277
SA688	316	292
SA688	316L	296
SA688 (XM-29)		418
SA705	630	375
SA705	631	382
SA705 (XM-12)		405
SA705 (XM-13)		402
SA705 (XM-25)		424
SA737 (XM-27)		422
ASTM		
A26	1064	40
A29	1005	6
A29	1006	6
A29	1008	6
A29	1010	10
A29	1012	10
A29	1015	11
A29	1016	13
A29	1017	13
A29	1018	13
A29	1019	13
A29	1020	16
A29	1021	18
A29	1022	18
A29	1023	20
A29	1025	20
A29	1026	20
A29	1030	21
A29	1034	21
A29	1035	21
A29	1038	26
A29	1038H	26
A29	1039	26
A29	1040	26
A29	1044	30
A29	1045	30
A29	1045H	30
A29	1046	30
A29	1050	31
A29	1055	34
A29	1059	34
A29	1060	34

UNITED STATES (continued)

Designation	AISI	Page
ASTM		
A29	1064	40
A29	1065	40
A29	1070	40
A29	1074	40
A29	1080	41
A29	1086	488
A29	1090	43
A29	1095	43
A29	1108	45
A29	1109	45
A29	1116	48
A29	1119	48
A29	1132	53
A29	1137	56
A29	1141	59
A29	1144	59
A29	1211	62
A29	1212	62
A29	1213	62
A29	12L13	65
A29	12L14	65
A29	1215	62
A29	1547	69
A29	1548	69
A29	15B48H	69
A29	9260	222
A36		483
A57	1064	40
A59	9260	222
A107	1117	48
A107	1118	48
A107	1141	59
A108	1008	6
A108	1010	10
A108	1015	11
A108	1016	13
A108	1017	13
A108	1018	13
A108	1020	16
A108	1030	21
A108	1035	21
A108	1040	26
A108	1117	48
A108	1118	48
A108	1137	56
A108	1141	59
A108	1144	59
A108	1211	62
A108	1212	62
A108	1213	62
A108	12L14	65
A108	1215	62
A131		484
A131(A)		484
A131 (AH32, DH32, EH32)		484
A131 (AH36, DH36, EH36)		484
A131 (B)		484
A131 (CS, DS)		484
A131 (D)		484
A131 (E)		484
A135		485
A139		486
A139 (B)		486
A139 (C)		486
A139 (D)		486
A139 (E)		486
A167	301	253
A167	302	258
A167	302B	262
A167	304	273
A167	304L	277
A167	305	281

UNITED STATES (continued)

Designation	AISI	Page
ASTM		
A167	308	283
A167	309	285
A167	309S	285
A167	310	288
A167	310S	288
A167	316	292
A167	316L	296
A167	317	301
A167	317L	303
A167	321	304
A167	347	311
A167	348	311
A176	403	317
A176	405	324
A176	409	327
A176	410	328
A176	410S	329
A176	430	353
A176	442	367
A176	446	369
A176 (XM-27)		422
A177	301	253
A181	1034	21
A182	304	273
A182	304L	277
A182	304N	280
A182	310	288
A182	316	292
A182	316L	296
A182	316N	300
A182	321	304
A182	347	311
A182	348	311
A182	430	353
A182 (XM-19)		407
A193	304	273
A193	316	292
A193	321	304
A193	347	311
A193	410	328
A193	501	371
A193 (grade B8S)		413
A194	303	265
A194	303Se	265
A194	304	273
A194	316	292
A194	321	304
A194	347	311
A194	410	328
A194	416	337
A194	416Se	337
A194	501	371
A194 (grade 8F)	303Se	265
A194 (grade 8S)		413
A213	304	273
A213	304L	277
A213	304N	280
A213	310	288
A213	316	292
A213	316L	296
A213	316N	300
A213	321	304
A213	347	311
A213	348	311
A227		487
A227M		487
A228	1086	488
A229		489
A229	1065	40
A230		489
A230	1064	40
A231		490
A231	6150	182
A232		490

UNITED STATES (continued)

Designation	AISI	Page
ASTM		
A240	302	258
A240	304	273
A240	304L	277
A240	304N	280
A240	305	281
A240	309S	285
A240	310S	288
A240	316	292
A240	316L	296
A240	316N	300
A240	317	301
A240	317L	303
A240	321	304
A240	348	311
A240	405	324
A240	410	328
A240	410S	329
A240	430	353
A240 (XM-8)	S43035	352
A240 (XM-19)		407
A240 (XM-21)	304HN	280
A240 (XM-27)		422
A240 (XM-29)		418
A249	304	273
A249	304L	277
A249	304N	280
A249	305	281
A249	309	285
A249	310	288
A249	316	292
A249	316L	296
A249	316N	300
A249	317	301
A249	321	304
A249	347	311
A249	348	311
A249 (XM-19)		407
A249 (XM-29)		418
A268	329	308
A268	405	324
A268	409	327
A268	430	353
A268	430Ti	352
A268	443	367
A268	446	369
A268 (XM-8)	S43035	352
A268 (XM-27)		422
A269	304	273
A269	316	292
A269	316L	296
A269	317	301
A269	321	304
A269	347	311
A269	348	311
A269 (XM-10)		415
A269 (XM-11)		415
A269 (XM-19)		407
A269 (XM-29)		418
A270	304	273
A271	304	273
A271	321	304
A271	347	311
A273	1026	20
A274	9840	231
A276		413
A276	302	258
A276	302B	262
A276	304	273
A276	304L	277
A276	304N	280
A276	305	281
A276	308	283
A276	309	285
A276	309S	285

UNITED STATES (continued)

Designation	AISI	Page
ASTM		
A276	310	288
A276	310S	288
A276	314	291
A276	316	292
A276	316L	296
A276	316N	300
A276	317	301
A276	321	304
A276	347	311
A276	348	311
A276	403	317
A276	405	324
A276	410	328
A276	414	332
A276	420	344
A276	430	353
A276	431	356
A276	440A	358
A276	440B	363
A276	440C	364
A276	446	369
A276 (XM-10)		415
A276 (XM-11)		415
A276 (XM-19)		407
A276 (XM-21)	304HN	280
A276 (XM-27)		422
A276 (XM-28)		420
A276 (XM-29)		418
A277		487
A283		491
A284		491
A284 (A)		491
A284 (B)		491
A284 (C)		491
A295	E50100	154
A295	E51100	154
A304	1330H	76
A304	1335H	81
A304	1340H	83
A304	1345H	87
A304	15B48H	69
A304	4027H	99
A304	4028H	99
A304	4037H	101
A304	4042H	102
A304	4047H	104
A304	4118H	105
A304	4130H	108
A304	4135H	110
A304	4137H	112
A304	4140H	113
A304	4142H	117
A304	4145H	118
A304	4147H	119
A304	4150H	120
A304	4161H	126
A304	4320H	127
A304	4340H	130
A304	4419H	137
A304	4620H	141
A304	4626H	144
A304	4718H	146
A304	4720H	147
A304	4815H	148
A304	4817	150
A304	4817H	150
A304	4820H	151
A304	50B40H	160
A304	50B44H	161
A304	5046H	162
A304	50B46H	162
A304	50B50H	164
A304	50B60H	165
A304	5120H	167

UNITED STATES (continued)

Designation	AISI	Page
ASTM		
A304	5130H	168
A304	5132H	170
A304	5135H	171
A304	5140H	172
A304	5150H	175
A304	5155H	177
A304	5160H	178
A304	51B60H	178
A304	6118	181
A304	6150H	182
A304	81B45H	185
A304	8617H	188
A304	8620H	189
A304	8622H	193
A304	8625H	194
A304	8627H	196
A304	8630H	197
A304	86B30H	197
A304	8637H	201
A304	8640H	202
A304	8642H	205
A304	8645H	206
A304	86B45H	206
A304	8650H	209
A304	8655H	212
A304	8660H	213
A304	8720H	215
A304	8740H	216
A304	8822H	218
A304	9260H	222
A304	94B15H	226
A304	94B17H	227
A304	94B30H	229
A304	E4340	130
A304	E4340H	130
A304	E9310H	223
A308		492
A311	1137	56
A311	1141	59
A311	1144	59
A312	304	273
A312	304L	277
A312	304N	280
A312	309	285
A312	310	288
A312	316	292
A312	316L	296
A312	316N	300
A312	317	301
A312	321	304
A312	347	311
A312	348	311
A312 (XM-19)		407
A312 (XM-29)		418
A313	302	258
A313	304	273
A313	305	281
A313	316	292
A313	631	382
A313 (XM-16)		426
A313 (XM-28)		420
A313 (XM-29)		418
A314	202	251
A314	302	258
A314	302B	262
A314	303	265
A314	303Se	265
A314	304	273
A314	304L	277
A314	305	281
A314	308	283
A314	309	285
A314	309S	285
A314	310	288
A314	310S	288

UNITED STATES (continued)

Designation	AISI	Page
ASTM		
A314	314	291
A314	316	293
A314	316L	296
A314	317	301
A314	321	304
A314	347	311
A314	348	311
A314	403	317
A314	405	324
A314	410	328
A314	414	332
A314	416	337
A314	416Se	337
A314	420	344
A314	430	353
A314	430F	349
A314	430F(Se)	349
A314	431	356
A314	440A	358
A314	440B	363
A314	440C	364
A314	446	369
A314	501	371
A314	502	371
A314 (XM-10)		415
A314 (XM-11)		415
A314 (XM-27)		422
A320	303	265
A320	303Se	265
A320	304	273
A320	316	292
A320	321	304
A320	347	311
A322	1330	76
A322	1335	81
A322	1340	83
A322	1345	87
A322	3140	95
A322	4023	97
A322	4024	98
A322	4027	99
A322	4028	99
A322	4037	101
A322	4042	102
A322	4047	104
A322	4118	105
A322	4130	108
A322	4137	112
A322	4140	113
A322	4142	117
A322	4145	118
A322	4147	119
A322	4150	120
A322	4161	126
A322	4320	127
A322	4340	130
A322	4419	137
A322	4615	140
A322	4620	141
A322	4626	144
A322	4718	146
A322	4720	147
A322	4815	148
A322	4817	150
A322	4820	151
A322	50B40	160
A322	50B44	161
A322	50B46	162
A322	50B50	164
A322	50B60	165
A322	5120	167
A322	5130	168
A322	5132	170
A322	5135	171
A322	5140	172

UNITED STATES (continued)

Designation	AISI	Page
ASTM		
A322	5150	175
A322	5155	177
A322	5160	178
A322	51B60	178
A322	6118	181
A322	6150	182
A322	81B45	185
A322	8615	187
A322	8617	188
A322	8620	189
A322	8622	193
A322	8625	194
A322	8627	196
A322	8630	197
A322	8637	201
A322	8640	202
A322	8642	205
A322	8645	206
A322	8650	209
A322	8655	212
A322	8660	213
A322	8720	215
A322	8740	216
A322	8822	218
A322	9255	220
A322	9260	222
A322	94B17	227
A322	94B30	229
A322	94B40	230
A322	9840	231
A322	E9310	223
A322	E51100	154
A322	E52100	154
A331	1330	76
A331	1335	81
A331	1340	83
A331	1345	87
A331	3140	95
A331	4023	97
A331	4024	98
A331	4027	99
A331	4028	99
A331	4037	101
A331	4042	102
A331	4047	104
A331	4118	105
A331	4130	108
A331	4137	112
A331	4140	113
A331	4142	117
A331	4145	118
A331	4147	119
A331	4150	120
A331	4161	126
A331	4320	127
A331	4340	130
A331	4419	137
A331	4615	140
A331	4620	141
A331	4626	144
A331	4718	146
A331	4720	147
A331	4815	148
A331	4817	150
A331	4820	151
A331	50B60	165
A331	5120	167
A331	5130	168
A331	5132	170
A331	5135	171
A331	5140	172
A331	5150	175
A331	5155	177
A331	5160	178
A331	51B60	178

UNITED STATES (continued)

Designation	AISI	Page
ASTM		
A331	6150	182
A331	8617	188
A331	8620	189
A331	8622	193
A331	8625	194
A331	8627	196
A331	8630	197
A331	8637	201
A331	8640	202
A331	8642	205
A331	8645	206
A331	8655	212
A332	8660	213
A331	8720	215
A331	8740	216
A331	8822	218
A331	9260	222
A331	94B17	227
A331	94B30	229
A331	E4340	130
A331	E52100	154
A355	4135	110
A358	304N	280
A358	309	285
A358	310	288
A358	316	292
A358	316N	300
A358	321	304
A358	347	311
A358	348	311
A358 (XM-29)		418
A366		492
A368	302	258
A368	304	273
A368	305	281
A368	316	292
A376	304	273
A376	304N	280
A376	316	292
A376	316N	300
A376	321	304
A376	347	311
A376	348	311
A387 (5)	501	371
A387 (5)	502	371
A401		492
A403	304L	277
A403	304N	280
A403	309	285
A403	310	288
A403	316	292
A403	316L	296
A403	316N	300
A403	317	301
A403	321	304
A403	347	311
A403	348	311
A403 (XM-19)		407
A409	304	273
A409	309	285
A409	310	288
A409	316	292
A409	317	301
A409	321	304
A409	347	311
A412	201	246
A412	202	251
A412 (XM-10)		415
A412 (XM-11)		415
A412 (XM-19)		407
A412 (XM-29)		418
A424		493
A429	201	246
A429	202	251
A429 (XM-19)		407

UNITED STATES (continued)

Designation	AISI	Page
ASTM		
A430	304	273
A430	304N	280
A430	316	292
A430	316N	300
A430	321	304
A430	347	311
A446		494
A453	651	391
A453	660	396
A457	651	391
A458	651	391
A463		494
A473	202	251
A473	302	258
A473	302B	262
A473	303	265
A473	303Se	265
A473	304	273
A473	304L	277
A473	305	281
A473	308	283
A473	309	285
A473	309S	285
A473	310	288
A473	310S	288
A473	314	291
A473	316	292
A473	316L	296
A473	317	301
A473	321	304
A473	347	311
A473	403	317
A473	405	324
A473	410	328
A473	410S	329
A473	414	332
A473	416	337
A473	416Se	337
A473	420	344
A473	430	353
A473	430F	349
A473	430F(Se)	349
A473	431	356
A473	440A	358
A473	440B	363
A473	440C	364
A473	446	369
A473	501	371
A473	502	371
A473 (XM-10)		415
A473 (XM-11)		415
A477	651	391
A478	302	258
A478	304	273
A478	304L	277
A478	305	281
A478	316	292
A478	316L	296
A478	317	301
A479		413
A479	302	258
A479	304	273
A479	304L	277
A479	304N	280
A479	310S	288
A479	316	292
A479	316L	296
A479	316N	300
A479	321	304
A479	347	311
A479	348	311
A479	403	317
A479	405	324
A479	410	328
A479	430	353

UNITED STATES (continued)

Designation	AISI	Page
ASTM		
A479 (XM-8)	S43035	352
A479 (XM-19)		407
A479 (XM-27)		422
A479 (XM-29)		418
A485		495
A485 (1)		495
A485 (2)		495
A485 (3)		495
A485 (4)		495
A492	302	258
A492	304	273
A492	305	281
A493	302	258
A493	304	273
A493	305	281
A493	321	304
A493	347	311
A493	384	315
A493	410	328
A493	430	353
A493	431	356
A493	440C	364
A493 (XM-7)		421
A500 (A, B)		497
A500 (A, B, C)		496
A501		496
A501		497
A505	4118	105
A505	4130	108
A505	4137	112
A505	4140	113
A505	4142	117
A505	4145	118
A505	4147	119
A505	4150	120
A505	4320	127
A505	4340	130
A505	4615	140
A505	4620	141
A505	4718	146
A505	4815	148
A505	4820	151
A505	5130	168
A505	5132	170
A505	5140	172
A505	5150	175
A505	5160	178
A505	6150	182
A505	8615	187
A505	8617	188
A505	8620	189
A505	8630	197
A505	8640	202
A505	8642	205
A505	8645	206
A505	8650	209
A505	8655	212
A505	8660	213
A505	8720	215
A505	8740	216
A505	9260	222
A505	E4340	130
A505	E51100	154
A505	E52100	154
A510	1005	6
A510	1006	6
A510	1008	6
A510	1010	10
A510	1012	10
A510	1015	11
A510	1016	13
A510	1017	13
A510	1018	13
A510	1019	13
A510	1020	16

UNITED STATES (continued)

Designation	AISI	Page
ASTM		
A510	1021	18
A510	1022	18
A510	1023	20
A510	1025	20
A510	1026	20
A510	1030	21
A510	1035	21
A510	1038	26
A510	1039	26
A510	1040	26
A510	1044	30
A510	1045	30
A510	1046	30
A510	1050	31
A510	1055	34
A510	1060	34
A510	1070	40
A510	1080	41
A510	1090	43
A510	1095	43
A510	1547	69
A510	1548	69
A511	302	258
A511	304	273
A511	304L	277
A511	305	281
A511	309	285
A511	309S	285
A511	310	288
A511	310S	288
A511	316L	296
A511	317	301
A511	321	304
A511	329	308
A511	347	311
A511	403	317
A511	405	324
A511	410	328
A511	414	332
A511	416Se	337
A511	430	353
A511	440A	358
A511	443	367
A511	446	369
A512	1025	20
A512	1030	21
A513	1016	13
A513	1017	13
A513	1018	13
A513	1019	13
A513	4130	108
A513	8620	189
A514		497
A514 (A)		497
A514 (C)		497
A514 (D)		497
A514 (E)		497
A514 (F)		497
A514 (G)		497
A514 (H)		497
A514 (J)		497
A514 (L)		497
A514 (M)		497
A514 (N)		497
A514 (P)		497
A515 (55)		491
A515 (60)		491
A517 (A)		497
A517 (C)		497
A517 (D)		497
A517 (E)		497
A517 (F)		497
A517 (G)		497
A517 (H)		497
A517 (J)		497
A517 (L)		497
A517 (M)		497
A517 (N)		497
A517 (P)		497
A519	1008	6
A519	1010	10
A519	1012	10
A519	1015	11
A519	1017	13
A519	1018	13
A519	1019	13
A519	1020	16
A519	1021	18
A519	1022	18
A519	1025	20
A519	1026	20
A519	1030	21
A519	1035	21
A519	1040	26
A519	1045	30
A519	1050	31
A519	1330	76
A519	1335	81
A519	1340	83
A519	1345	87
A519	3140	95
A519	4023	97
A519	4024	98
A519	4027	99
A519	4028	99
A519	4037	101
A519	4042	102
A519	4047	104
A519	4118	105
A519	4130	108
A519	4135	110
A519	4137	112
A519	4140	113
A519	4142	117
A519	4145	118
A519	4147	119
A519	4150	120
A519	4320	127
A519	4340	130
A519	4422	138
A519	4427	139
A519	4720	147
A519	4817	150
A519	4820	151
A519	50B40	160
A519	50B44	161
A519	5046	162
A519	50B46	162
A519	50B50	164
A519	50B60	165
A519	5120	167
A519	5130	168
A519	5132	170
A519	5135	171
A519	5140	172
A519	5150	175
A519	5155	177
A519	5160	178
A519	51B60	178
A519	81B45	185
A519	8630	197
A519	8637	201
A519	8640	202
A519	8642	205
A519	8645	206
A519	86B45	206
A519	8650	209
A519	8660	213
A519	8720	215
A519	8740	216
A519	8822	218
A519	9260	222
A519	94B15	226
A519	94B17	227
A519	94B30	229
A519	94B40	230
A519	9840	231
A519	E4340	130
A519	E9310	223
A519	E50100	154
A519	E51100	154
A519	E52100	154
A525		498
A525M		498
A526		498
A527		498
A528		498
A529		500
A534	4023	97
A535	4320	127
A535	4620	141
A535	4720	147
A535	4820	151
A535	E52100	154
A538 (A)		472
A538 (B)		474
A538 (C)		477
A544	1017	13
A544	1018	13
A544	1020	16
A544	1022	18
A544	1030	21
A544	1035	21
A544	1038	26
A545	1006	6
A545	1008	6
A545	1010	10
A545	1012	10
A545	1015	11
A545	1016	13
A545	1018	13
A545	1019	13
A545	1021	18
A545	1022	18
A545	1026	20
A545	1030	21
A545	1035	21
A545	1038	26
A546	1030	21
A546	1035	21
A546	1038	26
A546	1039	26
A546	1040	26
A547	1335	81
A547	1340	83
A547	4037	101
A547	4137	112
A547	4140	113
A547	4142	117
A547	4340	130
A548	1016	13
A548	1018	13
A548	1019	13
A548	1021	18
A548	1022	18
A549	1008	6
A549	1010	10
A549	1012	10
A549	1015	11
A549	1016	13
A549	1017	13
A549	1018	13
A554	301	253
A554	302	258
A554	304	273
A554	304L	277
A554	305	281
A554	309	285

UNITED STATES (continued)

Designation	AISI	Page
ASTM		
A554	309S	285
A554	309S(Cb)	285
A554	310S	288
A554	316L	296
A554	317	301
A554	347	311
A554	430	353
A554	430Ti	352
A564	630	375
A564	631	382
A564	632	384
A564	634	388
A564 (XM-12)		405
A564 (XM-13)		402
A564 (XM-16)		426
A564 (XM-25)		424
A565	422	347
A565	615	373
A567	661	398
A569		492
A570		500
A570 (A, B, C)		500
A570 (D, E)		500
A573		501
A573 (58)		501
A573 (65)		501
A573 (70)		501
A575	1008	6
A575	1010	10
A575	1012	10
A575	1015	11
A575	1017	13
A575	1020	16
A575	1023	20
A575	1025	20
A575	1044	30
A576	1008	6
A576	1010	10
A576	1012	10
A576	1015	11
A576	1016	13
A576	1017	13
A576	1018	13
A576	1019	13
A576	1020	16
A576	1021	18
A576	1022	18
A576	1023	20
A576	1025	20
A576	1026	20
A576	1030	21
A576	1035	21
A576	1038	26
A576	1039	26
A576	1040	26
A576	1044	30
A576	1045	30
A576	1046	30
A576	1050	31
A576	1055	34
A576	1060	34
A576	1070	40
A576	1080	41
A576	1090	43
A576	1095	43
A576	1547	69
A576	1548	69
A579	632	384
A579	634	388
A579 (grade 61)	633	386
A579 (grade 62)	631	382
A579 (grade 63)	431	356
A579 (grade 71)		472
A579 (grade 72)		474
A579 (grade 73)		477
A580		413

UNITED STATES (continued)

Designation	AISI	Page
ASTM		
A580	302B	262
A580	304	273
A580	304L	277
A580	305	281
A580	308	283
A580	309	285
A580	309S	285
A580	310S	288
A580	314	291
A580	316L	296
A580	317	301
A580	347	311
A580	348	311
A580	403	317
A580	405	324
A580	410	328
A580	414	332
A580	420	344
A580	430	353
A580	431	356
A580	440A	358
A580	440B	363
A580	440C	364
A580	446	369
A580 (XM-10)		415
A580 (XM-11)		415
A580 (XM-19)		407
A580 (XM-28)		420
A580 (XM-29)		418
A581	303	265
A581	303Se	265
A581	416	337
A581	416Se	337
A581	430F	349
A581	430F(Se)	349
A581 (XM-2)	303MA	271
A582	303	265
A582	303Se	265
A582	416	337
A582	416Se	337
A582	420F(Se)	346
A582	430F	349
A582	430F(Se)	349
A582 (XM-2)	303MA	271
A591		501
A592 (A)		497
A592 (F)		497
A599		502
A600	M1	465
A600	M2	465
A600	M3 (Class 1)	465
A600	M3 (Class 2)	465
A600	M4	465
A600	M6	465
A600	M7	465
A600	M10	465
A600	M30	465
A600	M33	465
A600	M34	465
A600	M36	465
A600	M41	465
A600	M42	465
A600	M43	465
A600	M44	465
A600	M46	465
A600	M47	465
A600	T1	463
A600	T2	463
A600	T4	463
A600	T5	463
A600	T6	463
A600	T8	463
A600	T15	463
A611		502
A619		503
A620		503

UNITED STATES (continued)

Designation	AISI	Page
ASTM		
A621		503
A622		503
A632	304	273
A632	304L	277
A632	310	288
A632	316L	296
A632	317	301
A632	348	311
A633	347	311
A638	660	396
A639	661	398
A642		498
A646	4130	108
A646	4140	113
A646	4340	130
A646	E52100	154
A651	304	273
A651	409	327
A651	430	353
A651	430Ti	352
A651	434	353
A651 (XM-8)	S43035	352
A659	1015	11
A659	1016	13
A659	1017	13
A659	1018	13
A659	1020	16
A659	1021	18
A659	1023	20
A666	201	246
A666	202	251
A666	301	253
A666	302	258
A666	304	273
A678		504
A678 (A)		504
A678 (B)		504
A678 (C)		504
A681	A2	445
A681	A3	445
A681	A4	445
A681	A5	445
A681	A6	445
A681	A7	445
A681	A8	445
A681	A9	445
A681	A10	445
A681	D2	449
A681	D3	449
A681	D4	449
A681	D5	449
A681	D7	449
A681	H10	457
A681	H11	457
A681	H12	457
A681	H13	457
A681	H14	457
A681	H19	457
A681	H21	457
A681	H22	457
A681	H23	457
A681	H24	457
A681	H25	457
A681	H26	457
A681	H41	457
A681	H42	457
A681	H43	457
A681	O1	440
A681	O2	440
A681	O6	440
A681	O7	440
A681	P2	453
A681	P3	453
A681	P4	453
A681	P5	453
A681	P6	453
A681	P20	453

UNITED STATES (continued)

Designation	AISI	Page
ASTM		
A681	P21	453
A681	S1	436
A681	S2	436
A681	S4	436
A681	S5	436
A681	S6	436
A681	S7	436
A682	1030	21
A682	1035	21
A682	1040	26
A682	1045	30
A682	1050	31
A682	1055	34
A682	1060	34
A682	1064	40
A682	1065	40
A682	1070	40
A682	1074	40
A682	1080	41
A682	1086	488
A682	1095	43
A686	W1	432
A686	W5	432
A688	304	273
A688	304L	277
A688	316L	296
A688 (XM-29)		418
A693	630	375
A693	631	382
A693	632	384
A693	633	386
A693	634	388
A693 (XM-12)		405
A693 (XM-13)		402
A693 (XM-16)		426
A693 (XM-25)		424
A699		504
A705	630	375
A705	631	382
A705	632	384
A705	634	388
A705 (XM-12)		405
A705 (XM-13)		402
A705 (XM-16)		426
A705 (XM-25)		424
A709		505
A710		506
A710 (A)		506
A710 (B)		506
A711	3140	95
A711	4135	110
A711	4720	147
A711	8660	213
A711	E9310	223
A711	E50100	154
A711	E51100	154
A711	E52100	154
A731 (XM-27)		422
B511	330	309
B512	330	309
B535	330	309
B536	330	309
B546	330	309
FED		
QQ-S-633 (C12L13)	12L13	65
QQ-S-635 (C1030)	1030	21
QQ-S-635 (C1035)	1035	21
QQ-S-635 (C1045)	1045	30
QQ-S-635 (C1050)	1050	31
QQ-S-637	1141	59
QQ-S-637	1215	62
QQ-S-637 (C1008)	1008	6

UNITED STATES (continued)

Designation	AISI	Page
FED		
QQ-S-637 (C1109)	1109	45
QQ-S-637 (C1116)	1116	48
QQ-S-637 (C1117)	1117	48
QQ-S-637 (C1118)	1118	48
QQ-S-637 (C1119)	1119	48
QQ-S-637 (C1132)	1132	53
QQ-S-637 (C1137)	1137	56
QQ-S-637 (C1144)	1144	59
QQ-S-637 (C1211)	1211	62
QQ-S-637 (C1212)	1212	62
QQ-S-637 (C1913)	1213	62
QQ-S-698 (C1008)	1008	6
QQ-S-698 (C1015)	1015	11
QQ-S-700 (C1025)	1025	20
QQ-S-700 (C1030)	1030	21
QQ-S-700 (C1035)	1035	21
QQ-S-700 (C1045)	1045	30
QQ-S-700 (C1050)	1050	31
QQ-S-700 (C1055)	1055	34
QQ-S-700 (C1065)	1065	40
QQ-S-700 (C1074)	1074	40
QQ-S-700 (C1080)	1080	41
QQ-S-700 (C1086)	1086	488
QQ-S-700 (C1095)	1095	43
QQ-S-763	202	251
QQ-S-763	302	258
QQ-S-763	304	273
QQ-S-763	304L	277
QQ-S-763	305	281
QQ-S-763	309	285
QQ-S-763	310	288
QQ-S-763	316	292
QQ-S-763	316L	296
QQ-S-763	317	301
QQ-S-763	321	304
QQ-S-763	347	311
QQ-S-763	403	317
QQ-S-763	405	324
QQ-S-763	410	328
QQ-S-763	414	332
QQ-S-763	420	344
QQ-S-763	430	353
QQ-S-763	440A	358
QQ-S-763	440B	363
QQ-S-763	440C	364
QQ-S-763	446	396
QQ-S-766	201	246
QQ-S-766	202	251
QQ-S-766	301	253
QQ-S-766	302	258
QQ-S-766	304	273
QQ-S-766	304L	277
QQ-S-766	309	285
QQ-S-766	310	288
QQ-S-766	316	292
QQ-S-766	316L	296
QQ-S-766	321	304
QQ-S-766	347	311
QQ-S-766	348	311
QQ-S-766	420	344
QQ-S-766	430	353
QQ-S-766	446	369
QQ-T-570	A2	445
QQ-T-570	A3	445
QQ-T-570	A4	445
QQ-T-570	A5	445
QQ-T-570	A6	445
QQ-T-570	A7	445
QQ-T-570	A8	445
QQ-T-570	A9	445
QQ-T-570	A10	445
QQ-T-570	D2	449
QQ-T-570	D3	449
QQ-T-570	D4	449
QQ-T-570	D5	449
QQ-T-570	D7	449

UNITED STATES (continued)

Designation	AISI	Page
FED		
QQ-T-570	H10	457
QQ-T-570	H11	457
QQ-T-570	H12	457
QQ-T-570	H13	457
QQ-T-570	H14	457
QQ-T-570	H19	457
QQ-T-570	H21	457
QQ-T-570	H22	457
QQ-T-570	H23	457
QQ-T-570	H24	457
QQ-T-570	H25	457
QQ-T-570	H26	457
QQ-T-570	H41	457
QQ-T-570	H42	457
QQ-T-570	H43	457
QQ-T-570	O1	440
QQ-T-570	O2	440
QQ-T-570	O6	440
QQ-T-570	O7	440
QQ-T-570	S1	436
QQ-T-570	S2	436
QQ-T-570	S4	436
QQ-T-570	S5	436
QQ-T-570	S6	436
QQ-T-590	M1	465
QQ-T-590	M2	465
QQ-T-590	M3 (Class 1)	465
QQ-T-590	M3 (Class 2)	465
QQ-T-590	M4	465
QQ-T-590	M6	465
QQ-T-590	M7	465
QQ-T-590	M10	465
QQ-T-590	M30	465
QQ-T-590	M33	465
QQ-T-590	M34	465
QQ-T-590	M36	465
QQ-T-590	M41	465
QQ-T-590	M42	465
QQ-T-590	M43	465
QQ-T-590	M44	465
QQ-T-590	M46	465
QQ-T-590	T1	463
QQ-T-590	T2	463
QQ-T-590	T4	463
QQ-T-590	T5	463
QQ-T-590	T6	463
QQ-T-590	T8	463
QQ-T-590	T15	463
QQ-W-412		490
QQ-W-412 (II)		492
QQ-W-423	302	258
QQ-W-423	304	273
QQ-W-423	305	281
QQ-W-423	310	288
QQ-W-423	316	292
QQ-W-423	321	304
QQ-W-423	347	311
QQ-W-423	410	328
QQ-W-423	416	337
QQ-W-423	420	344
QQ-W-423	430	353
QQ-W-461	1006	6
STD-66	202	251
STD-66	304	273
STD-66	310	288
STD-66	430	353
MIL SPEC		
MIL-C-24111	630	375
MIL-F-20138	304	273
MIL-S-862	302	258
MIL-S-862	303	265
MIL-S-862	303Se	265
MIL-S-862	304	273
MIL-S-862	304L	277

UNITED STATES (continued)

Designation	AISI	Page
MIL SPEC		
MIL-S-862	309	285
MIL-S-862	310	288
MIL-S-862	316	292
MIL-S-862	316L	296
MIL-S-862	317	301
MIL-S-862	321	304
MIL-S-862	347	311
MIL-S-862	403	317
MIL-S-862	405	324
MIL-S-862	410	328
MIL-S-862	416	337
MIL-S-862	416Se	337
MIL-S-862	420	344
MIL-S-862	430	353
MIL-S-862	430F	349
MIL-S-862	430F(Se)	349
MIL-S-862	431	356
MIL-S-862	440A	358
MIL-S-862	440B	363
MIL-S-862	440C	364
MIL-S-862	440F	366
MIL-S-862	440F(Se)	366
MIL-S-862	446	369
MIL-S-866	1016	13
MIL-S-866	8615	187
MIL-S-980	E52100	154
MIL-S-5000	E4340	130
MIL-S-5059	301	253
MIL-S-5059	304	273
MIL-S-5059	316	292
MIL-S-6049	8740	216
MIL-S-7420	E52100	154
MIL-S-7493 (A4615)	4615	140
MIL-S-7493 (A4620)	4620	141
MIL-S-8503	6150	182
MIL-S-11310 (CS1005)	1005	6
MIL-S-11310 (CS1006)	1006	6
MIL-S-11310 (CS1008)	1008	6
MIL-S-11310 (CS1010)	1010	10
MIL-S-11310 (CS1012)	1012	10
MIL-S-11310 (CS1017)	1017	13
MIL-S-11310 (CS1018)	1018	13
MIL-S-11310 (CS1020)	1020	16
MIL-S-11310 (CS1022)	1022	18
MIL-S-11310 (CS1025)	1025	20
MIL-S-11310 (CS1030)	1030	21
MIL-S-11310 (CS1040)	1040	26
MIL-S-11595 (ORD4150)	4150	120
MIL-S-11713 (2)	1070	40
MIL-S-16788 (C10)	1095	43
MIL-S-16974	1015	11
MIL-S-16974	1050	31
MIL-S-16974 (Gr. 1060)	1060	34
MIL-S-16974	1080	41
MIL-S-16974	1330	76
MIL-S-16974	1335	81
MIL-S-16974	1340	83
MIL-S-16974	3140	95
MIL-S-16974	4130	108

UNITED STATES (continued)

Designation	AISI	Page
MIL SPEC		
MIL-S-16974	4135	110
MIL-S-16974	4140	113
MIL-S-16974	4145	118
MIL-S-16974	4340	130
MIL-S-16974	8620	189
MIL-S-16974	8625	194
MIL-S-16974	8630	197
MIL-S-16974	8640	202
MIL-S-16974	8645	206
MIL-S-18411	1117	48
MIL-S-18411	12L13	65
MIL-S-18733	4135	110
MIL-S-20166 (CS1116)	1116	48
MIL-S-22141	E52100	154
MIL-S-22698 (B)		484
MIL-S-23195	304	273
MIL-S-23195	304L	277
MIL-S-23195	347	311
MIL-S-23195	348	311
MIL-S-23196	304	273
MIL-S-23196	304L	277
MIL-S-23196	347	311
MIL-S-23196	348	311
MIL-S-23495		484
MIL-S-23495 (C)		491
MIL-S-25043	631	382
MIL-S-46042	651	391
MIL-S-46049	1065	40
MIL-S-46049	1074	40
MIL-S-46409	1065	40
MIL-S-81506	630	375
MIL-S-81591	630	375
MIL-T-6845	304	273
MIL-T-8504	304	273
MIL-T-8506	304	273
MIL-W-46078	631	382
SAE		
J118	1034	21
J118	1059	34
J172		489
J217	631	382
J230	302	258
J403	1005	6
J403	1006	6
J403	1008	6
J403	1010	10
J403	1012	10
J403	1015	11
J403	1016	13
J403	1017	13
J403	1018	13
J403	1019	13
J403	1020	16
J403	1021	18
J403	1022	18
J403	1023	20
J403	1025	20
J403	1026	20
J403	1030	21
J403	1035	21
J403	1038	26
J403	1039	26
J403	1040	26
J403	1044	30
J403	1045	30
J403	1046	30
J403	1050	31
J403	1055	34
J403	1060	34
J403	1064	40
J403	1065	40
J403	1070	40
J403	1074	40

UNITED STATES (continued)

Designation	AISI	Page
SAE		
J403	1080	41
J403	1086	488
J403	1090	43
J403	1095	43
J403	1108	45
J403	1109	45
J403	1117	48
J403	1118	48
J403	1119	48
J403	1132	53
J403	1137	56
J403	1141	59
J403	1144	59
J403	1211	62
J403	1212	62
J403	1213	62
J403	12L13	65
J403	12L14	65
J403	1547	69
J403	1548	69
J404	1330	76
J404	1335	81
J404	1340	83
J404	1345	87
J404	4023	97
J404	4024	98
J404	4027	99
J404	4028	99
J404	4037	101
J404	4042	102
J404	4047	104
J404	4118	105
J404	4130	108
J404	4135	110
J404	4137	112
J404	4140	113
J404	4142	117
J404	4145	118
J404	4147	119
J404	4150	120
J404	4161	126
J404	4320	127
J404	4340	130
J404	4419	137
J404	4422	138
J404	4427	139
J404	4615	140
J404	4620	141
J404	4626	144
J404	4718	146
J404	4720	147
J404	4815	148
J404	4817	150
J404	4820	151
J404	50B40	160
J404	50B44	161
J404	5046	162
J404	50B46	162
J404	50B50	164
J404	5060	165
J404	50B60	165
J404	5120	167
J404	5130	168
J404	5132	170
J404	5135	171
J404	5140	172
J404	5150	175
J404	5155	177
J404	5160	178
J404	51B60	178
J404	6118	181
J404	6150	182
J404	81B45	185
J404	8615	187
J404	8617	188
J404	8620	189

UNITED STATES (continued)

Designation	AISI	Page
SAE		
J404	8622	193
J404	8625	194
J404	8627	196
J404	8630	197
J404	8637	201
J404	8640	202
J404	8642	205
J404	8645	206
J404	86B45	206
J404	8650	209
J404	8655	212
J404	8660	213
J404	8720	215
J404	8740	216
J404	8822	218
J404	9255	220
J404	9260	222
J404	94B15	226
J404	94B17	227
J404	94B30	229
J404	E4340	130
J404	E9310	223
J404	E50100	154
J404	E51100	154
J404	E52100	154
J405 (30201)	201	246
J405 (30202)	202	251
J405 (30301)	301	253
J405 (30302)	302	258
J405 (30302B)	302B	262
J405 (30303)	303	265
J405 (30303Se)	303Se	265
J405 (30304)	304	273
J405 (30304L)	304L	277
J405 (30305)	305	281
J405 (30308)	308	283
J405 (30309)	309	285
J405 (30309S)	309S	285
J405 (30310)	310	288
J405 (30310S)	310S	288
J405 (30314)	314	291
J405 (30316)	316	292
J405 (30316L)	316L	296
J405 (30317)	317	301
J405 (30321)	321	304
J405 (30330)	330	309
J405 (30347)	347	311
J405 (30348)	348	311
J405 (30384)	384	315
J405 (51403)	403	317
J405 (51405)	405	324
J405 (51409)	409	327
J405 (51410)	410	328
J405 (51414)	414	332
J405 (51416)	416	337
J405 (51416Se)	416Se	337
J405 (51420)	420	344
J405 (51420F)	420F	346
J405 (51420F-Se)	420F(Se)	346
J405 (51430)	430	353
J405 (51430F)	430F	349
J405 (51430F-Se)	430F(Se)	349
J405 (51431)	431	356
J405 (51434)	434	353
J405 (51436)	436	353
J405 (51440A)	440A	358
J405 (51440B)	440B	363
J405 (51440C)	440C	364
J405 (51440F)	440F	366
J405 (51440F-Se)	440F(Se)	366
J405 (51442)	442	367
J405 (51446)	446	369
J405 (51501)	501	371
J405 (51502)	502	371
J410c		237
J412	1005	6
J412	1008	6
J412	1010	10
J412	1012	10
J412	1015	11
J412	1016	13
J412	1017	13
J412	1018	13
J412	1019	13
J412	1020	16
J412	1021	18
J412	1022	18
J412	1023	20
J412	1025	20
J412	1026	20
J412	1030	21
J412	1034	21
J412	1035	21
J412	1038	26
J412	1039	26
J412	1040	26
J412	1044	30
J412	1045	30
J412	1046	30
J412	1050	31
J412	1055	34
J412	1059	34
J412	1060	34
J412	1064	40
J412	1065	40
J412	1070	40
J412	1074	40
J412	1080	41
J412	1086	488
J412	1090	43
J412	1095	43
J412	1108	45
J412	1109	45
J412	1116	48
J412	1117	48
J412	1118	48
J412	1119	48
J412	1132	53
J412	1137	56
J412	1141	59
J412	1144	59
J412	12L13	65
J412	12L14	65
J412	1215	62
J412	1330	76
J412	1335	81
J412	1340	83
J412	1345	87
J412	1547	69
J412	1548	69
J412	4023	97
J412	4024	98
J412	4027	99
J412	4028	99
J412	4037	101
J412	4042	102
J412	4047	104
J412	4118	105
J412	4130	108
J412	4135	110
J412	4137	112
J412	4140	113
J412	4142	117
J412	4145	118
J412	4147	119
J412	4150	120
J412	4161	126
J412	4320	127
J412	4340	130
J412	4419	137
J412	4422	138
J412	4427	139
J412	4615	140
J412	4620	141
J412	4626	144
J412	4718	146
J412	4720	147
J412	4815	148
J412	4817	150
J412	4820	151
J412	50B40	160
J412	50B44	161
J412	5046	162
J412	50B46	162
J412	50B50	164
J412	50B60	165
J412	5130	168
J412	5132	170
J412	5135	171
J412	5140	172
J412	5150	175
J412	5155	177
J412	5160	178
J412	51B60	178
J412	6150	182
J412	81B45	185
J412	8630	197
J412	8637	201
J412	8640	202
J412	8642	205
J412	8645	206
J412	86B45	206
J412	8650	209
J412	8655	212
J412	8660	213
J412	8740	216
J412	9255	220
J412	9260	222
J412	94B30	229
J412	E50100	154
J412	E51100	154
J412	E52100	154
J412 (1005)	1006	6
J412 (30330)	330	309
J412 (51410)	410	328
J414	1006	6
J414	1008	6
J414	1010	10
J414	1012	10
J414	1015	11
J414	1016	13
J414	1017	13
J414	1018	13
J414	1019	13
J414	1020	16
J414	1021	18
J414	1022	18
J414	1023	20
J414	1025	20
J414	1026	20
J414	1030	21
J414	1035	21
J414	1038	26
J414	1039	26
J414	1040	26
J414	1044	30
J414	1045	30
J414	1046	30
J414	1050	31
J414	1055	34
J414	1060	34
J414	1064	40
J414	1065	40
J414	1070	40
J414	1074	40
J414	1080	41
J414	1086	488
J414	1090	43

UNITED STATES (continued)

Designation	AISI	Page
SAE		
J414	1095	43
J414	1109	45
J414	1117	48
J414	1118	48
J414	1119	48
J414	1132	53
J414	1137	56
J414	1141	59
J414	1144	59
J414	12L14	65
J414	1547	69
J414	1548	69
J437	A2	445
J437	D2	449
J437	D3	449
J437	D5	449
J437	D7	449
J437	H11	457
J437	H12	457
J437	H13	457
J437	H21	457
J437	M1	465
J437	M2	465
J437	M3 (Class 1)	465
J437	M3 (Class 2)	465
J437	M4	465
J437	O1	440
J437	O2	440
J437	O6	440
J437	S2	436
J437	S5	436
J437	T1	463
J437	T2	463
J437	T4	463
J437	T5	463
J437	T8	463
J437 (W108)	W1	432
J437 (W109)	W1	432
J437 (W110)	W1	432
J437 (W112)	W1	432
J437 (W209)	W2	432
J437 (W210)	W2	432
J438	A2	445
J438	D2	449
J438	D3	449
J438	D5	449
J438	D7	449
J438	H11	457
J438	H12	457
J438	H13	457
J438	H21	457
J438	M1	465
J438	M2	465
J438	M3 (Class 1)	465
J438	M3 (Class 2)	465
J438	M4	465
J438	O1	440
J438	O2	440
J438	O6	440
J438	S1	436
J438	S2	436
J438	S5	436
J438	T1	463
J438	T2	463
J438	T4	463
J438	T5	463
J438	T8	463
J438 (W108)	W1	432
J438 (W109)	W1	432
J438 (W110)	W1	432
J438 (W112)	W1	432
J438 (W209)	W2	432
J438 (W210)	W2	432
J467	422	347
J467	H11	457
J467	H12	457

UNITED STATES (continued)

Designation	AISI	Page
SAE		
J467	H13	457
J467 (17-4PH)	630	375
J467 (17-7PH)	631	382
J467 (19-9DL)	651	391
J467 (A286)	660	396
J467 (AM-350)	633	386
J467 (AM-355)	634	388
J467 (Greek Ascoloy)	615	373
J467 (PH15-7-Mo)	632	384
J770	1330	76
J770	1335	81
J770	1340	83
J770	1345	87
J770	4023	97
J770	4024	98
J770	4027	99
J770	4028	99
J770	4037	101
J770	4042	102
J770	4047	104
J770	4118	105
J770	4130	108
J770	4135	110
J770	4137	112
J770	4140	113
J770	4142	117
J770	4145	118
J770	4147	119
J770	4150	120
J770	4161	126
J770	4320	127
J770	4340	130
J770	4419	137
J770	4422	138
J470	4427	139
J770	4615	140
J770	4620	141
J770	4626	144
J770	4718	146
J770	4720	147
J770	4815	148
J770	4817	150
J770	4820	151
J770	50B40	160
J770	50B44	161
J770	5046	162
J770	50B46	162
J770	50B50	164
J770	5060	165
J770	50B60	165
J770	5120	167
J770	5130	168
J770	5132	170
J770	5135	171
J770	5140	172
J770	5150	175
J770	5155	177
J770	5160	178
J770	51B60	178
J770	6118	181
J770	6150	182
J770	81B45	185
J770	8615	187
J770	8617	188
J770	8620	189
J770	8622	193
J770	8625	194
J770	8627	196
J770	8630	197
J770	8637	201
J770	8640	202
J770	8642	205
J770	8645	206
J770	86B45	206

UNITED STATES (continued)

Designation	AISI	Page
SAE		
J770	8650	209
J770	8655	212
J770	8660	213
J770	8720	215
J770	8740	216
J770	8822	218
J770	9255	220
J770	9260	222
J770	94B15	226
J770	94B17	227
J770	94B30	229
J770	E4340	130
J770	E9310	223
J770	E50100	154
J770	E51100	154
J770	E52100	154
J776	1038H	26
J776	1045H	30
J778	3140	95
J778	94B40	
J778	9840	231
J1268	1330H	76
J1268	1335H	81
J1268	1340H	83
J1268	1345H	87
J1268	4027H	99
J1268	4028H	99
J1268	4037H	101
J1268	4042H	102
J1268	4047H	104
J1268	4118H	105
J1268	4130H	108
J1268	4135H	110
J1268	4137H	112
J1268	4140H	113
J1268	4142H	117
J1268	4145H	118
J1268	4147H	119
J1268	4150H	120
J1268	4161H	126
J1268	4320H	127
J1268	4340H	130
J1268	4419H	137
J1268	4620H	141
J1268	4626H	144
J1268	4718H	146
J1268	4720H	147
J1268	4815H	148
J1268	4817H	150
J1268	4820H	151
J1268	50B40H	160
J1268	50B44H	161
J1268	5046H	162
J1268	50B46H	162
J1268	50B50H	164
J1268	50B60H	165
J1268	5120H	167
J1268	5130H	168
J1268	5132H	170
J1268	5135H	171
J1268	5140H	172
J1268	5150H	175
J1268	5155H	177
J1268	5160H	178
J1268	51B60H	178
J1268	6118H	181
J1268	6150H	182
J1268	81B45H	185
J1268	8617H	188
J1268	8620H	189
J1268	8622H	193
J1268	8625H	194
J1268	8627H	196
J1268	8630H	197
J1268	86B30H	197

UNITED STATES (continued)

Designation	AISI	Page
SAE		
J1268	8637H	201
J1268	8640H	202
J1268	8642H	205
J1268	8645H	206
J1268	86B45H	206
J1268	8650H	209
J1268	8655H	212
J1268	8660H	213
J1268	8720H	215
J1268	8740H	216
J1268	8822H	218
J1268	9260	222
J1268	94B15H	226
J1268	94B17H	227
J1268	94B30H	229
J1268	E4340H	130
J1268	E9310H	223
2317	A2317	88
2515	A2515	94
UNS		
G10050	1005	6
G10060	1006	6
G10080	1008	6
G10100	1010	10
G10120	1012	10
G10150	1015	11
G10160	1016	13
G10170	1017	13
G10180	1018	13
G10190	1019	13
G10200	1020	16
G10210	1021	18
G10220	1022	18
G10230	1023	20
G10250	1025	20
G10260	1026	20
G10300	1030	21
G10340	1034	21
G10350	1035	21
G10380	1038	26
G10390	1039	26
G10400	1040	26
G10440	1044	30
G10450	1045	30
G10460	1046	30
G10500	1050	31
G10550	1055	34
G10590	1059	34
G10600	1060	34
G10640	1064	40
G10650	1065	40
G10700	1070	40
G10740	1074	40
G10800	1080	41
G10860	1086	488
G10900	1090	43
G10950	1095	43
G11080	1108	45
G11090	1109	45
G11160	1116	48
G11170	1117	48
G11180	1118	48
G11190	1119	48
G11320	1132	53
G11370	1137	56
G11410	1141	59
G11440	1144	59
G12110	1211	62
G12120	1212	62
G12130	1213	62
G12134	12L13	65
G12144	12L14	65
G12150	1215	62
G13300	1330	76
G13350	1335	81

UNITED STATES (continued)

Designation	AISI	Page
UNS		
G13400	1340	83
G13450	1345	87
G15470	1547	69
G15480	1548	69
G31400	3140	95
G40230	4023	97
G40240	4024	98
G40270	4027	99
G40280	4028	99
G40370	4037	101
G40420	4042	102
G40470	4047	104
G41180	4118	105
G41300	4130	108
G41350	4135	110
G41370	4137	112
G41400	4140	113
G41420	4142	117
G41450	4145	118
G41470	4147	119
G41500	4150	120
G41610	4161	126
G43200	4320	127
G43400	4340	130
G43406	E4340	130
G44190	4419	137
G44220	4422	138
G44270	4427	139
G46150	4615	140
G46200	4620	141
G46260	4626	144
G47180	4718	146
G47200	4720	147
G48150	4815	148
G48170	4817	150
G48200	4820	151
G50401	50B40	160
G50441	50B44	161
G50460	5046	162
G50461	50B46	162
G50600	5060	165
G50601	50B60	165
G50986	E50100	154
G51200	5120	167
G51300	5130	168
G51320	5132	170
G51350	5135	171
G51400	5140	172
G51500	5150	175
G51550	5155	177
G51600	5160	178
G51601	51B60	178
G51986	E51100	154
G52986	E52100	154
G61180	6118	181
G61500	6150	182
G81451	81B45	185
G86150	8615	187
G86170	8617	188
G86200	8620	189
G86220	8622	193
G86250	8625	194
G86270	8627	196
G86300	8630	197
G86370	8637	201
G86400	8640	202
G86420	8642	205
G86450	8645	206
G86451	86B45	206
G86500	8650	209
G86501	86B50	209
G86550	8655	212
G86600	8660	213
G87200	8720	215
G87400	8740	216
G88220	8822	218

UNITED STATES (continued)

Designation	AISI	Page
UNS		
G92550	9255	220
G92600	9260	222
G93106	E9310	223
G94151	94B15	226
G94171	94B17	227
G94301	94B30	229
G94401	94B40	230
G98400	9840	231
H10380	1038H	26
H10450	1045H	30
H13300	1330H	76
H13350	1335H	81
H13400	1340H	83
H13450	1345H	87
H15481	15B48H	69
H40270	4027H	99
H40280	4028H	99
H40370	4037H	101
H40420	4042H	102
H40470	4047H	104
H41180	4118H	105
H41300	4130H	108
H41350	4135H	110
H41370	4137H	112
H41400	4140H	113
H41420	4142H	117
H41450	4145H	118
H41470	4147H	119
H41500	4150H	120
H41610	4161H	126
H43200	4320H	127
H43400	4340H	130
H43406	E4340H	130
H44190	4419H	137
H46200	4620H	141
H46260	4626H	144
H47180	4718H	146
H47200	4720H	147
H48150	4815H	148
H48170	4817H	150
H48200	4820H	151
H50401	50B40H	160
H50441	50B44H	161
H50460	5046H	162
H50461	50B46H	162
H50501	50B50H	164
H50601	50B60H	165
H51200	5120H	167
H51300	5130H	168
H51320	5132H	170
H51350	5135H	171
H51400	5140H	172
H51500	5150H	175
H51550	5155H	177
H51600	5160H	178
H51601	51B60H	178
H61180	6118	181
H61500	6150H	182
H81451	81B45H	185
H86170	8617H	188
H86200	8620H	189
H86220	8622H	193
H86250	8625H	194
H86270	8627H	196
H86300	8630H	197
H86301	86B30H	197
H86370	8637H	201
H86400	8640H	202
H86420	8642H	205
H86450	8645H	206
H86451	86B45H	206
H86500	8650H	209
H86550	8655H	212
H86600	8660H	213

UNITED STATES (continued)

Designation	AISI	Page
UNS		
H87200	8720H	215
H87400	8740H	216
H88220	8822	218
H92600	9260H	222
H93100	E9310H	223
H94151	94B15H	226
H94171	94B17H	227
H94301	94B30H	229
K01600		504
K01601		484
K01801		484
K01804		491
K01805		484
K01806		484
K02001		491
K02002		504
K02101		484
K02102		484
K02204		504
K02300		484
K02301		501
K02401		491
K02404		501
K02500		500
K02502		500
K02600		483
K02701		501
K02703		500
K02705		496
K03000		496
K03000		497
K03003		486
K03004		486
K03010		486
K03012		486
K06501		487
K06701		489
K07001		489
K08500		488
K10614		504
K11511		497
K11523		497
K11576		497
K11625		497
K11646		497
K11662		497
K11682		497
K11683		497
K11847		497
K11856		497
K11872		497
K15047		490
K15048		490
K15590		492
K19195		495
K19667		495
K19965		495
K19990		495
K20622		506
K20747		506
K21604		497
K21650		497
K63198	651	391
K66286	660	396
K92810		472
K92890		474
K93120		477
N08330	330	309
N08366		400
R30155	661	398
S13800		402
S14800		403
S15500		405
S15700	632	384
S17400	630	375

UNITED STATES (continued)

Designation	AISI	Page
UNS		
S17700	631	382
S20100	201	246
S20200	202	251
S20910		407
S21800		413
S21900		415
S21904		415
S24000		418
S24100		420
S30100	301	253
S30200	302	258
S30215	302B	262
S30300	303	265
S30323	303Se	265
S30345	303MA	271
S30400	304	273
S30403	304L	277
S30430		421
S30450	304N	280
S30452	304HN	280
S30500	305	281
S30800	308	283
S30900	309	285
S30908	309S	285
S30940	309S(Cb)	285
S31000	310	288
S31008	310S	288
S31400	314	291
S31600	316	292
S31603	316L	296
S31620	316F	299
31651	316N	300
S31700	317	301
S31703	317L	303
S32100	321	304
S32900	329	308
S34700	347	311
S34800	348	311
S35000	633	386
S35500	634	388
S38400	384	315
S40300	403	317
S40500	405	324
S40900	409	327
S41000	410	328
S41008	410S	329
S41400	414	332
S41600	416	337
S41623	416Se	337
S41800	615	373
S42000	420	344
S42020	420F	346
S42023	420F(Se)	346
S42200	422	347
S43000	430	353
S43020	430F	349
S43023	430F(Se)	349
S43035	S43035	352
S43036	430Ti	352
S43100	431	356
S43400	434	353
S43600	436	353
S44002	440A	358
S44003	440B	363
S44004	440C	364
S44020	440F	366
S44023	440F(Se)	366
S44200	442	367
S44300	443	367
S44600	446	369
S44625		422
S45000		424
S45500		426
S50100	501	371
S50200	502	371

UNITED STATES (continued)

Designation	AISI	Page
UNS		
T11301	M1	465
T11302	M2	465
T11304	M4	465
T11306	M6	465
T11307	M7	465
T11310	M10	465
T11313	M3 (Class 1)	465
T11323	M3 (Class 2)	465
T11330	M30	465
T11333	M33	465
T11334	M34	465
T11336	M36	465
T11341	M41	465
T11342	M42	465
T11343	M43	465
T11344	M44	465
T11346	M46	465
T11347	M47	465
T12001	T1	463
T12002	T2	463
T12004	T4	463
T12005	T5	463
T12006	T6	463
T12008	T8	463
T12015	T15	463
T20810	H10	457
T20811	H11	457
T20812	H12	457
T20813	H13	457
T20814	H14	457
T20819	H19	457
T20821	H21	457
T20822	H22	457
T20823	H23	457
T20824	H24	457
T20825	H25	457
T20826	H26	457
T20841	H41	457
T20842	H42	457
T20843	H43	457
T30102	A2	445
T30103	A3	445
T30104	A4	445
T30105	A5	445
T30106	A6	445
T30107	A7	445
T30108	A8	445
T30109	A9	445
T30110	A10	445
T30402	D2	449
T30403	D3	449
T30404	D4	449
T30405	D5	449
T30407	D7	449
T31501	O1	440
T31502	O2	440
T31506	O6	440
T31507	O7	440
T41901	S1	436
T41902	S2	436
T41904	S4	436
T41905	S5	436
T41906	S6	436
T41907	S7	436
T51602	P2	453
T51603	P3	453
T51604	P4	453
T51605	P5	453
T51606	P6	453
T51620	P20	453
T51621	P21	453
T72301	W1	432
T72302	W2	432
T72305	W5	432